Lower-Dimensional Systems and Molecular Electronics

NATO ASI Series

Advanced Science Institutes Series

A series presenting the results of activities sponsored by the NATO Science Committee, which aims at the dissemination of advanced scientific and technological knowledge, with a view to strengthening links between scientific communities.

The series is published by an international board of publishers in conjunction with the NATO Scientific Affairs Division

A	**Life Sciences**	Plenum Publishing Corporation
B	**Physics**	New York and London
C	**Mathematical and Physical Sciences**	Kluwer Academic Publishers
D	**Behavioral and Social Sciences**	Dordrecht, Boston, and London
E	**Applied Sciences**	
F	**Computer and Systems Sciences**	Springer-Verlag
G	**Ecological Sciences**	Berlin, Heidelberg, New York, London,
H	**Cell Biology**	Paris, Tokyo, Hong Kong, and Barcelona
I	**Global Environmental Change**	

Recent Volumes in this Series

Volume 248—Lower-Dimensional Systems and Molecular Electronics
edited by R. M. Metzger, P. Day, and G. C. Papavassiliou

Volume 249—Advances in Nonradiative Processes in Solids
edited by B. Di Bartolo

Volume 250—The Application of Charge Density Research to Chemistry and Drug Design
edited by George A. Jeffrey and Juan F. Piniella

Volume 251—Granular Nanoelectronics
edited by David K. Ferry, John R. Barker, and Carlo Jacoboni

Volume 252—Laser Systems for Photobiology and Photomedicine
edited by A. N. Chester, S. Martellucci, and A. M. Scheggi

Volume 253—Condensed Systems of Low Dimensionality
edited by J. L. Beeby

Volume 254—Quantum Coherence in Mesoscopic Systems
edited by B. Kramer

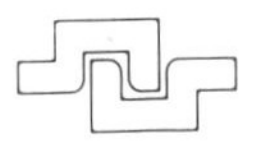

Series B: Physics

Lower-Dimensional Systems and Molecular Electronics

Edited by

Robert M. Metzger

The University of Alabama
Tuscaloosa, Alabama

Peter Day

Institut Laue-Langevin
Grenoble, France

and

George C. Papavassiliou

National Hellenic Research Foundation
Athens, Greece

Plenum Press
New York and London
Published in cooperation with NATO Scientific Affairs Division

Proceedings of a NATO Advanced Study Institute on
Lower-Dimensional Systems and Molecular Electronics,
held June 12-23, 1989,
at Hotel Spetses,
Spetses Island, Greece

Library of Congress Cataloging-in-Publication Data

NATO Advanced Study Institute on Lower-Dimensional Systems and Molecular Electronics (1989 : Hotel Spetses, Spetses Island, Greece)
Lower-dimensional systems and molecular electronics / edited by Robert M. Metzger, Peter Day, and George C. Papavassiliou.
p. cm. -- (NATO ASI series. Series B, Physics ; vol. 248)
"Proceedings of a NATO Advanced Study Institute on Lower-Dimensional Systems and Molecular Electronics, held June 12-23, 1989, at Hotel Spetses, Spetses Island, Greece"--T.p. verso.
"Held within the program of activities of the NATO Special Program on Condensed Systems of Low Dimensionality"--P.
Includes bibliographical references and indexes.
ISBN 0-306-43826-7
1. Organic superconductors--Congresses. 2. Molecular electronics--Congresses. 3. Polymers--Electric properties--Congresses. 4. Thin films, Multilayered--Electric properties--Congresses. 5. Crystals--Electric properties--Congresses. I. Metzger, R. M. (Robert M.), 1940- . II. Day, P. III. Papavassiliou, George C. IV. North Atlantic Treaty Organization. Scientific Affairs Division. V. Special Program on Condensed Systems of Low Dimensionality (NATO) VI. Title. VII. Series: NATO ASI series. Series B, Physics ; v. 248.
QC611.98.074N37 1989
537.6'23--dc20 91-3457
CIP

A Division of Plenum Publishing Corporation
233 Spring Street, New York, N.Y. 10013

Printed in the United States of America

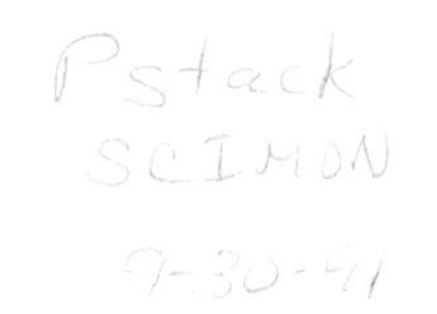

SPECIAL PROGRAM ON CONDENSED SYSTEMS OF LOW DIMENSIONALITY

This book contains the proceedings of a NATO Advanced Research Workshop held within the program of activities of the NATO Special Program on Condensed Systems of Low Dimensionality, running from 1985 to 1990 as part of the activities of the NATO Science Committee.

Other books previously published as a result of the activities of the Special Program are:

Volume 148 — INTERCALATION IN LAYERED MATERIALS
edited by M. S. Dresselhaus

Volume 152 — OPTICAL PROPERTIES OF NARROW-GAP LOW-DIMENSIONAL STRUCTURES
edited by C. M. Sotomayor Torres, J. C. Portal, J. C. Maan, and R. A. Stradling

Volume 163 — THIN FILM GROWTH TECHNIQUES FOR LOW-DIMENSIONAL STRUCTURES
edited by R. F. C. Farrow, S. S. P. Parkin, P. J. Dobson, J. H. Neave, and A. S. Arrott

Volume 168 — ORGANIC AND INORGANIC LOW-DIMENSIONAL CRYSTALLINE MATERIALS
edited by Pierre Delhaes and Marc Drillon

Volume 172 — CHEMICAL PHYSICS OF INTERCALATION
edited by A. P. Legrand and S. Flandrois

Volume 182 — PHYSICS, FABRICATION, AND APPLICATIONS OF MULTILAYERED STRUCTURES
edited by P. Dhez and C. Weisbuch

Volume 183 — PROPERTIES OF IMPURITY STATES IN SUPERLATTICE SEMICONDUCTORS
edited by C. Y. Fong, Inder P. Batra, and S. Ciraci

Volume 188 — REFLECTION HIGH-ENERGY ELECTRON DIFFRACTION AND REFLECTION ELECTRON IMAGING OF SURFACES
edited by P. K. Larsen and P. J. Dobson

Volume 189 — BAND STRUCTURE ENGINEERING IN SEMICONDUCTOR MICROSTRUCTURES
edited by R. A. Abram and M. Jaros

Volume 194 — OPTICAL SWITCHING IN LOW-DIMENSIONAL SYSTEMS
edited by H. Haug and L. Banyai

Volume 195 — METALLIZATION AND METAL-SEMICONDUCTOR INTERFACES
edited by Inder P. Batra

SPECIAL PROGRAM ON CONDENSED SYSTEMS OF LOW DIMENSIONALITY

Volume 198 MECHANISMS OF REACTIONS OF ORGANOMETALLIC COMPOUNDS WITH SURFACES
edited by D. J. Cole-Hamilton and J. O. Williams

Volume 199 SCIENCE AND TECHNOLOGY OF FAST ION CONDUCTORS
edited by Harry L. Tuller and Minko Balkanski

Volume 200 GROWTH AND OPTICAL PROPERTIES OF WIDE-GAP II–VI LOW-DIMENSIONAL SEMICONDUCTORS
edited by T. C. McGill, C. M. Sotomayor Torres, and W. Gebhardt

Volume 202 POINT AND EXTENDED DEFECTS IN SEMICONDUCTORS
edited by G. Benedek, A. Cavallini, and W. Schröter

Volume 203 EVALUATION OF ADVANCED SEMICONDUCTOR MATERIALS BY ELECTRON MICROSCOPY
edited by David Cherns

Volume 206 SPECTROSCOPY OF SEMICONDUCTOR MICROSTRUCTURES
edited by Gerhard Fasol, Annalisa Fasolino, and Paolo Lugli

Volume 213 INTERACTING ELECTRONS IN REDUCED DIMENSIONS
edited by Dionys Baeriswyl and David K. Campbell

Volume 214 SCIENCE AND ENGINEERING OF ONE- AND ZERO-DIMENSIONAL SEMICONDUCTORS
edited by Steven P. Beaumont and Clivia M. Sotomayor Torres

Volume 217 SOLID STATE MICROBATTERIES
edited by James R. Akridge and Minko Balkanski

Volume 221 GUIDELINES FOR MASTERING THE PROPERTIES OF MOLECULAR SIEVES: Relationship between the Physicochemical Properties of Zeolitic Systems and Their Low Dimensionality
edited by Denise Barthomeuf, Eric G. Derouane, and Wolfgang Hölderich

Volume 246 DYNAMICS OF MAGNETIC FLUCTUATIONS IN HIGH-TEMPERATURE SUPERCONDUCTORS
edited by George Reiter, Peter Horsch, and Gregory C. Psaltakis

PREFACE

This volume represents the written account of the NATO Advanced Study Institute "Lower-Dimensional Systems and Molecular Electronics" held at Hotel Spetses, Spetses Island, Greece from 12 June to 23 June 1989. The goal of the Institute was to demonstrate the breadth of chemical and physical knowledge that has been acquired in the last 20 years in inorganic and organic crystals, polymers, and thin films, which exhibit phenomena of reduced dimensionality.

The interest in these systems started in the late 1960's with lower-dimensional inorganic conductors, in the early 1970's with quasi-one-dimensional crystalline organic conductors, which by 1979 led to the first organic superconductors, and, in 1977, to the first conducting polymers. The study of monolayer films (Langmuir-Blodgett films) had progressed since the 1930's, but reached a great upsurge in the early 1980's. The pursuit of non-linear optical phenomena became increasingly popular in the early 1980's, as the attention turned from inorganic crystals to organic films and polymers. And in the last few years the term "molecular electronics" has gained ever-increasing acceptance, although it is used in several contexts. We now have organic superconductors with critical temperatures in excess of 10 K, conducting polymers that are soluble and processable, and used commercially; we have films of a few monolayers that have high in-plane electrical conductivity, and polymers that show great promise in photonics; we even have a few devices that function almost at the molecular level. All these results, and their exploitation, depend crucially on the dynamical interplay between synthetic chemists, physicists, and engineers.

The seventeen plenary lecturers, plus the three organizers of the Institute, endeavoured to cover all that was interesting and exciting in this vast subject. The seventy-eight participants added to the excitement of the Institute, not only by listening and discussing, but also by presenting short oral talks and posters, that either reinforced, or complemented the subjects covered in the plenary lectures. Given the eleven-day format, and budgetary constraints, some topics (notably inorganic lower-dimensional systems, and theory) were not covered in the plenary lectures, but were very well represented by the contributions of the participants.

The setting of the conference was in the inspiring landscape of Greece, where the noblest thoughts of Western man had originated twenty-seven centuries earlier. The sky and the sea were just as blue in 1989 as it must have been in 500 B.C., but instead of tragedies, comedies, and new ways of thinking about our very existence, the conferees discussed modern theory and experiments at the interface between chemistry, physics, and device engineeering. The Gordon conference format (open afternoons) gave senior and junior investigators many opportunities to mingle and think together, away from the hurly-burly of large international gatherings. It is also hoped that the Institute has suggested some new ideas, has created some new collaborations, has inspired all participants, and particularly the younger ones, to new efforts, and has laid the foundations for new and significant scientific discoveries.

This volume of 24 invited and 54 contributed papers appears, delayed by some tardy contributions, but enriched, it is hoped, by a lengthy subject index, which wishes to be a useful "catalogue raisonné" of the research results presented here. In the year that has elapsed since the conference, new results have appeared, one or two "records" have been broken, so the timeliness of the subject is at risk. However, the efforts made by all the contributors to provide a good overview of their respective fields, and the broad focus of the articles, should provide the reader with a useful work of reference.

Here is the place to thank the other Organizers for their help, the NATO Scientific Affairs Division for their support, Mr. George Karamalis of Hotel Spetses for his care and attention to the happiness of the conferees, and all who came to Spetses for their enthusiasm, professionalism, and congeniality.

Robert M. Metzger
Dept. of Chemistry
University of Alabama
Tuscaloosa, AL, USA

CONTENTS

PLENARY LECTURES: CHARGE-TRANSFER COMPOUNDS

DESIGN CONSTRAINTS FOR ORGANIC METALS AND SUPERCONDUCTORS.....1
D. O. Cowan, J. A. Fortkort, and R. M. Metzger

STRUCTURAL PROPERTIES OF MOLECULAR CHARGE-TRANSFER CONDUCTORS AND SEMICONDUCTORS FROM INFRARED AND RAMAN SPECTROSCOPY.....23
R. Bozio, A. Feis, D. Pedron, I. Zanon, and C. Pecile

PLENARY LECTURES: ION-RADICAL SALTS AND SUPERCONDUCTORS

REVIEW OF THE RADICAL-ION SALTS.....43
P. Delhaès

FRONTIERS OF ORGANIC SUPERCONDUCTORS.....67
G. Saito

RECENT DEVELOPMENTS IN ORGANIC SUPERCONDUCTORS.....85
D. Jérome

STRUCTURE-PROPERTY RELATIONSHIPS AS AN AID IN THE RATIONAL DESIGN OF β-$(ET)_2X$ ORGANIC SUPERCONDUCTORS.....91
J. M. Williams, H. H. Wang, A. M. Kini, M. A. Beno, U. Geiser, A. J. Schultz, K. D. Carlson, J. R. Ferraro, and M.-H. Whangbo

TRANSPORT PROPERTIES OF SINGLE CRYSTALS AND POLYCRYSTALLINE PRESSED SAMPLES OF $(BEDT\text{-}TTF)_2X$ SALTS AND RELATED COORDINATION POLYMERS.....97
D. Schweitzer, S. Kahlich, S. Gärtner, E. Gogu, H. Grimm, I. Heinen, T. Kutz, R. Zamboni, H. J. Keller, and G. Renner

SUPERCONDUCTIVITY IN MOLECULAR AND OXIDE LATTICES: A COMPARISON.....115
P. Day

OPTICAL STUDIES OF THE INTERPLAY BETWEEN ELECTRON-LATTICE AND ELECTRON-ELECTRON INTERACTIONS IN ORGANIC CONDUCTORS AND SUPERCONDUCTORS.....129
R. Bozio, M. Meneghetti, D. Pedron and C. Pecile

DESIGN AND SYNTHESIS OF POLYHETEROTETRAHETERAFULVALENES, METAL 1,2-DIHETEROLENES, AND THEIR LOW-DIMENSIONAL CONDUCTING AND SUPERCONDUCTING SALTS.....143
G. C. Papavassiliou

COMMUNICATIONS: ION-RADICAL CONDUCTORS

135 K CRYSTALLOGRAPHIC AND ELECTRONIC STRUCTURE OF $(TMTTF)_2SbF_6$ 163
T. Granier, B. Gallois, A. Fritsch, L. Ducasse, and C. Coulon

ELECTRON PARAMAGNETIC RESONANCE OF ORGANIC CONDUCTORS $(BEDT\text{-}TTF)_2X$ 169
M. Kurmoo, D. R. Talham, and P. Day

ORGANIC METALS FROM CHIRAL BEDT-TTF DONORS. 175
B.-M. Chen , F. Deilacher, M. Hoch, H. J. Keller, P.-J. Wu, S. Gärtner, S. Kahlich, and D. Schweitzer

NEW CONDUCTING SOLIDS BASED ON SOME SYMMETRICAL AND UNSYMMETRICAL π-DONORS. 181
G. C. Mousdis, V. C. Kakoussis, and G. C. Papavassiliou

TTF-DERIVATIVE RADICAL CATION SALTS WITH PLANAR TETRACYANOMETALLATE DIANIONS AND LARGE ACCEPTOR POLYMETALLATES. 185
L. Ouahab, S. Triki, D. Grandjean, M. Bencharif, C. Garrigou-Lagrange, and P. Delhaès

NEW SYNTHETIC METAL PRECURSORS: SUBSTITUTED TETRATHIOTETRACENE AND RELATED COMPOUNDS. 191
T. Maruo, M. Singh, and M. T. Jones

SPECTROSCOPIC INVESTIGATION OF $(2,5\text{-}DM\text{-}DCQNI)_2M$ MATERIALS IN BULK AND THIN FILM FORMS. 197
E. I. Kamitsos, G. D. Chryssikos, and V. Gionis

THE SERIES OF ORGANIC CONDUCTORS: $(PERYLENE)_x[M(mnt)_2]$. 205
V. Gama, R. T. Henriques, and M. Almeida

TRANSPORT AND MAGNETIC PROPERTIES OF $(PERYLENE)_2\ Au(i\text{-}mnt)_2$. 211
M. J. Matos, R. T. Henriques, and L. Alcácer

THE METALLOPORPHYRINS OF GROUP III_B AS PRECURSORS IN LOW-DIMENSIONAL MOLECULAR SOLIDS. 217
A. G. Coutsolelos and D. L. Ward

COMMUNICATIONS: ORGANIC SUPERCONDUCTORS

CRYSTAL STRUCTURES AND PHYSICAL PROPERTIES OF DMET SUPERCONDUCTORS. 223
K. Kikuchi, Y. Ishikawa, Y. Honda, K. Saito, I. Ikemoto, K. Murata, and K. Kobayashi

RADICAL CATION SALTS OF BEDT-TTF WITH METAL-THIOCYANATO ANIONS. 227
H. Müller, C.-P. Heidmann, H. Fuchs, A. Lerf, K. Andres, R. Sieburger, and J. S. Schilling

MAGNETOTRANSPORT IN $(BEDT\text{-}TTF)_2Cu(NCS)_2$: SHUBNIKOV-DE HAAS EFFECT AND UPPER CRITICAL FIELD. 233
C.-P. Heidmann, W. Biberacher, H. Müller, W. Joss, and K. Andres

RAMAN INVESTIGATIONS ON SINGLE CRYSTALS AND POLYCRYSTALLINE PRESSED SAMPLES OF ORGANIC SUPERCONDUCTORS. . 239
R. Zamboni, D. Schweitzer, and H. J. Keller

CALORIMETRIC STUDIES OF THE QUANTIZED MAGNETIC ORDERING IN THE ORGANIC METAL $(TMTSF)_2ClO_4$ 245
F. Pesty and P. Garoche

NONLINEAR ELECTRICAL TRANSPORT EFFECTS IN THE SPIN-DENSITY WAVE STATE OF THE ORGANIC CONDUCTORS $(TMTSF)_2X$ 251
S. Tomić

TWO NEW PHASES OF $(ET)_2I_3$ (β_d' AND λ_d) 257
L. Lu, B-H. Ma, H-M. Duan, S-Y. Lin, D-L. Zhang, X.-H. Wang, Y.-X. Yao, and D.-B. Zhu

COMMUNICATIONS: INORGANIC LOWER-DIMENSIONAL SYSTEMS

ONE-DIMENSIONAL LINEAR-CHAIN COMPLEXES OF PLATINUM, PALLADIUM, AND NICKEL 263
R. J. H. Clark

POLARONS AND SOLITONS IN HALOGEN-BRIDGED PLATINUM COMPLEXES 271
S. Kurita

SLIDING CHARGE-DENSITY WAVES IN INORGANIC CRYSTALS 277
A. Jánossy

SYNTHESIS, STRUCTURAL CHARACTERIZATION, AND CONDUCTIVITY OF THE FIRST LARGE ORGANIC ELECTRON INTERCALATES OF GRAPHITE 285
T. E. Sutto and B. A. Averill

SYNTHESIS, STRUCTURAL CHARACTERIZATION AND SUPERCONDUCTIVITY IN NOVEL NEUTRAL ORGANIC INTERCALATES OF TaS_2 AND TiS_2 289
T. E. Sutto and B. A. Averill

PLENARY LECTURES: CONDUCTING POLYMERS

RECENT PROGRESS IN CONDUCTING POLYMERS: CAN WE EXPECT POLYMERS WITH CONDUCTIVITY GREATER THAN COPPER AND STRENGTH GREATER THAN STEEL? 293
A. J. Heeger

THE POLYANILINES: RECENT ADVANCES IN CHEMISTRY AND PROCESSING 303
A. G. MacDiarmid and A. J. Epstein

ELECTRICAL, OPTICAL AND MAGNETIC PROPERTIES OF POLYANILINE PROCESSED FROM SULFURIC ACID AND IN SOLUTION IN SULFURIC ACID 317
Y. Cao, P. Smith, and A. J. Heeger

POLYANILINE: NEW PHYSICS IN AN OLD POLYMER 335
A. J. Epstein and A. G. MacDiarmid

VIBRATIONAL MOLECULAR SPECTROSCOPY OF CONDUCTING POLYMERS: A GUIDED TOUR 345
G. Zerbi

COMMUNICATIONS: CONDUCTING POLYMERS

ANELLATED DIPYRROLYL SYSTEMS: SYNTHESIS AND ELECTRICAL PROPERTIES OF THE POLYMERS 359
A. Berlin, S. Martina, G. Pagani, G. Schiavon, and G. Zotti

CONDUCTING POLYMERS FROM 3, 4-CYCLOALKYLTHIOPHENES. 363
A. Berlin, J. Rühe, and G. Wegner

ORIENTATION OF STRETCHED POLY(3-OCTYLTHIOPHENE) FILMS: VISIBLE AND INFRARED DICHROISM STUDIES. 369
G. Gustafsson and O. Inganäs

VIBRATIONAL PROPERTIES OF CONDUCTING POLYMERS WITH AROMATIC OR HETEROAROMATIC RINGS. 375
J.-P. Buisson, J. Y. Mevellec, S. Zeraoui, and S. Lefrant

STRUCTURAL INVESTIGATION AND ELECTRICAL PROPERTIES OF POLY(P-PHENYLENE) PREPARED IN TWO DIFFERENT WAYS. 381
H. Krikor, R. Mertens, P. Nagels, R. Callaerts, and G. Remaut

RAMAN CROSS-SECTIONS OF HIGHLY ORIENTED CIS POLYACETYLENE. 387
G. Lanzani, A. Piaggi, A. Borghesi, and G. Dellepiane

CHARACTERISATION OF POLY(P-PHENYLENE VINYLENE) [PPV] PREPARED BY DIFFERENT PRECURSOR ROUTES. 393
J. Martens, N. F. Colaneri, P. Burn, D. D. C. Bradley, E. A. Marseglia, and R. H. Friend

ELECTROCHEMICAL DOPING PROCESSES OF CONDUCTING POLYMERS STUDIED WITH IN SITU FTIR SPECTROSCOPY. 401
H. Neugebauer and N. S. Sariciftci

EVOLUTION OF ELECTRIC PERMITTIVITIES OF pTS AND pFBS DIACETYLENES DURING SOLID-STATE POLYMERIZATION. 407
M. Orczyk and J. Sworakowski

STRUCTURAL CHARACTERIZATION OF THIOPHENE-BASED MONOMERS, POLYMERS, AND OLIGOMERS, AS POWDERS, THICK AND THIN FILMS. 411
A. Bolognesi, M. Catellani, S. Destri, W. Porzio, C. Taliani, R. Zamboni, and S. Brückner

TRANSIENT PHOTOCONDUCTIVITY IN ORIENTED CONJUGATED POLYMERS. .415
J. Reichenbach, H. Bleier, Y. Q. Shen, and S. Roth

ENCAPSULATION OF CONDUCTING POLYMERS WITHIN ZEOLITES. 421
P. Enzel and T. Bein

LOW-DIMENSIONAL ELECTRICALLY CONDUCTIVE SYSTEMS. INTERCALATED POLYMERS IN V_2O_5 XEROGELS. 427
C. G. Wu, M. C. Kanatzidis, H. O. Marcy, D. C. DeGroot, and C. R. Kannewurf

GAMMA IRRADIATION OF POLY(PYRROLE)-COATED Pt ELECTRODES: THE EFFECT ON THE ELECTROCHEMICAL BEHAVIOR. 435
M. Arca, E. Arca, O. Güven, and A. Yildiz

COMMUNICATIONS: THEORY

BEYOND THE HUBBARD MODEL: SCREENED INTERACTIONS IN 1 D. 441
A. Painelli and A. Girlando

THEORETICAL CHARACTERIZATION OF THE ELECTRONIC STRUCTURE OF POLY(HETEROAROMATIC VINYLENES). 447
E. Ortí, M. C. Piqueras, R. Crespo, and F. Tomás

THEORETICAL DESIGN OF ORGANIC METALS BASED ON THE PHTHALOCYANINE MACROCYCLE. 455
E. Ortí, M. C. Piqueras, and R. Crespo

ELECTRICALLY CONDUCTIVE PHTHALOCYANINE ASSEMBLIES. STRUCTURAL AND NON-INTEGER OXIDATION NUMBER CONSIDERATIONS. . 461
F. Torrens, E. Ortí, and J. Sánchez-Marín

VALENCE-BOND TREATMENT OF 3/4-FILLED DIMERIZED CHAINS - EXTENDED HUBBARD RESULTS. 467
A. Fritsch and L. Ducasse

PLENARY LECTURES: LANGMUIR-BLODGETT FILMS

POSSIBLE NICHES FOR LANGMUIR BLODGETT FILMS. 473
G. G. Roberts

INVESTIGATIONS OF THE MICROSTRUCTURE OF LIPID INTERFACE FILMS. . . 491
M. Lösche

CHARGE-TRANSFER CONDUCTING LANGMUIR-BLODGETT FILMS. 503
M. Vandevyver

SOLID-STATE MOLECULAR ENGINEERING IN LANGMUIR-BLODGETT FILMS. .511
A. Ruaudel-Teixier

COMMUNICATIONS: LANGMUIR-BLODGETT FILMS

DEVELOPMENT OF NOVEL CONDUCTIVE LANGMUIR-BLODGETT FILMS: METALLIC PROPERTIES AND PHOTOCHEMICAL SWITCHING PHENOMENA. . . 519
T. Nakamura, H. Tachibana, M. Matsumoto, M. Tanaka, and Y. Kawabata

THE QUEST FOR HIGHLY CONDUCTING L-B FILMS. 527
K. Lerstrup, J. Larsen, P. Frederiksen, and K. Bechgaard

LANGMUIR-BLODGETT FILMS FROM DONOR-ACCEPTOR SUBSTITUTED POLYENES. 531
S. Hagen, H. Schier, S. Roth, and M. Hanack

LANGMUIR-BLODGETT FILMS OF A PYRROLE AND FERROCENE MIXED SURFACTANT SYSTEM. 537
L. Samuelson, A. K. M. Rahman, S. Clough, S. Tripathy, P. D. Hale, T. Inagaki, T. A. Skotheim, and Y. Okamoto

MEASUREMENT TECHNIQUES IN PYROELECTRIC LANGMUIR-BLODGETT FILMS. 549
M. W. Poulter, R. Colbrook, and G. G. Roberts

EFFECT OF LIGHT INTENSITY AND TEMPERATURE ON THE PHOTOVOLTAIC PARAMETERS OF CHLOROPHYLL b LANGMUIR-BLODGETT FILMS 557
A. Désormeaux and R. M. Leblanc

PLENARY LECTURES: NON-LINEAR OPTICS

PHOTONICS AND NON-LINEAR OPTICS - MATERIALS AND DEVICES. 563
P. N. Prasad

LANGMUIR-BLODGETT FILMS FOR NON-LINEAR OPTICS. 573
P. N. Prasad

ALL-OPTICAL SWITCHING USING OPTICAL FIBERS AND NON-LINEAR ORGANIC LIQUIDS. 579
L. Domash, P. Levin, J. Ahn, J. Kumar, and S. Tripathy

OPTICAL STUDIES OF AMPHIPHILIC MOLECULES WITH INTERESTING ELECTRO-OPTICAL AND NON-LINEAR OPTICAL PROPERTIES. 591
M. Lösche and G. Decher

COMMUNICATION: NON-LINEAR OPTICS

THE USE OF THE SURFACE PLASMON RESONANCE TECHNIQUE IN NON-LINEAR OPTICS. 605
S. J. Cooke and G. G.Roberts

PLENARY LECTURE: MOLECULAR DEVICES

REVIEW OF THE ORGANIC RECTIFIER PROJECT: LANGMUIR-BLODGETT FILMS OF DONOR-SIGMA-ACCEPTOR MOLECULES. 611
R. M. Metzger and C. A. Panetta

COMMUNICATIONS: MOLECULAR DEVICES

SOLID-STATE MICROELECTROCHEMICAL DEVICES: TRANSISTOR AND DIODE DEVICES EMPLOYING A SOLID POLYMER ELECTROLYTE. 627
D. R. Talham, R. M. Crooks, V. Cammarata, N. Leventis, M. O. Schloh, and M. S. Wrighton

POLYMER FIELD-EFFECT TRANSISTORS FOR TRANSPORT PROPERTY STUDIES. 635
J. Paloheimo, E. Punkka, H. Stubb, and P. Kuivalainen

MOLECULAR LINES. 643
L. L. Miller

MULTIFREQUENCY PHOTOCHROMIC MEMORY MATERIALS. 647
G. J. Ashwell, M. Szablewski, and A. P. Kuczynski

FUNCTIONALIZED CONDUCTING POLYMERS TOWARD MOLECULAR DEVICES. 653
T. Shimidzu

PROSPECTS FOR TRULY UNIMOLECULAR DEVICES. 659
R. M. Metzger

PARTICIPANTS 667

AUTHOR INDEX. 671

SUBJECT INDEX. 675

DESIGN CONSTRAINTS FOR ORGANIC METALS AND SUPERCONDUCTORS

Dwaine O. Cowan* and John A. Fortkort

Department of Chemistry
The Johns Hopkins University
Baltimore, Maryland 21218, USA

Robert M. Metzger

Department of Chemistry
The University of Alabama
Tuscaloosa, Alabama 35487-0336, USA

INTRODUCTION

Organic solids can now be made with only non-metallic elements (carbon, hydrogen, nitrogen, sulfur, etc.), which nevertheless can be metallic, or even superconducting. From the progress made since the synthesis of the first organic metal TTF-TCNQ in 1972-73 [1,2], design rules have emerged for new and technologically more interesting materials. These exotic compounds present synthetic and structure/property correlation challenges to the organic chemist, and have provided a veritable bonanza of interesting effects for the physicist (charge and spin density waves, Peierls transitions, Kohn anomalies, Mott-Hubbard insulators, magnetic field-induced Fermi surface instabilities, quantum Hall effects, and so on).

While many excellent reviews on the physics [3-15] and the chemistry [16-23] of organic metals and superconductors have been published, we will summarize here our own guidelines for the design of new organic conductors; these have evolved over the years [24-29]. We present twelve interlocking criteria for success, divided into: (a) molecular properties, (b) crystal structures and their stabilities, (c) molecular properties due to crystal packing, (d) other resultant solid-state properties. But before we do this, we review briefly certain concepts (band theory, the Hubbard Hamiltonian, and Mulliken charge-transfer theory) that are applicable to organic metals.

BAND THEORY

When two atoms are brought together to form a molecule, there is a concomitant mixing of atomic states to form molecular orbitals. Likewise, when several identical molecules are brought together to form a crystalline solid, the discrete and identical molecular energy levels are split by the solid-state interactions, first into 2 levels spaced 2 t apart (for a "dimer" of 2 molecules), then further (for a large number of molecules) into a series of energy levels so closely spaced, that they form a quasi-continuous band of width 4 t (Fig. 1a). The topmost filled energy level in this band is called the Fermi level, and has the corresponding Fermi energy E_F. Here the resonance or transfer integral t is a measure of the intermolecular interaction between neighboring molecules, and is very similar to the Mulliken charge-transfer integral, or the Hückel resonance integral β. One simple measure of t is to use Mulliken's approximation, and to assume that t is proportional to the intermolecular overlap integral S: $t = k\ S$. For typical organic solids (which are called narrow-band solids), t is of the order of a few hundred millielectron volts.

Lower-Dimensional Systems and Molecular Electronics
Edited by R.M. Metzger *et al.*, Plenum Press, New York, 1991

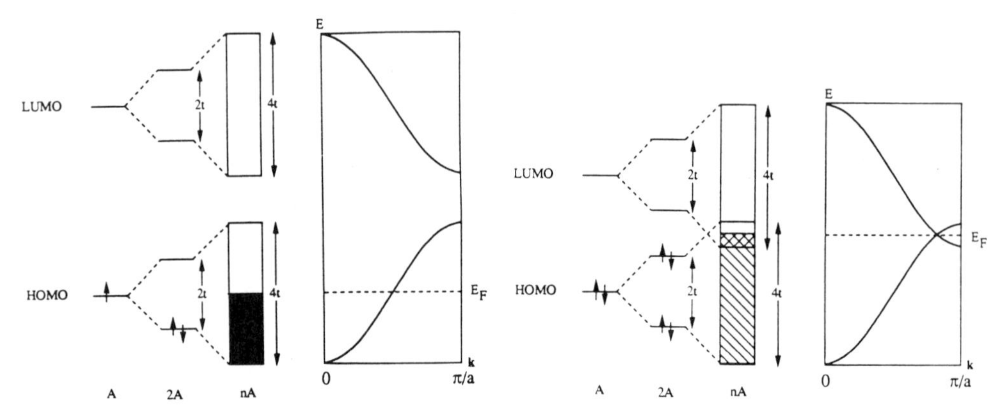

Fig. 1a. Partially filled bands formed via radicals

Fig. 1b. Partially filled bands formed via overlap

In considering the electronic properties of crystalline solids, one need only be concerned with the energy states near the Fermi level, since these are the states that affect the physical properties of the solid. In the organic charge-transfer salts we will be discussing, these states are derived from the Highest Occupied Molecular Orbitals (HOMOs) of the donor molecules, and Lowest Unoccupied Molecular Orbitals (LUMOs) of the acceptor molecules.

In solids the periodicity of the crystal lattice causes the appearance of a new set of quantum numbers, usually denoted by the wavevector **k** (units of 1/cm), which ranges from $\mathbf{k} = -\mathbf{k}_{max}$ to $\mathbf{k} = \mathbf{k}_{max}$ (in a one-dimensional solid of lattice parameter, or repeat distance a, kmax= π/a). In the free-electron approximation the energy as a function of k_x, k_y, k_z in one, two, and three dimensions is given by:

$$E_{\mathbf{k}} = h^2k_x^2/8\pi^2m^*, \quad E_{\mathbf{k}} = h^2(k_x^2 + k_y^2)/8\pi^2m^*, \quad E_{\mathbf{k}} = h^2(k_x^2 + k_y^2 + k_z^2)/8\pi^2m^*$$

respectively. On the right of Fig. 1a is shown the dispersion curve (from $\mathbf{k} = 0$ to $\mathbf{k} = \pi/a$), which depicts the allowed energy levels $E_{\mathbf{k}}$, filled up to $E_{\mathbf{k}} = E_F$ (and up to the Fermi wavevector $\mathbf{k} = \mathbf{k}_F$); the wavefunction at the "band origin" ($\mathbf{k} = 0$) is the all-even combination of HOMO wavefunctions, with coefficients of like sign from all contributing molecules; the wavefunction at the band edge ($\mathbf{k} = \pi/a$) will have coefficients with opposing signs from nearest-neighboring molecules. If the molecular wavefunctions are σ-type (symmetric) then the energy at $\mathbf{k} = 0$ will be minimum, and the energy at $\mathbf{k} = \pi/a$ will be a maximum (as shown for the HOMO-like band on the lower right of Fig. 1a). If the molecular wavefunctions are atomic π-like, then the band will be inverted, with the maximum energy at $\mathbf{k} = 0$ and the minimum energy at $\mathbf{k} = \pi/a$ (this situation is shown for the LUMO-derived band on the top right of Fig. 1a).

If the HOMO-derived band is only partially filled (i.e., there are more energy states in this band than electrons), then, due to the density of states, electrons are easily excited to higher energy levels, and metallic conductivity results. This is the case for certain charge-transfer salts, where the presence of open-shell cations, and/or anions, gives rise to partially filled bands. In theory, such partially filled bands may also be achieved by HOMO-LUMO overlap, that is, by decreasing the band gap between the HOMO and the LUMO, and by increasing the bandwidth (Fig. 1b). To the best of our knowledge, this second method of generating partially filled bands has not been achieved, perhaps because the HOMO-LUMO band gap is often several electron volts, while the bandwidths are frequently of the order of a few hundred millielectron volts.

If the uppermost band is completely full, then, for conduction to occur, electrons must be excited across the band gap, from what is called the valence band (built from the HOMOs) and into the conduction band (built from the LUMOs). Depending on the magnitude of the band gap, materials of this type exhibit either semiconducting or insulating properties.

The electrical conductivity of a solid σ (S/cm or Ω^{-1} cm^{-1}) is given by:

$$\sigma = N e \mu,$$

where N (cm^{-3}) is the concentration of charge carriers, e (Coulombs) is the electronic charge, and μ ($cm^2 V^{-1} s^{-1}$) is the mobility. It can be shown that, for narrow-band metals (such as organic metals), the conductivity is proportional to the square of the bandwidth:

$$\sigma \propto (4\,t)^2 .$$

The temperature dependence of the conductivity can differentiate between metals (for which $\sigma \uparrow$ as $T \downarrow$) and semiconductors (for which $\sigma \downarrow$ as $T \downarrow$). We next discuss a very useful way of looking at the physics of infinite one-dimensional chains of atoms or molecules, namely the Hubbard Hamiltonian.

HUBBARD HAMILTONIAN

In 1963 Hubbard [30-32] presented a Hamiltonian very useful in discussing the physics of one-dimensional chains:

$$H = U \Sigma_i\, n_{i,\sigma} n_{i,-\sigma} \;+ t\; \Sigma_{i,\sigma}\, (c_{i,\sigma}{}^{+} c_{i+1,\sigma} + c_{i+1,\sigma}{}^{+} c_{i,\sigma})$$

where the operator $c_{i,\sigma}{}^{+}$ creates a particle of spin σ (up or down) at site i, and $c_{i,\sigma}$ destroys a particle of spin σ (up or down) at site i; the operator $n_{i,\sigma}$ is a number operator defined by:

$$n_{i,\sigma} = c_{i,\sigma}{}^{+} c_{i,\sigma}$$

The important energies in this Hamiltonian are the <u>on-site energy U</u> (akin, in a very simple way, to the Hückel on-site integral α in simple Hückel molecular orbital theory) and the <u>off-site energy t</u> (same integral as in the bandwidth 4 t discussed above) These on-site (U) and off-site (t) energies are often used in discussions of organic metals.

MULLIKEN CHARGE-TRANSFER THEORY AND DA CRYSTALS

In 1952 Mulliken pointed out that, for traditional weak molecular complexes in solution, the ground and excited state wavefunctions are described by:

$$\Psi_G(D,A) = a\ \Phi_0\ (D\,.\,A\,) + \ b\ \Phi_{CT}(D^+A^-)$$
$$\Psi_E(D,A) = a\ \Phi_{CT}(\,D^+A^-) \ - \ b\ \Phi_0\ (D\,.\,A\,)$$

where a << b [33-35]. Thus, these "Charge-Transfer" (CT) complexes can retain the optical spectra of the neutral component molecules (D and A), yet exhibit a strongly allowed broad new transition (the CT band, for the transition $\Psi_G(D,A)$ <--> $\Psi_E(D,A)$, with energy $h\nu_{CT}$). The CT band energy $h\nu_{CT}$ is, to first order, approximated by the cost of ionizing the D,A pair, minus the Coulomb attraction energy $H = \Sigma_i \Sigma_j\, z_i\, z_j / (4\, \pi\, \varepsilon_o\, r_{ij})$ between the charges z_i and z_j on the D^+ and A^- ions, respectively, at the equilibrium internuclear distance r_{ij}. We will discuss the solid-state extensions of Mulliken's theory later.

Having reviewed some old theory, we now consider the strictly molecular properties desired for organic metals.

MOLECULAR PROPERTIES

<u>Criterion One: Stable open-shell (Free-radical) species</u>

The first property to be considered for any candidate molecule is the relative location of the HOMO and the LUMO. We are interested in highly conductive systems; this means that the"parent" neutral molecule, and the "daughter" anion radical or cation radical must be both thermodynamically and kinetically stable, and accessible (not too high in energy). Many relatively stable organic free radicals are known. There are, at present, four methods useful for stabilizing these species:

(1) One method is to use the steric hindrance of substituents. This prevents the approach of other radicals, and therefore prevents further reaction. DPPH (diphenylpicrylhydrazyl), nitroxide spin labels, or Yang's radical [36,37] are such species. However, because of the same steric hindrance, these molecules cannot stack well in the solid state, and therefore cannot interact well with their neighbors.

(2) A second method is to force the high spin density onto heteroatoms. This method is used for most of the known organic metals. This is shown in Fig. 2 for TTF, where the native system is, in organic chemists's parlance, a "7 π - 7 π" system; oxidation to the radical cation produces one resonance-stabilized 6 π electron structure; oxidation to the dication produces a 6 π - 6 π system. For TTF^+ more than 50% of the spin resides on the S atoms, while on $TCNQ^-$ only 25% of the spin density resides on the N atoms [38-40].

(3) A third method would be to to use a push-pull substituent effect [41] or the capto-dative or merostabilization method. An example of this is Kosower's radical [42] (1-ethyl-4-carbomethoxypyridinyl). This method has not yet been used to form organic metals.

C_2H_5—N•• •—$COOCH_3$

Kosower's radical

Phenalenium radical

(4) The fourth approach is to use derivatives of odd-alternant hydrocarbons, such as the phenalenium radical, in which the Hückel resonance energy is the same for the anion, the radical, and the cation, because the HOMO is a non-bonding MO [43,44]. However, the radicals seem too reactive.

Criterion Two: Reduce the HOMO-LUMO energy gap and increase the bandwidths

The goal to make good stable radicals can be rephrased in terms of band theory, and in terms of Fig. 1a above, in that the HOMO-LUMO energy gap should be reasonably small, and

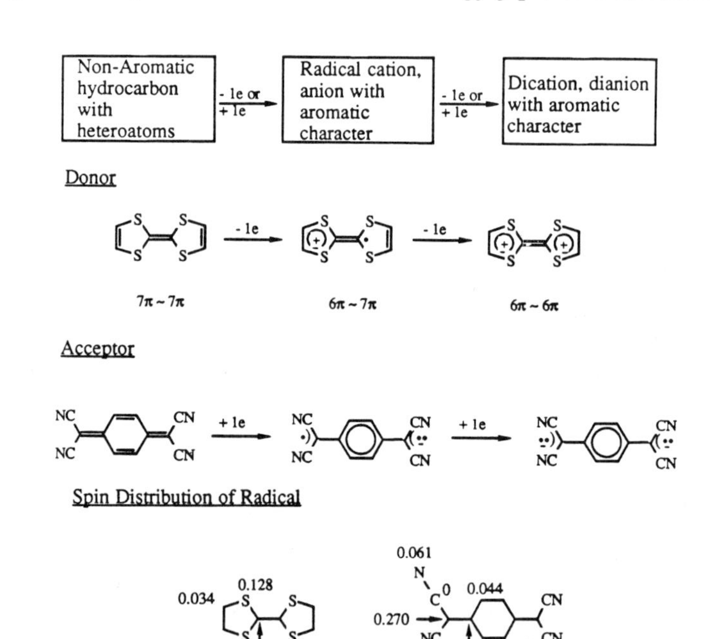

Fig. 2. Formation of stable radicals.

that, within the limits of reason, the metallic band be made as wide as possible. One can argue that the effective on-site repulsion U_{eff} should be made smaller than 4 t (obvious if you look at the black shaded region of Fig. 1a). Fig.1b shows what would happen if the HOMO and LUMO bands could be made to merge (as happens for the inorganic semimetal bismuth); this situation has not yet been encountered in organic metals, however.

Criterion Three: Use planar molecules with delocalized π molecular orbitals

Fig. 3 gives the structures of (i) six classes of good one-electron donors [TTF and friends; HMTTF and friends; TTT and friends; BEDT-TTF; perylene, and MPc (metal phthalocyanines)] and (ii) three classes of good one-electron acceptors (TCNQ, TNAP, and DCNQI). All these donors and acceptors are flat, or almost perfectly flat, molecules: this requirement seems almost self-evident if one wants to form a tight crystal lattice with good intermolecular overlap. It is conceivable that this requirement could be relaxed, but a good example of this has not been found. It is known that there are small non-planarities (e.g. in BEDT-TTF) but it is not clear whether this is critical for the solid-state properties.

Criterion Four: Cation and/or anion radical should be nominally divalent

It turns out also (for reasons explained better below) that, for electron donor molecules, the three states, derived from the emptying of the HOMO, must be kinetically and thermodynamically stable and accessible: neutral (charge $Z = 0$), cation radical ($Z = 1$) and dication ($Z = 2$) . Similarly, for electron acceptors, the three states derived from the filling of the LUMO must be stable and thermodynamically accessible: neutral (charge $Z = 0$), anion radical ($Z = -1$) and dianion ($Z = -2$).

In solution, one often considers the disproportionation reactions:

$$2\ \mathrm{TCNQ^{-}(soln)} \rightleftharpoons \mathrm{TCNQ^{0}(soln)} + \mathrm{TCNQ^{--}(soln)}$$

$$2\ \mathrm{TTF^{+}(soln)} \rightleftharpoons \mathrm{TTF^{0}\ (soln)} + \mathrm{TTF^{++}\ (soln)}$$

In a similar way, if one wants the electrons in a solid to hop from $TCNQ^0$ to $TCNQ^-$ to $TCNQ^0$, say, the intermediate state $TCNQ^{--}$ must be relatively easy to form. If these states are not accessible, then the motion of metallic electrons can only occur in a correlated, "Fröhlich" mode.

Criterion Five: The molecular polarizability can be important in reducing U

The molecular polarizability can be enhanced, e.g. by replacing the heteroatom S by Se or Te [45-58]. This should also help reduce the effective on-site energy U_{eff} [15,59,60], but often reduces the solubility of the chalcogen-containing molecule, a consequence which may hinder crystal growth. For example, the tellurium donors mentioned in Fig. 2 suffer from very low solubility, and the study of their complexes is still in progress.

Before we discuss the remaining design criteria, we introduce the known crystal structure types, then examine their thermodynamic stability, then discuss the Peierls instability, and finally, consider some appropriate theoretical models.

CRYSTAL PROPERTIES: STRUCTURE

There is a wealth of crystal lattice types of interest in this review. We can differentiate between solid CT complexes and ion-radical salts. The former are the solid-state equivalents of the Mulliken solution CT complexes, and are the "two-chain" crystals. The latter are the so-called "one-chain" compounds, or ion-radical salts, where the organic cation (anion) crystallizes with inorganic anions (cations) as counterions. The crystal structure types have been reviewed by Herbstein [61], Tanaka [62], Soos [63-65] and many others. Tables 1 and 2 update a classification introduced by Soos [63] and modified later by Wiygul *et al.* [66]. A few examples are shown diagrammatically in Fig. 4. The 1S lattices are the ion-radical salts; the 1M lattices are the CT crystals, the crystal equivalents of the solution CT complexes. The 2S lattices are the first organic metals found (TTF-TCNQ).

Tetrathiafulvalene (TTF): R= H, X= S
Tetraselenafulvalene (TSF): R= H, X= Se
Tetratellurafulvalene (TTeF): R= H, X= Te
Tetramethyltetrathiafulvalene (TMTTF): R= CH_3, X= S
Tetramethyltetraselenafulvalene (TMTSF): R= CH_3, X= Se

Hexamethylenetetrathiafulvalene (HMTTF): X= S
Hexamethylenetetraselenafulvalene (HMTSF): X= Se
Hexamethylenetetratellurafulvalene (HMTTeF): X= Te

Tetrathiatetracene (TTT): X= S
Tetraselenatetracene (TST): X= Se
Tetratelluratetracene (TTeT): X= Te

Bis(ethylenedithiolo)tetrathiafulvalene (BEDT-TTF)

Tetracyanoquinodimethane (TCNQ): R= H
Dimethyltetracyanoquinodimethane (DMTCNQ): R= CH_3

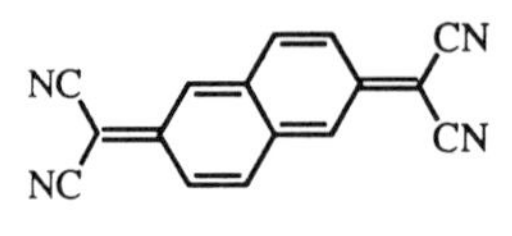

Tetracyanonaphthoquinodimethane (TNAP)

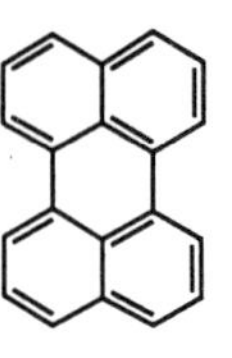

Perylene

N,N-Dicyanoquinonediimine (DCNQI)

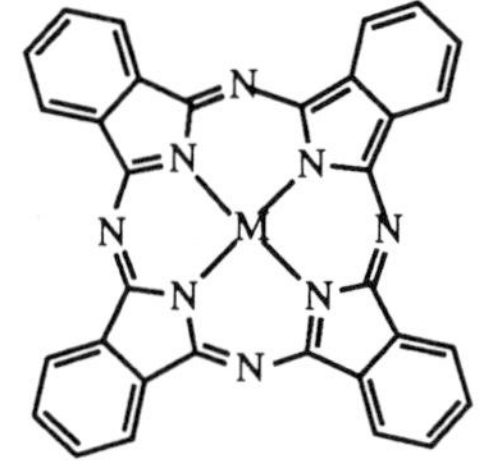

Metal Phthalocyanine [M(pc)]

Fig. 3. Good organic one-electron donors and acceptors: precursors to organic metals.

Table 1. The nABCD classification for organic donor-acceptor lattices (extended from [66]).

n = <u>Number of stacks</u>
- 1 = one-stack ion-radical salt
- 2 = two-stack CT crystal

A = <u>Stacking mode</u>
- S = segregated stacks
- M = mixed stacks
- L_2 = lone dimers

B = <u>Stoichiometry</u>
- U = unary (simple) 1:1 ratio donor:acceptor
- C = complex 1:n or n:1 ratio donor:acceptor
- B = Berthollitic, non-integer ratio

C = <u>Intermolecular distances</u>
- R = regular
- A_2 = alternating "diads"
- A_3 = alternating "triads"
- A_4 = alternating "tetrads"

D = <u>Ionicity, or valency</u>
- N = almost neutral
- P = partially ionic, mixed-valent
- D = partially ionic, discrete valencies
- I = almost ionic

CRYSTAL STABILITY - THERMODYNAMICS

A recurring argument is whether the highly conducting or superconducting crystals formed are kinetically or thermodynamically stable [67]. This was settled for TTF-TCNQ in favor of thermodynamic stability [68]. Wegner suggested that the kinetics of crystallization is just as important as the thermodynamic considerations [69]. For compounds which have several crystal polymorphs, belonging to different space groups and with different packings, as occurs often for $(BEDT\text{-}TTF)_2X$ salts, kinetic stability must, obviously, be very important (and the experimental lattice energy differences between the various forms must be very small indeed).

At this point it should be mentioned that at Hopkins we have been faced with the eternal question of who should grow single crystals: the organic bench chemist who made the molecules, or the physical chemist who needs the crystals. Our solution has been to make both sets of students and co-workers try their hand at growing crystals. After all, it is an art, and doubling the effort simply raises the chances for success.

CRYSTAL INSTABILITY: PEIERLS TRANSITION, CHARGE-DENSITY WAVES AND SPIN DENSITY WAVES

One-dimensional chains of equal objects are thermodynamically unstable to a static or dynamic distortion. In 1955 Peierls introduced the theorem [70], now named after him, which states that such a chain will distort spontaneously and dynamically to minimize the energy; the distortion will become static (i.e. permanent), rather than dynamic, if there is enough "off-chain coupling" with the lattice phonons. This Peierls theorem can be thought of as the

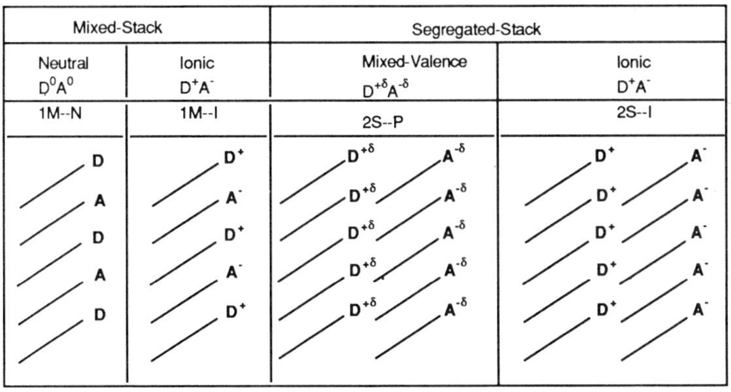

Mixed-Stack		Segregated-Stack	
Neutral D^0A^0	Ionic D^+A^-	Mixed-Valence $D^{+\delta}A^{-\delta}$	Ionic D^+A^-
1M--N	1M--I	2S--P	2S--I
D A D A D	D^+ A^- D^+ A^- D^+	$D^{+\delta}$ $A^{-\delta}$ $D^{+\delta}$ $A^{-\delta}$ $D^{+\delta}$ $A^{-\delta}$ $D^{+\delta}$ $A^{-\delta}$ $D^{+\delta}$ $A^{-\delta}$	D^+ A^- D^+ A^- D^+ A^- D^+ A^- D^+ A^-

Fig. 4. Schematic diagram of intermolecular overlap in mixed-stack (1M-N, 1MU-I) and segregated-stack (2SU-P and 2SU-I) structures.

Table 2. Examples of 1-chain and 2-chain compounds, with nABCD classification, ordered by degree of charge transfer and lattice type (adapted and extended from Ref. [66]).

Ionicity --> Cond.--> Magn.-->	N Insulators Diamagnetic	I Mott-Hubbard Insulators Paramagnetic (or Diamagnetic)	D Insulators Paramagnetic	P Metallic Conductors Paramagnetic
Chains: Segreg.	**1SURN** Anthracene	**1SURI** $TMPD^+ClO_4^-$ (above 186K) K^+ $TCNQ^-$ (above 400K) HMTTF-$TCNQF_4$	**$1SCA_3D$** Cs_2TCNQ_3	**$1SCA_2P$** $(TMTSF)_2PF_6^-$ (T_c(superc.)=1.5 K) β-$(BEDT\text{-}TTF)_2I_3$ (T_c(superc.)=8 K)
		$1SUA_2I$ $TMPD^+$ ClO_4^- (below 186K) K^+ $TCNQ^-$ (below 400K) TTF-$TCNQF_2$	**$1SCA_2D$** Rb^+TCNQ^-	**2SURP** TTF-TCNQ (above 58 K) TMTSF-TCNQ (black form) NMP-TCNQ TSF-TCNQ (above 40 K) HMTSF-TNAP TMTTF-TCNQ HMTTF-TCNQ
		2SURI HMTSF-$TCNQF_4$ HMTTF-$TCNQF_4$		
	2SURN none known	**2SBRI** $(TMTTF)_{1.3}(TCNQ)_2$		
Mixed	**1MURN** Anthr-TCNQ TTF-Chloranil (high T or low P) TMTSF-TCNQ (red form)	**1MURI** TMPD-TCNQ TTF-Chloranil (low T or high P) DBTTF-$TCNQF_2$ DBTTF-TCNQ		**2MURP** none known
Lone Dimers		**$1L_2UA_2I$** NBP-TCNQ		**$1L_2CA_2P$** δ-$(BEDT\text{-}TTF)_2I_3$ (no supercond.) κ-$(BEDT\text{-}TTF)_2Cu(NCS)_2$ (T_c(superc.)=11.4 K)

solid-state equivalent of the Jahn-Teller theorem [71]. (In a Jahn-Teller distortion a molecule with an incompletely filled degenerate HOMO is susceptible to a distortion that decreases the symmetry (point group) of the molecule and thereby removes the degeneracy). In a Peierls distortion, the lattice distortion costs some energy, which must be overcome by the energy gained by the reordering of states near E_F.

We show, in Fig. 5a, a one-dimensional chain of 4 electrons on 4 sites (atoms or molecules) spaced an equal distance (lattice period) b apart. At high temperature, if the on-site Coulomb repulsion U is small to medium, i.e. smaller than the bandwidth ($U<4t$), then the electrons are paired on two of the 4 sites (1/2-filled band), and the k-space states are occupied up to the Fermi wavevector $\pm k_F = \pm\pi/2b$, and Bragg or Umklapp scattering occurs between the $-k_F$ state and the $+k_F$ state (a difference of $2k_F$). Since the energy levels are only filled partially, up to the Fermi energy E_F, while the band extends out to π/b, there is metallic conduction.

As the temperature is lowered below the metal-to-insulator transition temperature T_{MI}, the lattice (thanks to the three-dimensionality of the lattice phonons) undergoes a static Peierls distortion to a dimerized lattice (the translation period becomes 2b) , and a gap E_g opens up in the energy level diagram (Fig. 5b). The valence band is now full, and the conduction band is empty: we now have a Peierls insulator or semiconductor, with an activation energy E_g for electrical conduction. Diffuse X-ray scattering "streaks" can be observed at $2k_F$, even in the high-temperature region (Fig. 5a), before the lattice dimerization of Fig. 4b occurs, because of dynamic "precursor" distortion charge-density waves (CDW), which become pinned (as static distortions) at or below T_{MI}: the $2k_F$ spots become narrow, but weak, "superlattice reflections".

If the on-site repulsion is large (U>4t), then two electrons cannot be accommodated on the same site, and the band fills up to $\pm\pi/b$ (Fig. 5c). The magnetic lattice period is twice the normal period, and one has a Mott-Hubbard semiconductor, in which Bragg or Umklapp scattering occurs at $4k_F$.

Another, lower-energy type of Peierls transition is possible for magnetic systems (Wurster's Blue Perchlorate, or alkali TCNQ salts). Then one talks about "spin-Peierls" transitions, whose X-ray signature is a $2k_F$ SDW (spin density wave) which, again, is dynamic and diffuse above the ordering temperature, and static (or "pinned") and sharpened below the transition temperature [72].

One can experimentally derive the degree of charge transfer (Z) [73-75] between donor and acceptor from the location of the superlattice X-ray spots (diffuse or sharp). Let the superlattice periodicity due to $2k_F$ or $4k_F$ scattering be λ, and let d be the regular (or high-temperature static) lattice spacing. Then it can be shown that:

$$\lambda / d = 2 / (j\, Z)$$

where j = 1 for $2k_F$ scattering and j = 2 for $4k_F$ scattering. We shall return later to comment on the difficulties created by the Peierls transition.

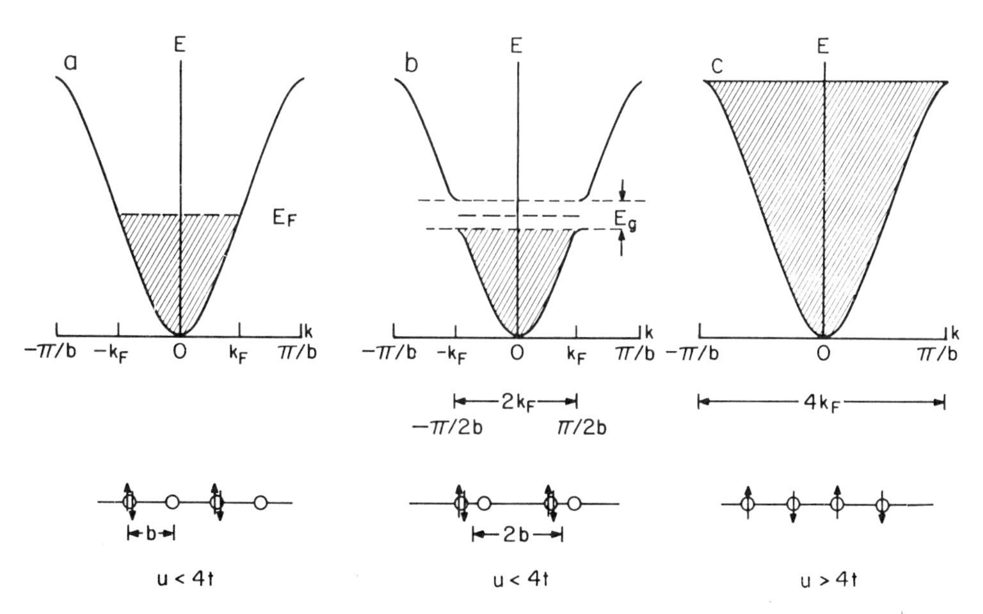

Fig. 5. Band structure for full charge transfer (four electrons on four sites: Z=1): (a) metal (b) Peierls semiconductor (c) Mott-Hubbard semiconductor

THEORIES FOR CHARGE-TRANSFER CRYSTALS AND ION-RADICAL SALTS

Most organic CT complexes crystallize such that the donor and the acceptor molecules alternate in the same stack (mixed stack or DA stack: the 1M--- lattice). However, in the 1950's and 1960's spectroscopy indicated that some crystals have an almost neutral (charge transfer $Z \approx 0$) ground state, while other CT crystals have an almost ionic ($Z \approx 1$) ground state. In 1965 McConnell, Hoffman, and Metzger [76] explained this by considering the competition between the cost of ionizing the donor-acceptor pair (ionization potential of the donor I_D minus the electron affinity A_A of the acceptor) and the electrostatic interaction (Madelung energy, E_M) of the ions once formed in the solid. The boundary between the neutral ($Z = 0$) and the ionic ($Z = 1$) ground states for the charge-transfer (CT) solids is discontinuous, and occurs when $I_D - A_A = E_M$. This theory did not provide an energy minimum for partial ionicity, or partial CT ($0 < Z < 1$) [because the energy is $Z(I_D-A_A) + Z^2E_M$ and $I_D-A_A > 0$ while $E_M < 0$], and thus could not explain the existence of partial ionicity in segregated-stack (2S) lattices such as TTF-TCNQ [77,78].

The addition of an overlap term by Soos [79] moved the boundary slightly, to a region between the almost neutral and the almost ionic states. Bloch resolved the dilemma of stabilizing the partially ionic state, by starting from a metallic or delocalized model, and obtaining several new terms in the cohesive energy equation [80]. In addition to the ionization energy $Z(I_D-A_A)$ and to the Madelung energy Z^2E_M, Bloch added three terms: one is $Z(Z-1)\langle U\rangle$, where $\langle U\rangle$ is the average value of the unscreened on-site Coulomb interaction U; the other two are Z-dependent contributions to the van der Waals energy. In partially published results for TTF-TCNQ, Bloch's estimates resolved the dilemma of stabilizing the partially ionic state, and matched the experimental cohesive energy of TTF-TCNQ, 2.4 eV [68].

Torrance [81] sought, and found, a neutral-ionic interface by mapping the crystalline $h\nu_{CT}$, not against I_D-A_A (for which not enough experimental data exist), but against the difference $\Delta E_{1/2}$ between the readily available solution oxidation and reduction potentials. A V-shaped plot was obtained, and a sharp break between neutral and ionic complexes was found at $\Delta E_{1/2} = 0.17$ V. Neutral CT complexes near the neutral-to-ionic boundary can be coerced into a phase change by the application of pressure or, in some cases, by lowering the temperature, because in most CT complexes there is a considerable volume contraction as the temperature is lowered [82-85]. For example, single crystals of TTF-chloranil exhibit a first-order phase transition from the "neutral" CT (Z=0.3) to the "ionic" state (Z=0.7) at a critical temperature (T_c) of 80 K; there is a "dimerization" of the TTF molecules, and also of the chloranil molecules, at that temperature.

Now, armed by theory and by known examples of insulators, semiconductors, and conductors, we can resume our discussion of the necessary criteria for success.

Criterion Six: Molecular components of appropriate or compatible size

A molecular property, intimately involved in the mode of crystallization, is the molecular size. This is Kitaigorodsky's familiar, but important argument, that "the key must fit the lock", i.e. that the packing forces in a crystal require that anions and cations be compatible for

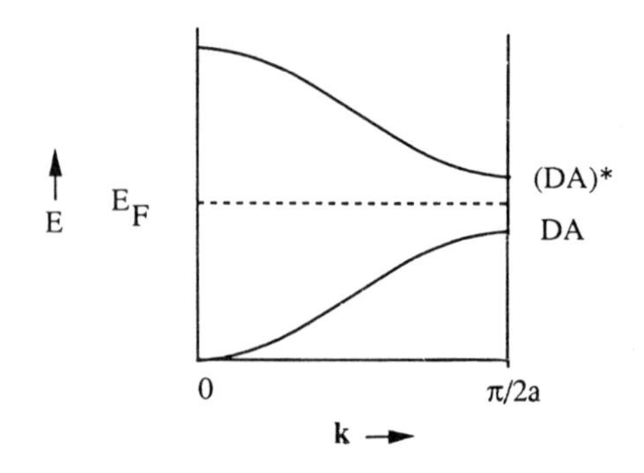

Fig. 6. Band structure for a DA (mixed-stack, 1M---) lattice

Table 3. Single crystal conductivity at room temperature of simple and complex TCNQ salts [88-90] (NMP= N-methylphenazinium): mixed valence helps!

Compound	Lattice	σ_{RT} (S/cm)
Li^+ $TCNQ^-$	1SURI	5 x 10^{-5}
Ph_3P^+Me $TCNQ^-$	1SURI	2 x 10^{-11}
Ph_3P^+Me $(TCNQ)_2^-$	$1SCA_4P$	2 x 10^{-2}
Quinolinium$^+$ $(TCNQ)_2^-$	$1SCA_4P$	100-200
NMP^+TCNQ^-	2SURP	200-400

efficient packing [86,87]. An illustrative example is provided by the $(TMTSF)_2X$ salts, to be described in detail later. For these compounds a linear correlation exists between the size of the tetrahedral anion and the metal-to-insulator transition temperature.

Criterion Seven: Segregated stacks of radical species

It is quite clear, from the experimental evidence, that the 1M lattices are electrical insulators, or, at best, semiconductors. Among the two-stack crystals we must seek the metals among those which also have partial charge transfer. One can understand why mixed-stack DA crystals (the 1M--- lattices) cannot be good conductors. If one constructs a band from $\Psi_{DA}(\mathbf{k} = 0)$ to $\Psi_{DA}(\mathbf{k} = \pi/a)$, using $\Psi_G(D,A)$ as the basis function, then the band filling is independent of CT, because the CT is already included in the basis function (see Fig. 6).

Criterion Eight: Fractional charge transfer ($0 < Z < 1$), or mixed-valence.

This is certainly one of the most important design considerations. Partial filling of the Brillouin zone is necessary for metallic conduction. In fact, even in the early TCNQ research at duPont in the 1960's, higher conductivities were achieved from complex TCNQ salts than with simple salts [88-90] (Table 3). The last two entries in Table 3 are really semiconductors with an activation energy for conduction, but the room-temperature resistivities are quite low. It is clear, however, that partial CT is essential; it is also clear that mixed valence (in the sense of the Robin-Day [91] class IIIB) must also be achieved, i.e. one cannot have discrete valences at discrete sites (as, e.g. in Cs_2TCNQ_3, which is a complex salt, but has low conductivity because it has a $1SCA_3D$ lattice).

The mixed-valence or partial charge transfer is, strictly, a bulk crystal effect, which is difficult to measure directly. As discussed above, Z can be obtained from X-ray or neutron diffraction measurements of the $2k_F$ and $4k_F$ scattering, if present. Other Z values have been estimated from X-ray diffraction measurements of certain bond lengths sensitive to Z [92], or from linear interpolations of shifts of IR- or Raman- active modes in the crystals, e.g. the Z-sensitive C-N stretch frequencies [93,94].

It is also difficult to predict Z from, say, I_D or A_A, or, more conveniently, from solution electrochemical $E_{1/2}$ data for the donor and acceptor molecules and ions. Already, these $E_{1/2}$ values can shift as much as 200 mV with solvent, electrolyte, or electrode material. If one attempts to correlate I_D (for the donors) with Z (for their complexes with TCNQ) [93,94], one must conclude that other factors (polarizability) play an important role (Table 4).

Table 4. Fractional charge transfer Z for TCNQ salts: no correlation with I_D [93,94].

Donor	I_D (eV)	Z
TTF	6.95	0.59
TSF	7.21	0.63
HMTTF	6.42	0.72
HMTSF	6.58	0.74

Despite such problems, several groups [95-98] have suggested that one could seek organic metals or superconductors by looking at ranges of solution $E_{1/2}$ values (Volts): Saito and Ferraris recommended [95] that metals be sought for:

$$-0.02 \leq E_{1/2}(D) - E_{1/2}(A) \leq 0.34$$

while Torrance, in a very beguiling and widely quoted article [97], suggests that one should not look for very good donors and very good acceptors (which tend to favor the 2MURI Mott-Hubbard insulator lattice), or for the very poor donors and acceptors (which will favor the 2MURN weak Mulliken CT lattice), but the "middle ground" of medium donors and acceptors. His somewhat fuzzy boundaries [98] present a useful terrain, in which to seek 2SURP lattices (Fig. 7).

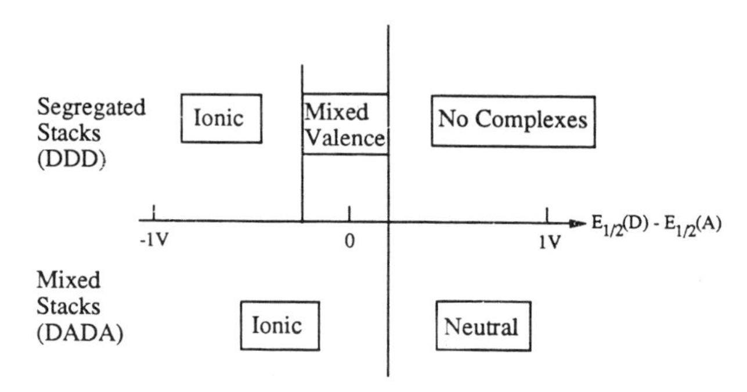

Fig. 7. Charge transfer Z, plotted against the difference between the electrochemical oxidation potential of the donor and the reduction potential of the acceptor [98].

Criterion Nine: Inhomogeneous charge and spin distribution

The next property, derived from a crystal packing effect, is again a molecular property. Most organic crystals stack in herringbone or otherwise slanted overlap in the crystalline state. Fig. 8 shows the crystal structure of the organic metal TTF-TCNQ [99], which shows an "interweaving herringbone motif" [100]. The overlap is not eclipsed, either between the neutral TTF^0 and the TTF^+ cations (which are crystallographically equivalent) in the donor stack, or between the neutral $TCNQ^0$ and the $TCNQ^-$ anions (also crystallographically equivalent) in the acceptor stack. The molecular charge [101] and spin [40,41] distribution is not uniform within the molecules, and the "ring over external bond" stacking motif helps reduce the Coulomb repulsions which would exist in an eclipsed stacking.

Criterion Ten: No Peierls distortion.

It is quite clear that one wants to avoid a metal-to-insulator transition if one wants to ultimately stabilize an organic superconductor. Nevertheless, most organic conductors do undergo a Peierls transition. The temperature dependence of the electrical conductivity of TTF-TCNQ (Fig. 9) is prototypical for the class of organic metals [1,2].

At high temperature the lattice is uniform; the conductivity increases with decreasing temperature (metallic behavior) until it reaches a maximum, then decreases (semiconducting behavior) when the temperature is decreased even more. For TTF-TCNQ the maximum conductivity (about 2×10^4 S/cm) is reached at about 60 K, which is the metal-insulator temperature (T_{MI}): the Peierls semiconducting state is more stable at lower temperatures.

For small or intermediate U (as in TTF-TCNQ) one can see extra Bragg reflections at both $2k_F$ and also at $4k_F$, while for TMTSF-TCNQ (small U) only $2k_F$ scattering is seen. For example, in TTF-TCNQ $\lambda / d = 3.39$ (the distortion is incommensurate with the original lattice), and $Z = 0.59$ is obtained. When the distortion is commensurate with the lattice (e.g. when Z is a rational fraction, 1/4, 1/3, 1/2, 3/4, or 1), then it usually requires less energy than an incommensurate distortion, which occurs at a higher temperature. For TTF-TCNQ, the situation is more complex: it is a two-chain material, and the bandwidth for the TTF chain is

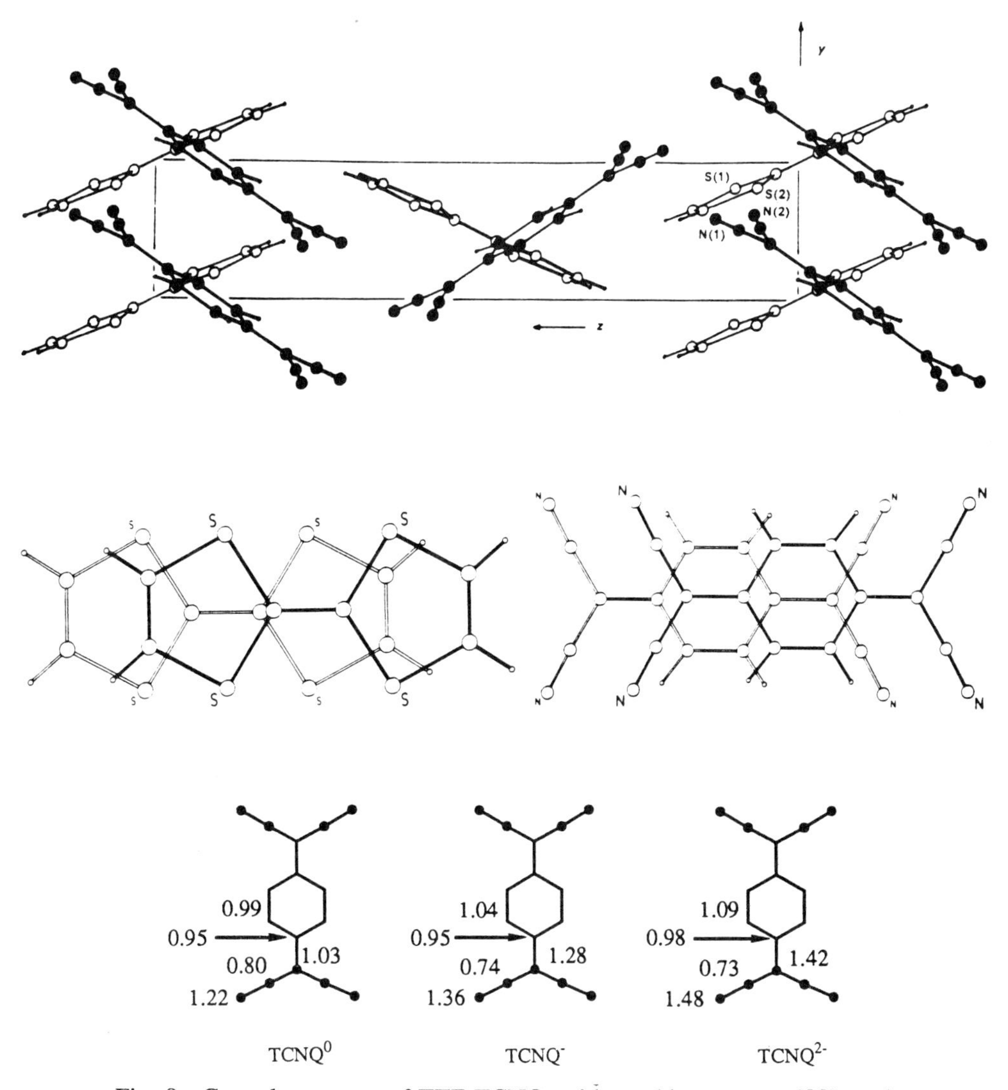

Fig. 8. Crystal structure of TTF-TCNQ, with stacking pattern [99], and charge distribution for the TCNQ moiety [101].

about 0.20 eV, while that of the TCNQ chain is about 0.45 eV [102,103]. Luckily, the selenium analog of TTF, TSF, forms with TCNQ a crystal isostructural with TTF-TCNQ [104,105]. Pure TSF-TCNQ maintains its metallic conductivity to about 40 K, and "alloy" crystals of composition $(TTF)_x(TSF)_{1-x}(TCNQ)$ have been grown [106,107]. From the combination of X-ray scattering [72-74], neutron scattering [108-110], and alloy studies [106,107] one sees that a CDW starts on the TCNQ chains at about 160 K; at 54 K there is coupling between CDW's on different TCNQ stacks, at 49 K a CDW starts on the TTF stacks (which could be driven by the CDW's on the TCNQ stacks), and by 38 K a complete Peierls state is observed. A slip of the TTF molecules along the long molecular axis of only 0.034 Å is the size of the structural modification involved [111].

Criterion Eleven: Strong or very weak interchain coupling to suppress Peierls transitions

It is easy to say that the Peierls transition should be suppressed. How does one do it? The relatively low-dimensional character of these molecular stacks can be a blessing or a curse. If one tries to prepare a new organic superconductor and observes, instead, a CDW or a SDW, that ultimately leads to a Peierls distortion and a semiconducting state at low temperature, one may feel beset by the dimensionality problem.

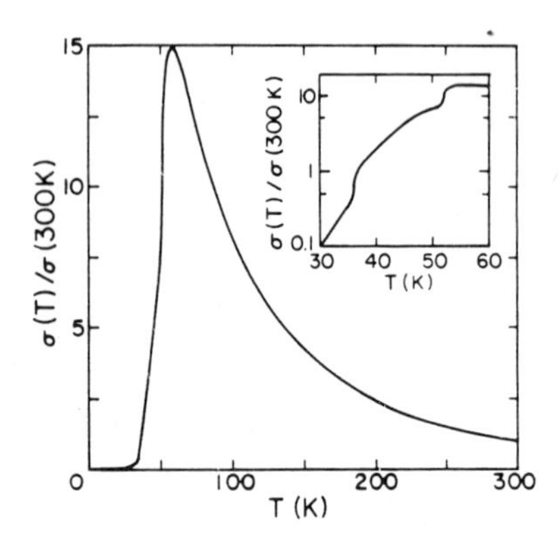

Fig. 9. Electrical conductivity of TTF-TCNQ [1]

Since long-range order at finite temperatures cannot occur in a strictly one-dimensional system, a phase transition (Peierls, superconducting, etc.) can take place only as a result of finite interchain coupling. Therefore, the actual transition temperature T_c is not the same as the "mean-field" temperature T_{MF} (which is the temperature, at which the transition would have occurred, if the interchain interactions had been equal to the intrachain interactions). If we define $T_{\perp}$ as the characteristic temperature for the interchain interactions, then T_c is approximated by the geometric mean $T_c = (T_{MF}\, T_{\perp})^{1/2}$. Thus, fluctuations or precursor events can be expected between T_c and T_{MF} [3, 14].

Some of the early Hopkins conductivity data for the series TMTTF-TCNQ, TTF-TCNQ (for whom the conductivity anisotropy is 500: 5 : <1), TMTSF-TCNQ, and HMTSF-TCNQ are shown in Fig. 10 [25, 26, 112-115]. From other (mainly magnetic) data it is surmised that

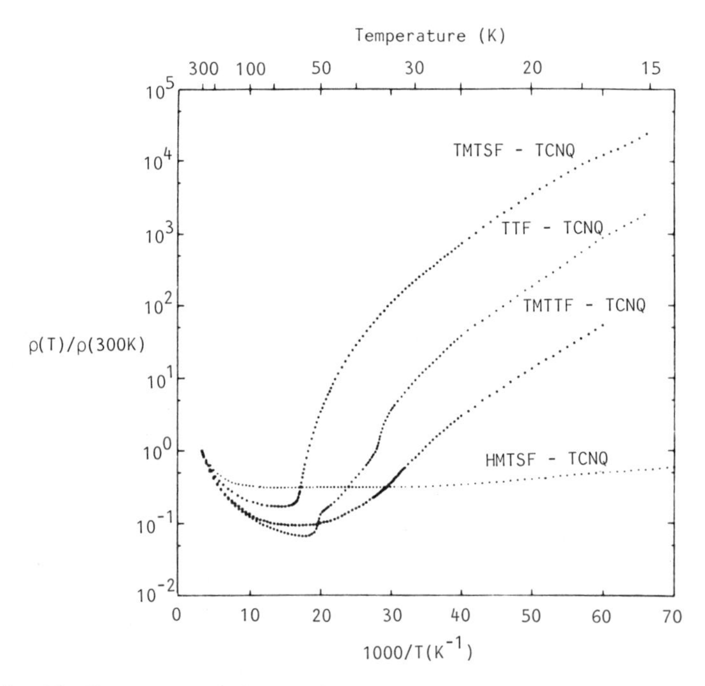

Fig. 10. Temperature-dependent resistivities of TTF-TCNQ, TMTTF-TCNQ, TMTSF-TCNQ, and HMTSF-TCNQ [26].

TMTTF-TCNQ is quite one-dimensional, and that HMTSF-TCNQ is the most two-dimensional of the series. Interestingly, in both cases the MI transition has been suppressed. In fact, HMTSF-TCNQ stays metallic (without going superconducting) down to 0.01 K. There are some stacking faults in the HMTSF-TCNQ lattice [116,117], and this disorder may also help to suppress the M-I transition. This interchain interaction can be increased, by altering the packing of the molecules, so that there is more equal interaction in two dimensions. An example of this can be found in the dimeric tiling found in the 10.4 K superconductor to be discussed later [κ-$ET_2Cu(NCS)_2$]. The interchain interactions can also be increased within a given stacking pattern, by increasing the interchain orbital overlap. An example of this can be found in the series TTF-TCNQ, TSF-TCNQ, and probably TTeF-TCNQ.

Criterion Twelve: Avoid disorder

The best early duPont TCNQ conductors (Table 3) quinolinium-TCNQ and NMP-TCNQ have unsymmetrical cations occupying centrosymmetric lattice sites. Nevertheless, the TCNQ

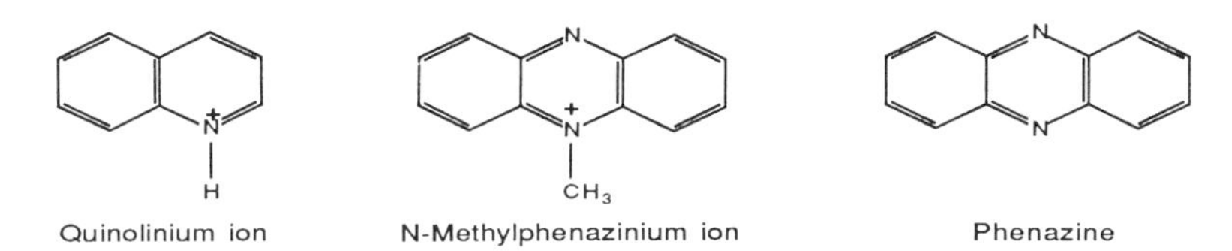

stacks are almost uniform (in NMP TCNQ the appearance of weak $2k_F$ streaks indicates that the charge transfer is $Z = 2/3$ [118]). The conductivity is not metallic, and does not become metallic for any solid solution with the centrosymmetric phenazine: $NMP_x(Phen)_{1-x}TCNQ$ [119]. Thus, some dipolar disorder (NMP^+) in these lattices permits relatively high conductivity. Decreasing the dipolar disorder by adding Phen• (while increasing the substitutional disorder and reducing the bandfilling) does not help achieve high conductivity: clearly the dipolar disorder pins the electrons, and conduction is via a hopping mechanism.

HOW BIG IS U, OR HOW BIG SHOULD IT BE?

The magnitude of the short-range Coulomb repulsion between conduction electrons on the same site (U_{eff}) can have a dramatic effect on the electrical and magnetic properties of the segregated-stack organic charge-transfer conductors. The unscreened cost of putting two electrons on a TCNQ molecule in the gas phase costs about 4 eV (and the same for putting two holes of the TTF molecule). If this value of U is not heavily screened in the solid, then U_{eff} will be much larger than the bandwidth 4 t (which is about 0.5 eV for TTF-TCNQ). Then the Mott-Hubbard criterion of Fig. 4c is met, and while one would have expected a nice $Z = 1$ half-filled band metal, one gets, instead, a filled-band insulator! Given the different X-ray signatures of large-U and small-U systems, one can show in Table 5 an interesting relationship between Z and U_{eff}.

Table 6 provides an abbreviated list of significant organic metals. At their conductivity maxima, some of these metals reach impressively high conductivities, but these high conductivity values are reached by compounds whose room-temperature conductivities are not so remarkable.

Table 5. Charge transfer Z and effective on-site repulsion energy U_{eff}

Charge transfer	U_{eff}	Signature	Example	Z
$0.50 < Z < 0.55$	large	$4k_F$	TMTSF-DMTCNQ	0.5
$0.50 < Z < 0.63$	intermediate	$2k_F + 4k_F$	TTF-TCNQ	0.59
$0.63 < Z < 0.80$	small	$2k_F$	HMTSF-TCNQ	0.74
$0.80 < Z < 1.0$	no examples	---	---	---
$Z = 1.0$	large	$4k_F$	HMTSF-$TCNQF_4$	1

Table 6. Selected organic metals.

Compound	Charge transfer[a] Z	Room-temperature conductivity σ_{RT} /Ω^{-1}cm^{-1}	Maximum conductivity σ_{max} /Ω^{-1}cm^{-1}	Temperature of max. cond.[b] T_{max} /K
TTF-TCNQ	0.59	500	2×10^4	59
TMTTF-TCNQ	0.65	350	5×10^3	60
HMTTF-TCNQ	0.72	500	2×10^3	75
TSF-TCNQ	0.63	800	1×10^4	40
TMTSF-TCNQ	0.57	1200	7×10^3	61
TMTSF-DMTCNQ	0.50	500	5×10^3	42
HMTSF-TCNQ	0.74	2000	7×10^3	(32)
TTeF-TCNQ	0.71	1800	2.5×10^4	< 4
HMTTeF-TCNQ	n.a.	550	9×10^2	(73)
HMTTeF-DMTCNQ	n.a.	460	1×10^3	(83)
HMTSF-TNAP	n.a.	2900	2×10^4	50
$(TTT)_2I_3$	0.50	1000	3×10^3	(60)
$(TST)_2I$	0.50	1500	1×10^4	(35)
$(TST)_2Cl$	0.50	2100	2×10^4	(26)
$(TMTSF)_2PF_6$	0.50	550	7×10^4	(18)
$(TMTSF)_2ClO_4$	0.50	650	5×10^5	(6)
$(BEDT\text{-}TTF)_2I_3$	0.50	30	1.5×10^4	(4.2)
$(Perylene)_2(PF_6)_{1.1}$	0.550	900	1×10^3	200
Ni(Pc)I	0.33	550	5×10^3	(25)
$H_2(Pc)I$	0.33	750	4×10^3	15
Cu(Pc)I	0.33	900	7×10^3	»(30)
$Ni(Pc)(BF_4)_{0.33}$	0.33	1000	4×10^3	80
$Ni(Pc)(ClO_4)_{0.42}$	0.42	700	1×10^3	(200)
$Cu(2,5\text{-}DM\text{-}DCNQI)_2$	n.a.	800	5×10^5	(3.5)

(a) fractional charge per cation, from Refs. [93,94] and references cited therein
(b) parentheses in this column indicate a broad conductivity maximum

ORGANIC SUPERCONDUCTORS.

TMTSF or Bechgaard salts.

Since we noted that, in TTF-TCNQ, the CDW which ultimately causes the M-I transition starts on the TCNQ chains, it seemed logical to discard the TCNQ stack altogether, and to concentrate on the donor stack alone. This is what Klaus Bechgaard did. Inspired by the fact that the HMTSF-TCNQ two-chain system stayed metallic down to 0.01 K, it seemed that selenium atoms would be helpful. In 1980 the first superconductor, $(TMTSF)_2PF_6$, was obtained by electrochemical crystallization [120] and was found to superconduct below the critical temperature $T_c = 0.9$ K at an applied pressure of 10 kbar [121-124]. In fact, a whole series of such salts $(TMTSF)_2X$ were grown (X = NO_3^-, BF_4^-, ClO_4^-, ReO_4^-, AsF_6^-, SbF_6^-) [120]. The stacking pattern of the $(TMTSF)_2X$ salts [125] is shown in Fig.11.

The overlap of the TMTSF π-orbitals (dominated by the Se atoms) in the crystallographic **a** direction yields a conduction band of width of about 1.4 eV [126,127], i.e. more than twice the bandwidth of TTF TCNQ. The Se-Se interactions along **b** are about 10 times smaller, and the interaction along **c** is even weaker. The anions form a sheet separating the TMTSF molecules in the **ab** plane. The Fermi surface of these salts consists of planes, which are

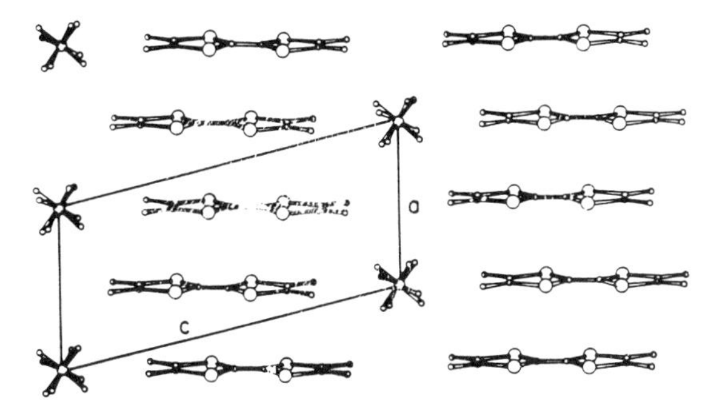

Fig. 11. Crystal structure of $(TMTSF)_2ClO_4$ [125].

slightly warped by the interchain coupling integral t_b along **b** (see Fig. 12). There is a danger that an external potential could induce perfect "nesting" of these warped planes, but, surprisingly, no precursor CDW has been found. Rather, a magnetic SDW can cause the transition to an insulating state [72], but this transition can be suppressed by applying hydrostatic pressure. The phase diagram for $(TMTSF)_2PF_6$ [11] is given in Fig. 13.

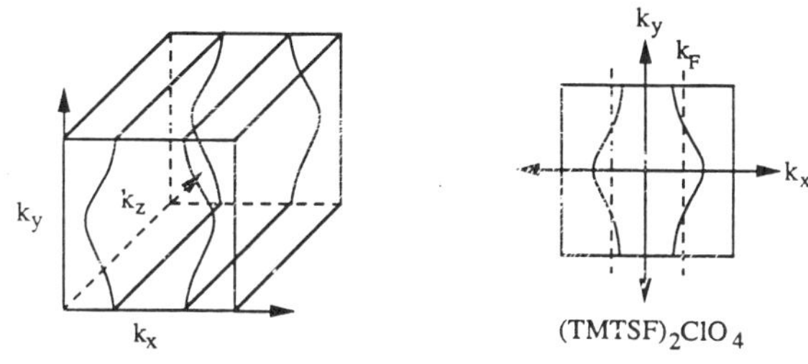

Fig.12. Schematic representation of 1-D Fermi surface with considerable interstack interactions

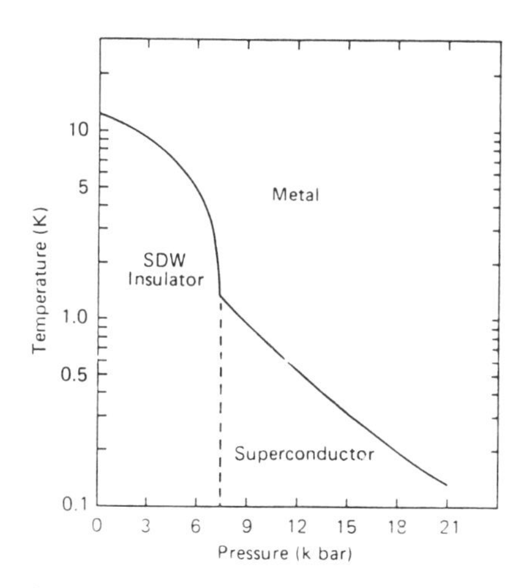

Fig.13. Phase diagram for $(TMTSF)_2PF_6$, redrawn from Ref. [11].

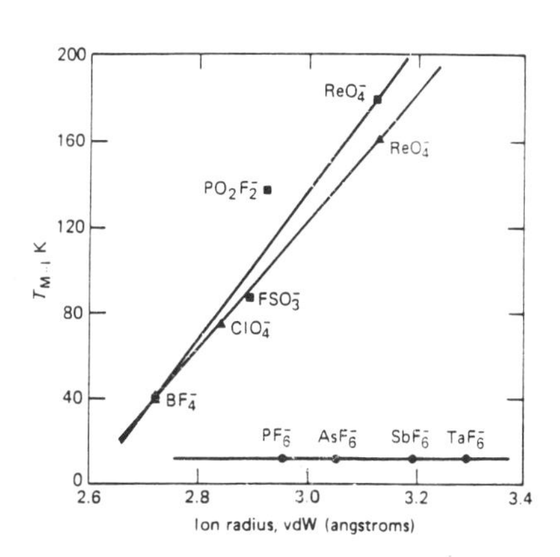

Fig.14. Metal-insulator phase transition temperature T_{MI} as a function of anion radius: ■ $(TMTSF)_2X$, T_h; ▲ $(TMTTF)_2X$, T_h; • $(TMTSF)_2X$, O_h (redrawn from Ref. [128]).

Table 7. Selected organic superconductors.

Compound	T_c /K	Pressure /kbar	Compound /K	T_c /kbar	Pressure
$(TMTSF)_2^+X^-$			**$(ET)_2^+X^-$**		
$X^- = PF_6^-$	0.9	10-12	$X^- = I_3^-$	1.6	none
$X^- = ClO_4^-$	1.2	none	$X^- = I_3^-$	8.0	1.5
$X^- = ReO_4^-$	1.3-1.5	12	$X^- = I_3^-$	2.5	none
$X^- = FSO_3^-$	2.5	>6	$X^- = IBr_2^-$	2.5-2.8	none
			$X^- = ReO_4^-$	1.5	5
$TTF[Ni(dmit)_2]_2$	1.6	7	$X^- = AuI_2^-$	4	none
			$X^- = Cu(NCS)_2^-$	10.4	none

These compounds are on the threshold of 2-dimensionality, and pressure is thought to increase the off-chain coupling enough, to prevent the "perfect nesting" distortion of the Fermi surface, and to thus enable high conductivity. Doping a pure $(TMTSF)_2X$ salt with S, in place of Se, depresses the critical temperature. Increasing the size of the anion decreases the off-chain Se-Se interactions. For the tetrahedral or pseudo-tetrahedral anions, the lattice orders at a fairly high metal-to-insulator temperature T_{MI}, while the octahedral anions never need to order, given their site symmetry. Therefore, it is interesting to find that T_{MI} depends on anion size for the tetrahedral and pseudo-tetrahedral anions, while there is no size effect on the ordering of the octahedral anions [128-131] (see Fig. 14).

BEDT-TTF or ET salts

The superconducting transition temperature of the Bechgaard salts was never very high. Saito found that a ReO_4^- salt of ET was metallic down to 1.4 K [132]. In 1982 the IBM group found superconductivity in ET_2ReO_4 at 1.5 K above 5 kbar pressure [133]. In the next few years, many salts of ET were prepared by electrochemical oxidation [133-136], and the present record T_c is 10.4 K for an undeuterated ET Cu isothiocyanate [137] (and 11 K for the fully deuterated donor [138]). The situation is reviewed [21, 22] in Table 7. There is a great diversity of stoichiometries and crystal phases [132]. The most common are ET_2X, ET_3X_2, ETX, and there are frequently α, β, γ, δ, ε, and ζ crystal forms, often several in the same batch. All of the metallic salts contain a two-dimensional network of ET molecules, loosely coupled by S...S interactions (probably of the TTF rings), in what we term the crystallographic **a b** plane. Band calculations indicate that the bandwidth in the interstack and intrastack directions are comparable in magnitude, so that we have a Fermi surface that is nearly isotropic in two dimensions [21]. Then we expect that phase transitions will not convert these metals to insulators. Indeed, the superconductor β-ET_2I_3 undergoes a phase transition to an incommensurately modulated structure at 200 K [21], but only small changes in the conductivity profile are seen, and the crystal remains metallic, until it turns into a low-temperature superconductor.

A final point to be made is the critical comparison of several superconductors (Table 8). The data are consistent with the notion (reinforced in recent high-temperature ceramic oxide superconductors) that the poorest (room-temperature) metals make the best superconductors.

Table 8. Comparison of room-temperature conductivity with superconducting critical temperature: there is an inverse correlation!

Compound	σ_{RT}/(S/cm)	T_c /K
$(TMTSF)_2\,ClO_4^-$	650	1.2
$(TMTSF)_2\,PF_6^-$	550	0.9 at 10 kbar
$(ET)_2\,ReO_4^-$	200	1.5 at 5 kbar
$(ET)_2\,I_3^-$	30	1.5, or 8 at 1.5 kbar
$(ET)_2\,IBr_2^-$	20	2.3-2.8
$(ET)_2\,Cu(NCS)_2$	14	10.4

CONCLUSION

We have tried in this article to describe the field of organic conductors and superconductors in a logical, *a priori* progression, where theory and experimental observations are closely interwoven. This is, of course, not the human, accidental, *a posteriori* order in which the field actually developed. Nevertheless, we can now see how the development of good organic metals and superconductors begins with recognizing the constraints that will lead to optimal solutions. In addition, we have tried to elucidate some of the more fascinating and dominant physical effects which give rise to these constraints.

ACKNOWLEDGEMENTS

The authors are grateful for the invaluable collaboration of colleagues and students whose names appear on many of the cited articles. One of us (DOC) would like to thank Drs. Aaron N. Bloch, Theodore O. Poehler, and Thomas J. Kistenmacher for their good humor and skillful sharing of the many burdens over the past 16 years. The generous support of the National Science Foundation for the recent work presented here (Grant DMR 86-15305) is very much appreciated.

REFERENCES

[1] J. Ferraris, D. O. Cowan, V. Walatka, and J. H. Perlstein, J. Am. Chem. Soc. 95: 498 (1973).
[2] L. B. Coleman, M. J. Cohen, D. J. Sandman, F. G. Yamagishi, A. F. Garito, and A. J. Heeger, Solid State Commun. 12: 1125 (1973)
[3] P. M. Chaikin and R. L. Greene, Physics Today 39: 24 (1986)
[4] J. Kommandeur, Nouv. J. Chim. 9: 341 (1985).
[5] T. J. Marks, Science 227: 881 (1985).
[6] M. R. Bryce and L. C. Murphy, Nature 309: 119 (1984).
[7] R. L. Greene and P. M. Chaikin, Physica B 126: 431 (1984).
[8] R. L. Greene and G. B. Street, Science 226: 651 (1984).
[9] F. Wudl, Acc. Chem. Res. 17: 227 (1984).
[10] B. M. Hoffman and J. A. Ibers, Acc. Chem. Res. 16: 15 (1983).
[11] D. Jérome and H. J. Schultz, Advan. Phys. 31: 299 (1982).
[12] J. B. Torrance, Acc. Chem. Res. 12: 79 (1979).
[13] A. J. Berlinsky, Contemp. Phys. 17: 331 (1976).
[14] J. S. Miller and A. J. Epstein, Prog.Inorg. Chem. 20: 1 (1976).
[15] A. F. Garito and A. J. Heeger, Acc. Chem. Res. 7: 232 (1974).
[16] D. O. Cowan and A. Kini in "The Chemistry of Organic Selenium and Tellurium Compounds", Vol. 2, edited by S. Patai (Wiley, New York, 1987) pages 463-494.
[17] J. M. Williams, H. H. Wang, T. J. Emge, U. Geiser, M. A. Beno, P. Leung, K. D. Carlson, R. J. Thorn, A. J. Schultz, and M.-H. Whangbo, Prog. Inorg. Chem. 35: 51 (1987).
[18] A. Krief, Tetrahedron, 42: 1209 (1986).
[19] M. R. Bryce, Ann. Reports B92: 377 (1985).
[20] J. M. Williams, Prog. Inorg. Chem. 33: 183 (1985).
[21] J. M. Williams, M. A. Beno, H. H. Wang, P. C.W. Leung, T. J. Emge, U. Geiser, and K. D. Carlson, Acc. Chem. Res. 18: 261 (1985).
[22] J. R. Ferraro and J. M. Williams, "Introduction to Synthetic Electrical Conductors" (Academic, San Diego, 1987).
[23] M. Narita and C. U. Pittman, Jr., Synthesis, 489 (1976).
[24] A. N. Bloch, D. O. Cowan, and T. O. Poehler in "Energy and Charge Transfer in Organic Semiconductors" edited by K. Masuda and M. Silver (Plenum, New York, 1974) page 167.
[25] D. O. Cowan, P. Shu, C. Hu, W. Krug, T. Carruthers, T. Poehler, and A. Bloch, in "Chemistry and Physics of One-Dimensional Metals" ed. by H. J. Keller (Plenum, New York, 1977) page 25.
[26] A. N. Bloch, T. Carruthers, T. O. Poehler, and D. O. Cowan, in "Chemistry and Physics of One-Dimensional Metals" ed.by H. J. Keller (Plenum, New York, 1977) page 47.

[27] D. O. Cowan, A. Kini, L.-Y. Chiang, K. Lerstrup, D. R. Talham, T. O. Poehler, and A. N. Bloch, Mol. Cryst. Liq. Cryst. 86: 1 (1982).
[28] D. O. Cowan and F. M. Wiygul, Chem. Eng. News 64: (29), 28 (1986).
[29] D. O. Cowan in "New Aspects of Organic Chemistry. I. Proceedings of the 4th International Kyoto Conference on New Aspects of Organic Chemistry" edited by Z. Yoshida, T. Shiba, and Y. Ohshiro (Kodansha, Tokyo, and VCH Verlagsgesellschaft, Federal Republic of Germany, 1989).
[30] J. Hubbard, Proc. Roy. Soc. (London) A276: 238 (1963).
[31] J. Hubbard, Proc. Roy. Soc. (London) A277: 237 (1963).
[32] J. Hubbard, Proc. Roy. Soc. (London) A281: 401 (1964).
[33] R. S. Mulliken, J. Am. Chem. Soc. 74: 811 (1952).
[34] R. S. Mulliken, J. Phys. Chem. 56: 801 (1954).
[35] R. S. Mulliken and W. B. Person in "Molecular Complexes: A Lecture and Reprint Volume" (Wiley, New York, 1969).
[36] N. C. Yang and A. J. Castro, J. Am. Chem. Soc. 82: 6208 (1960).
[37] K. Mukai, T. Mishina, and K. Ishizu, J. Chem. Phys. 66: 1680 (1977).
[38] F. Gerson, R. Heckendorn, D. O. Cowan, A. M. Kini, and M. Maxfield, J. Am. Chem. Soc. 105: 7017 (1983) and references cited therein.
[39] A. M. Kini, D. O. Cowan, F. Gerson, and R. Möchel, J. Am. Chem. Soc. 107: 556 (1985) and references cited therein.
[40] L. Cavara, F. Gerson, D. O. Cowan, and K. Lerstrup, Helv. Chim. Acta 69: 141 (1986).
[41] H. G. Viehe, Z. Janousek, R. Merenyi, and L. Stella, Acc. Chem. Res. 18: 148 (1985).
[42] E. M. Kosower, H. P. Waits, A. Teverstein, and L. C. Butler, J. Org. Chem. 43:`800 (1978).
[43] R. C. Haddon, Aust. J. Chem. 28: 2343 (1975).
[44] R. C. Haddon, S. V. Chichester, S. M. Stein, J. H. Marshall, and A. M. Mujsce, J. Org. Chem. 52: 711 (1987).
[45] F. Wudl and E. Aharon-Shalom, J. Am. Chem. Soc. 104: 1154 (1982).
[46] D. J. Sandman, J. C. Stark, and B. M. Foxman, Organometallics 1: 739 (1982).
[47] D. J. Sandman, J. C. Stark, G. P. Hamill, W. A. Burke, and B. M. Foxman, Mol. Cryst. Liq. Cryst. 86: 79 (1982).
[48] R. P. Shibaeva and V. F. Kaminskii, Cryst. Str. Commun. 10: 663 (1981).
[49] K. Lerstrup, D. Talham, A. Bloch, T. Poehler, and D. O. Cowan, J. Chem. Soc. Chem. Commun., 336 (1982).
[50] K. Lestrup, D. O. Cowan, and T. J. Kistenmacher, J. Am. Chem. Soc. 106: 8303 (1984).
[51] K. Lerstrup, M. Lee, D. O. Cowan, and T. J. Kistenmacher, Mol. Cryst. Liq. Cryst. 120: 295 (1985).
[52] K. Lerstrup, A. Bailey, R. McCullough, M. Mays, and D. O. Cowan, Synth. Met. 19: 647 (1987).
[53] A. B. Bailey, R. D. McCullough, M. D. Mays, D. O. Cowan, and K. A. Lerstup, Synth. Met. 27: 425 (1988).
[54] R. D. McCullough, G. B. Kok, K. A. Lerstrup, and D. O. Cowan, J. Am. Chem. Soc. 109: 4115 (1987).
[55] M. D. Mays, R. D. McCullough, D. O. Cowan, T. O. Poehler, W. A. Bryden, and T. J. Kistenmacher, Solid State Commun. 65: 1089 (1988).
[56] R. D. McCullough, M. D. Mays, A. B. Bailey, and D. O. Cowan, Synth. Met. 27: 487 (1988).
[57] M. D. Mays, R. D. McCullough, A. B. Bailey, D. O. Cowan, W. A. Bryden, T. O. Poehler, and T. J. Kistenmacher, Synth. Met. 27: 493 (1988).
[58] G. Saito, T. Enoki, M. Kobayashi, K. Imaeda, N. Sato, and H. Inokuchi, Mol. Cryst. Liq. Cryst. 119: 393 (1985).
[59] O. H. Le Blanc, J. Chem. Phys. 42: 4307 (1965).
[60] P. M. Chaikin, A. F. Garito, and A. J. Heeger, J. Chem. Phys. 58: 2336 (1973).
[61] F. H. Herbstein in "Perspectives in Structural Chemistry", Vol. IV, edited by J. D. Dunitz and J. A. Ibers (Wiley , New York, 1972), p.166.
[62] J. Tanaka and C. Tanaka, Mol. Cryst. Liq. Cryst. 126: 121 (1985).
[63] Z. G. Soos, Ann. Rev. Phys. Chem. 25: 121 (1974).
[64] Z. G. Soos, Isr. J. Chem. 23: 37 (1983).
[65] Z. G. Soos in "Organic and Inorganic Low-Dimensional Crystalline Materials", ed. by P. Delhaès and M. Drillon, NATO ASI Series B168: 47, (Plenum, New York, 1987).

[66] F. M. Wiygul, R. M. Metzger, and T. J. Kistenmacher, Mol. Cryst. Liq. Cryst. 107: 115 (1984).
[67] D. J. Sandman, Mol. Cryst. Liq. Cryst. 50: 235 (1979).
[68] R. M. Metzger, J. Chem. Phys. 66: 2525 (1977).
[69] G. Wegner, Angew. Chem. Intl. Ed. Engl. 20: 361 (1981).
[70] R. E. Peierls, "Quantum Theory of Solids" (Clarendon Press, Oxford, 1955) p. 108.
[71] H. A. Jahn and E. Teller, Proc. Roy. Soc. (London) A161: 220 (1937).
[72] J.-P. Pouget in "Organic and Inorganic Low-Dimensional Crystalline Materials" ed. by P. Delhaès and M. Drillon (Plenum, New York, 1987), p. 185 and references cited therein.
[73] S. K. Khanna, J.-P. Pouget, R. Comès, A. F. Garito, and A. J. Heeger, Phys. Rev. 16: 1468 (1977).
[74] F. Denoyer, R. Comès, A. F. Garito, and A. J. Heeger, Phys. Rev. Lett. 35: 445 (1975).
[75] S. Kagoshima, H. Anzai, K. Kajimura, and T. Ishiguro, J. Phys. Soc. Jpn. 39: 1143 (1975).
[76] H. M. McConnell, B. M. Hoffman, and R. M. Metzger, Proc. Natl. Acad. Sci. US 53: 46 (1965).
[77] R. M. Metzger and A. N. Bloch, J. Chem. Phys. 63: 5098 (1975).
[78] R. M. Metzger, Top. Curr. Phys. 26: 80 (1981).
[79] P. J. Strebel and Z. G. Soos, J. Chem. Phys. 53: 4077 (1970).
[80] A. N. Bloch and S. Mazumdar, J. Phys. (Paris) Colloque 44: C3-1273 (1983).
[81] J. B. Torrance, J. E. Vazquez, J. J. Mayerle, and V. Y. Lee, Phys. Rev. Lett. 46: 253 (1981).
[82] C. S. Jacobsen and J. B. Torrance, J. Chem. Phys. 78: 112 (1983).
[83] A. Girlando, R. Bozio, C. Pecile, and J. B. Torrance, Phys. Rev. B26: 2306 (1982).
[84] Y. Tokura, Y. Kaneko, H. Okamoto, S. Tanuma, T. Koda, T. Mitani, and G. Saito, Mol. Cryst. Liq. Cryst. 125: 71 (1985).
[85] S. Kagoshima, Y. Kanai, M. Tani, Y. Tokura, and T. Koda, Mol. Cryst. Liq. Cryst. 120: 9 (1985).
[86] A. I. Kitaigorodsky ,"Molecular Crystals and Molecules" (Academic, New York, 1973)
[87] A. I. Kitaigorodsky, "Mixed Crystals" (Springer, Berlin, 1984).
[88] L. R. Melby, R. J. Harder, W. R. Hertler, W. Mahler, R. E. Benson, and W. E. Mochel, J. Am. Chem. Soc. 84: 3374 (1962).
[89] W. J. Siemons, P. E. Bierstedt, and R. G. Kepler, J. Chem. Phys. 39: 3523 (1963).
[90] L. R. Melby, Can. J. Chem. 43: 1448 (1965).
[91] M. B. Robin and P. Day, Adv. Inorg. Chem. and Radiochem. 10: 247 (1967).
[92] S. Flandrois and D. Chasseau, Acta Crystallogr. B33: 2744 (1977).
[93] J. S. Chappel, A. N. Bloch, W. A. Bryden, M. Maxfield, T. O. Poehler, and D. O. Cowan, J. Am. Chem. Soc. 103: 2442 (1981).
[94] E. Kapmar and O. Neilands, Russian Chem. Rev. 55: 334 (1986).
[95] G. Saito and J. P. Ferraris, Bull. Chem. Soc. Jpn. 53: 2141 (1980).
[96] R. C. Wheland, J. Am. Chem. Soc. 98: 3926 (1976).
[97] J. B. Torrance, Acc. Chem. Res. 12: 79 (1979).
[98] J. B. Torrance in "Low-Dimensional Conductors and Superconductors" ed. by D. Jérome and L. G. Caron (Plenum, New York, 1986) page 113.
[99] T. J. Kistenmacher, T. Phillips, and D. O. Cowan, Acta Cryst. B30: 763 (1974).
[100] T. J. Kistenmacher, Ann. N. Y. Acad. Sci. 313: 333 (1978).
[101] H. Th. Jonkman and J. Kommandeur, Chem. Phys. Lett. 15: 496 (1972).
[102] A. J. Berlinsky, J. F. Carolan, and L. Weiler, Solid State Commun. 15: 795 (1974).
[103] D. R. Salahub, R. P. Messmer, and F. Herman, Phys. Rev. B13: 4252 (1976).
[104] E. M. Engler and V. V. Patel, J. Am. Chem. Soc. 96: 7376 (1974).
[105] E. M. Engler, B. A. Scott, S. Etemad, T. Penney, and V. V. Patel, J. Am. Chem. Soc. 99: 5909 (1977).
[106] S. Etemad, E. M. Engler, T. D. Schultz, T. Penney, and B. A. Scott, Phys. Rev. B17: 513 (1978).
[107] R. A. Craven, Y. Tomkiewicz, E. M. Engler, and A. R. Taranko, Solid State Commun. 23: 429 (1977).
[108] S. Megtert, R. Comès, C. Vettier, R. Pynn, and A. F. Garito, Mol. Cryst. Liq. Cryst. 85: 159 (1982).
[109] S. Megtert, R. Comès, C. Vettier, R. Pynn, and A. F. Garito, Solid State Commun. 37: 875 (1981).

[110] R. Comès, S. Shapiro, G. Shirane, A. F. Garito, and A. J. Heeger, Phys. Rev. Lett. 35: 1512 (1975).
[111] P. Coppens, V. Petricek, D. Levendis, F. K. Larsen, A. Paturle, G. Yan, and A. D. LeGrand, Phys. Rev. Lett. 59: 1695 (1987).
[112] A. N. Bloch, D. O. Cowan, K. Bechgaard, R. E. Pyle, R. H. Banks, and T. O. Poehler, Phys. Rev. Lett. 34: 1561 (1975).
[113] J. R. Cooper, M. Weger, D. Jérome, D. Lefur, K. Bechgaard, A. N. Bloch, and D. O. Cowan, Solid State Commun. 19: 749 (1976).
[114] K. Bechgaard, D. O. Cowan, and A. N. Bloch, J. Chem. Soc. Chem. Commun. 937 (1974).
[115] J. P. Ferraris, T. O. Poehler, A. N. Bloch, and D. O. Cowan, Tetr. Lett. 2553 (1973).
[116] T. E. Phillips, T. J. Kistenmacher, A. N. Bloch, and D. O. Cowan, J. Chem. Soc. Chem. Commun., 334, (1976).
[117] R. L. Greene, J. J. Mayerle, R. Schumaker, G. Castro, P. M. Chaikin, S. Etemad, and S. J. LaPlaca, Solid State Commun. 20: 943 (1976).
[118] J.-P. Pouget, R. Comès, A. J. Epstein, and J. S. Miller, Mol. Cryst. Liq. Cryst. 85: 203 (1982).
[119] J. S. Miller and A. J. Epstein, Angew. Chem. Intl. Ed. Engl. 26: 287 (1987) and references cited therein.
[120] K. Bechgaard, C. S. Jacobsen, K. Mortensen, H. J. Pedersen, and N. Thorup, Solid State Commun. 33: 1119 (1980).
[121] D. Jérome, A. Mazaud, M. Ribault, and K. Bechgaard, J. Physique Lett. 41: L95 (1980).
[122] K. Bechgaard, K. Carneiro, M. Olsen, F. Rasmussen, and C. S. Jacobsen, Phys. Rev. Lett. 46: 852 (1981).
[123] K. Bechgaard, K. Carneiro, F. Rasmussen, M. Olsen, G. Rindorf, C. S. Jacobsen, H. J. Pedersen, and J. C. Scott, J. Am. Chem. Soc. 103: 2440 (1981).
[124] D. U. Gubser, W. W. Fuller, T. O. Poehler, D. O. Cowan, M. Lee, R. S. Potember, L.-Y. Chiang, and A. N. Bloch, Phys. Rev. B24: 478 (1981).
[125] N. Thorup, G. Ringdorf, H. Soling, and K. Bechgaard, Acta Cryst.B37: 1236 (1981).
[126] P. M. Grant, Phys. Rev. Lett. 50: 1005 (1983).
[127] M.-H. Whangbo, J. M. Williams, M. A. Beno, and J. R. Dorfman, J. Am. Chem. Soc. 105: 645 (1983).
[128] T. J. Kistenmacher, Mol. Cryst. Liq. Cryst. 136: 361 (1986).
[129] T. J. Kistenmacher, Johns Hopkins Applied Physics Laboratory Technical Digest 7:(2), 142 (1986).
[130] J. M. Williams, M. A. Beno, J. C. Sullivan, L. M. Banovetz, J. M. Braam, G. S. Blackman, K. D. Carlson, D. L. Green, and D. M. Loesing, J. Am. Chem. Soc. 105: 643 (1983).
[131] T. J. Emge, J. M. Williams, P. C. W. Leung, A. S. Schultz, M. A. Beno, and H. H. Wang, Mol. Cryst. Liq. Cryst. 119: 237 (1985).
[132] G. Saito, T. Enoki, K. Toriumi, and H. Inokuchi, Solid State Commun. 42: 557 (1982).
[133] S. S. P. Parkin, E. M. Engler, R. R. Schumaker, R. Lagier, V. Y. Lee, J. C. Scott, and R. L. Greene, Phys. Rev. Lett. 50: 270 (1983).
[134] E. B. Yagubskii, I. F. Shchegolev, V. N. Laukhin, P. A. Konovich, M. V. Karstovnik, A. V. Zvarykina, and L. I. Buravov, JETP Lett. 39: 12 (1984).
[135] G. W. Crabtree, K. D. Carlson, L. N. Hall, R. T. Copps, H. H. Wang, T. J. Emge, M. A. Beno, and J. M. Williams, Phys. Rev. B30: 2958 (1984).
[136] K. Murata, M. Tokumoto, H. Anzai, H. Bando, G. Saito, K. Jajimura, and T. Ishiguro, J. Phys. Soc. Jpn. 54: 1236 (1985).
[137] H. Urayama, H. Yamochi, G. Saito, K. Nozawa, T. Sugano, M. Kinoshita, S. Sato, K. Oshima, A. Kawamoto, and J. Tanaka, Chem. Lett. 55 (1988).
[138] D. Schweitzer, K. Polychroniadis, T. Klutz, H. J. Keller, I. Hennig, I. Heinen, U. Haeberlen, E. Gogu, and S. Gärtner, Synth. Metals 27: A465 (1988).

STRUCTURAL PROPERTIES OF MOLECULAR CHARGE–TRANSFER CONDUCTORS AND SEMICONDUCTORS FROM INFRARED AND RAMAN SPECTROSCOPY

R. Bozio, A. Feis, D. Pedron, I. Zanon, and C. Pecile

Department of Physical Chemistry
University of Padova
2, Via Loredan
35131 Padova (Italy)

I. INTRODUCTION

In condensed-phase molecular systems, and particularly insulating molecular crystals, infrared and Raman spectroscopies are well-established tools for investigating molecular and crystalline structural properties.[1] On the other hand, in inorganic semiconductors and metals the same spectroscopic methods are widely used to evaluate electronic band structure parameters.[2] Organic charge transfer (*CT*) crystals, as molecular conductors and semiconductors, share some properties of both the above classes of solids, and, therefore, new spectroscopic phenomena are expected, and indeed observed for these materials.[3]

One basic reason is that in these solids some low-lying electronic excitations and some vibrational excitations, particularly intramolecular ones, come very close in energy. As a consequence, the coupling between electrons and vibrational degrees of freedom, which is always present in molecular solids, including the insulating ones, has in this case dramatic consequencies for the spectroscopic, as well as for other physical properties.

Besides these peculiar features, the spectrum of the electronic excitations also reveals unusual properties compared to those of ordinary metals and semiconductors. Most of the observed peculiarities derive directly from the structural properties of *CT* conductors such as the highly anisotropic character of the interactions (typical of quasi 1-D or 2-D systems), the comparatively low carrier density, the narrow bands, the easy localization of carriers.[4] The considerable interest devoted to the investigation of the optical properties is also motivated by the belief that they most dramatically reflect the genuine many–body properties introduced in the physics of *CT* crystals by the electron–electron correlations being comparable in magnitude to the one–electron interactions (intermediate regime).

Based on these considerations, one can easily understand that the spectroscopic studies of *CT* crystals and molecular conductors, and the companion linear and non-linear studies of conducting polymers,[5] have gradually evolved into entirely new chapters of solid-state spectroscopy. The scope of this paper is far from the one of giving a comprehensive account of the spectroscopy of *CT* crystals. We shall rather

Lower-Dimensional Systems and Molecular Electronics
Edited by R.M. Metzger *et al.*, Plenum Press, New York, 1991

draw, mainly from the results obtained in our group, to illustrate, both theoretically and experimentally, some of the novel spectroscopic phenomena encountered with these materials.

II. DIMER SPECTROSCOPY

Let us start by introducing some useful spectroscopic notions through a discussion of the simplest possible system containing most of the physical ingredients of CT organic molecular systems: a dimer of two open-shell conjugated molecules interacting through their π–electron systems. In describing the linear spectroscopic properties we shall use the macroscopic response functions, such as the complex dielectric function ($\hat{\varepsilon}(\omega)$), the complex refractive index ($\hat{N}(\omega)$) and the conductivity ($\hat{\sigma}(\omega)$). The definition of these quantities and the relation among them and with other experimentally measurable quantities such as the reflectance, $R(\omega)$, and the absorbance, $\alpha(\omega)$, as well as the dispersion relations connecting their real and imaginary parts (Kramers–Kronig relations) can be found in standard books. [1,2]

It is a well-known result that, for a dielectric molecular material, the dielectric function can be written using a Lorentz model [1]

$$\hat{\varepsilon}(\omega) = \varepsilon_\infty + \frac{4\pi N}{\hbar} \sum_n \frac{f_{n\circ}}{\omega_{n\circ}^2 - \omega^2 - i\omega\gamma_n} \tag{1}$$

where ε_∞ is a (real) core dielectric constant, N is the number density of molecules. The summation runs over all the low-lying excited states whose excitation frequencies, relaxation rates, and oscillator strenghts are $\omega_{n\circ}$, γ_n, and $f_{n\circ} = \omega_{n\circ}|\mu_{n\circ}|^2$ respectively; $\mu_{n\circ}$ is the matrix element of the electric dipole operator. In dealing with CT excitations, the latter is commonly expressed in the point-charge approximation and, for a molecular dimer, takes the form

$$\mu = -\frac{1}{2} ed(n_1 - n_2)$$

where d is the intermolecular spacing and $n_i = \sum_\sigma a_{i\sigma}^\dagger a_{i\sigma}$ is the electron number operator for molecule i = 1, 2 ($a_{i\sigma}^\dagger$ is the creation operator for an electron of spin σ). This is related to the fact that the model Hamiltonians used to describe the dimers usually consider one single orbital per molecular site.

A . Electronic spectra of symmetric dimers

Let us discuss first the simplest dimeric system, that is, a dimer of two identical molecules (hereafter a *symmetric dimer*) with only one electron in the frontier orbital. Denoting with t the electron transfer interaction energy, the model Hamiltonian can be reduced to

$$H_E = -t(a_1^\dagger a_2 + a_2^\dagger a_1) \tag{2}$$

with the following eigenvectors and eigenvalues

$$\begin{aligned} |\Psi_1\rangle &= \frac{1}{\sqrt{2}}(a_1^\dagger + a_2^\dagger)|0\rangle; \qquad & E_1 &= -t \\ |\Psi_2\rangle &= \frac{1}{\sqrt{2}}(a_1^\dagger - a_2^\dagger)|0\rangle; \qquad & E_2 &= t. \end{aligned} \tag{3}$$

Hence, we have a single charge transfer transition at the frequency $\omega_{CT} = \omega_{21} = 2t/\hbar$ with transition matrix element $\mu_{12} = -ed/2$.

The dielectric function is

$$\hat{\varepsilon}(\omega) = \varepsilon_\infty + 4\pi \frac{N_d}{\hbar^2} \frac{td^2e^2}{\omega_{CT}^2 - \omega^2 - i\omega\gamma_{CT}}. \tag{4}$$

N_d is the number density of dimers, that is $N/2$. It is seen that the oscillator strength of the CT transition, $f^{\circ}_{CT} = td^2e^2/\hbar$, linearly depends upon the electron transfer energy.

Let us switch to the case of a symmetric dimer with two electrons. Here the model Hamiltonian must include electron correlations. The simplest approximate way of doing this is by using a Hubbard model which includes an effective on-site electron correlation U.[6] Thus:

$$H_E = -t\sum_\sigma (a^\dagger_{1\sigma}a_{2\sigma} + a^\dagger_{2\sigma}a_{1\sigma}) + U(n_{1\uparrow}n_{1\downarrow} + n_{2\uparrow}n_{2\downarrow}). \tag{5}$$

The well-known eigenvalues and eigenvectors[6] are summarized in Table I.

TABLE I. Eigenvectors and eigenvalues of the Hubbard dimer Hamiltonian.

Basis states	Eigenvectors	Eigenvalues
$\lvert S_0\rangle = \frac{1}{\sqrt{2}}(a^\dagger_{1\uparrow}a^\dagger_{2\downarrow} + a^\dagger_{2\uparrow}a^\dagger_{1\downarrow})\lvert 0\rangle$	$\lvert 1\rangle = b_1\lvert S_0\rangle + c_1\lvert CT_+\rangle = A^\dagger_1\lvert 0\rangle$	$E_1 = E_-$
$\lvert CT_\pm\rangle = \frac{1}{\sqrt{2}}(a^\dagger_{1\uparrow}a^\dagger_{1\downarrow} \pm a^\dagger_{2\uparrow}a^\dagger_{2\downarrow})\lvert 0\rangle$	$\lvert 2\rangle = b_2\lvert S_0\rangle + c_2\lvert CT_+\rangle = A^\dagger_2\lvert 0\rangle$	$E_2 = E_+$
$\lvert T_1\rangle = a^\dagger_{1\uparrow}a^\dagger_{2\uparrow}\lvert 0\rangle$	$\lvert 3\rangle = \lvert CT_-\rangle = A^\dagger_3\lvert 0\rangle$	$E_3 = U$
$\lvert T_0\rangle = \frac{1}{\sqrt{2}}(a^\dagger_{1\uparrow}a^\dagger_{2\downarrow} - a^\dagger_{2\uparrow}a^\dagger_{1\downarrow})\lvert 0\rangle$	$\lvert 4\rangle = \lvert T_1\rangle = A^\dagger_4\lvert 0\rangle$	$E_4 = 0$
$\lvert T_{-1}\rangle = a^\dagger_{1\downarrow}a^\dagger_{2\downarrow}\lvert 0\rangle$	$\lvert 5\rangle = \lvert T_0\rangle = A^\dagger_5\lvert 0\rangle$	$E_5 = 0$
	$\lvert 6\rangle = \lvert T_{-1}\rangle = A^\dagger_6\lvert 0\rangle$	$E_6 = 0$
$b_{1,2} = [1 + (E_{1,2}/2t)^2]^{-\frac{1}{2}}$	$b^2_{1,2} + c^2_{1,2} = 1$	$E_\pm = \frac{1}{2}(U \pm \sqrt{U^2 + 16t^2})$

It is convenient to introduce creation (annihilation) operators $A^\dagger_n$ (A_n) for the eigenstates of the Hubbard dimer which render H_E diagonal, that is $H_E = \sum_n E_n A^\dagger_n A_n$. Based on these operators,the electric dipole operator is:

$$\begin{aligned}\mu &= -\frac{1}{2}ed(n_1 - n_2) = -ed[c_1(A^\dagger_1 A_3 + A^\dagger_3 A_1) + c_2(A^\dagger_2 A_3 + A^\dagger_3 A_2)] \\ &= \mu_{13}(A^\dagger_1 A_3 + A^\dagger_3 A_1) + \mu_{23}(A^\dagger_2 A_3 + A^\dagger_3 A_2).\end{aligned} \tag{6}$$

The corresponding matrix form is given in Table II.

TABLE II. Matrix representation of the operators μ, H_{emv}, H_{ep} for the symmetric dimer[a].

$$\mu \equiv \begin{vmatrix} 0 & 0 & \mu_{13} \\ & 0 & \mu_{23} \\ H.c. & & 0 \end{vmatrix}; \quad H_{emv} \equiv \begin{vmatrix} 0 & 0 & X_{13} \\ & 0 & X_{23} \\ H.c. & & 0 \end{vmatrix}; \quad H_{ep} \equiv \begin{vmatrix} Y_{11} & Y_{12} & 0 \\ & Y_{22} & 0 \\ H.c. & & 0 \end{vmatrix}$$

$$X_{mn} = \sum_i V_{mn}(i) q_i \qquad Y_{mn} = \sum_e P_{mn}(e) u_e$$

[a] Singlet eigenstates of the Hamiltonian H_E, Eq. (5), are used as the basis set. Definitions of the matrix elements are given in the text, Eqs. (6), (18), and (19).

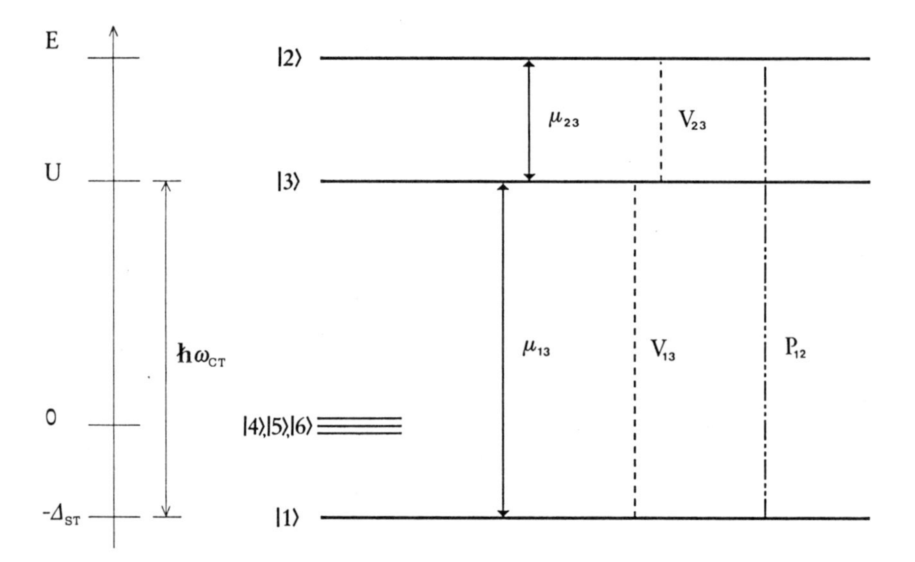

FIG. 1. Energy levels, transition dipoles, and electron-vibration (Herzberg–Teller) couplings in the symmetric Peierls–Hubbard dimer.

Figure 1 shows that the allowed dipole transitions are those from the ground state $|1\rangle$ and from the symmetric CT state $|2\rangle$ to the antisymmetric one $|3\rangle$. The single CT transition starting from the ground state occurs at a frequency

$$\omega_{CT} = \omega_{31} = \frac{1}{2\hbar}(U + \sqrt{U^2 + 16t^2}) \tag{7}$$

with matrix element $\mu_{CT} = \mu_{31}$. Thus one can see that for $U \gg 4t$ the location of the CT transition is directly related to the magnitude of the on-site correlation U.

The dielectric function is given by

$$\hat{\varepsilon}(\omega) = \varepsilon_\infty + 4\pi \frac{2N_d}{\hbar^2} \frac{1}{[1+(U/4t)^2]^{1/2}} \frac{td^2e^2}{\omega_{CT}^2 - \omega^2 - i\omega\gamma_{CT}}. \tag{8}$$

Comparison of Eqs. (8) and (4) shows that, beside the factor of 2, due to the fact that the number of electrons has doubled in respect to the previous case, there is a factor $[1+(U/4t)^2]^{-1/2}$, that accounts for a reduction of the oscillator strength with increasing on-site correlation U. This is a general phenomenon, not limited to the dimer model. Its origin has been discussed in quite general terms by Maldague[7] and by Mazumdar and Soos.[8]

The sum rule of oscillator strengths for transitions to all possible final states starting from the ground state $|0\rangle$ can be written as:

$$F_\circ = \sum_n f_{n\circ} = \frac{1}{2}\langle 0|[[\mu, H], \mu]|0\rangle. \tag{9}$$

In the point-charge approximation the dipole operator commutes with any Hamiltonian term containing electron number operators or their products. Therefore only the electron transfer term h_t contributes to the sum rule with its ground state expectation value, thus

$$F_\circ = -e^2d^2\langle h_t\rangle_\circ \tag{10a}$$

that is, in our case

$$F_\circ = e^2d^2t\langle\sum_\sigma(a_{1\sigma}^\dagger a_{2\sigma} + a_{2\sigma}^\dagger a_{1\sigma})\rangle_\circ. \tag{10b}$$

Stated in another way, each time there are terms in the Hamiltonian that tend to localize the electronic states, part of the oscillator strength is transferred to intramolecular electronic transitions (localized excitations, LE) that cannot be described by Hamiltonians considering one single orbital per site.

We can summarize this brief discussion of the electronic spectroscopy of dimers by noting that, in the dimer with two electrons, which is the simplest molecular cluster analog of an half-filled band system, the effect of the effective on-site correlation U is twofold: (i) it shifts the charge transfer band to higher energies; (ii) it reduces the oscillator strength of the CT transition. This behavior is shown in Fig. 2 for dimensionless quantities.

B. Vibronic spectra of symmetric dimers

Let us now introduce the electron–vibration interaction in the symmetric dimer, to account for the modulation of the site energies, and of the electron transfer integral, by the intramolecular vibrational modes and by the rigid molecular motions respectively. This leads us to the so-called Peierls–Hubbard model.[9]

The Hamiltonian now reads:

$$H = H_E + H_V + H_{EV} \tag{11}$$

where H_E is the Hubbard Hamiltonian defined before,

$$H_V = \sum_i \frac{\hbar\omega_i}{4}(P_{1i}^2 + P_{2i}^2 + Q_{1i}^2 + Q_{2i}^2) + \sum_e \frac{\hbar\omega_e}{4}(v_e^2 + u_e^2) \tag{12}$$

and

$$H_{EV} = \sum_i g_i(Q_{1i}n_1 + Q_{2i}n_2) - \sum_e \sum_\sigma g_e u_e (a^\dagger_{1\sigma}a_{2\sigma} + a^\dagger_{2\sigma}a_{1\sigma}) \tag{13}$$

Q_{ni} and ω_i are the dimensionless coordinate and the angular frequency of the ith intramolecular mode of the nth molecule. u_e and ω_e have corresponding meaning for the intermolecular (external) modes. As usual, the parameters $g_i = (\partial\varepsilon/\partial Q_i)_\circ$ and $g_e = (\partial t/\partial u_e)_\circ$ specify the strengths of the linear electron–molecular vibration (e–mv) and electron–intermolecular phonon (e–p) coupling respectively. ε is the potential energy surface of the frontier molecular orbital.

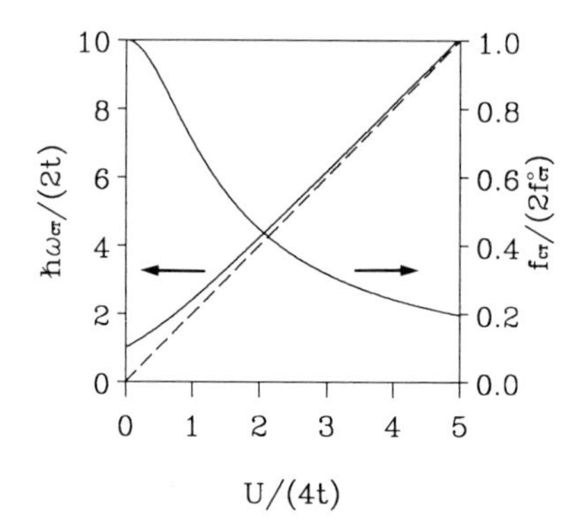

FIG. 2. Dimensionless CT transition energy and normalized oscillator strength *vs.* dimensionless on–site correlation energy in the symmetric Hubbard dimer. The dashed straight line is $\hbar\omega_{CT} = U$; f°_{CT} is the oscillator strength of the dimer with one electron.

Following the pioneering work of Rice,[10] once the intramolecular vibrational modes are properly symmetrized, by making in–phase (symmetric) s_i and anti–phase (anti–symmetric) q_i combinations, $s_i = 2^{-1/2}(Q_{1i} + Q_{2i})$; $q_i = 2^{-1/2}(Q_{1i} - Q_{2i})$, one immediately sees that there is complete decoupling of the symmetric intramolecular modes from the electronic degrees of freedom. This holds true as long as the coupling between CT and LE transitions is negligible. While in the model Hamiltonian this is granted by considering one single orbital per site, some caution is needed with real systems.

The e–mv and e–p coupling terms of the Hamiltonian,

$$H_{emv} = \sum_i \frac{1}{\sqrt{2}} g_i q_i (n_1 - n_2) \tag{14}$$

and

$$H_{ep} = -\sum_e g_e u_e \sum_\sigma (a^\dagger_{1\sigma}a_{2\sigma} + a^\dagger_{2\sigma}a_{1\sigma}) \tag{15}$$

can be rewritten using the creation (annihilation) operators $A_n^\dagger (A_n)$ for the eigenstates of the Hubbard dimer:[11]

$$H_{emv} = \sum_i V_{13}(i) q_i (A_1^\dagger A_3 + H.c.) + \sum_i V_{23}(i) q_i (A_2^\dagger A_3 + H.c.) \tag{16}$$

and

$$H_{ep} = \sum_e P_{11}(e) u_e A_1^\dagger A_1 + \sum_e P_{22}(e) u_e A_2^\dagger A_2 + \sum_e P_{12}(e) u_e (A_1^\dagger A_2 + H.c.) \tag{17}$$

(also reported in matrix form in Table II) with the following definition of the coupling constants:

$$V_{13}(i) = \sqrt{2} c_1 g_i \quad ; \quad V_{23}(i) = \sqrt{2} c_2 g_i \tag{18}$$

$$\begin{aligned} P_{12}(e) &= -2(b_1 c_2 + b_2 c_1) g_e \\ P_{11}(e) &= -4 b_1 c_1 g_e \quad ; \quad P_{22}(e) = -4 b_2 c_2 g_e . \end{aligned} \tag{19}$$

The terms which appear in the interaction Hamiltonians H_{emv} and H_{ep} can be recast in the language of molecular physics.[12] Those which are off–diagonal in the electronic eigenstates give rise to Herzberg–Teller couplings. They are responsible for non–Condon effects in vibronic absorption, emission and scattering spectroscopies.[12,13] Furthermore, the fact that several modes can couple with the same pair of states, either through the $V_{13}(i)$ (or $V_{23}(i)$) terms for intramolecular modes or through the $P_{12}(e)$ terms for intermolecular ones, bears the implication of frequency shifts and Duschinsky rotation, i.e. mode mixing, of the normal coordinates.[14] This is easily seen when one considers that the potential energy $E_\circ(Q)$, expanded to second order in the vibrational coordinates Q_i, contains products of the vibronic matrix elements. In fact

$$E_\circ(Q) = E_\circ(0) + \sum_i A_i^\circ Q_i + \frac{1}{2} \sum_{ij} B_{ij}^\circ Q_i Q_j \tag{20a}$$

where

$$\begin{aligned} B_{ij}^\circ &= \langle 0 | \frac{\partial^2 H}{\partial Q_i \partial Q_j} | 0 \rangle - \sum_n \frac{2}{\hbar \omega_{n\circ}} \langle n | \frac{\partial H}{\partial Q_i} | 0 \rangle \langle 0 | \frac{\partial H}{\partial Q_j} | n \rangle \\ &= F_{ij} - \sum_n \frac{2 V_{n\circ}(Q_i) V_{\circ n}(Q_j)}{\hbar \omega_{n\circ}} , \end{aligned} \tag{20b}$$

$|n\rangle$ is a crude adiabatic basis state, F_{ij} is a force constant matrix element. Note that, when second-order terms diagonal in the vibrational coordinate index are considered, the vibronic term of B_{ii}° is always negative, implying a tendency to downshift the mode frequency. Taking into account Eqs. (18) and (6), one can see that for intramolecular modes such a vibronically induced frequency shift is larger, the larger the e–mv coupling strength g_i, and the oscillator strength of the CT transition, the lower its excitation energy. On the other hand, terms such as $P_{11}(e)$ and $P_{22}(e)$, which are diagonal in the electronic state index, originate displacements of the vibrational coordinates, hence Franck–Condon factors and associated vibrational progressions in the spectra.[12,13]

According to the vibronic theory of infrared intensities,[12] the vibronic contribution to the transition dipole moment of the vibrational mode Q is related to

$\sum_n \mu_{no} V_{no}(Q)/\hbar\omega_{no}$. Therefore, if the ground state is connected to some excited state, both by a dipole allowed transition and by a Herzberg–Teller coupling, the vibrational mode taking part in such a coupling borrows oscillator strength from the electronic transition, and hence is vibronically enhanced in the infrared spectra. Comparing the matrix representations of the coupling Hamiltonians H_{emv} and H_{ep} with that of the dipole moment operator (Table II), it is easily seen that the above condition is fulfilled only for the anti-phase intramolecular modes q_i of the dimer (coupling terms $V_{13}(i)$). It is understood, of course, that these modes must correspond to molecular vibrations with non vanishing coupling constants g_i. As is well known, for non-degenerate orbitals and according to the Hellmann–Feynmann theorem, this occurs only with totally symmetric modes.

According to the above arguments, the very low frequency of the CT transitions is chiefly responsible for the overwhelming intensity of the vibronically enhanced absorptions in the infrared spectra of CT compounds.

Let us shortly digress by commenting on the fact that also the transition $|3\rangle \rightarrow |2\rangle$ is dipole-allowed and exhibits Herzberg–Teller coupling with the anti-phase intramolecular vibrations (the terms $V_{23}(i)$). Spectroscopically, this must imply the presence of vibronic structures in the CT absorption band analogous to false origins in the electronic molecular spectra.

Detailed calculations of the optical and infrared properties of the Peierls–Hubbard dimer with coupling to the intramolecular vibrations only have been reported following somewhat different schemes.[10,15,16] They have been recently reviewed and compared by Painelli and Girlando.[16] Without entering into technical details, we simply recall the well-known final result of the adiabatic, linear-response theory developed by Rice[10] and give an intuitive picture of the phenomena involved. The infrared conductivity along the direction joining the molecular centers in the dimer has been derived to be:

$$\hat{\sigma}(\omega) = -i\omega \frac{e^2 d^2}{4} N_d \frac{\chi(\omega)}{1 - \tilde{\chi}(\omega) D(\omega)} \tag{21}$$

where

$$\chi(\omega) = \frac{|\mu_{CT}|^2 2\omega_{CT}}{\omega_{CT}^2 - \omega^2 - i\omega\gamma_{CT}} \tag{22}$$

is the CT electronic susceptibility, $\tilde{\chi}(\omega) = \chi(\omega)/\chi(0)$ denotes its reduced form and

$$D(\omega) = \sum_i \frac{\lambda_i \omega_i}{\omega_i^2 - \omega^2 - i\omega\gamma_i} \tag{23}$$

is the phonon propagator in which the λ_i are dimensionless e–mv coupling parameters defined as: $\lambda_i = g_i^2 \chi(0)/\hbar\omega_i$. The expression for $\hat{\sigma}(\omega)$ describes the dimer charge transfer absorption, as well as a series of vibronically enhanced infrared absorptions due to the so called *dimer charge oscillations* all having the same polarization along the line joining the two molecules. Pictorically, the dimer modes corresponding to the anti-phase coupling of the vibrations of the individual molecules cause an alternating flow of electronic charge induced by the instantaneous inbalance of the site energies which are modulated by the vibrations.

The renormalized frequencies of the vibronic peaks are found as poles of $\hat{\sigma}(\omega)$ or, equivalently, as solutions of the equation $Re[\tilde{\chi}(\omega)D(\omega)] = 1$. The results confirm our expectations, based on the simple arguments of the vibronic theory, both for what concern the downshift of the frequencies and the dependence of the corresponding absorption intensities on the various model parameters.

We shall try to illustrate most of the things we have outlined up to this point with a single, if somewhat unconventional, example. We shall report and discuss the polarized infrared absorption spectra of a Langmuir–Blodgett (LB) film made out of the *CT* salt (octadecylphenanthrolinium^{++}) $(TCNQ^-)_2$ taken before and after treatment of the film with I_2 vapor.[17] One can easily obtain certain indications, from the features of the isothermal compression curve for the film at the air/water interface, from the dichroic behavior in the infrared, and from the insulating nature of the films, that their likely structure is as follows: phenanthrolinium dications intervene between $(TCNQ^-)_2$ dimers so as to effectively isolate them from each other; the molecular planes of both phenanthrolinium and $TCNQ^-$ lie on edge on the substrate. Treatment with I_2 yields oxidation of every other $TCNQ^-$ radical, leading to a structure of isolated mixed valence ($TCNQ^-$ $TCNQ^\circ$) dimers. Thus, this system enables us to compare the infrared spectra of isolated $(TCNQ^-)_2$ dimers with those of mixed-valence ($TCNQ^-$ $TCNQ^\circ$) dimers embedded in an ordered matrix that undergoes a minimum of structural changes in going from one situation to the other.

Comparison of the spectra before and after iodination (Fig. 3) shows that the broad absorption around 9000 cm^{-1} (1.1 eV), associated with the *CT* transition in the dimer with two electrons, shifts to 5000 cm^{-1} (0.6 eV) in the dimer with one single electron. The spectra in the mid-infrared region exhibit the typical absorptions due to the dimer charge oscillations (related to the totally symmetric modes a_g ν_3, ν_4, and ν_5 of TCNQ) at 1575, 1351, and 1177 cm^{-1} for the $(TCNQ^-)_2$ dimers and at 1567, 1347, and 1163 cm^{-1} for the mixed-valence dimers. As predicted, their polarization is the same as that of the *CT* absorptions, and their frequency shifts downward as the *CT* band moves to lower energy.

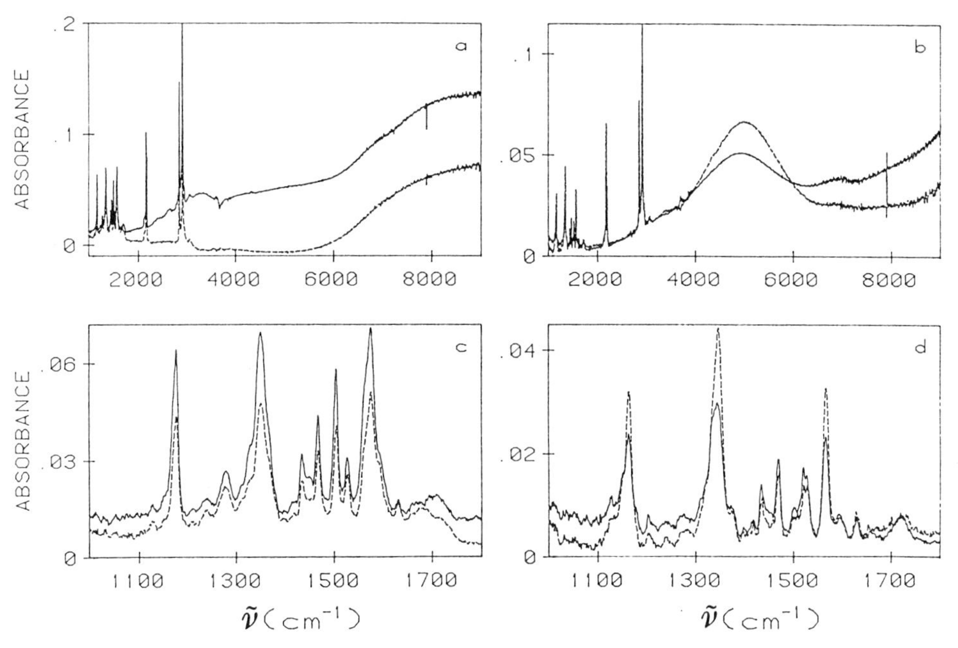

FIG. 3. Infrared spectra of LB films of (octadecylphenanthrolinium) $(TCNQ)_2$: (a) and (c) pristine form; (b) and (d) after treatment with I_2. Full and dashed curves are used for spectra obtained with electric vector parallel to the film plane and making 60° with it, respectively.

Comparison of the absorption intensities before and after the iodination process is made possible by using the C–H stretching absorptions of the long aliphatic chains (at 2851 and 2917 cm^{-1}) as internal standards. We have verified that their intensity is only a function of the number of layers in the LB film,and does not vary appreciably upon iodination. The peak intensities of the CT absorptions of the two samples turn out to be roughly in a 1:1 ratio. That means that in $(TCNQ^-)_2$ dimer the CT intensity is about half of that expected for the case of two uncorrelated electrons. Using the expression for the dielectric function given above,we get a ratio $U/t \approx 7$, which is consistent with the currently accepted values of $t \approx 0.15$ eV and $U \approx 1$ eV for insulating TCNQ salts. This analysis of the intensity data for the LB films is, however, a preliminary one. The changes in the dichroic behavior of the absorptions due to dimer charge oscillations show that there is an orientational rearrangement of the dimers upon iodination. Whether this induces relevant changes in the intradimer spacing d and in the transfer integral t cannot be said at the moment.

The availability of more detailed structural data and particularly of ESR studies would make this LB film an interesting model system for investigating further on the basics of dimer spectroscopy.

C. Resonance Raman Scattering of symmetric dimers

Let us now go back to considering the spectroscopic effects of the electron–intermolecular phonon (e–p) coupling. We first note that, for a symmetric dimer possessing inversion symmetry only, all three anti–phase combinations of the rigid translational motions of the two molecules belong to the gerade symmetry, and are therefore expected to be Raman-active. Furthermore, these modes modulate effectively the transfer integral,so that they exhibit non-zero coupling constant $g_e = (\partial t/\partial u_e)_o$.

In this section we shall discuss the enhancement of the scattering intensity when the Raman spectra are excited in resonance with the CT transition.[11] Among the many well established theories of resonance Raman scattering (RRS), Albrecht's vibronic model[18] is possibly best suited for discussing the underlying physics of the resonance enhancement. Using a first-order Herzberg–Teller expansion of the Born–Oppenheimer states, three different mechanisms are identified. The so–called A–type mechanism involves Franck–Condon factors, hence it implies displacements of the adiabatic potential minima. The B and C mechanisms are of a non–Condon type, with the former depending on a Herzberg–Teller coupling between excited states, the latter on a coupling involving the ground state.

For a Peierls–Hubbard dimer, consideration of the interaction scheme given above in Fig. 1 and Table II, shows that only the intermolecular phonon modes can be active in the A–type resonance enhancement. On the other hand, mechanisms B and C require that the two states, which are Herzberg–Teller coupled, are also individually coupled to a third state by allowed one–photon transitions. Note, in fact, that these mechanisms correspond to three–band terms of the theory of RRS from semiconductors,[19] whereas the A–type mechanism corresponds to two–band ones. The only coupling terms satisfaying the above condition are the $P_{12}(e)$. So, again, only the intermolecular phonons undergo resonance enhancement through a non–Condon mechanism. More specifically, since $P_{12}(e)$ couples the ground state with the symmetric CT state, it gives rise to C–type enhancement.

Actual calculation of the RRS spectra and excitation profiles of a PH dimer are not yet available, and we are currently working on that. One interesting qualitative feature is worth mentioning, though. The relative importance of the A–type and C–type enhancement mechanisms is determined by the coefficients $[P_{22}(e) - P_{11}(e)]$ and $P_{12}(e)$ respectively. Using the explicit expressions for these coupling coefficients in terms of the basic electronic interactions t and U (Eq. (19)), one can see that the C–type non–Condon mechanism is induced by the on–site Coulomb correlation U. Analysis of possible interference phenomena in the excitation profiles and in the polarization dispersion could therefore provide information on the electron correlations.

The main qualitative predictions of this model are strikingly demonstrated[11] by the low-temperature RRS spectra of crystalline (TTF)Br whose structure[20] consists of almost isolated $(TTF^+)_2$ dimers. The electronic absorption spectrum of polycrystalline (TTF)Br at 12 K is shown in Fig. 4a . The long wavelength absorption, displaying a doublet structure around 700 nm,is due to the *CT* transition of the dimers. The vertical bars indicate the laser wavelengths used to excite the Raman spectra. Those obtained with excitation wavelength at $\lambda_o = 530.9$ nm and at $\lambda_o = 647.1$ nm, resonant with LE and *CT* transition respectively, are shown in Fig. 4b.

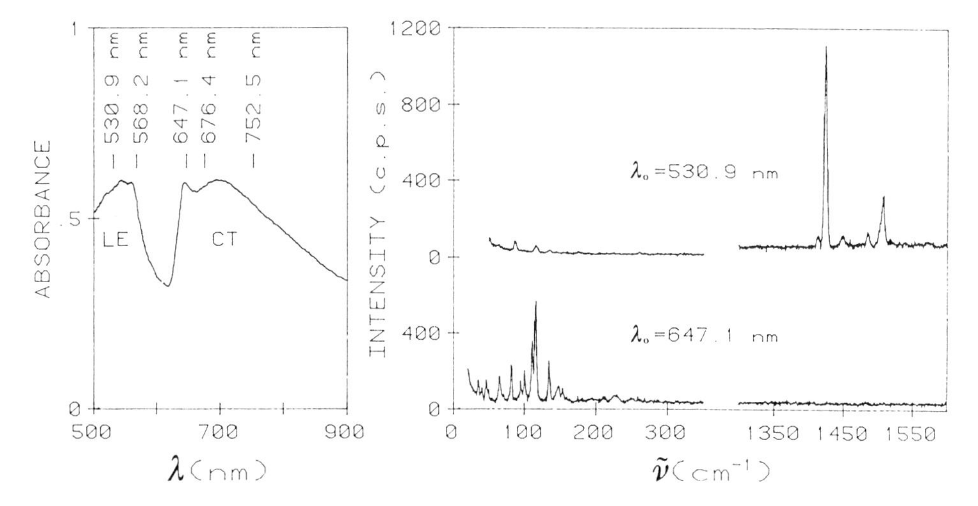

FIG. 4. Electronic absorption spectra (left) and resonance Raman spectra (right) of (TTF)Br powders at 20K. (After Ref. 11).

The fact that using excitation in the *CT* absorption a number of Raman bands appears in the intermolecular mode region,while the intramolecular bands practically disappear, can be understood on the basis of the dimeric model. The modulation of the transfer integral t by the intermolecular vibrations, particularly the anti-phase translational modes, provides an efficient mechanism for intensity enhancement at resonance with the *CT* transition. No such mechanism is operative for the intramolecular vibrational modes.

The $(TTF^+)_2$ dimer that we have used to test our spectral predictions appears to be well suited as a prototype system. In fact,the electron transfer interaction between TTF^+ cation radicals is strong,and the high symmetry of the dimer (effectively D_{2h}) prevents any mixing of the *CT* and the LE excitations. We emphasize that the selective RRS enhancement of the intramolecular phonon modes is a distinctive feature of the present model and test case. It occurs under the restrictive conditions that (i) the dimer is symmetric; (ii) the *CT* states do not mix with localized molecular excitations. When any of these two conditions is broken,the intramolecular modes can participate in the RRS enhancement.

D. Spectroscopy of unsymmetric dimers and donor–acceptor pairs

Let us now extend the arguments developed for the case of symmetric dimers to that of unsymmetric ones. When the asymmetry is due to the fact that the two molecules forming the dimer belong to chemically different species we are actually dealing with a donor–acceptor complex.[21]

At the same level of approximation used before for the symmetric dimer, but considering only the e–mv coupling, the Hamiltonian reads

$$H = H_E + H_{emv} + H_V \tag{24}$$

with

$$H_E = \Delta(n_1 - n_2) - t\sum_{\sigma}(a^{\dagger}_{1\sigma}a_{2\sigma} + a^{\dagger}_{2\sigma}a_{1\sigma}) + U_1 n_{1\uparrow}n_{1\downarrow} + U_2 n_{2\uparrow}n_{2\downarrow}, \tag{25}$$

$$H_{emv} = \sum_i g_{1i} n_1 Q_{1i} + \sum_j g_{2j} n_2 Q_{2j}. \tag{26}$$

H_V is the vibrational Hamiltonian for intramolecular modes having different frequencies on molecule 1 and 2. In Eq. (25) Δ is an effective one–electron energy difference between the two molecular sites.

The basic difference with respect to the previous case is that, since both the electronic site energies and the intramolecular vibrational frequencies are now different for the two moieties, neither the electronic wave functions,nor the vibrational coordinates couple to give symmetric and anti–symmetric combinations. One other difference, which,however,is not as fundamental as the previous one,is that, in dealing with donor–acceptor pairs, the electronic basis set is usually restricted by neglecting relatively high energy configurations in which the acceptor hosts two electrons while the donor is empty. It is easily seen that both the transition dipole moment and the linear e–mv interaction Hamiltonian can be expressed in terms of a single electron number operator (n_1 or n_2) and thus have isomorphous matrix representations. Therefore, it is obvious that both the donor and acceptor modes, having non–zero coupling constants,undergo the same spectroscopic phenomena described before for the anti–phase modes of the symmetric dimer. Namely, in infrared they borrow intensity from the *CT* transition and their frequency shifts downwards. Each mode can be observed both in infrared and in Raman since it is active in both spectroscopies.

One interesting feature which is worth emphasizing is that both the intensity of the *CT* absorption and of the vibronic peaks and the vibronic frequency shifts depend on the degree of ionicity ρ of the $D^{\rho+}A^{\rho-}$ pair as $\rho(1-\rho)$. Hence, the extent to which the above phenomena are observed is dependent on the degree of ionicity of the complex, being maximum when $\rho = 0.5$, and vanishing in the two limits $\rho = 0$ and $\rho = 1$.

III. SPECTROSCOPY OF HALF–FILLED SYSTEMS

The applicability of dimeric models for the interpretation of dimerized half–filled systems, such as ion-radical salts and donor–acceptor crystals, has been tested in numerous cases.[22–24] A theoretical rationale for such an applicability has been very recently provided by numerical calculation[25] of the optical properties of regular and dimerized strongly correlated half–filled systems. The results demonstrate that, with strong on–site correlation U and large dimerization, a strong absorption appears at exactly the same energy $\hbar\omega_{CT}$ given by Eq. (7) for the isolated dimer. Besides providing a qualitative understanding of the observed spectroscopic phenomena, application of the appropriate dimeric models for the fitting of the experimental data enables one to obtain estimates of some basic electronic and electron–vibration interaction parameters.

A. Selection rules and spectral predictions

Before giving a few examples of applications of dimeric models, and for the sake of comparison, it is worth to derive a few spectral predictions for the case of regular stacks.

Consider a regular segregated stack, i.e., one of regularly spaced identical molecules. Corresponding to each intramolecular mode, the system possesses one single phonon branch, with the $\vec{k} = 0$ mode corresponding to the in-phase motion of all the molecules. Analogy with the case of the in-phase dimer vibration shows that there is no coupling of this mode to the radical electrons hence no vibronic effect, that is, no frequency perturbation and intensity borrowing in infrared or resonance enhancement in Raman.

In a mixed (donor-acceptor) regular stack there are $\vec{k} = 0$ modes corresponding, to a first approximation, to vibrations of the donors only or the acceptors only. These modes are now coupled to the radical electrons, since they induce modulations of energy, hence charge, on alternating sites. However, since both the modes and the charge redistributions are of gerade symmetry, the frequency perturbation and intensity effects manifest themselves only in Raman scattering. It can be shown that in this case also the effects are greatest when $\rho = 0.5$.[16]

Let us now summarize our discussion of the spectroscopic properties related to the e-mv coupling in regular and dimerized stack systems. Based on the symmetry arguments just presented, and on the simple dimeric models, the overall picture sketched in Table III emerges. Note that, as a useful mnemonic, the various types of stacks can be simply characterized by associating them with a definite set of parity symmetry elements that, for the sake of illustration, are identified with inversion centers. The phonon modes generated in the stack by a single, totally symmetric (a_g), intramolecular mode are labelled A_g, either when the molecules sit on a center of symmetry, or when the motion of molecules related by the interchange symmetry in the unit cell occurs in-phase. The label A_u designates the corresponding anti-phase motion. The A symmetry is trivial, and poses no restriction on the shape of the phonon modes.

Thus, the regular segregated (*rs*) stack possesses inversion symmetry *on* the molecular sites and *between* them. The only crystal mode of A_g symmetry is Raman active and unaffected by the e-mv coupling.

The dimerized segregated (*ds*) stack has inversion symmetry only *between* molecular radicals. The A_g phonon mode is Raman-active and unperturbed, the anti-phase A_u mode is shifted in frequency and vibronically enhanced in infrared. Thus no frequency coincidence between infrared and Raman spectra is to be expected in this case.

In the regular mixed (*rm*) stack the only symmetry elements are the inversion symmetry points *on* which the molecules sit. Each donor or acceptor a_g mode gives rise to an A_g crystal mode, which is active only in Raman spectra at perturbed frequencies.

Finally, the dimerized mixed (*dm*) stack does not possess symmetry at all. Each mode is both infrared and Raman-active and perturbed by the vibronic interactions. Coincidence of frequencies between infrared and Raman spectra should therefore be expected for *dm* stacks.

Similar selection rules could be worked out for the effect of the e-p coupling but they are less useful as a tool for investigating structural properties. In fact, whereas the intramolecular frequencies and e-mv coupling constants are (to a good approximation) molecular parameters that carry over from one material to another, the intermolecular phonons and their coupling to electrons are peculiar to a specific structure, and should be investigated case by case.

Finally, we must emphasize that the selection rules outlined before are only valid for the case of dimerized and/or alternating (mixed) stack systems. More complicated structures such as trimerized, tetramerized or even incommensurate ones

TABLE III. Summary of selection rules for vibronic effects (related to e–mv coupling).

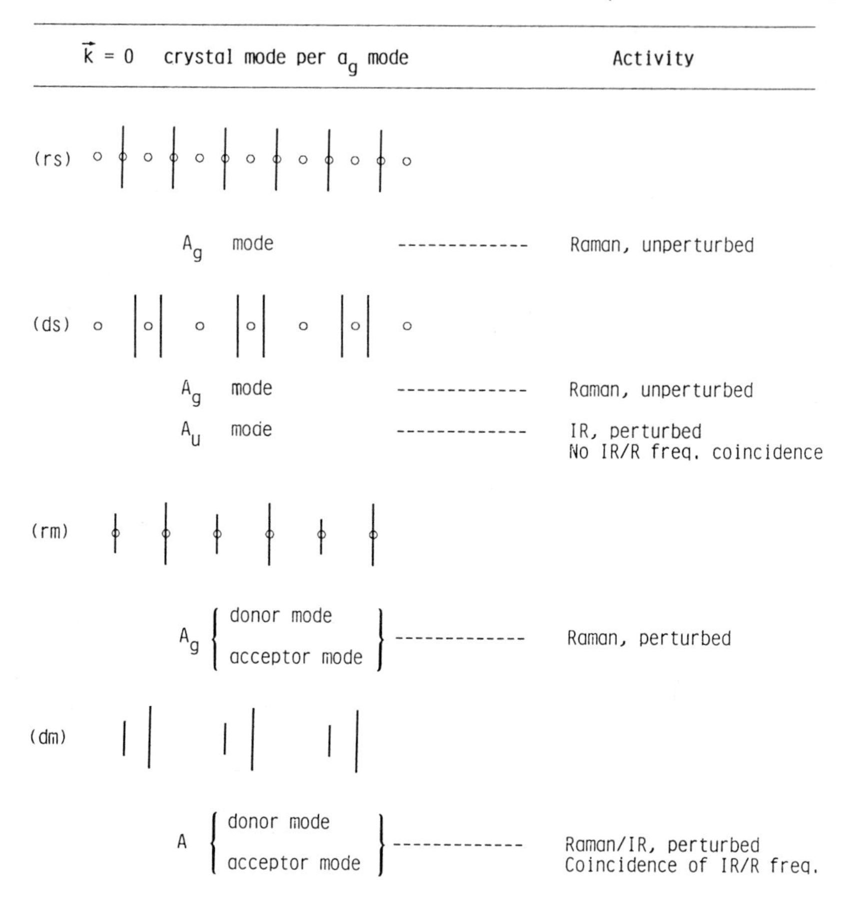

	$\vec{k} = 0$ crystal mode per a_g mode	Activity
(rs)	A_g mode	Raman, unperturbed
(ds)	A_g mode	Raman, unperturbed
	A_u mode	IR, perturbed No IR/R freq. coincidence
(rm)	A_g { donor mode, acceptor mode }	Raman, perturbed
(dm)	A { donor mode, acceptor mode }	Raman/IR, perturbed Coincidence of IR/R freq.

are characterized by a greater number of $\vec{k} = 0$ crystal modes. As a consequence one always has both infrared– and Raman–active vibronically perturbed modes and the selection rules are less clear-cut than those of Table III.

B. Applications to ion radical salts and CT crystals

The infrared spectra [26] of $RbTCNQF_4$ shown in Fig. 5 provide an illustrative example of the dramatic spectral changes displayed by half–filled ion radical stack systems when they undergo a structural transition from regular to dimerized.

The prominent new bands observed in the low-temperature infrared spectrum are due to the activation of dimer charge oscillations induced by the A_u crystal components of totally symmetric (a_g) intramolecular modes at frequencies Ω_i somewhat lower than those (ω_i) of the corresponding unperturbed A_g components. The latter can be measured directly in the Raman spectrum and the set of vibronic frequency shifts ($\omega_i - \Omega_i$) so derived allows one to evaluate the relative strength of the e–mv coupling constants g_i of the various a_g intramolecular modes. This is made possible by solving a set of linear coupled equations in the unknown quantities λ_i, i.e., the dimensionless coupling costants

$$\frac{\omega_{CT}^2 - \Omega_i^2}{\omega_{CT}^2} - \sum_i \frac{\lambda_i \omega_i^2(\omega_i^2 - \Omega_i^2)}{(\omega_i^2 - \Omega_i^2)^2 + \Omega_i^2\gamma_i^2} = 0. \tag{27}$$

These coupled equations are obtained from the equation $Re[\tilde{\chi}(\omega)D(\omega)] = 1$ by replacing for ω one of the experimentally observed vibronic frequencies Ω_i in turn.

The case of a mixed dimerized stack system is well exemplified by the room temperature infrared spectra of M_2P–TCNQ (M_2P = N,N'–dimethyl–phenazine). The polarized reflectance spectra[27] of a single crystal mosaic (Fig. 6a) are dominated by bands whose polarization is parallel to that of the CT band observed around 5000 cm^{-1}. The assignment of these bands to vibronic effects, related to totally symmetric modes of both the donor and the acceptor, is given in Table IV.

A detailed analysis of the normally infrared–active frequencies allows the evaluation of the degree of ionicity, which turns out to be $\rho = 0.5 \pm 0.1$. This is therefore one of the very rare cases of mixed stack compounds with intermediate ionicity.

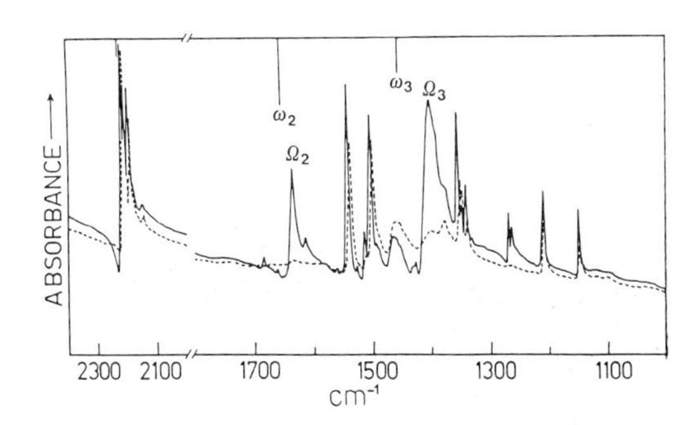

FIG. 5. Infrared spectra of $RbTCNQF_4$ at room temperature (dashed line) and at 20K (full line). (From Ref. 26)

The observation of vibronic absorptions in infrared is a consequence of the stack dimerization, while the coincidence of their frequencies with those observed in Raman, and shifted with respect to the bare ones, is related to the mixed nature of the stacks. The relevant vibronic intensities and the sizeable frequency shifts can be understood, based on the intermediate value of ρ.

The agreement between theory and measured data can be made quantitative through a spectral simulation[27] based on a mixed dimer model (Fig. 6b). The agreement is quite satisfactory, considering some uncertainty in the measured reflectance values, due to the small size of the crystals. The parameter values used in producing the theoretical fit and reported in Table IV provide an estimate of the basic interaction parameters, particularly the e–mv coupling constants for both the acceptor and the donor.

The related case of M_2P–$TCNQF_4$ offers us the possibility to demonstrate the effect of having the degree of ionicity close to the limit $\rho = 1$. Combination of

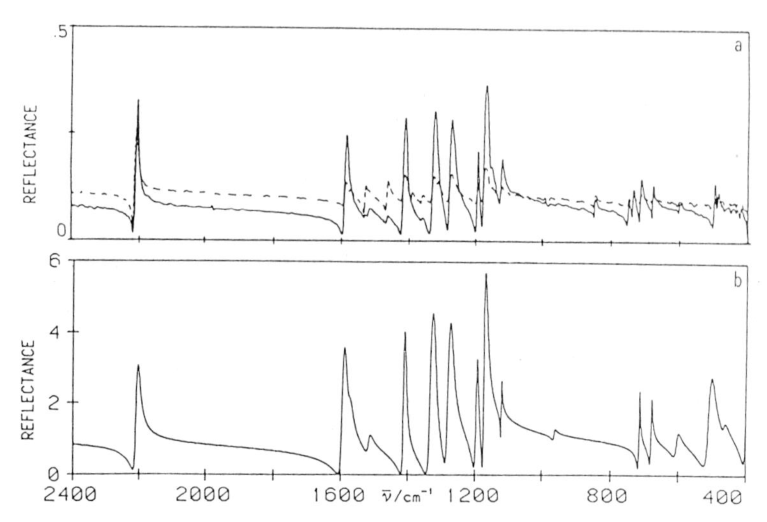

FIG. 6. Upper panel: polarized infrared reflectance spectra of a mosaic of single crystals of M_2P–TCNQ. Electric field vector parallel (full line) and perpendicular (broken line) to the stack axis. Lower panel: vibronic reflectance spectrum calculated on the basis of the dimer model. (From Ref. 27)

magnetic [28] and infrared spectroscopic [27] data show that this system undergoes a low temperature transition from a mixed regular to a mixed dimerized stack structure. The low temperature absorption spectrum of M_2P–$TCNQF_4$ shows the appearance of new bands,coincident in frequency with corresponding Raman bands,but their intensity is comparatively low,confirming the predicted scaling of all the vibronic effects in mixed stack compounds with the factor $\rho(1-\rho)$.

Finally, let us consider an example of how the spectroscopic phenomena and concepts we have been discussing can be fruitfully exploited to characterize some structural and electronic properties for cases in which, for the moment, other techniques are not viable or fail. Attempts at growing single crystals of TMPD–$TCNQF_4$ for structural investigations have been unsuccesful and, besides magnetic suseptibility measurements, [28,29] only powder infrared and Raman data [29] are available.

The 300 K powder spectrum (Fig. 7) clearly shows vibronic absorptions related to $TCNQF_4$ modes. Since their frequency is downshifted with respect to those measured in Raman (shown by bars in Fig. 7), this indicates a CT interaction typical of segregated dimerized systems. Further confirmation comes from the fact that, although analysis of the normally infrared active frequencies shows that the coumpound is fully ionic, still the vibronic intensities are rather strong, which would not be the case for mixed dimeric systems.

There is no indication on the packing of $TMPD^+$ radicals coming from the room temperature spectrum. The 15 K spectrum,instead,shows vibronic absorptions due to $TMPD^+$ modes,again at frequencies somewhat shifted from the Raman counterparts. The temperature dependence of their intensities, as well as of those of vibronic absorptions due to $TCNQF_4$, shows that the $TCNQF_4$ stacks are dimerized over the whole studied temperature range,whereas the TMPD stacks undergo a structural transition from regular to dimerized stacks at about 180 K.

The effective on–site electronic correlation parameter U (0.89; 1.56 eV), the hopping integral t (0.14; 0.11 eV), the magnetic gap Δ_{ST} (0.08; 0.03 eV) of both

TABLE IV. Dimer model parameters for the calculation of the vibronic spectra of M_2P-TCNQ.

TCNQ			M_2P		
$\omega_i(cm^{-1})$	$\gamma_i(meV)$	$g_i(meV)$	$\omega_i(cm^{-1})$	$\gamma_i(meV)$	$g_i(meV)$
2217	8	51	1613	15	21
1609	6.5	73	1576	18	16
1421	4	60	1519	15	26
1201	6	36	1346	7	66
964	8	11	1288	9	38
718	3	25	1176	3	26
608	15	30	1121	2	9
336	13	34	682	4	24
			523	18	55
			466	15	14
			406	10	39

$\omega_{CT} = 5000 cm^{-1}$ $\gamma_{CT} = 1600 cm^{-1}$ $\varepsilon_\infty = 1.3$ $N_d = 1.941 \cdot 10^{21} cm^{-3}$ $d = 3.22$Å

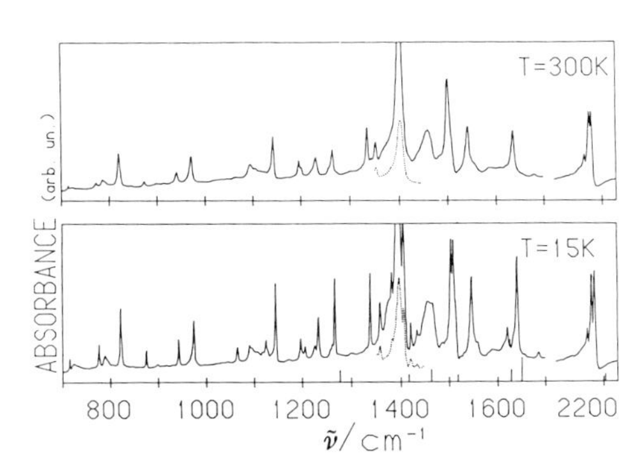

FIG. 7. Infrared spectra of Nujol mull of TMPD–$TCNQF_4$ at room temperature (upper panel) and at 15K (lower panel). (From Ref. 29)

stacks in the dimerized phase could be estimated[29] by using the dimer model and by comparison with known materials (values in parenthesis for $TCNQF_4$ and TMPD respectively).

IV. CONCLUSIONS

The fact that the *CT* crystals and molecular conductors exhibit peculiar spectroscopic phenomena has been recognized very early in this field. Part of them was rationalized rather soon by the pioneering work of Michael Rice[10,22] and connections with previously known concepts in the spectroscopy of molecular complexes have been pointed out.[16,30]

In this paper, illustrations of some peculiar aspects of the spectroscopy of *CT* crystals have been given with reference to simple dimeric models. Quantitative applications of such models are only appropriate for systems with localized electron states, such as insulating ion radical salts and donor–acceptor crystals. Analysis of spectroscopic data pertaining to these systems allows an accurate determination of some fundamental interaction parameters.

On the other hand, these models also provide a simple and intuitive picture for analogous spectroscopic phenomena, displayed by delocalized electron systems such as the organic molecular conductors with more complicated structures including incommensurate[31] and two–dimensional ones.[32]

Combined with some simple symmetry arguments applicable to infinite-stack systems with regular structures, the dimeric models provide us with a set of selection rules and spectral predictions that make infrared and Raman spectoscopy a powerful tool for structural investigations. These methods become rather unique if one considers the widespread availability of the required instrumentation and their applicability to disordered and amorphous materials like, e.g., thin films and Langmuir–Blodgett films which are more apt for applications than crystalline materials.

We must emphasize however that the straightforward application of the spectroscopic methods requires a previous extensive collection of basic molecular spectroscopic data. This has already been accomplished for most of the known key structures,[33] but would require further efforts as new structures emerge.

AKNOWLEDGMENTS

The authors wish to gratefully acknowledge the contribution of A. Girlando, M. Meneghetti, and A. Painelli in developing the research activity on which the present paper is based. Financial support by the Italian National Research Council and by the Ministry of Research and Universities is also acknowledged.

REFERENCES

1 J. C. Decius and R. M. Hexter, *Molecular Vibrations in Crystals*, (Mac Graw–Hill, New York, 1977).

2 F. Wooten, *Optical Properties of Solids*, (Academic Press, New York, 1972).

3 R. Bozio and C. Pecile, in *The Physics and Chemistry of Low Dimensional Solids*, ed. by L. Alcacer, (Reidel, Dordrecht, 1980), p. 165; R. Bozio, in *Molecular Electronics*, ed. by M. Borrisov, (World Scientific, Singapore, 1987), p. 666; C. S. Jacobsen, in Ref. 4, p. 243.

4 *Low Dimensional Conductors and Superconductors*, ed. by D. Jérome and L. G. Caron, (Plenum Press, New York, 1987).

5 See, e.g., other chapters in the present volume.

6 A. B. Harris and R. B. Lange, Phys. Rev. **157** 295 (1967).

[7] P. F. Maldague, Phys. Rev. B **16**, 2437 (1977).
[8] S. Mazumdar and Z.G. Soos, Phys. Rev. B **23**, 2810 (1981).
[9] Y. Toyozawa, J. Phys. Soc. Jpn. **50**, 1861 (1981); S. Weber and H. Büttner, J. Phys. C **17** , L–337 (1984); W. Schmidt and M. Schreiber, J. Chem. Phys. **86**, 953 (1987).
[10] M. J. Rice, Solid State Commun. **31**, 93 (1979).
[11] R. Bozio, A. Feis, I. Zanon and C. Pecile, J. Chem. Phys. **91**, 0000 (1989).
[12] G. Fischer, *Vibronic Coupling*, (Academic Press, New York, 1984).
[13] M. D. Frank–Kamenetskii and A. V. Lukashin, Sov. Phys. Usp. **18**, 391 (1976).
[14] G. J. Small, J. Chem. Phys. **54**, 3300 (1971).
[15] K. Prassides and P. N. Schatz, J. Phys. Chem. **93**, 83 (1989).
[16] A. Painelli and A. Girlando, J. Chem. Phys. **84**, 5655 (1986).
[17] M. Vandevyver, A. Ruaudel–Teixier, S. Palacin, J.–P. Bourgoin, A. Barraud, R. Bozio, M. Meneghetti, and C. Pecile, Mol. Cryst. Liq. Cryst., to be published.
[18] J. Tang and A. C. Albrecht, in *Raman Spectroscopy. Theory and Practice*, ed. by H. A. Szymanski (Plenum, New York, 1970), Vol. 2, Chap. 2, p. 33.
[19] A. Pinczuk and E. Burstein, in *Light Scattering in Solids*, ed. by M. Cardona, (Springer, Berlin, 1975), Chap. 2, p. 25.
[20] B.A. Scott, S. J. La Placa, J.B. Torrance, B. D. Silverman, and B. Welber, J. Am. Chem. Soc. **99**, 6631 (1977).
[21] A. Girlando, R. Bozio, C. Pecile, and J. B. Torrance, Phys. Rev. B **26**, 2306 (1982).
[22] M. J. Rice, N. O. Lipari, and S. Strässler, Phys. Rev. Lett. **39**, 1359 (1977).
[23] M. J. Rice, V. M. Yartsev, and C. S. Jacobsen, Phys. Rev. B **21**, 3437 (1980).
[24] A. Girlando, F. Marzola, C. Pecile, and J. B. Torrance, J. Chem. Phys. **79**, 1075 (1983).
[25] E. Y. Loh, Jr. and D. K. Campbell, Synth. Metals **27**, A499, (1988).
[26] M. Meneghetti and C. Pecile, J. Chem. Phys. **84**, 4149 (1986).
[27] M. Meneghetti, A. Girlando and C. Pecile, J. Chem. Phys. **83**, 3134 (1985).
[28] Z. G. Soos, H. J. Keller, K. Ludolf, J. Queckbörner, D. Wehe, and S. Flandrois, J. Chem. Phys. **74**, 5287 (1981).
[29] M. Meneghetti, R. Bozio, C. Bellitto, and C. Pecile, J. Chem. Phys. **89**, 2704 (1988).
[30] R. Bozio, A. Girlando, and C. Pecile, Chem Phys. **21**, 257 (1977).
[31] R. Bozio and C. Pecile, J. Phys. C **13**, 6205 (1980); Solid State Commun. **37**, 193 (1981).
[32] M. Meneghetti, R. Bozio, and C. Pecile, J. Physique **47**, 1377 (1986).
[33] Extensive vibrational analyses of key acceptor and donor structures and the corresponding radical ions are available, e.g., for (a) TCNQ: R. Bozio, I. Zanon, A. Girlando, and C. Pecile, J. Chem. Soc. Faraday Trans. 2 **74**, 235 (1978); I. Zanon and C. Pecile, J. Phys. Chem. **87**, 3657 (1983); (b) $TCNQF_4$: M. Meneghetti, and C. Pecile, J. Chem. Phys. **84**, 4149 (1986); (c) Chloranil and bromanil: A. Girlando, I. Zanon, R. Bozio, and C. Pecile, J. Chem. Phys. **68**, 22 (1978); (d) TTF: R. Bozio, I. Zanon, A. Girlando, and C. Pecile, J. Chem. Phys. **71**, 2282 (1979): (e) TMTTF and TMTSF: M. Meneghetti, R. Bozio, I. Zanon, C. Pecile, C. Ricotta, and M. Zanetti, J. Chem. Phys. **80**, 6210 (1984).

REVIEW OF THE RADICAL ION SALTS

P. Delhaès

Centre de Recherche Paul Pascal
Château Brivazac
33600 Pessac, FRANCE

INTRODUCTION

This review is divided into four major sections: (i) a general survey of π molecular conductors, including the chemistry and the structural classification of radical-ion salts; (ii) a summary of the physics of these electronic and magnetic low-dimensional systems; (iii) a discussion of the present "state-of-the-art" in these materials, with an emphasis on their mixed-valence character, and the impetus to create new compounds with specific low-temperature physical properties; (iv) a description of the current trends, which show an explosion in different directions, because one can get new materials with specific properties, by playing with the molecular organization in stable or metastable phases.

GENERAL SURVEY

The π charge transfer salts

The origins of molecular conductors based on π-electronic systems are in old studies, started a century ago, of molecular associations and complexes. It is now accepted that, when these associations give rise to a stable compound with a definite stoichiometry, thanks to an electron transfer between two partners (an electron donor and an electron acceptor), a stable complex is created. The first theory, which has rationalized these experimental results, was the intermolecular charge-transfer (CT) resonance theory presented by Mulliken in 1951-1952 [1]. This dimer model, which involves all types of electronic frontier orbitals, explains, on the one hand, the presence of either an almost neutral or an almost ionic ground state, and, on the other hand, the presence of a new intense CT band in the visible or near-IR range (this color change was the first experimentally observed signature of charge-transfer complexes (CTC)).

The next step occurred with the discovery of a particular class of conducting materials, involving cyclic planar molecules, where the π-type electrons play a dominant role. These CT salts are divided into two classes, the CTC and the radical-ion salts (RIS) [2]. In the first class (CTC), there is a real electron transfer from two neutral, closed-shall molecules. In the second class (RIS), following an oxidation or a reduction reaction, only one open-shell ion-radical is formed, which is associated with a closed-shell counter-ion, needed to preserve the charge equilibrium in the resulting solid. In both classes, the molecules involve π-electron-rich systems as strong donors (low ionization potential I_D) and rather poor π-systems as strong acceptors (high electron affinity E_A). A sketch of the two possible situations is presented (Fig. 1) using the molecular orbital description (a linear combination of atomic orbitals, which involves HOMO and LUMO orbitals, as explained below). In both cases, the energy of the CT excitation ($E_{CT} = h\nu_{CT}$) is lower than that of any intramolecular exciton mechanism [1]. In these bi- or polymolecular systems, the usual molecular constituents are flat radical-ions, with their structural arrangement in one preferred direction, together with a strong directivity of the

Lower-Dimensional Systems and Molecular Electronics
Edited by R.M. Metzger *et al.*, Plenum Press, New York, 1991

involved π-orbitals. In such a situation, the expected electronic and magnetic properties are very anisotropic, and they are more or less confined to one dimension.

To obtain conducting compounds, two major heuristic criteria, or rules of thumb, have been recognized as essential by several authors. Following Torrance [3], they are:

(i) a structural criterion, that requires segregated stacks or clusters of π donor/acceptor molecules

(ii) an energetic criterion, associated with an average charge density or a degree of charge transfer (ρ). This parameter, which characterizes the degree of ionicity, can be either an integer ($\rho = 0$ for a neutral compound, $\rho = 1$ or 2 for an ionic one) or a fraction (either $0 < \rho < 1$, or $1 < \rho < 2$). The degree of charge transfer is controlled by $| I_D - E_A | \propto \Delta E_{redox}$, and can be determined by cyclic voltammetry in solution [3]. If ρ is fractional, we can have a mixed-valence system and a partially filled electronic band: this is a necessary condition for a conducting state. (The physical background will be described in the next Section).

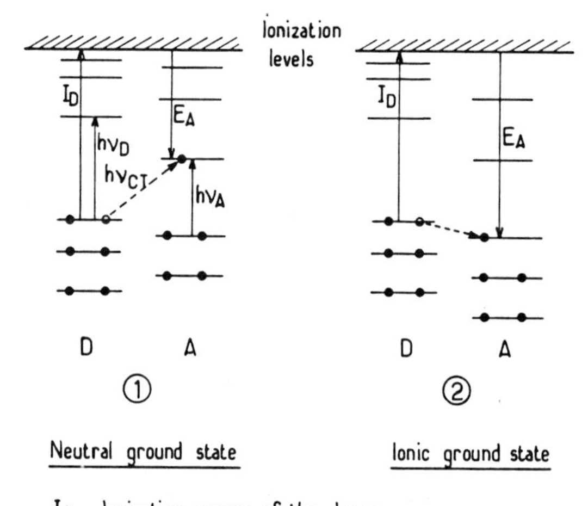

Fig. 1. Fundamental electronic transitions in charge-transfer salts.

Even if one uses these general rules of thumb, the fundamental question arises about how the molecular constituents (the ion-radicals) are related to their structural organization in the resulting solids. An answer to this question would allow us to predict a new series of materials, with selected physical properties.

Chemistry of the RIS

Restricting ourselves to the radical-ion salts, we will present a summary of the most investigated series of these compounds. One should remember that, historically, the first two series of organic conductors were the halides of condensed arenes [4], and the TCNQ salts [5].

Table 1. A general classification of radical-ion salts

1. Anion-Radical Salts	$(Y)_p (A)_n$ (solv.)	
π - acceptors A	. *Tetracyanoquinodimethane (TCNQ)* and analogs . *Dicyanoquinonediimines (DCNQI)* . *Dithiolates* . *quinones,* e.g. *chloranil*	
	Counter - ions Y	. *Metals* . *Quaternary ammoniums* . *Sulfoniums, phosphoniums*
2. Cation-Radical Salts	$(D)_n (X)_p$ (solv.)	
π - donors D	. *Arenes* . *Tetrathiotetracene (TTT)* . *Porphyrins and phthalocyanines* . *Tetrathiafulvalene (TTF) and derivatives*	
	Counter - ions X	. *Monovalent:* MX_6^-, MO_4^-, *Linear* ... . *Divalent:* MX_6^{--} ... . *Clusters*

As presented in Table 1, these salts are, usually, binary compounds. Sometimes, a third component is involved: it can be either a neutral solvent (solv.), or an ionic species (e.g.tri-iodide). When the stoichiometry and the counter-ion charge are defined, the average charge density is known ($\rho = p / n$ for singly charged species). Usually, we observe a radical-ion ($\rho \leq 1$) but, in a few particular cases, a diamagnetic di-ion (D^{++} or A^{--}) has been detected ($1 < \rho \leq 2$).

This means that the counter-ion characteristics are also essential for the synthesis and the physical properties of these salts. Their relevant parameters are:

(i) The geometry, shape and volume; centrosymmetry or non-centrosymmetry (in the last case, order-disorder phase transitions are possible).

(ii) The valence state; most of the usual ions are monovalent, but divalent or polyvalent ones are also used, as long as they are chemically stable under the experimental conditions. This characteristic implies an electrical charge redistribution (dipolar or quadrupolar moments).

(iii) The magnetic state; the counter-ions are usually diamagnetic, but they can be paramagnetic, if transition metals are involved.

All these characteristics will influence the structural organization and the physical properties.

To complete this survey of the main series of RIS, we present in Tables 2 and 3 the chemical structures of selected π acceptors and donors, with a special emphasis on the TTF series, which is, at present, the most investigated.

Finally, it should be emphasized that three steps are needed to obtain these compounds:

(i) Chemical synthesis of the neutral molecules (Tables 2 and 3)

(ii) Salt formation by direct oxidation, reduction, methathesis, photo-oxidation or electro-oxidation.

(iii) Growth of single crystals for structural and physical investigations.

Electrochemical crystal growth in the liquid phase is a very convenient method to get high-quality single crystals, as demonstrated by Bechgaard [6].

In the most common cases, the observed stoichiometric ratios (as defined in Table 1) are 2:1 and 3:2. Often, different ratios are obtained, depending upon the experimental conditions; in particular, for a given stoichiometry, a rich polymorphism is sometimes found. Outstanding examples are the BEDT salts with certain counter-ions (iodide, nitrate...), where five or six different phases have been found.

Table 2. Chemical lexicon of selected π acceptors.

Acceptors	Chemical structure	Abbreviation
7,7,8,8-tetracyanoquino-dimethane	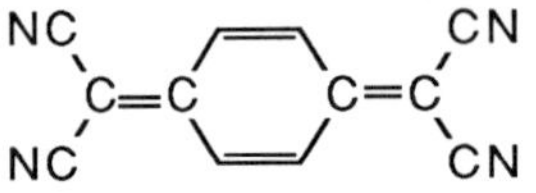 $(NC)_2C{=}C_6H_4{=}C(CN)_2$	TCNQ
dicyanoquinonediimine	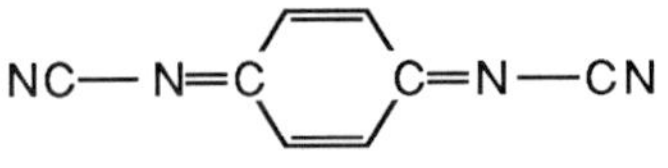 NC— N=C₆H₄=N—CN	DCNQI
metal 1,2-dithiolate type	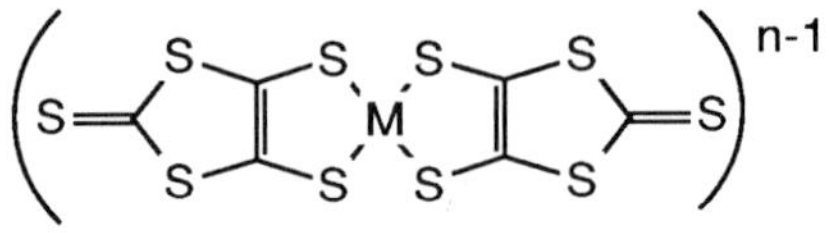 $(S{=}C_3S_4{=}M{=}C_3S_4{=}S)^{n-1}$	$M(dmit)_2$ M = Ni, Pd, Pt
haloanil (CA=chloranil)	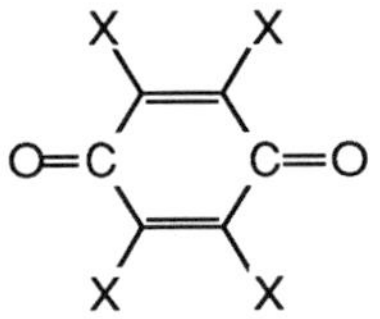 O=C₆X₄=O	CA X = Cl (Br, F)

A structural classification of RIS

The crystal structure of these molecular solids is a starting - point for understanding their physical properties. In the 1960's significant progress was made when it was shown, for the first time, that segregated stacks of either donors or acceptors could form, in spite of their mutual electrostatic repulsion [7]. It turns out that two kinds of stacks are possible, mixed or segregated, which were, in first approximation, associated with neutral or ionic compounds [8].

This naive picture is almost obsolete, because several hundreds of crystal structures have been solved, and a large variety of molecular organizations has been found. We propose, therefore, in Fig. 2, a general one-dimensional structural classification, which includes most, but not all, of the known crystal structures (in particular, sheets of dimer or tetramer stacks, which give rise to the 2-D electronic structure in BEDT salts, are not represented).

Except in a few cases, these compounds are formed by a self-complexation, which gives rise to more (or less) well-organized segregated stacks. One crucial parameter is the average charge density, which either is, or is not, identical on each radical ion. X-ray or neutron diffraction studies show that, if the crystallographic sites are identical, the degree of ionicity (ρ) must be uniform, but if they are different, this average charge sometimes becomes drastically different for the several (crystallographically unique) molecular sites. The concept of mixed valence is different in these two situations, because we are in presence of either delocalized or localized charges. This point will be developed further in the last Section of this review;. Here we note that, when ρ is an integer, the salts behave more or less as insulators. These results are verified in structural studies, which show changes in bond length and bond angles with the charge distribution; some useful empirical relationships between ρ and certain characteristic bond lengths have been established [9,10]. A second point concerns the multimer organization, which can take multiple forms, as seen in salts of TTF derivatives in the last few years. The regular stacking in a preferred direction can be changed, modified or even destroyed. In this last situation, we observe dimers or trimers, which are distributed in a given plane. The clear consequence will be an increased electronic

Table 3. Chemical lexicon of selected π donors.

Donors	Chemical structure	Abbreviation
pyrene	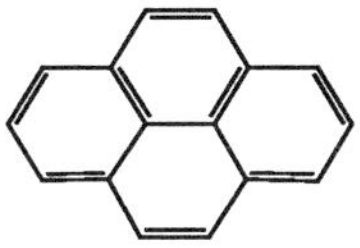	Py
perylene	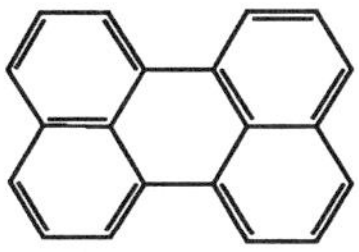	Per
tetrathiotetracene	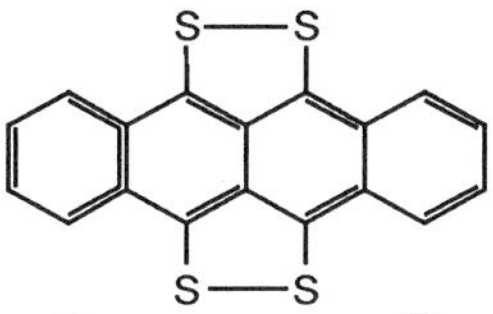 S—S / S—S	TTT (S→ Se: TST)
pyranylidenes	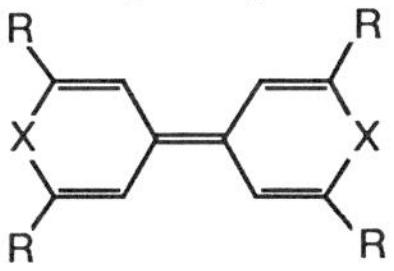 R, R, X, X, R, R	DIP (X = O, Se, Te)
porphyrins(benzo)	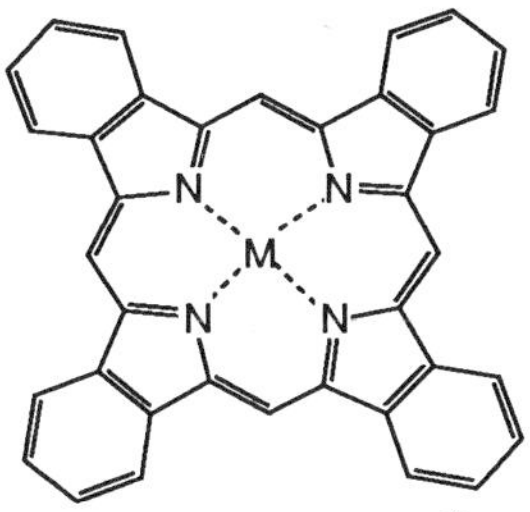 N, N, M, N, N	Por (M = Metal)
phthalocyanines	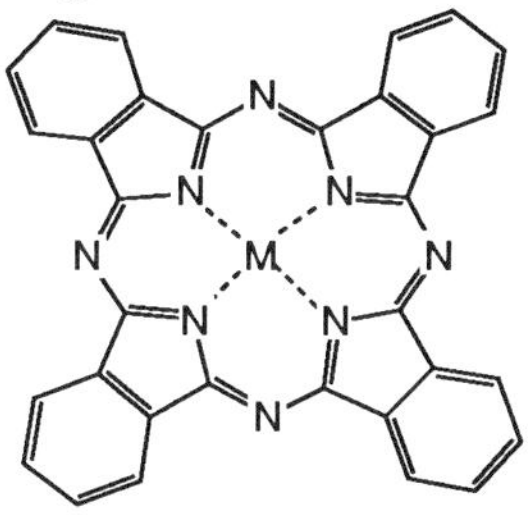 N, N, N, N, M, N, N, N, N	Pc (M = Metal)
tetrathiafulvalene	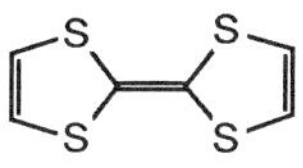 S, S, S, S	TTF (S → Se: TSF; S→Te: TTeF)
tetramethyltetrathiafulvalene	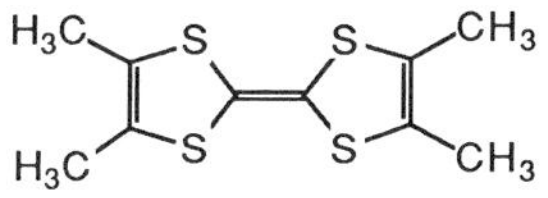 H_3C, S, S, CH_3, H_3C, S, S, CH_3	TMTTF (S → Se: TMTSF)
bis-ethylenedithiolo-tetrathiafulvalene	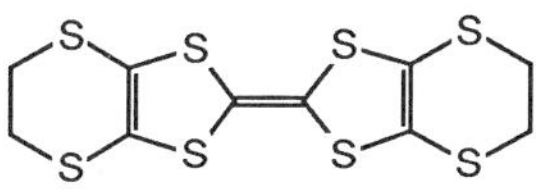 S, S, S, S, S, S, S, S	BEDT-TTF or ET (S→Se: BEDT-TSF or BEDS-TTF)
dimethyl(ethylenedithiolo)-tetrathiafulvalene	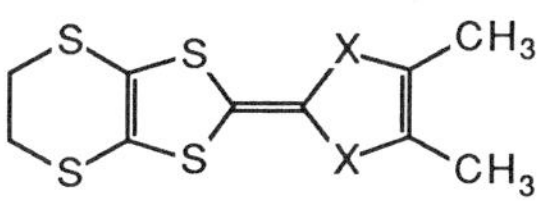 S, S, X, CH_3, S, S, X, CH_3	X = S: DIMET X=Se: DMET

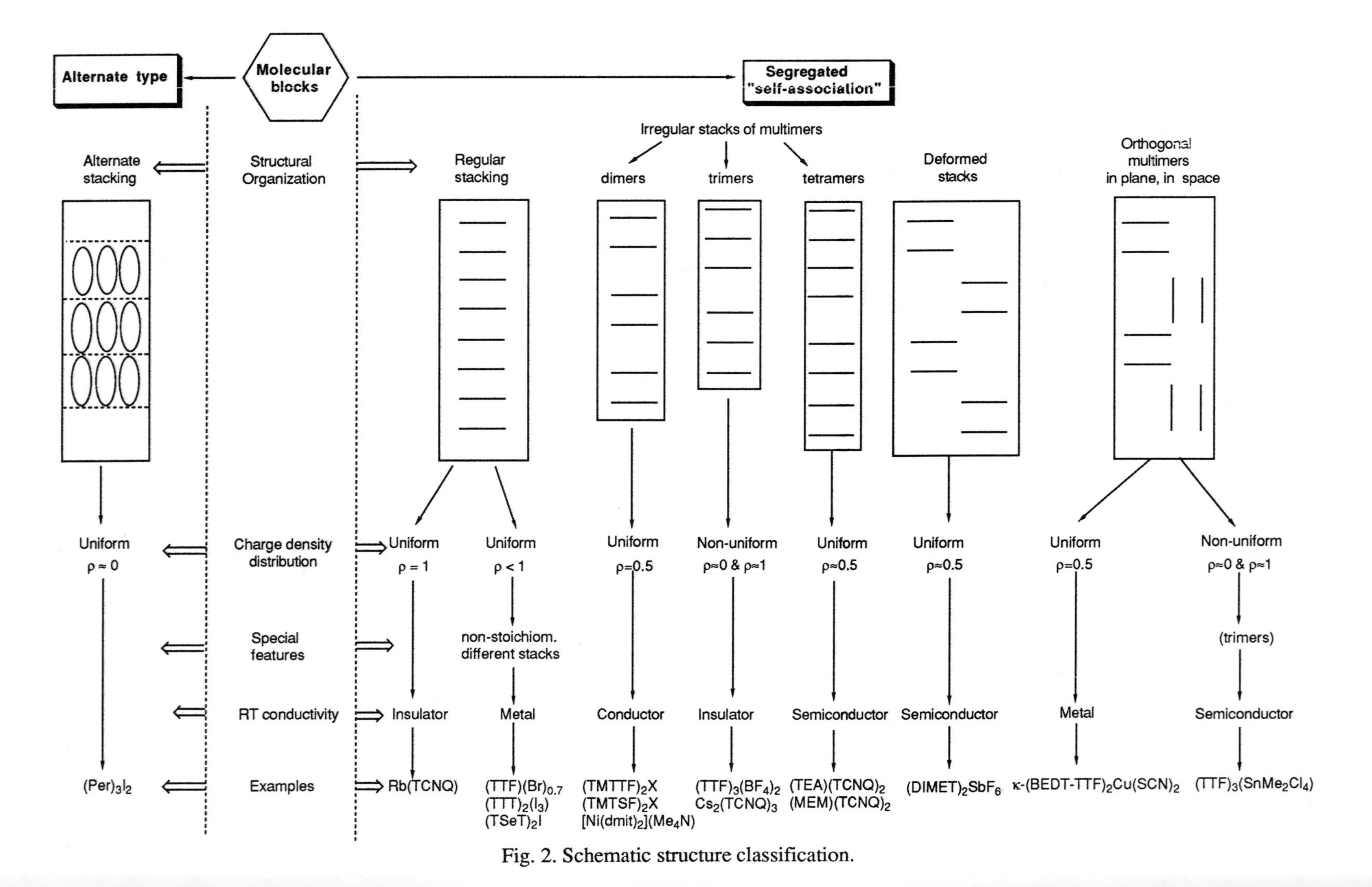

Fig. 2. Schematic structure classification.

or magnetic dimensionality in these new materials. Some other points are mentioned in Fig. 2, as, for example, the case of non-stoichiometric compounds, which give rise to incommensurable phases [11], or even to different kinds of stacks.

To complete the discussion about the structural organization, one should mention the theoretical calculations of the cohesive energy [12]. These calculations, based upon the crystal structures, have shown that, besides the strong Madelung energy, the covalent bonding, which is present for segregated stacks associated with a fractional or integer degree of ionicity, is also responsible for the thermodynamic stabilization. However, starting from the molecular blocks,these cohesive energy calculations are not able to predict a given structural organization, let alone explain the existence of several phases in many salts.

To conclude, we want to emphasize the following crystallographic points:

(i) The molecules are deformed, in accordance with their average charge. It means that there are strong interactions between the molecular vibrations and the electrons (or the holes), as described by a polaronic picture.

(ii) The lattice is soft: dimensional changes with the intensive variables T and P are observed, because the thermal dilation and the isothermal compressibility coefficients in these molecular solids are large.

(iii) Structural phase transitions are predicted at low temperatures, as, for example, a Peierls-type transition [13].

THE PHYSICAL BACKGROUND

One-dimensional electronic band model

There are two approximate starting points in quantum chemistry: the molecular orbital (MO) and valence bond (VB) methods. The MO theory derived from the Hückel treatment ignores the interactions between the electrons, whereas the VB theory forms the basis for the electron-paired chemical bond, and for the resonance concept [14]. In MO theory the electrons are delocalized (without any correlation), as contrasted to VB theory, where they are supposed to be localized.

To explain the origin of the electrical properties in these materials, we will start with the extended Hückel theory, as developed by Hoffmann [15], by taking into account all the valence electrons in a molecule.

A common approximation is to represent each monomer by its highest occupied molecular orbital (HOMO) or lowest unoccupied MO (LUMO), both with π symmetry (see Fig. 1). Where two such orbitals are interacting, they form the in-phase and out-of-phase combinations, which are separated by an energy called the dimer splitting [16], which is twice the transfer integral $t_{\|}$ (Fig.3). A generalization of this approach, for an infinite stacking, gives rise to a continuum of electronic states, which forms an energy band described, quantitatively, as the 1-D tight-binding approximation. This model, used for a one dimensional chain of equivalent molecules, gives the following energy dispersion relation in its simplest form:

$$E(\mathbf{k}) = \varepsilon - 2t_{\|} \cos (\mathbf{k.a}) \quad (1)$$

where ε is the site energy and $t_{\|}$ is the transfer integral, $t_{\|} = < \psi_i \mid \mathcal{H} \mid \psi_{i+1} >$. This transfer integral is proportional to the overlap integral $s = < \psi_i | \psi_{i+1} >$, which involves the Wannier function on two neighboring sites. The proportionality constant between s and $t_{\|}$ comes from the Mulliken-Wolfsberg-Helmholtz approximation; besides, $t_{\|}$ can be positive or negative, depending upon the topology of the orbital interactions [15]. This approximation is correct for a weak overlap integral ($s << 1$), and leads to small transfer integrals: we are in presence of very narrow electronic bands, compared with the usual atomic metals or semiconductors.

In molecular conductors, if we assume a strictly one-dimensional ideal system (no interrupted strands due to some impurities or disorder) with identical point sites (no internal degree of vibrations or librations), in a rigid lattice approximation (no modulation of the transfer integral by the lattice phonons), we observed a bandwidth, whose density of states N(E) (Fig. 3), is typically $\Delta = 4\, t_{\|} \leq 1$ eV [17].

The next problem is the band filling and the position of the Fermi level E_F at zero Kelvin. We have seen that the supposed uniform degree of ionicity ρ can be defined from the stoichiometry (cf. Table 2). It allows us to define the number of charge carriers per molecule, (N_e/N), and therefore, from the energy band dispersion $E(\mathbf{k})$, the wave vector $\mathbf{k}_F$ which is associated with the Fermi level position. It follows that every orbital can hold two electrons:

$$\rho = \frac{N_e}{N} = \frac{2a}{\pi} k_F \qquad (2)$$

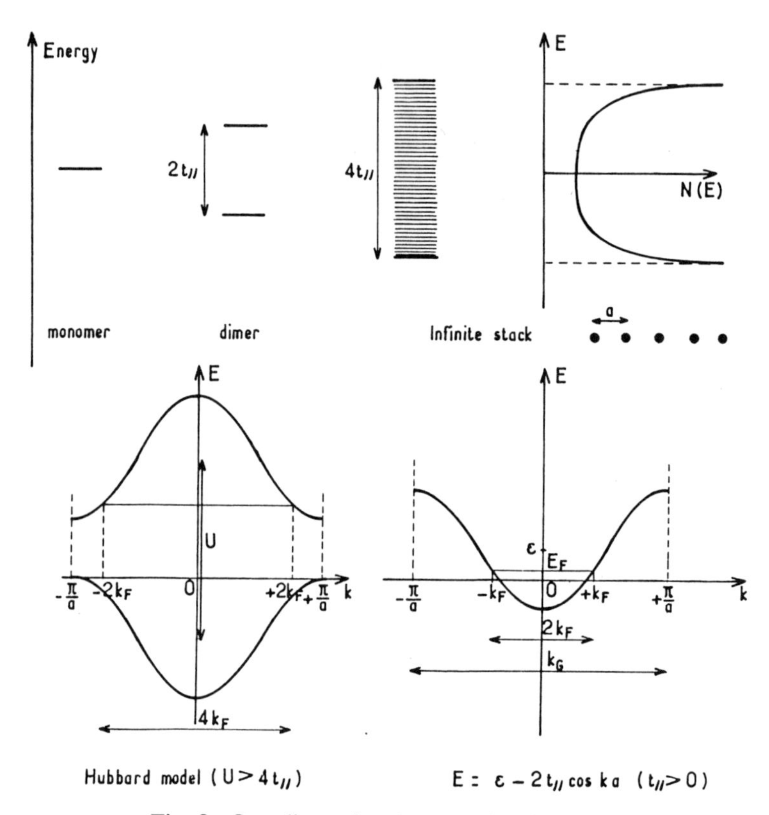

Fig. 3. One-dimensional energy band models

From basic concepts in solid state physics, we know that, if the Fermi energy lies within the band defined by the first Brillouin zone ($-\pi/a$, $+\pi/a$ with $k_G = 2\pi / a$ in the present case), a metallic behavior prevails if: $2 k_F \neq k_G$. Therefore, when $\rho = 1$, we must obtain a half-filled band, and a metallic compound. However, this is not the case for a compound such as K(TCNQ), which has a regular stacking of TCNQ molecules (see the structural classification, Fig. 2): a simple one-electron band calculation with independent electrons is not appropriate in this situation. One must go beyond this approximation, by introducing electronic correlations and electron-generalized phonons, to take into account the molecular degrees of freedom [18], and by considering transverse interactions, either of electrostatic or electronic origin.

Electron correlations: Hubbard models

It is currently admitted that the electronic correlations are the largest interactions in these 1-D narrow-electronic-band systems. The basic phenomenon is that it is energetically too costly for two electrons to occupy the same molecular orbital. When there is one electron per molecule ($\rho = 1$), the electrons (or holes) will tend to be localized, because any motion would necessarily force double occupancy of a given site. This is a Mott-Hubbard insulator [19]: with one electron per orbital, if $U >> t_{\|}$, the electronic band is filled from $-2k_F$ to $+2k_F$; for $\rho = 1$ the resulting Fermi wave vector is given by $4\,k_F = k_G$. A generalization of this argument is that the electron arrangement in the ground state is periodic, with a period depending on the degree of ionicity ρ: this is the Wigner lattice in 1-D as in 3-D systems [20].

A quantitative approach is to define these Coulomb interactions, not only the on-site interaction but also several near-neighbor interactions, which cannot be negligible, compared to the bandwidth. A valuable approximation is to consider the two first terms. The extended Hubbard hamiltonian [20] is, in second quantization notation (site representation):

$$H = \varepsilon \sum_{p,\sigma} n_{p,\sigma} - t_{\|} \sum_{p,\sigma} (C^{+}_{p+1,\sigma}\, C_{p,\sigma} + C^{+}_{p,\sigma} C_{p+1,\sigma}) + U \sum_{p} n_{p\uparrow}\, n_{p\downarrow} + V \sum_{p} n_{p+1} n_{p} \quad (3)$$

where $C^{+}_{p,\sigma}$ and $C_{p,\sigma}$ are the creation and anihilation operator, respectively, of an electron of spin σ on the site p, n_p is the occupation operator, ε and $t_{\|}$ are the one-electron site energy and transfer integral, respectively, as defined by Eq.1, and U, V are the on-site and next-neighbor Coulomb interactions.

These last two terms are essentially molecular characteristics, which include the charge distribution on a monomer and a dimer [21]. Nevertheless, the basic issue is the order of magnitude and the sign of U and V in these charge transfer salts. The first estimates give $U > V > 0$ with $U > 1$ eV [20,22], which is comparable to, or larger than, the bandwidth Δ (see the next Section).

Starting from this point, two main approaches have been developed [13]:

(i) The weak-coupling limit, where U, V are considered as perturbations of the one-electron model ($U, V < |t_{\|}|$). A vast literature has been developped ("g-ology" models) [23] which predicts different electronic instabilities and associated ground states.

(ii) The strong-coupling limit ($U, V >> |t_{\|}|$), where the strong electron-electron interactions lead to a charge localization on molecular sites or between them [20,22].

The easiest case to investigate is the zero-bandwidth or atomic limit; here, the standard Hubbard hamiltonian (with $V \equiv 0$) is equivalent to a Heisenberg spin hamiltonian [24]:

$$H_{sp} = J_{\|} \sum_{i} S_i S_{i+1} \quad \text{where } J_{\|} = [2(t_{\|})^2/U]\ \rho\ [1 - \sin(2\pi\rho)/2\pi\rho] \quad (4)$$

$J_{\|}$ is the exchange integral responsible for an AF coupling ($J_{\|} > 0$ in the present case). For an extended Hubbard model, it is necessary to use an effective U_{eff} [25] (instead of a bare U) which depends upon the band filling.

One drastic consequence of this model is the decoupling of charge and spin excitations, with the appearance of an energy gap along the energy axis. There is a split of the initial electronic band into two (U effect), or more (V effect) sub-bands, which are separated by a Hubbard gap (Fig. 3). The value of this gap has been calculated for a 1-D extended Hubbard model [26].

This explains the presence of an activation energy for the charge excitations (as detected by the electronic absorption and the temperature dependence of the electrical conductivity). From the spin excitations, we observe a Pauli-type paramagnetism, which is enhanced in the small-U approximation, and then progressively changes into the behavior of a regular magnetic chain, when the Hubbard terms become very large ("big U" approximation) [22].

For a non-regular chain, the situation becomes even more complex, because of the possible superposition of two different energy gaps, resulting, respectively, from a splitting in the energy scale, and from a reduction of the reciprocal lattice: these are the so-called Peierls-Hubbard models.

Departure from 1-D electronic structure: The transverse interactions

Two main, and different, causes for departure from 1-D physics have to be considered:

(i) The counter-ions can act as a perturbative external potential, through electrostatic interactions on the 1-D electronic structure. In absence of any structural change, they can modify either the site energy ε, or the transfer integral $t_{\|}$ values. There is diagonal and off-diagonal disorder, which give rise to site or bond charge localization. The most crucial effect is the site energy effect, which can change the mean charge distribution. As recognized a long time ago by Soos and Klein [27] for Cs_2TCNQ_3, crystallographically inequivalent sites lead to neutral and fully ionized molecules (Fig.2). In other cases, a less drastic situation could give rise to some weaker localization effect. Besides, when the counter-ion sublattice possesses extra degrees of freedom, the effect of new periodicities, due to the ordering of non-centrosymmetrical anions, as in TMTTF, TMTSF, and BEDT-TTF salts, induces new physical situations for the 1-D electronic chain [28].

(ii) The electronic transverse interactions. A strictly 1-D system (see Fig. 3) does not really exist, and a more realistic approach is to take into account the electronic interactions between more or less well-defined stacks. A simplified approach is to keep axial symmetry along the stacking axis. A two-dimensional tight-binding model, with a transverse transfer integral ($t_{\perp} \neq 0$), is sufficient to describe the physical properties [29]:

$$E(\mathbf{k}) = \varepsilon - 2\, t_{\|} \cos k_x a - 2\, t_{\perp} \cos k_y b \qquad (5)$$

This dispersion is for a rectangular lattice ($a \neq b$), with the longitudinal component as described in Eq. (1). Two sub-cases exist:

(a) The interchain coupling is small ($|t_{\perp}| << |t_{\|}|$). Inside the Brillouin zone, the planar Fermi surface of a strictly 1-D situation becomes modulated by a finite $t_{\perp}$, but the warped Fermi surface remains open in the transverse direction: This is a quasi-1-D model [30].

(b) The interchain coupling is efficient ($|t_{\perp}| < |t_{\|}|$). This is an anisotropic 2-D system with, eventually, a closed elliptic Fermi surface. A completely new situation arises; present theory studies the electronic correlations by means of 2-D Hubbard models [31]. A general phase diagram is not yet firmly established for such a situation.

1-D Instabilities and the different ground states

The underlying physical mechanism has been found by Peierls in 1955. A 1-D metal is unstable, at zero Kelvin, with respect to a periodic modulation of the wave vector $2k_F$, opening a gap at $\pm k_F$ in the reciprocal space: This is a metal-insulator phase transition.

For real systems, some interactions between chains are present, but, initially, the deviation from an ideal 1-D behavior is not drastically changed, except that phase transitions will occur at $T > 0$ K. Introducting the electronic correlations, as described in the Hubbard model, the theorists [23,32] have predicted different types of electronic instabilities and resulting ground states (the U and V values are now significant, when compared to the kinetic energy).

There are respectively:

(i) A charge-density wave (CDW), which is a periodic modulation of the charges with a periodic lattice distortion, resulting from an electron-phonon coupling (Peierls-Fröhlich approach)

(ii) A spin-density wave (SDW), which corresponds to a spin modulation state and an antiferromagnetic order (AF) below the Néel temperature.

(iii) The formation of Cooper pairs, which can give rise to a singlet or a triplet pairing superconducting state (SC). This last ground state possibility is an extension of Little's initial proposal about the prospects of excitonic superconductivity [33].

The main problem is the competition between these different instabilities. A departure from ideal 1-D situation, by increasing the transverse interactions, will suppress the Peierls state, as described by a mean-field approach [29]. Nevertheless, this description does not strictly apply to these low-dimensional systems, where large fluctuative regimes are present, as observed by diffuse X ray techniques [13,30].

For the experimentalist, the fundamental point is that the metallic state may not be stable down to very low temperatures. A phase transition might occur, which may or may not be a structural phase transition. In the first case, we will get an electronic Peierls distortion at 2 k_F for a free electron gas, or a magnetic analog at 4 k_F in the large-U approximation [28] (see Fig. 3). In the second case (AF and SC), there is a cooperative effect, which is manifested through different physical properties.

SUMMARY OF THE PHYSICAL PROPERTIES

A mixed-valence classification

The mixed-valence (MV) concept was first developed for inorganic compounds, and a classification scheme has been proposed by Robin and Day [34]. This classification is based on the degree to which the sites, occupied by the ions of different valency, could be distinguished in the ground state, and the consequent facility of transferring an electron from one site to another. It implies, therefore, that only materials of fractional degree of ionicity ρ can be mixed-valent. It is a means to classify many materials by their resulting physical properties, without taking account of the chemical type of compounds involved. A natural extension is to draw analogies between the Robin-Day classification and the π charge-transfer salts (see Fig. 2), based on their similar physical properties. This comparison is presented in Table 4; it should be noticed that we did not use class IIIA for clusters, and that the analogy for class I is rather loose [34].

The most famous examples belong to class IIIB, where all the molecular sites are identical in a continuous lattice, with the overlap integral $s \neq 0$ everywhere; in principle, the electrons or holes are free to move with no expenditure of extra energy. We will restrict the

Table 4. A comparative classification of mixed-valence (MV) molecular compounds.

Robin-Day classification [34]	π charge transfer salts	Physical properties
Class I	Alternate stacks; neutral ground state ($\rho \approx 0$) and MV excited states Ex.: $(Per)_3I_2$	Diamagnetic ground state insulator at low T; charge and spin excitations at $T \neq 0$
Class II	Segregated stacks with inequivalent lattice sites; localized MV system (charge ordering) Ex.: $(TTF)_3(BF_4)_2$, $Cs_2(TCNQ)_3$	Paramagnetic insulator or large-gap semiconductor. CT bands and vibronic effects
Class III B	Infinite array or sheet with equivalent lattice sites; delocalized states at the Fermi energy Ex.: $(TMTTF)_2X$, $(TMTSF)_2X$ $(BEDT\text{-}TTF)_2X$	Metallic compounds with plasma threshold in optical reflectivity and CT absorption bands with vibronic effects Pauli-type paramagnetism Magnetic or superconducting cooperative states at low T

Table 5. Relationships between the physical properties and the microscopic parameters.

Microscopic parameters	Physical properties
1) Transfer integrals	
- $t_{\|}$ longitudinal transfer integral	. electronic specific heat . optical reflectivity (plasma frequency) . thermoelectric power
- $t_{\perp}$ transverse transfer integral	. electrical conductivity anisotropy . ESR linewidths
2) Interaction terms	
- U intrasite Coulomb interaction - V intersite Coulomb interaction	. charge transfer absorption band (A and B type) . paramagnetic susceptibility
- Transverse electrostatic interactions	. X-ray diffuse scattering and phase transition temperatures
3) Electron-phonon interactions	
- Electron-molecular vibration coupling constant λ	. IR and Raman vibronic modes
- Electron-lattice phonon coupling constant	. Resonance Raman scattering

analysis of the physical properties to this class of materials only, with their new electronic, magnetic and optical properties due to the intermolecular CT [35]. It is difficult, but imperative, to obtain the microscopic parameters ($t_{\|}$, $t_{\perp}$, U, V, λ) from the different physical and structural properties, in order to establish a hierarchy between the different interactions involved [36].

A summary of these relationships is given in Table 5. We will not present all the details of these physical properties, but only the general trends and the most significant results, found in the last twenty years.

The electrical conduction

For the salt presenting a room-temperature (RT) conductive regime, it is useful to start from the 1-D tight-binding model. Within this one-electron picture (Fig. 2), independent of bandfilling, one obtains the mean free path Λ for the charge carriers:

$$\Lambda \; / \, a = \sigma_{\|} \, R \, A \, / \, (a \, z) \qquad (6)$$

where $R = \pi \hbar^2 / 2 \, e^2 = 6500 \, \Omega$ is a constant, A is the surface area of the cross section of the unit cell with the parameter a along the stacking axis, z is the number of stacks per unit cell, and $\sigma_{\|}$ is the longitudinal conductivity.

A d.c. conductivity critical value for $\Lambda / a \approx 1$ corresponds to $\sigma_{\|} \approx 300$ - $500 \, \Omega^{-1} \, cm^{-1}$ in these compounds. As emphasized by Weger [37] a few years ago, this threshold from diffuse to coherent longitudinal regime corresponds to $\Lambda \, k_F \approx 1$ (rule of Ioffe and Regel). This is the borderline between real metals, where the electronic motion is wavelike, and semiconductors, where the electronic motion is of a hopping type. From this basic argument, we can classify into three different groups the typical temperature dependences for the quasi-1-D mixed-valence salts with identical crystallographic sites, shown in Fig. 4:

(i) The semi-conducting salts with an apparent energy activation Δ_ρ. In general, these are strongly dimerized systems (Fig. 2), with a strong electronic localization, associated with a Hubbard-Peierls gap, as indicated by the optical absorption

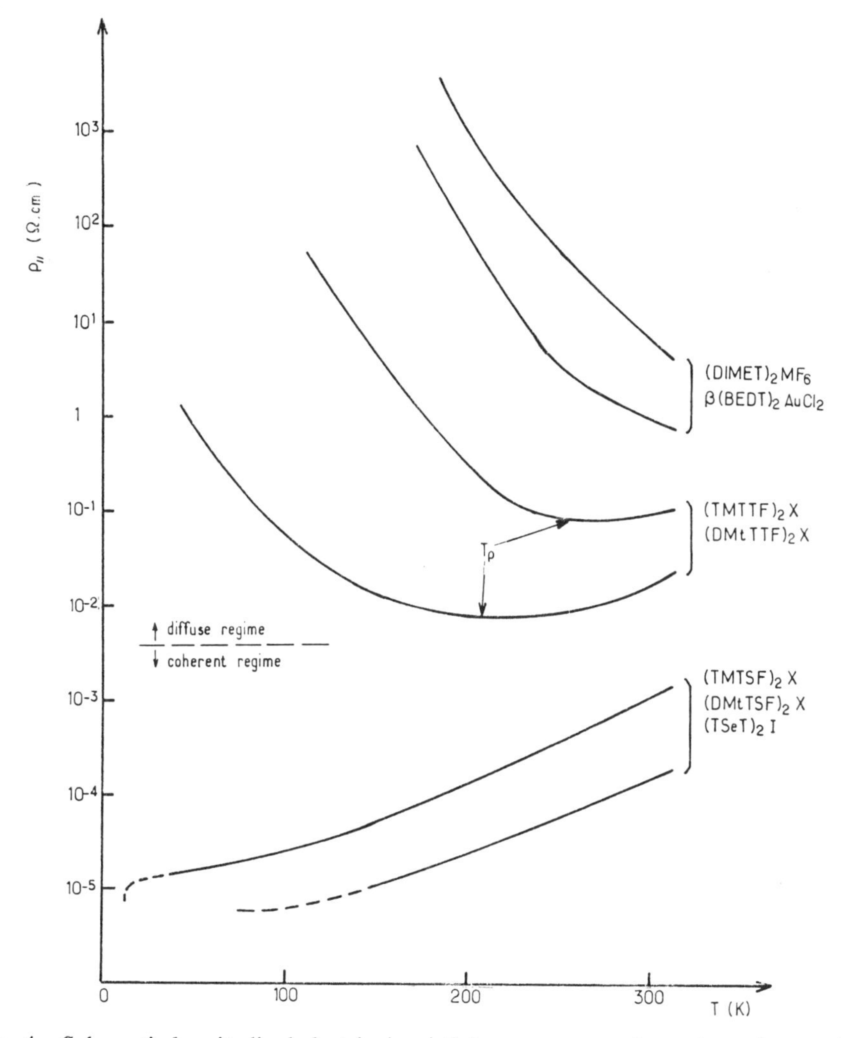

Fig. 4. Schematic longitudinal electrical resistivity temperature dependence for quasi-1-D RIS.

experiments [35]. The paramagnetic susceptibility is described by a Heisenberg model, Eq. (4). From optical absorption and magnetism, one can evaluate the longitudinal exchange integral $J_{\parallel} \approx t_{\parallel}^2/U_{eff}$ (for a non-regular stacking [35]).

(ii) The conducting salts, which are weakly dimerized or tetramerized stacks, with a moderate temperature dependence of Δ_ρ, and a broad minimum of $\rho(T)$ at a given temperature T_ρ [38]. The increase of resistivity below T_ρ is due to a gradual electronic localization, which freezes progressively the charge carriers. This effect has been recognized for the first time in a TCNQ salt [39], and is found in the TMTTF and DMtTTF (dimethyltrimethylene) salts. These compounds are characterized, on the one hand, by an enhanced Pauli paramagnetism ($\chi_{\rho exp}/\chi_{\rho calc} \approx 3-4$), and, on the other hand, by a relationship between the room-temperature d.c. electrical conductivity and the position of the mixed-valence CT band (labelled "A band" by Torrance) [22]. In Fig. 5 we have plotted Log $\sigma_{\parallel}$ versus the position of the CT band $\omega_{CT}{}^{max}$ and the characteristic frequency $\omega_{CT}{}^{DL}$ resulting from the fit based on a Drude-Lorentz model [40]: there is an optical gap of essentially Hubbard origin ($\Delta E_{CT} \approx V$), which might scale with the electrical gap activation energy.

(iii) The metallic salts. In this third class of compounds we observe a metallic dependence of the resistivity, down to very low temperatures: $\rho \propto T^n$, with $1 < n < 2$, depending on the scattering mechanism [41].

For the famous $(TMTSF)_2X$ salt series ($\sigma_{\parallel} \approx 500\ \Omega^{-1}\ cm^{-1}$ at 300 K), we observe a coherent regime along the stacking direction at low temperature ($\Lambda >> a$) (the thermoelectric power T dependence is also a good indicator of this regime change), but also a change in

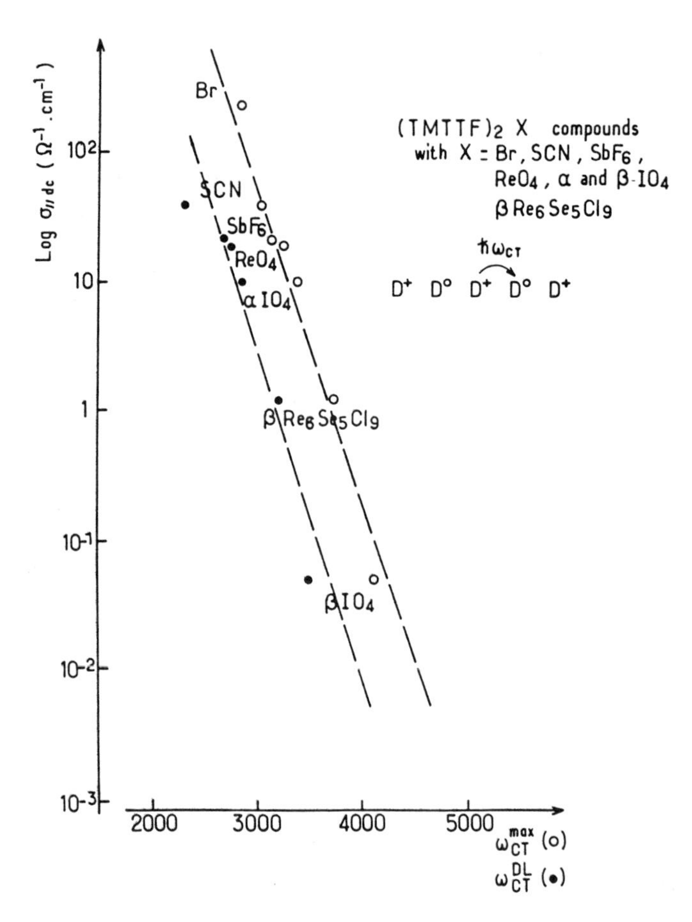

Fig. 5. Semi-logarithmic plot of RT electrical conductivity $\sigma_{\parallel}$ versus the optical activation energies [40].

electronic dimensionality. There are quasi-1-D systems, because $| t_{\parallel} | < | t_{\perp} |$) and experimentally the RT anisotropy $\sigma_{\parallel} / \sigma_{\perp} \approx$ 50 – 100, but, at low temperature, we have a cross-over from transverse diffuse regime to a coherent regime. This situation has been clearly demonstrated by polarized reflectivity experiments on single crystals, which show the apparition of a transverse plasma resonance at low temperature [42].

For these molecular metals, we still observe electronic correlations, whose presence is detected by an enhanced Pauli paramagnetism, and the analysis of the oscillator strength associated with the plasma frequency [42]. Besides, it will be shown in the following paragraphs that the largest RT electrical conductivities ($\sigma_{||} \geq 1000\ \Omega^{-1}cm^{-1}$) are not favorable for the appearance of a low-temperature cooperative state.

Another class of metallic salts is the quasi-2-D compounds, mainly the BEDT-TTF derivatives (β and κ phases). They have a moderate room-temperature conductivity ($\sigma_{||} \approx 10$ - $100\ \Omega^{-1}\ cm^{-1}$), rather lower than the usual minimum metallic value [19], with a small anisotropy ($\sigma_{||} / \sigma_{\perp} \leq 1$). They are 2-D electronic systems with a coherent regime at low temperature, in agreement with the electronic band description of a closed (or almost closed) Fermi surface, even in the presence of electronic correlations [43].

Phase transitions and observed ground states

We have briefly reviewed above the 1-D instabilities which have been predicted theoretically. With the synthesis of new series of molecular materials during these last twenty years, the experimentalists discovered most of these phase transitions which are specific of low-dimensional systems. Following a chronological order, we will next review the main experimental facts:

(i) A periodic lattice distortion with 1-D precursor effects at $2k_F$, and the occurrence of a metal-insulator transition, as predicted by Peierls, was observed by X-ray and neutron diffuse scattering in Krogmann's salt (potassium tetracyanoplatinate, or KCP: $K_2Pt(CN)_4Br_{0.3}$. $2.3H_2O$) and also in TTF-TCNQ during the seventies [44].

(ii) A collective state (AF, SC) occurs at low temperature, when the lattice distortion is prevented either by molecular engineering or by a physical restraint, such as pressure. The initial series of RIS was $(TMTTF/TMTSF)_2X$ in 1978-1982. The first organic singlet superconductors were discovered under hydrostatic pressure [30], and then, ultimately, without the application of any external pressure: $(TMTSF)_2ClO_4$ with $T_c = 1.2$ K [45]. During the same period, the competing AF ground state was found, e.g. in $(TMTTF)_2Br$ ($T_N = 14$ K).

A second, very promising series has been the $(BEDT\text{-}TTF)_2X$ salts. In these salts many phases form simultaneously, and only a few of them give a SC or AF low-temperature state. One of them, known as the β phase, has furnished several superconductors, in particular β-$(BEDT)_2I_3$ with $T_c = 8$ K [46]. Actually, the highest T_c was observed in a κ phase, which gives rise to a rather 2-D structural organization (Fig. 1): for κ-$(BEDT\text{-}TTF)_2Cu(SCN)_2$ $T_c = 11$ K [47].

As in the previous series, the AF state can also exist, but in different phases, as for example β-$(BEDT\text{-}TTF)_2AuCl_2$ for which $T_c = 27$ K [48]. In this series, we observe the cooperative phase transitions at higher temperatures, because we are in presence of more 2-D electronic or magnetic systems. There is a significant difference between these two series. For the TMTTF/TMTSF series there is, for one given phase stable under thermodynamic conditions, a competition between a spin-Peierls transition, a spin-density modulation (i.e. AF ordering) and a singlet SC state. This is nicely explained by Bourbonnais and Caron's model, which involves a transverse exchange interaction between chains in a quasi-1-D system [49]. Besides, novel phase transitions, induced by an external magnetic field, from a metallic state to a semi-metallic SDW state, have been observed [50]. Moreover, two new series of superconductors, based on unsymmetrical molecules $(DMET)_2X$ [51] and $(MDT)_2X$ [52], have been recently discovered. Except for the one case, $(DMET)_2Au(CN)_2$, we don't observe for the same thermodynamic phase a competition between the different ground states, because we are in presence of more 2-D systems, as in BEDT-TTF salts. During the same period, several series of symmetrical or unsymmetrical cation radical salts were found, based on the TTF backbone, which also present an AF state, with a Néel temperature up to 30 K [53].

The appearance of a low-temperature cooperative state in this series of compounds does not seem to be accidental, even if all the details about the mechanism of superconductivity are not completely understood. There is still high promise for future research of the rich polymorphism in these molecular materials, but the emerging fundamental question is: why are the TTF-type molecules almost exclusively able to provide organic superconductors?

Among the main series of molecules presented in Tables 2 and 3, there are, apparently, only two other significant cases, the DCNQI and $M(dmit)_2$ salts.

The synthesis of new anion radical salts from DCNQI-type molecules has been presented recently [54]. Several of these salts have metallic behavior down to liquid He^4 temperature; it has been suggested that one of them, $Cu(DMDCNQI)_2$, may have an AF ordered state around 5 K [55], but further experiments are necessary to confirm this point.

After the discovery of a SC state under hydrostatic pressure for a charge transfer complex $TTF[Ni(dmit)_2]_2$ (and its Pd analog) [56], it was also found that the radical anion salt $(CH_3)_4N[Ni(dmit)_2]_2$ is a superconducting compound at 5 K under an applied pressure of 7 kbar [57]. This is a crucial result, because, for the first time, a radical-ion salt different from the TTF analogs exhibits a SC state; it should open a route for synthesizing new series of interesting compounds.

Correlations between the molecular characteristics and the low-T physical properties

From a theoretical standpoint, it appears that two sets of requirements must be fulfilled for observing a low-temperature cooperative state (AF or SC):

(i) The presence of electronic correlations, with a delicate balance between the "molecular" parameters ($t_{\parallel}$, U, and V) and also weaker phonon-electron type interactions in a 1-D picture [36].

(ii) The presence of transverse electronic interactions, which give rise to a higher transition temperature [$k T_c \propto (J_{\parallel} J_{\perp})^{1/2}$] in a localized picture, with already defined exchange integrals ($J \propto t^2 / U$) [50]. From different physical properties, such as the spin susceptibility [58] and the CT optical absorption bands [40], these molecular parameters have been evaluated. Some typical estimates [36,40,58] are presented in Table 6 for TTF and TCNQ types of molecules.

Table 6. Estimates of extended Hubbard parameters.

Salts	U /eV solution dimer [36]	U/ eV solid state [40,58]	V /eV solid state [40,58]
TCNQ	1.3	1.4	0.4
TTF	1.7	1.7	0.6
TMTTF	----	1.3	0.3 - 0.4
DIMET	----	1.5	0.3
BEDT-TTF	----	1.2	0.4

These estimates, carried out in the atomic limit approximation of the extended Hubbard model, are, in principle, overestimated. Nevertheless, they confirm the general tendency: $U > 2V > 4 |t_{\parallel}|$, proposed by Mazumdar and Soos [59]. It must be also pointed out that the counter-ions, through their polarizability and shape, can modify the intrinsic U and V values [60].

The parameters should be compared with the bandwidth, i.e. with the longitudinal transfer integral $t_{\parallel}$ in pseudo-1-D systems. Since, in the presence of a weak dimerization, the Peierls gap is negligible (compared to the Hubbard gap), therefore, for regular or pseudo-regular stacking, the following requirement appears to be necessary:

$$0.5 \text{ eV} \leq (\Delta = 4\, t_{\parallel}) < 1 \text{ eV}$$

If the bandwidth is smaller than 0.5 eV, the disorder effects, static and dynamic, are larger and induce an Anderson-type localization [61]. If the bandwith is larger than 1 eV, the correlation effects are not pertinent any more. There are several examples of molecular conductors, which present a high RT electrical conductivity with a metallic dependence down to liquid helium temperature, without any cooperative ground state at low temperature:

$(DMtTSF)_2X$ (where DMtTSF = dimethyltrimethyleneTSF): $\sigma_{RT} \approx 1200\ \Omega^{-1}cm^{-1}$ [62];
$(TSeT)_2I$: $\sigma_{RT} \approx 2000\ \Omega^{-1}cm^{-1}$ [63].
Ni(Pc)I: $\sigma_{RT} \approx 550\ \Omega^{-1}cm^{-1}$ [64];
$Ni(Pc)(BF_4)_{1/3}$: $\sigma_{RT} \approx 1000\ \Omega^{-1}cm^{-1}$ [65].

For the tetraselenatetracene and phthalocyanine salts it appears that the electronic correlations are weak. In these molecules, larger than the TTF type, it turns out that $U < 4\ t_{||}$, and the delicate balance between the kinetic and the correlation energies seems to be destroyed (i.e. $U \approx 2V \approx 4\ t_{||}$). Besides, one must repeat that the transverse interactions have to be included, to account for the transition temperatures (see above), even if it is often delicate to estimate these parameters. As a conclusion, we can say that this approach may give clues as to why, among so many series of molecular conductors, only a few exhibit an AF or SC ground state.

PRESENT TRENDS

Cooperative phenomena

After the discoveries of these organic compounds, metals and low-temperature superconductors, the quest for higher superconducting transition temperatures, which could compete with the copper perovskite compounds, is a possible, but rather difficult task. One should note also that, at present, neither triplet superconductivity nor long-range ferromagnetic order have been found in molecular conductors. The molecular engineering to build up new structural arrangements is going towards an increasing electronic or magnetic dimensionality (2-D and even 3-D systems).

This trend goes further away from the peculiar physical properties which were predicted for 1-D electronic systems. There are the Peierls-Fröhlich CDW collective transport mechanism [30], and Little's excitonic mechanism for high-T_c superconductivity [33], in which an electronically polarizable entity is used instead of the ionic lattice. It is necessary to conclude that there is a divorce between the experimental results and these initial attractive concepts.

One different, but possible, way for the future is to prepare mixed-valence systems with higher degres of oxidation (or reduction) involving, therefore, diamagnetic di-ions, often called bipolarons. Indeed, it has been postulated that the degree of ionicity can be $1 < \rho < 2$ [25], and a few examples with $TCNQ^{--}$, TTF^{++} or $BEDT^{++}$ are known, but it is also possible to consider a new mixed-valence bipolaron system {± 2, 0}, which is not included in the usual classification (see Table 5). This situation has been theoretically investigated in inorganic systems by several authors, as in superconducting metal oxides [66], or in ligand-bridged mixed-valency systems [67]. For synthesizing such compounds, it appears that several conditions have to be fullfilled:

(i) The same potentials for the first two oxidation (or reduction0 waves, detected in solution by cyclic voltammetry (the energy to remove or to add an electron should be larger than the one necessary for removing or adding two electrons)
(ii) The same energy level for $D^{++}(A^{--})$ and $D^0(A^0)$ sites, which involves a soft lattice with a strong electron-lattice interaction.
(iii) A double charge hopping between neighbor sites is possible ($t_{||} \neq 0$).

Presently, some π donors or acceptors molecules could satisfy at least the first condition: there are DIP(S) and TTF-type molecules [68]. There is, therefore, the possibility of double valence fluctuating molecules, in order to get a bipolaronic conductor. Beside, Hirsch and Scalapino [69] point out that this mechanism could lead to tightly bound pairs coupled in real space, which undergo a Bose condensation to a high-T_c superconducting state. This idea is supported by the posssibility of obtaining "negative-U centers," as proposed initially by Anderson. This parameter, U, which is the intrasite Coulomb repulsion of the Hubbard hamiltonian defined in Eq. (3), is usually positive. For the present, there is no example of such $U_{eff} \approx 0$ in π-electron donor or π-electron acceptor molecules. We must be remember that, in some "g-ology" models [23], this concept of a bi-polaronic state was already invoked theoretically a few years ago [37].

Organic ferromagnets

The theoretical suggestions for a high-spin solid state involve triplet-state donors or acceptors, infinitely large alternant, or polyradical systems [70], or a superexchange mechanism [71]. The π charge-transfer salts are invoked by two of these mechanisms for observing a ferromagnetic molecular exchange. We will review the main steps, together with the principal experimental results.

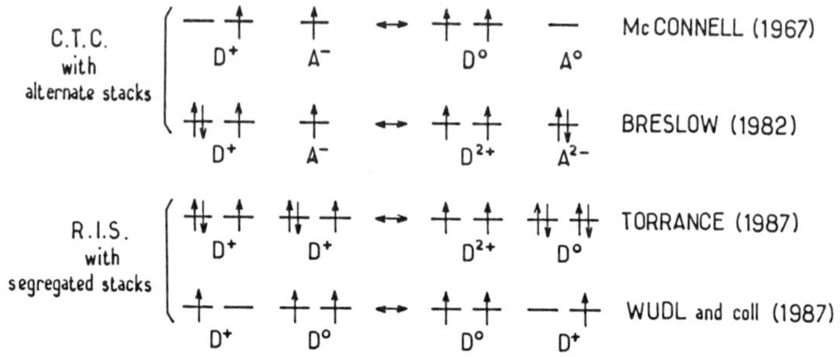

Fig. 6. Dimer models involving degenerate open shells (adapted from Lepage and Breslow [74]).

Existence of a triplet ground state. It is necessary to have a degenerate energy level, to apply the Hund's first rule, either by using a degenerate atomic orbital (d or f orbitals in transition metals), or resulting from a molecular symmetry [72] with a many fold-axis (C_3 or higher symmetry). In this last situation, with a doubly degenerate state, there is the possibility of a triplet state, as pointed out by McConnell several years ago [73]. As presented by Lepage and Breslow [74], several approaches have been proposed. They are summarized in Fig. 6 where we distinguish:

(i) The McConnell and Breslow models, for a π CTC with mixed stacks, on which we can have a neutral or a dication triplet state;

(ii) The more recent proposals by Torrance *et al.* [75] and Wudl and co-workers [76] for segregated stacks of RIS. These models could produce a conducting organic ferromagnet, because they use a mixed-valence state.

It must be noticed that, in Breslow and Torrance's approaches, triplet bipolarons are introduced, whereas, in the Wudl proposal, it is necessary to have a stable neutral biradical. To summarize the situation, for getting an organic ferromagnet one must attain two goals, which are, respectively, a stable triplet ground state ($S = 1$) in a non half-filled degenerate orbital [72], and a ferromagnetic spin-exchange interaction [i.e. $J < 0$ in Eq. (4)] between the molecular blocks which support this molecular triplet state. Long-range ferromagnetic interactions inside a given solid would be expected under these circumstances.

The first condition requires, therefore, a triplet state in solution. During these last few years, several molecules, with a C_3 or a higher-symmetry axis, have been synthesized [68]. In a few cases, evidence for a ground-state triplet has been obtained by low-temperature ESR experiments [74]. However, in most cases a Jahn-Teller distortion occurs, that lowers the molecular symmetry and splits the degenerate state [77,78].

Many attempts to prepare charge transfer compounds with high spin multiplicity have been unsuccessful; nevertheless, two series of experimental results are very promising. On one hand, there is the synthesis of the decamethylferrocene-tetracyanoethylene complex, which forms segregated stacks [72], and exhibits a paramagnetic-to-ferromagnetic phase transition at 4.8 K: this is the first molecular ferromagnet, but it involves the d orbitals of iron. On the other hand, there is the recent synthesis of a high spin density solid, resulting from a triplet state in the donor molecule hexamethyoxy-triphenylene (HMT) [78]. Starting from $(HMT)_2(TCNQF_4)$, an almost neutral CTC, it has been found that the introduction of arsenic pentafluoride, a strong oxidizing agent, induces a new magnetic state. Satellite ESR lines were observed, which are triplet exciton resonances, and a strong increase of the paramagnetic susceptibility at low temperature was seen, arising from spin triplets, which appears to be characteristic of a possible ferrimagnet. Nevertheless, most of the experimental work remains to be done in the field of ferromagnetic conductors.

Indirect exchange interaction. It is possible to consider a magnetic coupling between local moments and conduction electrons. This case has been considered in metals (s-d models); this situation can give rise to a ferromagnetic coupling through so-called RKKY oscillations [80]. This kind of magnetic interaction has been tentatively developed in 1-D materials, by using magnetic counter-ions, organized in clusters or in a magnetic chain. In this last case, one can wait for an interaction between an array of localized moments through an electronic chain.

One organo-mineral compound has been prepared, which meets this requirement: $(TTF)_6(MnCl_3)_3(MnCl_2)_3.13H_2O$ [81]. In this compound, there are Mn^{++} magnetic chains, as in TMMC (tetramethyl manganese chloride), and different TTF stackings; unfortunately, the magnetic interactions between them are weak, and the associated fluctuations below 30 K present an AF character, with a Néel temperature occurring around 1 K.

A more positive result was obtained with one another quasi 1-D conductor, a copper phthalocyanine iodide [82], where it has been shown that the interactions between Cu^{++} local moments arise from indirect exchange, involving the π charge carriers: below 2K a spin glass state is apparently present.

In order to succeed with this mechanism, it will be necessary to prepare compounds with a better overlap of the involved orbitals to get an effective indirect exchange mechanism.

Conductivity in organized media

There is a strong interest to prepare molecular conductors inside an anisotropic structural organization, which could be liquid crystals, mesomorphic polymers or Langmuir-Blodgett (LB) films.

Conducting liquid crystals. Several approaches have been tried, more or less successfully, based also on a 1-D stacking. One of the "firsts" has been to prepare a liquid crystal based on a phthalocyanine as an intrinsic semi-conductor [83]. One other attempt has been to use discotic liquid crystals, based on the triphenylene molecule oxidized by bromine [84] or by aluminium trichloride [8]; in the latter, the conductivity changes by six orders of magnitude (from $10^{-12}\ \Omega^{-1}\ cm^{-1}$ to $10^{-6}\ \Omega^{-1}\ cm^{-1}$), while the mesomorphic character is conserved.

The best result so far has been the synthesis of a CTC between TCNQ and DIP(S) with four alkyl chains [86], which has a respectable d.c. conductivity at room temperature ($\sigma = 0.7\ \Omega^{-1}\ cm^{-1}$). This solid phase undergoes a phase transformation at 123°C, and has a lamellar structure above this temperature, but with a lower electrical conductivity ($\sigma \approx 10^{-3}\ \Omega^{-1}\ cm^{-1}$) [87]. As shown in Fig. 7, the presumed structural organization is characterized by a stacking of the electroactive groups, which are called the cores, and their separation by alkane chains, as ocurs in the LB films (see next paragraph).

Finally, conducting or magnetic liquid crystals are potentially interesting materials, but, for preparing them, it is necessary to control the mesomorphic character, which can be modified or suppressed by the creation of new ionic and covalent bonds (these reinforce the cohesive energy).

Langmuir-Blodgett (LB) Films. This research field is of growing interest, because of the potential applications in molecular electronics [88]. From a fundamental point of view, the interest in creating conducting films is to try to orient the molecular organization and the associated physical properties in a different way, compared to classical crystal growth, but

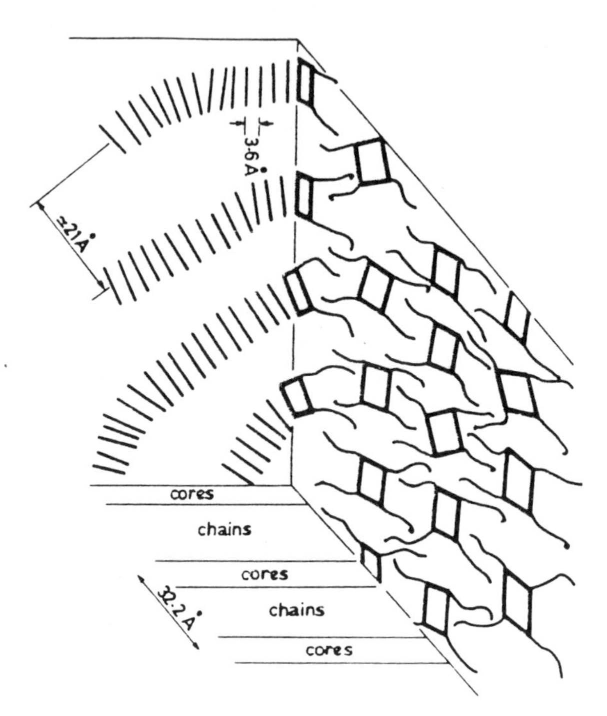

Fig.7. A schematic representation of the donor organization in the mesophase of a DIP(S)-TCNQ complex (the TCNQ molecules are not shown) (from P. Davidson *et al.* [84]).

rather analogous with the liquid crystals. In other words, it should be possible to create new metastable phases with the different spatial organization, imposed by the air/water interface of the Langmuir trough.

Presently, most of the experiments have been carried out with TCNQ, TTF and $M(dmit)_2$ - type electro-active groups, which have one on more aliphatic chains to be able to form a stable monolayer [90]. The most convincing results are the following:

(i) Conductive films based on a C_{18}-pyridinium TCNQ salt, doped with iodine [91].
(ii) The possible existence of Peierls-type fluctuations around 150 K in the CTC $(TMTTF)_3(C_{14}TCNQ)_2$ [92].
(iii) The realization of a metallic state, down to liquid He temperature, with a NR_4 $Au(dmit)_2$ salt (where R is an alkane chain). In these films, the possible presence of AF fluctuations has been envisaged around 20 K [93].

It appears, from these results, that the 1-D instabilities which have been first predicted, and then found, in the crystalline state, are also present in these LB films. However, it must be pointed out that, in several cases, there is no definite proof that these structural arrangements are molecular multilayers (rather than micro-crystals). The physical and structural studies are not developed enough to control the 2-D thermodynamics of a Langmuir monolayer, a necessary step for making these materials [94]. Presently, the conducting LB films are formed with 1-D stacking, as in crystals, and not a planar organization with the π donor or acceptor molecules parallel to the interface plane. Besides, the control of the mixed-valence situation *in situ*, or after the dipping process, by some redox process is not accomplished, in spite of a recent valuable improvement, which uses the electro-oxidation technique [95]. Finally, a good example that the present situation is not so evident is demonstrated by the realization of conducting LB films based on both CTC and RIS of a TTF derivative doped with iodine [96,97]. These films exhibit, after iodine oxidation, a decrease or an absence of the EPR signal from the radical-cation species. One possible way to explain this result is to suppose that we may be in presence of conducting iodine chains, or with some localized excitations, bipolarons or charged solitons, as in doped polymers. This speculative comment shows that the chemistry and the physics in these new systems is far from being understood, and needs careful study.

REFERENCES

[1] R. S. Mulliken and W. B. Person, "Molecular Complexes" (Wiley, New York, 1969).
[2] J. J. André, A. Bieber, and F. Gautier, Annales de Physique 1: 145-256 (1976).
[3] J. B. Torrance, Acc. Chem. Res. 12: 79 (1979).
[4] H. Inokuchi and H. Akamatsu, Solid State Physics 12: 93 (Academic, New York, 1961).
[5] W. R. Hertler, W. Malher, L. R. Melby, J. S. Miller, R. E. Putscher, and O. W. Vebster, Mol. Cryst. Liq. Cryst. (to appear, 1989).
[6] K. Bechgaard, Mol. Cryst. Liq. Cryst., 79: 1 (1982).
[7] F. H. Herbstein in "Perspectives in Structural Chemistry", Vol. IV, ed. by J. D. Dunitz and J. A. Ibers (Wiley, New York, 1971), p. 169.
[8] H. M. McConnell, B. M. Hoffman, and R. M. Metzger, Proc. Nat. Acad. Sci. U. S. 53: 46 (1965).
[9] S. Flandrois and D. Chasseau, Acta. Cryst. B33: 2744 (1977).
[10] T. C. Umland, S. Allie, T. Kuhlmann, and P. Coppens, J. Phys. Chem. 92: 6456 (1988).
[11] P. Delhaès in "The Physics and Chemistry of Low-Dimensional Solids", ed. by L. Alcácer , NATO ASI Series, C56: 281 (Reidel, Dordrecht, 1980).
[12] R. M. Metzger in "Crystal Cohesion and Conformational Energies", ed. by R. M. Metzger, Springer Top. Curr. Phys. 26: 91 (Springer, Berlin, 1981).
[13] J. P. Pouget in "Semiconductors and Semi-metals", Vol. 27, ed. by E. Conwell (Academic, New York, 1988), p.87.
[14] Z. G. Soos, J. Chem. Educ. 55(9): 547 (1978).
[15] R. Hoffmann, Rev. Mod. Phys. 60(3): 601 (1988).
[16] R. P. Messmer, Phys. Rev. B 15: 1811 (1977).
[17] P. M. Grant, Phys. Rev. B 26: 6888 (1982).
[18] M. Weger, in "The Physics and Chemistry of Low-Dimensional Solids", ed by L. Alcácer, NATO ASI Series, C56: 77 (Reidel, Dordrecht, 1980).
[19] N. F. Mott, "Metal-Insulator Transitions" (Taylor and Francis, London, 1974).
[20] J. Hubbard, Phys. Rev. B 17(2): 494 (1978).
[21] P. Pincus, in "Proc. 14th Latin Amer. School of Physics on Sel. Topics in Phys. Astrophys. and Biophys., Caracas, Venezuela, 10-28 Jul 1972," (Reidel, Dordrecht, 1973), pp.138-172.
[22] J. B. Torrance in "Chemistry and Physics of One-Dimensional Metals", ed. by H. J. Keller, NATO ASI Series B25: 137 (Plenum, New York, 1977).
[23] L. Sólyom, Adv. in Physics 28: 201 (1979).
[24] S. Mazumdar and S. N. Dixit, Phys. Rev. B 34: 3683 (1986).
[25] A. N. Bloch and S. Mazumdar, J. de Physique Colloque 44: C3-1273 (1983).
[26] J. Des Cloizeaux and M. Gaudin, J. Math. Phys., 7: 1384 (1966).
[27] Z. G. Soos and D. J. Klein, J. Chem. Phys. 55: 3284 (1971).
[28] J. P. Pouget in "Organic and Inorganic Low-dimensional Crystalline Materials", ed. by P. Delhaès and M. Drillon , NATO ASI Series B168: 185 (Plenum, New York, 1987).
[29] B. Horovitz, H. Gutfreund, and M. Weger, Phys. Rev. B 12: 3174 (1975).
[30] D. Jérome and H. J. Schulz, Adv. in Physics 31: 299 (1982).
[31] S. Tang and J. E. Hirsch, Phys. Rev. B, 37: 9546 (1988).
[32] V. J. Emery in "Highly Conducting 1D Solids", ed. by J. T. Devreese, R. P. Evrard, and V. E. Van Doren (Plenum, New York, 1979), p. 247.
[33] H. Gutfreund and W. A. Little in "Highly Conducting 1D Solids", ed. by J. T. Devreese, R. P. Evrard, and V. E. Van Doren (Plenum, New York, 1979), p. 305.
[34] M. B. Robin and P. Day, Adv. Inorg. Chem. and Radiochem. 10: 247 (1967); P. Day in "Low-Dimensional Cooperative Phenomena", ed. by H. J. Keller, NATO ASI Series B7: 191 (Plenum, New York, 1974).
[35] J. B. Torrance in "Low-Dimensional Conductors and Superconductors", ed. by D. Jérome and L. G. Caron, NATO ASI Series B155: 113 (Plenum Press, 1987).
[36] A. J. Heeger in "Highly Conducting 1D Solids", ed. by J. T. Devreese, R. P. Evrard, and V. E. Van Doren (Plenum, New York, 1979), p. 69.
[37] M. Weger, J. de Physique Colloque 39: C6- 1456 (1978).
[38] M. Coulon in "Organic and Inorganic Low-Dimensional Crystalline Materials", ed. by P. Delhaès and M. Drillon, NATO ASI Series B168: 201 (Plenum, New York, 1987).
[39] A. J. Epstein, S. Etemad, A. F. Garito and A. J. Heeger, Phys. Rev. B5: 952 (1972).

[40] P. Delhaès and C. Garrigou-Lagrange, Phase Transitions 13: 27 (1988).
[41] M. Kaveh and N. Wiser, Adv. in Phys. 33: 257 (1984).
[42] C. S. Jacobsen, in "Low-Dimensional Conductors and Semiconductors", ed. by D. Jérome and L. G. Caron, NATO ASI Series, B155: 253 (Plenum, New York, 1987).
[43] J. M. Williams, in "Progress in Organic Chemistry", ed. by S. J. Lippard, 33: 183 (Wiley, New York, 1985).
[44] R. Comès and G. Shirane in "Highly Conducting 1D Solids", ed. by J. T. Devreese, R. P. Evrard, and V. E. Van Doren (Plenum, New York, 1979), p. 17.
[45] K. Bechgaard, K. Carneiro, M. Olsen, F. B. Rasmussen, and C. S. Jacobsen, Phys. Rev. Lett. 46: 856 (1981).
[46] D. Schweitzer and H. J. Keller, in "Organic and Inorganic Low-Dimensional Crystalline Materials", ed. by P. Delhaès and M. Drillon, NATO ASI Series B168: 219 (Plenum, New York, 1987).
[47] H. Urayama, H. Yamochi, G. Saito, K. Nozawa, T. Sugano, M. Kinoshita, S. Sato, K. Oshima, A. Kawamoto, and J. Tanaka, Chem. Letters, 35 (1988).
[48] C. Coulon, R. Laversanne, J. Amiell, and P. Delhaès, J. of Phys.C 19: L753 (1986).
[49] D. Jérome, F. Creuset, and C. Bourbonnais, Nobel Symposium on Physics of Low-Dimensioal Systems, Granavallen, Sweden, June 1988 (to be published in Physica Scripta).
[50] P. M. Chaikin and R. L. Greene, Physics Today, p. 24 (May 1986).
[51] K. Murata, K. Kikuchi, T. Takahashi, K. Kobayashi, Y. Honda, K. Saito, K. Kanoda, T. Tokiwa, H. Anzai, T. Ishiguro, and I. Ikemoto, J. of Mol. Electronics 4: 173 (1988).
[52] G. Papavassiliou (Lecture at this NATO conference).
[53] C. Coulon, P. Vaca, T. Granier, and B. Gallois, Synth. Metals 27: B449 (1988).
[54] S. Hunig, A. Aumüller, P. Erk, H. Meixner, J. U. Von Schütz, H. J. Gross, U. Langohr, H. P. Werner, H. C. Wolf, Ch. Burscha, G. Klebe, K. Peters, and H. G. Von Schnering, Synth.Metals 27: B181 (1987).
[55] T. Nori, H. Inokuchi, A. Kobayashi, R. Kato, and H. Kobayashi, Synth.Metals, 27: B237 (1988).
[56] L. Brossard, M. Ribault, M. Bousseau, L. Valade, and P. Cassoux, Compt.Rend. Acad. Sci. Paris 302(II): 205 (1986).
[57] K. Kajita, Y. Nishio, S. Moriyama, R. Kato, H. Kobayashi, W. Sasaki, A. Kobayashi, H. Kim, and Y. Sasaki, Sol. St. Commun., 65: 361 (1988).
[58] S. Mazumdar and S. N. Dixit, Phys. Rev. B 34: 3683 (1986).
[59] S. Mazumdar and Z. G. Soos, Phys. Rev. B 33: 2810 (1981).
[60] J. S. Pedersen and K. Carneiro, Rep. Prog. Phys 50: 995 (1987).
[61] M. H. Cohen in "Organic Conductors and Semiconductors", ed.by L. Pál, G. Grüner, A. Jánossy, and J. Sólyom , Springer Lect. Notes in Phys. 65: 225 (Springer, Berlin, 1976).
[62] P. Delhaès, E. Dupart, J. P. Manceau, C. Coulon, D. Chasseau, J. Gaultier, J. M. Fabre, and L. Giral, Mol. Cryst. Liq. Cryst. 119: 269 (1985).
[63] P. Delhaès, C. Coulon, S. Flandrois, B. Hilti, C. W. Mayer, G. Rihs, and J. Rivory, J. Chem. Phys. 73: 1452 (1980).
[64] J. Martinsen, R. L. Greene, S. M. Plamer, and B. M. Hoffman, J.Am.Chem.Soc. 105: 677 (1983).
[65] T. Inabe, S. Nakamura, W. B. Liang, and T. J. Marks, J.Am.Chem.Soc. 107: 7224 (1985).
[66] L. J. De Jongh, J. de Chim. Phys. 85: 1105 (1988).
[67] P. Day in "Organic and Inorganic Low-Dimensional Crystalline Materials", ed. by P. Delhaès and M. Drillon, NATO ASI Series B168: 33 (Plenum, New York, 1987).
[68] Z. Yoshida and T. Sugimoto, Angew. Chem. Ind. Ed. Engl. 27: 1573 (1988).
[69] J. E. Hirsch and D. J. Scalapino, Phys. Rev. B 52: 5639 (1985).
[70] H. Iwamura, Pure Appl. Chem. 58: 187-196 (1986).
[71] T. P. Pradhakrisman, Z. G. Soos, H. Endres, and L. J. Azevedo, J. Chem. Phys. 85: 1126 (1986).
[72] J. S. Miller and A. J. Epstein, J. Am. Chem. Soc. 109: 3850 (1987).
[73] H. M. McConnell, Proc. Robert A. Welch Found. Conf. Chem. Res. 11: 164 (1967).
[75] J. B. Torrance, S. Oostra, and A. Nazzal, Synth. Metals 29: 709 (1987).
[76] E. Dormann, N. J. Nowak, K. A. Williams, R. O. Angus, and F. Wudl, J. Am. Chem. Soc. 109: 2594 (1987)

[77] J. S. Miller, D. A. Dixon, and J. C. Calabrese, Science 240: 1185 (1988).
[78] J. P. Morand, L. Brezinski, R. Lapouyade, C. Garrigou-Lagrange, J. Amiell, and P. Delhaès, Mol. Cryst. Liq. Cryst. , to appear (1989).
[79] L. Y. Chiang and D. P. Goshorn, Mol. Cryst. and Liq. Cryst., to appear (1989).
[80] A. J. Heeger, Solid State Physics 23: 283 (Academic, New York, 1969).
[81] M. Lequan, A. M. Lequan, C. Hauw, J. Gaultier, G. Macéno, and P. Delhaès, Synth. Metals, 19: 409 (1987).
[82] M. Y. Ogawa, B. M. Hoffman, S. Lee, M. Yudkowsky, and W. P. Halperin, Phys. Rev. Lett. 67: 1177 (1986).
[83] C. Piechocki, J. Simon, A. Skoulios, D. Guillon, and P. Weber, J. Am. Chem. Soc. 104: 5245 (1982).
[84] L. Y. Chiang, J. P. Stokes, C. R. Safinya, and A. N. Bloch, Mol. Cryst. Liq. Cryst. 125: 279 (1985).
[85] N. Boden, R. J. Bushby, J. Clements, M. V. Jesudason, P. F. Knowles, and G. Williams, Chem. Phys. Lett. 152: 94 (1988).
[86] V. Gionis, H. Strzelecka, M. Weber, R. Kormann, and L. Zuppiroli, Mol. Cryst. Liq. Cryst. 137: 365 (1986).
[87] P. Davidson, A. M. Levelut, H. Strzelecka, and C. Gionis, J. Phys. (Paris) Lettres 44: 823 (1983).
[88] A. Barraud, J. Chim. Phys. 85: 1121 (1988).
[89] M. Vandevyver, J. Chim. Phys. 85: 1033 (1988).
[90] T. Nakamura (Lecture at this NATO conference).
[91] J. Richard, M. Vandevyver, P. Lesieur, A. Ruaudel-Teixier, and A. Barraud, J. Chem. Phys. 86: 2428 (1987).
[92] K. Ikegami, S. Kuroda, K. Saito, M. Saito, and M. Sugi, Synth. Metals 27: B587 (1988).
[93] T. Nakamura, K. Kosima, M. Matsumoto, H. Tachibana, M. Tanaka, E. Manda, and Y. Kawabata, Chem. Letters, 369 (1989).
[94] A. Ruandel-Teixier (Lecture at this NATO conference).
[95] B. Tieke, A. Wegmann, Thin Solid Films, to appear (1989).
[96] J. P. Morand, R. Lapouyade, P. Delhaès, M. Vandevyver, J. Richard, and A. Barraud, Synth. Metals 27: B564 (1988).
[97] J. Richard, M. Vandevyer, A. Barraud, J. P. Morand, and P. Delhaès, J. Coll. Interf. Sci. 129: 254 (1989).

FRONTIERS OF ORGANIC SUPERCONDUCTORS

Gunzi Saito*

Institute for Solid State Physics
The University of Tokyo
Roppongi, Minato-ku
Tokyo 106, JAPAN

INTRODUCTION

Many investigations have been made on chemical, physical and structural properties of organic superconductors, since the discovery of the first organic superconductor $(TMTSF)_2PF_6$ by Jérome, Bechgaard *et al.* in 1980 [1]. As yet, they are not sufficient enough to yield a definitive picture of the mechanism of superconductivity, and to predict new organic superconductors. At present, more than 30 organic superconductors have been reported, and they have a maximum T_c (midpoint of resistance jump) in the range of 11.0 - 11.1 K (a T_c of 12.8 K was recently observed in one single crystal). The total number of organic systems is still less than 1%, and the maximum T_c is less than 1/10 of those of inorganic systems (Fig. 1). It is an urgent task for chemists to explore a variety of new organic superconductors, and to extend T_c to the range of 15 - 20 K, at least.

Here, an overview of the chemical and physical properties of the organic superconductor κ-(BEDT-TTF)$_2$Cu(NCS)$_2$, for which T_c is the highest in the organic systems, and of an organic metal, (BEDT-TTF)$_2$KHg(SCN)$_4$, will be described [2]. In particular, I make some remarks about the characteristics of κ-(BEDT-TTF)$_2$Cu(NCS)$_2$: 2-dimensional (2D) electronic and structural properties, inverse isotope effect, exceeded Pauli-limited H_{c2} value in the 2D plane, Shubnikov-de Haas oscillations and the Fermi surface, anisotropic thermopower, dc and ac magnetic susceptibilities, EPR, a big enhancement in the NMR T_1^{-1} well below T_c, and the energy gap. Finally, the future scope for T_c of organic superconductors will be discussed.

SEARCH FOR $Cu(NCS)_2$ AND $KHg(SCN)_4$ SALTS OF BEDT-TTF

There are five families of organic superconductors based on different organic moieties, namely the TMTSF (7 members), BEDT-TTF (about 13), DMET (7), MDT-TTF (1). and $M(dmit)_2$ (4) families. The BEDT-TTF family is the biggest, and T_c of it is the highest.

Before the $Cu(NCS)_2$ salt was discovered, several types of counter anions were exploited for BEDT-TTF superconductors: tetrahedral (ReO_4 [3]), linear trihalides (I_3 [4] and IBr_2 [5]), and metal halides (AuI_2 [6]), cluster anions ($Cl{\cdot}H_2O$ [7]), and polymeric anions ($Hg_{2.78}Cl_8$ [8], $Hg_{2.89}Br_8$ [9]). The linear anions, especially I_3, afforded several superconductors, such as iodine doped α- [10], low-T_c β- [4], high-T_c β- [11], γ- [12], θ- [13], and the κ-I_3 salt [14]. Among them, the β-phase showed a very curious behavior, and was investigated extensively. The T_c vs. pressure phase diagram of the β-I_3 salt, depicted in Fig. 2, indicates the coexistence of low-T_c (1.5 K) and high-T_c (8 K) phases at ambient pressure [11]. The β-I_3 salt was found to have an incommensurate superstructure below 175 K, and T_c = 1.5 K [15]. The superstructure is diminished by an application of a moderate pressure, and the salt

* Present address: Department of Chemistry, Kyoto University, Kyoto, JAPAN

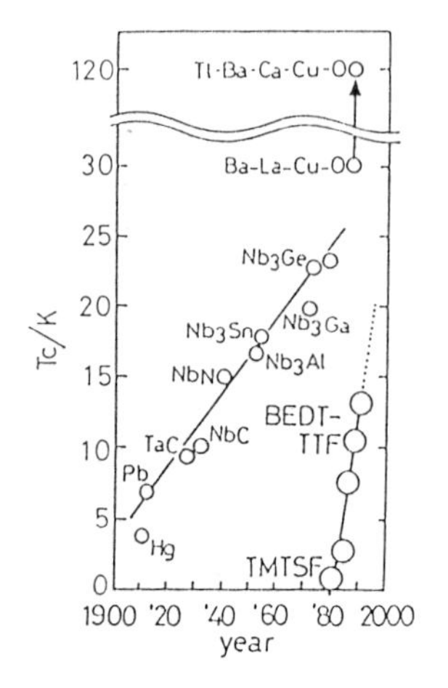

Fig. 1. History of organic vs. inorganic superconductors

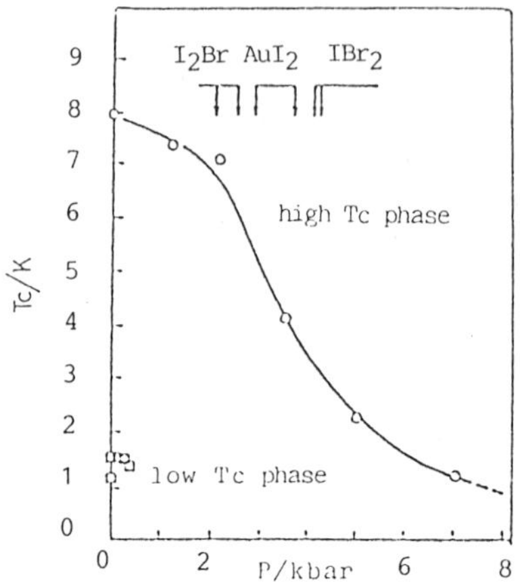

Fig. 2. T_c vs. pressure phase diagram of β-$(BEDT\text{-}TTF)_2I_3$ from Ref. [20].

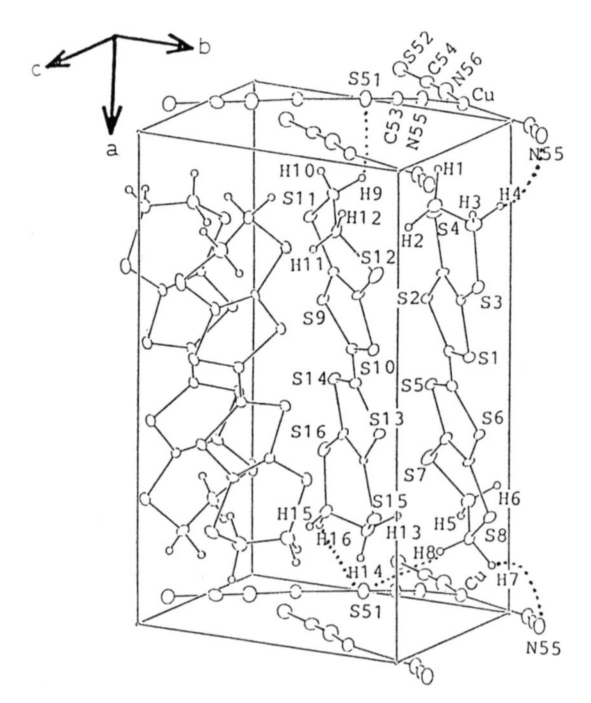

Fig. 3. Crystal structure of κ-$(BEDT\text{-}TTF\text{-}h_8)_2Cu(NCS)_2$ at 104 K (dextrorotatory form). Dotted lines indicate short atomic contacts between BEDT-TTF and anion.

becomes the high-T_c phase. Subtle conformational differences of the ethylene groups, and the resultant changes, both in anion-donor interactions and in ethylene ordering, are ascribed to the differences in T_c of these β-phase salts [16]. The superstructure is unstable below about 110 K and, hence, the high-T_c phase appears by annealing the salt for 20-40 hr at about 110 K [17].

Some of the reasons for the occurrence of multiphase BEDT-TTF salts are the conformational freedom of the ethylene groups of the BEDT-TTF molecule, and the cooperative interactions caused by the π-π (face-to-face) overlap and sulfur-sulfur (side-by-side) atomic contact; these cooperative interactions result in the 2D nature of BEDT-TTF charge-transfer (CT) complexes [18]. One of the polymorphs of β-I_3 salt (α-I_3), has been found to change into a stable 8 K superconductor (α_t-I_3) by tempering; this is considered to have a structure similar to that of high-T_c β-I_3 [19].

Analogous linear anions (IBr_2 and AuI_2) gave superconducting β-phase salts, and both of them are expected to be the high-T_c phase. But their T_c's are lower than that of the high-T_c β-I_3 salt. To explain how the size of the linear anion affects the T_c of the β-phase salts, an idea of "lattice pressure" has been presented [20]. The isostructural β-phase salt contracts by a use of smaller anions, and "lattice pressure" corresponds to this compression. The result is summarized in Fig. 2, where the β-IBr_2 salt (T_c = 2.7 K) and the β-AuI_2 salt (T_c = 3.4 - 4.9 K) correspond to lattice pressures of about 4 kbar and 3-3.7 kbar, respectively, compared with that of the β-I_3 salt. A plot of these lattice pressures and T_c's shows good agreement with the curve of the high-T_c β-I_3 salt in Fig. 2. An isostructural β-I_2Br salt, however, exhibits no superconductivity, due to the disordered orientation of the unsymmetrical anion. From these observations came a plan to obtain higher T_c in the β-phase salts: to find a β-phase salt of a symmetric linear anion with "negative" lattice pressure.

The length of the symmetric anion was found to be a critical factor in determining the T_c of a β-phase salt. T_c increases almost linearly with increasing the length of the anion; this trend has been discussed in terms of the density of states [21]. Although the explanations of this trend are still controversial, a molecular design can be derived from the trend as follows: the use of an anion longer than I_3 may yield a β-phase salt with T_c higher than 8 K, which just means the application of negative lattice pressure to the β-I_3 salt.

Since triiodide is the longest one among the symmetric linear polyhalides, we have searched for longer anions, among a variety of metal halides and metal pseudohalides, where metals are Cu, Ag, Hg, Ni, Pd, etc. and the pseudohalides are CN, OCN, SCN, SeCN, etc. In the following we describe only two BEDT-TTF complexes obtained in this process, namely superconducting $Cu(NCS)_2$ and metallic $KHg(SCN)_4$ salts; however, neither was found to be β-phase; furthermore, the anions were neither symmetric nor linear, but formed polymeric clusters.

CRYSTAL GROWTH

Crystal growth of the κ-$Cu(NCS)_2$ salt

Black, shiny single crystals with distorted-hexagon-shape (3 x 2 x 0.05 mm^3) of κ-$(BEDT\text{-}TTF)_2Cu(NCS)_2$ were prepared by the electrochemical oxidation of BEDT-TTF (prepared from CS_2 by conventional methods) in 1,1,2-trichloroethane (TCE), benzonitrile or THF in the presence of (1) CuSCN, KSCN and 18-crown-6 ether, (2) K(18-crown-6 ether) $Cu(NCS)_2$ or (3) CuSCN and TBA·SCN. For electrolytes (1) or (3), undissolved materials remained on the bottom of the cell during the course of electrocrystallization, but the precipitation did not affect the crystal growth. Crystals were grown in H cells (total volume *ca.* 20 mL) or modified H cells, where one cell compartment is an Erlenmeyer flask (total volume *ca.* 100 mL). The anode and cathode were separated by a medium-porosity frit.

For example, 30 mg of BEDT-TTF, 70 mg of CuSCN, 130 mg of KSCN and 200 mg of the crown were placed in the anode side, and the cell was evacuated, then filled with nitrogen or argon. 1090 mL of TCE (distilled under inert gas and eluted through Alumina Super I just before use under inert gas) was added, and stirred overnight under inert gas. The undissolved materials were precipitated, by standing, for one day, then electrolysis was carried out under a constant current of 1-5 μA. Crystal growth proceeded for 5-14 days, and the harvested crystals were washed with TCE, MeOH and THF, then dried under vacuum. An addition of a small amount of EtOH (1-2 wt. %) drastically accelerated the crystal growth, and a 4-5 mm length of crystals was achieved within a week [22].

The ESCA, EPR and crystal structure of this salt indicated the valency of the components to be +1/2 for BEDT-TTF, -1 for $Cu(NCS)_2$ and +1 for Cu. Even a partial substitution of Cu(I) in the salt by Cu(II) has not yet succeeded. The crystals were stable against light, air, moisture, and also thermally stable up to 190°C. Above 190°C the crystals started to lose weight, and decomposed at around 230°C with gas evolution.

Crystal growth of the $KHg(SCN)_4$ salt

The electrocrystallization of BEDT-TTF (57 mg) in the presence of $Hg(SCN)_2$ (319 mg), KSCN (98 mg) and the crown (328 mg) in CH_2Cl_2 (100 mL) under 1 μA gave long black crystals (2 x 0.25 x 0.03 mm^3) of $(BEDT\text{-}TTF)_3$ $[Hg(SCN)_3]$ after 18 days. The crystals were metallic (σ_{RT} = 23 S cm^{-1}) down to around 180 K, where a sharp metal-insulator transition occurred. The crystals obtained were not suitable for structural elucidation.

Using TCE as a solvent did not give any crystals, but adding 10 vol % of EtOH gave black thick plates (1.5 x 0.3 x 0.2 mm^3) of $(BEDT\text{-}TTF)_2KHg(SCN)_4$ after 35 days. The replacement of K ions in the salt with NH_4 ions was accomplished, by using NH_4SCN in the electrolysis, instead of KSCN.

CRYSTAL AND MOLECULAR STRUCTURES

Structure of the κ-$Cu(NCS)_2$ salt: [2a,c,e]

The crystal structures were determined at 298 K (R = 0.084) and 104 K (R = 0.042). The hydrogen atoms were included in the latter. There are two independent BEDT-TTF molecules in the asymmetric unit. The bond lengths of two independent BEDT-TTF molecules indicate that they have the same formal charge of +0.5. The BEDT-TTF molecules are slightly bent, with the dihedral angles between the least-squares planes of three tetrathioethylene moieties of 4.52°, 103.0°, 10.72° and 4.81° at 104 K. Two ethylene groups of a BEDT-TTF molecule deviate from the least-squares planes of the outer tetrathioethylene moieties. Although no significant conformational disorder was detected for the ethylene groups at 104 K, the thermal motion of one of the two ethylene groups is large at 298 K, which indicates a conformational disorder.

The crystal structure of this salt (Fig. 3) is very simiar to that of κ-$(BEDT\text{-}TTF)_2I_3$ [14]. Namely, two BEDT-TTF molecules form a dimerized pair, and the dimers are arranged over one another almost prependicularly. The interplanar spacing between the central tetrathioethylene moieties of a dimerized pair increases from 3.30 Å at 104 K to 3.38 Å at 298 K.

There are short intra- and interdimer sulfur-sulfur atomic contacts, so the donor molecules construct 2D conducting sheets in the **bc**-plane. The number of the short intra- and interdimer S..S contacts increases, on going from 298 K to 104 K, due to the subtle movement of BEDT-TTF molecules, but the overall features of the donor sheet at the two temperatures show only small differences. Every conducting layer is sandwiched by the insulating layers of anion $Cu(NCS)_2$ along the **a**-axis, which is very reminiscent of the Ginzburg model.

The lattice parameters show gradual changes, with lowering temperature, indicating that there is no distinct structural phase transition down to 104 K (Fig. 4). The temperature dilation of the **a**-axis is not linear with temperature. The **a**-axis remains constant down to 260-270 K, where it starts to expand gradually, with upper curvature down to 104 K, though the real separation of two anion layers (a sinβ) is almost unchanged.

Although the packing pattern of donor molecules and the lattice parameters of the crystals are almost the same for the κ-$Cu(NCS)_2$ and the κ-I_3 salts, their T_c's are considerably

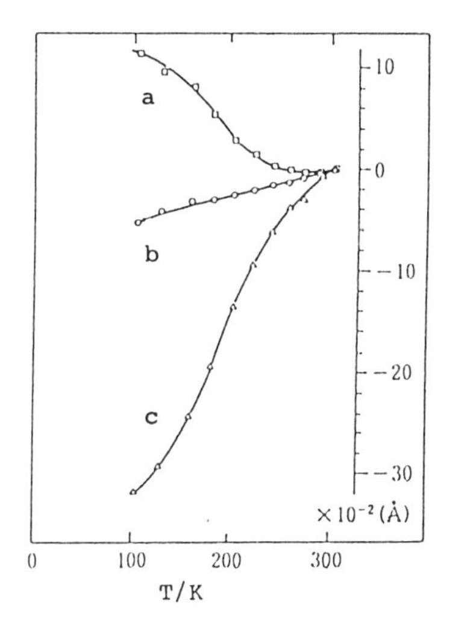

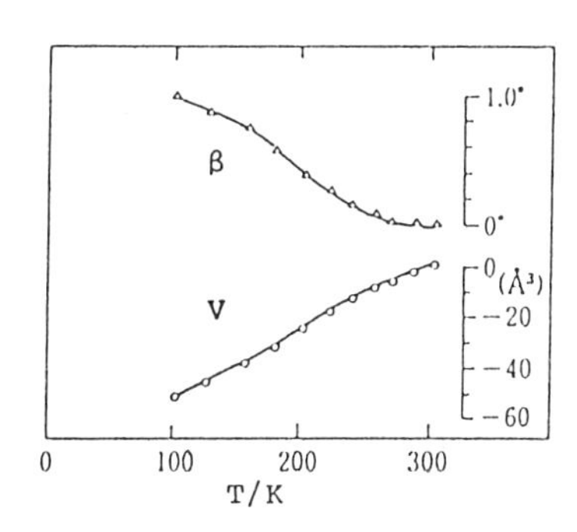

Fig. 4. Temperature dependences of lattice parameters of κ-$(BEDT\text{-}TTF\text{-}h_8)_2Cu(NCS)_2$.

Table 1. Crystal symmetry, unit cell parameters and T_c of κ-$(BEDT\text{-}TTF\text{-}h_8)_2Cu(NCS)_2$ and κ-$(BEDT\text{-}TTF\text{-}h_8)_2I_3$ [14].

Compound	$Cu(NCS)_2$ salt (298 K)	$Cu(NCS)_2$ salt (104 K)	I_3 salt
Crystal system	monoclinic	monoclinic	monoclinic
Space group	$P2_1$	$P2_1$	$P2_1/c$
a(Å)	16.248(5)	16.382 (4)	16.387(4)
b(Å)	8.440(2)	8.402(2)	8.466(2)
c(Å)	13.124(5)	12.833(4)	12.832(8)
β(°)	110.30(3)	111.33(2)	108.56(3)
V(Å³)	1688.0(9)	1645.3(7)	1687.6
Z	2	2	2
T_c (K)	10.4		3.6

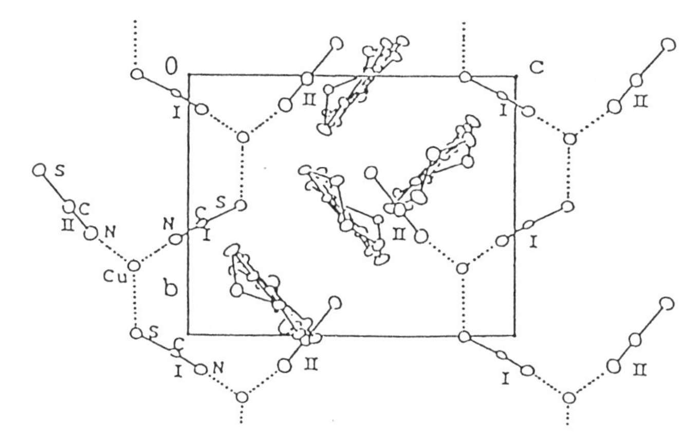

Fig. 5. Crystal structure of κ-$(BEDT\text{-}TTF\text{-}h_8)_2Cu(NCS)_2$ along the **a**-axis.

different (Table 1), which definitely indicates that there are no simple linear correlations between T_c and unit cell volume, contrary to an early prediction [21]. There are some structural differences between the two κ-phase salts, such as the anion structure and the interaction between the donor and anion sheets.

I_3 is a linear, symmetric anion, which is fairly isolated from other I_3's in the anion sheet of the β-I_3 salt. On the other hand, the $Cu(NCS)_2$ anion is neither symmetric nor linear, but asymmetrically bent, like a boomerang, where two crystallographically independent NCS groups (I and II) are almost linear (175° and 179° at 104 K) (Fig. 5). The repeating unit boomerang SCN(I)-Cu-NCS(II), is linked one after the other along the **b**-axis, to form a zig-zag 1D flat polymer. Both nitrogen and sulfur atoms of the NCS groups (I) coordinate to copper cations. On the other hand, it is the nitrogens of the NCS groups (II) which are directed towards Cu. Within a polymer, -SCN(I)-Cu- SCN(I)- forms an infinite zig-zag chain along the **b**-axis, and the other kind of ligand NCS(II) coordinates to the chain as a pendant. Thus the polymer is represented as --(SCN(I)-Cu---)$_n$---

|

NCS(II)

The thermal changes of angles associated with the ligand (II) are considerably larger than those of (I), indicating that the ligand (I) is under more rigid circumstance than (II), as expected. Every polymer lies in the same direction, to construct an insulating sheet in the **bc**-plane. As a consequence, the crystal does not have an inversion center, and hence is optically active, with the specific rotatory power of about 230° (25°C, 632.8 nm).

Taking into account the crystal symmetry ($P2_1$), the Fermi surface of the κ-$Cu(NCS)_2$ salt, calculated by extended Huckel MO, is composed of two different surfaces: one is a closed cylinder, and the other is a modulated 1D open surface (Fig. 6). A single big circular surface was proposed as a Fermi surface for the κ-I_3 salt, owing to its $P2_1/c$ symmetry [14].

Between the conducting and the insulating sheets, short atomic contacts were observed among the terminal ethylene groups of BEDT-TTF molecules and the N and S atoms of SCN(I) at low temperature (104 K). The meaningfully short atomic contacts of N..H (<2.6 Å) and S(anion)..H(<2.9 Å) are formed by using hydrogen atoms on both sides of the ethylene groups of the BEDT-TTF molecule. So, every 2D conducting layer of BEDT-TTF molecules is connected, by such atomic contacts, to the insulating anion layers along the **a**-axis. The role of these short C-H..N and C-H..S contacts in the superconductivity is not clear, though the substitution of hydrogen atoms of BEDT-TTF with deuterium atoms increased T_c by about 0.4 - 0.6 K *(vide infra)*. It seems important to elucidate the role of these atomic contacts in the transport properties of this compound, in the normal as well as in the superconducting state.

Structure of the $KHg(SCN)_4$ salt [2w]

The crystal data at 298 K and 104 K are summarized in Table 2. The BEDT-TTF molecules form a conducting sheet in the **ac**-plane, which is sandwiched by the insulating layers of the anion $KHg(SCN)_4$ along the **b**-axis (Fig. 7). There are three crystallographically independent BEDT-TTF molecules; one (A) is located in a general position, and the other two (B and C) are on the inversion center (Fig. 8). The bond lengths and angles of the BEDT-TTF molecules suggest that they have the same formal charge of about +0.5. The two kind of stacks (AA.. and BCBC.. along the **c**-axis) lie alternately along the **a**-axis, with their dihedral angles, as shown in Fig. 3. In addition to the face-to-face interactions of BEDT-TTF molecules along the stacking axis, several short S..S atomic contacts (3.435 - 3.60 Å) were observed along the side-by-side direction (**a**-axis). These features result in a complicated Fermi surface, calculated by the extended Huckel MO (Fig. 9), which contains both 2D closed and 1D open surfaces [23].

The total packing mode of BEDT-TTF molecules is much more similar to the α-I_3 salt [24] than that of the θ-I_3 salt [13], based on the periodicity along the **b**-axis, since every two donors are the repeating unit in the θ-phase, while each donor arranges equivalently, both in the α-phase and this salt. But the dihedral angles between the molecules in the neighboring columns are considerably different from those of the α-I_3 salt, the dihedral angles of which are 59.4° and 70.4° [24].

The anion arrangement of this salt is very curious. An anion layer consists of a triple-sheet ($y = 0, \pm 0.08$) parallel to (010). The bottom and top sheets ($y = \pm 0.08$) are made of a zig-zag arrangement of linear SCN groups. The middle sheet ($y = 0$) contains both K and Hg ions, where one of the K ions in a unit cell is electrostatically linked to four SCN groups in the bottom sheet, with nitrogen atoms, to form a pyramid (K..N = 2.85 - 2.89 Å, Fig. 10).

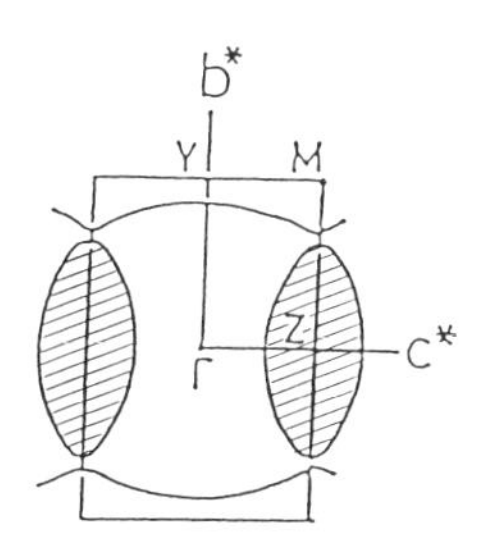

Fig. 6. Fermi surface of κ-(BEDT-TTF-h_8)$_2$Cu(NCS)$_2$ by the tight-binding method. Shaded regions indicate hole-like parts.

Table 2. Crystallographic data of (BEDT-TTF)$_2$[KHg(SCN)$_4$]

	298 K	104 K
Crystal System	triclinic	triclinic
Space group	$P\bar{1}$	$P\bar{1}$
a(Å)	10.082(10)	9.948 (2)
b(Å)	20.565(4)	20.505(11)
c(Å)	9.933(2)	9.833(4)
α(°)	103.70(2)	103.34(4)
β(°)	90.91(4)	90.53(3)
γ(°)	93.06(4)	92.80(3)
V(Å^3)	1997(2)	1949(1)
Z	2	2
D_x (g cm^{-3})	2.065	2.115
D_m (g cm^{-3})	2.069	----
Reflections used	7639	101100
R	0.0696	0.0418

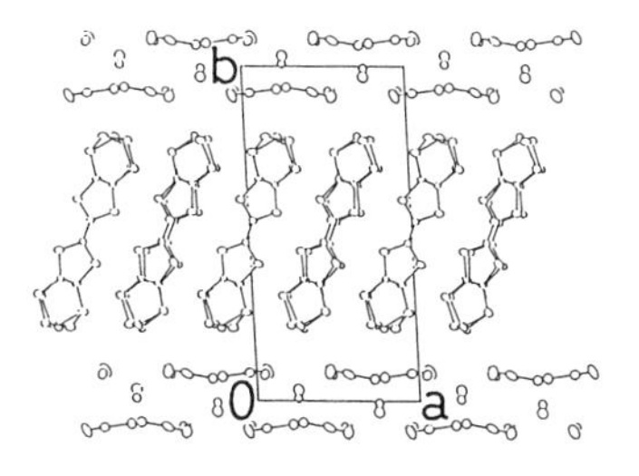

Fig. 7. Crystal structure of (BEDT-TTF)$_2$[KHg(SCN)$_4$] at 298 K along the **c** axis.

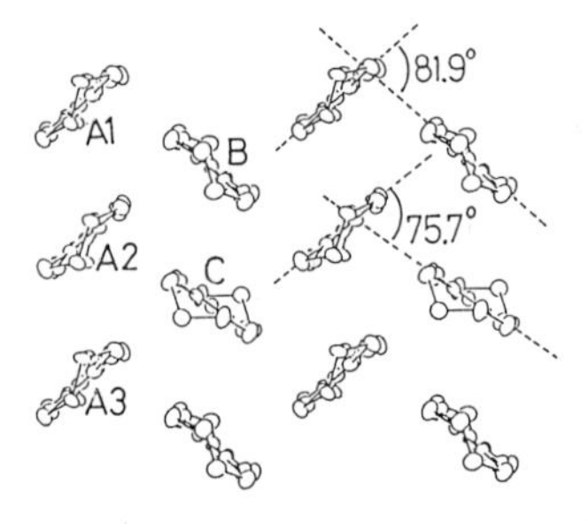

Fig. 8. Molecular stacking of BEDT-TTF in $(BEDT\text{-}TTF)_2[KHg(SCN)_4]$.

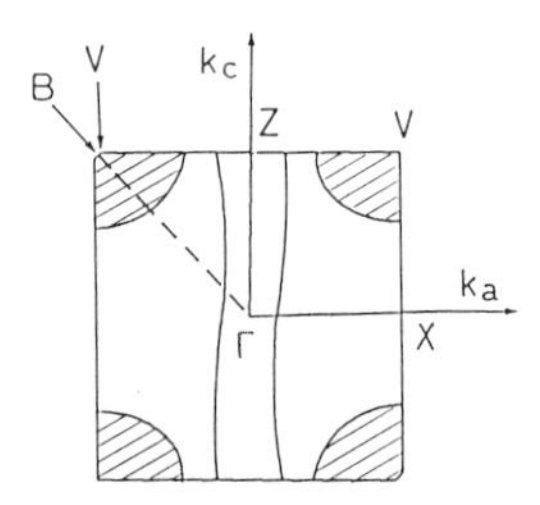

Fig. 9. Fermi surface of $(BEDT\text{-}TTF)_2[KHg(SCN)_4]$ calculated by the tight-binding method. Shaded regions indicate hole-like parts [23].

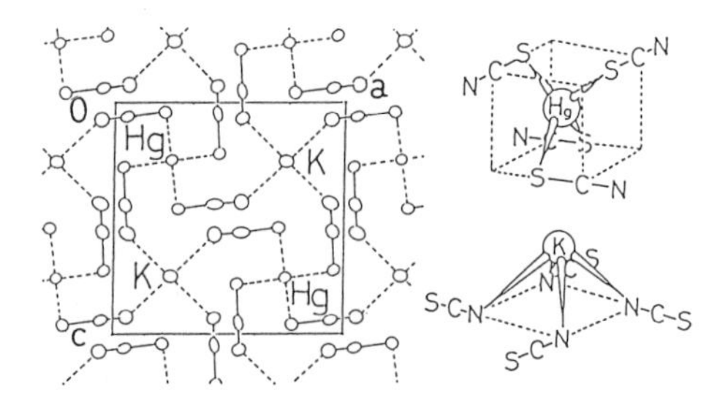

Fig. 10. Triple-sheet structure of anion $KHg(SCN)_4$ and schematic structures of its component clusters.

The other K ion forms a pyramid, in the same way, with the SCN ions in the top sheet. The calculated radius of N (1.33 - 1.37 Å), obtained by using the K..N distance and the ionic radius of K (1.52 Å), is just between the van der Waals (1.5 Å) and the covalent radius (0.70 Å) of N, indicating the ionic character of the K.N bond. The softer Hg ions are coordinated to four SCN ions, two in the bottom and two in the top sheet, with sulfur atoms, to form tetrahedra (Fig. 10). The Hg..S distances (2.55 - 2.57 Å) are little longer than the sum of covalent radii (2.52 Å) but considerably shorter than the sum of the ionic radii (2.95 Å). So, every SCN is linked to K^+ and Hg^{++} in this manner, to construct a thick (6.8 Å) network of infinite chains of:

```
                    NCS··          SCN··
                     |              |
 --SCN---K---NCS---Hg---SCN--
                     |              |
                    NCS··          SCN··
```

which spread in the **ac**-plane (Fig. 10). Similar structures have been reported for inorganic complexes, $M(II)Hg(SCN)_4$ (M(II) = Co, Zn, Cd) [25], which also contain Hg^{++} tetrahedrally coordinated to sulfur atoms.

The structure analysis at 104 K evidently clarified the existence of several short anion...ethylene contacts, such as S(anion)...H (2.79 - 2.96 Å), C(anion)...H (2.68 - 2.90 Å) and N(anion)...H (2.44-268 Å) contacts. The unit cell volume and lattice constants (except β, which is almost constant), decreased monotonically with temperature down to 104 K.

In the following, some physical properties of the κ-$Cu(NCS)_2$ and $KHg(SCN)_4$ salts, obtained in our group, will be described (see also Ref. [26] for other properties and the work of other groups on the former salt).

PHYSICAL PROPERTIES

Properties of the κ-$Cu(NCS)_2$ salt

Electrical Conductivity [2a,b,i,o,p,s]. The room-temperature conductivity along the b-axis is 10-40 S cm^{-1} by the four-probe method, and the conductivity ratio is $\sigma_{a*}: \sigma_b : \sigma_c$ = 1/600 : 1 : 1.2. The resistivity slightly decreased, down to 270 K, then it increased, with a maximum at around 90 K, followed by a rapid decrease of resistivity (Fig. 11). Some sample has been reported to show a monotonic decrease of resistivity, with a small temperature dependence, down to around 130 K, below which a sharp decrease of resistivity occurred [27]. T_c of the salt of BEDT-TTF-h_8 (H salt) is 10.2 - 10.4 K, and the T_c decreases with increasing pressure (-1.3 K/kbar). By the four-probe method, T_c of the salt of BEDT-TTF-d_8 (D salt) was observed at 10.8-11.0 K (a T_c of 11.1 K was recently claimed by Schweitzer *et al.* [28]). RF penetration depth measurements (contactless) also indicated that the T_c of the D salt is higher by 0.5-0.6 K than that of the H salt (Fig. 12). This isotope effect is just opposite to what one expects from the simple BCS theory:

$$T_c \propto \omega_D \propto M^{\alpha}, \alpha = -1/2$$

where ω_D is the Debye frequency, and M is the isotope mass. Even in a system with strong electron correlation, (organics may belong to this), the index α approaches zero, but does not exceed 0, as has been observed in transition metals. One of the explanations of this inverse isotope effect is that the superconductivity of this salt is, somehow, associated with the increase of the electron-phonon coupling in the D salt, compared to the H salt, just like the system of palladium-hydrogen and deuterium [29], through the short atomic interaction between the ethylene groups and the anion above mentioned.

Very recently, a T_c of 12.8 K was observed in only one deuterated single crystal (Fig. 13), out of several in one batch [30]. This superconductivity is overlapped with that of T_c = 11 K. The superconductivity is diminished by the application of a magnetic field of a few Tesla. It is not known whether this superconductor is a new phase of BEDT-TTF:Cu:SCN or not, since the crystal is still in a cryostat.

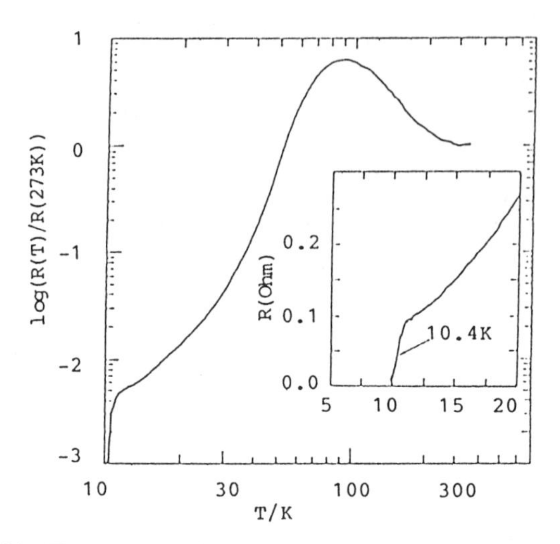

Fig. 11. Temperature dependence of the electrical resistance of κ-(BEDT-TTF-h_8)$_2$Cu(NCS)$_2$.

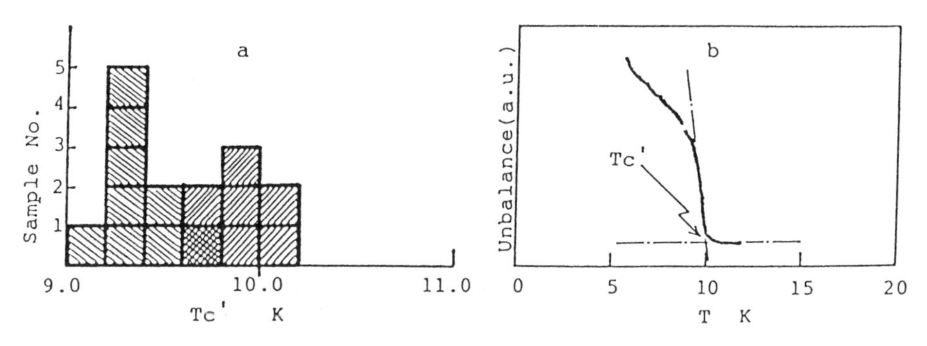

Fig. 12. (a) Scatter in T_c' for deuterated (▧) and undeuterated (▩) samples of κ-(BEDT-TTF)$_2$Cu(NCS)$_2$, determined by (b) RF penetration depth measurements.

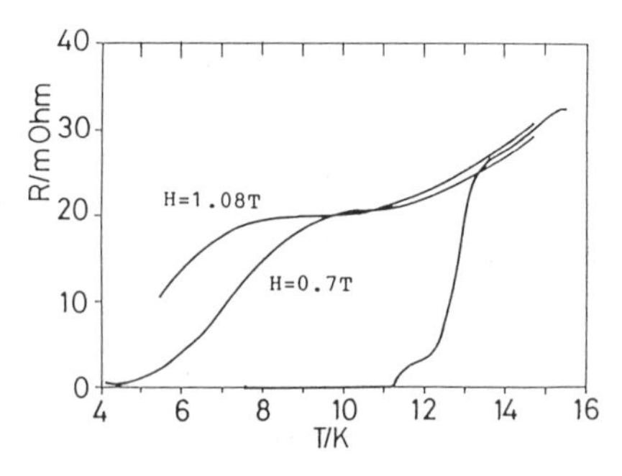

Fig. 13. Temperature dependence of the resistance of κ-(BEDT-TTF-d_8)$_2$Cu(NCS)$_2$. Only one crystal out of several in a batch showed this behavior [30].

Upper Critical Field [2b,i,p]. The early data of the temperature dependence of the upper critical field of the H salt (magnetic field < 13.5 Tesla) showed an inflection at around 9 K, and H_{c2} exceeded 13 Tesla at 5 K within the 2D plane (Fig. 14). H_{c2} measurements at higher magnetic fields, up to 24 Tesla, indicate that H_{c2}, estimated at 0.5 K in the **bc**-plane, exceeds the Pauli limiting value (H_p) [30]:

$$H_p(\text{kOe}) = 18.4\, T_c = \Delta_o / \sqrt{2}\, \mu_B$$

where $2\Delta_o$ is the energy gap and μ_B is the Bohr magneton. Along the **a***-axis, H_{c2} is very small, and increases steadily to about 0.8 Tesla at 5 K, then increases rather sharply at low temperatures. The exceeded H_{c2} value in the **bc**-plane for spin-pair breaking, and this upturned behavior along the **a***-axis are quite unusual, and we do not have any satisfactory interpretations at present.

The estimated anisotropy in the Ginzburg-Landau (GL) coherence lengths (which are calculated by using H_{c2} values near T_c and the relation:

$$H_{c2}{}^{i}(T) = \Phi_o / 2\pi\, \xi_j(T)\, \xi_k(T) = [\Phi_o / 2\pi\, \xi_j(0)\, \xi_k(0)]\, [(T_c - T) / T_c]$$

where $\Phi_o = h\, c / 2\, e$ is the flux quantum) is $\xi_{\|\mathbf{bc}} : \xi_{\mathbf{a}^*}$ = 182 Å: 9.6 Å = 19 : 1. It is noteworthy that the coherence length along the **a***-axis is less than the length of **a***. This situation is similar to that of the high-T_c oxide superconductors: for example, 3.8 - 6.3 Å (∥ **c**, **c** ≈ 11.6 Å) and 23 - 35 Å (in **ab** plane) at 0 K are reported in YBCO [31].

Shubnikov-de Haas Effect and Thermoelectric Power [2e,f,o]. Below 1 K and above 8 Tesla, Shubnikov-de Haas (SdH) oscillations were observed in the transverse magnetoresistance curve (Fig. 15). This is the first observation of a SdH signal in the organic superconductors, and is the conclusive evidence of the 2D nature of this compound. The oscillation of $\Delta(1/H) = 0.0015\ T^{-1}$ (for both H and D salts) corresponds to the area of the extremal orbit of $S = 6.37 \times 10^{14}\ cm^{-2}$ from:

$$S = 2\pi e / \Delta(1/H)\, \hbar c$$

Since the area of the first Brillouin zone is $1 / b\, c = 3.56 \times 10^{15}\ cm^{-2}$, that of the extremal orbit corresponds to 18% of the first Brillouin zone, which is quantitatively equal to that of the calculated cylindrical Fermi surface in Fig. 6.

Moreover, the calculated Fermi surface is in good qualitative agreement with the observed thermopower. The cylindrical Fermi surface is hole-like, and the modulated open Fermi surface is electron-like. This complicated nature of the Fermi surface may result in the anisotropic thermopower, which is positive along the **c**-axis, and negative along the **b**-axis, just as observed (Fig. 16), though a simple consideration of the chemical formula indicates that the carriers are only holes (carrier density $n = 1.18 \times 10^{21}\ cm^{-3}$). These results indicate the validity of the Fermi surface in Fig. 6 to some extent, in spite of the fact that the band calculation is based on a rather simple method. The thermopower became zero at around 10 K, indicating a superconducting transition. The temperature dependence of the SdH oscillation gave an effective mass at the Fermi level as $m / m^* = 3.5$, in accordance with those derived from the optical measurements (4.0 and 3.0 along the **b**- and **c**-axes, respectively) [2g,m]. A considerably smaller effective mass of $m / m^* = 2.4$ was obtained at 8 kbar.

Magnetic Susceptibility [2a,d,j,m]. The dc magnetic susceptibility of a polycrystalline sample (30 KOe) showed almost constant Pauli paramagnetism of $\chi_p = 4.1 - 4.6 \times 10^{-4}$ emu / mol between room temperature and about 10 K, which clearly indicates the metallic nature of this compound in this range (Fig. 17). The origin of a slight bump, which appeared at around 90 K, cannot be explained at present. From the room-temperature value, the density of states per formula unit for a single spin is calculated as $N(E_F) = 7.1\ eV^{-1}$ by using the relation:

$$\chi_p = 2\, \mu_B\, N_A N(E_F)\, [1 - (1/3)\, (m / m^*)^2]$$

and neglecting the Landau diamagnetism term (second term in the bracket), where N_A is the Avogadro constant. By using the optical data and the relation:

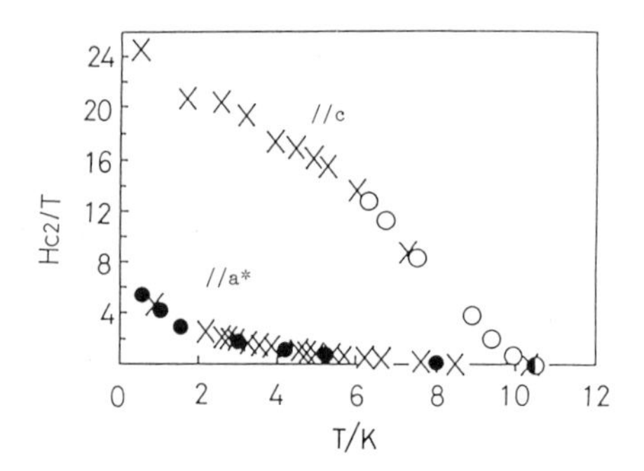

Fig. 14. Temperature dependence of the upper critical field of κ-(BEDT-TTF-h_8)$_2$Cu(NCS)$_2$ [30]. The circles are from Ref. [2b].

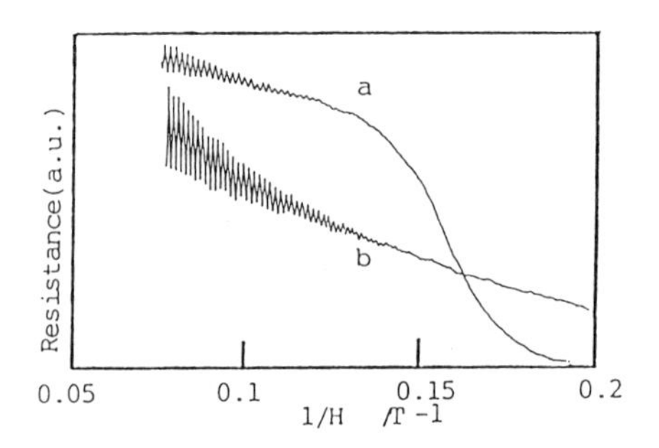

Fig. 15. Shubnikov-de Haas signals for κ-(BEDT-TTF-d_8)$_2$Cu(NCS)$_2$ (H, I // **a***) at (a) 1 bar, (b) 7 kbar.

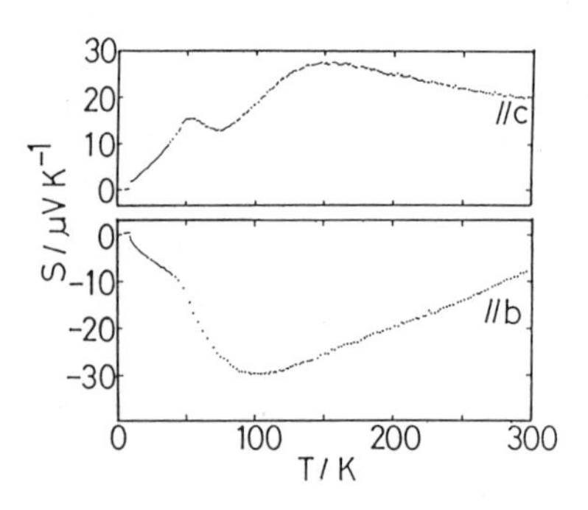

Fig. 16. Temperature dependence of thermoelectric power along the **b** and **c**-axes for κ-(BEDT-TTF-h_8)$_2$Cu(NCS)$_2$.

$$N(E_F) = (m_{b*}m_{c*})^{1/2} / (\pi \, \mathbf{a} \, \hbar^2)$$

where **a** is the lattice spacing, a comparable value of $N(E_F) = 7.5 \text{ eV}^{-1}$ was obtained, and the Fermi energy from the top of the band was estimated as 0.13 eV by using:

$$E_F = \pi \, \mathbf{a} \, n \, \hbar^2 / (m_{b*}m_{c*})^{1/2}.$$

A critical current density $J_c = 1060 \text{ A cm}^{-2}$ was estimated at 4.9 K and 50 Oe from the hysteresis curve of the dc magnetization.

The ac susceptibility of a polycrystalline sample showed that $T_c = 10.3 \pm 0.4$ K, and almost 100 % of the perfect diamagnetism ($-1/4\,\pi$ emu cm^{-3}) was observed below 7 K at 0.3 Oe (Fig. 17b). This is an unambiguous evidence for the bulk nature of the superconductivity of this material.

<u>EPR, NMR and Other Properties [2f,h,k,l,n]</u>. No signal assignable to Cu^{++} was observed in ESCA at room temperature and EPR down to 4 K. Due to the broading of the EPR main signal (Lorentzian), and to the appearance of a sharp additional signal (line width $\Delta H = 10 - 20$ G, $g = 2.0075$), which may be due to either defect or contamination with another phase, accurate estimations of g values and ΔH were not possible below 20 K (Fig. 18). The constant g values (2.0078 - 2.0070), down to 20 K, indicate that the molecular orientation of BEDT-TTF does not change substantially with respect to the crystal axes. The linewidth at room temperature was 61 G, and increased smoothly to about 80 G at 110 K. Then, ΔH increased more pronouncedly to 100 G down to 30 K, accompanied with a rapid increase to 120 G at 20 K. Since the electrical resistivity increases from 270 K to 90 K, the increase of ΔH in this region can be explained by an increase in the scattering rate of conduction electrons. The pronounced change of ΔH below 90 K, however, cannot be interpreted in terms of scattering rate, since ΔH does not follow the temperature dependence of the resistivity. So the EPR data are in contrast to the prediction of the Elliott formula [32]:

$$\Delta H \sim (\Delta g)^2 \tau_{\perp}^{-1}$$

where $\Delta g = g - 2.0023$, and $\tau_{\perp}$ is the interchain tunneling time, for the spin relaxation in 2D metals.

The EPR intensity, obtained with polycrystals, is almost constant (3.2×10^{-4} emu/mol at RT), in good agreement with the dc susceptibility for single crystal measurements.

A Korringa relation was observed between 77 K and 10 K in the ^1H NMR measurements on polycrystals (Fig. 19). The T_1T value is estimated as 1050 sec K, which is less than 2/3 of that of the low-T_c β-I_3 salt (1730 sec K, T_c = 1. 5 K). From the relation:

$$N(E_F) \propto (1/T_1T)^{1/2},$$

the $N(E_F)$ value of this salt is expected to be 1.28 times than that of the β-I_3 salt. By using the $N(E_F)$ value of 7.1 or 7. 5 eV^{-1} for the κ-$Cu(NCS)_2$ salt, a $N(E_F)$ value of 5.5 or 5.8 eV^{-1} was estimated for the β-I_3 salt, which is a little smaller than that obtained by static susceptibility measurements for the β-I_3 salt (6.5 eV^{-1}).

A big enhancement of the relaxation rate T_1^{-1} was noticed, with a peak at considerably lower temperature than T_c. The peak height is about 30 times that of the normal state at 3.28 KOe, and the peak position depends on the applied magnetic field. Such pronounced enhancement of T_1^{-1} is not explained by any known theory [33], and is unexpected for the usual superconductors known to date.

The superconducting gaps have been measured by a tunneling spectroscopic method on a $Cu(NCS)_2$ salt | Al_2O_3 | Au junction. No reliable spectra were obtained at temperatures above 4.2 K. Well below T_c, a superconducting energy gap, somewhat comparable to the BCS theory, was detected for one sample:

$$2\,\Delta_o = 4 \text{ meV}, \; 2\,\Delta H_o / k_B T_c = 4.5 \; (\textit{vs.}\ 3.52 \text{ for BCS ratio}).$$

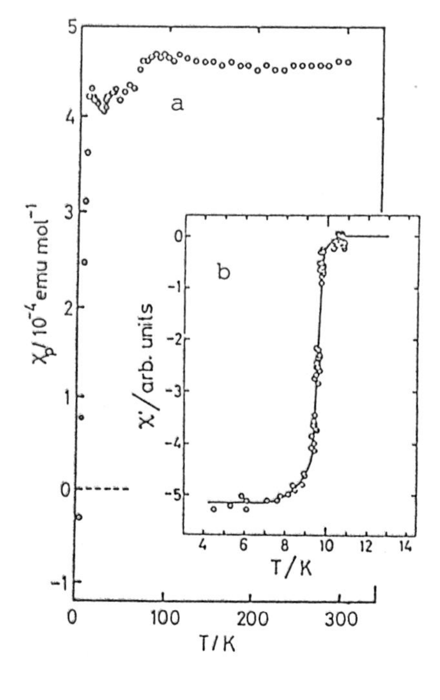

Fig. 17. Temperature dependence of the (a) dc ($H = 30$ kG) and (b) ac (H = earth magnetic field) susceptibilities of κ-(BEDT-TTF-h_8)$_2$Cu(NCS)$_2$ polycrystals.

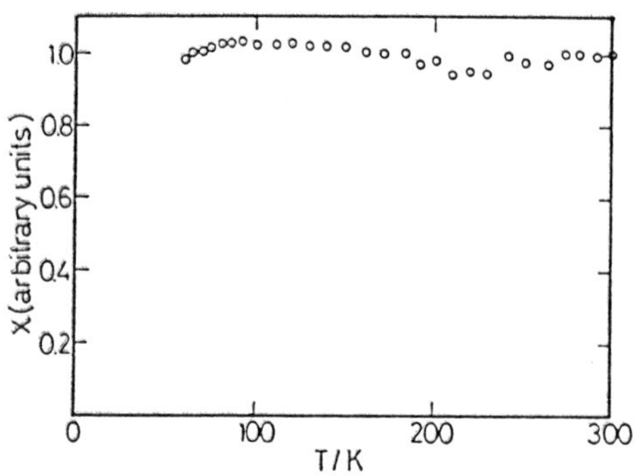

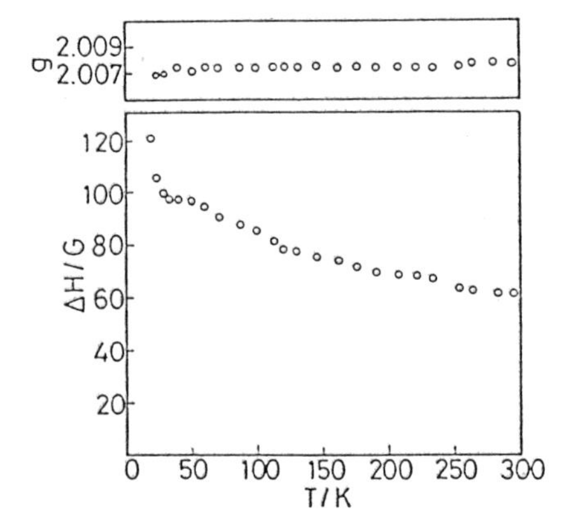

Fig. 18. Temperature dependences of the spin susceptibility χ for polycrystals, and the g value and linewidth ΔH for a single crystal of κ-(BEDT-TTF-h_8)$_2$Cu(NCS)$_2$.

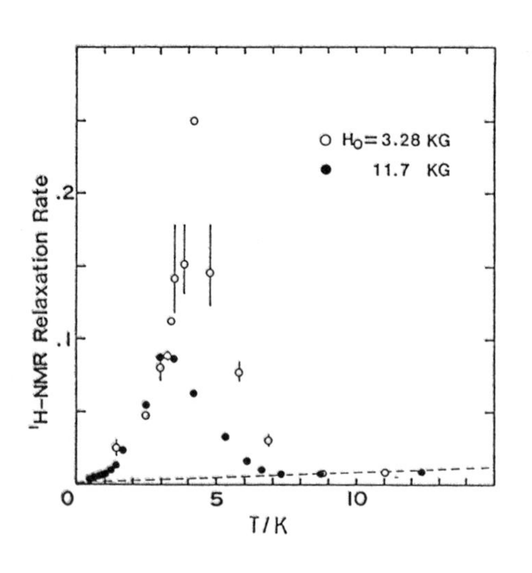

Fig.19. ^{1}H-NMR relaxation rate $1/T_1$ of κ-(BEDT-TTF-h_8)$_2$Cu(NCS)$_2$ at low temperatures. The broken line indicates the Korringa behavior determined at high temperatures.

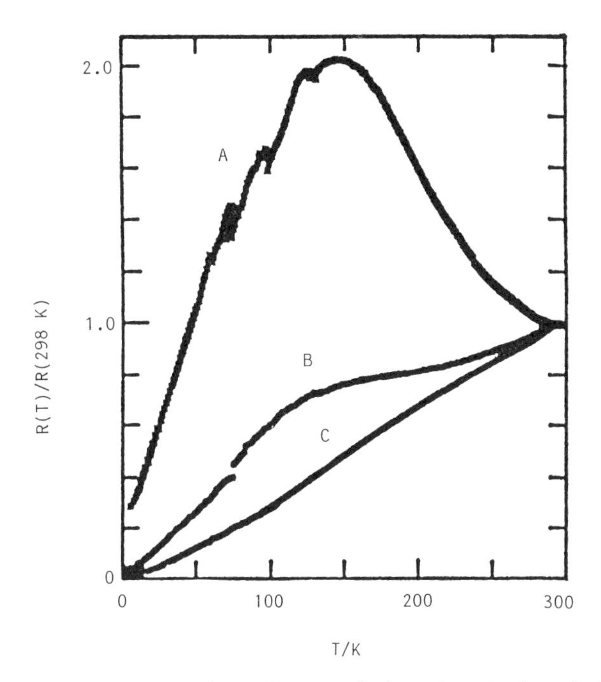

Fig. 20. Temperature dependence of the electrical resistivity of $(BEDT\text{-}TTF)_2[KHg(SCN)_4]$.

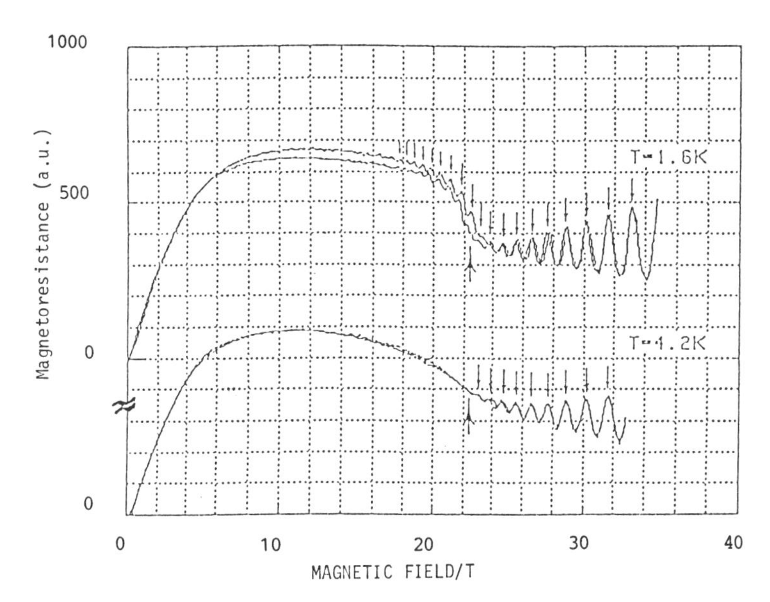

Fig. 21. Transverse magnetoresistance of $(BEDT\text{-}TTF)_2[KHg(SCN)_4]$. An unusual resistance decrease can be seen above 10 T, with a kink structure at 22 T indicated by the large arrow. The small arrows indicate Shubnikov-de Haas oscillations [34].

The other sample showed a much smaller gap, less than 1 meV. Therefore, at the moment, we may postulate that the superconducting gaps of this salt are anisotropic.

Preliminary data of optical properties [2g,m], lower critical fields [2j,m], and specific heat [2q] have already been reported. STM and μSR studies, by A. Kawazu *et al.* and Y. J. Uemura *et al.,* respectively, are now in progress.

Properties of the $KHg(SCN)_4$ salt [2w,23,34]

The room-temperature conductivities (20 - 100 S cm^{-1}) and the temperature dependences vary from sample to sample. Three typical behaviors of $R(T) / R_{RT}$ are depicted in Fig. 20. A big enhancement, or a weak shoulder of resistivity, was observed at around 130 K in Type A or B samples, respectively. A type C sample exhibited no anomaly above about 5 K, where the resistivity decreased more rapidly, but showed no zero resistance down to 1.8 K. In any case, the conductivity only improves about an order of magnitude from room temperature to 1.8 K, suggesting that the crystal contains some disorder.

The transverse magnetoresistance measurements on a Type A sample showed a saturation at around 10 Tesla, and a decrease up to about 20 Tesla [34]. SdH oscillations are superposed on this background magnetoresistance above 20 Tesla (Fig. 21). The oscillation period is 0.0015 T^{-1}. The band calculation, based on the extended Hückel MO, shows that the Fermi surface of this salt is composed of 2D closed (hole like) and 1D surfaces (Fig. 9) [23]. The area of the extremal orbit calculated from the SdH oscillation (16.5% of the first Brillouin zone) is comparable to that of the closed Fermi surface in Fig. 9 (19%). More precise measurements on the magnetoresistance and other properties are underway.

FUTURE SCOPE FOR T_c [2r,t]

At this stage, there are no reliable theories, and not enough physical parameters, to speculate on the T_c of organic superconductors. As an attempt to speculate on the future scope of T_c of BEDT-TTF superconductors, the T_c of several BEDT-TTF superconductors are plotted against the effective volume for one electron (V_{eff}) in Fig. 22. V_{eff} was calculated using the unit cell volume $V_{unitcell}$ (at RT), the anion volume V_{anion} (the anion is roughly approximated as a cylinder with the radius of the biggest atom in the anion), and the number of conduction carriers N in the unit cell: $V_{eff} = (V_{unitcell} - V_{anion}) / N$. Though Fig. 22 does not include every BEDT-TTF superconductor, and the calculated V_{eff} may contain uncertainty, there seems a linear relationship (except for low-T_c β-I_3) between T_c and V_{eff}, even among different phases. It is seen that the β-I_3 salt has bigger V_{eff} by about 10 Å^3, than the κ-I_3 salt. Therefore, if one could obtain a β-phase salt of $Cu(NCS)_2$, T_c of it might be about 13 K, as long as the relation holds. Also, by choosing an appropriate anion, one might extend the line to more than T_c =20 K within the family of BEDT-TTF salts. In the case of the $KHg(SCN)_4$ salt, there still may be some possibility of finding superconductivity, if we can improve the quality of the crystal, since the conductivity data suggest that the crystal has some disorder. But the expected T_c, derived from the correlation in Fig. 22, may be low, due to a small V_{eff} caused by the thick anion (V_{anion} is too big). Therefore, one may find high T_c with a big anion which forms a thin anion layer.

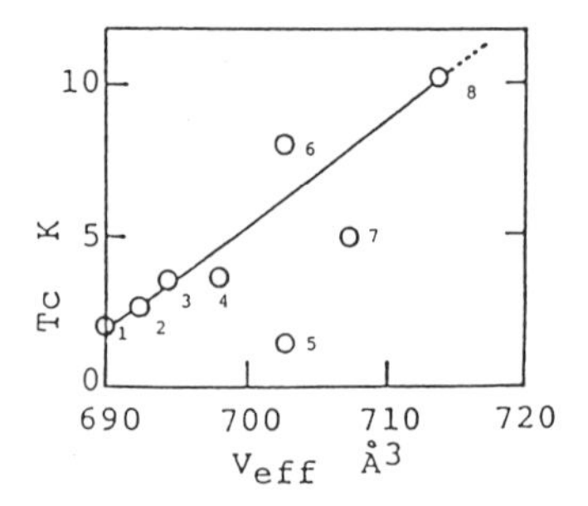

Fig. 22. Plot of T_c of BEDT-TTF superconductors *vs.* effective volume. 1: ReO_4, 2: β-IBr_2, 3: κ-I_3, 4: θ-I_3, 5: low-Tc β-I_3, 6: high-T_c β-I_3, 7: β-AuI_2, and 8: κ-$Cu(NCS)_2$ salts of BEDT-TTF.

However, there should be some limit to T_c in this treatment, since the larger the effective volume, the more dispersed BEDT-TTF molecules will be in a unit cell, which makes the crystal unstable. If one can control the band filling of organic superconductors of low molecular weight, just like a doped polymer or an inorganic oxide superconductor, V_{eff} may increase with small N without increasing the crystal instability.

ACKNOWLEDGEMENTS

The author wishes to thank the coauthors of Ref. [2] for exciting collaborations. In particular, the contributions of H. Mori (Urayama), H. Yamochi, M. Oshima, K. Omichi, K. Nozawa, T. Sugano, M. Kinoshita, S. Sato, Y. Ishida, T Yajima, and N. Miura (ISSP), K. Oshima (Okayama Univ.), T. Mori, H. Inokuchi, T. Inabe, and Y. Maruyama (IMS), T Takahashi (Gakushuin Univ.), J. Tanaka (Nagoya Univ.), T. Osada and S. Kagoshima (Tokyo Univ.) are invaluable. This work was partly supported by the Grant-in-Aid for Scientific Research from the Ministry of Education, Science and Culture, Japan (No. 63631006, 63604020, 63060004).

REFERENCES

[1] D. Jérome, A. Mazaud, M. Ribault, and K. Bechgaard, J. Phys. Lettres 41: L95 (1980).

[2] **For the κ-Cu(NCS)$_2$ salt:** (a) H. Urayama, H. Yamochi, G. Saito, K. Nozawa, T. Sugano, M. Kinoshita, S. Sato, K. Oshima, A. Kawamoto, and J. Tanaka, Chem. Lett. 55 (1988), (b) K. Oshima, H. Urayama, H. Yamochi, and G. Saito, J. Phys. Soc. Jpn.. 57: 730 (1988), (c) H. Urayama, H. Yamochi, G. Saito, S. Sato, A. Kawamoto, J. Tanaka, T. Mori, Y. Maruyama, and H. Inokuchi, Chem. Lett. 463 (1988), (d) K. Nozawa, T. Sugano, H. Urayama, H. Yamochi, G. Saito, and M. Kinoshita, *ibid.* 617 (1988), (e) K. Oshima, T. Mori, H. Inokuchi, H. Urayama, H. Yamochi, and G. Saito, Phys. Rev. B 37: 938 (1988), (f) H. Urayama, H. Yamochi, G. Saito, T. Sugano, M. Kinoshita, T. Inabe, T. Mori, Y. Maruyama, and H. Inokuchi, Chem. Lett. 1057 (1988), (g) T. Sugano, H. Hayashi, H. Takenouchi, K. Nishikida, H. Urayama, H. Yamochi, G. Saito, and M. Kinoshita, Phys. Rev. B 37: 9100 (1988), (h) T. Takahashi, T. Tokiwa, K. Kanoda, H. Urayama, H. Yamochi, and G. Saito, Physica C 153-155: 487 (1988), (i) K. Oshima, H. Urayama, H. Yamochi, and G. Saito, *ibid.* 154: 1148 (1988), (j) T. Sugano, K. Terui, S. Mino, K. Nozawa, H. Urayama, H. Yamochi, G. Saito, and M. Kinoshita, Chem. Lett. 1171 (1988), (k) Y. Maruyama, T. Inabe, H. Yamochi, and G. Saito, Solid State Commun. 67: 35 (1988), (1) H. Urayama, H. Yamochi, G. Saito, K. Oshima, S. Sato, T. Sugano, M. Kinoshita, A. Kawamoto, J. Tanaka, T. Inabe, T. Mori, Y. Maruyama, and H. Inokuchi, Synth. Met. 27: 393 (1988), (m) T. Sugano, K. Nozawa, H. Hayashi, K. Nishikida, K. Terui, T. Fukasawa, H. Takenouchi, S. Mino, H. Urayama, G. Saito, and M. Kinoshita, *ibid.* 27: 325 (1988), (n) T. Takahashi, T. Tokiwa, K. Kanoda, H. Urayama, H. Yamochi, and G. Saito, *ibid.* 27: 319 (1988), (o) K. Oshima, T. Mori, H. Inokuchi, H. Urayama, H. Yamochi, and G. Saito, *ibid.* 27: 165 (1988), (p) K. Oshima, H. Urayama, H. Yamochi and G. Saito, *ibid.* 27: 419, 473 (1988), (q) S. Katsumoto, S. Kobayashi, H. Urayama, H. Yamochi, and G. Saito, J. Phys. Soc. Jpn. 57: 3672 (1988); **for reviews**: (r) G. Saito, H. Urayama, H. Yamochi and K. Oshima, Synth. Met. 27: 331 (1988), (s) G. Saito, J. J. Appl. Phys., Series 1 165 (1988), (t) G. Saito, H. Urayama, H. Yamochi and K. Oshima in "Advances in Superconductivity," ed. by K. Kitazawa and T. Ishiguro, (Springer-Verlag, Berlin, 1989) p. 107, (u) H. Urayama, H. Yamochi, G. Saito and K. Oshima, *ibid.*, p. 113, (v) G. Saito and H. Urayama in "The Science of Superconductivity and New Materials," ed. by S. Nakajima (World Scientific Publ., Singapore, 1989), p. 60; **for the KHg(SCN)$_4$ salt:** (w) M. Oshima, H. Mori, G. Saito, and K. Oshima, Chem. Lett. 1159 (1989).

[3] S. S. P. Parkin, E. M. Engler, R. R. Schumaker, V. Y. Lee, J. C. Scott, and R. L. Greene, Phys. Rev. Lett. 50: 270 (1983).

[4] E. B. Yagubskii, I. F. Shchegolev, V. N. Laukhin, P. A. Kononovich, M. V. Karatsovnik, A. V. Zvarykina, and L. I. Buravov, Pis'ma Zh. Eksp. Teor. Fiz. 39:12 (1984).

[5] J. M. Williams, H. H. Wang, M. A. Beno, T. J. Emge, L. M. Sowa, P. T. Copps, F.

Behroozi, L. N. Hall, K. D. Carlson, and G. W. Crabtree, Inorg. Chem. 23: 3839 (1984).
[6] H. H. Wang, M. A. Beno, V. Geiser, M. A. Firestone, K. S. Webb, L. Nunez, G. W. Crabtree, K. D. Carlson, J. M. Williams, L. J. Azevedo, J. F. Kwak, and J. E. Schirber, *ibid.*, 24: 2466 (1985).
[7] T. Mori and H. Inokuchi, Solid State Commun. 64: 335 (1987).
[8] R. N. Lyubovskaya, R. B. Lyubovskii, R. P. Shibaeva, M. Z. Aldoshina, L. M. Goldenberg, L. R. Rozenberg, M. L. Khidekel, and Yu. F. Shulpyakov, Pis'ma Zh. Eksp Teor. Fiz. 42: 380 (1985).
[9] R. N. Lyuvobskaya, E. A. Zhilyaeva, A. V. Zvarykina, V. N. Laukhin, R. B. Lyubovskii, and S. I. Pestoskii, *ibid.* 45:416 (1987).
[10] E. B. Yagubskii, I. F. Shchegolev, V. N. Laukhin, R. P. Shibaeva, E. E. Kostyuchenko, A. G. Komenko, Yu V. Sushko, and A. V. Zvarykina, *ibid.* 40: 387 (1984).
[11] V. N. Laukhin, E. E. Kostyuchenko, Yu. V. Sushko, I. F. Shchegolev, and E. B. Yagubskii, *ibid.* 41:68 (1985); K. Murata, M. Tokumoto, H. Anzai, H. Bando, G. Saito, K. Kajimura, and T. Ishiguro, Solid State Commun. 54:1236 (1985).
[12] V. F. Kaminskii, T. G. Prokhoroea, R. P. Shibaeva, and E. B. Yagubskii, Pis'ma Zh. Eksp. Teor. Fiz. 41:15 (1984).
[13] H. Kobayashi, R. Kato, A. Kobayashi, Y. Nishio, K. Kajita, and W. Sasaki, Chem. Lett. 789 and 833 (1986).
[14] A. Kobayashi, R. Kato, H. Kobayashi, S. Moriyama, Y. Nishio, K. Kajita, and W. Sasaki, *ibid.* 459 (1987).
[15] P. C. W. Leung, T. J. Emge, M. A. Beno, H. H. Wang, J. M. Williams, V. Petricek, and P. Coppens, J. Am. Chem. Soc. 106: 7644 (1984).
[16] A. J. Schultz, H. H. Wang and J. M. Williams, *ibid.* 105: 7853 (1986).
[17] S. Kagoshima, Y. Nogami, M. Hasumi, H. Anzai, M. Tokumoto, G. Saito, and N. Mori, Solid State Commun. 69: 1177 (1989).
[18] G. Saito, T. Enoki, K. Toriumi, and H. Inokuchi, *ibid.* 42: 557 (1982).
[19] D. Schweitzer, P. Bele, H. Brunner, E. Gogu, U. Haeberlen, I. Hennig, I.Klutz, P. Swietlik, and H. J. Keller, Z. Phys. B-Condensed Matter 67: 489 (1987).
[20] M. Tokumoto, H. Bando, K. Murata, H. Anzai, N. Kinoshita, K. Kajimura, T. Ishiguro, and G. Saito, Synth. Met. 13: 9 (1986).
[21] J. M. Williams, M. A. Beno, H. H. Wang, U. W. Geiser, T. J. Emge, P. C. W. Leung, G. W. Crabtree, K. D. Carlson, L. J. Azevedo, E. L. Venturini, J. E. Shirber, J. F. Kwak, and M.-H. Wangbo, Physica 136B:371 (1986); T. J. Kistenmacher, Solid State Commun. 63: 977 (1987).
[22] T. Mishima, H. Kusuhara, Y. Ueba, K. Tada, and G. Saito, 56th Annual Meeting of Japan. Soc. of Applied Physics, April, 1989.
[23] T. Mori, Private communication.
[24] K. Bender, I. Hennig, D. Schweitzer, K. Dietz, H. Endres, and H. J. Keller, Mol. Cryst. Liq. Cryst. 108: 359 (1984).
[25] J. W. Jeffery, Nature 159: 610 (1947); R. W. G. Wyckoff, "Crystal Structures," 2nd Ed., Vols. I-V (Interscience, New York, 1963-65).
[26] Proc. Int. Conf. on Sci. and Tech. of Synth. Metals (ICSM '88), Synth. Met., 27 (1988) and 28 (1989).
[27] A. Ugawa, G. Ojima, K. Yakyshi, and H. Kuroda, Phys. Rev. B 38: 5122 (1988).
[28] D. Schweitzer, K. Polychroniadis, K. Klutz, H. J. Keller, I. Hennig, I. Heinen, U. Haeberlen, E. Gogu, and S. Gärtner, Synth. Met. 27: 465 (1988).
[29] J. E. Schirber and C. J. M. Northrup, Jr., Phys. Rev. B10: 3818 (1974).
[30] K. Oshima, R. C. Yu, P. M. Chaikin, H. Urayama, H. Yamochi, and G. Saito, in preparation.
[31] Y. Iye in "Studies of High Temperature Superconductors," ed. by A. V. Narlikar (NOVA Science Pub. Inc., 1989), p. 263.
[32] R. J. Elliott, Phys. Rev. 96: 266 (1954).
[33] Y. Hasegawa and H. Fukuyama, J. Phys. Soc. Jpn., 56: 877 (1987).
[34] T. Osada, R. Yagi, S. Kagoshima, N. Miura, M. Oshima, and G. Saito, in preparation.

RECENT DEVELOPMENTS IN ORGANIC SUPERCONDUCTORS

D. Jérome

Laboratoire de Physique des Solides (associé au CNRS)
Université Paris Sud
F- 91405 Orsay, FRANCE

INTRODUCTION

Superconductivity of radical ion salts is now well established in two families of organic conductors: the quasi-one-dimensional (Q-1-D) family, to which the prototype material $(TMTSF)_2PF_6$ belongs, with T_c in the one Kelvin range, and the two-dimensional series with T_c up to 10 K, which is illustrated by salts such as $(BEDT\text{-}TTF)_2X$ i.e. $(ET)_2X$ with $X = I_3$, AuI_2, $Cu(NCS)_2$, etc... We first recall the instabilities which are inherent to the $(TMTTF\text{-}TMTSF)_2X$ series, giving, as examples, the behaviour, under pressure, of the newly synthesized mixed sulfur-selenium compound $(TMDTDSF)_2PF_6$, and the discovery of a hidden spin density wave (SDW) phase in $(TMTSF)_2ReO_4$. Secondly, we present recent experimental results about the quantization of the Hall constant in the field-induced semimetallic states of $(TMTSF)_2PF_6$ under pressure, and the observation of giant magnetoresistance oscillations in the high-T_c phase of β-$(BEDT\text{-}TTF)_2I_3$, illustrating the two-dimensional character of the latter superconductor.

INSTABILITIES IN THE $(TMTSF)_2X$ SERIES

Superconductivity has always been a strong motivation for the research in organic conductors, since the original suggestion by Little [1] that organic matter could provide critical temperatures near ambient conditions. This driving force has proved to be justified, since superconductivity was indeed discovered in the 1 K range in 1980 [2], and significant improvements have been achieved in other materials discovered more recently [3,4]. However, the nature of the superconducting pairing in organic materials remains (as it is for high-T_c superconductors) an open question: phonon or non-phonon-mediated, singlet versus triplet pairing, etc. It becomes clear now that a better understanding of the context, in which the superconducting instability sets in, is a major step towards understanding organic superconductivity.

Superconductivity is only one among the various instabilities, which are observed at low temperature in the isostructural series, to which $(TMTSF)_2PF_6$ belongs. As shown in Figure 1, the electronic properties of the sulfur or selenium series are illustrated by a generalized phase diagram. The all-sulfur compound $(TMTTF)_2PF_6$ exhibits a marked charge localization below $T_\rho \approx 200$ K, which is attributed to strong on-site Coulomb repulsions [5] (1-D quantum antiferromagnet), and the onset of a 3-D spin singlet state (spin-Peierls, SP), accompanied by a lattice distortion below 19 K [6]. However, the Br salt of the same sulfur molecule exhibits a weaker charge localization below 100 K or so, and the onset of a spin-modulated state below 15-19 K, without any visible lattice distortion [7]. Quite a different behaviour is observed, instead, with the all-selenium compound $(TMTSF)_2ClO_4$, as no charge localization is visible above the superconducting transition at 1.2 K under ambient pressure.

The generalized phase diagram reveals two competitions between ground states: SP versus SDW on the one hand, and SDW versus superconductivity (SC) on the other.

Lower-Dimensional Systems and Molecular Electronics
Edited by R.M. Metzger *et al.*, Plenum Press, New York, 1991

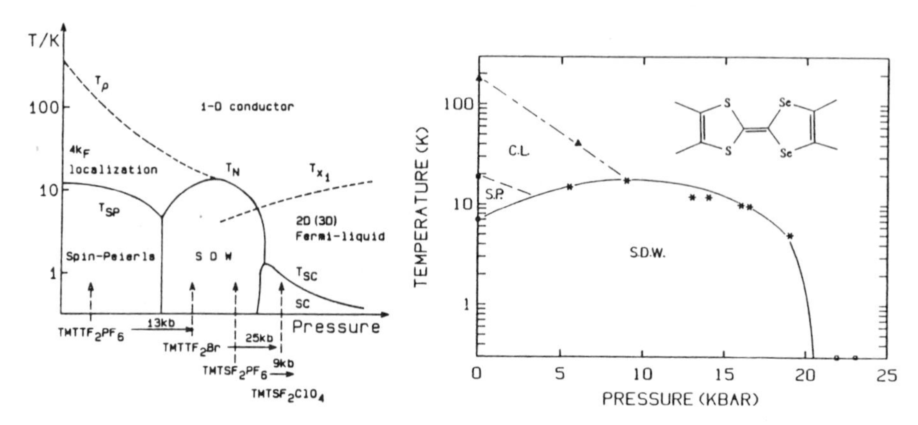

Fig. 1. Generalized phase diagram of $(TMTCF)_2X$ conductors (left) and phase diagram of $(TMDTDSF)_2PF_6$, showing the reentrance of SDW below the spin-Peierls state at 7 K, and the stabilization of a conducting ground state above 20 kbar (without superconductivity down to 0.4 K).

The charge localization and the SP-SDW competition are governed by the amplitude of the Umklapp scattering term g_3, which decreases under pressure, and also when moving from PF_6 to Br compounds. The SDW-SC competition is, very likely, governed by the interplay between the SDW phase transition temperature of the Q-1-D conductor and the small energy ($\approx t^2/t_{||}$) which characterizes the deviation of the Q-1-D Fermi surface from perfect nesting [8]. A recently synthesized S-Se hybrid molecule, TMDTDSF, has enabled a more thorough study of the SP-SDW competition [9]. As shown on the phase diagram of $(TMDTDSF)_2PF_6$, Figure 1, a spin-Peierls ordering is detected at 19 K, by its signature in the behaviour of the ESR spin susceptibility. However, the nature of the ground state, which is stabilized at 7 K, is clearly magnetic, according to a divergence of the nuclear spin relaxation rate, and the observation of antiferromagnetic resonance modes, instead of a Zeeman resonance, below 7 K. The conducting ground state becomes stable at low temperature above 20 Kbar, but no superconductivity has been detected so far, down to 0.4 K. At present, we can think of two possible interpretations for the absence of superconductivity: the pair-breaking effect of the disorder introduced by the non- symmetric organic molecule, or the weakness of the interchain coupling.

The intrinsic existence of the SDW-SC competition in the TMTSF series is also revealed by a recent investigation of the phase diagram of the organic superconductor $(TMTSF)_2ReO_4$. This latter system undergoes a superconducting transition at ($\approx$ 10 bar), which is necessary to stabilize the ordering of the non-centrosymmetric ReO_4 anions with the wave vector (0, 1/2, 1/2). Below the critical pressure, the stable (1/2, 1/2, 1/2) ordering doubles the lattice periodicity along the chain direction, and opens a gap at the Fermi level. The occurrence of this anion-ordered semiconducting state prevents us from studying the genuine instability of the Q-1-D organic stacks at low temperature. However, following a proper cooling procedure under pressure, the (0, 1/2, 1/2) configuration of the anions can be preserved at low temperature, even under ambient pressure [10]. Thus, another insulating state is revealed (Figure 2), below 15 K. It is assumed that this new insulating state is analogous to the SDW phase, which is stabilized in $(TMTSF)_2PF_6$ below the critical pressure.

In conclusion, the study of the $(TMTSF)_2PF_6$ and $(TMTSF)_2ReO_4$ phase diagrams has shown that the competition between antiferromagnetism and superconductivity is a firmly established character of Q-1-D superconductors, irrespective of the anion symmetry. The stabilization of superconductivity at a temperature higher than the 1 to 2 K range seems to be forbidden in Q-1-D conductors, such as $(TMTSF)_2X$, by the intrinsic competition with magnetic ordering. However, the borderline between the SDW and SC phases represents the optimum situation for superconductivity, at variance with high-T_c cuprate materials.

Another facet of this competition is revealed by the contribution of the low- frequency antiferromagnetic fluctuations to the nuclear spin-lattice relaxation, as shown in Figure 2. As far as $(TMTSF)_2ClO_4$ is concerned, the temperature dependence of $(T_1T)^{-1}$ scales with χ^2_s, in

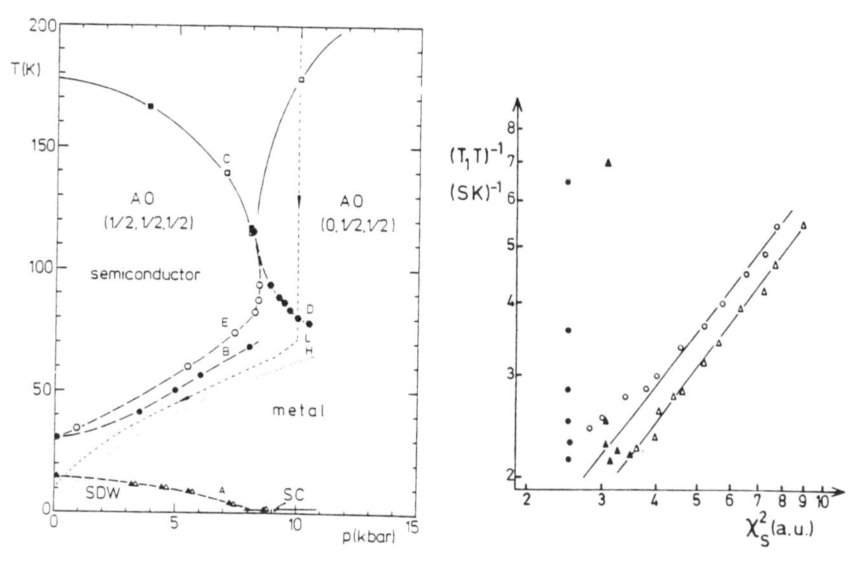

Fig. 2. Observation of a SDW phase in $(TMTSF)_2ReO_4$ competing with superconductivity, when a particular cooling procedure under pressure is followed (dotted line with arrows). Below 8 Kbar, the anions can be frozen at low temperature in the (0, 1/2, 1/2) configuration, allowing the conducting state to be stable down to the SDW or SC temperature (left). Temperature dependence of $(T_1T)^{-1}$ from ^{77}Se data plotted versus χ^2_s for $(TMTSF)_2PF_6$ (circles) and $(TMTSF)_2ClO_4$ (triangles). The departure from the straight line marks the onset of $2k_F$ low-temperature spin fluctuations, respectively at 100 and 30 K, in the PF_6 and ClO_4 salts.

the domain where uniform (q = 0) spin fluctuations are responsible for the temperature dependence of the spin susceptibility (T > 30K) [11]. Below 30 K, $2k_F$ spin correlations contribute predominantly to the relaxation rate, which then becomes nearly temperature-independent, down to the dimensionality cross-over at ≈ 8 K, where a renormalized Fermi liquid description is recovered. As shown in Figure 2, the effect of $2k_F$ spin correlations is even more pronounced in the case of $(TMTSF)_2PF_6$ (below 100 K) at ambient pressure, in agreement with the stronger tendency of that compound towards antiferromagnetism.

In conclusion, the proximity between magnetism and superconductivity is a remarkable feature of $(TMTSF)_2X$-like materials. In addition, NMR studies indicate that the long-range superconductivity order develops in a superconductivity background of strong antiferromagnetic fluctuations. Some theoretical approaches have used these experimental facts to propose a pairing mechanism in $(TMTSF)_2X$, based on the interchain exchange of antiferromagnetic spin fluctuations [12].

FERMIOLOGY IN ORGANIC CONDUCTORS

The open and quasi-planar nature of the Fermi surface of $(TMTSF)_2X$ materials is illustrated by the influence of the magnetic field on the stability of the conducting ground state. Figure 3 reports recent data of Hall effect and magnetoresistance obtained in $(TMTSF)_2PF_6$ under pressure [13]. The paramagnetic conducting state becomes unstable above 5T (T ≈ 0.5 K), against the formation of an antiferromagnetic semimetal, displaying a sequence of subphases. Each of these subphases is characterized by a field-independent Hall voltage. Furthermore, the Hall voltage is quantized, according to the law $R_H = h/Ne^2$ (N = integer). As shown in Figure 3, not only are the ratios of the plateaus given by successive integers, but the magnitude of the highest plateau between 14 and 18 T corresponds rather well to $h/2e^2$ = 12.9 kΩ/layer, the value expected for the quantum Hall effect (N = 1) in the presence of spin degeneracy. In addition, magnetoresistance data show well-defined peaks at the fields where the Hall voltage jumps to the next plateau. Above 18 T, the dramatic increase of the magneto-

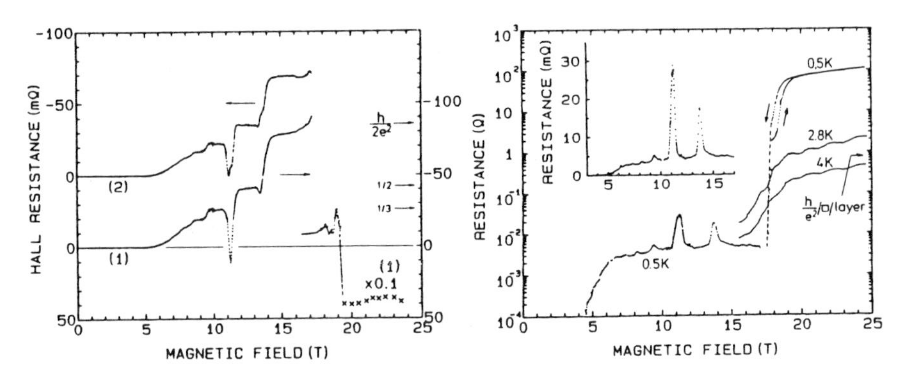

Fig. 3. Hall resistance versus magnetic field along the **c***axis for two $(TMTSF)_2PF_6$ samples (P ≈ 9 kbar, T = 0.5 K). The quantized values, $h/2ne^2$=12.9/n kΩ per molecular layer, are marked on the right for sample #1. The n = 0 phase is reached above 18 T (left). Magnetoresistance of the same sample up to 25 T (right; linear scale in the inset).

resistance, by about four orders of magnitude, suggests that the N = 0 state is attained. This state is likely to be similar to the SDW ground state observed under ambient pressure.

The ClO_4 compound behaves differently at very high fields. It exhibits a reentrance of the non-ordered phase. The ultra one-dimensionalization of the electron motion under large magnetic fields may be responsible for the reentrance phenomenon of $(TMTSF)_2ClO_4$.

Other spectacular Fermi surface effects, related to low-dimensionality, have also been observed in the $(ET)_2X$ series. In the $(ET)_2I_3$ material superconductivity can be stabilized at T_c = 8.1 K in the β_H phase, provided the cooling procedure avoids crossing a transition line, below which an incommensurate lattice modulation develops [14].

Figure 4 displays giant oscillations of the magnetoresistance, which exhibit a perfect (1/H) periodicity with the fundamental field H_o = 3730 T, and a beating phenomenon characterized by the much smaller field H_1 = 36.8 T [15]. The large amplitude and the beating phenomenon can both be understood in terms of a tube-like Fermi surface (FS) in a 2-D conductor. For such an anisotropic FS, the amplitude of the oscillations is enhanced by the factor $(m_c/m_a)^{1/2}$, where m_c and m_a are, respectively, the effective mass perpendicular and parallel to the conducting planes. The beating frequency is related to the warping of the tube, and thus to the interplanar coupling, namely $H_1/H_0 \propto t_a/t_c$. For the data in Figure 4, a ratio t_a/t_c= 140 is obtained. The giant magnetoresistance oscillations emphasize clearly the 2D nature of the FS of β-$(ET)_2X$ superconductors. The two-dimensionality is, very likely, responsible for both the absence of competition between superconductivity and antiferromagnetism (as the latter instability is favored by the nesting of quasi-planar surfaces), and for the enhancement of

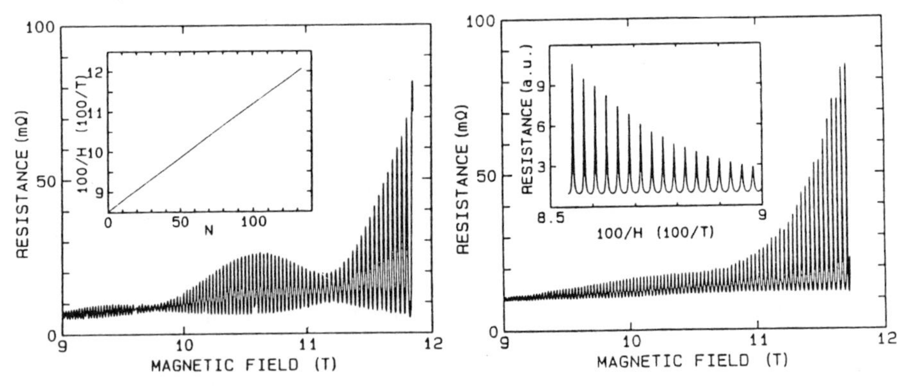

Fig. 4. Magnetoresistance of β_H-$(ET)_2I_3$ between 9 and 12 T at 0.38 K, and 1/H plot of the peak positions versus integer numbers (left inset). The oscillations become strongly anharmonic at high fields (right inset).

T_c above 8 K, as the negative effect of fluctuations is less severe for 2-D, than for Q-1-D ordering. Furthermore, band structure calculations point out that the Fermi energy of these 2-D conductors might be located close to a van Hove singularity in the density of states. We believe that such a situation should be considered for the interpretation of the high value of T_c, and for its very strong pressure dependence.

CONCLUSION

Organic superconductivity now exists in two families of organic compounds. In the first series, $(TMTSF)_2X$, the quasi-one-dimensionality manifests itself, first, in the competition between various ground states, which is governed by parameters such as the efficiency of Coulomb interactions, the amplitude of the interchain coupling, etc. and, secondly, in the remarkable field-induced spin density wave states.

The β and κ-phases of $(ET)_2X$ have proved to be two-dimensional conductors after the observation of magnetoresistance oscillations, which can exhibit remarkably large amplitudes in the case of β-$(ET)_2I_3$.

The nature of the pairing interaction remains an open question in organic superconductors. However, the existence of strongly developed antiferromagnetic fluctuations in $(TMTSF)_2X$ materials makes a non-phonon mediated pairing mechanism quite plausible.

ACKNOWLEDGEMENTS

I wish to acknowledge my collaborators, P. Auban, J. R. Cooper, W. Kang, S. Tomić and P. Wzietek who have contributed to the work at Orsay.

REFERENCES

[1] W. A. Little, Phys. Rev. 134 A: 1416 (1964).
[2] D. Jérome et al., J. Phys. Lett. Paris 41: L-95 (1980).
[3] V. N. Laukhin et al., JETP. Lett., 41: 81 (1985).
[4] H. Urayama et al., Chem Lett., 55 (1988).
[5] V. J. Emery et al., Phys. Rev. Lett. 48: 1039 (1982).
[6] J. P. Pouget et al., Mol. Cryst. Liq. Cryst.
[7] F. Creuzet et al., J. Phys. Lett. Paris 43: L-755 (1982).
[8] K. Yamaji, Mol. Cryst. Liq. Cryst. 119: 105 (1985).
[9] P. Auban et al., J. Physique Paris 50: 2727 (1989).
[10] S. Tomić and D. Jérome, J. Phys. Cond. Matter 1: 4451 (1989).
[11] C. Bourbonnais et al., Phys. Rev. Lett. 62: 1532 (1989).
[12] C. Bourbonnais and L. Caron, Europhys. Lett 6: 177 (1988).
[13] J. R. Cooper et al., Phys. Rev. Lett. 63: XXXX (1989).
[14] F. Creuzet et al., J. Phys. Lett. Paris 46: L-1079 (1985).
[15] W. Kang et al., Phys. Rev. Lett. 62: 2559 (1989).

STRUCTURE-PROPERTY RELATIONSHIPS AS AN AID IN THE RATIONAL DESIGN OF β-$(ET)_2X$ ORGANIC SUPERCONDUCTORS

Jack M. Williams, Hau H. Wang, Aravinda M. Kini, Mark A. Beno
Urs Geiser, Arthur J. Schultz, K. Douglas Carlson and John R. Ferraro

Chemistry and Materials Science Divisions
Argonne National Laboratory
Argonne, Illinois 60439, USA

Myung-Hwan Whangbo

Department of Chemistry
North Carolina State University
Raleigh, North Carolina 27695, USA

INTRODUCTION

The organic electron-donor molecule ET (Fig. 1) has been the current central focus of research in organic superconductors, because it forms numerous solid phases with anions, X^-,

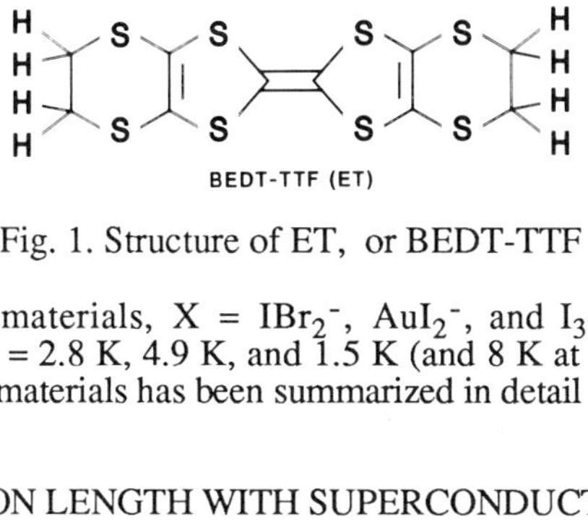

BEDT-TTF (ET)

Fig. 1. Structure of ET, or BEDT-TTF

of which the β-$(ET)_2X$ materials, X = IBr_2^-, AuI_2^-, and I_3^-, are ambient-pressure superconductors, with T_c's = 2.8 K, 4.9 K, and 1.5 K (and 8 K at p > 0.5 kbar), respectively [1]. The synthesis of these materials has been summarized in detail in the literature [1].

CORRELATION OF ANION LENGTH WITH SUPERCONDUCTING TRANSITION

An important feature of these materials, which allows the derivation of structure-property correlations, is that they are isostructural, with the anions residing within a cavity of ethylene

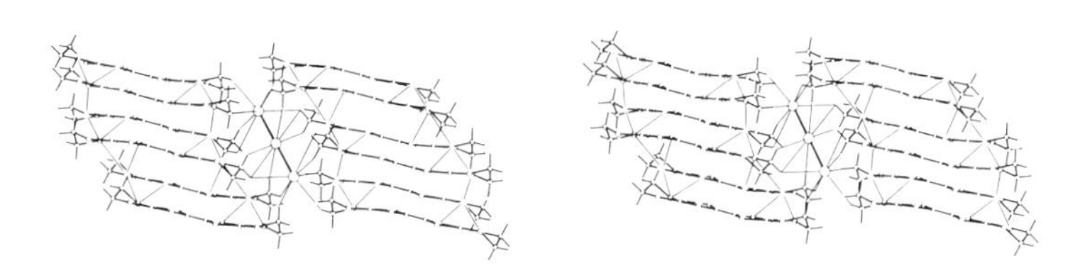

Fig. 2. Stereodiagram of ET layers

Table 1. Structural and conductivity data of β-$(ET)_2X$ salts [length of anion X, room-temperature unit cell volume V_c, and superconducting critical temperature T_c (or metal-to-insulator transition temperature T_{MI})].

X		Anion Length (Å)	V_c (Å^3 at 298 K)	T_c (K)
I_3^-	(I)(I)(I)	10.1	855.9	1.5
AuI_2^-	(I)(Au)(I)	9.4	845.2	5
I_2Br^-	(I)(I)(Br)	9.7	842.3	none
IBr_2^-	(Br)(I)(Br)	9.4	828.7	2.80
$ClIBr^-$	(Cl)(I)(Br)	9.0	821.3	?
ICl_2^-	(Cl)(I)(Cl)	8.7	814.3	T_{MI} = 22.1
$AuCl_2^-$	(Cl)(Au)(Cl)	8.1	800.7	$T_{MI} \cong 32$

group hydrogen atoms, formed by the "corrugated-sheet" layers of ET molecules, as shown in the stereodiagram of Fig. 2. The effect of varying the anion length is to change the unit cell volume in like fashion, and at the same time, alter the system of intra- and intermolecular S...S contact distances between ET molecular stacks. The linear anions, that yield β or β-like structures, are shown in Table 1, in which anion length, unit cell volume, and T_c (when appropriate) are given. Thus, over the range of anions given in Table 1, the electrical properties range from those with metal-insulator transitions in the shortest anions, to ambient-pressure superconductivity in the longer anions (IBr_2^-, AuI_2^-, and I_3^-). β-$(ET)_2I_3$ is novel, because modest pressures, of only 0.5 kbar, are sufficient to convert it to a new structure [2] [β*-$(ET)_2I_3$], with T_c = 8 K, which is the maximum found, to date, in β-$(ET)_2I_3$ - type superconductors. These observations (unit cell volume and T_c) form the basis of a useful structure-property correlation [1] for the isostructural salts, as shown in Fig. 3. From Fig. 3 we can see that anions longer than I_3^- could be expected to give new β-$(ET)_2I_3$ superconductors with T_c > 8 K!

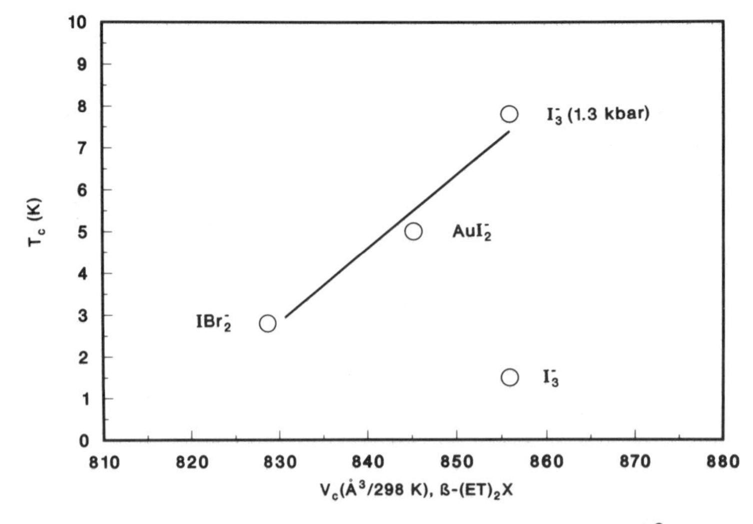

Fig. 3. Correlation of room-temperature unit cell volume of β-$(ET)_2I_3$ salts with superconducting critical temperature T_c

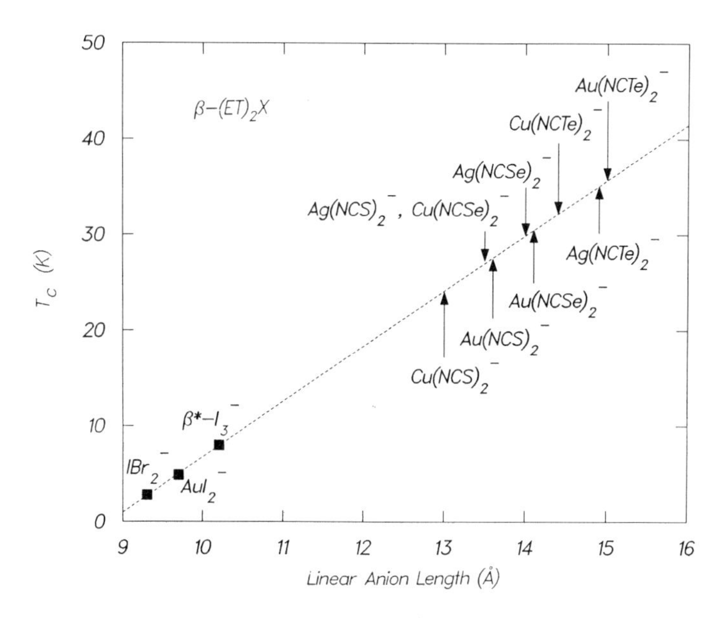

Fig. 4. Correlation of linear anion length in β-$(ET)_2I_3$ salts with superconducting critical temperature T_c.

useful structure-property correlation [1] for the isostructural salts, as shown in Fig. 3. From Fig. 3 we can see that anions longer than I_3^- could be expected to give new β-$(ET)_2I_3$ superconductors with $T_c > 8$ K!

However, triatomic anions containing terminal halide atoms that are longer than I_3^- are presently unknown, and new anionic species [1] such as $(NCS\text{-}M\text{-}SCN)^-$ and $(NC\text{-}M\text{-}CN)^-$, M = metal, are being investigated as the source of potential β-$(ET)_2I_3$ superconductors. A correlation, similar to that given in Fig. 3, but based on anion length, is shown in Fig. 4.

An inspection of Fig. 4 reveals that T_c's as high as ~ 40 K may be achieved in β-$(ET)_2I_3$ phase materials, if the linear anions depicted in Fig. 4 can be prepared and incorporated in the β-$(ET)_2I_3$ structure. Preparing such materials is one of the current goals of the organic superconductor research program at Argonne. Recently, the $Cu(NCS)_2^-$ system has led to a different structure type (vs. β), i.e., κ-$(ET)_2Cu(NCS)_2$, with $T_c = 10$ K, which contains a polymeric anion network [3-5], and one for which structure-property correlations do not yet exist.

Another very important factor, that must be considered, when attempting to prepare β-$(ET)_2I_3$ superconductors, and likely those derived from new organic electron-donor molecules yet to be prepared, is the nature of the interaction between the anion and the donor molecule. As shown in Fig. 2, the most significant anion-donor contacts in ET-anion systems are through $-CH_2...Y$ (Y = halide) hydrogen bonding interactions, and it is known the T_c's of the superconducting β-$(ET)_2I_3$ salts increase, as these interactions increase in "softness" [6]. This concept is derived from the BCS theory [7], in which electron-phonon coupling is essential for superconductivity, and the extent of this coupling is measured by the electron-phonon coupling constant λ. The value of λ is related to T_c through the McMillan equation [8], and the derived values for β-$(ET)_2I_3$ materials are shown in Table 2. From the Table it is clear that λ increases with T_c, and, since λ is proportional to $1/\langle\omega^2\rangle$, where $\langle\omega^2\rangle$ is the square of the phonon frequencies averaged over the phonon band, it is clear that a large value of λ (and correspondingly high T_c) is the result of a low-frequency phonon spectrum in the associated crystal lattice, i.e., when the lattice is "soft" toward certain vibrational modes, such as those involving translation, and the like. In the case of the β-$(ET)_2I_3$ system, the softness of the donor-anion interactions derives from the C-H...Y hydrogen bonding contacts, formed as a result of Coulomb interactions between $ET^{0.5+}$ donor molecules and X- triatomic

Table 2. Correlation of superconducting critical temperature T_c and the electron-phonon coupling constant, λ with short intermolecular contacts H...X- and H...H (as probes of the softness of the donor-anion interactions).

Salt	T_c (K)	λ	ET $-CH_2$ group configuration	H...X^- (Å)	H...H (Å)
β-$(ET)_2I_3$	1.4	0.37	eclipsed	2.842	2.477
			staggered	2.988	2.152
β-$(ET)_2IBr_2$	2.8	0.43	eclipsed	2.887	2.169
β-$(ET)_2AuI_2$	5.0	0.52	eclipsed	2.996	2.224
β*-$(ET)_2I_3$	8	0.62	staggered	3.014	2.261

softness of the donor-anion interactions derives from the C-H...Y hydrogen bonding contacts, formed as a result of Coulomb interactions between $ET^{0.5+}$ donor molecules and X- triatomic anions, which result from the anions residing in hydrogen "pockets", as shown in Fig. 2. In these hydrogen pockets, both short H...X- and H...H contacts occur (see Table 2), and the softness of the ET molecule motions can be correlated with the lengths of these short contacts. Thus, the longer these contact distances are, the softer the β-$(ET)_2I_3$ crystal lattice. Finally, these contacts depend upon whether the two ethylene groups in each ET molecule have an eclipsed or staggered arrangement (see Table 2). Only β*-$(ET)_2I_3$ has a completely staggered arrangement of the ethylene groups, and it is structurally unique [2,6] amongst the β-$(ET)_2I_3$ materials -- it also has the highest T_c of ~ 8 K. β*-$(ET)_2I_3$ has, in general, longer H...H and H...I contact distances than in β-$(ET)_2I_3$, which leads to a higher value of λ, and T_c, for the former salt. A general conclusion, regarding the softness of C-H...Y interactions in β-$(ET)_2I_3$ materials, is that softness increases, as hydrogen bonding ability decreases [9]. For these reasons, one would expect the softness of anion interactions involving halogen atoms to decrease as follows: I > Br > Cl > F. Thus, the search for new superconducting β-phase materials, and, in general, for new organic donor molecule salts, where C-H...Y interactions

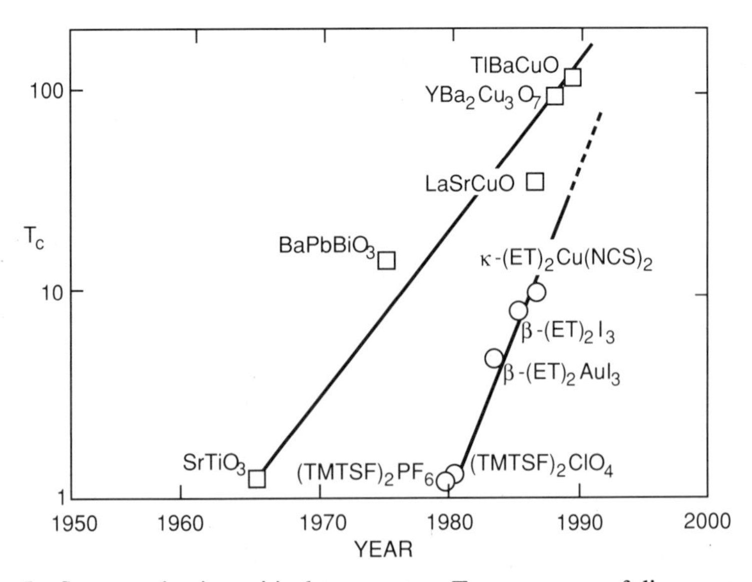

Fig. 5. Superconducting critical temperature T_c versus year of discovery, for inorganic and organic systems.

may be important, might prove more successful, if I or Br-containing anions are utilized. The relative softness of the interactions formed between donor molecules and $M(NCS)_2^-$, $M(NCSe)_2^-$, and $M(NCTe)_2^-$ anions is, as yet, largely unknown, and is the topic of much current research in the search for new ET molecule superconductors.

THE FUTURE OF ORGANIC SUPERCONDUCTORS

Finally, some perspective on the future of organic superconductors, relative to the recently discovered high-T_c copper oxide superconductors, can be gained from an inspection of Fig. 5. Clearly, since we now know that "high-T_c" materials exist, there is every reason to expect their eventual discovery in organic systems!

ACKNOWLEDGEMENT

Work at Argonne National Laboratory and North Carolina State University is supported by the U. S. Department of Energy, Office of Basic Energy Sciences, Division of Materials Sciences, Division of Materials Sciences under Contract W-31-109-ENG-38 and Grant DE-FG05-86-ER45259, respectively.

REFERENCES

[1] J. M. Williams, H. H. Wang, T. J. Emge, U. Geiser, M. A. Beno, P. C. W. Leung, K. D. Carlson, R. J. Thorn, A. J. Schultz, and M.-H. Whangbo, Prog. Inorg. Chem. 35: 51 (1987).
[2] A. J. Schultz, H. H. Wang, J. M. Williams, and A Filhol, J. Am. Chem. Soc. 108: 7853 (1986).
[3] H. Urayama, H. Yamochi, G. Saito, K. Nozawa, T. Sugano, M. Kinoshita, S. Sato, K. Oshima, A. Kawamoto and J. Tanaka, Chem. Lett. 55 (1988).
[4] S. Gärtner, E. Gogu, H. J. Keller, T. Klutz, and D. Schweitzer, Solid State Commun. 65: 1531 (1988).
[5] K. D. Carlson, U. Geiser, A. M. Kini, H. H. Wang, L. K. Montgomery, W. K. Kwok, M. A. Beno, J. M. Williams, C. S. Cariss, G. W. Crabtree, M.-H. Whangbo, and M. Evain, Inorg. Chem. 27: 965 (1988).
[6] M.-H. Whangbo, J. M. Williams, A. J. Schultz, T. J. Emge, and M. A. Beno, J. Am. Chem. Soc. 109: 90 (1987).
[7] J. Bardeen, L. N. Cooper, and J. R. Schrieffer, Phys. Rev. 106: 162 (1957).
[8] W. L. McMillan, Phys. Rev. 167: 331 (1968).
[9] J. R. Ferraro, H. H. Wang, J. Ryan, and J. M. Williams, Appl. Spectrosc. 41: 1377 (1987).

TRANSPORT PROPERTIES OF SINGLE CRYSTALS AND POLYCRYSTALLINE PRESSED SAMPLES OF $(BEDT\text{-}TTF)_2X$ SALTS AND RELATED COORDINATION POLYMERS

D. Schweitzer and S. Kahlich

3. Physikalisches Institut der Universität Stuttgart, Pfaffenwaldring 57
7000 Stuttgart 80, FEDERAL REPUBLIC OF GERMANY

S. Gärtner, E. Gogu, H. Grimm, I. Heinen, T. Klutz and R. Zamboni*

Max-Planck Institut für Medizinische Forschung
AG: Molekülkristalle, Jahnstraße 29
6900 Heidelberg, FEDERAL REPUBLIC OF GERMANY

H. J. Keller and G. Renner

Anorganisch Chemisches Institut der Universität Heidelberg
Im Neuenheimer Feld 270
6900 Heidelberg, FEDERAL REPUBLIC OF GERMANY

INTRODUCTION

Ten years ago, in 1979, superconductivity was observed for the first time in an organic metal [1]. Today, about 30 different organic metals are known, which become superconducting under pressure or ambient pressure. The organic superconductors with the highest transition temperatures are all radical salts of the donor bis(ethylenedithiolo)-tetrathiafulvalene (BEDT-TTF), namely at ambient pressure $(BEDT\text{-}TTF)_2Cu(NCS)_2$ (T_c = 10.4 K) [2] and α_t-$(BEDT\text{-}TTF)_2I_3$ (T_c = 8 K) [3] and under isotropic pressure β_H - $(BEDT\text{-}TTF)_2I_3$ (0.5 kbar, T_c = 7.5 K) [4,5]. The latter β_H-phase can even become superconducting at 8 K and ambient pressure, after a special pressure-temperature cycling procedure [6], i.e. pressurization up to 1 kbar at room temperature, and release of the helium gas pressure at temperatures below 125 K. Nevertheless, this superconducting state at 8 K in β_H-$(BEDT\text{-}TTF)_2I_3$ is only metastable [6,7], since warming up the crystal above 125 K, and cooling down again under ambient pressure, results only in superconductivity at 1.3 K, the so-called β_L - or β-phase.

In the following, some of the structural and physical properties of single crystals of the above-mentioned organic metals shall be briefly discussed. In a second part, some new results and observations, such as bulk superconductivity in polycrystalline pressed samples of the same materials, are presented. The remarkable fact of this finding is that it shows that organic superconductors can be used, in principle, for the preparation of electronic devices, such as SQUIDS, and might even be suitable for the preparation of superconducting cables. In addition, it indicates that the observation of superconductivity should also be possible in organic polymers. Therefore, in a final part, we discuss the possible conditions for superconductivity in organic polymers, and present our first investigations on a coordination polymer, in which such conditions might be fulfilled.

---

*On leave from Istituto di Spettroscopia Molecolare del C.N.R., Bologna, Italy

Lower-Dimensional Systems and Molecular Electronics
Edited by R.M. Metzger *et al.*, Plenum Press, New York, 1991

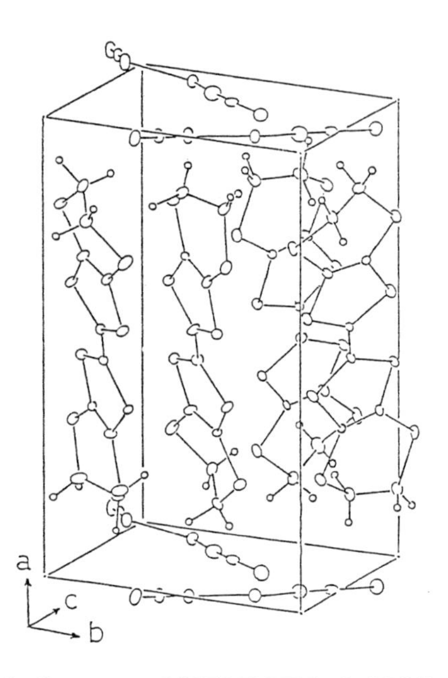

Fig. 1. Structure of $(BEDT\text{-}TTF)_2Cu(NCS)_2$.

SINGLE CRYSTALS

At the moment, the radical salt $(BEDT\text{-}TTF)_2Cu(NCS)_2$ is the organic metal which has the highest transition temperature, $T_c = 10.4$ K, into a stable superconducting state [2,8]. Single crystals, prepared from deuterated BEDT-TTF, have even higher transition temperatures of 11.1 K [2,9] or, in some cases, up to nearly 12 K. This is certainly not due to an isotopic effect, but probably due to a structural effect, since $(BEDT\text{-}TTF)_2Cu(NCS)_2$ undergoes phase transitions around 100 K and 50 K [8]. Unfortunately, the exact structure of $(BEDT\text{-}TTF)_2Cu(NCS)_2$ below 50 K, for both types of crystals, the protonated and deuterated one, is not yet known.

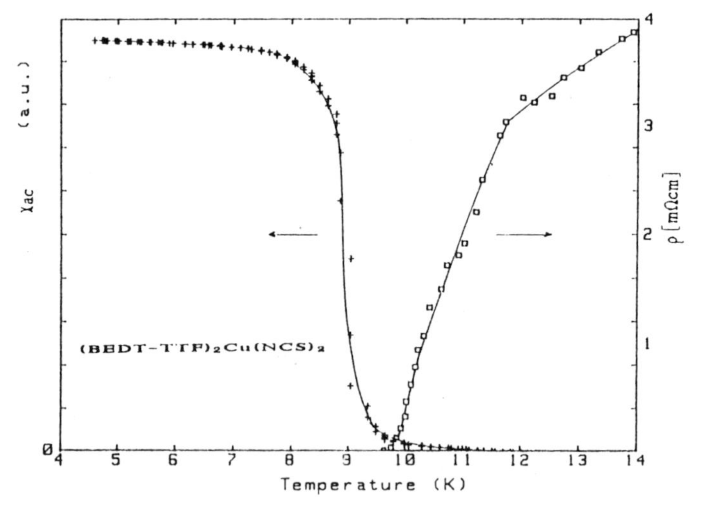

Fig. 2. **b**-axis resistivity versus temperature ρ and change in ac-susceptibility χ_{ac} with decreasing temperature in $(BEDT\text{-}TTF)_2\,Cu(NCS)_2$.

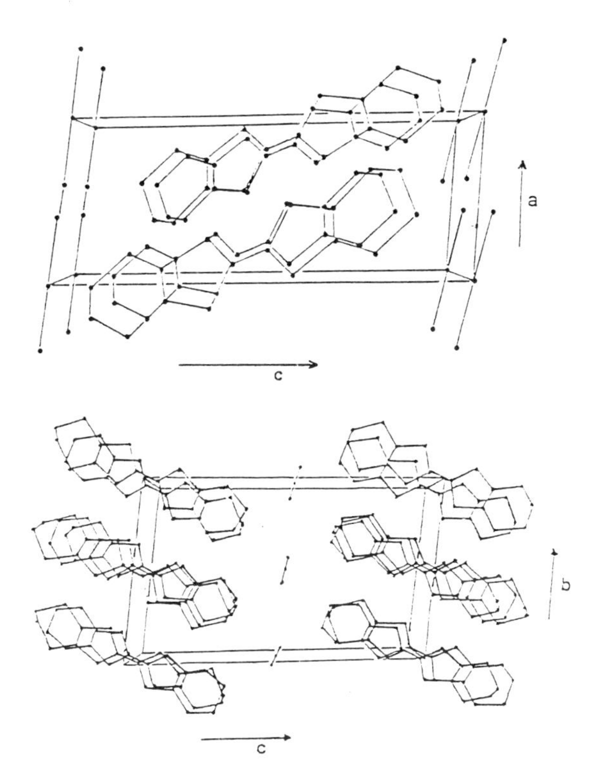

Fig. 3. Projections of the structure of (top) α-$(BEDT\text{-}TTF)_2I_3$ and (bottom) β-$(BEDT\text{-}TTF)_2I_3$ along the stacking axes.

$(BEDT\text{-}TTF)_2Cu(NCS)_2$ is a typical quasi-two-dimensional organic metal. The two-dimensional behavior is due to conducting sheets of dimerized BEDT-TTF molecules, which are separated by non-conducting anion layers. In the BEDT-TTF sheets the dimers are nearly orthogonally arranged with respect to each other [2,10], and the dimers do not form the usual face-to-face piling along the stacking axis (Fig. 1). The latter property increases the two-dimensional character of the molecular network. An important property of $(BEDT\text{-}TTF)_2Cu(NCS)_2$ is that the crystals show - compared to other organic superconductors - a relatively sharp transition into the superconducting state. This is observed, not only in resistivity (Fig. 2), but also in the ac-susceptibility measurement [8] (Fig. 2) and in the Meissner effect [2,11]. The two latter methods have shown that superconductivity in $(BEDT\text{-}TTF)_2Cu(NCS)_2$ is a nearly 100% bulk effect. Measurements of the upper critical fields have demonstrated the typical quasi-two-dimensional character of the crystals [12,13]. It could be shown that, at temperatures below 4 K, the superconducting properties can be described best in the picture of layered two-dimensional superconductors, as first discussed by Klemm *et al.* [14]. This picture of layered two-dimensional superconductors is also valid for the high-T_c copper oxide superconductors, and demonstrates the similarities of the principal structural buildup of both classes of materials. These similarities will be discussed in the following in some more detail, for the case of the organic superconductor α_t-$(BEDT\text{-}TTF)_2I_3$.

Crystals of α_t-$(BEDT\text{-}TTF)_2I_3$ are quasi-two-dimensional organic metals as well, which have a stable superconducting state at 8 K and ambient pressure [3]. The crystals can be prepared by tempering crystals of α-$(BEDT\text{-}TTF)_2I_3$ above 70°C for several days, resulting in a structural phase transition. This structural transformation was first reported by Baram *et al.* [15]. The structure of α_t-$(BEDT\text{-}TTF)_2I_3$ is very similar to the one of crystals of β-$(BEDT\text{-}TTF)_2I_3$, as was shown by several spectroscopic methods [3,15,16]. Nevertheless, its exact structure is not known yet, due to the fact that, after the structural transformation, the mechanically still very stable crystals have a mosaic type of build-up, so that a detailed X-ray crystal structure investigation has not been possible up to now. The mosaic type of transformation can be observed at 80°C under a polarization microscope, for very thin single crystals of α-$(BEDT\text{-}TTF)_2I_3$, showing that the resulting single crystalline areas of

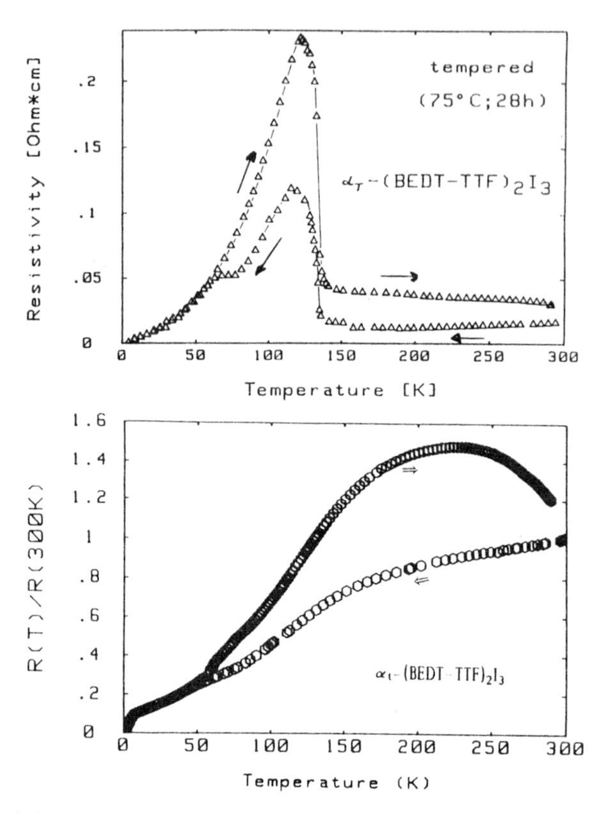

Fig. 4. Resistivity versus temperature: (a, top) for a not totally transformed α_t-(BEDT-TTF)$_2$I$_3$ crystal (tempering time 28 h at 75°C); (b, bottom) for a crystal of α_t-(BEDT-TTF)$_2$I$_3$ (tempering time 40 h at 75°C).

α_t-(BEDT-TTF)$_2$I$_3$ have typical diameters of about 10 μm. The mosaic-type build-up is a result of strong stress during the transformation, because the structures of α-(BEDT-TTF)$_2$I$_3$ and β-(BEDT-TTF)$_2$I$_3$ (Fig. 3) are quite different from each other. The BEDT-TTF molecules have to turn at least 35° around their long molecular axes during the transformation, since, in the α-phase, between BEDT-TTF molecules of non-equivalent neighboring stacks, dihedral angles of 59.4° and 70.4° exist [17], in contrast to β-(BEDT-TTF)$_2$I$_3$, where only one type of crystallographically equivalent stack is observed [18]. Nevertheless, even if the unit cell data of β-(BEDT-TTF)$_2$I$_3$ and α_t-(BEDT-TTF)$_2$I$_3$ and the arrangement of the molecules are the same, the structure of α_t-(BEDT-TTF)$_2$I$_3$ cannot be exactly the same as in β-(BEDT-TTF)$_2$I$_3$ (or, more correctly, as in β_H-(BEDT-TTF)$_2$I$_3$), since β_H-(BEDT-TTF)$_2$I$_3$ has only a metastable superconducting state at 8 K, while crystals of α_t-(BEDT-TTF)$_2$I$_3$ have a stable superconducting state at this temperature. The difference in the two structures exists, probably, only in the ordering of the terminal ethylene groups of the BEDT-TTF molecules. While in β_H-(BEDT-TTF)$_2$I$_3$ all the terminal ethylene groups have the so-called staggered form [19], in the crystals of α_t-(BEDT-TTF)$_2$I$_3$ these groups might have the so-called eclipsed form, or, even more plausibly, they might be ordered as in the α-phase, where both forms of ordering exist in neighboring stacks [17].

The time needed for the transformation of the crystals from α-(BEDT-TTF)$_2$I$_3$ to α_t-(BEDT-TTF)$_2$I$_3$ depends on the thickness of the crystals, and strongly on the temperature, which should be between 70°C and 100°C. The lower the temperature, the longer the period is for the total transformation, but the quality of the crystals is much better after a transformation at lower temperatures, than after a transformation at higher temperatures. At 75°C it takes about 3 days for the transformation to be completed. In Fig. 4a an example is shown, where the transformation is not complete. The usual metal-insulator transition of the α-phase at 135 K can still be partially observed. But, at lower temperatures, the crystal shows already a metallic behavior. In addition, above 60 K a strong dependence of the resistivity on the direction of the temperature cycle is observed. In the direction from lower to higher temperatures, the resistivity of the crystal is much higher, than by cooling the crystals down from room temperature. Nevertheless, after coming back to room temperature, the resistivity "relaxes" back to the starting value within several hours, and the cycle can be started again at the starting resistivity, resulting in exactly the same resistivity characteristics. It should be noticed that, below 60 K, the resistivity does not depend on the direction of the temperature cycle.

A similar "glassy" resistivity behavior can be observed also on fully transformed α_t-(BEDT-TTF)$_2$I$_3$ crystals. Fig. 4b shows the temperature dependence of the resistivity for a typical example of an α_t-crystal. By cooling down the crystal from room temperature, a typical metallic behavior is observed, and the usual phase transition at 135 K of the α-phase cannot be observed anymore. Above 60 K, again, the resistivity is larger during the warming-up cycle, than during the cooling-down procedure. It should be pointed out that, during the measurement, the temperature of the crystal was carefully controlled. In fact, it can be seen that the difference in resistivity between both cycles cannot be due to incorrect temperature measurements, since in the warming-up cycle the resistivity of the sample becomes 50% larger, compared to the starting value at 300 K. Again, at room temperature, it takes several hours for the resistivity to "relax" back to the starting value. A typical example of such a "relaxation" of the resistivity is shown in Fig. 5. At the moment, the reason for the different resistivity characteristics is not clear. It should be mentioned that such a behavior was not observed before, for instance, in crystals of α-(BEDT-TTF)$_2$I$_3$ or β-(BEDT-TTF)$_2$I$_3$. We assume that the α_t-(BEDT-TTF)$_2$I$_3$ crystals undergo a phase transition at around 60 K, since below 60 K the two resistivity characteristics are identical. By warming up the crystal above 60 K, the phase transition might be frozen in, and "relax" back to the starting structure at room temperature only after several hours. ESR experiments point to such an explanation as well. While the ESR linewidth (Fig. 6a) is, for both temperature cycles, identical within experimental error, the ESR signal amplitude shows, above 60 K, a tremendous dependence on the direction of the temperature cycle (Fig. 6b), indicating that the susceptibility, or in other words the density of states, is much smaller than in the warming-up cycle.

α_t-(BEDT-TTF)$_2$I$_3$ crystals have a stable superconducting state at 8 K. Fig. 7 shows the resistivity of an α_t-(BEDT-TTF)$_2$I$_3$ crystal below 11 K, as well as the increase in resonance frequency of an LC circuit, due to the exclusion of the rf-field by diamagnetic shielding currents (ac susceptibility) by lowering the temperature. The signal, which still increases on cooling, is, at 2 K, about 50 % of that expected for a perfect superconductor, and indicates a broad T_c distribution in the sample; nevertheless, superconductivity is a clear bulk effect here.

Fig. 8 shows the temperature dependence of the upper critical field for crystals of α_t-(BEDT-TTF)$_2$I$_3$, for the magnetic field perpendicular and parallel to the **c***-axis. It can clearly be seen that, for the region $T / T_c > 0.65$ ($T_c = 8$ K), the crystals behave as isotropic

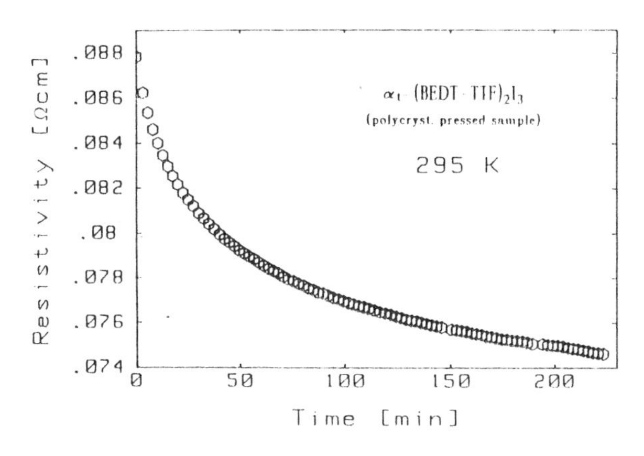

Fig. 5. "Relaxation" of the resistivity with time after a warming-up cycle of a α_t-(BEDT-TTF)$_2$I$_3$ crystal at 295 K (see text).

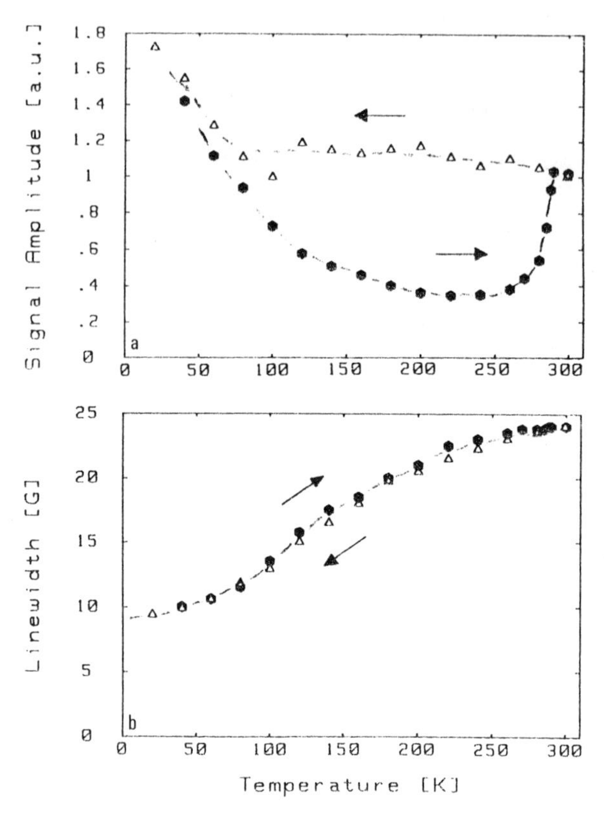

Fig. 6. Temperature dependence of the ESR signal amplitude (a: top) and the ESR linewidth (b: bottom) of a crystal of α_t-(BEDT-TTF)$_2$I$_3$ for a cooling-down and warming-up temperature cycle (see text).

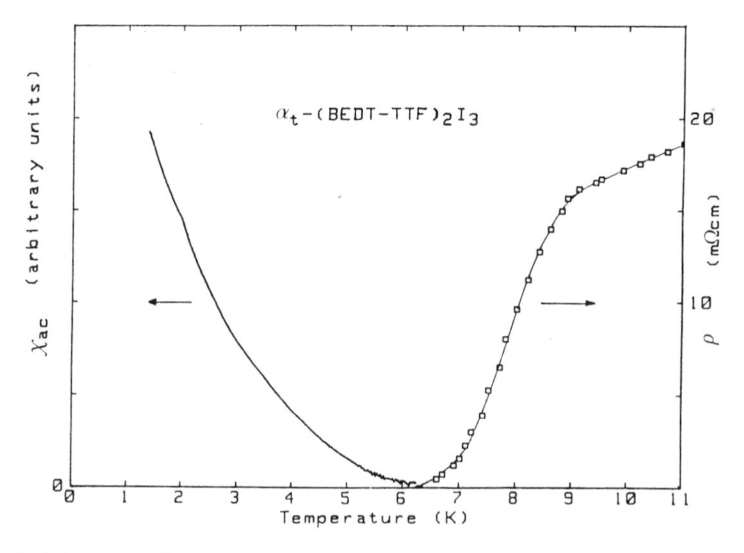

Fig. 7. Resistivity, and change in the ac-susceptibility by lowering the temperature, for a crystal of α_t-(BEDT-TTF)$_2$I$_3$ below 11 K.

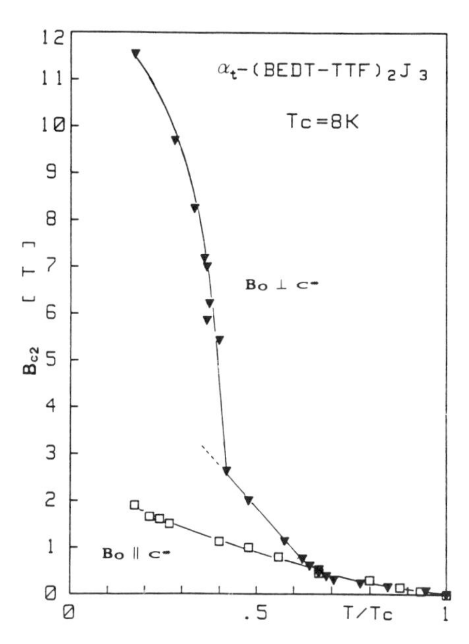

Fig. 8. Temperature dependence of the upper critical field $B_{c2}(T)$ for crystals of α_t-(BEDT-TTF)$_2$I$_3$, for the magnetic field perpendicular ($B_{c2\perp}$) and parallel ($B_{c2\parallel}$) to the **c***-axis.

three-dimensional superconductors, since for both directions the upper critical fields are identical. In the region $0.45 < T / T_c < 0.65$, an anisotropic three-dimensional behavior is found, while for $T / T_c < 0.45$, the material can be described as a typical two-dimensional layered superconductor [9,13]. By applying the anisotropic effective mass model of the Ginzburg-Landau (GL) theory, coherence lengths $\xi_{\parallel}(0) \approx 120$ Å and $\xi_{\perp}(0) \approx 20$ Å, and London penetration depths $\lambda_{\parallel}(0) = 4300$ Å and $\lambda_{\perp}(0) = 750$ Å were obtained. However, since the anisotropic GL theory is only valid sufficiently close to T_c, these values can give only the right order of magnitude. Nevertheless, the value $\xi_{\perp} \approx 20$ Å is comparable to the unit cell dimension in the **c**-direction. In addition, ^{13}C-NMR investigations [9,16,20] indicate that $\xi_{\perp}$ might even be only between 6 and 10 Å, since it was shown that the spin density (and therefore, to some degree, the charge density) is mainly located on the middle part of the BEDT-TTF molecules [20]. This means that the thickness of the superconducting layer is of the order of the length of the BEDT-TTF molecule, supporting nicely the picture of a Josephson-coupled two-dimensional layered superconductor, given by Klemm *et al.* [14].

As pointed out before, a close similarity exists between the organic superconductors and the high-T_c copper oxides. In both materials, short coherence lengths, of the order of the size of the unit cell, and large London penetration depths exist. These properties are due to structural features and the typical quasi-two-dimensional electronic conditions. Certainly, in the case of the organic superconductors, minor changes and differences in the structure, as, for instance, a different ordering of the terminal ethylene groups of the BEDT-TTF molecules, or a different symmetry of the anions, might influence strongly the superconducting transition temperature (for a review on BEDT-TTF radical salts see Ref. [21]). On the other hand, such small variations in the structure of organic materials open the chance to obtain a larger number of materials with different transition temperatures into a superconducting state; to vary these structural parameters might be a way to obtain higher transition temperatures, in the case of the organic materials as well.

Besides the typical quasi-two-dimensional structural features of the organic superconductors and the high-T_c copper oxides, there might exist another similarity between these two different classes of materials, namely a relatively strong carrier phonon interaction. In the case of the organic superconductors, this was demonstrated by tunneling measurements in the normal [22] and superconducting [23,24] states. The latter measurements provided the superconducting gap. The estimated linear electron-phonon coupling constant, $\lambda = 1 - 1.5$ [24,25], indicates a medium-to-strong interaction, while for the high-T_c superconductors this value might be above 3 [26], indicating a rather strong interaction.

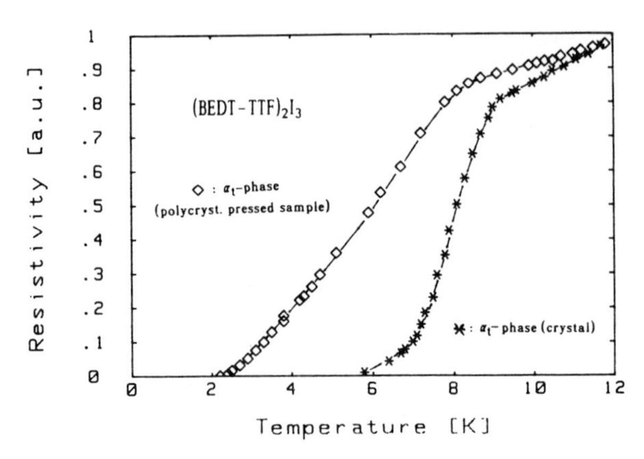

Fig. 9. Temperature dependence of the resistivity of a polycrystaline freshly pressed sample of α_t-(BEDT-TTF)$_2$I$_3$ (curve I, untempered after preparation), and of a sample of α_t-(BEDT-TTF)$_2$I$_3$ (curve II, tempered after the preparation for 3 days at 75°C).

Nevertheless, a principal difference exists between samples of the high-T_c copper oxide materials and the organic superconductors. The former are usually polycrystalline or granular, and single crystals are difficult to obtain; in the latter case, all physical investigations were done on single crystals. The electrochemical preparation of these organic metals results in single crystals suitable for physical investigations. But, since the organic superconductors have very short coherence lengths as well, it should be possible to prepare pressed samples from polycrystalline organic metals - as in the case of the "ceramics" - which should exhibit bulk superconductivity.

POLYCRYSTALLINE PRESSED SAMPLES OF ORGANIC METALS

Polycrystalline pressed samples of the size 4 x 1 x 0.5 mm^3 were prepared from grained single crystals (the resulting crystallites from the graining process had typical diameters of 0.5 - 10 μm) of the organic metals (BEDT-TTF)$_2$Cu(NCS)$_2$, α_t-(BEDT-TTF)$_2$I$_3$, α-(BEDT-TTF)$_2$I$_3$ and β-(BEDT-TTF)$_2$I$_3$, by applying a pressure of (3 to 10) x 10^3 kg cm^{-2} to the powder. These samples are mechanically very stable, and the temperature dependence of the resistivity can be measured easily by the usual four-point method. For the pressed samples of (BEDT-TTF)$_2$Cu(NCS)$_2$ and α_t-(BEDT-TTF)$_2$I$_3$, no bulk superconductivity could be found [27]. On the other hand, in the polycrystalline pressed sample of α_t-(BEDT-TTF)$_2$I$_3$ a metal-like behaviour was found over the whole temperature range between 300 and 1.3 K (see curve I in Fig. 9). This fact is, at least, worth noticing, since usually in polycrystalline pressed samples of organic conductors only a semiconducting behavior is found (this is especially valid for quasi-one-dimensional materials).

Since, from the single crystal investigations, it was known that small changes in the structure might suppress the superconductivity, it was assumed that, during the preparation of the polycrystalline pressed samples of (BEDT-TTF)$_2$Cu(NCS)$_2$ and α_t-(BEDT-TTF)$_2$I$_3$, phase transitions might occur. Therefore, newly prepared samples of both materials were tempered for 3 days at 75°C before the samples were provided with contacts and measured. While the sample of (BEDT-TTF)$_2$Cu(NCS)$_2$ again didn't show any superconductivity, in the case of α_t-(BEDT-TTF)$_2$I$_3$ a broad superconducting transition could be observed (see curve II in Fig. 9). In addition, it was noticed that the room-temperature conductivity (15 S cm^{-1}) was about a factor of 2 larger, than for the untempered polycrystalline pressed samples of α_t-(BEDT-TTF)$_2$I$_3$ (crystals of α_t-(BEDT-TTF)$_2$I$_3$ have, usually, room-temperature conductivities of 20-30 S cm^{-1} in the **ab**-plane). In the next step, samples from powdered α-(BEDT-TTF)$_2$I$_3$ material were made, and tempered for 3 days at 75°C: these polycrystalline pressed samples of α_t-(BEDT-TTF)$_2$I$_3$ showed a resistivity behaviour, and a broad superconducting transition, more or less identical to that of the sample, whose resistivity versus temperature is shown in Fig. 9 (curve II). Fig. 10 presents the temperature dependence of the resistivity (normalized at 12 K) for such a prepared sample of α_t-(BEDT-TTF)$_2$I$_3$, together with the resistivity of a crystal of α_t-(BEDT-TTF)$_2$I$_3$. In both samples, the onset of superconductivity is found near 9 K. In contrast to the crystal of α_t-(BEDT-TTF)$_2$I$_3$, the superconducting transition in the polycrystalline sample is much broader. Zero resistivity is found at 2.2 K, and the middle of the transition appears at about 5.5 K, while in the case of the crystals of α_t-(BEDT-TTF)$_2$I$_3$ zero resistivity appears already at 6 K, and the middle of the transition occurs at 8 K.

In order to ascertain, whether the superconductivity observed in resistivity is a bulk effect of the sample or not, the ac-susceptibility was measured at a frequency of 3 MHz, with an rf-field of about 0.2 Gauss. Fig. 11 shows the increase in resonance frequency of the LC-circuit, due to exclusion of the rf-field by diamagnetic shielding currents, together with the resistivity of a polycrystalline pressed sample of α_t-(BEDT-TTF)$_2$I$_3$. There is a clear evidence for an onset of diamagnetic shielding below 7 K. This behavior is in contrast with the observations usually made in single crystals of organic superconductors, where the temperature for the onset of diamagnetic shielding is observed at the temperature, at which the resistivity reaches zero. Here, in the polycrystalline sample, the onset for the diamagnetic shielding is far above the temperature at which the resistivity becomes zero. The signal (increase in resonance frequency), which still increases on cooling to 2 K, corresponds to about 50% of that for a perfect superconductor, indicating a clear bulk effect of the superconductivity in the polycrystalline sample of α_t-(BEDT-TTF)$_2$I$_3$. On the other hand, the large temperature range, in which the frequency shift of the resonance frequency is observed, shows that an inhomogenous distribution of superconducting transitions exists in the sample. This broad distribution might be due to the annealing procedure during the preparation of the α_t phase.

An interesting question is whether the upper critical field $B_{c2}(T)$ of the polycrystalline pressed sample of α_t-(BEDT-TTF)$_2$I$_3$ is similar to that of the crystals of α_t-(BEDT-TTF)$_2$I$_3$. Fig. 12 shows the first preliminary measurements of the upper critical field of a polycrystalline pressed sample of α_t-(BEDT-TTF)$_2$I$_3$, together with $B_{c2\parallel}$ of a crystal of α_t-(BEDT-TTF)$_2$I$_3$. The upper critical field of the polycrystalline pressed sample seems to be much higher than that of the crystals, but more extended investigations, up to higher magnetic fields, have to be done (it should be noticed that $T_c = 8$ K for the crystals, and $T_c = 5.5$ K for the pressed samples). Nevertheless, it would be not surprising if the polycrystalline pressed samples have higher upper critical fields than the crystals. During the pulverization process, and by applying a pressure of nearly 1 kbar to the powder during the preparation of the sample crystal, imperfections might be created in the crystallites, leading to additional pinning centers.

Since structural phase transitions during the preparation of the polycrystalline pressed samples of organic materials seem to play an important role, some additional investigations were done. Fig. 13 shows resistivity versus temperature data of a polycrystalline pressed sample of α-(BEDT-TTF)$_2$I$_3$, together with the resistivity characteristics of a single crystal of α-(BEDT-TTF)$_2$I$_3$. The polycrystalline pressed sample of α-(BEDT-TTF)$_2$I$_3$ still shows, in the temperature range between 300 K and 180 K, a metallic behaviour, but the sharp

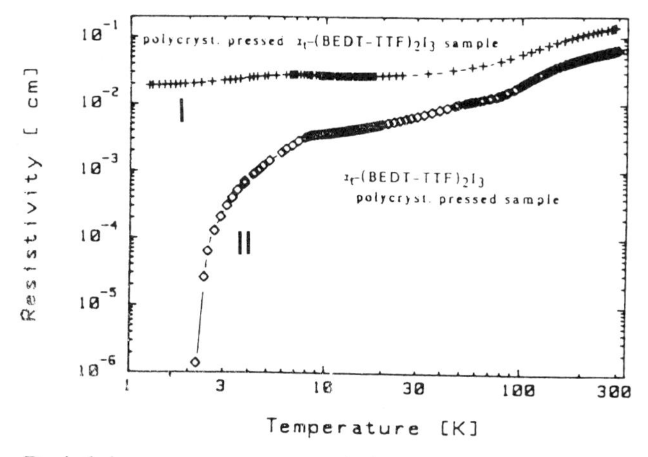

Fig.10. Resistivity versus temperature below 12K (normalized at 12 K) for a polycrystalline pressed sample of α_t-(BEDT-TTF)$_2$I$_3$ (prepared from α-(BEDT-TTF)$_2$I$_3$ and then tempered, see text) and for a crystal of α_t-(BEDT-TTF)$_2$I$_3$.

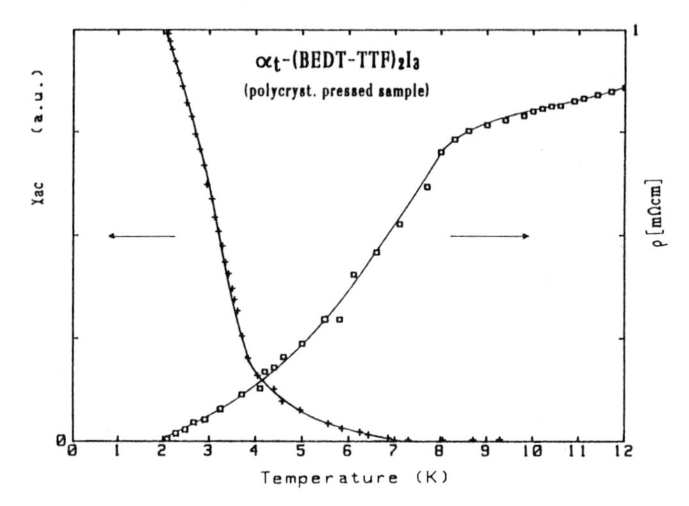

Fig.11. Resistivity below 12 K of a polycrystalline pressed sample of α_t-(BEDT-TTF)$_2$I$_3$ (right) and the change in ac susceptibility on cooling (left).

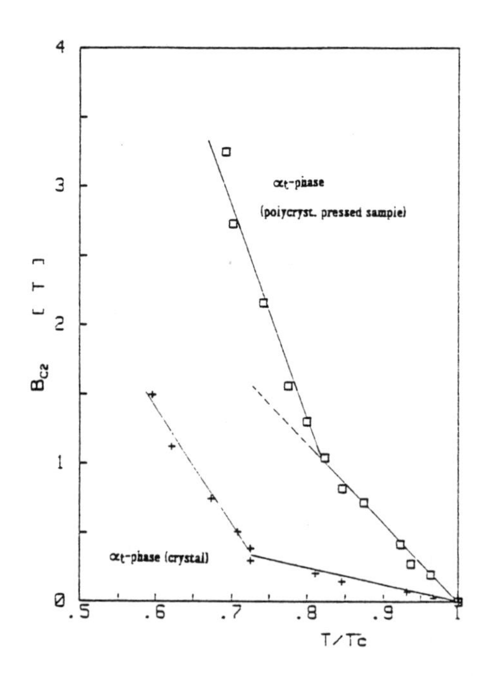

Fig.12. Upper critical field B_{c2} (T) of a polycrystalline pressed sample of α_t-(BEDT-TTF)$_2$I$_3$, and $B_{c2\parallel}$ (T) (see Fig.8) of a crystal of α_t-(BEDT-TTF)$_2$I$_3$.

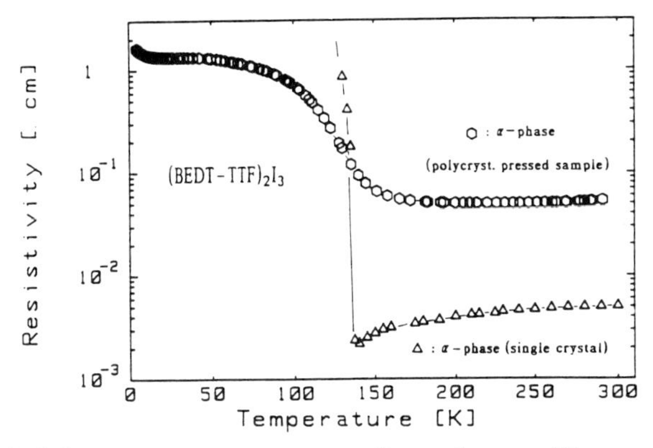

Fig. 13. Resistivity versus temperature of a polycrystalline pressed sample of α-$(BEDT\text{-}TTF)_2I_3$ and of single crystals of α-$(BEDT\text{-}TTF)_2I_3$.

metal-insulator transition, observed for the single crystal, cannot be seen any more. Instead, a very broad transition into a semiconducting state is found, and at 4 K a conductivity of about 0.5 S/cm is still observed, while the single crystals of α-$(BEDT\text{-}TTF)_2I_3$ are insulating at this temperature. This behaviour of the polycrystalline pressed sample of α-$(BEDT\text{-}TTF)_2I_3$ demonstrates again, that phase transitions might occur in these materials during the preparation of such pressed samples. This might also be of some importance for the preparation of pressed pellets for IR investigations of all organic conductors. In order to discriminate, whether the phase transition in the polycrystalline pressed sample of α-$(BEDT\text{-}TTF)_2I_3$ occurs already during the graining process, or later by applying the pressure of about 1 Kbar to the powder, the ESR linewidth of a powder of α-$(BEDT\text{-}TTF)_2I_3$, a polycrystalline pressed sample of α-$(BEDT\text{-}TTF)_2I_3$, and of the same pressed sample after tempering for 3 days at 75° C (α_t-$(BEDT\text{-}TTF)_2I_3$), was investigated. The results of these studies are compiled in Fig. 14. The ESR linewidth of the powder of α-$(BEDT\text{-}TTF)_2I_3$ corresponds, more or less, to the averaged ESR linewidths of a single crystal of α-$(BEDT\text{-}TTF)_2I_3$ [28]. Especially, the phase transition at 135 K can still be seen. In the case of the polycrystaline pressed sample of α-$(BEDT\text{-}TTF)_2I_3$, the linewidth at room temperature is smaller, and, as already observed for the resistivity, the sharper phase transition of the single crystal does not occur anymore. In the case of the tempered pressed sample, the ESR linewidth and its temperature dependence are more or less identical to those of crystals of α_t-$(BEDT\text{-}TTF)_2I_3$ or β-$(BEDT\text{-}TTF)_2I_3$ [3,28]. As a result of these ESR investigations, it can be said that, in the case of the polycrystalline pressed sample of α-$(BEDT\text{-}TTF)_2I_3$, the phase transition in the material is caused by the applied pressure during the preparation of the sample. A similar phase transition probably occurs also, when the samples are prepared directly from α_t-material. Therefore, polycrystalline pressed samples, which are prepared directly from the α_t-material, do not become superconducting if they are not tempered again after the preparation.

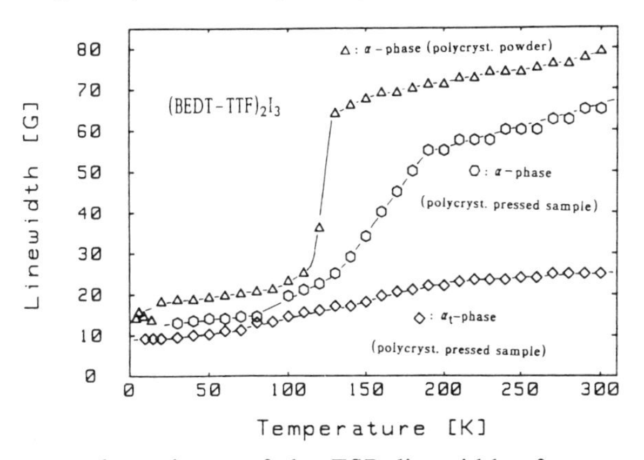

Fig. 14. Temperature dependence of the ESR linewidth of unpressed powder of α-$(BEDT\text{-}TTF)_2I_3$, of a polycrystalline pressed sample of α-$(BEDT\text{-}TTF)_2I_3$, and of the same sample after tempering (α_t-$(BEDT\text{-}TTF)_2I_3$).

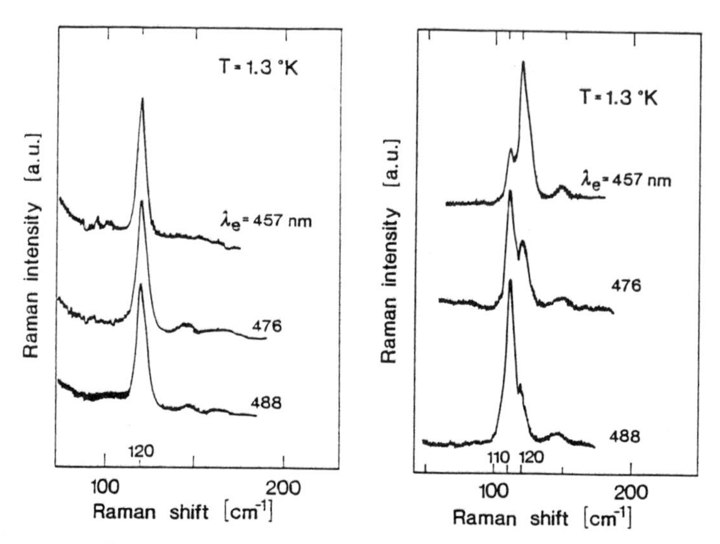

Fig. 15. Resonance Raman scattering, using different laser excitation frequencies at 1.3 K [30] of: (a: left) a polycrystalline pressed sample of α-(BEDT-TTF)$_2$I$_3$ (prepared from powdered α-crystals and then tempered), (b: right) of a polycrystalline pressed sample of α_t-(BEDT-TTF)$_2$I$_3$, where the surface of the pellet was polished with a razor blade (see text).

In order to obtain some more information about such phase transitions, which occur under pressure during the preparation of the polycrystallline pressed samples, resonance Raman investigations, in particular on the most intense vibrational symmetric stretching mode of the I$_3^-$ anions, were carried out. Earlier measurements on single crystals of α-, α_t- and β-(BEDT-TTF)$_2$I$_3$ [29] had shown that the resonance Raman spectra are very sensitive to the symmetry of the I$_3^-$ anions. The symmetric stretching mode of the linear symmetric I$_3^-$ anions usually is found at higher energy by about 10 cm^{-1}, compared to the asymmetric and non-linear I$_3^-$ anions [29]. In the resonance Raman spectra of the polycrystalline pressed sample of α-(BEDT-TTF)$_2$I$_3$, symmetric, linear and asymmetric I$_3^-$ anions are observed [30]. This indicates that the pressure, during the preparation of the samples, deforms the I$_3^-$ anions partially. A similar result can be observed for the polycrystalline pressed samples which were prepared directly from powdered α_t-crystals, and not tempered after the preparation. In contrast to this finding, the resonance Raman spectra of the tempered polycrystalline pressed samples of α_t-(BEDT-TTF)$_2$I$_3$ (which become superconducting) show only the stretching mode of the linear and symmetric I$_3^-$ anions (see Fig. 15a), indicating again the higher symmetry and higher order of the structure. In order to demonstrate how sensitive this structure is, in a further experiment, the surface of a sample was polished with a razor blade. Again, the resonance lines of the linear symmetric and asymmetric I$_3^-$ anions were observed (see Fig. 15b), while the resonance Raman spectra of the same sample, but just cut in the middle, had the same spectra, as that of the sample presented in Fig. 15a. A more detailed analysis of the resonance Raman studies is given in Ref. [30] in these proceedings.

The most surprising observation was made by measuring the temperature dependence of the resistivity of polycrystalline pressed samples of β-(BEDT-TTF)$_2$I$_3$ (in the following called β_p-(BEDT-TTF)$_2$I$_3$) [31]. Without tempering, the samples showed an onset of superconductivity at 9 K, zero resistivity at 3.2 K, and the middle of the resistive transition at 7.5 K (Fig. 16). This observation is surprising, because single crystals of β-(BEDT-TTF)$_2$I$_3$ show a rather sharp, stable superconducting transition, but only at 1.4 K [32] (onset temperature 1.6 K), and a metastable superconducting state at 8 K [4-7], as described before. Here, in the polycrystalline pressed samples of β_p-(BEDT-TTF)$_2$I$_3$, the superconducting state is stable, and a bulk effect of the sample, as seen from the measured change of the ac susceptibility (Fig. 16), corresponds at 2 K to about 50% of that expected for a perfect superconductor. Again, here, in the polycrystalline pressed sample of β_p-(BEDT-TTF)$_2$I$_3$, the onset for the diamagnetic shielding is far above the temperature, where the resistivity becomes zero, and the large temperature range, in which the frequency shift of the resonance frequency is observed, shows that an inhomogeneous distribution of superconducting transitions exists in the sample.

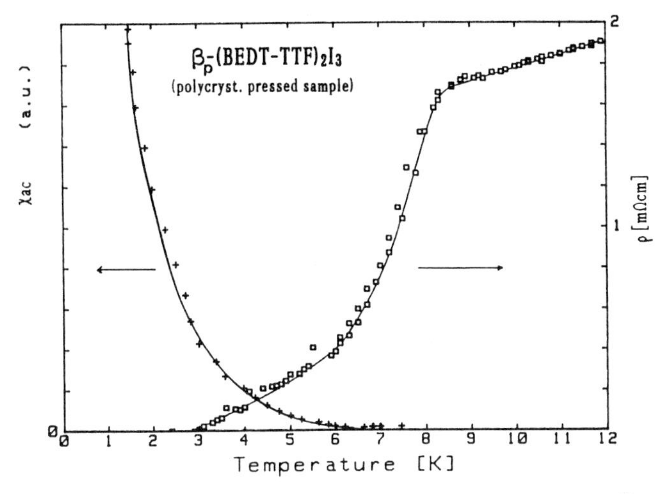

Fig. 16. Resistivity, and change in ac susceptibility versus temperature, of a polycrystalline pressed sample of β_p-(BEDT-TTF)$_2$I$_3$ (see text) below 12 K.

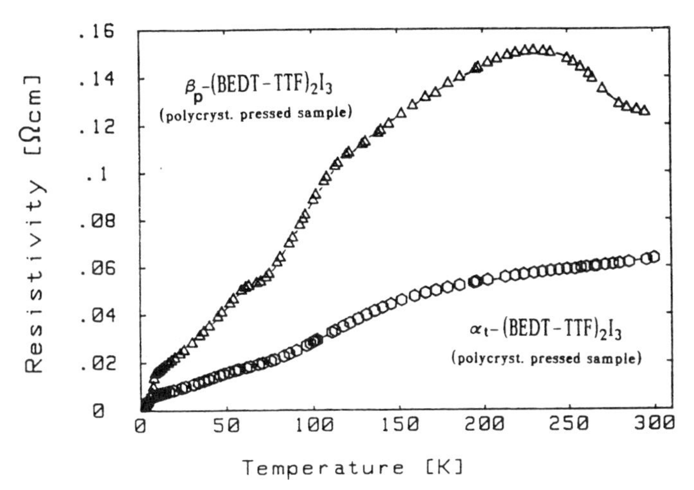

Fig. 17. Temperature dependence of the resistivity of a polycrystalline pressed sample of α_t-(BEDT-TTF)$_2$I$_3$ and β_p-(BEDT-TTF)$_2$I$_3$.

The room-temperature conductivity of the polycrystalline pressed sample of β_p-(BEDT-TTF)$_2$I$_3$ ranges between 5 and 10 S cm^{-1}, which is about a factor of 2 smaller, than in the pressed sample of α_t-(BEDT-TTF)$_2$I$_3$. Fig.17 shows the typical temperature dependence of the resistivity of a pressed sample of β_p-(BEDT-TTF)$_2$I$_3$ and of α_t-(BEDT-TTF)$_2$I$_3$. It can clearly be seen that the resistivity characteristics of β_p-(BEDT-TTF)$_2$I$_3$ differ remarkably from that of the pressed α_t-(BEDT-TTF)$_2$I$_3$. While the resistivity of the α_t-samples decreases upon cooling from room temperature (like in a metal), in the β_p-(BEDT-TTF)$_2$I$_3$ samples the resistivity first increases slowly on lowering the temperature. At around 220 K, the temperature at which an incommensurate structural modulation in single crystals of β-(BEDT-TTF)$_2$I$_3$ occurs [33], a maximum in resistivity is observed. Below 220 K, the resistivity decreases again. It should be mentioned that tempering of the polycrystalline pressed samples of β_p-(BEDT-TTF)$_2$I$_3$ for 1 to 3 days at 75°-90°C does not change the physical properties of the samples.

The resistivity of the polycrystalline sample of β_p-(BEDT-TTF)$_2$I$_3$ indicates, that the detailed structure is different from that of the polycrystalline pressed α_t-(BEDT-TTF)$_2$I$_3$, and different from single crystals of β-(BEDT-TTF)$_2$I$_3$, even if the unit cells are identical, and the molecular arrangement in the materials might be very similar. A possible difference might lay, again, in the ordering of the terminal ethylene groups of the BEDT-TTF molecules. The assumption, that both materials do not have identical structures, is supported by preliminary measurements of the upper critical fields B_{c2} of polycrystalline pressed samples of β_p-(BEDT-TTF)$_2$I$_3$. Fig. 18 shows the upper critical fields of pressed samples of β_p-(BEDT-TTF)$_2$I$_3$ and α_t-(BEDT-TTF)$_2$I$_3$. The B_{c2} of the β_p-sample is clearly higher than that of the α_t-sample. Measurements at higher magnetic fields are in preparation.

For both types of polycrystalline pressed samples, α_t-(BEDT-TTF)$_2$I$_3$ and β_p-(BEDT-TTF)$_2$I$_3$, again as in the case of the crystals of α_t-(BEDT-TTF)$_2$I$_3$, above 60 K a strong dependence of the resistivity on the direction of the temperature cycle is observed. Below 60 K, the resistivity is identical in both cycles. Fig. 19 compares the resistivity data for a temperature cycle for β_p-(BEDT-TTF)$_2$I$_3$. By warming up the sample, the resistivity becomes nearly twice larger than the starting room-temperature value. Just as for α_t-(BEDT-TTF)$_2$I$_3$, the resistivity of the sample of β_p-(BEDT-TTF)$_2$I$_3$ "relaxes" back to the starting value, at the end of the warming-up at room temperature, within a few hours. The explanation for this behaviour is probably the same, as for the crystals of α_t-(BEDT-TTF)$_2$I$_3$.

As in polycrystalline pressed α_t-(BEDT-TTF)$_2$I$_3$, in pressed samples of β_p-(BEDT-TTF)$_2$I$_3$ a pressure-induced structural phase transition plays an important role. As a consequence of the structural phase transition in β_p-(BEDT-TTF)$_2$I$_3$, the superconducting transition temperature is increased. This behavior re-emphasizes that organic superconductors might also be of interest for industrial applications, since polycrystalline materials are easier to use than single crystals. In addition, the discovery of bulk superconductivity in large pressed samples of crystallites of organic metals, of the typical diameter of 1 µm and below, indicates that the observation of superconductivity in conducting polymers should be possible as well.

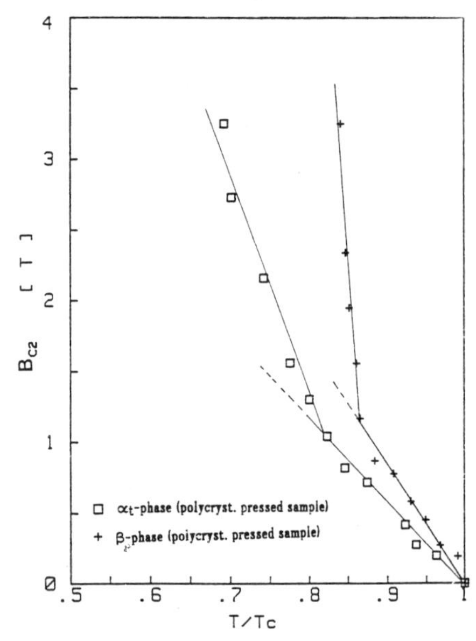

Fig. 18. Upper critical fields B_{c2}(T) of polycrystalline pressed samples of β_p-(BEDT-TTF)$_2$I$_3$ and α_t-(BEDT-TTF)$_2$I$_3$.

SUPERCONDUCTIVITY IN CONDUCTING POLYMERS?

The question arises, under which conditions superconductivity in conducting polymers might occur. We will discuss this question only in the picture of the usual electron-phonon interaction, and not consider other possible mechanisms or interactions. Investigations on single crystals of organic metals suggest that high symmetry is advantageous, in order to obtain superconductivity (it is not implicitly necessary to have such a high symmetry for an individual molecule, since it was demonstrated that also unsymmetrical donor molecules can form structures, which become superconducting [34], but, for instance, two or more molecules should form a symmetric unit, which repeats in the structure). In addition, it was shown that disorder in the anion channels [35], or disorder of molecular groups [23] (e.g. ethylene groups of BEDT-TTF molecules) might lead to lower superconducting transition temperatures, or might even suppress the superconductivity totally [36,37]. Therefore, structures of high symmetry are favorable for superconductivity. On the other hand, such conditions are difficult to achieve in organic conducting polymers. Nevertheless, we have seen that, in the case of a medium or even strong electron-phonon coupling, as in the single crystals of the BEDT-TTF radical salts, the coherence lengths can be of the order of only 10 Å; this means that it would only be necessary to have ordered and symmetrically structured areas, in such conducting polymers, of diameters of 10-30 Å.

Such symmetric units might be connected by other, non-symmetric or disordered polymer units, which are not superconducting. They might have even a length of the order of 100 Å. Due to a proximity effect and/or a Josephson coupling between the symmetric and ordered units, it still would be possible to obtain bulk superconductivity in the polymer. Therefore, besides small, symmetric, and ordered polymer areas, the main condition for superconductivity would be a medium to strong electron-phonon coupling, and therefore short coherence lengths. In the case of weak electron-phonon coupling, and long coherence lengths, superconductivity in conducting polymers presumes large ordered symmetric polymer areas, or long ordered chains. In the case of medium or strong electron-phonon coupling and short coherence lengths, the chains in the polymer must not be very long. But it would be advantageous, if such chains would be coupled in two, or even three, dimensions.

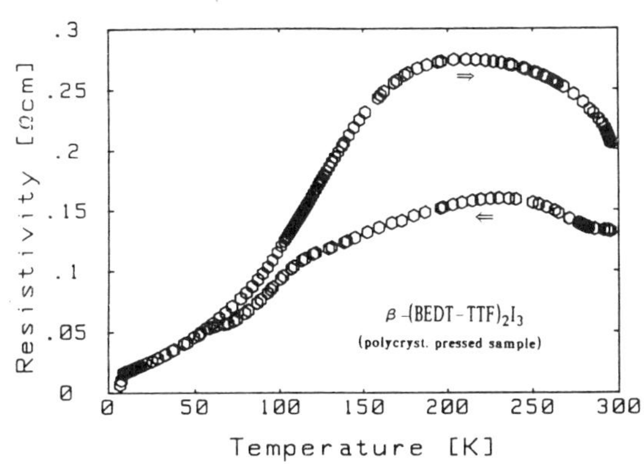

Fig.19. Temperature dependence of the resistivity of a polycrystalline pressed sample of β_p-(BEDT-TTF)$_2$I$_3$ in a cooling-down and a warming-up temperature cycle.

Since BEDT-TTF radical salts are environmentally quite stable, we synthesized metallic coordination polymers [38] with BEDT-TTF related ligands.

METALLIC COORDINATION POLYMERS

One main problem, hampering the technical applications of organic conducting polymers up to now, is their environmental and thermal instability. Metal ions can be introduced to stabilize polymeric backbones. Because of the enormous coordinating ability of sulfur to many transition metal ions, a polymeric carbon-sulfur backbone, similar to BEDT-TTF and TTF units, was used in order to obtain coordination polymers.

As reported earlier [39-42] CS_2 can be reacted to thiapendione, which can be converted into ethylenetetrathiolate (TT), by cleavage with strong chemical bases. Additionally, thiapendione can be "dimerized" to bis(1,3-dithiole-2-one)tetrathiafulvalene, which can be converted into tetrathiolatotetrathiafulvalene (TTF-TT), again by a cleavage reaction with strong bases. Both tetrathiolates react with transition metal ions, like nickel(II) or copper(II), to form coordination polymers which are remarkably inert, chemically as well as thermally [38]. The reactions, and their final products, are schematically summarized in Fig. 20. Fig. 21 shows a schematic picture of the possible coordination between different chains.

In the case of nickel (M = Ni) m = 0 and p = 0. The polynuclear compound contains, formally, nickel(IV) without counterions. Using copper(II) as metal ion, a Cu(III)/Cu(I) mixed-valence solid with m = 1 and p = 1 is obtained.

The compounds, which are obtained as powders, can be pressed to compact pellets, which show conductivities up to 150 S cm^{-1}. The electrical conductivity of some of these materials varies only slightly, in the temperature range between 4.2 K and 400 K (Figs. 22,

Fig. 20. Schematic description of the synthesis of "polymer" metal tetrathiolates, starting from CS_2 (M = Ni, Cu, ...).

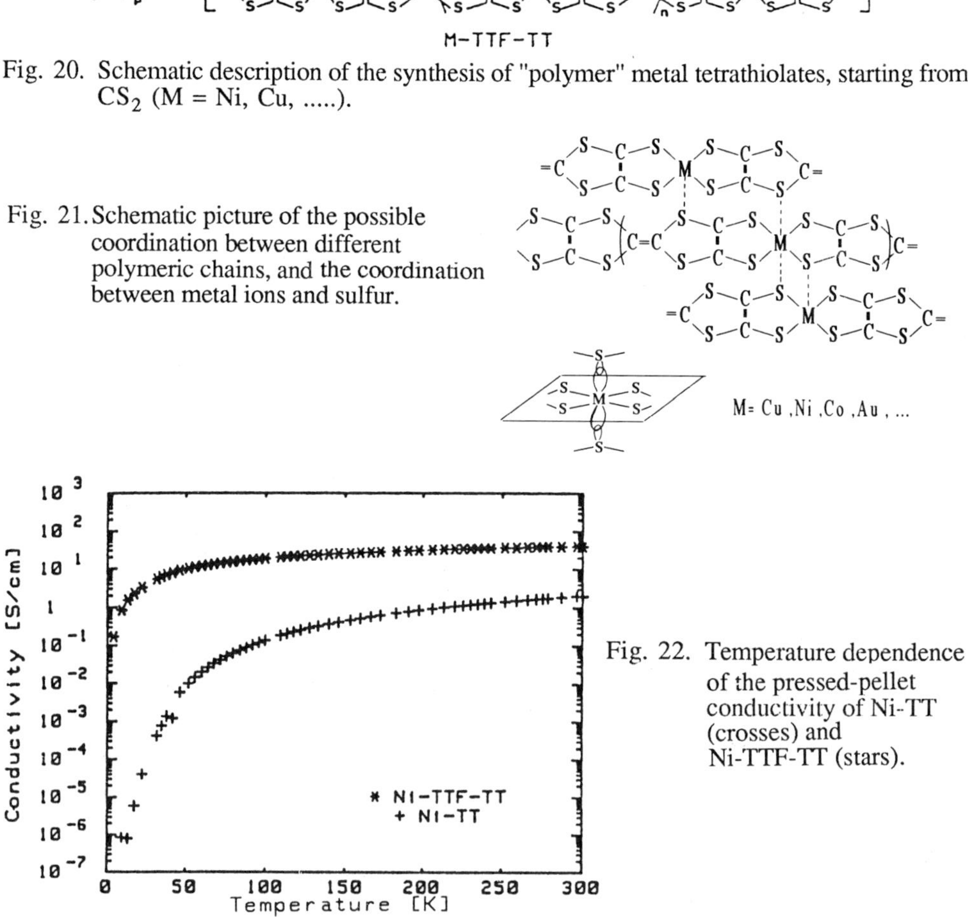

Fig. 21. Schematic picture of the possible coordination between different polymeric chains, and the coordination between metal ions and sulfur.

Fig. 22. Temperature dependence of the pressed-pellet conductivity of Ni-TT (crosses) and Ni-TTF-TT (stars).

23). Thermopower measurements indicate a metallic character of these solids [42,43]. Principally, the conductivity of the isolated TT-derivatives is lower (2-7 S cm^{-1}), compared with the TTF-TT compounds (40-150 S cm^{-1}).

A comparison between equivalent nickel and copper compounds shows that the copper species usually conduct better than the nickel derivatives, which are much more stable environmentally, compared to the copper compounds. The higher stability of the nickel species

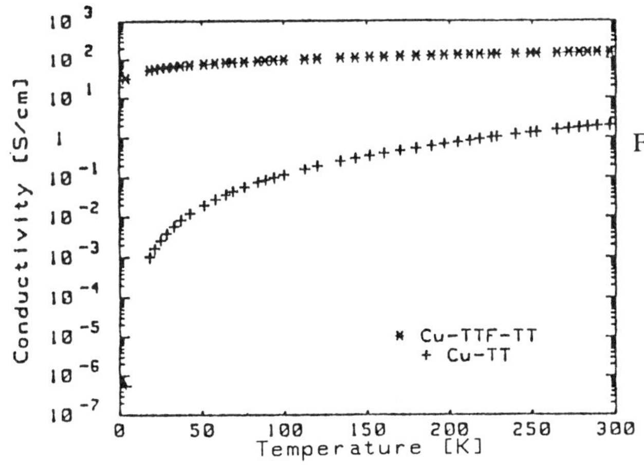

Fig. 23. Temperature dependence of the conductivity of Cu-TT (crosses) and Cu-TTF-TT (stars).

may be explained by the surprising fact, that these materials are obtained in an electrically neutral "undoped" state. The conducting mechanism is not clear yet. Nevertheless, the environmental stability of all these coordination polymers is at least 100 times better, than polypyrrole. Both polymers have been investigated unaged and aged (24 - 120 days at 80°C in humid air) [38,44]. It is remarkable that Cu-TTF-TT has a room temperature conductivity of about 100 S cm^{-1}, and nearly no temperature dependence of the conductivity down to 4.2 K (Fig. 22). On the other hand, it should be stressed that, up to now, none of the investigated coordination polymers become superconducting. Further investigations, and several chemical variations, such as doping of these coordination polymers, are necessary.

ACKNOWLEDGEMENT

We gratefully acknowledge the financial support of this work by the Deutsche Forschungsgemeinschaft and the European Economic Community (contract ST 2J-0315-C).

REFERENCES

[1] D. Jérome, A. Mazaud, M. Ribault, and K. Bechgaard, J. Physique Lett. 41: L95 (1980).
[2] H. Urayama, H. Yamochi, G. Saito, K. Nozava, T. Sugano, M. Kinoshita, S. Saito, K. Oshima, A. Kawamoto, and J. Tanaka, Chem. Lett. 55 (1988).
[3] D. Schweitzer, P. Bele, H. Brunner, E. Gogu, U. Haeberlen, I. Hennig, T. Klutz, R. Swietlik, and H. J. Keller, Z. Phys. B - Condensed Matter 67:489 (1987).
[4] K. Murata, M. Tokumoto, H. Anzai, H. Bando, G. Saito, K. Jajimura, and T. Ishiguro, J. Phys. Soc. Japan 54: 1236 (1985).
[5] V. N. Laukhin, E. E. Kostynchenko, Yu. Y. Sushko, I. F. Shegolev, and E. B. Yagubskii, Soviet Physics JETP Lett. 41: 81 (1985).
[6] F. Creuzet, G. Creuzet, D. Jérome, D. Schweitzer, and H. J. Keller, J. Physique Lett. 46: L1079 (1985).
[7] F. Creuzet, D. Jérome, D. Schweitzer, and H. J. Keller, Europhys. Lett. 1: 461 (1986).
[8] S. Gärtner, E. Gogu, I. Heinen, H. J. Keller, T. Klutz, and D. Schweitzer, Solid State Commun. 65: 1531 (1988).
[9] D. Schweitzer, K. Polychroniadis, T. Klutz, H. J. Keller, I. Henning, I. Heinen, U. Haeberlen, E. Gogu and S. Gärtner, Synth. Metals 27: A465 (1988).
[10] H. Urayama, H. Yamochi, G. Saito, S. Sato, A. Kawamoto, J. Tanaka, T. Mori, Y. Maruyama, and H. Inokuchi, Chem. Lett. 463 (1988).
[11] S. Klotz, J. S. Schilling, S. Gärtner, and D. Schweitzer, Solid State Commun. 67: 1981 (1988).
[12] K. Oshima, H. Urayama, H. Yamochi, and G. Saito, J. Phys. Soc. Japan 57: 730 (1988).
[13] E. Gogu, Ph.D. Thesis, Universität Heidelberg, 1989.

[14] R. A. Klemm, A. Luther, and R. M. Beasley, Phys. Rev. B 12: 877 (1975).
[15] G. O. Baram, L. I. Buravov, L. C. Degtariev, M. E. Kozlov, V. N. Laukhin, E. E. Laukhina, V. B. Drischenko, K. L. Pokhodnia, M. K. Scheinkmann, R. P. Shibaeva, and E. B. Yagubskii, JETP Lett. 44: 2913 (1986).
[16] I. Hennig, U. Haeberlen, I. Heinen, D. Schweitzer, and H. J. Keller: Physica C 153-155: 493 (1988).
[17] K. Bender, I. Hennig, D. Schweitzer, K. Dietz, H. Endres, and H. J. Keller, Mol. Cryst. Liq. Cryst. 108: 359 (1984).
[18] T. Mori, A. Kobayashi, Y. Sasaki, H. Kobayashi, G. Saito, and H. Inokuchi, Chem. Lett. 957 (1984).
[19] A. J. Schultz, H. H. Wang, J. M. Williams, and A. Filhol, J. Am. Chem. Soc. 108: 7853 (1986).
[20] I. Hennig, Ph. D. Thesis, Universität Heidelberg 1989.
[21] D. Schweitzer and H. J. Keller in "Organic and Inorganic Low Dimensional Crystalline Materials," ed. by P. Delhaès and M. Drillon, NATO ASI Series B168: 219 (Plenum, New York, 1987); J. M. Williams, H. H. Wang, T. J. Emge, U. Geiser, M. A. Beno,
[22] A. Nowak, M. Weger, D. Schweitzer, and H. J. Keller, Solid State Commun. 60: 199 (1986).
[23] M. E. Hawley, K. E. Grey, B. R. Terris, H. H. Wang, K. D. Carlson, and J. M. Williams, Phys. Rev. Lett. 57: 629 (1986).
[24] A. Nowak, U. Poppe, M. Weger, D. Schweitzer, and H. Schwenk: Z. Phys. B. Condensed Matter 68: 41 (1987).
[25] M. Weger, K. Bender, T. Klutz, D. Schweitzer, F. Groß, C. P. Heidmann, Ch. Probst, and K. Andres, Synth. Metals 25: 49 (1988).
[26] M. Surma, Physica C 153-155: 243 (1988).
[27] D. Schweitzer, S. Gärtner, H. Grimm, E. Gogu, and H. J. Keller, Solid State Commun. 69: 843 (1989).
[28] B. Rothaemel, L. Forró, J. R. Cooper, J. S. Schilling, M. Weger, P. Bele, H. Brunner, D. Schweitzer, and H. J. Keller, Phys. Rev. B 34: 704 (1986).
[29] R. Swietlik, D. Schweitzer, and H. J. Keller, Phys. Rev. B 36: 6881 (1987).
[30] R. Zamboni, D. Schweitzer, and H. J. Keller in "Lower Dimensional Systems and Molecular Electronics" ed. by R. M. Metzger, P. Day and G. C. Papavassilliou, (Plenum, New York, in press).
[31] D. Schweitzer, E. Gogu, H. Grimm, S. Kahlich, and H. J. Keller, Angew. Chem. Adv. Mater. 101: 977 (1989).
[32] E. B. Yagubskii, I. F. Shegolev, V. N. Laukhin, P. A. Kononovich, M. V. Kartsovnik, A. V. Zwarykina, and L. I. Biwavov, Sov. Phys. JETP Lett. 39: 12 (1984).
[33] P. C. W. Leung, T. J. Emge, M. A. Beno, H. H. Wang, and J. M. Williams, J. Am. Chem. Soc. 106: 7644 (1984).
[34] K. Kikuchi, Y. Honda, Y. Ishikawa, K. Saito, I. Ikemoto, K. Murata, H. Anzai, T. Ishiguro, and K. Kobayashi, Solid State Commun. 66: 405 (1988).
[35] H. Endres, H. J. Keller, R. Swietlik, D. Schweitzer, K. Angermund, and C. Krüger, Z. Naturforsch. 41a: 1319 (1986).
[36] T. J. Emge, H. H. Wang, M. A. Beno, P. C. W. Leung, M. A. Firestone, H. C. Jenkins, J. D. Cook, K. D. Carlson, J. M. Williams, E. I. Venturini, L. J. Azevedo, and J.E. Schirber, Inorg. Chem. 24: 1736 (1985).
[37] H. Endres, M. Hiller, H. J. Keller, K. Bender, E. Gogu, I. Heinen, and D. Schweitzer, Z. Naturforsch. 40b: 1664 (1985).
[38] G. Renner, Ph.D. Thesis, Universität Heidelberg 1987.
[39] H. Poleschner, W. John, G. Kempe, E. Hoyer, and E. Fanghänel, Z. Chem. 15:345 (1978).
[40] G. E. Holdcroft and A. E. Underhill, Synth. Metals 10: 437 (1985).
[41] R. Vicente, J. Ribas, P. Cassoux, and L. Valade, Synth. Metals 13: 265 (1986).
[42] H. J. Keller, T. Klutz, H. Münstedt, G. Renner, and D. Schweitzer, in: "Electronic Properties of Conjugated Polymers" ed. by H. Kuzmany, M. Mehring and S. Roth, (Springer, Heidelberg, 1987).
[43] T. Klutz, Diplomarbeit, Universität Heidelberg, 1987.
[44] K. Bender, E. Gogu, I. Hennig, D. Schweitzer, and H. Münstedt, Synth. Metals 18: 85 (1987).

P. C. W. Leung, K. D. Carlson, R. J. Thorn, A. J. Schultz, and M. H. Whangbo, Prog. Inorg. Chem. 35: 51 (1987).

SUPERCONDUCTIVITY IN MOLECULAR AND OXIDE LATTICES: A COMPARISON

Peter Day

Institut Laue-Langevin
156X
38042 Grenoble Cedex, France

INTRODUCTION

Superconductivity is, perhaps, the most dramatic of the various collective ground states that can be formed by a Fermi gas composed of electrons in the periodic potential of a crystalline lattice. Zero resistance is readily manifested, and the concomitant property of flux exclusion, shown up as perfect diamagnetism, is a necessary observable. In the context of the present Advanced Study Institute, one should recall the simple fact that, to be a superconductor, a material must first be metallic. On the other hand, by no means all metals are superconductors, either because there is competition from other ground states, like charge density wave and spin density wave, or because the electronic band width is too large, and vibronic interaction too weak. The challenge, therefore, is to define, and then optimize, the electronic and structural factors that determine this competition. Structure types providing paradigms for conventional superconductivity have traditionally been three-dimensional continuous lattice in type, from the close-packed structures of elements, such as Hg and Nb, to the more elaborate, but still largely isotropic, arrays of the A15 compounds, like Nb_3Sn. Recent developments, however, have brought superconductivity into contact with molecular and inorganic solid-state chemistry, with an enlargement in the variety of structure types that one would expect, when synthetic chemistry comes to bear on a problem and, in one instance at least, dramatic enhancement of critical temperature.

The new superconductors were born out of quite distinct fields of chemistry, and quite different methods of preparation, crystallization and sample characterization have to be used in each case. Nevertheless, there are some tantalizing similarities between the two classes of structure, and between some of their physical properties. That makes comparison between them a very suitable topic for this meeting, which is otherwise entirely occupied by molecular materials and devices. Before setting out that comparison, however, it will be instructive to set both categories, molecular and oxide, against the background of the evolution of our knowledge of superconductors, since the discovery of the phenomenon by Kammerlingh Onnes in 1910.[1]

EVOLUTION OF SUPERCONDUCTORS

Whilst the critical temperature of the transition from metallic to superconducting state is far from being the only parameter of interest, the

Lower-Dimensional Systems and Molecular Electronics
Edited by R.M. Metzger *et al.*, Plenum Press, New York, 1991

history of superconductivity can be charted, quite simply, in terms of this single observable. The evolution of the highest known T_c with time has been plotted quite frequently, but a change of scale on the temperature axis serves to emphasise the progress made with different classes of materials, and hence the importance of venturing into new categories of substances. In Figure 1 the temperature scale is logarithmic, to embrace the widest range of temperature. The study of superconductivity started at the limit of the liquid helium range, with experiments on what are structurally, the simplest materials, the elements. After the original discovery in Hg (T_c 4.15K), the highest known T_c nearly doubled in three years (Pb, T_c 7.2K in 1913)[2] With passing time, the whole set of elements was exhausted, yielding a maximum possible T_c of 9.2K for Nb.

Exploration of binary phases (of which, of course, there are vastly more) then got under way, with a further sharp increase, followed by more gradual progress over a 40-year time span, culminating in the A15 compounds, with Nb_3Ge at T_c = 23.2K (1973)[3] remaining unchallenged, until the arrival of the mixed valency copper oxide phases in 1986.

Meanwhile, however, superconductivity had already been found in pure oxide phases (to be distinguished from the T_c enhancement seen by the surface layer oxidation of e.g. Nb metal). The first oxide perovskite phase in the field was $SrTiO_{3-x}$[4] which, having an extremely small carrier density (x small), had T_c well below 1K. During the later 1960's and early 1970's, however, further perovskite-based systems were discovered, in which the electrons near the Fermi surface were either of d or s character. Among the former, one could mention W bronzes, and for the latter, the mixed-valency Bi compounds $BaPb_{1-x}Bi_xO_3$[5] which, with the spinel $LiTi_2O_4$, first took T_c for oxides above 10K.[6] All of these oxides are what inorganic chemists would call 'mixed valency',[7] i.e. on a formal electron count, assigning a closed shell of 8 electrons to O^{2-} and integral oxidation states to the A-subgroup cations, the remaining transition or post-transition element has an average oxidation state that is non-integral. The same is also true, of course, of the high-T_c copper oxides which, in the top right-hand corner of Figure 1, represent the current 'state of the art'.

The final group of materials in Figure 1 are those most relevant to the present meeting, namely the molecular crystals. Although one polymeric material, polysulphurnitride ($(SN)_x$) was found to be a superconductor in 1974 (T_c=0.35K),[8] no others have followed. The earliest ambient-pressure organic superconductor was bis(tetramethyl-tetraselenofulvalene) perchlorate $(TMTSF)_2ClO_4$, otherwise known as Bechgaard's salt, in 1981 with T_c=1.2K[9], and all the other increases in T_c, up to the current limit of 10.5K[10], originate from, broadly, the same class of compound. All are charge transfer salts of stoichiometry A_2B, with A being a tetra-chalcogeno(S or Se)-fulvalene molecule, substituted in the four peripheral carbons by methyl or bis-ethylene-dithio groups, and B a uninegative anion. This is not to say that other types of molecular salts, or other stoichiometries, have not been found to superconduct, but merely that they have not exhibited the highest T_c's in this class. For example, the A_3B_2 salt $(BEDT\text{-}TTF)_3Cl_2.2H_2O$ becomes superconducting under pressure (14 kbar, 5.1K)[11] and the metal complex salt $(Ni(d\ mit)_2)_2$ likewise (6 kbar, 1.6k).[12]

What Figure 1 shows very clearly is the way that, for every category of material, initial progress (as measured by the rate of increase of T_c with time) has always been fast, but that advance has subsequently slowed, as further examples are discovered, so that qualitative improvement is replaced by merely quantitative. Nevertheless, even in this second phase of consolidation, percentage improvements may still be crucial, as when the evolution from Nb_3Sn (T_c=18.1K) to Nb_3Ge (T_c=23.2K) first took T_c above the boiling point of liquid hydrogen. Viewed from this perspective, progress on the oxides and the molecular superconductors has been equally spectacular, granted that T_c's in the latter remain an order of magnitude lower than in the former. It is worth noting, though, that before the discovery of the

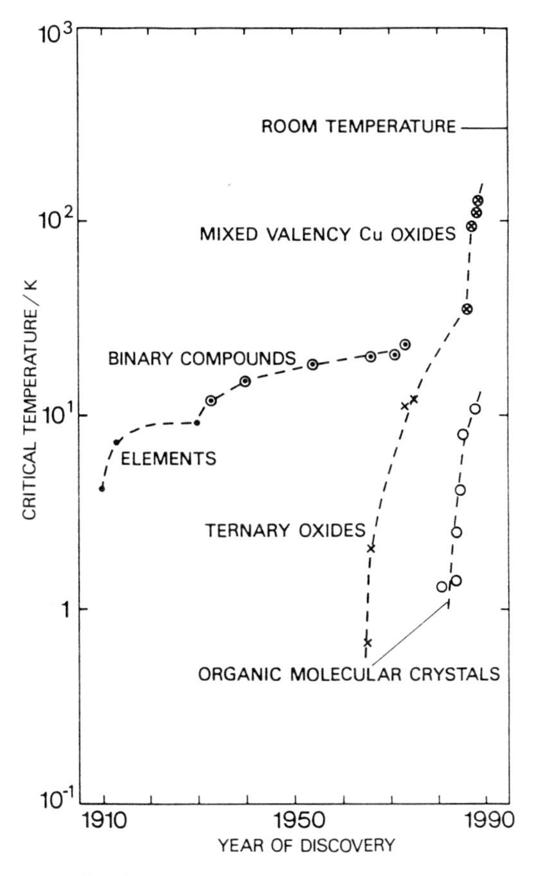

Fig. 1. How the highest known superconductivity T_c has increased from 1910 to 1989. Note the families of materials.

cuprates, a T_c of 10.5K, as found in 1988 in $(BEDT-TTF)_2Cu(NCS)_2$,[10] would have been in the upper quartile of all 1400 known superconductors. The time derivative of T_c for both classes is comparable, and well in excess of those in times past for the elements and binaries. Furthermore, neither shows any sign of saturating yet!

STRUCTURAL FEATURES OF MOLECULAR AND OXIDE SUPERCONDUCTORS

Although the detailed structural chemistry of the molecular and copper oxide superconductors is naturally quite different, there are broader features that they have in common, sufficient to make a comparison worthwhile.

Molecular Superconductors

Given that the dominant structural feature in conducting molecular crystals is the columns formed by plane-to-plane stacking of flat, or nearly flat, conjugated molecules, the physics of the earlier examples of the type were dominated by the properties of a near one-dimensional Fermi surface, and in particular by the various instabilities to which such a surface is subject.[13] A global view of the different kinds of behaviour of charge transfer salts is given in Figure 2, which shows the variation of resistivity with temperature from room temperature to 1K. Specific resistivity varies at least 11 orders of magnitude over this range.

The notable feature of the earliest examples of charge transfer salts is the existence of insulating, rather than metallic, behaviour at low temperature. Taking the TCNQ salts first, Torrance[14] distinguished four categories of behaviour, depending on whether the chemical stoichiometry was simple (1:1) or complex (1:2, 2:3 etc.) and whether the extent of charge transfer was complete or not. In the first category (e.g. K(TCNQ)) there is formally one unpaired electron per anion, and the compound is a Mott insulator. Similarly, $TEA(TCNQ)_2$ is a salt in the sense that the charge on the TEA (triethylammonium) cation is precisely +1, but now the number of unpaired electrons per TCNQ is 1/2. As is well known, the one-dimensional electron gas is unstable to either a charge density wave (CDW), or a spin density wave (SDW) state, the former via the Peierls distortion[15]. The 1:2 stoichiometry means that a commensurate CDW is formed by simple dimerization with intradimer spacing of 3.22Å, and interdimer spacings of 3.32 and 3.34Å,[16] leading to an insulating ground state. In the case of (TTF)(TCNQ), the situation is more complicated, since, in this segregated stack compound, both TTF and TCNQ conduction bands have to be considered. A massive volume of work on this compound[17] has led to the conclusion that there is partial charge transfer between the stacks, amounting to 0.59 electrons per formula unit. The metal-insulator transition at 54K is understood to arise from a Peierls distortion on the TCNQ stack. Further phase transitions follow at 48 and 38K, the former being the corresponding Peierls distortion on the TTF stack, and the latter arising from interstack interaction.

Progressing to the $(TMTSF)_2X$ series, one might have thought that there would also be a Peierls distortion because, as in $TEA(TCNQ)_2$, there is, again, 1/2 an unpaired electron per molecule (although, this time, on the cation stack) and the resulting CDW would be commensurate. However, the structures of the $(TMTSF)_2X$ salts are unusual, in that they are triclinic ($P\bar{1}$), with each unit cell containing two TMTSF molecules (Figure 3).[18] The TMTSF stacks are zig-zag and very slightly dimerized, with S...S intermolecular spacings of (3.964, 4.021) versus (3.874, 3.934)Å. However, although the valence band would be 3/4 filled, on the assumption of uniform stacking, the triclinic space group, and slight dimerization, finally lead to a half-filled band. The final ambient-pressure ground state at low temperature is, now, not a CDW, but an SDW, where the X^- ion is octahedral (PF_6^-, AsF_6^-),[19] but, eventually, superconducting in the ClO_4^- salt.[9] Thus $(TMTSF)_2ClO_4$ was the first-ambient pressure organic molecular super-

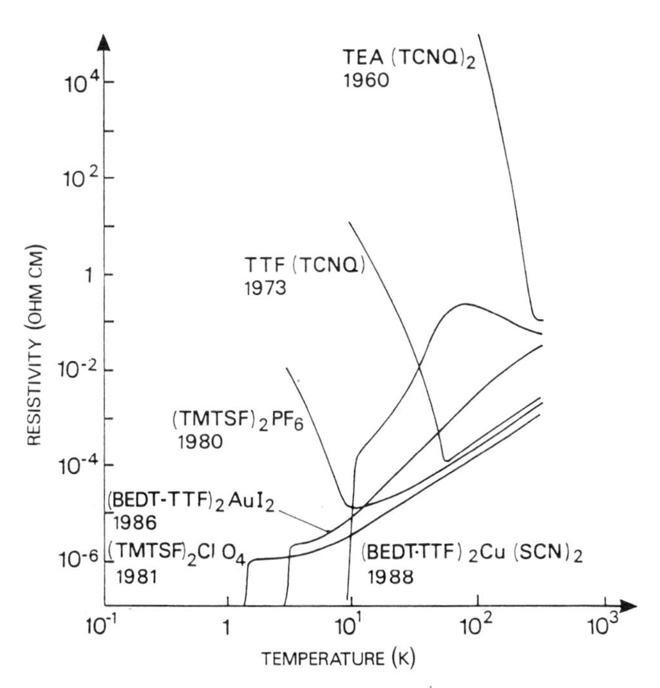

Fig. 2. Variation of resistivity of representative molecular charge transfer salts with temperature.

conductor, although the SDW state in $(TMTSF)_2PF_6$ is suppressed by pressure to give a superconducting state (T_c=1.0K) at 8 kbars.[20]

An important consideration, in deciding what compounds to synthesise, in order to stabilise the superconducting state versus the CDW or SDW in molecular lattices, is the effective dimensionality. A first priority is to make the lattice less purely one-dimensional, so as to make it stable to Peierls distortion. In that context, the compound (HMTSF)(TCNQ) was a significant early example.[21] HMTSF (hexa-methylene-tetraselenofulvalene) was synthesised to explore the consequence of introducing more bulky substitution of the TSF moieties, with the expectation that the system would be rendered more purely one-dimensional. Yet, it was found that, unlike TMTSF-TCNQ, which like TTF-TCNQ has a major metal-insulator transition near

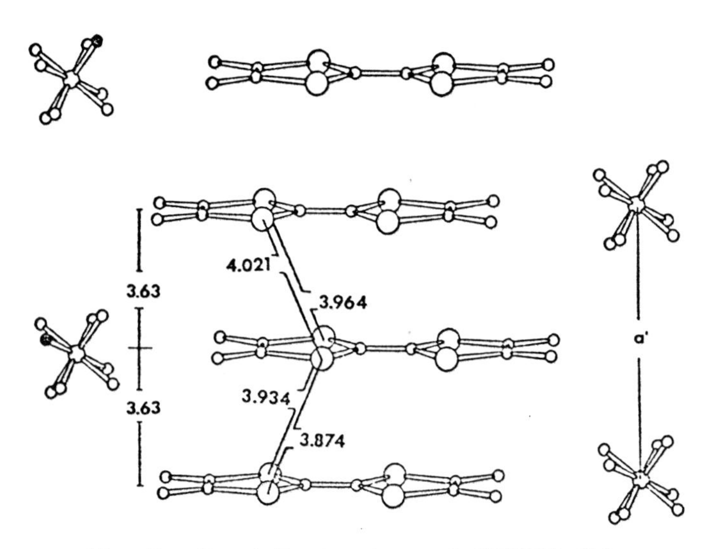

Fig. 3. Crystal structure of $(TMTSF)_2ClO_4$.

60K, the hexamethylene-substituted complex retains metallic conductivity down to very low temperatures (Figure 4). The reason for this striking difference in behaviour lies in the crystal structure (Figure 5).[22] So, far from decreasing the interaction between stacks, the bulky substituents induce a change in the packing, such that a terminal N of a TCNQ is brought within van der Waals contact (3.18Å) of an Se from TMTSF. A suitable strategy for maintaining metallic conduction down to temperatures where superconductivity has a chance to take over is, thus, to promote such close non-bonded contacts. It is of interest here to compare the transfer integrals along and between stacks in (TTF)(TCNQ) and superconducting $(TMTSF)_2ClO_4$. In the former, the transfer integrals between TTF molecules along and between stacks (t_{FF} : $t_{FF'}$) are in the ratio 500:1, and between TCNQ's (t_{QQ} : $t_{QQ'}$) are 25:1.[23] On the other hand, the corresponding ratio (t_{SS} : $t_{SS'}$) in the superconductor is only 10:1.[24] In both, the intrastack transfer integrals are of the same order of magnitude (0.1 eV).

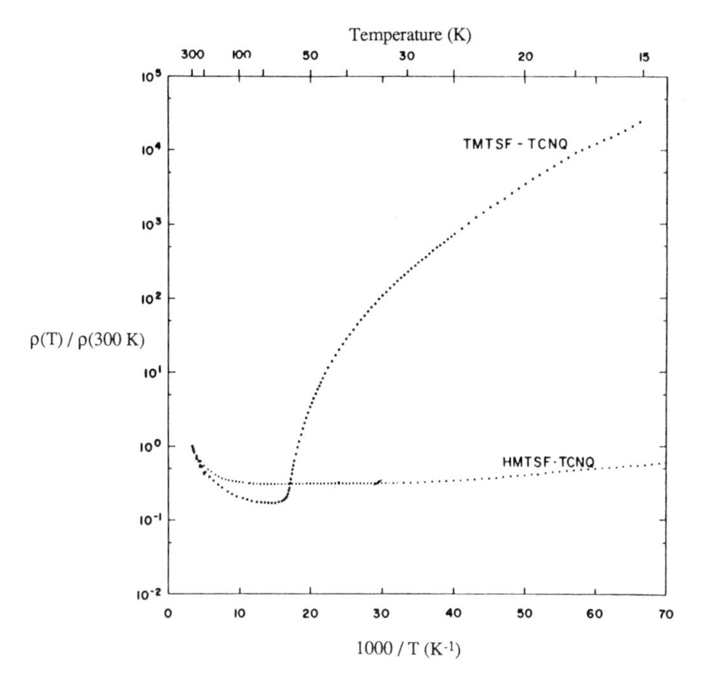

Fig. 4. Temperature variation of resistance of (a) (TMTSF)(TCNQ) and (b) (HMTSF)(TCNQ).[21]

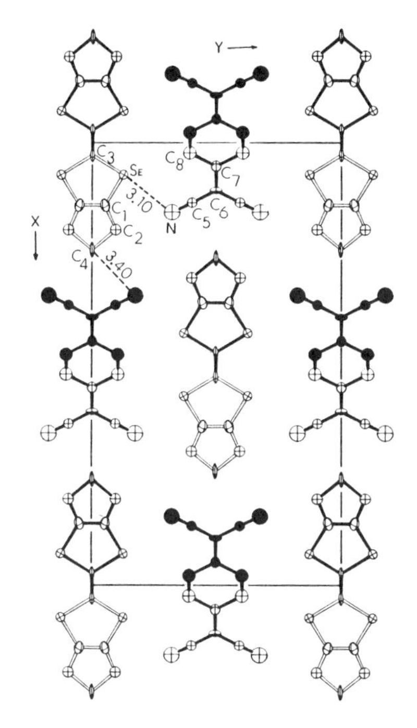

Fig. 5. Crystal structure of (HMTSF)(TCNQ).[22] Note the S...N contact.

Evidence that chalcogen-chalcogen interactions are effective in stabilising the superconducting state came with the discovery of the superconductivity in the BEDT-TTF salts, which contain four additional chalcogen atoms per molecule. The network of interstack interactions is very extensive. An example is shown in Figure 6.[25] The ratio of intra- to interstack transfer integrals is then still smaller (~5:1). In the so-called β-series of salts with linear mononegative triatomic anions (I_3^-, IBr_2^-, AuI_2^- etc.), the structure approximates much more closely to two-dimensional, since the anions are segregated quite clearly into layers, separated by 'blocks' of the organic material. The conductivity anisotropy, parallel and perpendicular to these blocks, is of the order of 500.

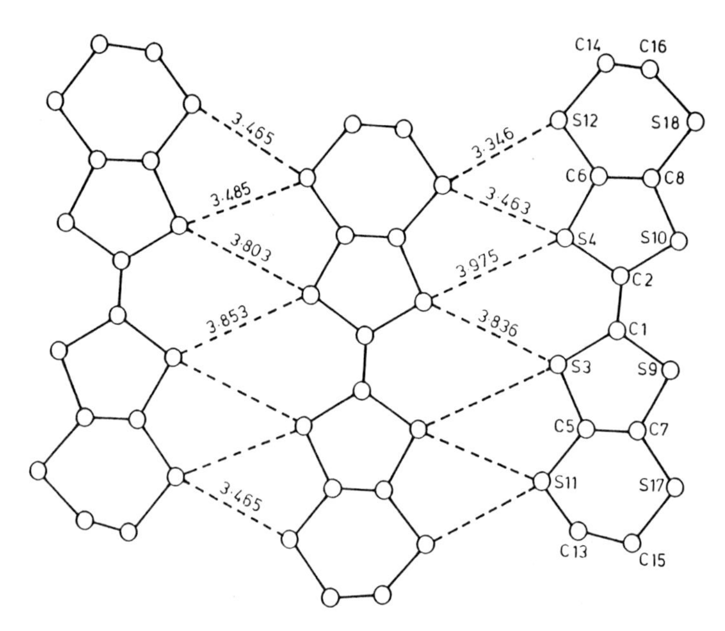

Fig. 6. Interstack S...S interactions in $(BEDT\text{-}TTF)_2AuBr_2$.[25]

With the β-$(BEDT\text{-}TTF)_2X$ series, superconducting T_c's have climbed from 1.4K (I_3^-),[26] 2.5K (IBr_2^-)[27] and 4.2 (AuI_2^-)[28] to 8K (high pressure form of the I_3 salt).[29] However, the highest presently known T_c among the $(BEDT\text{-}TTF)_2X$ is not a β-salt, but belongs to the κ-series.[10] In the latter, there are no stacks of cations, but only face-to-face pairs with the planes of the nearest neighbour dimers orthogonal to each other (Figure 7). With a segregation of the anions into layers, similar to that found in the β-series, the layers of organic cations are all but isotropic, as is the conductivity within that plane. Thus, overall, there is a clear evolution in the metallic charge transfer salts from the strongly one-dimensional TTF-TCNQ, with a transition at low temperatures to a CDW state, towards more and more two-dimensionally isotropic structures in the $(TMTSF)_2X$, β-$(BEDT\text{-}TTF)_2X$ and finally κ-$(BEDT\text{-}TTF)_2Cu(NCS)_2$, having superconducting ground states with higher and higher T_c.

Oxide Superconductors

Turning to the ternary and quaternary mixed-valency copper oxide class of superconductors, first uncovered by Bednorz and Müller in 1986,[30] the immediate feature of structural comparison with the charge transfer salts concerns dimensionality. Five distinct categories of copper oxide superconductors are known, as follows[31]:

			T_c/K
A.	$La_{2-x}M^{II}_{x}CuO_{4-y}$	M^{II} = Ca, Sr, Ba	20 - 40
B.	$LnBa_2Cu_3O_{7-y}$	Ln = Y, all lanthanides except Ce, Pr	70 - 95
C.	$Ln_{2-x}M^{IV}_{x}CuO_{4-y}$	Ln = Pr, Nd; M^{IV} = Ce, Th	20
D.	$Bi_2Sr_2Ca_{n-1}Cu_nO_{2n+4}$		80 - 110
E.	$Tl_2Ba_2Ca_{n-1}Cu_nO_{2n+4}$		80 - 125

Category A, the first to be discovered, has the K_2NiF_4 structure, with layers of corner-linked CuO_6 octahedra, tetragonally elongated perpendicular to the layers. The structure of the category B compounds is that of an anion-deficient perovskite, with vacancies ordered in such a way, that layers of square pyramidal CuO_5 are formed by joining their equatorial corners. Further oxide vacancies occur in chains, so that one-dimensional arrays are built up by sharing opposite corners of square-planar CuO_4. The category C phases are tetragonal, and can be considered as derived from the K_2NiF_4 structure by rotating alternate layers by 45°, and removing the apical O from the CuO_6 units. Each of the categories D and E consists of a homologous series of phases. Layers having the stoichiometry M_2O_2 with (on average) the NaCl structure are separated by from one to four layers of corner-sharing square CuO_4 units. The outer layers of the latter form the equatorial parts of CuO_5 square pyramids, while the inner layers are square planar.

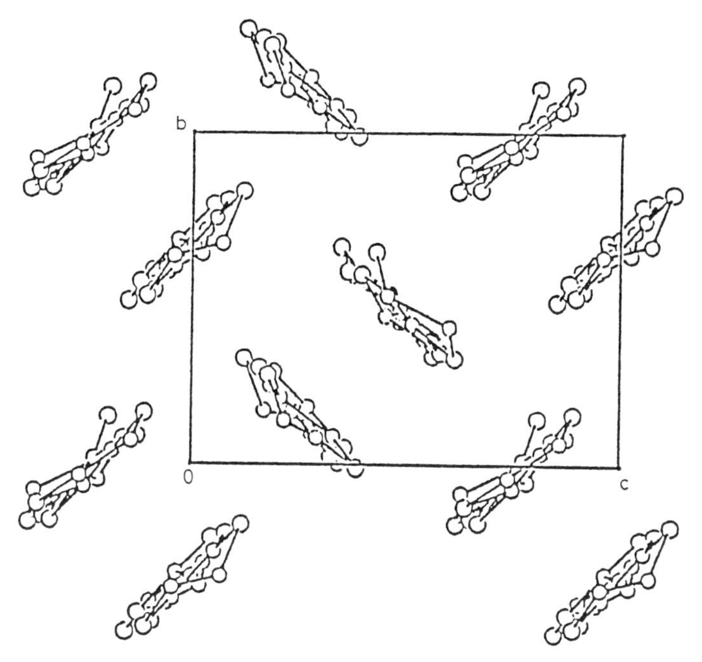

Fig. 7. Orthogonal BEDT-TTF dimers in $(BEDT\text{-}TTF)_2Cu(NCS)_2$[10].

The only structural feature shared by all five of the copper oxide categories is the presence of square-planar CuO_4 units linked through all four of their corners, to form an infinite layer with the overall stoichiometry CuO_2. The oxygen coordination number of the Cu atoms in these layers is sometimes strictly fourfold, as in C and the middle layers in the higher homologues of D and E. It may also be five, as in B or in the outer layers in D and E, or six, as in A. However, even when one or two oxide ions are coordinated to the Cu perpendicular to the plane of the layers, the axial Cu-O bond distance is always much greater than the equatorial. Hence the physical properties are very strongly two-dimensional. Recall that in the 10.5K molecular superconductor κ-(BEDT-TTF)$_2$Cu(NCS)$_2$, the transport within the layers of BEDT-TTF is also almost isotropic, but with an anisotropy of at least 100 parallel and perpendicular to the layers[10].

In four of the five categories of copper oxide superconductors, there is, also, a small breaking of the fourfold symmetry of the CuO_2 layer, the structural mechanism being different in each case. Only in the electron superconductor C is the tetragonal symmetry preserved. The separation of the CuO_2 layers is achieved either by large individual cations (e.g. Ln^{3+}, Ba^{2+}) with 8- or 9-fold oxygen coordination (A,B,C) or by complete layers with 1:1 cation-anion stoichiometry, and average (though not local) NaCl structure (D,E). In the 1:2:3 series (category B) there is the further complication of the one-dimensional CuO_4 chain parallel to the b-axis which, in the early days of the high T_c superconductors, was thought to be the origin of the enhancement in T_c from category A to B. However, the subsequent discovery of the Bi and Tl phases, with even higher T_c's and no Cu chains, indicated that this structural feature was not crucial to high T_c.

POLYMORPHISM AND PHASE TRANSITIONS

Both the high-T_c copper oxides and the molecular charge transfer salts give evidence of lattice instabilities, which may have a bearing on the occurrence and mechanism of the superconductivity.

Molecular charge transfer salts

Particularly in the (BEDT-TTF)$_2$X series, polymorphism is rife, as witness the nomenclature established for the various phases ($\alpha, \alpha', \beta, \beta', \beta'', \kappa$ etc). The extreme sensitivity of the electronic properties of the ground state to the fine details of intermolecular packing can be exemplified by the fact that superconductivity has been found inthe β and κ series, while the α phases, which differ by the introduction of a zig-zag alternation in the molecular stacks, are Mott-Hubbard insulators[25]. However, it is noteworthy that, in the case of the I_3 salt, heating crystals of the α-phase above 70°C for 24 hours transforms them, via a solid-state rearrangement, to the β'' phase with T_c = 8K![32]

A particularly interesting phenomenon occurs in the 10.5K superconductor κ-(BEDT-TTF)$_2$Cu(NCS)$_2$, whose resistivity first increases on cooling below room temperature, and passes through a maximum around 100K before becoming metallic in its temperature dependence (see Figure 2). Similar resistivity maxima have been seen in related compounds, such as (DMET)$_2$AuBr$_2$, which becomes superconducting under pressure (1.5 kbar, 1.6K)[33]. In the Cu(NCS)$_2$ salt, application of a modest hydrostatic pressure (~ 1 kbar) serves to sharpen the maximum, but, at higher pressures, it progressively diminishes, till above 5 kbar it disappears completely, leaving a metallic temperature dependence from room temperature down to the superconducting T_c (Figure 8)[34]. The presumption is, that there is a structural phase transformation, largely uncorrelated between the layers at ambient pressure, with the ambient-pressure low-temperature phase being stabilised over the whole temperature range above 5 kbar. High pressure crystallographic studies are in progress.

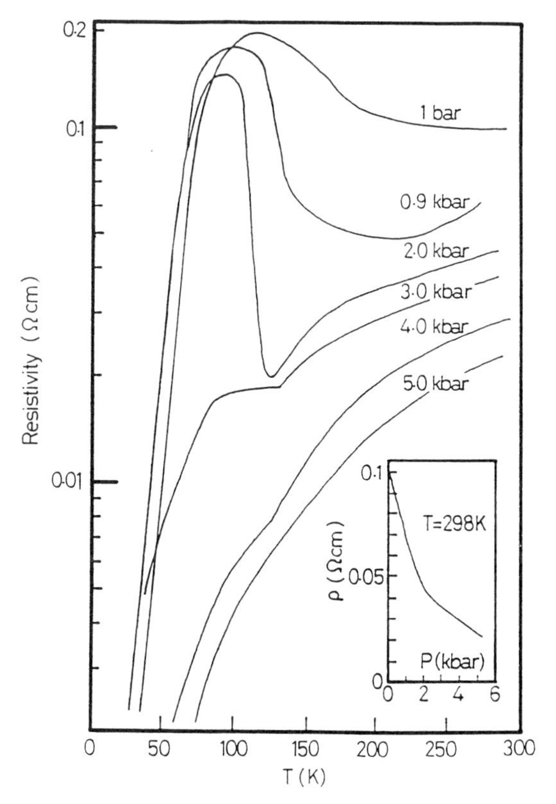

Fig. 8. Pressure dependence of the normal state resistivity of κ-$(BEDT\text{-}TTF)_2Cu(NCS)_2$[34]

Oxide Superconductors

When superconductivity was found in the $La_{2-x}Ba_xCuO_4$ system, it was thought, at first, that the exceptionally high T_c was due to coupling of the conduction electrons with a high density of low-energy phonons, arising from a soft mode associated with a structural phase transition from orthorhombic to tetragonal. Thus, La_2CuO_4 itself is orthorhombic (Cmca)(O) at room temperature, with a transition to the tetragonal (T) (I4/mmm) K_2NiF_4 structure at 540K. In the orthorhombic phase, the CuO_2 layers are puckered along [110] so that the Cu-O-Ca angle decreases from 180° to 173°. On doping with M^{II}, the O → T transition temperature falls, until, above x ~ 0.20, only the T phase occurs at all temperatures. It was originally thought to be significant that this is just the composition, beyond which superconductivity is no longer found. However, careful studies of the temperature dependence of the structural parameters, by high-resolution powder neutron diffraction, showed that, for the compositions that optimize the superconductivity (x ~ 0.15), both La:Sr[35] and La:Ba[36] compounds were already orthorhombic at T_c, e.g. in $La_{1.85}Sr_{0.15}CuO_4$, T_{OT} was 180K while T_c was 36K. An orthorhombic-tetragonal transition also occurs in the 1:2:3 (category B) phases, not as a function of temperature though, but rather as a function of either oxygen deficiency or disorder.

MAGNETIC PROPERTIES

A striking similarity between the charge transfer salts and oxides is the importance of electron correlations, and hence the proximity of superconducting and magnetic states in the phase diagrams, as a function of band filling or structural modifications. This arises from the fact that both types of compound are narrow-band systems.

Charge Transfer Salts

The superconducting β-$(BEDT\text{-}TTF)_2X$ are Pauli paramagnets in their normal states, but the α-salts with the same X have the strongly enhanced and temperature-dependent susceptibility, characteristic of a low-dimensional antiferromagnet. The magnitude of the susceptibility is exactly what one would expect for one unpaired electron per $(BEDT\text{-}TTF)_2$ dimer [37]. An even closer concurrence of magnetism and superconductivity was found in the early work of Jérome and his colleagues on $(TMTSF)_2PF_6$[20]. At ambient pressures, this compound is metallic down to 10K, where it undergoes a transition to a spin density wave (antiferromagnetic) state, and the resistivity increases again (Figure 2). If pressure is applied, though, the SDW state is suppressed, and the compound becomes superconducting. Jérome's chapter in the present volume contains much more detail on this topic.

Given that the molecular superconductors are salts, there is the further option of including magnetic ions into the lattice, by incorporating them into the anion X^-. Thus, TMTSF and BEDT-TTF salts of complex anions containing 3d elements have been prepared, such as $(BEDT\text{-}TTF)_2FeCl_4$ and $(BEDT\text{-}TTF)_3CuCl_4.H_2O$[38]. Unfortunately, no superconducting examples have been found yet, so it has not been possible to test for the coexistence of superconductivity and magnetism in this class of compound.

Oxide Superconductors

The facile transformation from SDW to superconductivity occurs in the copper oxide series, not as a function of pressure, like the TMTSF case, but as a function of band filling. Thus, La_2CuO_4, with a half-filled band based on Cu $3d(x^2\text{-}y^2)$ and O 2p, is antiferromagnetic, with $T_N \sim 270K$. On doping the La^{3+} site with Sr^{2+}, and so bringing about a corresponding decrease in the Cu-O band filling, T_N decreases very rapidly, so that for $x > \sim 0.06$ both T_N and the staggered moment tend to zero. It is significant that, for $0 < x < 0.05$, there is no detectable variation in the periodicity of the SDW, which remains [1/2 1/2 0][34]. Hence, one can talk of an antiferromagnetic state. Still, the limiting value of the moment for $x = 0$ ($\sim 0.5\mu_B$) is strongly reduced, by comparison with what one would expect for a localized $Cu^{2+}(3d^9)$ configuration. This is because of the very extensive Cu-O covalent mixing: indeed, the electrons at the Fermi surface have more O(2p) than Cu(3d) character. Rather analogous results are also found in the 1:2:3 series (category B), where the Cu-O band filling is systematically varied, by changing the occupancy of the O between the Cu forming the chains parallel to the b-axis, and breaking up the -Cu-O-Cu-O-Cu-. In the limit, e.g. in $YBa_2Cu_3O_6$, the latter Cu are all linearly coordinated, in a fashion characteristic of $Cu(I)(3d^{10})$, rather than square-planar, as expected for $Cu(III)(3d^8)$. The moments reside in the CuO_2 planes, and $YBa_2Cu_3O_6$ is a two-dimensional antiferromagnet[40]. With increasing x in $YBa_2Cu_3O_{6+x}$, the moment and T_N fall, and are replaced by superconductivity.

An alternative method for introducing localized magnetic moments into the Cu oxide superconductors is similar to the method used in the organics, though, in this case, by substituting the cations rather than anions. Thus, in category B, all Ln^{3+}, except Pr and Ce, yield superconductors, but the Ln moments do not order until far below T_c. Then they order antiferromagnetically, with no great effect on the superconductivity. The Ln(4f) magnetic electrons have only negligible interaction with those of the Cu-O planar band. In contrast, substitution of Cu by elements of the 3d block depresses T_c markedly.

CONCLUSION

As well as presenting a brief overview of the molecular and oxide superconductors themselves, this chapter has attempted to place both groups of compounds in the context of superconducting phases in general. It is highly probable that both groups have yet to achieve their maximum T_c. There are also some similarities between them, that may point the way to a further evolution of structures: dimensionality provides a common theme, and the occurrence of antiferromagnetic phases, differing only marginally in electronic and crystal structure from the superconducting ones, is a further point of similarity.

REFERENCES

1. H. Kammerlingh Onnes, Akad. van Weterschappen (Amsterdam), 14:113, 818 (1911).
2. H. Kammerlingh Onnes, Leiden Comm. 13:133d (1913).
3. J. R. Gavaler, Appl. Phys. Lett. 23:480 (1973).
4. J. E. Schooley, W. R. Hasler and M. L. Cohen, Phys. Rev. Lett. 12:474 (1964).
5. A. W. Sleight, J. L. Gillson and F. B. Bierstedt, Sol. St. Comm. 17:27 (1975).
6. D. C. Johnston, H. Prakash, W. H. Zachanasen and R. Viswathan, Mat. Res. Bull. 8:777 (1973).
7. M. B. Robin and P. Day, Adv. Inorg. Chem. & Radiochem. 10:248 (1967).
8. R. L. Greene, G. B. Street and L. J. Slater, Phys. Rev. Lett. 34:577 (1975).
9. K. Bechgaard, K. Carneiro, F. B. Rasmussen, M. Olsen, G. Rinsdorf, C. S. Jacobsen, H. J. Pedersen and J. C. Scott, J. Amer. Chem. Soc. 103:2440 (1981).
10. H. Uragama, H. Yamochi, G. Saito, K. Nozawa, T. Sugamo, M. Kinoshita, S. Sato, K. Oshima, A. Kawamoto and J. Tanaka, Chem. Lett. 55 (1988).
11. T. Mori and H. Inokuchi, Sol. St. Comm. 64:335 (1987); M. Kurmoo, M. J. Rosseinsky, P. Day, P. Auban, W. Kang, D. Jerome and P. Batail, Synth. Met. 27:A425 (1988).
12. M. Bousseau et al. J. Amer. Chem. Soc. 108:1908 (1986).
13. For reviews, see e.g. "The Physics and Chemistry of Low Dimensional Solids", L. Alcacer, ed., D. Reidel Publ. Co. (1980).
14. J. B. Torrance, Ann. N.Y. Acad. Sci. 313:210 (1978).
15. R. E. Peierls, "Quantum Theory of Solids", Clarendon Press, (1964).
16. J. Jaud, D. Chasseau, J. Gaultier and C. Hauw, C. R. Acad. Sci. (Paris) C278:769 (1974).
17. See summaries in "Physics & Chemistry of One-Dimensional Metals", H. J. Keller, ed., Plenum Press (1977).
18. N. Thorp et al, J. Physique, Colloques C3, 44:1017 (1983).
19. D. Jerome and H. J. Schultz, Adv. Phys. 31:299 (1982).
20. D. Jerome, A. Mazaud, M. Ribault and K. Bechgaard, J. Physique Lett. 41:L95 (1980).
21. A. N. Bloch, D. O. Cowan, K. Bechgaard, R. E. Pyke and R. H. Banks, Phys. Rev. Lett. 34:1561 (1975).
22. T. E. Phillips, T. J. Kistenmacher, A. N. Bloch and D. O. Cowan, Chem. Commun. 334 (1976).
23. V. K. S. Shante et al, Bull. Amer. Phys. Soc. 21:287 (1976).
24. P. M. Grant, Phys. Rev. B26:6888 (1982).
25. D. Talham, M. Kurmoo, P. Day, A. Obertelli, I. Parker, R. H. Friend, A. Stringer and J. A. K. Howard, Sol. St. Commun. 61:459 (1987).

26. E. B. Yagubskii et al, JETP Lett. 39:12 (1984).
27. J. M. Williams et al, Inorg. Chem. 23:3839 (1984).
28. J. M. Williams et al, Physica B+C (1985).
29. V. N. Lanklin et al, JETP Lett. 41:81 (1985).
30. J. G. Bednorz and K. A. Müller, Z. Phys. B64:189 (1986).
31. Since there is a vast literature on high T_c Cu oxides, we do no include a list here. For recent reviews concentrating on structures with many references, wee e.g. J. M. Williams et al, Acc. Chem. Rés. 21:1 (1988); C. N. R. Rao and B. Raveau, ibid., 22:106 (1989).
32. G. O. Barem et al, Pis'ma Zh. Eksp. Teor. Fiz. 44:293 (1986).
33. K. Kikuchi et al, J. Phys. Soc. Jap. 56:3436 (1989).
34. I. D. Parker, R. H. Friend, M. Kurmoo and P. Day, J. Phys.: Cond. Matt. in press
35. P. Day, M. J. Rosseinsky, K. Prassides, W. I. F. David, O. More and A. Saper, J. Phys. C: Sol. St. Phys. 20:L729 (1987).
36. D. McK. Paul et al, Phys. Rev. Lett. 58:1976 (1987).
37. S. D. Obertelli, R. H. Friend, D. R. Talham, M. Kurmoo and P. Day, Synth. Mat. 27:A375 (1988); idem, J. Phys.: Cond. Matt., in press.
38. T. Mallah, C. Hollis, S. Bott, P. Day and M. Kurmoo, Synth. Mat. 27:381 (1988); idem J. Chem. Soc., Dalton Trans., in press.
39. M. J. Rosseinsky, K. Prassides and P. Day, to be published.
40. P. Burlet et al, Physica C 153:1115 (1988).

OPTICAL STUDIES OF THE INTERPLAY BETWEEN ELECTRON- LATTICE AND ELECTRON-ELECTRON INTERACTIONS IN ORGANIC CONDUCTORS AND SUPERCONDUCTORS

R. Bozio, M. Meneghetti, D. Pedron, and C. Pecile

Department of Physical Chemistry
University of Padova
2, Via Loredan
35131 Padova, ITALY

INTRODUCTION

In another paper [1] in this volume we have discussed some basic notions of the spectroscopy of charge transfer (CT) crystals and molecular conductors, mainly based on very simple dimeric models. The practical applicability of these models is limited to dimerized systems with localized electron states. This situation most often occurs for half-filled systems with intermediate to strong electron correlations.

In this paper we shall be mostly concerned with the spectroscopy of conducting molecular materials characterized by delocalized electron states. More specifically, we shall be dealing with quarter-filled systems, typified by the well-known families of organic conductors and superconductors, namely, the Bechgaard salts: $(TMTSF)_2X$ and their sulfur analogs $(TMTTF)_2X$ (X is an inorganic anion). Furthermore, our emphasis will not be in the application of spectroscopic methods for investigating the crystal structural properties. These properties are, rather, assumed to be known at the outset, as it is the case for the above-mentioned materials, and the investigation of the optical properties is used as one of the most powerful means of linking the macroscopic physical behavior with the microscopic properties. Here our main concern is on obtaining some insight on the interplay between electron-lattice (i.e. coupling of electrons with phonons and with static crystal potentials) and electron-electron interactions, a subject of much current interest for CT conductors, as well as for conducting polymers [2].

BACKGROUND

When discussing the optical properties of delocalized electron systems, one needed elementary reference case is that of a regular tight-binding chain. Consider a chain of N molecules with regular lattice spacing d, and a single orbital per molecular site, with nearest-neighbor transfer integral t. For a one-dimensional band, filled with N/2 electrons or holes, i.e. a quarter-filled band, use of the well-known energy eigenvalues $\varepsilon_k = 2\,t\,\cos(k\,d)$ in a self-consistent calculation [3] of the dielectric function $\hat{\varepsilon}(\omega)$ yields:

$$\hat{\varepsilon}(\omega) = \varepsilon_\infty - 4\pi \frac{N}{\hbar^2} \frac{2\sqrt{2}}{\pi} \frac{t\, d^2 e^2}{(\omega^2 + i\,\omega\gamma)} \tag{1}$$

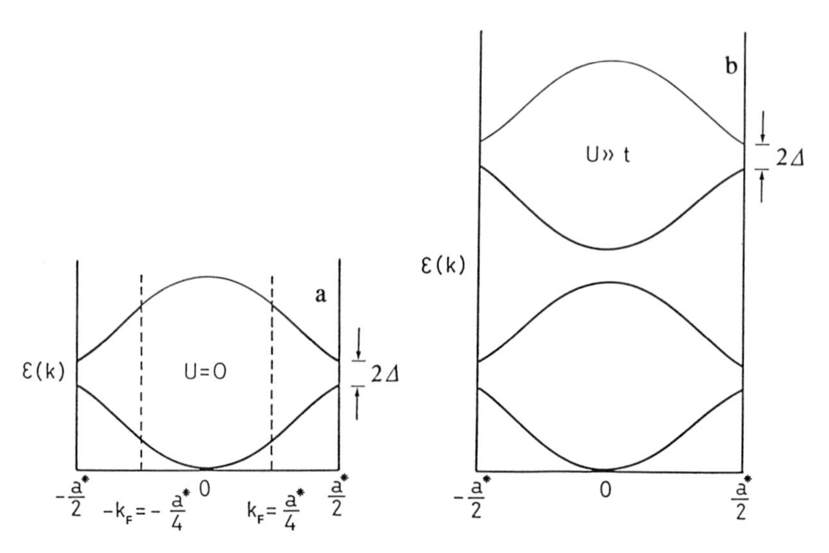

Fig. 1. Schematic band structure for a dimerized tight-binding chain: (a) U = 0 and (b) U >> t. Here a* = π/d.

It is seen that, compared to a collection of non-interacting dimers, each with one electron, (cf. Eq. (4) of Ref. [1]) the oscillator strength is increased by a factor $4\sqrt{2}/\pi \approx 1.8$. Remember, in fact, that the sum rule is $F_o = -e^2 d^2 \langle h_t \rangle_o$ (cf. Eq. (10) of Ref. [1], and Ref. [4]), and we have additional terms in the electron transfer Hamiltonian h_t, accounting for the delocalization of electrons over the whole chain.

The dimerization of the chain opens a gap at half-zone, that is, away from the Fermi surface. The band structure in the folded Brillouin zone now appears as shown in Fig. 1a, and the intraband contribution to the dielectric function becomes:

$$\hat{\varepsilon}(\omega) = \varepsilon_\infty - 4\pi \frac{N\sqrt{2}}{\pi\hbar^2} \frac{(4t^2 - \Delta^2)}{(4t^2 + \Delta^2)^{1/2}} \frac{d^2 e^2}{(\omega^2 + i\omega\gamma)} \tag{2}$$

where Δ is the amplitude of the dimerization gap. Eq. (2) and Fig. 2 show that, as the gap increases, while keeping constant the average transfer integral t, the oscillator strength decreases, corresponding to a transfer of intensity to the interband transitions. Note also that, for dimerization gap amplitudes on the order of those commonly found in quarter-filled CT conductors (≤ 10% of the total bandwidth), the loss of intraband oscillator strength is only a few percent. One can therefore conclude that small-amplitude chain dimerizations do not affect appreciably the spectra of quarter-filled systems of uncorrelated electrons.

The calculation of the optical properties of correlated electron systems is still a challenging problem in many-body physics, since it requires knowledge of the excited-state wave functions, in addition to the ground-state wavefunction. No general solution is available for a one-dimensional system with arbitrary coupling strengths and band fillings. With a few exceptions [4-6], the available perturbative calculations deal with a half-filled Hubbard system [7]. Furthermore, they appear to be of limited relevance for a discussion of the optical properties of molecular CT conductors since, by a number of experimental indications, these materials fall in the so-called *intermediate regime:* that means that the direct electron-electron correlations are neither negligible, nor overwhelming, with respect to the one-electron interactions. Exact solutions, in the numerical sense, have been provided by the valence-bond approach, applied to finite-size rings and chain segments with $1 > \rho \geq 0.5$ (ρ being the number of electrons per molecule) and arbitrary values of the interaction parameters [4,8].

Besides some uncertainty, inherent in extrapolating results for finite-size systems to infinite ones, these calculations did not consider the effect on the optical properties of possible chain distortions, nor did they include the coupling of electrons to inter- and intra-molecular phonons. More recent efforts [9] have been focusing on half-filled systems, motivated by the current interest on linear and non-linear excitations in conducting polymers.

It is useful, for the ensuing discussion, to keep in mind one well-known elementary result of the Hubbard model [10]. When the on-site correlation energy is much greater than the bandwidth ($U >> 4t$), each one-electron band splits into a band of singly occupied states at lower energy, and a band of doubly occupied states at energy approximately U above. Correspondingly, the Fermi wave-vector shifts to twice its value for the independent electron case. Thus, as shown in Fig. 1b, an opening of a dimerization gap in a strongly correlated quarter-filled system would empty the density of states at the Fermi surface. Therefore, one can easily understand that, in the presence of strong on-site correlations, the effect of a stack dimerization on the optical properties of a quarter-filled system is much greater, than in the opposite limit of negligible correlations.

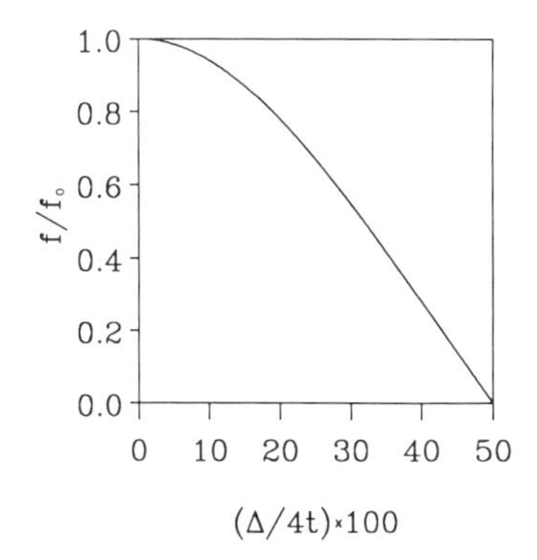

Fig. 2. Intraband contribution to the oscillator strength of a quarter-filled dimerized chain vs. the dimerization gap Δ. f_o is the oscillator strength for a regular chain.

One should also mention that a more realistic modeling of a quarter-filled system would require the inclusion of at least the nearest-neighbor Coulomb repulsion terms in addition to the on-site ones. That means switching from the original Hubbard to the extended Hubbard model Hamiltonian:

$$H_{EH} = -t \sum_{n,\sigma}(a_{n,\sigma}{}^{\dagger} a_{n+1,\sigma} + h.c.) + U \sum_{n} n_{n\uparrow} n_{n\downarrow} + V \sum_{n} n_n n_{n+1}. \qquad (3)$$

It is known that the excitation spectrum of a quarter-filled extended Hubbard system with $U >> 4t$ exhibits a gap of order V, even in the absence of dimerization [8].

The argument, just briefly outlined, show that, in a quarter-filled system, the presence of strong on-site correlations leads to deviations of the optical properties from the Drude behavior [3] expected for independent electrons. Such deviations are not as large, as those predicted for a half-filled system, and, in particular, do not relate directly to U itself, but, rather, to dimeric lattice distortions and, possibly, to the nearest-neighbor repulsion V.

To the best of our knowledge, all the quarter-filled molecular CT conductors whose spectral properties have been measured (whether classified as metals or not) exhibit marked deviations form the Drude model [11]. The most apparent among them are: (i) a shift of the conductivity maximum from zero frequency, and the associated appearance of optical gaps or pseudo-gaps, (ii) the appearance of structures in the spectrum, attributable to electron-molecular vibration and electron-phonon coupling; (iii) deviations from the simple partial sum rule.

In the last few years, we have endeavored to extract some basic information on the interplay between electron-electron and electron-lattice interactions from the optical spectra of slightly dimerized quarter-filled molecular conductors characterized by narrow optical gaps [12-15]. The $(TMTSF)_2X$ and $(TMTTF)_2X$ salts, on which we have mainly focused so far, do belong to this category, since their isomorphous room-temperature structures are characterized by a lattice dimerization, whose amplitude varies from salt to salt. In general, the dimerization is greater in the sulfur-containing materials. For instance, the infrared conductivity spectra [14,16], polarized along the stack direction, show increasing deviations from the Drude behavior in the order $(TMTSF)_2PF_6 < (TMTTF)_2Br < (TMTTF)_2PF_6$. In particular, as the d.c. conductivity decreases, in going from the first to the last of the above compounds, the infrared conductivity peak shifts to higher frequencies, while, simultaneously, the structure in the mid-infrared region becomes increasingly pronounced.

A SPECTROSCOPIC VIEW OF THE INTERACTIONS IN THE $(TMTSF)_2X$ AND $(TMTTF)_2X$ SALTS

A key point of our approach is to investigate quantitatively the apparent scaling between the location of the conductivity maximum and the intensity of the structures induced by the coupling with the molecular vibrations [1]. This is made possible, on the one hand, by collecting high-quality, high-resolution polarized reflectance spectra in the range from the near to the far infrared and, on the other hand, by using, for the data analysis, a dimerized molecular chain model we have previously developed [13]. While kept at a fairly low level of complexity, such a model includes enough structural and electronic features, as to allow a meaningful comparison with the optical data of real systems.

Dimerized molecular chain model

The model was originally developed for 1-D molecular chain systems carrying general two-fold commensurate charge density waves, that is, bond-order waves and charge density waves centered between, and on the sites, respectively, both with arbitrary amplitude [13]. Structurally, this corresponds to including a lattice dimerization or a site alternation, or both. It should be noted that, by virtue of their crystal symmetry, the Bechgaard salts can only bear a bond-order wave, since all molecular sites are equivalent. In the following, we shall then briefly outline the model in the corresponding reduced form.

The Hamiltonian we consider is as follows:

$$H = H_E + H_{EV} + H_V \tag{5}$$

$$H_E = -t \sum_{n,\sigma} (a_{n+1,\sigma}{}^{\dagger} a_{n,\sigma} + \text{h.c.}) + \Delta_b \sum_{n,\sigma} (-1)^n (a_{n+1,\sigma}{}^{\dagger} a_{n,\sigma} + \text{h.c.}) \tag{6}$$

where

$$\Delta_b = B_x + 2\,(\partial t/\partial u)_o u_o. \tag{7}$$

H_E is the one-electron Hamiltonian of a molecular chain exhibiting a lattice dimerization, wnose amplitude is specified by the longitudinal displacement u_o of the molecules from their location in the undistorted chain. The electrons are also subjected to a potential, say from the inorganic anion chains and from some mean-field electron correlation, of period twice the regular chain spacing d, acting on the bonds (i.e. between sites) with an amplitude B_x.

The linear coupling of the electrons to an acoustic phonon, and to the intramolecular modes, is expressed in the usual way [1] by:

$$H_{EV} = \sum_{i}\sum_{n,\sigma} g_i q_{n,i} a_{n,\sigma}^\dagger a_{n,\sigma} - \sum_{n} (\partial t / \partial u)_o (u_{n+1} - u_n)(a_{n+1,\sigma}^\dagger a_{n,\sigma} + h.c.) \qquad (8)$$

where the linear electron-molecular vibration (e-mv) coupling constants g_i have been introduced. H_V is the vibrational Hamiltonian in the harmonic approximation.

Diagonalization of the Hamiltonian H_E yields the two tight-binding bands shown schematically in Fig. 1a:

$$E_{k,n} = (-1)^{n+1}(\varepsilon_k^2 + |\Delta_k|^2)^{1/2} \qquad (n = 0,1) \qquad (9)$$

where $\Delta_k = i\, \Delta_b \sin(k\, d)$.

The parameters u_o, $(\partial\, t / \partial\, u)_o$, B_x and the amplitude $2\,\Delta_b$ of the gap, that opens at the boundaries of the reduced zone, are all connected in a self-consistent way, so that knowledge of any two of them, say from structural and spectroscopic data, uniquely determines the other two [13]. This allows us to estimate the individual contributions of the phonons and of the static potential to the total gap.

The interband and vibronic contributions to the infrared properties of the model system have been calculated in the adiabatic, linear-response approximation. Two possible schemes for the occupation of the band states by a number of electrons or holes $N_p = N/2$ (corresponding to half a carrier per molecule) have been considered: (i) the case of regular fermion particles with spin, where the lower band only is half filled, (ii) that of spinless fermions, with the lower band completely full. Although no electron correlation term is explicitly included in the Hamiltonian, the latter case represents the situation that is attained, when the on-site correlation of an Hubbard model is $U >> t$.

Actual calculations for the two limit cases above include also Drude-like intraband contributions, with a plasma frequency calculated self-consistently from the band structure [3]. As already discussed (cf. Eq. (2)), in the $U = 0$ limit this is the dominating contribution to the optical properties. However, intraband terms are also important in the large-U limit, where they account for the relevant contribution of the carriers thermally excited across the narrow gaps at high temperatures.

Experimental procedures

When high-quality quantitative optical data are needed for organic conductors that absorb strongly in the infrared [11], the only viable technique consists in measuring the polarized reflectance at normal (or near-normal) incidence from single crystals. The minimum required sample dimension is typically in the order of 1 mm for the visible or near infrared, and 4 mm for the middle- to far-infrared measurements.

In the usual procedure, each measurement on the sample is followed immediately by a measurement of the reflectance from a metallic mirror taken under the same instrumental conditions. This allows one to compensate for possible fluctuations in the spectrometer response. The crystalline sample is then coated with a thin layer of gold, and its reflectance is measured again. The final data are obtained by normalizing the spectrum of the sample to that obtained after coating. In this way, one can compensate for the loss of reflection intensity, due to surface roughness and irregularities.

By using suitable extrapolations [3] of the experimental data, beyond the upper and lower frequency limits of the measurements, and carrying out a Kramers-Kronig analysis, it is possible to obtain the dispersion of the real parts of the dielectric function and of the conductivity. The latter function is particularly suited for a comparison with the results of theoretical models, and for the determination of physical parameters by model fittings.

Experimental results and data analysis

Let us consider comparing the experimental stack-axis polarized reflectance and conductivity spectra with calculations in the two opposite limits of large and vanishing U. We choose $(TMTSF)_2PF_6$ because it is representative of the Bechgaard salts with metallic behavior. The influence of the interstack interactions is thought to be minimized by considering the optical data at room temperature; also the calculations refer to room temperature.

Input data for the calculation were the known room-temperature structure [17] and vibrational studies of TMTSF [18]. As suggested by the structural data, the dimerization amplitude u_o was taken to be 0.03 Å. The parameters, whose value is adjustable and determined by the fitting of the experimental data, are: the transfer integral t, the gap amplitude Δ_b, the e-mv coupling constants g_i and the core dielectric constant ε_∞. Use of the self-consistent relations mentioned before allows us to determine the e-p parameter $(\partial t / \partial u)_o$ and the percentage contribution of the static potential B_x to the gap Δ_b.

The experimental and calculated reflectance spectra in the range 20-10,000 cm^{-1} and conductivity spectra in the range 100-5000 cm^{-1} are shown in Fig. 3. One notices immediately that the $U = 0$ calculations fail to reproduce several features of the observed data. In particular, the flattening of the conductivity spectrum below about 1000 cm^{-1} is missed, and the rich vibronic structure is completely absent in the calculated spectrum for $U = 0$. This is because the vibronic structures derive only from the coupling to interband transitions. With a small amplitude gap and a relatively large electron transfer integral, the latter transitions appear as a weak hump in reflectance and conductivity around 8000 cm^{-1}. Higher-order e-mv coupling to the intraband transitions, leading to the so-called Holstein effect [19], are not included in our calculation, but are expected to give rise to structures certainly weaker than the observed ones.

As shown in Fig. 3, the calculation for the large-U case allows a satisfactory fit of both the conductivity and the reflectance spectra. Note that the plasma edge in reflectance, the conductivity levels in the infrared, and the e-mv structure are simultaneously fitted to a reasonable level of accuracy, although further refinement of the parameter values is still possible. The parameters used are collected in Table 1. The corresponding picture of the electronic structure exhibits a total bandwidth of 1 eV, with a small gap of amplitude $2\,\Delta_b = 66$ meV, associated with a bond-order wave. The gap opens at the Fermi surface, which is shifted

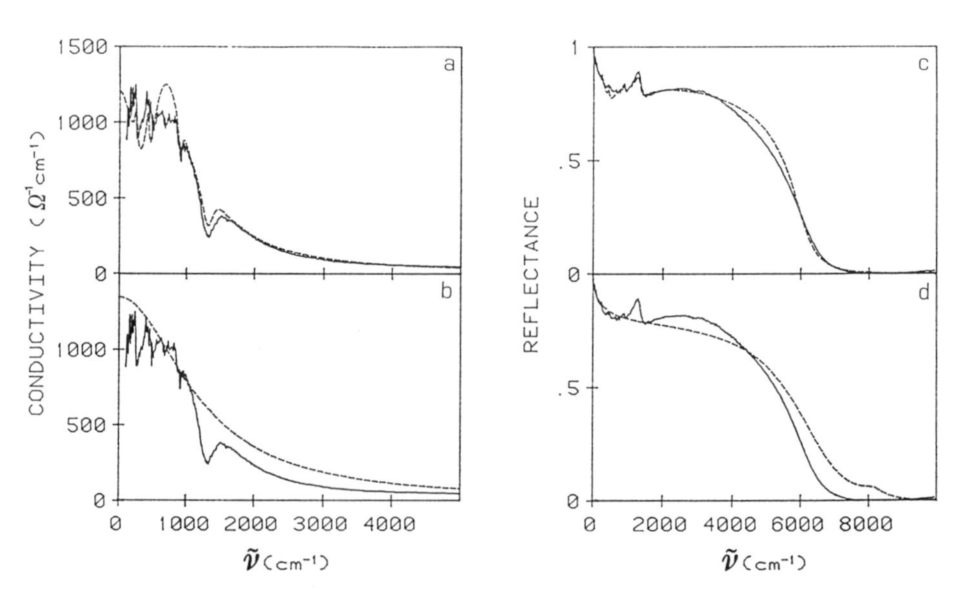

Fig. 3. Experimental (full line) and calculated (dashed line) infrared conductivity and reflectance spectra of $(TMTSF)_2PF_6$ at room temperature. Calculated spectra shown in panels (a) and (c) refer to the larger U case, whereas those of panels (b) and (d) refer to the $U = 0$ case.

to the reduced-zone boundaries for spinless fermions. The experimentally observed d.c. conductivity is easily accounted for, in terms of the contribution from thermally excited carriers.

Consistent with our previously stated aim, we should now check the scaling between the location of the conductivity maximum and the amplitude of the vibronic structures, by comparing the results for systems with different degrees of dimerization.

Measurements and calculations have been performed for the series of compounds including $(TMTSF)_2ClO_4$, $(TMTSF)_2PF_6$, $(TMTTF)_2Br$ and $(TMTTF)_2PF_6$ (Fig. 4). The dimerization parameter, taken from crystallographic data [17,20-22] as the difference between the two alternating interplanar distances, is reported in Table 2. For the sake of comparison with our fitting parameters, we have also reported the average value of the intrastack transfer integral, calculated [23,24] by quantum-chemical methods using different orbital basis sets.

The first apparent feature of the large-U calculations is their ability to reproduce the increasing amplitude of the vibronic structures, as the conductivity maximum shifts to higher frequencies in the compounds that have larger dimerization. It should be noted that the same set of e-mv coupling constants have been used for all the TMTSF salts and, with some expected modifications, also for the TMTTF salts.

Let us first consider the values assumed by the fitting parameters. The values of the estimated average transfer integral t scale properly with those of the calculated ones, and with

Table 1. Values of the parameters used in the calculation of the T = 300 K spectra of $(TMTSF)_2PF_6$ for the U —> ∞ case.

$t = 0.25$ eV $\Delta_b = 33$ meV $\varepsilon_\infty = 2.00$ $\Gamma_{intra}{}^a = 300\ cm^{-1}$ $\Gamma_{inter}{}^a = 300\ cm^{-1}$

mode no.	bare freq./cm^{-1}	g_i /meV	$\gamma_i{}^a$ / cm^{-1}
ν_3 (a_g)	1599	20	20
ν_4 (a_g)	1469	100	20
ν_7 (a_g)	1060	10	20
ν_8 (a_g)	920	30	20
ν_9 (a_g)	452	30	20
ν_{10} (a_g)	282	30	20

[a] Γ_{intra}, Γ_{inter} and γ_i are the constant damping factors for intraband and interband transitions and for the vibrational modes, respectively.

Table 2. Structural and electronic interaction parameters for $(TMTSF)_2X$ and $(TMTTF)_2X$ salts.

				This work				
Compound	d^a /Å	$u_o{}^a$ /Å	$t_{calc}{}^b$ /eV	t /eV	Δ_b /meV	B_x /%	$(\partial t/\partial u)_o$ /(eV/Å)	ε_∞
$(TMTSF)_2ClO_4$	3.635	0.01	0.24, 0.37	0.27	33	88	0.19	2.30
$(TMTSF)_2PF_6$	3.645	0.03	0.23, 0.36	0.25	33	1	0.55	2.00
$(TMTTF)_2Br$	3.515	0.03	0.13, 0.24	0.20	43	69	0.22	2.00
$(TMTTF)_2PF_6$	3.570	0.10	0.11	0.17	84	0	0.42	2.00

[a] From Refs. [17, 20-22].
[b] From Ref. [24] (left values) and Ref. [23] (right values).

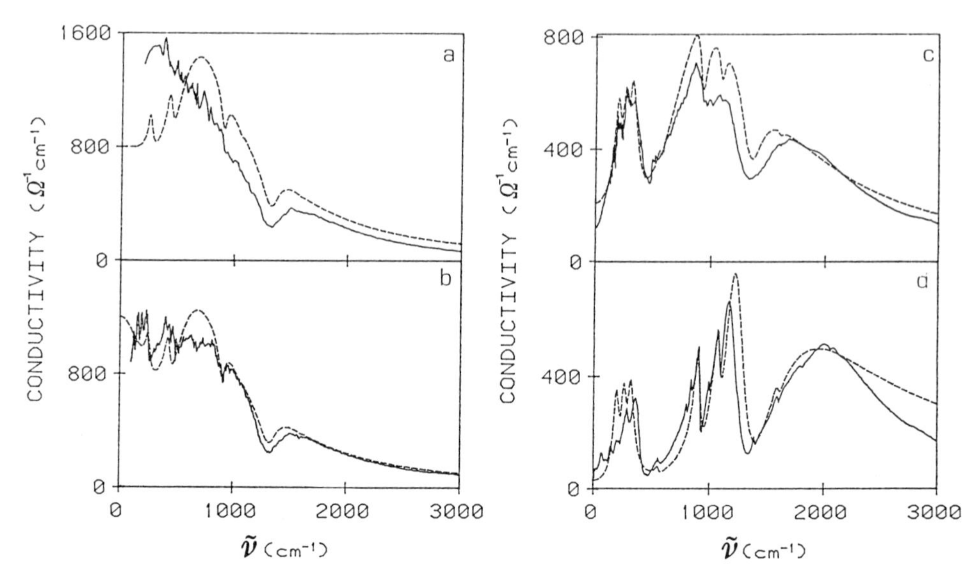

Fig. 4. Room-temperature conductivity spectra of: (a) $(TMTSF)_2ClO_4$, (b) $(TMTSF)_2PF_6$, (c) $(TMTTF)_2Br$, and (d) $(TMTTF)_2PF_6$. Full line: experimental data; dashed line: calculated spectra for the large-U case.

the average interplanar spacing d = a / 2. The estimated values for the TMTTF salts are possibly a little larger than expected. The width of the gap Δ_b displays the right tendency to increase, as the dimerization amplitude u_o increases. However, there is not a simple linear relation between the two quantities. This might be explained by the fact that, given a pair of values for u_o, from the structure and for Δ_b from the spectra, the contribution of the static potential B_x, possibly ascribed to the anions, is determined, and turns out to vary widely from salt to salt. We do not mean, however, that this, and the corresponding large variations in the value of the e-p coupling parameter $(\partial t/\partial u)_0$, carry any real physical implication. To a large extent, it is rather thought to reflect the uncertainties in parametrizing the dimerization amplitude in such complicated structures.

As for a more detailed comparison between theory and experiment, we cannot expect, of course, that all the features of the measured spectra can be accurately reproduced with a model as simple and rough as the one we have used for our calculations. For instance, in the case of $(TMTSF)_2ClO_4$, in order to reproduce the amplitudes of the observed vibronic structures, we have to use a gap parameter that causes the calculated conductivity maximum to be higher in frequency than the observed one. This kind of discrepancy is much less pronounced in the case of $(TMTSF)_2PF_6$, although both the gap parameter and the transfer integral are very similar to those of the ClO_4^- salt.

This might be explained by the fact that the deviations from an ideal 1-D behavior are stronger for the ClO_4^- than for the PF_6^- salt. Experimental support for that comes from the structural data [17,20] (interstack lattice parameter **b** = 7.678 Å (ClO_4^-); 7.711 Å (PF_6^-)), and from the observation of a larger transverse plasma frequency in the low-temperature spectra of the former salt [16]. The presence of relevant interchain interactions would affect the spectra in two ways. First, it would make the thermodynamic gap smaller than the optical one, thus enhancing the intraband contributions at high temperatures. Second, it would imply additional interband transitions if, as suggested [24], the band structure is not accurately represented by a rectangular model.

This, as well as other difficulties that might be encountered in carrying the analysis to an increasing degree of sophistication, do not affect the achievement of our primary goal. We have, in fact, shown, that the observed scaling between the location of the conductivity maximum and the amplitude of the vibronic structures cannot be explained, unless we account for the existence of a narrow gap at the Fermi surface. The amplitude of this gap increases with increasing degree of dimerization of the stacks. Its pronounced effects on the optical spectra of a 1/4-filled system can be modeled, by assuming that the double occupancy of the molecular sites is effectively forbidden, corresponding to the strong correlation limit U >> t.

The question now is: how strong is strong correlation? To get some hints on this problem we have paralleled our analyses, based on the infinite dimerized chain model, by the somewhat complementary approach of using models based on finite-size molecular clusters.

Tetramer model analysis: a complementary approach

In the preceding paper [1] we have used the symmetric dimer with two electrons as the simplest molecular cluster analog of a 1/2-filled band system. For 1/4-filled systems the trivial choice is that of a tetramer with two electrons, as a system of minimal dimension that still allows one to include electron correlations. The extended Hubbard Hamiltonian for such a tetrameric model can easily be solved numerically, for arbitrary values of the interaction parameters, to obtain exact eigenvalues and eigenvectors, from which the optical properties can be calculated [25]. To apply a tetramer model for the analysis of optical data of Bechgaard salts we must, on the one hand, choose a system with a tetramerized structure and, on the other hand, make some tailoring on the original extended Hubbard model, in order to incorporate in it some important features of the real structure. In addition, we have to include the e-mv coupling in the model Hamiltonian. Details of the model and of the calculation of the infrared and Raman properties, as well as on the experimental results, will be given elsewhere. The chosen system is $(TMTSF)_2ReO_4$ in its low-temperature phase, below the anion ordering transition [26]. Direct structural evidence [27] shows that, in this phase, the TMTSF stacks are tetramerized.

In defining our tetramer model, sketched in Fig. 5, we have accounted for the pattern of interactions suggested by the structural data. Thus, different transfer integrals t and t' have been considered for the interaction of B-B and A-B pairs, respectively, with t > t' (cf. Fig. 5). In addition, a difference Δ in the energy of sites A and B has been introduced, to account for the fact that the anion-donor interaction is stronger for the central pair of sites. The latter feature is expected to induce a localization of positive charge on the central molecules, whose intramolecular vibrational frequencies and linear e-mv coupling constants will, therefore, be different from those of the end ones. To have this feature directly built into our model, we have included the e-mv coupling, both at the linear and the quadratic levels.

In addition to calculating the contributions to the optical conductivity of the CT bands and of the vibronically induced absorptions, we have also calculated the renormalized frequencies of the Raman-active tetramer modes. In fact, it can easily be seen that, for each e-mv coupled

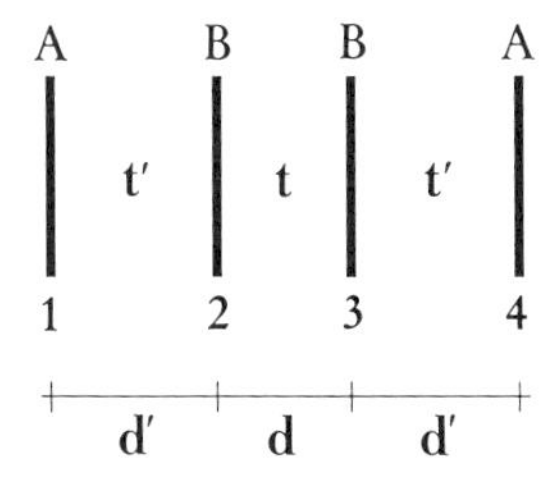

Fig. 5. Schematic drawing of a tetramer unit of four chemically identical molecules. A and B moieties are not equivalent by symmetry, and have different site energies. t and t' are transfer integrals; d and d' denote the distances between molecular planes.

mode of TMTSF, there are two IR-active and two Raman-active modes, corresponding to the mixing of the *ungerade* and *gerade* tetramer coordinates, respectively. All of these modes are perturbed by the vibronic coupling. Comparison of the calculated Raman frequencies with recent high-quality resonance Raman data [28] will allow us to check further the reliability of our analysis.

The stack-axis infrared conductivity spectrum, obtained from a Kramers-Kronig transformation of polarized reflectance data, is shown in Fig. 6 for the frequency range 500-2500 cm^{-1}. The conductivity, calculated according to the tetramer model, is plotted using the ordinate scale indicated on the right side. The latter choice is because the calculated conductivity is smaller than the observed one, by more than a factor of two. This reflects an intrinsic limitation of the cluster models. In fact, in these models the oscillator strength and the energy of the lowest CT transition are both strongly affected by the transfer interaction inside the cluster. In an infinite chain, made up of interacting clusters, the oscillator strength depends on the average transfer interaction [1,4], whereas the excitation energy is mostly affected by the difference between the interactions inside a cluster and those between clusters. In our calculation, we have chosen to reproduce the observed frequencies, and not the oscillator strengths.

The peak frequencies of both the vibronic and CT absorptions are in excellent agreement with the observed ones. The relative values of the integrated intensities are also well reproduced, although the width of the calculated CT band is narrower than the observed one and, correspondingly, the peak conductivity is higher. This is likely related to the fact that the interaction between tetramers in the TMTSF chain spreads each tetramer state into a band [26].

With the parameters used in the calculation, the predicted doublet structure of the vibronic bands becomes evident only for the TMTSF mode with strongest coupling, namely, the ν_4 mode, whose bare frequency is at 1469 cm^{-1} for the molecule with a formal charge of +0.5. The calculated Raman-active frequencies of the tetramer modes, corresponding to the TMTSF modes ν_3, ν_{28}, and ν_4, are shown in Fig. 7, as vertical bars superimposed on the low-temperature Raman spectrum [28] of $(TMTSF)_2ReO_4$. They have been calculated using the same set of parameters as for the IR spectra, and the rather good agreement with the experiment makes us more confident of the estimated parameter values.

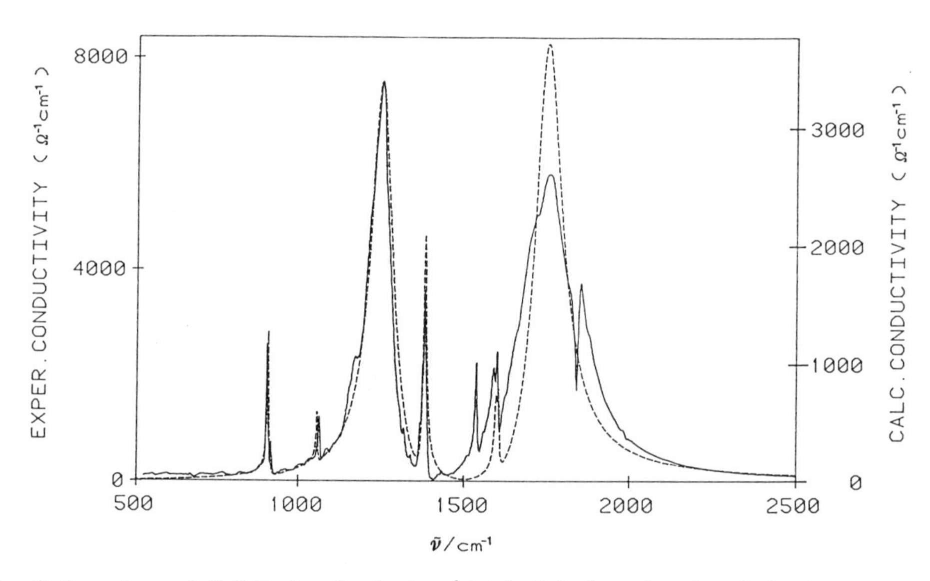

Fig. 6. Experimental (full line) and calculated (dashed line) stack-axis polarized conductivity spectra of $(TMTSF)_2ReO_4$ at 120 K. Note the different scales used for the two spectra.

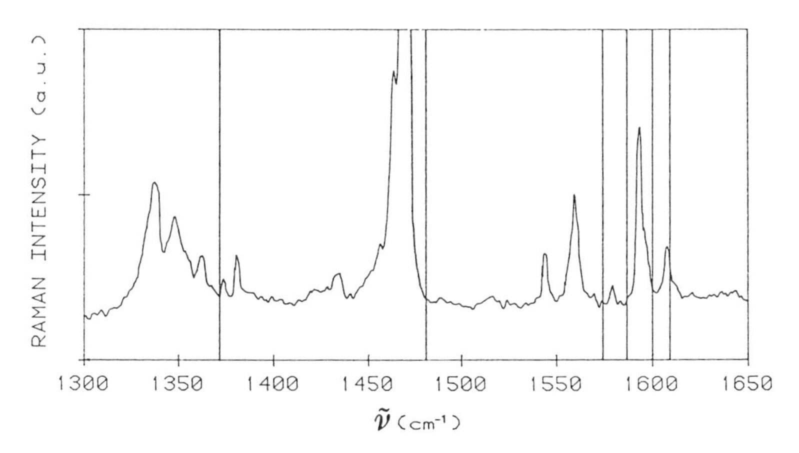

Fig. 7. Experimental polarized Raman spectra (electric vector of the incident and scattered radiation perpendicular to the stack axis) of $(TMTSF)_2ReO_4$ at T = 30 K, below the anion ordering transition (from Ref. [28]). Calculated Raman frequencies are indicated by vertical lines. Some of the bands observed in the region around 1350 cm^{-1} are due to methyl group vibrations.

The latter are collected in Table 3. With the exception of the transfer integrals, which must be considered as effective parameters for the reasons discussed above, the other interactions have assumed values, which are well within an acceptable range suggested by the current knowledge of the system. In particular, the relative magnitude of the correlation parameters U and V is reasonable, and their absolute value, when compared to that commonly accepted for the average transfer integral (t = 0.25 - 0.30 eV), characterizes $(TMTSF)_2ReO_4$ at low temperature as an extended Hubbard system in the intermediate regime.

Analysis of the eigenvectors of the electronic Hamiltonian allows us to gain a deeper understanding of the observed spectral features, as well as of the other CT transitions (the total number is four) that result from the calculation at frequencies higher than those reported in Fig. 6. The data in Table 4 show that the two lower-energy CT transitions, calculated at 0.19 and 0.50 eV, correspond to a spreading of the charge almost evenly on the four sites, whereas, in the ground state, the central sites bear a charge of +0.72, and the end ones, +0.28. The double occupancy of the sites has a very small weight in these low-energy states, and changes

Table 3. Values of the parameters used in the calculation of the T = 120 K conductivity spectra of $(TMTSF)_2ReO_4$.[a]

Δ = 55 meV		t = 173 meV		t' = 127 meV
U = 600 meV		V = 140 meV		Γ_{CT} = 90 cm^{-1}
mode	ω_i /cm^{-1}	ω_i^+/cm^{-1}	g_i/meV	γ_i/cm^{-1}
ν_3 (a_g)	1625	1573	12	6
ν_{28}(b_{1u})	1617	1558	6	8
ν_4 (a_g)	1539	1399	56	6
ν_7 (a_g)	1057	1064	5	4
ν_8 (a_g)	916	924	14	4
ν_{10}(a_g)	266	298	30	6

[a] ω_i and ω_i^+ are the vibrational frequencies of the i-th total symmetric mode of TMTSF and TMTSF$^+$. g_i is the linear e-mv coupling constant. Γ_{CT} and γ_i are the damping factors, used for CT transitions and for the vibrational modes, respectively.

Table 4. CT transitions $S_{1g} \rightarrow S_{iu}$: energies, transition moments and occupation numbers.

	S_{1g}		S_{1u}		S_{2u}		S_{3u}		S_{4u}	
E (eV)	0.0		0.19		0.50		0.89		1.12	
$\langle S_{1g} \| n_3 - n_2 \| S_{iu} \rangle$	--		0.09		0.86		0.43		0.03	
$\langle S_{1g} \| n_4 - n_1 \| S_{iu} \rangle$	--		0.69		0.14		0.09		0.04	
	$\langle n_i \rangle$	$\langle n_{i\uparrow} n_{i\downarrow} \rangle$	$\langle n_i \rangle$	$\langle n_{i\uparrow} n_{i\downarrow} \rangle$	$\langle n_i \rangle$	$\langle n_{i\uparrow} n_{i\downarrow} \rangle$	$\langle n_i \rangle$	$\langle n_{i\uparrow} n_{i\downarrow} \rangle$	$\langle n_i \rangle$	$\langle n_{i\uparrow} n_{i\downarrow} \rangle$
Site 1	0.28	0.00	0.49	0.02	0.46	0.04	0.15	0.10	0.90	0.84
Site 2	0.72	0.11	0.51	0.03	0.54	0.12	0.85	0.80	0.10	0.04
$\langle \sum_{i=1}^{4} n_i n_{i+1} \rangle$	0.53		0.40		0.40		0.08		0.11	

very little in the first two transitions. It is seen that also the nearest-neighbor correlation V contributes very little to their transition energies. The latter is, therefore, dominated by one-electron interaction terms, such as t , t', and Δ. Conversely, the two CT transitions at higher energy are characterized by a large increase in the double occupancy, either of the central pair of sites ($\hbar\omega_{CT3}$ = 0.89 eV) or the end ones ($\hbar\omega_{CT4}$ = 1.12 eV), and are therefore dominated by the interaction U, with an important contribution from Δ for the 1.12 eV transition. Their oscillator strength, as well as that of the second CT transition, is much smaller than that of the first one.

The analysis of the infrared-active modes of the tetramer shows that their vibronic intensity is almost totally derived from the coupling to the two lowest energy CT transitions. One can therefore summarize the above findings as follows. If one limits oneself to considering the excitation spectrum, including the effects of the e-mv coupling, up to an energy lower than the on-site correlation energy U, then all its features can be modeled with the use of one-electron interactions only, *provided that the double occupancy of the sites is effectively excluded.*

CONCLUSIONS

The observation of vibronic structures in the infrared spectra of organic CT conductors is one of the most sensitive probes for the existence of gaps originated by electron charge density modulations, that is CDW, whether induced by the electron-phonon coupling, by some static lattice potential, or by electron correlations. This is true, *provided* that the gap opens at the Fermi level. The fundamental reason for that is, that these vibronic structures derive their intensity from interband transitions. Therefore, as such transitions shift to higher energies and lose oscillator strength, when the gap moves apart from the Fermi energy, the amplitude of the vibronic structures decreases dramatically. For investigating quantitatively the relation between the frequency location of the infrared conductivity maximum, due to interband transitions and the amplitude of the vibronic structures, one needs to use an infinite-chain model. However, severe difficulties arise, when electron correlations in the intermediate regime have to be accounted for.

We believe this paper has been successful in demonstrating the fruitfulness of combining two different analyses of the spectroscopic data of organic CT conductors. As a first application, we have reported and analysed polarized reflectance and conductivity data for a series of Beechgaard salts, including $(TMTSF)_2ClO_4$, $(TMTSF)_2PF_6$, $(TMTTF)_2Br$, $(TMTTF)_2PF_6$ at room temperature, as well as $(TMTSF)_2ReO_4$ at 120 K, i.e., below its anion-ordering transition.

Comparing the experimental data at room temperature and the calculations using the dimerized chain model [13], we conclude that the observed trends in the spectra can only be explained, if we assume a narrow gap at the Fermi level in all of the above compounds . The amplitude of this gap increases with increasing degree of dimerization, that is, in the order $(TMTSF)_2ClO_4 \approx (TMTSF)_2PF_6 < (TMTTF)_2Br < (TMTTF)_2PF_6$. Coincidence between a dimerization gap and the Fermi level can only occur, for quarter-filled systems, if the Fermi wave vector is shifted to twice the value ($k_F = \pi / 4$ d) expected for independent electrons or, in real space, if the double occupancy of the molecular sites is effectively forbidden. The above situation is usually associated to the presence of a strong on-site electron correlation, or, U >> 4 t.

At this stage, the use of molecular cluster models, whose electronic (extended Hubbard) Hamiltonian is exactly soluble with arbitrary values of the interaction parameters, becomes relevant. Application of a tetramer model to the fitting of the conductivity spectra of $(TMTSF)_2ReO_4$ at 120 K has allowed us to locate the appropriate range of interaction parameters for this material. With these parameter values, the participation of doubly occupied sites to the low-lying electronic states is small, and its change upon electronic excitation is even smaller. One can, therefore, argue that the spectra, observed up to an energy smaller than the correlation energy U, can be understood considering only one-electron interactions, and restricting the electronic basis set only to those states that do not include doubly occupied sites.

Similar arguments should provide a rationale for the success in reproducing the main features of the spectra of Bechgaard salts at high temperature, by using calculations carried out for the dimerized chain model in the strong-correlation limit. In other words, it is not required that the on-site correlation energy be much larger than the one-electron bandwidth, in order for the low-energy spectra to behave, as if the double occupancy was effectively excluded.

The general picture of the Bechgaard salts at high temperature, that is offered by the results of the optical studies reported here, is one of correlated narrow-gap semiconductors, and is applicable both to the TMTTF and TMTSF salts. This view is related to the concept of a 4 k_F charge localization previously proposed [29,30] for the TMTTF salts alone. We note, however, that our results do not imply that the same picture can be maintained for the $(TMTSF)_2X$ salts at lower temperature. It however emphasizes the similarities between the sulfur and selenium-containing compounds. The strength of the electron correlations seems not to differ in an important way between the two subclasses. More relevant factors are the larger dimerization and weaker two-dimensional character of the $(TMTTF)_2X$ salts. This is well illustrated by the changes in the behavior of these materials accompanying the structural changes induced by the application of pressure [30].

ACKNOWLEDGEMENTS

The authors wish to thank Prof. R. M. Pick for making available his Raman data prior to publication and for stimulating discussions. Valuable technical help from C. Ricotta, F. Marzola, and M. Zanetti is gratefully acknowledged. Financial support by the Italian National Research Council and by Ministry of Research and Universities is also acknowledged.

REFERENCES

[1] R. Bozio, A. Feis, D. Pedron, I. Zanon, and C. Pecile, this volume, and references therein.
[2] "Proceedings of the International Conference on Science and Technology of Synthetic Metals (ICSM'88)", ed. by M. Aldissi, Synth. Metals 27: 1-99 (1988).
[3] F. Wooten, "Optical Properties of Solids", (Academic Press, New York, 1972).
[4] P. F. Maldague, Phys. Rev. B 16: 2437 (1977).
[5] W. Gasser and R. Höfling, Phys. stat. sol. (b) 92: 91 (1979).
[6] S. K. Lyo and T. Holstein, Phys. Rev. B 14: 5137 (1976).
[7] K. Kubo, J. Phys. Soc. Jpn. 31: 30 (1971); S. K. Lyo and J.-P. Gallinar, J. Phys. C 10: 1693 (1977); S. K. Lyo, Phys. Rev. B 10: 1854, 5835 (1978); J.-P. Gallinar, J. Phys. C 12: L335 (1979).
[8] S. Mazumdar and Z. G. Soos, Phys. Rev. B 23: 2810 (1981).
[9] E. Y. Loh, Jr. and D. K. Campbell, Synth. Metals 27: A499, (1988).
[10] J. Hubbard, Proc. Roy. Soc. London Ser. A 276: 238 (1963); 277: 237 (1964); 281: 401 (1964); 285: 542 (1965).

[11] C. S. Jacobsen in "Low Dimensional Conductors and Superconductors", ed. by D. Jérome and L. G. Caron, (Plenum Press, New York, 1987), p. 253; R. Bozio in "Molecular Electronics", ed. by M. Borissov, (World Scientific, Singapore, 1987), p. 666.
[12] R. Bozio, M. Meneghetti, D. Pedron, and C. Pecile, Synth. Metals 27: B109, B129 (1988).
[13] R. Bozio, M. Meneghetti, and C. Pecile, Phys. Rev. B 36: 7795 (1987).
[14] R. Bozio, M. Meneghetti, C. Pecile, and F. Maran, Synth. Metals 19: 309 (187); *Ibid.* 20: 393 (1987).
[15] R. Bozio, M. Meneghetti, and C. Pecile, J. Chem. Phys. 76: 5785 (1982).
[16] C. S. Jacobsen, D. B. Tanner, and K. Bechgaard, Phys. Rev. B 28: 7019 (1983).
[17] N. Thorup, G. Rindorf, H. Soling, and K. Bechgaard, Acta Cryst. B37: 1236 (1981).
[18] M. Meneghetti, R. Bozio, I. Zanon, C. Pecile, C. Ricotta, and M. Zanetti, J. Chem. Phys. 80: 6210 (1984).
[19] P. B. Allen, Phys. Rev. B 3: 305 (1971).
[20] G. Rindorf, H. Soling, and N. Thorup, Acta Cryst. B38: 2805 (1982).
[21] J. L. Galigné, B. Liautard, S. Peytavin, G. Brun, J. M. Fabre, E. Torreilles, and L. Giral, Acta. Cryst. B34: 620 (1978).
[22] B. Liautard, S. Peytavin, G. Brun, and M. Maurin, J. Physique 43: 1453 (1982).
[23] P. M. Grant, J. Physique Colloq. 44: C3-847 (1983).
[24] L. Ducasse, M. Abderrabba, J. Hoarau, M. Pesquer, B. Gallois, and J. Gaultier, J. Phys. C. 91: 3805 (1986).
[25] V. M. Yartsev, Phys. stat. sol . (b) 126: 501 (1984).
[26] A preliminary report on some of the results discussed here is found in: R. Bozio, M. Meneghetti, D. Pedron, and C. Pecile, Synth. Metals 27: B109 (1988).
[27] G. Rindorf, H. Soling, and N. Thorup, Acta. Cryst. C40: 1137 (1984).
[28] R. M. Pick, private communication; J. Breitenstein, Ph.D. Thesis, Université Paris VII, 1988.
[29] V. J. Emery, R. Bruinsma, and S. Barisic, Phys. Rev. Lett. 48: 1039 (1982).
[30] D. Jérome, F. Creuzet, and C. Bourbonnais, Physica Scripta T27: 130 (1989); D. Jérome, this volume.

DESIGN AND SYNTHESIS OF POLYHETEROTETRAHETERAFULVALENES, METAL 1,2-DIHETEROLENES, AND THEIR LOW-DIMENSIONAL CONDUCTING AND SUPERCONDUCTING SALTS

G.C. Papavassiliou

Theoretical and Physical Chemistry Institute, National Hellenic Research Foundation, 48, Vassileos Constantinou Ave., Athens 116/35, Greece

INTRODUCTION

Most of the known organic and metalloorganic crystalline conductors (and superconductors) have been based on the tetrathiafulvalene derivatives, transition-metal 1,2-dithiolenes and selenium- or tellurium- analogs [1]-[5]. The synthesis and physicochemical properties of these compounds reported by 1988 are summarized in a number of excellent review articles [6]-[12]. Recent work, concerning synthesis and physicochemical properties of these and similar compounds, is described in refs [13]-[56] and in references cited therein. Generally, 2-thioxo-1,3-dithioles and selenium or tellurium analogs have been used as starting materials for preparation of both tetraheterafulvalenes and metal 1,2-diheterolenes. For a systematic investigation, we have divided the simplest 2-thioxo-1,3-dithioles in classes, based on the number and the nature of the additional heteroatoms to the thioxo-dithiole group (see [41]). Table 1 shows examples of the simplest 2-thioxo-1,3-dithioles divided in classes: Class 0 contains compounds without additional heteroatoms; Class 1 contains compounds with a sulfur heteroatom in an additional ring or chain; Class 2 contains compounds with a nitrogen heteroatom in an additional ring or chain; and so on. One can design similar classification tables for the selenium, tellurium, oxygen and nitrogen analogs , as well as for their self- or cross-coupling products (polyheterotetraheterafulvalenes) and for the corresponding metal 1,2-heterolenes and similar compounds [57]-[60]. The addition of groups and heteroatoms to the tetraheterofulvalene core or to the metal 1,2-diheterolene core play an important role in the chemical and physicochemical properties of these compounds (see for example [6]). Also, they play an important role in the crystal structure and physical properties of the corresponding charge transfer complexes (CTC), cation radical salts (CRS) and cation deficient metal 1,2-diheterolenes (CDMD) (see for example [4]).

During the last ten years, a number of low-dimensional conducting crystals, based on oxygen-, sulfur-, selenium- and nitrogen- containing molecules have been investigated in our institute, and in a collaboration of our institute with others [20], [37]-[41], [61]-[98]. In this paper we report on new tetraheterafulvalenes and metal 1,2-diheterolenes, and discuss general synthetic methods which have been developed and applied recently for the preparation of this kind of compounds. Also,

Table 1. Examples of simple 2-thioxo-1,3-dithioles (X=S)

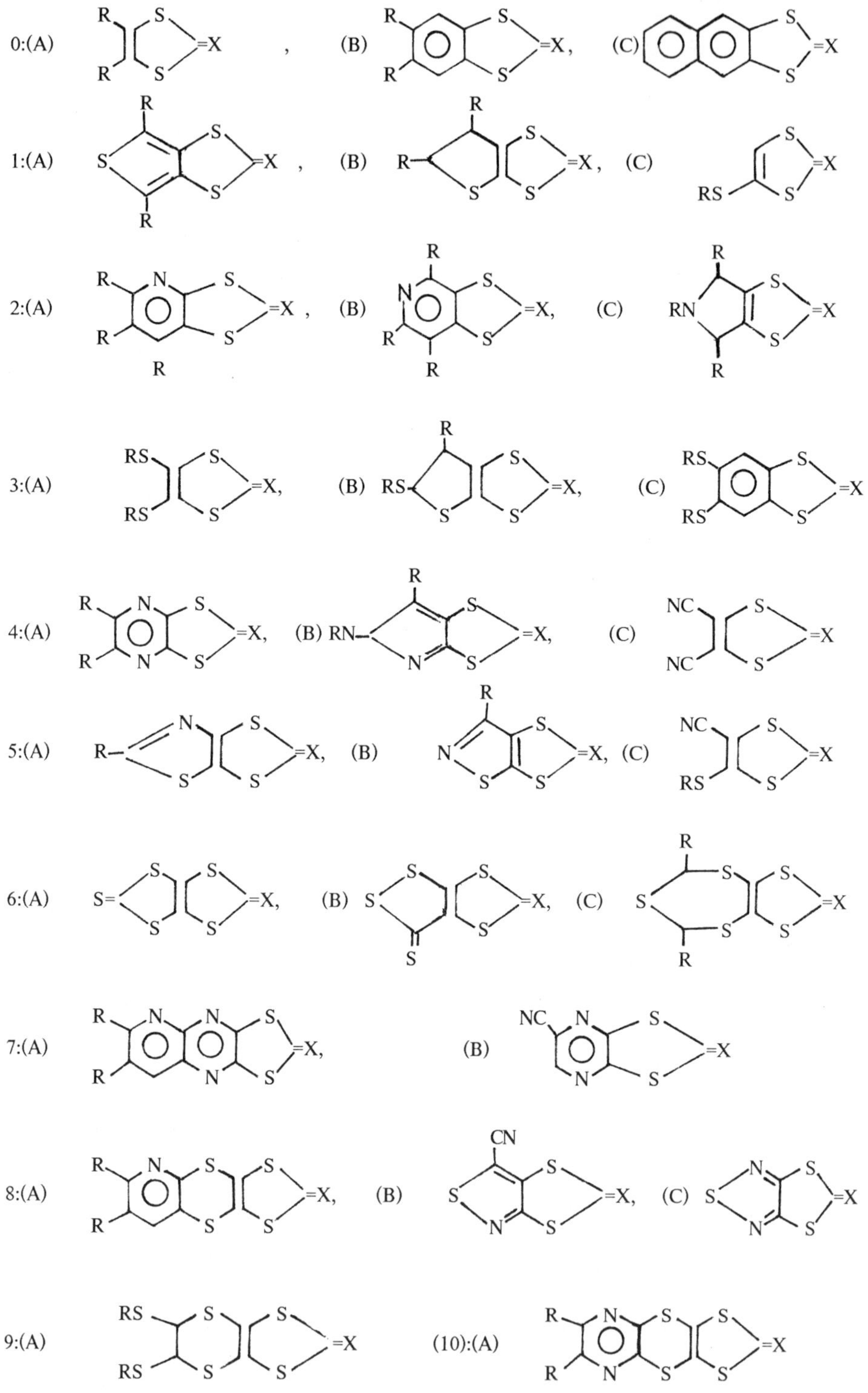

0:(A) , (B) =X, (C) =X

1:(A) , (B) =X, (C) =X

2:(A) , (B) =X, (C) =X

3:(A) =X, (B) =X, (C) =X

4:(A) =X, (B) =X, (C) =X

5:(A) =X, (B) =X, (C) =X

6:(A) =X, (B) =X, (C) =X

7:(A) =X, (B) =X

8:(A) =X, (B) =X, (C) =X

9:(A) =X (10):(A) =X

R=H,CH_3,CF_3,etc, 2R=CH_2, CH_2-CH_2 etc

we briefly discuss the synthesis, structure and properties of some conducting crystalline salts based on these compounds.

DESIGN OF NEW COMPOUNDS

Starting from the compounds given in Table 1, and selenium, tellurium, oxygen, and nitrogen analogs, one can design the preparation of a large number of tetraheterofulvalenes (π-donors) (all possible binary combinations of compounds of Table 1 as well as O, Se, N and Te analogs), and metal 1,2-diheterolenes (π-acceptors), and a very large number of their CTC, CRS and CDMD, which means a very large number of low-dimensional conducting crystals. The design of new tetraheterafulvalenes and metal 1,2-diheterolenes depends on the kind of products needed, namely, crystalline conductors or superconductors [1]-[4], conducting Langmuir-Blodgett films [16], [30],[31],[46], ferromagnetic organic metals [58],[60], materials for nonlinear optics [60], etc. At present, our interest is in the preparation of crystalline conductors and superconductors. Consequently, before we design and prepare new tetraheterafulvalenes and metal 1,2-diheterolenes, we should take into account all the current literature on organic conductors and superconductors based on already known compounds.

It is well known that compounds based on 2-thioxo-1,3-dithioles of Class 0 and Class 1 gave 1-dimensional conductors (see for example [5],[8]). However, 1-dimensional metallic systems are unstable and exhibit a metal-to-semiconductor transition at lower temperatures. The dimensionality can be increased by incorporating sulfur, selenium, tellurium or nitrogen atoms into the π-donor or π-acceptor molecule [4], [25],[14],[87],[88],[55]. By this modification, a superconducting ground state was found to be stabilized in salts with varying degrees of anisotropy. Thus, some salts of the type $(TMTSF)_2X$ ($X=PF_6$, ClO_4 etc) based on the (selenium containing) π-donor molecules of tetramethyltetraselenafulvalene (TMTSF), were found to be quasi-1-dimensional superconductors (see for example [2]), while the salts of the type β-$(BEDTTTF)_2X$($X=I_3$,AuI_2,IBr_2, etc) based on the (sulfur containing) π-donor molecules of bis(ethylenedithio)tetrathiafulvalene (BEDTTTF), were found to be 2-dimensional superconductors (see for example [1]-[4]). The compound (Me_4N) $[Ni(dmit)_2]_2$ based on the (sulfur containing) π-acceptor molecules of bis(4,5-dimercapto-1,3-dithiole-2-thione)nickelate $[Ni(dmit)_2]$ was found to be a 2-dimensional superconductor, while (TTF) $[Ni(dmit)_2]_2$ was found to be a quasi-1-dimensional superconductor (see ref.[1] and refs [33]-[35] cited therein). The salts $(DMET)_2X$ ($X=I_3$, IBr_2 etc) and $(MDTTTF)_2AuI_2$ based on the unsymmetrical π-donor molecules of dimethyl(ethylenedithio)diselenadithiafulvalene (DMET) and methylenedithiotetrathiafulvalene (MDTTTF) were found to be 2-dimensional superconductors (see ref [94] and refs therein, [98]-[100]). The salts $(DMtTSF)_2X$ ($X=ClO_4$, PF_6,AsF_6) [101], $(EDTTTF)_2IBr_2$, and $(EDTDSDTF)_2$ IBr_2 [94], [98] based on the unsymmetrical π-donor molecules of dimethyl(trimethylene)tetraselenafulvalene (DMtTSF), ethylenedithiotetrathiafulvalene (EDTTTF) and ethylenedithiodiselenadithiafulvalene (EDTDSDTF) were found to remain metallic down to very low temperatures. Also, some compounds of the type $(BEDO\text{-}TTF)_xX$ based on the symmetrical donor bis(ethylenedioxo)tetrathiafulvalene (BEDO-TTF) were found to be metallic down to low temperatures [19]; BEDO-TTF is better donor than BEDTTTF.The compounds TMTSF, BEDTTTF, DMET, MDTTTF, EDTTTF, DMtTSF, and $M(dmit)_2$, which gave good conductors or superconductors, were found to be fairly good π-donors and π-acceptors [6],[47], [101], [102], respectively. The half-wave oxidation potentials of π-donors showed that the donor capacity decreases from tetrathiafulvalene to the corresponding tetraselenafulvalene and increases again for the corresponding tetratellurafulvalene [6],[14]. The π-donor capacity of tetraheterafulvalenes and π-acceptor capacity of metal 1,2-diheterolenes depends also on the nature of additional groups (chains or rings). Groups such as alkyls, cycloalkyls and thioalkyls are electron-donating groups, while groups such as trifluoromethyl, nitriles and esters are electron-withdrawing substituents. Addition of methyl groups, for example, increases the donor capacity in TTF, while addition of nitrile groups decreases the donor capacity in TTF (see for example [6]). The addition of a benzene- or vinylenedithio-group to the TTF-core increases the extent of its π-electron system, and the electron-donor capacity of the obtained (planar) TTFs then decreases, compared with the unsubstituted TTF [6], [36]. The addition of a pyrazine-(or a pyridine-, etc.) group decreases the π-electron donor capacity, but increases the ability for the formation of intermolecular contacts (and consequently of 2-or 3-dimensional systems) in the salts of (planar) TTFs [38],[73],[87], [88]. Similar effects have been discussed for the π-acceptor molecules of metal 1,2-diheterolenes. It was observed that, in metal 1,2-diheterolenes with a small cation (such as Li, Na, Me_4N,) the intermolecular contacts are strong and consequently good conductors (cation-deficient metal 1,2-diheterolenes) can be obtained (see [44],[47],[50],[53],[55],[63],[64],[83],[86] and refs therein).

Tetraheterafulvalenes, such as bis(ethylenediseleno)tetrathiafulvalene, bis (methylenediseleno) tetrathiafulvalene, tetratellurafulvalenes, which are rich in selenium or tellurium hetero-atoms, showed two oxidation potentials, with values close to those of BEDTTTF, and a few of them gave highly conducting complexes (see [14],[25],[32] and refs therein). However, the low solubility in common organic solvents of these π-donors seriously hampers crystal-growing efforts, and often results in poor crystal quality [14] [32].

Taking into account the above statement, we suggest the design and synthesis of symmetrical or unsymmetrical (planar) tetraheterafulvalenes poor in selenium or tellurium heteroatoms, and (small cation-) metal 1,2-diheterolenes fairly rich in sulfur, sellenium, (tellurium) or nitrogen heteroatoms, to obtain good conductors or superconductors.

2-THIOXO-1,3-DITHIOLES AND SELENIUM ANALOGS

2-Thioxo-1,3-dithioles and selenium analogs have been usually used as starting materials for the syntheses of both tetraheterafulvalenes and metal 1,2-diheterolenes. In this section, we describe methods of preparation and some properties of the following compounds, (1)-(18):

CH_3OOC, CH_3OOC / S, S / =S, (1)

CH_3OOC, H / S, S / =S, (3)

CH_3OOCCH_2S, CH_3OOCCH_2S / S, S / =S (5)

CH_3OOC, CH_3OOC / Se, Se / =Se, (2)

CH_3OOC, H / Se, Se / =Se, (4)

CH_3OOCCH_2Se, CH_3OOCCH_2Se / S, S / =S (6)

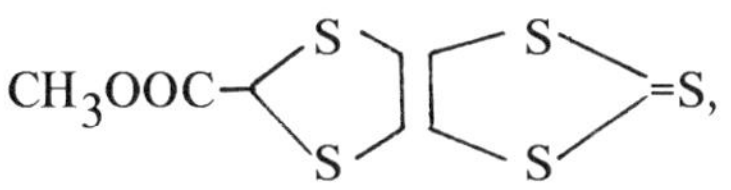

(a):R=CH_3, (b):2R=CH_2, (c):2R=CH_2CH_2, (d):2R=$CH_2CH_2CH_2$
(e):2R=$CH(CH_3)CH_2$, (f):2R=$CH(CH_3)CH(CH_3)$, (g):2R= CH=CH

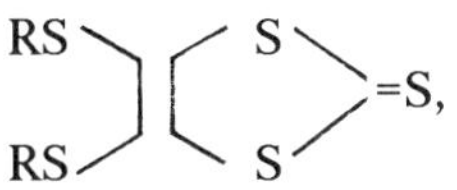

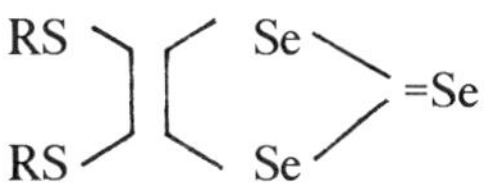

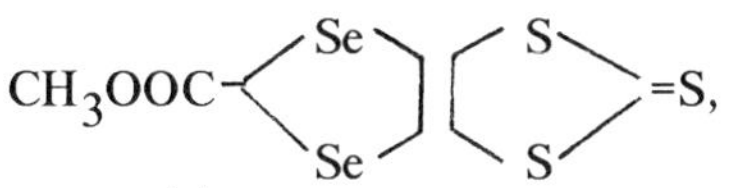

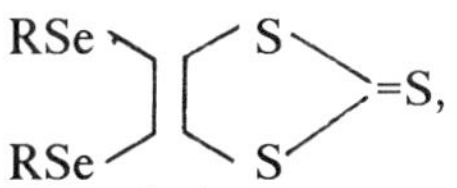

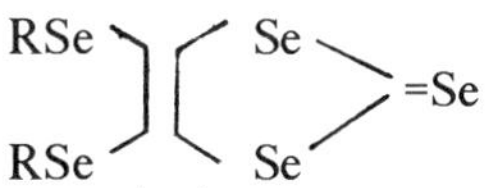

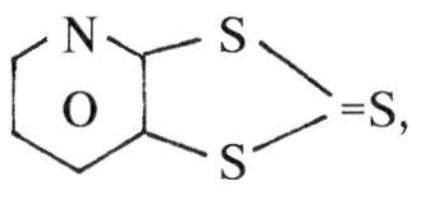

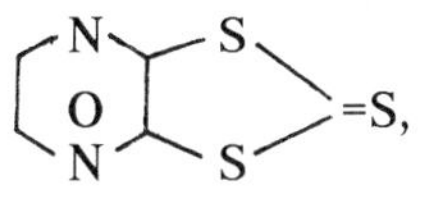

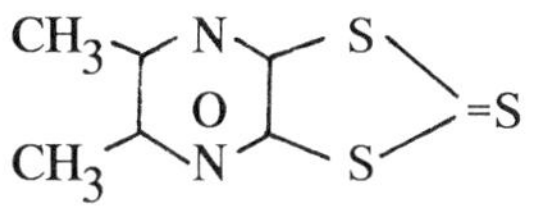

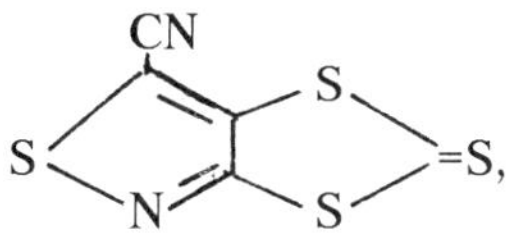

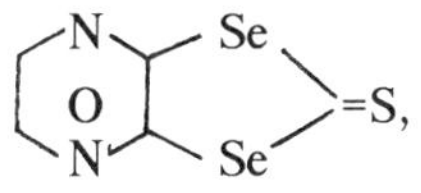

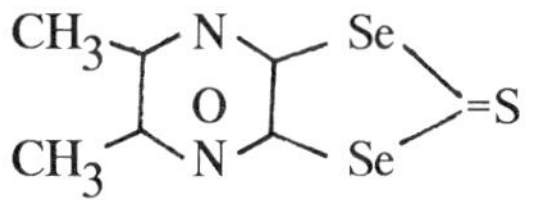

4,5-Bis(methylcarboxy)-1,3-dithiole-2-thione (1) and its selenium analog (2) were known years ago (see ref [11] cited in ref [37] here and ref [30] cited in ref [20] here). They have been prepared by treatment of ethylene trithiocarbonate and selenium analog, respectively, with dimethyl acetylenedicarboxylate in toluene at reflux temperature. Methylcarboxyl-1,3-dithiole-2-thione(3)[103], and selenium-analog (4) [104], have been prepared by treatment of ethylene trithiocarbonate and selenium analog, respectively, with methyl propiolate in toluene at reflux temperature for several hours followed by silica gel column chromatography separation (C_6H_6-cyclohexane 1:1).

Compounds (5) [40], (6) [104], (7) [39], (8) [104], (9a)-(9f) [76], [79], (10a)-(10f) [25],[29], [41],[43] , [104], (11a)-(11f) [105], [106], (12a)-[12f) [105]-[107] have been prepared by treatment of the corresponding tetrabutyammonium salt of zinc 1,2-dithiolates and selenium analogs (19A)-(19D) [79],[41] [106] or the Hg-[65] or Ni-[25] analogs with the corresponding alkyl halide ([79],[39]-[41],[106] and refs therein). Disodium 4,5-dithiolates and selenium analogs (20A)-(20D) should be used instead of (19A)-(19D) for the preparation of (9g), (10g), (11g) and (12g) (see refs [5],[6] cited in ref [20] here). Compounds (9f), (10f) can be also prepared from (20A), (20B), respectively, by a two-step sequence (treatment with the sulfate diester of (R,R)-butane-2,3-diole in MeOH and cyclization on heating in THF) (see ref[8] cited in ref [39] here). Compounds (20A)-(20D) can be prepared from the corresponding zincates (19A)-(19D) according to Scheme 1 [53], [63], [106]. However, compounds (19c) and (20c) have not been isolated yet in a pure form, [106]. For the preparation of the corresponding compounds (11) in a pure form, the method reported in ref [105] has been used. This method has been applied only in the case of 2R=CH_2CH_2, according to Scheme 2[105] ,[106],[22].Also, it has been applied for the preparation of (9c) as an alternative method [22].

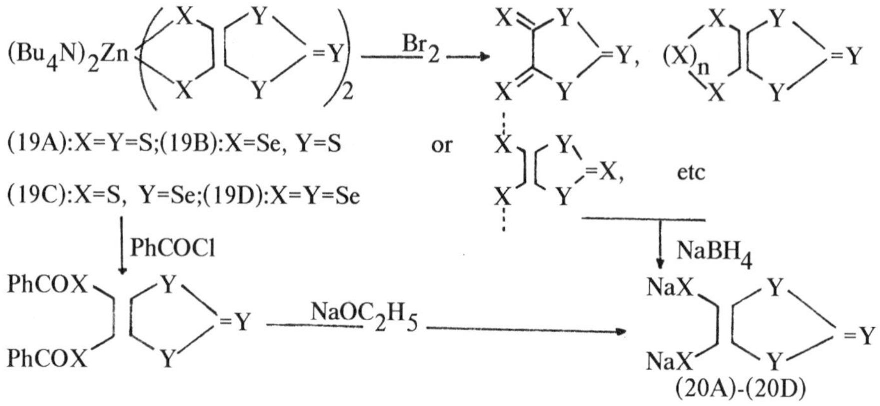

Scheme 1

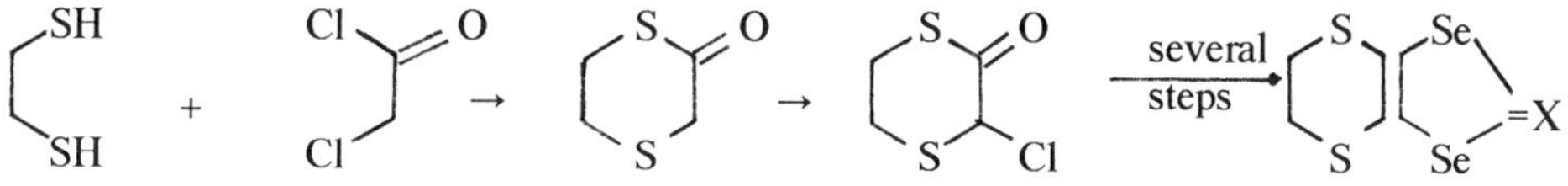

Scheme 2

Compound (13) has been prepared from 2-chloro-3-nitropyridine by a four-step sequence (treatment with NaHS, reduction with tin(II)chloride, diazotization with nitrous acid and treatment with carbon disulfide at 220°C)[72],[85]. Compound (14) has been prepared by treatment of disodium 5-cyanoisothiazoledithiolate with thiophosgene [83], [86]. Compounds (15)-(17) have been prepared by a two-step sequence (treatment with KHS or NaHSe and reaction with thiophosgene) [78],[82]. Reactions with phosgene yielded the corresponding -2-ones [78],[82],[83],[86]. Compounds (15),(17) have been prepared also, by treatment of 2,3-dichloropyrazine and 2,3-dichloro-5,6-dimethylpyrazine, respectively, with potassium trithiocarbonate [78].

Compounds (1)-(18) are soluble in common organic solvents (dichloromethane, chloroform, benzene, etc), and they can be transformed easily to the 2-oxo-1,3-dithioles and 2-oxo-1,3-diselenoles, by treatment with mercury acetate in CH_2Cl_2-CH_3COOH or $CHCl_3$-CH_3COOH mixtures (see for example [39]-[42]).They can be also converted to the corresponding selenones by a three-step sequence: reaction with methyl fluorosulfonate, treatment with morpholine, and reaction with hydrogene selenide (see ref [110] cited in ref [7] here). Compounds (7)-(10) can be converted to the corresponding 1,2-dithiolium salts by a three-step sequence reaction with sulfur dichloride, reduction with sodium borohydride, and reaction with fluoroboric acid) according to Scheme 3 [39]:

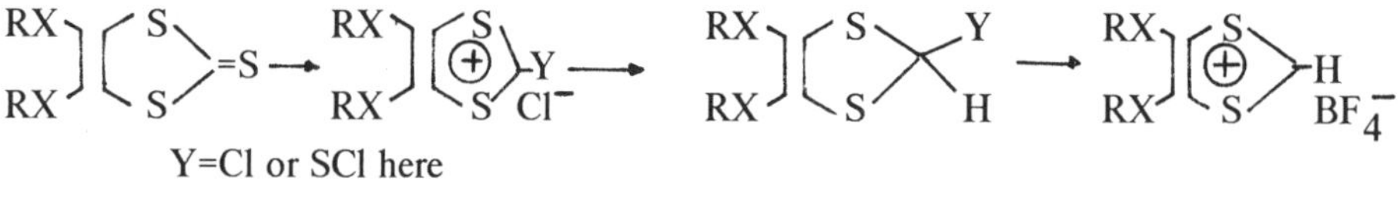

Scheme 3

These salts can be prepared by a four-step sequence (methylation with dimethyl sulfate, treatment with fluoroboric acid, reduction with sodium borohydide and reaction with fluoroboric acid) [6],[39].

Compounds (1)-(18) should be purified, by gel column chromatography and /or recrystallization, before use. In the mass spectra of fairly pure compounds (9),(10), the presence of some decomposition products similar to those obtained by bromine-oxidation of (19)(see Scheme 1) and their isomers [66] has been observed. In the case where 2R=CH_2, the contents have been found to be lower than those of the cases where 2R=CH=CH, CH_2CH_2. Also, compounds (9),(10) give several products in alkaline medium, depending on the nature of R and the conditions [21],[34],[45], [108] as it is shown in Scheme 4:

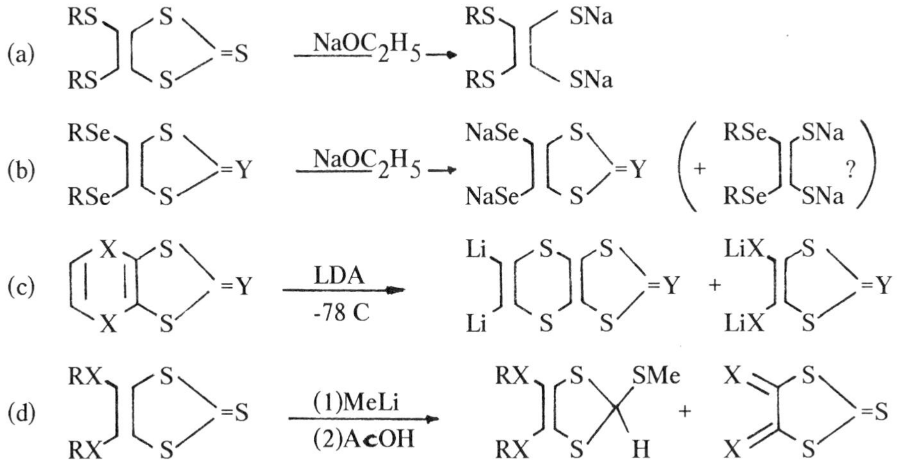

(a) RS, RS, S, S, =S —$NaOC_2H_5$→ RS, RS, SNa, SNa

(b) RSe, RSe, S, S, =Y —$NaOC_2H_5$→ NaSe, NaSe, S, S, =Y (+ RSe, RSe, SNa, SNa ?)

(c) X, X, S, S, =Y —LDA, -78 C→ Li, Li, S, S, S, S, =Y + LiX, LiX, S, S, =Y

(d) RX, RX, S, S, =S —(1)MeLi, (2)AcOH→ RX, RX, S, S, SMe, H + X, X, S, S, =S

Scheme 4

It has been elucidated that the products obtained from the routes (b)-(d) are not due to the presence of decomposition products in the samples of (9),(10).

POLYHETEROTETRAHETERAFULVALENES

A number of symmetrical polyheterotetraheterafulvalenes have been prepared by self-coupling reaction of the corresponding 2-thioxo-1,3-dithioles and selenium analogs via triethyl phosphite or other trivalent derivatives of phosphorous, in yields depending on the nature of the additional group and the reaction conditions (see for example [6], [7], [9]). It has been observed that 2-oxo-1,3-dithiols and 2-oxo-1,3-diselenoles give, in most cases, higher yields than the corresponding 2-thioxo- and 2-selenoxo-compounds, and the yield of cross-coupling products is higher than those of the self-coupling byproducts. Sometimes, coupling of 1,2-dithiolium salts (Scheme 3) or selenium analogs via triethylamine has been applied for the preparation of the corresponding tetraheterafulvalenes (see for example [6], [7], [39]). A number of sulfur [24], selenium [24], [32], [35] and tellurium [24]-containing symmetrical polyheterotetraheterafulvalens have been prepared from the corresponding simplest tetraheterofulvalenes (TTF, TSF, TTeF) by a three-step sequence: lithiation with lithium diisopropylamide (LDA), chalcogene-addition and reaction with alkyl halide. A similar procedure has been applied for an oxygen-containing tetraheterafulvalene [32]. Some sulfur-[7], [78], selenium-[14], tellurium-[14] and nitrogen-containing [78] tetraheterafulvalenes have been prepared by treatment of the corresponding dialkali dithiolates, as well as selenium- and tellurium- analogs with tetrachloroethylene.

In this section we report methods for preparation of some unsymmetrical tetraheterafulvalenes of the following formulas, (I)-(XII):

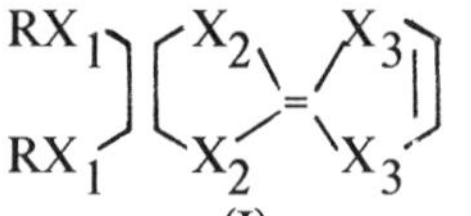

(I) 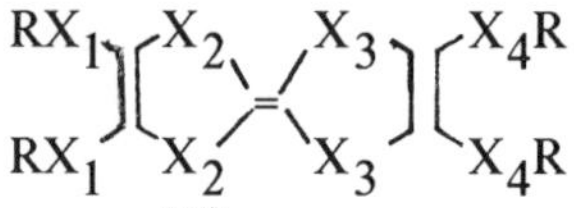 (II) 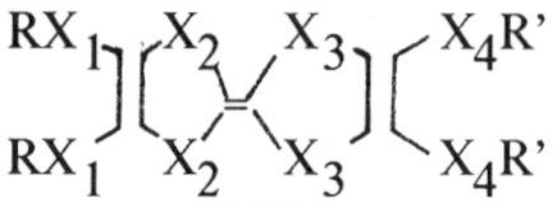 (III)

(A): $X_1=X_2=X_3=S$, $X_4=Se$ (B): $X_1=X_2=X_4=S$, $X_3=Se$
(C): $X_1=X_4=Se$, $X_2=X_3=S$ (D): $X_1=X_3=X_4=Se$, $X_2=S$
(E): $X_1=X_3=S$, $X_2=X_4=Se$ (F): $X_1=X_4=S$, $X_2=X_3=Se$
(G): $X_1=X_2=Se$, $X_3=X_4=S$ (H): $X_1=X_2=X_3=Se$, $X_4=S$

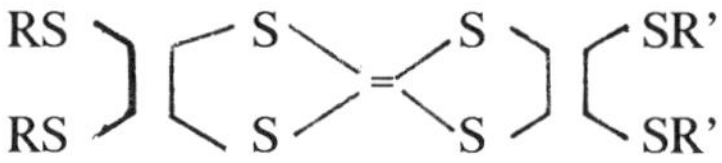

(IV), 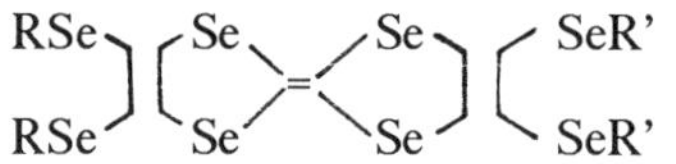 (V)

(a): R,R'=CH_3, (b): 2R,2R'=CH_2, (c): 2R, 2R'=CH_2CH_2, (d): 2R, 2R'=$CH_2CH_2CH_2$
(e): 2R,2R'=$CH(CH_3)CH_2$, (f): 2R, 2R'=$CH(CH_3)CH(CH_3)$, (g): 2R, 2R'=CH=CH

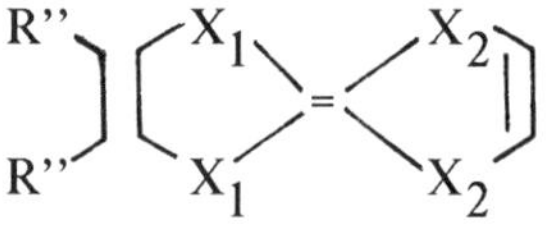

(VI), 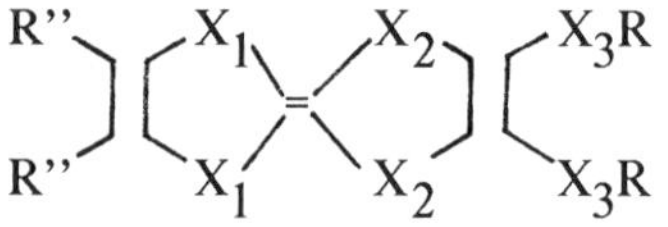 (VII),

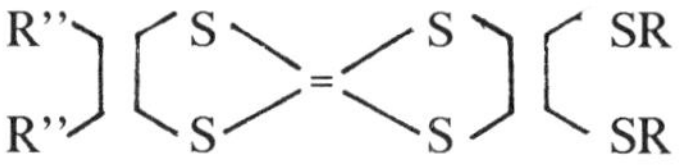

(VIII), 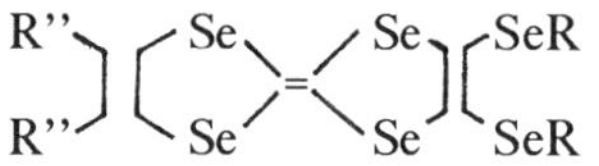 (IX)

(A): $X_1=X_2=S$, $X_3=Se$, (B): $X_1=Se$, $X_2=X_3=S$, (C): $X_1=S$, $X_2=X_3=Se$,
(D): $X_1=X_2=Se$, $X_3=S$

(a''): 2R''=pyridine-, (b''): 2R''=5-cyanoisothiazole-, (c''): 2R''=pyrazine- (d''): 2R''=dimethylpyrazine-, (e''): 2R''=isothiazole-group fused to the diheterafulvalene ring.

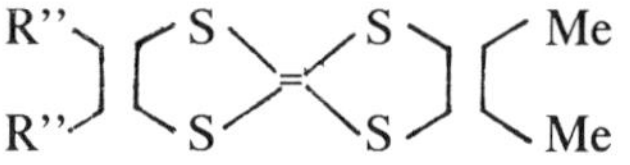

(X) 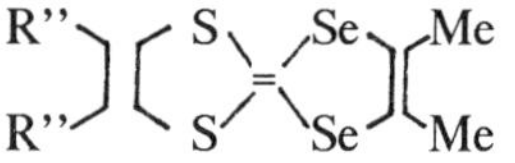 (XI) 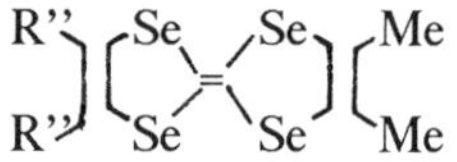 (XII)

A number of them have been prepared by cross-coupling reaction of materials (9)-(18), or the corresponding 2-oxo-1,3-dithioles and selenium-analogs, via triethyl phosphite. Because of the widely varying polarity of the cross-coupling products and the self-coupling byproducts, the separation by gel column chromatography is in most cases very easy (see [38],[74],[75],[78],[82],[83],[85],[86],[109]). Compounds (I)-(III) can be also prepared by this method, but the chromatography separation of the products is difficult (see refs [48], [59] cited in ref [38] here, [21]). To avoid these separation difficulties, a two- or three-step procedure, using one unfunctionalized component and one component containing groups of highly polarity such as $COOCH_3$, CN, has been applied (see [7], [20], [39]-[41] and refs therein).

The functional groups should be eliminated, or modifed, in the last step of the reaction procedure. Compounds of the type (I), for example, have been prepared from (9)-(12) and (1) - (4) according to the following Scheme 5 [20],[37], [41],[106]:

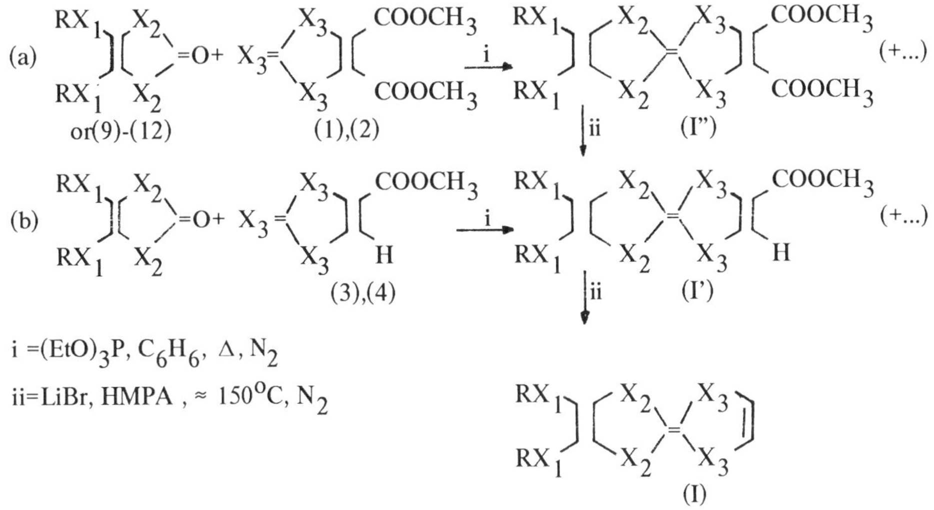

(a) RX_1, X_2 =O + X_3= X_3 COOCH$_3$ —i→ RX_1 X_2 X_3 COOCH$_3$ (+...)
or(9)-(12) (1),(2) (I")

(b) RX_1 X_2 =O + X_3= X_3 COOCH$_3$ H —i→ RX_1 X_2 X_3 COOCH$_3$ H (+...)
(3),(4) (I')

ii ↓ (I)

i =$(EtO)_3P$, C_6H_6, Δ, N_2

ii=LiBr, HMPA , ≈ 150°C, N_2

Scheme 5

It has been found that the yield of compounds of the type (VI), prepared by same procedure, is higher than that of the corresponding procedure using unfunctionalized components (:vinylene trithiocarbonate, vinylene triselenocarbonate [110] and (13)-(18)) [20], [38] [109]. Compounds similar to (I) and (VI) can be prepared by the same methods, using more complicated starting materials of Classes 9 [34], 10[78] (Table 1) and selenium analogs [36]. Also, using (5)-(8) and (9)-(12), tetraheterafulvalenes of the type (II)-(V) have been prepared by similar procedures (Scheme 6) [20], [39], [40], [106] [109]. Isothiazole-containing tetrathiafulvalenes of the type (VI) and (VIII) have been prepared, from the corresponding 5-cyanoisothiazole--containing tetrathiafulvalenes, by elimination of CN-group [86],

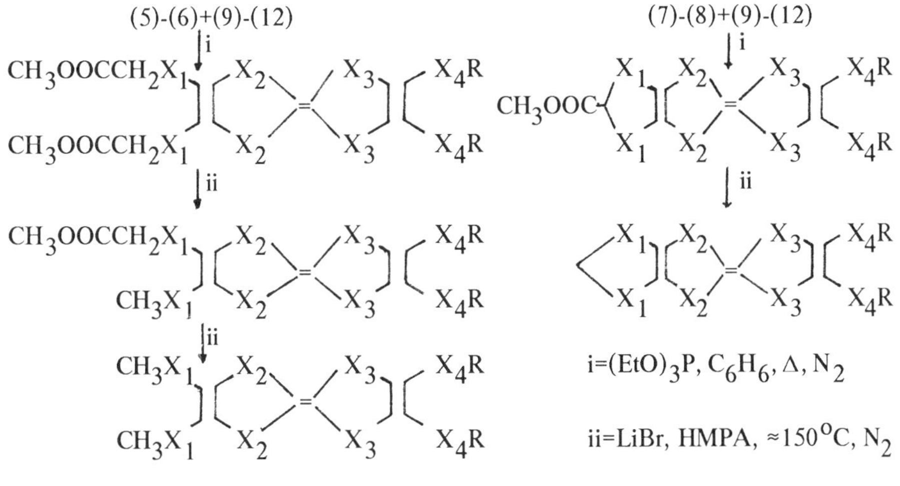

(5)-(6)+(9)-(12) —i→ $CH_3OOCCH_2X_1$ / $CH_3OOCCH_2X_1$ X_2 X_3 X_4R —ii→ $CH_3OOCCH_2X_1$ / CH_3X_1 X_2 X_3 X_4R —ii→ CH_3X_1 / CH_3X_1 X_2 X_3 X_4R

(7)-(8)+(9)-(12) —i→ CH_3OOC X_1 X_2 X_3 X_4R —ii→ X_1 X_2 X_3 X_4R

i=$(EtO)_3P$, C_6H_6, Δ, N_2

ii=LiBr, HMPA, ≈150°C, N_2

Scheme 6

[109]. Compounds of the type (III)-(V) have been prepared from 2-thioxo-1,3-dithioles by the trithioorthoformate procedure [21]: thiophilic addition of MeLi, acidification with AcOH, treatment with n-BuLi, addition of a second 2-thioxo-1,3-dithiole, followed by methylation with MeI, and thermolysis in organic solvents. They could be also prepared by a three-step procedure (lithiation with LDA, chalcogene-addition, treatment with alkylhalide) [20], which has been applied for the preparation of some symmetrical tetraheterafulvalenes [24],[32]. A number of subproducts, similar to those of Scheme 4, produced on the application of the last two procedures, are due to the alkaline reaction media. Also, impurities are due to the thermal isomerization of (20)[66],[79]. Sometimes, high-performence liquid chromatography, followed by recrystallization, is needed for a complete purification of the required tetraheterafulvalenes. The solutions of TTFs in organic solvents ($CHCl_3$, CH_2Cl_2 etc) are more or less unstable. It was observed, for example, that the NMR spectra of compounds (I)[41] in $CDCl_3$ do not show the band at ca δ: 6.33 (CH=CH) some days after the preparation of the solutions even in absence of oxygen . This means that a partial or complete transformation of TTTs to undesirable products takes place on standing of their solution for long time.

METAL 1,2-DIHETEROLENES

A number of metal 1,2-dithiolenes, mainly corresponding to 1,3-dithioles of Class 0 (Table 1), and selenium-, tellurium-, etc- analogs, have been described in refs [10]-[12] and refs therein. Recent work has been described in refs [48], [52], [54]. In this section we describe methods of preparation and properties of cation-rich salts of some metal 1,2-dithiolenes and selenium analogs, having ligands with additional heteroatoms (S,Se,N) to an additional ring or chain. These are compounds of the following formulas, (XIII)-(XXII),

$Z_xM(X_1, X_1; R_1, R_2)_2$ (XIII), $Z_xM(X_1, X_1; R', R')_2$ (XIV)

Z=Li,Na,etc; Bu_4N,Me_4N, etc; x=2,1; M=Zn,Hg,Ni, Pd etc
(A):X_1=S, R_1=R_2=CN, (B): X_1=S, R_1=CN, R_2=SCH_3
(C):X_1=Se, R_1=R_2=CN, (D):X_1=S, R_1+R_2= S-S-C(=S)
(A'):X_1=S, 2R'=pyridine-, (B'):X_1=S, 2R'=5-cyanoisothiazole-, (C'):X_1=S,2R'=pyrazine-group fused to the metal 1,2-dithiolate ring and similar with X_1=Se.

$Z_xM(S, S; S, S >=X)_2$ (XV):X=S , $Z_xM(Se, Se; Se, Se >=X)_2$ (XVII):X=Se

$Z_xM(X_1, X_1; X_2, X_2 >=X)_2$ (XVII), $Z_xM(X_1, X_1; X_2R, X_2R)_2$ (XVIII)

(A):X_1=Se, X_2=X=S, (B):X_1=S, X_2=X=Se, (C):X_1=X_2=S, X=Se, (D):X_1=X_2=Se, X=S.

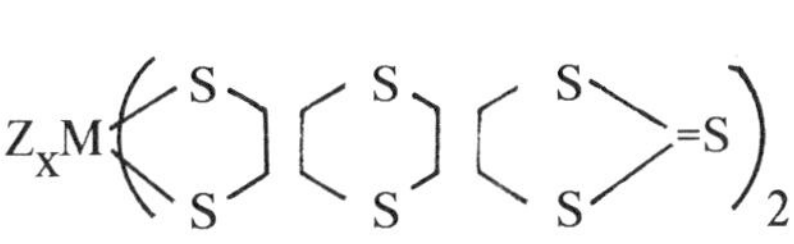

(A): $X_1=X_2=X_3=S$, $X_4=Se$, (B): $X_1=X_2=X_4=S$, X_3Se etc
(a): $R=CH_3$, (b): $2R=CH_2$, (c): $2R=CH_2CH_2$ etc
(i): $2R=>=O$, (ia): $2R=>=S$, etc

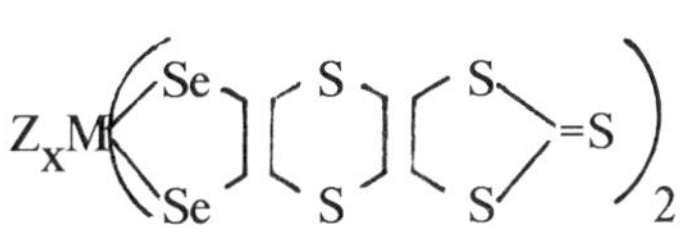

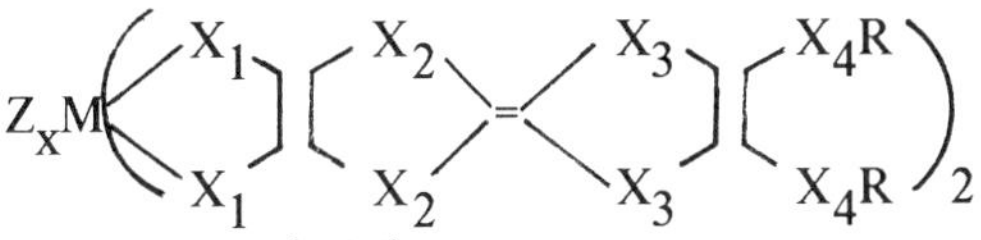

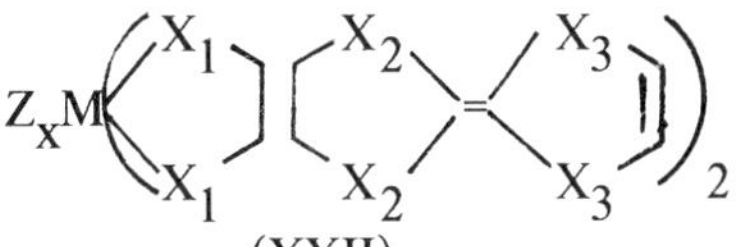

Compounds of the type (XIII) have been prepared from the corresponding disodium dithiolates, or 1,3-dithioles [11], [27],[79], and Se-analogs [111], by methods reported in refs [10] - [12]. Also, compounds of the type (XIV) have been prepared by similar methods [72], [78], [83], [85], [86], [112],[113]. Compounds of the type (XV) have been prepared by chemical or electrochemical reduction of carbon disulfide in dimethylformamide (DMF), followed by reaction with transition-metal salts according to Scheme 7 (see [61]-[65], [78] and refs cited therein). Selenium-

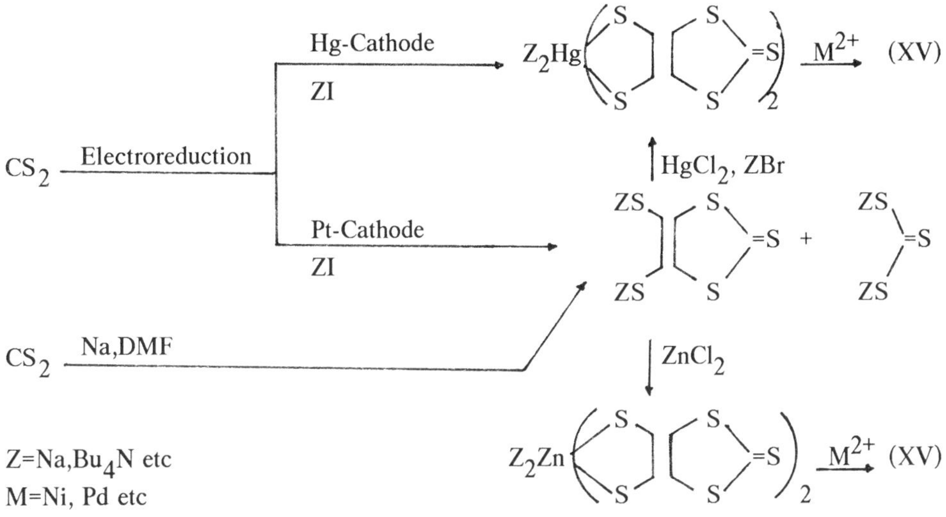

Z=Na, Bu_4N etc
M=Ni, Pd etc

Scheme 7

analogs (XVI) have been prepared from carbon diselenide by similar methods (see [53], [62]-[65],[107]). Using CSSe, instead of carbon disulfide or carbon diselenide, a number of alloys have been obtained [53]. Compounds (XV), (XVII A) and (XVI) have been prepared from vinylene trithiocarbonate and vinylene triselenocarbonate, respectively, by a three-step procedure according to Scheme 8 (see [25],[29],[41],[106]). Attempts to prepare (XVII B)-(XVIID) and similar compounds

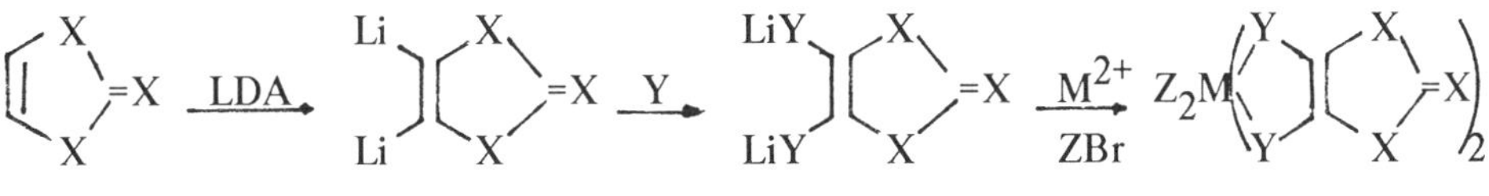

Scheme 8

from the corresponding vinylene trichalcogenocarbonate [114] by this method have been unsuccessful. Instead of them, a number of alloys with symmetrical or unsymmetrical molecules have been obtained [104]. The reason is that, in alkaline medium (LDA), a rearrangement of sulfur and selenium atoms takes place [104]. 1,3-Thiaselenole-2-thione, for example, can be isomerized to 1,3-dithiole-2-selenone, and vice versa [104],[114]. Treatment of the resulting alloys (M=Zn) with alkyl halides, followed by chromatographic separation, a number of compounds (9), (21), (22) etc have been obtained [104]. The isolation of these compounds shows the presence of the corresponding ligand in the considered alloy. Compounds(XV) and (XVIIA), with X=O, [49], [50],[64] instead of X=S, can be prepared by same procedure (Scheme 8) using 1,3-dithiole-2-one and 1,3-diselenole-2-one, respectively, instead of vinylene trithiocarbonate. Some compounds with formulas (XV)-(XVII) can be prepared from 4,5-vinylenedithio-1,3-dithiole-2-thione, 4,5-vinylenediseleno-1,3-dithiole-2-thione, etc by a two-step sequence: reaction with LDA (see Scheme 4c) followed by treatment of a transition metal salt with ZBr. A few of them can be prepared by an easier procedure from the ethylene-analogs (see Scheme 4b) [108]. Some compounds with formula (XVIII) have been prepared by treatment of the corresponding 1,3-dithioles and selenium-analogs with potassium ethoxide or potassium hydroxide, followed by treatment with a transition-metal salt in presence of ZBr [47] [76]. However, this method can not be applied for the preparation of (XVIIIB) (see Scheme 4b). Instead, the compounds of the type (XVIIA) have been obtained [108]. Compounds of the type (XIX) can be isolated from the reaction products of 4,5-vinylenedithio-1,3-dithiole-2-thione with LDA, after sulfurization with sulfur, and treatment with a transition-metal salt in the presence of ZBr, according to the following Scheme 9 [34] , [36],[45] ,[108]:

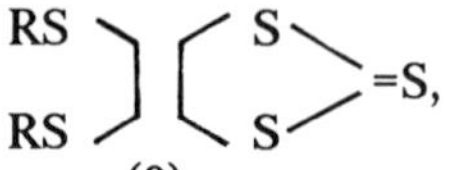

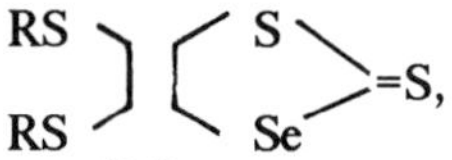

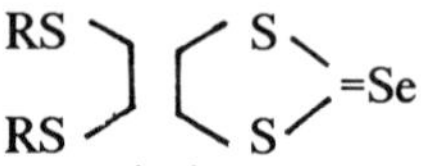

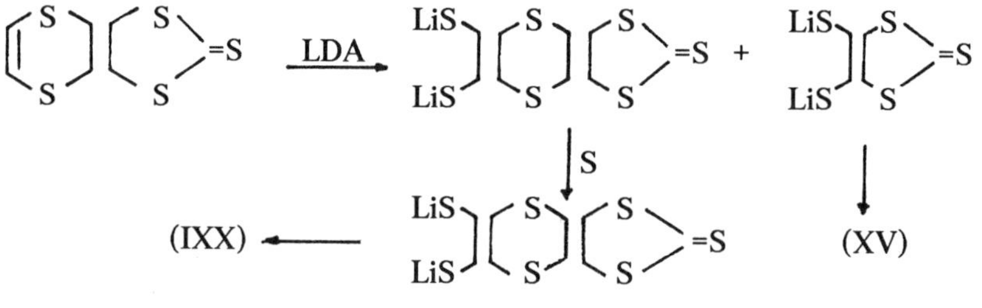

Scheme 9

Compounds of the type (XX) can be prepared by a similar procedure. However, in both of the last cases, the separation of (XIX) and (XX) from the byproducts (XVIII) is difficult [108]. Similar procedures have been applied for the preparation of (XXI), (XXII) and similar compounds (see [76], [77],[108],[115]) from the corresponding compounds of the type (I), (II) etc.

Purification of compounds (XIII)-(XXII) and conversion from a cation Z to an other Z', or from a metal to another one, can be performed by a three-step sequence: treatment with benzoyl chloride, reaction with sodium ethoxide and treatment with a transition metal salt in presence of ZBr or Z'Br. All the compounds of the type (XIII)-(XXII) can be transformed to the corresponding neutral compounds (x=0) after treatment with an oxidizing reagent (e.g. Br_2) or by electrooxidation.

Using compounds of the type (I)-(XXVII), a number of charge-transfer complexes (CTC), cation-radical salts (CRS), and cation deficient metal 1,2-diheterolenes (CDMD) have been prepared [20], [37]-[41], [61]-[99], [108], [109], by chemical (direct reaction or diffusion) or electrochemical procedures, some of which are outlined in the following Scheme 10 :

(a) $xD + TCNQ \rightarrow D_xTCNQ$ (x=0,5,1,2;D=donor)

(b) $xD + I_3^- \rightarrow D_xI_3$ (x=1,2 etc)

(c) $ZX + X' \rightarrow ZX' + X$; $ZX + KTCNQ + TCNQ \longrightarrow Z(TCNQ)_2 + KX$

(d) $Z_2X \rightarrow Z_1X \rightarrow Z_xX (1>x>0)$

(e) $ZX + (TTF)X' \rightarrow Z(TTF) + ZX' (Z=cation)$

Scheme 10

The salts, obtained by these methods, have been found have in several varying chemical compositions, crystal structures and physical properties, which depend in most cases on the counter ion and the conditions of crystallization. Table 2 gives the formulas of some salts, and information on their corresponding composition and

Table 2. Examples of some conducting and superconducting salts

	Salt	Behavior	Refs
(1)	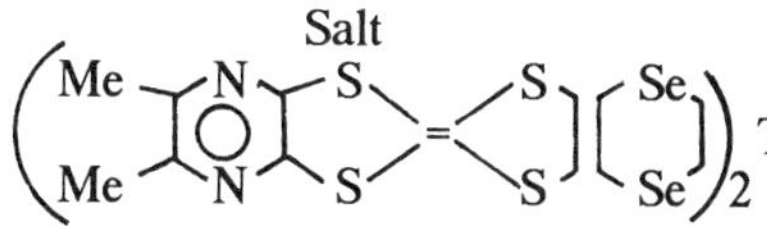 TCNQ	semiconducting	[109]
(2)	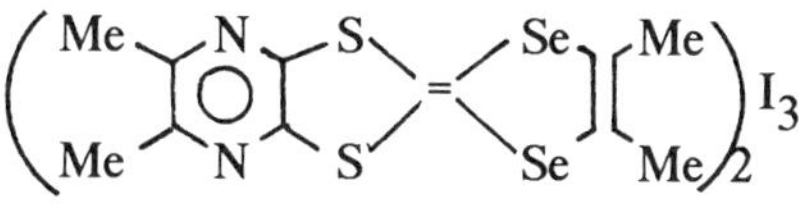 I_3	metallic,MIT at ca 120K	[89]
(3)	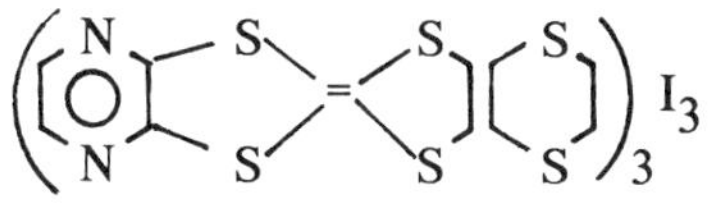 I_3	metallic,MIT at ca 60K	[80]
(4)	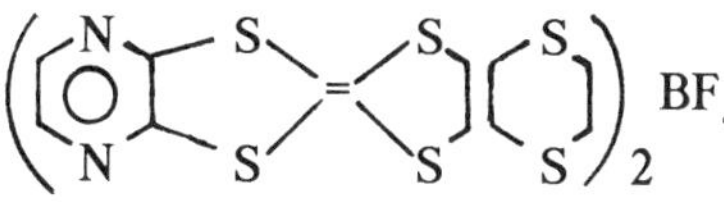 BF_4	metallic,MIT at ca 180K	[84]
(5)	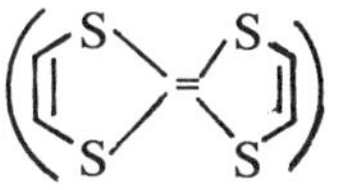 $Ni(dmit)_2$	semiconducting	[61],[62]
(6)	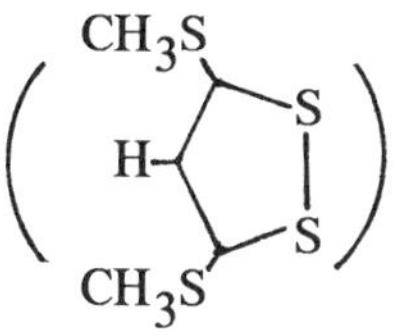 $[Ni(dmit)_2]_2$	semiconducting	[76]
(7)	$K_2[Ni(dmit)_2]_5$	semiconducting	[44],[63]-[65]
(8)	$Cs[Pd(dmit)_2]_2$	metallic, MIT at 60K	[44],[63][65]
(9)	$(Me_4N)_x[Pd(dmit)_2]$	metallic MIT at ca 280K	[98]

Compound	Anion	Behavior	Ref.
(10)	$Ni(dcit)_2$	metallic, MIT at ca 125 K	[98]
(11)	I_{3+x}	semiconducting	[93]
(12)	AuI_2	metallic, MIT at 60 K	[41],[98]
(13)	IBr_2	metallic down to 1.35 K	[94]
(14)	IBr_2	metallic down to 1.35K	[94]
(15)	AuI_2	superconducting, T_c at ca5K	[98],[94]

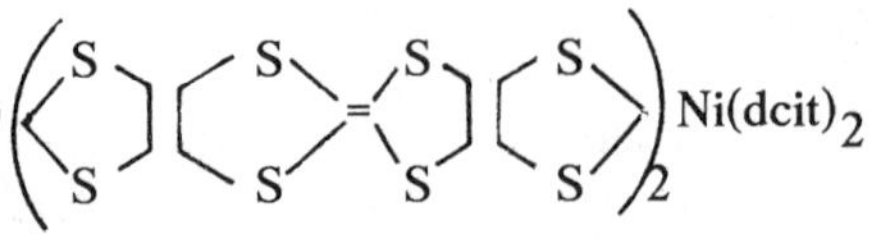

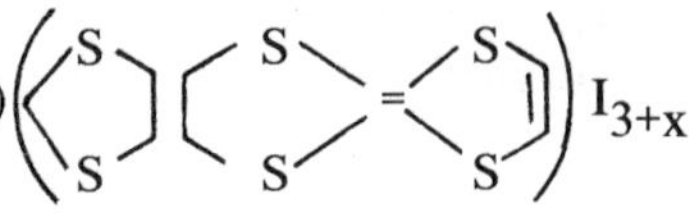

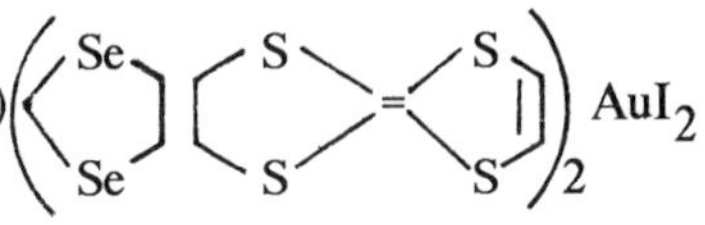

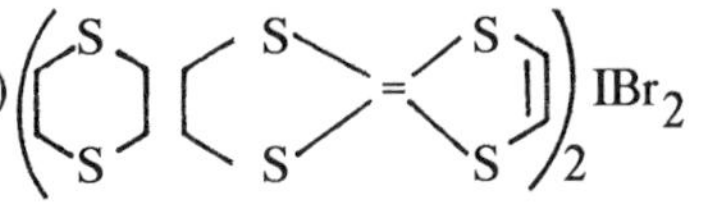

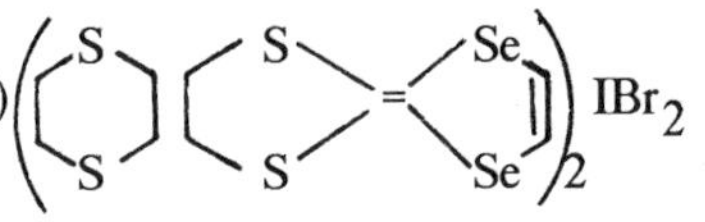

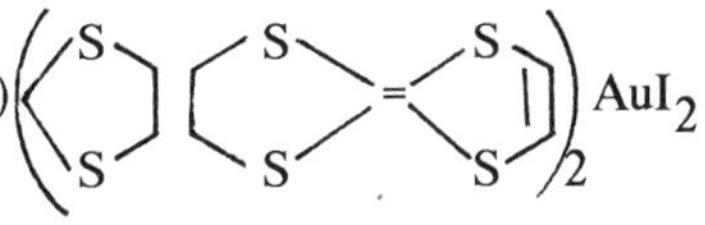

$Ni(dmit)_2$: S=… =S, $Ni(dcit)_2$: …

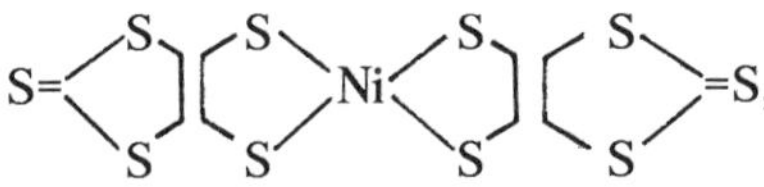

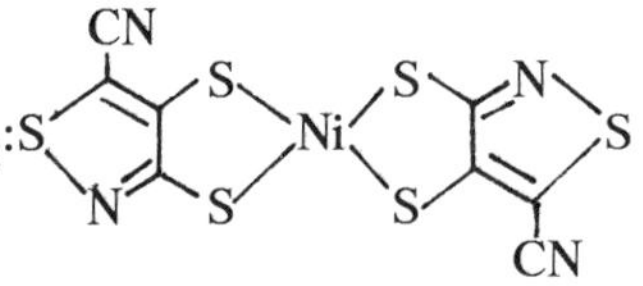

electrical behavior. Crystal structure solutions, which have been performed in a number of salts, showed that there are strong intermolecular interactions due to the S--S, S--N, Se--Se, Se---S, Se---N contacts [95]-[97],[91],[87],[88] [80], [73]. Compound(1) behaves as a semiconductor, similar to those reported in ref [14b] and ref [15] cited therein, which are compounds of tellurium and selenium:(DMHM-TTeF)$_x$TCNQ, (CpHMT-TSF)$_x$TCNQ, (DMHM-TSF)$_2$TCNQ. Compound (3) and similar compounds, reported in refs [75],[80], [87], have a composition 3:1. According to a treatment given in ref [116], in the case of salts with stoichiometry 2:1, 4:3 and 1.1:1, the anionic chains (A) provide an external potential on the organic stack (D) which corresponds to a Fermi wave vector q^A with values $4k_F^D$, $2k_F^D$ and $\frac{4}{3}k_F^D$, respectively [116]. In the case of 3:1 salts, it should be $q^A \geq 6k_F^D$. Compound (6) and similar compounds [66], [69], [76], [79] are semiconductors under ambient pressure, and are of the same stoichiometry as (Me_4N) $[Ni(dmit)_2]_2$ (:superconductor under pressure). Conductivity measurements under high pressure are required for this kind of salts. Compounds (13) and (14) which are isostructural with the β-(BEDTTTF)$_2$X superconductors [4], are metallic down to 1.35 K.

These compounds have a residual conductivity $\sigma_{1.35K}$>10000 Ω^{-1} cm^{-1} [98]. According to the treatment reported in refs [117], [3], the superconductivity in organic systems requires a minimum residual conductivity of 6000 Ω^{-1} cm^{-1}, a value consistent with the theoretical estimation based on the Anderson localization. Compounds (13) and (14) could be candidates for superconductivity as they fulfill the above conditions. This means that conductivity measurements are required at lower temperatures. Compound (15) consists of unsymmetrical molecules of methylene dithiotetrathiafulvalene (MDTTTF) [20], [37]. Compound (15) and compounds (16),(17),(18) are almost isostructural at room temperature[94],[97], [99],[100], [118],[119]. They are superconducting crystals, which consist of tightly linked dimers almost orthogonally arranged, to form 2-dimensional S---S, or Se---S networks (see [94], [119] and refs therein).

κ-( )$_2$ I_3, (16) κ-( )$_2$ $Cu(NCS)_2$ (17)

κ-( )$_2$ $AuBr_2$ (18)

Fig.1 shows the temperature (T) dependence of resistivity (ρ) of superconductors (15) [94], (16)[119] , (17) [118] and (18) [99]. One can observe that the resistivity of (15) decreases smoothly with decreasing temperature down to superconducting transtition temperature. This observation indicates that the crystal structure

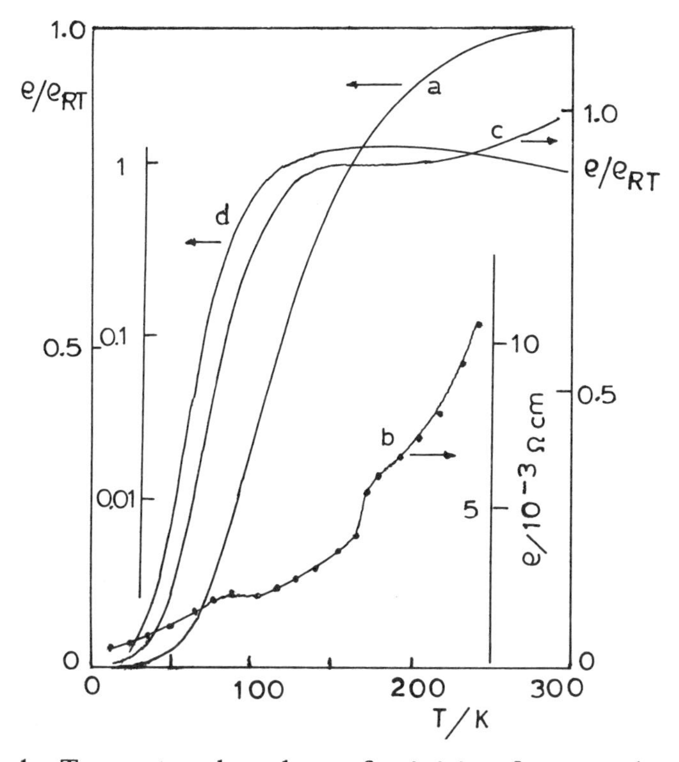

Fig. 1. Temperature dependence of resistivity of superconductors: (15)(a), (16)(b), (17)(c), and (18)(d) for $T_c < T \leq T_{RT}$.

remains the same as that at room temperature. These findings from the considered compound (15) firmly dispel the following two notions [8]:

(1) symmetrical molecules are essential for designing new conductors and superconductors,

(2) the molecules must be packed in regular stacks in the crystals of conductors and superconductors

ACKNOWLEDGEMENTS

I want to express my deepest gratitude to my coworkers and colleagues who have participated to the experimental part of this work (synthesis and crystal structure) particularly C.Mayer, J.Zambounis, G.Mousdis,V.Kakoussis,V.Gionis, K.Kobayashi, A.Terzis and E.Kamitsos. Also, I am grateful to B.Hilti for communicating results of the conductivity measurements before publication.

REFERENCES

1. H.Inokuchi "New Organic Superconductors" Angew.Chem.,Int.Ed.Engl. 27, 1747 (1988).
2. T.Ishiguro "Superconductivity in Organic Charge Transfer Salts" Physica C, 153-155, 1055 (1988).
3. M.Tokumoto "Experimental Study on the Superconductivity in Organic Metals, , β-$(BEDTTTF)_2X$" Researches of the Electrochemical Laboratory 1-1-4, Umezono, Tsukuba, Ibaraki, Japan (1988).
4. J.M.Williams, H.H.Wang, T.J.Emge, U.Geiser, M.A.Beno, P.C.W.Leung, K.D.Carlson, R.J.Thorn, A.J.Schultz, and W.H.Whangbo, "Rational Design of Synthetic Metal Superconductors" Progr.Inorg.Chem. 35, 51 (1987).
5. L.Alcacer and H.Novais "Linear Chain 1,2-Dithiolene Complexes" in "Extended Linear Chain Compounds" Ed.J.S.Miller, Plenum, N.Y. 3, 319 (1983).
6. G.Schukat, A.M.Richter, and E.Fanghanel "Synthesis, Reactions, and Selected Physico-chemical Properties of 1,3- and 1,2-Tetrachalcogenefulvalenes" Sulfur Reports, 7, 155 (1987).
7. A.Krief "Syntheses of Tetraheterafulvalenes and of Vinylene Triheterocarbonates- Strategy and Practice" Tetrahedron 42, 1209 (1986).
8. R.N.Lyubovskaya "Organic Metals and Superconductors Based on Tetrathiofulvalene Derivatives" Russ. Chem.Rev. 52, 736 (1983).
9. M.Narita and C.U.Pittman, Jr., "Preparation of Tetrathiafulvalenes (TTF) and their Selenium Analogs-Tetraselenafulvalenes (TSeF)" Synthesis 489 (1976).
10. U.T.Mueller-Westerhoff and B.Vance "Dithiolenes and Related Species, Including Dithiolene-like Ligands" Comprehensive Coord.Chem. 2, (chapt.15, part 5) , Ed.G.Wilkinson, Pergamon Press (1987).
11. R.P.Burns and C.A. Mc Auliffe, "1,2-Dithiolene Complexes of Transition Metals" Adv.Inorg.Chem.Radiochem. 22, 303 (1979).
12. S.Alvarez, R.Vicente, and R.Hoffmann, "Dimerization and Stacking in Transition-Metal Bisdithiolanes and Tetrathiolates", J.Am.Chem.Soc.107, 6253 (1985).
13. K.Lerstrup, I.Johannsen, and M.Jorgensen, Synth.Metals 27, B9 (1988).
14. R.D.McCullough, M.D.Mays, A.B.Bailey, and D.O.Cowan, Synth.Metals 27, B487(1988); A.B.Bailey, R.D.McCullough, M.D.Mays, D.O.Cowan, and K.A.Lerstrup, Synth.Metals 27, B425 (1988);D.O.Cowan, this volume.

15. S.Bittner, A.Maradpour, and P.Krief, Synthesis 132 (1989).
16. A.S.Dhindsa, M.R.Bryce, J.P.Lloyd , M.C.Petty, K.Kobayashi and H.Tukada, J.Chem.Soc., Chem.Commun., 1391 (1988); A.S.Dhindsa, M.R.Bryce, J.P.LLoyd, and M.C.Perry, Thin Solid Films 165, L97(1988);Synth.Metals 27, B563 (1988);R.M.Metzger, this volume.
17. Y.A.Jackson, C.L.White, M.V.Lakshmikantham and M.P.Cava, Tetrahedron Lett. 28, 5635 (1988).
18. J.M.Fabre, A.K.Gouasmia, L.Giral and D.Shasseau, Tetrahedron Lett., 29, 2185 (1988).
19. O.G.Safier, D.V.Nazarov, V.V.Zorin, and D.L.Rakhmankulov, Khim.Geterotsikl.Soedin, 61, 852 (1988); M.Ishida, K.Sugiuva, K.Takagi, H.Hiraoka, and S.Kato, Chem.Lett. 1705 (1988); M.E.Jung, R.B.Blum, B.J.Gaede, and M.R.Gisler, Heterocycles 28, 93 (1988);T.Suzuki, H.Yamochi, G.Srdanov, K.Hinkelmann and F.Wudl, J.Am.Chem.Soc. 111, 3108 (1989) and ref.[13] cited therein.
20. G.C.Papavassiliou, G.A.Mousdis, S.Y.Yiannopoulos, V.C.Kakoussis and J.S.Zambounis, Synth.Metals 27, B373 (1988).
21. A.M.Kini, B.D.Gates, S.F.Tykto, T.J.Allen, S.B.Kleinjan, H.H.Wang, L.K.Montgomery, M.A.Beno, and J.M.Williams, Synth.Metals 27, B445 (1988).
22. J.Larsen and C.Lenoir, Synthesis 134 (1989).
23. J.Larsen and K.Bechgaard, J.Org.Chem. 52, 3285 (1987); K.S.Varma, J.Evans, S.Edge, A.E.Underhill, G.Bojesen, and J.Becher, J.Chem. Soc., Chem. Commun. 257 (1989).
24. N.Iwasawa, H.Urayama, H.Yamochi, G.Saito, K.Imaeda, T.Mori, Y.Maruyama, H.Inokuchi, T.Enoki, Y.Higughi, and N.Yasuoka, Synthetic Metals 27, B463 (1988).
25. P.J.Nigrey, B.Morosin, and E.Duesler, Synth.Metals 27, B481 (1988); P.J.Nigrey, B.Morosin, and J.F.Kwak, in "Novel Superconductivity" Ed.S.A.Wolf and V.Z.Kresin, Plenum, 171 (1987).
26. Ya.N.Kreitsberga, K.A.Bolodis, Z.L.Kraupsha, and O.Ya.Neiland, Zh.Org.Chim. 24, 1243 (1988).
27. W.Kolling, M.Augustin, and R.Ihrke, Synthesis 655 (1987).
28. R.R.Schumaker, E.Dupart, J.P.Morand, C.Coulon and P.Delhaes, to be published.
29. P.J.Nigrey, J.Am.Chem.Soc. 53, 201 (1988); Synth.Metals 27, B365 (1988).
30. P.Morand, R.Lapouyade, P.Delhaes, M.Vanderyver, J.Richard, and A.Barraud, Synth.Metals 27, 569 (1988).
31. A.Otsuka, G.Saito, T.Nakamura, M.Matsumoto, K.Kawabata, K.Honda, M.Koto, and M.Kurahashi, Synth.Metals 27, B575 (1988).
32. S.-Y. Hsu and L.Y.Chiang , Synth.Metals 27, B651 (1988);G.Seitz and P.Imming, Arch.Pharm. 321, 757 (1988).
33. W.Chen, M.P.Cava, M.A.Takassi, and R.M.Metzger , J.Am.Chem.Soc.110, 7903 (1988).
34. T.Nogami, Jpn.J.Appl.Phys. Series 1, p.197 (1988).
35. S.Rajeswari, Y.A.Jackson, and M.P.Cava, J.Am.Chem.Soc., 110, 1089 (1988).
36. H.Nakano, T.Nogami, and Y.Shirota, Bull.Chem.Soc.Jpn. 61, 2973 (1988).
37. G.C.Papavassiliou, J.S.Zambounis, G.A.Mousdis, V.Gionis and S.Y.Yiannopoulos, Mol.Cryst.Liq.Cryst. 156, 269 (1988).
38. G.C.Papavassiliou, V.Gionis, S.Y.Yiannopoulos, J.S.Zambounis, G.A.Mousdis, K.Kobayashi, and K.Umemoto, Mol.Cryst. Liq.Cryst. 156, 277 (1988).

39. G.C.Papavassiliou, G.A.Mousdis, S.Y.Yiannopoulos and J.S.Zambounis, Chem.Scripta 28, 3654 (1988),and unpublished work for Se-analogs..
40. G.C.Papavassiliou, V.C.Kakoussis, G.A.Mousdis, and J.S.Zambounis, Chem.Scripta, in press and unpublished work for Se-analogs.
41. G.C.Papavassiliou, V.C.Kakoussis, J.S.Zambounis, G.A.Mousdis, and C.W.Mayer, Chem.Scripta, in press and unpublished for S and Se-analogs.
42. R.Kato, H.Kobayashi, H.Kim, A.Kobayashi, Y.Sasaki, T.Mori, and H.Inokuchi, Synth.Metals 27, B359 (1988).
43. H.H.Wang et al, Chem.Materials, 1, 140 (1989).
44. R.A.Clark and A.E.Underhill, Synth.Metals 27, B515 (1988); R.A.Clark, A.E.Underhill, I.D.Parker and R.H.Friend, J.Chem.Soc., Chem.Commun. 228 (1989).
45. C.Le Coustumer, N.Bennasser, and Y.Mollier, Synth.Metals 27, B523 (1988).
46. T.Nakamura, H.Tanaka, M.Matsumoto, H.Tochibana, E.Manda, and Y.Kawabata, Synth.Metals 27, B601 (1988);T.Nakamura et al Chem.Lett. 367 (1989).
47. T.Nakamura, T.Nogami, and Y.Shirota, Bull.Chem.Soc.Jpn 60, 3447 (1987).
48. I.Tabushi, K.Yamamura, and H.Nonoguchi, Chem.Lett. 1373 (1987).
49. R.Vicente, J.Ribas, S.Alvarez,A.Segui,X.Solans,andM.Verdaguer, Inorg.Chem. 26, 4004 (1987); R.M.Olk et al , Z.anorg.Allg. Chem. 456, 131 (1988).
50. R.Vicente, J.Ribas, C.Faulman, J.-P.Legros, and P.Cassoux, C.R.Acad.Sci.Paris 305, Series II, 1055 (1987).
51. R.Vicente, J.Ribas, X.Solans, M.Font-Altaba, A.Mari, P.deLoth, and P.Cassoux, Inorg.Chim.Acta, 132, 229 (1987).
52. D.M.Giolando, T.B.Rauchfuss and A.L.Rheingold, Inorg.Chem.26, 1636(1987).
53. R.M.Olk, W.Dietzsth, J.Mattusch, J.Stach, C.Nieke, E.Hoyer, W.Meiler and W.Robien, Z.anorg.Allg.Chem. 544, 199 (1987); ref [13] cited therein.54. H.Shiozaki, H.Nakazumi, Y.Nakado, and T.Kitao, Chem.Lett., 2393 (1987).
55. A.Kobayashi, H.Kim, Y.Sasaki, S.Moriyama, Y.Nishio, K.Kajita, W.Sasaki, R.Kato, and H.Kobayashi, Synth.Metals 27, B339 (1988).
56. J.Amzil, J.-M.Catel, G.LeCoustumer, Y.Mollier, and J.-P.Sauve, Bull.Soc.Chem.France No1, 101 (1988); see also ref [66] here.
57. R.R.Schumaker, S.Rajeswari, M.V.Joshi, M.P.Cava, M.A.Takassi, and R.M.Metzger, J.Am.Chem.Soc., 111, 308 (1989)
58. Z.Yoshida and T.Sugimoto, Angew.Chem.Int.Ed.Engl. 27, 1573 (1988).
59. U.Zoller, Tetrahedron, 44, 7413 (1988).
60. R.Gompper and H.-U.Wagner, Angew.Chem.Int.Ed.Eng. 27, 1437 (1988).
61. G.C.Papavassiliou, Z.Naturforsch, 36b, 1200 (1981).
62. G.C.Papavassiliou, Mol. Liq.Cryst., 86, 159 (1982).
63. G.C.Papavassiliou , Z.Naturfors. 37b, 825(1982).
64. G.C.Papavassiliou, J.Physique C3, 1257 (1983).
65. G.C.Papavassiliou, A.M.Cotsilios and C.S.Jacobsen,J.Mol.Struct., 115, 41 (1984).
66. G.C.Papavassiliou, Chem.Scripta 25, 167(1985).
67. G.C.Papavassiliou, S.Y.Yiannopoulos and J.S.Zambounis, Mol.Cryst.Liq.Cryst. 120, 333 (1985).
68. G.C.Papavassiliou, E.I.Kamitsos and J.S.Zambounis, Mol.Cryst.Liq.Cryst. 120, 315 (1985).
69. G.C.Papavassiliou, J.S.Zambounis, A.E.Underhill, B.Kaye and H.P.Geserich,Mol.Cryst. Liq.Cryst. 134, 53 (1986).
70. A.E.Underhill, B.Kaye, G.C.Papavassiliou and S.Y.Yiannopoulos, Mol.Cryst. Liq.Cryst. 134, 59(1986).

71. G.C.Papavassiliou, H.P.Geserich, S.Y.Yiannopoulos and J.S.Zambounis, J.Mol.Structure, 143, 215(1986).
72. G.C.Papavassiliou, Chim.Chron. (New Series), 15, 161 (1986).
73. A.Terzis, A.Hountas and G.C.Papavassiliou, Acta Crystallor. C42, 1584(1986).
74. G.C.Papavassiliou, S.Y.Yiannopoulos, and J.S.Zambounis, J.Chem.Soc.Chem.Commun. 820, (1986).
75. G.C.Papavassiliou, S.Y.Yiannopoulos, and J.S.Zambounis , Physica 143B, 310 (1986).
76. G.C.Papavassiliou, J.S.Zambounis and S.Y.Yiannopoulos, Physica 143 B, 307 (1986).
77. G.C.Papavassiliou, J.S.Zambounis and A.G.Mousdis, Abstr. of Papers of XXIV Int.Conf.Coord.Chem., August 24-29, 1986, Athens, Greece; Chim.Chron. New Series (Special Issue) August 1986, p.246; see also ref [76] here.
78. G.C.Papavassiliou, S.Y.Yiannopoulos, and J.S.Zambounis, Chem.Scripta, 27, 265 (1987).
79. G.C.Papavassiliou, J.S.Zambounis, and S.Y.Yiannopoulos, Chem.Scripta 27, 265 (1987); and unpublished work.
80. G.C.Papavassiliou, A.Terzis, A.E.Underhill, H.P.Geserich, B.Kaye, A.Hountas and S.Y.Yiannopoulos, Synth.Metals, 19, 703 (1987).
81. A.Terzis, E.I.Kamitsos, V.Psycharis, J.S.Zambounis, J.Swiatek and G.C.Papavassiliou, Synth.Metals, 19, 481 (1987).
82. G.C.Papavassiliou, S.Y.Yiannopoulos, J.S.Zambounis, K.Kobayashi, and K.Umemoto, Chem.Lett., 1279 (1987).
83. G.C.Papavassiliou, G.A.Mousdis, V.Gionis, J.S.Zambounis and S.Y.Yiannopoulos, Z.Naturfors. 42b, 1050 (1987).
84. G.C.Papavassiliou, A.E.Underhill, B.Kaye, H.P.Geserich, A.Terzis and S.Y.Yiannopoulos, Material Science, 13, 185 (1987).
85. G.C.Papavassiliou, S.Y.Yiannopoulos and J.S.Zambounis, in “Organic and Inorganic Low Dimensional Crystalline Materials” Ed.P.Delhaes, and M.Drillon, Plenum, 168, 305 (1987);ref [8] cited therein.
86. G.C.Papavassiliou, G.Z.Mousdis, J.S.Zambounis and S.Y.Yiannopoulos, in Organic and Inorganic Low Dimensional Crystalline Materials” Ed.P.Delhaes and M.Drillon, Plenum 168, 301 (1987).
87. V.Psycharis, A.Hountas, A.Terzis and G.C.Papavassiliou, Acta Cryst. C., 44, 125 (1988).
88. A.Terzis, V.Psycharis, A.Hountas and G.C.Papavassiliou, Acta Cryst. C.44, 128 (1988).
89. K.Kikuchi, H.Kamio, K.Saito, G.C.Papavassiliou, K.Kobayashi, I.Ikemoto and S.Y.Yiannopoulos, Bull.Chem.Soc.Jpn., 61, 741 (1988).
90. J.S.Zambounis, E.I.Kamitsos, A.Terzis and G.C.Papavassiliou, J.Mol.Structure, 174, 189 (1988).
91. A.Terzis, A.Hountas, and G.C.Papavassiliou, Sol.St.Commun., 66, 1161 (1988).
92. J.S.Zambounis, E.I.Kamitsos, M.A.Karakassides and G.C.Papavassiliou, Proc.XIth Int.Conf.Raman Spectrosc. (London 5-9 Sept. 1988), Ed.R.J.H.Clark and D.A.Long, J.Wiley and Sons, 425 (1988).
93. A.Terzis, A.Hountas, A.E.Underhill, A.Clark, B.Kaye, S.Y.Yiannopoulos, G.Mousdis and G.C.Papavassiliou, Synth.Metals, 27, B97 (1988).
94. G.C.Papavassiliou, G.A.Mousdis, J.S.Zambounis, A.Terzis, A.Hountas, B.Hilti, C.W.Mayer and J.Pfeiffer, Synth.Metals, 27, B379 (1988).

95. A.Terzis and G.C.Papavassiliou, H.Kobayashi, A.Kobayashi, Acta Cryst. in press.
96. A.Hountas, A.Terzis, G.C.Papavassiliou, B.Hilti and J.Pfeiffer, Acta Cryst. in press;A.Terzis, A.Hountas, G.C.Papavassiliou, B.Hilti and J.Pfeiffer, Acta Cryst., in press.
97. A.Hountas, A.Terzis, G.C.Papavassiliou, B.Hilti, M.Burkle, C.W.Mayer and J.S.Zambounis, Acta Cryst., in press.
98. B.Hilti et al to be published; M.M.Freund and B.Hilti, Eur.Conf.Abst.; 9th Gen.Conf.Condensed Matter, (6-9 March 1989, Nice, France) 13A, I-61.
99. K.Kikuchi, Y.Ishikawa, K.Saito, I.Ikemoto, and K.Kobayashi, Synth.Metals, 27, B391 (1988) and refs therein.
100. A.M.Kini, M.A.Beno, D.Son, H.H.Wang, K.D.Carlson, L.C.Porter, U.Welp, B.A.Vogt, J.M.Williams, D.Jung, M.Evain, M.-H.Whangbo, D.L.Overmyer and J.E.Schirber, Sol.St.Commun., 69, 503 (1989).
101. J.M.Fabre, L.Giral, E.Dupart, C.Coulon, J.P.Manceau and P.Delhaes, J.Chem.Soc., Chem.Commun., 1477 (1983).
102. C.Mayer, J.Zambounis, and G.C.Papavassiliou, unpublished: E_1, E_2 of MDTTTF, EDTTTF etc are lower but close to those of BEDTTTF: see also refs [39],[58] cited in ref [38] here.
103. N.Plarac, L.W.J.Still, M.S.Chauham and D.M.McKinnon, Can.J.Chem., 53, 836 (1975); refs [32], [65] cited in ref[6] here.
104. G.A.Mousdis, V.C.Kakoussis and G.C.Papavassiliou, unpublished work: (4) is a red solid, mp=138°C; (6) is a yellow solid, mp=45°; (8) is a greenish -yellowsolid,mp=171°C;(21b)mp=152°C;(21c)mp=116°C;(22b)mp=176°C;(22c) mp=138°C see also ref [109] herein.
105. R.R.Schumaker, V.Y.Lee, and E.M.Engler, J.Physique C3 44, 1139 (1983).
106. G.C.Papavassiliou, V.C.Kakoussis and G.A.Mousdis, to be published, see also ref [40] herein.
107. V.Y.Lee, E.M.Engler, R.R.Schumaker, and S.S.P.Perkin J.Chem.Soc.,Chem.Commun. 236 (1983)
108. G.C.Papavassiliou and G.A.Mousdis unpublished work.
109. G.A.Mousdis, V.K.Kakoussis and G.C.Papavassiliou, this volume and unpub lished work.
110. H.Poleschner, Z.Chem. 26, 138 (1986).
111. H.Poleschner and E.Fanghanel, J.Prakt.Chem. 324, 691 (1982).
112. G.C.Papavassiliou and S.Y.Yiannopoulos, unpublished work.
113. S.Lahner, Y.Wakatsuki, and H.Kish, Chem.Ber. 120, 1011 (1987).
114. M.V.Lakshmikantham and M.P.Cava, J.Org.Chem. 45, 2632 (1980).
115. E.M.Engler, V.V.Patel and R.R.Schumaker, Tetrahedron Lett. 22, 2035 (1981); R.R.Schumaker and E.M.Engler, J.Am.Chem.Sol., 102, 6652 (1980).
116. J.Pouget, in "Organic and Inorganic Low-Dimensional Crystalline Materials", Ed.P.Delhaes and M.Drillon , Plenum 168, 185 (1987).
117. M.Tokumoto, H.Anzai, K.Kurata, K.Kajimura and T.Ishiguro, Jpn.J.Appl.Physics 26, suppl. 26-3, 1977 (1987) ; Synth.Metals 27A, 251 (1988).
118. A.Ugawa, G.Ojima, K.Yokushi and H.Kuroda, Synth. Metals 27, A445 (1988) and refs therein;G.Saito, this volume.
119. H.Kobayashi, R.Kato, A.Kobayashi, S.Moriyama, Y.Nichio, K.Kajita and W.Sasaki, Synt.Metals 27, A283 (1980);R.Kato et al Chem. Lett. 507 (1987)and refs therein.

135 K CRYSTALLOGRAPHIC AND ELECTRONIC STRUCTURE OF $(TMTTF)_2SbF_6$

T. Granier*, B. Gallois*, A. Fritsch#, L. Ducasse#, and C. Coulon§

Laboratoires de *Cristallographie et de #Physicochimie Théorique
et §Centre de Recherche Paul Pascal
Université de Bordeaux I
F-33405 Talence, FRANCE

INTRODUCTION

Extensive studies of $(TMTCF)_2X$ salts (where C=T (for sulfur), S (for selenium), and X is a small inorganic anion, i.e. PF_6^-, AsF_6^-, SbF_6^-, ClO_4^-, ...) have been carried out since the discovery of superconductivity in the first organic compound $(TMTSF)_2PF_6$ at low temperature (T = 1.2 K) and high pressure (P = 10 kbar) [1]. An important result of these studies has been the building of a unified (T, P) phase diagram by Caron and Bourbonnais [2] (Fig. 1), which relates the different physical behaviors of the sulfur and selenium compounds: $(TMTSF)_2X$ salts generally evolve at low temperature from a metallic state to an antiferromagnetic (AF) ground state, which is destroyed under moderate pressure, to give rise to superconductivity. Sulfur compounds, on the other hand, exhibit an electronic localization around 120-200 K [3], and reach at lower temperature either a Spin-Peierls (SP) or an AF ground state [4].

Differences between S and Se-based series are at present interpreted with the help of theoretical models, in which major parameters have a structural origin: electronic dimerization of the organic stacks, as well as interchain coupling, appear to be relevant parameters in the driving forces of competing instabilities.

As far as octahedral anions are concerned, the TMTTF salts are close to the boundary between the SP and the AF phases: the PF_6 compound reaches the SP ground state at 1 bar and 15 K, and crosses the SP-AF boundary at a pressure of 10 Kbars, while SbF_6 is AF at

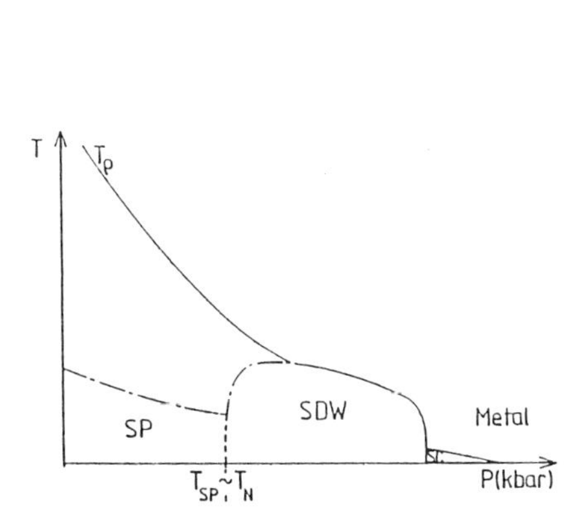

Figure 1. Unified (T, P) phase diagram of $(TMTTF)_2X$ and $(TMTSF)_2X$ from Ref. [2].

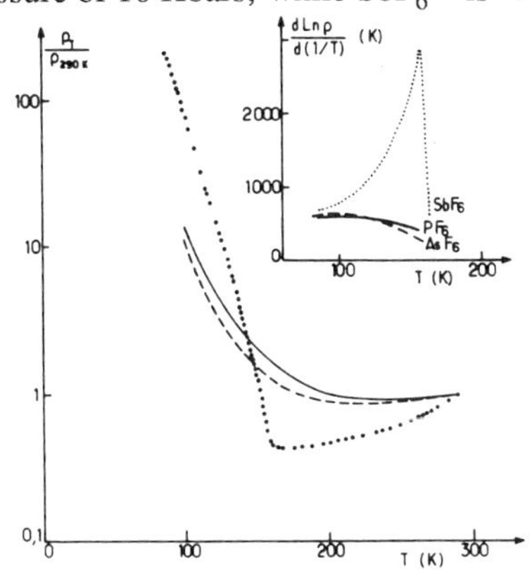

Figure 2. Normalized resistivity of $(TMTTF)_2PF_6$, $(TMTTF)_2AsF_6$ and $(TMTTF)_2SbF_6$ from Ref. [3].

Lower-Dimensional Systems and Molecular Electronics
Edited by R.M. Metzger *et al.*, Plenum Press, New York, 1991

ambient pressure and for a temperature of 6 K. Therefore, we expect the crossover to be induced by small differences in the structures. Our previous structural studies at low temperature (4 K) on the $(TMTTF)_2PF_6$ and AsF_6 salts [5] have raised several unanswered questions regarding the validity of the different theoretical models: in these latter, parameters, such as the electronic dimerization [6] or T_ρ, the temperature of maximum conductivity [2] play an important role. From this point of view, it is clear that the study of the structure of $(TMTTF)_2SbF_6$ at low temperature is necessary to bring new data. Unfortunately, due to the small size of the SbF_6 crystals, it was not possible to determine by neutron diffraction the 4 K structure of this compound. Nevertheless, we carried out an X-ray diffraction study at a temperature of 135 K. Although this temperature is far from 4 K, we believed that these structural data would be important for the following reasons:

- an anomaly in the conductivity measurements of $(TMTTF)_2SbF_6$ is observed at 154 K, in contrast with smooth variations observed in $(TMTTF)_2PF_6$ and $(TMTTF)_2AsF_6$ [3] (see Figure 2).

- when lowering temperature, major structural changes occur within the first 200 K, and a structure determination at around 100 K could shed some light on the structural evolution of this salt at lower temperature.

In the following, we describe structural data of $(TMTTF)_2SbF_6$ at 135 K, and compare the results with those of the PF_6 and AsF_6 structures at 4 K: transfer integral calculations are discussed along with the influence of the anion on the organic stacks in terms of coulombic potential interactions.

STRUCTURAL DATA AND TRANSFER INTEGRAL

Experimental details

X-ray diffraction experiments were performed on an Enraf-Nonius CAD-4 four- circle diffractometer. The crystal cooling to 135 K was achieved by cold nitrogen gas flow, surrounded by a jacket of dry nitrogen gas at room temperature, to prevent frost from growing around the sample. The X-ray radiation wave length was $\lambda = 1.5418$ Å, the crystal size was 0.12 x 0.30 x 0.60 mm^3. The unit cell parameters were determined by least-squares refinement of 21 reflections. 1796 independent reflections were collected (range: $0 \leq h \leq 6$; $-7 \leq k \leq 7$; $-13 \leq l \leq 13$), using the ω-2θ technique. Data reduction was performed with the $\sigma(I)/I$ algorithm of Lehmann and Larsen [7]. Intensities were corrected for Lorentz and polarization factors and empirical absorption [8]. Full-matrix least-squares refinement of F_o was performed using SHELX76 [9] and the X-ray room-temperature atomic coordinates as starting values [10]. The final reliability factor was $R = 0.0365$.

Structure evolution

The results at 135 K are very similar to the room-temperature structure: Figure 3 shows the projection of the crystal packing along the C=C central double bond of the TMTTF molecules. However, subtle differences must be noticed: these can be revealed through structural parameters, which may account, on average, for intermolecular interactions and their variations upon applied constraint: these parameters, defined previously [11, 12], are the spherical coordinates (R, θ, ϕ) of the centers of mass (middle of C=C central double bond) of neighboring TMTTF molecules, relative to the TMTTF origin molecule. In the next section, we present the calculated transfer integrals (t) for the intrachain (S) and interchain (I) interactions (labelled in Figure 3), in relation with their corresponding (R, θ, ϕ) parameters (Table 1). Then, we discuss the possible influence of the anion potential on the organic stacks.

Transfer integrals

The transfer integrals are evaluated, using the dimer splitting approximation [11], within the framework of the extended Hückel method [13]. The calculation of the t's is then reduced to the determination of the energy gap between the two highest occupied molecular orbitals (HOMO) of the monocation dimer. These orbitals are the + and - combinations of the HOMO's of the monomer. The sign of the integrals is given by the relative energies of the two

Table 1. Structural parameters (R in Å, θ and φ in degrees) relative to I's (interchain interactions).

		I_1	I_2	I_3	I_4
R	300 K	8.33	6.89	7.66	8.94
	135 K	8.12	6.74	7.61	8.57
θ	300 K	38.0	-18.5	30.4	28.0
	135 K	36.0	-15.4	27.5	25.8
φ	300 K	-11.7	15.0	14.0	36.6
	135 K	- 8.1	10.3	17.8	33.1

combinations, E_+ and E_-, and one can formally write $2\,t = E_+ - E_-$. The relative position and orientation of each molecule are of great importance for the magnitude and the sign of the transfer integrals; in the stack, the bonding orbital is the negative combination, and always leads to positive values of t, but interchain interactions may show a stabilization of the positive combination (yielding negative t), as molecules may lay face-to-face in very close molecular planes.

The variation of the transfer integrals is strongly dependent on the modification of the organic packing upon cooling. Previous studies on the AsF_6 and PF_6 salts [5] have shown two main effects: an increase of the intrastack integrals, together with a reduction of the dimerization t_{S1} - t_{S2}; and also an increase of the interchain integrals, as the relative position of the stacks is modified. The distinct evolutions of each interchain integral have been related to the R, θ, φ temperature dependence. We give in Table 2 the transfer integrals of the PF_6, AsF_6 and SbF_6 salts at room temperature, at 4 K for the PF_6 and AsF_6 salts, and at 135 K for the SbF_6 salt.

Intrachain interactions

The evolution of the SbF_6 salt seems, at first sight, very surprising, and is quite different from the two other salts. The change in the t_{S1} integral doesn't follow the reduction of the R parameter, and is reduced by 7% , while t_{S2} is increased by more than 20%; the corresponding R variation being -0.06 and -0.12 Å. Such a discrepancy in the compression of the successive intermolecular spaces is strong evidence of the influence of the closer anions on the S_1 interaction, which will be discussed in the next section. The small reduction of t_{S1} is quite unusual, and is related to a small sliding of the 1 molecule in the direction perpendicular to the

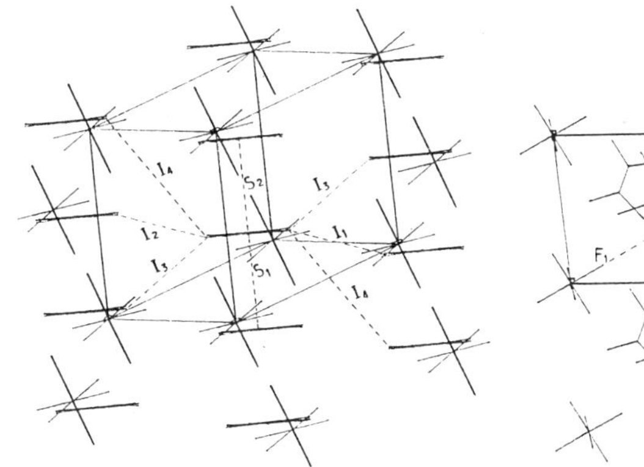

Figure 3. Projection of the structure along the C=C central double bond of molecule I. The SbF_6 anions project inside the dimer.

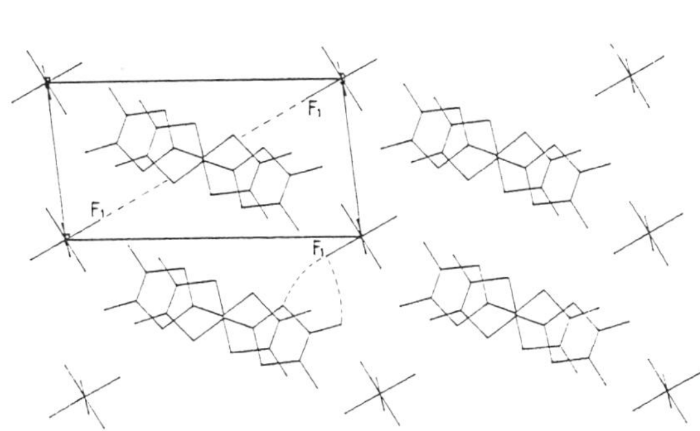

Figure 4. Projection of the structure along the a axis. F_1 points to the center of the dimer.

Table 2. Intrachain and interchain transfer integrals (meV) for PF_6, AsF_6 and SbF_6 salts.

		S_1	S_2	I_1	I_2	I_3	I_4
PF_6	300 K	137.0	93.0	1.3	0.0	12.3	0.7
	4 K	146.0	125.0	-4.5	-10.6	16.9	1.3
AsF_6	300 K	144.0	84.0	0.0	7.8	13.9	0.0
	4 K	145.0	130.0	-6.4	-14.3	15.9	1.6
SbF_6	300 K	137.8	85.0	2.8	10.3	12.2	0.3
	135 K	128.0	103.0	-0.8	-2.3	9.9	0.0

molecular axis (in the molecular plane). Nevertheless, the subsequent decrease of the dimerization of the t's is not as pronounced, as in the AsF_6 and PF_6 salts. However, the structural determination was not carried out at the same temperature, which limits the quantitative comparison of the data.

Interchain interactions

The interchain transfer integrals are known to be strongly dependent on the (R, θ, ϕ) parameters [14]. Their low values, and the fact that the corresponding molecules are closer to the compensation area, where the sign of t is changing, prevent us from extrapolating the low-temperature integrals, since small changes (for example, a further compression of the S_2 interdimer space) may lead to large differences of the t_I values at 135 K.

I_2 and I_2 both cross the compensation area, but the amplitude of the variation is not sufficent to yield larger absolute values at 135 K. This is related to the values of R, which are larger by approximately 0.12 Å than in the two other salts: for similar ϕ variations, a larger R yields a "slower" crossing of the compensation area. A similar result holds for the I_3 interaction, which is closer to this area than in the PF_6 and the AsF_6 compounds, and thus more sensitive to slight changes. In this case, Table 2 shows an increase of ϕ and decrease of θ, but with small changes of R; this gives rise to a decrease of the transfer integral. Therefore, the interchain interactions are weakened as temperature is lowered. The electronic system is thus, paradoxically, more one-dimensional at 135 K. Since further significant changes may occur at lower temperature, one cannot conclude on the details of the electronic structure at lower T.

Anion-cation interaction

The importance of the anion size effect on the organic stacks is related to the metal-fluorine bond length increase (1.60 Å for P-F, 1.73 Å for As-F and 1.86 Å for Sb-F). In the crystal structure, the anions are located in the cavities left by the zig-zag type stacking of the organic molecules. From Figure 3, it can be seen that the projection of the anions lies inside the TMTTF dimers, the contraction of which is thus limited upon cooling. On the other hand, the interdimer distance may diminish more easily: this accounts for the small variation in t_{S1}, compared to t_{S2}, when the sample is cooled down. However, the shortest contacts are those between the TMTTF molecules and the MF_6^- anions, which alternate along the **c-b** diagonal (see figure 4). The shortest distances (smaller than the sum of the van der Waals radii) observed are:

F-S:	PF_6	(4 K): 3.14 Å	F-C:	PF_6	(4 K): 3.15 Å
	AsF_6	(4 K): 3.13 Å		AsF_6	(4 K): 3.14 Å
	SbF_6	(135 K): 3.15 Å		SbF_6	(135 K): 3.14 Å

It then seems that the SbF_6 salt has reached the compactness observed in the two other salts at 4 K. This needs to be discussed more in detail, since these contacts point in the direction of the diagonal (**c** - **b**) of the unit cell: this lattice direction does not contract upon cooling. This is illustrated in Table 3: the greatest variation of (**c** - **b**) is observed for $(TMTTF)_2PF_6$, while no signifcant change is noticed for the other salts. This stiffness speaks for strong coulombic

Table 3. Anion-cation interaction: d = F_1-dimer center distance; δ = (metal-F_1, **c-b**) angle; |**c-b**| = diagonal of the (**b, c**) plane

		d(Å)	δ (°)	|**c** - **b**|
PF_6	300 K	5.63	5.8	14.81
	4 K	5.56	3.3	14.69
AsF_6	300 K	5.54	8.9	14.80
	4 K	5.47	1.3	14.80
SbF_6	300 K	5.42	8.6	15.01
	135 K	5.44	1.0	15.04

observed for the SbF_6 salt at all temperatures; the variations, as a function of temperature, are the combination of two effects: contraction of the lattice ,on the one hand, rotation of the metal-F_1 bond about the center of inversion, on the other hand; this rotation is illustrated by the angle δ between the metal - F_1 bond and the diagonal **c** - **b** (Table 3); it should be noticed that δ varies greatly upon cooling, and that in the case of SbF_6, the Sb-F_1 vector points almost exactly towards the center of the dimer.

These considerations could be of some importance, due to the charge distribution on the SbF_6^- anion. Molecular orbital calculations performed by Teramae et al [15] have shown that fluorine atoms are more negatively charged in the SbF_6^- anion, than in the PF_6^- and AsF_6^-. One may, then, conclude that the coulombic potential, due to the F_1 atom, could perturb significantly the intradimer charge distribution. Recently, Laversanne *et al* [16] have pointed out that the contribution of the anions could induce a non-negligible renormalization of the dimerization of the t's. Nevertheless, no reliable calculation has yet been able to confirm this. Moreover, there still remains some uncertainty on the validity of t_{S2}-t_{S1} as the relevant parameter of the electronic localization [2, 6]. Parameters related to the charge density between the molecules, in particular, may be more accurate, and the role of the anion potential should then be a major feature of the model.

The structural study on $(TMTTF)_2SbF_6$ did not allow us to draw any conclusion on the evolution of t at lower temperature. On the other hand, we have demonstrated a stronger influence of the anions on the organic packing, which could be further used, when improving the models describing the electronic localization.

ACKNOWLEDGEMENT

We would like to thank R. Laversanne for providing us the $(TMTTF)_2SbF_6$ crystals.

REFERENCES

[1] D. Jérome, A. Mazaud, M. Ribault, and K. Bechgaard, J. Physique Lett. 41: L95 (1980).
[2] L. G. Caron and C. Bourbonnais, Physica 453: 1438 (1986).
[3] R. Laversanne, C. Coulon, B. Gallois, J. P. Pouget, and R. Moret , J. Physique Lett. 45: L-393 (1984).
[4] L. G. Caron, F. Creuset, B. Buto and C. Bourbonnais, Synth. Metals 27: B123 (1988).
[5] T. Granier, B. Gallois, L. Ducasse, A. Fritsch, and A. Filhol, Synth. Metals 24: 343 (1988).
[6] V. J. Emery, R. Bruinsma, and S. Barisic, Phys. Rev. Lett. 48: 1039 (1982).
[7] M. S. Lehmann and K. Larsen, Acta Cryst. A30: 580 (1974).
[8] A. C. T. North, D. C. Phillips, and F. S. Matthews, Acta. Cryst. A24: 351 (1968).
[9] G. M. Sheldrick, SHELX76, Program for crystal structure determination, Univ. of Cambridge (1976).
[10] B. E. Amrani, Thesis, Université de Bordeaux (1986).

[11] L. Ducasse, M. Abderrabba, J. Hoareau, M. Pesquer, B. Gallois, and J. Gaultier, J. Phys. C.: Solid State Phys. 19: 3805 (1986).
[12] L. Ducasse, M. Abderrabba, and B. Gallois, J. Phys. C: Solid State Phys. 18: L947 (1985).
[13] R. Hoffmann, J. Chem. Phys. 39: 1397 (1963).
[14] M. Abderrabba, Thesis, Université de Bordeaux (1987).
[15] H. Teramae, K. Tanaka, and T. Yamabe, Solid State Comm. 44: 431 (1982).
[16] R. Laversanne, C. Coulon, and J. Amiel, Europhys. Lett. 2: 401 (1986).

ELECTRON PARAMAGNETIC RESONANCE OF ORGANIC CONDUCTORS $(BEDT\text{-}TTF)_2X$

M. Kurmoo, D. R. Talham§, and P. Day

Inorganic Chemistry Laboratory
University of Oxford
South Parks Road
Oxford OX1 3QR, UNITED KINGDOM

INTRODUCTION

The electron paramagnetic resonance (EPR) of five different structural classes [β, β', β", α' and κ] of the organic conductors, $(BEDT\text{-}TTF)_2X$, X = linear anion, is examined. The experimental data show a strong dependence of the peak-to-peak linewidths, spin susceptibilities and microwave conductivities on the crystal structures, and thus, the electronic band structures. The electron spin relaxation times do not obey the Elliott relation in all cases. The g-value shift is very small, and is temperature-independent, except when $T<T_{SDW}$ for β'-$(BEDT\text{-}TTF)_2AuCl_2$. The spin susceptibility is weakly temperature-dependent for the superconducting β and κ phases, and slightly more for the conducting β" phase; it increases with increasing temperature. The two insulating phases β' and α' show Bonner-Fisher type one-dimensional antiferromagnetic behavior, with transitions to a spin-density wave and a spin-Peierls ground state, respectively. These observations are finally correlated with conductivity and reflectivity measurements, and are discussed in terms of the calculated band structures.

The study of the EPR of conducting radical cation salts of the TTF family, and its synthetic analogs, can be an invaluable source of information in understanding these unusual materials. It has so far been useful in (a) identifying new phases (mainly from the magnitude of the linewidth), and classifying them into their different structural types, (b) sorting out good conductors from poor ones from the asymmetry of the lineshape, (c) checking sample purity and (d) identifying spin or charge-density wave and spin-Peierls phase transitions [1]. Although there are some general trends and typical behaviors of the various types, there is still a lack of theoretical interpretation of the electron spin relaxation mechanism and of the reason for the enhanced spin susceptibility.

It is clear that the linewidth of the EPR signal is dependent on the spin-orbit coupling constants of the chalcogen atom (S or Se). For example, the S-containing salts (e.g. TMTTF, tetramethyltetrathiafulvalene) have smaller linewidths than the Se analogs (e.g. TMTSF, tetramethyltetraselenafulvalene) [2].

The materials examined here are salts of the sulphur-based donor, $(BEDT\text{-}TTF)_2X$, where BEDT-TTF or ET is bis(ethylenedithio)tetrathiafulvalene, and X are linear inorganic anions of length 7.8 - *ca.* 10.3 Å [3]. This donor has been shown to crystallize in different crystal structures with the same anion and under the same conditions. For example, there are five different crystal phases for X = $Ag(CN)_2$ [4]. The EPR linewidths vary from 1 to 150 Oersted at room temperature, depending on the crystal structure. On the other hand, the

§Present address: Department of Chemistry, Univ. of Florida, Gainesville, FL 32611, USA.

behaviour, within a series of compounds belonging to the same crystal modification, is similar, with only minor differences in g-values and linewidths [5,6]. To understand these differing magnitudes, precise knowledge of the band structures is needed, and of the contribution of the spin-orbit coupling from each band at the Fermi surface [7]. The lack of this information requires that one carries out an empirical study of the five structural classes of BEDT-TTF salts defined below.

We present EPR data on representative examples of all five, as follows (underlined materials are used as examples in the following discussions):

A: β-(BEDT-TTF)$_2$X — X = <u>AuI_2</u>, IBr_2, and I_3
B: β''-(BEDT-TTF)$_2$X — X = <u>$AuBr_2$</u>, ICl_2 and AuIBr
C: β' -(BEDT-TTF)$_2$X — X = <u>$AuCl_2$</u> and ICl_2
D: α'-(BEDT-TTF)$_2$X — X = <u>$Ag(CN)_2$</u>, $Au(CN)_2$, $AuBr_2$ and $CuCl_2$
E: κ-(BEDT-TTF)$_2$X — X = <u>$Ag(CN)_2.H_2O$</u>, $Cu(NCS)_2$ and I_3

RESULTS AND DISCUSSION

Crystal Structures and Electronic Band Structures

The crystal structures and simple electronic band structures of all these classes of compounds have been determined previously [3]. The salient feature of the crystal structure is an array of alternating layers of organic molecules and anions. The arrangements of the BEDT-TTF molecules in the layers determine the band structures, and thus the physical properties.

Class A compounds consist of face-to-face stacks in a steplike side-by side arrangement. The band structure consists of a 2D-closed Fermi surface [~ 50% of the Brillouin zone (BZ)]. For Class B, the BEDT-TTF molecules form stacks inclined at about 60° to the molecular plane; in the transverse direction they are coplanar to within 5°. The band structure gives both an open 1D Fermi surface and some small pockets (~ 10% of the BZ). For Class C, a stack of strongly dimerized face-to-face pairs is formed; the side-by-side arrangement is steplike. The band structure gives a purely 1D open Fermi surface. The structure of the Class D compounds is, again, stacks of highly dimerized pairs; the molecules of each pair are rotated with respect to each other by ~ 30°. The band structure consists of face-to-face dimerized pairs, and each pair is rotated about the long axis of the molecule by ~90°. This band structure shows a 2D closed Fermi surface (~20% of the BZ) [8].

Peak-to-Peak Linewidth

The range of the room temperature linewidths, for rotation of the crystals with respect to the static field along three orthogonal axes, are given in Table 1, and their temperature dependencies are shown in Fig. 1. For the conducting β, β'', and semiconducting β' phases,

Table 1. EPR and Conductivity Data

Compounds	Peak-to-peak RT linewidth H_{pp} Oersted	Slope of log H_{pp} vs. log T, α^a and T range -- and /K	EPR g-value --	Type of spin susceptib. $(\chi_{spin})^b$ --	Room-temperature conductivity σ_{RT} S/cm
β-ET_2AuI_2	14-20	+0.6 (4- 300)	2.003-2.011	Pauli	10-30 (T_c = 3-5 K)
β''-ET_2AuBr_2	35-63	+1.85 (30-300)	2.002-2.013	Linear T	~500 (metallic)
β'-ET_2AuCl_2	5- 8	+1 (30-300)	2.003-2.010	BF/SDW	10^{-2} (E_a = 0.1 eV)
α-$ET_2Ag(CN)_2$	21-34	-1 (4- 70)	2.003-2.009	BF/SP	10^{-2} (E_a = 0.3 eV)
β-$ET_2Ag(CN)_2.H_2O$	58-84	-0.3 (50-300)	2.003-2.010	Pauli	~20 (metallic)

a Slope of the fit: log H_{pp} = α log T (in parentheses the temperature range)
b Type of paramagnetic susceptibility: BF = Bonner-Fischer type one-dimensional antiferromagnet, SDW = Spin density wave, SP = Spin Peierls.

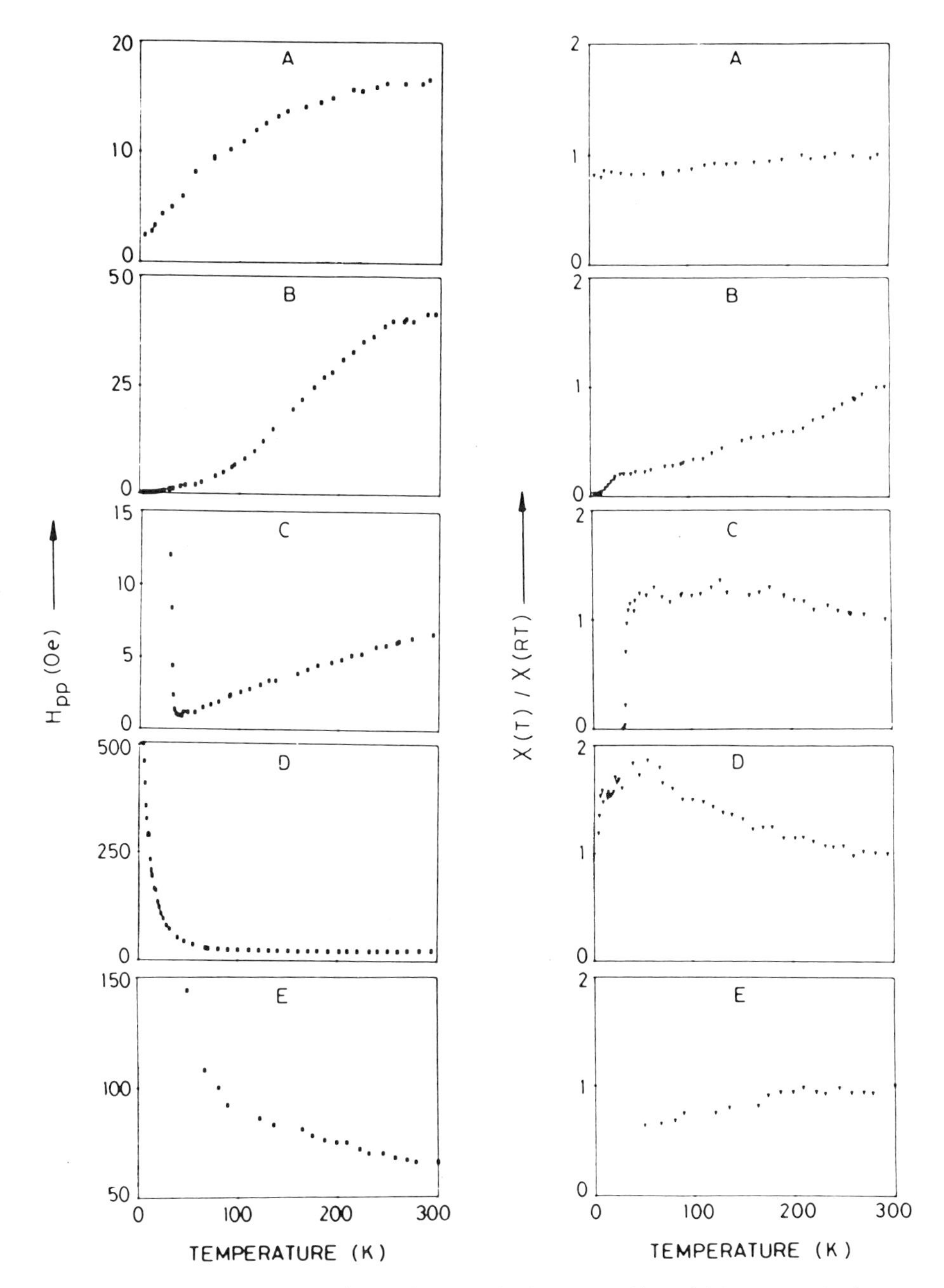

Fig. 1. Temperature dependence of the EPR linewidth H_{pp} and the normalised spin susceptibility $\chi(T)/\chi(RT)$ of:
(A) β-$(BEDT\text{-}TTF)_2AuI_2$ (B) β''-$(BEDT\text{-}TTF)_2AuBr_2$
(C) β'-$(BEDT\text{-}TTF)_2AuCl_2$ (D) α'-$(BEDT\text{-}TTF)_2Ag(CN)_2$ and
(E) κ-$(BEDT\text{-}TTF)_2Ag(CN)_2.H_2O$

an increase in linewidth is observed with temperature, while for the α' and κ phases a decrease is obtained. In both the β" and the β' phases there is a change in dH_{pp}/dT at the phase transitions at ~ 30 K. The nature of the phase transition in the β" phase is not clear, as no structural or resistive anomaly is observed. That in the β'-phase is consistent with the onset of a SDW, being accompanied by an increase in g-value and a sharp drop in spin susceptibility. This inference has been confirmed by antiferromagnetic resonance experiments [9]. There is no correlation between the magnitudes of the linewidths and either the dimensionalities of the crystals or the electronic band structures. The Elliott relation [10], $H_{pp} = (\Delta g)^2 \tau^{-1}/\gamma$, where $\Delta g = g - 2.0023$ measures the spin-orbit coupling, and γ is the gyromagnetic ratio, is not obeyed for all classes. The temperature dependence of the linewidths for the β and β" phases is consistent with an increase in τ^{-1} with increasing temperature, as also observed in the resistivity measurements [11], but agreement between the two kinds of measurement is not always good. Fits to a power law for the temperature dependence of the linewidths are given in Table 1.

g-Value

The g-values (Table 1) are close to the free electron value of 2.0023, indicating very small spin-orbit coupling (ξ) [7]. Although small, the contribution of ξ from each band at the Fermi surface will add to the width of the EPR signal. In all cases, only a single Lorentzian line is observed. The g-value is temperature-independent, within the accuracy of the experiment. For β'-$(BEDT\text{-}TTF)_2AuCl_2$ an increase in g-value is observed at the phase transition, characteristic of a transition to a SDW ground state.

Spin Susceptibility

The normalised spin susceptibilities of the five compounds are shown in Fig.1. For the superconducting phases (β and κ) a weak temperature dependence is observed, while the conducting β" exhibits a much larger T-dependence and a transition at T ≈ 30 K, where the susceptibility drops exponentially. The two semiconducting salts (α' and β') show one-dimensional antiferromagnetic behavior of the Bonner-Fisher type, with χ_{max} at ~100 K. The former shows an exponential decrease in susceptibility at 7 K, and the latter a sharp drop to zero susceptibility at T ≈ 30 K, which are associated with a SP and a SDW transition, respectively [5,9]. The static susceptibilities, χ_{ems}, are only known for the β, α' and κ phases [5,11,12]. The excess molar values are (3.4-4.6) x 10^{-4}, (9.0-9.2) x 10^{-4} and 4.6 x 10^{-4} emu mole^{-1}, respectively. The temperature dependence of χ_{ems} agrees with that measured by EPR. The χ_{ems} values correspond to a density of states at the Fermi energy of 5-8 states/(eV spin formula unit) at room temperature for the conducting β and κ phases, assuming no enhancement (i.e. $\chi_{ems} = 2\mu_B N(E_F)$), while for the α' phase the magnitude of χ_{ems} is close to the value calculated for a 1/2 spin per dimer. It is clear that a relatively small adjustment in the size of the π electron bandwidths is sufficient to switch the BEDT-TTF salts from a metal and superconductor to an insulator of the Mott-Hubbard type.

Microwave Conductivity

Microwave conductivities have only been obtained for the β and β" phases, because in the other cases one is not in the anomalous skin depth regime. The calculated conductivity for the β phase increases monotonically to low temperatures, by a factor of ~10^3 from 100 K to 4 K, while the β" phase shows a constant value from room temperature to ~50 K. Below 50 K the conductivity increases to a maximum at ~30 K, where the anomaly in susceptibility is observed. This, however, contrasts with the resistive measurements, which indicate the transition to be at a lower temperature (~8 K) [13].

CONCLUSION

We have shown that very different magnitudes of linewidth and susceptibility can be obtained for the organic conductors $(BEDT\text{-}TTF)_2X$, belonging to different structural classes. The temperature dependence of these quantities also differs, due to the difference in the electronic band structures. Precise information about the spin-orbit coupling in all the bands at the Fermi surface is needed to understand the differing magnitude of the linewidths. Recent Shubnikov-de Haas and de Haas-van Alphen measurements show the complexity of the band structures at the Fermi surface [14,15]. Certainly, much theoretical work is needed to give an insight in the electron spin relaxation. To this end, we are currently studying the effect of magnetic anions on the conduction electrons [16].

ACKNOWLEDGEMENTS

This work was supported by grants from the Science and Engineering Research Council and the University of Oxford.

REFERENCES

[1] Proceedings of the International Conferences on Science and Technology of Synthetic Metals (Kyoto), Synth. Metals 19 (1986); (Santa Fe), Synth. Metals 27 (1988).
[2] F. J. Adrian. Phys. Rev. B33: 1537 (1986), and references therein.
[3] J. M. Williams, H. H. Wang, T. J. Emge, U. R. Geiser, M. A. Beno, P. C. W. Leung, K. D. Carlson, R. J. Thorn, and A. J. Schultz, Prog. Inorg. Chem. 35: 51 (1987).
[4] M. Kurmoo, D. R. Talham, K. L. Pritchard, P. Day, A. M. Stringer, and J. A. K. Howard, Synth. Metals 27: A177 (1988).
[5] S. D. Obertelli, R. H. Friend, D. R. Talham, M. Kurmoo, and P. Day, J. Phys. C: Solid State Phys. (1989) in press.
[6] E. L. Venturini, J. E. Schirber, H. H. Wang, and J. M. Williams, Synth. Metals 27: A243 (1988).
[7] Y. Yafet, in "Solid State Physics", Vol. 14, Ed. F. Seitz and D. Turnbull (Academic, New York, 1963) p.1.
[8] (a) M. Kurmoo and L. Ducasse, unpublished, (b) T. Mori and H. Inokuchi, J. Phys. Soc. Japan 57: 3674 (1988).
[9] C. Coulon, R. Laversanne, J. Amiell and P. Delhaès, J. Phys. C: Solid State Phys. 19: L753 (1986).
[10] R. J. Elliott, Phys. Rev. 89: 689 (1953); 96: 226 (1954).
[11] I. D. Parker, S. D. Obertelli, R. H. Friend, D. R. Talham, M. Kurmoo, P. Day, J. A. K. Howard, and A. Stringer, Synth. Metals 19: 185 (1987); M. Weger, K. Bender, T. Klutz, D. Schweitzer, F. Gross, C. P. Heidmann, Ch. Probst, and K. Andres, Synth. Metals 25: 49 (1988).
[12] S. Klotz, J. S. Schilling, S. Gärtner, and D. Schweitzer, Solid State Commun. 67: 981 (1988).
[13] M. Kurmoo, D. R. Talham, P. Day, I. D. Parker, R. H. Friend, A. M. Stringer, and J. A. K. Howard, Solid State Commun. 61: 459 (1987).
[14] F. L. Pratt, A. J. Fisher, W. Hayes, J. Singleton, S. J. R. M. Spermon, M. Kurmoo, and P. Day, Phys. Rev. Lett. 61: 2721 (1988).
[15] I. D. Parker, D. D. Pigram, R. H. Friend, M. Kurmoo, and P. Day, Synth. Metals 27: 387 (1988).
[16] T. Mallah, C. Hollis, S. Bott, P. Day, and M. Kurmoo, Synth. Metals 27: 381 (1988) ; J. Chem. Soc. (Dalton Trans.) (1989) submitted.

ORGANIC METALS FROM CHIRAL BEDT-TTF DONORS

Ben-ming Chen,* F. Deilacher, M. Hoch, H. J. Keller, and Pei-ji Wu*

Anorganisch-Chemisches Institut
Universität Heidelberg, Im Neuenheimer Feld 270
D-6900 Heidelberg, GERMAN FEDERAL REPUBLIC
*Guests from Academia Sinica, Institute of Chemistry, Beijing, CHINA

S. Gärtner, S. Kahlich, and D. Schweitzer**

Max-Planck-Institut für Med. Forschung
AG.: Molekülkristalle, Jahnstr. 29
D-6900 Heidelberg, GERMAN FEDERAL REPUBLIC
**3. Physikalisches Institut der Universität Stuttgart, Pfaffenwaldring 75
D-7000 Stuttgart 80, GERMAN FEDERAL REPUBLIC

STATUS QUO ANTE

The chemistry and physics of solids with layered structures has been of central interest in the solid state sciences for several decades [1]. One of the many scientifically and technically important aspects connected with these materials is the possibility of pronounced two-dimensional delocalized interatomic and/or intermolecular electronic interactions, which lead to unusual magnetic and electric properties in the bulk. Especially remarkable are a variety of phase transitions, which occur in selected materials of this type at different temperatures. Typical, and very well-known, examples of such specimens are graphite, the binary sulfides MoS_2 and TaS_2, together with several of their intercalation compounds, ternary or quarternary chalcogenides and/or halides of the transition elements, and so on. Inorganic ternary systems, composed of anionic transition metal oxide layers and different kinds of countercations located between these anion sheets, are especially well-suited for systematic variations in composition and - as a consequence thereof - in solid state properties. Varying stages of band fillings can be achieved "chemically," depending on the number of electrons per metal ion in the different sheets. The systematic work on these materials culminated in the discovery of superconductivity above 40 K by Müller and Bednorz in layered copper oxide structures [2].

With the exception of graphite-related compounds - which consist mainly of carbon- there were no comparable classes of electronically collective interacting layered organic materials. Most of the two-dimensional organic solids - which means layered matrices made up mainly of the main group elements carbon, nitrogen, oxygen, hydrogen and sulfur - are diamagnetic and insulating. The electrically conducting and/or magnetically interesting organic charge transfer solids, on the other hand, contain columns or chains of interacting molecules or atoms. This arrangement results in pronounced "one-dimensional" physical properties, with the well-known phase transitions at comparatively high temperatures.

STATUS QUO

Up to now, there is one organic two-dimensional "exception:" the cation radical salts of the donor BEDT-TTF. Numerous BEDT-TTF salts - metallic, semiconducting and/or insulating - have been prepared in the last few years [3,4], after Saito *et al.* firstly pointed to the typical two-dimensional molecular interactions in a radical salt of this donor [5]. Almost all of the solids, obtained so far by electrochemical oxidation of BEDT-TTF, in the presence of suitable counter anions, crystallize in two-dimensionally interacting arrays. The typical arrangement consists of sheets of anions - normally diamagnetic and not interacting with each other, nor with the BEDT-TTF partners - and layers of BEDT-TTF cations, in which the molecules can be oriented in quite different ways. There are often columns, or pairs of cations, with strong intermolecular S-S-interactions. Independent of these differences in molecular arrangement, the main feature of all these materials is a layered structure. The different orientations of the cation radicals in a layer with respect to each other, and in relation to the counter anion lattice, leads, in most cases, to strong two-dimensional, in other cases, to typical one-dimensional electronic interactions. Nevertheless, the general structure in most of the materials is of the layered type. This variety of $[\text{BEDT-TTF}]_y^{n+}X^{n-}$ structures shows that this material establishes a new class of two-dimensional organic matrices.

THE PLAN

The versatility of this organic counterpart of the well-known inorganic layered solids opens the possibility for systematic "chemical" variation of band fillings and electronic interactions. The redox properties of the different lattice elements can be varied systematically by chemical methods, since the BEDT-TTF salts are open to synthetic variations in both types of layers very similar to the famous inorganic ternary oxides. So far, most of the recent work to vary the physical properties by chemical action concentrated on the preparation of BEDT-TTF salts with a wide variety of anions. Fewer efforts were concerned with a systematic variation of the BEDT-TTF moieties [6]. One major problem arises in the pursuit of the latter goal: the substitution of the peripheral hydrogen atoms could lead to a manifold of chiral molecules.

THE PROJECT

For a beginning, we decided to substitute the hydrogen atoms of the peripheral ethylene groups of BEDT-TTF systematically by methyl groups. The preparation of these substituted BEDT-TTF donors is by no means trivial. The hydrogen methyl group exchange leads to different configurational isomers and, especially, to a variety of chiral molecules.

First, we concentrated our efforts on two symmetrical molecules, DIET and TMET.

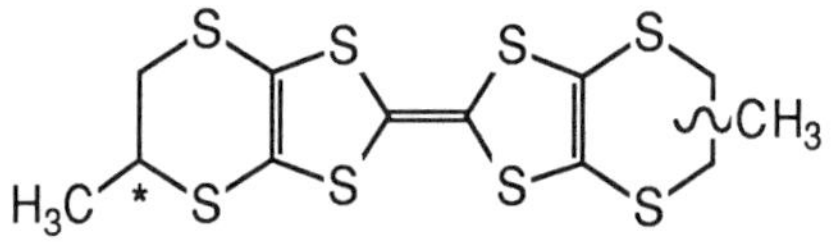

DIET

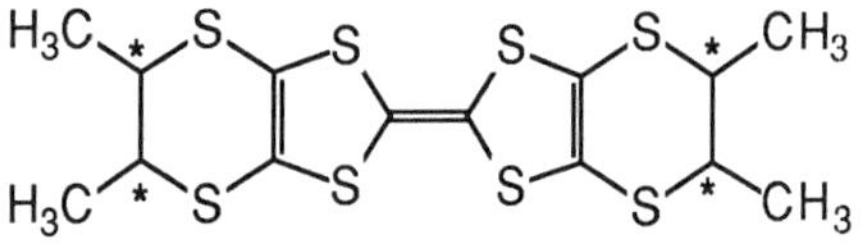

TMET

DIET can be arranged in a "trans" - or a "cis" - form.

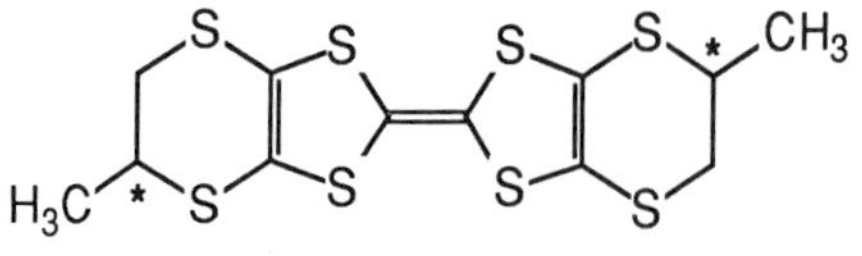

"trans"-DIET

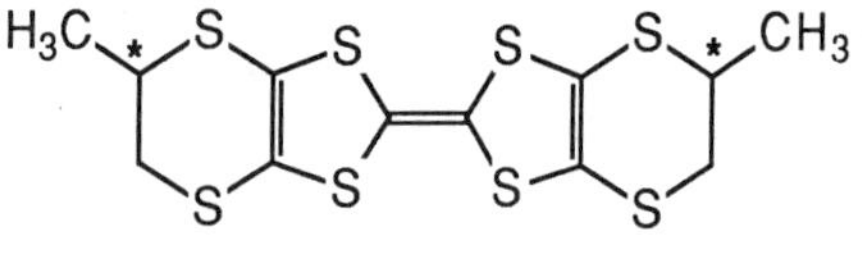

"cis"-DIET

Each of these DIET isomers could occur in two enantiomers and an additional diastereomer. In the case of the tetramethyl derivative TMET, two pairs of enantiomers and three diastereomers could result from the preparation.

Chemistry

The synthetic route to the different donors is in imitation of the usual procedure to obtain BEDT-TTF. A complex of the dithiolate I

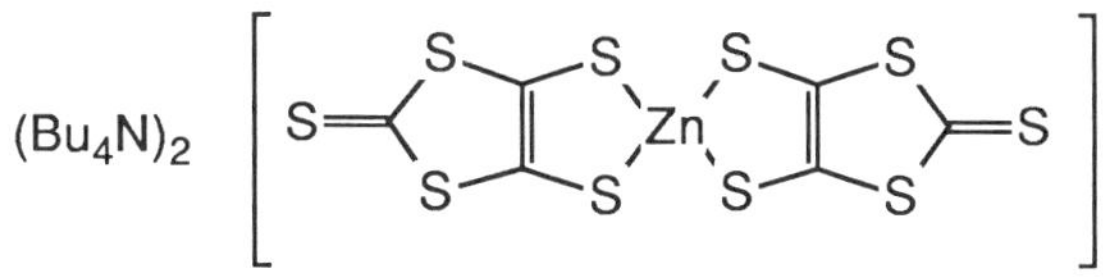

I

is reacted with 1,2-dibromopropane and 2,3-dibromobutane, respectively, to yield the thiones II and III

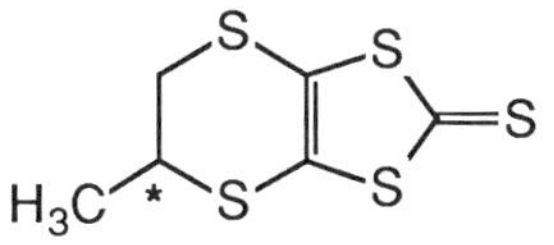

II 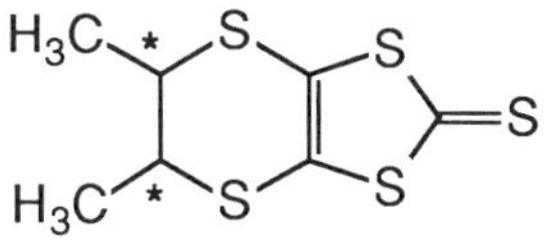 III

which are converted to the ketones IV and V

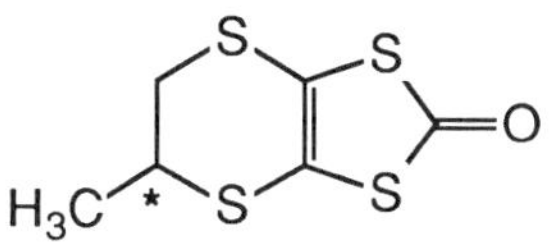

IV 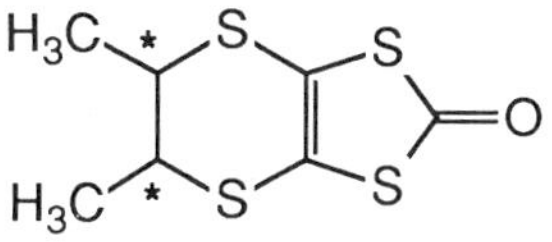 V

in order to be self-coupled by action of $P(OR)_3$. A mixture of DIET and TMET isomers results. DIET has been obtained earlier by a similar procedure [6], but no details concerning its structure have been reported so far. TMET was isolated starting from a sulphate diester earlier, and some of its radical salts have been described [7].

In order to investigate the basic chemical and structural properties of the donors, we decided to electrocrystallize mixtures of the isomers of the donor, obtained by starting with racemic precursors. DIET salts crystallize with more ease at the electrode, compared to the TMET materials. The radical salts of the latter are very soluble in polar solvents, making the crystallization more difficult. Polar solvents - on the other hand - are necessary to dissolve the supporting electrolyte. In the case of DIET this difficulty is less pronounced. So far, we obtained well characterized "racemic" DIET solids with FSO_3^-, ClO_4^-, HSO_4^-, NO_3^-, I_3^-, $[AuI_2]^-$, ReO_4^-, PF_6^-, IO_4^-, $SO_3CF_3^-$ and BF_4^- as counter anions.

X-ray Characterization

Structural work on these materials is complicated by the fact that the crystals contain mixtures of different isomers, and are therefore disordered. Nevertheless, it could be shown that the donor DIET is obtained solely as the trans-isomer. Surprising is the comparatively long "central" C=C bond (1.4 Å) of the DIET donor.

A typical example for the structural problems in the radical salts is given in Fig. 1. It shows the molecular structure of $(DIET)_2(FSO_3)_3$. All the indicated atomic positions are statistically occupied. The same is valid for the isostructural $(DIET)_2(BF_4)_3$. Nevertheless, the structure elucidation clearly reveals the 2:3 composition of both compounds, which results in an average +1.5 charge on each DIET cation. It is interesting to note that the central C=C bond distances of 1.322(35) Å in the former, and of 1.292(26) Å in the latter material, are appreciably shortened with respect to these C=C distances in the neutral donor. Relating to recently proposed models [3,8], this trend would not be indicative of positive charges at all,

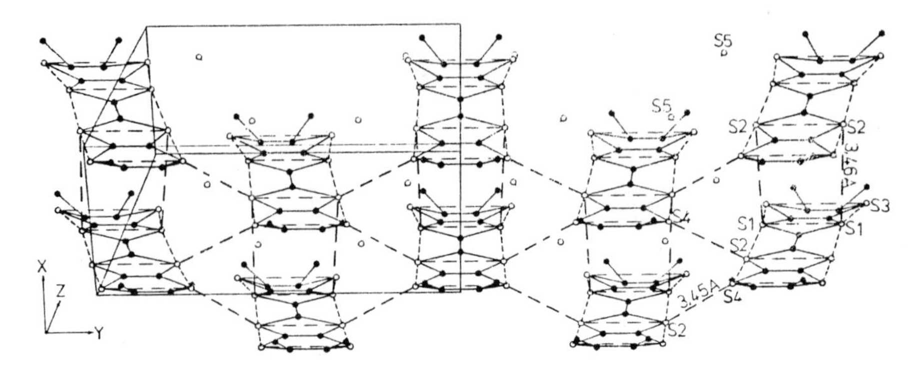

Fig. 1. Crystal structure of $(DIET)_2(FSO_3)_3$.

which is in contradiction to the observed stoichiometry. Furthermore, the crystal structure clearly reveals three-dimensional interactions between the radical cations. This finding might explain the "smearing out" of sharp phase transitions.

Physical Properties

Fig. 2 shows the temperature dependence of the electrical resistivity of $(DIET)_2(BF_4)_3$ as a typical example. The compound behaves as a metal at room temperature and below. At lower temperature, a very broad phase transition is indicated in the resistivity data. Fig. 3 summarizes results of the thermopower measurement on this material at different temperatures. The latter data clearly reveal a phase transition at around 50 K. Similar results have been obtained from DIET salts with other anions like HSO_4^- or NO_3^-, as shown in Fig. 4 for the compound $(DIET)_x(NO_3)_y$, the full structure of which could not be solved so far.

CONCLUSIONS

We have shown that is is possible to obtain organic metals from chiral molecules. The X-ray results prove an oxidation number of +1.5 for the radical cations in at least two cases. To our best knowledge, these are the first examples of BEDT-TTF-related radical cation salts

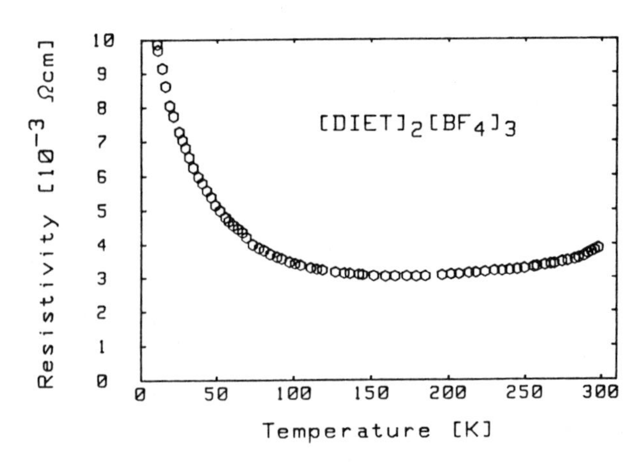

Fig. 2. Temperature dependence of the electrical resistivity of $(DIET)_2(BF_4)_3$.

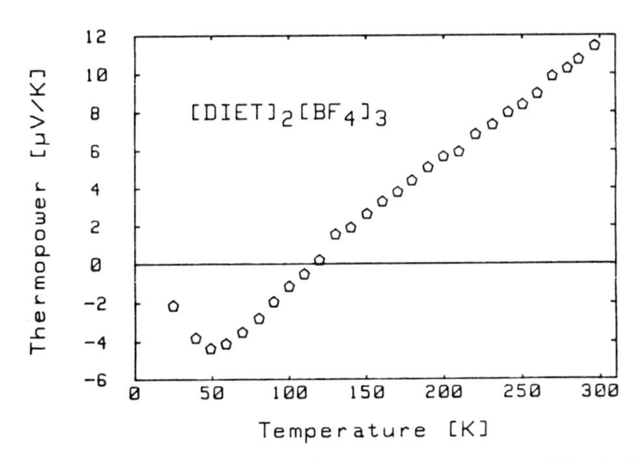

Fig. 3. Temperature dependence of the thermopower of $(DIET)_2(BF_4)_3$.

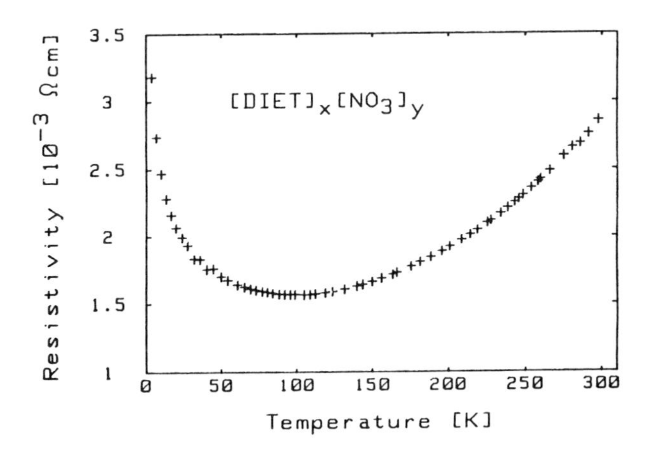

Fig. 4. Temperature dependence of the resistivity of $(DIET)_x(NO_3)_y$.

with this oxidation number. As expected, we obtained statistically disordered crystals, containing the two different enantiomers in a "racemic" mixture. The broad smeared-out phase transitions are probably due to this disorder, or may be caused by the three-dimensional interactions which have been observed, so far, only once in a BEDT-TTF radical salt. We will crystallize the metallic compounds using "optically pure" isomers, in the hope to be able to isolate chiral metals.

REFERENCES

[1] See for example in: "Physical and Chemistry of Materials with Layered Structures," Vol. 1-5 (D. Reidel, Dordrecht, 1976).

[2] K. A. Müller and J. G. Bednorz, Z. Phys. B64: 189 (1986).

[3] J. M. Williams, H. H. Wang, T. J. Emge, U. Geiser, M. A. Beno, P. C. W. Leung, K. D. Carlson, R. J. Thorn, A. J. Schultz, and M.-H. Whangbo, Progr. Inorg. Chem. 30: 51 (1987).

[4] Proceedings of the ICSM 88, Santa Fe, Synth. Metals 27 (1989).

[5] G. Saito, T. Enoki, K. Toriumi, and H. Inokuchi, Solid State Commun. 42: 557 (1982).

[6] G. C. Papavassilou, J. S. Zambounis, and S. Y. Yiannopoulos, Chem. Scripta 27: 261 (1987), and references cited therein.
[7] A. Karrer, J. D. Wallis, J. D. Dunitz, B. Hilti, C. W. Mayer, M. Bürkle, and J. Pfeiffer, Helv. Chim. Acta 70: 942 (1987).
[8] T. C. Umland, S. Allie, T. Kuhlmann, and P. Coppens, J. Phys. Chem., in print.
[9] U. Geiser, H. H. Wang, L. E. Gerdom, M. A. Firestone, L. M. Sowa, J. M. Williams, and M.-H. Whangbo, J. Am. Chem. Soc. 107: 8305 (1985).

NEW CONDUCTING SOLIDS BASED ON SOME SYMMETRICAL AND UNSYMMETRICAL π-DONORS

G.C. Mousdis, V.C. Kakoussis and G.C. Papavassiliou

Theoretical and Physical Chemistry Institute
National Hellenic Research Foundation
48, Vassileos Constantinou Ave., Athens
11635, Greece

INTRODUCTION

Recently, conducting and syperconducting crystals based on unsymmetrical π-donor molecules have been described [1]-[9]. In this paper we describe the preparation and preliminary studies of new conducting salts based on some unsymmetrical π-donors (1)-(13) (Table 1) [10]-[16] as well as on the following symmetrical π-donors: TTF (14), TMTSF (15), (BMDTTTF) (16), (BEDTTTF)(17), (BVDTTTF) (18) [17], (BPDTTF) (19) [18], (BDMETTF) (20) [19], (BMDSTTF) (21) [20], (BEDSTTFR) (22) [20] and (BTMDSTTF) (23) [20].

EXPERIMENTAL

The preparation of 1,3-dithiole-2-ones and selenium-analogs, precursors of unsymmetrical π-donors, is described in refs [10]-[16] and in the following paragraphs.

Preparation of 4,5-bis(methylacetate) diseleno-1,3-dithiole-2-thione (I)

This compound was prepared by a method similar to that reported for the preparation of its sulfur analog [13]: Treatment of bis(tetrabutylammonium) bis(1,3-dithiole-2-thione-4,5-diselenolato)-zincate [12] with methyl bromoacetate in boiling acetone, followed by cooling at -10°C and recrystallization of the resulting precepitate from boiling cyclohexane, gave a yellow crystalline solid ; mp=45°C; UV (CH_3CN):377, 280, 227 nm.

Preparation of 4,5-(methylacetatediseleno)-1,3-dithiole-2-thione (II)

This compound was prepared by a method similar to that reported for the preparation of its sulfur analog [14]: Treatment of bis(tetrabutylammonium) bis(1,3-dithiole-2-thione-4,5-diselenolato) -zincate [12] with methyl dichloroacetate in boiling acetone, followed by cooling at -10°C, and recrystallization of the resulting precipitate from boiling ethanol, gave a greenish-yellow solid; mp=171 °C; UV(CH_3CN): 418, 274 nm.

Lower-Dimensional Systems and Molecular Electronics
Edited by R.M. Metzger *et al.*, Plenum Press, New York, 1991

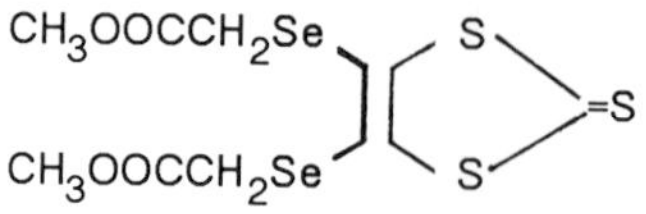

CH_3OOCCH_2Se / CH_3OOCCH_2Se (I) CH_3OOC (II)

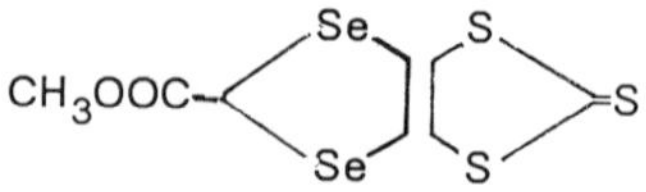

Preparation of 4,5-bis(methylacetate)diseleno-1,3-dithiole-2-one and 4,5-(methylacetatediseleno) -1,3-dithiole-2-one.

These compounds were prepared by treating the corresponding -2-thiones (I) or (II) with mercury acetate in a mixture of dichloromethane-acetic acid at refluxing temperature (see [12]-[14]). 4,5-Bis(methylacetate)diseleno-1,3-dithiole-2-one is a white solid; mp=18-20°C; UV(CH_3CN) 264 nm, and 4,5-(methylacetatediseleno)-1,3-dithiole-2 one is a white-yellow solid; mp=97 °C; UV(CH_3CN): 307 nm.

Table 1. Molecular formulas of unsymmetrical π-donors

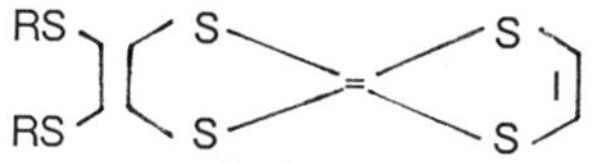

(1a)-(1g) 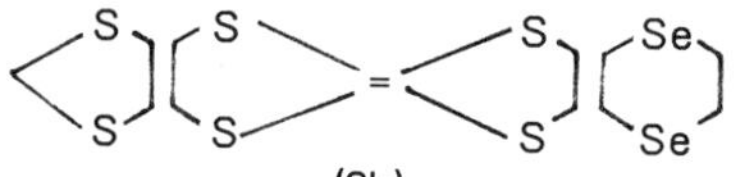 (8b)

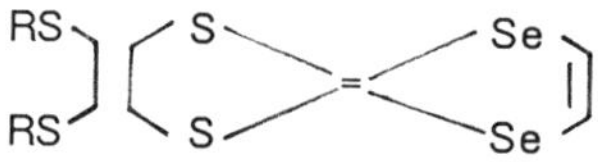

(2b)-(2c) 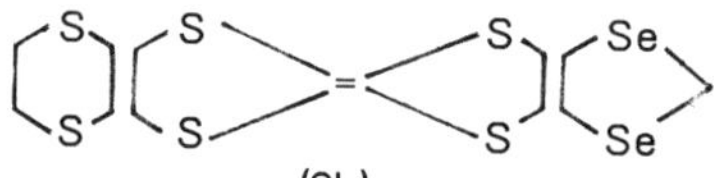 (9b)

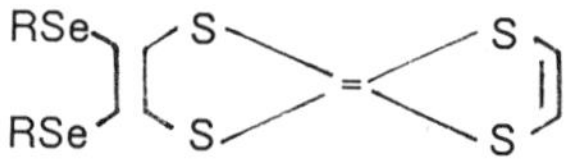

(3b)-(3d) 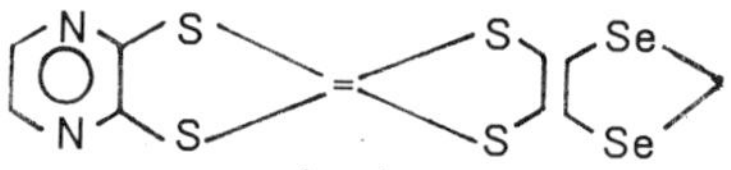 (10b)

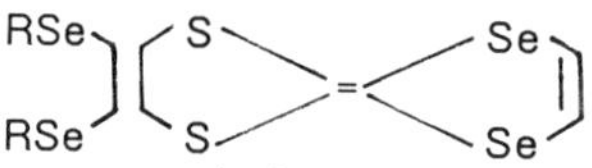

(4c) 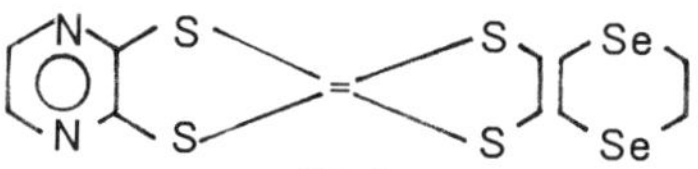 (10c)

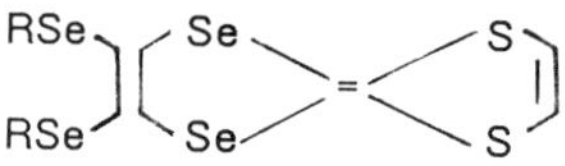

(5b)-(5c) 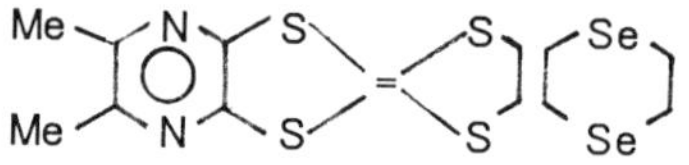 (11b)

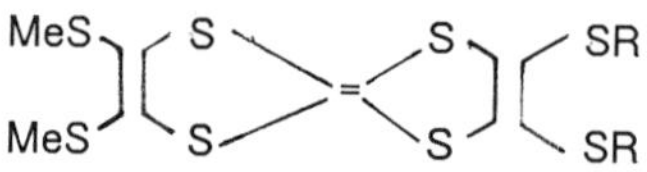

(6b)-(6g) 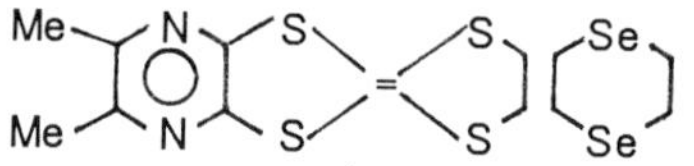 (11c)

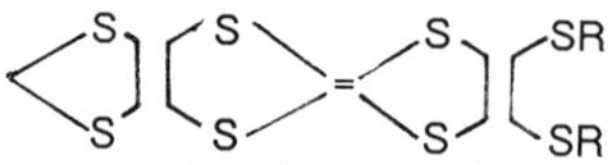

(7a)-(7f) 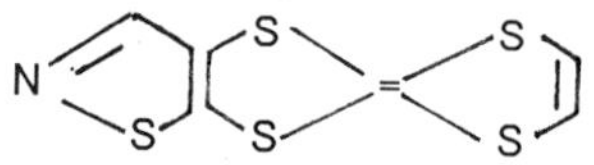 (12)

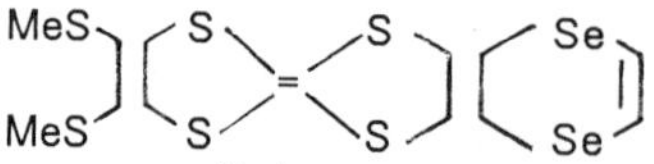

(8a) 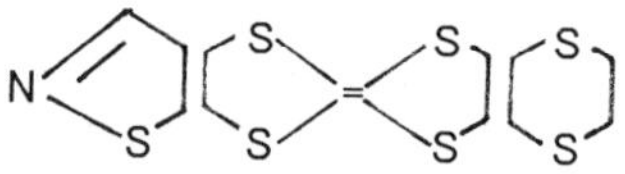 (13c)

(a):R=CH_3, (b):2R=CH_2, (c):2R=CH_2CH_2, (d):2R=$CH_2CH_2CH_2$, (e):2R=$CH(CH_3)CH_2$, (f):2R=$CH(CH_3)CH(CH_3)$, (g):2R=CH=CH

Table 2. Salts, their method of preparation and their appearance.

$(1b)_2AuBr_2$	EL	black plates	$(6f)_xIBr_2$	CH	black needles
$(1b)_xNi(dcit)_2$	EL	black μ-crystals	$(6f)_xTCNQ$	CH	black needles
$(1c)_2AuBr_2$	EL	black plates	$(6d)_xI_3$	CH	golden μ-crystals
$(1c)_2Ag(CN)_2$	EL	black needles	$(6d)_xIBr_2$	CH	black plates
$(1c)_xNi(dcit)_2$	EL	black μ-crystals	$(7c)_xIBr_2$	EL	bronze plates
$(1g)_xAuI_2$	EL	black needles	$(7c)_xI_3$	EL	bronze-blackplates
$(1g)_xIBr_2$	EL	black needles, plates	$(8a)_xI_3$	CH	black μ-crystals
$(1g)_xNi(dcit)_2$	EL	black μ-crystals	$(8a)_xIBr_2$	CH	black μ-crystals
$(2b)_xIBr_2$	CH	bronze needles, plates	$(9b)_xTCNQ$	CH	black μ-crystals
$(2c)_xIBr_2$	EL	black needles, plates	$(9b)_xI_3$	CH	needles
$(3b)_xTCNQ$	CH	black needles	$(10b)_xI_3$	CH	dark-copper
$(3b)_xI_3$	CH	copper plates	$(10b)_xIBr_2$	CH	dark-copper
$(3b)_xIBr_2$	CH	bronze needles	$(10b)_xBr_3$	CH	brown chips
$(3b)_xAuI_2$	CH	black plates	$(10c)_xI_3$	CH	golden needles
$(3c)_xTCNQ$	CH	black μ-crystals	$(10c)_xIBr_2$	CH	brown μ-crystals
$(3c)_xI_3$	CH	black μ-crystals	$(10c)_xBr_3$	CH	golden brown
$(3c)_xIBr_2$	CH	golden μ-crystals	$(11b)_xIBr_2$	EL	black plates
$(3c)_xIBr_2$	EL	black, golden plates	$(11b)_xI_3$	CH	golden chips
$(3d)_xTCNQ$	CH	black μ-crystals	$(11b)_xIBr_2$	CH	copper chips
$(3d)_xI_3$	CH	needles, μ-crystals	$(11b)_xI_3$	EL	bronze plates
$(3d)_xIBr_2$	CH	black μ-crystals	$(11c)_2TCNQ$	CH	dark blue plates
$(3d)_xI_3$	EL	black needles, μ-crystals	$(11c)_xI_3$	CH	dark bronze
$(4c)_xI_3$	CH	bronze needles	$(11c)_xIBr_2$	CH	black plates
$(5c)_xTCNQ$	CH	black μ-crystals	$(11c)_xIBr_2$	EL	black plates
$(5c)_xI_3$	CH	golden needles	$(11c)_xAuI_2$	EL	black plates
$(5c)_xI_3$	EL	black needles, plates	$(12)_xTCNQ$	CH	brown μ-crystals
$(5c)_xIBr_2$	CH	black needles	$(13c)_xBr_3$	CH	black μ-crystals
$(5c)_xAuI_2$	EL	black μ-crystals	$(14)_x$-$(15)_xNi(dcit)_2$	EL	black μ-crystals
$(6e)_xIBr_2$	CH	black μ-crystals	$(16)_xNi(dcit)_2$	EL	red-black chips
$(6g)_xIBr_2$	CH	black μ-crystals	$(17)_x$-$(23)_xNi(dcit)2$	EL	black μ-crystals

CH= chemical method, EL= electrochemical method

$Ni(dcit)_2$ =

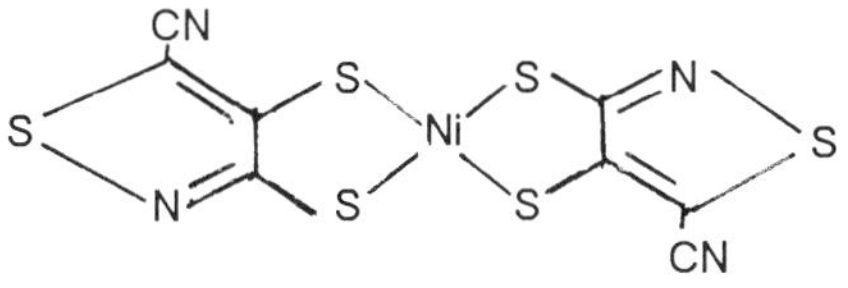

The unsymmetrical π-donors (1)-(13) were prepared from the corresponding precursors by methods reported in refs. [10]-[16]. A number of salts based on the π-donors (1)-(23) was prepared by chemical or electrochemical procedures. These salts, their method of preparation, and their appearance are summarized in Table 2.

RESULTS AND DISCUSSION

Conductivity measurements on polycrystalline compressed pellets or on single crystals performed at room temperature in the salts of Table 2 showed that these are conducting salts. Some of them, such as $(1b)_2AuBr_2$, $(1b)_xNi(dcit)_2$, $(1c)_2AuBr_2$, $(1g)_xIBr_2$, were found to be metallic down to low temperatures. These preliminary results indicate that the compounds are worthy of further studies.

REFERENCES

1. G.C.Papavassiliou, G.A.Mousdis, J.S.Zambounis, A.Terzis, A.Hountas, B.Hilti, C.W.Mayer and J.Pfeiffer , Synth.Metals 27 B379 (1988).
2. A.Terzis, A.Hountas, A.E.Underhill, A.Clark, B.Kaye, B.Hilti, C.Mayer, J.Pfeiffer, S.Y.Yiannopoulos, G.A.Mousdis and G.C.Papavassiliou, Synth.Metals 27 B97 (1988).
3. A.Terzis, A.Hountas, and G.C.Papavassiliou, Sol.St. Commun., 66, 1161 (1988).
4. A.Hountas, A.Terzis, G.Papavassiliou, B.Hilti and J.Pfeiffer, Acta Cryst., C, in press.
5. A.Hountas, A.Terzis, G.C.Papavassiliou, B.Hilti, N.Bürkle, C.W.Mayer, and J.S.Zambounis, Acta Cryst. C, in press.
6. A.Terzis, A.Hountas, G.C.Papavassiliou, B.Hilti and J.Pfeiffer, Acta Cryst., in press.
7. B.Hilti et al , to be published.
8. A.M.Kini et al ,Sol.St.Commun., 69, 503 (1989).
9. A.Kobayashi, R.Kato, and H.Kobayashi, to be published.
10. G.C.Papavassiliou, V.C.Kakoussis and G.A.Mousdis to be published:(5c)(orange, mp=210°C), (8a) (yellow, mp=105°C), (9b)(orange-brown, mp=189 ° C),(10b) (orange-red, mp=244°C), (10c) (orange mp=209°C), (11b) (yellow-orange, mp=240°C), (11c)(yellow, mp>240°C), (13)(yellow, mp=197°C) have been pre pared by methods similar to those reported in refs [11]-[16] herein.
11. G.C.Papavassiliou, this volume.
12. G.C.Papavassiliou, V.C.Kakoussis, J.S.Zambounis, and G.A.Mousdis, Chem.Scripta, in press (1989).
13. G.C.Papavassiliou, V.C.Kakoussis, G.A.Mousdis, J.S.Zambounis and C.W.Mayer, Chem.Scripta, in press (1989).
14. G.C.Papavassiliou, G.A.Mousdis, S.Y.Yiannopoulos, and J.S.Zambounis, Chem.Scripta 28, 365 (1988).
15. G.C.Papavassiliou, G.A.Mousdis, S.Y.Yiannopoulos, V.C.Kakoussis and J.S.Zambounis, Synth.Metals 27, B373 (1988).
16. G.C.Papavassiliou, J.S.Zambounis, G.A.Mousdis, V.Gionis and S.Y.Yiannopoulos, Mol.Cryst.Liq.Cryst., 156, 269 (1988).
17. K.Inone et al Chem.Lett. 781 (1986): K.S.Varma et al Physica 143B, 321 (1986).
18. G.C.Papavassiliou et al Chem.Scripta, 27, 261 (1987).
19. J.D.Wallis et al Helv.Chim.Acta 69, 69 (1986).
20. P.J.Nigrey et al Synth.Metals, 27, B481 (1988).

TTF-DERIVATIVE RADICAL CATION SALTS WITH PLANAR TETRACYANOMETALLATE DIANIONS AND LARGE ACCEPTOR POLYMETALLATES

L. Ouahab, S. Triki and D. Grandjean

Laboratoire de Chimie du Solide et Inorganique Moleculaire
U.R.A. 254 CNRS, Université de Rennes I
35042 Rennes cedex, FRANCE

M. Bencharif

Institut de Chimie, Université de Constantine
Constantine, ALGERIA

C. Garrigou-Lagrange and P. Delhaès

Centre de Recherche Paul Pascal
Chateau Brivazac
33600 Pessac, FRANCE

INTRODUCTION

In the search for new conducting molecular materials, several attempts have been made, during the last few years, to assemble blocks of organic and inorganic molecules. For the particular class of radical-ion salts, it has been demonstrated that the counter-ions are playing a key role; there is no reason, therefore, to restrict oneself to diamagnetic monoanions.

Several studies used anion modifications based on (sometimes paramagnetic) polyanions, and obtained new structural arrangements of these large molecular blocks, associated with powerful organic donors derived from tetrathia(selana)fulvalene [TCF, C=S, Se]. These previous studies have included large clusters such as $Mo_6Cl_{14}^{--}$ [1,2] $Re_6Se_5Cl_9^-$ [1,3], $Ta_6Cl_{18}^{n-}$ [4]. In particular, the first molecular perowskite $[(TTF)_3Re_6Se_5Cl_9(Cl)]$ [1,3] has been discovered among these series. The latter series are characterized by the conflicting tendency of the organic entities to stack, and of the inorganic anions to give compact salts, derived from the CsCl or $BaTiO_3$-type of structures, for instance. In order to attenuate this structural duality, we have chosen anions which are known to be responsible for magnetic or electronic properties in organic systems:

(i) The tetracyanometallate complexes $[M(CN)_4^{--}$ with M=Pt(II), Ni(II)]. These anions are known to form partially oxidized one-dimensional $Pt(CN)_4$ stacks, and to give a highly conducting salt [5]. To clarify our goal in this organic-inorganic mixing, let us remember the concluding question of one of our papers [6] "Is it possible to stabilize together mixed-valence systems of organic and inorganic entities (especially in the case of the tetracyanoplatinate dianions) in the same solid?" In the course of our attempts to answer this question, we have obtained a large variety of new radical-cation salts [6,7]. Recently, a $(BEDT\text{-}TTF)_4Pt(CN)_4$ salt, with a different unit cell, has been reported by Shibaeva *et al.* [8], and the corresponding $Ni(CN)_4$ salt is described by Tanaka *et al.* [9] Additionally, Schweitzer and co-workers [10] have also reported a new organic metal with square platinates $[Pt(C_2O_4)_2]^{--}$.

(ii) The polyoxometallates, which are MO_6-condensed octahedra (M = Mo, W, Nb, V). Known for over a century, these polyoxometallates are still of interest, because of their high electron acceptor capabilities [11]. They offer the possibility of making "Organic Donor - Inorganic Acceptor charge transfer (CT) salts", which have not been investigated yet, as far as we know. We have focused our work on the Lindquist (M_6O_{19}, M = Mo, W) [12] and the Keggin ($XM_{12}O_{40}$, X = P, Si, ...; M = Mo, W) polyanions [13].

We report here the X-ray crystal structures and the optical properties of several of these salts, after a general presentation in Table 1. We limit our discussion to the salts which show interesting physical possibilities.

RESULTS AND DISCUSSION

Planar tetracyanometallate compounds

$(TTF)_5Pt(CN)_4.2\ CH_3CN$ (1): In this compound (Fig. 1a), the unit cell contains two independent $Pt(CN)_4(TTF)_{2.5}$ blocks [6]. The TTF molecules of each independent unit form centrosymmetric isolated pentamers. The $Pt(CN)_4^{--}$ dianions are separated by the disordered acetonitrile molecules. The structural features of the TTF molecules show that the central molecule of the pentamer is neutral, while the other four are fully oxidized, indicating mixed valence in the isolated pentamers. This new type of organization has particular optical properties. The room-temperature electronic absorption spectrum (Fig. 2) exhibits the two

Table 1. Room-temperature conductivity and unit cell parameters.

Salts[a]	σ[b]	**a**(Å)	**b**(Å)	**c**(Å)	α(°)	β (°)	γ (°)	S.G.[c]	Z[c]	Ref
1 $Pt_2(TTF)_5A_2$	-	11.697[d]	15.005	17.020	78.33	75.54	72.98	$P\bar{1}$	2	[6]
2 $Pt(TMTSF)_3$	0.25	7.957	10.817	12.834	85.85	86.58	76.17	$P\bar{1}$	1	[6]
3 $Ni(TMTTF)_2$	Ins	8.186	8.574	10.361	92.54	97.47	98.14	$P\bar{1}$	1	[7]
4 $Ni(TMTTF)_3$	0.55	7.800	10.579	12.495	85.42	86.83	77.16	$P\bar{1}$	1	[6]
5 $Ni(TMTSF)_3$	0.43	7.932	10.887	12.868	86.91	87.07	75.52	$P\bar{1}$	1	[6]
6 $Pt(ET)_4$[e]	12.	9.721	11.127	16.552	76.90	81.52	62.88	$P\bar{1}$	1	[6]
7 $Ni(ET)_4$	2.04	9.699	11.140	16.430	77.18	82.13	63.01	$P\bar{1}$	1	[6]
8 $(TTF)_3Mo_6$	0.0014	9.942	10.417	10.601	72.33	78.77	63.01	$P\bar{1}$	1	[16]
9 $(TTF)_3W_6$	0.0005	9.965	10.503	10.634	71.93	78.63	63.38	$P\bar{1}$	1	[16]
10 $(TMTSF)_3W_6.2D$	0.03	11.589	19.385	13.681	90.	99.53	90.	$P2_1/c$	1	[16]
11 $E_2(TTF)_6PW_{12}$	0.03	15.563	19.498	14.178	90.	90.	90.	Cmmm	1	[18]
12 $(TMTSF)_3PW_{12}$	Ins	11.729	18.453	16.207	90.	98.08	90.	$P2_1/n$	1	
13 $(ET)_3PW_{12}$	Ins	12.687	13.493	13.621	77.93	74.17	62.93	$P\bar{1}$	1	

a $Pt=Pt(CN)_4$, $Ni=Ni(CN)_4$, $Mo_6=Mo_6O_{19}^{--}$, $W_6=W_6O_{19}^{--}$, $PW_{12}=PW_{12}O_{40}^{---}$; TMTTF= tetramethyltetrathiafulvalene, TMTSF=tetramethyltetraselenafulvalene, ET=BEDT-TTF= Bis(ethylenedithiolo)tetrathiafulvalene, $A=CH_3CN$, $D=C_3H_7ON$, $E=(C_2H_5)_4N$.
b conductivity at 300 K in S/cm, Ins = insulator.
c S.G.=space group; Z=number of formula units per unit cell.
d the estimated standard deviations on the last digit range between 3 and 6.
e This compound has also been recently reported [8], but with a unit cell which is twice the one given here (**a**=11.002, **b**=17.906, **c**=16.625Å, α=77.28, β=84.17, γ=81.26°) [8].

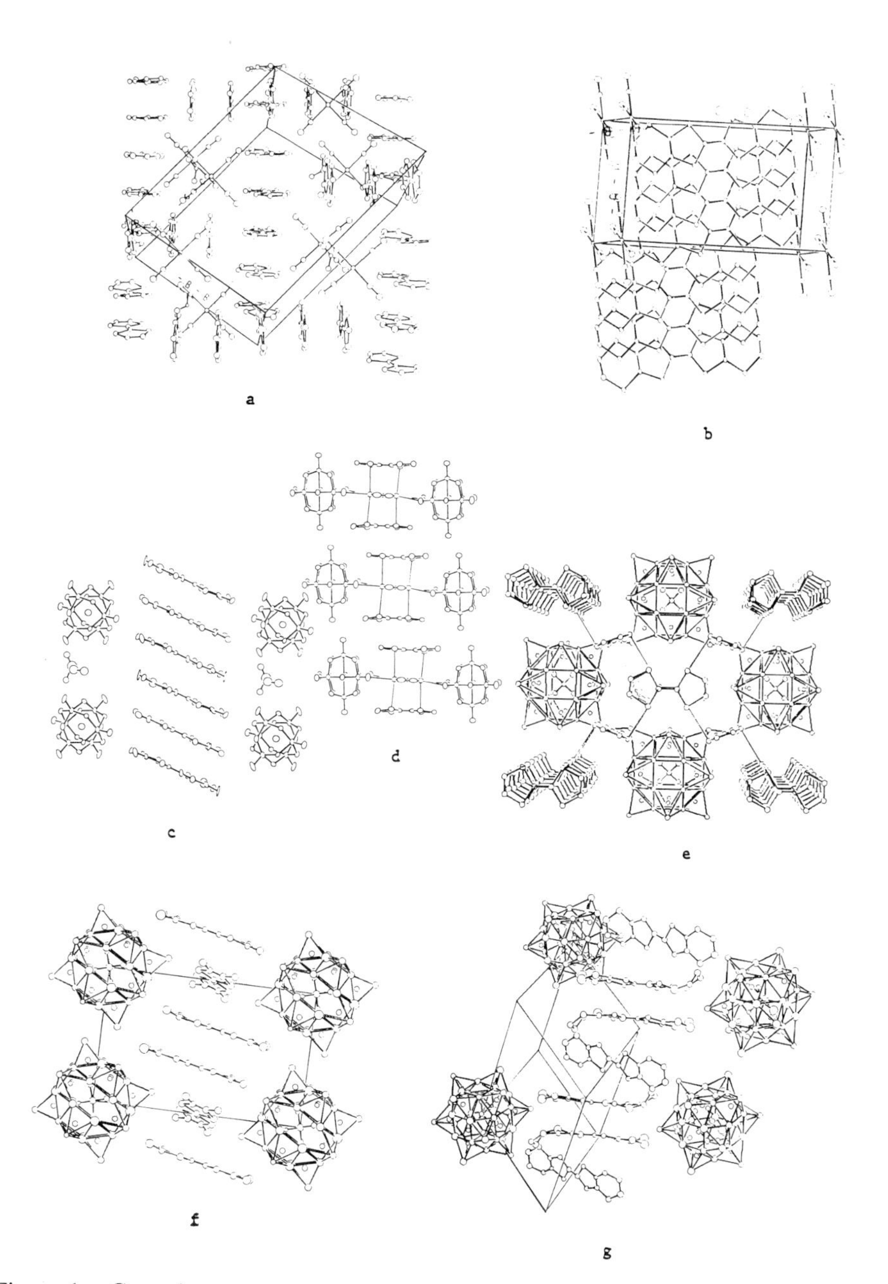

Figure 1. Crystal structures at room temperature of: (a) $(TTF)_5Pt(CN)_4.2\ CH_3CN$, (b) $(BEDT\text{-}TTF)_4Pt(CN)_4$, (c) $(TMTSF)_3W_6O_{19}.2$ DMF, (d) $(TTF)_3W_6O_{19}$, (e)$E_2(TTF)_6PW_{12}O_{40}$, (f) $(TMTSF)_3PW_{12}O_{40}$, (g) $(BEDT\text{-}TTF)_3PW_{12}O_{40}$

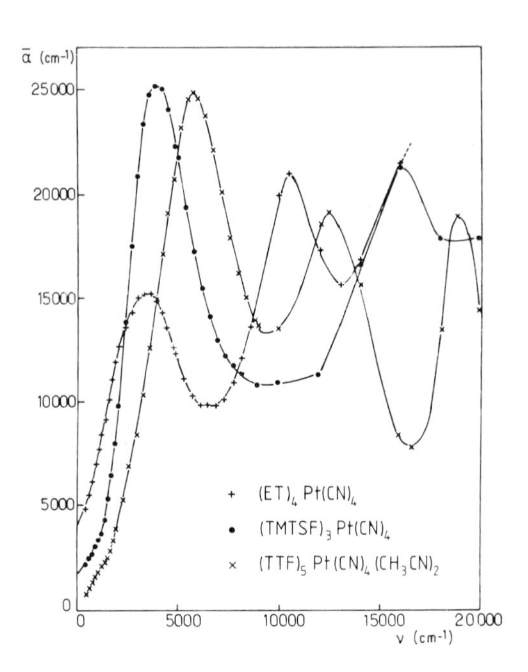

Figure 2. Tetracyanometallate salts: mean electronic absorption coefficient ($\bar{\alpha}$), determined on a KBr pellet at room temperature, versus the energy (in wavenumbers).

usual charge transfer (CT) bands, respectively characteristic of a mixed-valency system (the "A peak" around 5700 cm^{-1}) and of a doubly occupied site (the "B peak" just above 11,000 cm^{-1}) [14]. Additionally, the IR absorption spectrum shows the presence of several vibronic lines between 1300 and 1400 cm^{-1} [a_g mode ($\nu_{C=C\ central}$)] which are under current examination [15].

(BEDT-TTF)$_4$M(CN)$_4$, M=Pt, Ni (6,7): In these compounds [6], the "ET" molecules form tetramerized slipped stacks (Fig. 1b), already observed in other β-type "ET" salts, which present a tendency to a 2-D electronic structure. These salts exhibit also the two characteristic electronic (CT) absorption bands (Fig. 2) associated with strong vibronic IR modes, which are temperature-dependent [15]. Indeed, we have shown that there is a variation of the electron-molecular vibration (e-mv) coupling effect, associated with a "Peierls-like" phase transition around 200 K for both salts. These results are confirmed by the investigation of the electrical and the magnetic properties of the $Ni(CN)_4$ salt, which indicate a metal-insulator transition in the same temperature range [9]. Nevertheless, low-temperature structural investigations are needed to fully characterize this phase transition.

Polyanion compounds

(D)$_3$M$_6$O$_{19}$, M = Mo, W; D = TTF, TMTSF (8-10) [16]: In the compounds **8** and **9** (Fig. 1d) the TTF molecules form stacks of trimers, as in the tetracyanometallates **2, 4, 6**. The intra-trimer overlap is a "criss-cross" type, and the inter-trimer overlap is a "deformed ring-double bond" type. Compound **10** (Fig. 1c) crystallizes with two DMF solvent molecules. In this salt, the TMTSF molecules form stacks of trimers, with intra- (and inter) -trimer ring-double bond overlaps, identical to those observed in the classical $(TMTSF)_2X$ series.

The preliminary electronic absorption spectra for this class of salts are presented in Fig. 3. We observe, as previously, a strong "A band" which is the sign of a mixed-valence system, accompanied by a "B band", more or less intense, depending on the particular type of compound. A comparison between the TTF and TMTSF salts presented in Figs. 2 and 3 allows us to confirm that the position of these CT bands are mainly characteristic of the molecules, almost independent of the counter-ions, but indicative of strong coulombic interactions in the organic blocks [17]. Moreover, strong vibronic IR modes are detected, and

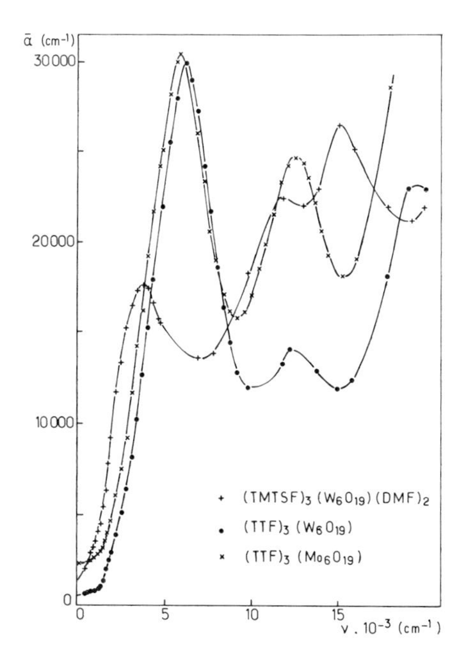

Figure 3. Polymetallate salts: mean electronic absorption coefficient ($\bar{\alpha}$), determined on a KBr pellet at room temperature, versus the energy, expressed in wave numbers.

are under current study, in relationship with the semi-conducting character indicated in Table 1. In particular, $(TTF)_3W_6O_{19}$ presents a T-dependence of the absorption spectra, which could indicate the presence of a structural change at low temperature.

$(D)_xPW_{12}O_{40}$ [D = TTF, x = 6 (**11**) [18]; D = TMTSF, x = 3 (**12**), D = "ET", x = 3 (**13**)]: Two different situations occur. Firstly, in the case of the TTF salt **11**, the donor molecules overlap in an eclipsed fashion, to form regular stacks, lying in a channel formed by four inorganic anions and four isolated TTF molecules (Fig. 1e). The conductivity and magnetism measurements reveal that this material is semi-conducting, and contains one unpaired electron located on the anion [19]. The RT electronic absorption spectra have also shown the presence of a CT "A band" around 4000 cm^{-1}, with a shoulder around 5500 cm^{-1}. This case could constitute a material, where the mixed-valence states coexist on the organic and inorganic moieties, but further study is necessary to confirm this point. Finally, the TMTSF (**12**) and the "ET" (**13**) salts, which present fully ionized trimers (Figs. 1f, 1g) are insulators, as is compound **3** (see Table 1). This situation is confirmed by the absence of any CT absorption band in the lower IR range, as described previously (see Figs. 2 and 3) but only a strong "B-type band" around 11,000 cm^{-1}. For these salts we have not yet been able to stabilize a partially oxidized state.

CONCLUDING REMARKS

In this research field, where organic and inorganic blocks are combined to prepare new materials with sophisticated physical properties, we have briefly presented our recent results, concerning two new series of radical-cation salts. Their analysis, and their comparison with other, similar results with the same kind of inorganic clusters, lead us to the two following observations:

(i) The stabilization of a mixed-valence organic system can be accomplished with large polyanions, as demonstrated by the presence of a characteristic CT band (the so-called "A peak"). However, the equivalency between the molecular sites is a crucial parameter. In trimers or pentamers we often observe a charge-ordering effect, whereas a homogeneous charge distribution is detected for dimers or tetramers. This empirical rule indicates that, to get a metallic compound, it should be better to have an even number of similarly charged molecules, to create the basic organic arrangement, as, for example, in the superconducting κ–$ET_2Cu(SCN)_2$.

(ii) The spatial organization of these blocks is a second necesary step to build new materials. As indicated in the Introduction, there is a conflict between the organic and inorganic entities. On one hand, the radical-cations are trying to develop stacks, because there is a gain of covalent energy through the overlapping of π-orbitals, whereas, on the other hand, the polyanions attempt to develop their own order. Because of the influence of their steric effect (volume and shape), combined with the charge distribution effect (electrostatic long-range interactions), any kind of structural arrangement can result. A rich structural variety should be present, but without any valuable predictive rule. Nevertheless, there is a considerable promise in this approach, because materials with electronic or magnetic dimensionality ranging from zero to 3-D is predicted, i.e. from mixed-valence isolated clusters (for example $(TTF)_5Pt(CN)_4.2CH_3CN$) to a real 3-D conducting compound (π-orbital overlaps in all space).

REFERENCES

[1] L. Ouahab, Thèse de Doctorat d'Etat, 1985, Rennes, France.
[2] L. Ouahab, P. Batail, C. Perrin, and C. Garrigou-Lagrange, Mater. Res. Bull. 21: 1223 (1987).
[3] P. Batail, L. Ouahab, A. Penicaud, C. Lenoir, and A. Perrin, C. R. Acad. Sci. Paris, Serie II 304: 111 (1987).
[4] A. Slougui, L. Ouahab, C. Perrin, D. Grandjean, and P. Batail, Acta Cryst. C45: 388 (1989); A. Slougui, Thèse de Magister, 1989, Constantine Algerie.
[5] (a) K. Krogmann, Angew. Chem. Int. Ed. Engl. 8: 35 (1969);
(b) for a review, see: J. S. Miller and A. J. Epstein, Prog. Inorg. Chem. 20: 1 (1976).
[6] L. Ouahab, J. Padiou, D. Grandjean, C. Garrigou-Lagrange, P. Delhaès, and M. Bencharif, J. Chem. Soc. Chem. Commun. in press.
[7] M. Bencharif and L. Ouahab, Acta Cryst. C44: 1514 (1988).
[8] R. P. Shibaeva, R. M. Lobkovskaya, V. E. Korotkov, N. D. Kushch, E. B. Yagubskii and M. K. Makova, Synth. Metals, 27: A457 (1988).
[9] M. Tanaka, H. Takeuchi, M. Sano, T. Enoki, K. Lsuzuki, and K. Imaeda,Bull. Chem. Soc. Japan, 1989, 62: 1432 (1989).
[10] S. Gärtner, I. Heinen, D. Schweitzer, B. Nuber, H. J. Keller, Synth. Metals, in press.
[11] (a) M. T. Pope and G. M. Varga, J. Inorg. Chem., 1966, 5: 1249 (1966).
(b) M. T. Pope and E. Papaconstantinou, J. Inorg. Chem 6: 1147 (1967).
(c) G. M. Varga, E. Papaconstantinou, and M. T. Pope, J. Inorg. Chem 9: 662 (1970).
(d) J. P. Launay, J. Inorg. Nucl. Chem. 18: 807 (1976).
[12] I. Lindquist, Arkiv Kemi, 5: 247 (1953).
[13] J. F. Keggin, Proc. Roy. Soc. London, Ser. A, A144: 75 (1934).
[14] C. Garrigou-Lagrange, E. Dupart, J. P. Morand and P. Delhaès, Synth. Metals, 27: B537 (1988).
[15] C. Garrigou-Lagrange, L. Ouahab, D. Grandjean and P. Delhaès, Synth.Metals, in press.
[16] S. Triki, L. Ouahab, J. Padiou, and D. Grandjean, J. Chem. Soc. Chem. Commun. in press.
[17] P. Delhaès and C. Garrigou-Lagrange, Phase Transitions, 13: 27 (1988).
[18] L. Ouahab, M. Bencharif, and D. Grandjean, C. R. Acad. Sci. Paris, Série II 307: 749 (1988).
[19] L. Ouahab *et al.,* to be submitted.

NEW SYNTHETIC METAL PRECURSORS: SUBSTITUTED TETRATHIOTETRACENE AND RELATED COMPOUNDS

Toshio Maruo, Megh Singh and M. Thomas Jones

Department of Chemistry
University of Missouri-St. Louis
St. Louis, MO 63121
USA

INTRODUCTION

The electron donor tetrathiotetracene (TTT), Fig. 1, forms charge transfer salts which, at room temperature, possess some of the highest electrical conductivities of any known synthetic metal[1]. For example, a conductivity of ca 10^3 S/cm is observed for $(TTT)_2I_3$ at room temperature. As the temperature of the complex is decreased, the conductivity increases[2] until ca 35 K, at which point it decreases dramatically[2]. While not firmly established, it is believed that the latter behavior is the result of a Peierls transition, which allows alternating pairwise interactions to occur between adjacent TTT molecules and cations.

It has been demonstrated that the introduction of a substituent or substituents into otherwise very symmetrical donors and acceptors (i.e., D_{2h}) can suppress the temperature at which the Peierls transition occurs. Such is the case of 2-fluorotetraselenotetracene[3], wherein the conductivity of the salt, $(2\text{-FTSeT})_2Br$, increases smoothly from room temperature to 4.2 K.

One of the problems in working with tetrathiotetracene and its tetrachalcogeno relatives is that synthesis is

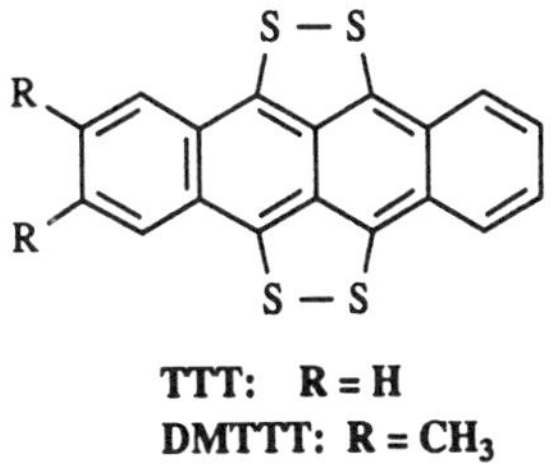

TTT: R = H
DMTTT: R = CH_3

Fig.1. Structure of Dimethyltetrathiotetracene and Relatives

Lower-Dimensional Systems and Molecular Electronics
Edited by R.M. Metzger *et al.*, Plenum Press, New York, 1991

complicated, because of their relative insolubility, which makes purification difficult. Thus, we asked the question of whether it would be possible to synthesize substituted tetrachalcogenotetracenes, which are more soluble, and which possess lower symmetry such that their salts would resist the tendency to undergo Peierls transitions at lower temperatures.

We decided to attempt the synthesis of dimethyl-TTT (DMTTT), Fig. 1, and related chalcogeno derivatives. The methyl group was selected as the substituent, because it was expected to impart improved solubility to the parent compound, because of its size, and because it was expected to enhance the electron-donating ability of DMTTT relative to TTT. The two methyl groups were placed such that the symmetry of the donor is C_{2v}.

SYNTHESIS OF DIMETHYLTETRATHIOTETRACENE

The series of reactions for the synthesis of DMTTT are outlined in the scheme (See Fig. 2).

4,5-Dimethylphthalic anhydride (I) and 1,4-naphthalenediol (II) were mixed with B_2O_3 and heated at 190°C with stirring, to give 2,3-dimethyl-6,11-dihydroxy-5,12-naphthacenequinone (III). (III) was refluxed with PCl_5 and $POCl_3$, followed by the partial dehydrochlorination with NaI in DMF, to give 2,3-dimethyl-5,6,11,12-tetrachlorotetracene (IV). (IV) was heated with Na_2S_2 in HMPA at 100°C for 12 h, to give crude DMTTT (V). (V) was washed with H_2O, MeOH, sublimed and recrystallized from monochlorobenzene, to give very dark green crystals with yield of 20%. The physical properties for the new compounds are (I) colorless crystals (benzene), m.p. 206°C, IR (KBr, cm^{-1}) 1850, 1770, 1615, 1460, 1395, 1345, 1290, 1260, 1200, 1110, 1060, 1015, 910, 895, 740, 635, and 530, ^1H ($CDCl_3$) δ 2.43 (s, 6H, 2xCH_3), 7.40 (s, 2H, aromatic), ^{13}C ($CDCl_3$) δ 20.81, 126.16, 129.23, 146.56, and 163.13; (III)

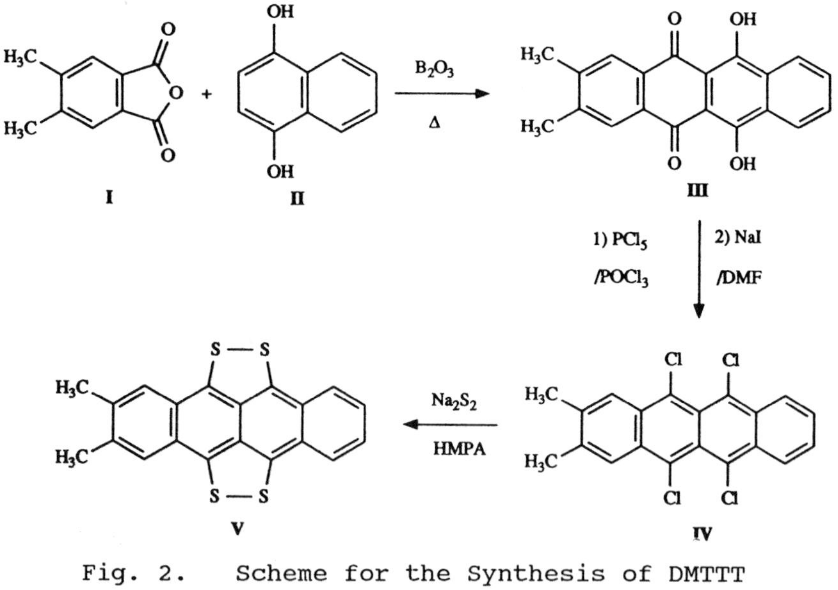

Fig. 2. Scheme for the Synthesis of DMTTT

orange red needles (nitrobenzene), m.p. 322-4° C, IR (KBr, cm^{-1} 3450 (br, OH) 1630, 1585, 1512, 1470, 1275, 1020, 875, and 740, ^1H ($CDCl_3$) δ 2.46 (s, 6H, $2xCH_3$), 7.80 (m, 2H, aromatic), 8.28 (s, 2H, aromatic), 8.45 (m, 2H, aromatic); (IV) dark red needles (chloroform), m.p. 212-3° C, IR (KBr, cm^{-1} 1460, 1278, 1122, 752 (C-Cl), and 658, ^1H ($CDCl_3$) δ 2.52 (s, 6H, $2xCH_3$), 7.58 (m, 2H, aromatic), 8.32 (s, 2H, aromatic), 8.58 (m, 2H, aromatic); (V) very dark green fine needles, anal. calcd for $C_{20}H_{12}S_4$: C, 63.12; H, 3.18; S, 33.70: found C, 62.45; H, 3.29; S, 33.11 %, m.p. 300-5°C (melts with dec.), IR (KBr, cm^{-1}) 1622, 1605, 1470, 1440, 1315, 1300, 1232, 1150, 840, 735, and 680.

PROPERTIES OF DIMETHYLTETRATHIOTETRACENE (DMTTT)

The electronic absorption spectrum of DMTTT in benzene displays three absorption bands at 470, 635 and 693 nm which compare with those observed at 472, 642, and 701 nm for TTT in benzene[4].

Cyclic voltammetry traces for TTT and DMTTT have been obtained, and the results are summarized in Table 1. Two reversible one-electron steps are observed for each compound. As expected, the trace for DMTTT is shifted to more negative potentials relative to TTT.

PROPERTIES OF DMTTT CATION RADICAL IN SOLUTION

Reaction of dilute solutions of DMTTT with trifluoroacetic acid in such solvents as methylene chloride and tetrahydrofuran yields the cation radical of DMTTT. ESR spectra taken from dilute degassed solutions show 11 of the expected 13 approximately equally spaced absorption lines, separated by 0.63 Gauss, and a g-value equal to 2.00778. These values fall within the range expected for the cation radical of DMTTT, and agree quite well with those reported[5] for the TTT cation radical in solution.

Table 1. Redox Potentials of Dimethyltetrathiotetracene and Tetrathiotetracene[a,b]

Compound	E_{pa}^1	E_{pc}^1	E_{pa}^2	E_{pc}^2
TTT	0.26V	0.18V	0.67V	0.59V
DMTTT	0.18V	0.11V	0.60V	0.54V

[a]Measured in benzonitrile solution
[b]0.10 M $(n\text{-}C_4H_9)_4NPF_6$ was the supporting electrolyte and the reference electrode was Ag/AgCl.

The electronic absorption spectrum of $DMTTT^+$ in benzene with trifluoroacetic acid displays three absorption bands at 463, 503 and 591 nm, which compare well with those observed for TTT^+ in dimethylsulfoxide[6].

PROPERTIES OF (DIMETHYLTETRATHIOTETRACENE)X (X = Cl or I)

While the ESR spectra of TTT_2I_3 and $DMTTT_2I_3$ consist of single and more symmetrical lines, the ESR spectra of the corresponding 1:1 chloride salts, prepared by photooxidation of TTT and DMTTT under the same conditions as described by Masson et al[6] for the preparation of TTTCl, are characteristic of solid state anisotropic g-tensor spectra. The analysis of these spectra are shown in Table 2 in comparison with the g-tensors reported by Jones et al[5] for TTT^+ in frozen solution.

The ESR spectrum for DMTTTI is a single line which is more symmetrical than that of DMTTTCl. Its overall spectral width, however, is comparable, as shown in Table 2.

Compressed pellet electrical conductivity of DMTTTI was measured at room temperature using a four-probe technique. Electrical contacts were made with conductive silver paste (GC Electronics) using 1 micron gold wires. The room temperature conductivity is observed to be 1×10^{-3} S/cm. This compares very well with that reported for TTTI[1].

PROPERTIES OF $(DIMETHYLTETRATHIOTETRACENE)_2I_3$

Reaction of DMTTT with I_2 in nitrobenzene, under the same conditions as described by Kaminskii et al (See ref. 7) for the preparation of TTT_2I_3, yields a paramagnetic solid, which is dark green with a golden luster. Its ESR spectrum consists of

Table 2. Comparison of g-Values and ESR Spectral Widths for 1:1 Salts

	$TTT^{+,a}$	$TTTCl^b$	$DMTTTCl^b$	$DMTTTI^b$
g_1	2.0026	2.0037	2.0028	2.0039^c
g_2	2.0094	2.0093	2.0084	2.0076
g_3	2.0114	2.0108	2.0112	2.0118^c
<g>	2.0078	2.0079	2.0074	2.0078
$Width/G^c$	14.3	11.6	13.8	12.8

[a]See reference 5.
[b]This work.
[c]This represents the spectral width between points of maximum and minimum amplitude (i.e., the distance in Gauss between g_1 and g_3 for the anisotropic spectra).

a single, almost symmetrical line of width 21.8 Gauss with a g-value of 2.0082 at crossover. This compares with the ESR spectrum of TTT_2I_3 prepared under similar conditions, where the spectral width is 22.5 Gauss and the g-value at crossover is equal to 2.0077[6].

FUTURE WORK

Plans for future work include the preparation of a larger number of charge transfer salts based on DMTTT and a broader range of physical studies to better characterize these new materials.

Preliminary attempts to prepare dimethyltetraselenotetracene (DMTSeT), according to the scheme outlined here for the synthesis of DMTTT, appear to have been successful. However, sufficient quantities have not been obtained to date to fully characterize the product, which we believe to be DMTSeT. Our results in this area will be described more fully at a later date.

ACKNOWLEDGEMENTS

The partial support of this study by the University of Missouri through the Missouri Research Assistance Act and of Toshio Maruo through a Mallinckrodt Research Fellowship is gratefully acknowledged. Also acknowledged is the help of Dr. David Pipes of the Mallinckrodt Co. who kindly made the cyclic voltammetry measurements. Also gratefully acknowledged is the help of Professor Bernard Feldman of the Department of Physics at the University of Missouri-St. Louis who made the room-temperature electrical conductivity measurements.

REFERENCES

1. L. I. Buravov, G. I. Zvereva, V. F. Kaminskii, L. P. Rosenberg, M. L. Khidekel, R. P. Shibaeva, I. F. Shchegolev, and E. B. Yagubskii, J. Chem. Soc. Commum. 720 (1976); L. C. Iset and E. A. Perez-Albuerne, Solid State Comm. 21:433 (1977); I. F. Shchegolev and E. B. Yagubskii, "Cation-Radical Salts of Tetrathiotetracene and Tetraselenotetracene: Synthetic Aspects and Physical Properties" in Linear Extended Chain Compounds, Vol. 2, pp. 385-434. edited by J. S. Miller, Plenum Press, NY (1982); B. Hilti, C. W. Mayer, E. Minder, K. Hauenstein, J. Pfeiffer, and M. Rudin, Chemica 402:56 (1986).

2. B. Hilti, C. W. Mayer, Helv. Chim. Acta 61:502 (1978).

3. B. Hilti, C. W. Mayer, G. Rihs, H. Loeliger, and P. Paltzer, Mol. Cryst. Liq. Cryst. 120:267 (1985).

4. E. A. Perez-Albuerne, H. Johnson, Jr., D. J. Trevoy, J. Chem. Phys. 55:1547 (1971).

5. M. T. Jones, S. Jansen, L. A. Acampora, and D. J. Sandman, J. de Phys. 44:C1159 (1983).

6. M. Masson, P. Delhaes and S. Flandrois, Chem. Phys. Lett.

76:92 (1980). These authors report ESR parameters of 22.5 Gauss for the width and 2.0080 for the g-value at crossover.

7. V. F. Kaminskii, M. L. Khidekel, R. B. Lyubovskii, I. F. Shchegolev, R. P. Shibaeva, E. B. Yagubskii, A. V. Zvarykina, and G. I. Zvereva, Phys. stat. sol (a). 44:77 (1977).

SPECTROSCOPIC INVESTIGATION OF $(2,5\text{-DM-DCNQI})_2M$ MATERIALS IN BULK AND THIN FILM FORMS

E.I. Kamitsos, G.D. Chryssikos and V. Gionis

Theoretical and Physical Chemistry Institute
National Hellenic Research Foundation
48, Vass. Constantinou Ave., Athens 11635
Greece

INTRODUCTION

Recently, a new class of electron acceptors based on N,N'-dicyanoquinonediimine (DCNQI), and a number of their charge-transfer complexes and anion radical salts have been prepared (1-5). Among them the compound $(2,5\text{-DM-DCNQI})_2Cu$ has attracted special interest, because it exhibits very high electrical conductivity (up to 5×10^5 S cm^{-1} below 10 K) and retains its metallic behavior down to 1.3 K without metal-insulator transition (3). Anion radical salts with other counterions exhibit lower conductivities and are semiconductors, or undergo a Peierls transition in the temperature range 100-150 K (4).

To understand such differences in transport properties, various studies have been undertaken, including structural investigations (6-9), pressure effects (10-12), XPS, ESR and susceptibility measurements (5,7,13). These studies have led to the quite general agreement that the properties of the copper salts can be understood on the basis of the mixed-valence state of copper, *i.e.* a formal oxidation state of +1.3 . Thus, it has been suggested that a mixing of copper $3d_{xy}$ orbital with the $p\pi$ LUMO of DCNQI takes place, causing the suppression of the one-dimensional character of the copper-DCNQI compound. Recently, evidence for such an interaction

Lower-Dimensional Systems and Molecular Electronics
Edited by R.M. Metzger *et al.*, Plenum Press, New York, 1991

was reported by Yakushi et al. (14) on the basis of the optical reflectance spectra of $(2,5\text{-DM-DCNQI})_2M$ (M=Cu, Na) compounds. It was shown that the Cu-compound exhibits an additional absorption at ca. 12,000 cm^{-1}, assigned to a charge-transfer between Cu ions and DCNQI anions.

It has been demonstrated, in the case of TCNQ-based materials, that vibrational spectroscopy is a useful tool in studying the state of the acceptor species, the knowledge of which is essential in understanding the materials properties (15). In this paper we wish to report on the Raman, infrared and uv-vis spectra of $(2,5\text{-DM-DCNQI})_2M$ (M=Cu, Ag, Na) materials. Thin film formability of the Cu- and Ag-compounds has been also investigated, and preliminary results are reported here.

EXPERIMENTAL

2,5-DM-DCNQI was synthesized according to the published procedure (1). The anion radical salts of Cu and Na were prepared upon reduction of the acceptor by the corresponding metal iodide, while silver powder was used as the reducing agent for the preparation of the Ag-salt (4). These reactions were carried out in CH_3CN, and resulted in the powders of the anion radical salts. For spectral measurements, free-standing pellets were pressed from the prepared compounds.

Infrared spectra were recorded in the reflectance mode on a Fourier-transform Bruker IFS 113v spectrometer. The appropriate sources, beam splitters and detectors were utilized to cover effectively the 30-5,000 cm^{-1} spectral range. Reflectance spectra in the nir-vis-uv region (4,000-40,000 cm^{-1}) were recorded on a Varian 2390 spectrometer. Raman spectra were measured on a Ramanor HG 2S Jobin-Yvon spectrometer. To avoid sample decomposition, care was taken to maintain the incident laser power as low as possible (<20 mW), and have the sample in the spinning pellet form.

Thin films of $(2,5\text{-DM-DCNQI})_2M$ (M=Cu, Ag) were prepared by the vacuum deposition and thermal treatment technique employed previously for analogous TCNQ-based materials (16,17). Thus, succesive layers of acceptor and metal were vacuum-deposited on the appropriate substrate to achieve a 2:1 DCNQI:metal molar ratio. The so-obtained thin films were heat-treated at *ca* 150^0 C for a few minutes, to result in light-blue films for M=Ag and goldish films for M=Cu. Various substrates, such as quartz, KBr or CaF_2 were used for transmission spectroscopic measurements covering the mid-infrared to uv spectral range.

RESULTS AND DISCUSSION

Reflectance Spectra

The reflectance spectra of the compounds studied are shown in Figure 1 in an extended spectral range. The infrared region alone is shown as insert in the same Figure. The spectra of Cu- and Na-salts are quite similar to those reported by Yakushi et al. in the 750-25,000 cm^{-1} range (14). In the high-frequency region both Na- and Ag-salts show bands at *ca*. 13,000 and 28,000 cm^{-1}, as well as the onset of absorption at *ca*. 38,000 cm^{-1}. We note that a dilute solution of (2,5-DM-DCNQI)$_2$Na in CH_3CN exhibits bands at 29,400 and 41,500 cm^{-1}, arising presumably from localized or intramolecular excitations of the charged DCNQI species, in analogy with TCNQ compounds (18). Thus, the bands of the solid state spectra at *ca*. 28,000 and 38,000 cm^{-1} are assigned to intramolecular excitations of the DCNQI

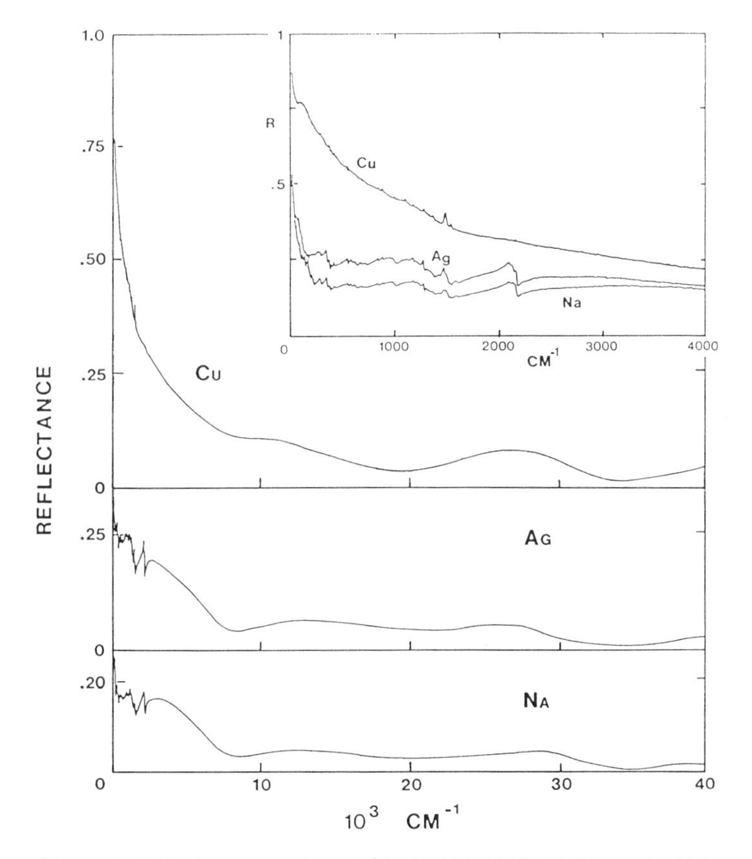

Figure 1. Reflectance spectra of (2,5-DM-DCNQI)$_2$M (M=Cu,Ag,Na) from far-infrared to ultraviolet spectral region. The infrared region alone is shown in the insert.

anion, termed LE_1 and LE_2 respectively (19). The lower-energy band at *ca*. 13,000 cm^{-1} is assigned to the charge- transfer (CT) excitation between charged DCNQI neighbours along the same stack, denoted CT_1 (19). The same feature was assigned by Yakushi et al. to the lowest intramolecular excitation of $DCNQI^{\dot{-}}$ (LE_1), on the basis of the similarity of this band with the triplet (15,000 , 16,500 and 17,900 cm^{-1}) observed in a solution of 2,5-DM-$DCNQI^{\dot{-}}$. However, we have measured this triplet only in concentrated solutions of $(2,5\text{-DM-DCNQI})_2Na$, indicating that this should be of a charge-transfer, rather than of a localized type (20).

In analogy to the above, the band of the Cu-salt at *ca*. 27,000 cm^{-1} and the onset of absorption at *ca*. 40,000 cm^{-1} are assigned to localized excitations LE_1 and LE_2, respectively. The broad feature at *ca*. 10,500 cm^{-1} results from contributions of both CT_1 type and charge-transfer between copper ion and DCNQI species (14), the latter involving the $3d_{xy}$ orbital of Cu^+ and the LUMO orbital of $DCNQI^{\dot{-}}$ (7). A CT transition of the same nature was measured for the first time in CuTCNQ (*ca*. 10,000 cm^{-1}) and was attributed to CT from the full d orbitals of Cu^+ to LUMO of $TCNQ^{\dot{-}}$ (19).

The infrared conductivity obtained by Kramers-Kronig analysis of the reflectance spectra is shown in Figure 2. The spectra of Na- and Ag-salts show broad but well-defined maxima at *ca*. 4,000 and 3,500 cm^{-1} respectively. These are assigned to CT excitations between a charged and a neutral DCNQI neighbour along the 1-D stacks (CT_2) (18, 19). In contrast to this behaviour, the spectrum of the Cu-compound exhibits a very broad absorption, covering effectively the entire infrared region, suggesting the delocalized nature of interactions in this compound, compared to the Na- and Ag-counterparts.

Another pronounced difference in the spectrum of the Cu-salt is observed in the $C \equiv N$ stretching region : both Na- and Ag-salt spectra are characterized by the presence of a strong triplet (*ca*. 2160, 2130 and 2105 cm^{-1}), while that of the Cu-salt shows a weak CN absorption in the same region. This effect may well be due to the presence of the broad and strong CT absorption, which effectively masks the $C \equiv N$ absorption of the Cu-compound. In addition, a considerable effect on the $C \equiv N$ stretching could be expected, because of the strong tetrahedral coordination of copper to the nitrogen atoms of the CN groups. This was shown to be manifested by the short Cu-N distance (4,7) and the inter-chain charge-transfer through the d_{xy} orbital of copper (14).

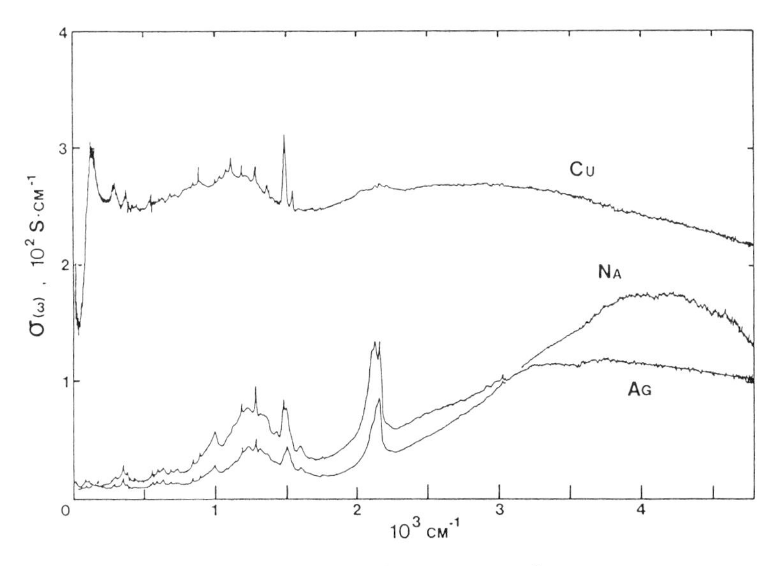

Figure 2. Infrared conductivity of $(2.5\text{-DM-DCNQI})_2M$ compounds obtained by Kramers-Kronig analysis of reflectance spectra.

In the lower frequency region (500-2000 cm^{-1}) all spectra exhibit similar features, with the noticeable presence of a broad envelope centered at *ca*. 1250 cm^{-1}. In the far-infrared, the spectrum of the Cu-salt shows a strong band at *ca*. 140 cm^{-1}, which is not present in the rest of the spectra. The origin of this feature can not be identified at present with certainty, but it can be tentatively assigned to the Cu-N vibration, resulting from the strong coordination of Cu.

Raman Spectra

It was demonstrated for TCNQ compounds that the reduction of the acceptor molecule causes significant vibrational frequency shifts, which are mostly pronounced for the totally symmetric modes (21-23). Thus, the Raman spectra of the three anion radical salts of 2,5-DM-DCNQI have been measured and are presented in Figure 3. The spectrum of the neutral acceptor is also included for comparison. Even though no vibrational analysis for the neutral or charged acceptor has been reported so far, assignments of the main Raman bands can be proposed on the basis of the work on TCNQ compounds (21,22).

The spectrum of the neutral species shows the strongest bands at 2170, 1635 and 1502 cm^{-1}. The main activity of the rest of the spectra is also exhibited in the same frequency regions. Clearly, the first band can be assigned to the C≡N

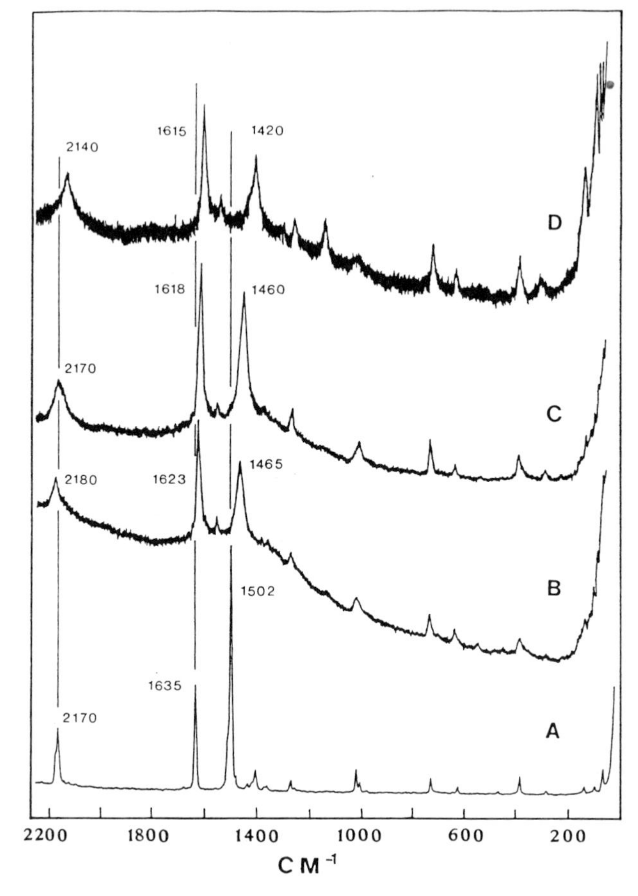

Figure 3. Raman spectra of $(2,5\text{-DM-DCNQI})_2M$ materials in pressed pellet forms (488.0 nm excitation): (A) 2,5-DM-DCNQI; (B) M=Na; (C) M=Ag; (D) M=Cu.

stretching vibration (ν_2), and the second one to the C=C stretching vibration of the ring (ν_3), (the notation used for TCNQ compounds is accepted in this work). The strongest feature at 1502 cm^{-1} is analogous to the ν_4 mode of TCNQ compounds and can be assigned to the C=N stretching vibration (24).

While the spectra of all salts presented in Fig.3 show a considerable broadening in the region of the ν_2 mode, only the spectrum of the Cu-salt exhibits a clear frequency downshift by 30 cm^{-1}, compared to that of the neutral acceptor. A similar downshift of the ν_2 mode has been observed in TCNQ compounds (21,22), and found to increase with negative charge on the acceptor molecule (degree of charge-transfer) (23). It should be noted though, that a frequency decrease of the C$\equiv$N stretching vibration can also result via a π-back-bonding mechanism, through which electron density is transfered from metal to CN group (25). The opposite mechanism, *i.e.* σ-donation from CN to metal, results in an increase of the CN stretching frequency (25).

The ν_3 and ν_4 modes of the salts appear downshifted, compared to those of the neutral acceptor. The greatest frequency shift is demonstrated by the ν_4 mode of the Cu-salt. It is noted that the frequency of the ν_4 mode of TCNQ compounds was found to change linearly with negative charge on TCNQ-species, *i.e.* the degree of charge-transfer, ρ (23). Assuming that a similar effect holds for DCNQI-compounds, then the Raman spectra of Fig.3 suggest that the negative charge on DCNQI is greater in the Cu-compound, compared to the Na- and Ag-analogs. This result is in agreement with the previous findings, that is the existence of copper ion in a mixed-valence state with a +1.3 average oxidation number.

Spectroscopic studies (ir, uv-vis) of the Cu- and Ag- compounds in thin film forms have also been conducted in the context of this work. The results indicate that the thin film materials are of the same nature with their bulk analogs, and exhibit pronounced orientational effects on various substrates. This work will be reported in detail elsewhere.

REFERENCES

1. A.Aumüller and S.Hünig, Angew. Chem. 96: 437 (1984); Angew. Chem. Int. Ed. Engl. 23: 447 (1984); Liebigs Ann. Chem. 142: 165 (1986).
2. A.Aumüller, E.Hädicke, S.Hünig, A.Schätzle, J.U. von Schütz, Angew. Chem.96: 439 (1984); Angew. Chem. Int. Ed. Engl. 23: 449 (1984).
3. A Aumüller, P.Erk, G.Klebe, S.Hünig, J.U. von Schütz and H-P. Werner, Angew. Chem. Int. Ed. Engl. 25: 740 (1986).
4. S.Hünig, A. Aumüller, P.Erk, H.Meixner, J.U. von Schütz, H.J.Gross, U.Langohr, H-P. Werner, H.C.Wolf, C.Burschka, G.Klebe, K.Peters and H.G. von Schnering, Synthetic Metals, 27: B181 (1988)
5. J.U. von Schütz, M.Bair, H.J.Gross, U.Langohr, H.P.Werner, H.C.Wolf, D.Schmeiber, K.Graf, W.Gopel, P.Erk, H.Meixner and S.Hünig, Synthetic Metals, 27: B249 (1989).
6. R.Kato, H.Kobayashi, A.Kobayashi, T.Mori, H.Inokuchi, Chem.Lett, 1579 (1987)
7. A.Kobayashi, R.Kato, H.Kobayashi, T.Mori, H.Inokuchi, Solid State Comm. 64: 45 (1987).

8. R.Kato, H.Kobayashi, A.Kobayashi, T.Mori and H.Inokuchi, Synthetic Metals 27: B263 (1988).
9. R.Moret, Synthetic Metals 27: B301 (1988)
10. T.Mori, K.Imaeda, R.Kato, A.Kobayashi, H.Kobayashi and H.Inokuchi, J. Phys. Soc. Jpn. 56: 3429 (1987).
11. S.Tomic, D.Jérome, A.Aumüller, P.Erk, S.Hünig and J.U. von Schütz, J. Phys. C. Solid State Phys. 21: L203 (1988).
12. S.Tomic, D.Jérome, A.Aumüller, P.Erk, S.Hünig and J.U.von Schütz, Synthetic Metals 27: B281 (1988).
13. T.Mori, H,Inokuchi, A.Kobayashi, R.Kato and H.Kobayashi, Synthetic Metals 27: B237 (1988).
14. K.Yakushi, G.Ojima, A.Ugawa and H.Kuroda, Chem. Lett. 95 (1988).
15. For a review article see R.Bozio and C.Pecile, in "The Physics and Chemistry of Low-Dimensional Solids", Ed. by L.Alcácer (Reidel, Boston, 1980).
16. E.I.Kamitsos and W.M.Risen, Solid State Comm. 45: 165 (1983) and 42:561 (1982).
17. E.I.Kamitsos, Mol.Cryst.Liq.Cryst. Inc. Nonlin. Opt. 161: 335 (1988)
18. K.Kamaràs, G.Grüner and G.A.Sawatzky, Solid State Comm. 27: 1171 (1978).
19. E.I.Kamitsos, G.C.Papavassiliou and M.A.Karakassides, Mol.Cryst. Liq. Cryst., 134: 43 (1986).
20. V.Gionis et al, to be published.
21. A.Girlando and C.Pecile, Spectrochim. Acta 29: 1859 (1973).
22. R.Bozio, A.Girlando and C.Pecile, J.Chem. Soc. Faraday Trans. II71: 1237 (1975) and II74: 235 (1978).
23. E.I.Kamitsos and W.M.Risen, J.Chem. Phys. 79: 5808 (1983).
24. N.B.Colthup, L.H.Daly and S.E.Wiberley, in "Introduction to Infrared and Raman Spectroscopy", Academic Press, New York (1975).
25. K.Nakamoto, in "Infrared and Raman Specta of Inorganic and Coordination Compounds", J.Wiley, New York (1978).

THE SERIES OF ORGANIC CONDUCTORS: (PERYLENE)$_x$ [M(mnt)$_2$]

V. Gama, R. T. Henriques and M. Almeida

Dept. Quimica, Instituto de Ciências e Engenharia Nucleares
Laboratório Nacional de Engenharia e Tecnologia Industrial
P-2686 Sacavém Codex, PORTUGAL

INTRODUCTION

The family of organic conductors perylene$_2$[M(mnt)$_2$] (mnt = maleonitriledithiolate) has been studied for more than fifteen years, since the Cu and Ni compounds were reported to be semiconductors by Alcácer and Maki [1,2]. Subsequent work paid special attention to the Au, Pt and Pd compounds [3-9], that exhibit quasi one-dimensional metallic behaviour down to a temperature T_{MI}, where a metal-insulator transition occurs. These compounds crystallize in space group $P2_1/c$ [3,8], and the **b** axis is the high-conductivity direction ($\sigma_{\parallel} / \sigma_{\perp} \approx 10^3$).

In these solids there are two segregated stacks, the perylene stack (responsible for hole conduction) [3,6,7] and the M(mnt)$_2$ stack (which, in the case of Pt and Pd, has localized magnetic moments: S = 1/2 per unit). For the Pt and Pd cases, there is evidence of spin-exchange interactions between the itinerant spins on the perylene stacks and the localized spins on the M(mnt)$_2$ stacks [6,9]; there is also evidence for the existence of a spin-Peierls transition at T_{MI} [4, 6].

In order to gain more insight into this class of compounds, and to study the nature of the low-temperature phase transitions, we reinvestigated the Ni and Cu analogues, and prepared new compounds, based on Fe and Co.

SAMPLE PREPARATION

Single crystals of Per$_x$[M(mnt)$_2$] (M = Cu, Ni, Pt, Au, Co, Fe) can be prepared either by electrocrystallization from dichloromethane solutions containing perylene and the corresponding tetraalkylammonium salt of $[M(mnt)_2]^-$, or by perylene oxidation with iodine. The first method usually gives better results, and has been more extensively used, but, for M = Pd, crystals could only be obtained by iodine oxidation.

The crystals obtained by these methods are long and needle-shaped, with typical dimensions $\leq$ 4 x 0.1 x 0.05 mm^3, depending on the compounds, with the exception of M = Co, where most of the crystals are significantly thicker ($\approx$ 5 x 0.5 x 0.5 mm^3) and M = Fe, where crystals 10 x 1 x 0.1 mm^3 can be obtained easily.

RESULTS AND DISCUSSION

Systematic transport measurements, performed on crystals obtained in the same preparation, revealed, in most cases, the existence of more than one phase.

The first phase, which we shall call the α-phase, is metallic (for M = Cu, Ni, Co, Fe) at room temperature, with transport properties comparable to those previously reported for the

Lower-Dimensional Systems and Molecular Electronics
Edited by R.M. Metzger *et al.*, Plenum Press, New York, 1991

Au, Pd and Pt analogues. A second phase (β-phase), for M = Cu, Ni, Co, is clearly semiconducting at room temperature, as originally reported for the Ni and Cu compounds. Finally, a third phase, with rather different behaviour, has metallic properties at high temperature: this was observed only for M = Co, and is associated with a different stoichiometry, $Per[Co(mnt)_2].xCH_2Cl_2$, with $x \approx 0.5$, as found by elemental analysis. The presence of CH_2Cl_2 was confirmed by SEM-EDS, that indicated the presence of Cl incorporated in the structure. In all other cases (α and β phases) the elemental analyses are consistent with the stoichiometry $(Per)_2[M(mnt)_2]$, confirmed, in some cases, by the crystal structure [8].

The electrical resistivity ρ and thermoelectric power of the α and β phases are shown in Figs. 1 and 2, respectively.

α-Phases

The α-Ni crystals have properties very similar to the Pd and Pt analogues. The room-temperature conductivity is ≈ 700 S/cm. The resistivity decreases with temperature, reaches a minimum at $T_\rho \approx 50$ K, and there is a metal-insulator transition at $T_{MI} = 25$ K, corresponding to the maximum of $d\ln\rho / d(1/T)$. At low temperatures, $d\ln\rho / d(1/T)$ indicates an energy gap ≥ 15 meV. Preliminary data for α-Cu (not shown in Fig. 1) show a similar behaviour down to 65 K.

The α-phases with M = Co, Fe present a relatively lower room-temperature conductivity, ≈ 150 S/cm, with metal-insulator transitions at higher temperatures (73 and 58 K, respectively) and also slightly higher activation energies at low temperatures, indicative of gaps of approximately 60 and 50 meV, respectively.

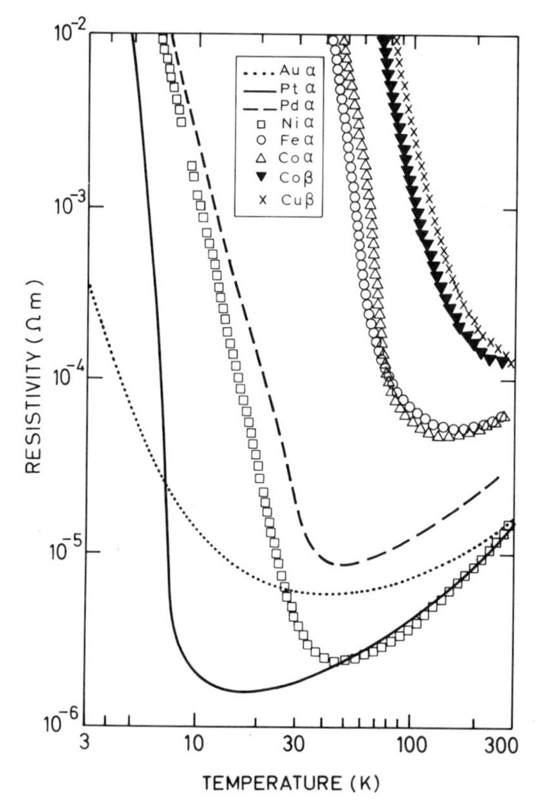

Fig. 1. Temperature-dependent resistivity of $(Per)_2[M(mnt)_2]$ salts.

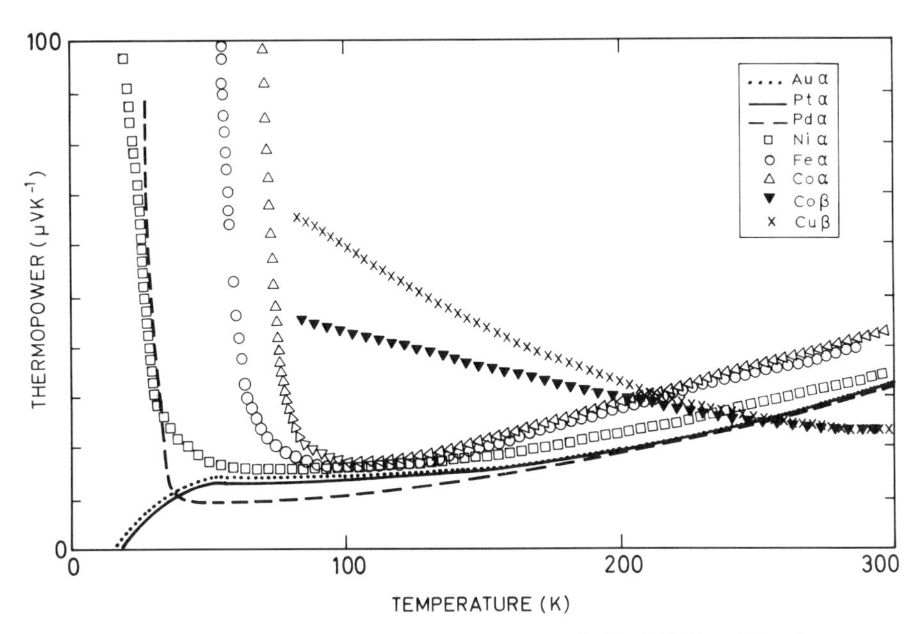

Fig. 2. Temperature-dependent thermopower of $(Per)_2[M(mnt)_2]$ salts.

In the metallic regime, all the α-phases have a very similar thermopower (≈ 35 μV K^{-1} for Au, Pt, Pd and Ni, and ≈ 45 μV K^{-1} for Fe and Co at room temperature), that decreases upon cooling, as expected in metals, and in agreement with the resistivity data. At T_{M-I}, dS / d(1/T) also shows an anomaly, and, at lower temperatures, the thermopower varies at 1/T, as expected in a semiconductor. The positive sign of thermopower in the metallic state is indicative of hole-type conduction, consistent with the stoichiometry $(Per)_2M(mnt)_2$, and with conduction in a 3/4-filled band of perylene chains. From the linear regime at high temperatures, and as was previously done for the Au, Pt, and Pd [7] analogues, it is possible to estimate the bandwidth to be ≈ 0.6 eV for M = Ni (result similar to the one already obtained for M=Au, Pt, Pd) [7] and ≈ 0.5 eV for M = Co, Fe.

The differences in the behaviour of α-Co and α-Fe from the other α-type compounds may be attributed to differences in the structure. All known solids based on $M(mnt)_2$ with M = Co, Fe are dimerized as $[M(mnt)_2]_2^{2-}$. Mössbauer studies in the α-Fe compound [9] showed that the Fe bisdithiolate is dimerized, even at room temperature, without any significant change between room temperature and 10 K. Preliminary X-ray studies indicate that the unit cell parameter of the α-Fe compound is doubled, when compared with the corresponding Au, Pd, Pt and Ni compounds.

β-Phases

A semiconducting behaviour was found in the β-phases of Co and Cu. In agreement with earlier studies [1] for M = Cu and Ni, the conductivity was found to be ≈ 50-80 S/cm at room temperature, and thermally activated in the temperature range 100 - 300 K. In these compounds dlnρ / d(1/T) increases smoothly from room temperature down to 150 K, where it reaches a constant value, corresponding to an energy gap of ≈ 90 meV. This semiconducting behaviour is in agreement with the thermopower measurements shown in Fig. 2. At room temperature S = 22 μV K^{-1}, increasing as T decreases. Further X-ray diffraction studies, now in progress, will, hopefully, clarify the structural differences between the α and the β phases.

$Per[Co(mnt)_2].0.5(CH_2Cl_2)$

The temperature-dependent electrical resistivity and thermopower of this compound are represented in Fig. 3. As expected from its 1:1 stoichiometry, this compound exhibits a rather different behaviour from all the other members of the series.

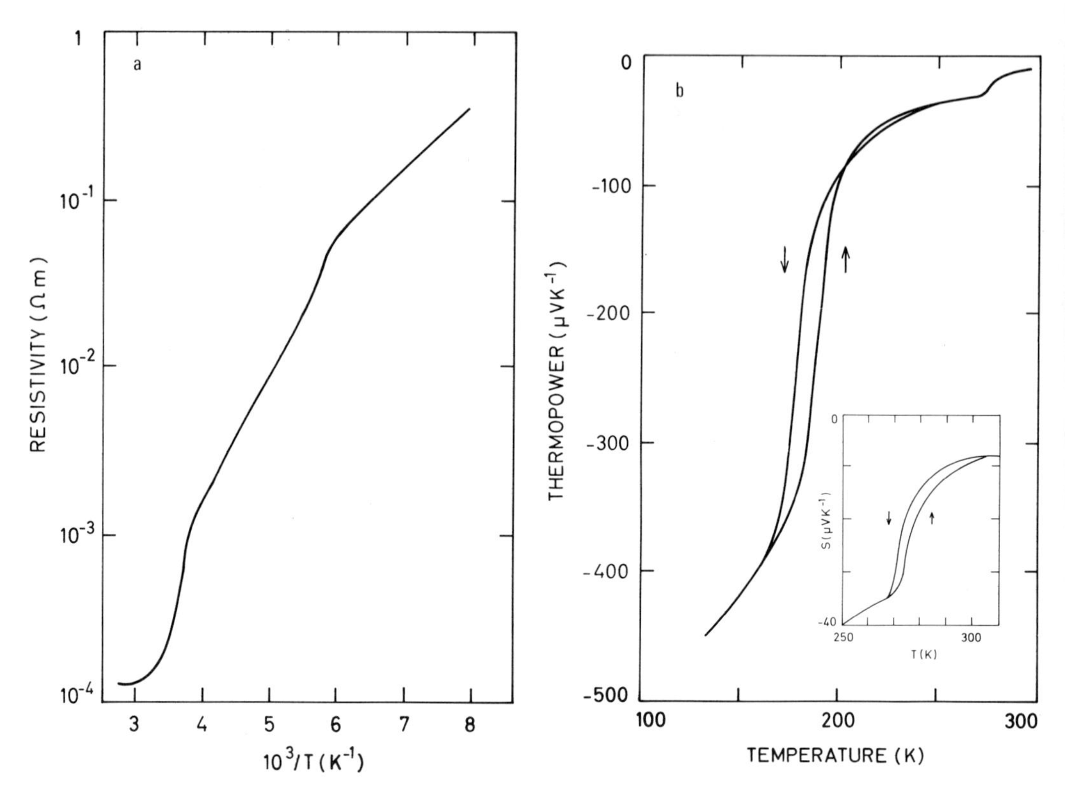

Fig. 3. Temperature-dependent resistivity (a) and thermopower (b) of (Per)[Co(mnt)$_2$]. 0.5(CH$_2$Cl$_2$).

Table 1. Properties of different phases of (Per)$_x$[M(mnt)$_2$] (n = number of d electrons, S = spin, σ_{RT} = room-temperature conductivity, S_{RT} = room-temperature thermopower, $T_{M\text{-}I}$ = metal insulator phase transition temperature (except for Co with x = 1, where the two phase transitions are reported), Δ = low-temperature energy gap derived from resistivity measurements).

Compound x	M	n	S	$(d\rho_{RT}/dT)$	σ_{RT} (S/cm)	S_{RT} (μV K^{-1})	$T_{M\text{-}I}$ (K)	Δ (meV)
x=2	α-Au [3]	8	0	>0	700	32		
	α-Cu	8	0	>0	≈700	22	<60	
	β-Cu	8	0	<0	80	22	——	90
	α-Pt	7	1/2	>0	700	32	7	≈10
	α-Pd	7	1/2	>0	300	32	28	
	α-Ni	7	1/2	>0	700	35	25	≤15
	β-Ni [1]	7	1/2	<0	50		——	102
	α-Co	6	0	>0	150	42	73	60
	β-Co	6	0	<0	80	22	——	90
	α-Fe	5	1/2	>0	150	42	58	50
x=1								
	Co	6	0	>0 (T>345)	50	-10	272-277 178-188	350 300

For $Per[Co(mnt)_2].0.5(CH_2Cl_2)$, the metallic phase is restricted to a small temperature range at higher temperatures ($T > 270$ K). Two phase transitions with large hysteresis are observed: one from a metallic to an insulator state at 272-277 K, and another between two semiconducting states at 178-188 K. At the semiconductor-to-semiconductor phase transition, the activation energy decreases slightly, from 0.175 eV (in the higher temperature phase) to 0.15 eV (in the lower temperature phase). This phase transition is, most probably, associated with the freezing of the dichloromethane molecules incorporated in the structure. The small and negative thermopower at room temperature, and the high metal-insulator transition temperature denote a different electronic structure (half-filled band) in this compound, as a consequence of its stoichiometry. One should mention that a 1:1 compound with M = Pt has also been described by Shibaeva *et al* [10], who used slightly different preparative conditions, and obtained poor semiconducting properties ($\sigma_{RT} \approx 10^{-3}$ S/cm), as a consequence of a mixed-stack arrangement.

CONCLUSIONS

The present results are summarized in Table 1, where the most important transport parameters are indicated. They show a rich variety of metallic and semiconducting behaviours in the compounds of this family, due to their multiple phases.

ACKNOWLEDGEMENTS

The authors acknowledge the collaboration of M. Matos and E. B. Lopes in the transport measurements and fruitful discussion with Prof. L. Alcácer. This work is partially supported by the EEC under contract ESPRIT Basic Research Action 3121.

REFERENCES

[1] L. Alcácer and A. H. Maki, Electrically conducting metal dithiolate-perylene complexes, J. Phys. Chem. 78: 215 (1974).

[2] L. Alcácer and A. H. Maki, Magnetic properties of some electrically conducting perylene-metal dithiolate complexes, J. Phys. Chem. 20: 158 (1978).

[3] L. Alcácer, H. Novais, F. Pedroso, S. Flandrois, C. Coulon, D. Chasseau and J. Gaultier, Synthesis, structure and preliminary results on electrical and magnetic properties of $(perylene)_2$ $[Pt(mnt)_2]$, Sol. State Commun. 35: 945 (1980).

[4] R. T. Henriques, L. Alcácer, J. P. Pouget and D. Jérome, Electrical conductivity and X-ray diffuse scattering study of the family of organic conductors $(perylene)_2M(mnt)_2$, (M=Pt, Pd, Au), J. Phys. C 17: 5197 (1984).

[5] L. Alcácer, Metal-insulator transitions in the perylene-dithiolate conductors, Mol. Cryst. Liq. Cryst. 120: 221 (1985).

[6] R. T. Henriques, L. Alcácer, M. Almeida and S.Tomic, Transport and magnetic properties of the family of perylene-dithiolate conductors, Mol. Cryst. Liq. Cryst. 120: 237 (1985).

[7] R. T. Henriques, M. Almeida, M. J. Matos, L. Alcácer and C. Bourbonnais, Thermoelectric power of the $(perylene)_2M(mnt)_2$ family, Synth. Met. 19: 379 (1987).

[8] A. Domingos, R. T. Henriques, V. Gama, M. Almeida, A. Lopes Vieira and L. Alcácer, Crystalline structure/transport properties relationship in the $(perylene)_2M(mnt)_2$ family (M = Au, Pd, Pt, Ni), Synth. Met. 27: B411 (1988).

[9] J. C. Waerenborgh, J. M. P. Cabral, L. C. J. Pereira, V. Gama and R. T. Henriques, Mössbauer spectroscopy study of $Per_2Fe(mnt)_2$, to be published.

[10] R. P. Shibaeva, V. F. Kaminskii, M. A. Simonov, E. B. Yagubskii and E. E. Kostiosenko, Structure of electrically conducting complexes from radical cations. XIII. Radical cation salt of perylene with platinum (II) bis (dicyanoethylene-1,2-dithiolate), $(C_{20}H_{12})$ $[Pt(S_4C_4(CN)_4]$, Kristallografia: 30, 488 (1985).

TRANSPORT AND MAGNETIC PROPERTIES OF $(PERYLENE)_2Au(i\text{-}mnt)_2$

Manuel J. Matos and Rui T. Henriques

Dept. Quimica, Instituto de Ciências e Engenharia Nucleares
Laboratório Nacional de Engenharia e Tecnologia Industrial
P-2686 Sacavém Codex, PORTUGAL

Luis Alcácer

Dept. Quimica, Instituto Superior Tecnico
P-1097 Lisboa Codex, PORTUGAL

INTRODUCTION

Perylene-based compounds of the family $(Per)_2M(mnt)_2$ have been studied in our laboratory for several years, after they were first reported by Alcácer and Maki [1].

One of the main interests in this family of compounds lies in the possibility of having high electrical conductivity and, simultaneously, a magnetic susceptibility, arising predominantly from localized magnetic moments in the $M(mnt)_2$ stack, for some of the members of the family (e.g. M = Pt, Pd) [2,3]. On the other hand, it is possible to synthesize compounds, for which the $M(mnt)_2$ is diamagnetic (e.g. M = Au), and, consequently, to estimate the contribution to the magnetic susceptibility of the charge carriers only, and to explore the influence of the non-conducting stack on the metal-insulator transition. So far, $Per_2Au(mnt)_2$ was the only member extensively studied in such conditions. It reveals striking differences in the resistivity profile and in the structural instabilities, relative to M = Pt and Pd [4]. Other perylene-based conductors with diamagnetic countrions, $Per_2X.ySolv$ (X= AsF_6, PF_6, Solv = CH_2Cl_2, THF) have a narrower temperature range for metallic conduction, as $T_{M\text{-}I} \approx 200$ K [5,6]. However, the EPR signal for these systems has very narrow linewidths [7] comparable to those found for other hydrocarbon radical-cation conductors like $(Fluoranthrene)_2PF_6$, where $\Delta H_{pp} \approx 10$ mGauss [8], whereas for $Per_2Au(mnt)_2$ it is $\approx$ 0.5 Gauss [2].

In order to get insight into the above problems, we decided to investigate the behavior of one member (M = Au) of the family of compounds of general formula $(Per)_2M(i\text{-}mnt)_2$, in which the ligand **i-mnt** (**1**) is the geometrical isomer of **mnt** (**2**).

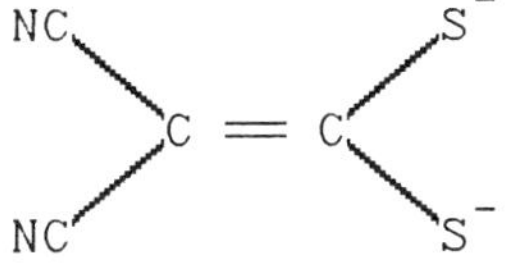

1

i-mnt =iso-maleonitriledithiolate

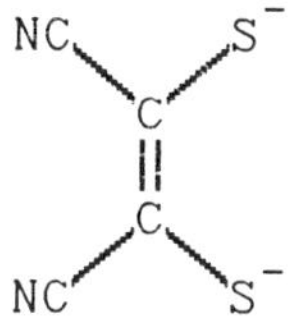

2

mnt = maleonitriledithiolate

Lower-Dimensional Systems and Molecular Electronics
Edited by R.M. Metzger *et al.*, Plenum Press, New York, 1991

SYNTHESIS

The ligand iso-maleonitriledithiolate was synthesized as the potassium salt, according to the procedure of Gomper and Topel [9]. Its gold complex was obtained by the reaction of potassium tetrachloroaurate with $K_2(i\text{-mnt})_2$ in the presence of tetra-n-butylammonium bromide [10].

Crystals of $Per_2Au(i\text{-mnt})_2$ were obtained in the anode compartment of an electrochemical cell containing a dichloromethane solution of perylene and $(TBA)Au(i\text{-mnt})_2$ in stoichiometric proportions, with platinum electrodes, and using a galvanostatic method, in which the current density ranged from 0.5 to 5 $\mu A/cm^2$. The crystals, obtained after 3 to 5 days, have the shape of small, elongated platelets, with average dimensions of 4 x 0.04 x 0.02 mm^3.

TRANSPORT PROPERTIES.

We previously reported preliminary data on this compound [11], in which a shallow minimum in the resistivity was observed at different temperatures for different samples, this being attributed to different concentrations of defects and/or impurities. Subsequent work with better-quality samples is reported here.

The electrical resistivity, measured in single crystals along the chain axis by a lock-in technique (88 Hz) shows a metallic behavior ($d\rho / dT > 0$) in a wide range of temperature (see Fig. 1), with high values of electrical conductivity at room temperature ($\sigma_{RT} \approx 10^3$ S/cm). At low temperature, the resistivity rises with decreasing temperature. It is not yet clear if there is a structural transition, or only a gradual loss of electronic degrees of freedom.

The thermoelectric power is positive, indicating that the majority of charge carriers are holes (Fig. 2). This agrees with the $Per_2^+Au(i\text{-mnt})_2^-$ formula, assuming that conduction occurs in the stack of the perylene molecules, corresponding to a 3/4-filled band. Using the expression derived by Chaikin *et al.* [12] for the thermopower in the tight-binding approximation, and neglecting the energy dependence of the scattering time, a bandwidth of 0.6 ± 0.02 eV can be estimated from the slope of the linear part of the S(T) line in the high-temperature range.

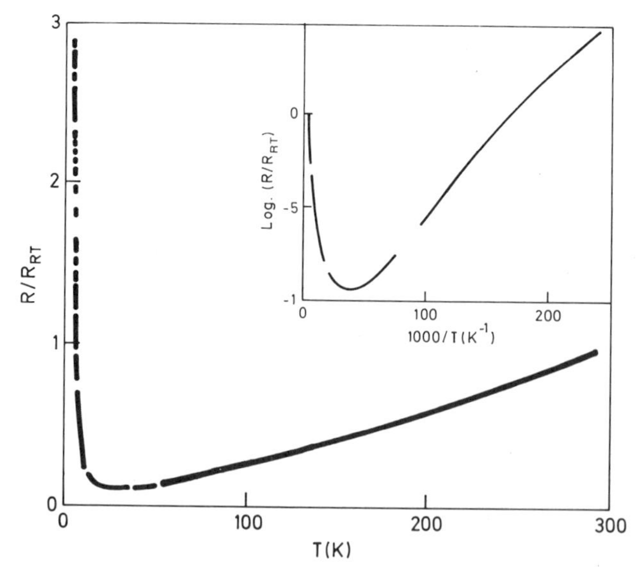

Fig.1. The normalized resisitivity as a function of temperature. In the inset, the same data, in a logarithmic scale versus reciprocal temperature.

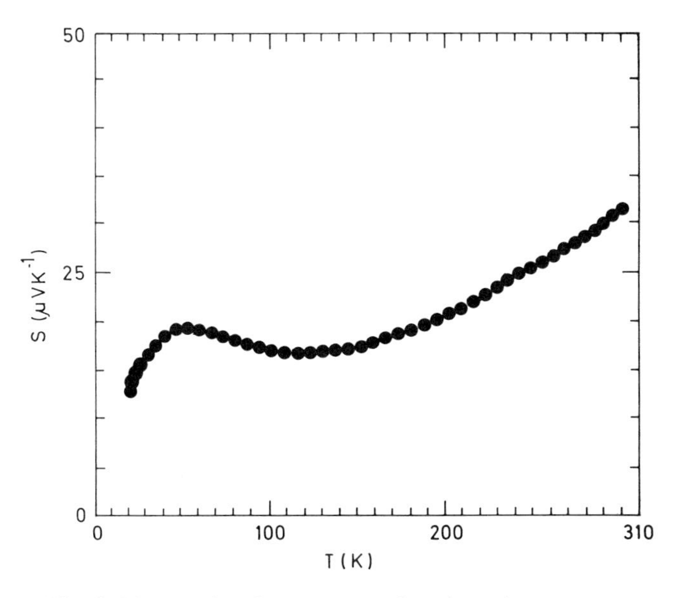

Fig. 2. Thermoelectric power as a function of temperature.

MAGNETIC PROPERTIES

Electron paramagnetic resonance (EPR) studies in a conventional X-band spectrometer showed the following:

(i) The magnetic susceptibility (proportional to the integrated EPR intensities) is almost independent of the temperature in a wide range ($50 < T < 300$ K), and shows a Curie tail below 50 K (Fig. 3b). This kind of behavior is expected for itinerant conduction electrons (Pauli-type susceptibility), with a small concentration of paramagnetic impurities and/or localization of charge carriers at low temperature.

(ii) The dependence of the linewidth (ΔH_{pp}) on temperature (Fig. 3a) does not follow the Elliott relation for elemental metals ($\Delta H_{pp} \alpha \rho$) [13]. The inapplicability of the Elliott relation, which also occurs in other molecular compounds with metal-like conduction, has been ascribed to the unidimensionality of their electronic structure [14].

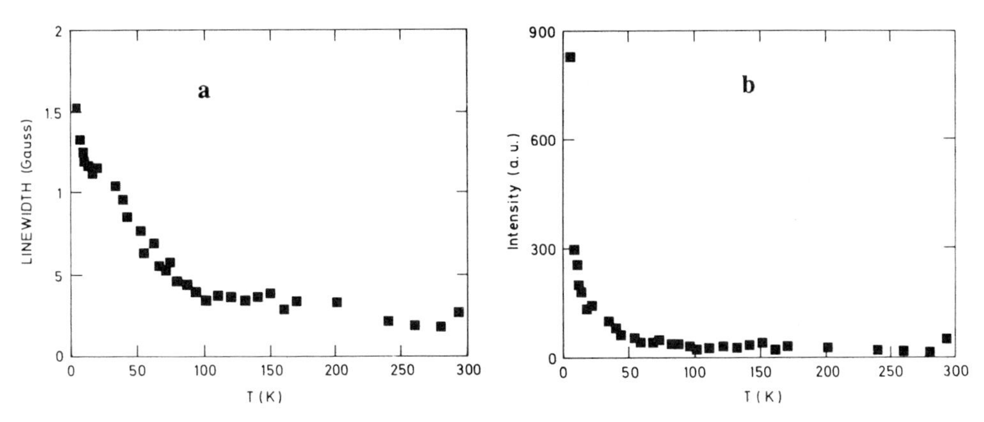

Fig.3. Temperature dependence of the EPR signal: (a) peak-to-peak linewidth (b) integrated intensity.

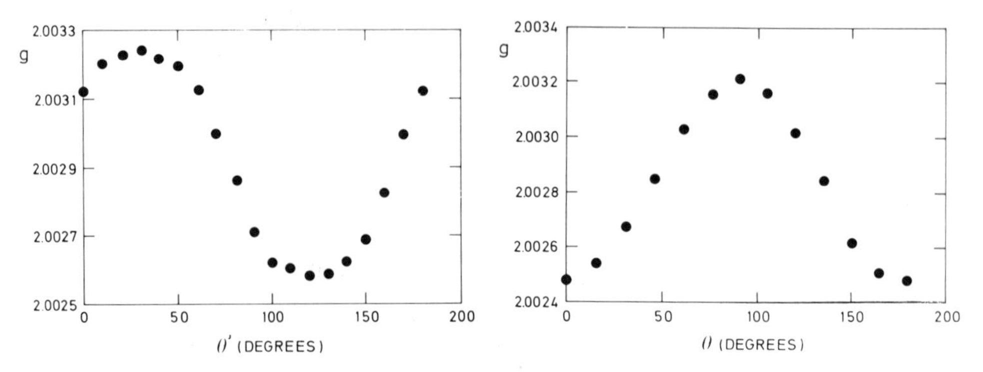

Fig. 4. The anisotropy of g values for vertical (θ') and horizontal rotation (θ).

(iii) The EPR signal has to peak-to-peak width (ΔH_{pp}) less than 0.5 Gauss, similar to the one found for the $Au(mnt)_2^-$ salt, but smaller, than that for other conductors based on perylene with diamagnetic counter-ions (for example PF_6^-, AsF_6^-).

(iv) The g values, although anisotropic (see Fig. 4), are close to the free electron value (g = 2.0023), as expected for an EPR signal due only to the perylene radicals.

CONCLUSIONS

$Per_2Au(i\text{-}mnt)_2$ presents quasi-unidimensional metallic properties similar to those observed in $Per_2Au(mnt)_2$. Thermopower and EPR indicate that conduction takes place in the perylene chains. This is in agreement with the general observation, that the slight change, from the ligand **mnt** to the very similar ligand **i-mnt**, almost does not modify the physical properties.

Changes in the transition metal atom of the complex are expected to have larger effects on the physical properties [15]. Attempts to prepare compounds, where such effects should be present (namely with Pt, Ni and Cu), failed, due to the instability of those complexes in the +3 formal oxidation state (lack of resonance stabilization in this structure) [10].

The determination of the crystal structure, and structural instabilities, by X-ray diffraction and X-ray diffuse scattering, respectively, are underway, in order to clarify the nature of the metal-insulator transition, and to allow comparisons with the **mnt** analog.

ACKNOWLEDGEMENTS

The authors are grateful to Manuel Almeida and Vasco da Gama for their encouragement and fruitful discussions. This work is partly supported by EEC under contract ESPRIT Basic Research Action 3121.

REFERENCES

[1] L. Alcácer, and A. H. Maki., J. Phys. Chem. 78: 215 (1974).
[2] R. T. Henriques, M. Almeida, L. Alcácer, and S. Tomic, Mol. Cryst. Liq. Cryst. 120: 237 (1985).
[3] L. Alcácer, Mol. Cryst. Liq. Cryst. 120: 221 (1985).
[4] R. T. Henriques, L. Alcácer, J. P. Pouget, and D. Jérome, J. Phys. C 17: 5197 (1984).
[5] H. J. Keller, D. Nöthe, H. Pritzkow, D. Wehe, M. Werner, P. Koch, and D. Schweitzer, Mol. Cryst. Liq. Cryst. 17: 93 (1980).
[6] D. Schweitzer, I. Henig, K. Bender, H. Endres, and H. J. Keller, Mol. Cryst. Liq. Cryst. 120: 213 (1985).

[7] D. Meenenga, K. P. Dinse, D. Schweitzer, and H. J. Keller, Mol. Cryst. Liq. Cryst. 120: 243 (1985).
[8] H. Eichele, M. Schwoerer, C. Krohnke, and G. Wegner, Chem. Phys. Lett. 77: 311 (1981).
[9] R. Gomper and W. Topel, Angew. Chem. 74: 2861 (1962).
[10] B. G.Verden, E. Billig, and H. B. Gray, Inorg. Chem. 5: 78 (1966).
[11] M. J. Matos, R. T. Henriques, and L. Alcácer, Synth. Metals.19: 389 (1987).
[12] P. M. Chaikin, R. L. Greene, S. Etemad, and E. Engler, Phys. Rev. B 13: 1627 (1976).
[13] R. J. Elliott, Phys. Rev. 96: 266 (1954).
[14] M. Weger, J. Physique Colloque, 39: 1456 (1978).
[15] V. Gama, M. Almeida, and R. T. Henriques, These Proceedings.

THE METALLOPORPHYRINS OF GROUP IIIB AS PRECURSORS IN LOW-DIMENSIONAL SOLIDS

A. G. Coutsolelos*[a] and D. L. Ward[b]

[a] Department of Chemistry, Laboratory of Inorganic Chemistry
University of Crete, P. O. Box 1470
71110 - Iraklion, Crete, GREECE

[b] Department of Chemistry, Michigan State University
East Lansing, MI 46824, USA

INTRODUCTION

The search for highly conducting low-dimensional materials has been at the forefront of chemical research in the last decade. There are two criteria which have to be satisfied, and which appear naturally from the theories of mixed valence: (i) the metal-ligand units should occupy crystallographically similar sites and (ii) they should be mixed-valence units.

Recently, Wynne and co-workers [1,2] and Marks and co-workers [3-6] described the synthesis and characterization of inorganic-organic polymers, in which a metallic or pseudometallic element alternates with a linear-chain bridging atom, like oxygen or fluorine. The metallic or pseudometallic element is usually the central atom of a phthalocyanine system, and the bridge-stacked polymeric structure is rigid. These derivatives are electrical conductors after iodine oxidation.

We report, in this work, the synthesis, physicochemical characterization, and x-ray analysis of substituted porphyrin complexes of the general type M(Por)F (M = Ga, In and Tl, Por = porphyrin), as precursors of new low-dimensional molecular solids. We also give preliminary data on their doping with iodine, and their electrical conductivity.

EXPERIMENTAL

Synthesis of precursor porphyrins

The (polymeric) precursor porphyrins, [Ga(Por)F], were prepared and purified, following established procedures [7]. [In(Por)F] and [Tl(Por)F] were prepared by reaction of NaF with the corresponding chloro-metalloporphyrins. The porphyrins used were octamethylporphyrin (OMP), octaethylporphyrin (OEP), and tetraphenylporphyrin (TPP).

Iodine doping

The precursor porphyrin Ga(Por)F is used to form partially oxidized (iodine-doped) low-dimensional solids. The Ga(Por)F starting material (8.85×10^{-5} mmol) was dissolved under inert atmosphere in freshly distilled toluene (30 mL), when Por = OEP or TPP, and in 1,2,4-trichlorobenzene, when Por = OMP. This solution was introduced into an "H-tube" [24]. At the same time, 30 mL of I_2 (7.33×10^{-4} mmol) in heptane were added. The

Lower-Dimensional Systems and Molecular Electronics
Edited by R.M. Metzger *et al.*, Plenum Press, New York, 1991

apparatus was heated at 80 °C for one week, and then cooled by 3 °C per day, until room temperature was reached, affording dark-violet crystals. The yield was about 60-70%, based on the porphyrin complex. The iodination reaction is reversible: the iodine can be completely removed by heating the materials in vacuo at 100°C.

Spectroscopy

Products were characterized by a number of different spectroscopic methods. Infrared spectra were recorded on a Perkin-Elmer Model 298 or 580 B spectrophotometer. Samples were prepared as 1% dispersions in CsI or Nujol mulls. ^1H nmr spectra were recorded on a Varian Model FT-80 spectrometer: solutions of 5 mg of complex in 0.6 mL $CDCl_3$ were used,with tetramethylsilane as internal reference. Electronic absorption spectra were recorded on a Perkin-Elmer Model 330 spectrometer, using 5 x 10^{-3} mol/L dichloromethane solutions. Raman spectra were recorded on samples cooled by liquid nitrogen, on a DILOR Model RTI 30 spectrometer, using an excitation wavelength of 514.5 nm. ESR spectra of the powders in $MgSO_4$ were recorded at X-band frequency with a Varian Model E-4 spectrometer. Signal g values and intensities were calibrated with a standard diphenylpicrylhydrazyl (DPPH) sample (g = 2.0036).

X-ray structure determination

To verify (or disprove) the polymeric structure of the series of In (III) and Tl (III) porphyrins, the X-ray crystal structure of Tl(TPP)F was determined. A suitable crystal of Tl(TPP)F was obtained, by slow diffusion of pentane into a solution of the compound in methylene chloride. Preliminary Weissenberg photographs along the c axis revealed a four-fold symmetry, and from the systematic absences of h0l (h+l = 2n), 0k0 (k=2n), and from the subsequent least-squares refinement, the space group was determined to be $P2_1/n$.

The structure was solved by the Patterson heavy-atom method, which revealed the position of the Tl atom. The remaining atoms were located in successive difference Fourier syntheses; the hydrogen atoms were included in the refinement, but constrained to ride on the atom to which they are bonded. The structure was refined by full-matrix least-squares, where the function minimized was $\Sigma_w (|F_o|-|F_c|)^2$, and the weight w is defined as $4F_o^2/\sigma(F_o^2)^2$.

Conductivity measurements

Conductivities of polycrystalline pressed pellets were determined by van der Pauw method at the University of Nancy (France).

RESULTS AND DISCUSSION - PRECURSOR PORPHYRINS

Infrared spectroscopy

The determination of the stoichiometric formulas M(Por)F by mass spectrometry could give information about the axial ligand (fluorine), but there is but a small intensity for the molecular ion, due to the character of the M-F bond. The axial conformation for the ligand was, therefore, verified by infrared spectroscopy, by comparing the precursor complexes M(Por)Cl and the final compound M(Por)F. For all the compounds, we observed a shift of the metal-halogen bond to higher values, when going from chlorine to fluorine as axial ligands. This trend agrees with the observed results for similar complexes of Ga, In, and Tl [8].

^1H-NMR Data

^1H-NMR spectra of the M(Por)X complexes were obtained.at ambient temperature. The spectra of M(Por)X give poor structural information, since the resonance lines corresponding to the *meso* and *methyl* protons are two singlets, which appear in the same region, whatever the nature of X. The bridge-stacked polymeric structure of Ga(OMP)F was demonstrated by EXAFS spectroscopy [9]: the same arrangement may be postulated for Ga(OMP)Cl and Ga(OMP)I in the solid state [7]. We have also obtained the ^1H-NMR spectra of the gallioctaethylporphyrin Ga(OEP) systems: we observe an ABR_3 multiplet for the ethyl groups of Ga(OEP)Cl, which can be explained only by the presence of a non-octahedral monomeric structure [10,11]. In contrast, the resonance signal of the methylenic protons of Ga(OEP)F is a quadruplet, and these protons have very similar shifts (see Fig. 1). These observations may be consistent wih a polymeric structure for Ga(OEP)F, similar to that of Ga(OMP)F, the other complexes being pentacoordinate.

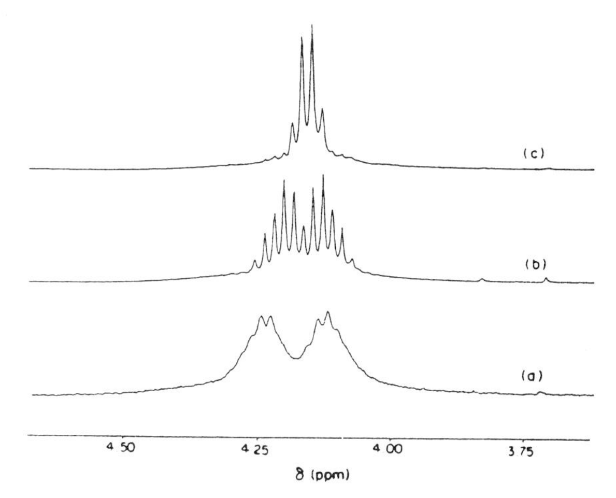

Fig. 1. ^{1}H NMR signal of A and B methylenic protons of (a) Ga(OEP)I, (b) Ga(OEP)Cl and (c) Ga(OEP)F.

At room temperature, the resonances of the *ortho* phenyl protons are broad for Ga(TPP)Cl, and split into a doublet for Ga(TPP)F. This reflects the well-known phenomenon of the restricted rotation of phenyl rings [12-17]. Even if the rate of phenyl ring rotation is not very well established, it is clear that the chemical shift difference is higher, when the axial ligand is the more electronegative halide (X = F). It thus appears that the metal is not coplanar with the porphyrin. The above assumptions were confirmed by the x-ray crystal structure of Ga(TPP)Cl [7].

For In(Por)F complexes, the ^{1}H-nmr data indicate that a five-coordinate coordination sphere can be proposed. Similar results have been obtained for the series of thallium (III) porphyrins. The *meso* and pyrrole protons give a doublet for Tl(OEP)F and Tl(TPP)F. The same spectral morphology is observed for Tl(Por)Cl, and results in a coupling between the

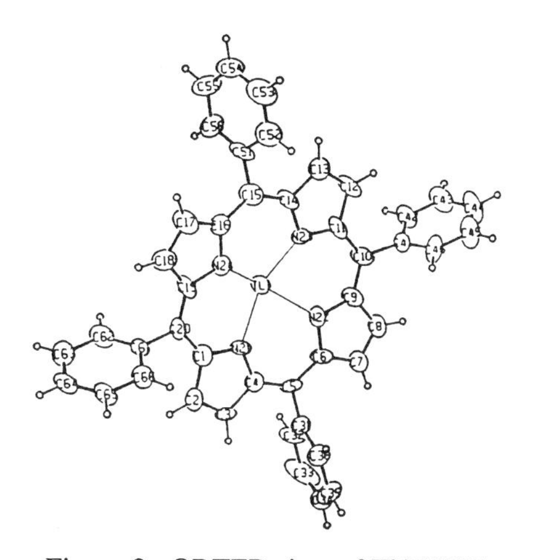

Figure 2. ORTEP view of Tl(TPP)F.

Table 1. Bond distances (Å) in Tl(TPP)F.

Atom1	Atom2	Distance	Atom1	Atom2	Distance	Atom1	Atom2	Distance
Tl1	F1	2.441(6)	C8	C9	1.43(3)	C41	C46	1.38(3)
Tl1	N21	2.221(12)	C9	C10	1.42(2)	C42	C43	1.36(2)
Tl1	N22	2.223(14)	C10	C11	1.42(3)	C43	C44	1.34(3)
Tl1	N23	2.181(12)	C10	C41	1.50(2)	C44	C45	1.33(3)
Tl1	N24	2.225(14)	C11	C12	1.43(2)	C45	C46	1.38(3)
N21	C1	1.40(2)	C12	C13	1.32(3)	C51	C52	1.42(3)
N21	C4	1.36(2)	C13	C14	1.44(2)	C51	C56	1.37(3)
N22	C6	1.37(2)	C14	C15	1.42(2)	C52	C53	1.39(3)
N22	C9	1.39(2)	C15	C16	1.42(2)	C53	C54	1.37(3)
N23	C11	1.36(2)	C15	C51	1.51(3)	C54	C55	1.36(3)
N23	C14	1.39(2)	C16	C17	1.41(2)	C55	C56	1.34(3)
N24	C16	1.39(2)	C17	C18	1.39(3)	C61	C62	1.35(3)
N24	C19	1.39(2)	C18	C19	1.41(3)	C61	C66	1.38(3)
C1	C2	1.49(3)	C19	C20	1.39(2)	C62	C63	1.46(3)
C1	C20	1.38(3)	C20	C61	1.50(3)	C63	C64	1.40(3)
C2	C3	1.32(3)	C31	C32	1.37(3)	C64	C65	1.35(3)
C3	C4	1.43(2)	C31	C36	1.37(3)	C65	C66	1.37(2)
C4	C5	1.41(2)	C32	C33	1.41(3)			
C5	C6	1.42(2)	C33	C34	1.38(4)			
C5	C31	1.46(3)	C34	C35	1.37(3)			
C6	C7	1.41(3)	C35	C36	1.37(3)			
C7	C8	1.35(2)	C41	C42	1.40(3)			

porphyrin protons and the ^{203,205}Tl metal [18]. This interaction is weaker for the bimetallic Tl(Por)M(L) derivatives, than for the Tl(Por)Cl complexes, and also shows that the $Co(CO)_4$ ligand is the only metallate group to possess some electron-withdrawing ability [19]. Furthermore, this interaction is the same, when changing OEP for TPP, or changing axial ligands, Cl for F.

UV-visible spectra

M(Por)X exhibit electronic absorption spectra belonging to the "normal" class [20]. They show an intense band in the 395-420 nm region, and two bands between 500 and 600 nm (α- and β- bands). One extra band [Q(2,0)] is observed in the visible region, and another one appears blue-shifted, relative to the Soret band. As previously reported by Gouterman *et al.* [21] for $Sn(OEP)X_2$ (with X = F, Cl, Br, or I), the shifts in the transition energies and in the relative intensities correlate with changes in the electron density in the porphyrin ring.

Crystal and molecular structure determination of Tl(TPP)F

Figure 2 is an ORTEP view of Tl(TPP)F, and Table 1 gives the bond lengths and angles [23]. The bond distances and angles in the porphyrin group are statistically equal to those usually found in other crystal structure of porphyrins. A detailed discussion of the polymeric structure of Tl(TPP)F will be given separately [23].

RESULTS AND DISCUSSION-PARTIALLY OXIDIZED FLUOROGALLIOPORPHYRINS

The precursor porphyrin Ga(Por)F is polymeric in the solid state, and should be written as $[Ga(Por)F]_n$. Infrared spectra for the unoxidized and oxidized Ga(Por)F were compared: the new vibrational frequencies in the latter are in good agreement with the proposed formula $[Ga(Por)FI_x]_n$. The vibrations in the range 1625-1190 cm^{-1} are due to the vibrators C_{meso}-H and $C_{\beta\text{-pyrrole}}$-H of the ring. We can also observe the frequencies due to the aspiration of the porphyrin ring, in the range 965-650 cm^{-1}. For the absorption frequencies of Ga-F, a large displacement (Δ_v=41 cm^{-1}) was observed for the case Por = OEP, while Δ_v= 9 and 8 cm^{-1}, respectively for Por = TPP, OMP. On the basis of these results, it appears that molecular iodine has been incorporated, but its fate in the solid must be ascertained.

Raman spectroscopy should help to characterize the nature of the incorporated anionic units (I_3^- or I_5^-). Fig. 3 shows representative solid-state resonance Raman spectra in the polyiodide scattering region for the $[(Ga(Por)FI_x]_n$ materials. There is no evidence of free I_2

(expected at 200 cm^{-1}). The characteristic symmetric stretching fundamental of I_3^- at 109.6-111.8 cm^{-1}, and the stretching fundamental of I_5^- at 165.2-170.0 cm^{-1} have been observed for all the complexes. The relative intensity of these two frequencies varied with the changes in the porphyrin ring. The ratio of I_3^-/I_5^- depends of the ease of reduction of the initial compound, and on the resultant morphology of the iodine chain.

Electronic absorption spectra (nujol mulls) of solid samples of $[Ga(Por)FI_x]_n$ confirm the existence of I_3^- and I_5^- anionic units.

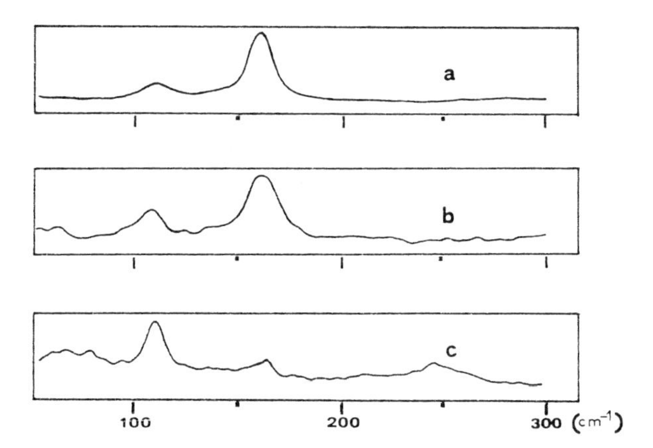

Fig. 3. Solid-state resonance Raman spectra (514.5 nm excitation) of: (a) $[Ga(OEP)FI_x]_n$, (b) $[Ga(TPP)FI_x]_n$, (c) $[Ga(OMP)FI_x]_n$.

All of the $[Ga(Por)FI_x]_n$ materials exhibit EPR spectra (Fig.4), which show a single, narrow, nearly symmetric absorption, with g values very close to the free-electron value, and to g values for related materials [3]. We can observe, for Por = OMP, a isotopic hyperfine structure (Fig. 4a). This coupling, due to the super-hyperfine interaction of the free electron with the four N-atoms, is about 17 Gauss, and has been observed for other porphyrin series [25].

Compressed-pellet conductivity measurements reveal conductivities of the order of 10^{-8}-10^{-6} S/cm for these materials. The conductivity increases after iodination of the starting material, and also increases with temperature (i.e., the $[Ga(Por)FI_x]_n$ compounds are semiconductors).

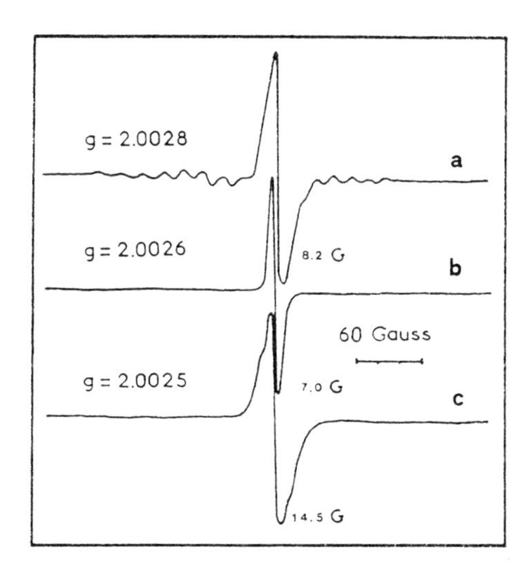

Fig. 4. EPR spectra of (a) $[Ga(OMP)FI_x]_n$, (b) $[Ga(TPP)FI_x]_n$, (c) $[Ga(OEP)FI_x]_n$.

By variation of crystallization conditions, we have obtained single crystals of these materials. A x-ray study of a representative of this family is underway.

REFERENCES

[1] R. S. Nohr and K. J. Wynne, J. Chem. Soc., Chem. Commun. **1981**, 1210.
[2] P. Brant, D. C. Weber, S. G. Haupt, R. S. Nohr, and K. J. Wynne, J. Chem. Soc., Dalton Trans. 289 (1989).
[3] T. J. Marks, J. Coatings Technol. 48: 53 (1976).
[4] T. J. Marks and D. W. Kalina, in Extended Linear Chain Compounds, Vol. 1, ed. by J. S. Miller (Plenum Press, New York, 1982), p. 197.
[5] C. W. Dirk and T. J. Marks, Inorg. Chem. 43: 4325 (1984).
[6] E. Ciliberto, K. A. Doris, W. J. Pietro, D. M. Reisner, D. E. Ellis, I. Fragala, F. M. Herbstein, M. A. Ratner, and T. J. Marks, J. Am. Chem. Soc. 106: 7748 (1984).
[7] A Coutsolelos, R. Guilard, D. Bayeil, and C. Lecompte, Polyhedron 5: 1157 (1986).
[8] K. Nakamoto in Infrared and Raman Spectra of Inorganic and Coordination Compounds 3rd Edition, (Wiley-Interscience, New York, 1977).
[9] J. Goulon, P. Friant, C. Goulon-Ginet, A. Coutsolelos, and R. Guilard, Chem. Phys. 83: 367 (1984).
[10] C. A. Busby and D. Dolphin, J. Magn. Reson. 23: 211 (1976).
[11] R. J. Abraham and K. M. Smith, Tetrahedron Lett. 36: 335 (1971).
[12] A. Laurie, W. Shryer, C. Lorbereaux, and S. S. Eaton, J. Org. Chem. 45: 4296 (1980).
[13] S. S. Eaton, D. M. Fishwild, and G. R. Eaton, Inorg. Chem. 17: 1542 (1978).
[14] S. S. Eaton and G. R. Eaton, J. Am. Chem. Soc. 99: 6594 (1977).
[15] S. S. Eaton and G. R. Eaton, J. Am. Chem. Soc 97: 3660 (1975).
[16] D. V. Behere, R. Birdy, and S. Mitra, Inorg. Chem. 21: 386 (1982)
[17] J. Huet, A. Gaudemer, C. Boucly-Goester, and P. Boucly, Inorg. Chem. 21: 3413 (1982).
[18] T. R. Janson and J. J. Katz in The Porphyrins, Vol. IV, ed. by D. Dolphin (Academic Press, New York, 1978), ch. I.
[19] R. Guilard, A. Zrineh, M. Ferhat, A. Tabard, P. Mitaine, C. Swistak, P. Richard, C. Lecompte, and K. M. Kadish, Inorg. Chem. 27: 697 (1988).
[20] M. Gouterman, in The Porphyrins, Vol III, ed. by D. Dolphin, (Academic Press, New York, 1978), p. 1.
[21] M. Gouterman, F. P. Schwartz, and P. D. Smith, J. Chem. Phys. 59: 676 (1973).
[22] M. Gouterman, F. P. Schwartz, and P. D. Smith, J. Chem. Phys. 61: 3491 (1974).
[23] A. Coutsolelos and C. Ward, to be published.
[24] A. Coutsolelos, Thèse d'Etat, University of Dijon, France, 1985.
[25] J. Fajer and M. S. Dais, in The Porphyrins, Vol. IV, ed. by D. Dolphin (Academic Press, New York, 1979), p. 197.

CRYSTAL STRUCTURES AND PHYSICAL PROPERTIES OF DMET SUPERCONDUCTORS

Koichi Kikuchi, Yoshimitsu Ishikawa, Yoshiaki Honda, Kazuya Saito and Isao Ikemoto

Department of Chemistry, Faculty of Science
Tokyo Metropolitan University
Fukazawa, Setagaya-ku
Tokyo 158, JAPAN

Keizo Murata

Electrotechnical Laboratory
Tsukuba
Ibaraki 305, JAPAN

Keiji Kobayashi

Department of Chemistry, College of Arts and Science
The University of Tokyo
Komaba, Meguro-ku
Tokyo 153, JAPAN

INTRODUCTION

Since the first discovery of superconductivity in an organic material [1], a number of organic superconductors have been found in the TMTSF and BEDT-TTF families. However, there are many differences in the crystal structures and physical properties between the two families. Recently we have discovered seven organic superconductors in the family of the unsymmetrical donor DMET [2-6]. These new superconductors are the first ones based on an unsymmetrical molecule. They are also useful for understanding the organic superconductors systematically, because the DMET salts seem to link the TMTSF and BEDT-TTF families: DMET has halves of the structures of TMTSF and BEDT-TTF. Indeed, the investigation of the temperature dependence of the resistivity revealed that DMET superconductors can be classified into three groups: one group is similar to the TMTSF salts, another is similar to the BEDT-TTF salts [7,8]. In this report we summarize the crystal structures and the physical properties of DMET superconductors.

DMET

The temperature dependence of resistivities of five typical DMET salts is shown in Fig. 1 [8]. Among them, three show superconductivity. $(DMET)_2Au(CN)_2$ shows a resistivity upturn at 28 K at ambient pressure, and needs some pressure to become a superconductor [2]. The insulating phase is SDW [9,10] and the T-P phase diagram (Fig. 2) [11] is similar to that of the $(TMTSF)_2X$ salts. However, the pressure range, where both superconductivity and

Lower-Dimensional Systems and Molecular Electronics
Edited by R.M. Metzger *et al.*, Plenum Press, New York, 1991

SDW are observed, is very large. So $(DMET)_2Au(CN)_2$ is a suitable material to investigate the relation between SDW and superconductivity. The other phase transition, at high temperature, is observed through the measurements of electrical resistivity, thermopower and heat capacity [12,13]. The pressure regions, where this and SDW transitions occur, concide with each other. $(DMET)_2AuI_2$ and $(DMET)_2AuCl_2$ show a similar temperature dependence of resistivity [5]. Therefore, these salts can be classified into the $(DMET)_2Au(CN)_2$ group.

$(DMET)_2I_3$ exhibits a metallic behavior down to low temperature, and shows superconductivity at ambient pressure with $T_c = 0.47$ K [4]. This temperature dependence of resistivity is similar to that of β-$(BEDT\text{-}TTF)_2X$. $(DMET)_2IBr_2$ also belongs to the same group as $(DMET)_2I_3$.

$(DMET)_2AuBr_2$ salt belongs to another group. This salt show two similar types of temperature dependence of resistivity [6]. Both exhibit a semiconductor-like behavior between 180 K or 150 K and room temperature, and a metallic behavior at lower temperature. Such a semiconductor-to-metal transition is also observed in $(BEDT\text{-}TTF)_2Cu(NCS)_2$. One type of $(DMET)_2AuBr_2$ shows superconductivity at 1.9 K at ambient pressure. The other shows superconductivity only under pressure.

The crystal structure of $(DMET)_2Au(CN)_2$ is shown in Fig. 3 [14]. A columnar structure is observed, as in $(TMTSF)_2X$ salts, but the normal to the DMET molecule is largely tilted from the stacking axis. Within a stack, there are two types of overlap between the neighboring molecules. One type has two Se-Se contacts shorter than the sum of the van der Waals radii. The other has four shorter Se-S contacts. Between stacks, we can find some short interstack contacts. There are one shorter Se-Se and two shorter S-S contacts between the neighboring molecules. Therefore $(DMET)_2Au(CN)_2$ has some two-dimensional character.

The crystal structure of $(DMET)_2I_3$ is very similar to that of $(DMET)_2Au(CN)_2$ [15,16]. However, these salts are different from each other, because $(DMET)_2Au(CN)_2$ is similar to $(TMTSF)_2X$, while $(DMET)_2I_3$ is similar to β-$(BEDT\text{-}TTF)_2X$. The difference in properties

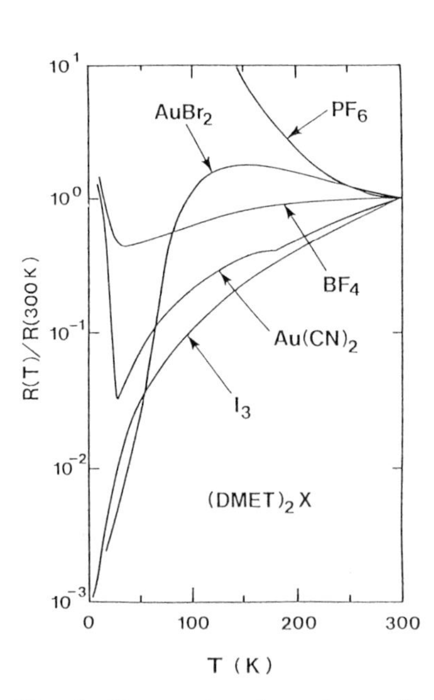

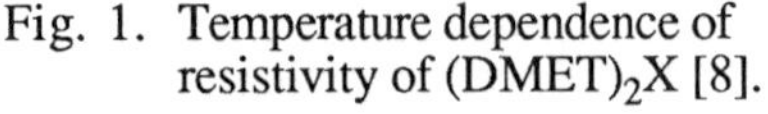
Fig. 1. Temperature dependence of resistivity of $(DMET)_2X$ [8].

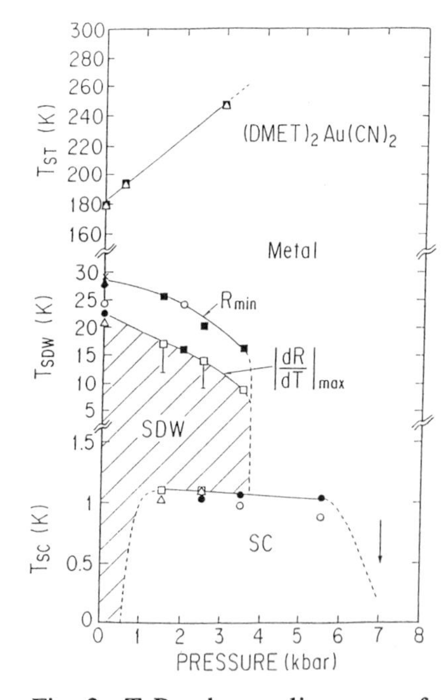

Fig. 2. T-P phase diagram of $(DMET)_2Au(CN)_2$ [11].

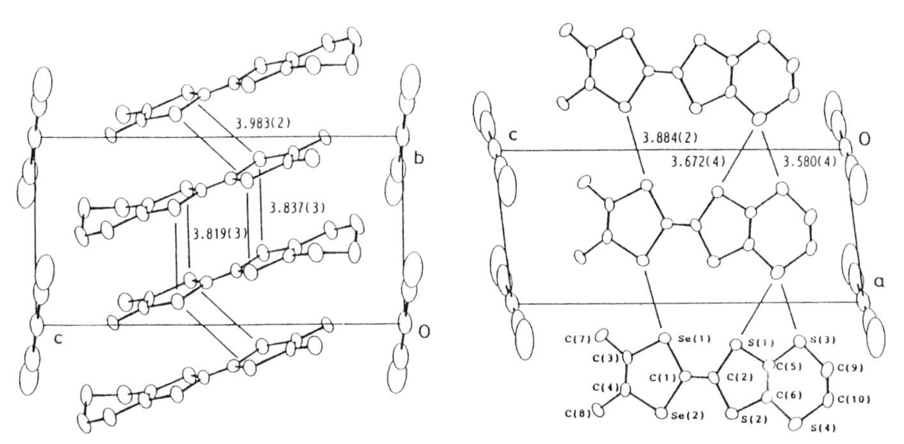

Fig. 3. Crystal structure of $(DMET)_2Au(CN)_2$ [14].

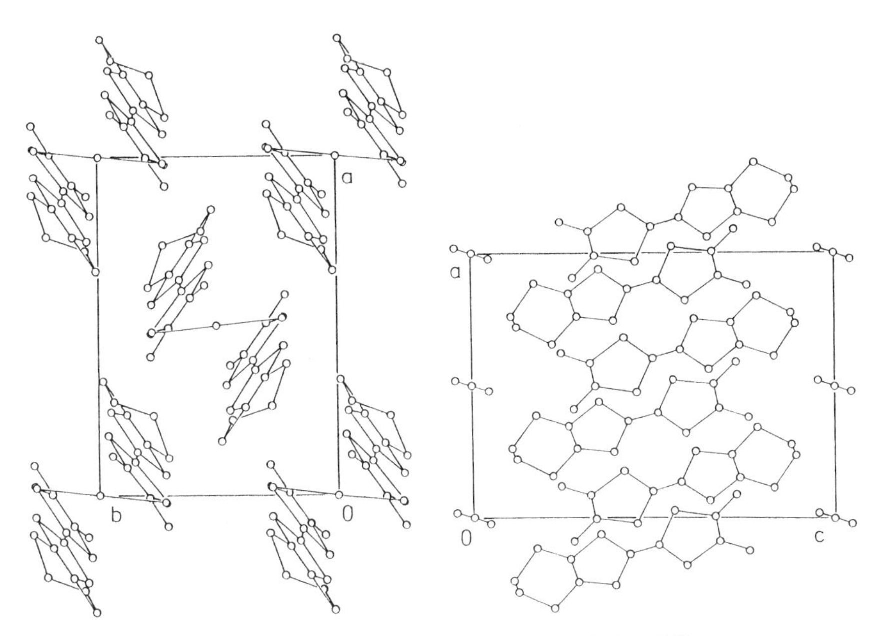

Fig. 4. Crystal structure of $(DMET)_2AuBr_2$ [6].

Table 1. Comparison of distances of short interstack atomic contact in DMET superconductors.

Compound	Se...Se/Å	S...S/Å	S...S/Å
$(DMET)_2Au(CN)_2$	3.884	3.580	3.672
$(DMET)_2AuI_2$	3.832	3.552	3.611
$(DMET)_2I_3$	3.784	3.504	3.604
$(DMET)_2IBr_2$	3.774	3.494	3.588

between the TMTSF and the BEDT-TTF salts has generally been explained to be related to their difference in dimensionality. The TMTSF salts have less two-dimensionality than the BEDT-TTF salts, so the TMTSF salts have SDW and superconducting phases. On the other hand, the electronic systems of BEDT-TTF salts are simple, being only metallic and/or superconducting. We compare the distance of short interstack contacts of $(DMET)_2X$ (X = $Au(CN)_2$, AuI_2, I_3, IBr_2), to consider their dimensionality. Their distances are given in Table 1 [14,16,17]. The distances in I_3 and IBr_2 salts are shorter than the corresponding distances in $Au(CN)_2$ and AuI_2 salts. This suggests that the $Au(CN)_2$ and AuI_2 salts have less two-dimensionality than the I_3 and IBr_2 salts. This fact is consistent with their electrical properties. That is, the less-two-dimensional $Au(CN)_2$ and AuI_2 salts have SDW and superconducting phases, and the more-two-dimensional I_3 and IBr_2 salts have a simple electronic system.

The crystal structure of $(DMET)_2AuBr_2$ is shown in Fig. 4 [6]. The structure is not columnar, in contrast to other DMET superconductors, and consists of sheets of DMET molecules and $AuBr_2$ anions. Thus, this crystal is a two-dimensional material. The structure of the $AuBr_2$ salt is very similar to that of $(BEDT\text{-}TTF)_2Cu(NCS)_2$. This is consistent with the similarity of their electrical properties.

REFERENCES

[1] D. Jérome, A. Mazaud, M. Ribault, and K. Bechgaard, J. Phys. (Paris) Lett. 41: L95 (1980).

[2] K. Kikuchi, M. Kikuchi, T. Namiki, K. Saito, I. Ikemoto, K. Murata, T. Ishiguro, and K. Kobayashi, Chem. Lett. 931 (1987).

[3] K. Kikuchi, K. Murata, Y. Honda, T. Namiki, K. Saito, K. Kobayashi, T. Ishiguro, and I. Ikemoto, J. Phys. Soc. Jpn. 56: 2627 (1987).

[4] K. Kikuchi, K. Murata, Y. Honda, T. Namiki, K. Saito, K. Kobayashi, T. Ishiguro, and I. Ikemoto, J. Phys. Soc. Jpn. 56: 3436 (1987).

[5] K. Kikuchi, K. Murata, Y. Honda, T. Namiki, K. Saito, K. Kobayashi, T. Ishiguro, and I. Ikemoto, J. Phys. Soc. Jpn. 56: 4241 (1987).

[6] K. Kikuchi, Y. Honda, Y. Ishikawa, K. Saito, I. Ikemoto, K. Murata, H. Anzai, T. Ishiguro, and K. Kobayashi, Solid State Commun. 66: 405 (1988).

[7] K. Murata, K. Kikuchi, T. Takahashi, K. Kobayashi, Y. Honda, K. Saito, K. Konoda, T. Tokiwa, H. Anzai, T. Ishiguro, and I. Ikemoto, J. Mol. Electron. 4: 173 (1988).

[8] K. Kikuchi, K. Saito, I. Ikemoto, K. Murata, T. Ishiguro, and K. Kobayashi, Synth. Metals 27: B269 (1988).

[9] K. Kanoda, T. Takahashi, T. Tokiwa, K. Kikuchi, K. Saito, I. Ikemoto, and K. Kobayashi, Phys. Rev. B38: 39 (1988).

[10] Y. Nogami, M. Tanaka, S. Kagashima, K. Kikuchi, K. Saito, I. Ikemoto, and K. Kobayashi, J. Phys. Soc. Jpn. 56: 3738 (1987).

[11] Y. Honada, K. Murata, K. Kikuchi, K. Saito, I. Ikemoto, and K. Kobayashi, Solid State Commun., in press.

[12] K. Saito, H. Kamio, K. Kikuchi, K. Kobayashi, and I. Ikemoto, J. Phys.: Condensed. Matt., in press

[13] K. Saito, H. Kamio, Y. Honda, K. Kikuchi, K. Kobayashi, and I. Ikemoto, submitted.

[14] K. Kikuchi, Y. Ishikawa, K. Saito, I. Ikemoto, and K. Kobayashi, Acta Cryst. C44: 466 (1988).

[15] M. Z. Aldoshina, L. O. Atovmyan, L. M. Goldenberg, O. N. Krasochka, R. N. Lyubovskaya, R. B. Lyubovskii, and M. L. Khidekel, Dokl. Akad. Nauk SSR 289: 1140 (1986).

[16] K. Kikuchi, Y. Ishikawa, K. Saito, I. Ikemoto, and K. Kobayashi, Synth. Metals 27: B391 (1988).

[17] Y. Ishikawa, K. Kikuchi, K. Saito, I. Ikemoto, and K. Kobayashi, Acta Cryst. C45: 572 (1989).

RADICAL CATION SALTS OF BEDT-TTF WITH METAL-THIOCYANATO ANIONS

H. Müller[+], C.P. Heidmann[+], H. Fuchs[+], A. Lerf[+], K. Andres[+]
R.Sieburger[++] and J.S. Schilling[++]

\+ Walther-Meissner-Institut, 8046 Garching FRG
++ Sektion Physik LMU München, 8000 München FRG

ABSTRACT

The synthesis, characterization and physical properties of two superconducting types of $(ET)_2Cu(NCS)_2$ and of the semiconductor $(ET)Ag_x(SCN)_2$ ($x \simeq 1,5$) are reported. Conductivity measurements at ambient and applied pressure, as well as Raman and ESR spectroscopy were performed. Moreover, the results of electrocrystallization experiments with $(n\text{-}Bu_4N)_2Hg(SCN)_4$ as counterion are reported.

INTRODUCTION

In the past years, numerous organic superconductors have been obtained with various donor molecules, such as BEDT-TTF[+], TMTSF[++], DMET (1) and MDT-TTF (2), BEDT-TTF being the most successful, especially in combination with halides (I_3^-, IBr_2^-, Cl^-) and metal halides (AuI_2^-, $Hg_3Cl_8^{2-}$, $Hg_3Br_8^{2-}$). During our research on conducting and superconducting salts of ET, we have made also several attempts to synthesize salts with metal-thiocyanato (and metal-selenocyanato) counterions, as well as with the cyanate and thiocyanate anions themselves (3,4).

Recently, Urayama et al. (5) reported on superconductivity in $(ET)_2Cu(NCS)_2$, the organic superconductor with the highest T_c(10.4) known so far. In this paper, we present our results concerning the synthesis of two superconducting types of $(ET)_2Cu(NCS)_2$. Furthermore, we report on the synthesis of radical cation salts of ET, using dithiocyanatoargentate(I) and tetrathiocyanatomercurate (II) anions, and their physical properties.

\+ BEDT-TTF is Bis-(ethylenedithiolo)-tetrathiafulvalene, or "ET" for short
++ TMTSF is Tetramethyltetraselenafulvalene

Lower-Dimensional Systems and Molecular Electronics
Edited by R.M. Metzger *et al.*, Plenum Press, New York, 1991

EXPERIMENTAL

Electrocrystallization

with $Cu(NCS)_2^-$ anion yielded two superconducting types of $(ET)_2Cu(NCS)_2$, denoted by 'A' and 'B' respectively, when carried out according to previously published procedures (6).

with $Ag(SCN)_2^-$ anion produced black, rhombic crystals (typ. dimensions: 1.8 x 1.1 x 0.44 mm^3) in an electrolyte containing ET (1 mmol/l) and a suspension of AgSCN, KSCN and 18-crown-6 (10 mmol/l each) as anion source (Pt electrodes; I = 1 $\mu A/cm^2$). As solvents, either tetrahydrofuran (THF)/acetonitrile (AC) (2.5:1) or dichloromethane (CH_2Cl_2) were used.
Crystals obtained by both methods are very similar, and were characterized by ESR and evaluation of unit cell data. Elemental analysis revealed a composition $(ET)Ag_x(SCN)_2$ ($x \simeq 1.5$).
The orthorhombic unit cell parameters are $\underline{a}$ = 4.204(1) Å, $\underline{b}$ = 11.597(2) Å, $\underline{c}$ = 40.125(9) Å, $\alpha=\beta=\gamma=90^o$, V = 1956.4(6) $Å^3$ in the space group Pbca (Z=4; $\rho_{calc.}$ (for $(ET)Ag_{1.5}(SCN)_2$ = 2.08g/cm^3), $\rho_{meas.}$ = 2.15g/cm^3).
The room-temperature ESR line-widths of single crystals were found to be 36 - 40 G, and did not show a significant angular dependence.

with $Hg(SCN)_4^{2-}$ anion generated black, thin, extremely fragile needles (typical size: 1.4 x 0.06 x 0.006 mm^3) in a solution containing ET (1mmol/l) and $(n\text{-}Bu_4N)_2Hg(SCN)_4$ (10mmol/l) at current densities of 3 - 5 $\mu A/cm^2$ (Pt electrodes). As solvents, either 1,1,2-trichloroethane (TCE), or a mixture of chlorobenzene (CB)/AC (4:1) were used. At the end of electrolysis, always a greenish-yellow deposit ($(SCN)_x$) on the anode, and sometimes at the crystal surface could be observed. Very low yields (about 1 - 2 mg for 50 ml cell volume) impeded elemental analysis, and thus these data, and the results of microprobe analysis (standard: $(ET)_4Hg_3I_8$) on single crystals differ somewhat. For crystals generated in TCE, we propose a formal stoichiometry $(ET)_4Hg(SCN)_4(TCE)$, while for crystals obtained in CB/AC a composition $(ET)_2Hg_x(SCN)_4$ ($x \simeq 1.4 - 1.5$) is derived from analytical data available.

Conductivity measurements

Four-terminal conductivity measurements were made with gold wires attached to previously evaporated gold pads on the crystals with silver paste. Measuring currents (dc) were 0.1 - 100 μA, depending on sample size and conductivity behaviour. Hydrostatic pressure could be applied in a He-gas pressure cell.

RESULTS AND DISCUSSION

We have synthesized samples of the organic superconductor $(ET)_2Cu(NCS)_2$ using different solvents and anion-sources; i.e. CH_2Cl_2/KSCN/CuSCN/Dibenzo-18-crown-6 for 'A' and TCE/18-crown-6/KSCN for 'B' (6).
Fig.1 shows the temperature dependence of resistivity of a sample prepared by our procedure (type 'A'), in comparison with one obtained by literature methods (5) (type 'B').

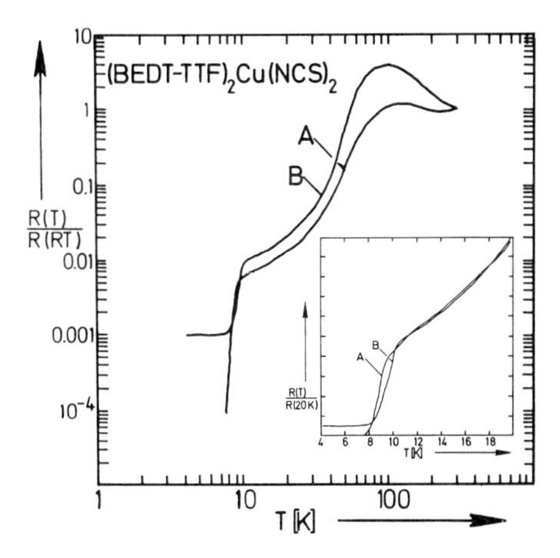

Fig.1. Resistivity of 'A' and 'B' samples of $(ET)_2Cu(NCS)_2$ normalized to r.t.values; insert: low temperature resistivities normalized to 20 K values

From a starting value of 100 mΩ·cm (σ = 10S/cm) at room temperature, the resistivity passes for (type 'A') through a shallow minimum to a maximum at about 130 K with $R_{max}/R_{RT} \simeq 0.95 - 1.3$ depending on the sample. In contrast, type 'B' crystals show activated resistivity (corresponding to a gap of about 50 meV) down to ≃ 170 K. Below, the resistivity levels to pass through a maximum at 100 K ($R_{max}/R_{RT} \simeq 3$; comparable to literature values). Upon further cooling, a steep decrease in resistivity followed by the onset of superconducting transition at 9.4 K ('A') and 10.3 K ('B') is observed.
In both cases, the resistive transition is about 2 K wide (T_{cmid} = 8.7 K for 'A'; T_{cmid} = 9.4 K for 'B').
Surprisingly, in all type 'B' samples (compare Fig.1 insert) measured so

far, the resistance remained finite below the superconducting transition, and showed ohmic behaviour (linear I-V characteristic), whereas in type 'A' samples the resistance falls to zero within experimental accuracy. However, there is no doubt that superconductivity is a volume property in both cases, since this has been established by low field dc magnetization measurements (6). When comparing the overall temperature dependence of the resistivity, we note that the usual definition of the residual resistance ratio RRR = R_{RT}/R_{onset}, gives a higher value for the 'A' samples (≃180 for 'A' vs. ≃100 for 'B'). However, from other observations (morphology, X-ray, Shubnikov-de Haas amplitude) type 'B' crystals appear to be of superior quality. Thus, the ratio R_{max}/R_{onset} (≃400 for 'B' vs. ≃240 for 'A') seems to be the more appropriate choice as a measure of crystal quality.

Though 'A' and 'B' differ markedly in morphology, physical and chemical characteristics (composition, ESR line-with (72 G for 'A'), the unit cell data, Raman spectra, Shubnikov-de Haas frequency) are very similiar for both types. Remarkable differences in conductivity behaviour, T_c, critical fields (preliminary results), however, suggest that we are dealing with two different types of $(ET)_2Cu(NCS)_2$.

We may tentatively attribute the differences between 'A' and 'B' samples to an "alloying" effect in our 'A' samples (i.e. SCN^- anions are incorporated in addition to, or in place of, $Cu(NCS)_2^-$ anions). Further evidence is, however, necessary to clarify this point.

Electrocrystallization with the $Ag(SCN)_2^-$ anion in various solvents always produced - in contrast to $Cu(NCS)_2^-$ - a compound with complex anion stoichiometry: $(ET)Ag_x(SCN)_2$ (x ≃ 1.5).+++

X-ray studies indicate a "doubling" of the a-axis. The alternating appearance of sharp reflexions with extremely broad ones points out the existence of a disorder, too.

Fig.2 shows the temperature dependence of resistivity measured independently at samples of $(ET)Ag_x(SCN)_2$ (x ≃ 1.5) from the same batch at ambient pressure (upper curves), and under application of hydrostatic pressure (2.6 kbar; lower curve). $(ET)Ag_x(SCN)_2$ shows (activated) semiconducting behaviour with σ(290 K) = 1S/cm and E_{gap} ≃ 185 meV at room temperature.

An anomalous upturn in resistivity is observed below 215 K, followed by a region were resistivity is again simply activated (E_{gap} ≃ 230 meV) down to the lowest temperatures measured. Under the application of pressure (2,6

+++ A compound $(ET)Ag_x(SCN)_2$ (x ≃ 1.6), recently reported by Williams et al. (7), is similiar to the salt originally synthesized by us (3,6); unit cell data and stoichiometry are closely related, r.t. conductivity, however, differs widely from our results.

kbar) the 'upturn' shifts to 195 K. Both the room- and low-temperature gaps are reduced by about 10% under applied pressure, to 167 and 210 meV, respectively.

Using $(n\text{-}Bu_4N)_2Hg(SCN)_4$ as counterion in various solvents, we obtained two kinds of thin, extremely fragile crystals in very low yields, which could not be characterized very well: $(ET)_2Hg_x(SCN)_4$ ($x \simeq 1.4 - 1.5$) and $(ET)_4Hg(SCN)_4(TCE)$. Unfortunately, it has so far been impossible to grow any single crystals of sufficient quality for X-ray studies. Samples prepared in CB/AC (Fig.3) show activated semiconducting behaviour with σ(290K) = 3S/cm and $E_{gap} \simeq 84$ meV.

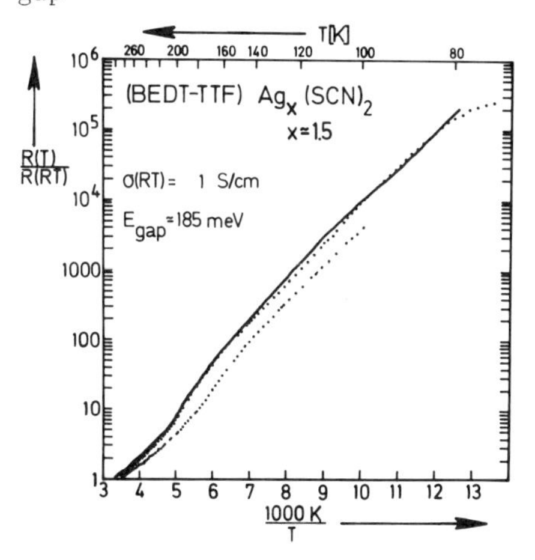

Fig.2. Normalized resistivities of $(ET)Ag_x(SCN)_2$ ($x \simeq 1{,}5$); Samples from the same batch measured independently at ambient (solid and upper dotted curve) and applied pressure (2.6 kbar, lower dotted curve)

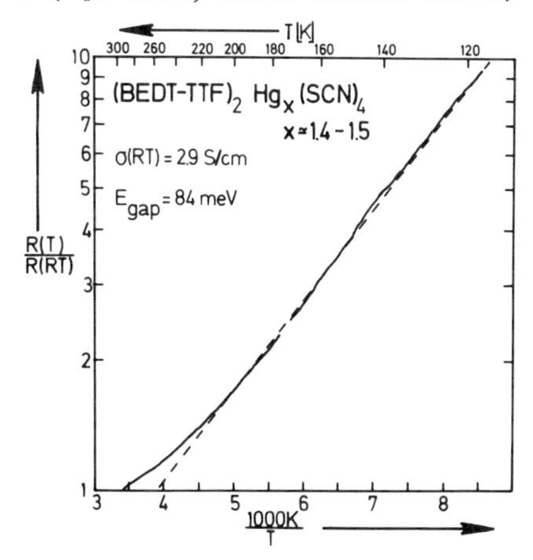

Fig.3. Normalized resistivity of $(ET)_2Hg_x(SCN)_4$ ($x \simeq 1{,}4-1{,}5$); the broken line corresponds to activated behaviour with $E_{gap} \simeq 84$ meV.

ACKNOWLEDGEMENT

We wish to thank J. Riede, München, for X-ray studies, R. Wagner and E. Dormann, Bayreuth, for ESR measurements.

We gratefully acknowledge financial support by Stiftung Volkswagenwerk, Hannover.

REFERENCES

1 K. Kikuchi, M. Kikuchi, T. Namiki, K. Saito, I.Ikemoto, K. Murata, T. Ishiguro and K. Kobayashi, Chem. Lett. 931 (1987)

2 G.C. Papavassilou, G.A. Mousdis, S.Y. Yiannopoulos and J.S. Zambounis, Chemica Scripta 28, 365 (1988)

3 H. Fuchs, PhD thesis, Univ. München, 1987

4 H. Müller, Diploma thesis, TU München, 1987

5 H. Urayama, H. Yamochi, G. Saito, K. Nozawa, T. Sugano, M. Kinoshita, S. Sato, K. Oshima, A. Kawamoto and J. Tanaka, Chem. Lett. 55 (1988)

6 H. Veith, C.P. Heidmann, H. Müller, H.P. Fritz, K. Andres and H. Fuchs, Synth. Met. 27, A 361 (1988)

7 U. Geiser, M.A. Beno, A.M. Kini, H.H. Wang, A.J. Schultz, B.D. Gates, C.S. Carriss, K.D. Carlson and J.M. Williams, Synth. Met. 27, A 235 (1988)

MAGNETOTRANSPORT IN $(BEDT\text{-}TTF)_2Cu(NCS)_2$: SHUBNIKOV-DE HAAS EFFECT AND UPPER CRITICAL FIELD

C.-P. Heidmann[+], W. Biberacher[+,++], H. Müller[+], W. Joss[++] and K. Andres[+]

+ Walther-Meissner-Institut, D-8046 Garching, FRG
++ Max-Planck-Institut für Festkörperforschung, Hochfeld-Magnetlabor, F-38042 Grenoble, France

ABSTRACT

The transverse magnetoresistance of $(BEDT\text{-}TTF)_2Cu(NCS)_2$ was measured at temperatures down to 0.3K and in magnetic fields up to 24T. Shubnikov-de Haas oscillations with a single frequency of 597±7 T were found consistently for samples from six different batches. The angular dependence of the fundamental SdH frequency shows a $1/\cos\theta$ behaviour up to the highest angle measured (69°), which indicates a highly two-dimensional electronic structure of this compound (cylindrical Fermi Surface), and is in agreement with previous band structure calculations.

For H ∥ b the upper critical field H_{c2} exceeds the Chandrasekhar-Clogston limit by 25%. A finite slope of $H_{c2}(T)$ is found even at $T/T_c<0.1$.

INTRODUCTION

The combination of the organic donor BEDT-TTF[+] with a wide selection of inorganic anions has yielded a huge number of conducting radical cation salts with a remarkable variety of structures, among them many metallic and even superconducting systems. From the anisotropy of their electrical conductivity /1/ and superconducting critical fields /2,3,4/ the latter are mostly classified as (quasi) two-dimensional metals. Although one may thus have anticipated the existence of closed portions of the Fermi Surface in these salts, direct evidence, such as the observation of the de Haas-van Alphen or the Shubnikov-de Haas effect has not been available until quite recently for β_L and β_H-ET_2I_3 /5,6/, β-ET_2IBr_2 /5,7/, β-ET_2AuI_2 /8/, β"-ET_2AuBr_2 /9/ and $ET_2Cu(NCS)_2$ /10/.

Besides being the organic superconductor with the highest T_c (10.4 K) known to date /11/, $ET_2Cu(NCS)_2$ shows a unique structural feature in that it is lacking any preferential orientation within its cation layer such as a stacking or ribbon direction. Instead, dimers of ET molecules are packed in a chequerboard manner with their molecular planes roughly perpendicular to each other, thus facilitating close S···S contacts between neighbouring molecules in all directions /11/.

+ BEDT-TTF is bis(ethylenedithiolo)tetrathiafulvalene or "ET"

Lower-Dimensional Systems and Molecular Electronics
Edited by R.M. Metzger *et al.*, Plenum Press, New York, 1991

We have studied the transverse magnetoresistance of $ET_2Cu(NCS)_2$ in order to explore its H-T phase diagram to lower temperatures and higher fields than before and especially to test whether the "two-dimensional" structure in real space is reflected in the electronic structure.

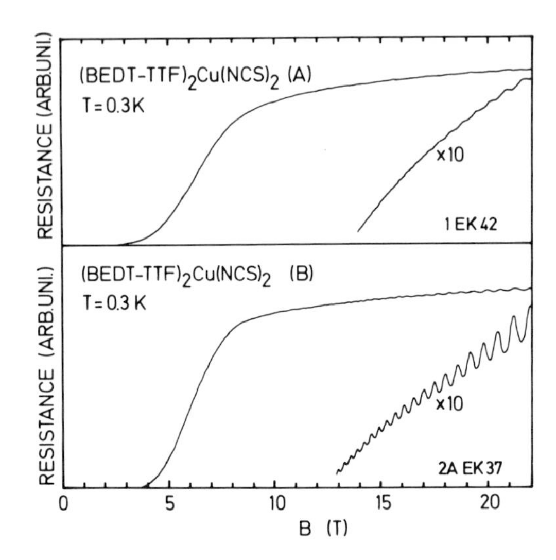

Fig. 1. Magnetoresistance of a type A (top) and a type B (bottom) sample for fields normal to the crystal bc-plane; oscillatory contribution shown on an expanded scale

EXPERIMENTAL

Two types of samples, denoted by "A" and "B" respectively, were synthesized by electrocrystallization as described elsewhere /12/. Four contacts, grouped along the crystal b axis, were made with silver paste to previously evaporated gold pads, resulting in contact resistances below a few ohms at helium temperature. A constant low-frequency ac current (100 μA - 1 mA, 90 Hz) and lock-in detection were used for the resistance measurements. The samples were mounted on a holder that allowed them to be rotated about two orthogonal axes while immersed in liquid ^3He at a temperature between 0.3 and 1.3 K. Magnetic fields up to 22 T were generated in a polyhelix type magnet at the high-field facility in Grenoble. Data points equidistant in 1/H were taken while the magnetic field was swept between 22 T and 14 T and stored in a desk-top computer for further processing (Fourier transformation).

RESULTS AND DISCUSSION

The resistance as a function of magnetic field at a constant temperature of 0.3 K is shown in Fig. 1 both for an A and a B sample. After a steep increase at the superconducting transition, the resistance tends to saturate at higher fields, and an oscillating contribution appears above 10 T.

The Fourier spectrum of these oscillations ($H \parallel a^*$), shown in Fig. 2, consists of a sharp symmetric peak centered at a frequency of about 600 T, and a smaller peak at around 1200 T, obviously the second harmonic in the spectrum of an anharmonic oscillation. The oscillations being thus periodic in 1/H, we are confident that we observe in fact the Shubnikov-de Haas effect. An observed frequency F is then related to an extremal cross-section S of the Fermi Surface normal to the magnetic field direction by $S=(2\pi e/\hbar c)F$ /13/, and thus geometric information about the Fermi Surface can be obtained from the angular dependence $F(\Theta)$. The result for the fundamental peak frequency in $ET_2Cu(NCS)_2$ is shown in Fig. 3.

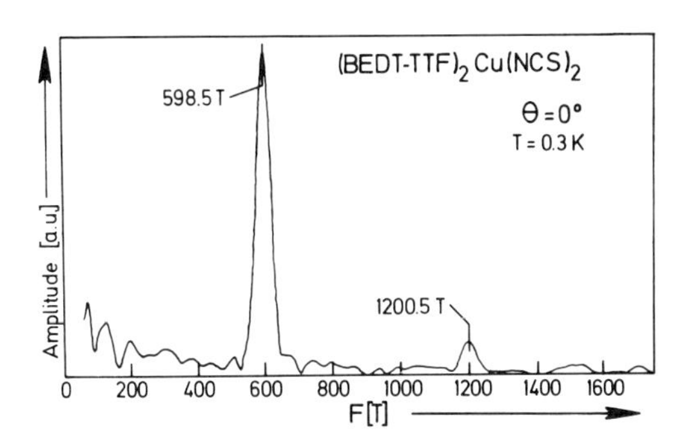

Fig.2. Fourier spectrum of magnetoresistance (Shubnikov-de Haas) oscillations for fields normal to the crystal bc-plane

From a previous observation of the Shubnikov-de Haas effect, as well as from a tight-binding band structure calculation /10/, a cylindrical Fermi Surface was inferred, in which case the angular dependence takes the form $F(\Theta)=F_o/\cos\Theta$, where F_o is the peak frequency for $H \parallel a^*(\Theta=0)$, and so this expression was fitted to the experimental data with F_o and a (trivial) constant angular offset as the adjustable parameters. Theory (solid line in Fig. 3) and data are in agreement, within experimental error, up to the highest angle (69°) and for all samples measured. A mean value of 597 ± 7 T is obtained for F_o (corresponding to 15% of the Brillouin Zone cross section) for samples from six different electrocrystallization batches, including two crystals of type A, crystals of type B synthesized with ET from three different sources, as well as a deuterated crystal. We note that the scatter in F_o values is only slightly larger than the resolution (channel separation, 5T) of the Fourier transform.

Our result is in disagreement with the value of around 660 T reported by Oshima et al. /10/ and it is currently not clear how this discrepancy can be resolved.

Both from the observation of a single frequency in the Fourier spectrum and its angular dependence we conclude that the Fermi Surface of $ET_2Cu(NCS)_2$ comprises a closed part that can be described to a very good approximation as a cylinder with axis along a* (without noticeable warping). This means that the band structure shows very small dispersion along a* ,if any, and the electron motion is effectively confined to the cation layers.

As far as the dependence of the oscillation amplitude on angle, temperature, and field is concerned, we first note that in all our experiments the amplitude reaches its maximum value at an angle of 15-20° between magnetic field and a* axis, independent of the axis of rotation (b or c). This phenomenon can also be seen in Fig. 3 of ref. /14/. We can, therefore, exclude trivial explanations like misaligned or twinned crystals.

The temperature dependence of the SdH amplitude for H $\parallel$a* was used to estimate the cyclotron mass of the orbit by the standard procedure /13/. We find a value of $m/m_o \simeq 3.2$, which agrees well with previously published data. Using this value, a Dingle temperature of T_D=0.7 K is obtained from the field dependence of the amplitude (type B crystal).

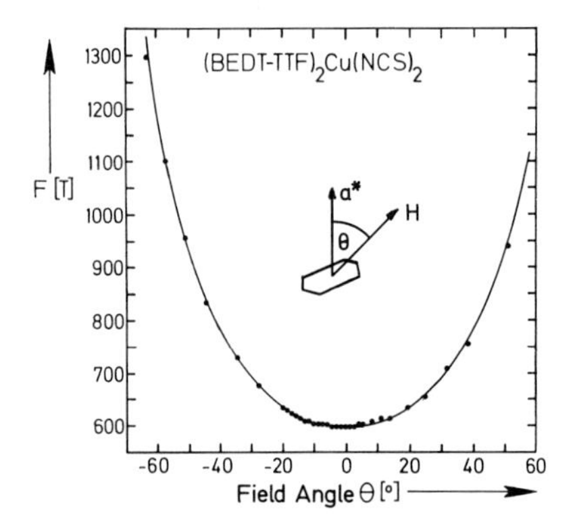

Fig. 3. Angular dependence of the SdH frequency at 0.3 K dots - experimental values, solid line - theoretical curve for a cylindrical Fermi Surface

Upper critical field values H_{c2} were determined from plots of the magnetoresistance such as shown in Fig. 1 using the midpoint definition of H_{c2} for fields parallel to the a* and b axes (the inflection point had to be taken where the transition could not be observed completely).

For H $\parallel$ b, the upper critical field reaches 20.2 T at 1.17 K, and 21.6 T at 0.33 K. The superconducting transition temperature T_c for this sample, as determined in a separate run prior to the magnetoresistance measurements, was 9.4 K (midpoint), with an onset at 10.3 K. Therefore H_{c2b} exceeds the paramagnetic Chandrasekhar-Clogston limit given by 1.84 T/K$\cdot T_c$ [K] = 17.3 T. Of special interest is also the large temperature dependence of H_{c2} ($\sim$1.6 T/K) at values of T/T_c even below 0.1.

For H $\parallel$a* the 1.2 K value of H_{c2} (4.1 T) agrees with published data, while the 0.3 K value of 6.0 T lies on the extrapolation of this curve (Fig. 5 in /4/), indicating the absence of saturation at low temperatures for this direction as well.

Crystals of $ET_2Cu(NCS)_2$ synthesized by different electrocrystallization procedures showed differences in conductivity behaviour, T_c and other properties, and were therefore classified as type A and B /12,15/. From our present results, however, both types appear to be very similar. As can be seen in Fig. 1, their overall magnetoresistance behaviour, as well as the values of the upper critical field (midpoint of the superconducting transition), are nearly identical. Moreover, type A and B samples show exactly the same Shubnikov-de Haas frequency, which strongly suggests that their electronic structure is identical. We therefore attribute the smaller SdH amplitude found in type A samples to an enhanced scattering rate, due to a higher concentration of lattice defects in these crystals.

SUMMARY

The electronic structure of $ET_2Cu(NCS)_2$ was established to be of two-dimensional nature by the observation of Shubnikov-de Haas oscillations as a function of angle at 0.3 K and in magnetic fields between 14 T and 22 T. The SdH frequency, for magnetic fields perpendicular to the cation layers, was found consistently to be 597±7 T for samples from six different batches. This result agrees qualitatively with the published band structure, but is about 10% smaller than the value derived from a previous study in magnetic fields below 13.5 T.

The upper critical field exceeds the Chandrasekhar-Clogston limit by 25% at 0.3 K and fields parallel to the cation layers (b axis), and shows an anomalously large temperature dependence at low temperatures.

ACKNOWLEDGEMENT

We gratefully acknowledge financial support by Stiftung Volkswagenwerk, Hannover, and by Deutsche Forschungsgemeinschaft.

NOTE ADDED

In a high-field run (24 T) we have very recently found high frequency Shubnikov-de Haas components at 3245 T and two higher frequencies, spaced by about the value of the low SdH oscillation frequency observed in these crystals. This might indicate that we are observing larger orbits due to magnetic breakdown, as was discussed in /10/.

REFERENCES

1. L.I. Buravov, M.V. Kartsovnik, P.A. Kononovich, V.N. Laukhin, S.I. Pesotskii and I.F. Shchegolev,
   Sov. Phys. JETP 64 (1986) 1306
2. C.-P. Heidmann, K. Andres and D. Schweitzer,
   Physica 143B (1986) 357
3. M. Tokumoto, H. Anzai, H. Bando, G. Saito, N. Kinoshita, K. Kajimura and T. Ishiguro,
   J. Phys. Soc. Jpn. 54 (1985) 1669
4. K. Oshima, H. Urayama, H. Yamochi and G. Saito,
   Physica C153 (1988) 1148
5. K. Murata, N. Toyota, Y. Honda, T. Sasaki, M. Tokumoto, H. Bando, H. Anzai, Y. Muto and T. Ishiguro,
   J. Phys. Soc. Jpn. 57 (1988) 1540

6. W. Kang, G. Montambaux, J.R. Cooper, D.Jérome, P. Batail and C. Lenoir, Phys. Rev. Lett. 62 (1989) 2559
7. M.V. Kartsovnik, V.N. Laukhin, V.I. Nizhanovskii and A.A. Ignat'ev, JETP Lett. 47 (1988) 362
8. I.D. Parker, D.D. Pigram, R.H. Friend, M. Kurmoo and P. Day, Synth. Met. 27 (1988) A387
9. F.L. Pratt, A.J. Fisher, W. Hayes, J. Singleton, S.J.R.M. Spermon, M. Kurmoo and P. Day, Phys. Rev. Lett. 61 (1988) 2721
10. K. Oshima, T. Mori, H. Inokuchi, H. Urayama, H. Yamochi and G. Saito, Phys. Rev. B38 (1988) 938
11. H. Urayama, H. Yamochi, G. Saito, K. Nozawa, T. Sugano, M. Kinoshita, S. Sato, K. Oshima, A. Kawamoto and J. Tanaka, Chemistry Letters (1988) 55
12. H. Müller, C.-P. Heidmann, H. Fuchs, A. Lerf, K. Andres, R. Sieburger and J.S. Schilling, (these Proceedings)
13. D. Shoenberg, *Magnetic Oscillations in Metals*, Cambridge University Press, 1984
14. K. Oshima, T. Mori, H. Inokuchi, H. Urayama, H. Yamochi and G. Saito, Synth. Met. 27 (1988) A165
15. H. Veith, C.-P. Heidmann, H. Müller, H.P. Fritz, K. Andres and H. Fuchs, Synth. Met. 27 (1988) A361

RAMAN INVESTIGATIONS ON SINGLE CRYSTALS AND POLYCRYSTALLINE PRESSED SAMPLES OF ORGANIC SUPERCONDUCTORS

R. ZAMBONI[1] and D. SCHWEITZER
Max-Planck-Institut für Medizinische Forschung, AG Molekülkristalle, Heidelberg, FRG

H. J. KELLER
Anorganisch-Chemisches Institut, Universität Heidelberg, FRG

INTRODUCTION

The investigation of molecular vibrations is a powerful tool, which can increase our knowledge on structures and on electron molecular vibrations, which are due to charge oscillation between dimerized molecules, coupled with totally symmetric intramolecular modes [1-3]. Raman scattering studies account for totally symmetric vibrations. In addition, Raman spectroscopy can take advantage of resonant effects. In fact, when resonant conditions are fulfilled, selected molecular vibrations are obtained, as well as information on the electronic manifold involved in the resonance process.

In this paper, we report on Raman investigations of polycrystalline pressed materials of superconducting tempered α-(BEDT-TTF)$_2$I$_3$, called, in the following, α_t-(BEDT-TTF)$_2$I$_3$, and on untempered α-(BEDT-TTF)$_2$I$_3$.

Furthermore we report on the polarized resonant Raman scattering in the superconducting regime of (BEDT-TTF)$_2$Cu(NCS)$_2$ single crystals.

EXPERIMENTAL

α-(BEDT-TTF)$_2$I$_3$ crystals were grown in a THF-solution by electrochemical methods, as previously reported [4]. The samples for Raman investigations have been prepared by grinding single crystals, and by applying a pressure of 10^4kg/cm^2 to the powder. The superconducting α_t-(BETT-TTF)$_2$I$_3$ [5] samples were obtained, by tempering the pressed α-samples at 75 °C for several days.

Single crystals of (BEDT-TTF)$_2$Cu(NCS)$_2$ have been obtained by electrochemical preparation [6]. The largest crystals obtained in this way (10x2x0.1 mm), have been selected and oriented in a patchwork of about 10 mm^2. The Raman spectra have been carried out with the lines of an Ar$^+$ ion laser (Coherent Innova 90) in a 90° scattering configuration. The laser beam was filtered with a monochromator, and a proper interference filter, in order to prevent plasma lines.

The laser power was maintained below 10 mw in focused conditions for pressed pellets of α_t-(BEDT-TTF)$_2$I$_3$ and $\alpha-$(BEDT-TTF)$_2$I$_3$, but not completely focused on single crystals of (BEDT-TTF)$_2$Cu(NCS)$_2$, in order to avoid sample damage. For (BEDT-TTF)$_2$I$_3$ materials, the conditions of a single Raman scan, with a resolution of the double monochromator of 0.5 cm^{-1}, were sufficient for a good signal-to-noise ratio. For the (BEDT-TTF)$_2$Cu(NCS)$_2$, ten spectra were accumulated, with a resolution of the monochromator of about 3 cm^{-1}, in order to achieve an acceptable signal-to-noise ratio, and the Raman shift between 40 cm^{-1} and 3500 cm^{-1} has been investigated.

---

1 On leave from Istituto di Spettroscopia Molecolare del C.N.R., Bologna, Italy

RESULTS AND DISCUSSION

The $(BEDT\text{-}TTF)_2Cu(NCS)_2$ single crystal.

The $(BEDT\text{-}TTF)_2Cu(NCS)_2$ is up to now the ambient pressure organic superconductor with the highest critical temperature [7,8] (T_c = 10.4 K).

In addition, this compound shows a nearly isotropic electrical and optical conductivity in the *bc* plane,and a very distinctive structure. The latter consists in a bidimensional sheet of dimerized (BEDT-TTF) molecules where one dimer is nearly orthogonal to the neighbouring ones [7], and they do not form the usual face-to-face piling along the stacking axis. The $Cu(NCS)_2$ counter anion builds a sheet, where the copper coordinates with a sulfur atom and two nitrogen ones, to form a coordination polymer [7,10].

Fig. 1 shows the polarized Raman scattering at 1.3K from the (100) crystal surface, with the laser beam polarized along the *c* axis, which is the long molecular axis of the (BEDT-TTF)[7].

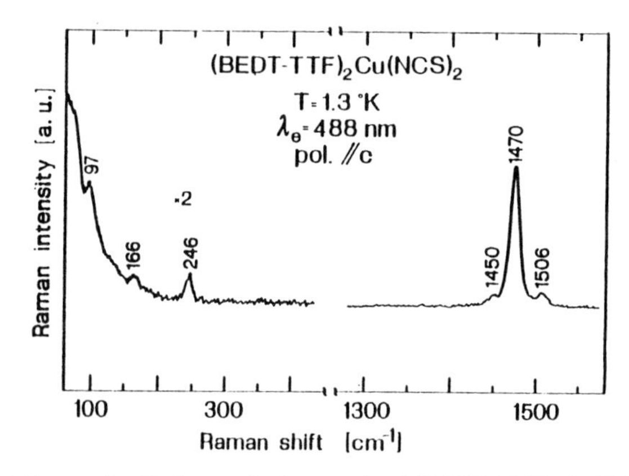

Fig. 1 Resonant Raman Scattering on single crystals of $(BEDT\text{-}TTF)_2Cu(NCS)_2$. The laser beam is polarized along the major axis of the (BEDT-TTF) molecules.

The spectrum consists of a few bands of very low intensity. The most intense one is at 1470 cm^{-1}. In fig. 2 we show details of this range, varying the polarization of the laser beam.

As we can see, the 1470 cm^{-1} band totally disappears, and a band at 1492 cm^{-1} appears, when the beam polarization rotates perpendicular to the long molecular axis of the (BEDT-TTF). We can assign the 1470 cm^{-1} band to the totally symmetric C=C stretching vibration of the central carbons, and the 1492 cm^{-1} band to the C=C symmetric stretching vibration of the ring.

Note the consistency of the Raman data as a function of beam polarization; in fact, at 45° with respect to the major axis of the (BEDT-TTF), the intensity of the 1470 cm^{-1} is nearly one half.

The relatively high intensity of the band at 1470 cm^{-1} is due to a strong coupling of this vibration to the electronic transition at 20000 cm^{-1} (see Fig. 3) because, in this resonance Raman experiment, we excite at about 20500 cm^{-1} (488 nm). This electronic transition is probably a bonding-antibonding one, which is mainly localized on the rigid and planar TTF fragment of the (BEDT-TTF) molecule. The weak bands, at 1450 cm^{-1} and 1506 cm^{-1}, are certainly connected with C=C vibrations, as results from IR and Raman scattering on deuterated samples of (BEDT-TTF) neutral molecules and α-$(BEDT\text{-}TTF)_2I_3$ crystals [10-12].

The authors in ref. (11) found a band at 1511 cm^{-1} in the Raman spectrum of the neutral (BEDT-TTF), and assigned this band to an antisymmetric b_{1u} vibration of the ring C=C stretching, which is allowed in Raman by departure from molecular symmetry. In our data, we found a band at 1506 cm^{-1} in both beam polarizations. This finding could support the previous assignment, as well as the strong asymmetric environment surrounding of the BEDT-TTF molecules [7].

In addition, we found very weak features at 774 cm^{-1}(b_{1u} C-S stretching [11]) and 1037 cm^{-1} (C-C-H bending totally symmetric vibration [3]). The signal-to-noise ratio of these bands is very poor. The most resolved and intense band, in the lower part of the spectrum, is the one at 246 cm^{-1}. Authors [12] reported on Raman scattering at room temperature with the 514.5 nm line, on (BEDT-TTF) Br powder, and they found a weak band at 256 cm^{-1}, which was attributed to a b_{2u} mode that became Raman-active.

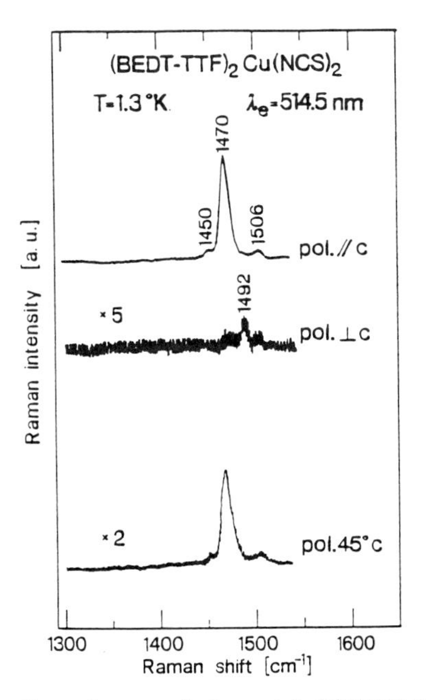

Fig. 2. Resonant Raman Scattering on a single crystal of $(BEDT\text{-}TTF)_2Cu(NCS)_2$ in the C=C stretching frequency range with different laser beam polarization.

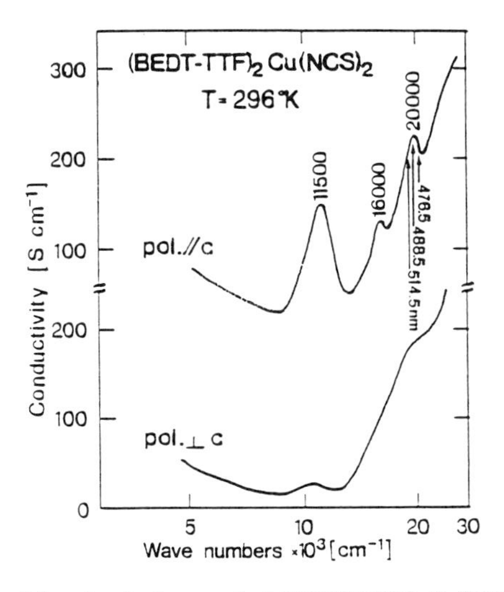

Fig. 3. Optical conductivity of a single crystal of $(BEDT\text{-}TTF)_2Cu(NCS)_2$. The polarization is taken parallel and perpendicular to the major axis of (BEDT-TTF) molecules.

In order to extend this attribution to our polarized 246 cm^{-1} band (see fig. 1) we would have to explain the relative high intensity of an antisymmetric mode in resonance Raman scattering conditions. Antisymmetric vibrational scattering may be activated by a mixing of excited electronic states with vibrational modes. These modes can gain Raman intensity via a Herzberg-Teller term, which can become dominant in resonance with a partially allowed transition, vibronically coupled to a nearby strongly allowed one [13,14]. If this is the case, the 246 cm^{-1} and the 1506 cm^{-1} bands are particularly active in the vibronic mixing. The band at 166 cm^{-1} (see fig. 1) is assigned to a totally symmetric skeleton vibration [12] along the major axis of the BEDT-TTF molecules. The 97 cm^{-1} band could be a lattice mode.

Nevertheless, the Raman scattering intensity of this material is very low. Antiresonance conditions, and de-enhancement of Raman intensity could account for this low intensity [15] but, as we can see in fig. 3, the strong absorption of the $(BEDT\text{-}TTF)_2Cu(NCS)_2$ is certainly a reason for covering the Raman scattering intensity. In addition we cannot exclude a totally symmetric Cu-NCS vibration for the 246 cm^{-1} band [16].

α-$(BEDT\text{-}TTF)_2I_3$ and α_t-$(BEDT\text{-}TTF)I_3$ polycrystalline pressed materials.

The superconductivity in organic materials is very sensitive to any kind of disorder, or structural changes. Recently resonant Raman spectra on single crystals of α_t- and α-phases of $(BEDT\text{-}TTF)_2I_3$ have been reported [17-19]. Resonance conditions with the $\sigma_g \rightarrow \sigma_u$ electronic transition of the I_3^- anion are fulfilled with the usual lines of the Ar^+-ion laser, and a typical resonant Franck-Condon-like Raman scattering has been recorded [18,19].

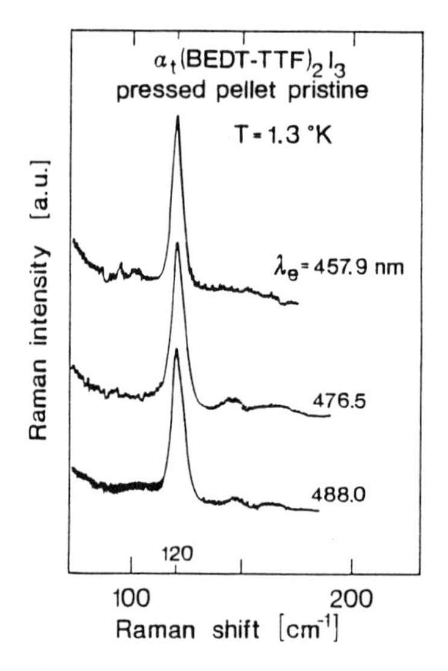

Fig. 4. Resonant Raman Scattering on the tempered polycrystalline pressed pellet of α-$(BEDT\text{-}TTF)_2I_3$ with different laser frequencies.

In particular, the most intense vibrational mode, coupled with the $\sigma_g \rightarrow \sigma_u$ electronic transition, is the symmetric stretching vibration of the I_3^-. Nevertheless, the recent discovery of bulk superconductivity at ambient pressure in polycrystalline pressed samples [5] stimulates the Raman investigation on polycrystalline pressed materials.

Fig. 4 shows the Raman scattering at 1.3K on pressed pellets of α_t-$(BEDT\text{-}TTF)_2I_3$, prepared as reported in the experimental section. The intense mode at 120 cm^{-1} is assigned to the linear I_3^- symmetric stretching. It remains structureless, and in the same frequency position, varying the laser frequency and the sample temperature up to room temperature.

Fig. 5a shows the Raman scattering at 1.3K on pressed pellets of non superconducting α-(BEDT-TTF)I_3. The I_3^- stretching vibration shows a peak at 110 cm^{-1} and a shoulder at 120 cm^{-1}, when working with the 514.5 nm line. The shoulder at 120 cm^{-1} becomes dominant by varying the laser frequency from 514.5 nm to 413.5 nm. A similar behaviour is obtained by working with the laser frequency at 514.5 nm, and varying the sample temperature (see fig. 5b). This splitting observed here is, probably, due to a partial departure from linearity of the I_3^- anion, which causes a splitting of the symmetric mode [18]. The drastic change in relative intensity between the two modes varying the laser frequency, as well as changing the sample temperature, is due to the tuning inside the electronic manifold of the I_3^- anion.

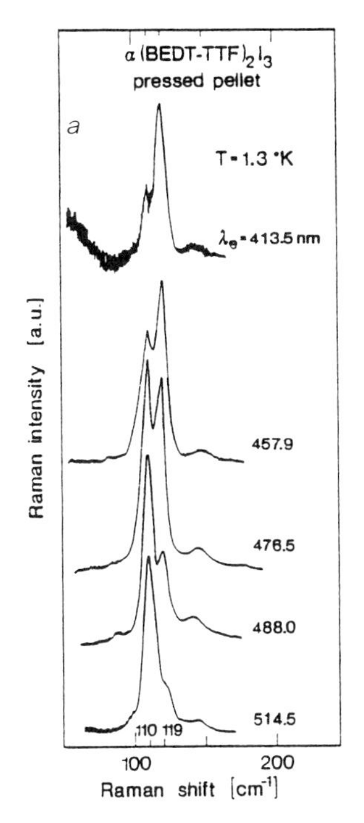

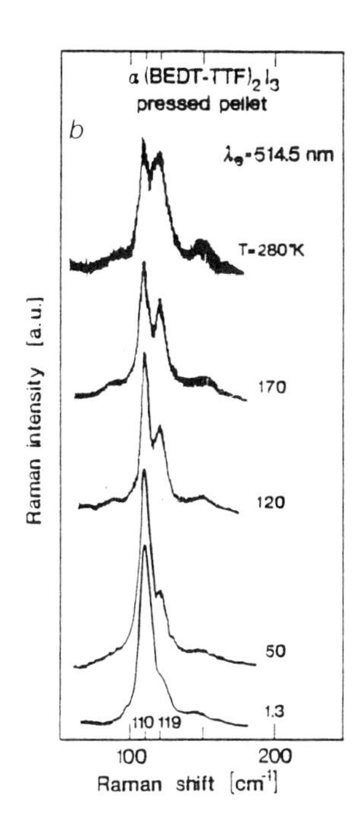

Fig. 5 a) Resonant Raman Scattering on the untempered polycrystalline pressed pellet of α-(BEDT-TTF)$_2I_3$ with different laser frequencies at 1.3K

b) Resonant Raman Scattering on the untempered polycrystalline pressed pellet of α-(BEDT-TTF)$_2I_3$ at a fixed laser frequency, varying the sample temperature.

If this is the case, in the α-phase of (BEDT-TTF)$_2I_3$ the superconductivity could be correlated with the linearity of I_3^- anion, that means with the order in the system. In fact, in the superconducting α_t-(BEDT-TTF)$_2I_3$, the I_3^- shows only one, symmetric stretching mode (see fig. 4). Nevertheless, the grinding and pressure process, during the preparation of the pellets, could cause strains in the material, and different atomic charge distributions resulting in non-linear and asymmetric anions.

In fig. 6 the Raman spectrum of a α_t-(BEDT-TTF)I_3 sample shows such a splitting of the I_3^- anion. In this case, the surface of the tempered pellet has been polished with a razor blade. The appearance of the splitting of I_3^- anion means that the pressure, applied on the surface of the pellet by the razor blade, is sufficient to recreate the conditions of the untempered material, as well as to destroy the superconducting transition [5].

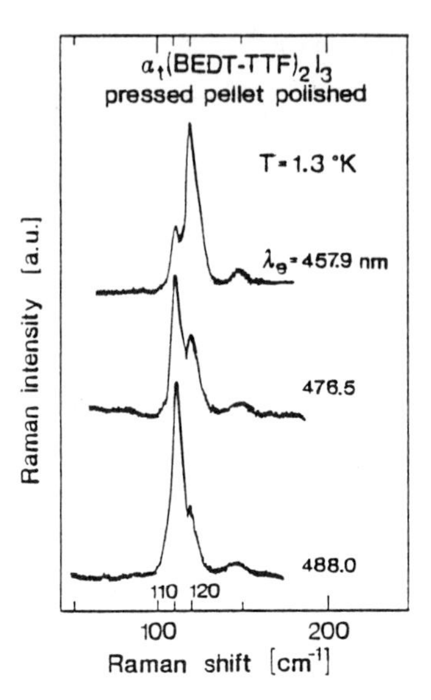

Fig. 6. Resonant Raman Scattering on the tempered pressed pellet of α-(BEDT-TTF)$_2$I$_3$ with different laser frequencies at 1.3K. The surface of the pellet was polished with a razor blade.

ACKNOWLEDGMENTS

We thank profs. C. Taliani, R. Tubino and dr. G. Ruani for useful discussions. We gratefully acknowledge the financial support of the European Economic Community (contract ST2J-0315-C)

REFERENCES

1. M. J. Rice, Phys. Rev. Lett. 37:36 (1976).
2. M. J. Rice, N. O. Lipari and S. Strassler, Phys. Rev. Lett., 39:1359 (1977).
3. M. Meneghetti, R. Bozio and C. Pecile, J. de Physique, 47:1377 (1986).
4. K. Bender, I. Henning, D. Schweitzer, K. Dietz, H. Endres and H. J. Keller, Mol. Cryst. Liq. Cryst., 108:359 (1984).
5. D. Schweitzer, S. Gärtner, H. Grimm, E. Gogu and H. J. Keller, Sol. St. Comm. 69:843 (1989).
6. D. Schweitzer, K. Polychroniadis, T. Klutz, H. J. Keller, I. Henning, I. Heinen, U. Haeberlen, E. Gogu and S. Gärtner, Synth. Met., 27:A465 (1988).
7. H. Urayama, H. Yamuchi, G. Saito, S. Sato, A. Kawamoto, J. Tanaka, T. Mori, Y. Maruyama and H. Inokuchi, Chem. Lett., 463 (1988).
8. S. Gärtner, E. Gogu, I. Heinen, H. J. Keller, T. Klutz and D. Schweitzer, Sol. St. Commm. 65:1531 (1988)
9. T. Sugano, H. Hayashi, H. Takenouchi, K. Nishikida, H. Urayama, H. Yamochi, G. Saito and M. Kinoshita, Phys. Rev., B37:9100 (1988).
10. R. Zamboni, D. Schweitzer, H. J. Keller and C. Taliani, Z. Naturforsch., 44a:429 (1989).
11. M. E. Kozlov, K. I. Pokhodnjia and A. A. Yurchenko, Spectrochim. Acta, 43A:323 (1987).
12. M. E. Kozlov, K. I. Pokhodnjia and A. A. Yurchenko, Spectrochim. Acta, in press.
13. T. G. Spiro and P. Stein, Ann. Rev. Phys. Chem., 28:501 (1977).
14. A. R. Gregory, W. H. Henneker, W. Siebrand and M. Z. Zgierski, J. Chem. Phys., 63:5475 (1975).
15. T. Stein, V. Miskowski, W. H. Woodruff, J. P. Griffin, K. G. Weiner, B. P. Gaber and T. G. Spiro, J. Chem. Phys., 64:215 (1976).
16. Y. M. Bosworth, R. J. H. Clark and P. C. Turtle, J.C.S. Dalton, 2027 (1975).
17. D. Schweitzer, E. Gogu, I. Henning, T. Klutz and H. J. Keller, Ber. Bunsenges. Phys. Chem., 91:890 (1987).
18. R. Swietlik, D. Schweitzer and H. J. Keller, Phys. Rev., B36:6881 (1987).
19. S. Sugai and G. Saito, Sol. St. Comm., 58:759 (1986).

CALORIMETRIC STUDIES OF THE QUANTIZED MAGNETIC ORDERING IN THE ORGANIC METAL $(TMTSF)_2ClO_4$

François Pesty and Pierre Garoche

Laboratoire de Physique des Solides
Associé au CNRS, U.P.S., Bât 510
91405 Orsay Cedex (FRANCE)

ABSTRACT

We review a calorimetric experiment performed on the Spin Density Wave induced by the magnetic field. Our investigations have been concerned with the cascade of weak first-order transitions which separate SDW phases, as well as with the complex second-order line separating the normal metal from the SDW states. Three major features arise from these results. First, the normalized specific heat jump at the metal-insulator transition exhibits an oscillatory behavior as a function of the magnetic field, and takes a value, at high field, far above the BCS prediction of 1.43. Second, the threshold field for the appearance of the SDW phases is shifted toward high fields,as the cooling rate through the anion-ordering transition is lowered. Third, for the well-ordered samples, a new set of transition lines appears in the low-field range: we obtain a treelike phase diagram with a self-similar character.

INTRODUCTION

In the past decade, various experimental techniques[1-3] have been used to study in detail the Spin Density Wave induced by the magnetic field (FISDW). The cascade of phase transitions was first characterized at intermediate fields, from transport[4-8] as well as from thermodynamic[9,10] measurements. The "standard" model[11-15] accounts fairly well for the observed field-induced transitions. They are believed to result from an electronic low-dimensional instability favored by the topology of the quasi-two dimensional open orbit Fermi surface. In a magnetic field applied along the c^* axis, the electronic motion is indefinite in the longitudinal direction a, but is bounded in the transverse direction b. This motion becomes more and more one dimensional as the field is increasing, favoring the instability and leading to the cascade of FISDW transitions.

However the richness of the FISDW was not exhausted: a striking result has been the discovery of a reentrance of the metallic state at around 30 teslas[16-18], an unexpected behavior within the standard model. Few models have been proposed to explain such a behavior. One of them[19] takes explicitely into account interactions between electrons, instead of merely considering the nesting properties of the Fermi surface, as in the standard model.

For the $(TMTSF)_2ClO_4$ compound, an another puzzling property is the influence of the cooling rate on the low temperature ground states[20-21]. In particular,for the FISDW, a very well ordered sample exhibits a complex Hall effect with negative plateaus[21]. In order to clarify the new aspect of the phase diagram of the perchlorate salt, we have focussed on studying this last cooling rate effect. Several calorimetric methods have been used,so as to get complementary information on the thermodynamics of the FISDW.

Lower-Dimensional Systems and Molecular Electronics
Edited by R.M. Metzger *et al.*, Plenum Press, New York, 1991

First, an anomalous Shubnikov-de Haas effect was reported in the $(TMTSF)_2PF_6$ compound under pressure[4]. Specific heat measurements subsequently revealed a surprising metal-insulator transition induced by a moderate magnetic field of 6 teslas, in the $(TMTSF)_2ClO_4$ salt[22]. The electronic density of states at the Fermi level was then found to be magnetic field dependent[23]. To check such a surprising behavior, it was necessary to improve our experimental apparatus, in order to increase significantly its signal-to-noise ratio.

Using a new sample-holder with a very low heat capacity (less than 1 nJ/K at 0.3 K), prepared by using a microlithographic process and a thin-film technique, we were able to measure accurately the heat capacity of a single crystal of $(TMTSF)_2ClO_4$ [10,24]. By means of the ac calorimetry technique[25], we have built up a phase diagram from two specific heat measurement techniques: either at fixed field and varying temperature, or at fixed temperature and sweeping field.

All these results allow us to describe precisely the transition line that separates the normal metal from the FISDW phases, up to 10 teslas[26] (Fig. 1). The 10 to 30 teslas range has recently been studied by calorimetric experiments using a similar technique[27]; these data give additional information on the surprising reentrance of the metallic phase at high field.

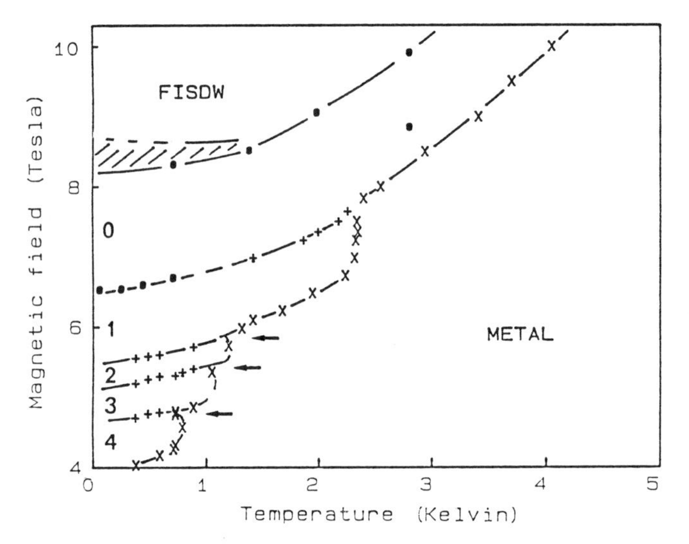

Fig. 1. Phase diagram of the slowly cooled $(TMTSF)_2ClO_4$. Cross and plus symbols represent calorimetric data. The dots represent magnetization data[9]. Arrows indicate the partial reentrances of the metal into the FISDW.

The transition line which marks the appearance of the FISDW is not monotonic: the mean positive slope indicates that the critical temperature for the transition from the metal to the SDW increases as the field increases, which shows that the SDW is stabilized by the magnetic field. Nevertheless, at the limit between two subphases, a negative slope is associated with a partial reentrance of the metallic phase into subphases (arrows on Fig. 1). In between a given quantized SDW phase, for a particular field B_{max} , the transition temperature has a maximum value: $T_c(B_{max})$. But when the field is increased,or decreased,from B_{max} toward a phase boundary, T_c reaches a minimum at the multicritical point, i.e. at the metallic reentrance. This reentrance illustrates the competition between order parameters of adjacent SDW subphases.

Furthermore, this line is always related to a second-order phase transition, though a non-mean-field transition behavior is observed just above the reentrance points[28]. Along this transition line the specific heat jump ranges from 30% to 400% of the BCS prediction. This strongly suggests that the weak coupling limit is not a convenient description for the quantized magnetic ordering, and that an intermediate coupling model must be developed.

At the multicritical points associated with the partial reentrances of the metal, the specific heat jump displays a discontinuity, from a low value on the low field side, to a large value on the high field one (arrows on Fig. 2). This oscillatory behavior has been tentatively explained within the quantized nesting model[29], but only a qualitative description has been achieved, due to the quoted limitation of the weak coupling limit.

In addition, our direct calorimetric measurements as a function of the magnetic field lead us to describe the cascade of weak first-order transitions separating the SDW subphases[10,30,31]. Figure 3 shows the magnetocaloric effect recorded for two cooling rates. The anomalies correspond to the temperature variation ΔT of the sample assembly, weakly coupled to a thermal bath, when the field is swept up and down. ΔT is directly proportional to the thermal coefficient of the magnetization, $(\partial M/\partial T)_B$. The direction of heat exchange flow indicates an entropy decrease as the magnetic field is raised. It corresponds to paramagnetic jumps followed by diamagnetic relaxations, in good agreement with magnetization data[9].

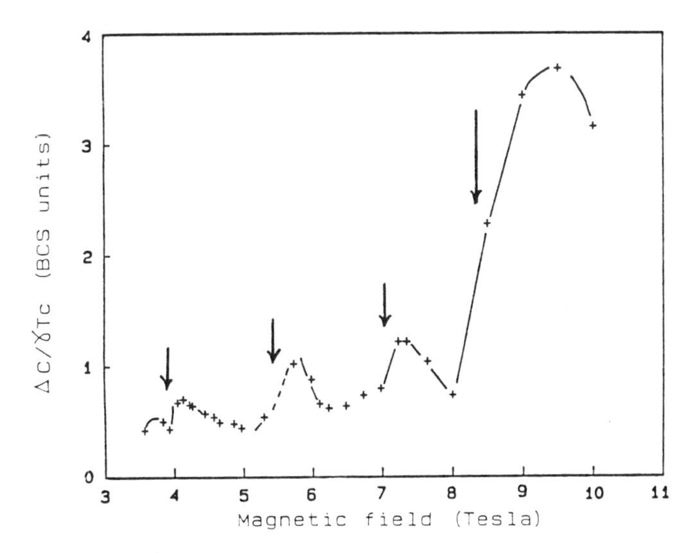

Fig. 2. Oscillatory behavior of the normalized specific heat jump at the metal-FISDW transition, plotted as a function of the magnetic field. Notice the jumps of the jump (arrows), and at high field the strong change in the behavior.

In the low-field range the weak first order transitions are fully reversible, but above 8 teslas we observe the strong hysteresis associated with the formation of a metastable SDW phase[30]. Finally, the direct measurement of the specific heat as a function of the field showed no variation of the density of state at the Fermi level, for the normal metallic state below the FISDW threshold field[10].

EFFECTS OF THE COOLING RATE

Figures 3 and 4 illustrate, for the $(TMTSF)_2ClO_4$ compound, the two main effects of the cooling rate. The first effect concerns the shift of the threshold field, as displayed on Fig. 3: the threshold fields, corresponding to quenched and relaxed states, are represented by solid and dashed arrows, respectively. This shift is responsible for the discrepancy we observed between the threshold fields, as determined by specific heat[10] or by magnetocaloric effect measurements [31,32]. In fact, Figure 3 leads to a first conclusion: in order to construct a coherent phase diagram, all the experiments are to be performed for the same cooling rate, at the same time. If not, as is clearly shown on Fig. 3, distinct threshold fields can be wrongly attributed to the very same transition line.

The second effect of the cooling rate is far more spectacular: it concerns the arborescence of the phase diagram, that is observed in the very well ordered samples[32]. Figure 4 shows the splitting of the transition lines observed by $(\partial M/\partial T)_B$ measurements. It proceeds by dichotomy, in an iterative process. A two-stage splitting is exemplified by the $(\partial M/\partial T)_B$ anomaly number 3: this anomaly is narrow at 625 mK (a), broader as the temperature is lowered to 545 mK (b), it begins to separate at 470 mK (c), and is fully split at 425 mK (d). The first anomaly on Fig. 4 gives an example of a three-stage process: it is already split in (c), and it gives rise in (d) to at least four transition lines. The result is a phase diagram with more than twenty phases within the 4 to 6 tesla range.

We have recently developed an experimental device to measure simultaneously the specific heat C_B and the thermal coefficient of the magnetization, $(\partial M/\partial T)_B$, as functions of the magnetic field. Our results allow us to draw a fully consistent picture of the cooling effect[33]: at high temperature (say, 0.7 K), there is a unique,coherent,set of transition lines. They give rise to a unique set of anomalies for the Hall effect, the magnetization, the specific heat, and the magnetocaloric effect. As the temperature is decreased, these lines split into a new set of lines (arrows on Fig. 4), leading to a treelike phase diagram. Correspondingly, new plateaus appear on the Hall effect curves[21].

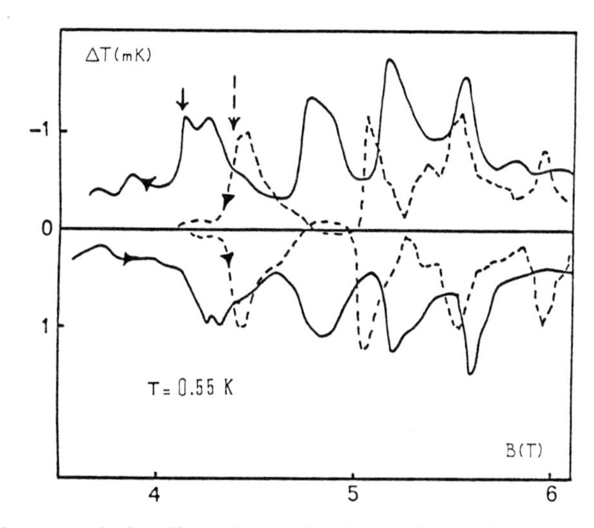

Fig. 3. Magnetocaloric effect observed at increasing and decreasing fields, for two cooling rates: 2 K/hour (dashed line), and about 1 K/minute (solid line). It is worth noting the strong shift of the threshold field for the two series of anomalies.

This behavior can be understood in the framework of the quantized nesting model[34]: the integer quantization results from an interference effect between the SDW periodicity and the cyclotron period[11-15]. The wave vector $\mathbf{Q}=(Q_{//}, \pi/b)$ of the SDW obeys an integer condition: the Fermi level must lay in between one of the gaps. These gaps are separated from the Peierls gap by a multiple of the cyclotron energy $\hbar\omega_c$, and the longitudinal component of the wave vector reads:

$$Q_{//} = 2k_F + n^*\hbar\omega_c/v_F$$

where n^* is a whole number, k_F the Fermi wave vector and v_F the Fermi velocity.

When the lattice periodicity is taken into account, higher-order gaps are opened near the Fermi level of the metal. The condition,that the Fermi level lies in the middle of one of these gaps, now reads[34]:

$$m^*Q_{//} = m^*2k_F + n^*h\omega_c/v_F$$

where m is an odd integer number. As is indicated by the evolution of the specific heat jump at the metal-FISDW transition[28], the situation of the Bechgaard salts corresponds to the intermediate coupling regime. These gaps are thus large enough to stabilize new subphases, corresponding to a rational quantization of the wave vector. The treelike phase diagram predicted by such a model, with a self-similar character, is in a good qualitative agreement with the phase diagram we obtain from the $(\partial M/\partial T)_B$ measurements in the $(TMTSF)_2ClO_4$ salt.

CONCLUSION

To summarize, the first result of these calorimetric investigations on the MFISDW states concerns the complex second-order line separating the metal from the magnetic subphases. This line displays a series of multicritical points, corresponding to partial reentrances of the metal into the SDW phases. The specific heat jump oscillates along this line, and at about 10 teslas its magnitude deviates strongly from the BCS prediction.

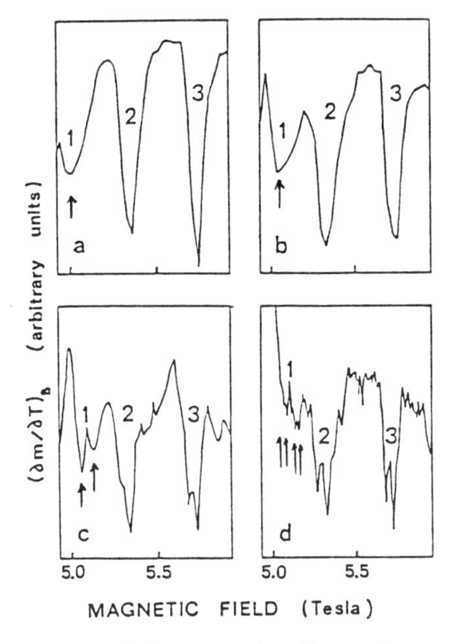

Fig. 4. Splitting process of the transition lines, observed by $(\partial M/\partial T)_B$ measurements in a well-ordered (1.5 K/hour) 2 mg single crystal of $(TMTSF)_2ClO_4$. Figures a, b, c and d correspond respectively to 625, 545, 470 and 425 mK.

Secondly, we observe two major effects of the cooling rate. They lead to two kinds of phase diagrams, according to the sample preparation: either in the quenched or in the relaxed state. The threshold fields between SDW subphases are shifted toward high fields, as the cooling rate below the anion ordering temperature is lowered. Moreover, a qualitatively new diagram arises when the sample is prepared in a very well ordered state: the transition lines split into sublines, in an iterative process. The resulting phase diagram has a treelike structure, with a self-similar character.

We would like to thank M. Héritier for number of fruitful discussions.

REFERENCES

1. For a review of experimental and theoretical results on the Bechgaard salts, see: Proc. of ICSM'88, Santa Fe, Syn. Metals 27-29 (1988-89)
2. M. Ribault, in Low Dimensional Conductors and Superconductors, Eds D. Jérome and L.G. Caron (NATO ASI Plenum Press) 155 (1986) 199 (experimental review)
3. P.M. Chaikin, M.Ya Azbel, M.J. Naughton, R.V. Chamberlin, X. Yan, S. Hsu and L.Y. Chiang, Syn. Metals 27 (1988) B163 (recent review, focussing on the high-field behavior)
4. J.F. Kwak, J.E. Shirber, R.L Greene and E.M. Engler, Phys. Rev. Lett. 46 (1981) 1296
5. K. Kajimura, H. Tokumoto, K. Murata, T. Ukashi, H. Anzai, T. Ishiguro and G. Saito, J. Phys. (Paris) 44 (1983) C3-1059
6. P.M. Chaikin, Mu-Yong Choi, J.F. Kwak, J.S. Brooks, K.P. Martin, M.J. Naughton, E.M. Engler and R.L. Greene, Phys. Rev. Lett. 51 (1983) 2333
7. M. Ribault, D. Jérome, J. Tuchendler, C. Weyl and K. Bechgaard, J. Phys. (Paris) Lett. 44 (1983) L953;
   B. Piveteau, L. Brossard, F. Creuzet, D. Jérome, R.C. Lacoe, A. Moradpour and M. Ribault, J. Phys. C19 (1986) 4483
8. H. Schwenk, S.S.P. Parkin, R. Schumaker, R.L. Greene and D. Schweitzer, Phys. Rev. Lett. 56 (1986) 667
9. M. J. Naughton, J.S. Brooks, L.Y. Chiang, R.V. Chamberlain and P.M.Chaikin, Phys. Rev. Lett. 55 (1985) 969
10. F. Pesty, P. Garoche and K. Bechgaard, Phys. Rev. Lett. 55 (1985) 2495
11. L. P. Gor'kov and Lebed', J. Phys. (Paris) Lett. 45 (1984) L-433
12. M. Héritier, G. Montambaux and P. Lederer, J. Phys. (Paris) Lett. 45 (1984) L943, and ibid 46 (1985) L831
13. K. Yamaji, Syn. Metals 13 (1986) 29
14. M.Ya. Azbel, P. Bak and P.M. Chaikin, Phys. Rev. A34 (1986) 1392
15. A. Virosztek, L. Chen and K. Maki, Phys. Rev. B34 (1986) 3371
16. J.P. Ulmet, A. Khmou, P. Auban and L. Bachere, Solid State Commun. 58 (1986) 753
17. T. Osada, N. Miura and G. Saito, Physica 143B (1986) 403
18. M.J. Naughton, R.V. Chamberlin, X. Yan, S.-Y. Hsu, L.Y. Chiang, M.Ya. Azbel and P.M. Chaikin, Phys. Rev. Lett. 61 (1988) 621
19. V.M. Yakovenko, Zh. Eksp. Teor. Fiz. 93 (1987) 627 (Sov. Phys. JETP 66 (1987) 355), and Phys. Rev. Lett. 61 (1988) 2276
20. P. Garoche, R. Brusetti and K. Bechgaard, Phys. Rev. Lett. 49 (1982) 1346
21. M. Ribault, Mol. Cryst. Liq. Cryst 119 (1985) 91
22. P. Garoche, R. Brusetti D. Jérome and K. Bechgaard, J. Phys (Paris) Lett. 43 (1982) L147
23. R. Brusetti, P. Garoche and K. Bechgaard, J. Phys. C16 (1983) 3535
24. P. Garoche and F. Pesty, J. Mag. Mag. Mat. 54-57 (1986) 1418
25. P.F. Sullivan and G. Seidel, Phys. Rev. 173 (1968) 679; F. Pesty and P. Garoche, to be published
26. F. Pesty, G. Faini and P. Garoche, J. Appl. Phys. 63 (1988) 3061
27. J.S. Brooks, N.A. Fortune, M.J. Graf, P.M. Chaikin, L.Y. Chiang and S. Hsu, Syn. Metal 27 (1988) B29
28. F. Pesty, P. Garoche and M. Héritier, Proc. of the MRS 1989 Fall Meeting, Boston, USA (Nov. 1989)
29. G. Montambaux, J. Phys. C20 (1987) L327
30. M. Ribault, F. Pesty, L. Brossard, B. Piveteau, P. Garoche, J. Cooper, S. Tomic, A. Moradpour and K. Bechgaard, Physica 143B (1986) 393
31. G. Faini, F. Pesty and P. Garoche, J. Appl. Phys. 63 (1988) 3058
32. G. Faini, F. Pesty and P. Garoche, J. Phys. (Paris) 49 (1988) C8-807
33. F. Pesty, P. Garoche and M. Héritier, Proc. of Int. Symp. on Phys. and Chem. of Organic Supercond. (ISSP-ISOS), Tokyo, Japan (August 1989)
34. M. Héritier, in Low Dimensional Conductors and Superconductors, Eds D. Jérome and L.G. Caron (NATO ASI Plenum Press) 155 (1986) 243

NONLINEAR ELECTRICAL TRANSPORT EFFECTS IN THE SPIN-DENSITY WAVE STATE OF THE ORGANIC CONDUCTORS $(TMTSF)_2X$

Silvia Tomić

Institute of Physics of the University
POB 304, 41001 Zagreb, Yugoslavia
Laboratoire de Physique des Solides
Bât 510, 91405 Orsay, France

INTRODUCTION

Various highly anisotropic organic and inorganic materials are ideal systems to study collective transport phenomena. Depending on the material and the applied pressure, there is usually a phase transition to a superconducting (SC), a charge-density wave (CDW), or a spin-density wave (SDW) ground state at low temperatures[1]. Translational modes of these ground states couple to an applied electric field, leading so to a collective transport. While novel phenomena associated with the dynamics of CDW are well established by a broad range of experimental techniques, and also theoretically understood, not much is known about the dynamics of the spin-density wave ground state. According to Lee, Rice and Anderson, similar behaviour might be expected for a SDW state, because collective transport should not depend on the nature of the underlying interaction mechanism: electron-electron, rather than electron-phonon, as in the CDW case[2]. The SDW model systems are some members of the $(TMTSF)_2X$ family, in which the SDW nature of the ground state (critical temperatures are about 10K) has been firmly established by various magnetic measurements, like nuclear magnetic resonance, antiferromagnetic resonance, static susceptibility and electron spin resonance[1,3,4,5].

The purpose of this paper is twofold. First, to discuss some essential theoretical aspects of the SDW dynamics and second, to review recent results of electric-field-dependent experiments, performed in the spin-density wave ground state of two organic conductors $(TMTSF)_2PF_6$ and $(TMTSF)_2NO_3$.

THEORETICAL BACKGROUND

Spin-density wave ground state, as well as the charge-density wave one, is established below a well-defined transition temperature (T_c), which is related to a single-particle gap (2Δ) through the BCS expression[1]

$$2\Delta / k_B T_c \simeq 3.5 . \quad (1)$$

Important feature of these ground states is the modulation of the charge and spin in the CDW and the SDW state, respectively. In the former state, the charge density is given by

Lower-Dimensional Systems and Molecular Electronics
Edited by R.M. Metzger *et al.*, Plenum Press, New York, 1991

$$\Delta\rho = \rho_0 + \rho_1 \cos(2k_F x + \Phi) , \tag{2}$$

where ρ_0 is the total electronic charge and ρ_1 and Φ are the amplitude and phase of the condensate. In the SDW state, the spin density is given by

$$\Delta s = s_0 + s_1 \cos(2k_F x + \Phi) , \tag{3}$$

where s_0 is the total spin and, s_1 and Φ the amplitude and phase of the SDW condensate. The spin density (3) is accompanied with a periodic charge density, which is for a given spin direction γ :

$$\Delta\rho_\gamma = \frac{1}{2}\rho_0 + \rho_1 \cos(2k_F x + \frac{\pi}{2}\gamma + \Phi) , \tag{4}$$

where $\gamma = \pm 1$ for spin-up and spin-down bands, respectively.

If a finite external electric field is applied, the collective mode translates with a finite velocity, and carries a current

$$j_{CDW} = \frac{\rho_0 \rho_c e}{2k_F} \frac{d\Phi}{dt} , \tag{5}$$

where the collective density ρ_c equals zero at T_c and unity at T=0K[6]. As the renormalized total charge $\rho_0 \rho_c$ couples to the electric field (and not the amplitude ρ_1), it follows that the SDW translational mode can carry a current as well as the CDW does. Moreover, such a motion of the collective mode would lead to superconductivity, as originally proposed by Fröhlich[7]. However, in real materials this motion is restricted by various pinning mechanisms, which, basically, depend on the modulation amplitude of charge for both CDW and SDW mode. For the latter, according to the relation (4), in the approximation up to the first order in the electron-impurity interaction, the pinning should be negligibly weak, because the total density is constant

$$\Delta\rho = \rho_0 . \tag{6}$$

However, recent theories, developed independently by several authors, suggest that there is a broad range of SDW pinning energies which could easily attain the values found for a CDW, or could be even larger[8,9,10]. The starting idea is that nonmagnetic impurities will pin the SDW, by inducing a distortion of the total electron density near the impurity sites. This should happen because the up- and down-spin components of the charge density deform differently, to yield the net local charge. Basically, the problem is reduced to the coupling of the nonmagnetic impurities to the second-order harmonic CDW that coexist with the SDW. In the framework of the phenomenological model by Maki and Virosztek (ref.9.), the density oscillation, up to the second order, is given by

$$\Delta\rho = \rho_0 + \pi^2 N_0 \Delta \tanh \frac{\Delta(T)}{2T} \cos(4k_F x + 2\Phi) \tag{7}$$

and the coupling between impurities and the sliding SDW mode is expressed by

$$H = (\frac{\pi}{2} N_0 V)^2 \Delta(T) \tanh \frac{\Delta(T)}{2T} \sum_i \cos(4k_F x + 2\Phi) , \tag{8}$$

where N_0, $\Delta(T)$ and V are the electronic density of states, the SDW gap and the impurity potential, respectively. The electric field, required to depin the SDW, can be estimated by simply comparing the total electric field energy with the pinning energy per unit volume. In the strong-pinning limit, the SDW threshold field (E_T) is given by

$$E_T(0) = \frac{2k_F}{e} \frac{n_i}{n_e} (\pi N_0 V)^2 \Delta(0) , \tag{9}$$

where e, n_e and n_i are the electronic charge, the electron concentration and the impurity concentration, respectively. The temperature dependence is

$$\frac{E_T(T)}{E_T(0)} = \frac{\Delta(T)}{\Delta(0)} \tanh \frac{\Delta(T)}{2T} \frac{\rho_0}{\rho_c(T)} . \tag{10}$$

From this equation, $E_T(T)/E_T(0)$ increases monotonically from unity at T=0K to 1.33 at $T=T_c$. This behaviour differs from one observed in the CDW materials, where E_T exhibits a divergence at $T=T_c$, and a minimum slightly below T_c, which results from an increase in E_T at low temperatures, due to phase fluctuations.

Hence, one should expect to observe similar properties of the SDW current-carrying state to ones of the CDW: nonlinear current-voltage characteristics, accompanied by broad and narrow band noise, with sharp threshold fields, frequency-dependent conductivity, interference effects between the ac voltage generated in the sample, and an external rf field, hysteresis and memory effects etc.

EXPERIMENTAL RESULTS AND DISCUSSION

In what follows, we will review recent experiments performed to check one of the properties of the SDW current-carrying state: dc electric conductivity which increases above a finite threshold field[11,12,13]. We have investigated two materials: $(TMTSF)_2 NO_3$ and $(TMTSF)_2 PF_6$, with the SDW transition temperatures of 11 and 11.5K, and the SDW single-particle gaps of about 16 and 32K, respectively. As far as the frequency-dependent conductivity is concerned, the results obtained by G.Grüner et al., clearly show the existence of a collective mode in the SDW state of $(TMTSF)_2 PF_6$, with a pinning frequency of about 30GHz, and with a relaxation time and effective mass comparable with the parameters which characterize the metallic state[14]. Here, we will not comment on experiments performed in the last ten years, which, generally, have failed to identify nonlinear effects because of technical problems related to contacts and heating effects (see ref.11. and refs. therein).

The electric-field-dependent conductivity observed in $(TMTSF)_2 NO_3$ and $(TMTSF)_2 PF_6$ is shown in Fig.1. In the normal state, i.e. at temperatures above the SDW transition, the conductivity stays constant in the whole field range measured (up to about 0.7V/cm). However, in the SDW state, the conductivity is constant, until a threshold field is reached, above which the conductivity increases. Values of the threshold field, measured at 4.2K are 40 and 7.5 mV/cm for the NO_3 and PF_6 compound, respectively. The sharpness of the threshold field was checked by continuous current measurements (see insert of Fig.1.b.) and by dynamic resistance measurements (Fig.2.). The value of the threshold field is temperature-independent below $T_c/2$, and, as observed for the PF_6 compound, increases close to T_c (Fig.1.b.).

Furthermore, an amount of impurities, which is large enough to broaden the SDW transition, but does not affect T_c, strongly increases E_T: E_T=140mV/cm at 1.7K for the PF_6 compound. In addition, the excess conductivity is smaller in samples with a lower resistivity ratio $\rho(RT)/\rho(min)$, (Fig.3.).

Finally, the excess current, associated with the field-dependent conductivity (eq.(5)), can be expressed as a function of electric field (Fig.4.)

$$j_{ex} = (\sigma - \sigma_0)E , \tag{11}$$

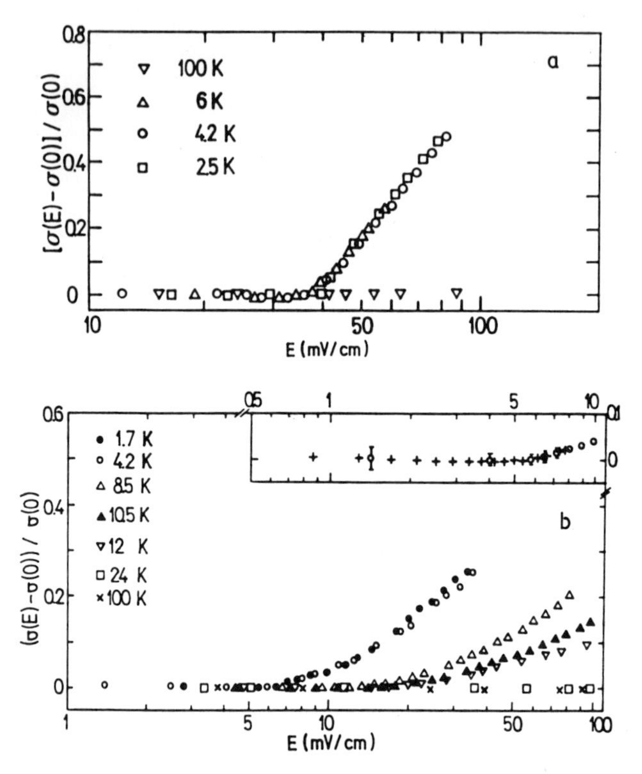

Fig.1. Non-ohmic conductivity $\sigma(E)-\sigma(E\to 0)/\sigma(E\to 0)$ versus logarithm of electric field (E) at various temperatures for (a) $(TMTSF)_2\,NO_3$ and (b) $(TMTSF)_2\,PF_6$.

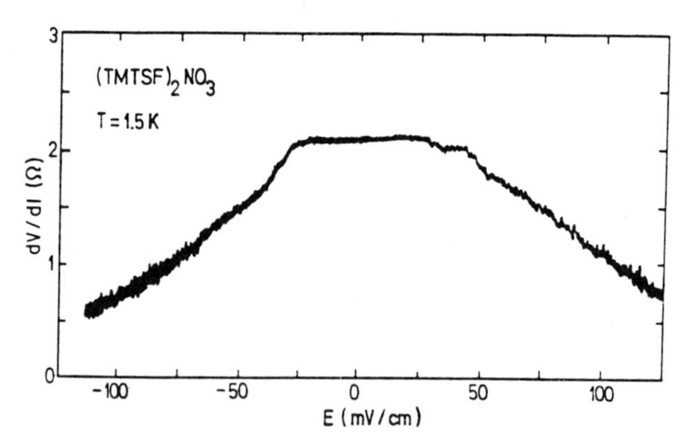

Fig.2. Dynamic resistance (dV/dI) versus electric field (E) for $(TMTSF)_2\,NO_3$.

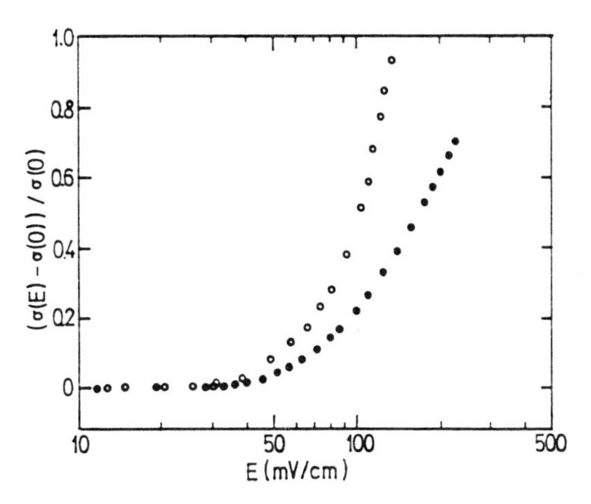

Fig.3. Non-ohmic conductivity for two samples of $(TMTSF)_2NO_3$ at 1.5K with different resistivity ratio (rr). Open and close circles for rr$\simeq$ 170 and 60, respectively.

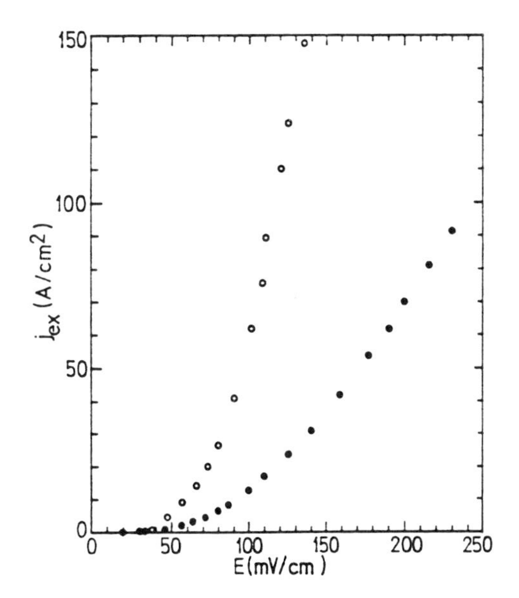

Fig.4. Excess current (j_{ex}) versus electric field (E) for two samples of $(TMTSF)_2NO_3$ with different resistivity ratio (rr). Open and close circles for rr$\simeq$170 and 60, respectively.

where $\sigma=\sigma(E)$ and σ_0 is the low-field ohmic conductivity.

The observation of non-ohmic conductivity, below the temperature which corresponds to the onset of three-dimensional SDW order, strongly suggests that the nonlinearity is associated with the appearance of SDW. Indeed, it is difficult to explain the low threshold fields, at which deviations from Ohm's law are observed by models based on a single-particle picture. First, we note that the threshold fields are far too low, for Zener breakdown to be relevant.

Second mechanism, which must be considered, is related to so-called hot-electron effects. If the inelastic mean free path is sufficiently long, an electron may attain a large kinetic energy, equal to the voltage drop across the distance between two scatterings. When this energy becomes comparable to the single-particle gap, electrons can be excited across the gap, and such a relaxation process is a non-ohmic one. For $(TMTSF)_2X$ salts, the electron mean free path at low temperatures is estimated to be about 14000Å. Thus, the maximum energy from the electric field is 0.06K, what is much smaller than $k_B T \simeq 10K$. Therefore, electron heating effects can be also ruled out as the source of substantial nonlinearity.

The observed values of threshold fields are close to ones predicted by theoretical model of Maki and Virosztek (ref.9.). According to the expression (9) and taking $N_0 V=0.1$, $\Delta(0) \simeq 10^{-3}$ eV and n_i/n_e a few ppm, one gets $E_T(0) \simeq 10mV/cm$. Furthermore, the temperature dependence of E_T is also qualitatively in accordance to a theoretical prediction (eq.(10)). However, we find a stronger increase of E_T above $T_c/2$, namely $E_T(T_c)/E_T(1.7K) \simeq 2.5$, if compared to a theoretically expected value of 1.33. Note that the latter value is calculated for a strong-pinning limit.

CONCLUSION

Two essential features of the sliding CDW mode seem to be rather well established for the SDW mode, too. Non-ohmic electrical transport sets in above a finite threshold field, and a collective mode contribution is evident at finite frequencies in the spin-density wave ground state. These features are also in qualitative agreement with theoretical models. However, important differences are observed, like the temperature dependence of the threshold field and the relaxation time and effective mass of the collective mode. Further experiments are required to establish details of this behaviour, as well as to search for other nonlinear effects, like broad and narrow-band noise, hysteresis and memory effects etc.

ACKNOWLEDGMENTS

We wish to acknowledge constant and close collaboration with D.Jérome and J.R.Cooper and fruitful discussions with K.Maki and J.Voit.

REFERENCES

1. See, for example, "Low-Dimensional Conductors and Superconductors", ed. by D.Jérome and L.G.Caron, NATO ASI, Series B, Vol.155, Plenum, New York (1987).
2. P.A.Lee, T.M.Rice, and P.W.Anderson, Solid State Commun.14:703 (1974).
3. K.Mortensen, Y.Tomkiewicz, T.D.Schultz, and E.M.Engler, Phys.Rev.Lett. 46:1234 (1981).
4. W.M.Walsh, F.Wudl, E.Aharon-Shalom, L.W.Rupp, J.M.Vandenberg, K.Andres, and J.B.Torrance, Phys.Rev.Lett.49:885 (1982).
5. P.Baillargeon, C.Bourbonnais, S.Tomić, P.Vaca, and C.Coulon, Synthetic Metals 27:B83 (1988).
6. P.A.Lee and T.M.Rice, Phys.Rev.B19:3970 (1979).
7. H.Fröhlich, Proc.Roy.Soc.London A223:296 (1954).
8. P.F.Tua and J.Ruvalds, Phys.Rev.B32:4660 (1985).
9. K.Maki and A.Virosztek, preprint (1988).
10. I.Tütto and A.Zawadowski, Phys.Rev.Lett.60:1442 (1988).
11. S.Tomić, J.R.Cooper, D.Jérome, and K.Bechgaard, Phys.Rev.Lett.62:462 (1989).
12. S.Tomić, J.R.Cooper, W.Kang and D.Jérome, to be submitted.
13. S.Tomić and J.R.Cooper, unpublished results.
14. G.Grüner, Synthetic Metals 29:F453 (1989).

TWO NEW PHASES OF $(ET)_2I_3$ ($\beta_{d'}$ and λ_d)

Li Lu, Bei-hai Ma, Hong-min Duan, Shu-yuan Lin, Dian-lin Zhang
Xiao-hong Wang*, You-xin Yao*, and Daoben Zhu*

Institutes of Physics and *Chemistry
Chinese Academy of Sciences
Beijing, CHINA

INTRODUCTION

We present our study of the new phases, β_d'- and λ_d-(BEDT-TTF)$_2$I$_3$, which were recently synthesized in our laboratories by a diffusion method. Both of the new organic crystals show a phase transition in their transport properties, and further investigation indicates that the transition is related to the disorder, that possibly exists in these crystals. However, the mechanism of the two phase transitions seems not to be the same.

The radical cation salts (BEDT-TTF)$_2$I$_3$ have drawn much attention, since one of them was found to be an organic superconductor. This series of salts can appear in many modifications, known as the α-, β-, θ-, κ-phases. Under ambient pressure, the α-(BEDT-TTF)$_2$I$_3$ phase undergoes a metal-insulator phase transition at 135 K [1], while the β-, θ-, and κ- (BEDT-TTF)$_2$I$_3$ become superconductors below ~ 1.3 K, 3.6 K and 3.6 K, respectively [2-5]. After some particular pressure and temperature processing, the β-phase shows superconductivity at ambient pressure, up to a temperature as high as ~ 8.1 K [6]. Unlike these phases, which are usually synthesized by an electrochemical method, two new phases, called β_d'- and λ_d-(BEDT-TTF)$_2$I$_3$, were synthesized recently by D. Zhu and co-workers by a diffusion method [7,8].

The β_d'-phase has a distorted hexagonal plate-shaped morphology, with cleavage angles of 109°, 121°, and 130°, respectively. Its crystal structure is very similar to that of the β-phase, but the arrangement of I_3^- chains is a little different (Fig. 1) [7,9].

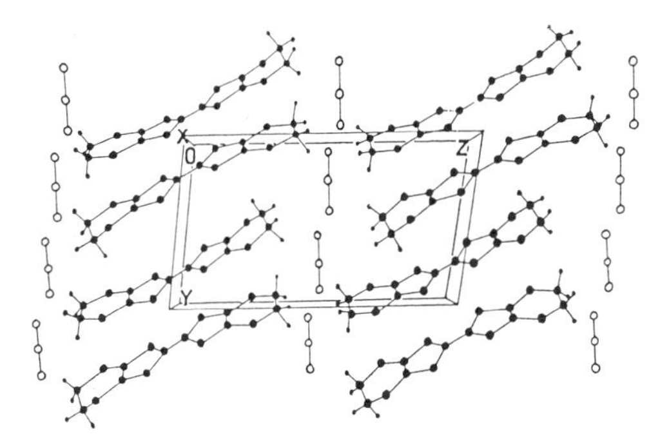

Fig. 1. Crystal structure of β_d'-(BEDT-TTF)$_2$I$_3$. The figure shows full occupancy of the iodine sites; in reality, only half the sites are randomly occupied.

Lower-Dimensional Systems and Molecular Electronics
Edited by R.M. Metzger *et al.*, Plenum Press, New York, 1991

From the room-temperature X-ray diffraction data, the unit cell of the β_d'-phase should contain two complete I_3^- chains, if all the iodine site occupation factors were unity, but, actually, the site occupation factors of I1, I2, and I3 in each I_3^- chain are only 0.5, 0.34, and 0.66, respectively (where I1 represents the middle site of the chain). Therefore, a total of three iodine atoms occupy, randomly, six sites in the unit cell.

In contrast to the hexagonal-shaped β_d'-phase, another new phase, called the λ_d-phase, is also found in the products of diffusion synthesis [7]. The λ_d-phase is a rectangular plate-shaped crystal. The determination of the detailed structure is still in progress. Preliminary results show that the lattice parameters of the λ_d-phase are similar to that of the θ-phase [3,4] with Z = 4. However, the new phase is characterized by its peculiar transport behavior (see below). Table 1 shows the structure parameters of the β_d'- and λ_d-phases, in comparison with the other isomers. It can be seen that the reduced unit cell volume in the λ_d-phase is a little larger than that of the θ-phase, indicating either the existence of excess iodine atoms, or the misarrangement of the iodine chain. In the following, we will present experimental results for the two new phases.

Table 1. Structural parameters of the $(BEDT\text{-}TTF)_2I_3$ family.

	a/Å	b/Å	c/Å	α/deg.	β /deg.	γ /deg.	V(Å^3)/Z
α	10.785	9.172	17.39	82.08	96.92	95.62	845.15
β	6.609	9.083	15.267	85.63	95.62	70.22	852.2
γ	16.387	8.466	12.832		108.56		844
θ	10.676	9.928	34.220		98.39		846.5
β_d'	6.466	9.257	15.273	98.28	89.75	108.91	854.9
λ_d	10.686	9.947	34.31		98.51		901.7

β_d'-PHASE

We have studied the resistance and thermopower behavior of the β_d'-phase [9]. From the temperature-dependent resistance curve, one can see clearly a metal-semiconductor phase transition at about 140 K (Fig. 2), whereas on the temperature-dependent thermopower curve, only a very small (but distinct) kink appears at the same temperature (Fig. 3). To interpret these seemingly conflicting phenomena, a two-energy band model was used [9]. In this model, the conductivity is due to a combination of the two bands, and the thermopower is due to a competition of the two bands. From the room-temperature X-ray diffraction, which shows the iodine atoms arranged randomly, we speculate that, at room temperature, there should be an energy gap at the Fermi surface, but that the random arrangement of iodine atoms smears the gap. Thus, the crystal stays metallic at room temperature.

However, there is evidence that, when temperature goes down to ~140 K, the iodine atoms undergo a disorder-to-order transition, and hence the smearing of the energy gap is removed, resulting in the metal-to-semiconductor phase transition. In fact, from a Raman scattering investigation [10], the structure of the 121 cm^{-1} peak (4579 Å excitation) shows a relatively rapid change near ~ 140 K, as shown in Fig. 4. At room temperature the 121 cm^{-1} peak is unusually wide, compared with its isomers, and even has a multiplet structure, but at 100 K, the peak becomes sharp and clean. Comparing with the Raman investigation of other materials containing similar I_3^- chains [11-13], the 121 cm^{-1} peak is assigned to the vibrations of the I_3^- chain. Therefore, the changes in the 121 cm^{-1} peak (Fig. 4) may be due to the ordering of the iodine atoms.

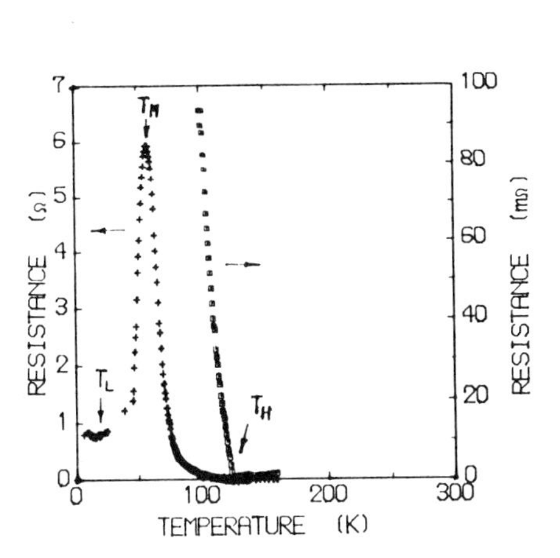

Fig. 2. The temperature-dependent resistance of β_d'-phase shows a metal-semiconductor transition at 140 K and a maximum at ~60 K. Estimated energy gap ≈ 0.07 eV.

Fig. 3. The anisotropic thermopower of β_d'-(BEDT-TTF)$_2$I$_3$ in the **ab**-plane. A small kink in S_a can be identified at ~ 150 K.

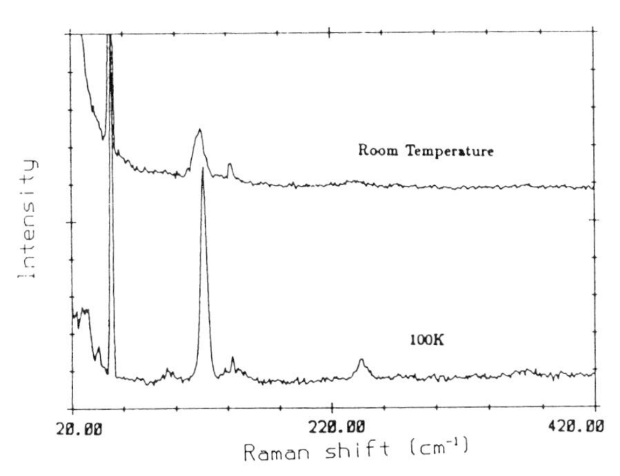

Fig. 4. Raman spectra of β_d'- (BEDT-TTF)$_2$I$_3$ crystal (with 4579 Å excitation).

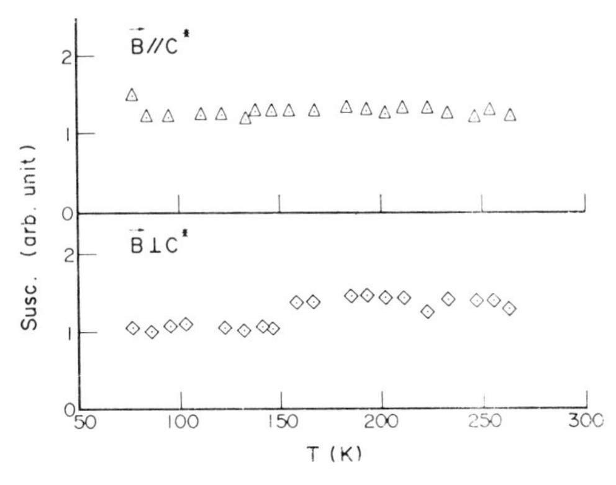

Fig. 5. The temperature dependence of the spin susceptibility, with the static field parallel and perpendicular to the **ab**-plane.

Fig. 5 shows the temperature-dependent EPR spin susceptibility of the β_d'-phase [14]. The susceptibility is related to the density of (spin) free electrons. For $\mathbf{B}\perp\mathbf{c}$, the susceptibility has a sudden decrease, when temperature reaches ~ 150 K, which may correspond to a sudden shrinkage of the Fermi surface at that temperature. This phenomenon is also consistent with our assumption, that the ordering of iodine atoms removes the impurity levels, and thus the energy gap recurs.

λ_d-PHASE

The λ_d-phase has a relatively higher room temperature conductivity [8]. Estimated from our measurement, σ_{RT} falls in the range of 5 x 10^2 ~ 5 x 10^3 S cm^{-1} for all samples. Fig. 6 shows the temperature-dependent resistance curve, and Fig. 7 shows the influence of thermal cycles on its transport behavior. It can be seen that: (i) there exists a transition temperature, below which the resistivity increases almost linearly; (ii) it is somewhat unstable in both the transition temperature and the behavior (slope), above and below the transition. In contrast to the resistance measurement, the thermopower of the λ_d-phase is rather stable [8]. Fig. 8 shows the temperature-dependent thermopower of the sample. These particular

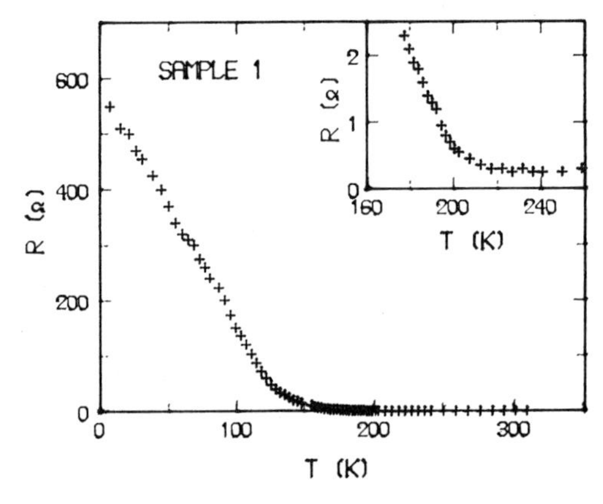

Fig. 6. Temperature dependence of the resistance of λ_d- (BEDT- TTF)$_2$I$_3$ in the **ab**-plane.

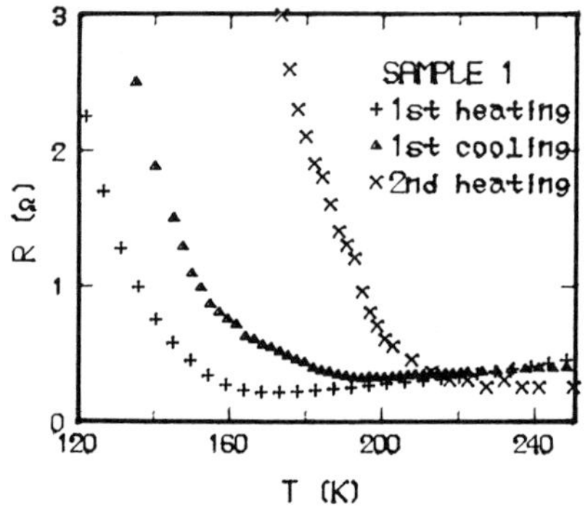

Fig. 7. Influence of thermal cycle on the resistance of λ_d- (BEDT- TTF)$_2$I$_3$

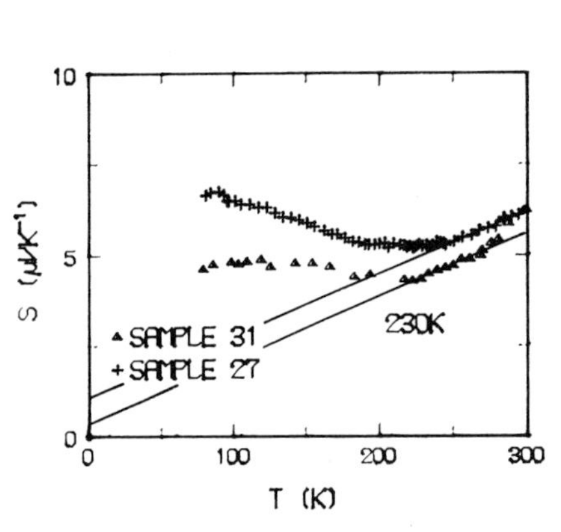

Fig. 8. The temperature-dependent thermopower of λ_d-(BEDT-TTF)$_2$I$_3$ in **ab**-plane.

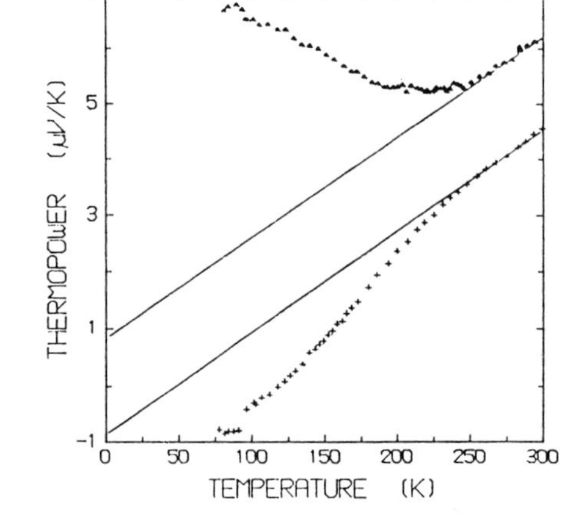

Fig. 9. Annealing changes the thermopower of λ_d-(BEDT-TTF)$_2$I$_3$ (lower curve).

Table 2. The room temperature conductivity σ_{RT}, thermopower S_{RT}, and average transfer integral t_S of β_d'- and λ_d- $(BEDT\text{-}TTF)_2I_3$ and other isomers. All data are for directions in the **ab**-plane.

phase	$\sigma_{RT}(S\ cm^{-1})$	$S_{RT}(\mu V\ K^{-1})$	$t_S(eV)$
α	60 ~ 250	35	0.2
β	20 ~ 70	15 ~ 25	0.12
κ	40 ~ 150		
θ	30 ~ 100		
β_d'	100 ~ 1000	10 ~ 30	0.12
λ_d	500 ~ 5000	4 ~ 6	0.6

transport properties indicate, that the transition may not originate from a band gap formation; the carrier scattering process in the sample may play an important role in the resistance measurement, and cause the steep rise of resistance at low temperature.

Since the resistive behavior is sensitive to thermal cycling, the disorder might easily be changed by a slight heat treatment. We annealed a sample at 80 - 90°C for 24 hours, and found large changes in the thermopower at low temperatures (Fig. 9). At high temperature the value of S_{RT} and the slope of the S-T curve are nearly the same as before annealing, indicating that the band structure above the transition temperature is unchanged by annealing. Below the transition temperature, the resistance decreases, and the thermopower deviates from a straight line in the direction opposite to that before annealing (Fig. 9). The annealing may change the disorder, and influence the transport properties below the phase transition.

CONCLUSION

From the results obtained thus far, we speculate that, in both the β_d'- and λ_d-phase, disorder may play an important role. Table 2 lists the room-temperature conductivity, thermopower, and average transfer integral of the two new phases, and the other known phases. The unusually large transfer integral t_S of the λ_d-phase (estimated from the slope of the S-T curve) does not conflict with its high conductivity, as shown in Fig. 10. However, further work is needed, especially for the λ_d-phase.

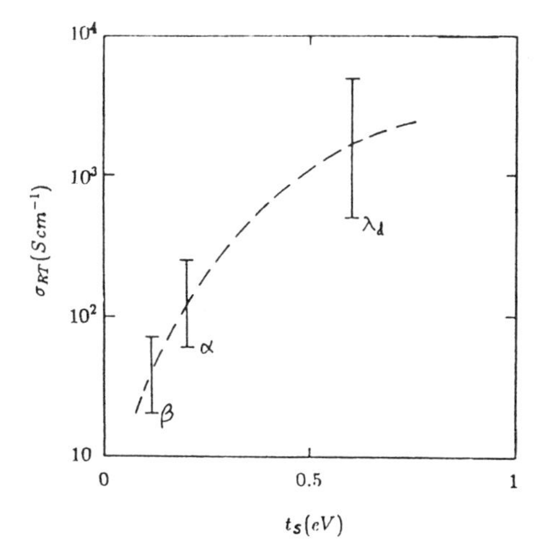

Fig.10. Relation between the transfer integral t_S and the room-temperature conductivity σ_{RT}.

REFERENCES

[1] K. Bender, I. Hennig, D. Schweitzer, K. Dietz, H. Endres, and H. J. Keller, Mol. Cryst. Liq. Cryst. 108: 359 (1984).
[2] E. B. Yagubskii, I. F. Shchegolev, V. N. Laukhin, and L. I. Buravov, JETP Lett. 39: 12 (1984).
[3] H. Kobayashi, R. Kato, A. Kobayashi, Y. Nishio, K. Kajita, and W. Sasaki, Chem. Lett. 789, 833 (1986).
[4] A. Kobayashi, R. Kato, H. Kobayashi, S. Moriyama, Y. Nishio, K. Kajita, and W. Sasaki, Chem. Lett. 2017 (1986)
[5] R. Kato, H. Kobayashi, A. Kobayashi, S. Moriyama, Y. Nishio, K. Kajita, and W. Sasaki, Chem. Lett. 507 (1987).
[6] F. Creuzet, D. Jérome, D. Schweitzer, and H. J. Keller, Europhys. Lett. 1: 461 (1986).
[7] Min-xie Qian, Xiao-hong Wang, Yu-lan Zhu, Daoben Zhu, Li Lu, Bei-hai Ma, Hong-min Duan, and Dian-lin Zhang, Synth. Metals 27: A227 (1988).
[8] Li Lu, Bei-hai Ma, Hong-min Duan, Shu-yuan Lin, Dian-lin Zhang, Xiao-hong Wang, and Daoben Zhu, Synth. Metals 27: A311 (1988).
[9] Li Lu, Bei-hai Ma, Hong-min Duan, Shu-yuan Lin, Dian-lin Zhang, You-xin Yao, Min-xie Qian, Yu-lan Zhu, Xiao-hong Wang, and Daoben Zhu, Synth. Metals 31: 37 (1989).
[10] Lu Li, Liu Jing-qian, Ma Bei-hai, Zhang Dian-lin, Wang Xiao-hong, and Daoben Zhu, Synth. Metals 31: 45 (1988).
[11] S. Sugai and G. Saito, Solid State Commun. 58: 759 (1986); Synth. Metals 19: 231 (1987).
[12] M. Mizuno, J. Tanaka, and I. Harada, J. Phys. Chem. 85: 1789 (1981).
[13] R. Swietlik, D. Schweitzer, and H. J. Keller, Phys. Rev. B36: 6881 (1987).
[14] Ma Bei-hai, Lu Li, Zhang Dian-ling, Wang Xiao-hong, and Zhu Daoben, Solid State Commun. 68: 433 (1988).

ONE-DIMENSIONAL LINEAR-CHAIN COMPLEXES OF PLATINUM, PALLADIUM AND NICKEL

Robin J.H. Clark

Christopher Ingold Laboratories
University College London
20 Gordon Street
LONDON WC1H 0AJ, U.K.

INTRODUCTION

Potassium tetracyanoplatinate, $K_2Pt(CN)_4Br_{0.30}.3H_2O$ (KCP) and its analogues (Fig. 1a) are, on account of their being one-dimensional conductors, the best studied linear-chain complexes of platinum. However, several other kinds of linear-chain complex have been the subject of much recent attention. In particular, halogen-bridged linear-chain complexes of the platinum group, despite their having much lower conductivities than KCP, have proved to be of immense interest, primarily spectroscopically.[1,2] Two types of such chain complex will be discussed in this article, *viz.* the Wolffram's red (WR) type, where WR = $[Pt^{II}(C_2H_5NH_2)_4][Pt^{IV}(C_2H_5NH_2)_4Cl_2]Cl_4.4H_2O$ (Fig. 1c), and the type formed from the barrel-shaped complex ion $[Pt_2(pop)_4]^{4-}$, where pop = diphosphite =

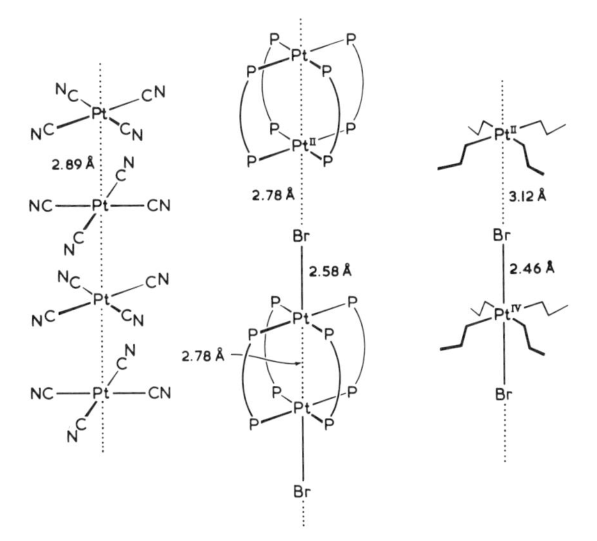

Fig. 1 Structure of (a) KCP, $K_2Pt(CN)_4Br_{0.30}.3H_2O$, (b) $K_4[Pt_2(H_2P_2O_5)_4Br].3H_2O$ at 19 K, and (c) Reihlen's green, $[Pt(C_2H_5NH_2)_4][Pt(C_2H_5NH_2)_4Br_2]Br_4.4H_2O$.

Lower-Dimensional Systems and Molecular Electronics
Edited by R.M. Metzger *et al.*, Plenum Press, New York, 1991

band maxima of all these complexes vary in the order Cl > Br > I, and Pd^{II}/Pt^{IV} > Ni^{II}/Pt^{IV} > Pt^{II}/Pt^{IV} > Pd^{II}/Pd^{IV} > Ni^{II}/Ni^{IV}. This result implies that the valence electrons are most delocalised for the Ni^{II}/Ni^{IV} complexes, an implication which is consistent with

WR-TYPE CHAIN COMPLEXES

Complexes of the WR sort, $[PtL_4][PtL_4X_2]Y_4$, have chain structures as indicated below:

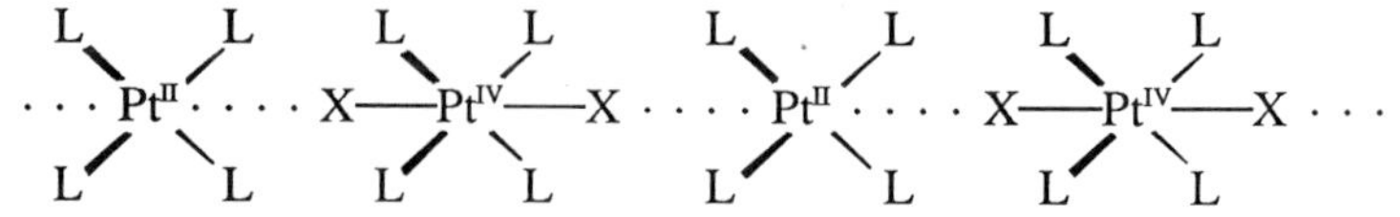

```
   L     L     L     L          L     L     L     L
    \   /       \   /            \   /       \   /
 · · · PtII · · · · X——PtIV——X · · · · PtII · · · · X——PtIV——X · · ·
    /   \       /   \            /   \       /   \
   L     L     L     L          L     L     L     L
```

where L, a neutral equatorial ligand, is usually an amine such as NH_3, CH_3NH_2, or $C_2H_5NH_2$ and X = Cl, Br or I. Bidentate ligands LL can also form complexes of this general structural type, where LL can be 1,2-diaminoethane (en), 1,2-diaminopropane (pn), 1,3-diaminopropane (tn), *etc.* Moreover, both terdentate ligands, LLL, such as diethylenetriamine (dien) and N-methyldiethylenetriamine (Medien), and quadridentate ligands, LLLL, such as 1,4,8,11-tetraazacyclotetradecane (cyclam) will also form complexes of this general type.[3] WR consists of Pt^{II} and Pt^{IV} units, each of which bears a +2 charge, and hence four anions are required for electroneutrality. In addition to this type of complex, however, chain complexes of the same basic structure can also be formed, in which the equatorial amine ligands are replaced by one, two, three, or four halide ions, with the result that the overall charge must be balanced by two, zero, two, or four counter ions, *e.g.* as in $[Pt^{II}(Medien)I][Pt^{IV}(Medien)I_3]I_2$, $[Pt^{II}(en)Cl_2][Pt^{IV}(en)Cl_4]$, $K_2[Pt^{II}(NH_3)Cl_3][Pt^{IV}(NH_3)Cl_5].2H_2O$ and $K_4[Pt^{II}I_4][Pt^{IV}I_6]$, respectively. The counter ions may be Cl^-, Br^-, I^-, ClO_4^-, HSO_4^-, BF_4^-, NO_3^-, K^+ or Cs^+ *etc.*; they play an important role, often along with water of crystallisation, in helping to hold the chains together by hydrogen bonding.

The intense colours of linear-chain complexes, which differ markedly from those of their constituent platinum(II) and (IV) entities, are caused by intervalence transitions $Pt^{II} \rightarrow Pt^{IV}$, which are polarised in the chain direction; they occur in the regions 25,000-18,200 cm^{-1} (X = Cl), 23,600-14,300 cm^{-1} (X = Br), and 20,600-7,500 cm^{-1} (X = I). Extensive research has established that the shorter the $Pt^{II}\cdots Pt^{IV}$ chain distance, the lower in wavenumber is the intervalence transition. This is consistent with the idea that shortening of the chain repeat unit is primarily brought about by a shortening of the $Pt^{II}\cdots X$, rather than the Pt^{IV}-X distance, resulting in more central bridging by the chain halide. Central halide bridging is known also to be critical in bringing about increased chain conductivity.

The Raman spectra of halogen-bridged mixed-valence complexes of platinum at resonance with the $Pt^{II} \rightarrow Pt^{IV}$ intervalence band, are very intense. In particular, the symmetric X-Pt^{IV}-X chain stretching mode (ν_1), together with long overtone progressions $\nu_1\nu_1$, are tremendously enhanced at resonance. The overtone progressions reach as far as $17\nu_1$ in some cases (Fig. 2),[4] implying a very substantial (0.05-0.10 Å) change in Pt^{IV}-Cl bond length on changing from the ground to the intervalence state.[2] The ν_1 values (for ν_0 at resonance with the intervalence transition) range from 309.1-297.8 cm^{-1} (X = Cl), 175.7-172.0 cm^{-1} (X = Br) and 122.3-114.2 cm^{-1} (X = I). Recent synthetic, spectroscopic and structural work has led to the characterisation of structurally analogous palladium complexes, as well as mixed-metal derivatives of the sort $[Ni^{II}(en)_2][Pt^{IV}(en)_2Cl_2][ClO_4]_4$, $[Pd^{II}(en)_2][Pt^{IV}(en)_2Cl_2][ClO_4]_4$, and analogous 1,2-diaminopropane complexes.[5-7] Mixed-valence nickel(II,IV) complexes of the same structural type have been thought to exist for some years (most recently on the basis of EXAFS work),[8] but the characterisation of these has been uncertain, until the excellent

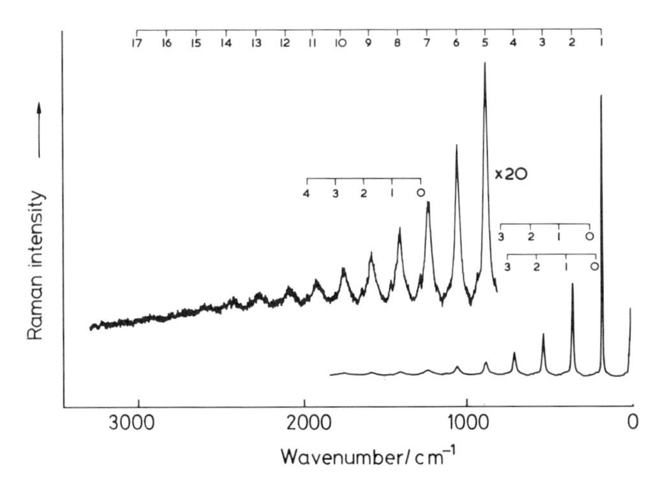

Fig. 2 Resonance Raman spectrum of $[Pt(pn)_2][Pt(pn)_2Br_2][Cu_3Br_5]_2$, pn = 1,2-propanediamine, in a KBr disc at *ca.* 80 K with 568.2 nm excitation. (Ref. 4).

recent work of Toriumi, Yamashita and others.[9] The wavenumbers of the intervalence $H_2P_2O_5^{2-}$ (Fig. 1b). The last, a chain bromide, has a chain conductivity ($\sigma_{\parallel}$) of *ca.* 10^{-3} Ω^{-1} cm^{-1}, which is intermediate between that of Reihlen's green (WR, but with bromide in the chain rather than chloride, $\sigma_{\parallel} \approx 10^{-9}$ Ω^{-1} cm^{-1}) and that of KCP ($\sigma_{\parallel} \approx 10^{3}$ Ω^{-1} cm^{-1}). the comparatively high (for a halogen-bridged complex) chain conductivity of the nickel complexes.

The most remarkable of the nickel complexes is $Ni^{III}(chxn)_2Br_3$, chxn = cyclohexanediamine, for which the bridging bromide is required by symmetry to bridge centrally, thus indicating that the complex should be formulated as one of nickel(III).[10] What feature of this complex makes it possible for it not to undergo the expected Peierls distortion of a symmetric linear chain is not clear, unless it is exceptionally strong coupling of cation chains to anions. Interconversion from nickel(II,IV) to nickel(III) chain complexes is easily brought about by moisture and, in some cases, reversed by hydrogen halide. Raman spectroscopy of this complex shows no evidence of a ν_1 band (or its overtone bands), consistent with the behaviour expected for a centrally bridged halide; a weak band appearing at near-resonance conditions (λ_0 = 752.5 nm, maximum of the intervalence transition 10,300 cm^{-1}) at 213 cm^{-1} may be ν_1, activated by defects in the chain.[11]

A summary of the types of M^{II}/M^{IV} linear-chain complexes of the +2,+2 charge type now known is given in Fig. 3.

For certain complexes of this sort - particularly those for which X = Br or I - the wavenumber of ν_1 appears, under conditions of only moderate resolution, to increase as that of the exciting line (ν_0) increases.[12-16] For any particular set of complexes, the shift in ν_1 increases in the order Cl < Br < I, and is largest for those, in which the bond length ratio $r(Pt^{IV}\text{-}X)/r(Pt^{II}\text{-}X)$ is nearest to unity, and in which the $Pt^{II}\cdots Pt^{IV}$ distance is least. Careful, high resolution studies of these complexes, as single crystals at low temperatures, have now shown that the apparent dispersion in ν_1 is caused by the fact, that the ν_1 band consists of many components, those of higher wavenumber being resonance-enhanced with lines of higher ν_0. It is possible that each component of ν_1 is related to a chain segment of a different correlation length (defined as the length over which the oxidation states

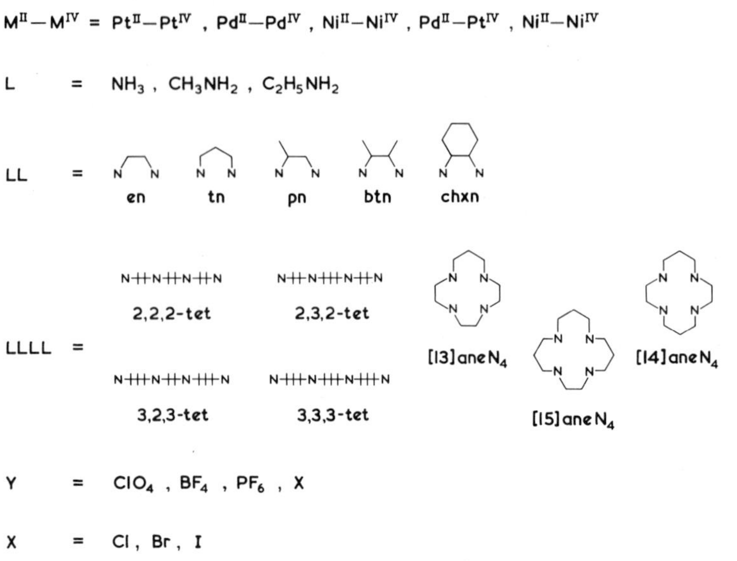

$M^{II}-M^{IV}$ = $Pt^{II}-Pt^{IV}$, $Pd^{II}-Pd^{IV}$, $Ni^{II}-Ni^{IV}$, $Pd^{II}-Pt^{IV}$, $Ni^{II}-Ni^{IV}$

L = NH_3 , CH_3NH_2 , $C_2H_5NH_2$

LL = en, tn, pn, btn, chxn

LLLL = 2,2,2-tet, 2,3,2-tet, 3,2,3-tet, 3,3,3-tet, [13]aneN_4, [14]aneN_4, [15]aneN_4

Y = ClO_4 , BF_4 , PF_6 , X

X = Cl , Br , I

Fig. 3 Summary of known M^{II}/M^{IV} linear-chain complexes of the types $[ML_4][ML_4X_2]Y_4$, $[M(LL)_2][M(LL)_2X_2]Y_4$ and $[M(LLLL)][M(LLLL)_4X_2]Y_4$.

alternate II and IV regularly and without defects of any sort). Thus, each segment is thought to have a specific wavenumber, absorption band maximum, and excitation profile maximum. Since shorter segments are known to give rise to absorption band maxima of higher wavenumber and higher ν_1 values, this possible explanation requires the existence of a certain small distribution of discrete, but relatively short, correlation lengths in each chain; each chain length can thus be made, in turn, to come into resonance as ν_0 is changed, leading (where the resolution is only moderate) to the apparent dispersion of the unresolved ν_1 band with change of ν_0.

Somewhat analogous behaviour has also been observed for certain bands (*e.g.* ν(C=C) and ν(C-C) at ~1450 and ~1060 cm^{-1}, respectively) of trans-polyacetylene, $(CH)_x$ and $(CD)_x$;[17-21] these also exhibit a dependence on excitation wavenumber, which has been attributed to the existence of segments, within each chain, of different conjugation lengths. Such segments are also expected to have different vibrational wavenumbers, absorption band maxima, and excitation profile maxima; if the component bands are not resolved, the composite Raman bands, likewise, appear to show dispersion with change of ν_0.

The better understanding of the nature and properties of the segments giving rise to the discrete components to ν_1 - thought to be closely linked to the properties of both solitons and polarons[22,23] - is important to the understanding of the structures, spectral properties and conductivity mechanisms for both Wolffram's red type complexes, as well as for trans-polyacetylene. In this context, iodine doping of $[Pt(en)_2][Pt(en)_2I_2](ClO_4)_4$ leads to an increase in the conductivity of over 10^7. Moreover, the absorbance of two midgap bands and the e.s.r. intensity in dilute doped crystals increase with dopant concentration. Low levels of halogen doping give the same effects on optical absorption and e.s.r. spectra, as does photoexcitation at a wavenumber greater than, or equal to, that of the intervalence band. Doping-induced states are located on single chains, and have both charge and spin, suggesting that it is polarons which are formed on doping.[24]

Many other studies on these intriguing complexes are in hand, particularly related to the understanding of their luminescence,[24] and its dependence on pressure.

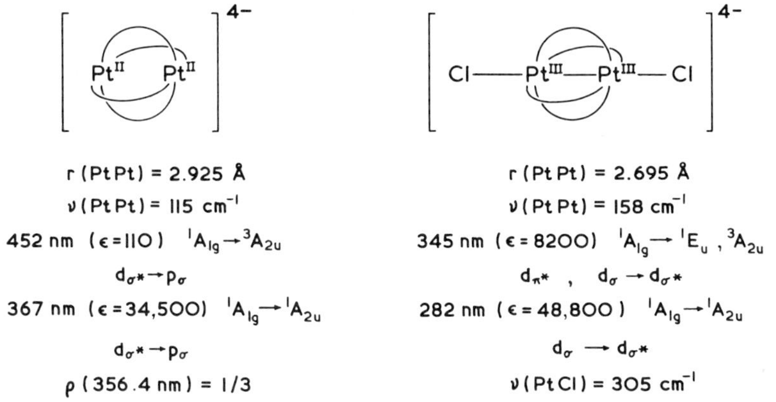

$$\left[\text{Pt}^{II} \quad \text{Pt}^{II} \right]^{4-} \qquad \left[\text{Cl}-\text{Pt}^{III}-\text{Pt}^{III}-\text{Cl} \right]^{4-}$$

r (PtPt) = 2.925 Å	r (PtPt) = 2.695 Å
ν (PtPt) = 115 cm^{-1}	ν (PtPt) = 158 cm^{-1}
452 nm (ε = 110) $^1A_{1g} \rightarrow {}^3A_{2u}$	345 nm (ε = 8200) $^1A_{1g} \rightarrow {}^1E_u, {}^3A_{2u}$
$d_{\sigma^*} \rightarrow p_\sigma$	d_{π^*} , $d_\sigma \rightarrow d_{\sigma^*}$
367 nm (ε = 34,500) $^1A_{1g} \rightarrow {}^1A_{2u}$	282 nm (ε = 48,800) $^1A_{1g} \rightarrow {}^1A_{2u}$
$d_{\sigma^*} \rightarrow p_\sigma$	$d_\sigma \rightarrow d_{\sigma^*}$
ρ (356.4 nm) = 1/3	ν (PtCl) = 305 cm^{-1}

Fig. 4 Spectroscopic and structural properties of the complex ions $[Pt_2(H_2P_2O_5)_4]^{4-}$ and $[Pt_2(H_2P_2O_5)_4Cl_2]^{4-}$.

DIPHOSPHITE COMPLEXES

The two barrel-shaped ions $[Pt_2(H_2P_2O_5)_4]^{4-}$ and $[Pt_2(H_2P_2O_5)_4X_2]^{4-}$, the properties of which (for X = Cl) are given in Fig. 4, may be co-crystallised to give a second type of chain complex of the form:

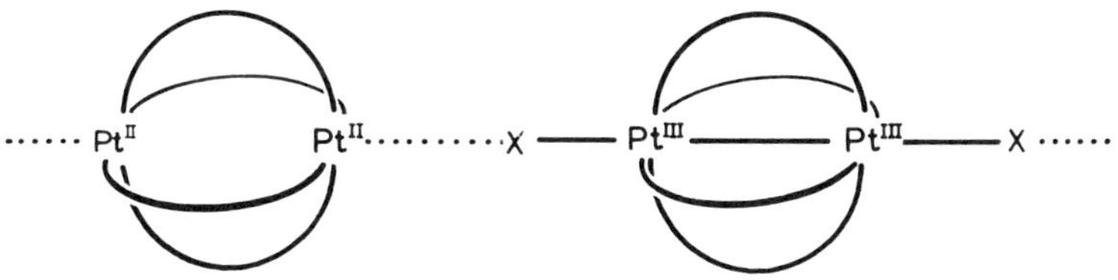

$$\cdots\cdots \text{Pt}^{II} \quad \text{Pt}^{II} \cdots\cdots\cdots\cdots \text{X} — \text{Pt}^{III} —— \text{Pt}^{III} —— \text{X} \cdots\cdots$$

where  ⌒ represents the diphosphite phosphorus-donor ligand

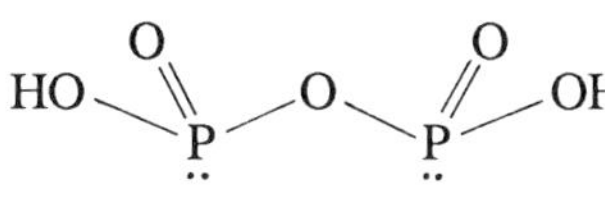

HO–P(=O)(:)–O–P(=O)(:)–OH, $H_2P_2O_5^{2-}$ (pop)[26] or HO–P(=O)(:)–CH_2–P(=O)(:)–OH (pcp)[27]. Such complexes form as golden metallic crystals, with chain conductivities ($\sigma_{//}$) about six orders of magnitude greater than that of Wolffram's red salts with the same bridging ligand.[28-32] A more detailed view of one repeat unit along the chain is given in Fig. 5, which also shows the two half-chloride ions detected along the chain. The conductivity of $K_4[Pt_2(H_2P_2O_5)_4Br].3H_2O$ is thermally activated, with a band gap of ~0.08 eV.

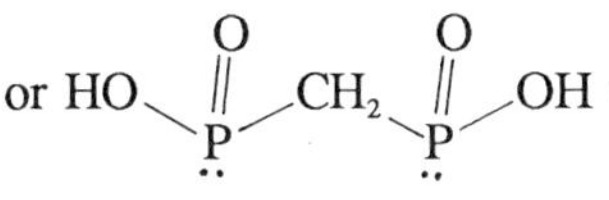

The Raman spectrum of $K_4[Pt_2(H_2P_2O_5)_4Cl].3H_2O$, at resonance with the intervalence band, is dominated by a band at 291 cm^{-1} (λ_0 = 647.1 nm), which is assigned to the symmetric Pt-Cl stretching mode of the chain.[30] Its value is a little lower, on account of bridging, than that found (305 cm^{-1}) for the discrete di-platinum(III) species, $[Pt_2(H_2P_2O_5)_4Cl_2]^{4-}$. Six-membered progressions in the 291 cm^{-1} band are detected in the

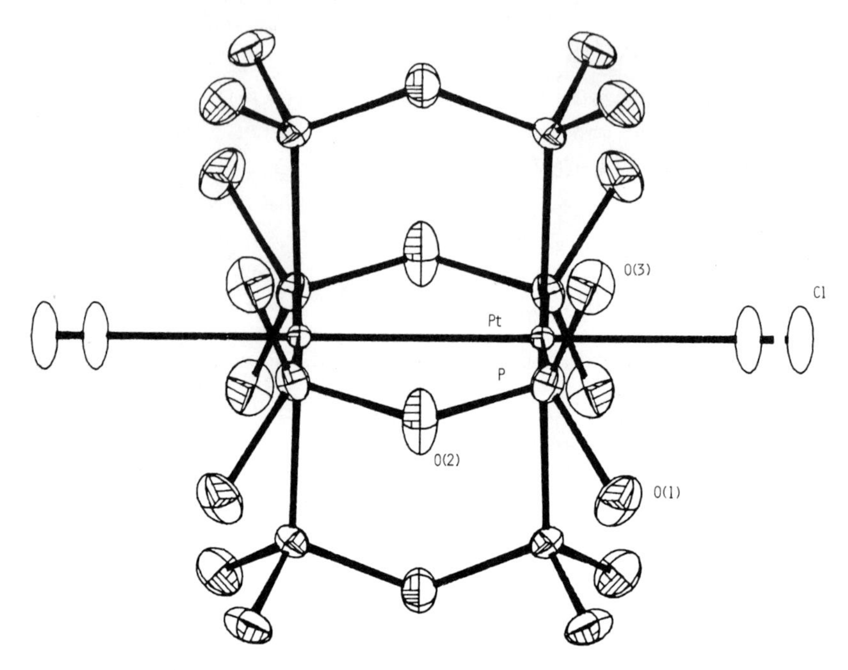

Fig. 5 Chain repeat unit in $K_4[Pt_2(H_2P_2O_5)_4Cl].3H_2O$.

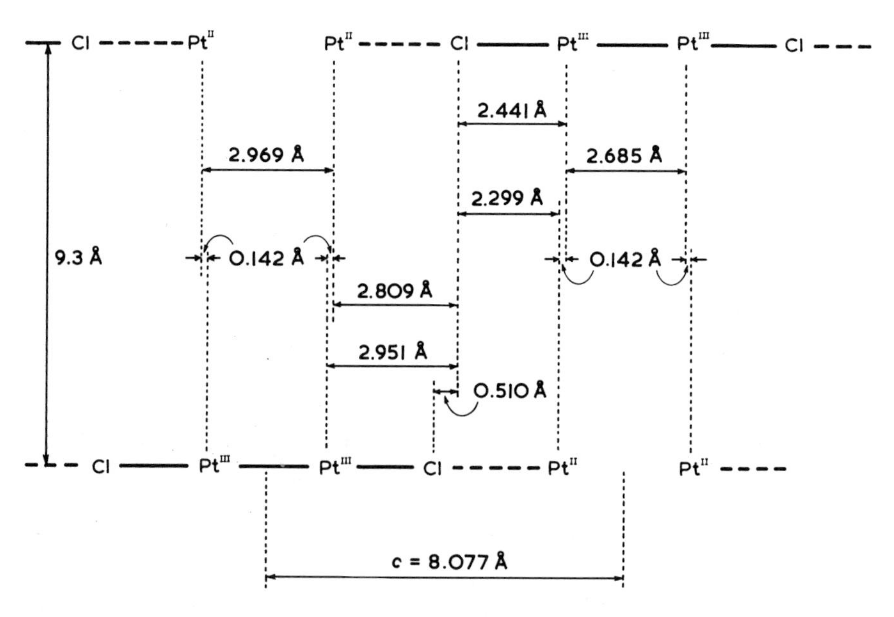

Fig. 6 $(AABCCB)_n$ chain structure of $K_4[Pt_2(H_2P_2O_5)_4Cl].3H_2O$ at 22 K.

Table 1. Analogy between Wolffram's Red salts and Chain POP Complexes; Platinum-Halogen (PtX) Data (X = Cl or Br)

	$[Pt(etn)_4][Pt(etn)_4X_2]X_4.4H_2O$		$K_4[Pt_2(H_2P_2O_5)_4X].3H_2O$			
r(PtCl)/Å	2.26	0.87	2.37 [a]	0.51	2.299 2.441 [b]	0.142
	3.13		2.88		2.809 2.951	0.142
ν_1(ClPtCl)/cm^{-1}	312.3 (to ν_1=16 at res)		291 (to ν_1=6 at res)			
r(PtBr)/Å	2.46 3.12	0.66	2.58 2.78	0.20		
ν_1(BrPtBr)/cm^{-1}	177.0 (to ν_1=11 at res)		210 223			

[a,b] Assumes equivalent (a) or two different (b) PtPt bond lengths.

resonance Raman spectrum implying (a) that the chloride ion is not centrally bridging and (b) that the principal structural change undergone by the ion on excitation to the intervalence state is - as for Wolffram's red complexes - substantial, and along the Pt-Cl coordinate; it thus implies that the platinum ion valences approximate to ...II,II...III,III... along the chain. Analogous ν(PtX) progressions in the resonance Raman spectrum of the corresponding bromide and iodide could not be detected; instead, a progression in the Pt-Pt stretching mode is observed. This implies that the bridging group is, in these cases, much nearer to the central position, than is the case for a bridging chlorine atom, and that the platinum ions are close to being in the +2.5 oxidation state (class IIIA species, in the parlance of the chemistry of mixed-valence species).[3] These conclusions are fully in accord with those drawn on the basis of X-ray diffraction studies of the complexes at *ca.* 20 K.[32] The relationship between the chains for $K_4[Pt_2(H_2P_2O_5)_4Cl].3H_2O$, an $(AABCCB)_n$ structure, is shown in Fig. 6.

CONCLUSION

Both the degree of enhancement of the band assigned to the X-Pt^{IV}-X symmetric stretch (ν_1), as well as the length of the overtone progressions in ν_1 observed at resonance with the intervalence band, are related to the extent of structural change attendant upon excitation to the intervalence state.[2] Spectacularly detailed band progressions are observed for these complexes, yielding an immense amount of valuable spectroscopic information. Of particular interest is the comparison between the Wolffram's red type complex and the diphosphite complexes; the latter have chain conductivities ~10^6 times greater than the former, owing to the more nearly central positioning of chloride or bromide in the chain for diphosphite complexes (Table 1). Work to characterise further linear-chain complexes, involving the diphosphite pcp, is currently in progress, as is work concerned with reversing the Peierls distortion of such complexes by the action of pressure to 4 GPa.[33] Our principal concern has been with understanding the implications of all this information, and with attempting to understand the relationships between structure, spectroscopy, conductivity, and bonding in chain complexes likely to be of interest as new materials. In this context, the chain nickel complexes are potentially very important, since they involve a cheaper metal, and have the higher chain conductivities.

REFERENCES

1. R.J.H. Clark, in "Advanced in Infrared and Raman Spectroscopy", ed. R.J.H. Clark and R. E. Hester, Wiley/Heyden, Chichester, vol.11, pp.95-132 (1984).
2. R.J.H. Clark and T.J. Dines, *Angew Chem.*, 25:131 (1986).
3. R.J.H. Clark, *Chem. Soc. Rev.*, 13:219 (1984).
4. R.J.H. Clark, M. Kurmoo, H.J. Keller, B. Kepler, and U. Traeger, *J. Chem. Soc. (Dalton),* 2498 (1980).
5. R.J.H. Clark and V.B. Croud, *Inorg. Chem.*, 24:588 (1985); 25:1751 (1986).
6. R.J.H. Clark, V.B. Croud, and R.J. Wills, *Inorg. Chem.*, 27:2096 (1988).
7. R.J.H. Clark, V.B. Croud, R.J. Wills, P.A. Bates, H.M. Dawes, and M.B. Hursthouse, *Acta Cryst.*, B45:147 (1989).
8. J. Evans, J.T. Gauntlett, and W. Levason, *Inorg. Chem.*, 27:4523 (1988).
9. M. Yamashita, Y. Nonaka, S. Kida, Y. Hamaue, and R. Aoki, *Inorg. Chim. Acta,* 52:43 (1981).
10. K. Toriumi, Y. Wada, T. Mitani, S. Bando, and M. Yamashita, *J. Am. Chem. Soc.,* (1989).
11. R.J.H. Clark and D.J. Michael, unpublished work.
12. R.J.H. Clark and M. Kurmoo, *J. Chem. Soc. Faraday Trans. II,* 79:519 (1983).
13. R.J.H. Clark and V.B. Croud, *J. Phys. C: Solid State Phys.*, 19:3467 (1986).
14. M. Tanaka and S. Kurita, *J. Phys. C: Solid State Phys.*, 19:3019 (1986).
15. S.D. Conradson, R.F. Dallinger, B.I. Swanson, R.J.H. Clark, and V.B. Croud, *Chem. Phys. Lett.*, 135:463 (1987).
16. R.J.H. Clark and D.J. Michael, *J. Mol. Struct.*, 189:173 (1988).
17. H. Kuzmany and P. Kroll, *J. Raman Spectrosc.*, 17:89 (1986).
18. Z. Vardeny, J. Ornstein, and G.L. Baker, *Phys. Rev. Lett.*, 50:2032 (1983).
19. Z. Vardeny, E. Ehrenfreund, O. Brafman, and B. Horovitz, *Phys. Rev. Lett.*, 51:2326 (1983).
20. R. Tiziani, G.P. Brivio, and E. Mulazzi, *Phys. Rev. B,* 31:3019 (1985).
21. A.W. Tarr and W. Siebrand, submitted for publication.
22. N. Kuroda, M. Sakai, and Y. Nishina, *Phys. Rev. Lett.*, 58:2122 (1987).
23. S. Kurita, M. Haruki, and K. Miyagawa, *J. Phys. Soc. Japan,* 57:1789 (1988).
24. M. Haruki and S. Kurita, *Phys. Rev. B,* 39:5706 (1989).
25. H. Tanino, N. Koshizuka, K. Kobayashi, M. Yamashita, and K. Hoh, *J. Phys. Soc. Japan,* 54:483 (1985).
26. R.P. Sperline, M.K. Dickson, and D.M. Roundhill, *J. Chem. Soc., Chem. Commun.*, 62 (1977).
27. C. King, R.A. Auerbach, F.R. Fronczek, and D.M. Roundhill, *J. Am. Chem. Soc.*, 108:5626 (1986).
28. C.-M. Che, W.P. Schaefer, H.B. Gray, M.K. Dickson, P. Stein, and D.M. Roundhill, *J. Amer. Chem. Soc.*, 104:4253 (1982).
29. R.J.H. Clark and M. Kurmoo, *J. Chem. Soc. (Dalton),* 579 (1985).
30. M. Kurmoo and R.J.H. Clark, *Inorg. Chem.*, 24:4420 (1985).
31. R.J.H. Clark, M. Kurmoo, H.M. Dawes, and M.B. Hursthouse, *Inorg. Chem.*, 25:409 (1986).
32. L.G. Butler, M.H. Zietlow, C.-M. Che, W.P. Schaefer, S. Sridhar, P.J. Grunthaner, B.I. Swanson, R.J.H. Clark, and H.B. Gray, *J. Amer. Chem. Soc.*, 110:1155 (1988).
33. B.I. Swanson, M.A. Stroud, S.D. Conradson, and M.H. Zietlow, *Solid State Comm.*, 65:1405 (1988).

POLARONS AND SOLITONS IN HALOGEN-BRIDGED PLATINUM COMPLEXES: EFFECTS OF PHOTO-EXCITATION AND HALOGEN-DOPING ON OPTICAL ABSORPTION AND ESR SPECTRA

Susumu Kurita

Laboratory of Applied Physics
Faculty of Engineering, Yokohama National University
Hodogaya, Yokohama 240, JAPAN

INTRODUCTION

In recent years, much attention has been given to one-dimensional insulating systems, where various kinds of fascinating phenomena, like spin density wave(SDW), charge density waves(CDW) etc. are expected, depending on the relative magnitudes of the electron-electron and the electron-phonon interactions. In addition to the interest in their ground states, these systems with a degenerate structure offer a possibility of studying nonlinear excitations of kink-soliton as well as polarons, bipolarons etc.. The studies of such nonlinear excitations have been done in conjugated polymers intensively,[1] and it appears that such excitations play important roles in the optical and electrical properties of one-dimensional materials. However, the conjugated polymers are only one case of weak electron-phonon interaction. Experimentally, it is quite difficult to get a single crystal for such a polymer. Moreover, the magnitude of the interactions between the neighbouring chains is unknown. These prevent the studies of nonlinear excitation states. Then, it is important to find materials, which grow as single crystals easily, and provide new model systems for studying the problems of nonlinear excitations. As one of such materials, we propose the family of halogen-bridged mixed-valence transition metal complexes, especially platinum complexes.

The halogen-bridged mixed-valence transition metal complexes have a doubly degenerate linear chain structure,

$$\cdots —X—M^{4+}—X—M^{2+}—X—M^{4+}—X—M^{2+}— \cdots$$

here X refers to the halogen atom (Cl,Br,I) and M to the transition metal(Ni,Pt,Pd). Figure 1 shows a chain structure of $[Pt(en)_2][Pt(en)_2Cl_2](ClO_4)_4$, (here en denotes ethylenediamine) which is a typical halogen-bridged platinum complex[2,3], and is mainly discussed in this paper. In this chain, the chlorine ions are located, not at the midpoint between the Pt^{2+} and Pt^{4+} ions, but at a point closer to the Pt^{4+} ion. This periodically alternating valence of Pt ions can be regarded as a charge density wave with the commensurability index of 2. Moreover, the Pt-X chain is surrounded by the neutral ligands of ethylenediamine, so that the interaction between the different Pt-X chains is supposed to be fairly small. These facts mean that the halogen-bridged mixed-valence transition metal complex qualifies a good one-dimensional compound with doubly degenerate structure (CDW).

As above mentioned, the displacement of halogen atoms from the midpoints between neighbouring metal ions brings on the doubly degeneracy, that is, the Peierls state, in the chain, so that the breathing mode of halogen

Lower-Dimensional Systems and Molecular Electronics
Edited by R.M. Metzger *et al.*, Plenum Press, New York, 1991

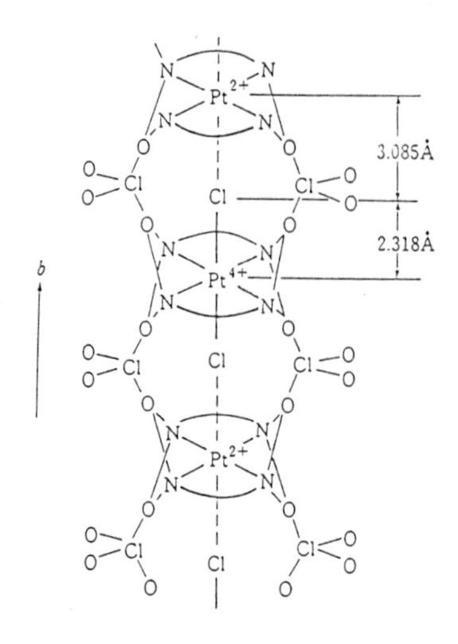

Fig.1. Chain structure of $[Pt(en)_2][Pt(en)_2Cl_2](ClO_4)_4$.

$\cdots Pt^{2+} \cdots\cdots X—Pt^{4+}—X \cdots\cdots Pt^{2+} \cdots\cdots X—Pt^{4+}—X \cdots$

		2		4		2		4		
(G)	2	4	2	4	2	4	2	4	2	4
(S)	2	4	3 S^0	2	4	2	3 S^0	4	2	4
(S)	2	4	4 S^+	2	4	2	2 S^-	4	2	4
(P)	2	4	3 P^+	4	2	4	2	3 P^-	2	4

Fig.2. Schematic illustration of quasi-one-dimensional chain of halogen-bridged Pt complex. Simplified notations 2, 3, and 4 denote Pt^{2+}, Pt^{3+}, and Pt^{4+}, respectively. Halogens and ligand molecules are omitted.

vibration, centered at Pt^{4+} ion modulates valence electrons of Pt ions strongly. This strong electron-phonon interaction has been seen experimentally by resonant Raman scattering, where 17 overtones of breathing vibration are observed[4,5]. Baeriswyl and Bishop[6] have calculated intrinsic defect states (solitons, polarons, bipolarons, and excitons) in the strong CDW limit, and demonstrated that they are strongly localized in the M-X chain, in contrast to the relatively extended defect states in a conjugated polymer with a weak electron-phonon interaction. Such defect states of the Pt complexes may be depicted as in Fig. 2. The ground state of Pt-X chain is represented by the simplified notation of 2424, where 2, 3 and 4 denote Pt^{2+}, Pt^{3+} and Pt^{4+}, respectively, omitting intervening halogen ions. Using this notation, the ground state, kink-solitons(S^0, S^+, S^-) and polarons(P^+, P^-) can be represented by the configurations (G), (S), (S') and (P) in Fig.2. As it is clear from the figure, polarons have both spin and charge, but solitons cannot have both at same time. These are summarized as follows[6].

		Spin	Charge	Formation Energy
Polaron	P^+	1/2	e	$3E_g/8$
	P^-	1/2	-e	$3E_g/8$
Soliton	S^0	1/2	0	$3E_g/8$
	S^+	0	e	$E_g/4$
	S^-	0	-e	$E_g/4$
Bipolaron	BP^+	0	2e	$E_g/2$
	BP^-	0	-2e	$E_g/2$

Both solitons and polarons have their characteristic absorption bands below the band gap energy[6]. Then, for the identification of the nonlinear excited states, the optical absorption and the electron spin resonance spectra must be studied, to get information about the midgap states and the spin, respectively. Studies of the electrical properties are also needed to get information about the charge. In this report, the excited states in single crystals of $[Pt(en)_2][Pt(en)_2Cl_2](ClO_4)_4$ are studied by the experimental methods mentioned above, and the photo-induced excited state in this material is shown to be polarons, which are also produced by halogen-doping.

EXPERIMENTAL PROCEDURES

All measurements were done by using single crystals prepared by the method described previously[7]. To study photo-excitation effects, light from an Ar ion laser was focussed on the sample surface, with the light intensity of 0.24 mW/mm^2. The change in the absorption spectra(photo-induced absorption) with the total excitation light intensity was measured from the time dependence of spectra under the constant irradiation. Halogen doping was carried out by exposing the crystals in a glass vessel to the halogen vapor, that is, chlorine gas of 1 atom was directly injected into the glass vessel at room temperature. The contents of doped chlorine was not measured directly. The absorption change by the doping was obtained by measuring the absorption spectra in proper time intervals.

The absorption spectra were measured by a double pass method. The sample in a glass vessel was fixed in the optical system during photo-excitation, or halogen-doping, so that the experimental configuration was kept in same condition and the sample was not exposed to air.

ESR measurements were carried out at X band with a modulation field of 2 gauss at 100 kHz. The resonant magnetic field was calibrated using a proton NMR gaussmeter. All samples were loaded into the quartz ESR tube in such a way that the chain axes were directed perpendicular to the resonant magnetic field. For the effects of halogen-doping, ESR measurement was done at room temperature. For the effects of photo-excitation, the ESR signals were measured after irradiation of laser light at 77K.

If the doping time was less than 48 hours, the samples maintained their crystal properties. In this study, doping was done for less than 24 hours.

RESULTS AND DISCUSSION[8,9,10]

For the light polarized parallel to the chain axis (E//b), a weak absorption band (A band) is commonly observed for the halogen-bridged platinum complexes on the low-energy side of the charge transfer (CT) absorption band (electron transfer from Pt^{2+} to Pt^{4+})[7,11]. On the crystals of $[Pt(en)_2][Pt(en)_2Cl_2](ClO_4)_4$, a weak additional absorption band B is also seen between the A band and the CT band at low temperatures. They are very likely some intrinsic excited states mentioned above because they are commonly seen. At first, the effects of light irradiation upon the A and B absorption bands were investigated. The results at nitrogen temperature are shown in Fig. 3; the photo-excitation was done by the light of 2.41eV

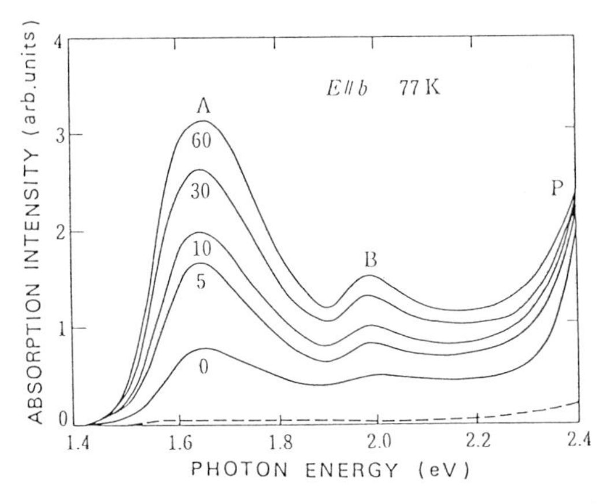

Fig.3. Photo-induced absorption spectra of $[Pt(en)_2][Pt(en)_2Cl_2](ClO_4)_4$ single crystal at 77K. The spectra were measured for the light polarized parallel(E//b solid lines) and perpendicular to the chain(E⊥b, dashed line). The numbers are the time of photo-excitation (minutes) with the constant irradiation of the laser light(2.41eV, 0.24mW/mm^2).

with the intensity of $0.24 mW/mm^2$. P is the absorption edge of the CT band. The curves are absorption spectra, measured after laser irradiation for 0, 5, 10, 30 and 60 minutes, respectively. The absorption bands are observed only for the polarization E//b. The photo-induced absorption bands A and B are stable, and do not decrease with time, as long as the samples are kept at 77K. A small thermal bleaching was recognized after the annealing time of 4 hours at 173K. It should be noted that the spectral shape of the A and B bands, observed in the annealing process, is similar to those in the photo-induced process, which suggests that the A and B bands originate from same defect state. The defect state is on the single chain, because the A and B bands are active only for the light polarized parallel to the chain.

Photo-induced ESR spectra were measured in order to see whether the photo-induced defect states have spin or not. The results are shown in Fig.4, where the magnetic field was applied parallel to the chain axis. The photo-excitation was done by a Hg lamp with a visible-light-cut filter. The light intensity on the surface of sample was not measured. The curve 0 in the figure is the spectrum before the photo-excitation was made, which coincides with the spectrum observed by Kawamori et al.[12]. The curves 5, 20 and 45 are the spectra measured after the constant irradiation of light of 5, 20 and 45 minutes. This result shows that the photo-induced states have a spin. When these photo-induced states, which are on the single chain, are related to the nonlinear excitation ones, they should be neutral solitons or polaron pairs (P^+-P^-), because the photo-induced states must be neutral (see table). To determine whether the photo-induced state is pairs, neutral solition pairs, or polaron pairs, it is very helpful to consider that both the absorption and the ESR spectra before the photo-excitation are similer in shape to photo-induced ones as seen in Figs. 3, 4. This fact suggests that the photo-induced states are the same as defect ones (which as-grown crystals happen to have).

For the neutral soliton to grow in the crystal, the disorder in the segments of Pt^{2+} and Pt^{4+} must occur. and at the same time Pt^{2+} or Pt^{4+} ions change their charge to Pt^{3+}, as seen in Fig.2(S)[13]. The formation energy of the neutral soliton is higher than that of charged one, so that the charged soliton may be more favorable, if solitons grow in the process of crystallization.

For the polaron formation, excess charge must be introduced on the chain. Such charge may be produced by excess halogens which are taken up in the crystal when the material is synthesized. This may happen with high probability, because the excess halogens have a chance to remain in the solution of Pt^{4+} ions[7]. It can be verified easily, by a halogen doping experiment,

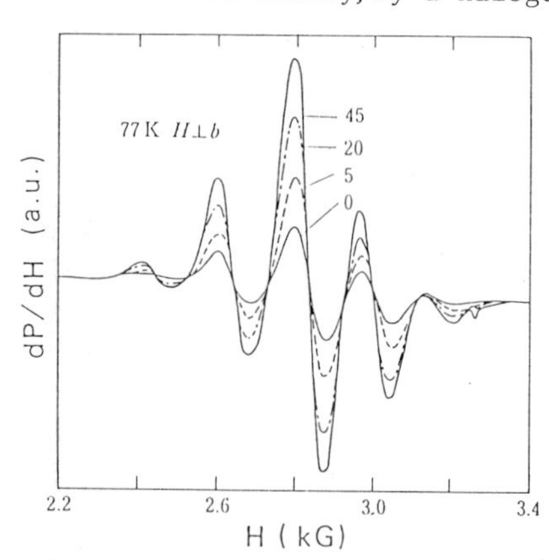

Fig.4. Photo-induced ESR spectra of $[Pt(en)_2][Pt(en)_2Cl_2](ClO_4)_4$ at 77K. The excitation was done by the Hg lamp with the VIS-cut filter. Curve 0 is the spectrum before the excitation was made, and 5, 20 and 45 are those after the excitation for 5, 20 and 45min., respectively.

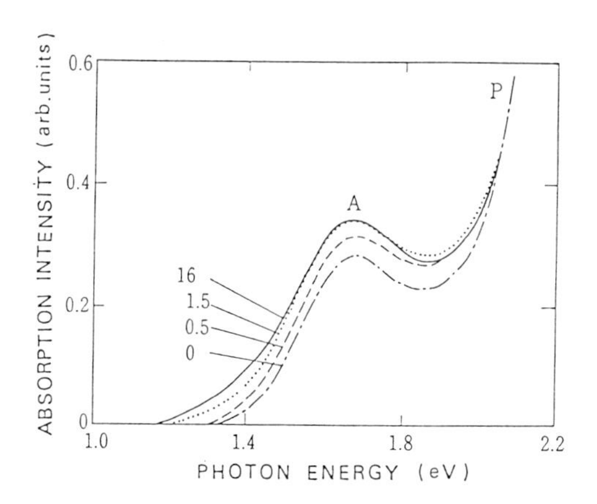

Fig.5. Chlorine-doping-induced absorption spectra of $[Pt(en)_2][Pt(en)_2Cl_2](ClO_4)_4$ for the polarization parallel to the chain axis at room temperature. The numbers are the reaction time in hours with the chlorine gas of 1 atom.

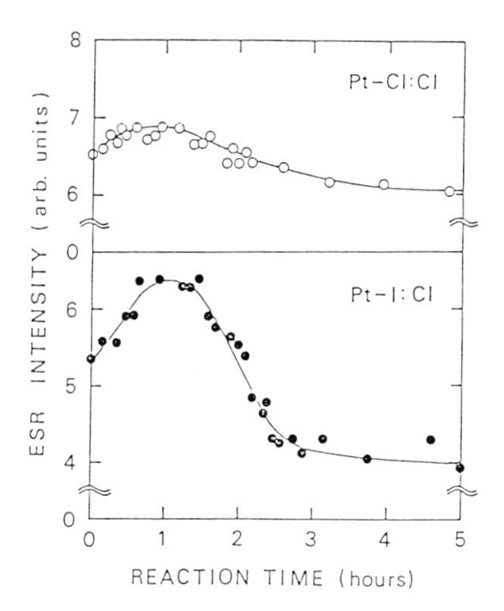

Fig.6. Chlorine-doping-induced ESR intensities of $[Pt(en)_2][Pt(en)_2X_2](ClO_4)_4$ (X=Cl, I) as a function of the reaction time.

whether the defect states are polarons or not. If the A and B band absorption and the ESR intensity increase with the halogen-doping, the defect will be inferred to be a polaron. Figure 5 shows the halogen-doping effect on the optical absorption spectra of the A band at room temp-ature. The absorption spectra were measured as a function of reaction time with chlorine gas. The A and B bands do not separate at room temperature. As seen in the figure, the absorption intensity increases with halogen-doping at light doping levels. This can also be seen in the ESR spectra (Fig.6). The intensity of ESR increases with the reaction time if the reaction time is shorter than one hour. These results show that the photo-induced states are a polaron pair. The decrease of the ESR intensity for doping for more than one hour will be discussed later.

Let us consider the mechanism of polaron formation by photo-excitation: when the charge-transfer transition is induced by light, a d_{z^2} electron on a Pt^{2+} ion transfers to the d_{z^2} orbitals of the neighbouring Pt^{4+} ions. Then, two Pt^{3+} ions are produced in the chain. Such a CT excited state can return to the original ground state, by emitting a photon. The strong luminescence[4,14] indicates that this process is the dominant relaxation channel. However, the CT exciton will have some chance to produce a pair of P^+ and P^- states, either by succesive transfer of an electron, or by simultineous transfer of two electrons to the neighbouring ions(Fig.2(P)). The energy, needed for this electron transfer can be supplied by the excess energy when photon excites the electron beyond the band gap. The polaron mass in this crystal has been estimated to be several hundred times as much as that of free electron[15]. Polaron (and the other excitation species) in the chain are far less mobile than in trans-polyacetylene. This explains why the photo-induced states(polarons) are stable at low temperature.

Nasu and Mishima calculated the ground and excited states with the one-dimensional extended Peierls Hubbard model[16,17]. From the adiabatic energy diagram, they have shown that the CT state, created by a photon energy higher than the band gap, relaxes to the excited states of the self-trapped exciton, and then is partially dissociated into a separated polaron pair but that the path to soliton pair, from the photo-excited state, has a potential barrier to be surmounted. This explains why the photo-excited state relaxes to a polaron pair even though the formation energy of polaron is higher than that of soliton(see table).

The halogen doping produces positive polarons on the chains. At the heavy doping level, the density of positive polarons is so high, that the interaction between the polarons will change two polarons into the positive soliton pair(bipolaron), which is more stable than two polarons.

$$P^+ + P^+ \rightarrow S^+ + S^+ \text{ (bipolaron)}$$

As a result of this reaction, the spin density decreases and a new absorption band appears at the low-energy side of the A band (Fig. 4 and Ref. 10) as the halogen-doping increases. This should not happen for the photo-induced polarons, because the photo-induced states are a polaron pair (P^+-P^-) and they will recombine at high density.

REFERENCES

1. See, for example, Y. Lu, "Solitons and polarons in conducting polymeres" World Scientific Singapore (1988).
2. M. Yamashita, K. Toriumi and T. Ito, Acta Crystallogr., C41:876 (1985).
3. M. Yamashita, M. Matsumoto and S. Kida, Inorg. Chem. Acta, 31:L381 (1978).
4. M. Tanaka, S. Kurita, Y. Okada, T. Kojima and Y. Yamada, Chem. Phys., 96:343 (1985).
5. M. Tanaka and S. Kurita, J. Phys. C: Solid State Phys., 19:3019 (1986).
6. D. Baeriswyl and A. R. Bishop, J. Phys., C21:339(1988).
7. M. Tanaka, S. Kurita, T. Kojima and Y. Yamada, Chem. Phys., 91:257 (1984).
8. S. Kurita, M. Haruki and K. Miyagawa, J. Phys. Soc. Jpn., 57:1789 (1988).
9. S. Kurita and M. Haruki, Synth. Metals, 29:F129 (1989).
10. M. Haruki and S. Kurita, Phys. Rev., B39:5706 (1989).
11. M. Matsushita, N. Kojima, T. Ban and I. Tsujikawa, J. Phys. Soc. Jpn., 56:3808 (1987).
12. A. Kawamori, R. Aoki and M. Yamashita, J. Phys. C:Solid State Phys., 18:5487 (1985).
13. N. Kuroda, M. Sakai, Y. Nishina, M. Tanaka and S. Kurita, Phys. Rev. Lett., 58:2122 (1987).
14. H. Tanino and K. Kobayashi, J. Phys. Soc. Jpn. 52:1446 (1983).
15. Y. Onodera, J. Phys. Soc. Jpn., 56:250 (1987)
16. K. Nasu and A. Mishima, Rev. Solid State Science, 2:539 (1988).
17. A. Mishima and K. Nasu, Phys. Rev. B39:5758 (1989), ibid B39:5763 (1989).

SLIDING CHARGE DENSITY WAVES IN INORGANIC CRYSTALS

András Jánossy

Central Research Institute for Physics
H-1525 Budapest 114, POB 49, Hungary

I. INTRODUCTION

Some quasi-one dimensional materials have a curious low-temperature structure: along the chains of highly overlapping molecules or atoms, both the electronic density and the atomic positions are periodically modulated with respect to the high-temperature uniform structure. The periodic lattice distortion (PLD) and the charge density wave (CDW) wavelength, determined by the band filling, is incommensurate in these materials to the lattice constant along the chains, consequently in ideal systems the phase of the modulation is not coupled to the lattice. Here an "ideal system" means an infinite crystal with no lattice defects. If boundaries and impurities were of no importance, the CDW (together with the PLD) would "float" within the crystal.

The weak interchain coupling orders the modulated structure three-dimensionally at the Peierls transition temperature, T_p . The force due to an applied electric field, within regions where the CDW is coherent, is counterbalanced by the pinning force of impurities or other defects. For high-purity crystals the application of small electric fields may depin the CDW from the lattice, and the modulated structure slides as a whole.

Fröhlich[1] was the first to describe in 1954, theoretically, this type of collective electron motion. Bardeen[2] suggested twenty years later that it may contribute to the conductivity of the organic charge transfer salt TTF-TCNQ above T_p, and later proposed[3] that this "Fröhlich mode" gives rise to the non-linearities of the conductivity in $NbSe_3$ observed by Monceau et al.[4] below T_p .

Since then sliding of CDW's has been found in diverse materials[5], most of them inorganic, like TaS_3 or the blue bronzes $A_{0.3}MoO_3$ (A = K, Rb, Tℓ). In TTF-TCNQ the non-linear conductivity has been observed[6] in the temperature range below T_p where the CDW is incommensurate.

The non-linear conductivity appearing above a threshold electric field has become a fairly convincing sign for the sliding of the CDW, especially since it was shown[7] by X-ray diffraction that the PLD amplitude is not changed when the CDW is depinned. The most unambiguous test[8], however, is by NMR, which probes the motion of the CDW's by the fluctuations of the hyperfine field at some nuclei nearby or on the chains.

A number of reviews[9] have been published on the subject and in this short paper we shall deal with only a limited number of experiments aimed

Lower-Dimensional Systems and Molecular Electronics
Edited by R.M. Metzger *et al.*, Plenum Press, New York, 1991

at the understanding of the sliding motion. We take the "blue bronzes" as a model system, for which a large body of structural, transport, and NMR data is available.

We first deal with the structure of the CDW, in particular with observations of the incommensurability. This is followed by the description of the sliding state in view of NMR and voltage noise experiments. Finally some experimental evidence is presented for topological defects of the CDW, which may help to overcome the restoring forces at the interface between sliding and pinned regions.

II. CHARACTERIZATION OF THE STATIC CDW OF BLUE BRONZES

For the understanding of the dynamics it is important to know the main structural properties of the pinned CDW. In the simplest case one may neglect completely the (incommensurate) lattice potential, and the CDW is a plane wave with a constant phase ϕ:

$$\rho = \rho_o + \rho_1 \cos(\underline{q}\underline{r} + \phi) \quad (1)$$

where ρ_o is the uniform density and ρ_1 is the amplitude of the CDW. The wavevector, $\underline{q}$, is determined by the band filling, and $q_b = 2k_F$ for a one-dimensional chain along the lattice vector $\underline{b}$. If q_b is very close to low-order commensurability (i.e. to $q_c = (n/m)(2\pi/b)$, where n and m are small integers), the lattice potential may modulate the CDW and instead of (1), it will consist of commensurate regions, with $q = q_c$ and ϕ optimized to the lattice potential, interrupted by discommensurations in which ϕ varies rapidly. So far CDW sliding has been observed only in incommensurate systems; the possibility that it occurs in a commensurate one is remote, since the periodic lattice potential pins the CDW very effectively. In a nearly incommensurate system the charge would be carried by the motion of discommensurations, such a system, however, has not yet been found. In incommensurate systems the pinning to defects is essential; the threshold electric field increases with defect concentration. The CDW phase coherence length may be determined by the impurities or by crystal boundaries. The dynamics may be affected by boundaries between pinned and unpinned regions for systems with long phase coherence lengths.

The blue bronzes $K_{0.3}MoO_3$ and $Rb_{0.3}MoO_3$ have almost identical physical properties. Chains of MoO_6 octahedra form a quasi-1D metal at room temperature. The one-dimensional CDW fluctuations are well observed[10] by neutron and X-ray spectroscopy. Below T_p = 180 K, the Peierls transition temperature, a gap opens at the Fermi momentum, k_F, in the one-dimensional electronic spectrum. The CDW becomes static with a 3D ordered structure and a wavevector of $\underline{q} = (0, q_b, 1/2)$ in reciprocal lattice units. Along the chains $q_b = 2k_F$, and the band filling is such that the wavelength is incommensurate to the lattice constant. The CDW is coupled to a periodic lattice distortion, which has components both longitudinal and transverse to $\underline{b}$. The Peierls transition is of second order and the displacements may be characterized by a complex order parameter, $ne^{i\phi}$. The CDW amplitude is proportional to n, while its phase is given by ϕ. By lowering the temperature below T_p, n first increases rapidly and then saturates. Below 100 K, n, is practically constant.

The wavevector q_b depends somewhat on temperature; it approaches closely the commensurate value of q_c = 3/4 as the temperature is decreased. Some studies[11] reported a lock-in to a commensurate phase at 100 K. This is ruled out by the form of the ^{87}Rb NMR spectrum[12]. The NMR resonance frequency of the Rb sites at R_i is modulated by the CDW and in the local approximation for a plane wave, up to second order in the CDW amplitude:

$$\delta\nu(R_i) = \nu_1 \cos(\underline{q}\,\underline{R}_i) + \nu_2 \cos^2(\underline{q}\,\underline{R}_i) \quad (2)$$

where ν_1 and ν_2 depend on the electric field gradient tensor and are proportional to n and n^2, respectively. For an incommensurate wave $\cos(\underline{q}\,\underline{R}_i)$ takes all values between -1 and $+1$, and the NMR spectrum consists of a continuum[13]

$$g(\nu) = K[1-(\nu/\nu_1)^2]^{-1/2} \quad \text{for} \quad |\nu| < \nu_1$$
$$g(\nu) = 0 \quad \text{for} \quad |\nu| > \nu_1 \quad (3)$$

Here ν is measured from the unperturbed Larmor frequency, and $\nu_2 << \nu_1$ is assumed, as is valid for $Rb_{0.3}MoO_3$. K is a normalizing constant. On the other hand the spectrum of a commensurate system consists of a small number of discrete lines. The NMR spectrum of $Rb_{0.3}MoO_3$ clearly corresponds to an incommensurate CDW. It is rather surprising that the lattice potential does not modulate the CDW measurably, despite the closeness of q_b to $3/4$. At 100 K $q_b - q_c \approx 0.0025$; nevertheless the CDW is well approximated by a plane wave.

Commensurate regions are found[14], however, by the ESR of Mo^{5+} defects below 50 K, where the spectrum consists of narrow lines superimposed on a

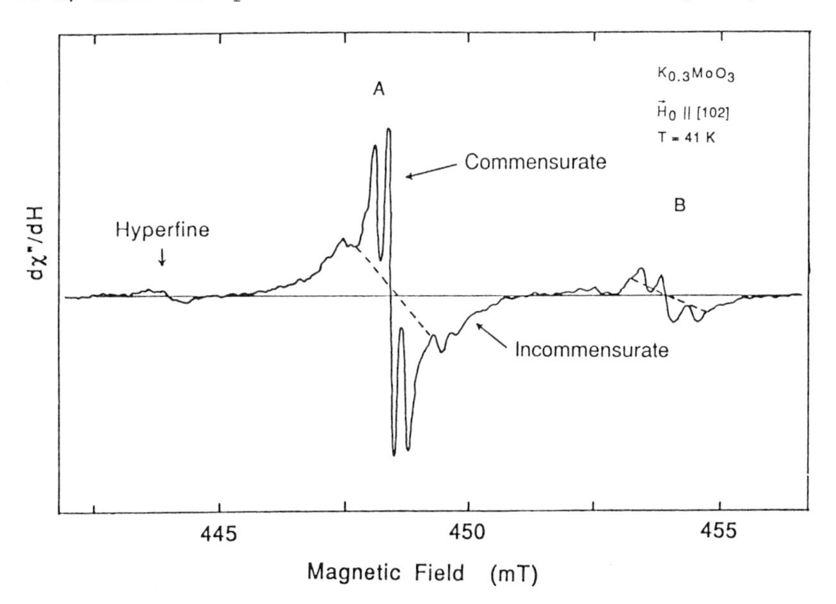

Fig. 1. ESR absorption derivative of $K_{0.3}MoO_3$ at T = 41 K. Lines A and B correspond to two types of defects. The dotted line indicates a decomposition into an inhomogeneously broadened component and a set of narrow lines due to regions of incommensurate and commensurate CDW, respectively. (Reproduced from Ref. 14.)

the NMR: the CDW modulates the Mo^{5+} gyromagnetic-factor. The transition to a commensurate CDW is incomplete down to 4.2 K, although the commensurate regions increase gradually with decreasing temperature. The formation of a discommensuration lattice can not be ruled out, although the spectra indicate, rather, that commensurate and discommensurate regions are well separated.

broad line (Fig. 1). The explanation for these spectra is the same as for Barisić[15] suggested that the screening of charged impurities may drive regions locally commensurate. At low temperatures, where the conduction electron density is small, the screening length is long, and by evenly distributing the screening charge, so that locally $q_b = q_c$, a single impurity may drive a region of many hundred wavelengths commensurate. Further ESR work and ^{87}Rb NMR study at low temperatures would be highly desirable to settle the question.

The pinning of the CDW to lattice defects may be strong or weak, depending on whether the phase ϕ at the impurity is determined solely by the potential of this impurity, or by many surrounding ones. One expects charged defects to pin the CDW efficiently. For strongly pinning defects only a single narrow EPR line is expected in both the commensurate and incommensurate phase. The ESR study[14] is at variance with this explanation. One observes a range of resonance frequencies for the incommensurate regions, and at least six (out of the eight possible) lines for the commensurate regions, showing that the pinning of the CDW to the Mo^{5+} defects is predominantly weak.

The CDW phase coherence length measured by X-ray diffraction increases[16] rapidly below T_p. For high-quality crystals at about 100 K the longitudinal coherence becomes longer than 1 μm. At temperatures below 100 K the transverse coherence decreased[16,17] by applying a current for a while.

III. THE SLIDING CDW STATE

A CDW depinned from lattice imperfections by an applied electric field is simply described by eq. (1), but with a phase $\phi(t)$ continuously increasing with time. It is observed first of all, as an excess current, usually called "nonlinear" to distinguish it from the "linear" current carried by the normal electron excitations across the gap (Fig. 2). The nonlinear

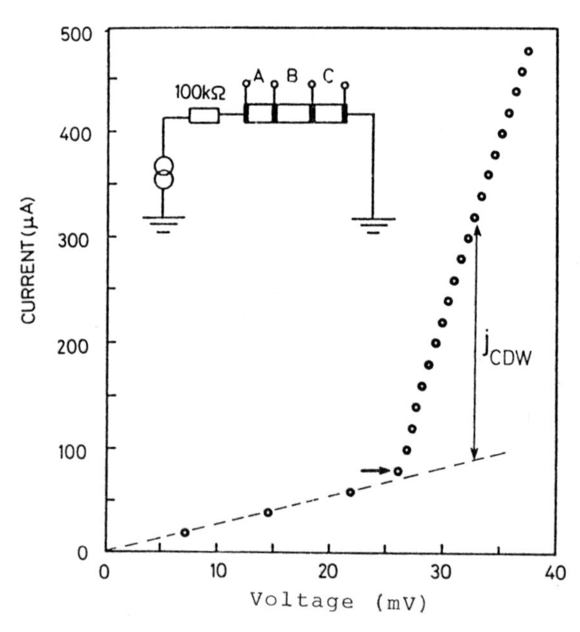

Fig. 2. Current-voltage characteristics of segment B of a $Rb_{0.3}MoO_3$ crystal at T = 77 K measured in the configuration shown in the inset. The voltage pulses of Fig. 4 appear at the current indicated by the arrow. (From Ref. 25.)

current may be expressed by the drift velocity of the condensate, v_d:

$$j_{CDW} = ne\, v_d = \frac{ne}{q_b}\frac{d\phi}{dt} . \quad (4)$$

A direct way to measure[18] the drift velocity is by NMR. The hyperfine field at the ^{87}Rb nuclei, described by eq. (2) for the static case, now oscillates with a frequency of

$$\nu_d = \frac{1}{2\pi}\frac{d\phi}{dt} \quad (5)$$

since the PLD, which moves together with the CDW, corresponds to a coherent temporal oscillation of the atoms. If ν_d is constant in time and homogeneous in space, the NMR spectrum is [19]:

$$G(\nu_1, \nu_d) = \sum_{p=0}^{\infty} J_p^2(\nu_1/\nu_d)\ \delta(\nu - \nu_2/2 - p\nu_d) \quad (6)$$

neglecting dipolar and other broadening. (J_p is the Bessel function of order p). In contrast to the broad static spectrum, for $\nu_d > \nu_1$ most of the intensity is in the narrow central line $p = 0$. This motionally narrowed central line and the $p = \pm 1$ sidebands at $\pm\nu_d$ have been observed[8] in blue bronzes (Fig. 3). Although in real systems ν_d is inhomogeneous, its spatial average is still proportional to the full nonlinear current flowing in the sample.

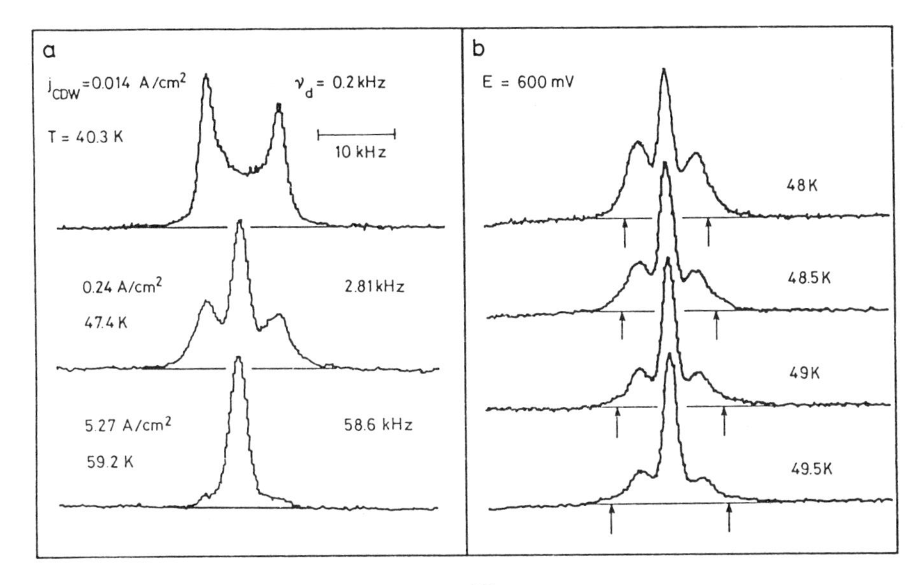

Fig. 3. Motional narrowing of the ^{87}Rb NMR of a $Rb_{0.3}MoO_3$ crystal under current. (a). The applied field was held constant and j_{CDW} is increased by increasing the temperature. The average local-field oscillation frequencies, $\bar{\nu}_d$, are indicated. For large currents only the $p = 0$ central component is observed. (b). Evidence for the $p = \pm 1$ NMR sidebands. The arrows indicate the positions of the sideband maxima, as expected from the experimental noise spectra. (Spectra of Fig. 3. (a) and (b) are taken with different crystals.) (Reproduced from Ref. 18.)

It has been shown[20,9], for a number of sliding CDW systems, that the drift velocity varies periodically in time in a constant electric field. The frequency of the related voltage noise ν_n is found to be proportional to the CDW current. A phenomenological model[21], in which the CDW is considered rigid, and is described by a single coordinate moving in a potential with a periodicity equal to the CDW wavelength λ, provides a simple guide to the phenomenon. In such a potential, which could arise from a single impurity, for a constant total current, the voltage along the sample oscillates periodically. The period $2\pi/\nu_n$ corresponds to a displacement by λ of the CDW. Except for $E \approx E_t$, the modulation (with frequency ν_n) of the drift velocity v_d is small. ν_n is measured by a spectrum analyser and ν_d (related directly to v_d) by the motional narrowing of the NMR. In agreement with the measurement of the CDW current/noise frequency ratio, the simultaneous measurement[18] of the NMR lineshape and voltage noise under current leads to $\nu_n = \nu_d$.

The "impurity potential" models neglect the problem of boundaries. Ong, Verma, and Maki[22] and independently Gor'kov[23] suggested that boundaries between pinned and unpinned regions are the source of voltage noise. In the theory of Ong, Verma, and Maki, CDW phase vortices generated near the electrode contacts move transverse to the current. The vortices serve to convert the CDW current into normal current which flows into the pinned regions neighbouring the contacts.

Lee and Rice[24] pointed out several years ago that the CDW may not become depinned simultaneously in the whole crystal but only within CDW phase dislocation loops. The sample is divided into conducting and pinned regions extending along the conducting axis if the pinning and/or the electric field is inhomogeneous. The CDW can not, however, slip simultaneously at the full boundary of such regions; the slip occurs only at vortices or dislocation loops, which propagate along the crystal. The loops are supposed to be perpendicular to the conducting axis. Within the loop each chain carries a charge of 2e, in excess to that on pinned chains outside. The charge is more or less localized to the sheet of the loop, and the CDW current propagates in pulses along the crystal. The loops must be created and annihilated at the ends of the crystal as envisaged by Ong, Verma, and Maki.

Very recently evidence has been found[25] for the propagation of voltage pulses in blue bronze. Fig. 4 shows the simultaneous oscilloscope traces of the voltage of three segments of a sample to which a current just above threshold has been applied. Voltage pulses appear in all three segments, but with a delay increasing from one end to the other. This delay which is found to decrease with applied electric field is inversely proportional to the propagation velocity and increases rapidly with decreasing temperature. For higher fields the frequency of pulses increases and they merge into a quasi-periodic voltage oscillation. The phase of the oscillation shifts from segment to segment, showing that the CDW voltage noise propagates with a finite velocity.

The "sources" and "sinks" of the pulses where the dislocation loops are created and annihilated have been distinguished by experiments[26] in which the segments were selectively perturbed by a periodic external field. Perturbation of the segment in which the source is found, changes the rate of voltage pulses; perturbation of other segments does not influence the pulses.

In the blue bronzes two types of depinning have been observed: at temperatures above about 30 K with a relatively small threshold field[27] of the order of $E_t \approx 0.1$ V/cm and at lower temperatures with E_t in the order of

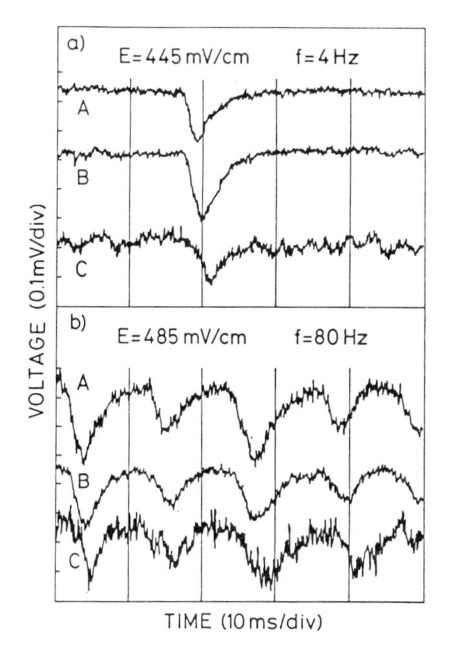

Fig. 4. Propagation of quasi-periodic voltage noise in a $Rb_{0.3}MoO_3$ crystal. The experimental set up is indicated in Fig. 1. By slightly increasing the current the voltage pulses of (a) merge into oscillations of (b). The voltages of the three segments A, B, C are recorded simultaneously; the delay corresponds to a propagation from A towards C. (Ref. 25.)

10 V/cm[28]. At low temperatures the normal current is negligible (at 4.2 K the normal resistance is unmeasurably high) and, since charge neutrality must be maintained, the CDW current has to be uniform along the sample. In contrast, at high temperatures the normal excitations allow an inhomogeneity of the CDW current along the conducting direction and - as we have shown above - this is indeed the case.

This work was supported by the Hungarian Academy of Sciences grant AKA 86-292 and an OTKA grant.

REFERENCES

1. H. Fröhlich, Proc. Roy. Soc. London, Ser A223, 296 (1954).
2. J. Bardeen, Solid State Comm. 13, 357 (1973).
3. J. Bardeen, in Highly Conducting One-Dimensional Solids, edited by J.T. Devreese, R.P. Evrard and V.E. van Doren; p. 375, Plenum Press New York, 1979.
4. P. Monceau, N.P. Ong, A.M. Portis, A. Meerschant and J. Rouxel, Phys. Rev. Lett. 37, 602 (1976).
5. See e.g. Proceedings of the International Conference on Charge Density Waves in Solids, Budapest, Hungary, 1984, edited by Gy. Hutiray and J. Sólyom; (Lecture Notes in Physics, Vol. 217, Springer Verlag--Berlin; 1985).

6. R.C. Lacoe, H.J. Schulz, D. Jérome, K. Bechgaard, I. Johannsen, Phys. Rev. Lett. 55, 2351 (1985).
7. R.M. Fleming, D.E. Moncton, and D.B. Mc Whan, Phys. Rev. B18, 5560 (1978).
8. J.H. Ross, Jr., Z. Wang, and C.P. Slichter, Phys. Rev. Lett. 56, 663 (1986); K. Nomura, K. Kume, and M. Sato, J. Phys. C19, L289 (1986); P. Ségransan, A. Jánossy, C. Berthier, J. Marcus and P. Butaud, Phys. Rev. Lett. 56, 1854 (1986).
9. N.P. Ong, Can. J. Phys. 60, 757 (1981); G. Grüner, Comments Solid State Phys. 10, 173 (1983); P. Monceau, D. Riedel, Publishing Company Holland, p. 139 (1985); J.C. Gill, Contemp. Phys. 27, 37 (1986).
10. J.P. Pouget, C. Escribe-Filippini, B. Hennion, R. Currat, A.H. Moudden, R. Moret, J. Marcus and C. Schlenker, Molec. Crystals Liqu. Crystals, 121, 111 (1985); M. Sato, H. Fujishita and S.Hoshino, J. Phys. C16, L877 (1983).
11. R.M. Fleming, L.F. Schneemeyer and D.E. Moncton, Phys. Rev. B31, 899 (1985).
12. K. Nomura, K. Kume and M. Sato, Solid State Commun. 61, 33 (1987).
13. R. Blinč, Phys. Report 79, 331 (1981); R. Blinč, P. Prelovsek, V. Rutar, J. Seliger and S. Zumer in Incommensurate Phases in Dielectrics, Eds. R. Blinč and A.P. Levanyuk, Elsevier Science Publishers, BV 1986 Ch. 4.
14. A. Jánossy, G.L. Dunifer and J.S. Payson, Rev. B38, 1577 (1988).
15. S. Barišič and I. Batistič, to be published
16. R.M. Fleming, R.G. Dunn and L.F. Schneemeyer, Phys. Rev. B31, 4099 (1985).
17. T. Tamegai, K. Tsutsumi, S. Kagoshima, Y. Kanni, H. Tomozawa, M. Tani, Y. Nogami and M. Sato, Solid State Commun. 56, 13 (1985).
18. A. Jánossy, C. Berthier, P. Ségransan and P. Butaud, Phys. Rev. Lett. 59, 2348 (1987); P. Butaud, A. Jánossy, P. Ségransan and C. Berthier, to be published in Phys. Rev. B.
19. M. Kogoj, S. Zumer and R. Blinč, J. Phys. C17, 2415 (1984).
20. R.M. Fleming and C.C. Grimes, Phys. Rev. Lett. 42, 1423 (1979).
21. G. Grüner, A. Zawadowski and P.M. Chaikin, Phys. Rev. Lett. 46, 511 (1981); P. Monceau, J. Richard, and M. Renard, Phys. Rev. B25, 931 (1982).
22. N.P. Ong, G. Verma and K. Maki, Phys. Rev. Lett. 52, 663 (1984).
23. L.P. Gor'kov, Pisma Zh. Eksp. Teor. Fiz. 38, 76 (1983); JETP Lett. 38, 87 (1983).
24. P.A. Lee and T.M. Rice, Phys. Rev. B19, 3970 (1979).
25. T. Csiba, G. Kriza and A. Jánossy, Europhysics Lett. 9, 163 (1989).
26. T. Csiba, G. Kriza and A. Jánossy, preprint to be published
27. J. Dumas, C. Schlenker, J. Marcus, R. Buder, Phys. Rev. Lett. 50, 757 (1983)
28. G. Mihály and P. Beauchène, Solid State Commun. 63, 911 (1987).

SYNTHESIS, STRUCTURAL CHARACTERIZATION, AND CONDUCTIVITY OF THE FIRST LARGE ORGANIC ELECTRON INTERCALATES OF GRAPHITE

Thomas E. Sutto and Bruce A. Averill

Department of Chemistry
University of Virginia
Charlottesville, Va 22901, USA

INTRODUCTION

The intercalation of graphite with various alkali metal and rare-earth oxides has been well characterized [1]. However, the synthesis of organic intercalates of graphite has, up to now, been limited to solvent inclusion of small-ring systems. Reported here are the first intercalates of graphite with large aromatic ring systems that are used to form organic metals.

The layers of graphite possess not only a high degree of conductivity within the plane, but these layers can also act as both electron donors, as in the case of potassium intercalation, or as electron acceptors, as is the case of most rare-earth oxide intercalates [1]. This aspect of graphite, as well as the metal-like conductivity of the planes, makes it an ideal host species for the intercalation of various organic ring systems that are good one-electron donors or acceptors.

SYNTHESIS

The use of sonication in enhancing intercalation reactions has been previously reported [2]. In an attempt to adapt this technique to graphite (Crystalline Graphite, Aesar, 300-mesh), the nitrate-graphite intercalate was prepared as an intercalation precursor, by refluxing graphite in fuming nitric acid for 48 hours. The x-ray data on the material indicated that graphite nitrate had indeed been formed.

The precursor was then placed in a solution of (2 : 2 : 1) - (DME : DMF : H_2O), and sonicated for 4 minutes. This suspension was then added to a previously dissolved solution containing the organic species to be intercalated. The materials intercalated were: tetracene (TT), 1,3,10,12-tetrachlorotetracene (TCT), perylene (Per), tetracyano-tetrathiafulvalene (TCN-TTF), and 2,3-dithio-dipyrazine (DTDP): these are good electron donors, which are expected to donate electrons to the graphite planes. Intercalation reactions were run at 60 °C for one week. The products were isolated by filtering the reaction mixture, and washing six times each with water and ethanol. The material was then washed with acetone, until the effluent was clear.

X-ray data were taken on a Scintag powder diffractometer from $\theta = 5 - 60°$, in steps of 0.01°, using Cu K_α radiation. Resistivity data were collected on pressed pellets at room temperature, using a two-probe d.c. method. Resistivity measurements were normalized to those of graphite.

Lower-Dimensional Systems and Molecular Electronics
Edited by R.M. Metzger *et al.*, Plenum Press, New York, 1991

RESULTS

The most interesting aspect of these intercalation reactions is the speed with which they occurred. The intercalation of graphite with TCT occurred within ten minutes. TCT was insoluble in the solution medium used, however, at the end of a ten-minute period, the only remaining material was red graphite. In general, this pattern held true for the other intercalates; the reactions seemed to occur quickly, and with the exception of the DTDP intercalate, the resulting graphite intercalates were brightly colored species.

Chemical analysis of the TT intercalate indicates that the ratio of TT to carbon in graphite was between 1 : 60 and 1 : 70. Since the chemical analysis determined only the hydrogen content of TT, the exact stoichiometry of the material is questionable. The nitrogen content was found to be less than 0.5%, indicating that the nitrate was completely displaced, and that little or no solvent inclusion occured.

The X-ray diffraction patterns were extremely clean. The most prevalent line was the 001 reflection. This allowed us to examine the **c**-axis expansion easily. The reflections with non-zero h or k, needed to determine the **a** and **b** parameters, were relatively few, but the **a** and **b** axes were obtained, with relatively large standard deviations of 0.3 Å. The unit cell dimensions are given in Table 1.

Table 1. Unit cell parameters of graphite-organic intercalates.

Compound	a/Å	b/Å	c/Å
$TT_{1.0}C_{60}$	8.52	15.51	24.014
TCT_xC_y	5.71	15.37	19.128
Per_xC_y	11.35	13.03	19.566
$DTDP_xC_y$	8.64	12.96	13.011
$TCN\text{-}TTF_xC_y$	8.10	15.67	14.314

The large values for the **c**-axis indicate that the materials intercalate into graphite vin stages, with most of them orienting parallel to the graphite layers. The precursor, graphite nitrate, is known to be staged as $\text{-C-C-NO}_3\text{-C-NO}_3\text{-C-C-NO}_3\text{-C-}$, where each -C- indicates a carbon layer. Using the same type of fitting pattern to analyze the new organic intercalates, one finds that the staging for the TT intercalate is -C-C-C-TT-C-C-TT-C-C-C-. The intercalation staging for the Per intercalate is similar to that of the nitrate, i.e. -C-C-Per-C-Per-C-C-. The staging for the DTDP and the TCN-TTF intercalates are the same third-stage type of intercalation, which is -C-C-C-Guest-C-C-C-. The TCT intercalate also has a third-stage intercalation. However, the distance between the layers of graphite, as well as the small a-axis, indicate that TCT is orienting perpendicular to the graphite planes, not parallel.

The resistivities and normalized resistivities are given in Table 2. The data indicate that the organic intercalates decrease the conductivity of the material. The TCN-TTF gave the lowest resitivity, while the TCT intercalate yielded the largest. However, since these were unoriented pressed pellets, the actual conductivity in the planes may be considerably larger than reported here.

DISCUSSION

One of the more interesting structural findings is that the TCT intercalate is oriented perpendicular to the graphite planes. This is most likely due to the preference of the chlorine atom to interact with the graphite layers at an angle of 180° from its bond to the TT molecule. Otherwise, the overlap of the chlorine would be at the unlikely angle of 90°.

In light of this, the resistivity data can be empirically interpreted. The presence of the neutral, flat aromatic rings between the graphite layers most likely decreases the interlayer interaction. The presence of the perpendicular TCT molecule would cause the greatest

Table 2. Resistivities of graphite-organic intercalates.

Compound	Resistivity (S/cm)	Normalized Resisitivty
Graphite	1.33	1.00
$TT_{1.0}C_{60}$	9.46	7.11
TCT_xC_y	11.85	8.91
Per_xC_y	11.51	8.65
$DTDP_xC_y$	5.28	3.97
$TCN\text{-}TTF_xC_y$	3.47	2.61

separation, and thus, cause the largest decrease in inter-layer interactions. Therefore, the TCT intercalate has the highest resistivity. For the TCN-TTF and the DTDP intercalates, the heterocyclic ring systems most likely does not cause as big a decrease in the inter-layer interactions, and these therefore have the lowest resistivity.

CONCLUSION

The intercalation of graphite with large organic systems has been investigated. Numerous physical properties still need to be investigated, including thin-film resistivity, and magnetic susceptibility probes of the Pauli susceptibility of these materials.

Since these few compounds have been synthesized, the intercalation of graphite with numerous other guest species seems possible. These include intercalation of numerous organic metals, as well as other heterocyclic systems. It is hoped that the potential for these types of reactions will give rise to new classes of intercalates, and possible even graphite polymers.

REFERENCES

[1] M. S. Whittingham and A. Jacobson, Editors, "Intercalation Chemistry" (Academic Press, New York, 1982) pages 31, 44.

[2] K. Chatakondu, M. Green, M. Thompson, and K. Suslick, The enhancement of intercalation reactions by ultrasound, J. Chem. Soc. Chem. Commun. 900 (1987).

SYNTHESIS, STRUCTURAL CHARACTERIZATION, AND SUPERCONDUCTIVITY IN NOVEL NEUTRAL ORGANIC INTERCALATES OF TaS_2 AND TiS_2

Thomas E. Sutto and Bruce A. Averill

Department of Chemistry
University of Virginia
Charlottesville, Va 22901, USA

INTRODUCTION

The intercalation of TaS_2 and TiS_2 with large aromatic organic guests is presented here. Recently, it has been reported that sonication of the sodium intercalate of these materials allows for a rapid intercalation of many large molecules [1]. Therefore, we attempted to introduce large, neutral aromatic compounds, such as pentacene and tetracene, into these metal dichalcogenides, in the hope of observing novel structural and physical characteristics.

SYNTHESIS

The sodium intercalates were prepared via a salt reaction in solution. X-ray data confirmed that the sodium intercalate had been formed, with the highest, 001 reflection corresponding to the hydrated $Na_{0.33}TaS_2$ or $Na_{0.33}TiS_2$. The sodium intercalate was then placed in an acetonitrile-water solution, and sonicated for 1.5 minutes. This suspension was then added to a previously prepared solution of the guest species. The guest species intercalated into TaS_2 and TaS_2 were: tetracene (TT), 1,3,10,12-tetrachlorotetracene (TCT), perylene (Per), and pentacene (Pen). Intercalation reactions were allowed to run for 20 days at 60 °C. The materials were washed six times in succession with water , then ethanol, then acetonitrile. The samples were finally dried in an active vacuum for 24 hours.

X-ray data were obtained with a Scintag powder diffractometer in the Bragg angle range $\theta = 5 - 70°$, in 0.01° increments, using Cu K_α radiation. Magnetic susceptibility data were taken on a S.H.E. SQUID, with the powdered sample placed in a magnetically neutral Mylar bucket. Resistivity measurements were made on pressed pellets at 296 K, with a two-probe d.c. method.

RESULTS

Visual inspection of the samples showed that the intercalated materials took on the color of the guest species. The tetrachlorotetracene intercalates were bright red. The tetracene intercalates were bright orange, while the perylene intercalates appeared yellow. However, the pentacene intercalates had a blue-black color.

The best fit to the observed X-ray data yielded orthorhombic unit cells for all the intercalates (Table 1); however, the standard deviation for these fits was around 0.2 Å. Cell expansion along the **c**-axis indicates that these planar organic species are going into the material at an angle of nearly 45°. This can be deduced from a simple geometric consideration of the guest species. In the case of tetracene, which is 13.9 Å long and 7.2 Å wide, the cell

Lower-Dimensional Systems and Molecular Electronics
Edited by R.M. Metzger *et al.*, Plenum Press, New York, 1991

parameters are **a** = 8.918 Å, **b** = 10.071 Å, and **c** = 24.270 Å. The nearest fit to the **a** axis is the width of tetracene, while the **b** axis could correspond to the projected length of the tetracene molecule onto the planes of the host species, if the molecule is at an angle of 44.1°. If the molecule is placed in that configuration, then the overall **c**-axis expansion corresponds to approximately double the depth of a single tetracene molecule at a 44.1° angle, indicating that the stacking of the guest species is bilayered, at angles of about 45° to the layers of the metal dichalcogenide.

Table 1. Unit cell parameters of metal dichalcogenide-organic intercalates.

Compound	a/Å	b/Å	c/Å
Per_xTaS_2	15.181	8.047	12.502
TT_xTaS_2	8.918	10.052	24.270
TCT_xTaS_2	10.107	9.909	24.620
Pen_xTaS_2	8.886	14.110	28.054
Per_xTiS_2	15.498	10.071	12.122
TT_xTiS_2	9.189	10.302	23.062
TCT_xTiS_2	10.310	10.155	24.620
Pen_xTiS_2	9.389	13.348	29.149

Resistivity measurements were made on pressed pellets of the material: therefore, it could not be determined whether conductivity in one direction had increased. All samples were normalized to their respective starting materials (either TaS_2 or TiS_2). The resisitivity was high, in the order of 100 to 400 Ω cm for all samples. This is probably due to the very small particle sizes that arise from sonication.

Low-temperature susceptibility checks were run on all samples. Pen_xTaS_2 was shown to undergo a superconducting transition at 7.02 K in a 5 G field. Increasing the field strength to 350 G was sufficient to quench any superconductivity, which is due, again, to the small particle size of the sonicated material. All other samples showed a weak paramagnetic signal.

DISCUSSION

The most striking feature of these materials is the existence of superconductivity in the Pen_xTaS_2 material. The most likely reason for this is the donating of electrons in the π cloud of the pentacene molecule to the metal-sulfur layers, similar to the effect in the octadecylamine intercalate, which has a slightly lower superconducting transition at 4 K [2]. The lack of superconductivity in Pen_xTiS_2 is, most likely, due to the same reason for which neither TT_xTaS_2 nor TT_xTiS_2 superconduct. That is, that the amount of electron density that the guest species can donate to the host layers in the tetracene intercalates and in Pen_xTiS_2 is not great enough to cause superconductivity. Since titanium is considered to be more strongly electron-withdrawing than tantalum, it seems likely that hexacene of heptacene intercalation of TiS_2 should give rise to a similar form of superconductivity.

Also interesting is that these reactions were not done as cation-exchange reactions. Instead, a neutral organic species is substituted for the hydrated sodium. The ability to intercalate neutral species is not only shown to be feasible, but also facile.

CONCLUSION

The intercalation of metal dichalcogenides with large flat aromatic species, in the neutral form, has been demonstrated to give rise to superconductivity, as well as interesting structural features. These reactions can be considered as opening the door to the intercalation of various organic metals into metal dichalcogenides, in the hope of achieving interesting electronic properties, possibly even superconductivity, at temperatures even higher than those found for the unintercalated organic metals.

REFERENCES

[1] K. Chatakondu, M. Green, M. Thompson, and K. Suslick, The enhancement of intercalation reactions by ultrasound, J. Chem. Soc. Chem. Commun. 900 (1987).

[2] M. S. Whittingham and A. Jacobson, Editors, "Intercalation Chemistry" (Academic Press, New York, 1982) pages 236-237.

RECENT PROGRESS IN CONDUCTING POLYMERS: CAN WE EXPECT POLYMERS WITH CONDUCTIVITY GREATER THAN COPPER AND STRENGTH GREATER THAN STEEL?

Alan J. Heeger

Department of Physics
Institute for Polymers and Organic Solids
University of California, Santa Barbara, California 93106

I. INTRODUCTION

Although improvement of solid state properties through higher quality materials is a general goal of materials science, until recently there had been little optimism that this would be successful for the electrical conductivity of polymers. The reason for this is that,in order to achieve "metallic" behavior, doping to a relatively high level is required; the resulting charged impurities might then be expected to cause scattering and localization. Thus, the need for doping would negate any improvements toward macromolecular chain perfection.

Recent studies have shown that this is not the case in polyacetylene; improvements in synthesis and orientation have resulted in electrical conductivities as high as 10^5 S/cm.[1-4] Furthermore, the absence of a metallic temperature dependence with resistivity decreasing as the temperature is lowered[1-4] implies that the measured conductivity is still limited by material imperfections. Consequently, it is quite clear that the intrinsic electrical conductivity of polyacetylene and, by implication, of other conducting polymers, may be significantly greater than that of copper.

Why is the high concentration of (partially) disordered charged impurities so ineffective in scattering the conduction electrons? The answer is that,in a very fundamental sense, conducting polymers are self-organizing systems. The stiff conjugated chains self-consistently force the dopant counter-ions to go into channels or planes within the structure. The details of this self-organization have been demonstrated in an _in-situ_ study of the evolution of the structure of polyacetylene during electrochemical doping.[5] Because of this self-organization, the charged dopant ions are spatially removed from the quasi-one-dimensional conduction path, and resistive back-scattering is suppressed.[6] The situation is analogous to that which occurs in artificially layered semiconductors (quantum well heterostructures) where the high-mobility carriers are confined to layers spatially separated from the alternate layers which contain the dopant ions. This effect is enhanced for conducting polymers by a combination of the anisotropic screening and the quasi-one dimensional nature of the transport[6] (only the $2k_F$ Fourier component of the scattering potential is important). The conclusion is that in high quality chain-aligned materials, scattering from the charged counter-ion impurities can essentially be ignored.[6] Other limitations (molecular weight, interchain packing and order, sp^3 defect density, etc) can be removed by a combination of improvements in synthesis and by better processing; i.e. by the methodology of materials science.

A principal goal of the field of conducting polymers is, therefore, to strive for advances in materials quality that will enable the exploration of the intrinsic electrical

Lower-Dimensional Systems and Molecular Electronics
Edited by R.M. Metzger *et al.*, Plenum Press, New York, 1991

properties. With knowledge of these intrinsic properties, we will be in a position to begin to understand in detail the fundamental transport processes in this class of materials.

II. CONDUCTING POLYMERS: "DIRTY" CONDUCTORS OR TRUE METALS WITH DELOCALIZED CARRIERS?

The electrical conductivity (σ) of a metal can be expressed as a product of the number of carriers per unit volume (n) times the carrier mobility (μ):

$$\sigma = ne\mu \tag{1}$$

where e is the electronic charge. To optimize the conductivity, one wants to achieve the highest carrier density, and one wants these carriers to have high mobility. In the context of nearly free electron theory,

$$\mu = e\tau/m^* \tag{2}$$

where τ is the mean scattering time and m^* is the carrier effective mass. Since the carrier densities are of the order of 10^{21} or greater, and since these carriers go into relatively broad energy bands, the Pauli exclusion principle demands that they form a degenerate Fermi gas. Thus, the typical carrier velocity is the Fermi velocity, v_F, of the order of 10^8 cm/s. Consequently, the mean distance between scattering events (the "mean free path") is given by

$$\lambda = v_F\tau. \tag{3}$$

The shortest possible mean free path is one lattice constant (a); in the limit where $\lambda \sim a$, the transport is better described as due to *hopping of localized carriers*, rather than *scattering of free carriers*. If we consider the hopping limit (mean free path of about one carbon-carbon distance), $\lambda \sim 10^{-8}$ cm, then $\tau \sim 10^{-16}$ seconds, and $\mu \sim 0.2$ cm^2/V-s.

To get a moderately high density of carriers per unit volume is relatively easy. For example in polyacetylene, a doping level of 10% per carbon can be achieved, corresponding to a carrier concentration of about 4×10^{21} cm^{-3}. This is a typical number for conducting polymers such as polyacetylene, poly(paraphenylenevinylene), poly(paraphenylene), etc. The addition of side chains, as in the poly(3-alkylthiophenes), reduces the maximum carrier density somewhat, simply because of the smaller fractional volume occupied by the conjugated backbones. For the emeraldine salt of polyaniline, there is one carrier per (B-NH-B-NH)$^{+\cdot}$ repeat unit; again $n \approx 5 \times 10^{21}$.

We conclude that, in the hopping limit, the conductivity of a typical conducting polymer would be given by Eq. (1), with $n \approx 10^{21}$ cm^{-3} and $\mu < 0.2$ cm^2/Vs; i.e. $\sigma < 30$ S/cm. For maximum doping levels, the smaller conductivities that are often reported in systems such as polypyrrole, polyaniline, etc. imply that the carriers are truly localized and that the transport is via thermally-activated hopping, as for example by means of the famous phonon-assisted variable-range hopping mechanism.[7]

This elementary analysis of the electrical conductivity demonstrates that there are two important classes of heavily doped polymers:

(1) "Dirty" conductors with electrical conductivities of the order of 10 S/cm or less: Such systems, can be achieved with relative ease; *all that is required is a moderately high density of carriers.* Carrier delocalization is neither required nor implied by such values. In such cases, the heavily doped polymers are highly disordered systems, which can be considered as examples of the "Fermi glass" concept.[8] Indeed, π-conjugation is not even needed to achieve this low level of conductivity in polymers.[9]

(2) True metals in which the carrier mean free path is at least a few lattice constants: This only becomes relevant for those polymers in which the electrical conductivities are in excess of several hundred S/cm. In such systems, the molecular weight is sufficiently high, interchain order is sufficiently good, and

the defect density is sufficiently low that delocalization occurs leading to "free" metallic carriers with mean free paths much greater than a carbon-carbon repeat unit.

Note that the same "polymer" could fit into both categories, depending upon how it was synthesized and processed. Obvious examples are conjugated systems made by the precursor polymer route: e.g. polyacetylene made by the Durham route[10] and polyphenylenevinylene (and its derivatives).[11] If the amorphous precursor polymers are simply converted to the conjugated polymers, these are disordered systems, dominated by localization and hopping, and with electrical conductivities of a few S/cm. If processed during conversion, so as to achieve significant chain extension and chain alignment, these become true metals with delocalized carriers and electrical conductivities of several thousands of S/cm.

III. THE INTRINSIC ELECTRICAL CONDUCTIVITY: MEAN FREE PATH LIMITED BY PHONON SCATTERING

When the conjugation length is sufficiently large, the conductivity will reach the intrinsic level, and the electronic mean free path will no longer be limited by static defects and imperfections. In this "clean" limit, the electron mean free path is determined by phonon scattering; i.e. by scattering from the deviations from the regular periodic atomic spacings due to thermal motion of the atoms about their ideal equilibrium configurations.

The intrinsic back-scattering rate resulting from thermal phonons has been calculated:[6]

$$(1/\tau_{ph}) \approx [8\alpha^2/M\omega_o t_o] \exp-(\hbar\omega_o/k_BT) \tag{4}$$

where ω_o is the frequency of the $2k_F$ phonon, α is the electron-phonon coupling constant, M is the mass of the repeat unit (e.g. the CH mass in polyacetylene) and $t_o \approx 2.5$-3 eV is the π-electron transfer matrix element ($4t_o$ is the π-band width). On substitution of τ_{ph} into Eq. (1) and (2) with the effective mass appropriate for a nearly half-filled one-dimensional energy band, $m^*=(\pi\hbar/aV_F)$, one obtains $\sigma_{\|}$:

$$\sigma_{\|} = (e^2/4\pi\hbar a)na^3(M\omega_o t_o^2/\alpha^2\hbar)\exp(\hbar\omega_o/k_BT) \tag{5}$$

A similar exponential temperature dependence, $\exp(\hbar\omega_o/k_BT)$, is expected for all quasi-one-dimensional polymers; the important parameters are the carrier density (n), the bandwidth ($4t_o$), and $\hbar\omega_o/k_BT$. For polyacetylene, $\alpha \approx 4.1$eV/Å and $\hbar\omega_o \approx 0.12$ eV. Using these values, we estimate a room temperature value for the intrinsic conductivity of metallic trans-$(CH)_x$ which is 2 x 10^6 S/cm, about four times greater than that of copper.

In the intrinsic regime, the temperature dependence will be dominated by the exponential factor. Since $\hbar\omega_o/k_BT \sim 4$-5 at room temperature, decreasing the temperature to 150K would lead to an increase in $\sigma_{\|}$ by more than a factor of 50-100! This large value for $\hbar\omega_o/k_BT$ originates in the stiffness of the carbon-carbon bond. However, since even the best materials produced to date, with $\sigma_{\|}$(300K) in the range (0.2 -1)x10^5 S/cm, show no indication of this strong temperature dependence,[1-4] it is clear that the intrinsic room temperature conductivity of polyacetylene and all other conducting polymers is still severely limited by the quality of the material.

IV. AN ANALYSIS OF LIMITING FACTORS: WHAT ARE THE REQUIREMENTS?

Dramatic improvements in the physical properties of polymers through chain extension and chain alignment are not unfamiliar in polymer science. For example, ordinary polyethylene is useful in a wide variety of low-value applications, but it is certainly not considered a high-performance material. On the other hand, ultra-high molecular weight polyethylene which has been chain-aligned through gel-spinning and subsequent tensile drawing is one of the strongest materials known;[12] in this case the measured strength approaches the intrinsic theoretical limit set by the strength of the carbon-carbon bond.

long as

$$x \gg (t_o/t_{3d}). \tag{7}$$

Thus for any value of t_{3d}, the transport becomes three dimensional when the material is sufficiently good that[6]

$$L/a|_{crystalline} \gg t_o/t_{3d}. \tag{8}$$

In the case where there is good chain extension and good chain alignment, but when that alignment is nematic (i.e. with random interchain phase along the chain), the criterion is more severe;[6]

$$L/a|_{nematic} \gg (t_o/t_{3d})^2. \tag{9}$$

This is a much more stringent condition than expressed in Eq. (8). For example, in polyacetylene,[13] $(t_{3d}/t_o) \approx 0.03$, so that in the case of coherent interchain motion Eq. (8) would require $L/a \gg 30$ whereas in the incoherent case, the chain-defect concentration would have to be well below 10^{-3}, or $L/a \gg 1000$! Note that in the first case, Eq. (8), X-ray diffraction would yield Bragg spots along the chain direction, whereas in the second case, Eq. (9), the scattering would be in the form of sheets. In reality, with finite longitudinal structural coherence lengths we expect the physical condition to be somewhere in between those expressed in Eq. (8) and (9).

Even if (t_{3d}/t_o) is reduced by extending the interchain spacings through addition of side-chains to the conjugated backbone (as in the poly-(3-alkylthiophenes)), it is still possible to satisfy the inequalities (8) or (9). However, since L/a would have to be correspondingly greater, the implied difficulties for the polymer scientist would be that much greater!

When the appropriate inequality is satisfied, the transport is that of an anisotropic three dimensional metal, and the conductivity is given by Eqs. (1) and (3) with $\lambda \approx L$;

$$\sigma_{||} \approx (ne^2/m^*)(L/v_F). \tag{10}$$

In this case, the conductivity should increase in proportion to the mean distance between chain interruptions, cross links, sp^3 defects, etc. If the mean defect scattering time, $\tau_{def} = (L/v_F)$, becomes sufficiently long, that phonon scattering limits the mean free path, then Eq. 5 becomes valid, and the system is in the clean and intrinsic transport limit.

An analogous argument can be constructed for understanding the requirements for achieving the intrinsic strength of a polymer.[14] If one imagines a fiber like that shown in Figure 1, what is the requirement that the chains not slip with respect to one another, such that the ultimate strength is determined by that of the carbon-carbon bond? If E_o is the energy required to break the carbon-carbon bond and E_{3d} is the weaker interchain bonding energy (from Van der Waals forces and hydrogen bonding for saturated polymers, or from interchain transfer, t_{3d} for conjugated polymers), then the requirement is coherence over a length L such that $L/a \gg E_o/E_{3d}$. In this limit the large number (L/a) of weak interchain bonds add coherently, such that the polymer yields by breaking a carbon-carbon bond. Note that for a conjugated polymer, this criterion is less stringent, since t_{3d} is typically much greater than the energy expected from Van der Waals forces or from hydrogen bonding. Moreover, the increased bond-order due to the π-bonds will increase E_o over that expected for a saturated polymer. Thus, chain-aligned conjugated polymers should have exceptional mechanical properties at lower chain lengths than their saturated counterparts.

The need for high molecular weight conducting polymers that can be processed into the chain aligned configuration is two-fold. Clearly the molecular weight (and chain perfection) must be sufficient to satisfy the appropriate inequality, Eq. (8) or (9). More important, however, is the need for sufficiently high molecular weight to enable the processing that will yield the chain extended and chain aligned material.[12b]

The question before us in the field of conducting polymers is whether or not we can reach the same level: Can we make conducting polymers with sufficient quality that the mean free path is limited by the intrinsic scattering from thermal vibrations of the lattice (phonons)?

The principal problem is that of localization; quasi-one-dimensional electronic systems are especially prone to localization of electronic states due to disorder. The origin of this tendency to localize is easy to understand. Consider an array of polymer chains, each with a few defects, as shown schematically in Figure 1. We imagine the defect density (e.g. sp^3 sites) to be such that the typical distance, L, between defects is many lattice sites, but is still considerably less than the physical molecular weight. A carrier on a typical chain will then move with the Fermi velocity, until it comes to such a defect, at which point it back-scatters and moves in the opposite direction, until it back-scatters from the next defect on the same chain, etc. This multiple resonant scattering localizes the electronic wave-function; the resulting "conjugation length" is much less than the chain length. The result is localization, with carrier transport limited by phonon-assisted hopping.[7] Note that in this case, the conductivity is inherently small, and would go to zero as $T \rightarrow 0$, in contrast to the behavior of a metal.

To avoid the localization inherent to one-dimensional systems, one must have the possibility of interchain charge transfer. In the case of relatively high molecular weight and relatively few sp^3 defects, even weak interchain coupling (t_{3d}<0.1 eV) is sufficient to avoid one-dimensional localization. The problem is essentially three-dimensional, so long as there is a high probability that an electron will have diffused to another chain between scattering events. If τ is the backscattering lifetime and $P(t)$ is the probability that an electron which was initially on a specific chain will still be on that chain a time t later, then this criterion is[6]

$$P(\tau) \ll 1. \tag{6}$$

For scattering off chain-breaks separated by a mean distance L, $\tau = L/v_F = (a/v_F)x$, where $x = L/a$.

For well-ordered crystalline material, in which the chains have precise phase order, the interchain diffusion is a coherent process. In this case, the inequality (6) is satisfied so

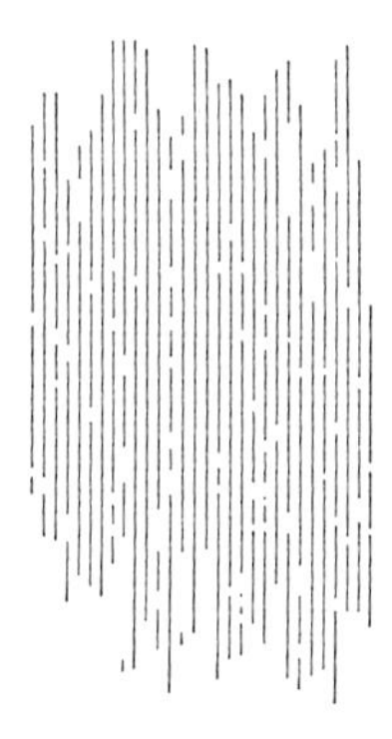

Figure 1. An array of aligned polymer chains, each with a few defects, as in an oriented fiber. The defects might be either chain ends or sp^3 defects, etc. which interrupt the π-conjugation.

V. IS "METALLIC" POLYACETYLENE A METAL?

The analysis given in the previous sections assumes that at heavy doping concentrations, polyacetylene (and other conducting polymers) have the electronic structure of metals. Only in a few cases has this been checked in detail. For example, the existence of a temperature-independent Pauli susceptibility has been established for polyacetylene[15] and polythiophene,[16] indicative of a metallic system with a finite density of states at the Fermi surface. For polyaniline, a Pauli contribution to the susceptibility has been inferred,[17] but it only dominates in the most crystalline material.[18]

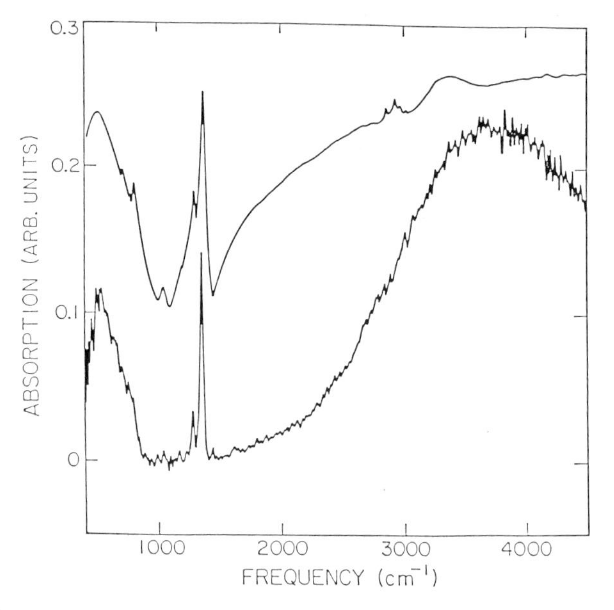

Figure 2. A direct comparison of the infrared absorption spectrum of the heavily doped sample (at 80 K) with that of the photoinduced absorption spectrum of pristine *trans*-$(CH)_x$ (at 80 K).

Nevertheless, even for polyacetylene, the electronic structure is not that of a *simple* metal in which the bond-alternation and the π-π^* gap have gone to zero; there are infrared active vibrational modes (IRAV) and a pseudo-gap.[19] This is indicated by the spectra[19] in Figure 2, which demonstrate the remarkable similarity between the doping-induced absorption found with heavily doped *trans*-$(CH)_x$, and the photoinduced absorption spectrum observed in the pristine semiconductor containing a very few photoexcitations. Not only are the same IRAV mode spectral features observed, they have almost identical frequencies.

The "metallic" regime in polyacetylene is characterized by three important aspects of the data:[19]

a) The IRAV modes are absorption bands; there is no indication of Fano-like anti-resonances studied in the one-dimensional charge-transfer salts.[20] In addition, since there is no indication of free-carrier Drude absorption, the intraband contribution to $\sigma(\omega)$ must be in the far-infrared[21] below 450 cm^{-1}. Therefore, there is a gap in the excitation spectrum (or a pseudo-gap where α is small but nonzero) with magnitude of about 1500 cm^{-1} ($\approx$0.2 eV).

2) The intensities of the IRAV modes increase linearly with the dopant concentration, with essentially the same slope as observed at more dilute concentrations.[22] This implies that all the doping-induced charges are involved, and that the IRAV in the "metallic" state are not due to a small number of residual inhomogeneities (or nonuniformities) in the charge distribution.

3) Since the IRAV mode frequencies are essentially identical with those observed with photoexcitation, the pinning of the π-electron charges which cause the IRAV (i.e. all the charges, see 2) has virtually disappeared.

These three conclusions are not consistent with the excitation spectrum of the simple metal which would result if the Peierls' gap had been reduced to zero (there would be no gap and no IRAV modes).

As noted above, the free carrier contribution which extrapolates to the measured $\sigma(0)$ must be in the far-IR below 450 cm^{-1}.[15] Nevertheless, most of the π-electron oscillator strength remains in the broad absorption band above 0.2 eV. An alternative which appears to be in agreement with thc essential experimental facts is that polyacetylene is an example of a polaronic metal.[23] The polaron lattice with a half-filled polaron band is certainly consistent with the observed susceptibility. In the case of a polaron lattice,[23] the IRAV modes are expected, and would be red-shifted from the Raman modes, provided that the pinning is weak. Although the intensity of the IRAV modes was initially calculated to be much too weak, this calculation ignored the effect of the counter-ions[24]; the counter-ion Coulomb potentials may lead to sufficient nonuniformity in the charge density, to yield the observed IRAV mode intensities.[25] For the polaron lattice, $\sigma(\omega)$ would have two contributions with a "gap" in between:

(i) a free-carrier contribution, corresponding to the mobile carriers in the lattice of polaron-like distortions;

(ii) an interband contribution: for hole polarons the transition is from the filled π band to the Fermi level in the lower polaron band; for electron polarons, the transition is from the Fermi level in the upper polaron band to the empty π^* band.

The concept of the polaronic metal has been applied to polyaniline as well,[26] and may be a more general feature of the metallic state in conducting polymers. In the context of the discussions presented in Section I through Section IV, the polaronic metal is indeed a "metal" and the analysis given is appropriate. In particular, the importance of the self-organization of the doped polymer (to avoid counter-ion scattering) and the need to avoid localization etc. are of clear importance.

VI. DISCUSSION AND CONCLUSIONS

Although the ability to dope conjugated polymers (and thereby change their conductivity by many orders of magnitude) is well-known, disorder introduced by the same doping process had been previously thought to render such materials intrinsically "dirty" conductors. Because of the self-organizing capability of the conjugated chains, and because of the anisotropic screening in such systems,[6] this is not the case. While the dopant potential will certainly lead to an inhomogeneous charge distribution (e.g. a polaron with its mean charge density peaked in the vicinity of the dopant counter-ion), so long as there is reasonably good local order, this potential causes almost no backscattering. Thus, the electrical conductivity of conducting polymers is not necessarily limited by scattering from the dopant ions.

There are certainly other sources of disorder which are present in real conducting polymers; e.g. there are strains, there are gentle bends, twists, kinks and jogs, etc. of the polymer backbone, and there are regions where neighboring polymer chains are imperfectly aligned. Although these have not been treated explicitly, none of these sources of disorder are expected to be seriously limiting, since the conductivity is only sensitive to the component of the disorder potential at $2k_F$. Only sharp, localized defects on the polymer chain (e.g. chain ends or cross-links etc) will have a significant effect on the conductivity. This is in accord with recent experimental results[1-4] which demonstrate increases in conductivity (to values in excess of 10^5 S/cm) as the number of sp^3 defects is reduced, even though the structural coherence length, as obtained from x-ray scattering, is only ~ 100 Å.

For a tangled or cross-linked amorphous polymer, metallic transport with $\lambda >> a$ cannot be expected (nor can high stiffness and strength). Such systems will always fall into the category of "dirty" conductors with modest conductivities. Nevertheless, since a wide range of conductivity levels can be useful, both classes of materials can potentially be important in applications. High-molecular weight, chain-aligned conducting polymers will be the high-performance polymers of the field. For more routine requirements, conducting polymers from the "dirty" conductor category may be sufficient. Note, however, that composites made up of only a small fractional volume of (phase-separated) connected regions of the high-performance component in the medium of a second polymer may yield conductivities that are of broad interest. However, it is from materials of the high performance category that we will learn in more detail about the metal-physics of conducting polymers, and about the various contributions to the scattering, which limit the mean free path, and thereby limit the electrical transport. To produce such materials, we must strive for advances in the synthesis of high molecular weight polymers, and we must develop novel methods of processing these polymers into the chain-aligned configuration.

Based upon these results, there is good reason to expect that conducting polymers can be made as high-performance materials with electrical conductivities significantly greater than those even the best conventional metals. To achieve these high conductivities, methods must be developed to synthesize oriented conducting polymers with high molecular weight and with a high degree of chain perfection. However, it is not necessary that the material be crystalline or even well-ordered, since the electronic mean free path can be much greater than the structural coherence length as measured in a scattering experiment.

Since the phonon frequencies are higher than in conventional metals, and since only the $2k_F$ phonon is relevant, the intrinsic conductivity should be very high at room temperature, and it will grow exponentially as the temperature is lowered. These high phonon frequencies are a direct manifestation of the strength of the carbon-carbon bond, which ultimately offers the promise of making such polymers truly high performance materials with conductivities significantly greater than copper and with strengths greater than steel by an order of magnitude.

Acknowledgement: This paper was stimulated by an EPRI workshop in Baltimore (March 1989) and was prepared under support from EPRI. Many of the specific results were drawn from earlier work. In particular, I thank Prof. S. Kivelson for many important discussions and many detailed results. Any insight on the relationship of electrical and mechanical properties in polymers is due to my interactions with Prof. Paul Smith.

REFERENCES

1. H. Naarmann and N. Theophilou, Synth. Met. 22:1 (1987).
2. N. Basescu, Z.-X. Liu, D. Moses, A.J. Heeger, H. Naarmann and N. Theophilou, Nature (London) 327:403 (1987).
3. T. Schimmel, W. Reiss, G. Denniger, J. Gmeiner, M. Schwoerer, H. Naarmann and N. Theophilou, Sol. State. Commun. 65:147 (1988).
4. N. Theophilou, D.B. Swanson, A.G. MacDiarmid, A. Chakraborty, H.H.S. Javadi, R.P. McCall, S.P. Treat, F. Zuo, and A.J. Epstein, Synth. Met. 28:D35 (1989).
5. M. Winokur, Y. B. Moon and A. J. Heeger, Phys. Rev. Lett. 58:2329 (1987).

6. S. Kivelson and A.J. Heeger, Synth. Met. 22:371 (1988).
7. N.F. Mott and E.A. Davis, "Electronic Processes in Non-Crystalline Materials", Clarendon Press, Oxford (1979).
8. a. P.W. Anderson, Comments on Solid State Physics 2:193 (1970).
   b. L. Fleischman and P.W. Anderson, Phys. Rev.B 21:2366 (1980).
   c. H. Kamimura, Electron-Electron Interactions in Disordered Materials in: "Modern Problems in Condensed Matter Science", V.M. Agranovich, and A.A. Maradudin Editors, North Holland, Amsterdam (1985); Vol. X, p. 555.
9. M. Thackur, Macromolecules 21:1379 (1988).
10. W.J. Feast, "Handbook on Conducting Polymers" Vol. 1, p.1, Marcel Dekker, Inc., New York and Basel (1986)
11. a. D.R. Gagnon, J.D. Capistron, F.E. Karasz, R.W. Lenz and S. Antoun, Polymers 28:567 (1987); D.R. Gagnon, F.E. Karasz, E.L. Thomas, and R.W. Lenz, Synth. Met. 20:85 (1987).
   b. T. Momii, S. Tokito, T. Testsui and S. Saito, Chem. Lett. 1201 (1988)
   c. S. Yamada, S. Tokito, T. Tetsui and S. Saito, J. Chem. Soc. Chem. Comm. 1448 (1987).
12. a. P. Smith and P.J. Lemstra, J. Mater. Sci. 15:505 (1980).
   b. P. A. Irvine and P. Smith, Macromolecules 19:204 (1986).
13. a. P.M. Grant and I. Batra, J. Phys. (Paris) Colloq. 44: C3-437 (1983).
   b. P. Vogl and G. Leising, Synth. Met. 28:D209 (1989).
14. Y. Termonia, P. Meakin and P. Smith, Macromolecules 19:154 (1986).
15. a. T.-C. Chung, F. Moraes, J.D. Flood and A.J. Heeger, Phys. Rev. B 29:1341 (1984).
   b. F. Moraes, J. Chen, T.-C. Chung and A.J. Heeger, Synth. Met. 11:271 (1985).
   c. J. Chen and A.J. Heeger, Phys. Rev. B 33:1990 (1986); *ibid*, Synth. Met. 24, 311 (1988).
16. F. Moraes, D. Davidov, M. Kobayashi, T.-C. Chung, J. Chen, A.J. Heeger and F. Wudl. Synth. Met. 10:169 (1985).
17. J.M. Ginder, A.F. Richter, A.G. MacDiarmid and A.J. Epstein, Solid State Commun. 63:97 (1987).
18. C. Fite, Y. Cao and A.J. Heeger, Sol. St. Commun. (in press).
19. Y.H. Kim and A.J. Heeger, Phys. Rev. B (in press).
20. a. M.J. Rice, L. Pietronero and P. Breusch, Sol. State Commun. 21:757 (1977).
   b. Y.H. Kim, M. Nowak, Z.G. Soos and A.J. Heeger, J. Phys. C 21:L503 (1988).
21. H.S. Woo, D.B. Tanner, N. Theophilou and A.G. MacDiarmid, Meeting of the American Physical Society, St. Louis (March, 1989); the results presented in this report indicated an increase in $\sigma(\omega)$ toward the measured $\sigma(0)$ at frequencies below 100 cm^{-1}.
22. a. A.J. Epstein, H. Rommelmann, R. Bigelow, H.W. Gibson, D.M. Hoffman and D.B. Tanner, Phys. Rev. Lett. 50:1866 (1983).
   b. X.Q. Yang, D.B. Tanner, M.J. Rice, H.W. Gibson, A. Feldblum and A.J. Epstein, Sol. St. Commun. 62:335 (1987); *ibid*, Mol. Cryst. Liq. Cryst. 117:267 (1985).
   c. X.Q. Yang, D.B. Tanner, G.A. Arbuckle, A.G. MacDiarmid and A.J. Epstein, Synth. Met. 17:277 (1987).
   d. D.B. Tanner, G.L. Doll, A.M. Rao, P.C. Eklund, G.A. Arbuckle and A.G. MacDiarmid, Proc. ICSM '88, Synth. Met. 28:D141 (1989).
23. S. Kivelson and A.J. Heeger, Phys. Rev. Lett. 55:308 (1985).
24. a. H.Y. Choi and E.J. Mele, Phys. Rev. B 34:8750 (1986).
   b. J.C. Hicks, J. Tinka Gammel, H.-Y. Choi and E.J. Mele, Synth. Met. 17:57 (1987).
25. S. Kivelson, Private communication.
26. S. Stafstrom, J.L. Brédas, A.J. Epstein, H.S. Woo, D.B. Tanner, W.S. Huang and A.G. MacDiarmid, Phys. Rev. Lett. 59:1464 (1987).

THE POLYANILINES: RECENT ADVANCES IN CHEMISTRY AND PROCESSING

Alan G. MacDiarmid

Department of Chemistry
University of Pennsylvania
Philadelphia, Pa 19104-6323, U.S.A.

Arthur J. Epstein

Department of Physics and Department of Chemistry
The Ohio State University
Columbus, OH 43210-1106, U.S.A.

Until relatively recently, "polyaniline", probably the oldest known synthetic organic polymer, consisted of an ill-defined class of materials, obtained by the chemical or electrochemical oxidative polymerization of aniline. Early studies were fraught with problems of uncertain compositions, and it was not until the mid-1980's, with the advent of better characterized materials, that significant physical studies became possible.

The current phase of polyaniline research was entered in 1980,[1] with several groups following suit during the following decade. The fascination of the conducting polymer community with this complex, challenging, and potentially technologically important polymer system can be gauged by the fact that it has generated almost 500 publications and patents in the last three years (since Jan. 1986). It is, undoubtedly, at the present time, the most rapidly growing area in the conducting polymer field.

The term "polyaniline", as commonly employed today[2,3] refers to a class of polymers consisting of up to 1,000 or more (ring-N-) repeat units, which can be considered as being derived from a polymer, the base form of which has the generalized composition $[(C_6H_4-NH-C_6H_4-NH)_Y(C_6H_4-N=C_6H_4=N)_{1-Y}]_x$ and consists of alternating reduced, $-C_6H_4-NH-C_6H_4-NH-$ and oxidized, $-C_6H_4-N=C_6H_4=N-$ repeat units. In principle, "y" can be varied continuously from one, to give the completely reduced polymer, $[C_6H_4-NH-C_6H_4-NH-C_6H_4-NH-C_6H_4-NH]_x$, to zero, to give the completely oxidized polymer, $[C_6H_4-N=C_6H_4=N-C_6H_4-N=C_6H_4=N]_x$. The imine ni-

Lower-Dimensional Systems and Molecular Electronics
Edited by R.M. Metzger *et al.*, Plenum Press, New York, 1991

trogen atoms, in any of the species, can be protonated, in whole or in part, to give the corresponding salts, the degree of protonation of the polymeric base depending on its oxidation state, and on the pH of the aqueous acid.

The terms "leucoemeraldine", "emeraldine" and "pernigraniline", used in the following discussion will refer to the different average oxidation states of the polymer where y = 1, 0.5 and 0 respectively, either in the base form, e.g. emeraldine base or in the protonated salt form, e.g. emeraldine hydrochloride.[2,3] It seems highly likely that the true average emeraldine oxidation state, where y is exactly equal to 0.5, may never have been synthesized from aniline. The term "emeraldine", in the following discussion, will therefore refer to an average oxidation state, where y is approximately equal to 0.5, in the generalized formula of the polyaniline bases given above.

SYNTHESIS OF POLYANILINE

The partly protonated emeraldine hydrochloride salt can be synthesized easily, as a partly crystalline black-green precipitate (dark green by transmitted light), by the oxidative polymerization of aniline, $(C_5H_5)NH_2$, in aqueous acid media by a variety of oxidizing agents, the most commonly used being ammonium peroxydisulfate, $(NH_4)S_2O_8$, in aqueous HCl.[2,4] It can be deprotonated by aqueous ammonium hydroxide, to give an essentially amorphous black-blue (dark blue by transmitted light) "as-synthesized" emeraldine base powder, with a coppery, metallic glint having an average oxidation state corresponding approximately to that of the ideal emeraldine oxidation state. The ^{13}C[5] and ^{15}N NMR[6,7] spectra of emeraldine base are consistent with its being composed, principally, of alternating oxidized and reduced repeat units.

Polyaniline salts may also be deposited by "*in situ* adsorption polymerization" in a few minutes, as strongly adhering films on a variety of substrates, such as natural and synthetic fibers and textiles,[8] plastic, glass, silver chloride pellets etc.,[9] by immersing the substrate in a freshly mixed acidic aqueous solution of aniline and oxidizing agent, such as ammonium peroxydisulfate. It is believed that a reactive intermediate, possibly an oligomeric radical cation of aniline, is first adsorbed, which subsequently polymerizes.[8]

Polyaniline salts of the approximate average emeraldine oxidation state may also be conveniently synthesized by electrochemical oxidation of aniline in aqueous acid media, as thin films on metal or conducting glass electrodes.[10] Under certain potential/current conditions, the polymer is formed as a matt of ~2,000Å fibrils.[10]

NON-OXIDATIVE "DOPING" OF THE EMERALDINE OXIDATION STATE

Protonic Acid Doping

Until recently, all known conducting polymers were doped to the highly conducting, frequently metallic, state by partial oxidation or partial reduction of the pi system of the polymer, e.g.,[11]

$$[CH]_x + (0.105x)I_2 \longrightarrow [CH^{+0.07}(I_3)^{-0.07}]_x \quad (1)$$

Thus the number of electrons associated with the polymer were either decreased or increased, respectively, during the doping process. The emeraldine base form of polyaniline was the first well-established example[2,3,12,13] of the "doping" of an organic polymer to a highly conducting regime, by a process in which the number of electrons, associated with the polymer, remain unchanged during the doping process. This is accomplished by treating emeraldine base with aqueous protonic acids, and is accompanied by a 9 to 10 order of magnitude increase in conductivity (to 1 - 5 S/cm; compressed powder pellet), reaching a maximum in ~ 1M aqueous HCl, with the formation of the fully protonated emeraldine hydrochloride salt, viz.,

$$\left[\left(\text{C}_6\text{H}_4\text{-NH-C}_6\text{H}_4\text{-NH}\right)\left(\text{C}_6\text{H}_4\text{-N=C}_6\text{H}_4\text{=N}\right)\right]_x \xrightarrow{2x\ HCl} \left[\left(\text{C}_6\text{H}_4\text{-NH-C}_6\text{H}_4\text{-NH}\right)\left(\text{C}_6\text{H}_4\text{-NH}^+\text{=C}_6\text{H}_4\text{=NH}^+\right)\right]_x\ (Cl^-)_2 \quad (2)$$

The crystal structure of the emeraldine hydrochloride polymer differs, according to the past history of the emeraldine base, from which it is produced.[14,15] The degree of protonation, and resulting conductivity, can be readily controlled by changing the pH of the dopant acid solution.[2,3,4] The process is reversed[2,3,16] when the protonated polymer is treated with aqueous alkali. The above fully protonated polymer is, generally, referred to as "50% protonated", since only 50% of all the nitrogen atoms have undergone protonation. It was originally believed, that preferential protonation of imine sites occurred exclusively.[2,3,4] However, more recent studies[16,17] show that, depending on the pH of the acid, some amine sites are also protonated, to give $(-NH_2-)^+$ groups before all the imine sites are protonated.

It is well known that other strong protonic acids, such as $R-SO_3H$ (R= organic group), besides HCl, dope polyaniline according to equation 2. It has recently been found,[18] when a solution of emeraldine base in concentrated H_2SO_4 is heated, that $-SO_3H$ groups replace a hydrogen atom on the (C_6H_4) rings, thus making R-(polyaniline chains). The proton from the SO_3H groups then protonates the -N= groups, resulting in a "self-doped" polymer (σ~0.5S/cm; 4-probe, compressed pellet), in which the protonic acid (dopant) is part of the polymer itself!, viz., $\left[\left(\text{C}_6\text{H}_4\text{-NH-C}_6\text{H}_3(SO_3^-)\text{-NH}^{+\cdot}\right)\left(\text{C}_6\text{H}_4\text{-NH-C}_6\text{H}_3(SO_3^-)\text{-NH}^{+\cdot}\right)\right]_x$.

If the fully protonated, i.e. ~50% protonated, emeraldine base should have the above dication, i.e. bipolaron, constitution, as shown in equation 2, it would be diamagnetic. However, extensive early magnetic studies[19] show that it is strongly paramagnetic, and that its Pauli (temperature independent) magnetic susceptibility increases linearly with the extent of protonation. These observations, and other earlier studies,[3,12,13] show that the protonated polymer is a polysemiquinone radical cation, one resonance form consisting of two separated polarons being $\left[\left(C_6H_4-\overset{H}{N}-C_6H_4-\underset{Cl^-}{\overset{H}{\overset{+\cdot}{N}}}\right)\left(C_6H_4-\overset{H}{N}-C_6H_4-\underset{Cl^-}{\overset{H}{\overset{+\cdot}{N}}}\right)\right]_x$.

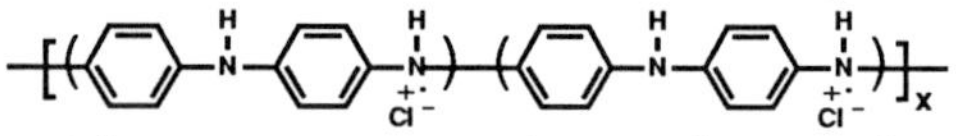

It can be seen, from the alternative resonance form, where the charge and spin are placed on the other set of nitrogen atoms, that the overall structure is expected to have extensive spin and charge delocalization.

"<u>Pseudo-protonic Acid"Doping</u>

It has been found that a number of substances will add to the imine nitrogen in emeraldine base in an analogous manner to protons[20,21] viz.,

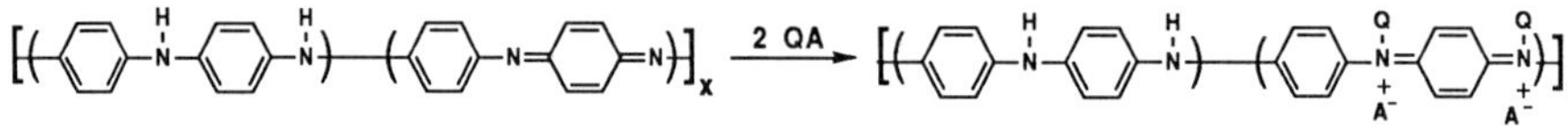

$$\left[\left(C_6H_4-\overset{H}{N}-C_6H_4-\overset{H}{N}\right)\left(C_6H_4-N=C_6H_4=N\right)\right]_x \xrightarrow{2\,QA} \left[\left(C_6H_4-\overset{H}{N}-C_6H_4-\overset{H}{N}\right)\left(C_6H_4-\underset{A^-}{\overset{Q}{\overset{+}{N}}}=C_6H_4=\underset{A^-}{\overset{Q}{\overset{+}{N}}}\right)\right]_x \quad (3)$$

where, for example, QA = $(CH_3)_2SO_4$ i.e. "$(CH_3)^+(CH_3SO_4)^-$. or $(RSO_2)_2O$, i.e. $(RSO_2)^+(RSO_3)^-$.[22] It is clearly apparent, when the same group is not attached to each nitrogen atom in the polymer, that the charge will be less evenly distributed along the polymer chain. This is consistent with the smaller Pauli susceptibility of the polymer doped with $(CH_3)_2SO_4$, and with its smaller conductivity ($\sigma \sim 10^{-2}$S/cm).

OXIDATIVE "DOPING" OF POLYANILINE

Leucoemeraldine base, the completely reduced form of polyaniline base, can be controllably oxidatively doped to the highly conducting regime, analogously to $(CH)_x$, viz.,

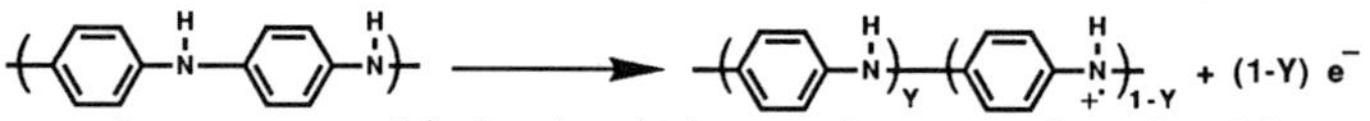

$$\left(C_6H_4-\overset{H}{N}-C_6H_4-\overset{H}{N}\right) \longrightarrow \left(C_6H_4-\overset{H}{N}\right)_Y\left(C_6H_4-\overset{H}{\overset{+\cdot}{N}}\right)_{1-Y} + (1-Y)\,e^- \quad (4)$$

This can be accomplished either electrochemically or chemically.[10,23,24]

Electrochemical oxidation and reduction of polyanilines has been extensively studied[10,23,24] by cyclic voltammetry techniques, usually in aqueous acid solution, using either electrochemically synthesized polyaniline films, or chemically synthesized powders. This is a convenient method for the electrochemical doping of extremely small amounts of leucoemeraldine, but it is not well suited for the synthesis of large quantities of oxidatively doped polymer in a known oxidation state.

The chemical oxidative doping of leucoemeraldine, as a preparative synthetic method for making relatively large quantities of the doped polymer, has recently been investigated,[25,26,27] although it is not as simple as the oxidative doping of conventional conducting polymers, due to spontaneous deprotonation, which can occur, as the oxidation state of the polymer increases to values of y less than ~0.5, with concomitant reduction in its base strength.

Leucoemeraldine base powder can be oxidized, for example, as a suspension in a solution of chlorine in CCl_4, or by solutions of $(NO)^+(PF_6)^-$, $FeCl_3$ or $SnCl_4$ in organic solvents, and by oxygen, or H_2O_2, as a suspension in aqueous acid media.[25,26,27] The analogy between oxidative doping of $(CH)_x$ and leucoemeraldine can be seen from equations 5 and 6

$$(CH)_x + (xy)\,(NO)^+(PF_6)^- \longrightarrow [(CH)^+_y\,(PF_6)^-_y]_x + (xy)\,NO \quad (5)$$

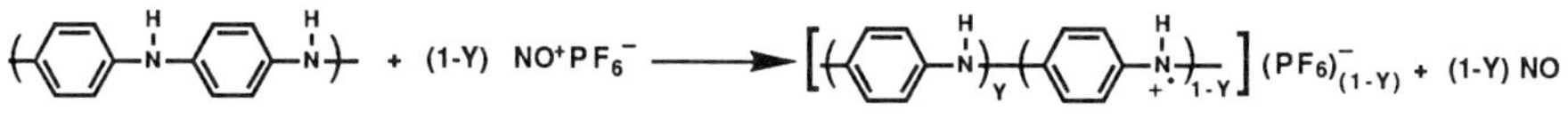

$$+(1-Y)\,NO^+PF_6^- \longrightarrow [\ \ ]\,(PF_6)^-_{(1-Y)} + (1-Y)\,NO \quad (6)$$

When oxygen or H_2O_2 are used in the doping process, they oxidize the leucoemeraldine, while the anion of the acid in the aqueous solution provides the necessary counter-anion for the doped polymer. This is a convenient method for introducing different dopant anions, e.g.,

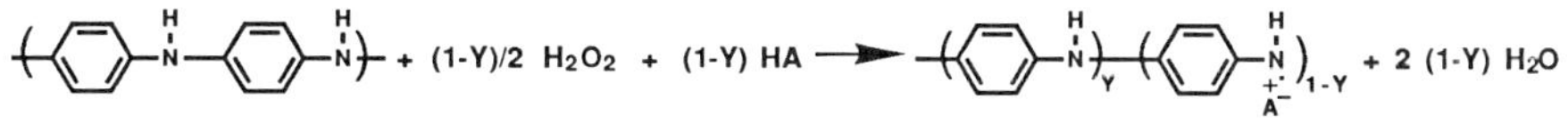

$$+ (1-Y)/2\,H_2O_2 + (1-Y)\,HA \longrightarrow \ \ + 2\,(1-Y)\,H_2O \quad (7)$$

(Where $A^- = Cl^-$, HSO_4^-, $H_2PO_4^-$, etc)

The doping level and/or oxidation state of the resulting polymer can be controlled, within limits, according to the stoichiometric ratio of leucoemeraldine base and oxidant employed.

The fundamental difference between the oxidative and non-oxidative doping methods for synthesizing the highly conducting form of polyaniline is illustrated diagrammatically below for emeraldine hydrochloride:

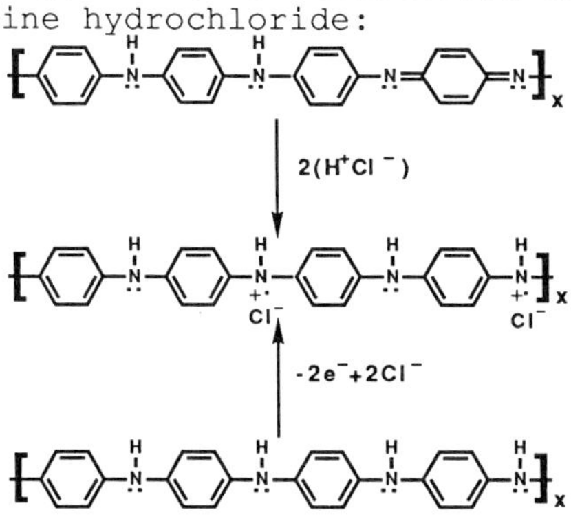

$2(H^+Cl^-)$

$-2e^- + 2Cl^-$

LEUCOEMERALDINE AND PERNIGRANILINE OXIDATION STATES

Synthesis

Leucoemeraldine, the completely reduced form of polyaniline, $[\ \ ]_x$, was first synthesized in

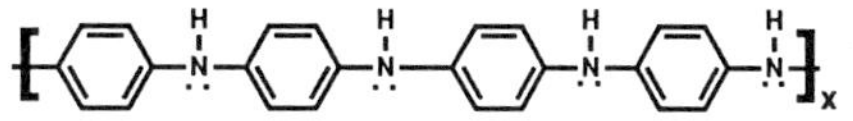

1910.[28,29] It can be conveniently prepared as an analytically pure, off-white powder, by the reduction of emeraldine base,[28,29] the most commonly used reducing agents being phenyl hydrazine or hydrazine.[7] Similarly to emeraldine base, free-standing films can be cast from NMP solutions.[7] The surfaces of the particles in the powder, or of the film, are oxidized relatively slowly by air, turning blue; however ^{13}C NMR studies[5] show that the bulk of the polymer is still in the leucoemeraldine oxidation state.

Pure pernigraniline, the completely oxidized form of polyaniline, $[\text{-C}_6\text{H}_4\text{-N=C}_6\text{H}_4\text{=N-C}_6\text{H}_4\text{-N=C}_6\text{H}_4\text{=N-}]_x$, has recently been synthesized for the first time.[30] It was reported in 1910 [28,29] to be formed, in an impure state, by the oxidation of emeraldine base, but that it was unstable, and rapidly "disproportioned", especially when wet. Synthesis of the analytically pure, dark purple, partially crystalline powder has been accomplished by the controlled oxidation of emeraldine base with m-$Cl(C_6H_4)C(O)OOH/N(C_2H_5)_3$ in NMP containing a trace of $CrCl_3$. Free-standing, lustrous, copper-colored films can be cast from this solution, and subsequently leached in a methanol/acetone mixture to remove excess oxidizing agent.

The fact, that the two extreme members of polyaniline bases — the completely reduced and completely oxidized — have been synthesized, should now make it easier to interpret future systematic studies of intermediate members of the series.

<u>Reaction of Leucoemeraldine with Pernigraniline</u>

It has recently been observed[31] that the completely reduced base form of polyaniline, (leucoemeraldine), spontaneously reduces, and is itself oxidized, when its solution in NMP is mixed with an NMP solution of the completely oxidized base form of polyaniline, (pernigraniline). This reaction may be studied by "titrating" a solution of the leucoemeraldine with a solution of pernigraniline, and by following the course of the reaction by the changes in the electronic spectrum of the reaction mixture. When equimolar solutions of the reactants are mixed, this redox reaction is found to approach equilibrium after ~20 hours at room temperature, to produce the emeraldine base, viz.,

$$[\text{-C}_6\text{H}_4\text{-NH-C}_6\text{H}_4\text{-NH-C}_6\text{H}_4\text{-NH-C}_6\text{H}_4\text{-NH-}]_x + [\text{-C}_6\text{H}_4\text{-N=C}_6\text{H}_4\text{=N-C}_6\text{H}_4\text{-N=C}_6\text{H}_4\text{=N-}]_x \longrightarrow 2[(\text{-C}_6\text{H}_4\text{-NH-C}_6\text{H}_4\text{-NH-})(\text{-C}_6\text{H}_4\text{-N=C}_6\text{H}_4\text{=N-})]_x \quad (8)$$

No definitive evidence has been found, in preliminary studies, for the existence of oxidation states intermediate between the leucoemeraldine and emeraldine oxidation states, at least when the reaction is performed in solution. This is qualitatively consistent with cyclic voltammetry/spectroscopic studies of oligomers, and of solid polyaniline, which have suggested that polyaniline may exist, locally, in only three discrete oxidation states, although the average oxidation state of the bulk polymer may be intermediate between them.[32]

PROCESSING AND SELECTED PHYSICAL PROPERTIES

Structural Modification by Processing

"As-synthesized" emeraldine base is soluble in N-methyl pyrrolidinone, NMP[33], and in concentrated H_2SO_4.[28,29,34] However, the term "soluble" must be used with caution, since it is not clear how much of the polymer in, for example, a viscous ~20 weight % NMP "solution", is in "true" solution.[33,35] It has been known for some time that emeraldine base is readily solution-processible[33,36] and that it may be cast as free-standing, flexible, coppery-colored films from its solutions in NMP. These films can be doped with ~1M aqueous HCl to give the corresponding flexible, lustrous purple-blue films (σ~1-4S/cm) of emeraldine hydrochloride.[33]

Exhaustive extraction at room temperature under nitrogen of "as-synthesized" essentially amorphous emeraldine base, first with tetrahydrofuran, THF, then with NMP results in the removal of ~20 weight% of oligomeric material and impurities.[37,38,39] The resulting "processed" emeraldine base powder has an excellent elemental analysis, and is up to ~50% crystalline (orthorhombic).[14,15] It is insoluble in NMP and in concentrated sulfuric acid, under the same experimental conditions, in which the "as-synthesized" emeraldine base powder is "soluble" in these solvents. Increased crystallinity, not unexpectedly, tends to result in decreased solubility.

It is interesting to note that the above, partly crystalline, emeraldine base powder can be repeatedly interconverted to its amorphous form, and then back to its crystalline form. For example, protonation with HCl, to give the corresponding emeraldine hydrochloride (σ~18 S/cm.; 4-probe compressed pellet), followed by deprotonation with NH_4OH, results in the amorphous form. Treatment of this amorphous form with THF, in which it is essentially insoluble, results in some solvation of the polymer, giving it sufficient mobility to rearrange, at least in part, back to the more thermodynamically stable, partly crystalline form.

A solution of the amorphous "processed" emeraldine base in NMP exhibits a bimodal molecular weight distribution (by g.p.c; monodispersed polystyrene standard), the maximum molecular weight fraction corresponding to approximately 325,000,[35,37] consistent with earlier studies.[37,39] Lower molecular weights, in concentrated H_2SO_4 solution of emeraldine base, synthesized in a slightly different manner, have also been reported.[34] Partly crystalline films of the insoluble protonated polymer are formed, when thin layers of these concentrated H_2SO_4 solutions are exposed to air.[34]

Recent studies suggest, that it is possible to controllably convert emeraldine base from a pseudo one-dimensional polymer to a two-or three-dimensional material.[30] On heating to ~300° in vacuum, it undergoes an exothermic reaction, believed to be caused by cross-linking of the polymer chains by phenazine groups, to give a material having the same elemental composition, and which is still electroactive.[30]

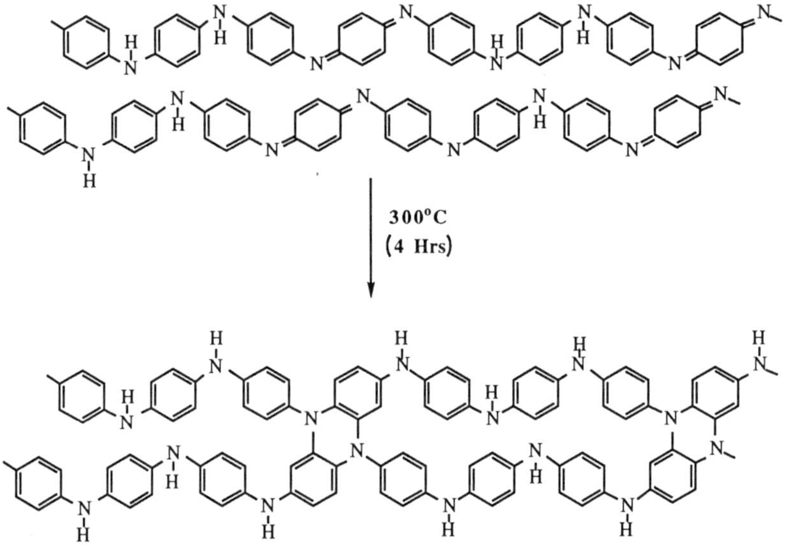

300°C
(4 Hrs)

(9)

It seems likely that cross-linked polyanilines represent a new important class of material in the polyaniline system.

Oriented Films

Oriented, partly crystalline emeraldine base films are obtained by simultaneous heat treatment and stretching of films, formed from "as-synthesized" emeraldine base containing a plasticizer, such as NMP.[40,41] Samples are observed to elongate by up to 400%, when held above the glass transition temperature [≥~110°C].[40,41] The resulting films have an anisotropic x-ray diffraction, and optical response, with a misorientation of only a few degrees.[15] Oriented polyaniline may also be obtained by evaporating (in the present of a heat lamp) a solution of emeraldine base in NMP on polyethylene, polyacetylene or other substrates, while the substrate is being mechanically stretched.[42]

Oriented Fibers

Fibers of emeraldine base can be formed, by drawing a ~20 weight % "solution" of any form of amorphous emeraldine base in NMP in a water/NMP solution.[35,38] They can be thermally stretch-oriented up to 400%, in a similar manner, to emeraldine base films.[35] X-ray diffraction studies show directional enhancement of the Debye-Scherrer rings. They differ from the oriented, partly crystalline fibers, prepared from solutions of emeraldine base in concentrated sulfuric acid[34] by coagulation in water, in that they consist of undoped polymer. Doping with 1M aqueous HCl results in a significant increase in the conductivity parallel to the direction of stretching (to 40 - 170 S/cm; Av., 85 S/cm),[35] similar to that observed with fibers formed from concentrated H_2SO_4 solution.[34] If desired, the emeraldine base "solution" in NMP may also be drawn in aqueous HCl, which results in direct formation of the doped fiber.

The thermally stretch-oriented base fibers have a tensile strength of ~330 MPa,[43] which falls within the tensile strength range of many commercial fibers, e.g., polyamides such as Nylon 66, etc..[44] Doping of such fibers with 1M HCl reduces the tensile strength to ~185/MPa, possibly caused by the somewhat reduced degree of observed crystallinity.[35] These early results suggest that further work will result in tensile strengths at least equal to that of commercial fibers.

SUBSTITUTED POLYANILINES

As seen from the preceding discussion, the base form of a polyaniline polymer can exist in a large number of different average oxidation states, each of which can, in principle, be protonated to a variety of different levels. The richness of both the chemistry, and associated physics, of the polyanilines can be further extended by substitution of hydrogen atoms on the ring, or on the amine nitrogen atoms, giving a potentially vast series of bases, each of which are potentially capable of protonic or pseudo-protonic doping to a variety of different levels, with concomitant increase in conductivity.

Ring Substitution

Substituted polyanilines of the type

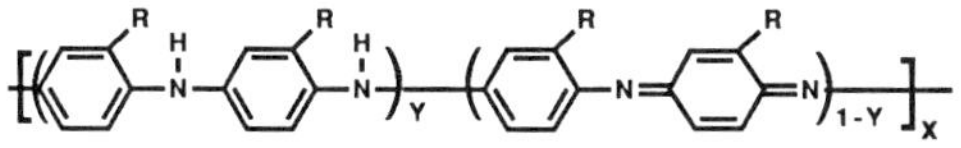

where R= $-CH_3$, $-C_2H_5$, $-OCH_3$, $-OC_2H_5$ have been synthesized by chemical and/or electrochemical oxidative polymerization of the corresponding substituted aniline monomers[9,45,46,47] by essentially identical experimental procedures to those used for the synthesis of the corresponding parent polyanilines. They are, usually, synthesized in their hydrochloride salt form, and are, then, depronated by aqueous alkali to the base form which can be subsequently re-pronated by an appropriate aqueous acid. The hydrochloride salts (R = $-OCH_3$ or R = $-OC_2H_5$) are soluble in water after wetting with methanol. Fibers of the $-OC_2H_5$ base form can be drawn from CH_2Cl_2 solution.

The compressed pellet conductivities of powders of the polymers, after protonation in 1M HCl, fall in the range 10^{-1} to 10^{-3} S/cm, significantly less than that of the parent emeraldine hydrochloride (~1-5 S/cm) synthesized in a similar manner.[9,45,46,47] This is, qualitatively, consistent with the larger π-π^* transition observed in the substituted polyaniline bases, as compared to emeraldine base, and with their cyclic voltammograms, both these observations suggesting reduced conjugation in the substituted polymers.

These observations can be rationalized on the basis of the steric effects of the substituents having a greater effect on the above properties, than their electronic effects.[45]

Substitution at Nitrogen

The polyanilines differ from heterocyclic conducting polymers, such as polypyyrole, etc., in that their molecular orbitals involving nitrogen must be directly involved in their electronic transport process. In, for example, polypyyrole, conduction could occur exclusively via the carbon chain backbone. The effect of changing substituents on the nitrogen is, therefore, of fundamental importance in understanding the chemistry and physics of the polyanilines. In principle, the hydrogen atoms on nitrogen in the fully reduced leucoemeraldine form of polyaniline can be replaced, in whole or in part, by a variety of organic and/or inorganic groups.

Polymeric material may be obtained by the electrochemical, or chemical, oxidative polymerization of $(C_6H_5)N(CH_3)H$, using methods similar to those employed for the parent and ring-substituted polyanilines.[21,46,47] Elemental analysis and magnetic studies[21] of the chemically-synthesized polymer (σ~10^{-4} S/cm; compressed powder pellet) are consistent with a dominant bipolaronic structure,

$$[(C_6H_4-N(CH_3)-C_6H_4-N(CH_3))_{0.56}(C_6H_4-N(CH_3)-C_6H_4-N^+H(CH_3)Cl^-)_{0.30}(C_6H_4-N^+(CH_3)Cl^-=C_6H_4=N^+(CH_3)Cl^-)_{0.14}\cdot(H_2O)_{0.90}]_x$$

On heating in vacuo, deprotonation occurs, with the loss of water and HCl, to give a composition (σ~10^{-8}S/cm; compressed powder pellet) by elemental analysis consistent with

$$[(C_6H_4-N(CH_3)-C_6H_4-N(CH_3))_{0.86}(C_6H_4-N^+(CH_3)Cl^-=C_6H_4=N^+(CH_3)Cl^-)_{0.14}]_x \; .$$

CONCLUSIONS

Although extensive advances have been made, during the last four years, in obtaining an understanding of the fundamental electronic, magnetic, spectroscopic, transport and theory, relating to this large and diverse area of conducting polymers, it is apparent that the field has not yet been clearly delineated. The fact that aniline is a relatively inexpensive compound, that its polymerization can be readily accomplished by simple chemical procedures, and that some of its conducting derivatives apparently exhibit good thermal and environmental stability strongly suggest it may have many good technological uses. The richness of the chemistry, electrochemistry, and physics of the polyanilines indicates they will continue to serve as a focus for challenging interdisciplinary research in the future.

ACKNOWLEDGEMENTS

The authors wish to thank Dr. R.K. Kohli and Mr. E.M. Scherr, in particular, for their invaluable assistance in preparing the manuscript. Most of the work reported by the authors was supported by the Defense Advanced Research Projects Agency through a contract monitored by the office of Naval Research.

References

1. A.F. Diaz and J.A. Logan, J Electroanal Chem. 111:111 (1980).
2. J-C. Chiang and A.G. MacDiarmid, Synth. Met. 13:193 (1986).
3. A.G. MacDiarmid, J-C. Chiang, A.F. Richter and A.J. Epstein, Synth. Met. 18:285 (1987).
4. A.G. MacDiarmid, J-C. Chiang, A.F. Richter, N.L.D. Somasiri and A.J. Epstein in: "Conducting Polymers", ed. L. Alcacér (Reidel Publications, Dordrecht, 1987).
5. S. Kaplan, E.M. Conwell, A.F. Richter and A.G. MacDiarmid, Synth. Met. 29:E235 (1989).
6. A.F. Richter, A. Ray, K.V. Ramanathan, S.K. Manohar, G.T. Furst, S.J. Opella, A.G. MacDiarmid and A.J. Epstein, Synth. Met. 29:E243 (1989).
7. J. Masters, S.K. Manohar, A. Ray, A.G. MacDiarmid and A.J. Epstein, unpublished results; A. Ray Ph.D. dissertation, University of Pennsylvania, (1989).
8. R.V. Gregory. W.C. Kimbrell and H.H. Kuhn, Synth. Met. 28:C-823 (1989).
9. S.K. Manohar, A.G. MacDiarmid and A.J. Epstein, Bull. Am. Phys. Soc. 34:582 (1989); S.K. Manohar, A.G. MacDiarmid and A.J. Epstein, 1989, unpublished observations.
10. W-S. Huang, B.D. Humphrey and A.G. MacDiarmid, J. Chem. Soc. Faraday Trans. 1 82:2385 (1986).
11. A.G. MacDiarmid and A.J. Heeger, Synth. Met. 1:101 (1979/80).
12. P.M. McManus, S.C. Yang and R.J. Cushman, J. Chem. Soc. Chem. Commun. 1556 (1985).
13. G.E. Wnek, Synth. Met. 15:213 (1986).
14. M.E. Jozefowicz, R. Laversanne, H.H.S. Javadi, A.J. Epstein, J.P. Pouget, X. Tang and A.G. MacDiarmid, Phys. Rev. B, Rapid Commun. 39:12958 (1989).
15. A.J. Epstein and A.G. MacDiarmid, in: "Electronic Properties of Conjugated Polymers", eds. H. Kuzmany, M. Mehring and S. Roth, Springer-Verlag, Berlin (1989) in press.
16. C. Menardo, M. Nechtschein, A. Rousseau, J.P. Travers and P. Hany, Synth. Met. 25:311 (1988).
17. A. Ray, A.F. Richter, A.G. MacDiarmid, A.J. Epstein, Synth. Met. 29:E151 (1989).

18. J. Yue, A.J. Epstein and A.G. MacDiarmid, to be published (1989).
19. A.J. Epstein, J.M. Ginder, F. Zuo, R.W. Bigelow, H.S. Woo, D.B. Tanner, A.F. Richter, W-S. Huang and A.G. MacDiarmid, Synth. Met. 18:303 (1987); J.M. Ginder, A.F. Richter, A.G. MacDiarmid and A.J. Epstein, Solid State Commun. 63:97 (1987).
20. M. Angelopoulos, S.P. Ermer, S.K. Manohar, A.G. MacDiarmid and A.J. Epstein, Mol. Cryst. Liq. Cryst. 160:223 (1988).
21. S.K. Manohar, A.G. MacDiarmid, K.R. Cromack, J.M. Ginder and A.J. Epstein, Synth. Met. 29:E349 (1989).
22. R.E. Cameron, S.K. Preto Clement, A.L. Landis, Bull. Am. Phys. Soc. 34:582 (1989).
23. E.W. Paul, A.J. Ricco and M.S. Wrighton, J. Phys. Chem. 89:1441 (1985).
24. E.M. Genies, M. Lapkowski and J.F. Penneau, J. Electroanal. Chem. 249:97 (1988).
25. A. Ray, G.E. Asturias, D.L. Kershner, A.F. Richter, A.G. MacDiarmid and A.J. Epstein, Synth. Met. 29:E141 (1989).
26. J.G. Masters, D.L. Kershner, A.G. MacDiarmid and A.J. Epstein, unpublished observations, (1989).
27. X. Tang, D.L. Kershner, J.G. Masters, Y. Sun, A.G. MacDiarmid and A.J. Epstein, Bull. Am. Phys. Soc. 34:583 (1989).
28. A.G. Green and A.E. Woodhead, J. Chem. Soc. Trans. 97:2388 (1910).
29. A.G. Green and A.E. Woodhead, J. Chem. Soc. Trans. 101:1117 (1912).
30. Y. Sun, A.G. MacDiarmid and A.J. Epstein, unpublished observations, (1989).
31. J.G. Masters, Y. Sun, A.G. MacDiarmid and A.J. Epstein, unpublished observations (1989).
32. L.W. Shacklette, J.F. Wolfe, S. Gould and R.H. Baughman, J. Chem. Phys. 88:3955 (1988).
33. M. Angelopoulos, G.E. Asturias, S.P. Ermer, A. Ray, E.M. Scherr, A.G. MacDiarmid, M. Akhtar, Z. Kiss and A.J. Epstein, Mol. Cryst. Liq. Cryst. 160:151 (1988).
34. A. Andreatta, Y. Cao, J-C. Chiang, A.J. Heeger and P. Smith, Synth. Met. 26:383 (1988).
35. X. Tang, A.G. MacDiarmid and A.J. Epstein, unpublished observations, (1989).
36. M. Angelopoulos, A. Ray, A.G. MacDiarmid and A.J. Epstein, Synth. Met. 21:21 (1987).
37. A.G. MacDiarmid, G.E. Asturias, D.L. Kershner, S.K. Manohar, A. Ray, E.M. Scherr, Y. Sun, X. Tang and A.J. Epstein, Polymer Preprints 30-1:147 (1989).
38. X. Tang, E. Scherr, A.G. MacDiarmid and A.J. Epstein, Bull. Am. Phys. Soc. 34:583 (1989).
39. X. Tang, A.G. MacDiarmid, A.J. Epstein and Y. Wei, unpublished observations, (1989).
40. K.R. Cromack, M.E. Jozefowicz, J.M. Ginder, R.P. McCall, A.J. Epstein, E.M. Scherr and A.G. MacDiarmid, Bull. Am. Phys. Soc. 34:583 (1989).
41. E.M. Scherr, A.G. MacDiarmid and A.J. Epstein, unpublished observations, (1989).
42. N. Theophilou, A.G. MacDiarmid, D. Djurado, J.E. Fischer and A.J. Epstein, in: "Electronic Properties of Conjugated

Polymers," eds. H. Kuzmany, M. Mehring and S. Roth, Springer-Verlag, Berlin, (1989) in press.
43. X. Tang, C.H. Hsu, C.N. Barry, A.G. MacDiarmid and A.J. Epstein, to be published (1989).
44. H.R. Allcock and F.W. Lampe, "Comtemporary Polymer Chemistry", Prentice-Hall, Inc., Engelwqod Cliffs, NJ; (1981), p. 537; J. Brandrup and E.H. Immergut, "Polymer Handbook", John Wiley & Sons, New York (1975), p. V-84.
45. Y. Wei, W.W. Focke, G.E. Wnek, A. Ray and A.G. MacDiarmid, J. Phys. Chem. 93:495 (1989); M. Leclerc, J. Guay and L.H. Dao, Macromolecules 22:649 (1989); J.M. Ginder, A.J. Epstein and A.G. MacDiarmid, Synth. Met., in press, (1989).
46. S.K. Manohar, A.G. MacDiarmid and A.J. Epstein, Bull. Am. Phys. Soc. 34:583 (1989).
47. N. Comisso, S. Daolio, G. Mengoli, R. Salmaso, S. Zecchin and G. Zotti, J. Electroanal. Chem. 255:97 (1988); J. Guay, Polym. Commun. 33:149 (1989); L.H. Dao, M. Leclerc, J. Guay and J.W. Chevalier, Synth. Met. 29:E-377 (1989).

ELECTRICAL, OPTICAL AND MAGNETIC PROPERTIES OF POLYANILINE PROCESSED FROM SULFURIC ACID AND IN SOLUTION IN SULFURIC ACID

Y. Cao, P. Smith and A.J. Heeger

Institute for Polymers and Organic Solids
University of California, Santa Barbara 93106

I. INTRODUCTION

Although polyaniline has been investigated extensively since the beginning of this century,[1] it has recently attracted considerable interest as a conducting polymer.[2] Among conducting polymers, polyaniline is unique in that its electronic structure and electrical properties can be reversibly controlled both by charge transfer doping (to vary the oxidation state of main chain) and by protonation.[3] The wide range of associated electrical, electrochemical and optical properties, coupled with good stability, make polyaniline attractive as an electronic material for potential use in a variety of applications.

Throughout the extensive literature, polyaniline has been generally categorized as an intractable material. Recently, however, two groups[4,5] have reported methods to dissolve and process polyaniline without changing of the structure of the polymer (in N-methylpyrrolidinone[4] or in concentrated sulfuric[5] and other strong acids[5]). In contrast to alternative methods for achieving solubility through preparation of substituted[6] polyaniline or through the synthesis of graft or block copolymers,[7] the resulting films[4,5] and fibers[5] are highly conductive after processing.

The use of concentrated acids[5] has specific advantages in that both the salt and the base form of polyaniline can be completely dissolved at room temperature, with polymer concentrations ranging from extremely dilute to more than 20% (w/w), in concentrated protonic acids such as H_2SO_4, CH_3SO_3H, and CF_3SO_3H. Although there has been considerable progress toward the development of soluble conducting polymers (such as substituted polythiophenes and substituted polyparaphenylene-vinylene), polyaniline in sulfuric acid is the first example of a stable, concentrated solution of a conjugated polymer in which the material is converted by the solvent to the conducting form, and from which the polymer can be processed directly into the metallic salt, with no need for subsequent doping.[8]

The solubility of this conducting polymer opens the way to processing pure, partially crystalline polyaniline, or composites of polyaniline, with the other commercial polymers into fibers and films, etc. In addition, this solubility enables extensive characterization of polyaniline as a macromolecular system (e.g. viscosity in solution as a probe of molecular weight, etc.) and of polyaniline as a conducting polymer (e.g. optical studies of spin-cast films as a probe of electronic structure of the salt or base forms). The latter is the subject of this paper.

The structure of the semiconducting emeraldine base form of polyaniline is

$$[(1A)(2A)]_n \qquad (1)$$

Lower-Dimensional Systems and Molecular Electronics
Edited by R.M. Metzger *et al.*, Plenum Press, New York, 1991

where

$$(1A) = (B—NH—B—NH—) \quad (2a)$$

and

$$(2A) = (B—N{=}Q{=}N—). \quad (2b)$$

In the above, B denotes a C_6H_4 ring in the benzenoid form and Q denotes a C_6H_4 ring in the quinoid form. The existence of one quinoid ring (out of four) in the emeraldine base has been well established in the literature.[9] The emeraldine base can be fully reduced to leucoemeraldine, $(1A)_n$, a large-bandgap insulator with π-π^* transition at nearly 4 eV. Upon protonation of $[(1A)(2A)]_n$ to the emeraldine salt, there is a structural change (with no change in the number of electrons) leading to a half-filled band and a metallic state (described as a polaronic metal) of the form[10-13]

$$[1S]^{\cdot}(A^-)_n = [B—NH—B—NH^+—]^{\cdot}_n(A^-)_n \quad (3)$$

where A^- is the counter ion (e.g. ClO_4^-, Cl^-, HSO_4^-etc.), and the $[\ ]^{\cdot}$ denotes one unpaired electron per formula unit. Further oxidation of $[1S]^{\cdot}(A^-)_n$ is expected to yield the fully-oxidized copolymer with alternating quinoid and benzenoid monomer units:

$$(B—NH^+{=}Q{=}NH^+—)_n \quad (4)$$

with an H-atom bonded to each nitrogen, and with one counterion for each charge on the chain. Since the $(-N^+{=}Q{=}N^+-)$ unit is a charged bipolaron, the polymer structure in (4) is that of a charged bipolaron lattice with a two-fold degenerate ground state. Note that (4) can in principle be obtained directly from $(1A)_n$ by charge-transfer doping.

In the following sections, we briefly review recent progress in the structure and properties of the various forms of polyaniline with emphasis on the use of sulfuric acid as a means of characterizing and processing the polymer.

II. STRUCTURE[14]

Polyaniline (emeraldine salt) can be recovered in *partially crystalline* form from solutions in sulfuric acid.[5] This is demonstrated by the wide-angle X-ray diffraction pattern of such a polyaniline film, shown as the upper solid curve in Figure 1a. Essentially identical

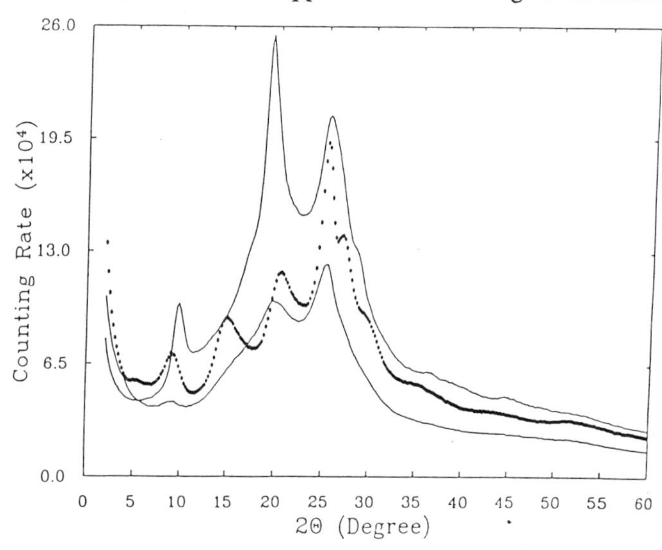

Fig. 1a. X-ray scans for $[1S]^{\cdot}(HSO_4^-)_n$ (upper solid curve), $[1S]^{\cdot}(Cl^-)_n$ (dotted curve), and $[1S]^{\cdot}(ClO_4^-)_n$ (lower solid curve). For the upper curve, the counting rate is counts per five-minute internal; for the other two curves, the counting rate is counts per four-minute interval.

data were obtained by X-ray scans from the film and powder samples. In addition, we found that the as-synthesized emeraldine salts, $[1S]^{\cdot}(Cl^{-})_n$ (dotted curve) and $[1S]^{\cdot}(ClO4^{-})_n$ (lower solid curve), are partially crystalline. The data indicate that the $[1S]^{\cdot}(HSO4^{-})_n$ samples and $[1S]^{\cdot}(Cl^{-})_n$ sample are more highly crystalline than the $[1S]^{\cdot}(ClO4^{-})_n$ sample.

The intensity of the various reflections and their clear definition above the background amorphous scattering indicate that the $[1S]^{\cdot}(A^{-})_n$ materials have significant crystallinity. For the $[1S]^{\cdot}(HSO4^{-})_n$ and $[1S]^{\cdot}(Cl^{-})_n$ samples, as many as eight reflections can be identified. The relatively narrow widths of the diffraction peaks (full width at half maximum as narrow as approximately 1°) indicate relatively long-range structural coherence.

Figure 1b compares the corresponding wide angle X-ray scans for the $[1S]^{\cdot}$ material (as precipitated from sulfuric acid) and the emeraldine base prepared by carefully compensating the same material by exposure to aqueous ammonia solution. Although compensation reduces the crystallinity, the resulting emeraldine base shows X-ray reflections which clearly indicate structural order.

Attempts to establish the details of the interchain packing and crystal structure through comparison of the calculated structure factors with the data in Figures 1a and 1b are currently underway. However, some more general aspects of the structure can be inferred directly from the data. The coherence lengths parallel and perpendicular to the chains (based on the narrowest lines at $2\theta \approx 10^{o}$ and 19°, respectively) are comparable. From the Scherrer formula,

$$\zeta \approx 2\pi\lambda/\delta(2\theta)$$

where $\delta(2\theta)$ is the full width at half-maximum, we find $\zeta \approx 10$ nm.

The similarities between the molecular structure of the emeraldine base and those of poly(p-phenylene sulfide),[15a] poly(p-phenylene oxide),[15b] and even poly(p-phenylene terephthalamide)[15c] suggest that a similar interchain packing might be expected. Indeed, the two strongest equatorial peaks in Figure 1b (at $2\theta=19.5^{o}$ and $2\theta=22.8^{o}$) correspond to the similar peaks observed in all three materials.

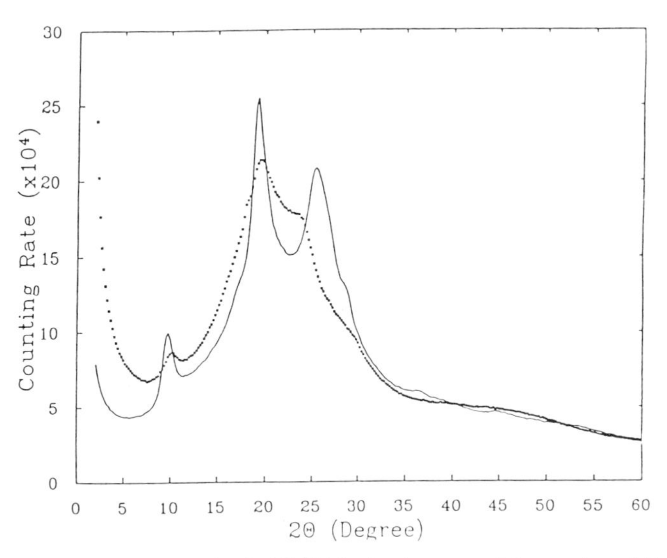

Fig. 1b. X-ray scans for $[1S]^{\cdot}(HSO_4^{-})_n$ (upper solid curve) and the $[(1A)(2A)]_n$ emeraldine base; the counting rate is counts per five-minute internal.

III. SPECTROSCOPY

By utilizing the partial solubility of polyaniline in dimethylformamide (DMF), dimethylsulfoxide (DMSO) and tetrahydofuran (THF), and the more complete solubility in N-methylpyrrolidinone (NMP), films of chemically polymerized polyaniline have been prepared for spectroscopic studies.[11] With DMF, DMSO and THF, it is necessary to remove insoluble materials in order to cast films from solution. Since we have recently found that this soluble part has low molecular weight, and usually consists of only about 20% of the total mass,[16] there is a need to extend the spectroscopic studies, to evaluate the effect of molecular weight on the principal absorption features.

Spectroscopic measurements have provided information on the electronic structure of the insulating emeraldine base form of polyaniline[17-19] and the evolution of the electronic band structure as a function of the degree of protonation. On the basis of magnetic susceptibility experiments,[10-13,20] it was proposed that the electronic structure changed from a semiconductor in the emeraldine base (the structure given in equation 1) to a metal in the protonated emeraldine salt (the structure given in equation 3), as a result of the formation of a polaron band resulting from an internal redox reaction[17,21] described as proton-induced spin unpairing.[20] Upon protonation, the absorption at 2 eV in the emeraldine base polymer disappears, and two new absorptions appear, centered at 1.5 and 2.9 eV. Using the results of band structure calculations,[17] the latter have been assigned[17] to excitations from the highest and second-highest occupied energy bands to the partially filled polaron band. The long tail extending into the infrared has been assigned to intraband free carrier excitations.[12,13]

Recently, the results of photoinduced absorption studies (i.e. excitation spectroscopy) of the emeraldine base polymer have been reported.[22-24] Photoinduced absorptions were observed[23,24] at 0.9eV, 1.4eV and 3.0 eV, accompanied by photo-induced bleaching at 1.8eV and at energies above 3.5 eV. The 1.4 and 3.0 eV photoinduced absorptions have been interpreted, by analogy with the absorption spectrum of the emeraldine salt, to the photoproduction of polarons in emeraldine base.[24] Stafstrom et al[25] concluded that positive and negative polarons, as well as positive bipolarons account very well for the 1.4 and 3.0 eV photoinduced absorptions. No specific explanation has been proposed for the 0.9 eV photoinduced peak.

We have recently published the results of a comprehensive spectroscopic study of the emeraldine salt of polyaniline in dilute solution in concentrated sulfuric acid and in the form of thin films spin-cast from sulfuric acid solutions. Three spectral features are observed, at 1 eV (with a tail extending deeper into the infrared), at 1.5 eV, and at 3 eV. In the films, the relative intensities of the 1eV and 1.5 eV absorption bands were found to be strongly dependent on both the molecular weight and the protonation level; the 1.0 eV absorption is strongest in the fully protonated emeraldine salt with the highest molecular weight. We also present a method for preparation and stabilization of the fully oxidized polyaniline charged bipolaron lattice form of polyaniline (the structure in equation 4) and give the first spectroscopic characterization of this novel alternating copolymer. We analyze the available data in comparison with the results of band structure calculations[17,18] to provide the basis for an understanding of the electronic structure of the four principal forms of polyaniline: the fully reduced leucoemeraldine, $(1A)_n$; the partially oxidized emeraldine base, $[(1A)(2A)]_n$; the oxidized and fully protonated emeraldine salt, $[1S]^{\cdot}(A^-)_n$; and the fully oxidized bipolaron lattice, $(\text{-B-NH}^+\text{=Q=NH}^+\text{-})_n$.

A. Experimental Methods and Techniques

The polyaniline samples used for the spectroscopic studies were synthesized according the method described in detail previously.[16] The oxidizing agent was prepared by dissolving 23.0 g (0.11 mole) of ammonium persulfate (Aldrich) into 250 ml 1.5M aqueous HCl solution. This was slowly added (while stirring vigorously) to 250 ml of aqueous HCl solution of the same molarity that contained 20 ml (0.22 mole) of aniline. During the entire 3 hour period of addition, the temperature of the polymerizing mixture was carefully maintained at 0°C (to within 1°C). After the oxidant was added, the reaction mixture was left stirring at $0^{o}C$ for an additional hour. The precipitated polyaniline was recovered from the polymerization vessel, filtered, and then washed with distilled water until the washing liquid was completely colorless. In order to remove oligomers and other organic by-products, the precipitate was washed with several portions of methanol until the methanol solution was colorless. Finally, the material was washed twice with ethyl ether and subsequently dried at room temperature for 48 hours in dynamic vacuum, until constant mass was reached. The pristine polyaniline salt was converted to the base form by treatment with 3% aqueous NH_4OH solution for two hours, followed by washing with distilled water, methanol and ethyl ether. Polyaniline synthesized in this way has been thoroughly characterized in terms of viscosity (as a measure of molecular weight), structure and magnetic properties, as well as electrical conductivity.

The emeraldine salt films were spin-cast from a viscous 3% (w/w) sulfuric acid solution of as-polymerized polyaniline onto sapphire substrates at spin rates of $1.5\text{-}2.0 \times 10^3$ cpm during a period of 2-3 minutes. After spinning, the film on the substrate was left in air for 1 hour in order to complete the precipitation of the polyaniline. The resulting thin film was then washed several times in distilled water to remove any residue of free sulfuric acid and subsequently dried in dynamic vacuum. Such polyaniline emeraldine salt films appear homogeneous, and they adhere tightly to the substrate. Typical film thicknesses were 5000Å as determined by Dektak measurements. In order to convert a film to the base form, it was treated with 3% aqueous NH_4OH solution, followed by washing with distilled water and drying in dynamic vacuum. Fully reduced leucoemeraldine films with structure $(1A)_n$ were prepared by dipping an emeraldine base film into phenylhydrazine for several hours under nitrogen,followed by thorough washing with deoxygenated acetone.

Polyaniline films with different degrees of protonation were prepared by dipping the films (on the substrate) for 30 minutes into a large excess of previously prepared HCl solution, the pH of which was preset and determined by a Fisher Model 955 pH meter. The films were again dried in dynamic vacuum. Since we found that the spectra did not change after 30 minutes in the HCl solutions, this is sufficient time for the thin film to come to equilibrium with the solution.

Solution spectra were obtained from samples prepared by diluting a homogeneous 3% solution of polyaniline in sulfuric acid by 97% H_2SO_4 to appropriate concentrations. Unless otherwise stated, the solution spectra were obtained at the dilute concentrations of 10^{-4} mole/liter.

The inherent viscosity of the polyaniline base (in concentrated sulfuric acid) was used as an indicator of the molecular weight. Measurements of the inherent viscosity of the polyaniline emeraldine base (obtained by exposure to aqueous NH_3 solutions and subsequently washed) were carried out using an Ubbelohde viscometer at $25^{o}C$ using 0.1% w/w polymer solutions in H_2SO_4. The polyaniline batch synthesized for this study had an inherent viscosity of 1.0 dL/g at $25^{o}C$ (0.1 wt% of the emeraldine base in H_2SO_4). The material was fractionated for studies of the molecular weight dependence. Samples with η_{in}=0.18 dL/g were made from the low molecular weight fraction extracted from the same batch; i.e. that fraction which was soluble in THF. Samples with η_{in}=1.6 dL/g sample were made from that fraction which was insoluble in THF (after the elimination of THF-soluble part by extraction). Note that the fraction which was insoluble in THF was soluble in concentrated sulfuric acid. Thus, the fraction which is insoluble in common organic solvents is not crosslinked, but simply has a relatively high molecular weight.

The fraction of the sample which gives η_{in}=1.6 dL/g has a molecular weight (as estimated from the Mark-Houwink relations[26]) in the range from about 15,000 (rigid chain limit) to about 60,000 (flexible chain limit). Based on these estimates of the molecular weight, the concentration (10^{-4}mole/liter) used for the solution spectroscopic studies is dilute; i.e. involving negligible interchain interaction.

Electronic absorption spectra (from ≈0.3 eV to ≈ 6.0 eV) were recorded with a Perkin-Elmer Lambda 9 UV/VIS/IR spectrophotometer. Mid-IR spectra were obtained with a Perkin-Elmer 1330 Infrared Spectrophotometer.

B. Experimental Results

a. *Molecular weight (viscosity)*

The absorption spectra of polyaniline films cast from sulfuric acid are strongly dependent on the molecular weight (viscosity). Figure 2 shows spectra of three thin films spin-cast from sulfuric acid solution, and subsequently treated by 0.5M HCl solution to achieve full protonation. To study the effect of molecular weight on the absorption spectrum, we used emeraldine base samples fractionated from the same preparation batch (polymerization and compensation) as described in the previous section. This procedure prevents any uncertainty in the viscosity value, and avoids slight variations in the preparation procedure.

The absorption spectrum of the low molecular weight sample (η_{in}=0.18 dL/g) shown in Figure 2 is quite similar to spectra for the emeraldine salt reported previously;[11,17] there are two peaks, centered at 1.5 and 2.9 eV. The minimum in absorption near 2.0-2.4 eV is not as deep or well-defined as the corresponding minimum in the spectra from higher molecular weight samples, implying excess absorption near 2 eV; in addition, there is residual interband absorption near 4 eV. Since both of these features are characteristic of the emeraldine base, the spectrum of the low molecular weight material implies incomplete conversion to the emeraldine salt. The fact that these features remained after protonation for the η_{in}=0.18 dL/g sample suggests that the low molecular weight (protonated) material is not fully metallic, consistent with the low electrical conductivity of this sample (after precipitation from sulfuric acid, we find σ= 4 S/cm for η_{in}= 0.18 dL/g, compared with σ=33 S/cm for η_{in}=1.6 dl/g). In comparison with previously reported spectra, an important difference in the spectrum in Figure 1 for the low molecular weight sample is the clear shoulder at 1.0 eV.

Figure 2 shows that, upon increasing the molecular weight through η_{in}=1.0 dL/g to η_{in}=1.6 dL/g, the intensity of the 1 eV feature increases, until the infrared absorption becomes a broad band, with a long tail extending into the deep infrared, consistent with the metallic character of these samples. In addition, for the highest molecular weight sample, there is a single asymmetric absorption band, which peaks at 3.0-3.2 eV, with no residual interband absorption near 4 eV from the emeraldine base.

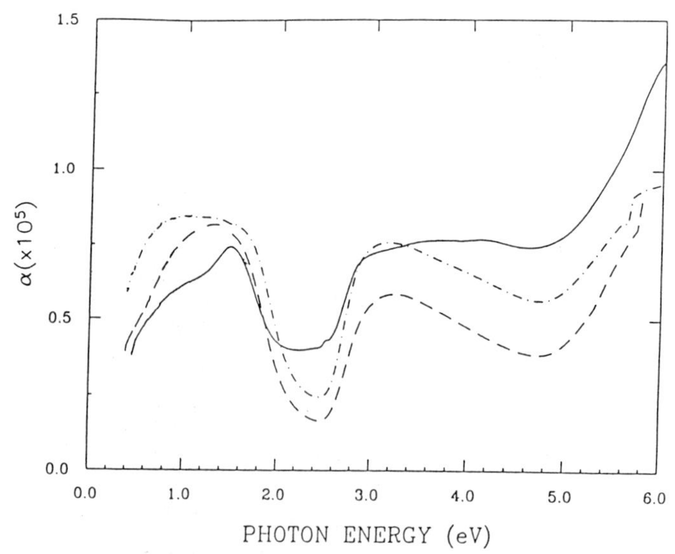

Fig. 2. Optical absorption spectra of films of the fully protonated emeraldine salt $[1S]^{\cdot}(A^{-})_n$, spin-cast from sulfuric acid; the absorption coefficient is plotted as a function of photon energy. The three spectra were obtained from films with different molecular weights, as inferred from viscosity measurements (see text).

——————— η_{in}=0.18 dL/g
— — — η_{in}=1.0 dL/g
-- -- -- η_{in}=1.6 dL/g

Since the films used in earlier studies[11,17] were cast from DMF or DMSO, it is not surprising that only the 1.5 eV peak had been reported. The fraction (~20%) of polyaniline emeraldine base which is soluble in these solvents has the same viscosity (and hence approximately the same low molecular weight) when redissolved in sulfuric acid as the THF-soluble emeraldine base used to obtain the solid curve on Figure 1. Moreover, unlike the homogeneous films obtained by spin-casting from sulfuric acid, in our experience the thin films cast from organic solvent solution are more granular, leading to decreased spectral resolution due to stronger scattering.

In Figure 3, we show the solution spectra of samples prepared with different viscosity. We present the data as a function of wavelength to emphasize the subtle changes found in the near infrared region. Comparing Figures 2 and 3, one finds that the two near infrared features appear at almost the same positions, 1.0 and 1.5 eV, respectively, while their relative intensities are inverse in the solid and solution spectra. In the dilute solution spectra, the intensity of 1.0 eV feature was always much higher than that of the 1.5 eV absorption; in fact, the 1.5 eV absorption is completely absent in the high molecular weight material, while the 1.0 eV absorption (with a long tail extending farther into the infrared) completely dominates. This result implies that the 1.0 eV absorption band arises from an intrachain excitation, while the 1.5 eV transition is due to interchain charge separation. Note that the 3.0 eV peak in the solid film spectra was red-shifted to about 2.5 eV in the solution spectra; i.e. opposite to the usual solid state/solution effect.

b. *Degree of protonation*

Figure 4 shows a series of absorption spectra of polyemeraldine spin-cast from solutions with η_{in}=1.6 dL/g, and subsequently equilibrated with aqueous HCl solutions of different pH. Chiang et al[27] have calibrated the relationship between the pH of the solution

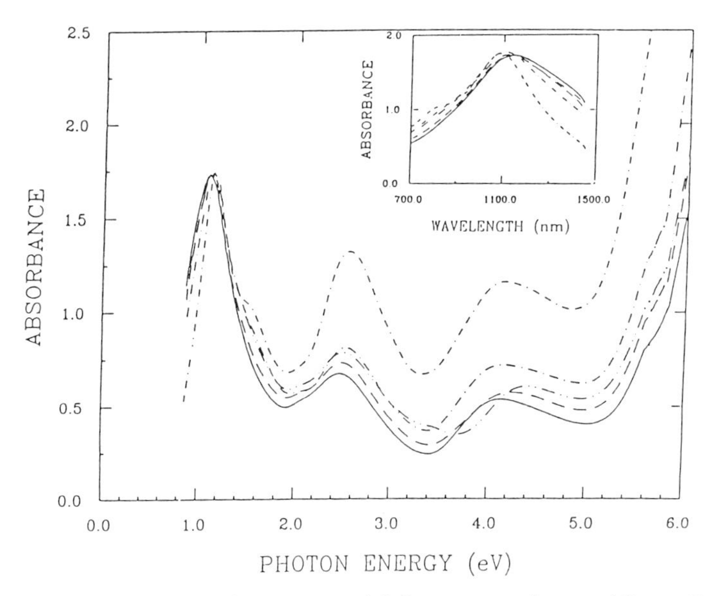

Fig. 3. Optical absorption spectra of fully protonated emeraldine salt, $[1S]^{\cdot}(A^{-})_n$, in sulfuric acid solution (97%); the absorbance is plotted as a function of photon energy. The IR region is shown as a function of wavelength in more detail in the inset. The spectra were obtained from solutions with different molecular weight as inferred from viscosity measurements (see text).

— — · — — · — —	η_{in}=0.18 dL/g
— · — · —	η_{in}=1.0 dL/g
— ·· — ·· —	η_{in}=1.45 dL/g
— — —	η_{in}=1.65 dL/g
————	η_{in}=2.27 dL/g

in which emeraldine base samples were equilibrated and the fractional protonation; the lower pH gives the higher degree of protonation. We demonstrated previously that treatment of emeraldine base with aqueous solution of pH=1 resulted in essentially complete protonation of the emeraldine salt.[16] As can be seen from Figure 4, at pH=5 (where the protonation begins), the band at 2 eV shifted slightly to lower energy and the intensity decreased, while a new shoulder appeared at about 1.4 eV, indicative of partial protonation. The simultaneous existence of the 2 eV and 1.4 eV bands suggests phase segregation of fully protonated and unprotonated domains. At pH=3, although the protonation level is still very low (the protonated fraction[25] is less than 0.05) the 2 eV absorption has vanished completely, while in the near-infrared region only a single, almost symmetric peak is observed centered at about 1.4 eV. Further increase of protonation level (by equilibration in pH as low as 0.15) leads to an important red shift and broadening of this peak. Noting the existence of the two features in this spectral region (as described in previous section), the data indicate that the intensity of the lower energy (intrachain) feature increases with increasing level of protonation, qualitatively consistent with an increase in the number of free carriers. The complete absence of the well-defined absorption at 4 eV (characteristic of the π-π^* transitions in leucoemeraldine and in the emeraldine base) is also consistent with metallic behavior; the oscillator strength has shifted into the infrared.

c. *Leucoemeraldine and the Emeraldine Base*

The spectra of leuoemeraldine, $(1A)_n$, and the emeraldine base, $[(1A)(2A)]_n$, have been extensively studied elsewhere.[28,29] We include the spectra of these two important forms of polyaniline obtained from films spin-cast from sulfuric acid (see Figure 5) in order that they can be directly and quantitatively compared with those of the emeraldine salt and the bipolaron lattice. Leucoemeraldine is an insulator with a large band gap. The π-π^* transition shows an onset at approximately 3 eV, with a peak at 3.7-3.8 eV. The emeraldine base shows two principal absorption bands (see Figure 5), with maxima at 2 eV and at 3.9 eV, respectively.

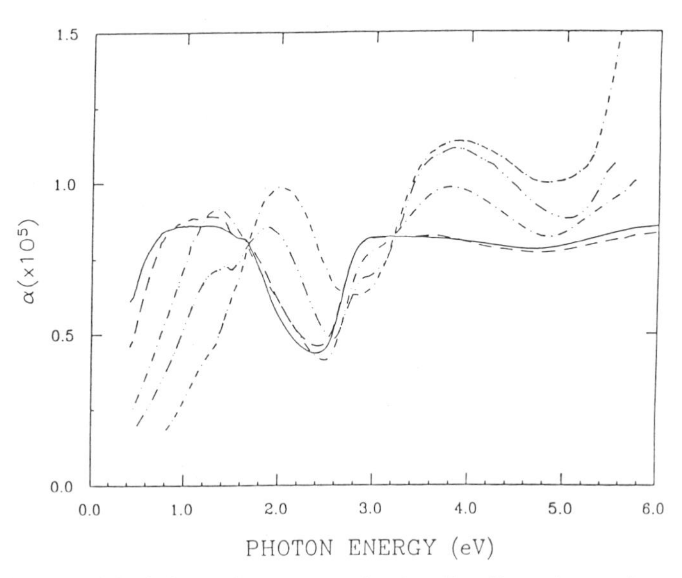

Fig. 4. Optical absorption spectra of polyaniline film spin-cast from sulfuric acid and subsequently exposed to media with different pH to control the protonation level. The absorption coefficient is plotted as a function of photon energy. The film was made from relatively high molecular weight material with η_{in}=1.6 dL/g.

— — · — — · — —	pH>10
—— ·· —— ·· ——	pH=5
—— · —— · ——	pH=3
—— —— ——	pH=2
————————	pH=0.15

We found that when fully reduced leucoemeraldine film is exposed to air, the 2 eV peak appeared quickly, within the first several hours. The intensity continued to increase for the first two days, saturated during the fourth day, and remained almost constant during several more days of exposure. The relative intensity of the 2 eV to 4 eV absorption peaks of this auto-oxidized and air-stabilized polyaniline was much smaller than that of the polyemeraldine base. Comparing with the spectra of the various oxidation states of octaaniline,[30] we conclude that the degree of oxidation for this sample is near to y=0.25 (corresponding protoemeraldine).

d. *Fully-oxidized polyaniline in sulfuric acid*

Asturias et al[31] noted that overoxidized polyaniline will gradually degrade toward the emeraldine structure (in 80% acetic acid solution) due to the hydrolysis of -C=N- double bonds). We find, however, that fully oxidized polyaniline is stable in concentrated sulfuric acid (97%).

Figure 6 shows the spectral changes that occur as a result of the overoxidation of polyaniline in sulfuric acid. After taking the initial spectrum, a small drop of a solution of 2.5 g of $(NH_4)_2S_2O_8$ oxidant dissolved in 100 g of 97% sulfuric acid was added into a quartz cell containing a $3x10^{-4}$ mole/liter solution of polyaniline in concentrated sulfuric acid. After vigorously shaking the quartz cell for one minute, the absorption spectrum was recorded; the spectrum was then recorded again at 15 minutes and at 10 hours after addition of oxidant. As can be seen from Figure 6, the spectrum changed almost immediately after addition of oxidant; the characteristic features of the emeraldine salt disappeared, and an intense transition with onset below 2 eV (peak at 2.2 eV) was observed. In addition, the absorption near 4 eV increased, and developed into a well-defined peak. The new absorption near 2 eV is characteristic of quinoid rings in the polyaniline chain. No significant changes in the spectrum were observed after 40 minutes; the oxidation-induced changes were stable and remained without any change at least for several days.

That this procedure does indeed yield the fully oxidized charged bipolaron lattice was demonstrated by nuclear magnetic resonance (NMR). ^{13}C NMR spectra of the poly(emeraldine) salt before and after oxidation in sulfuric acid solution are shown in Figure 7. For the emeraldine salt in solution (Fig. 7a), we observe only two lines: a more intense line at 129.7 ppm, attributable to the proton-bonded carbons in the benzenoid rings, and a weaker line at 144.7 ppm, attributable to the nitrogen-bonded carbons in the benzenoid rings. The fact that the peaks corresponding to carbons in quinoid rings

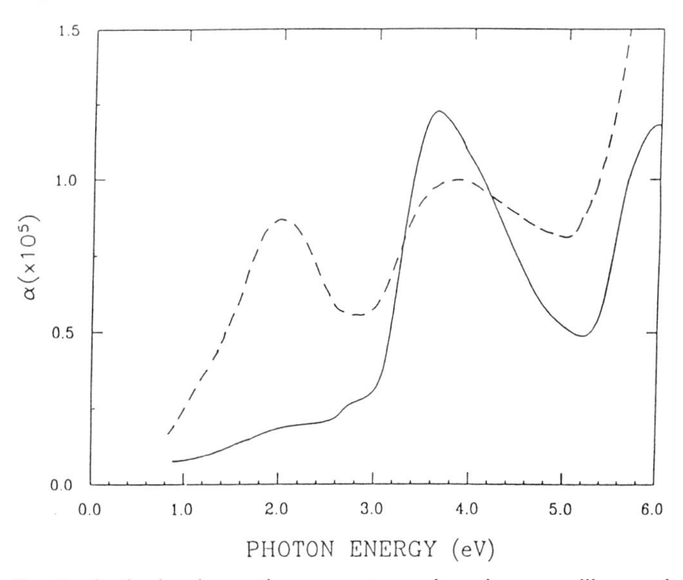

Fig. 5 Optical absorption spectra of spin-cast films of leucoemeraldine and of the emeraldine base.

——————— $(1A)_n$

— — — $[(1A)(2A)]_n$

completely disappear (in comparison with the spectra obtained for the emeraldine base and salt in the solid state[9a,32]) demonstrates that all benzene rings become equivalent in sulfuric acid solution. After subsequent oxidation (as described in the previous paragraph), two additional features appear: one line at around 136 ppm, corresponding to proton-bonded carbons in quinoid rings, and a second line at 161.5 ppm, corresponding to the nitrogen-bonded carbons in quinoid rings. The downfield shift to 161.5 ppm, from 157 ppm in the neutral emeraldine base,[9d,33] suggests that the fully oxidized polyaniline [i. e. after treatment with $(NH_4)_2S_2O_8$] in sulfuric acid must be protonated.

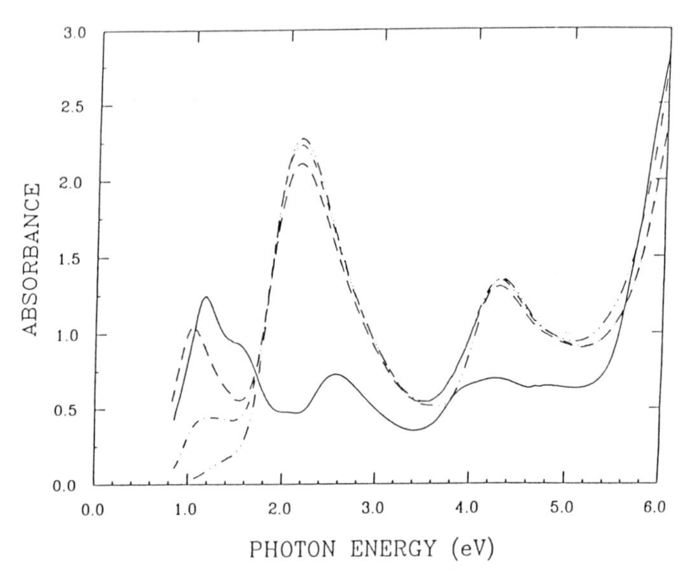

Fig.6. Solution optical absorption spectra showing the formation of the fully oxidized bipolaron lattice form of polyaniline.

———— initial emeraldine base
— — — 1 minute after addition of ammonium persulfate
— — — 5 minutes after addition of ammonium persulfate
—— —— 10 hours after addition of ammonium persulfate

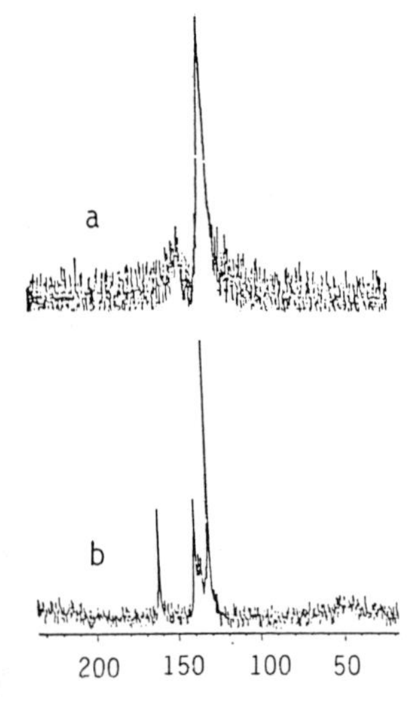

Fig. 7. ^{13}C NMR spectra of the emeraldine salt before (7a) and after (7b) oxidation in sulfuric acid

Comparing the positions and relative intensities of the 2 eV and 4 eV peaks in Figure 6 with the data obtained from the fully-oxidized tetramer[9b] and octamer[30] suggests a formal positive charge on each nitrogen in the protonated structure in sulfuric acid solution (i.e. a doubly charged bipolaron associated with each quinoid unit). These results imply that the polyaniline was fully-oxidized to the copolymer with alternating quinoid and benzenoid monomer units, with the structure given in (4). The absence of any sign of free-carrier absorption in the infrared indicates complete conversion of the polaronic metal to the doubly charged bipolaron lattice after oxidation of the protonated emeraldine salt.

The stabilization of the bipolaron lattice structure provides an opportunity for studies of this new degenerate ground-state polymer, at least in solution. However, although the bipolaron lattice form appears to be stable in concentrated sulfuric acid, hydrolysis occurs when the solution is exposed to moisture from the atmosphere. As a result, attempts toward recovering (4) in the pure solid state have thus far been unsuccessful. Efforts in this direction are continuing in our laboratory.

C. Discussion of Spectroscopic Data

The molecular weight dependence of the $[1S]^{\cdot}(A^{-})_n$ absorption spectrum demonstrated in Figure 1 is important, for it emphasizes the need for long, uninterrupted chains in order to obtain the intrinsic electronic structure. Short chains (low molecular weight) appear to affect the electronic structure in two ways. First, they lead to localization of the carriers, as inferred from the loss of the infrared free-carrier absorption (at 1 eV and below in Figure 1). This is consistent with the inferences obtained from magnetic susceptibility studies;[34] only in well-ordered partially crystalline samples does one clearly observe the temperature-independent Pauli spin susceptibility expected for delocalized Fermions. Second, the residual 2 eV and 4 eV absorption present in the low molecular weight samples (after exposure to pH=0.15) indicate incomplete protonation. This may be due to the relative importance of end effects in the low molecular weight oligomers which are soluble in THF and DMF.

The available data provide the basis for an understanding of the electronic structure of the four principal forms of polyaniline: the fully reduced leucoemeraldine, $(1A)_n$; the emeraldine base, $[(1A)(2A)]_n$; the oxidized and fully protonated emeraldine salt, $[1S]^{\cdot}(A^{-})_n$; and the fully oxidized bipolaron lattice, $(\text{-B-NH}^{+}\text{=Q=NH}^{+}\text{-})_n$.

For leucoemeraldine, the absorption spectrum (Figure 4) consists of a single asymmetric absorption with maximum at 3.7-3.8 eV, and with a shape consistent with that expected for the joint density of states for an intrachain transition in a quasi-one dimensional band structure. This is the π-π^* transition of the $(1A)_n$ conjugated polymer. For the emeraldine base structure there are two transitions, as inferred from the data shown in Figure 4. The close agreement of the energies of the ultraviolet absorption peaks for the two materials, $(1A)_n$ and $[(1A)(2A)]_n$, implies that they are of the same origin.

Valence Effective Hamiltonian (VEH) calculations of the band structures of the $(1A)_n$ and $[(1A)(2A)]_n$ conjugated polymers have been carried out by Boudreaux et al;[18] their results are reproduced in Figures 8a and 8b. As shown in Figure 8a, the π-π^* transition in leucoemeraldine is from band b (the highest occupied molecular orbital or HOMO), to the flat band above the Fermi level. From Figure 8a, the onset of absorption is estimated as about 4 eV. The VEH band structure[18] of the emeraldine base is reproduced in Figure 8b. The high-energy transition observed in the emeraldine base is from band b to the flat band 4 eV higher in energy. The 2 eV absorption, characteristic of the existence of quinoid rings in the emeraldine base, can be assigned to the transition from band b to band b'; however, the observed energy ($\approx$2 eV) is higher than that inferred from the calculations (approximately 0.8 eV). These assignments of the optical transitions are in agreement with the results of photoconductivity measurements, which indicate the photoproduction of free carriers for both 2 eV and 4 eV pumping.

From the data of Figures 1 and 2, we conclude that the absorption spectrum of partially crystalline films of the high molecular weight and fully protonated emeraldine salt, i.e. $[1S]^{\cdot}(A^{-})_n$, has three important spectral features:

(i) an intrachain absorption band at 1 eV (with a long tail into the deep infrared)
(ii) an intrachain absorption band at 3 eV (with onset at approximately 2.5 eV),

and

(iii) an interchain absorption at 1.5 eV.

The 1 eV absorption appears to be dominated by the free carrier intraband absorption (the Drude-like tail extending into the deep infrared). Although there may be a weak interband contribution (with a finite energy gap) as well, this is uncertain because of the strength of the metallic free carrier absorption.

These features of the $[1S]^{\cdot}(A^-)_n$ spectrum can also be compared with the results of electronic band structure calculations. The single-chain VEH band structure for the $[1S]^{\cdot}(A^-)_n$ polaron metal[17] is reproduced in Figure 8c. Since the highest occupied band (band a) is half-filled, the intraband Drude-like contribution is an obvious feature, to be expected for the $[1S]^{\cdot}(A^-)_n$ polaronic metal. The 3 eV intrachain absorption must then arise from vertical transitions from the Fermi level in band a to the higher-energy flat bands, which in the VEH band structure are about 4 eV above E_F. Although allowed, the vertical transition from band b to the Fermi level in band a would be weak, because of the small joint density of states. This transition could, however, contribute to the background absorption below 2 eV.

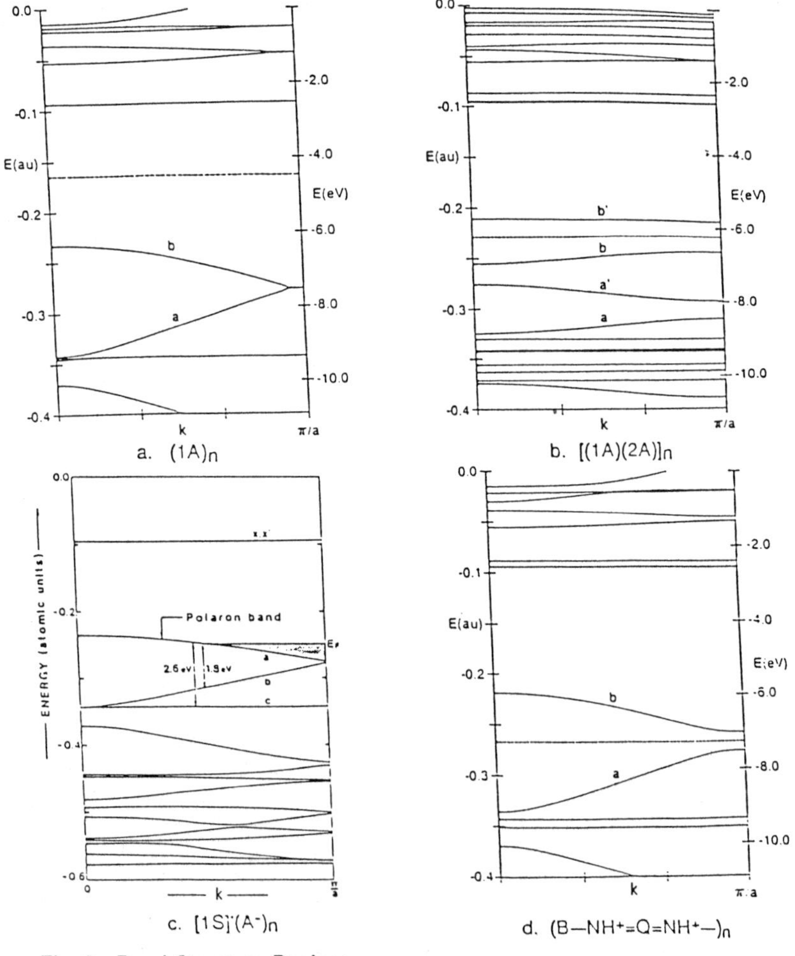

Fig. 8. Band Structure Designs

a. $(1A)_n$ = (B—NH—B—NH—); (from Boudreaux et al.[18])
b. $[(1A)(2A)]_n$ =[(B—NH—B—NH—)(B—N=Q=N—)]$_n$; (from Boudreaux et al.[18])
c. $[1S]^{\cdot}_n$ = [B—NH—B—NH$^+$—]$^{\cdot}_n$; (from Stafstrom et al[17])
d. (B—NH$^+$=Q=NH$^+$—)$_n$, (from Boudreaux et al[18])

The electronic absorption spectrum of the fully oxidized bipolaron lattice with the $(\text{-B-NH}^+\text{=Q=NH}^+\text{-})_n$ structure consists of two bands, an intense band at 2.2 eV and a weaker band at 4.3 eV; the latter being the residual π-π^* transition of leucoemeraldine, $(\text{-B-NH})_n$. Both are asymmetric, consistent with the expected joint density of states for an intrachain transition in a quasi-one dimensional band structure. The VEH band structure calculated[18] for the $(\text{-B-NH}^+\text{=Q=NH}^+\text{-})_n$ bipolaron lattice is reproduced in Figure 8d. Although there is no change in the size of the unit cell, when compared with $[1S]^{\cdot}(A^-)_n$, the half-filled polaron band is emptied upon oxidation, and gaps are increased at the symmetry points (zone edge and zone center) because of the benzenoid-quinoid bond alternation. Thus, we assign the 2.2 eV absorption to the transition between the HOMO (band a in Figure 8d) and the lowest unoccupied band (band b in Figure 8d); the experimental value is considerably larger than that inferred from the direct transition at the zone edge in Figure 8d (approximately 0.5 eV). Note that band b corresponds to the half-filled metallic band in $[1S]^{\cdot}(A^-)_n$. Since the lowest interband absorption energy (2.2 eV in the bipolaron semiconductor) presumably arises from the benzenoid-quinoid bond alternation, the smaller calculated value may imply a larger bond alternation. The 2.2 eV absorption can also be viewed as a transition from the valence band of leucoemeraldine to a bipolaron band, formed in the gap as a result of oxidation to the bipolaron lattice. In this context, we note that the experimental results indicate only a single bipolaron band in the gap. This asymmetry was noted earlier[17] and arises fundamentally from the higher electronegativity of nitrogen over that of carbon and nitrogen.

We conclude that the available band structure calculations provide the basis for a semiquantitative understanding of the absorption spectra. There are, however, quantitative discrepancies. In addition, the importance of the 1.5 eV interchain transition in $[1S]^{\cdot}(A^-)_n$ indicates that more extensive three-dimensional band calculations will be required for a complete understanding of the electronic structure of the polaronic metal.

The similarity of the photoinduced absorption spectrum[22-24] with that of the metallic $[1S]^{\cdot}(A^-)_n$ suggests that the photoinduced absorption may arise from photogenerated polarons according to the following scheme:

$$(\text{-B-N-B-N-B-N=Q=N-}) \xrightarrow[\hbar\omega]{} (\text{-B-N-B-N-B-N}^+\text{=Q=N}^+\text{-})$$

$$\downarrow$$

$$(\text{-B-N-B-N}^{+\cdot}\text{-B-N-B-N}^{+\cdot}\text{-})$$

Although consistent with the photoinduced infrared active vibrational (IRAV) mode spectrum, this scheme must be verified by photoinduced electron spin resonance measurements, since the mechanism implies the photogeneration of unpaired spins via charge induced spin unpairing (i.e. directly analogous to the proton-induced spin unpairing).

IV. MAGNETIC SUSCEPTIBILITY[34b]

The discovery[5] that, when in solution in concentrated sulfuric acid, polyaniline is in the protonated form, $[\text{B-NH-B-NH-}]^{+\cdot}{}_n$, and that it is recovered as the partially crystalline salt, $[\text{B-NH-B-NH-}]^{+\cdot}{}_n(HSO_4^-)_n$, from solutions in sulfuric acid (by precipitation in water or methanol) has opened the way to a more complete characterization of the polymer, and to studies directed toward the determination of the intrinsic properties of the ordered material. For example, the temperature independence of the spin susceptibility[34] of the more highly ordered crystalline material above 125K is consistent with the Pauli spin susceptibility expected for a metal, with a density of states at the Fermi level estimated as 1 state per eV per formula unit (two rings).

The solubility in concentrated sulfuric acid allows the investigation of the novel state of metallic chains of polyaniline in solution in sulfuric acid. Since the material should be fully protonated when in solution in sulfuric acid, this would be the first example of a metallic polymer as a "liquid metal". On the other hand, when in solution the interchain interactions are clearly much weaker than in the solid state. As a result, the electronic structure is expected to be more nearly one-dimensional, and the possibility of a Peierls' transition with the formation of an energy gap in the excitation spectrum might be anticipated.[35,36] In this case, the metallic chains would distort to form a superlattice structure with an associated energy gap in the excitation spectrum. The existence of such a gap should be evident from the magnitude of the magnetic susceptibility. Alternatively, unless the $[B\text{-}NH\text{-}B\text{-}NH\text{-}]^{+\cdot}{}_n$ chains are rigid rods when in solution, disorder would be expected to play a significant role (and perhaps to even suppress the Peierls' instability).[37]

Solutions of emeraldine salt were prepared by dissolving an appropriate amount of the polymer in 97% sulfuric acid.[5,16] The mixture was left overnight at room temperature under moderate stirring, to yield a homogeneous solution. Part of the solution was transferred into an ESR tube under nitrogen and sealed under vacuum. Using this method, solutions of emeraldine salt in sulfuric acid with concentrations 3 wt%, 5 wt%, 7 wt% and 10 wt% were prepared.

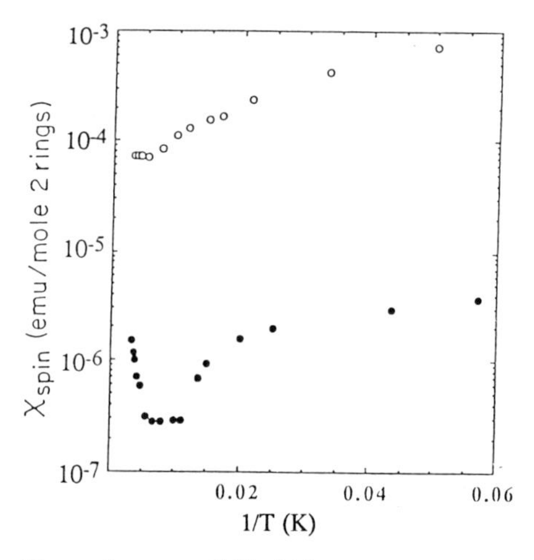

Fig. 9. Magnetic suceptibility(χ) versus temperature (T); black points, emeraldine salt in 10% (w/w) solution in acid; open circles, solid emeraldine salt.

The magnetic susceptibility was determined by means of ESR measurements (9.5 Ghz). The magnitude of χ was obtained by comparison of the results of double integration of the signal from the sample and from a calibrated (National Bureau of Standards) ruby sample as reference. Since the reference is in the cavity with the sample, this method automatically compensates for changes in Q arising from loss due to the ionic conductivity of the sulfuric acid. The measured susceptibility for all the different concentration solutions was in the range of $(2\pm1)\times10^{-6}$ emu/mole (2 rings) at room temperature. Since the ESR signal originates from the polyaniline in solution, the magnitude of χ is given in units appropriate to the amount of emeraldine salt present in the solution. Because of the low Q of the cavity (due to ionic conductivity of the sulfuric acid) and the small number of unpaired spins, we were unable to detect the ESR signal at concentrations less than 3 wt%.

Figure 9 shows the temperature (T) dependence of the susceptibility in the case of the solution with 10 wt% polyaniline in sulfuric acid. The temperature dependence of the peak-to-peak linewidth (ΔHpp) of the ESR absorption line is presented in Figure 10. All the solutions which were studied (with different polymer concentrations) exhibited the same dependences of χ and ΔHpp versus T as those shown in Figure 9 and Figure 10. The magnetic susceptibility of the $[B\text{-}NH\text{-}B\text{-}NH\text{-}]^{+\cdot}{}_n(X^-)_n$ salt obtained by precipitation from sulfuric acid (and subsequently treated with HCl to ensure complete protonation) is also shown for comparison in Figure 9.

As the temperature is lowered (Figure 9), the susceptibility drops from 1.5×10^{-6} at room temperature to 3×10^{-7} emu/mole (2 rings) below 200K, at which point χ is reduced from that of the solid $[B\text{-}NH\text{-}B\text{-}NH^{+}\text{-}]^{\cdot}_{n}(X^{-})_{n}$ salt by more than a factor of 200. The linewidth of the ESR signal in the frozen glass is approximately an order of magnitude greater than that of the solid $[B\text{-}NH\text{-}B\text{-}NH^{+}\text{-}]^{\cdot}_{n}(X^{-})_{n}$ salt. It is temperature-independent below 200 K, and begins to narrow above 200 K.

As noted above, because of the reduced interchain electron transfer interaction for conjugated chains in solution, the electronic structure is expected to be more nearly one-dimensional. Thus, the possibility of a Peierls' transition with the formation of an energy gap in the excitation spectrum might be anticipated. The susceptibility results shown in Figure 9 suggest that this is indeed the case for polyaniline in solution in sulfuric acid. Such a large decrease in χ cannot be accounted for in the context of the half-filled band expected for $[B\text{-}NH\text{-}B\text{-}NH\text{-}]^{+\cdot}_{n}$. To reduce χ by such a large factor would require an increase in the band width by the same factor; i.e. to a band width of more than 200 eV!

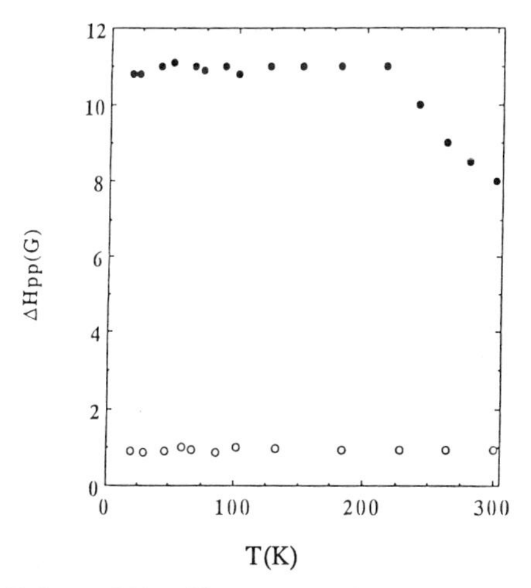

Fig. 10. ESR line width ΔHpp versus the temperature T; black points, emeraldine salt in 10% (w/w) solution in acid; open circles, solid emeraldine salt.

The small density of states, ~0.004 states per eV per formula unit (two rings), indicates the formation of an energy gap at the Fermi surface, consistent with a Peierls' transition. This residual value may arise from a sum of two contributions:

(i) A small fraction of the polyaniline is not in solution, but is suspended as fine particles;

(ii) States in the gap as a result of disorder in the $[B\text{-}NH\text{-}B\text{-}NH\text{-}]^{+\cdot}_{n}$ chains.

The small measured density of states sets an upper limit on the sum of these two contributions.

The magnetic susceptibility data imply, therefore, that polyaniline in sulfuric acid forms a true solution,rather than a suspension of fine particles; at least about 99.6% of the polymer is in molecular solution (and that estimate is a lower limit). This is confirmed by the measured linewidth (Figure 10). There is no indication of even a small fraction of the material with a linewidth of about 1-2 Gauss; i.e. that of the solid $[B\text{-}NH\text{-}B\text{-}NH\text{-}]^{+\cdot}_{n}(X^{-})_{n}$ salt. Note that this conclusion is valid at low temperatures (e.g. 100K) implying that even in the glass of frozen sulfuric acid, the $[B\text{-}NH\text{-}B\text{-}NH\text{-}]^{+\cdot}_{n}$ chains are molecularly dispersed and in matrix isolation.

Unless the $[B\text{-}NH\text{-}B\text{-}NH\text{-}]^{+\cdot}{}_n$ chains are rigid rods when in solution, disorder would be expected to play a significant role, and, perhaps, to even suppress the Peierls' instability. It is known, however, that for conjugated chains in solution, the conformation is determined by a competition between the increased entropy associated with the many degrees of freedom of a coiled or worm-like chain, and the increased electronic energy associated with the breaking of π-bonds.[38] Moreover, even if the insulating conjugated chain is coiled or worm-like, the introduction of free carriers leads to the self-consistent formation of straight-chain segments in the vicinity of the carriers, i.e. conformons.[39] Thus, for the $[B\text{-}NH\text{-}B\text{-}NH\text{-}]^{+\cdot}{}_n$ chains, which have one carrier per formula unit, one is perhaps not surprised to find evidence of a rod-like conformation. That this must indeed be the case is implied by the small value of the residual susceptibility for chains in matrix isolation in the frozen glass. The weak disorder is not sufficient to suppress the Peierls transition, nor is it sufficient to yield a density of localized states within the energy gap greater than about 0.004 states per eV per formula unit (two rings). Moreover, the increase in $\chi(T)$ at low temperatures can be used as an indication of disorder-induced localized states. Even this contribution is extremely small, compared with that observed in the partially crystalline material cast from sulfuric acid. We conclude that in solution, the $[B\text{-}NH\text{-}B\text{-}NH\text{-}]^{+\cdot}{}_n$ chains have a relatively long persistence length (and a correspondingly long localization length), implying that the chain conformation is probably rod-like.

The order of magnitude increase in the ESR linewidth of the $[B\text{-}NH\text{-}B\text{-}NH\text{-}]^{+\cdot}{}_n$ chains in the frozen matrix is consistent with the interpretation that the residual susceptibility arises from localized states in the gap. The onset of narrowing at temperatures above 200 K may imply thermal excitation to more delocalized states, or it may simply reflect the onset of motion of the $[B\text{-}NH\text{-}B\text{-}NH\text{-}]^{+\cdot}{}_n$ chains as the frozen glass begins to melt.

Although the small magnitude of the magnetic susceptibility implies the existence of a gap in the excitation spectrum, it is not possible to directly determine the magnitude of the gap from the data. The up-turn in $\chi(T)$ at temperatures above 250K (and the corresponding onset of motional narrowing as inferred from the linewidth) may indicate the onset of 1d fluctuations as T approaches the mean-field Peierls temperature (T_c^{MF}). However, one must be careful with such an analysis, since this is the temperature range where the sulfuric acid is undergoing the transition from frozen glass to liquid. To investigate this possibility in more detail, we monitored the microwave loss by measuring the full width at half maximum of the cavity as a function of temperature. The results imply that for the polyaniline containing sample (10 wt %), the melting occurs between 220K and 230K, and that the melting point is suppressed somewhat by the dissolution of the polymer in the acid. One might argue that the increase in ionic conductivity could partially shield the $[B\text{-}NH\text{-}B\text{-}NH\text{-}]^{+\cdot}{}_n$ chains inside the solution from the microwave field (and thereby underestimate χ of the solution). However, since there is no step in χ at the melting point, such shielding effects are not important. On the other hand, the increase in χ continues for temperatures at least up to 325K; i.e. 100K above the solid-liquid transition. Our preliminary conclusion is, therefore, that the increase in χ is not an artifact of the melting transition, suggesting that T_c^{MF} may be somewhat greater than room temperature.

V. SUMMARY and CONCLUSION

The results of structural, spectroscopic and magnetic studies of the various forms of polyaniline carried out recently at UC-Santa Barbara have been briefly reviewed. Although polyaniline can be prepared in partially crystalline form, the molecular weight dependence of the $[1S]^{\cdot}(A^-)_n$ absorption spectrum is particularly important, for it demonstrates the need for long, uninterrupted chains in order to obtain the intrinsic electronic structure.

From the spectroscopic data, we conclude that high molecular weight and fully protonated emeraldine salt, $[1S]^{\cdot}(A^-)_n$, has three important spectral features; an <u>intra</u>chain absorption band at 1 eV (with a long tail into the deep infrared), an <u>intra</u>chain absorption band at 3 eV (with onset at approximately 2.5 eV), and an <u>inter</u>chain absorption at 1.5 eV. The 1 eV absorption appears to be dominated by the free carrier "metallic" intraband absorption.

We have re-analyzed the available data in comparison with the results of band structure calculations, to provide the basis for an understanding of the electronic structure of the four principal forms of polyaniline: the fully reduced leucoemeraldine, $(1A)_n$; the partially oxidized emeraldine base, $[(1A)(2A)]_n$; the oxidized and fully protonated emeraldine salt, $[1S]^{\cdot}(A^-)_n$; and the fully oxidized bipolaron lattice, $(\text{-B-NH}^+\text{=Q=NH}^+\text{-})_n$.

Studies of the magnetic susceptibility of the emeraldine salt in the solid state and in solution in sulfuric acid have demonstrated that in sulfuric acid, the $[\text{B-NH-B-NH-}]^{+\cdot}{}_n$ chains are molecularly dispersed in a genuine solution (as opposed to a suspension of fine particles). The indications of weak disorder imply that the chains have a relatively long persistence length (and a correspondingly long localization length), suggesting that the chain conformation is probably rod-like.

The magnetic susceptibility of protonated $[\text{B-NH-B-NH-}]^{+\cdot}{}_n$ chains in solution was found to be decreased, with respect to χ in the solid state, by more than two orders of magnitude. This decrease was interpreted as indicative of a Peierls transition; in solution the metallic chains apparently distort to form a superlattice, which leads to the opening of a characteristic energy gap in the excitation spectrum.

Acknowledgment: The spectroscopic studies were supported by the Office of Naval research under N00014-83-K-0450; the synthesis and solution processing were supported by the National Science Foundation through an MRG grant (NSF DMR-87-03399); X-ray measurements and analysis were supported by the Office of Naval Research, and the ESR studies were supported by the National Science Foundation (NSF DMR85-21392).

REFERENCES

1a. A.G. Green and A.E. Woodhead, J. Chem. Soc. 97:2388 (1910); 101:1117 (1912);
1b. R. Wilstetter and S. Dorogi, Ber. 42:2147 (1909) and 42:4118 (1909).
2. A.G. MacDiarmid, J.-C. Chiang, M. Halpern, S.-L. Mu, N.L.D. Somasori, W.Wu, and S.I. Yaniger, Mol. Cryst. Liq. Cryst. 121:173 (1985).
3. W.R. Salaneck, I. Lundstrom, W.-S. Huang and A.G. MacDiarmid, Synth. Met. 13:291 (1986).
4. M. Angelopoulos, G.E. Asturias, S.P. Ermer, A. Ray, E.M. Scherr, A.G. MacDiarmid, M. Akhtar, Z. Kiss, and A.J. Epstein, Mol. Cryst. Liq. Cryst. 160:151 (1988).
5. A. Andereatta, Y. Cao, J.C. Chiang, P. Smith, A.J. Heeger, Synth. Met. 26:383 (1988).
6. A. Ray and A.G. MacDiarmid, Synth. Met. (in press; Proceedings of ICSM '88).
7. S. Li, H. Dong and Y. Cao, Synth. Met. (in press; Proceedings of ICSM '88).
8. A special case is the early report of solubility of undoped polyphenylenesulphide in a mixture of SbF_3 and SbF_5; see J. Frommer, R.E. Elsenbaumer and R.R. Chance, ACS Symp. Ser. 242:44 (1984). However, the resulting solution is unstable. Moreover, after casting, the resulting films are no longer reversibly soluble in the same solvent, implying either a change in structure or significant crosslinking (or both).
9a. T. Hjertberg, W.R. Salaneck, I. Lundstrom, N.L.D. Somasiri, A.G. MacDiarmid, J. Polym. Sci., Polym. Lett. Ed. 23:503 (1985).
9b. Y. Cao, S. Li, Z. Xue, and D. Guo, Synth. Met. 16:305 (1986).
9c. A. Ray, G.E. Asturias, D.L. Kershner, A.F. Richter, A.G. MacDiarmid, and A.J. Epstein, Synth. Met. (in press).
9d. S. Kaplan, E.M. Conwell, A.F. Richter and A.G. MacDiarmid, J. Am. Chem. Soc. (in press); *ibid* Macromolecules (in press).
9e. A.F. Richter, A. Ray, K.V. Ramanathan, S.K. Manohar, G.T. Furst, S.J. Opella, A.G. MacDiarmid,and A. J. Epstein (in press).
10a. A. G. MacDiarmid, J. C. Chiang, A. F. Richter and A. J. Epstein, Synth. Met. 18:285 (1987).
10b. A.G. MacDiarmid, J.C. Chiang, W.-S. Huang, B.D. Humphrey and N.L.D. Somarisi, Mol. Cryst. Liq. Cryst. 121:181 (1985).

11. A.J. Epstein, J.M. Ginder, F. Zuo, R.W. Bigelow, H.-S. Woo, D.B. Tanner, A.F. Richter, W. S. Huang and A.G. MacDiarmid, Synth. Met. 18:303 (1987).
12. J.M. Ginder, A.F. Richter , A.G. MacDiarmid and A.J. Epstein, Solid State Comm. 63:97 (1987).
13. A.J. Epstein, J.M. Ginder, F. Zuo, H.-S. Woo, D.B. Tanner, A.F. Richter, M. Angelopoulos, W.-S. Huang and A. G. MacDiarmid, Synth. Met. 21:63 (1987).
14. Y. Moon, Y. Cao, P. Smith and A.J. Heeger, Polymer Commun. (in press).
15a. B.J. Tabor, E.P. Magre and J. Boon, Eur. Polym. J. 7:1127 (1971).
15b. J. Boon and E.P. Magre, Makromol. Chem. 126:130 (1969).
15c. M. G. Northold, Eur. Polym. J. 10:799 (1974).
16. Y. Cao, A. Andretta, A.J. Heeger, P. Smith, Polymer (in press).
17. S. Stafstrom, J.L. Brédas, A.J. Epstein, H.S. Woo, D.B. Tanner, W.S. Huang and A.G. MacDiarmid, Phys. Rev. Lett. 59:1464 (1987).
18. D.S. Boudreaux, R.R. Chance, J.F. Wolf, L.W. Shacklette, J.L. Brédas, B.Themans, J.M. André, R. Silbey, J.Chem. Phys. 85:4584 (1986).
19. E.M. Conwell, C.B. Duke, A. Paton, L. Jeyadev, J.Chem. Phys. 88:331 (1988).
20. F.Wudl, R.O. Angus, F.L. Lu, P.M. Allemand, D.J. Vachon, M. Nowak, Z.X Liu and A.J. Heeger, J. Amer. Chem. Soc. 109:3677 (1987).
21. G. Wnek, Synth. Met. 15:213 (1986).
22. A.J. Epstein, J.M. Ginder, M.G. Roe, T.L. Gustafson, M.Angelopoulos and A.G. MacDiarmid, Mat. Res. Soc. Symp. 109:317 (1988).
23. G. Roe, J.M. Ginder, P.E. Wigen, A.J. Epstein, M. Angelopoulos, A.G. MacDiarmid, Phys. Rev. Lett. 60:2798 (1988).
24. Y.H. Kim, C. Foster, J. Chiang, A.J. Heeger, Synth. Met. 26:49, (1988); Synth. Met. (in press; Proceedings of ICSM '88).
25. S. Stafstrom, B. Sjogen, and J.L. Brédas, Synth. Met. (in press; Proceedings of ICSM '88).
26a D.G. Baird and J.K. Smith, J. Polym. Sci., Polym. Chem. Ed. 16:61 (1978).
26b. K. Kamide and M. Yukio, Kobunshi Ronbunshi, 35:467 (1978).
27. J.-C. Chiang and A.G. MacDiarmid, Synth. Met. 13:193 (1986).
28a. P.M. McManus, R.J. Cushman, and S.C. Yang, J. Phys. Chem. 91:744 (1987).
28b. E.M. Genies and M. Laporski, J. Electrochem. 220:67 (1987).
29. S.D. Phillips, G. Yu, Y. Cao and A.J. Heeger, Phys. Rev. B (in press).
30. F.-L. Lu, F. Wudl, M. Nowak and A.J. Heeger, J. Amer. Chem. Soc. 108:8311 (1986).
31. G.E. Asturias, A.G. MacDiarmid and A.J. Epstein, Synth. Met., (in press, Proceedings of ICSM '88).
32. F. Devreux, G. Bidan, A.A. Sayed, C. Tsintavis, J. Physique 46:1595 (1985).
33. C. Menardo, M. Nechtschein, A. Rousseau, T. P. Travers, and P. Hany, Synth. Met. (in press).
34a. C. Fite, Y. Cao, and A.J. Heeger, Sol. State Commun. (in press).
34b. C. Fite, Y. Cao, and A.J. Heeger, Phys. Rev. Lett. (submitted)
35. R.E. Peierls, "Quantum Theory of Solids," Oxford University Press, London (1955), p. 108.
36. J.T. Devrees, R.P. Evrard and V.E. Van Doren (eds.), "Highly Conducting One Dimensional Solids," Plenum, New York (1978).
37. E.J. Mele and M.J. Rice, Phys. Rev. B 23:5397 (1981).
38. D.R. Spiegel, P.A. Pincus and A.J. Heeger, Polymer Commun. 29:264 (1988).
39. P. Pincus, G. Rossi, and M. E. Cates, Europhys. Lett. 4:41 (1987).

POLYANILINE: NEW PHYSICS IN AN OLD POLYMER

A. J. Epstein

Departments of Physics and Chemistry
The Ohio State University
Columbus, Ohio 43210-1106, USA

A. G. MacDiarmid

Department of Chemistry
University of Pennsylvania
Philadelphia, Pennsylvania 19104-6323, USA

INTRODUCTION

The study of conducting polymers began as a new field with the introduction of doped polyacetylene in 1977 [1]. Polyacetylene (Fig. 1a) has a backbone consisting of carbon atoms with alternating single and double bonds. The p_z wave function of each carbon atom overlaps to form the π-band, which is split into the filled valence (π) band and empty conducting (π*) band, due to the dimerization of the chain. Extensive studies of undoped, and p and n-doping of polyacetylene, have shown the important excitations to be neutral and charged solitons [2,3]. Because motion of a soliton distortion of the bond dimerization involves small excursions of the carbon atoms, the effective mass of solitons is small, of order the electron mass [4].

The subsequently studied polythiophene (Fig. 1b) polypyrrole (Fig. 1c), and similar systems, also have electronic structures based upon overlap of the p_z orbitals of the bonded carbon atoms [2,3]. The heteroatom in the chain (S for polythiophene, NH for the polypyrrole), to the simplest approximation, does not participate in the electronic structure [5]. Again, the bond length alternation leads to a (Peierls) gap in the electronic energy spectrum, with an additional contribution to the energy gap, due to the difference in energies of the two possible phases of bond alternation. Higher-order effects aside, these polymers are flat and planar, and charge-conjugation symmetric, as is polyacetylene. Because of the difference in energy of the two phases of bond alternation, the stable charge excitations are positive and negative polarons and bipolarons. Ignoring the role of Coulomb repulsion, these polarons and

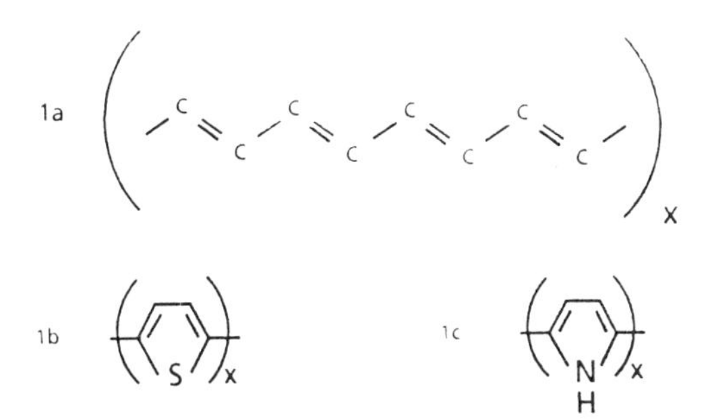

Fig.1. (a) Trans-polyacetylene; (b) polythiophene; (c) polypyrrole

Lower-Dimensional Systems and Molecular Electronics
Edited by R.M. Metzger *et al.*, Plenum Press, New York, 1991

bipolarons have two energy levels in the gap, symmetrically placed in the energy gap. Again, since the movements of a polaron or bipolaron involve only small movements of a carbon atom, the effective masses of these charge defects are of order an electron mass [6].

THE POLYANILINES

The polyaniline family of polymers differs substantially from these earlier studied polymer ststems [7-9]. The polymer backbone consists of alternating C_6H_4 rings and nitrogen atoms, in a variety of different oxidation and protonation levels (Fig. 2). Three stable, insulating forms are the leucoemeraldine or polyphenyleneamine (Fig. 2a), emeraldine or polyphenyleneamineimine (Fig. 2b), and pernigraniline or polyphenyleneimine (Fig. 2c). The presence of both carbon and nitrogen covalently linked within the chain makes this material an "A-B" polymer [10], with an absence of charge conjugation symmetry. A metallic phase of the polymer, emeraldine salt (Fig. 2d), can be obtained, through either oxidation of the leucoemeraldine, or protonation of the emeraldine base polymers [11,12]. Electronic structure calculations show that the energy gap in the leucoemeraldine and emeraldine base forms are due to the presence of the C_6 rings, and not due to the presence of a Peierls [13-15] distortion, while a Peierls distortion may be present in the pernigraniline form of the polymer [16,17].

Improvements in the quality of the materials available [18,19] has enabled extensive studies in the physics of the system, revealing a number of unusual properties. Extensive study of the magnetic and crystalline properties and the crystallinity, as a function of protonation of the emeraldine base form of the polymer, shows the growth of a metallic phase in the crystalline regions, and a spinless electronic state in the amorphous regions of the polymer [20]. Charge transport studies have shown the metallic nature of the islands formed, and the conduction via charging-energy-limited tunneling between the islands [21]. Extensive optical and photoinduced optical studies of this material [22-30] have shown unusual time dynamics, reflecting upon the important role of ring torsion angles in the physical properties of the ring-containing polymers [15]. Analysis of photoinduced optical spectra shows that the photoexcited charges are orders of magnitude more massive, than those in the earlier studied

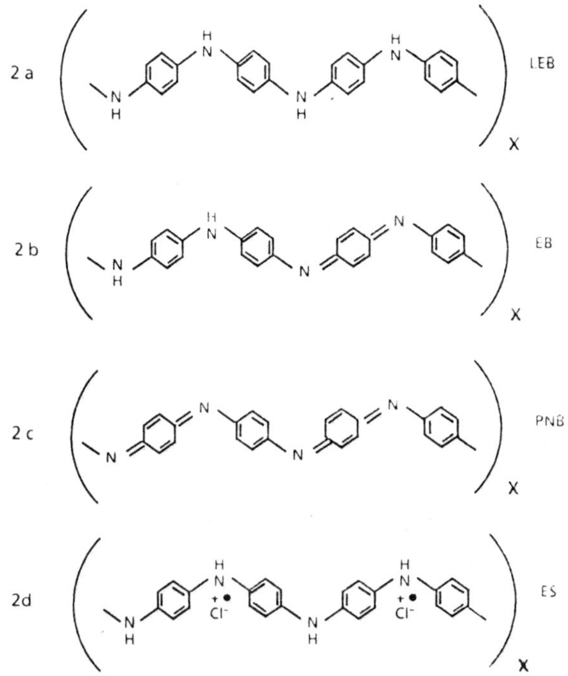

2 a LEB

2 b EB

2 c PNB

2d ES

Fig. 2. (a) Leucoemeraldine base (LEB or polyphenyleneamine); (b) emeraldine base (EB or polyphenyleneamineimine); (c) pernigraniline base (PB or poly-phenyleneimine); (d) emeraldine salt (ES).

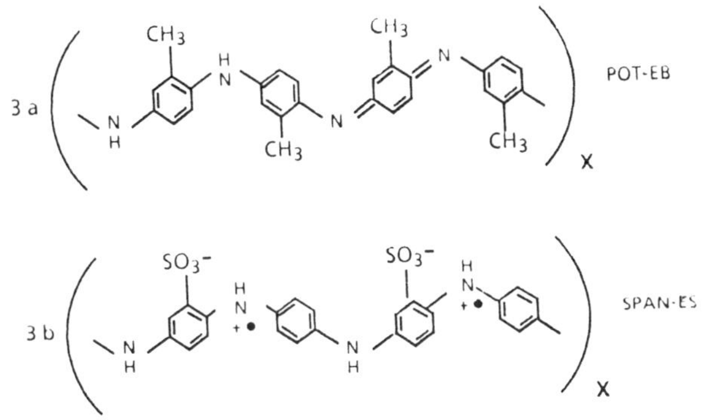

Fig. 3. (a) Polyorthotoluidene (emeraldine base form); (b) sulfonated polyaniline (self-doped salt form).

polyacetylene and polythiophene-type systems [24]. New models have been developed to describe the physics of these systems, in terms of the role of ring torsion angles on the electronic structure and on the self-localization of charge [15].

These phenomena, combined with the processibility and chemical flexibility of these materials, have lead to the suggestion of new applications for these types of systems, for use as erasable optical information storage [31], microwave absorbers [32], and nonlinear optical active media [26]. The chemical structure of the polyaniline family allows for easy derivatization at the ring and nitrogen positions. This allows for critical study of the roles of ring torsion angles and interchain separation. For example, careful study of the monomethyl ring-substituted system, polyorthotoluidene, (Fig. 3a) demonstrates the important role of localization, that occurs upon forced separation of the chains [28], while sulfonation at the phenyl ring results in a self-doped polymer, soluble in aqueous media [34]. This latter system holds particular promise from the point of processibility and speed of electrochemical response.

ELECTRONIC STRUCTURE

The band structure of the various forms of polyaniline has been calculated by several groups [13-15]. The electronic structure is sensitive to the overlap integral between the nitrogen p_z orbital and the p_z orbitals on the carbon atoms adjacent to nitrogen [35]. This overlap, in turn, depends on the angle, at which the C_6H_4 rings lie with respect to the plane of the nitrogen atoms, with increasing bandwidths for the phenyl rings lying in the plane of the nitrogen atoms. The steric repulsion between adjacent phenyl rings is substantial, for rings lying in the plane [15] (Fig. 4), hence there is a tendency for the rings to lie at alternate angles $\pm\Psi$ to the nitrogen plane, to reduce the steric energy. The electornic band structure, calculated for leucoemeraldine base with $\Psi = 56°$, is shown in Fig. 5, assuming a repeat of only a single phenyl NH group. There are seven energy bands, corresponding to six p_z energy levels of C_6 and the one p_z energy level of N. Considering the presence of one p_z electron for each carbon atom and two p_z electrons for the nitrogen, the lower four energy bands are completely filled, with the Fermi level lying in the ~ 3.8 eV gap. This energy gap has its origin in the electronic structure of the C_6 rings, and not in the presence of a Peierls distortion. Oxidation of the leucoemeraldine base to the emeraldine salt level partially depopulates the upper band. Assuming a repeat distance of two rings and two NH's, the Brillouin zone is halved, with the upper energy band being only half occupied for the emeraldine salt state. Further oxidation to the pernigraniline state leads to the uppermost band of Fig. 5 being only half full. It has been a suggestion that a Peierls distortion, driven by the electron coupling to the bond lengths [16] and ring torsion angles [17], occurs, to give a Peierls gap in the pernigraniline form.

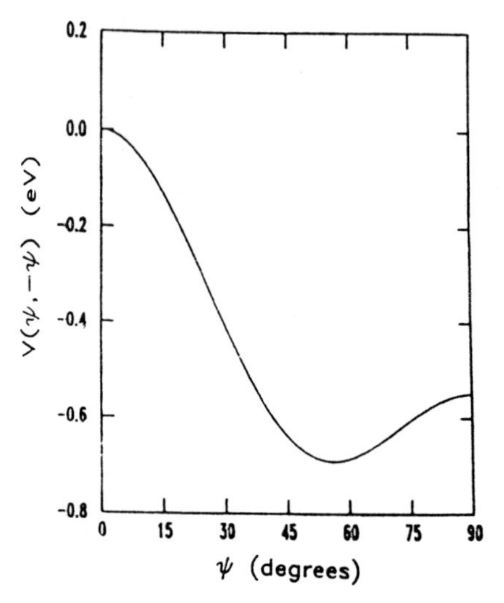

Fig. 4. Approximate energy per site versus ring twist angle Ψ for leucoemeraldine base, assuming equal near-neighbor angles.

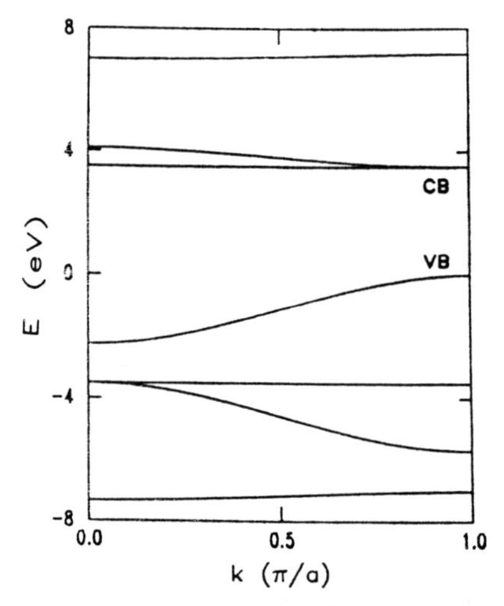

Fig. 5 Calculated band structure for a leucoemeraldine base within the extended zone scheme.

STRUCTURE

The x-ray diffraction studies [20,36-38] of the polyaniline family of polymers shows a multiplicity of structures available, dependent upon the preparation and treatment conditions of the polyanilines. Polymers precipitated as salt, either through chemical or electrochemical methods, are substantially crystalline. For samples with counterion Cl^-, its structure is termed ES-I. It is monoclinic. A schematic structure [20] associated with ES-I is shown in Fig. 6. Dedoping of ES-I leads to amorphous EB-I. Class II materials include those prepared from EB-I by exhaustive extraction, first with THF, then with NMP. This structure is up to fifty percent crystalline, and has an orthorhombic space group (Fig. 6). Doping of EB-II with HCl leads, initially, to protonation of only the noncrystalline regions. In the higher protonation levels, the crystalline form is doped, to form orthorhombic ES-II (Fig. 6). While there are few free parameters in fitting the EB-II structure, the latitude in setting angles, halide position, and central chain position leads to some ambiguity in the ES-I and ES-II structures. The coherence length of the ordered regions is as large as ~ 50 Å perpendicualr to the chains, and

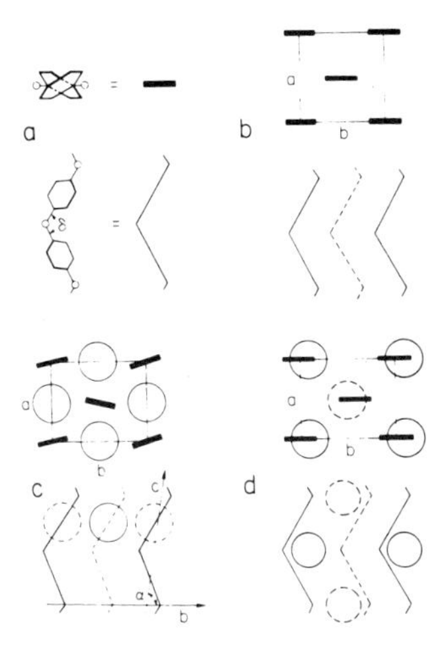

Fig. 6. Projection along chain axis and side view of proposed structure for (a) polyaniline chain, (b) EB-II, (c) ES-I, and (d) ES-II. Large circles represent Cl^- anions. Note that the exact placement of the central chain along the b- and c- axes is uncertain for ES-I and ES-II.

~150 Å in the chain direction for oriented emeraldine base, i.e., the coherence lengths are similar to those of polyacetylene [39].

The magnetic susceptibility was first measured as a function of protonation for the Class I family of materials [12]. Results showed a quasi-linear increase in Pauli susceptibility, which led to the proposal of the formation of metallic islands of the emeraldine salt in the emeraldine base form with protonation. X-ray diffraction studies of the ES-I series, as a function of protonation, gives a quasi-linear increase in the crystalline salt fraction with protonation, supporting the formation of crystalline metallic islands with increasing protonation level. Electron spin resonance studies of the Class II materials, as a function of protonation level, show a dramatically different behavior [20]. Initial protonation of the base EB-II leads to a spinless material, until compositions are achieved, such that essentially all of the amorphous regions are fully doped. At that point, there is an increase in the Pauli susceptibility, corresponding to the formation of crystalline ES-II (Fig. 7). Preliminary studies show that the Pauli susceptibility is essentially the same for the ES- and ES-II structures.

The absence of Pauli susceptibility in the protonated amorphous regions of ES-II correlates with the absence of Pauli susceptibility in the tetramer oligomer salt of aniline, which shows a spinless state, analogous to a bipolaron state being formed [40]. Similarly, recent electron spin resonance studies of emeraldine salt in solution show the presence of a spinless state in solution [41]. Hence, it appears that, in the presence of only short-range order, the spinless-type bipolaron state is stabilized, as compared with stabilization of the metallic polaron lattice in the crystalline forms. This variation in the magnetic state and electronic structure of the emeraldine salt, depending upon local order, may explain the variation in magnetic

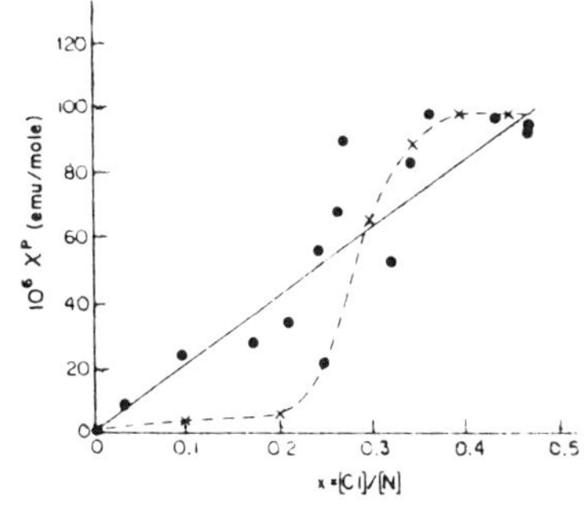

Fig. 7. χ_p versus x for EB-I (●, solid line) and EB-II (x, dashed line). The solid and dashed lines are drawn as guides to the eye.

properties reported by different groups, especially the electron spin resonance on very thin electrochemically synthesized polyaniline [42].

CHARGE TRANSPORT STUDIES

Four-probe dc conductivity of ES-I, as a function of protonation level, showed [21] a conductivity proportional to $\exp[-(T_0/T)^{1/2}]$. The slope of the conductivity vs $T^{1/2}$ curves is independent of dopant concentration for $x \equiv [Cl]/[N] \approx 0.3$. Given the behavior of susceptibility vs. protonation level for the ES-I family, the conductivity is best understood in terms of charging-energy-limited tunneling among the small metal islands [43]. For $[Cl]/[N] > 0.3$, it appears that the barriers between the islands remain the same, while the number of pathways increases. Together with the electric field dependence of the conductivity, T_0 can be used to estimate [21] the separation among the metallic islands as ~100 Å, in accord with later x-ray diffraction studies [20]. The temperature-dependent thermoelectric power, as a function of protonation level, shows a clear crossover in behavior, as a function of protonation level [21]. Analysis of the data is consistent with effective medium theory for a metallic phase embedded in a nonmetallic phase [44].

The ability to stretch-orient the emeraldine polymer, as well as the ability to form stretch-oriented fibers of the emeraldine, has enabled the study of conductivity in anisotropically oriented samples [18]. Preliminary studies of oriented ES-II films show that the conductivity increases approximately fifty-fold to $\sigma(295\text{ K}) \approx 8$ S/cm with $\sigma_{\parallel} / \sigma_{\perp} \approx 2.5$. The temperature dependence of the conductivity is nearly the same as that of the unoriented samples, with a small increase in anisotropy at low temperatures. These results point to a continuing presence of barriers in the oriented ES-II polymers.

For low protonation levels, the charge transport differs [45]. The dc conductivity varies from 10^{-10} S/cm for $x = 0$, to 10^{-6} S/cm for $x = 0.08$, and is proportional to $\exp[-(T_0/T)^{1/2}]$, with T_0 decreasing with increasing x. The temperature-dependent audio frequency f (10^1-10^5 Hz) conductivity varies as f^s, with $s \approx 0.9$ for $x = 0$, decreasing with increasing x. For frequencies greater than 10^3 Hz, the dielectric constant agrees with the T-independent dielectric constant measured by microwave techniques. At lower frequencies and high temperatures, the dielectric constant increases. A Cole-Cole analysis shows the presence of, primarily, a single, thermally activated relaxation process in these materials, with a dispersion in relaxation rates. These results support hopping of charge among positively charged polaron and bipolaron or neutral defect (polaron) states in the emeraldine polymer at the low protonation levels.

The microwave conductivity is larger than the dc conductivity by many orders of magnitude for the base, but approaches that of dc for more conducting emeraldine salts [32]. For lowly protonated emeraldine polymers, the microwave frequency dielectric constant is small, and almost temperature-independent. For higher protonation levels, the dielectric constant is linear in T, with deviations observed at maximum protonation level. The dielectric constant increases monotonically with protonation, for intermediate and higher-level protonated emeraldine salts. The temperature dependence of microwave conductivity differs from that of dc, especially at lower temperatures. In general, the data support the phase segregation of the protonated polymer into "metallic islands" and insulating background, with localization prevailing at low temperatures. In view of the presence of barriers within the metallic islands and the increase of coherence length of the charge carriers with temperature, the phrase "textured metallic islands" was introduced [32] to describe the delicate role of temperature on the interplay between localization and delocalization. From the low-temperature dielectric constant, it was estimated that the "metallic" conductivity within the islands exceeds several hundred S/cm, even at low temperatures. The detailed frequency and composition behavior of the conductivity, as a function of moisture content, are consistent with the role of charging-energy-limited tunneling among small granular polymeric metallic particles, with the barrier being affected by the presence of the moisture [46,47].

EFFECTS OF SUBSTITUENTS ON LOCALIZATION

The role of localization in quasi-one-dimensional systems has long been of interest. As an electron in an ideal one-dimensional system is localized to a finite length, the importance of interchain coupling has been recognized [48]. The addition of a methyl group to the ring of polyaniline, to form polyorthotoluidene (Fig. 3a) enables the study of the effects of substituents on the electronic structure and transport mechanisms in this polymer. Preliminary studies [33] show that the optical spectra are blue-shifted, supporting the important role of increased steric repulsion, in leading to increased ring torsion angle, and, consequently, to narrower energy bands. Magnetic susceptibility studies shows a decrease in the Pauli susceptibility and an increase of the Curie term, reflecting increased localization. The dc conductivity is decreased, from that of the emeraldine salt form, by three orders of magnitude, with a much stronger temperature dependence, consistent with, primarily, hopping in one dimension, while the thermopower is no longer metallic. The microwave conductivity and dielectric constant also decrease dramatically, reflecting the localization of charge. These results support the idea that the combination of decreased interchain bandwidth, and decreased interchain transfer integral, leads to effective localization of charge within single chains. Hence, unless the local structure of derivatized polymers allows for the close approach of the backbones, to give sufficient interchain coupling, derivatization of the polymers will, likely, lead toa substantial reduction in electrical properties.

SPECTROSCOPY

The optical spectroscopy of polyaniline reflects the evolution of electronic structure with change in oxidation level (ratio imine to amine). Earlier studies of solutions [49] of emeraldine base and emeraldine salt, as well as solid emeraldine base and salt, show the presence of a π-to-π^* gap of ~3.8 eV in emeraldine base, together with a broad absorption, centered at ~2 eV, contributing to the formation of an exciton upon charge transfer excitation from the highest occupied energy levels on the benzenoid groups, to the lowest unoccupied orbital centered on the quinoid groups [11,14,50]. Spectroscopy of the leucoemeraldine base form shows the presence of only the π-to-π^* transition at ~3.8 eV, while the emeraldine salt form has no 2 eV absorption, and instead two new absorptions appear, centered at 1.4 eV and 2.9 eV, assigned to the optical absorptions of the polaron energy band [11,14,29].

PHOTOINDUCED SPECTROSCOPIES

Extensive photoinduced absorption spectroscopy has been carried out, with time resolutions ranging from picoseconds through seconds, for polyaniline polymers in the emeraldine base oxidation state, as well as selected studies in the leucoemeraldine base and pernigraniline base forms. Steady-state (millisecond) photoinduced absorption spectroscopy of emeraldine base shows the bleaching of the 2 eV exciton transition, as well as the π–to-π^* (3.8 ev) transition, and the formation of three new photoinduced absorptions at 0.9, 1.4, and 3.0 eV [22,27,47]. The monomolecular decay kinetics of the 1.8 eV subgap bleaching is consistent with the model of optically excited localized molecular exciton. The 1.4 eV and 3.0 eV peaks have the same power and time dynamics, and are associated with the polaron transitions, as earlier identified in the optical absorption of the emeraldine salt. The low-energy, i.e., 0.9 eV, peak increases in relative intensity at low temperatures, as compared with the 1.4 eV absorption. For very long time (greater than minutes) photoinduced spectroscopy, the 0.9 eV absorption is nearly unobservable, as compared with the 1.4 eV absorption [23].

Photoinduced infrared absorption studies show that the photoinduced infrared modes are much weaker in intensity, than the photoinduced electronic transition [23], in contrast to the behavior of polyacetylene [51,52] and polythiophene [6]. Analysis, within the amplitude mode formalism [53], indicates that the polarons are massive, > or ~60 m_e, while use of bond order [54] or Holstein [55] polaron formalisms leads to an even larger estimate of the mass of the polarons. Photoexcitation, into either the exciton peak or the π-to-π^* peak of emeraldine base, produces essentially identical long-lived photoinduced infrared modes [29]. The long-lived photoinduced spectra of the leucoemeraldine base are much weaker. Similarly, the photoinduced infrared absorptions in the pernigraniline base are much weaker, than the photoinduced electronic transitions, again indicating massive photoinduced defects [30].

A model for the effects of electron-lattice coupling via ring rotation on the electronic structure of leucoemeraldine base has been developed [15]. The charge defect states in leucoemeraldine and other phenyl ring-containing polymers, particularly hole polarons, are associated with localized distortions in the ground-state ring torsion angle toward a planar coformation, in constrast with the bond-alternation defects present in polyacetylene and related polymers. It has been shown that this self-localized positive polaron, P^+, leads to a single energy level in the gap, split off by ~ 0.7 eV from the valence band edge, leading to two absorptions, one at 0.7 eV and the second at ~3 eV (from the P^+ level to the conduction band). It is anticipated that the P^+ can, in turn, be trapped near a quinoid group, leading to a formation of P^+_Q, whose electronic transitions are at ~1.4 eV and 3.0 eV. The negative polaron, in contrast, is not expected to be trapped by ring torsion angles, because of the absence of charge density in the conduction band on the carbon atoms adjacent to the nitrogen sites [29]. Hence, the P^- is expected to be stabilized only by bond distortion, and not by ring torsion angle distortion. Hence, for the pure leucoemeraldine base, the P^- is anticipated to be only weakly localized. On the other hand, the presence of a quinoid or imine group in the polymer chain will act as a trap for the P^- to form a P^-_Q, the energy level of which is expected to be approximately in the middle of the π-to-π^* gap [29].

Additional relevant excitations include the formation of a charge-transfer exciton, which has negative charge centered on a quinoid, with +e distributed on the phenyl groups on either side of the quinoid. The exciton is expected to exist in excited states, EX* and also in long-lived metastable states, due to metastable ring conformations, EX†. Fig. 8 summarizes a proposed decay scheme for electrons and holes photoexcited from the ground state for pure leucoemeraldine base (with the presence of a few quinoid groups) and also for emeraldine base. The formation of trap charge at P^+_Q and P^-_Q is reflected in the long lifetimes observed in the photoinduced spectroscopies. The presence of long-lived photoinduced absorption changes, associated with trap-charge defect states, has led to the proposal of utilizing polyaniline's derivatives as a medium for erasable optical information storage [31].

The pernigraniline form of the polymer may undergo a Peierls dimerization. It has earlier been pointed out that dimerization can involve changes in the bond lengths [16], as expected from infrared spectroscopy, showing the presence of quinoid groups within the pernigraniline system [30]. Extension of the model of electron/phonon coupling, *via* ring torsion angles [15], leads to the expectation that any Peierls gap in pernigraniline contains a substantial component in its order parameter, due to alternation of the ring torsion angles [17]. The presence of a significant ring torsion angle order parameter is expected to lead to a dramatic

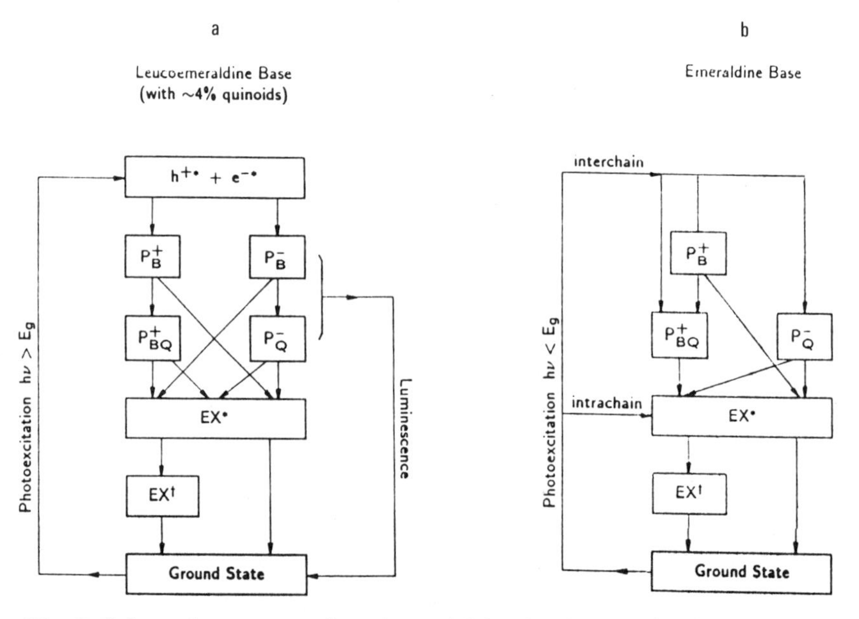

Fig. 8. Schematic representation of a model for the photoproduction and time-dependent relaxations of defect states in (a) leucoemeraldine base for above the π - π^* gap photoexcitation and (b) emeraldine base for photoexcitation in the "exciton" band.

increase in the mass of any charge defects excited in the pernigraniline polymer, in agreement with preliminary experiments [30].

SUMMARY

The polyaniline family of polymers provides a flexible backbone, which is easily chemically modified, and that can be used to study physical phenomena in a broad range of electronic states. Though some similarities exist in the properties of the polyanilines, as compared with polyacetylene and polythiophene, and earlier studies of other polymers, there are very substantial differences in the physical phenomena of these systems, based upon the key role of the ring torsion angle, and its coupling to electronic energy states, as well as the presence of nitrogen atoms within the chain, the presence of nitrogen atoms within the chain, and the absence of charge-conjugation symmetry. As a result, new phenomena are observed, and many unusual combinations of polymer properties and electronic properties exist.

This work is supported in part by the Defense Advanced Research Projects Agency through a contract monitored by the U. S. Office of Naval Research.

REFERENCES

[1] H. Shirakawa, E. J. Louis, A. G. MacDiarmid, C. K. Chiang, and A. J. Heeger, J. Chem. Soc. Chem. Commun. 578 (1977).
[2] See, for example, "Handbook of Conducting Polymers", Vols. 1 and 2, ed. by T. Skotheim, (Dekker, New York, 1986).
[3] See, for example, Proceedings of the International Conferences on Synthetic Metals, Santa Fe, NM, June 1988 [Synth. Met. 27 (1988); 28, 29 (1989)]; Kyoto, Japan, June, 1986 [Synth. Met. 17-19 (1987)]; Abano Terme, Italy, June 1984 [Mol. Cryst. Liq. Cryst. 117-121 (1985)].
[4] W. P. Su, J. R. Schrieffer, and A. J. Heeger, Phys. Rev. B 22: 2099 (1980).
[5] J. L. Brédas, B. Thémans, J. G. Fripiat, J. M. André, and R. R. Chance, Phys. Rev. B 29: 6761 (1984).
[6] Z. Vardeny, E. Ehrenfreund, O. Brafman, A. J. Heeger, and F. Wudl, Synth. Met. 18: 183 (1987).
[7] A. G. MacDiarmid and A. J. Epstein, J. Chem. Soc., Faraday Trans., in press.
[8] A. J. Epstein and A. G. MacDiarmid, in "Proc. Winter School on Electronic Processible Polymers, Kirchberg, Austria", ed. by H. Kuzmany, M. Mehring and S. Roth (Springer, Berlin, 1989), in press.
[9] A. G. MacDiarmid and A. J. Epstein, these Proceedings.
[10] M. J. Rice and E. J. Mele, Phys. Rev. Lett. 49: 1455 (1982).
[11] A. J. Epstein, J. M Ginder, F. Zuo, R. W. Bigelow, H. S. Woo, D. B. Tanner, A. F. Richter, W. S. Huang, and A. G. MacDiarmid, Synth. Met. 18: 303 (1987).
[12] J. M. Ginder, A. F. Richter, A. G. MacDiarmid, and A. J. Epstein, Solid State Commun. 67: 97 (1987).
[13] D. S. Boudreaux, R. R. Chance, J. F. Wolf, L. W. Shacklette, J. L. Brédas, B. Thémans, J. M. André, and R. Silbey, J. Chem. Phys. 85: 4584 (1986).
[14] S. Stafstrom, J. L. Brédas, A. J. Epstein, H. S. Woo, D. B. Tanner, W. S. Huang, and A. G. MacDiarmid, Phys. Rev. Lett. 59: 1464 (1987).
[15] J. M. Ginder, A. J. Epstein, and A. G. MacDiarmid, to be published.
[16] M. C. dos Santos and J. L. Brédas, Phys. Rev. Lett. 62: 2499 (1989); see also M. C. dos Santos and J. L. Brédas, Synth. Met. 29: E321 (1989).
[17] J. M. Ginder and A. J. Epstein, to be published.
[18] K. R. Cromack, M. E. Jozefowicz, J. M. Ginder, R. P. McCall, A. J. Epstein, E. Scherr, and A. G. MacDiarmid, Bull. Am. Phys. Soc 34: 583 (1989), and to be published.
[19] X. Tang, E. Scherr, A. G. MacDiarmid, and A. J. Epstein, Bull. Am. Phys. Soc. 34: 583 (1989), and to be published.
[20] M. E. Jozefowicz, R. Laversanne, H. H. S. Javadi, A. J. Epstein, J. P. Pouget, X. Tang, and A. G. MacDiarmid, Phys. Rev. B 39: 12958 (1989).
[21] F. Zuo, M. Angelopoulos, A. G. MacDiarmid, and A. J. Epstein, Phys. Rev. B 36: 3475 (1987).

[22] M. G. Roe, J. M. Ginder, P. E. Wigen, A. J. Epstein, M. Angelopoulos, and A. G. MacDiarmid, Phys. Rev. Lett. 60: 2789 (1988).
[23] R. P. McCall, J. M. Ginder, M. G. Roe, G. E. Asturias, E. M. Scherr, A. G. MacDiarmid, and A. J. Epstein, Phys. Rev. B 39: 10174 (1989).
[24] R. P. McCall, M. G. Roe, J. M. Ginder, T. Kusumoto, A. J. Epstein, G. E. Asturias, E. M. Scherr, and A. G. MacDiarmid, Synth. Met. 29: E433 (1989).
[25] Y. H. Kim. S. D. Phillips, M. J. Nowak, D. Spiegel, C. M. Foster, G. Yu, J. C. Chiang, and A. J. Heeger, Synth. Met. 29: E291 (1989).
[26] J. M. Ginder, A. J. Epstein, and A. G. MacDiarmid, Synth. Met. 29: E395 (1989).
[27] M. G. Roe, J. M. Ginder, R. P. McCall, K. R. Cromack, A. J. Epstein, T. L. Gustafson, M. Angelopoulos, and A. G. MacDiarmid, Synth. Met. 29: E425 (1989).
[28] M. G. Roe, J. M. Ginder, T. L. Gustafson, M. Angelopoulos, A. G. MacDiarmid, and A. J. Epstein, Phys. Rev. B 40 (15 Aug. 1989).
[29] R. P. McCall, J. M. Ginder, H. J. Ye, J. M. Leng, S. K. Manohar, J. G. Masters, G. E. Asturias, A. G. MacDiarmid, and A. J. Epstein, to be published.
[30] J. M. Ginder, R. P. McCall, J. M. Leng, H. J. Ye, S. K. Manohar, Y. Sun, and A. G. MacDiarmid, to be published.
[31] R. P. McCall, J. M. Ginder, and A. J. Epstein, to be published.
[32] H. H. S. Javadi, K. R. Cromack, A. G. MacDiarmid, and A. J. Epstein, Phys. Rev. B 39: 3579 (1989); H. H. S. Javadi, Microwave Journal 162 (Feb.1989).
[33] Z. Wang, K. R. Cromack, H. H. S. Javadi, J. Yue, J. M. Ginder, A. J. Epstein, A. Ray, S. K. Manohar, and A. G. MacDiarmid, Bull Am.Phys. Soc. 34: 824 (1989), and to be published.
[34] J. Yue, A. J. Epstein, and A. G. MacDiarmid, to be published.
[35] C. B. Duke, A. Paton, E. M. Conwell, W. R. Salaneck, and I. Lundström, J. Chem. Phys. 86: 3414 (1987).
[36] J. P. Pouget, M. Jozefowicz, X. Tang, A. G. MacDiarmid, and A. J. Epstein, to be published.
[37] W. Fosong, T. Jinsong, W. Lixiang, Z. Hongfand, and M. Zhishen, Mol. Cryst. Liq. Cryst. 160: 175 (1988).
[38] Y. B. Moon, Y. Cao, P. Smith, and A. J. Heeger, Polymer Commun. 30: 196 (1989).
[39] P. Robin, J. P. Pouget, R. Comès, H. W. Gibson, and A. J. Epstein, Phys. Rev. B 27: 3938 (1989).
[40] H. H. S. Javadi, S. P. Treat, J. M. Ginder, J. F. Wolf, and A. J. Epstein, J. Chem. Phys. Solids, in press.
[41] C. Fite, Y. Cao, and A. J. Heeger, submitted.
[42] K. Mizoguchi, M. Nechtschein, J. P. Travers, and C. Menardo, Phys. Rev. Lett. 63: 66 (1989).
[43] B. Abeles, P. Sheng, M. D. Coutts, and Y. Arie, Adv. Phys. 24: 407 (1975); P.Sheng, B. Abeles, and Y. Arie, Phys. Rev. Lett. 31: 44 (1973).
[44] I. Webman, J. Jortner, and M. H. Cohen, Phys. Rev. B 16: 2959 (1977).
[45] F. Zuo, M. Angelopoulos, A. G. MacDiarmid, and A. J. Epstein, Phys. Rev. B 39: 3570 (1989).
[46] H. H. S. Javadi, F. Zuo, M. Angelopoulos, A. G. MacDiarmid, and A. J. Epstein, Mol. Cryst. Liq. Cryst. 160: 225 (1988).
[47] H. H. S. Javadi, M. Angelopoulos, A. G. MacDiarmid, and A. J. Epstein, Synth. Met. 26: 1 (1988).
[48] See, for example, Yu. A. Firsov, in "Localization and Metal-Insulator Transition," ed.by H. Fritzsche and D. Adler (Plenum, New York, 1985).
[49] F. Zuo, R. P. McCall, J. M. Ginder, M. G. Roe, J. M. Leng, A. J. Epstein, G. E. Asturias, S. P. Ermer, A. Ray, and A. G. MacDiarmid, Synth. Met. 29: E445 (1989).
[50] C. B. Duke, E. M. Conwell, and A. Paton, Chem. Phys. Lett. 131: 82 (1986).
[51] G. B. Blanchet, C. R. Fincher, T. C. Chung, and A. J. Heeger, Phys. Rev. Lett. 50: 1938 (1983).
[52] N. Theophilou, D. B. Swanson, A. G. MacDiarmid, A. Chakraborty, H. H. S. Javadi, R. P. McCall, S. P. Treat, F. Zuo, and A. J. Epstein, Synth. Met. 28: D35 (1989).
[53] E. Ehrenfreund, Z. Vardeny, O. Brafman, and B. Horovitz, Phys. Rev. B 36: 1535 (1987).
[54] M. J. Rice and E. J. Mele, Phys. Rev. Lett. 49: 1455 (1982).
[55] T. Holstein, Ann Phys. 8: 325 (1959); 8: 343 (1959).
[56] S. Stafstrom, B. Sjorgren, and J. L. Brédas, Synth. Met. 29: E219 (1989).

VIBRATIONAL MOLECULAR SPECTROSCOPY OF CONDUCTING POLYMERS: A GUIDED TOUR

Giuseppe Zerbi

Dipartimento Chimica Industriale
Politecnico
Piazza L. Da Vinci 32
Milano, ITALY

INTRODUCTION

For the understanding of the structural properties of polyconjugated systems in their pristine (insulating) and doped (conducting) states, many physical techniques have been used. Researchers in this field can be classified in two main groups. "Problem-oriented" people try to collect data from all techniques, sometimes accepting gross simplifications, or avoiding a detailed specific analysis, in favour of an overall view of the problem, in an attempt to reconcile and correlate data from different techniques, all aiming at a specific target. "Technique-oriented" people often object to such quick conclusions, and try to dig out from each physical technique all possible detailed information, in order to assess critically the validity and limits of the structural data obtained. Disagreement may often occur between the two groups.

Among the physical techniques, used in the field of conducting polyconjugated polymers and of photoactive polyconjugated organic molecules, vibrational infrared and Raman spectroscopy, as well as lattice dynamical calculations, have been used as a source of useful and relevant data [1].

In this paper, we wish to introduce the reader and the student into the way technically-oriented spectroscopists would approach the understanding of the experimental data obtained from the materials mentioned above. This paper can be taken as a guided tour in the vibrational spectroscopy of polyconjugated materials. Polyacetylene (PA) is taken as a case study. The same procedures can be applied to other polyconjugated molecules. A few experimental results will be presented, and others will be referred to, together with the references to relevant fundamental publications in the field.

VIBRATIONAL SPECTROSCOPY

The three main quantities, which can be derived from an experiment in infrared absorption spectroscopy, or Raman scattering, are: (i) vibrational frequencies, (ii) integrated intensities and (iii) band shapes [1]. Each of these quantities can provide independent information on the local and/or overall dynamics and electrical properties of molecules. In this paper we shall not discuss case (iii), since it has not yet been applied extensively to the case of conducting polymers.

Lower-Dimensional Systems and Molecular Electronics
Edited by R.M. Metzger *et al.*, Plenum Press, New York, 1991

Frequency spectroscopy

Molecular dynamics of organic molecules or polymers as one-dimensional crystals is treated, generally, in the harmonic oscillator approximation. Indeed, the vibrational amplitudes about the equilibrium geometry are very small, and only at high vibrational quantum numbers does anharmonicity play a non-negligible role. In the discussion which follows, we shall treat the dynamics of the systems in terms of chemical bonds and valence angles. Let us describe the normal vibrations Q_i of a molecule by a set of internal coordinates R, suitably chosen to represent bond stretching and angle bendings or torsions [2]. In what follows, we shall use matrix notation. R's can be defined as linear combinations of the cartesian displacement coordinates x by the relation:

$$\mathbf{R} = \mathbf{B}\ \mathbf{x} \tag{1}$$

wher **B** is a matrix of geometrical coefficients. The R's are used for expressing the intramolecular vibrational potential that determines the fundamental frequencies of the system. Let the harmonic vibrational molecular potential V be written as

$$2V = \tilde{\mathbf{R}}\ \mathbf{F}_R\ \mathbf{R} \tag{2}$$

where $\mathbf{F}_R$ is the matrix of valence force constants $((F_R)_{ij} = \partial^2 V/\partial R_i\, \partial R_j)_o)$. Let the purely vibrational kinetic energy of the system be defined as

$$2T = \tilde{\dot{\mathbf{R}}}(G_R)^{-1}\dot{\mathbf{R}} \tag{3}$$

$$\mathbf{G}_R = \mathbf{B}\mathbf{M}^{-1}\tilde{\mathbf{B}} \tag{4}$$

where $\mathbf{M}^{-1}$ is the diagonal matrix of the inverse of the atomic masses.

The solution of the dynamical problem leads to the solution of the eigenvalue equation

$$\mathbf{G}_R\, \mathbf{F}_R\, \mathbf{L}_R = \mathbf{L}_R\, \Lambda \tag{5}$$

where Λ is the diagonal matrix of the eigenvalues, with $\lambda_i = 4\ \pi^2 c^2 \nu^2{}_i$; ν_i (in cm^{-1}) is the vibrational frequency of the i-th normal mode, and $(L_R)_i$ is the corresponding eigenvector, which gives the vibrational displacements. The description of the normal modes, in terms of R coordinates, is given by:

$$\mathbf{R} = \mathbf{L}_R\, \mathbf{Q} \tag{6}$$

Eqs. 1 through 6 treat the case of a finite molecule, consisting of N masses with 3N-6 degrees of vibrational freedom. The treatment can be extended to molecular systems with translational periodicity in one or three dimensions. For these systems, intramolecular or intermolecular coupling between oscillators generates phonon waves, which can be described with the corresponding wave vector **k.**

Lattice dynamical treatments [3] allow to rewrite eq. 5 in the following way

$$\mathbf{G}_R(\mathbf{k})\ \ \mathbf{F}_R(\mathbf{k})\ \mathbf{L}_R(\mathbf{k}) = \mathbf{L}_R(\mathbf{k})\ \Lambda(\mathbf{k}) \tag{7}$$

With the use of eq. 7, phonon dispersion curves $\nu_i(\mathbf{k})$ can be calculated [4]; phonons with $\mathbf{k} = 0$ are the infrared and/or Raman active modes to be matched with the experiments. From the data obtained from eq. 7, the one- and multi-phonon density of vibrational states $g^i(\nu)$ can be calculated. When translational periodicity of long chains is lost, a suitable modification of eq. 5 can be introduced, in order to calculate $g^1(\nu)$ for disordered chains [5].

The peculiarity of the physics of conducting polymers is that they can be considered one-dimensional crystals, since intermolecular forces (van der Waals type) are much weaker than the intramolecular forces (covalent forces). This concept is immediately translated in eq. 7, by the fact that the **k** vector has only one component for 1-D crystals, and phonons propagate only longitudinally along the chain backbone [3,4].

Intensity Spectroscopy

The experimental datum of vibrational intensity has been generally neglected by most of the workers, because of the lack of a suitable model of interpretation. Its importance has been recently realized for the understanding of the electronic properties of molecules, and the attention of many workers is being re-focussed at this field [6].

We consider, here, the case of infrared absorption spectroscopy. Raman intensities have been treated in a parallel way [6]. In the double harmonic approximation (where the forces and the changes in the total molecular dipole moment are linear with atomic displacements) the integrated intensity A_i of the i-th normal mode is related to molecular properties by the relation:

$$A_i = (N \pi g_i / 3 c^2 v_i) \; | \partial M / \partial Q_i |^2 \tag{8}$$

where N is Avogadro's number, c the speed of light, g_i the degeneracy of the i-th mode, and **M** is the total molecular electric dipole moment) [2].

The relevant quantity appearing in eq. 8, namely $\partial \mathbf{M} / \partial Q_i$, can be separated into the contributions from the various vibrational internal coordinates during Q_i. One can write

$$\partial \mathbf{M} / \partial Q_i \doteq \Sigma_t (\partial \mathbf{M} / \partial R_t) (\partial R_t / \partial Q_i) \tag{9}$$

and from eq. 6

$$\partial \mathbf{M} / \partial \mathbf{Q}_i = \Sigma_t (\partial \mathbf{M} / \partial R_t)(L_R)_{ti} \tag{10}$$

thus showing the close connection between the data from intensities (i.e. electronic properties) and molecular dynamics [7].

The importance of the intensity datum has been re-discovered, when the schools of Gribov in Russia [8] and later of Gussoni in Milano [7,9-12] have presented and developed the "optical theory of valence". The important concept of such a theory lies in the fact, that the instantaneous **M** can be expressed as a sum of bond dipole moments μ_k directed at any instant along the bonds

$$\mathbf{M} = \Sigma_k \, \mu_k \, \mathbf{e}_k \tag{11}$$

The infrared intensity, as measured by $\partial \mathbf{M} / \partial Q_i$, can be expressed as

$$\partial \mathbf{M} / \partial Q_i = f\,(\mu_k^\circ, (\partial \mu_k / \partial R_t), (\partial R_t / \partial Q_t)) \tag{12}$$

The first two terms in eq. 12 are known as "electro-optical parameters." If the instantaneous molecular dipole moment is expressed in terms of effective electrical charges, localized on the atoms, $\mathbf{M} = \Sigma_\alpha q_\alpha \mathbf{r}_\alpha$, the total dipole moment change can then be expressed as a function of

$$\partial \mathbf{M} / \partial Q_i = f\,(q_\alpha^\circ, \partial q_\alpha / \partial R_t, \partial R_t / \partial Q_t) \tag{13}$$

where q_α° measures the equilibrium charge of atom α, and $\partial q_\alpha / \partial R_t$ measures the charge flux on atom α, when the t-th internal coordinate is activated. Many measurements, correlations and applications are available in the literature [10-12], and we shall use this theory also for polyacetylene.

Vibrational Potentials

The calculation and the correlative understanding of harmonic and anharmonic force constants in molecules has been a very active field in the last 25 years. The vibrational assignment of innumerable molecules and their isotopic derivatives, the study of ro-vibronic parameters, the application of group theory and chemical correlations have allowed to introduce, into grand least-squares fitting procedures [2,13,14] by computers, many experimental data, from which seemingly reliable sets of vibrational force constants have been derived. This statement is certainly true for molecules which contain covalent σ bonds, or isolated double and triple bonds. This fact can be understood, since, in a network of σ bonds, electronic interactions are generally limited to the first neighbours, and no long-range intramolecular forces exist. The problem becomes much more difficult, when networks of delocalized (conjugated), highly polarizable, bonds exist in the system, and long-range forces become dominant.

For σ-bonded covalent 1-D lattices (such as n-alkanes and poly-ethylene) very reliable force fields were obtained, with a fitting procedure of the observed frequencies of n-alkanes with increasing chain length [13]. The observed frequencies were easily assigned to $\mathbf{k}\neq0$ phonons, and experimental phonon dispersion curves were constructed, thus reaching a goal never yet attained by any neutron scattering study.

In the case of PA, several authors have thought to use the same method, and used the observed frequencies for conjugated oligoenes of various lengths, and decided to use experimental $\mathbf{k}=0$ modes and/or $\mathbf{k}\neq0$ modes of such molecules to increase the number of experimental data, in order to obtain long-range interactions. As already pointed out [15], but never noticed or understood by several authors, the vibrational frequencies of each oligoene molecule form a particular set of dispersion curves, which are different from those of the other oligomers in the homolog series. This can be proven conceptually [15,16] and numerically [17]. The range of interaction (extent of electron/phonon coupling), and its role in the physics of these systems is matter of active study, both experimentally and theoretically, and should be determined in some way.

From the viewpoint of lattice dynamics, the whole vibrational force field, and hence the extent and distance of long-range interactions, could be calculated from *"ab initio"* quantum chemical methods. Vibrational force constants have already been calculated successfully for small molecules, and critical reviews on the success and limitations of such calculations are already available [18]. Attempts have been made also in the case of polyconjugated molecules, and of oligoenes as models of PA [17,19]. The problem is not simple, since reliable results could, in principle, be achieved with the use of *"ab initio"* quantum chemical methods, with large basis sets, and with the inclusion of electron correlations. Such calculations require an extemely long computing time, and an enormous computer memory; by necessity, one must reduce the size of the systems to be studied, just the opposite of what one is aiming at for these systems. Clementi and his school have published the results on the largest and longest calculations ever made, and have treated the case of a molecule with 12 double bonds [19]. Generally, one is forced to use simpler and faster calculations, using semi-empirical methods, such as MNDO. In such a situation, one does not aim at a quantitative result, but limits oneself to accept a trend of values, which may explain some physical phenomena [17].

The consequences of such a lack of good estimates of the coupling terms between double bonds is immediately reflected in the shape of the phonon dispersion curves [20]. It has been shown, that the curvature of the dispersion curve of the C-C skeletal stretching in polyenes is positive, and is a function of the extent of electron-phonon coupling [16]. The real shape of the phonon branches throughout the first Brillouin zone is yet unknown, either theoretically or experimentally.

PRISTINE POLYACETYLENE

Trans-PA in the pristine (i.e. insulating) state is known to be obtained from a solid-state thermal isomerisation of cis-PA synthesized in various ways [21], whose discussion is beyond the scope of this paper. The mechanism of thermal isomerisation is not yet known. Time-dependent FTIR vibrational spectra have been recorded at time intervals of two seconds; a set of spectra is reported in Fig. 1 for an experiment at 145°C [22,23].

It is observed that the C-H out-of-plane deformation modes shift during isomerisation, as shown in Fig. 2. From the very long chain of cis-PA (γ_{C-H} = 735 cm^{-1}), p electrons become more and more localized within shorter chains, which are necessarily found at the end of the reaction (γ_{C-H} = 743 cm^{-1}). The case of trans-PA is opposite. The original short chains (γ_{C-H}= 1013 cm^{-1}) generate a long and highly delocalized system of p electrons (γ_{C-H} = 1008 cm^{-1}). Qualitatively, the isomerisation develops through three time domains: (i) cis-PA: 0-30, 30-130, 130-200 secs; (ii) trans-PA: 0-30, 30-60, 60-200 secs. The time evolution of the

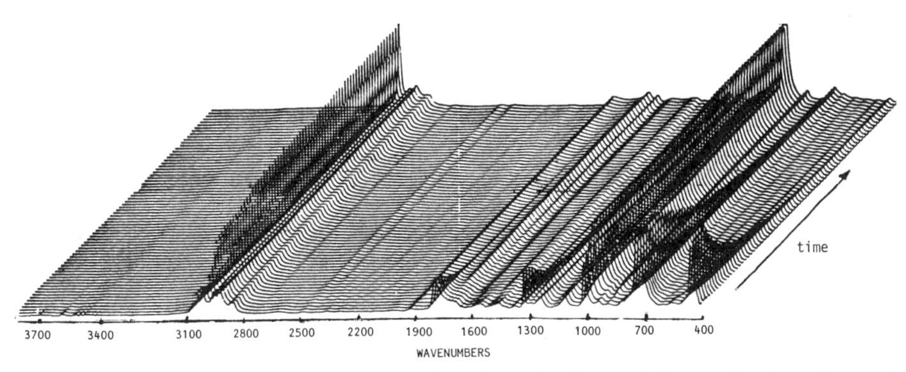

Fig. 1. Time-dependent FTIR spectrum, showing the thermal isomerization from cis to trans-polyacetylene (T = 145°C). One spectrum every two seconds.

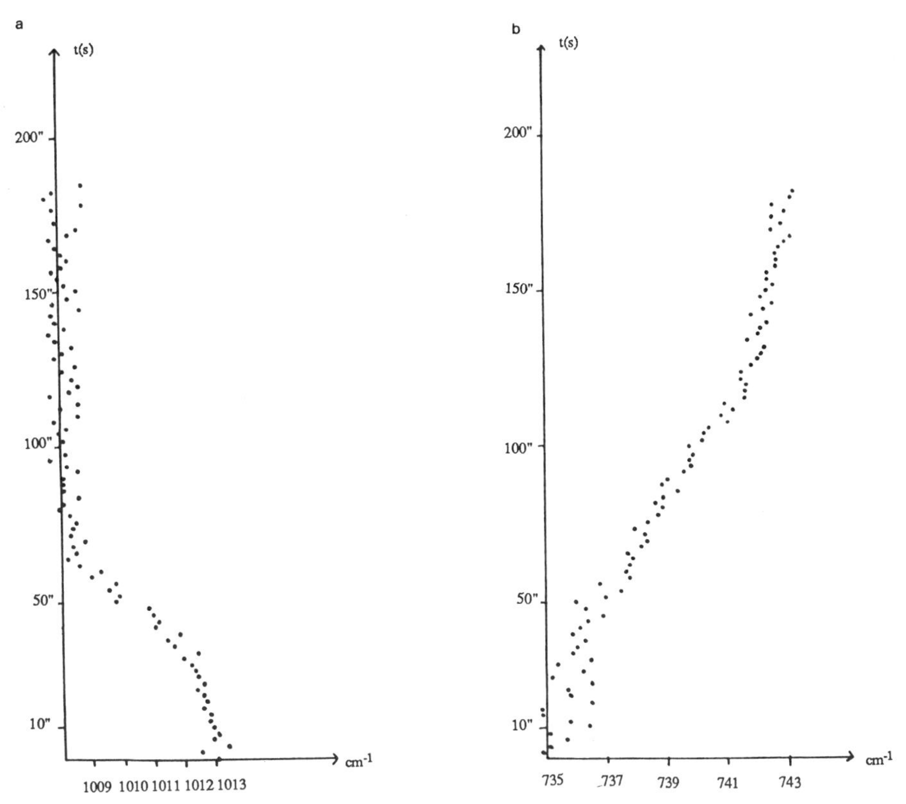

Fig. 2. Evolution with time of the frequencies of the C-H out-of-plane motions of cis (a) and trans (b) polyacetylene for a thermal isomerization at 145°C

frequencies shows that the collectivity of the p electron delocalisation develops with time in different ways, thus indicating that the reaction proceeds stepwise, through the formation of chain segments, and not in a single-step unwinding of the cis-chain.

The infrared spectra of the starting cis-PA and the final trans-PA are compared in Fig. 3. The infrared spectrum in polarized light of a stretch-oriented and highly oriented sample of trans-PA is given in Fig. 4 [24].

The Raman spectra of trans-PA exhibit several peculiar characteristics, which will be discussed later in this paper. We just mention here that the Raman spectrum shows, in general,

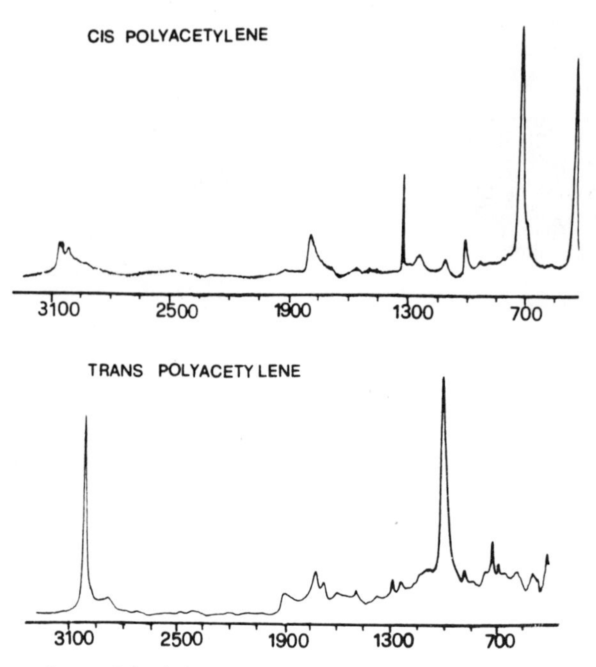

Fig. 3. Comparison of the infrared spectra of cis and trans-polyacetylene (notice the different scales of the abscissae).

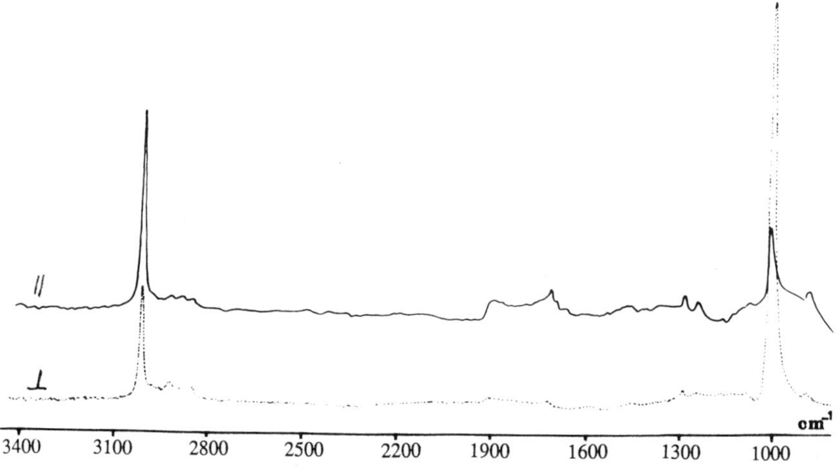

Fig. 4. Infrared spectrum in polarized light of a highly oriented film of trans-polyacetylene (Assoreni): electric vector perpendicular (dotted) and parallel (full line) to the stretching direction

three main lines: ν_1 in the range 1500 cm^{-1} (strong and sharp); ν_2 near 1320 cm^{-1} (very weak); ν_3 in the range 1000 cm^{-1} (strong) [25].

Chain dimerisation (bond alternation)

The first step is to derive the structure of trans-PA from the vibrational IR and Raman spectra. We consider trans-PA as a 1-D crystal, since no indication is found in IR and Raman spectra of sizeable intermolecular interactions, even if, as presently known from structural studies, the crystalline orthorhombic unit cell contains two molecular chains, just as in the case of polyethylene. The molecular models which could be considered are given in Fig. 5. First, one has to establish the existence of chain dimerisation, i.e. the existence of bond alternation along the chain (structures B or C).

Classical group theory helps in this kind of study [26]. If the chain is undimerized, it belongs to the D_{2h} point group, and the k=0 phonons are distributed in the following irreducible representations: in-plane modes: $2A_g(R) + 2B_{2g}(R) + 1B_{1u}(IR) + 1B_{3u}(IR)$; out-of-plane modes: $1B_{1g}(R) + 1B_{1u}$ (IR). If the chain is dimerized (structures A or B), the point group is D_{2h}, with $4A_g(R) + 2B_u(IR)$ in-plane modes and $1B_g(R)$ and $1A_u(IR)$ out-of-plane modes. Based on the number of bands, IR spectroscopy cannot distinguish between structures C or A, since both predict three IR-active modes.

Let us focus on the vibrations of the chain backbone, neglecting high-frequency C-H stretches. Group theory predicts $1A_g + 2B_{1g}$ modes for structure C below 2000 cm^{-1}; structure A should show $3A_g$ modes. Again, from simple Raman spectroscopy, one could not decide between the two structures. However, for polyconjugated polymers, because of p-electron delocalisation, the energy gap E_g between valence and conduction bands (HOMO-LUMO gap) is relatively small, and the optical excitation energies fall in the visible spectrum. In this case, any Raman spectrum with excitation in the visible is recorded in resonance conditions. We neglect, so far, the discussion on Raman dispersion. From the theory of resonance Raman spectroscopy [27], it is known that the totally symmetric A_g modes are strongly and selectively enhanced, with respect to the other B_g modes. On these grounds, structure C should show only one Raman line below 2000 cm^{-1}, while model A (or B) should show three lines. Experimentally, two strong and one weak Raman lines (ν_1, ν_3, and ν_2 respectively) are observed, thus ruling out the existence of the undimerized chain.

The analysis of the infrared spectrum completes the picture of the structure. Let m_{pa} and m_{pr} indicate the transition moments parallel and perpendicular to the chain axis, i.e. to the stretching direction. In the infrared spectrum in polarized light of a stretch-oriented sample, one predicts (i) for a fully undimerized chain, m_{pr} for C-H stretching and m_{pa} for in-plane C-H wagging and out-of-plane C-H deformation, and (ii) for a dimerized chain, the C-H stretching motion should show both pa and pr components, since, upon stretching the C-H bond, the flux of electrons from C=C and C-C bonds must be different, thus tilting the transition moment along the chain axis [24]. This is indeed observed experimentally [24], as seen in Fig. 4.

Electron delocalisation (conjugation)

Once we have proven the existence of bond alternation, we must prove the existence of p-electron delocalisation (i.e. conjugation). In this case, we use vibrational intensities (see section on "Intensity Spectroscopy" above), and compare the data derived from butadiene, the shortest trans-diene. Let us consider the electro-optical parameters which describe the change

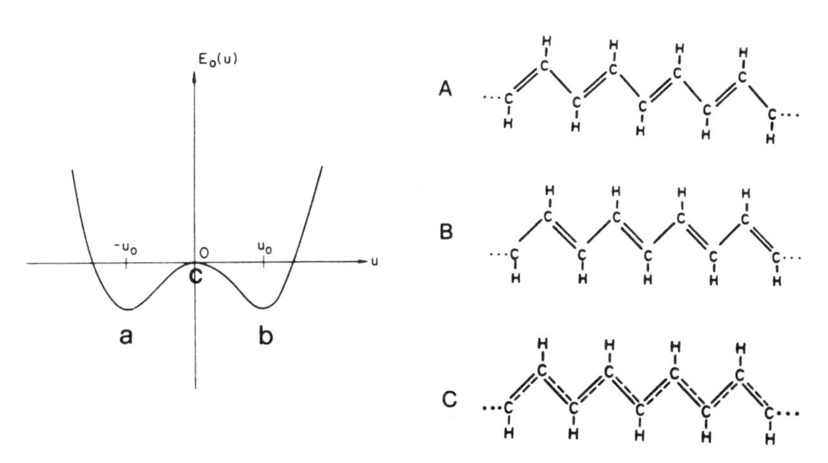

Fig. 5. Electronic energy vs. dimerisation coordinate u in trans-polyacetylene; the two equal minima correspond to the A and B structures. The top of the barrier C would correspond to the undimerized chain C.

of the dipole moment of the C=C bond upon stretching (d) or wagging (w) the C-H bond. The numerical values obtained are the following:

Table 1. Experimental Electro-optical Infrared Intensity Parameters for Butadiene and trans Polyacetylene.

	butadiene	trans-PA
$\partial M_{C=C}/\partial d$ (D/A)	-0.068	-0.217
$\partial M_{C=C}/\partial w$ (D/rad)	-0.260	-0.400

The charge flux along the C-C skeleton increases with conjugation length, thus proving (from infrared) the existence of delocalisation [28].

Effective atomic point charges and bond distance in C-H bonds

With the methods mentioned above (section on "Intensity Spectrscopy") the static atomic point charge on the H atom, as measured from IR intensities, is $q_H{}^\circ = 0.13$ e, rather high, if compared with that measured for polyethylene and n-alkanes ($q_H{}^\circ = 0.04$ e). Moreover, using IR frequency / bond length correlations [29] the bond lengths measured from the C-H stretching frequency are the following: Polyethylene, n-alkanes $d_{C-H} = 1.096$ A; butadiene $d_{C-H} = 1.088$ A; trans-PA $d_{C-H} = 1.090$ A [28], in good agreement with IR intensity / bond length correlations [9].

RAMAN SPECTRA OF PRISTINE TRANS-PA

Raman dispersion

It is presently a known fact that, while the Raman frequencies of each individual short finite-chain oligoene do not show "dispersion" (i.e. change of Raman frequencies with change of the wavelength of the exciting radiation), any sample of trans-PA shows, remarkably, the phenomenon of "Raman dispersion." This fact has been the subject of extensive and highly controversial theoretical and experimental studies by many authors, but we shall limit ourselves to the basic facts, and to the consequences in structural and dynamical studies of trans-PA [25].

The experimental fact about "dispersion" is that ν_1 and ν_3 frequencies and intensities depend on the exciting frequency, while ν_2 does not. The conjugation length dependence of the Raman lines has also been observed with short oligoenes, with increasing the number of conjugated double bonds, thus indicating that conjugation length is the determining factor. It is presently accepted, at least conceptually, that the observed Raman dispersion originates from the fact that any sample of polyconjugated polymer consists of a distribution of conjugation lengths, hence of band gaps E_g. Each excitation frequency selects the corresponding gap, produces a $\pi-\pi^*$ transition, and exhibits the resonance Raman spectrum of a particular fraction of polyene chains within the distribution of conjugation lengths [25,30]. On this basis, innumerable, controversial theoretical and experimental attempts have been made to derive, from the Raman spectrum, the distribution of chain lengths.

Amplitude modes and effective conjugation coordinate

Having established a correlation between Raman dispersion and conjugation length, a better explanation of the experimental observation is required. The problem has been successfully and thoroughly studied by Horowitz *et al.*, in terms of the amplitude mode theory (AMT) [30]. The dispersion of the three Raman modes is accounted for, in the AMT, by a parameter λ, which expresses the electron-phonon coupling in chains of a given conjugation length. Another parameter, α, "the pinning potential" of the electron onto the chain, is defined when the chain is doped. AMT has been reformulated, in molecular terms, with the introduction of the "effective conjugation coordinate" (ECC), $\mathcal{R}$ [25,31], which has a more direct meaning for the understanding of molecular properties.

The $\mathcal{R}$ coordinate is defined as linear combination of the C-C and C=C skeletal internal coordinates of stretching, which are mostly involved in p-electron conjugation. It has been shown, that the coordinate $\mathcal{R}$ (i) is unique in a given polyconjugated molecule [32] (ii) is totally symmetric, thus coupling only with totally symmetric normal modes, (iii) enters only in the A term of the resonance Raman (RR) scattering theory by Albrecht, (iv) selectively gives strong RR intensity enhancement to those Raman modes, which are coupled with it, their intensity being proportional to $|L_{\mathcal{R}}|^2$, (see eq. 5) [16,25,31].

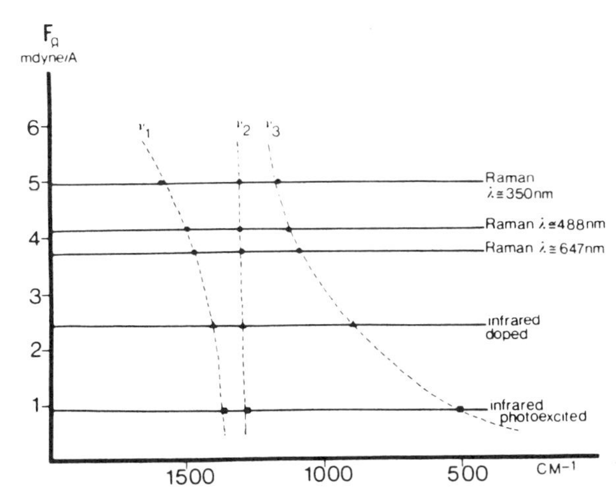

Fig. 6. ν vs. $F_{\mathcal{R}}$ for the A_g normal modes of trans-polyacetylene. The experimental values are: (●) Raman shifts for different excitation wavelengths; (▲) infrared doping-induced bands; and (■) infrared bands induced by photoexcitation.

When lattice dynamical calculations are made, one can define an effective conjugation force constant $F_{\mathcal{R}}$ expressed, for trans-PA, as

$$F_{\mathcal{R}} = (K_{C=C} + K_{C\text{-}C} - 4\Sigma_n^N f^{(n)}_{C\text{-}C/C=C} + 2\Sigma_n^N f^{(n)}_{C=C/C\text{-}C} + 2\Sigma_n^N f^{(n)}_{C=C/C\text{-}C})/2 \qquad (14)$$

where n indicates the distance of interaction (n=1 for first neighbour, etc.) and N indicates the maximum distance of interaction. N is, as yet, unknown, and only *"ab initio"* calculations (discussed above) could provide some information. From quantum theory and numerical calculations, it is known that interactions of the type $f^{(n)}_{C\text{-}C/C=C}$ must be positive, while interactions involving C=C/C=C and C-C/C-C pairs are negative. Thus, the value of $F_{\mathcal{R}}$ decreases with increasing conjugation. The $F_{\mathcal{R}}$ dependence of the vibrational frequencies of A_g species is obtained by use of eq. 5. The results are reported in Fig. 6. It has been shown that the Raman intensities are proportional to $|L_{\mathcal{R}}|^2$ in each mode. The plot of $|L_{\mathcal{R}}|^2$ vs. $F_{\mathcal{R}}$ is given in Fig. 7.

ECC theory accounts for Raman dispersion, and defines a unique parameter $F_{\mathcal{R}}$. The numerical value of $F_{\mathcal{R}}$ can be obtained from theory, if each individual term of eq. 14 is evaluated from quantum chemical calculations and, as a whole, experimentally from the Raman spectrum, if the plot of Fig. 6 is used.

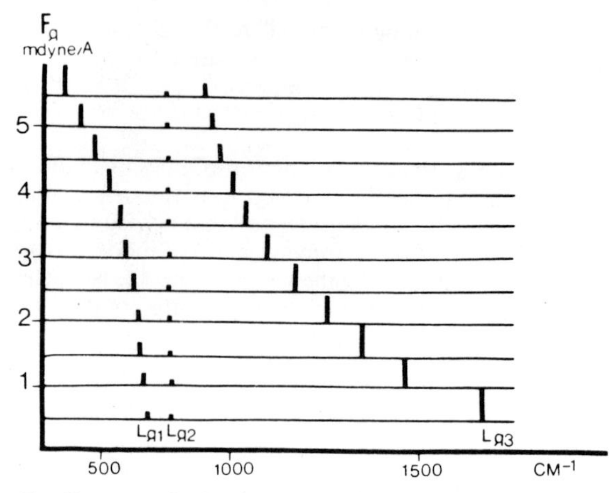

Fig. 7. Content of $\mathcal{A}$ in the normal modes of the plot in Fig. 6.

$F_\mathcal{A}$ vs electronic properties

It has been shown, from simple quantum chemical calculations (MNDO), that $F_\mathcal{A}$ is linearly related to molecular electronic properties, such as E_g, ionisation potential I_p and bandwidth, Fig. 8 [33]. It then becomes possible, in principle, to correlate directly Raman vibrational frequencies with the above electronic properties. The theoretical linear relation has also been verified experimentally, for the case of E_g measured from the UV-visible spectrum, and $F_\mathcal{A}$ derived from Raman experiments, Fig. 9 [33].

It has to be pointed out, that the plots of Figs. 8 and 9 are also very useful in the case of biologically relevant molecules, such carotenes, carotenoids, etc. Moreover, similar theoretical plots have also been obtained for other polymers, such as polypyrroles and polythiophenes [33].

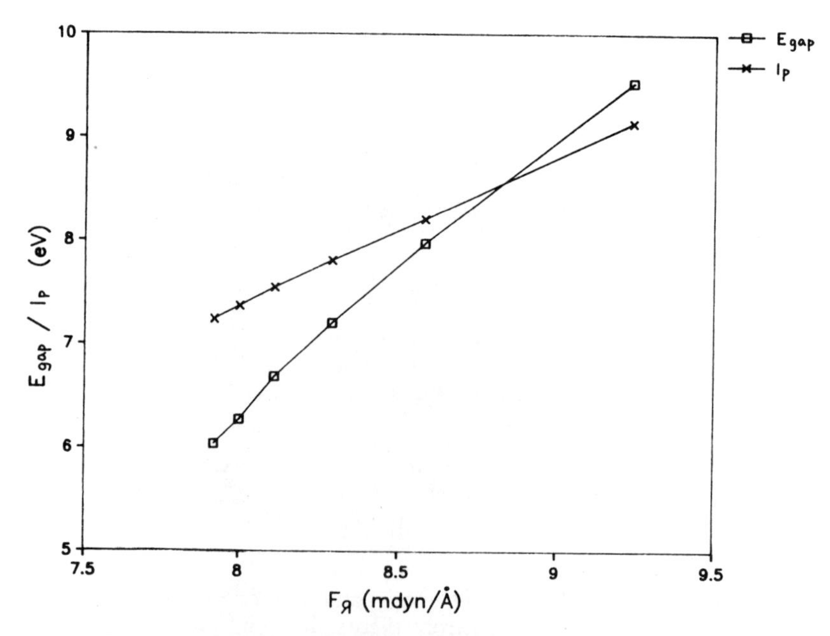

Fig. 8. Relation between MNDO calculated effective conjugation force constants $F_\mathcal{A}$ and MNDO-calculated E_g and I_p.

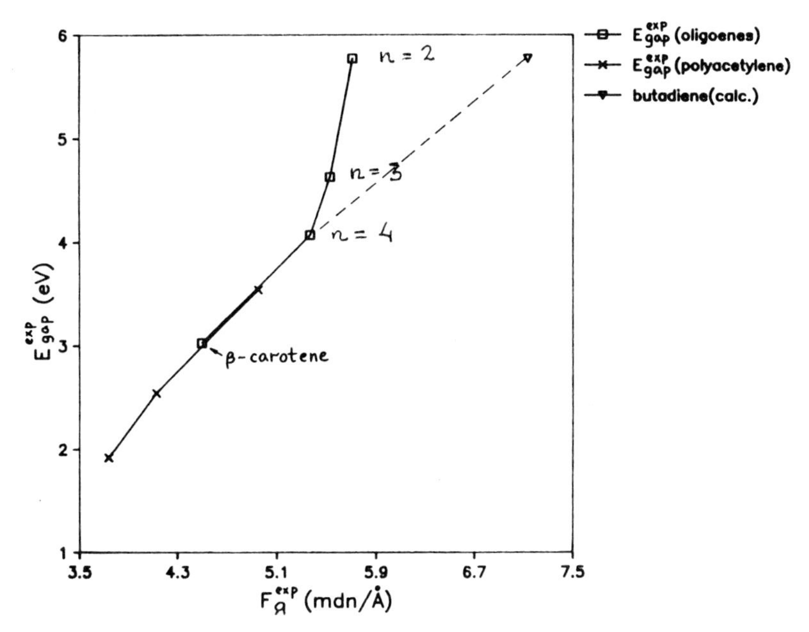

Fig. 9. Relation between "experimental" $F_{\mathcal{R}}$ (from Raman spectra of oligoenes) and experimental E_g (from UV/visible spectra). The dotted line connects the "experimental" $F_{\mathcal{R}}$ value of octatetraene with the $F_{\mathcal{R}}$ value for butadiene derived from careful force field calculations. The discrepancy for short chains is due to end effects.

VIBRATIONAL SPECTRA OF DOPED POLYACETYLENE

It has been shown that, upon doping, three new bands appear in the infrared, with such a very large absorption coefficient that, even for weak doping, they dominate the whole IR spectrum [34,35]. The doping-induced bands show the following characteristics in the frequencies ν_i and the intensities I_i:

(i) for pristine trans-PA the I_i's show the pattern: ν_1 strong, ν_2 weak, ν_3 strong and generally very broad; the pattern is similar to the Raman lines of pristine PA, even with different frequencies.

(ii) the ν_i's show dispersion with effective conjugation length.

(iii) the ν_i's are substantially independent from p- or n-doping.

(iv) ν_i's are isotopically dependent (D or ^{13}C), and show a different pattern for perdeutero-PA.

(v) the absorption coefficients of ν_1, ν_2, and ν_3 decrease with decreasing average effective conjugation length (this happens in doped and segmented PA, i.e. with sp^3 defects).

As to the Raman spectrum, most of the published studies show that the spectrum of doped samples is practically the same as that of pristine materials, with upper shifts of ν_1 and ν_3 bands. If the concentration of doping is increased, the spectrum changes drastically (e.g. at 12% for iodine doping), indicating a sort of phase transition [36]. It has been shown that, on the basis of the ECC theory, the Raman spectrum of the doped species originates from the undoped part of the material. At the maximum doping level, the Raman spectrum still identifies the existence of segments of very short effective conjugation length, either arising from long distorted chains, or from short transplanar chains, which, for some reason, cannot be reached by the dopant molecule.

The conjugational defect

Upon doping, a charge-transfer bond is formed between the polymer chain and the p or n dopant, and the electronic structure of the chain is strongly perturbed, with the generation of conjugational defects. The problem, whether doped sites are randomly distributed within the polymer chain, or whether they combine into "1-D clusters", is yet unsolved experimentally.

It has been shown, that the strong intensity of the IR spectrum cannot be reproduced by introducing only large charge fluxes during vibrations. The important term to be introduced is the large equilibrium bond dipole moment of C=C and C-C bonds [37]. While, in the pristine PA, all skeletal bonds are apolar, doping removes (injects) charges, and induces a very large polarity of the bonds. The 1-D lattice becomes strongly polarized (as also predicted from quantum chemistry [38]) and the vibrational transition moment becomes extemely large. It follows that the symmetry of the molecule is lowered; the originally only Raman-active A_g modes become IR active, and very strong in IR, because of lattice polarisation [39].

Eq. 14 has to be used for the infrared spectrum of doped PA. The strong perturbation by the dopant changes the value of either the diagonal $K_{C=C}$ and K_{C-C} terms, or of the interaction terms, thus affecting the values of $F_{\mathcal{R}}$. The whole electronic perturbation can be represented by $F_{\mathcal{R}}$, as a change of the "effective conjugation" within the doped chain. What is important to stress is that one can still define a $\mathcal{R}$ coordinate, and that the observed and calculated frequencies do not arise from localized defects, but from collective phonons, due to the vibrational coupling of the doped site with the undoped part of the chain [39].

On this basis, the plot of ν vs. $F_{\mathcal{R}}$ (Fig. 6) keeps its validity also for the IR spectra of doped PA, and the doping-induced bands can be accounted for, in frequencies and intensities, as the modes which contain, to various extents, the $\mathcal{R}$ mode, just as in the case of the Raman spectra of pristine polyacetylene.

Based on the fact, that doping-induced bands come from collective phonons in the doped 1d-lattice, phonon dispersion curves have been calculated for different values of $F_{\mathcal{R}}$ [39]. For long, but finite, doped chains, $k \neq 0$ phonons become the so-called "translational modes". $k \neq 0$ phonons may form a progression of bands, with rapidly decreasing infrared intensity, just like the case of long, but finite, n-alkanes. These modes are generally called in the literature "shape modes".

An important conclusion follows, namely that solitons, as well as polarons and bipolarons, cannot be distinguished in the IR spectrum, since all produce strong polarisation of the 1d-lattice, and all produce vibrational coupling, as described by eq. 14. The naming of the doping-induced infrared bands as "soliton bands" is, then, misleading [20, 37].

Charge flux and domain size

Let us define the electro-optical parameter [40]

$$A = \partial M^i / \partial R_i + \Sigma_k \, \partial M_k / \partial R_i \quad (i = \text{CC bonds in the defect region}) \qquad (15)$$

which describes the charge flux along the chain skeleton. A has been fitted to the infrared intensities of PA and perdeutero-PA. A increases with increasing conjugation length, and is small, if the chain is short. From the study of segmented doped PA [41], we derive A=12 DA^{-1}, corresponding to an average electrical perturbation of ~18 CC bonds [40], in agreement with quantum chemistry [38]. The A value of the charge flux from eq. 15 is extremely large, and measures a large polarizability and/or mobility of the p-electrons in the doped domains within the chain.

PHOTOEXCITED POLYACETYLENE

Illumination by a suitable laser radiation in the visible generates conjugational defects, whose vibrational spectra have been recorded [42]. Under photoexcitation, PA becomes a photoconductor with a fast decay time. The nature of the photoexcited species is not yet identified, even if theories have been proposed. In the first, elementary, step of the excitation, within the conduction band a very short-lived species is generated, which decays rapidly into other states with longer lifetimes.

The infrared spectrum of the, as yet, unidentified excited electronic state has been recorded, and shows a pattern very similar to the doping-induced spectrum. ν_1 and ν_2 modes are almost superimposable in the doped and photoexcited material, ν_3 occurs at frequencies lower that the corresponding doping-induced mode, and is isotope-dependent [42]. Photoexcitation of "trans in cis" PA shows extra small peaks, assigned to "shape modes" [43].

The interpretation of the photoexcited spectrum uses ECC again directly, as in the cases of pristine and doped PA. The photoinduced species polarizes the 1d-lattice, lowers the symmetry, and gives dipole strength to the originally totally symmetric modes. From the values of the observed frequencies, the corresponding $F_{\mathcal{R}}$ indicates a much longer conjugation length, or a larger effective conjugation [25,39].

A practical outcome of the spectroscopic properties of photoexcited PA is that the ν_3 infrared-active mode is very sensitive to conjugation length, and can be used as a probe for the relative measure of the length of the delocalisation path in unknown samples.

CONCLUSIONS

The guided tour, presented here, for the stepwise analysis of the vibrational data recorded for trans-PA, can be and has been, equally applied for the interpretation of the spectra and the structural analysis of other polyconjugated polymers such as polyparaphenylene [44], polypyrrole [45], polythiophene [46] and their alkyl derivatives [47]. For these substances, more experimental data are needed. Moreover, additional theoretical problems must be considered, due to the fact that the electron delocalisation depends on the torsional angle between the aromatic rings. Moreover, these systems cannot sustain solitonic excitation, but only polarons or bipolarons can be formed with the generation of "quinoid structures."

AKNOWLEDGEMENT

The various steps discussed in this guided tour are the results of many enlightening discussions, intelligent work and careful calculations by Dr. M. Gussoni, Dr. C. Castiglioni, Dr. B. Tian and Dr. J. T. Lopez-Navarrete.

REFERENCES

[1] G. Zerbi, in "Advances in Infrared and Raman Spectroscopy", vol. 11 (R. J. H. Clark and R. E. Hester, eds.) (Wiley-Heyden, New York, 1984) p. 301.
[2] E. B. Wilson, J. C. Decius, and P. Cross, "Molecular Vibrations", (McGraw-Hill, New York, 1955); S. Califano, "Vibrational States" (Wiley, New York, 1976).
[3] L. Piseri and G. Zerbi, J. Mol. Spectrosc. 26: 254 (1968); J. Chem. Phys. 48: 3561 (1968).
[4] G. Zerbi and L. Piseri, J. Chem. Phys. 49: 3840 (1969).
[5] G. Zerbi, in "Lattice Dynamics and Intermolecular Forces" (S. Califano ed.) (Academic Press, New York, 1975); M. Tasumi and G. Zerbi, J. Chem. Phys. 48: 3813 (1968).
[6] M. Gussoni, in "Vibrational Intensities in Infrared and Raman Spectroscopy", (W. B. Person and G. Zerbi eds.) (Elsevier, Amsterdam, 1984) pp. 96-121; *ibid.* pp. 221-238.
[7] M. Gussoni, in "Advances in Infrared and Raman Spectroscopy" Vol. 6 (R. J. H. Clark and R. H. Hester eds.) (Heyden, London, 1976) p. 61-126.
[8] L. A. Gribov, "Intensity Theory for Infrared Spectra of Polyatomic Molecules", (Consultants Bureau, New York, 1964).
[9] M. Gussoni, C. Castiglioni, and G. Zerbi, J. Phys. Chem. 88: 600 (1984).

[10] M. Gussoni, C. Castiglioni, and G. Zerbi, J. Chem. Phys. 80: 1377 (1984).
[11] P. Jona, M. Gussoni and G. Zerbi, J. Chem. Phys. 75: 1677 (1981).
[12] C. Castiglioni, M. Gussoni, and G. Zerbi, J. Chem. Phys. 82: 3534 (1985).
[13] J. H. Schachtschneider and R. G. Snyder, Spectrochim. Acta. 19: 17 (1961).
[14] R. G. Snyder and G. Zerbi, Spectrochim. Acta 23A: 391 (1967).
[15] G. Zerbi and G. Zannoni, J. de Physique 44C: 3273 (1983).
[16] G. Zerbi, C. Castiglioni, J. T. Lopez-Navarrete, B. Tian, and M. Gussoni, Synth. Met. 28: D359 (1989).
[17] J. T. Lopez-Navarrete and G. Zerbi, Synth. Met., in press.
[18] P. Pulay, G. Fogarasi, G. Pongor, J. E. Boggs, and A. Vargha, J. Am. Chem. Soc. 105: 7073 (1983).
[19] J. Dupuis and E. Clementi, J. Chem. Phys. 89: 4989 (1988).
[20] G. Zannoni and G. Zerbi, J. Mol. Struct. 100: 485 (1983).
[21] W. J. Feast, in "Handbook of Conducting Polymers" , Vol. 1 (T. A. Skotheim ed.) (Dekker, New York, 1986) pp. 1-43.
[22] G. Zerbi in "Advances in Applied Fourier Transform Infrared Spectroscopy" (M. W. Mackenzie ed.) (J. Wiley, New York, 1988) p. 247-322.
[23] G. Zerbi and G. Dellepiane, Gazzetta Chim. Ital. 117: 591 (1987).
[24] C. Castiglioni, G. Zerbi, and M. Gussoni, Solid State Comm. 56: 863 (1985).
[25] For a collection of references see: C. Castiglioni, M. Gussoni, J. T. Lopez-Navarrete, Solid State Comm. 65: 625 (1988).
[26] G. Zannoni and G. Zerbi,. Solid State Comm. 48: 871 (1983).
[27] A. C. Albrecht, J. Chem. Phys. 34: 1476 (1961).
[28] C. Castiglioni, M. Gussoni, M. Miragoli, G. Lugli, and G. Zerbi, Spectrochim. Acta 41A: 371 (1985).
[29] D. C. McKean, Chem. Soc. Rev. 7: 399 (1978).
[30] B. Horovitz, Solid State Comm. 41: 792 (1982).
[31]. C. Castiglioni, J. T. Lopez-Navarrete, G. Zerbi, and M. Gussoni, Solid State Comm. 65: 625 (1988); G. Zerbi, C. Castiglioni, J. T. Lopez-Navarrete, B. Tian, and M. Gussoni, Synth. Met. 28: D359 (1989); M. Gussoni, C. Castiglioni, and G. Zerbi, Synth. Met. 28: D375 (1989).
[32] G. Zerbi, C. Castiglioni, B. Tian, and M. Gussoni, Proceedings of the International Winterschool on Electronic Properties of Polymers and Related Compounds, Kirchberg 1989 (H. Kuzmany, M. Mehring and S. Roth, eds.) Springer Verlag, in press.
[33] J. T. Lopez-Navarrete, B. Tian and G. Zerbi, Solid State Comm. submitted.
[34] A. J. Heeger, in "Handbook of Conducting Polymers" Vol. 2 (T. A. Skotheim ed.) (Dekker, New York, 1986) pp. 729-756.
[35] I Harada, Y. Furukawa, M. Tasumi, H. Shirakawa, and S. Ikeda, J. Chem. Phys. 73: 4746 (1980).
[36] J. Tanaka, Y. Saito, M. Shimizu, and M. Tanaka, Synth. Met. 17: 307 (1987).
[37] C. Castiglioni, M. Gussoni, M. Miragoli, and G. Zerbi, Mol. Cryst. and Liquid Cryst. 121: 287 (1985).
[38] J. L. Brédas in "Handbook of Conducting Polymers" Vol. 2 (T. A. Skotheim ed.) (Dekker, New York, 1986) pp. 825-857 and 859-913.
[39] G. Zerbi, C. Castiglioni, J. T. Lopez-Navarrete, T. Bogang, and M. Gussoni, Synth. Met. 28: D359 (1989).
[40] G. Zerbi, C. Castiglioni, S. Sala, and M. Gussoni, Synth. Met. 17: 93 (1987).
[41] Y. Furukawa, T. Arakawa, H. Takeuchi, I. Harada, and H. Shirakawa, J. Chem. Phys. 81: 2907 (1984).
[42] J. Orenstein, in "Handbook of Conducting Polymers" , Vol. 2 (T. A. Skotheim ed.) Dekker, New York, 1986) p. 729.
[43] See refs. 16 and 25.
[44] C. Castiglioni, M. Gussoni, and G. Zerbi, Synth. Met. 29: E1 (1989).
[45] B. Tian and G. Zerbi, Proceedings of the International Winterschool on Electronic Properties of Polymers and Related Compounds, Kirchberg, 1989, (H. Kuzmany, M. Mehring, S. Roth eds.) Springer Verlag, in press.
[46] J. T. Lopez-Navarrete and G. Zerbi, Synth. Met. 28: C15 (1989); J. Chem. Phys., submitted.
[47] B. Tian, G. Zerbi, J. Ruhe, and G. Wegner, Proceedings of the International Conference on Solid State Chemistry, Como, 1989.

ANELLATED DIPYRROLYL SYSTEMS: SYNTHESIS AND ELECTRICAL PROPERTIES OF THE POLYMERS

Anna Berlin,[a] Stefano Martina,[a] Giorgio Pagani,[a] Gilberto Schiavon,[b] and Gianni Zotti[b]

[a]Dipartimento di Chimica Organica e Industriale dell'Università and Centro CNR Speciali Sistemi Organici, via C. Golgi 19, 20133 Milano, ITALY

[b]CNR, Istituto di Polarografia ed Elettrochimica Preparativa, Corso Stati Uniti 4, Padova, ITALY

INTRODUCTION

Functionalization of heterocyclic monomers with medium-length alkyl chains[1,2] has provided a route to the obtainment of (at least partially) soluble polymers. To minimize interference of the functionalization with the planarity and regularity of the polymer, we have approached the problem starting from monomers constituted by two heterocycles separated by a conjugatively unsaturated spacer:[3] this can be a ring onto which the terminal heterocycles are anellated and it is intended to be the site of functionalization. In view of the merits of polypyrrole,[4] we have focused our attention on systems (1)-(4), in which two terminal pyrrole units are

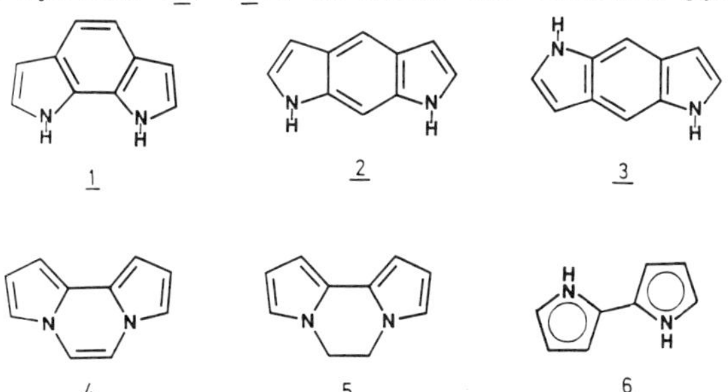

anellated to a central aromatic ring. Heterocycles (1)-(4) are all isomers and constitute a body of structural and electronic variation of the same skeleton. A further electronic variation is provided by saturation on the double bond of the ethenylic fragment joining the two nitrogen atoms in (4): for this reason also the dihydro compound (5) was included in the series. Results are reported on the oxidative electrochemical polymerization of the above monomers.

Lower-Dimensional Systems and Molecular Electronics
Edited by R.M. Metzger *et al.*, Plenum Press, New York, 1991

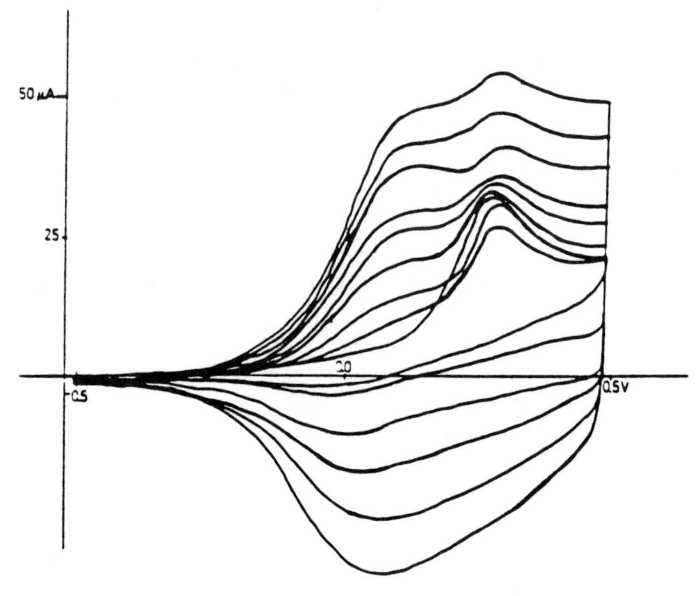

Figure 1

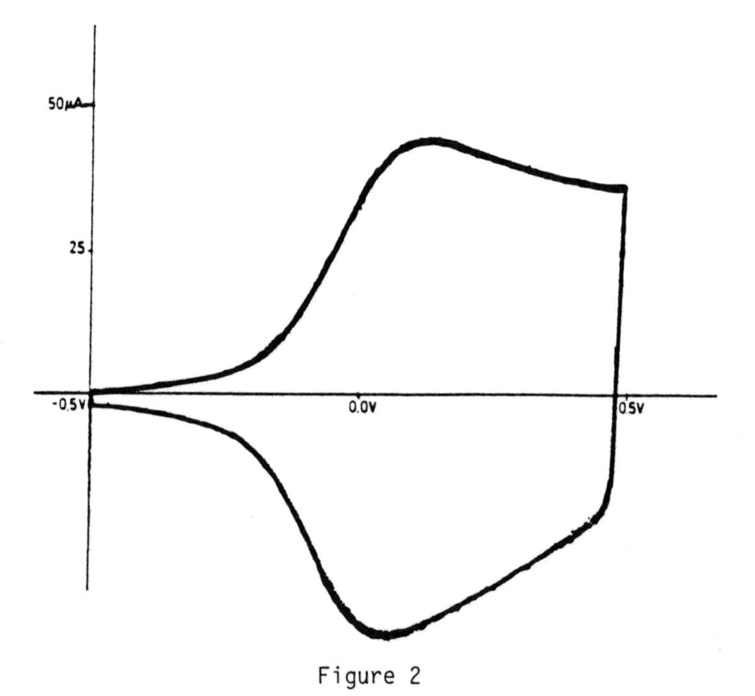

Figure 2

RESULTS

Synthesis of the monomers

Access to compounds (1)-(3) has been recently described[5] starting from the proper dinitroxylenes, while preparation of the dipyrrolopyrazine (4) requires dehydrogenation of the corresponding reduced dipyrrolyl derivative (5).[6]

Electrochemical polymerization

Electrochemical experiments were performed at room temperature in nitrogen degassed acetonitrile solutions 0.5-1x10^{-2} M in substrate and 0.1 M in tetraethylammonium perchlorate (TEAP). The reference electrode was $Ag/AgClO_4$ (0.1 M in acetonitrile) (ca. 0.3 V vs SCE).

The cyclic voltammograms of all the substrates show an irreversible oxidation peak followed by a fast chemical reaction: Figure 1 shows the case for compound (4). Repetitive cycling with a starting potential of -0.6 V and a switching potential usually 0.1 V beyond the peak, produces immediately and progressively the build up of a reversible process. The electrodes covered by thin films of the conductive polymer, upon transfer into blank acetonitrile solution, show the reversible redox cycles observed in the filming solution (Figure 2).

Table. Electrochemical and Conductivity Data

Monomer	Ep/V[a]	E_o/V[b]	Doping level[c]	σ/Scm^{-1}
1	0.40	0.15	0.4	5
2	0.30	0.00	0.5	3
3	0.27	-0.10	0.6	0.2
4	0.23	0.06	0.5	0.2
5	0.17	-0.16	0.56	2.5
6	0.25	-0.50	0.6	100

[a] Oxidation peak potential of the monomer
[b] Polymer redox potential
[c] Counterion units per monomer

Oxidation peak potentials of the monomers Ep and polymer redox potentials are reported in the Table, together with the doping level of the oxidized polymers and conductivity data. Microanalytical data of the oxidized polymers provide the number of counterion units per monomer (doping level), numbers that are in very good agreement with the number of electrons reversibly exchanged per monomer in the redox cycle (ca. 0.4-0.6 electrons).

DISCUSSION

Compound (5) may be looked at as a dipyrrolyl system bridged at the two nitrogen atoms with an ethylene fragment: indeed, compound (5) has almost the same Ep value as dipyrrolyl (6). One interesting aspect is that, while N-substitution of pyrrole causes a dramatic drop in the conductivity of the corresponding polymer (e.g. poly-N-methylpyrrole = 10^{-3} Scm^{-1}), N-substitution does not seem to be particularly detrimental in decreasing the conductivity of a N,N'-bridged dipyrrolyl polymer. It is possible that in the polymer of (5) the ethylene fragment joining the two nitrogen atoms has smaller steric requirements than two N-methyl groups in poly-N-methylpyrrole, thus allowing in the former polymer an increased coplanarity.

In analogy with compound (5), compounds (1) and (4) can be looked at as dipyrrolyl systems functionalized with an ethenylic bridge between positions ß,ß'- and N,N'- respectively. If so, it should be concluded that such anellations increase in both cases the E_0 values. In conclusion, the neutral polymers derived from (1) and (4), thanks to their higher stability towards air, may be exploited as models for performing structural studies instead of the extremely air sensitive and unstable polypyrrole.

REFERENCES

1. M. Sato, S. Tanaka and K. Kaeriyama, J. Chem. Soc. Chem. Commun., 1346 (1986).
2. J. Rühe, T.A. Ezquerre and G. Wegner, Synth. Met., 28, C177 (1989).
3. A. Berlin, S. Bradamante, R. Ferraccioli, G.A. Pagani and F. Sannicolò, J. Chem. Soc. Perkin 1, 2631 (1987).
4. "Handbook of Conducting Polymers", Vol. 1, T.A. Skotheim, ed. Marcel Dekker (1986).
5. A. Berlin, S. Bradamante, R. Ferraccioli, G.A. Pagani and F. Sannicolò, J. Chem. Soc. Chem. Commun., 1176 (1987).
6. U. Berger, F. Dreier, Tetrahedron, 39, 2065 (1983).

CONDUCTING POLYMERS FROM 3,4-CYCLOALKYLTHIOPHENES

Anna Berlin,[a] Jürgen Rühe,[b] and Gerhard Wegner[b]

[a] Centro CNR Speciali Sistemi Organici, via C. Golgi 19
20133 Milano, ITALY

[b] Max Planck-Institut für Polymerforschung, Ackermannweg 10
D-6500, Mainz, FEDERAL REPUBLIC OF GERMANY

INTRODUCTION

The conduction behaviour of poly(3,4-cycloalkylpyrrole)perchlorates is controlled by certain topological parameters correlated with the size of the alkyl substituents.[1,2] We were interested whether similar electrical properties are obtained in poly(3,4-cycloalkylthiophene)s.

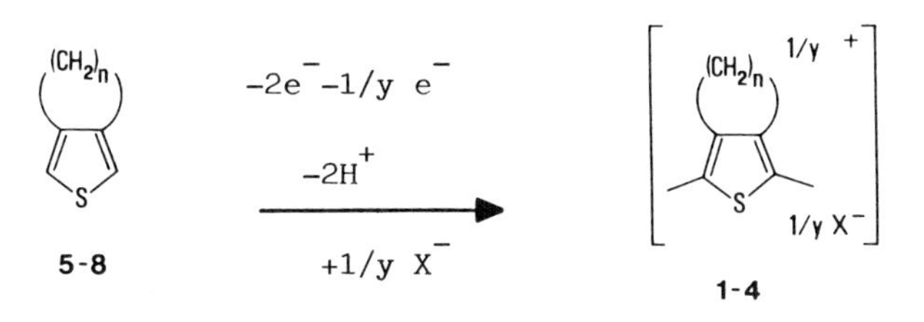

$X^- = ClO_4^-$; PF_6^-; $CF_3SO_3^-$; $FeCl_4^-$

n =	3	4	5	10
	(1)	(2)	(3)	(4)
	(5)	(6)	(7)	(8)

Lower-Dimensional Systems and Molecular Electronics
Edited by R.M. Metzger *et al.*, Plenum Press, New York, 1991

Therefore we have synthesized thiophenes with a fused alkyl ring in the 3 and 4 position of the thiophene moiety and polymerized them either electrochemically or by chemical oxidation. Here the electrochemical, electrical and spectroscopic properties of the obtained polythiophene derivatives are presented.

RESULTS

Synthesis of the monomers

Cyclopenta[c]thiophene (5), cyclohexa[c]thiophene (6) and cyclohepta-[c]thiophene (7) were prepared according to literature prescriptions;[3-5] cyclododeca[c]thiophene (8) was prepared following the same route described for (7)[5] with some modifications.

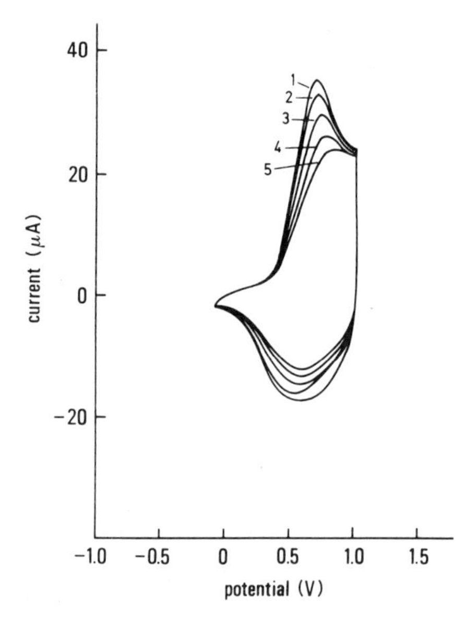

Fig 1. Cyclic voltammogram of poly(cyclohexa[c]thiophene); reference Ag/AgCl; sweep rate 100 mV/s; electrode surface 7.85×10^{-3} cm^2; (1)=cycle 1; (2)=cycle 3; (3)=cycle 50; (4)=cycle 100; (5)=cycle 300.

Synthesis of the polymers

Polymeric salts of the poly(3,4-cycloalkylthiophene)s were obtained by electropolymerization in a divided compartment cell at 25°C in acetonitrile or nitrobenzene as solvent. The monomer and electrolyte concentrations were 0.01 and 0.1 mol/l, respectively. The optimum current density under galvanostatic conditions was found to be 2 mA/cm^2. Chemical polymerizations were performed using iron trichloride in acetonitrile or chloroform as oxidizing agent.

Table 1. Electrochemical data[a]

monomer	Ea^{p}(monomer)[b]	Ec^{p}(polymer)[b]	Ea^{p}(polymer)[b]
(5)	1.80	0.45	0.64
(6)	1.70	0.60	0.73
(7)	1.78	1.00[c]	1.20[c]
(8)	1.76	1.10[c]	1.30[c]
thiophene (9)	1.90	0.75	0.90
3-methylthiophene (10)	1.65	0.53	0.65

[a] Reference Ag/AgCl; sweep rate 100 mV/s; polymers deposited at 23°C.
[b] Peak potential of the anodic (a) or cathodic (c) current in volts.
[c] Only poorly reversible.

Electrochemical properties

The cyclic voltammograms of all the monomers show an irreversibile oxidation peak, followed by a fast chemical process, which is attributable to the polymerization; the electrodes covered by 1 μm film, upon transfer into blank solution, show the reversible redox cycles attributable to the electroactivity of the polymers.

Electrochemical data of the 3,4-cycloalkylthiophenes (5)-(8), thiophene (9), 3-methylthiophene (10) and of the corresponding polymers are presented in the Table. From these data it must be noted that the oxidation potential of the polymers increases with increasing ring size of the substituents.

Polymers (1) and (2) can be reduced and reoxidized several hundred times, while (3) and (4) are stable for only a few cycles. In Figure 1 the case of (2) is presented. The electroactivity of the polymeric films is lost if potentials higher than +2.0 V vs. Ag/AgCl are applied ("overoxidation").

Electrical properties

As shown by molecular models, due to the bulky substituents, adjacent thiophene moieties of the same chain cannot be arranged in a plane, but have to be twisted. Each individual polythiophene backbone is cylindrically surrounded by a layer of methylene groups, the thickness of which is determined by the size of the alkyl ring fused to the thiophene units in the 3,4 position.

Therefore, adjacent polymer backbones in the bulk material must be separated from each other by a distance R which is related to the size of the substituent. The relevant distances R have been evaluated from space filling models.[2] According to the hopping theory of localized charge carriers,[6] the conductivity, at sufficiently high temperatures, can be described by

$$\sigma = \sigma_o \exp(-2\alpha R) \exp(-E_A / kT) \qquad 1$$

α being the inverse localization length, and E_A the activation energy for a thermally activated hopping process. From this equation a linear dependence of $\log\sigma$ on the hopping distance R should be expected at a given temperature.

The logarithm of the room temperature conductivity of the investigated polythiophene derivatives (1)-(4) shows this linear dependence on the minimum interchain distance R as shown in Figure 2

If the dependence of E_A on R is known, the localization length $1/\alpha$ can be calculated according to eq. 1. The E_A values for different polymers are obtained from the slopes of $\log\sigma$ vs. 1/T plots which are found to be linear at relatively high temperatures.

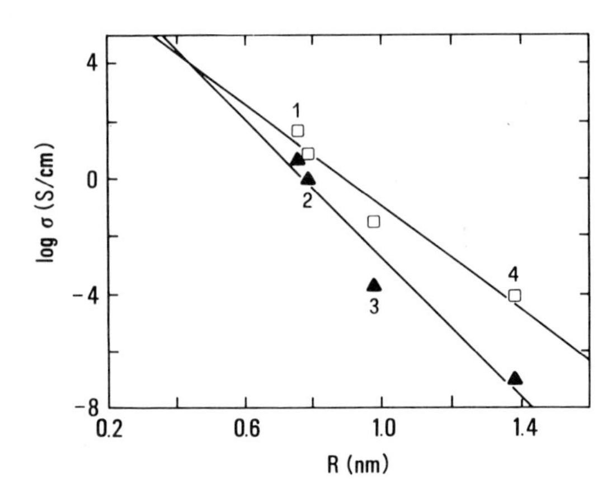

Fig. 2 Logarithm of room temperature d.c. conductivity of the investigated polymeric salts as a function of R; $FeCl_4^-$ salts (triangles) and PF_6^- salts (squares).

Empirically, E_A is found to depend linearly on R, that is $E_A \sim BR$ with the proportionality constant B=0.55±0.08 eV/nm for polythiophenes-$FeCl_4^-$ and B=0.15±0.03 eV/nm for the corresponding polypyrroles-ClO_4^-. Taking into account the empirical linear dependence of E_A on R, eq 1 gives:

$$\log\sigma \sim -R\,(2\alpha + B/KT)$$

The term $2\alpha + B/KT$ can be identified as the slope of the straight line. Following this treatment, polythiophenes-$FeCl_4^-$ data give an α^{-1} value of 0.38±0.16 nm, very close to the value found for the corresponding polypyrroles-ClO_4^- (0.36 nm)

CONCLUSION

Analogously to the case of 3,4-cycloalkylpolypyrroles, the conduction in 3,4-cycloalkylpolythiophenes can be understood as a hopping of localized charge carriers between adjacent chains separated by poorly conducting ("insulating") regions of methylene groups. The probability for the hopping process is controlled by the substituent size which represents the minimum interchain distance. These results emphasize the importance of interchain charge-carrier hopping processes for the macroscopic conduction behaviour of polymers.

REFERENCES

1. J. Rühe, Ch. Krönke, T. A. Ezquerra, F. Kremer and G. Wegner, Ber. Bunsengesell Physik. Chem., 9, 885 (1987).
2. T. A. Ezquerra, J. Rühe and G. Wegner Chem. Phys. Lett., 144, 194 (1988).
3. D. W. H. Mac Dowell, T. B. Frame and D. L. Ellison, J. Org. Chem., 32, 1227 (1967).
4. R. Meyer, H. Kleinert, S. Richter and K. Gewald, J. Prakt. Chem., 20, 244 (1963).
5. S. Hauptmann and E. M. Warner, J. Prakt. Chem, 314, 499 (1972).
6. N. F. Mott and E. A. Davis "Electronic process in non-crystalline materials", Clarendon Press, Oxford, 1979.

ORIENTATION OF STRETCHED POLY(3-OCTYLTHIOPHENE) FILMS: VISIBLE AND INFRARED DICHROISM STUDIES

G. Gustafsson and O. Inganäs

Department of Physics
University of Linköping
Linköping, SWEDEN

INTRODUCTION

Conducting polymers are considered to be quasi-one-dimensional materials. To be able to study the intrinsic properties of these materials, such as intra- and inter-chain electron motions, it is important to reach a high degree of orientation of the polymer chains. Orientation is also important for achieving a high conductivity in these materials, since an increased order also increases the overlap of orbitals between chains, and thus increases the hopping probability.

Orientation of the polymer chains can be obtained by different methods. Highly oriented films of polyacetylene have been made by performing polymerization in a liquid crystal [1,2]. It has also been shown that it is possible to attain highly oriented films of polyacetylene [3] and poly(p-phenylene vinylene) [4] by stretching their precursors. Kaneto *et al.* [5] have oriented polythiophene by stretching a film, prepared electrochemically on In-Sn oxide (ITO)-coated poly(ethyleneterephthalate) films. The processability of the recently developed poly(3-alkylthiophenes) [6-8] makes this class of polymers suitable for stretching. Yoshino *et al.* [9] have shown that it is possible to orient poly(3-alkylthiophene), both as free-standing films and in a blend with an elastomer.

EXPERIMENTAL

Poly(3-octylthiophene) (P3OT) was prepared by the method described elsewhere [10], utilizing $FeCl_3$ as a catalyst. P3OT was used in two configurations: (a) free- standing film: the film thickness after stretching was around 2 μm; (b) film spun on a polyethylene (PE) foil.

These structures were mechanically stretched at room temperature to a stretch ratio l/l_0 where l is the final length, and l_0 is the initial length. Infrared spectra were taken with a Bruker Model IFS 113v Fourier transform spectrometer equipped with a wire grid polarizer. Optical absorption spectroscopy was done using a Perkin-Elmer Lambda 9 spectrophotometer. The light beam was polarized with a Glan-Thompson polarizing prism.

RESULTS AND DISCUSSION

Degree of orientation

The degree of orientation is often expressed in the so called orientation function f defined as [11]:

Lower-Dimensional Systems and Molecular Electronics
Edited by R.M. Metzger *et al.*, Plenum Press, New York, 1991

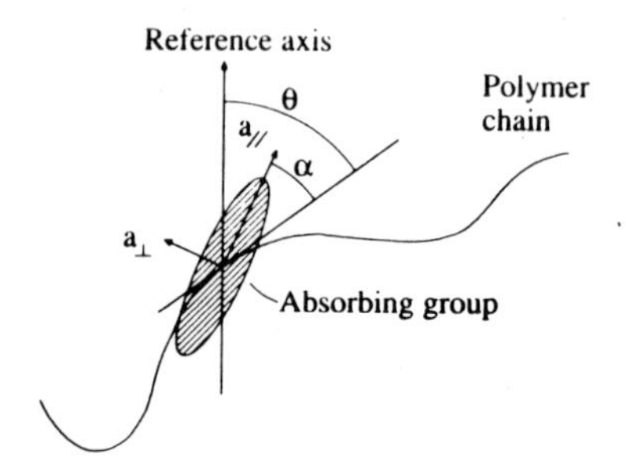

Figure 1. Illustration of the coordinates of equations (1) to (3), from Ref. [11].

$$f = \frac{3\langle\cos^2\theta\rangle_{av} - 1}{2} \qquad (1)$$

where θ is the angle between the chain axis and the stretch direction, and the quantity $\langle\cos^2\theta\rangle_{av}$ represents the average value of $\cos^2\theta$, taken over all chains in the sample (Fig. 1).

If we let $A_{\|}$ and $A_{\perp}$ designate the absorbances for light polarized with the electric vector parallel and perpendicular to the stretching direction, the dichroism $D = A_{\|} / A_{\perp}$ can be related to the orientation function by

$$\frac{D-1}{D+2} = \frac{a_{\|} - a_{\perp}}{a_{\|} + 2a_{\perp}} f_o f = Cf \qquad (2)$$

$a_{\|}$ and $a_{\perp}$ are the absorbances parallel and perpendicular to the principal axis of absorption, and f_o is the orientation factor for the principal axis of absorption, defined by

$$f_o = \frac{3\langle\cos^2\alpha\rangle_{av} - 1}{2} \qquad (3)$$

where α is the angle between the principal axis of absorption and the chain axis.

In infrared absorption, the principal axis of absorption usually corresponds to the transition moment direction **M** for the actual vibration, i.e. $a_{\perp} = 0$ and $C = f_o$.. In the optical regions the absorption process involves electronic motions and, in general, there is a perpendicular contribution to the absorbance i.e. $a_{\perp} \neq 0$ and $C \neq f_o$.

Infra-red absorption measurements

The infrared spectra of a free standing film of P3OT, stretched to a ratio $l / l_o = 2.5$, for light polarized parallel ($E_{\|}$) and perpendicular ($E_{\perp}$) to the stretch direction, is shown in Fig. 2. In both spectra, well-defined interference fringes appear, which indicate the macroscopic uniformity of the film. The difference of the period of these fringes reflects the anisotropy of the refractive index in the material. The greatest dichroism in the spectra is observed for the bands at 3050 cm^{-1}, 1510 cm^{-1} and 820 cm^{-1}. These bands are assigned to thiophene ring CH stretch, C=C stretch and thiophene ring CH out-of-plane bend, respectively. The thiophene ring CH out-of-plane bending mode (820 cm^{-1}) should have a transition moment **M** perpendicular to the chain axis, i. e. $\alpha = 90°$ and $f_o = -1/2$. The dichroic ratio D for this peak is 0.27. Using this value, Eqs. (1) and (2) give

$$f = 0.64 \text{ and } \theta = 29° \qquad (\theta = 54.7° \text{ for random orientation}).$$

The band at 1510 cm^{-1}, which has a dichroic ratio D = 7.3, is assigned to a C=C stretch

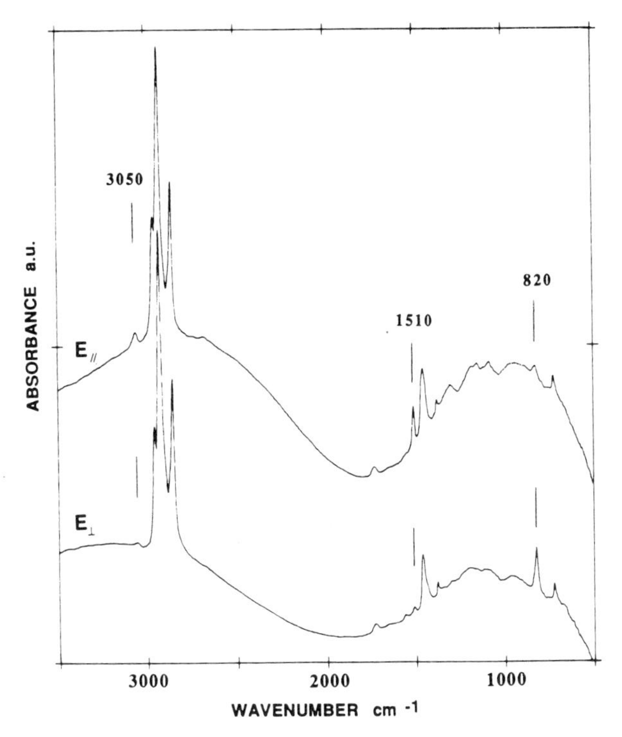

Figure 2. Infrared spectra of a stretched, free-standing film of poly(3-octylthiophene), for light polarized with the electric vector parallel ($E_{||}$) and perpendicular ($E_{\perp}$) to the stretch direction. $l / l_0 = 2.5$.

vibration. The direction of the transition moment of this band is uncertain, but if we assume that **M** is directed parallel to the chain, we get a lower limit for the orientation function:

$$f \geq 0.68$$

and an upper limit for the angle θ:

$$\theta \leq 28°$$

A better stretch ratio, and thus also a higher degree of orientation, can be obtained by casting a solution of P3OT on a polyetylene (PE) foil, and stretching the structure as a whole. This method is also preferred, since films with a thickness suitable for UV/VIS/NIR spectroscopy can be used. Fig. 3 shows the band at 1510 cm^{-1} of a P3OT/PE structure, which was stretched to a final stretch ratio $l / l_0 = 3$. The measured dichroic ratio is D = 14, which gives

$$f \approx 0.81 \text{ and } \theta \approx 21°$$

Optical absorption measurements

Additional information about the orientation can be obtained, by studying the dichroism of the optical absorption. Fig. 4 shows the optical absorption spectra of the same film, as in Fig. 3, for light polarized parellel ($E_{||}$) and perpendicular ($E_{\perp}$) to the stretch direction. Upon stretching, a large anisotropy in the optical absorption is induced. This appears not only as a difference in the intensities of the absorption parallel and perpendicular to the stretch direction, but also as a difference in the peak positions. This gives rise to different colors when observing the sample through a polarizer. By orienting the polarizer parallel to the stretch direction of the sample, a deep red-violet color appears. Rotating the polarizer by 90°, the

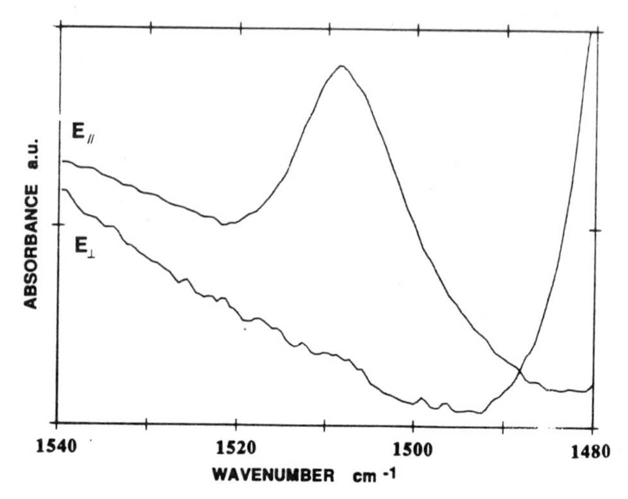

Figure 3. The 1510 cm^{-1} (C=C stretch) band of a stretched poly(3-octylthiophene)/ polyethylene structure, for infrared light polarized parallel ($E_{||}$) and perpendicular ($E_{\perp}$) to the stretch direction. The stretch ratio $l / l_0 = 3$.

color of the sample changes to bright yellow. As can be seen in Fig. 4, the dichroic ratio peaks at a value of 20, for energies around 2.1 eV. The absorption at these energies is caused by segments of chains with long conjugation lengths, which means that the degree of orientation calculated from this value does not represent the polymer as a whole, but only regions with long straight-chain segments. If we, to a first approximation, neglect interchain and intrachain π - π* transitions perpendicular to the chain axis, i.e. $a_{\perp} = 0$ and $\alpha = 0$, we get the result

$$f = 0.86 \text{ and } \theta = 18°$$

If there are significant contributions from interchain and/or intrachain π - π* transitions, the orientation function f is larger.

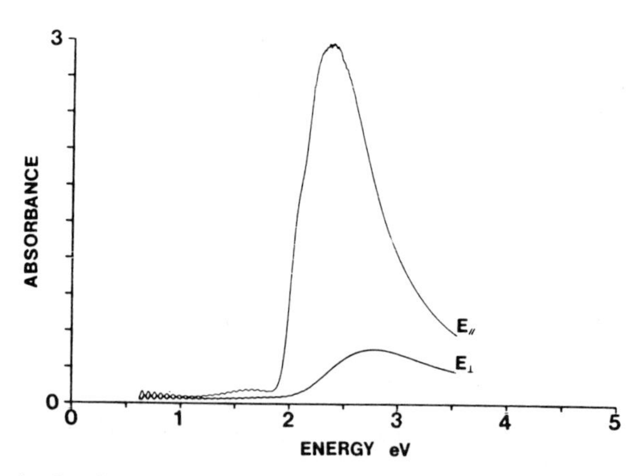

Figure 4. Optical absorption spectra of a stretched poly(3-octylthiophene)/polyethylene structure, for light polarized parallel ($E_{||}$) and perpendicular ($E_{\perp}$) to the stretch direction. The stretch ratio $l / l_0 = 3$.

Structure of oriented poly(3-octylthiophene)

Poly(3-octylthiophene) is best described as a partially crystalline polymer. In unconjugated, saturated polymers, crystallites are described as lamellae, in which the chain must fold in order to be accommodated within the thin lamellae. However, in the conjugated polymers it is believed that the chains are too rigid to fold. A fringe-micellar- like morphology is probably a more appropriate description for this class of polymers. This morphology consists of crystalline and amorphous regions, where the crystalline regions are built up by segments from different chains. In conducting polymers there should be one, possibly nematic, phase of stiff chain segments with long conjugation lengths, which interacts to form some kind of crystallites, and another region with coiled chains, with shorter conjugation lengths, which are connecting the crystallites.

In the optical dichroism measurements presented above, there is a nonzero absorption of light polarized perpendicular to the stretch direction. There can be at least two contributions to this absorption: (i) a nonzero $a_{\perp}$, (ii) a misalignment of the chains. The mis-aligned chains must preferentially be those chains, or segments of chains, which have short conjugation lengths, compared to the aligned chains, since the absorption maximum of the perpendicular absorption is shifted 0.3 - 0.4 eV to higher energy, compared to the parallel absorption. If we make the assumption that contributions of type (i) are much smaller than those of type (ii), then the optical measurements show that, upon stretching P3OT, segments of chains with long conjugation lengths are more preferentially oriented along the stretch direction, than segments with shorter conjugation lengths, which are more randomly oriented. The assumption made above is supported by measurements of the anisotropy of the absorption coefficient in polyacetylene [12] which shows that the transverse absorption coefficient is a factor of 25 lower, than the parallel absorption coefficient.

Optical anisotropy of doped P3OT

The optical anisotropy of a similar structure as in Fig. 4, but doped with $FeCl_3$ after stretching, is shown in Figure 5. The peak at 3.4 eV, appearing in both spectra, is the absorption from the anion, $FeCl_4^-$. The orientation function, evaluated from the maximum of the dichroism (at 2.1 eV) of the optical spectra of the film in the neutral state, is $f = 0.75$. As can be seen, a very high doping level can be obtained, even after the orientation process. This shows that doping ions can be intercalated easily into the oriented polymer. The large anisotropy observed indicates, that a high degree of orientation is preserved, even after doping.

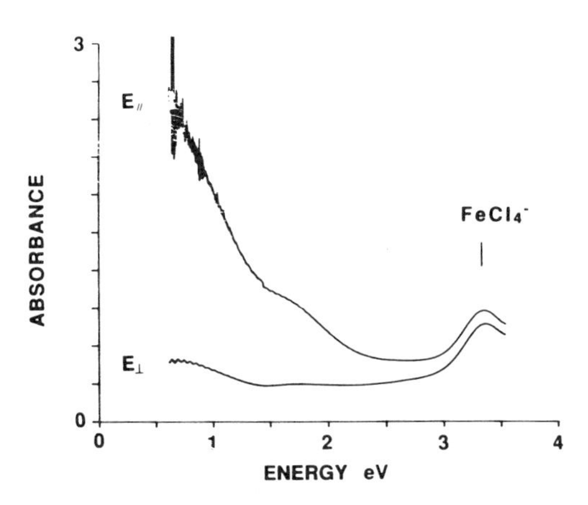

Figure 5. Optical absorption spectra of a doped stretched poly(3-octylthiophene)/ polyethylene structure, for light polarized parallel ($E_{\parallel}$) and perpendicular ($E_{\perp}$) to the stretch direction. The poly(3-octylthiophene) was doped by immersing the stretched structure as a whole in a $FeCl_3/CHCl_3$ solution. The stretch ratio $l / l_0 = 3$.

CONCLUSIONS

Poly(3-octylthiophene) has been oriented, by stretching a free-standing film and by stretching a sandwich structure, polyethylene/poly(3-octylthiophene). The infrared dichroism studies indicate that P3OT is highly oriented at a rather low draw ratio:

$$f = 0.81 \text{ and } \theta = 21° \text{ at } l / l_0 = 3$$

The optical dichroism studies support a model for a stretch-oriented P3OT, in which phases of chains with long conjugation lengths, e.g. crystallites, are oriented preferentially, compared to phases with shorter conjugation length. Optical absorption measurements on a doped film of oriented P3OT show that the stretched polymer can be doped into the metallic regime, and that it is highly ordered even after doping.

ACKNOWLEDGEMENTS

We are grateful to Dr. J. Laakso and Dr. J.-E. Österholm at Neste Oy for preparing poly(3-octylthiophene), and to B. Liedberg for fruitful discussions. The work was supported by the Nordic Fund for Industrial and Technological Development.

REFERENCES

[1] K. Akagi, S. Katayama, H. Shirakawa, K. Araya, M. Mukoh, and T. Narahara, Synth. Met. 17: 241 (1987).
[2] A. Montaner, M. Rolland, J. L. Sauvajol, L. Meynadier, R. Almairac, J. L. Ribet, M. Galtier, and C. Gril, Synth. Met. 28: D19 (1989).
[3] G. Leising, Polym. Bull.11: 401 (1984).
[4] D. D. C. Bradley, R. H. Friend, H. Lindenberger, and S. Roth, Polymer 27: 1709 (1986).
[5] K. Kaneto, F. Uesugi, and K. Yoshino, Solid State Commun. 65: 783 (1988).
[6] M. Sato, S. Tanaka and K. Kaeriyama, J. Chem. Soc. Chem. Commun. 1346 (1986).
[7] K. Y. Yen, G. G. Miller and R. L. Elsenbaumer, J. Chem. Soc. Chem. Commun. 1346 (1986).
[8] S. Hotta, S. D. D. V. Rughooputh, A. J. Heeger, and F. Wudl, Macromolecules 20: 212 (1987).
[9] K. Yoshino, M. Onoda, and R. Sugimoto, Jpn. J. Appl. Phys. 27: L2034 (1988).
[10]. J.-E. Österholm, J. Laakso, P. Nyholm, H. Isotalo, H. Stubb, O. Inganäs, and W. R. Salaneck, Synth. Met. 28: C435 (1989).
[11] Y. Shindo, B. E. Read, and R. S. Stein, Makromol. Chemie 118: 272 (1968).
[12] P. D. Townsend and R. H. Friend, Synth. Met. 17: 361 (1987).

VIBRATIONAL PROPERTIES OF CONDUCTING POLYMERS WITH AROMATIC OR HETEROAROMATIC RINGS

J. P. Buisson, J. Y. Mevellec, S. Zeraoui and S. Lefrant

Laboratoire de Physique Cristalline, I. P. C. M.
Université de Nantes
2 rue de la Houssinière
44072 Nantes Cedex 03, FRANCE

INTRODUCTION

Since the discovery that unsaturated polymers can exhibit specific properties, after doping with oxidizing or reducing agents, a large number of studies have been carried out in many different domains [1]. Although most of the attention has been given to polyacetylene, because of its simple structure, its instability to ambient atmosphere considerably reduces the possibility of its use in technical applications. Therefore, more stable materials, like polyaromatics, have been studied in detail. More recently, other polymers, like polyarylene vinylenes, which, in addition, exhibit good mechanical properties and, in some cases, high optical quality, have been the subject of intense investigation [2-5].

In this paper, we present a comparative study of polyaromatics or polyheteroaromatics by means of Resonant Raman Scattering (RRS). Although a larger number of these compounds exhibit a very intense fluorescence, hiding, in some cases, the Raman signal, our capability to choose excitation lines from the near U.V. ($\lambda_L = 351.1$ nm) to the near infrared ($\lambda_L = 752.2$ nm) allows us to detect most of the Raman peaks. Associated with infrared absorption measurements, a complete knowledge of the vibrational modes leads to the establishment of dynamical models, with a determination of force constants, which can be compared from one compound to the other. Unlike for polyacetylene, these force constants can be determined in a more accurate way, since it is possible to use oligomers as model compounds. Notice also that, in our study, the "frequency" dispersion, which has been observed in polyacetylene [6-9], and discussed in the frame of different models [10-12], is discussed in terms of π-electron delocalization. Finally, we present and discuss results obtained in RRS after doping some samples, and it is shown that the modifications, which are observed in the Raman spectra, are consistent with the existence of a quinoid structure along the polymeric chains.

RESULTS

Polyparaphenylene

In Fig. 1, we present Raman spectra of polyparaphenylene samples, undoped and doped with AsF_5, together with the deuterated analog, undoped and doped. All samples are obtained by a "Kovacic" synthesis [13], and therefore exhibit a strong fluorescence signal, when excited in the visible range. Therefore both $(C_6H_4)_x$ and $(C_6D_4)_x$, when undoped, have been investigated with $\lambda_L = 363.8$ nm. The main Raman bands are experimentally observed at 1220, 1280 and 1598 cm^{-1} for $(C_6H_4)_x$, and 894, 1262 and 1568 cm^{-1} for $(C_6D_4)_x$. Detailed experimental data on these polymers have been published elsewhere [14]. Notice that, upon doping with electron acceptors (here AsF_5), modifications in the Raman spectra are observed.

Lower-Dimensional Systems and Molecular Electronics
Edited by R.M. Metzger *et al.*, Plenum Press, New York, 1991

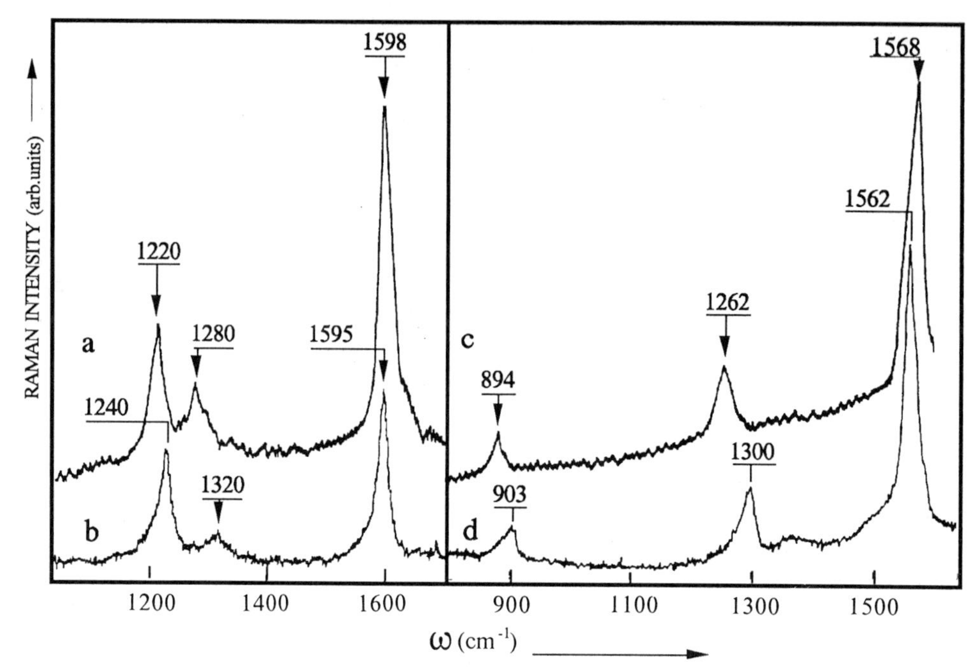

Fig.1. Raman spectra of polyparaphenylene at T = 300 K.
a) Undoped $(C_6H_4)_x$.λ_L = 363.8 nm. b) AsF_5-doped $(C_6H_4)_x$.λ_L = 514.5 nm.
c) Undoped $(C_6D_4)_x$.λ_L = 363.8 nm. d) AsF_5-doped $(C_6D_4)_x$.λ_L = 514.5 nm.

In $(C_6H_4)_x$, they consist in the appearance of two bands at 1240 and 1320 cm^{-1}, while the high-frequency peak remains unshifted. A similar situation is seen in $(C_6D_4)_x$ doped with AsF_5, since the Raman spectrum is composed of bands at 903, 1300 and 1562 cm^{-1}. Experimental spectra, reported in Fig. 1 (curves b and d), have been taken with λ_L = 514.5 nm, since, after doping, the fluorescence is completely annealed. Notice also that the new bands, which are observed, are strongly enhanced when the excitation laser line goes towards the red range, indicating a resonance behavior in this region.

In order to interpret these experimental data, we carried out calculations, in which we assumed a planar geometry for the polymer (neglecting the steric effects, which induce an angle between the phenyl rings). Then the point group is D_{2h}, and the in-plane and out-of-plane vibrations can be separated. If we only consider the in-plane vibrations, the k = 0 vibrational modes are classified as follows: $5A_g + 5B_{1g} + 4B_{2u} + 4B_{3u}$. In the valence force field calculations that we carried out, a total set of 26 force constants are needed (15 intra-ring and 11 inter-ring parameters), but three of them play, in this case, a more important role. They are those involving the carbon-carbon bonds, F_t2, $F_{t'}2$ and F_R2, which are defined in the following scheme:

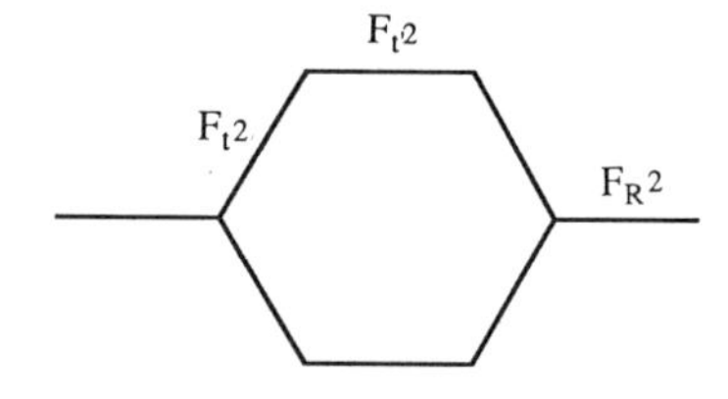

The calculated frequencies are obtained by performing a fit of the experimental Raman and infrared modes, by minimizing the differences by a least-squares method. In addition, in order to obtain a more accurate fit, we started the force constants determination by considering

Table 1. Experimental and calculated Raman and I. R. frequencies (cm^{-1}) in hydrogenated and deuterated PPP. We report only the frequencies smaller than 2000 cm^{-1}.

Symmetry	$(C_6H_4)_x$ Experimental	$(C_6H_4)_x$ Calculated	$(C_6D_4)_x$ Experimental	$(C_6D_4)_x$ Calculated
A_g	805	803	---	776
	1220	1184	894	863
	1280	1290	1262	1253
	1598	1601	1568	1572
B_{1g}	410	418	---	383
	618	616	---	593
	--	1343	---	1044
	--	1652	---	1620
B_{2u}	--	1118	---	855
	--	1343	---	1268
	1400	1412	1329	1347
B_{3u}	1000	1005	816	811
	--	1044	977	977
	1482	1490	1355	1364

benzene, as well as all oligomers from biphenyl to sexiphenyl. In Table 1, have collected the experimental and calculated frequencies, resulting from a fit considering both hydrogenated and deuterated biphenyl and PPP. More detailed results can be found in Ref. [14]. Note that, in this case, F_t2 is found equal to $F_{t'}2$ (Table 2), since we deal with a benzenoid structure. In order to understand the changes in the Raman spectra of AsF_5-doped PPP and AsF_5-doped

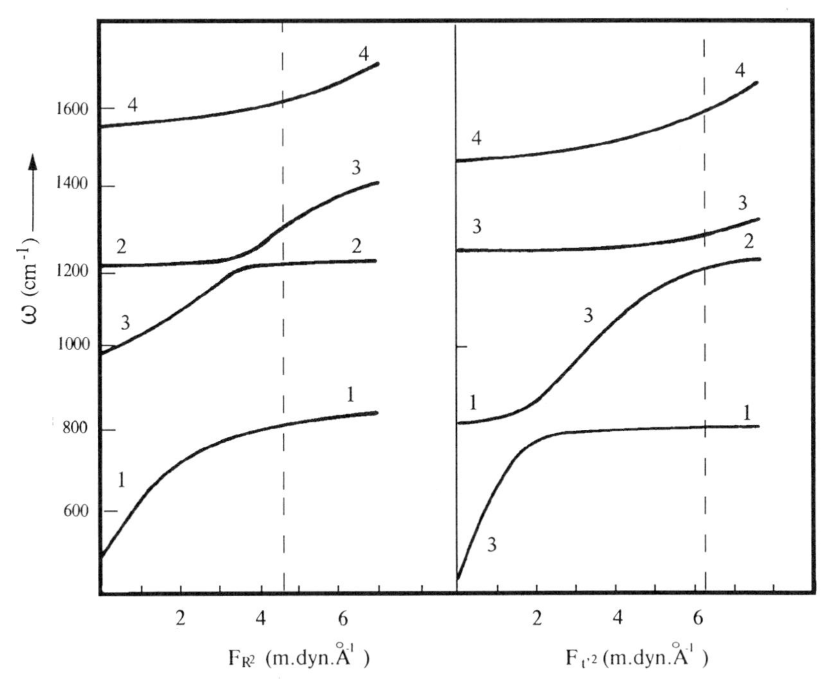

Fig. 2. Calculated frequencies as a function of F_R2 and $F_{t'2}$ force constants. The vertical dashed lines indicate the values used in our calculations for undoped polymer vibrational modes.

Table 2. Experimental and calculated values of the three main Raman mode frequencies (cm^{-1}) for undoped and doped $(C_6H_4)_x$ and $(C_6D_4)_x$. We indicate also the three main force constants (mdyn $Å^{-1}$), which are modified after doping the material.

	F_R2	$F_{t'}2$	F_t2		$(C_6H_4)_x$			$(C_6D_4)_x$		
Undoped	4.62	6.21	6.21	Exp.	1220	1280	1598	894	1262	1568
				Cal.	1184	1290	1601	863	1253	1572
Doped	5.05	7.53	4.38	Exp.	1240	1320	1595	903	1300	1562
				Cal.	1212	1320	1600	868	1308	1566

deuterated PPP (theory predicts the appearance of a quinoid character along the polymer chain upon doping [15]), we studied the values of the calculated frequencies, as a function of F_R2, F_t2 and $F_{t'}2$, respectively (for each variation, the other parameters are those fitting the undoped PPP modes). As an example, we show in Fig. 2 such variations in the case of PPP, for the two force constants F_R2 and $F_{t'}2$. These curves are of interest, especially in the case of F_R2, which represents the force constant between the benzene rings. It is shown, in particular, that the mode labelled 3 on Fig. 2a, observed at 1280 cm^{-1} in PPP (F_R2 = 4.62 mdyn $Å^{-1}$), is derived from the breathing A_g mode of benzene (F_R2 = 0), observed epxerimentally at 992 cm^{-1}. Then, we tried a new fit for the modified spectra of the doped samples (PPP and deuterated PPP), by allowing the three parameters F_R2, F_t2 and $F_{t'}2$ to move, while the others were kept constant.

In Table 2, we present the results of the calculations. We find that, to get a good fit of the modified Raman spectra, F_R2 and $F_{t'}2$ have increased, while F_t2 has decreased. This result is consistent with a quinoid structure on the doped chain, although the F_R2 value (5.05 mdyn $Å^{-1}$) is smaller than usually obtained for a double carbon-carbon bond.

Polyarylene-Vinylenes

A detailed experimental RRS study of PPV has been performed and presented in Ref. [16]. We recall the main features of the Raman spectra. For λ_L = 676.4 nm, we observe Raman peaks at 966, 1174, 1330, 1550, 1586 and 1628 cm^{-1}. Upon doping with $FeCl_3$, the modifications which occur, consist essentially of the disappearance of the triplet, which is now unresolved, with a prominent peak at 1567 cm^{-1} and shoulders at 1530 and 1590 cm^{-1}.

A theoretical treatment, similar to the one described for PPP, was applied to this polymer. It is somewhat an intermediate case between PPP and trans-polyacetylene, since the vinylene group consists in an alternation of single and double carbon-carbon bonds. Also, we used trans-stilbene as a model compound, in order to obtain a better fit for PPV. The point group in this case in C_{2h}, and if we restrict the calculations to in-plane vibrations, once the translation modes are withdrawn, the Raman and IR-active modes are classified as folows: 14 A_g + 12 B_u. Again, among the 56 force constants, which are needed to establish the dynamical model, in turns out that only F_D2, F_R2,F_t2 and $F_{t'}2$ are the most important (F_D2 is the force constant of the C=C bond of the vinylene group, and the others are identical to those described in PPP). In the case of undoped PPV, we have assumed a benzenoid structure for the benzene ring on the chain, i.e. we have taken $F_{t'}2 = F_t2$. Notice that this value is very close to the one determined in PPP (F_R2 = 4.8 mdyn $Å^{-1}$, $F_t2 = F_{t'}2$ = 6.32 mdyn $Å^{-1}$, F_D2 = 7.06 mdyn $Å^{-1}$). For the $FeCl_3$-doped PPV, a new fit for the Raman spectrum is obtained, if we let the four force constants move and keep the others unchanged. We find that F_R2 goes from 4.8 to 6.00 mdyn $Å^{-1}$, F_t2 from 6.32 to 5.74 mdyn $Å^{-1}$, $F_{t'}2$ from 6.32 to 6.50 and F_D2 from 7.06 to 4.48 mdyn $Å^{-1}$. Although the increase of $F_{t'}2$ is not very large, the behavior is in good agreement with a quinoid-like structure along the chain.

In the case of polythienylene-vinylene (PTV) and polyfurylene-vinylene (PFV), a complete experimental RRS study has been carried out, as a function of the excitation wavelength. Spectra have been presented, in Ref. [17], for both compounds, which show rather similar behavior. In particular, going from $\lambda_L = 676.4$ to $\lambda_L = 457.8$ nm, a small band shift is observed. This is an indication that a slight π-electron delocalization occurs on the chain, unlike for PPP or PPV. We have carried out preliminary calculations in a first step, to account for the in-plane vibrational modes. If we take the geometries published by Eckhart *et al.* [18], the polymers are planar, and the point group is D_{2h}. The normal group analysis gives: $9A_{1g} + 9B_{3g} + 9B_{1u} + 9B_{2u}$, with one B_{1u} and one B_{2u} translation modes. The different Raman modes seen in PTV at 932, 1045, 1212, 1287, 1354, 1410 and 1586 cm^{-1} have been the subject of a preliminary assignment, whose details can be found in Ref. [17].

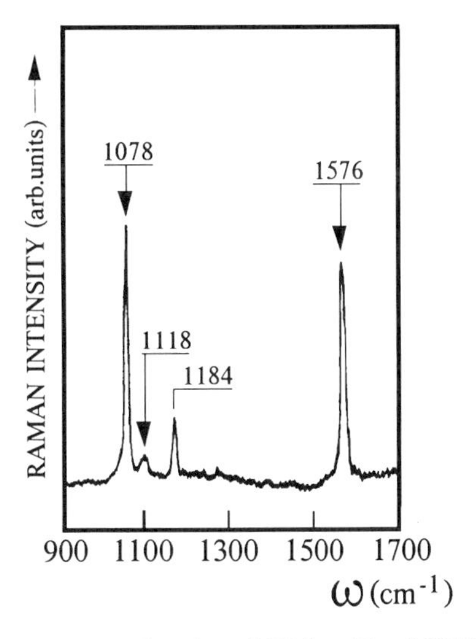

Fig. 3. Raman spectrum of undoped PPS at T = 300 K. $\lambda_L = 676.4$ nm.

Other Polymers

In a similar way, a preliminary investigation of other polymers containing phenyl rings has been carried out, and a Raman spectrum is shown in Fig. 3. We observe in PPS, for example, Raman bands at 1078, 1118, 1184 and 1576 cm^{-1}. Calculations are in progress, in order to interpret these modes, and correlate the vibrations to those observed in the polymers described above.

CONCLUSION

We have presented Resonant Raman Spectra of a series of conducting polymers with aromatic or heteroaromatic rings. First of all, by building a dynamical model, we are able to assign IR and Raman-active vibrational modes. In our procedure, the use of oligomers or model compounds allows a better determination of the force constants. We found that those corresponding to the phenyl rings are very close from one compound to the other. In general, a very good fit is obtained, except for the Raman mode at 1220 cm^{-1} in PPP, for which calculations give 1184 cm^{-1}. This is a special behavior when phenyl rings are linked together, since, experimentally, it is found at 1174 cm^{-1} in PPV, and 1184 cm^{-1} in PPS. Further studies are in progress to know if this discrepancy comes from the assumption of a planar geometry in this case.

The Raman spectra of doped PPP and doped PPV are well interpreted, if modifications in some force constants are made. They concern, more specially, the intra-ring and inter-ring carbon-carbon bonds. The changes are consistent with a quinoid-like structure along the chain.

Finally, a slight "frequency dispersion" is observed in heteroaromatic polymers, and is not in compounds with phenyl rings, indicating a stronger π-electron delocalization in the former case.

ACKNOWLEDGEMENTS

The help of J. C. Ricquier in the preparation of the manuscript is greatly appreciated. Samples of PPV, PFV and PTV were provided by Allied-Signal Inc., and some of the experimental data were taken in this company. Many thanks to Dr. H. Eckhardt for a very fruitful collaboration.

REFERENCES

[1] See, for example, the Proceedings of the Interantional Conference on Science and Technology of Synthetic Metals, Santa Fe, NM (ICSM'88), Synth. Met. 27, 28 and 29 (1989).
[2] R. L. Elsenbaumer, K. Y. Jen, G. C. Miller, H. Eckhardt, L. W. Shacklette, and R. Jow, Springer Series in Solid State Sciences 76: 400 (1987).
[3] D. D. C. Bradley, J. Phys. D.: Appl. Phys. 20: 1389 (1987).
[4] I. Murase, T. Ohnishi, T. Noguchi, and M. Hirooka, Polym. Comm. 25: 327 (1984).
[5] J. D. Capistrau, D. R. Gagnon, S. Antoun, R. W. Lenz, and F. E. Karasz, ACS Polym. Preprints 25: 282 (1984).
[6] D. Fitchen, Mol. Cryst. Liq. Cryst. 83: 95 (1982).
[7] S. Lefrant, J. Phys. (Paris) Coll. 44: C3-247 (1983).
[8] E. Mulazzi, G. P. Brivio, E. Faulques, and S. Lefrant, Sol. St. Commun. 46: 851 (1983).
[9] H. Kuzmany, E. A. Imhoff, D. B. Fitchen, and A. Sarhangi, Phys. Rev. B26: 7109 (1984).
[10] Z. Vardeny, E. Ehrenfreund, O. Brafman, and B. Horovitz, Phys. Rev. Lett. 51: 2326 (1983).
[11] G. P. Brivio and E. Mulazzi, Phys. Rev. B30: 676 (1984).
[12] S. Lefrant, E. Faulques, G. P. Brivio, and E. Mulazzi, Solid St. Commun. 53: 583 (1985).
[13] O. Kovacic and A. Kyriakis, J. Am. Chem. Soc. 85: 454 (1983).
[14] J. P. Buisson, S. Krichene, and S. Lefrant, Synth. Met. 21: 229 (1987).
[15] J. L. Brédas, J. Chem. Phys. 22: 3808 (1985).
[16] S. Lefrant, E. Perrin, J. P. Buisson, H. Eckhardt, and C. C. Han, Synth. Met. 29: E91 (1989).
[17] J. Y. Mevellec, J. P. Buisson, S. Lefrant, H. Eckhardt, and K. Y. Jen, Synth. Met. (in press, 1989).

STRUCTURAL INVESTIGATION AND ELECTRICAL PROPERTIES OF POLY(P-PHENYLENE) PREPARED IN TWO DIFFERENT WAYS

Hilde Krikor, Robert Mertens, Piet Nagels, Robert Callaerts
and Germain Remaut

R.U.C.A., University of Antwerp
B-2020 Antwerpen, Belgium

ABSTRACT

Poly(p-phenylene) was prepared by the oxidative cationic polymerization of benzene, and by a recently discovered, route based on a thermal aromatization of high molecular poly(cyclohexa-1,3-dienyl-5,6-diacetate), resulting in the formation of poly(p-phenylene) and acetic acid. An investigation using infrared spectroscopy, thermogravimetric analysis and gel permeation liquid chromatography revealed, however, that the thermal treatment of the diacetate polymer resulted in a scission of the polymer chains. The electrical conductivity of both types of polymers was measured after doping with $FeCl_3$.

1. INTRODUCTION

The synthetic route most commonly applied to produce poly(p-phenylene) is based on a polycondensation reaction. The Kovacic method[1] consists of a direct coupling of benzene molecules using a Lewis acid catalyst system, and is similar to a Friedel-Crafts reaction. This process is known as an oxidative cationic polymerization. Poly(p-phenylene), prepared in this way, is an insoluble, low molecular weight product, with a degree of polymerization that does not exceed 15 phenylene units. Therefore, the product is more correctly defined as an oligomer rather than a polymer. Moreover, crosslinking yielding a branched polymer cannot be completely avoided. Another route to arrive at an all-para linked polymer is the dehalogenation of para-dihalobenzenes, according to an Ullmann reaction which requires metals such as copper. Again, short chain products are obtained (10-12 repeat units).

Recently, Ballard and co-workers[2] discovered a novel path to prepare linear polyphenylene, based on the polymerization of derivatives of 5,6-dihydroxycyclohexa-1,3-diene. The advantage of this method is that the diene polymer remains soluble in a variety of solvents, because of the presence of solubilizing groups, such as ester functions. Because of this, molecular weight determinations can be made before the final pyrolysis to remove the attached groups. Ballard et al. reported that the final product is a high molecular weight poly(p-phenylene).

It is known that the electrical conductivity of a doped polymer is markedly dependent on its chain length. In the present paper, we report on a structural investigation of poly(p-phenylene) prepared according to the Kovacic and Ballard methods. The polymers were doped with $FeCl_3$ and their dc electrical conductivity was measured.

2. EXPERIMENTAL PROCEDURES

The first synthesis of poly(p-phenylene) used in this study was the direct polymerization of benzene, using a mixture of aluminium chloride and cupric chloride (2/1 molar ratio), as described in detail by Kovacic et al.[1]. The reaction temperature was maintained at 35°C for two hours. The product, under the form of a powder, was purified by repeated washings in hot hydrochloric acid. Traces of the catalyst system are difficult to remove. Before use, the powder was annealed in argon atmosphere at 400°C for 24 hrs. This thermal treatment eliminates lower molecular weight fractions and chlorine and, moreover, increases the degree of crystallinity, as was evidenced by X-ray powder diffraction measurements. A weight loss of approximately 2 % was found. Elemental analysis yielded: C : 89.95; H : 4.7; Al : 0.006; Cu : 0.114 and Cl : 0.236 wt %. The C/H atomic ratio was 1.60, which gives a higher carbon content than the limiting theoretical value equal to 1.5.

The second synthesis used to prepare poly(p-phenylene) was the method recently introduced by Ballard et al.[2]. The starting product is 5,6-dihydroxycylohexa-1,3-diene (DHCD), which is commercially available. The preparation proceeds in three steps: 1) alkylation or acylation of the two alcohol groups; 2) polymerization of these derivatives and 3) elimination of the groups inplanted on the cyclohexene units, accompanied by an aromatization of the polymer. The reaction scheme is as follows:

OH, OH ========> (1) $OCOCH_3$, $OCOCH_3$ polymerization ========> (2)

[$OCOCH_3$, $OCOCH_3$]$_n$ pyrolysis =======> (3) [ ]$_n$ + 2 CH_3COOH

To synthesize a diacetate DHCD-DA according to step (1), the diol was dissolved in pyridine, cooled at - 10°C, and mixed with CH_3COCl. Hereby, we followed the procedure described in detail by Ballard et al. The reaction did not yield the expected ester, as was proved by a structural analysis using ^{13}C-NMR. We found that the use of acetyl chloride resulted in the formation of phenyl acetate. In Ballard's description of the preparation procedure, the acylation of DHCD was accomplished with the help of acetic anhydride. We also found that the use of this reagent resulted in the formation of cyclohexa-1,3-dienyl-5,6-diacetate.

The monomer was polymerized at 70°C, using benzoyl peroxide as initiator. The polymers of the DHCD derivative can be aromatized by a thermal elimination of acetic acid. Aromatization in solution was tried, by refluxing the precursor polymers under nitrogen atmosphere in N-methylpyrrolidone at 175°C and in 2,4,6-collidin at 172°C for 24 hours. Infrared spectra of the products, recorded after heating, showed strong absorption bands typical for ester groups and, hence, it was concluded that this procedure was unsuccessful. Aromatization of poly(DHCD-DA) in the solid state was followed by thermogravimetric analysis. The weight loss was measured as function of temperature at a fixed heating rate of 5°C/min under nitrogen atmosphere. The TGA thermogram yielded a weight loss of 70 % on heating to 320°C. This value corresponds well with the theoretically expected weight loss, due to the elimination of two acetic acid molecules per monomeric unit.

$FeCl_3$ doping of the Kovacic and Ballard polymers was accomplished using the method described by Pron et al.[3]. The iron content was determined by atom emission spectroscopy-inductive coupled plasma (AES-ICP).

3. STRUCTURAL ANALYSIS

3.1. Poly(p-phenylene) synthesized according to Kovacic

An analysis, including differential scanning calorimetry (DSC), X-ray diffraction and infrared spectroscopy, yielded similar results to those already reported in the literature by other authors.

Differential scanning calorimetry demonstrated that the polymer is thermally stable up to 450°C. The X-ray diffraction pattern of the annealed powder exhibited rather broad peaks, indicating that the material is only partially crystalline. Three X-ray reflections, corresponding to d-spacings of 4.53, 4.00 and 3.20 Å, were observed. The most intense one at 4.53 Å represents the distance between phenyl groups in the direction of the polymer chain.

The infrared spectrum of the undoped polymer, recorded between 1600 and 400 cm^{-1} is shown in Figure 1a. The strong absorption band occurring at 810 cm^{-1}, can be attributed to the out-of-plane carbon-hydrogen vibrations for adjacent hydrogen atoms on para-disubstituted rings and, hence, gives good evidence for the para configuration. The position of this para band shifts to longer wavelength as the number of phenyl rings linked together increases[1]: p-terphenyl, 837 cm^{-1} : p-quinquephenyl, 818 cm^{-1}. The less intense absorption bands at 765 and 695 cm^{-1} are associated with the out-of-plane vibrations of the hydrogen atoms on the terminal phenyl groups. The two other IR lines at 1485 and 1403 cm^{-1} can be assigned to the asymmetric and symmetric C=C stretching vibrations.

Figure 1b shows the IR spectrum of $FeCl_3$-doped poly(p-phenylene) in the 1600-400 cm^{-1} region. The spectrum is characterized by a large number of absorptions, and is in good agreement with the IR spectra observed by Iqbal et al.[4] on AsF_5 and PF_5-doped poly(p-phenylene). The most important features are the appearance of two broad and strong absorptions at 1560 and 1190 cm^{-1}, and the disappearance of the 765 and 695 cm^{-1} bands. In addition, new absorptions are observed at 1360, 1275 and 1110 cm^{-1}. Iqbal et al. have interpreted the IR absorptions to be due to defect absorption. They concluded that, at the defect site itself, localized charge oscillation would make the defect modes strongly IR active. The nature of the defects in poly(p-phenylene) is still a matter of speculation. Contrary to trans-polyacetylene, this polymer is not expected to form solitons. Via a theoretical study Brédas et al.[5] have shown that polaron defect formation occurs at low doping levels, whereas at higher doping levels the defects interact to give a spinless quinoid bipolaron.

3.2. Poly(p-polyphenylene) synthesized according to Ballard

^{13}C-NMR spectra were recorded on solutions of DHCD, DHCD-DA and poly(DHCD-DA) in $CDCl_3$ (50/50 %). The results are summarized in Table 1.

Table 1. ^{13}C-NMR resonances in DHCD(1), DHCD-DA(2) and poly(DHCD-DA)(3) (in ppm relative to TMS)

	(1)	(2)	(3)
$-CH_3$		20.7	21
C•			40
C*	67.6	67	69.1
C°	124.6; 129.4	125.7; 126.3	128
- O-CO		170	170.4

The most important feature is the appearance of a resonance peak at 40 ppm after polymerization, characteristic for the presence of sp^3 hybridized C atoms between two cyclohexene units. The presence of ester groups is revealed by the NMR peaks at 67 and 170 ppm in the spectrum of the DHCD-DA monomer.

Since the DHCD-DA polymer is soluble in organic solvents, the molecular weight of the material was determined by gel permeation liquid chromatography (GPLC) in tetrahydrofuran solution using a refractive index and a U.V.-254 nm detector. Styragel columns (100, 500, 10^3, 10^4, 10^5 and 10^6 Å) were calibrated with standard polystyrenes. The mean value of the numeric molecular weight was equal to 24,757, corresponding to a polymerization degree n ≅ 126. The polydispersion degree amounted to 2.8.

Figure 2 represents the IR spectra of poly(DHCD-DA) and its aromatized derivative after thermal treatment at 320°C for 2 hours. The three broad and strong absorption bands at 1750, 1250 and 1050 cm^{-1}, due to C = O stretching, O-CO stretching and $-O-CH_3$ stretching vibrations observed in the diacetate polymer (Fig. 2a), completely disappeared after heating (Fig. 2b). New absorptions appeared at 810, 767 and 700 cm^{-1}, which, as mentioned earlier, are characteristic of poly(p-phenylene). The two bands at 767 and 700 cm^{-1}, assigned to monosubstituted benzene, are recorded with a higher intensity, compared to the ones observed in the IR spectrum of the Kovacic polymer (Fig. 1a). This finding casts some doubts on the preservation of the original chain of the precursor polymer during the thermal aromatization process.

In order to avoid the chain scission during aromatization, the poly(DHCD-DA) polymer was subjected to a milder heating procedure. The powder was heated at a fixed heating rate of 5°C/min to 320°C, and immediately cooled down to room temperature. From TGA analysis a weight loss of 23.3 % was calculated, corresponding to an aromatization degree of 37.9 %. This incomplete aromatization also followed from the IR spectrum. The 1750, 1250 and 1050 cm^{-1} bands were still present as medium intensity bands. The 810, 767 and 700 cm^{-1} bands, appearing in the spectrum, are indicative for the presence of phenylene units. Since the product was still almost completely soluble in tetrahydrofuran, the GPLC technique was again applied to determine the molecular weight distribution. The mean numeric molecular weight decreased to a value of 1,166, with a broader polydispersion degree of 6.14. Guinier-Hägg X-ray diffraction technique yielded d-spacings of 4.49, 3.93 and 3.18 Å.

The main conclusion is that chain scission seems to occur during the aromatization process at 320°C in the solid state. Indeed, the molecular

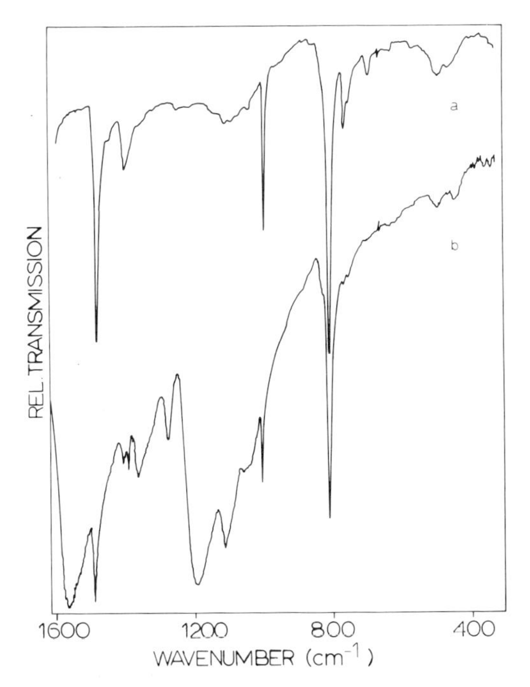

Fig. 1. Infrared spectra of poly(p-phenylene) synthesized according to Kovacic et al.[1]
(a) undoped; (b) $FeCl_3$ doped (9 wt % Fe)

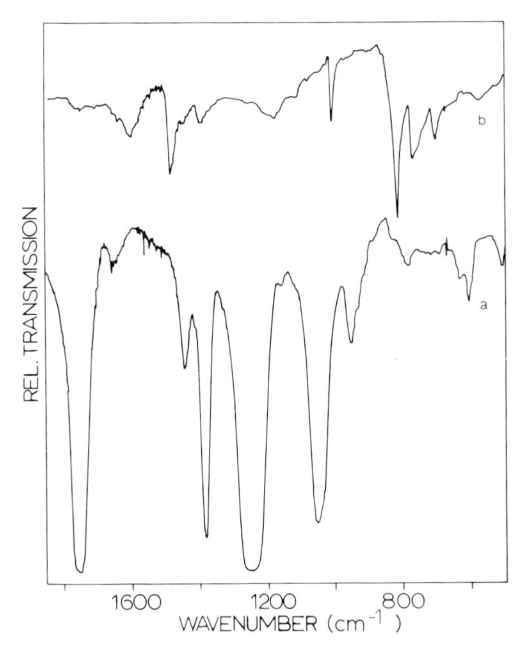

Fig. 2. Infrared spectra of poly(p-phenylene) synthesized according to Ballard et al.[2]
(a) poly(cyclohexa-1,3-dienyl-5,6-diacetate) (DHCD-DA) ;
(b) poly(p-phenylene)

weight is reduced significantly. This finding is not directly in contradiction with the work of Ballard et al. who reported the preparation of high molecular weight poly(p-phenylene). These authors mainly studied the polymerization and aromatization of another DHCD derivative, having dimethoxycarbonyl, instead of diacetate, as function. Ballard et al. reported that the polymer of this dimethylcarbonate derivative, DHCD-DMC, can be aromatized at 148°C in a N-methylpyrrolidone solution, yielding high molecular weight products. The eliminated acid in this case is methyl carbonic acid, which is an easier leaving group. The elimination of the ester functions in the diacetate requires higher temperatures.

4. ELECTRICAL PROPERTIES

4.1. Poly(p-phenylene) synthesized according to Kovacic

Undoped poly(p-phenylene) is a good insulator, with a room temperature conductivity σ_{RT} of the order of 10^{-14} $ohm^{-1} cm^{-1}$. Figure 3 represents σ_{RT} as a function of the Fe content. The conductivity rapidly increases by many orders of magnitude in the range $0 < Fe\ wt\ \% < 1$, followed by a slight increase to a saturation value of 3.9 $ohm^{-1} cm^{-1}$ for 9.2 wt % Fe.

4.2. Poly(p-phenylene) synthesized according to Ballard

The dc measurements performed on films prepared from a solution of 50 mg poly(DHCD-DA) and 1 ml CH_2Cl_2, thermally treated at 320°C for 2 hours under argon atmosphere, yielded low conductivity values. Further investigation is needed on films of better quality.

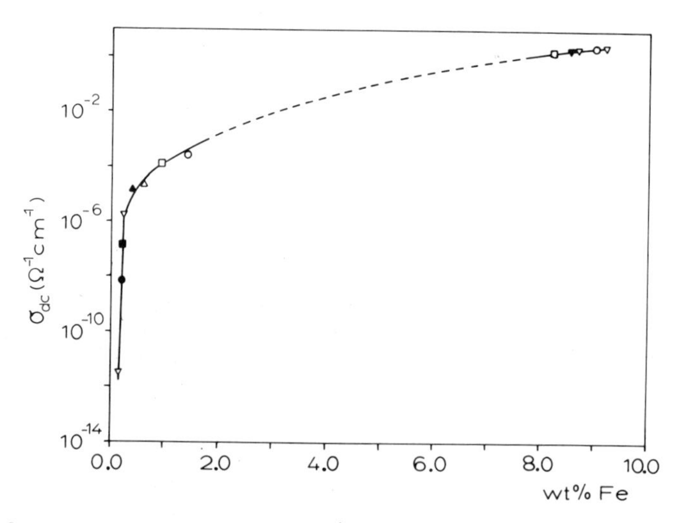

Fig. 3. Room temperature electrical conductivity of $FeCl_3$ doped poly(p-phenylene) as a function of the wt % Fe

REFERENCES

1. P. Kovacic and J. Oziomek, J. Org. Chem. 29:100 (1964).
2. D. Ballard, A. Courtis, I. Shirley and S. Taylor, Macromolecules 21:294 (1988).
3. A. Pron, D. Billaud, I. Kulszewicz, C. Budrowski, J. Przyluski and J. Survalski, Mat. Res. Bull. 16:1229 (1981).
4. Z. Iqbal, H. Bill and R. Baughman, J. de Physique C23:761 (1983).
5. J. Brédas, R. Chance and R. Sibley, Phys. Rev. B 26:5843 (1982).

RAMAN CROSS-SECTIONS OF HIGHLY ORIENTED CIS-POLYACETYLENE

G. Lanzani *#, A. Piaggi #, A. Borghesi #, and G. Dellepiane *

Dipartimento di Fisica "Alessandro Volta"
Università di Pavia
Pavia, ITALY

* Istituto di Chimica Industriale
Università di Genova
Genova, ITALY

INTRODUCTION

It is well known that resonance Raman scattering, obtained by exciting wavelengths close to the lowest allowed optical transition of semiconducting polymers, provides a great amount of information concerning structural, optical and transport properties. In particular, the investigation on highly oriented polyacetylene (HOPA) [1-3] has expanded considerably in recent years. Its simple one-dimensional morphology provides direct information on the electronic properties of the polymeric chain.

All the Raman data previously reported did not take into account the true values of the optical function anisotropy of these materials. This is due to intrinsic difficulties in measuring the fundamental optical constants of these samples. In fact, it is expected that large corrections, due to reflectivity and absorption, should be applied to the observed Raman intensities. Simple models have been proposed [1,3] to roughly estimate these effects, using approximated values of the optical functions, deduced from theoretical models [2] or from measurements on similar samples [1,3].

Our report is the first to present intensity corrections of the Raman spectra on *cis* -rich HOPA [1], deduced from the optical functions measured on the same sample.

EXPERIMENTAL PROCEDURE AND RESULTS

The polarized reflectivity, in the energy range between 0.06 and 0.6 eV, is shown in Fig.1. The measurements on highly oriented (draw ratio 7) *cis* -polyacetylene, (polymerized with a procedure described elsewhere [5]), were performed by a Bruker 113v Fourier transform spectrometer, equipped with a KRS-5 polarizer.

A Perkin-Elmer 330 automatic spectrophotometer was used in the energy range between 0.5 and 4.2 eV, with dichroic sheet polarizers in the sample and reference beams. We obtained the complex refractive index $\tilde{n}(\omega) = n(\omega) + i\,k(\omega)$, shown in Fig. 2, by Kramers-Kronig (KK) analysis on the experimental reflectivity spectra for both light polarizations. The integral, required in the KK analysis, calls for the reflectivity over an infinite range of energies, so it was necessary to introduce an extrapolation procedure in the high-energy region. In our analysis, a tail [7] $R(\omega)=R(\omega_2)\cdot(\omega_2/\omega)^3$ was used beyond the last experimental energy ω_2. Such extrapolation is expected to give quite good results, if the parameters can be adjusted to reproduce the value of some optical functions, independently known at one or more selected energies.

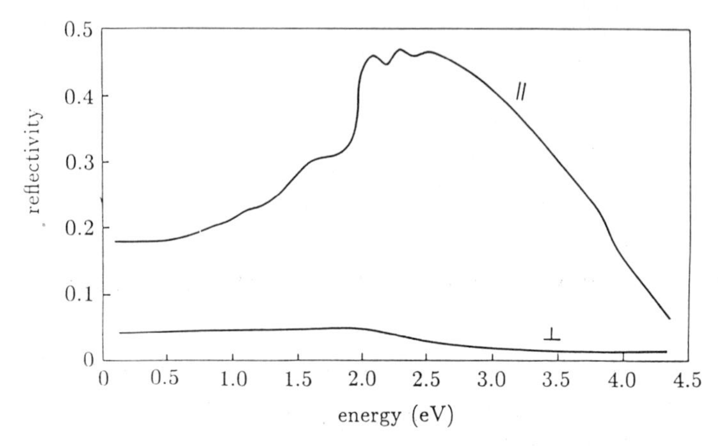

Fig. 1. Optical reflectivity measured with light parallel and perpendicular to the stretching direction.

Fig 3 shows the polarized resonance Raman spectra on the same sample type. The scattering geometry is reported in Fig. 4. The four scattering configurations are labeled with a standard notation A(BC)D, where A represents the direction of the incident beam, B the polarization of incident beam, C the polarization of scattered beam and D the direction of scattered beam. The exciting wavelength was 614.9 nm.

The X(ZZ)Y configuration exhibits three fundamental, resonantly enhanced, Raman bands at 909, 1250 and 1520 cm^{-1}, typical of the *cis*- polyacetylene. Moreover, the two Raman bands of *trans* - polyacetylene, at about 1080 cm^{-1} and 1450 cm^{-1}, are visible in the spectra.

Ref. 1 gives a detailed analysis of the spectra. Here we discuss the effect of the corrections on the Raman intensities, due to the high anisotropy of this material. In fact, true

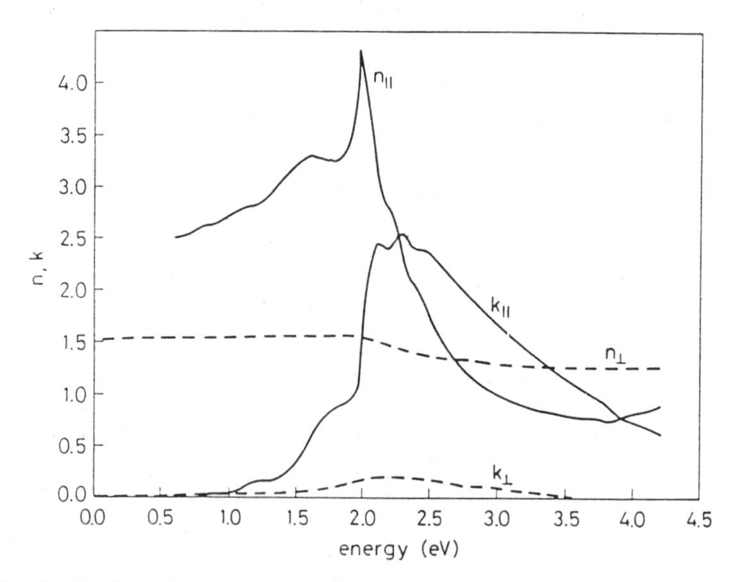

Fig. 2. Real and imaginary part of the refractive index, as obtained by KK analysis on experimental spectra for both light polarizations.

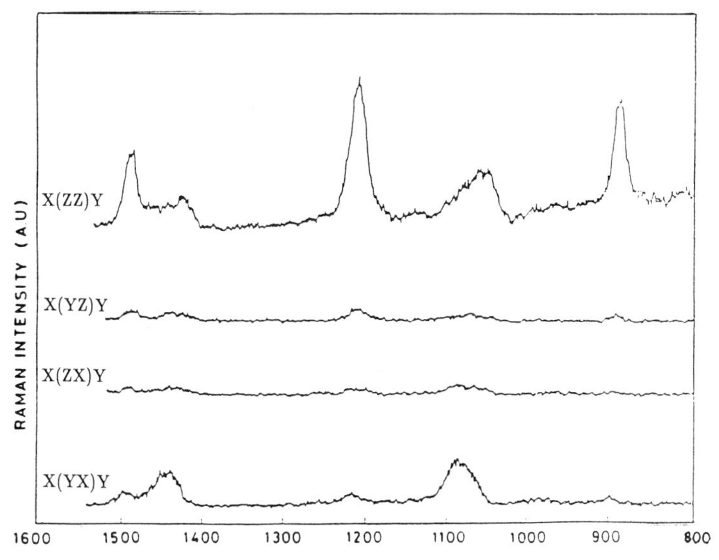

Fig. 3. Resonance Raman scattering of HOPA film in the four scattering configurations. Exciting laser: $\lambda = 614.9$ nm. T ~ 80 K.

Raman cross-sections can be obtained, from the measured intensities in the various scattering configurations, only if optical reflectivity, absorption coefficient, and their anisotropies are known.

The reflectance measurements, performed at near-normal incidence, cannot be used directly to correct the Raman data, because these were obtained with an incident angle of 70°. As a consequence, we considered the angular dependence of the reflectivity, using the Fresnel formulas in the case of a complex refractive index [6]. The measured value of reflectivity, for light polarization parallel to the stretching direction, was 0.43 at the wavelength used in the Raman experiment. However, this value should be raised to 0.75 for an incidence angle of 70°. The value of the reflectivity, for light polarization perpendicular to the stretching direction, changes from 0.05 to 0.04, when varying the incidence angle form 0 to 70°.

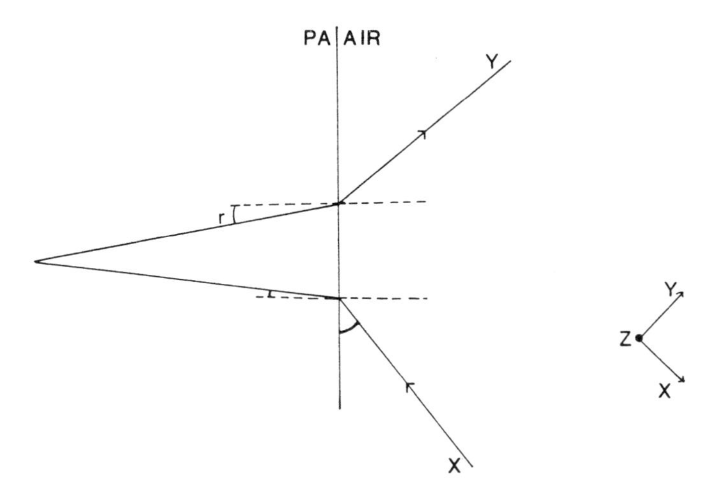

Fig. 4. Raman scattering geometry, viewed along the chain axis of the stretched polymer (Z axis). X is the direction of the incident laser beam, and Y the direction of collection optics.

Table 1. Optical constants of the four different configurations.

	INCIDENT			SCATTERED		
	n	k	R(70°)	n	k	R(0°)
ZZ	4.15	1.55	0.476	4.15	1.55	0.43
ZX	4.15	1.55	0.476	1.55	0.17	0.05
YZ	1.55	0.17	0.042	4.15	1.55	0.43
YX	1.55	0.17	0.042	1.55	0.17	0.05

It has been shown [1] that, if we only consider the light scattered in the XY plane with sample thickness >> $1/\alpha_i$, the true Raman intensity $I(\lambda_s)$ can be calculated with the observed intensity $S(\lambda_s$ by the relation:

$$I(\lambda_s) = K \cdot S(\lambda_s)$$

with

$$K = I_o (1 - R_i)(1 - R_s)\{\alpha_i + \alpha_s [(1 - \cos^2\beta_i/n_i^2)/(1 - \sin^2\beta_i/n_s^2)]^{1/2}\}^{-1}$$

I_o is the incident laser intensity; R and α are sample reflectivity and absorption coefficient, respectively; n is the refractive index, and β_i is the angle between the incident beam and the sample surface; i and s refer to the incident and scattered beam, respectively. We used the proper values of n_i and n_s obtained by KK analysis; the values of α_i and α_s were deduced from the values of $k_{\|}$ and $k_{\perp}$, applying the well-known relation $\alpha = 4\pi k/\lambda$. The values of R_i and R_s were inserted as obtained above. All the previous optical parameters were evaluated at the wavelength of the exciting laser.

The correction factors K for each configuration are:

$$K[X(ZZ)Y] = I_o(1 - R_{\|}(70°))(1 - R_{\|}(0°))\{\alpha_{\|} + \alpha_{\|}[(1 - \cos^2\beta_i/n_{\|}^2)/(1 - \sin^2\beta_i/n_{\|}^2)]^{1/2}\}^{-1}$$

$$K[X(ZX)Y] = I_o(1 - R_{\|}(70°))(1 - R_{\perp}(0°))\{\alpha_{\|} + \alpha_{\perp}[(1 - \cos^2\beta_i/n_{\|}^2)/(1 - \sin^2\beta_i/n_{\perp}^2)]^{1/2}\}^{-1}$$

$$K[X(YZ)Y] = I_o(1 - R_{\perp}(70°))(1 - R_{\|}(0°))\{\alpha_{\perp} + \alpha_{\|}[(1 - \cos^2\beta_i/n_{\perp}^2)/(1 - \sin^2\beta_i/n_{\|}^2)]^{1/2}\}^{-1}$$

$$K[X(YX)Y] = I_o(1 - R_{\perp}(70°))(1 - R_{\perp}(0°))\{\alpha_{\perp} + \alpha_{\perp}[(1 - \cos^2\beta_i/n_{\perp}^2)/(1 - \sin^2\beta_i/n_{\perp}^2)]^{1/2}\}^{-1}$$

where $\|$ ($\perp$) refers to light polarization parallel (perpendicular) to the stretching direction.

The calculated values, using the parameters listed in Table 1, and normalized to the X(ZZ)Y configuration, are:

$$\frac{K[X(YX)Y]}{K\{X(ZZ)Y]} = 0.014$$

$$\frac{K[X(ZX)Y]}{K\{X(ZZ)Y]} = 0.33$$

$$\frac{K[X(YZ)Y]}{K\{X(ZZ)Y]} = 0.22$$

We obtained the true *cis* -polyacetylene Raman cross-sections, reported in Table 2, by multiplying the integrated observed Raman intensities by these correction factors. Moreover, corrected Raman intensity ratios for the X(YX)Y configuration have been derived for the C–C mode of *trans*- polyacetylene, which peaks around 1100 cm^{-1}. Given the dispersion of the

Table 2. Corrected Raman intensities, normalized to X(ZZ)Y. The observed values are reported in brackets.

(cm^{-1})	X(ZZ)Y	X(ZX)Y	X(YZ)Y	X(YX)Y
909	1		0.01(0.07)	~0.001(0.06)
1250	1	0.01(0.03)	0.01(0.06)	~0.001(0.06)
1540	1	0.01(0.04)	0.02(0.12)	<0.001(0.015)

Table 3. Corrected intensity ratios for the trans band.

(cm^{-1})	X(YX)Y/X(ZZ)Y
1052	0.001
1089	0.01
1100	0.02

Raman bands in *trans*- polyacetylene [2,3], Table 3 reports the corrected Raman intensities for three different values of the Raman shift, namely 1052, 1089, and 1100 cm^{-1}.

CONCLUSIONS

From Table 2 we can conclude that the main effect of the correction factors is the reduction of the intensity in X(YX)Y configuration. The angular dependence of the reflectivity accounts for the intensity difference observed between configurations X(ZX)Y and X(YZ)Y; in fact, this difference disappears in the corrected values.

The extinction of the *cis*- polyacetylene peak intensities, observed upon changing the scattering configuration, indicates almost perfect uniaxial alignment of the *cis* chain. The Raman band corrected intensity for *trans*- polyacetylene confirms a different behavior, observed with changes in scattering configurations. The depolarization ratio increases drastically, as the frequency increases towards the high-energy side of the 1100 cm^{-1} Raman band. This could be considered evidence that the short *trans* -segments in *cis* -polyacetylene sample are strongly misaligned.

REFERENCES

[1] G. Lanzani, S. Luzzati, R. Tubino and G. Dellepiane, "Polarized resonance Raman scattering of cis-polyacetylene", J. Chem. Phys. (1989).

[2] G. Masetti, E. Campani, G. Gorini, R. Tubino, P. Piaggio and G. Dellepiane, "Polarized Raman scattering of highly oriented trans-polyacetylene", Chem. Phys. 108: 141 (1986).

[3] E. Falques, E. Rzepka, S. Lefrant, E. Mulazzi, G. Brivio, and G. Leising, "Polarized resonant Raman spectra of fully oriented trans-polyacetylene: Experiments and theory", Phys. Rev. B 33: 8622 (1986).

[4] D. Comoretto, A. Borghesi, G. Dellepiane, G. Lanzani, G. F. Musso, A. Piaggi and R. Tubino, "Optical properties of highly oriented fibrous polyacetylene", to be published.

[5] G. Lugli, U. Pedretti and G. Perego, "Highly Oriented Polyacetylene", J. Polymer Sci. Polym. Lett. Ed. 23:129 (1985).

[6] M. Born and E. Wolf, "Principles of Optics," Pergamon Press, Oxford (1965).

[7] F. Stern, "Solid State Physics," Academic Press, New York (1968).

CHARACTERISATION OF POLY(P-PHENYLENE VINYLENE) [PPV] PREPARED BY DIFFERENT PRECURSOR ROUTES

J. Martens, N. F. Colaneri, P. Burn, D. D. C. Bradley,
E.A. Marseglia and R.H. Friend

Cavendish Laboratory, Cambridge CB3 OHE, U.K.

INTRODUCTION

Studies on the physical properties of conjugated polymers have been hampered in the past by difficulties in purification and sample fabrication, arising from the intractibility associated with the presence of a rigid chain backbone. The use of a precursor-route synthesis, in which the processing is performed for a non-conjugated precursor polymer (that can subsequently be converted to the conjugated material through a thermal treatment) has greatly improved the possibilities for the preparation of samples in a variety of forms, including the homogeneous thin films required for optical and electro-optical processing applications.

Poly(p-phenylene vinylene) [PPV] is the simplest member of a family of poly(arylene vinylene) polymers that can be prepared via such a precursor route[1]. The initially described literature synthesis of this material involves preparation of a sulphonium polyelectrolyte precursor polymer, which can be cast from either water or methanol solution, to form films of varying thickness[2]. The conversion of this material to the conjugated PPV is achieved by heating in-vacuo, and requires the elimination of one dialkyl sulphide and one hydrogen halide group per repeat unit. It is experimentally observed that the conversion does not proceed smoothly, and that the formation of an alkyl mercaptan side group, through elimination of a methyl halide moiety, competes with the desired straightforward elimination[3,4]. As a result of this, the temperature required to achieve full conversion (as judged by complete loss of the C-S stretching mode from the infrared spectrum) is raised to 300°C. Moreover, the presence of such mercaptan units appears to frustrate crystallinity at lower conversion temperatures, and is likely to have other consequences for the state of order of the polymer chains.

More recently, an alternative precursor has been proposed, in which the dialkyl sulphonium moiety is replaced by a tetrahydrothiophenium group[5]. The elimination of this leaving group proceeds without noticeable side reactions, and can be effected at significantly lower temperatures, full conversion being achieved at 200°C. In this paper we report spectroscopic and diffraction measurements, undertaken to study the effects of this simplified conversion reaction upon the resultant conjugated polymer films.

Lower-Dimensional Systems and Molecular Electronics
Edited by R.M. Metzger *et al.*, Plenum Press, New York, 1991

SYNTHESIS

The tetrahydrothiophenium precursor polymer was prepared by the following process:

I

II

1) 0.4 M ^{-}OH

2) 0.4 M H^{+}

3) Dialysis H_2O

0.4 M in methanol

A methanolic solution of the precursor monomer (I) was reacted with aqueous hydroxide for one hour at 0°C, followed by neutralization with aqueous acid. Dialysis of the resultant reaction mixture against water, to remove inorganic salts, followed by removal of the solvent and dissolution of the residue in methanol, afforded a viscous solution of the precursor polymer (II). The unoriented fully conjugated polymer, denoted in this paper as PPV[B], was then formed by thermal treatment at 300°C for one hour under high vacuum. The dialkyl sulphonium precursor polymer was prepared following the standard synthetic route,[2] and then converted to PPV[A] at 300°C in a similar manner as PPV[B].

SPECTROSCOPY OF UNORIENTED FILMS

Significant differences can be seen in the electronic and vibrational absorption bands of PPV prepared via the two different precursor routes, as has been reported in Ref. 5. Figure 1 shows the UV-visible absorption spectra of thin films of PPV[A], synthesized by the dialkyl sulphonium salt precursor route (broken line), and of PPV[B] from the tetrahydrothiophenium salt precursor route (solid line). These bands are mainly due to π to π^* transitions of π electron states delocalised along the polymer chain, and of π states localised within a single phenylene ring.[1] The peak position of the lowest-energy absorption can lie in the range 2.97 to 3.50 eV, depending on casting and conversion conditions, and, as we see here, on the method of synthesis. This is because the shape and peak position of this absorption band is determined by the distribution of conjugation lengths in the sample. Thus, the more complete conversion of PPV[B] to its fully conjugated form is evident from the ~0.1 eV reduction of the interband transition energy to about 2.4 eV. Figure 2 shows the spectrally resolved luminescence of PPV[A] (broken line) and PPV[B] (solid line) following excitation with an argon ion laser (lasing all lines in the ultraviolet). Consistent with the data of Figure 1, the emission of PPV[B] appears nearly a tenth of an eV lower in energy than that of PPV[A]. Further details of this luminescence behaviour can be found in Ref. 6.

Differences also appear in the infrared absorption spectra of samples prepared by the two precursor routes. Figure 3 shows the infrared spectra of PPV[A] (lower curve) and PPV[B] (upper curve) following thermal elimination at 300°C for one hour.

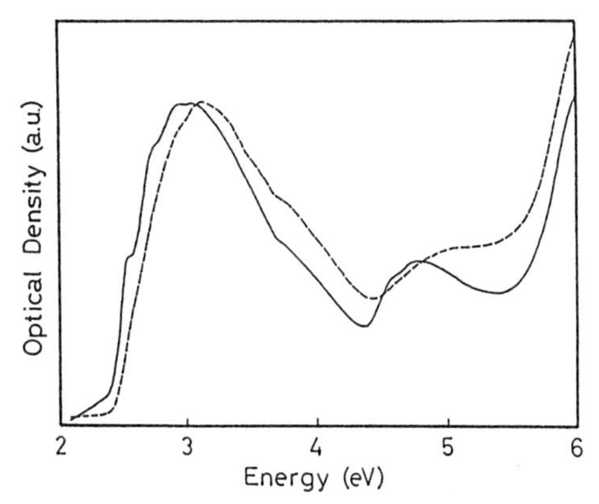

Fig. 1 Ultraviolet and visible absorption spectra of PPV prepared by two different precursor routes by heating to 300°C for one hour. The absorption edge of the material prepared from a tetrahydrophenium salt precursor (solid line) appears nearly 0.1 eV lower than that prepared from a sulphonium salt precursor (broken line), and shows more pronounced phonon structure.

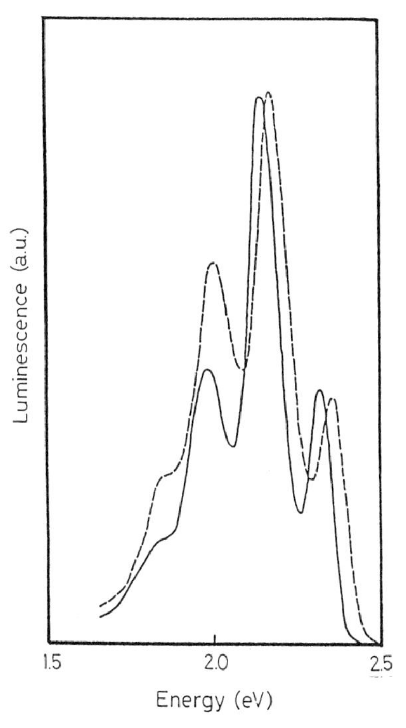

Fig. 2 Photoluminescence spectra at 80 K of the two samples whose absorption is displayed in Figure 1. The emission results from irradiation of the samples with the ultraviolet lines of an argon ion laser (363 nm, 351 nm, and 334 nm). The solid curve is the emission of a sample of PPV[B], and the broken line is that of a sample of PPV[A].

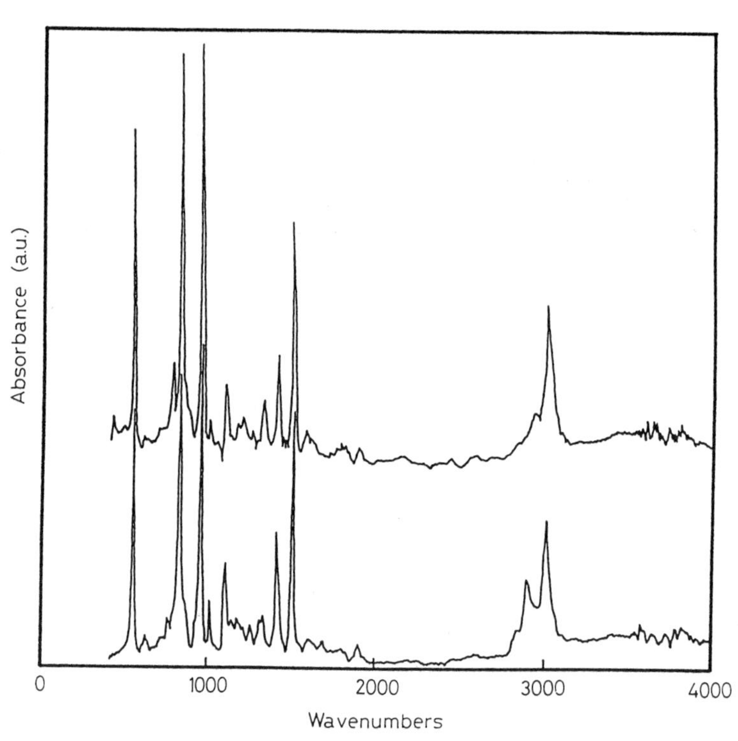

Fig. 3 Infrared absorption spectra of unoriented free standing films of samples of PPV[A] (lower curve) and PPV[B] (upper curve).

In the lower spectrum, a weak feature appears at 2920 cm^{-1}, which is absent from the upper spectrum. This transition, which slowly disappears from the spectrum of PPV[A] during prolonged thermal elimination, is indicative of incomplete conversion of the dialkyl sulphonium precursor to PPV[A]. Further evidence to this effect is provided by considering the ratio of the oscillator strength of the very strong mode at 837 cm^{-1} to that of the weak mode at 1013 cm^{-1}. The ratio of the strengths of these two transitions, associated with the out-of-plane and in-plane phenylene CH bending modes, respectively, is closely related to the degree of π electron delocalization in the material. The greater value of this ratio for the spectrum of PPV[B] is consistent with its lower bandgap. Finally, we note that the weak feature near 784 cm^{-1} grows in intensity, and shifts toward higher energy between the lower and upper spectra in Figure 3. This behaviour has been correlated with an increase in crystallinity in previous studies of samples of PPV[A][1].

PREPARATION OF ORIENTED FILMS OF PPV

Electron diffraction photographs of the as-grown unoriented PPV[A] and PPV[B] described above showed that there were no gross structural differences in the two forms. In both cases, sharp Debye-Scherrer rings were seen, showing that both types of film have an isotropic distribution of well-defined crystallites.

Films made from the tetrahydrothiophenium precursor were slowly heated up to 140°C, in order to start the elimination process. At 60°C the films began to stretch, and for about 10 mins it was possible to stretch the sample at the same rate. After this time stretching was no longer possible and the application of further stress only led to tearing of the films. In this way, films of a draw ratio of up to 11 could be obtained. The partly converted films were then heated to 300°C for 4 hours to complete the elimination. During the processing the films were kept as far as possible under vacuum or argon. The thickness of the films was between 600 and 2000 Å, making them suitable for electron microscopy. For x-ray studies much thicker samples were needed and it was necessary to glue several layers of the stretched film together. This was carried out under a microscope to ensure sufficient alignment of the whole sample. We report here on preliminary diffraction measurements on these oriented films of PPV[B].

ELECTRON DIFFRACTION

Oriented films of PPV[B] were examined, using a JEOL 2000EX electron microscope operating at 200kV, and the diffraction patterns were recorded on Kodak TMAX400 films. This high-sensitivity film material, and the high voltage used, made it possible to record the diffraction pattern as a single exposure photograph, instead of the multi-exposure photographs of PPV[A] previously reported[7]. Thus we were able to detect more reflections out to higher order, and in fact, reflections up to the 14th order were observed. A typical electron diffraction pattern, obtained from a fully converted film of PPV[B], is shown in Figure 4, with a draw ratio of eleven. In this electron diffraction pattern there are six measurable equatorial reflections perpendicular to the fibre axis, and the corresponding d-spacings are listed in Table 1. These equatorial reflections do not show any arcing, which indicates that the crystallites in the sample are well oriented along the polymer axis. Tilting of the sample about the c-axis did not change the positions and the intensities of the reflections, showing that there is no preferred orientation of the crystallites about the c-axis. From the spacing of the layer lines the c axis was calculated to be 6.4 Å. This distance can

Table 1
Observed and Calculated Equatorial d Spacings

d_{obs} electron diffraction	d_{obs} x-ray diffraction	d_{calc}	h k l
4.3	4.327	4.298	1 1 0
-	4.000	4.01	2 0 0
3.131	3.183	3.15	2 1 0
2.411	2.427	2.426	1 2 0
2.132	-	2.149	2 2 0
1.835	1.839	1.843	3 2 0
-	-	1.866	4 1 0
-	1.640	1.660	1 3 0
1.578	1.564	1.575	4 2 0
		1.530	5 1 0
		1.563	2 3 0

d_{calc} for $a = 8.02$
$b = 6.07$
$\alpha = 123°$

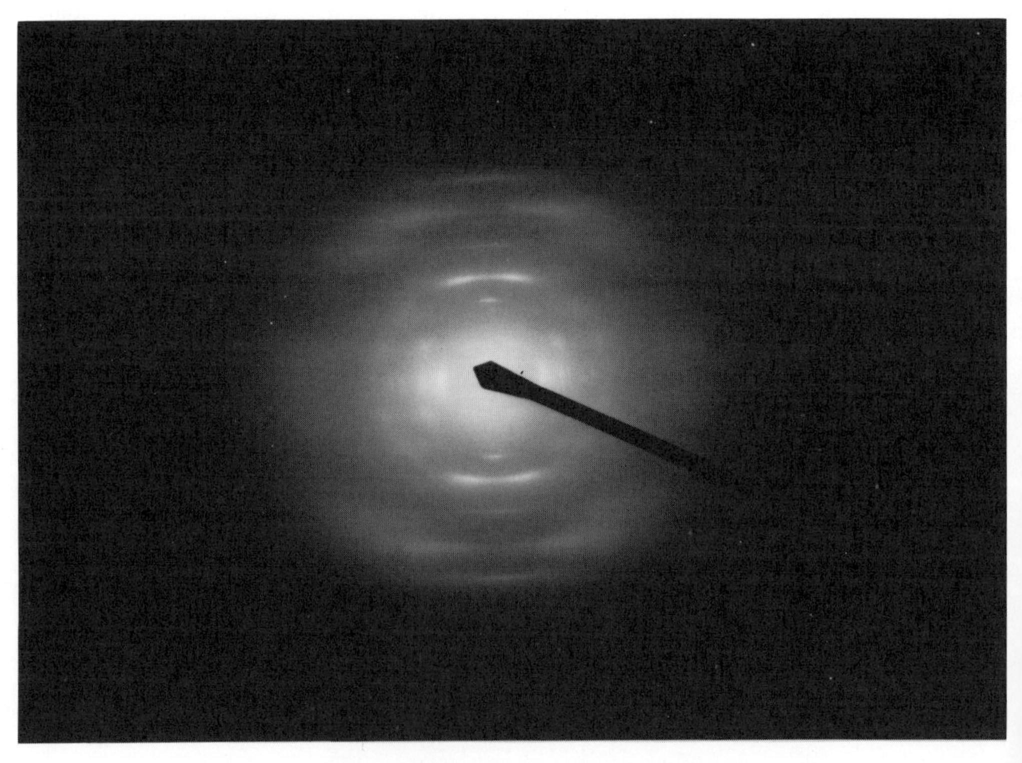

Fig. 4 Electron diffraction pattern of PPV[B] (L/L_o=11)

be identified with the chain repeat distance; thus the c-axis of the polymer is aligned along the polymer chain,and also along the stretch direction,as expected.

X-RAY DIFFRACTION

Measurements on the same PPV[B] films were also carried out on a four-circle-diffractometer using Cu-Kα radiation, with the fibre axis (c axis) oriented perpendicular to the x-ray beam. A 2-θ scan was used to record the reflections along the equator and the d spacings deduced from this are shown in Table 1. The unit cell derived from both the x-ray and electron measurements is monoclinic, with $a = 8.02$, $b = 6.07$, $c = 6.4$ Å and $\alpha = 123^{\circ}$, and the space group is pgg, in good agreement with the results obtained for PPV[A] by us[8] and by Granier et al[7].

CONCLUSION

We have reported measurements on films of poly(p-phenylene vinylene) PPV prepared by two different precursor routes, and have found that,although the overall unit cell of the polymer remains the same,the materials prepared by the tetrahydrothiophenium precursor method, PPV[B], are of higher quality, and are more fully converted from the precursor polymer. Our findings are in excellent agreement with those reported by Stenger-Smith et al[5]. In particular, we find that this method leads to films with higher local order along the polymer chain, as evidenced by a lower π to π^* gap seen in optical absorption and photoluminescence, although there is no overall increase in the degree of crystallinity, or ordering of as-grown samples. Further experiments are in progress to investigate the detailed structure of oriented films, to look for any possible changes within the unit cell.

REFERENCES

1. Bradley, D. D. C. J. Phys. D: Appl. Phys. **20** (1987) 1389
2. Wessling, R.A. J Polym. Sci., Polym. Symp. **72** (1986) 55
3. Tokito, S., Tsutsui, T., Saito, S. and Tanaka, R. Polymer Commun. **27** (1986) 333
4. Gagnon, D. R., Capistran, J. D., Karasz, F. E., Lenz, R. W. and Antoun, S. Polymer **28** (1987) 567
5. Strenger-Smith, J. D., Lenz, R. W. and Wegner, G. Polymer **30** (1989) 1048
6. Friend, R. H., Bradley, D.D.C., and Townsend, P.D. J. Phys. D: Appl. Phys. **20** (1987) 1367
7. Granier, T., Thomas, E.L., Gagnon, D.R., Karasz, F.E. and Lenz , R.W. J. Polm. Sci: Part B: Polm. Phys. **24** (1986) 2793
8. Bradley, D. D. C., Friend R.H., Hartmann, T. Marseglia E.A.,Sokolowski,M. M. and Townsend,P.D. Synth. Met. **17** (1987) 473

ELECTROCHEMICAL DOPING PROCESSES OF CONDUCTING POLYMERS STUDIED WITH IN SITU FTIR SPECTROSCOPY

H. Neugebauer

Institut für Physikalische Chemie, Universität Wien, Vienna, Austria

N. S. Sariciftci

Institut für Festkörperphysik, Universität Wien, Vienna, Austria

INTRODUCTION

In situ FTIR spectroscopy has proved to be very useful for the investigation of the reaction mechanisms of electrochemical reactions and structural changes of substances involved in these reactions[1]. Two different methods are used: external reflection absorption spectroscopy and internal reflection spectroscopy. The application of these methods to the study of the electrochemical doping process of polypyrrole, and the comparison of the results are described in this contribution.

EXPERIMENTAL

Polypyrrole (PPy) was synthesized electrochemically in an electrolyte containing 0.5 M $LiClO_4$ and 0.2 M pyrrole in CH_3CN by applying a constant potential of +800 mV on the electrode[2] using the standard three electrode technique (Potentiostat Jaissle 1001 TNC, sweep generator Prodis 1/14 I, reference electrode SCE; all potential values in this paper refer to this electrode). The polymerization was stopped after one minute. The polymer-covered electrode was washed with CH_3CN, and transfered into the spectroelectrochemical cell. The electrolyte for the in situ experiments was 0.5 M $LiClO_4$ in CH_3CN. After the electrochemical reduction of the polymer, by applying a potential of -700 mV on the electrode, the potential was altered between -700 mV and +100 mV with a sweep rate of 1 mV/s. With this procedure, the non-conducting reduced form of PPy was converted into the conducting oxidized form[2] ("doping process"). During this potential sweep, FTIR spectra were recorded consecutively[3] (LPS-FTIRS technique) using a Nicolet 60 SX FTIR spectrometer equipped with a KBr/Ge beamsplitter and an MCT detector. The spectral resolution was 4 cm^{-1}. For each spectrum, a number of 128 interferograms was coadded before performing the Fourier transformation. Due to the registration time for the collection of 128 interferograms (about 38 seconds), each spectrum covers a range of about 38 mV in the potential sweep. The spectra were related to the spectrum at -700 mV. In this way, only the spectral changes during the electrochemical doping process, relative to the reduced form of PPy, are shown.

Lower-Dimensional Systems and Molecular Electronics
Edited by R.M. Metzger *et al.*, Plenum Press, New York, 1991

EXTERNAL REFLECTION ABSORPTION SPECTROSCOPY

The setup for external reflection absorption spectroscopy[4,5,6] is shown in Fig. 1. A PPy-covered platinum disc electrode is pressed against a ZnSe window, yielding a distance of some μm between the electrode and the window. The electrochemical current of the oxidation process has to pass the thin electrolyte layer. Since only parallel-polarized light interacts with substances near a reflecting metal surface, the IR light was polarized in this way. The IR beam permeates the window, the electrolyte and the polymer and is reflected at the Pt surface. Only this part of the radiation (*a* in Fig. 1) contains information on the polymer absorption. Reflections at the window and the polymer surface (*b*, *c* and *d* in Fig. 1), which also reach the detector, can lead to disturbing spectral features and have to be eliminated or corrected[7].

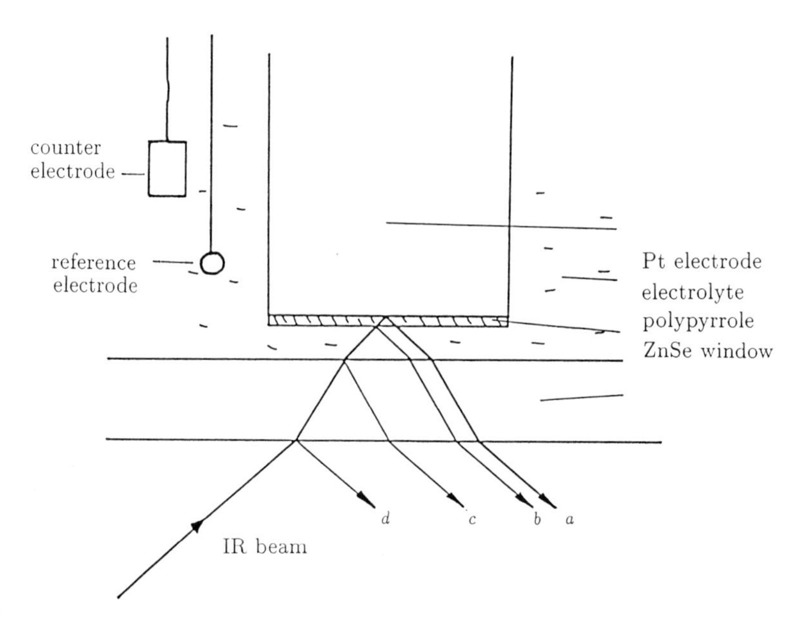

Fig. 1. Setup for external reflection absorption spectroscopy

The spectral range between 7000 and 500 cm^{-1} is accessible using a ZnSe window and an MCT detector. The spectral changes in the range between 2300 cm^{-1} and 500 cm^{-1} during the electrochemical doping of PPy, using external reflection absorption spectroscopy, are shown in Fig. 2. Several bands appear with increasing intensity during the doping process. The bands at 1546, 1288, 1217, 1038, 924 and 792 cm^{-1} can be attributed to polymer vibrations. The strong increase in intensity originates from high dipole moment changes of the vibrations, due to the occurence of positive charges on the polymer chain during the doping process. A more detailed description of the origin and the behaviour of these vibrations is given by Zerbi and Bogang[8,9]. A broad absorption at wave-numbers higher than 1800 cm^{-1}, and extending into the near IR, comes from an electronic transition (of polaronic or bipolaronic type[10,11]) in connection with the increasing conductivity of the polymer.

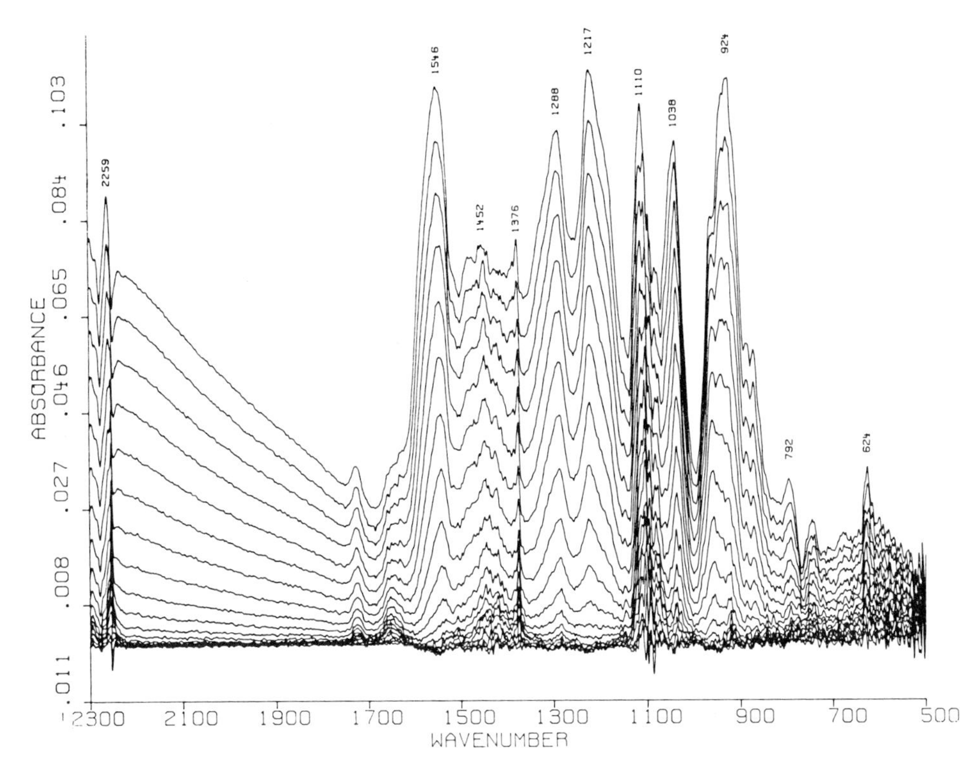

Fig. 2. Spectral changes during the electrochemical doping process of PPy using external reflection absorption spectroscopy.

In addition to the polymer bands, spectral features at 2259, 1452 and 1376 cm^{-1} can be seen. Using external reflection absorption spectroscopy, absorption bands of the electrolyte cannot be compensated completely, leading to these disturbing effects. The bands at 1110 and 624 cm^{-1} are bands of the ClO_4^--ion. The electrochemical current of the oxidation process causes the migration of these ions from the surrounding bulk electrolyte into the thin electrolyte layer, resulting in an increase in the concentation and in IR absorption of this species.

INTERNAL REFLECTION SPECTROSCOPY

The setup for internal reflection spectroscopy[12,13] is shown in Fig. 3. The electrode consists of a germanium reflection element, covered on one side with PPy in contact with the electrolyte solution. The IR beam is totally reflected at the interface between the reflection element and the PPy layer. The penetration depth of the IR beam into the optical thinner medium (polymer and electrolyte) at each total reflection is in the order of some μm. Spectral information on this small region is available without disturbances by the bulk electrolyte. Since the counter electrode and the reference electrode can be mounted in an advantageous arrangement, no electrochemical limitations, connected with a thin electrolyte layer, occur using this method. Due to the limited stability of Ge in aqueous electrochemical systems, the water content of the electrolyte solution has to be very low.

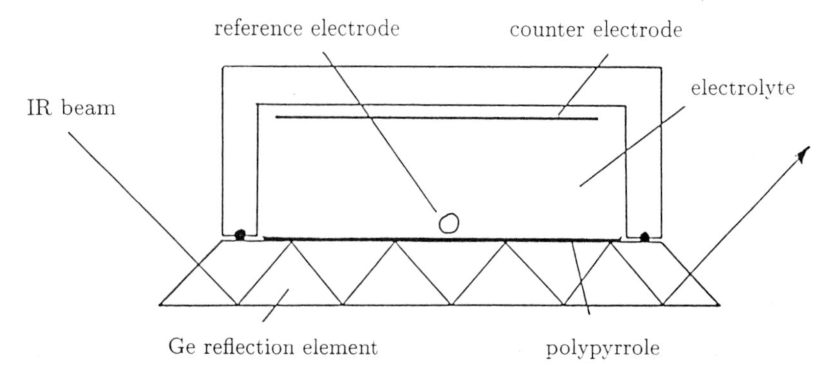

Fig. 3. Setup for internal reflection spectroscopy

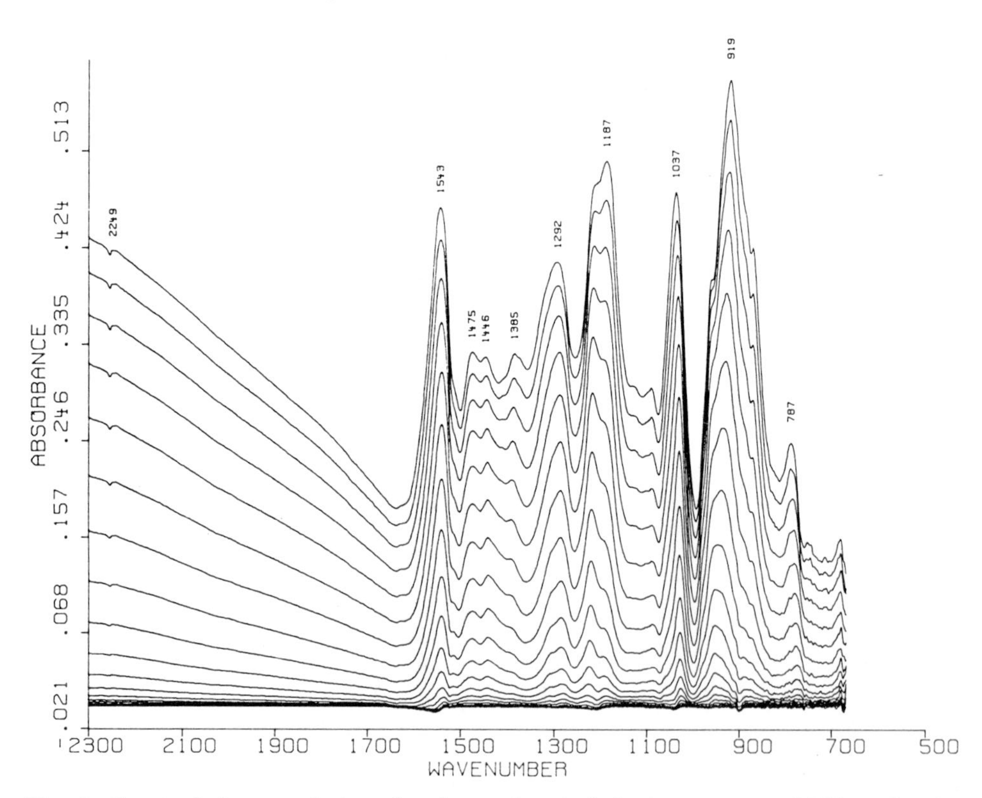

Fig. 4. Spectral changes during the electrochemical doping process of PPy using internal reflection spectroscopy.

Germanium is transparent in the spectral range between 5100 cm^{-1} and 670 cm^{-1}. The spectral changes in the range between 2300 cm^{-1} and 670 cm^{-1}, during the electrochemical doping of PPy, using internal reflection spectroscopy, are shown in Fig. 4. Besides small features at 2249 cm^{-1}, no electrolyte incompensation effects can be seen. The same increasing polymer absorption bands, as seen with external reflection absorption spectroscopy (Fig. 2), appear. In addition, three more polymer bands at 1475, 1446 and 1385 cm^{-1} can be detected. Because of electrolyte incompensation effects, these bands are not seen using external reflection absorption spectroscopy. The spectral resolution of 4 cm^{-1} and the wavelength dependence of the penetration depth[14] give rise to small differences in frequencies and in the relative intensities of the IR bands. At the position of the ClO_4^--absorption band (around 1100 cm^{-1}) only a very weak band can be seen. The S/N ratio of the spectra is better compared with external reflection absorption spectroscopy, indicating a higher sensitivity of internal reflection spectroscopy.

CONCLUSION

Both methods — external reflection absorption spectroscopy and internal reflection spectroscopy — can be used to obtain in situ spectral information on the electrochemical doping process of polypyrrole.

The main problem using external reflection absorption spectroscopy shows to be the thin electrolyte layer. The absorption bands of the electrolyte cannot be compensated completely, resulting in spectral disturbances. Besides difficulties with the ohmic resistance for the electrochemical current, concentration changes in the thin electrolyte layer occur during electrochemical reactions.

Internal reflection spectroscopy is limited on the one hand by the electrochemical stability of the reflection element (dissolution of Ge in aqueous electrolytes at anodic potentials), and on the other hand by the limited transparency of the reflection element. Furthermore, the intensities of the absorption bands depend on the penetration depth of the IR beam and therefore on the wavelength of the IR radiation.

SUMMARY

External Reflection Absorption Spectroscopy:

Advantages:	Disadvantages:
Extended spectral range	complex IR signal
electrochemically stable materials	hindered electrochemistry
	electrolyte incompensation effects
	ion migration effects

Internal Reflection Spectroscopy:

Advantages:	Disadvantages:
electrochemistry not hindered	limited stability of the reflection element
no electrolyte incompensation effects	limited transparency of the reflection element
high sensitivity	absorption intensities wavelength dependent

REFERENCES

1. A.Neckel, Mikrochim.Acta III : 263 (1987)
2. A.F.Diaz, J.I.Castillo, J.A.Logan and W.Y.Lee, J.Electroanal.Chem 129 : 115 (1981)
3. H.Neugebauer, G.Nauer, N.Brinda-Konopik and G.Gidaly, J.Electroanal.Chem. 122 : 381 (1981)
4. A.Bewick, in: "Trends in Interfacial Electrochemistry", A.F.Silva, ed., Reidel, Dordrecht (1986) p.331
5. H.Neugebauer, A.Neckel, N.S.Sariciftci and H.Kuzmany, Synthetic Metals 29 : E185 (1989)
6. N.S.Sariciftci, M.Bartonek, H.Kuzmany, H.Neugebauer and A. Neckel, Synthetic Metals 29 : E193 (1989)
7. H.Neugebauer, N.S.Sariciftci, H.Kuzmany and A.Neckel, Springer Series in Solid State Sciences, in press
8. G.Zerbi, this conference
9. T.Bogang and G.Zerbi, Synthetic Metals 28 : C1 (1989)
10. J.L.Brédas, B.Thémans and J.M.André, Phys.Rev.B 27 : 7827 (1983)
11. J.H.Kaufman, N.Colaneri, J.C.Scott and G.B.Street, Phys.Rev.Lett. 53 : 1005 (1984)
12. H.Neugebauer, G.Nauer, A.Neckel, G.Tourillon, F.Garnier and P.Lang, J.Phys. Chem. 88 : 652 (1984)
13. H.Neugebauer, A.Neckel and N.Brinda-Konopik, Springer Series in Solid State Sciences 63 : 227 (1985)
14. N.J.Harrick, "Internal Reflection Spectroscopy", Wiley, New York (1967)

EVOLUTION OF ELECTRIC PERMITTIVITIES OF pTS AND pFBS DIACETYLENES DURING SOLID-STATE POLYMERIZATION

Maciej Orczyk and Juliusz Sworakowski

Institute of Organic and Physical Chemistry
Technical Unversity of Wroclaw
50-370 Wroclaw, POLAND

INTRODUCTION

Some substituted diacetylenes (R–C≡C–C≡C–R, hereinafter referred to as DA) constitute a class of crystals polymerizable in the solid state [1]. One may thus obtain large single crystals, built of polydiacetylene (PDA) chains parallel to one another. A pronounced anisotropy of the systems, due to the presence of long π–conjugated chains, resulted in an interest concerning electrical and optical properties of DA [2,3]. Much less attention has been paid to studies of their dielectric properties [4-6].

In this work, we present results of dielectric measurements of two diacetylenes: pTS (R = CH_2–O–SO_2–ϕ–CH_3) and pFBS (R = CH_2–O–SO_2–ϕ–F). At room temperature, and down to ca. 200 K, both crystals are isomorphous (monoclinic $P2_1/c$, Z = 2), with almost identical crystallographic packings [7-9]. The monomer of pTS exhibits two phase transitions (at ca. 200 K and ca. 160 K), the phase between them being incommensurate [10]. The low–T phase remains $P2_1/c$, with four molecules per unit cell [8]. Poly-pTS displays only one phase transition, at ca. 190 K, directly to the low–T phase, which is isomorphous with that of monomer [11]. On the other hand, spectroscopic investigations of polymerized pFBS [12] did not show any evidence of a phase transition, from troom temperature down to 4 K.

We report on the determination of the complete tensor of relative electric permittivity in single crystals of pTS, as a function of temperature and polymer contents. Since the dielectric measurements prove to be a suitable method, enabling one to follow the second- order phase transition in pTS, we compare the dielectric behavior of monomer and polymer of pTS and pFBS, looking for a possible signature of a phase transition in the latter crystals.

EXPERIMENTAL

The diacetylenes were synthesized by Dr. M. Bertault. The material was then purified, and the crystals grown according to the procedure described by Bertault *et al.* [13]. The samples employed in the measurements were cut in the desired directions with a wire saw, and then polished on tissues soaked with ethyl acetate. The thicknesses of the samples were ca. 0.5 mm, their electrode area amounting to about 20 mm^2. As–grown samples, used for the measurements in the direction perpendicular to the **bc** plane, were usually twice larger.

The electric permittivities (ε) were calculated from capacitance measurements carried out at 1 kHz in the temperature range 80–333 K, in the dark and under ambient nitrogen pressure. The samples were polymerized *in situ* at 333 K, the polymer amount being determined directly from the time–conversion dependences [13, 14].

We performed measurements of the electric permittivity of pTS crystals in 7 independent crystallographic directions. The experiments on pFBS crystals were carried out in two orthogonal directions: the direction of growing polymer chains (**b**), and normal to the largest natural cleavage plane (**a'**).

RESULTS AND DISCUSSION

Temperature dependence of the ε tensor in polymerizing pTS

The principal components of the permittivity tensor were calculated according to the procedure described by Nye [15]. As is shown in Fig. 1, only one of them (ε_2) substantially increases during polymerization, seemingly following the polymer content [13]. Up to ca. 19 hours of polymerization (corresponding to a polymer content of about 8%), ε_2 was found to vary to a minor degree (2–6% increase). In the autocatalytic region of polymerization, we observed a pronounced increase of ε_2 (about 50% at room temperature). Annealing for periods longer than 24 hours did not result in any further changes in the permittivity.

The components ε_1 and ε_3 were found to vary to a minor degree upon polymerization (ca. 2% decrease).

The phase transitions in both pTS and poly-pTS clearly manifest themselves on the $\varepsilon_2(T)$ dependences (cf. Fig. 1). The other two principal components vary to a much lesser degree: independently of the conversion to polymer, ε_3 is a weak function of temperature, and ε_1 remains practically constant over the entire temperature range under investigation.

Fig. 2 shows two cross–sections of the permittivity ellipsoid in pTS crystals, determined at room temperature in monomer and in fully polymerized crystals. The ellipsoid undergoes a substantial elongation along the **2** axis upon polymerization, while the cross–section perpendicular to this axis, nearly circular for the monomer, remains practically unchanged. The symmetry of the pTS crystals allows the principal axes 1 and 3 of the tensor to be arbitrarily oriented with respect to the crystal axes **a** and **c**. It is, however, interesting to note that, within the experimental error (±3°), these two principal axes of the tensor practically coincide with the **c** and **a'** crystallographic directions, respectively (cf. Fig. 3).

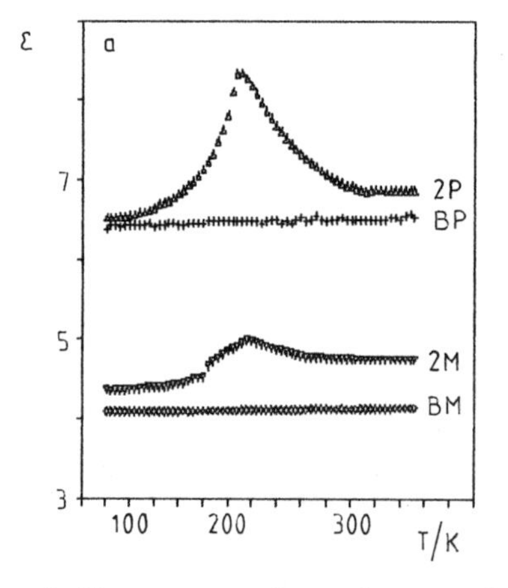

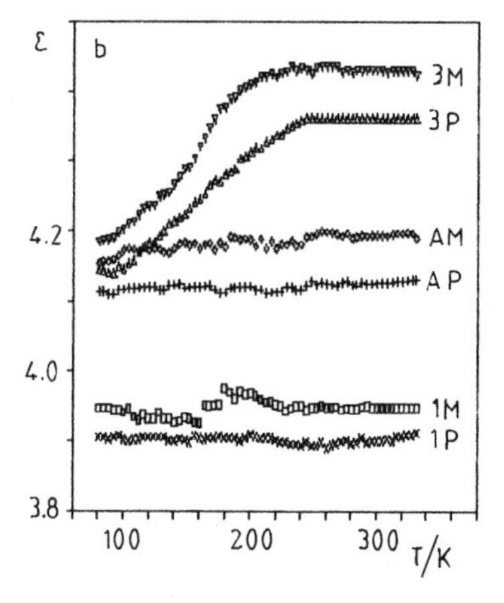

Fig. 1. Temperature dependences of the principal components of the electric permittivity tensor in monomer and fully polymerized pTS (ε_i). The numbers correspond to the indices of the components, the symbols "M" and "P" refer to monomer and polymer, respectively. Results obtained for pFBS are also shown for comparison: curves "A" and "B" refer to the measurements carried out along the **a'** and **b** directions of pFBS, respectively. Note the different scales of ε in Figs. 1a and 1b.

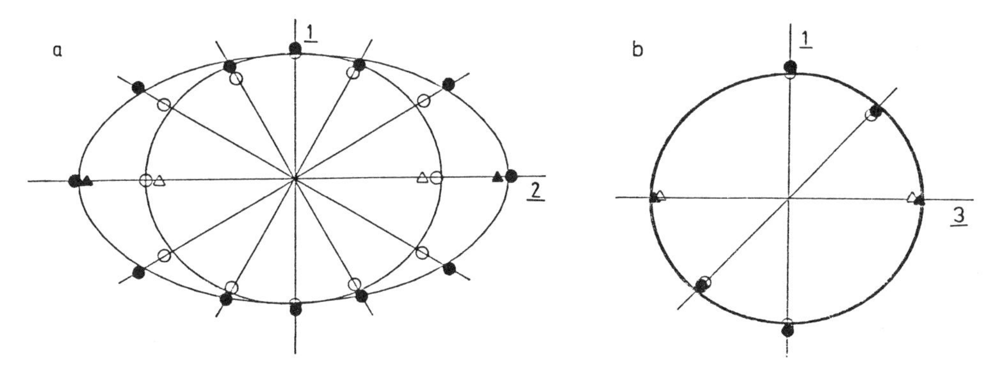

Fig. 2. Cross-sections of the ellipsoid of permittivity measured at 295 K for monomer and polymer of pTS: normal to the **3** axis (a) and normal to the **2** axis (b). (○), (●): the experimental data points for monomer and polymer, respectively. The results obtained for pFBS (Δ) and poly–pFBS (▲) are shown for comparison.

Dielectric measurements of pFBS crystals

Contrary to pTS, permittivities of the monomer and polymer pFBS crystals are practically independent of temperature over the entire range under study, exhibiting no evidence of a phase transition. Far from the region of phase transitions in pTS, and in particular at low temperatures, the electric permittivities in monomers and in polymers of both crystals are very similar. This fact, as well as the increase of ε_2 during polymerization, very similar in both crystals, indicate that changes of the permittivities (except for the transition region) are mainly due to increasing lengths of the polymer backbones, resulting in the increase of the polarizability of their π–electron systems. Basing on the above, as well as on the structural similarities of both crystals, one may suppose that the electric permittivities measured in the **b** and **a'** directions of pFBS correspond to the principal components ε_2 and ε_3 of the tensor, as is the case in pTS.

It is reasonable to assume that the phase behavior of pFBS, different from that of pTS, is primarily associated with a difference of some parameter(s) characterizing their side groups, the most obvious one being a difference of the effective sizes of the methyl group and fluorine atom. The van der Waals volume increments amount to 23.5 Å^3 and 9.6 Å^3 for CH_3 and F,

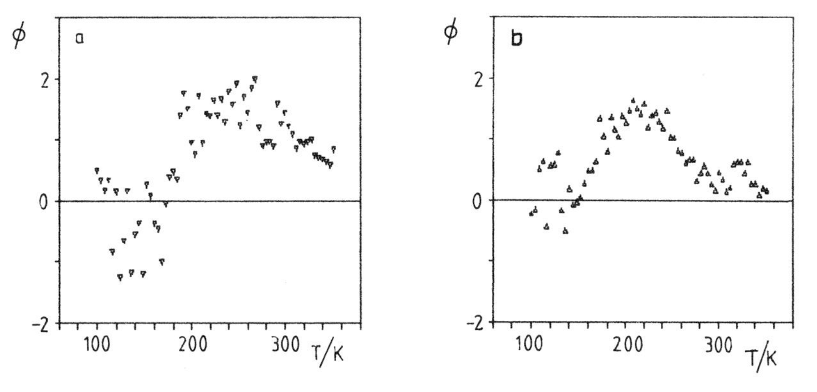

Fig. 3. Temperature dependences of the angle ϕ between ε_1 and the **c** crystallographic axis in pTS (a) and poly–pTS (b).

respectively [15]; however, one should bear in mind that the substituents are only small parts of larger moieties. Thus, it is difficult to regard the difference in the geometrical sizes of fluorine and methyl as a main factor governing the phase behavior of both DA crystals.

It seems conceivable that the differences in the behavior of pTS and pFBS (and probably also in the kinetics of thermal polymerization [14]) are connected with different energies of the electrostatic dipole–dipole interactions of the side groups. The electric moment of the tosyl group amounts to ca. 4 Deybe units [16]. Replacing the methyl group by the highly electronegative fluorine atom results in a decrease of this value by ca. 1.8 D [17]. Such a significant difference in the dipole moments is likely to influence their mutual interactions, which are responsible for the presence (or absence) of the phase transitions.

ACKNOWLEDGEMENTS

Thanks are due to Dr. M. Bertault for the gift of pTS and pFBS. The research reported in this paper was supported by the Polish Academy of Sciences within the Programme CPBP 01.14.

REFERENCES

[1] G. Wegner, Topochemical polymerization of monomers with conjugated triple bonds, Makromol. Chem. 154: 35 (1972).
[2] H. J. Cantow, ed, "Polydiacetylenes" (Springer, Berlin, 1984).
[3] D. Bloor, and R. R. Chance, eds., "Polydiacetylenes: synthesis, structure and electronic properties", (M. Nijhoff, The Hague, 1985).
[4] G. F. Lipscomb, A. F. Garito, and T. S. Wei, An apparent ferroelectric transition in an organic diacetylene solid, Ferroelectrics 23: 161 (1980).
[5] R. Nowak, J. Sworakowski, B. Kuchta, M. Bertault, M.Schott, R. Jakubas, and H. A. Kolodziej, Dielectric properties of single crystals of the substituted diacetylene pTS: effect of the solid-state polymerization and phase transitions, Chem. Phys. 104: 467 (1986).
[6] R. Zielinski, and J. Kalinowski, The electric permittivity and phase transition in the bis-(toluene sulphonate) of 2,4-hexadiyne-1,6-diol polymer single crystal, J. Phys. C: Solid State Phys. 20: 177 (1987).
[7] D. Kobelt, and E. F. Paulus, Poly[1,2-bis(p-tolylsulfonyloxymethylen)-1-buten -3-inylen], Acta Cryst. B30: 232 (1974).
[8] J. P. Aimé, J. Lefebvre, M. Bertault, M. Schott, and J. O. Williams, Studies of a polymerizable crystal: I. Structure of monomer pTS (bis-p-toluenesulphonate of 2,4-hexadiyne-1,6–diol) by neutron diffraction at 120 and 221 K, J. Phys. (Paris) 43: 307 (1982).
[9] J. P. Aimé, M. Schott, M. Bertault, and L. Toupet, Study of reactive diacetylenes: Structure of monomer and polymer pFBS crystals, Acta Cryst. B44: 617 (1988).
[10] J. N. Patillon, P. Robin, P. A. Albouy, J. P. Pouget, and R. Comès, X-ray study of the incommensurate phase of bis-(p-toluenesulphonate) of 2,4-hexadiyne-1,6-diol as a function of polymer conversion., Mol. Cryst. Liq. Cryst. 76: 297 (1981).
[11] V. Enkelmann, The crystal structure of the low-temperature phase of poly[1,2-bis(p-tolylsulphonyloxy-methylene)-1-butene-3-ynylene], Acta Cryst. B33: 2842 (1977).
[12] R. R. Chance, K. C. Yee, R. H. Baughmann, H. Eckhardt, and C. J. Eckhardt, Electronic spectra of two polydiacetylene isomorphs, J. Polym. Sci.: Polym. Phys. Ed 18: 1651 (1980).
[13] M. Bertault, M. Schott, M. J. Brienne, and A. Collet, Studies of a polymerizable crystal. II. Microcalorimetric study of polymerization and polymorphism of pTS, Chem. Phys. 85: 481 (1984).
[14] V. Enkelmann, The solid-state polymerization, physical properties and crystal structures of diacetylene mixed crystals, Makromol. Chem. 184: 1943 (1983).
[15] A. I. Kitaigorodsky, "Molecular crystals and molecules", (Academic Press, New York, 1973).
[16] Landolt-Börnstein, "Zahlenwerte und Funktionen", (Springer, Berlin, 1951).
[17] V. I. Minkin, O. A. Osipov, and Y. A. Zhdanov, "Dipolyne momenty v organicheskoi khimiyi", (Izdat. Khimiya, Leningrad, 1968).

STRUCTURAL CHARACTERIZATION OF THIOPHENE-BASED MONOMERS, POLYMERS, AND OLIGOMERS AS POWDERS, THICK, AND THIN FILMS

A. Bolognesi, M. Catellani, S. Destri, W. Porzio, C. Taliani*, R. Zamboni*, and S. Brückner#

Istituto di Chimica delle Macromolecole del C.N.R.
Milano, ITALY

*Istituto di Spettroscopia Molecolare del C.N.R.
Bologna, ITALY

#Istituto di Chimica, Università di Udine
Udine, ITALY

INTRODUCTION

Since the field of electrically conducting materials (or, more generally, electronic materials) has been enriched by the discovery of organic polymers and organic charge-transfer complexes, the need for accurate and detailed structural data has been felt more and more acutely. The very few structural data available for polymers were, in fact, the basis of most attempts to describe the phenomena, occurring during the conduction process, in terms of relatively simple models. These attempts allowed confident predictions of new molecular architectures, and reliable interpretations of the electronic properties of polymeric systems. Notable among these are the theories of the charge transport, based on the notion of soliton and polaron-bipolaron excitations [1-3].

In spite of the great number of polymers, prepared chemically or electrochemically from heteroaromatic compounds [4], the available structural data are few and often approximate, limiting the accuracy and the reliability of calculations aiming to design new molecules. However it must be stressed that, for polymeric systems, one must generally expect a reduced three-dimensional (3-D) order, due to the presence of chemical defects (sp^3 hybridization, chain ends and anything which interrupts the π-conjugation) and to preparation conditions that are unfavorable to 3-D packing.

Finally, the recent discovery of a high non-linear optical response, in both polymers and oligomers based on heterocyclic compounds, has renewed the efforts to find and clarify the relationships between structure and properties.

The aim of this contribution is to show the promise of structural analysis, when it is applied to the various steps of the production of materials with electrical and/or non-linear optical properties.

Lower-Dimensional Systems and Molecular Electronics
Edited by R.M. Metzger *et al.*, Plenum Press, New York, 1991

Table 1. C-C ring bond distance differences (δr), from crystal structure determinations of thiophene-based monomers, and corresponding solution oxidation potentials (OP, Volts versus SCE)[a]

Compound		δr (Å)	OP(Volts)	Ref.
I	Thiophene	0.053	2.1	[5]
II	Thiophtene	0.050	1.6	[6]
III	Dithieno[3,2-b: 2',3'-d]thiophene	0.043	1.2	[7]
IV	Dithieno[3,4-b: 3',4'-d]thiophene	0.033	1.05	[8]
Dithieno[3,2-b: 2',3'-d]thiophene:TCNQ (CT complex)[b]		0.0285	---	[9]

(a) δr is the minimum difference observed between intra-ring C-C bond distances ["long" (single-bond character) and averaged "short" (double-bond character)].

(b) TCNQ=tetracyanoquinodimethane. This charge-transfer (CT) complex is quoted to show the effect of CT mixing on the C-C distances (see text).

MONOMERS

In the designing new molecular architectures, one preliminary objective is finding the relationships among the factors which affect the electronic properties of the final materials. Specifically, the properties one may wish to control in the starting monomer are: (i) symmetry, to influence the HOMO bandwidth of the polymer, and hence the mobility of the charge carriers, (ii) π-electron conjugation, which should be maintained in the polymer backbone, (iii) oxidation or reduction potential, to favor desired reactions for polymer synthesis, (iv) positional selectivity, to prevent the formation of defects that may interrupt the π-conjugation.

We have studied the x-ray crystal structure geometries of a thiophene-based series of monomers, and we have correlated these features with the molecular oxidation potentials. In Table 1 are reported the oxidation potentials (OP) and δr values, defined as the minimum difference observed between "long" (single) and mean "short" (double) C-C distances within a monomer ring. This parameter δr is indicative of the "degree of aromaticity", which influences the "quinoid character" of the ground-state description in the resulting polymer. Table 1 shows that δr decreases, as the "quinoid character" increases. The shortest δr value is for a charge-transfer complex, where the "quinoid" contribution should be dominant. OP values also decrease correspondingly, going fron thiophene to fused-ring monomers, confirming the direct relationship between OP and "aromaticity" in heterocyclic monomers.

Of course, it can be argued that one cannot disentangle the contributions of the steric stress and of the electronic and symmetry factors, which together determine the resulting crystalline geometries. However, we only aim to follow trends in the "aromaticity" of the compounds of Table 1, and to correlate these trends with the oxidation potentials.

Moreover, we note that monomer IV in Table 1 satisfies also the last requirement listed above, i.e. the high positional selectivity, since only the α-positions are available for polymerization.

POLYMERS

Similar considerations are applicable to the corresponding polymers. Although the factors already mentioned (preparation conditions, chemical defects, distribution of molecular weights), which are responsible for chain aggregation, also reduce the 3-D order, nevertheless structural analysis can yield non-trivial information.

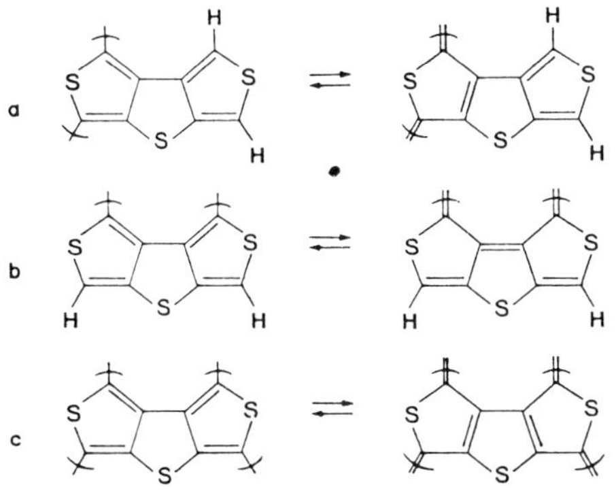

Fig. 1. Resonance valence-bond ground state structures of DTT', which assume the several possible sites of polymerization.

As an example, we discuss a structural study, by the Rietveld [10] method, of neutral polythiophene [11]. This polymer is, sofar, the most ordered polyheterocycle, even though no oriented films have been obtained up to now, and so the only available samples are thermally annealed powders. Interestingly, aside from the expected (and found) co-planarity between adjacent rings, the detected 3-D order can be interpreted in terms of a discrete number of preferred arrangements, each of them shifted with respect to another along the chain axis, and co-present in the crystallites. This hypothesis accounts for the relative doping ease of polythiophene.

By contrast, ClO_4^--doped polydithieno[3,4-d: 3',4'-b]thiophene (DTT') displays only one diffraction peak (d spacing = 3.53 Å), but we have inferred for this polymer a 3-D structure, corroborated by molecular modelling analysis [12]. In this model, the polymer chains consists of about eight residues, slightly bent ($\leq 10°$) with respect to each other, and packed at a distance d from the next chain; these crystalline islands are randomly surrounded by the counterions (ClO_4^-). A similar situation is inferred for other thiophene-based polymers, which also present a single diffraction line [13].

For DTT' one can envision three different chain propagation directions (enchainments), all of which preserve the conjugation pathway (Fig. 1). The enchainment of Fig. 1(c), with four chain propagation points, should be ruled out, considering the steric hindrance of adjacent residues. The other two configurations in Fig. 1(a,b), involve only two attack points, and were the object of molecular modelling calculations [12]. The one developing onto a thiophene residue, allows for an effective π-overlap in the polymer backbone, and hence is largely to be preferred.

OLIGOMERS

Thiophene oligomers received recent attention, becuase of their non-linear optical properties. Detailed structural investigations in this new topic can really help to clarify the complex phenomena, which govern fast optical response in polyheterocyclic systems. Unfortunately, efforts to obtain single crystals of these compounds have been unsuccessful, so it is necessary to undertake powder diffraction analysis.

This has been performed for quaterthienyl (QT). QT crystallizes in microscopic needles (less than 25 μm thick), which aggregate like sheaves of wheat. QT is monoclinic, space group $P2_1/n$, with refined unit cell parameters **a** =5.02(2) Å, **b** = 30.63 (8) Å, **c** = 4.85 (2) Å,

β = 105.8(5)°. Assuming an inversion center in the middle of the molecule, one obtains a calculated density of 1.50 g cm^{-3} (for Z = 2), which matches very well the value observed by the gradient density method (1.50 g cm^{-3}). An effective molecular axis overlap between adjacent molecules is inferred. A complete and detailed Rietveld analysis [10] of QT is in progress, and will be reported elsewhere.

THIN FILMS

Diffraction methods, specifically modified for thin-film studies, are just now developing, and are giving information on layer spacing, on mean domain dimensions, and perhaps on chain packing. Several diffraction techniques have been used to characterize Langmuir-Blodgett (LB) films, using electron-, x-ray, synchrotron, and neutron radiation. Suitable diffraction geometries (grazing incidence, Seeman-Bohlin focussing) are required to enhance the signal-to-noise ratio, and position-sensitive detectors can help to detect weak signals.

We have tested LB films (40 layers) of cadmium arachidate, and we have been able to detect 20 diffraction orders along the layer direction. This allows for an accurate determination of both the interplanar spacing (55.13±0.02 Å) and the mean domain coherence (more than 7 bilayers).

Other systems, such as polydiacetylenes, and also polyphthalocyanines, have been examined; recent investigations on thiophene oligomers and polymers, in forms of thin films, are in progress.

CONCLUSIONS

We have applied high-precision diffraction analysis to monomers, oligomers, and polymers based on thiophene residues, in the form of single crystals, of powders, of cast thick films, and even as thin films prepared using the Langmuir-Blodgett technique. Preliminary results encourage us to perform further detailed studies on diffraction techniques suitable for well-ordered thin films, which are of interest for non-linear optics and for electronic applications.

REFERENCES

[1] W. P. Su, J. R. Schrieffer, and A. J. Heeger, Phys. Rev. B22: 2099 (1980).
[2] A. R. Bishop, D. K. Campbell, and K. Fesser, Mol. Cryst. Liq. Cryst. 77: 253 (1981).
[3] J. L. Brédas, R. R. Chance, and R. S. Silbey, Mol. Cryst. Liq. Cryst. 77: 319 (1981).
[4] C. Taliani, in "Molecular Electronics", M. Borissov, Ed. (World Scientific Publishers, Singapore, 1987), p. 394.
[5] B. Bak, D. Christiansen, L. Hansen-Nygaard, and J. Rastrup-Andersen, J. Mol. Spectr. 7: 58 (1961).
[6] E. G. Cox, R. J. J. H. Gillot, and G. A. Jeffrey, Acta Cryst, 2: 356 (1949).
[7] F. Bertinelli, P. Palmieri, C. Stremmenos, G. Pelizzi, and C. Taliani, J. Phys. Chem. 87: 2317 (1983).
[8] M. Catellani, S. Destri, and W. Porzio, Acta Cryst. C44: 545 (1988).
[9] D. Zobel and G. Ruban, Acta Cryst. B39: 638 (1983).
[10] H. M. Rietveld, Acta Cryst. 22: 151 (1967).
[11] S. Brückner and W. Porzio, Makromol. Chem. 189: 961 (1988).
[12] A. Bolognesi, M. Catellani, S. Destri, D. R. Ferro, W. Porzio, C. Taliani, R. Zamboni, and P. Ostoja, Synth.Metals 28: C527 (1989).
[13] R. Lazzaroni, C. Taliani, R. Danieli, R. Zamboni, P. Ostoja, W. Porzio, and J. L. Brédas, Synth. Metals 28: C515 (1989).

TRANSIENT PHOTOCONDUCTIVITY IN ORIENTED CONJUGATED POLYMERS

J. Reichenbach, H. Bleier, Y.Q. Shen, S. Roth

Max-Planck-Institut für Festkörperforschung
D-7000 Stuttgart 80, Heisenbergstr. 1, FRG

INTRODUCTION

A powerful technique for probing the underlying physics in a material is the investigation of the behaviour of photo-excited states, which are produced by absorption of light. Electroactive conjugated polymers exhibit a wide variety of photoexcitations, such as solitons, polarons, bipolarons, and excitons[1].

Fast transient photoconductivity experiments, which measure the product of the number of charge carriers and their mobility, can serve as a local probe, and provide information on charge transport along the polymer chains, as well as on the recombination kinetics of charged excitations.

In this paper we report on the photoconductivity of fully oriented Durham/Graz-polyacetylene and of fully converted, highly oriented polyphenylenevinylene (PPV) films. Though the two materials are different in several aspects - e.g. polyacetylene is a degenerate ground state system, whereas PPV is the non-degenerate one, which has important consequences for the kind of possible excitations - they both show similar behaviour, when looking at the transient photocurrent response.

EXPERIMENTAL

Oriented free-standing trans-$(CH)_x$ films (about 2 μm thick) have been prepared by the Durham/Graz-method, i.e. by stretch-orienting a precursor polymer during thermal conversion (120°C) to $(CH)_x$[2,3]. In a similar way, PPV has been fully converted from a water soluble precursor polymer by heating at 300°C[4].

The samples were mounted on a quartz-substrate, onto which two gold electrodes had been evaporated, separated by a gap of 200 μm. Sample preparation and handling was performed without any exposure to air, and the photoconductivity experiments were carried out under vacuum of $5 \cdot 10^{-6}$ bar.

Lower-Dimensional Systems and Molecular Electronics
Edited by R.M. Metzger *et al.*, Plenum Press, New York, 1991

For the transient photocurrent measurements a Dye laser pumped by 500 ps pulsed Nitrogen laser was used for the optical excitation. The incident light was linearly polarized parallel and perpendicular to the stretch-direction. The bias field was applied parallel to the c-axis, therefore the photocurrent reflects the transport properties along the chains. Details of the photoconductivity apparatus can be found in Ref. 5.

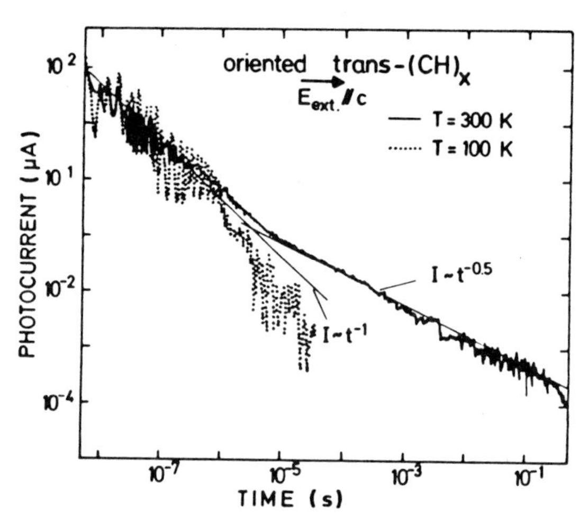

Fig. 1. Decay of the photocurrent after 500 ps excitation investigated from nanoseconds to seconds at 300 K (full line) and at 100 K (dotted line) for oriented trans-$(CH)_x$

RESULTS AND DISCUSSION

Fig. 1 shows the decay of the photocurrent in trans-$(CH)_x$ over 8 orders of magnitude in time on a log-log plot for two different temperatures. There are two components, that can be clearly distinguished - a fast one decaying nearly linearly in time, and a slow one with $I \sim t^{-0.5}$. For the case of PPV films, the behaviour looks quite similar. There the fast component in the short time region decays as $I \sim t^{-0.9}$, whereas the slow component, which dominates after 1 μs, follows a power law with an exponent of -0.6.

For both polymers the temperature dependence of these two components is quite different. The fast picosecond response is nearly temperature-independent from room temperature down to 10 K, the slow component varies strongly with temperature, and can be frozen out by cooling the samples (see also the dotted curve in Fig. 1).

For the intensity dependence of the peak photocurrent there is a linear relationship in the case of trans-$(CH)_x$ (see Fig. 2), whereas in the case of PPV a different behaviour has been observed (see Fig. 3). In the low-intensity region we see a linear dependence, which changes to a sublinear relationship for higher intensities. This suggests, that the recombination of some of the charge carriers follows bimolecular recombination kinetics. When looking at the decay kinetics at different light intensities (Fig. 4) this process is clearly confirmed; the higher the intensity the faster the

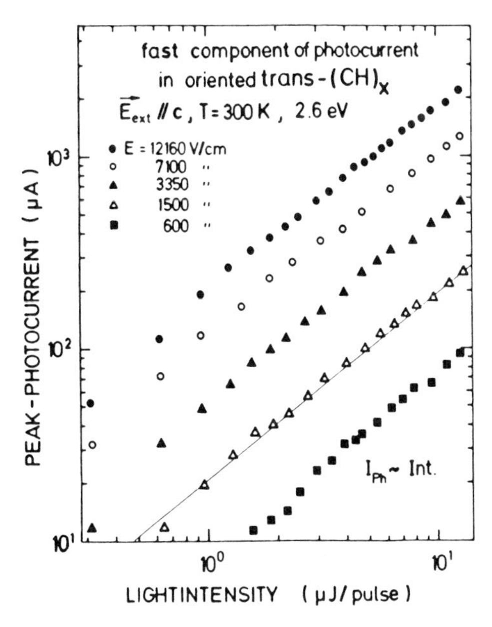

Fig. 2. Intensity dependence of the fast component on a log-log plot at different electric fields for trans-$(CH)_x$. The slope of 1.0 clearly demonstrates the linear behaviour between the peak photocurrent and the light intensity.

photocurrent decays. Such a mechanism has already been observed by Friend et al.[6] for the neutral polaron-exciton states in PPV. The efficiency of the (non-radiative) bimolecular processes was seen to increase with the conjugation length. This is fully consistent with the observation, that the intensity dependence becomes linear for partially converted PPV films with shorter conjugation length. The shorter conjugation decreases the mobility of the carriers and, as a consequence, the probability for bimolecular recombination is therefore lowered.

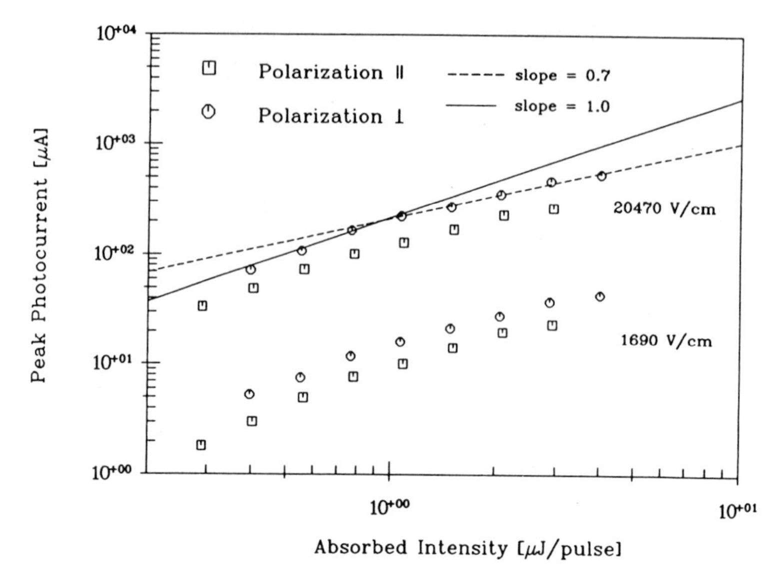

Fig. 3. Intensity dependence of the peak photocurrent in an oriented PPV film at 300 K for excitation light parallel and perpendicular to the chain orientation at different bias fields.

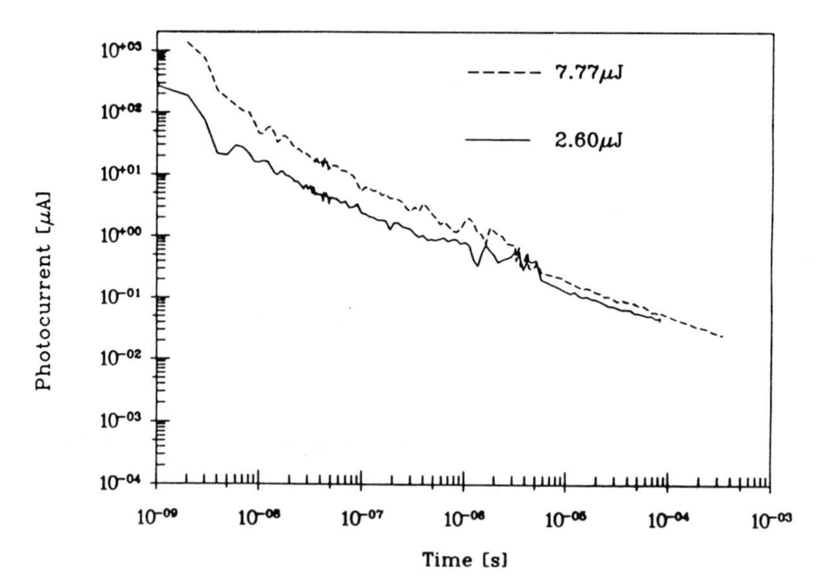

Fig. 4. Decay of the transient photocurrent in oriented PPV films excited by perpendicularly polarized light at two different intensities.

Looking at the anisotropy of the photoresponse with respect to the polarization of the incident light, we find that the photocurrent in fully converted PPV is higher for the polarization perpendicular to the c-axis than for the parallel case. This result can be explained by the proposed quadratic recombination mechanism, since in the parallel case the photoexcited charge carrier density is about 7 times higher than for a perpendicular polarization. Therefore, the probability for bimolecular recombination is increased. Additional evidence for this interpretation comes from results obtained in partially converted films, where the linear intensity dependence demonstrates the strongly decreasing influence of the bimolecular recombination channel. At the same time the anisotropy with respect to the polarization of the incident light disappears.

The peak photocurrent in highly oriented trans-$(CH)_x$ with light polarized perpendicular to the chain orientation is about two times higher than that excited by light polarized parallel to the chains[7]. Since there is no experimental evidence for bimolecular recombination in trans-$(CH)_x$[8] (the decay of the photosignal obeys the same power law, when changing the carrier density by more than two orders of magnitude[5]), this result has to be explained by an increased probability to separate the photoexcited electron-hole pairs to different chains, thereby avoiding geminate recombination. Charge carriers produced by intra-chain excitation recombine immediately (geminately) or convert quickly into neutral soliton pairs[9], which do not contribute to the photocurrent.

The absence of bimolecular recombination in trans-$(CH)_x$ may probably be explained by the short conjugation length and the disorder in this material compared to partially converted PPV. With (hypothetical) much more crystalline samples such a bimolecular mechanism should also be expected.

ACKNOWLEDGEMENT

We would like to thank D.D.C. Bradley and all our colleagues from Polymer Hill for their help and valuable discussions. Support of Deutsche Forschungsgemeinschaft (SFB 329) is gratefully acknowledged.

REFERENCES

1. For a review see e.g.
   S. Roth, H. Bleier: Adv. in Physics 36, 385 (1988)
   A.J. Heeger, S. Kivelson, J.R. Schrieffer, W.P. Su: Rev. Mod. Phys. 60, 781 (1988)
2. J.H. Edwards, W.J. Feast: Polymer 21, 595 (1980)
3. G. Leising: Polym. Bull. 11, 401 (1984)
4. J.D. Capistran, D.R. Gagnon, S. Anton, R.W. Lenz, E.F. Karasz: A.C.S. Polymer Preprints 25, 282 (1984)
5. H. Bleier, S. Roth, Y.Q. Shen, D. Schäfer-Siebert, G. Leising: Phys. Rev. B 38, 6031 (1988)
6. R.H. Friend, D.D.C. Bradley, P.D. Townsend: J. Phys. D: Appl. Phys. 20, 1367 (1987)
7. H. Bleier: in "Electronic Properties of Conducting Polymers" (Kirchberg III), H. Kuzmany, M. Mehring, S. Roth (eds.), Springer Series in Solid State Sciences, Heidelberg 1989, in print
8. Z. Vardeny, E. Ehrenfreund: Proceedings of the "Conference on Transport and Relaxation Processes in Random Materials", Gaithersburg 1985
9. S. Kivelson, W.K. Wu: Phys. Rev. B 34, 5423 (1986)

ENCAPSULATION OF CONDUCTING POLYMERS WITHIN ZEOLITES

Patricia Enzel and Thomas Bein*

Department of Chemistry, University of New Mexico
Albuquerque, NM 87131, USA

INTRODUCTION

A great deal of current research efforts are aimed at the design and understanding of conducting and semiconducting structures at sub-micrometer dimensions. The term 'molecular electronics' describes the ultimate reduction of electronic circuitry to the molecular level.[1] Beyond the development of concepts,[2] a major challenge in this area is to create isolated, addressable molecular units that function as useful electronic components. We study the *encapsulation of conducting polymers within the crystalline channel systems of zeolite hosts* as a promising approach to isolated, well-defined chains of molecular conductors. The molecular-size channels of these hosts limit the dimensions of the polymer chains to molecular dimensions. The conducting polymers polypyrrole and polythiophene have previously been studied in larger scale host structures such as layered FeOCl and V_2O_5.[3] Polypyrrole fibrils with diameters between 0.03 and 1 μm at 10 μm length have been synthesized in Nucleopore membranes.[4]

Zeolites are open-framework aluminosilicates with pore sizes between 0.3 and 1.2 nm, and exchangeable cations compensating for the negative charge of the framework.[5,6] Zeolite Y, mordenite (MOR) and zeolite A were used in this study (Figure 1). Zeolite Y is composed of interconnected "sodalite" cages, and mordenite features a twelve-ring channel system. Both structures have an open pore size of about 0.7 nm. The structure of zeolite A is based upon sodalite cages interconnected via double four-rings, with pore-openings of about 0.4 nm.

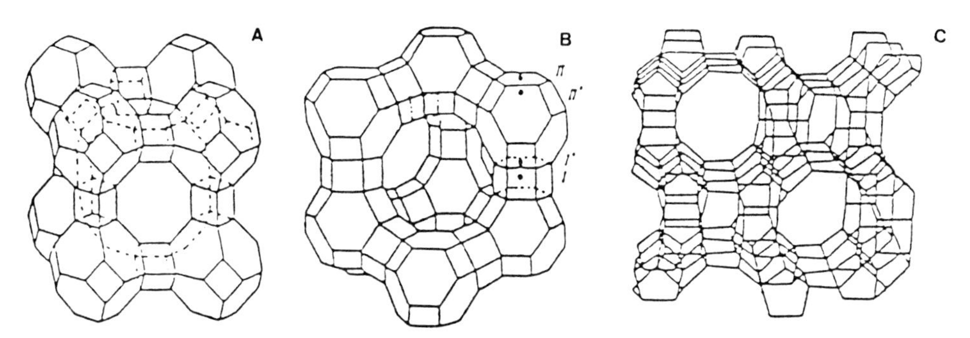

Figure 1. Zeolite structures. Three-dimensional channel systems of zeolite A (A), zeolite Y (B), and the pseudo one-dimensional channel system of mordenite (C).

Table 1. Composition of zeolite/polymer samples.

Samples	Aniline[a] PANI	Pyrrole[a] PPy	Thiophene, 3MTh[a] PTh, P3MTh
$Na_{56}Y$	43/uc, orange[b]		
$H_{46}Na_{10}Y$	31/uc, deep blue		
Na_8MOR	5/uc, light orange[b]		
H_8MOR	3/uc, blue		
$Na_{56}Y$-V[c]		41/uc, white	37 and 31/uc, white
$Na_{56}Y$-H[d]		7/uc, white	white
$Fe_{12}Na_{32}Y$-V[c]		39/uc, dark turquoise	29 and 25/uc, dark turquoise
$Cu_{15}Na_{26}Y$-V[c]		50/uc, dark blue	35 and 32/uc, dark blue
$Cu_{15}Na_{26}Y$-H[d]		6.5/uc, dark blue	5.4 and 5.7, dark blue
Na_8MOR-V[c]		2/uc, white	2 and 2/uc, white
Na_8MOR-H[d]		1/uc, white	white
Fe_3Na_2MOR-V[c]		1/uc, turquoise	1 and 1/uc, grey-green
$Cu_{2.5}Na_3MOR$-V[c]		0.8/uc, blue-grey	1.5 and 1/uc, blue-grey
$Cu_{2.5}Na_3MOR$-H[d]		0.5/uc, blue	blue
$Cu_8Na_{80}A$-V[c]		0.3/uc, light blue	0.2/uc, light blue

a- Monomer loadings and color of the resultant products; uc = zeolite unit cell. The loading levels were determined gravimetrically (vapor loadings) or spectroscopically (solution experiments).
b- Unknown oxidation products, no polymer was detected.
c- Vapor phase loadings; samples were saturated with monomer vapor at 295 K.
d- Hexane solution loadings, adjusted to achieve approximately optimum reaction stoichiometries.

We recently succeded in forming intrazeolite polyaniline (PANI)[7,8], polypyrrole (PPy)[9], polythiophene (PTh) and poly(3-methylthiophene) (P3MTh)[10] by oxidative polymerization inside the cavities of different zeolites, as demonstrated by vibrational, ESR, and electronic absorption data. It was observed that the dimensionality and pore size of the host determine the polymerization rates and intrazeolite products. This communication compares the above zeolite/polymer systems and discusses evidence for polymerization inside the host channel structures.

EXPERIMENTAL

Zeolite host materials were derived from the sodium and ammonium forms of zeolite Y(LZ-Y52,LZ-Y62; Alfa), Na-mordenite (MOR; LZ-M5; Union Carbide), and zeolite A (Alfa 5A). The ammonium form of MOR was obtained by refluxing Na-MOR two times in excess of 0.1 M NH_4Cl for 12 h. Cu(II) and Fe(II) ions were introduced into the zeolites via ion exchange with 0.1 M $Cu(NO_3)_2$ and 0.1 M $FeSO_4$, respectively. All zeolites were degassed in an oxygen stream followed by evacuation (620 K, 10^{-5} Torr). This treatment generates the acidic forms in the case of ammonium zeolites, and oxidizes the Fe(II) zeolites to Fe(III) zeolites, respectively. The resulting zeolite cation contents per unit cell are $Na_{56}Y$, $H_{46}Na_{10}Y$, $Cu_{15}Na_{26}Y$, $Fe_{12}Na_{32}Y$, Na_8MOR, H_8MOR, $Cu_{2.5}Na_3MOR$, Fe_3Na_2MOR, and $Cu_8Na_{80}A$. Intrazeolite PANI was synthesized in the acidic zeolite forms by analogy with the chemical polymerization of aniline in acidic solution.[11] Aniline was loaded into the degassed zeolites from hexane solution (Table 1) and subsequently reacted with an aqueous solution of $(NH_4)_2S_2O_8$ at a ratio of 4:1 intrazeolite aniline:oxidant. Intrazeolite Cu(II) and Fe(III) ions served as oxidants for the oxidative polymerization reaction of pyrrole[12] and thiophenes.[13] Pyrrole, thiophene and 3-methylthiophene monomers were loaded into the degassed zeolites from either the vapor phase in small quartz reactors, or from zeolite suspensions in water and hexane (Table 1). Bulk polymers were synthesized according to published procedures.[11,12,13]

RESULTS AND DISCUSSION

The monomer-loaded zeolite samples display dramatic color changes from white to different hues of blue and green when (a) aniline in different acidic zeolite forms is treated with the oxidant, or when (b) pyrrole or thiophene monomers are admitted into Cu(II)/Fe(III)-containing zeolites Y or MOR from the vapor phase or from hexane (or other hydrocarbon) solutions (Table 1). These color changes correspond to those observed in bulk synthesis reactions.[11-13] No reaction is observed with the zeolite sodium forms, indicating that the polymerizations proceed only in the presence of intrazeolite protons and/or appropriate oxidants. No polymer formation is detected in zeolite Cu(II)A (pore size 0.4 nm, smaller than pyrrole or thiophene). This is

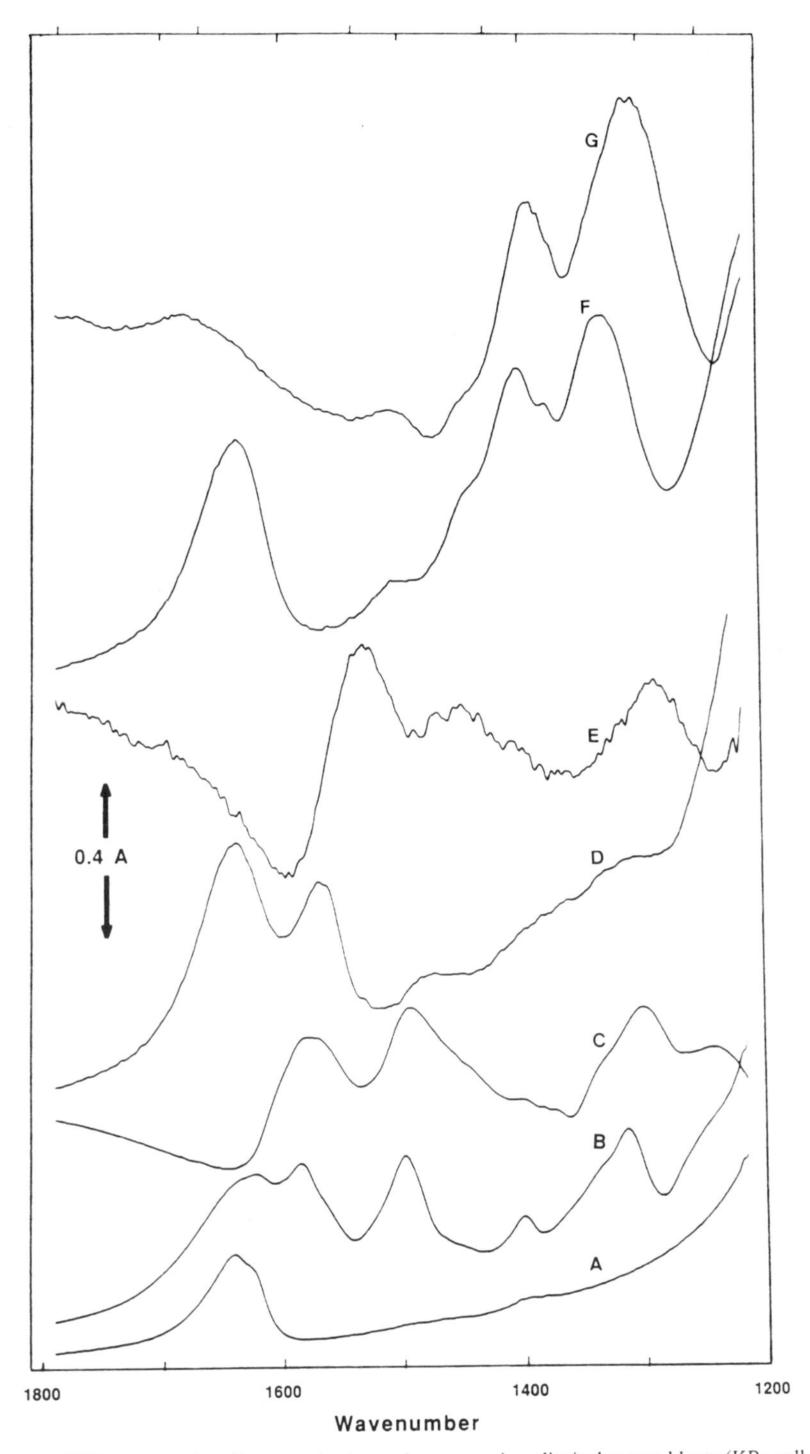

Figure 2. FTIR spectra of zeolites, conducting polymers, and zeolite/polymer adducts (KBr pellets, 4 cm^{-1} resolution). $Na_{56}Y$ (A), $H_{46}Na_{10}Y$/PANI (B), PANI (C), $Fe_{12}Na_{32}Y$-V/PPy (D), PPy (E), $Cu_{15}Na_{26}Y$-V/P3MTh (F), P3MTh (G). For sample names, see Table 1.

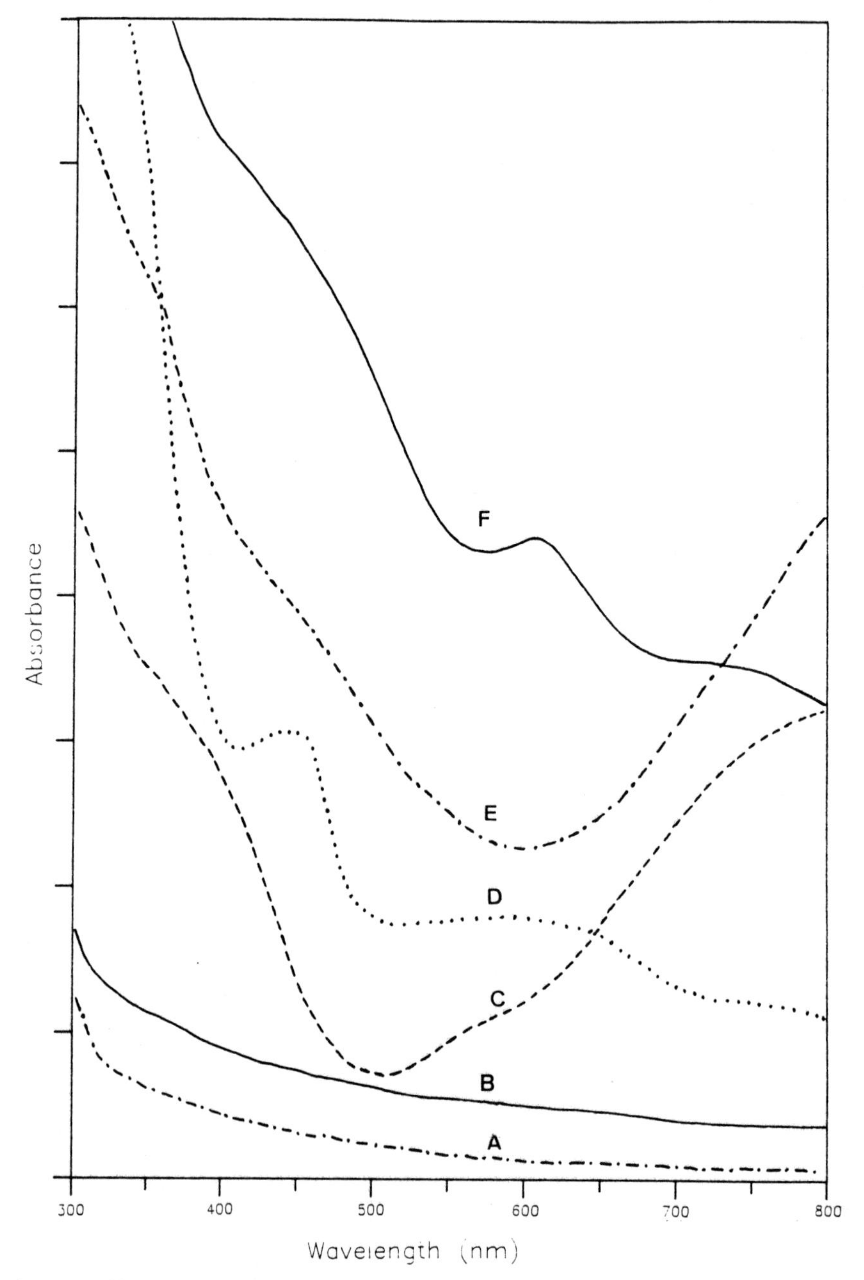

Figure 3. Electronic absorption spectra (samples dispersed in glycerol) of zeolites and zeolite/polymer adducts. Na_8MOR (A), $Na_{56}Y$ (B), $H_{46}Na_{10}Y$/PANI (C), Fe_3Na_2MOR-V/PPy (D), $Fe_{12}Na_{32}Y$-V/PPy (E), $Cu_{15}Na_{26}Y$-V/P3MTh (F). For sample names, see Table 1.

consistent with the inability of the monomers to diffuse into the zeolite cavities where the majority of the oxidant ions are located. In contrast to trends observed in bulk synthesis reactions, polar solvents such as water do not favor the intrazeolite polymerization of pyrrole and thiophenes, probably because the intrazeolite metal ions are screened by the polar solvent molecules. Based upon the small surface capacity of the zeolite crystals relative to the total pore volume (ca. 0.2 monomer molecules per unit cell of zeolite Y, for 1 μm crystals), and the high monomer uptake (Table 1), it is evident that most of the monomer molecules will reside in the pore system of the zeolite host. Hence, it can be concluded that most of the polymer is also formed inside the zeolites. It should be noted that in the zeolite/PPy and PTh systems derived from vapor phase-saturated zeolites, the excess of monomer vs. oxidant ions will prevent 100% conversion to the polymer. However, since the monomers have no visible absorption and much smaller absorption coefficients in the IR than the polymers, no spectroscopic interference is expected. No deposition of polymer on the surface of the zeolite crystals was detected in scanning electron micrographs, while zeolite samples deliberately coated with polymer showed distinct bulk polymer features.

Intrazeolite polymers show mid-IR bands similar to those typical for the bulk polymers PANI[14], PPy[15], and P3MTh[16], and a characteristic tail of the electronic excitation corresponding to free carrier absorption (Figure 2). Certain shifts are observed between IR frequencies of the bulk polymers and the zeolite/polymer adducts, suggesting some interaction of the polymer chains with the host. The intrazeolite polymers could be recovered by dissolution of the zeolite host with HF. IR spectra of the recovered products are comparable to those of chemically synthesized bulk materials.

The electronic absorption spectra of the zeolite/polymer samples (Figure 3) show absorption bands that are related to different electronic transitions in the bulk polymers. The zeolite/PANI samples display a weak band at 560 nm (2.2 eV) associated with quinone diimine structures of polymer at high oxidation levels. Shoulders at 400 nm (3.1 eV) and the broad features in the red (<1.8 eV) are associated with radical cations supporting *polarons* as charge carriers.[17] The electronic spectra of intrazeolite polypyrrole are influenced by the nature of the zeolite host. The spectrum of sample $Fe_{12}Na_{32}Y$-V/PPy (Figure 3E) shows a shoulder at ca. 450 nm (2.7 eV) and at energies lower than 650 nm (1.9 eV). The red absorption of zeolite Y samples is generally stronger than that in mordenite. In addition, sample Fe_3Na_2MOR-V/PPy shows a feature between 520 and 700 nm (ca. 2 eV, Figure 3D). Absorption maxima at 2.3 and 0.7 eV have been observed with electrochemically formed, highly doped polypyrrole; the higher energy band shifted to lower values at lower oxidation levels.[18] Thus, we assign the bands at 2.7, 2.0 eV and that in the near infrared to *bipolaron* absorptions typical for PPy at different oxidation levels, and conclude that the oxidation level of PPy in zeolite Y is high, while in MOR probably a bimodal distribution of PPy at high and intermediate oxidation levels is present. Intrazeolite PTh and P3MTh show bands at 450 nm (2.8 eV), 600 nm (2 eV) and absorption extending into the near infrared (< 1.7 eV, Figure 3F). Transitions at about 2.5 eV have been associated with interband excitations of neutral P3MTh[19] which decrease in intensity with progressive oxidation. The additional bands at 2 eV and in the red/near IR are assigned to *bipolaron* transitions typical of intermediate oxidation levels (in bulk P3MTh: 1.6 and 0.6 eV). It would not be unexpected if the decoupled intrazeolite polymer chains had a different electronic structure (and band positions) than the corresponding bulk material. Additional bands, observed for some of the zeolite/polymer samples, can be attributed to different chain-lengths and/or oxidation levels of the constrained polymers in the zeolites channels.

ESR data of the intrazeolite PANI show the presence of ca. 0.0025 Curie-type spins per aniline loaded (comparable to bulk spin densities if the lower polymer content in the zeolites is considered), with g-values (g=2.0034) similar to bulk polymers.[20] ESR spectra of intrazeolite PPy confirm the low spin count expected for bipolaron formation and g-values (2.0027) characteristic of polypyrrole.[21] Large linewidths in both cases (8-10 G) could indicate strong dipolar interactions with the zeolite host.

No bulk conductivity is observed in pressed wafers of the zeolite/polymer samples (detection limit: 10^{-8} S/cm). However, pressed pellets of zeolite samples deliberately coated with a thin film of polymer show conductivities of about 10^{-6} S/cm, compared to 1-10 S/cm for bulk polymers. The products recovered after dissolution of the zeolite/polymer adducts in HF have conductivities at the order of 0.001 S/cm. These observations indicate that the polymers do not coat the zeolite crystals and that the intrazeolite polymer chains are probably decoupled from each other.

The location of the polymer phase is further illustrated by the polymerization rates in the zeolite hosts which are orders of magnitude slower than in bulk chemical solution syntheses (no reaction in A, MOR < Y << solution). The oxidant and/or the monomers have to diffuse into the channels of the zeolites in order to reach the intrazeolite reaction partners. These diffusion and pore volume limitations would not have been observed if the polymers had only formed on the crystal surfaces.

CONCLUSION

We have demonstrated that oxidative polymerization reactions leading to conducting polymers can be carried out within the channel systems of zeolites. Acidic zeolite forms are required to synthesize intrazeolite polyaniline by analogy to the oxidative coupling of aniline in acidic solutions. The presence of intrazeolite oxidants such as Cu(II) and Fe(III) ions is fundamental for the polymerization of pyrrole, thiophene and 3-methylthiophene. The degree of polymer chain oxidation and probably the chain lengths are influenced by the dimensionality of the zeolite channels.
This is the first approach towards the encapsulation and stabilization of *molecular wires* in well-defined hosts of molecular dimensions.

ACKNOWLEDGMENT

Instrumentation used in this work was acquired with funding from the National Science Foundation. We thank L. Zuppiroli and F. Beuneu (Ecole Polytechnique, Palaiseau) for the ESR studies.

REFERENCES

1 F. L. Carter, Ed. *Molecular Electronic Devices II*, Marcel Dekker, New York, 1987.
2 J. J. Hopfield, J. N. Onuchic, B. N. Beratan, *Science* 1988, **241**, 817.
3 (a) M. G. Kanatzidis, L. M. Tonge, T. J. Marks, H. O. Marcy and C. R. Kannewurf, J. Am. Chem. Soc. 1987, **109**, 3797. (b) M. G. Kanatzidis, M. Hubbard, L. M. Tonge, T. J. Marks, H. O. Marcy and C. R. Kannewurf, *Synth. Met.*, 1989, **28**, C89. (c) M. G. Kanatzidis, C.-G. Wu, C. R. Kannewurf and H. O. Marcy, Paper INOR134 presented at the 197th ACS National Meeting, Dallas, April 1989.
4 (a) R. M. Penner, C. R. Martin, *J. Electrochem. Soc.*, 1986, **133**, 2206. (b) Z. Cai, C. R. Martin, J.Am. Chem. Soc., in press.
5 D. W. Breck, *Zeolite Molecular Sieves,* R. E. Krieger, Malabar, Fl, 1984.
6 R. Szostak, *Molecular Sieves. Principles of Synthesis and Identification,* Van Nostrand Reinhold, New York, 1989.
7 T. Bein and P. Enzel, *Synth. Met.* , 1989, **29**, E163.
8 P. Enzel and T. Bein, J. Phys. Chem., *to be published.*
9 T. Bein and P. Enzel, J. Am. Chem. Soc., *submitted.*
10 P. Enzel and T. Bein, J. Chem. Soc., Chem. Comm., *to be published.*
11 A. G. MacDiarmid, J. C. Chiang, A. F. Richter, N. L. D. Somasiri, and A. J. Epstein, in *Conducting Polymers* , L. Alcácer, Ed., Reidel Publications, Dordrecht, The Netherlands, 1986, 105.
12 (a) R. E. Myers, *J. Electron. Mat.*, 1986, **15**, 61. (b) S. P. Armes, *Synth. Met.*, 1987, **20**, 365. (c) S. Rapi, V. Bocchi and G. P. Gardini, *Synth. Met.* 1988, **24**, 217.
13 (a) N. Mermilliod-Thevenin and G. Bidan, *Mol. Cryst. Liq. Cryst.*, 1985, **118**, 227. (b) M. B. Inoue, E. F. Velazquez and M. Inoue, *Synth. Met.*, 1988, **24**, 223.
14 (a) L. W. Shacklette, J. F. Wolf, S. Gould, and R. H. Baughman, *J. Chem. Phys.* 1988, **88**, 3955. (b) N. S. Sariciftci, M. Bartonek, H. Kuzmany, H. Neugebauer, and A. Neckel, *Synth. Met.*, 1989, **29**, E193. (c) Y. Furukawa, F. Ueda, Y. Hyodo, I. Harada, T. Nakajima, and T. Kawagoe, *Macromolecules*, 1988, **21**, 1297.
15 (a) K. G. Neoh, T. C. Tau, and E. T. Kang, *Polymer*, 1988, **29**, 553. (b) M. Zagorska, A. Pron, S. Lefrant, Z. Kucharski, J. Suwalski and P. Bernier, *Synth. Met.*, 1987, **18**, 43.
16 (a) J. E. Osterholm, P. Sunila and T. Hjertberg, *Synth. Met.*, 1987, **18**, 169. (b) H. Neugebauer, G. Nauer, A. Neckel, G. Tourillon, F. Garnier and P. Lang, *J. Phys. Chem.*, 1984, **88**, 652.
17 (a) A. J. Epstein, and A. G. MacDiarmid, *Mol. Cryst. Liq. Cryst.* 1988, **160**, 165. (b) S. H. Glarum and J. H. Marshall, *J. Phys. Chem.*, 1988, **92**, 4210.
18 J. H. Kaufman, N. Colaneri, J. C. Scott, K. K. Kanazawa and G. B. Street, *Mol. Cryst. Liq. Cryst.* 1985, **118**, 171.
19 (a) G. Dian, G. Barbey and B. Decroix, *Synth. Met.*, 1986, **13**, 281. (b) N. Colaneri, M. Nowak, D. Spiegel, S. Hotta and A. J. Heeger, *Phys. Rev. B*, 1987, **36**, 7964.
20 H. H. S. Javadi, R. Laversanne, A. J. Epstein, R. K. Kohli, E. M. Scherr and A. G. MacDiarmid. *Synth. Met.*, 1989, **29**, E439.
21 J. C. Scott, P. Pfluger, M. T. Krounbi and G. B. Street, *Phys. Rev. B*, 1983, **28**, 2140.

LOW-DIMENSIONAL ELECTRICALLY CONDUCTIVE SYSTEMS
INTERCALATED POLYMERS IN V_2O_5 XEROGELS

Chun-Guey Wu[(a)], Mercouri G. Kanatzidis*[(a)], Henry O. Marcy[(b)], Donald C. DeGroot[(b)] and Carl R. Kannewurf[(b)]

(a) Department of Chemistry and Center for Fundamental Materials Research, Michigan State University, East Lansing MI 48824 (b) Department of Electrical Engineering and Computer Science, Northwestern University, Evanston, IL 60208

INTRODUCTION

During the last decade, sol-gel derived materials have attracted increasing attention from the chemical, materials science and physics communities, due to their demonstrated, as well as potential, value for the development of advanced structural and electronic ceramics[1]. Vanadium oxide xerogels, in particular, are unique among sol-gel generated materials, because they are electron semiconductors[2], they possess low-dimensional layered structure, and they are capable of host-guest intercalation chemistry with a variety of species[3] (i.e. alkylamines, alcohols, sulfoxides, tetrathiafulvalene (TTF), benzidine, alkali ions etc). The interlayer spacing is 11.55Å[4]. The driving force for this intercalation chemistry depends on the substrate and involves cation exchange, acid-base or redox chemistry. $V_2O_5 \cdot nH_2O$ (**I**) xerogels (n~1.5-2.0) are good oxidizing species, and participate in redox chemistry with a variety of substrates. It is well known that organic molecules, such as aniline , pyrrole and thiophene yield, upon chemical or electrochemical oxidation, technologically promising, robust, electrically conductive polymers.[5] Progress in elucidating the real structure of these materials, thus far, has been hampered by their invariably amorphous nature. The lack of structural data inhibits, to a certain degree, the development of theoretical models aimed at understanding charge transport in these materials. Furthermore, owing to their limited processability, oriented specimens for important anisotropic studies are lacking. An exciting development in the chemistry of these materials was the intercalation of conductive polymers into layered inorganic host structures such as FeOCl[6,7] and V_2O_5[8]. This can be accomplished by the *in-situ* intercalation/polymerization of the monomers, to yield layered materials containing monolayers of conductive polymers inserted in the intralamellar space of the host. When intercalated, the polymers can be thought of as being ordered, at least perpendicular to the stacking direction. These materials are interesting, because they are molecular composites of two electrically active, but chemically diverse, components: organic conductive polymers and inorganic metal-oxide bronzes.[9] In principle, such systems may exhibit unique electrical properties, which may not be possible with either component separately.[10] Recently, we reported preliminary results on the intercalative polymerization of aniline in V_2O_5 xerogels.[4] Here we report preliminary chemical, physical and electrical data on the intercalative polymerization products of pyrrole and 2,2'-bithiophene with V_2O_5 xerogels, as well as new information on the polyaniline/V_2O_5 system.

RESULTS AND DISCUSSION

Synthesis and Spectroscopy

The reaction of neat pyrrole or aniline with a H_2O-preswollen V_2O_5 xerogel is extremely fast, and results in intercalative polymerization of these monomers, to yield products of various stoichiometries, depending on the monomer/xerogel ratio used. If excess monomer is used, amorphous products are obtained. This occurs because of over-reduction of the V_2O_5, which causes major structural changes in the framework.[3d] The reaction with pyrrole is particularly difficult to control, due to its high reactivity with V_2O_5 xerogels. The resulting black shiny products have a layered structure when dilute acetonitrile solutions of pyrrole are refluxed with V_2O_5 xerogel films. The interlayer distance of the products is larger than that of the pristine xerogel, which is consistent with intercalation. Although thiophene does not react with V_2O_5 xerogel, 2,2'-bithiophene does (in refluxing CH_3CN solution) to yield $(Pbth)_{0.40}V_2O_5{\cdot}nH_2O$ (Pbth=polybithiophene, $C_8H_4S_2$). Both free-standing films and powders of these materials can be obtained, depending on the initial state of the starting xerogel. Films of these materials are quasi flexible. Table (I) summarizes the different intercalation compounds, and their corresponding interlayer separation, synthesized thus far. The X-ray powder diffraction patterns of $(C_6H_4NH)_{0.44}V_2O_5\ 0.5H_2O$ **(II)** and (I) are shown in figure 1. No *(hkl)* or *(hk0)* reflections are observed in the X-ray patterns of (I) and (II), confirming the random orientation (turbostratic structure) of the V_2O_5 slabs perpendicular to the stacking axis. This quasi-crystallinity, the high surface area and the high porosity of the V_2O_5 xerogels is essential for the success of these intercalative polymerization reactions. For example, if crystalline V_2O_5 is used instead, no intercalation occurs.

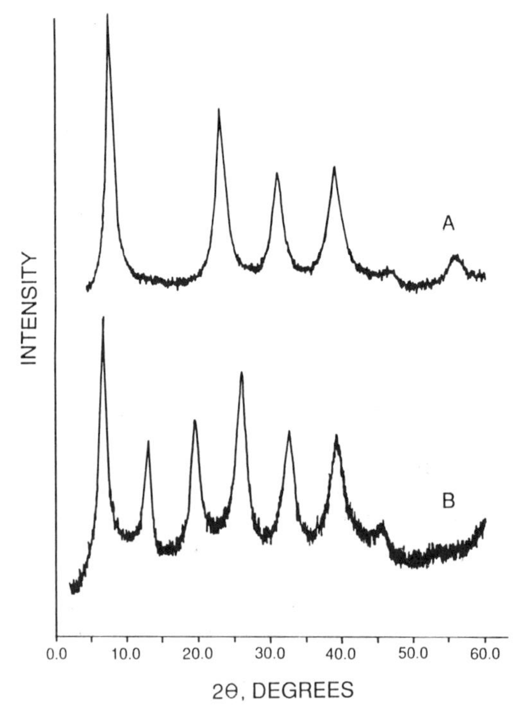

Figure 1. Typical X-ray diffraction diagram taken from films of $V_2O_5{\cdot}nH_2O$ (top) and $(C_6H_4NH)_{0.44}V_2O_5\ 0.5H_2O$ (bottom).

The formation of polypyrrole, polyaniline and polythiophene inside the host's gallery space is evident in the FT-IR spectra of these compounds, which clearly show characteristic vibrations of the corresponding polymers in the 1000-1600 cm^{-1} region (for the case of (V) see figure 2). The FT-IR spectra of these compounds compare favorably with those of chemically and electrochemically prepared (conventional) conductive polymers.[11] There is no indication from the IR spectra that monomers, dimers, or trimers might also be present. Furthermore, the corresponding conductive polymers can indeed be isolated from these materials by simply dissolving the host layers of vanadium oxide with aqueous 50% HCl or 2% NaOH solutions.[12] Such a treatment is not known to cause polymerization of the monomers. As in conventional conductive polymers, the degree of polymerization of the intercalated polymers in these materials is currently unknown.

Table I. $(Polymer)_xV_2O_5{\cdot}nH_2O$ Materials Prepared by Intercalative Polymerization Reactions of Monomers with $V_2O_5{\cdot}nH_2O$.

Formula	d-Spacing(Å)	Designation	Starting Monomer
$(C_6H_4NH)_{0.44}V_2O_5\ 0.5H_2O$	13.94(2)	(II)	Aniline
$(C_6H_4NH)_{0.21}V_2O_5\ 1.32H_2O$	13.92(2)	(III)	Aniline
$(2,6\text{-}Me_2C_6H_2NH)_{0.44}V_2O_5\ 0.47H_2O$	14.55(2)	(IV)	2,6-dimethylaniline
$(C_4H_3N)_{0.64}V_2O_5 1.5H_2O$	14.30(2)	(V)	Pyrrole
$(C_4H_2NMe)_{0.46}V_2O_5 2.39H_2O^*$	14.30(2)	(VI)	N-Methylpyrrole
$(C_4H_3N)_{0.26}V_2O_5 1.9H_2O^*$	16.60(2)	(VII)	Pyrrole
$(C_8H_4S_2)_{0.40}V_2O_5 1.5H_2O$	14.26(2)	(VIII)	2,2'-Bithiophene

* The difference in interlayer distance in these two phases may reflect the varying amounts of intercalated water as well as differences in the disposition and orientation of the polypyrrole chains in the intralamellar space.

During intercalative polymerization, the vanadium oxide framework is reduced to form V^{4+} centers. The unpaired electrons in these d^1 centers are fairly localized (small polarons), and hop between the vanadium sites, thus giving the layers a finite *thermally activated* n-type conductivity.[2] Assuming all the electrons removed by oxidation from the monomers are transferred to V_2O_5, the degree of reduction of the V_2O_5 framework should be $2x+\delta$ (according to eq(1)). Therefore the V_2O_5 framework qualifies as a bronze, and the $(Polymer)_xV_2O_5$ materials can be viewed as polymer bronzes. Alternatively, from a conductive-polymer viewpoint, the $[V_2O_5]^{(2x+\delta)-}$ portion of the compound can be regarded as the *dopant.*

$$x\,C_6H_5\text{-}NH_2 + V_2O_5.nH_2O \longrightarrow \left\{\left[\left(C_6H_4\text{-}NH\right)_n\right]^{\delta+}_x V_2O_5.nH_2O\right\} + 2xH^+ + 2xe + \delta \quad (1)$$

The reaction of toluidine (4-methylaniline) with $V_2O_5{\cdot}nH_2O$ is facile and results in a dark-blue material. However, after treating this material with HCl or NaOH solutions to dissolve the $V_2O_5{\cdot}nH_2O$ matrix, no polymer was found, which is consistent with the inability of toluidine to polymerize oxidatively. The oxidation product in this case is a soluble, neutral, oligomeric, organic species and the dark-blue material is the corresponding hydrogen bronze $H_xV_2O_5{\cdot}nH_2O$. By contrast, and as expected, 2,6-dimethylaniline (DMAN) *does* intercalatively polymerize to yield $\{poly(DMAN)\}_{0.40}V_2O_5{\cdot}0.50H_2O$ (IV), which possesses a slightly larger interlayer distance of 14.55Å. Indeed a black-polymer was isolated from (IV), which resembles, spectroscopically and electrically, an authentic sample of {poly(DMAN)}HCl.[13] By analogy, the use of 2,5-dimethylpyrrole did not result in intercalative polymerization either, while N-methylpyrrole did.[13]

The electron paramagnetic resonance (EPR) spectra of (II) through (VIII) show broad (ΔH_{pp}~250 G) isotropic signals centered around g~1.96, and arise primarily from d^1 vanadium centers. The typical narrow EPR signals, arising from static defects in conducting polymers, are not observed in these materials, suggesting that the unpaired spins on such static defects are magnetically coupled to the proximal unpaired spins in the V_2O_5 layers.

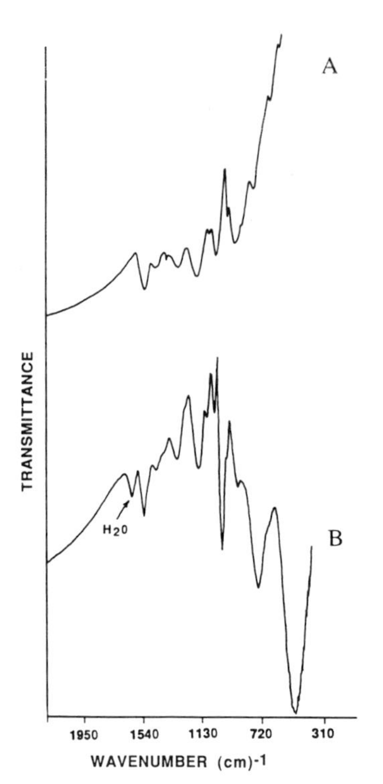

Figure 2. Fourier Transform Infra-Red spectra of bulk $(Ppy)Cl_{0.35}$, (A) and $(C_4H_3N)_{0.64}V_2O_5 1.5H_2O$, (B). The strong peaks observed below 1000 cm^{-1} arise from the V_2O_5 layers.

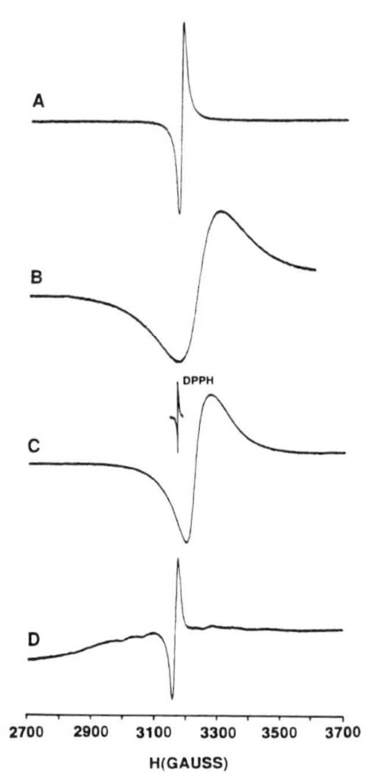

Figure 3. Room temperature X-band EPR spectra of (a) a bona-fide sample of chemically prepared polyaniline hydrochloride (conductive form) (b) $(C_6H_4NH)_{0.44}V_2O_5nH_2O$ (II) (c) $Na_{0.40}V_2O_5nH_2O$ (d) polymer extracted from (II); the distortion of the baseline at low field is due to residual paramagnetic vanadium impurities. Spectrometer frequency, 9.06 GHz. Microwave Power, 20mW. Gain, $4x10^2$. Modulation frequency, 100 KHz.

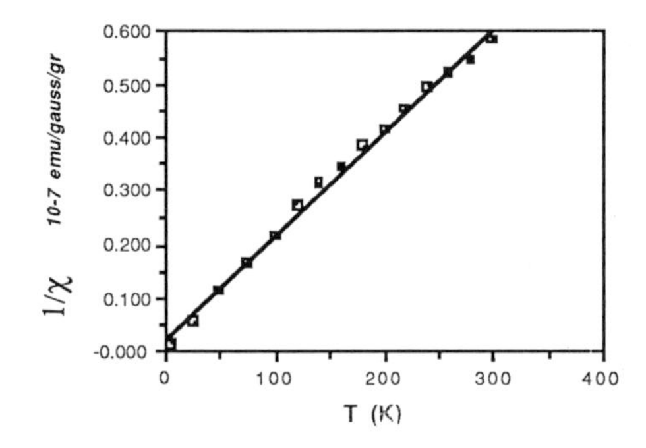

Figure 4. Magnetic susceptibility data for $(C_6H_4NH)_{0.44}V_2O_5 0.5H_2O$ showing the paramagnetic nature of this compound.

Therefore, the magnetic behavior of these polymer bronzes is dominated by the V_2O_5 network.

Figure 3 shows the EPR spectra of protonated polyaniline, (II), $Na_{0.40}V_2O_5nH_2O$ and polyaniline hydrochloride extracted from (II). $Na_{0.40}V_2O_5nH_2O$ was prepared as a model compound to represent the $V_2O_5nH_2O$ portion (reduced) of the

$(Polymer)_xV_2O_5nH_2O$ materials. The broad signal observed for (II) arises from interacting V^{4+} centers. The absence of the typical polyaniline signal in (II), suggests the presence of antiferromagnetic exchange between unpaired density on the polymer backbone and V_2O_5. The EPR signal of polyaniline[14] is recovered only after the polymer is separated from the host matrix, figure 4d. These data show these materials are pure phases, and not mixtures of polymer and V_2O_5 phases. The EPR signals of (II) and $Na_{0.40}V_2O_5nH_2O$ are similar, consistent with reduction of the V_2O_5 network in (II). Consistent with the EPR data, preliminary variable temperature magnetic susceptibility data of (II) (see figure 4) and (V) suggest that these compounds are paramagnetic, obeying Curie-Weiss law. A detailed analysis of the magnetic and EPR data is in progress.

Charge Transport Measurements

The electrical properties of these materials are strongly dependent on the polymer/V_2O_5 ratio. Figure 5 shows the electrical conductivity of (II), (V) and (VIII) in a wide temperature range. The decrease in conductivity with temperature for all samples, indicates thermally activated behavior, as is typical for most conductive polymers[5] and V_2O_5 bronzes.[9] Thus far, we have found that compounds with high polymer/V_2O_5 ratio show higher conductivities than corresponding samples with low ratio (less than 0.3). Since both the polymer and the V_2O_5 portions in these materials are potentially capable of charge transport, the question, arises as to which would be the preferred charge transport medium. This is an important question since the type and mechanisms of charge transport will be different through the various chemical components. For example, if the V_2O_5 portion is the preferred conduction medium, then n-type semiconductivity is expected. However, if the polymer section dominates the conductivity, then p-type (for polypyrrole and polythiophene) and n-type (for aniline) metal-like behavior should be observed. An interesting situation may arise, at an appropriate stoichiometry, when both charge transport media (organic and inorganic) are equally favorable. Unlike electrical conductivity measurements, thermoelectric power measurements could provide answers to these questions. Preliminary thermoelectric power data obtained from these materials indicate that, in compounds with high polymer/V_2O_5 ratio, the charge transport is primarily through the polymer section; whereas, in compounds with low polymer/V_2O_5 ratio, the inorganic network is the primary conductor. The results of the thermoelectric power studies (in a limited temperature range) are shown in table II.

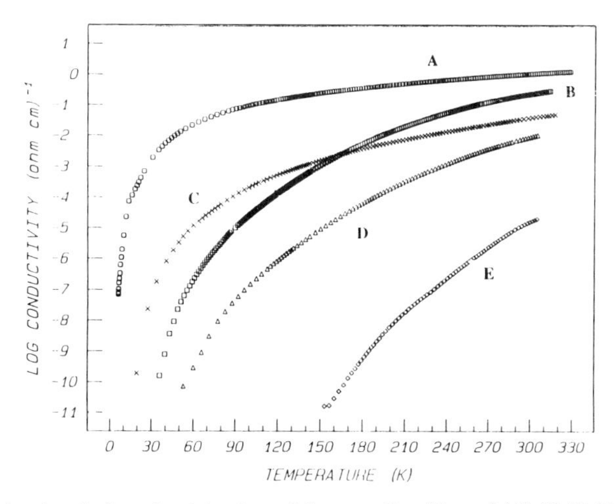

Figure 5. Four-probe electrical conductivity data of free standing films of (A) $(C_4H_3N)_{0.64}V_2O_5 \cdot 1.5H_2O$, (B) $(C_8H_4S_2)_{0.40}V_2O_5 \cdot 1.5H_2O$, (C) $(C_6H_4NH)_{0.44}V_2O_5 \cdot 0.5H_2O$ and (D) $Na_{0.40}V_2O_5 \cdot nH_2O$ and (E) pristine $V_2O_5 \cdot nH_2O$

The dependence of electrical properties on the amount of intercalated conductive polymer probably reflects the variation in the average polymer interchain distance in the high and low polymer-content materials. Low polymer/V_2O_5 fractions will result in long interchain distances,thus making interchain electron (or hole) hopping less favorable. In such a case charge transport through the polypyrrole is inhibited and conductivity from the V_2O_5 network takes over,giving rise to n-type semiconducting characteristics. This is supported by the fact that alkali metal bronzes of these xerogels such as $Na_{0.40}V_2O_5nH_2O$ (n~1.50-2.0) exhibit conductivity and thermoelectric power very similar to those of (III) and (VII).[13]

Table II. Thermoelectric Power Data and Conductivity-type for $(Polymer)_xV_2O_5 \cdot nH_2O$ Materials

Formula	Seebeck coeff* μV/K	Carrier type	σ $\Omega^{-1}cm^{-1}$ at 300 K	Behavior
$(C_6H_4NH)_{0.44}V_2O_5\ 0.5H_2O$	-30	n	0.1	metal-like
$(C_6H_4NH)_{0.21}V_2O_5\ 1.32H_2O$	-120	n	10^{-3}	semiconductor
$(C_4H_3N)_{0.64}V_2O_51.5H_2O$	+15	p	0.5	D. S.***
$(C_4H_2NMe)_{0.46}V_2O_52.39H_2O$	~-300	n	10^{-4}	semiconductor
$(C_4H_3N)_{0.26}V_2O_51.9H_2O^*$	-180	n	10^{-3}	semiconductor
$(C_8H_4S_2)_{0.40}V_2O_51.5H_2O$	-5+-20**	p	0.2	metal-like(T>210K)

* Value at room temperature

** The behavior of this material is unusual. The Seebeck coefficient is slightly negative at 300 K but quickly crosses to positive values and remains positive down the lowest temperature measured (~100 K)

*** Degenerate semiconductor

Acknowledgement. Financial support from the Center of Fundamental Materials Research for Michigan State University is gratefully acknowledged. At Northwestern University, support was provided by ONR; this work made use of Central Facilities supported by NSF through the Materials Research Center.

REFERENCES

1) (a) Brinker, C.J.; Clark, D.W.; and Ulrich, D.R.(Eds) in "Better Ceramics Through Chemistry III", *Mat. Res. Soc. Symp. Proc.* **1986,** Vol. 73, references therein. (b) Gross, M.E.; Jasinsky, J.M.; Yates, J.T. Jr. (Eds) in "Chemical Perspectives of Microelectronic Materials" *Mat. Res. Soc. Symp. Proc.* **1989,** Vol. 131, and references therein.

2) (a) Bullot, J.; Gallais, O.; Gauthier, M.; Livage, *J. Appl. Phys. Lett.* **1980**, 36,986-988. (b) Sanchez, C.; Babonneau, F.; Morineau, R.; Livage, J.; Bullot, J. *Philos. Mag. [Part] B* **1983**, 47, 279-290. (c) Bullot, J.; Cordier, P.; Gallais, O.; Gauthier, M.; Livage, J. *J. Non-Cryst. Solids* **1984**, 68, 123-134.

3) (a) Aldebert, P.; Baffier, N.; Legendre, J.-J.; Livage, J. *Revue Chim. Min.* **1982,** 19, 485-495. (b) Lemordant, D.; Bouhaouss, A.; Aldebert, P.; Baffier, N. *J. de Chim. Physique* **1986,** 83, 105-113. (c) Aldebert, P.; Baffier, N.; Gharbi, N.; Livage, J. *Mat. Res. Bull.* **1981,** 16, 949-955. (d) Masbah, H.; Tinet, D.;Crespin, M.; Erre, R.; Setton, R.; Van Damme, H. *J. Chem. Soc., Chem. Commun.* **1985**,935-936

4) (a) Legendre, J.-J.; Livage, J. *J. Colloid Interface Sci.* **1982,** 94, 75-83. (b) Legendre, J.J.; Aldebert, P.; Baffier, N.; Livage, J. *J. Colloid Interface Sci.* **1982,** 94, 84-89. (c) Aldebert, P.;Haesslin, H.W.; Baffier, N.; Livage, J. *J. Colloid*

Interface Sci. **1983**, 98, 478-483. (d) Gharbi, N.; Sanchez, C.; Livage, J.; Lemerle, J.; Nejem, L.; Lefebvre, J. *Inorg. Chem.* **1982,** 21, 2758-2765.

5) (a) For a recent review of the field see: "Proceedings of the International Conference on Science and Technology of Synthetic Metals (ICSM'88)" Aldissi, M. (ed) *Synth. Met.* **1989**, 27-29, references therein. (b) Skotheim, T.A. (ed) in "Handbook of Conductive Polymers", Marcel Dekker: New York, Vol. 1,2, 1986 (c) Marks, T. J. *Science*, **1985**, 227, 881-889

6) (a) Kanatzidis, M. G.; Tonge, L. M.; Marks, T. J.; Marcy, H. O.; Kannewurf, C. R. *J. Am. Chem. Soc.* **1987**, 109, 3797-3799 (b) Kanatzidis, M. G.; Marcy, H. O.; McCarthy, W. J.; Kannewurf, C. R.; Marks, T. J. *Solid State Ionics* **1989** in press (c) Kanatzidis, M. G.; Hubbard, M.; Tonge, L. M.; Marks, T. J.; Marcy, H. O.; Kannewurf, C. R. *Synth. Met.* **1989**, 28, C89-C95

7) Wu, C.-G.; Kanatzidis, M.G.; Marcy, H.O.; DeGroot, D.C.; Kannewurf, C.R. submitted for publication

8) (a) Wu, C.-G.; Kanatzidis, M.G.; Marcy, H.O.; Kannewurf, C.R. *J. Am. Chem. Soc.* **1989**, 111, 4139-4141 (b) Wu, C.-G.; Kanatzidis, M.G.; Marcy, H.O.; DeGroot, D.C.; Kannewurf, C.R. *Polym. Mat. Sci. Eng.* **1989**, 61, in press.

9) (a) Hagenmuller, P. in "Non-stoichiometric Compounds, Tungsten Bronzes, Vanadium Bronzes and Related Compounds" Bevan, D.J.; Hagenmuller, P. (eds), Vol. 1, Pergamon Press, Oxford 1973

10) Day, P. *Phil. Trans. R. Soc. Lond. A* **1985**, 314, 145-158

11) (a) Street, G.B. in ref 5b Vol. 1, pp 280.(b) Vachon, D.; Angus, R. O. Jr; Lu, F. L.; Nowak, M.; Liu, Z. X.; Schaffer, H.; Wudl, F.; Heeger, A. J. *Synth. Met.* **1987,** 18, 297-302

12) TGA experiments under flowing argon, done on all samples, show that after an initial loss of water a thermally resistant component exists with no appreciable volatility up to 350 ^{o}C. The final decomposition residue is crystalline orthorhombic V_2O_5.

13) Wu, C.-G.; Kanatzidis, M.G.; Marcy, H.O.; Kannewurf, C.R. to be submitted for publication.

14) Javadi, H.H.S.; Laversanne, R.; Epstein, A.J.; Kohli, R.K.; Scherr, E.M.; MacDiarmid, A.G. *Synth. Met.* **1989**, 29, E439-E444

GAMMA IRRADIATION OF POLY(PYRROLE)-COATED Pt ELECTRODES: THE EFFECT ON THE ELECTROCHEMICAL BEHAVIOR

Meral Arca, Emin Arca*, Olgun Güven, and Attila Yildiz

Departments of Chemistry and *Chemical Engineering
Hacettepe University
Beytepe, Ankara, 06532, TURKEY

INTRODUCTION

The effect of gamma irradiation on poly(pyrrole) (PPy)-coated electrodes was investigated by cyclic voltammograms, taken with ferrocene and benzoquinone systems, in the positive and negative potential regions of the electrochemical window of PPy, respectively. It was observed that gamma irradiation does not change the response of the electrode in the positive potential region, whereas cyclic voltammograms in the negative potential region are deformed almost instantly, when the electrodes are exposed to air after irradiation. ESR studies showed that the physisorption of oxygen on the irradiated PPy is accelerated. Furthermore it was determined by IR spectroscopy that some oxygen is chemically bonded in the structure of the polymer, which is exposed to air following irradiation.

Radiation processing with gamma rays is a commonly used technique in the field of polymer research. The interaction of highly energetic gamma rays with the polymer skeleton causes some important changes in the chemical and physical properties of the polymer. These changes depend, not only on the polymer, but also on the temperature, the medium in which irradiation is carried out, as well as the duration and the dose of irradiation. Some of the chemical changes are: chain breakage, cross-linking, bond formation, formation of gaseous products, and immobilization of reactive radical intermediates in the polymer matrix. Gamma irradiation is also commonly used to prepare graft copolymers in a controlled manner. Physical properties, like mechanical strength, color, conductivity and crystallinity of the polymer may also be changed by radiation processing.

Gamma irradiation was used [1] to prepare non-conductive polymer network-coated graphite and platinum electrodes, using poly(dimethyldiallyl ammonium chloride), poly(acrylonitrile) and poly(ethylenimine). The conducting polymer polythiophene was doped with SF_5^-, by irradiating the sample with gamma rays in an SF_6 atmosphere [2]. In the course of our attempts [3] to obtain a radiation-induced graft copolymer of pyrrole and 4-vinylpyridine, we have detected important, radiation-induced, changes in the physical and chemical properties of poly(pyrrole) itself.

We report here the result of an electrochemical study of the effect of gamma irradiation on a poly(pyrrole)-coated Pt electrode. The effect of the irradiation on the chemical structure of poly(pyrrole) was followed by changes in its vibrational spectrum. The results were interpreted, together with ESR spectroscopic data obtained with irradiated and nonirradiated samples of poly(pyrrole).

Lower-Dimensional Systems and Molecular Electronics
Edited by R.M. Metzger *et al.*, Plenum Press, New York, 1991

EXPERIMENTAL

Poly(pyrrole) films were prepared on a Pt electrode, by the electro-oxidation of freshly distilled pyrrole in nonaqueous acetonitrile, containing 0.1 M tetraethylammonium tetrafluoroborate as the supporting electrolyte. All electrochemical studies were carried out in a dry and oxygen-free nitrogen atmosphere. Acetonitrile was distilled in five steps. It was distilled over CaH_2 (2 g/L), followed by a second distillation over $AlCl_3$ 10 g//L). The third distillation was carried out in the presence of $KMnO_4$ (10 g/L) and Li_2CO_3 (10 g/L). $KHSO_4$ (15 g/L) was the reagent used for the fourth distillation. Finally, the distillate was redistilled over CaH_2 (2 g/L), and stored under nitrogen. The acetonitrile, obtained after these procedures, showed no measurable absorbance, down to 196 nm.

The supporting electrolyte was prepared through the reaction of tetrafluoroboric acid with tetraethylammonium bromide, and was recrystallized from a methyl alcohol: petroleum ether mixture. It was kept under nitrogen atmosphere after drying in vacuum.

Electrochemical measurements were carried out in a three-electrode cell, with separate compartments for the reference electrode (Ag | AgCl) and for the counter electrode (Pt spiral); Pt disk electrode (area A = 0.44 cm^2) was used as the working electrode. All poly(pyrrole) films were prepared by the controlled-potential electrolysis of the monomer at 0.9 volt. The thickness of the films was controlled by counting the charge passed during the anodic electrolysis. For electrochemical work, 0.8 μm thick poly(pyrrole) films, deposited on Pt, were used. For spectroscopic investigations, 5 μm thick films were deposited, and stripped off the Pt surface. Ferrocene (BDH) was used as a depolarizer in the positive potential region (0.0 to +1.0 Volt). For the negative potential region (0.0 to -1.0 Volt), benzoquinone (BDH) was used.

A PAR Model 173 Potentiostat and a PAR Model 179 Digital Coulometer were used for controlled- potential electrolysis experiments. The potentiostat was connected to a PAR Model 175 Universal Programmer and a Houston Instruments Model 2000 X-Y Recorder for cyclic voltammetry experiments. A Perkin-Elmer Model 1710 FTIR Spectrophotometer and Varian Model E-9 ESR Spectrometer were used in spectroscopic measurements. For the ESR spectra taken in vacuum, the samples were filled in ESR tubes, pumped down to 10^{-4} torr for one hour, and sealed on a vacuum line. The ESR spectra were taken at room temperature. A Gammacell Model 220 ^{60}Co gamma ray source was used for irradiation.

RESULTS AND DISCUSSION

Table 1 summarizes the cyclic voltammetric results obtained with ferrocene solutions on Pt/PPy electrode, before and after irradiation with gamma rays. The results obtained, after the exposure of this electrode to air following irradiation, are also included in Table 1. The cyclic voltammograms do not change their shape, neither with irradiation nor with exposure to air after irradiation. The ΔE_p values vary between 90 and 160 mV, the anodic peak current changes linearly with the square root of the potential scan rate, and the peak current ratios are close to unity under all conditions.

Table 2 shows the results of the cyclic voltammetry experiments carried out with benzoquinone solutions. It is seen that the cyclic voltammetric response is also well defined, before and after gamma irradiation. As soon as the electrode is exposed to air, however, important deformations in the cyclic voltammetric peaks are observed (Figs. 1,2). A very well-defined peak of ferrocene was still observed with this electrode, while the cyclic voltammetric peak of benzoquinone vanishes irreversibly.

Comparison of the finger-print region of the IR spectra of irradiated and nonirradiated PPy samples (Fig. 3) reveals the appearance of new sharp bands at 1080 and 1130 cm^{-1}. The same bands were observed with the films prepared with intentionally added water to acetonitrile during electropreparation of the polymer [4]. These bands were assigned to a C = C - O - C stretching vibration. It is clear that oxygen is induced to chemically bind to the polymeric structure, after the irradiated electrodes are exposed to air.

Table 1. Cyclic voltammetric results of 5 x 10^{-3} M ferrocene solution on Pt/PPy before, and after, irradiation with gamma rays (A: 0.44 cm^2, film thickness: 0.8 μm, monomer concentration: 5.5 x 10^{-2} M, irradiation dose: 0.3 MRad).

Electrode	Scan rate (mV/s)	E_p^a (mV)	E_p^c (mV)	ΔE_p (mV)	i_p^a (μA)	i_p^c (μA)	i_p^a/i_p^c
Pt/PPy before irradiation	10	425	335	90	240	235	1.02
	20	430	330	100	325	325	1.00
	50	450	325	125	485	490	0.99
	100	460	310	150	675	680	0.99
Pt/PPy after irradiation	10	425	335	90	240	240	1.00
	20	430	330	100	320	325	0.98
	50	450	320	130	485	480	0.99
	100	465	310	155	675	680	0.99
after 1 min. exposure to air following irradiation	10	425	330	95	240	240	1.00
	20	435	325	100	325	325	1.00
	50	450	350	130	485	490	0.99
	100	460	310	150	680	680	1.00
after 10 min. exposure to air following irradiation	10	425	335	90	240	245	0.98
	20	430	330	100	325	320	1.02
	50	450	325	125	485	490	0.99
	100	464	305	160	675	685	0.99

$PPy^+ClO_4^-$ gives a sharp ESR line in vacuum, which broadens significantly when the sample is exposed to oxygen [5]. The original sharp signal can be obtained, if the sample is pumped down again, and the broadening of the signal has been attributed to the spin-flip relaxation process, as the physisorbed oxygen diffuses through the polymeric matrix [5]. We have also observed this behavior for the $PPy^+BF_4^-$ samples prepared from dry acetonitrile. In vacuum, the peak-to-peak width of the ESR line was 0.7 G; after the uptake of oxygen, the

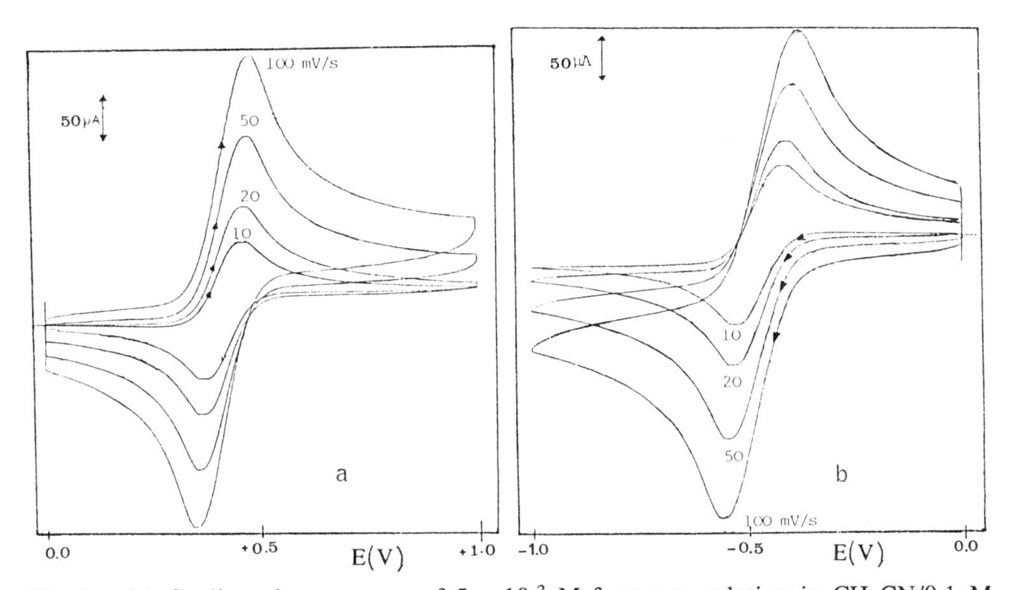

Fig. 1. (a) Cyclic voltammogram of 5 x 10^{-3} M ferrocene solution in CH_3CN/0.1 M Et_4NBF_4 on Pt | PPy electrode before irradiation
(b) Cyclic voltammogram of 5 x 10^{-3} M benzoquinone solution in CH_3CN/0.1 M Et_4NBF_4 on Pt | PPy electrode before irradiation

Table 2. Cyclic voltammetric results of 5 x 10^{-3} M benzoquinone solution on Pt/PPy before and after irradiation with gamma rays (A: 0.44 cm^2, film thickness: 0.8 μm, monomer concentration: 5.5 x 10^{-2} M, irradiation dose: 0.3 MRad).

Electrode	Scan rate (mV/s)	$-E_p^a$ (mV)	$-E_p^c$ (mV)	ΔE_p (mV)	i_p^a (μA)	i_p^c (μA)	i_p^a/i_p^c
Pt/PPy before irradiation	10	545	440	105	200	210	0.95
	20	555	435	120	270	270	1.00
	50	565	430	135	480	485	0.99
	100	575	415	160	630	635	0.99
Pt/PPy after irradiation	10	545	440	105	200	210	0.95
	20	555	430	125	270	275	0.98
	50	565	430	135	480	485	0.99
	100	570	415	155	630	630	1.00
after 1 min. exposure to air following iradiation	10	585	425	160	160	170	0.94
	20	600	415	185	210	230	0.91
	50	615	385	230	305	335	0.92
	100	630	360	270	400	455	0.88
after 10 min. exposure to air following irradiation	10	615	400	215	100	120	0.83
	20	630	380	250	135	170	0.79
	50	640	340	300	220	265	0.83
	100	650	300	350	270	330	0.82

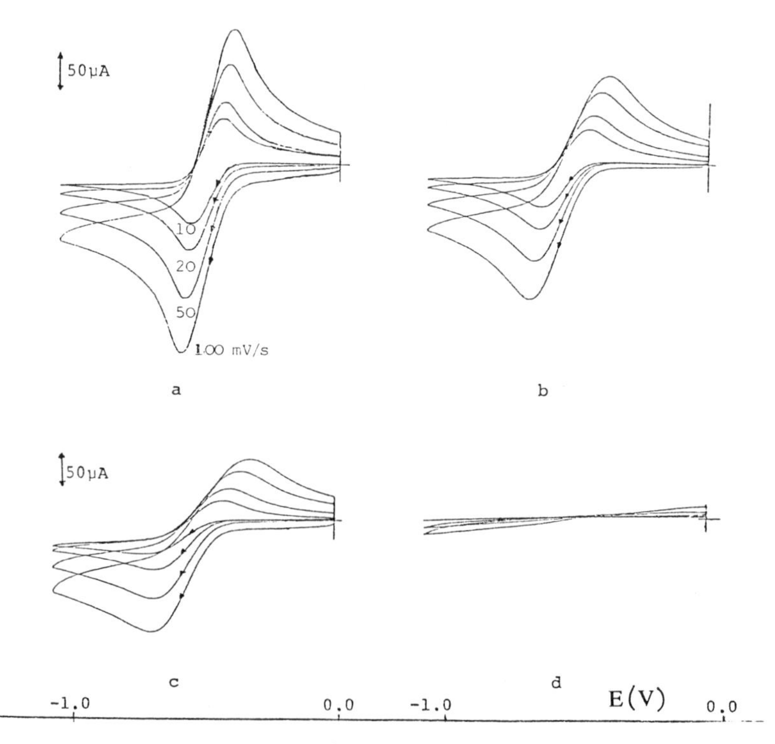

Fig. 2. Cyclic voltammograms of 5 x 10^{-3} M benzoquinone solution on a Pt | PPy electrode irradiated with gamma rays (area 0.44 cm^2, film thickness=0.8 μm, monomer concentration 5.5 x 10^{-2} M, irradiation dose 0.3 MRad)
(a) Right after irradiation in vacuum, (b) After 1-minute exposure to air, (c) After 10-minute exposure to air, (d) after 24-hour exposure to air

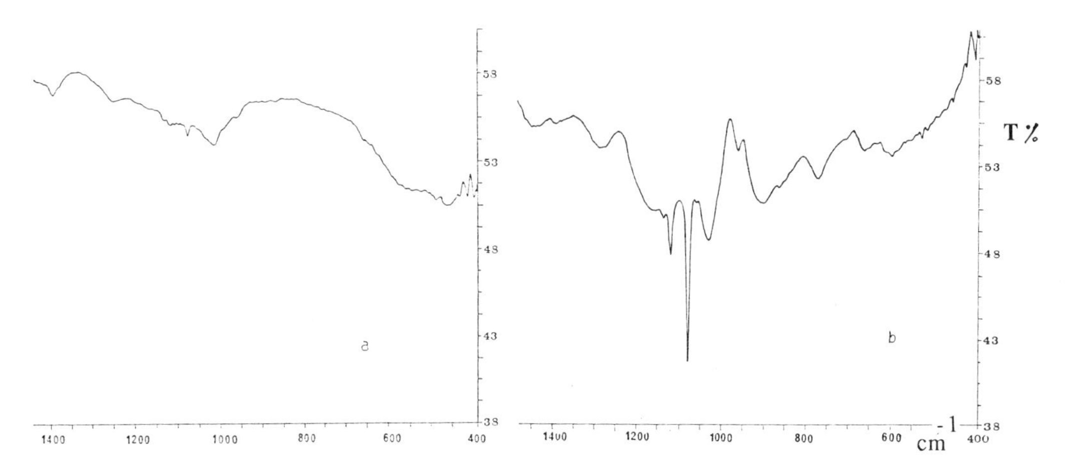

Fig. 3. FTIR spectra of (a) non-irradiated, (b) irradiated samples of PPy (irradiation dose 0.3 MRad)

peak-to-peak linewidth increased to 8 G. Upon the removal of physisorbed oxygen, the broad line disappears, and the sharp line with 0.7 G linewidth returns. The intensity of the broad signal increases with time, and saturation in the signal intensity is reached after a few days.

The ESR behavior of the gamma-irradiated samples of $PPy^{+}BF_4^{-}$ also resembles that of the non-irradiated sample, as far as the broadening of the signal is concerned. Firstly, the line intensity of the ESR signal increases more than twice with irradiation [6]. Secondly, the oxygen uptake is accelerated, upon exposure to air following irradiation; and a saturation in the intensity of the broad line is reached after 40 minutes.

CONCLUSION

Gamma-irradiation of PPy induces changes in its chemical structure, and probably forms some oxypolypyrrole structure, as has been proposed in the literature [7]. Furthermore, gamma irradiation creates more new sites for the physisorption of oxygen molecules. It is probably the latter effect, which is responsible for the deformation of the cyclic voltammetric peaks of benzoquinone, taken with an irradiated electrode which is exposed to air. The presence of oxygen, in the pores of the polymer in a physisorbed state, prevents benzoquinone from reaching the Pt surface for reduction. The pore diffusion of the depolarizer, at the negative potentials applied to the PPy electrode, is hindered in a very short time, if the electrode is irradiated and allowed to interact with oxygen. The direct electron-transfer process, from ferrocene to polymer chain in the positive potential region, is not affected at all by radiolysis of the polymer, up to 0.5 MRad of irradiation.

ACKNOWLEDGEMENT

The electrochemical instruments used in this research were dontated to A. Yildiz by the Alexander von Humboldt Foundation. The authors thank Y. Kaptan for his valuable help in obtaining ESR data.

REFERENCES

[1] E. S. De Castro, E. W. Huber, D. Villaroel, C. Galiatsatos, J. E. Mark, W. R. Heineman, and P. T. Murray, Electrodes with Polymer Network Films Formed by Gamma Irradiation Cross-Linking, Anal. Chem. 59: 134 (1987).

[2] S. Hayashi, S. Takeda, K. Kaneto, K. Yoshino, and T. Matsuyama, Radiation- Induced Effect in Conducting Polymers, Synth. Met. 18: 591 (1987).
[3] M. Arca, E. Arca, A. Yildiz, and O. Güven, Preparation of an Electroactive Copolymer by Radiation-Induced Grafting of n-Vinyl-4-Pyridine onto Polypyrrole, Radiat. Phys. Chem. 31: 647 (1988).
[4] M. Arca, Ph.D. Thesis, Investigation of the Structure of Polypyrrole by Electrochemical and Spectroscopic Techniques, and the Effect of Gamma Rays on Polypyrrole, Hacettepe University, Ankara, Turkey, 1986.
[5] J. C. Scott, P. Pfluger, M. T. Krounbi , and G. B. Street, Electron Spin Resonance Studies of Pyrrole Polymers: Evidence for Bipolarons, Phys. Rev. B28: 2140 (1983).
[6] M. Arca, E. Arca, Y. Kaptan, A.Yildiz, and O. Güven, The Effect of Gamma Rays on the Conducting Properties of Polypyrrole, 6th Tihany Symp. on Radiation Chemistry, Balatonszéplack, Hungary (1986).
[7] R. A. Bull, F. F. Fan and A. J. Bard, Electrochemical Behavior at Polypyrrole- Coated Platinum and Tantalum Electrodes, J. Electrochem. Soc. 129: 1009 (1982).

BEYOND THE HUBBARD MODEL: SCREENED INTERACTIONS IN 1D

Anna Painelli

Centro Studi Stati Molecolari Radicalici ed Eccitati del CNR
Department of Chemical Physics, Padova University
Padova, ITALY

Alberto Girlando

Institute of Chemical Physics, Parma University
Parma, ITALY

INTRODUCTION

A central role in the understanding of the fundamental, and rather peculiar, properties of conducting polymers has been played by very simple, yet very elegant, one-electron theories, like SSH [1]. However, in recent years, several features have been experimentally observed, which cannot be rationalized in terms of one-electron theories, unambiguously pointing towards the need of including electron-electron (e-e) interactions in complete models for conducting polymers [2]. Hubbard models, which explicitly account for e-e interactions, are widely and successfully applied to investigate the properties of quasi one-dimensional (1D) charge-transfer (CT) salts [3], but they cannot be transferred to polymers without due modification. In fact, conducting polymers have larger bandwidths than CT salts. Moreover, the range of the interelectronic potential may be very different in the two kinds of systems, due to the different screening properties of the materials [4]. In the next Section we briefly discuss how e-e interactions can be modeled for 1D systems with different bandwidth and screening properties. Then we focus on half-filled systems, and investigate their instabilities in terms of the proposed models.

SETTING THE STAGE: ELECTRON-ELECTRON INTERACTIONS IN 1D

The problem of modeling e-e interactions in a system, whose one-electron hamiltonian is written in terms of the tight-binding picture, has recently been extensively discussed [5-7], at least in the two extreme cases of a poorly screened, long-range interelectronic potential, and of a completely screened, δ-function interelectronic potential. The fundamental ingredients of the discussion are: (i) the observation that in a tight-binding picture the overlap between orbitals on adjacent sites (S) is a small but non-negligible quantity, and can be conveniently chosen as an expansion parameter; (ii) the orthogonalization, up to a given order in S, of the starting atomic or molecular tight-binding basis. Skipping the details of the treatment, we summarize briefly the main results as follows.

In the case of a long-range potential, or, more precisely, when the range of the potential is much larger than the range of the on-site wavefunctions, only the site-diagonal elements of the matrix of the interelectronic potential have to be retained in the model. Up to the first order in S, one ends up with the well known extended Hubbard model (EHM):

$$H = t\,\Sigma_{i,\sigma}(a_{i,\sigma}^{+}a_{i+1,\sigma} + \mathrm{H.c.}) + U\,\Sigma_{i}\, a_{i,\alpha}^{+}a_{i,\beta}^{+}a_{i,\beta}a_{i,\alpha} + \Sigma_{i>j}(-1)^{i+j}V_{ij}n_i n_j \qquad (1)$$

Lower-Dimensional Systems and Molecular Electronics
Edited by R.M. Metzger *et al.*, Plenum Press, New York, 1991

where $a_{i,\sigma}$ ($a_{i,\sigma}^+$) is the Fermi annihilation (creation) operator of one electron with spin σ on site i, n_i is the usual electron number operator for site i, and t is the (negative) CT integral between nearest-neighbor sites. U and V_{ij} represent the repulsive interactions between two electrons residing on the same site and on i and j sites, respectively. At higher orders in S, CT interactions, beyond the nearest neighbor-one, have to be added to the previous hamiltonian, whereas the two-electron part does not change. This sharp result is independent of the detailed shape of the on-site orbitals.

In the case of a δ-function potential, the situation is more complex. Up to the first order in S, and for rather large S values ($S \geq 0.2$), the system may be described in terms of the simple Hubbard model (SHM):

$$H = t\, \Sigma_{i,\sigma}(a_{i,\sigma}^+ a_{i+1,\sigma} + \text{H.c.}) + U\, \Sigma_i\, a_{i,\alpha}^+ a_{i,\beta}^+ a_{i,\beta} a_{i,\alpha} \tag{2}$$

At smaller S values, a non-site diagonal term has to be introduced:

$$H = t\, \Sigma_{i,\sigma}(a_{i,\sigma}^+ a_{i+1,\sigma} + \text{H.c.}) + U\, \Sigma_i\, a_{i,\alpha}^+ a_{i,\beta}^+ a_{i,\beta} a_{i,\alpha}$$
$$+ 2\, X\, \Sigma'_{i,\sigma}(a_{i,\sigma}^+ a_{i+1,\sigma} + \text{H.c.})(a_{i,\sigma'}^+ a_{i,\sigma'} + a_{i+1,\sigma'}^+ a_{i+1,\sigma'}) \tag{3}$$

The X term accounts for the interaction between electrons residing on a site and on the adjacent bond. It is important to notice that X is a <u>negative</u> quantity [7], indicating <u>effective attraction</u> between site and bond electronic densities. This rather surprising result originates from the orthogonalization of the on-site wavefunctions. Extending the treatment up to the S^2 order [7], the situation becomes even more complicated. In fact, whereas at fairly large S values ($S \geq 0.3$) SHM is regained, at smaller S ($0.25 \leq S \leq 0.3$), the negative X term has to be added to the model, together with a <u>positive</u> next-nearest neighbor CT interaction. Finally, at even smaller S, other e-e interaction terms become non-negligible, and the complete model becomes rather intractable. In the case of a δ-function potential, the results depend on the actual choice of the on-site wavefunctions. The above results have been obtained by choosing the 2p Slater atomic orbitals of the carbon atom, so, strictly speaking, they are relevant only to polyacetylene and related systems. On the other hand, due to the fact that atomic or molecular orbitals always show an exponential decay with the site distance, we believe that the above discussion can be rather safely extended to the vast majority of 1D materials.

INSTABILITIES IN 1D HALF-FILLED SYSTEMS: THE VALENCE BOND VIEW

Having reviewed the models relevant to 1D systems with very long and very short range interelectronic potentials, we are now in the position of looking at their properties. In this paper we consider 1D half-filled systems, and investigate the effect of e-e interactions on their instabilities. For the sake of simplicity, we limit attention to models derived in order S, so we start with SHM (Eq. 2) and add either the X interaction (short range potentials) or V_{ij} terms (long range potentials). In the latter case, we consider an unscreened Coulomb potential, so that $V_{ij} = V/d_{ij}$, where V is the interaction between two electrons on nearest neighbor sites, and d_{ij} is the distance between i and j sites, measured in units of the intersite distance. In both cases we are dealing with interacting electrons, and it is convenient to adopt a real-space approach, like the valence bond (VB) one [8]. The VB basis functions, usually called diagrams, are the electronic configurations of the system. The diagrams are eigenstates of the total spin perator and, being interested to the ground state, we just consider the singlet subspace [9]. A singlet diagram may be graphically represented as a combination of dots, crosses and sticks, representing empty and doubly occupied sites, and singlet pairing between two electrons on different sites, respectively [8]. As an example:

$$|\cdot\, x - \widehat{x\, \cdot}\rangle = (1/2)\, a_{2\alpha}^+ a_{2\beta}^+ (a_{3\alpha}^+ a_{4\beta}^+ - a_{3\beta}^+ a_{4\alpha}^+)(a_{5\beta}^+ a_{8\beta}^+ - a_{5\beta}^+ a_{8\alpha}^+) a_{6\alpha}^+ a_{6\beta}^+ |0\rangle,$$

where $|0\rangle$ is the vacuum state, and the sites are numbered from left to right. It is not difficult to realize that, with the VB basis, the site-diagonal e-e interaction terms (U and V_{ij}) give rise to diagonal elements in the hamiltonian matrix, whereas the t and X terms give rise to off-diagonal elements. This fact makes the VB approach particularly useful, as it allows us to investigate the instabilities of the system, and to understand the effect of the various e-e interaction terms without need of any explicit calculation.

For a half-filled 1D chain one can consider two extreme kinds of VB diagrams: The "all-cross" diagrams, where there are no singlet pairs (no sticks), but half of the sites are doubly occupied (crosses) and half are empty (dots), and the "all-stick" diagrams, where there are neither empty nor doubly occupied sites (no dots, no crosses), but just singlet pairs. Starting with SHM, the diagonal energy of the all-stick diagrams is zero, whereas that of the all-cross diagrams is N U/2, N being the number of sites in the chain. In the infinite-stack limit, the energy of the all-cross diagrams is therefore infinite ($U \neq 0$). Thus, any finite off-diagonal interaction, either t or X, cannot induce a mixing of the two kinds of diagrams. Therefore, the all-cross diagrams cannot contribute to the ground state of an infinite chain described by SHM (Eq. 2) or by the (t, U, X) model (Eq. 3).

Adding diagonal V_{ij} interactions to SHM, to get EHM, the energy of the all-cross diagrams, depends on the distribution of crosses and dots, being lowest when crosses and dots regularly alternate. The energy difference between these diagrams and the all-stick diagrams may be written as $N\gamma$, where $\gamma = (U - V\alpha)/2$ represents the balance between on-site and intersite e-e interactions, $\alpha = -(NV)^{-1}\Sigma'_{i,,j}(-1)^{i+j}V_{ij}$ being the generalized Madelung constant [10]. Once more, the energy difference between the two extreme kinds of VB diagrams is infinite (if $\gamma \neq 0$), and they cannot be mixed by any off-diagonal element. But this time, depending on the relative magnitude of U and V_{ij} interactions, either the all-stick or the all-cross diagrams will contribute to the ground state. Actually, due to the different spin statistics of all-stick and all-cross diagrams, the interface between the regions, where the ground state has no contribution from one of the two diagrams, is expected to occur at a critical γ (γ^*) different from zero [10].

We are now in the position of understanding the instabilities of the system. When the all-stick diagrams have infinite energy, actually all the diagrams with an infinite number of sticks have infinite energies, so that only diagrams with an infinite number of crosses and dots can contribute to the ground state. Moreover, the diagrams with an infinite number of adjacent doubly occupied sites (xx) also have infinite energy ($V\alpha \neq 0$). The only diagrams which contribute to the ground state, therefore, contain domains of infinite extension, described either as |...x·x·x·...> or |...·x·x·x...>. Since an infinite number of electrons have to be transferred to go from the |...x·x·x·...> configuration to the |...·x·x·x...> one, such a transformation requires an infinite time: a system prepared in one state has zero probability to transform to the other state. In different words, the system spontaneously breaks the site-equivalence symmetry (the inversion center on the middle of each bond), giving rise to site-charge density wave (site-CDW) condensation [10]. Analogously, it can be shown that, when all-cross diagrams have infinite energies, the ground state of the system is described in terms of diagrams with all-stick domains of infinite extension, leading to an intrinsic instability with respect to both bond-CDW and spin density wave (SDW) condensation [10,11].

INSTABILITIES IN HALF-FILLED SYSTEMS: VALENCE BOND CALCULATIONS

VB is a very convenient approach to the problem of interacting electrons, also due to the fact that its numerical implementation [8], the diagrammatic VB technique, allows one to exactly solve the electronic structure of finite-size systems. The finite-size results, which can be extrapolated to the infinite-stack limit, provide useful tests of the theories, and also allow one to get quantitative information about the behavior of the investigated systems.

To investigate the instabilities of 1D half-filled systems for both long- and short-range interelectronic interactions, we consider finite-size rings, described in terms of either the EHM or the (t, U, X) model. The choice of rings allows us to introduce the complete symmetry of the infinite stack: the translational invariance of the sites, the inversion center on each site (i) and in the middle of each bond (i'). To investigate period 2-site- and bond-CDW instabilities, we introduce a fictitious 2-sites unit cell, limiting attention to the zero-wavevector singlet subspaces relevant to our discussion [10]: A_1, totally symmetric; A_2, antisymmetric with respect to i; B_1, antisymmetric with respect to i'. Since both SHM and EHM are symmetric with respect to electron-hole (e-h) exchange, we also introduce e-h symmetry in our calculations. On the other hand, the X term breaks the e-h symmetry in the (t, U, X) model. It is easy to realize that the last term of Eq. 3 is neither symmetric nor antisymmetric with respect to e-h exchange. However, the symmetric and antisymmetric parts may be separated as follows:

$$H = (t + 2X) \Sigma_{i,\sigma}(a_{i,\sigma}{}^{+}a_{i+1,\sigma} + H.c.) + U \Sigma_i a_{i,\alpha}{}^{+}a_{i,\beta}{}^{+}a_{i,\beta}a_{i,\alpha}$$

$$+ 2X \Sigma'_{i,\sigma}(a_{i,\sigma}{}^{+}a_{i+1,\sigma} + H.c.)(a_{i,\sigma'}{}^{+}a_{i,\sigma'} + a_{i+1,\sigma'}{}^{+}a_{i+1,\sigma'} - 1) \quad (4)$$

obtaining the interesting result that the symmetric part simply renormalizes t. For both EHM and the (t, U, X) model, calculations have been performed for rings with 4, 6, 8, 10 sites and for U = 1, 4, 12 (here and henceforth $\sqrt{2}\,|\,t\,| = 1$). In EHM $V\alpha$ is varied from 0 to 2 U, whereas in the (t, U, X) model X is varied from 0 to -0.05 U, in agreement with our estimates [7].

Some results relevant to EHM are reported in Fig. 1. In particular we show, as a function of γ / U, the energy difference between the lowest singlets in the A_1 - A_2 subspaces (left panel) and in the A_1 - B_1 subspaces (right panel), as extrapolated for the infinite stack. Even if these results are rather preliminary, the extrapolations are sufficiently accurate for our aims, as can be inferred from the insets in the same figure, where we show, as an example, the results for finite-size rings with U = 4. We observe that, in most cases, the results of 4 n and 4 n + 2 rings tend to the same limit from opposite directions, giving upper and lower boundaries for the infinite stack. More specifically, for 4 n + 2 rings the lowest singlet has always A_1 symmetry, whereas the lowest singlet of 4 n rings has A_2 symmetry in the high-γ regime, and B_1 symmetry at low γ. It is not difficult to convince oneself that, in the infinite stack limit, there is a critical γ, so that for $\gamma < \gamma^*$ the A_1 and B_1 lowest singlets are degenerate, while there is a finite A_1 - A_2 energy gap. By increasing γ beyond γ^*, the A_1 - B_1 gap opens, while the A_1 - A_2 gap closes. These results agree with the picture proposed in the previous Section. In fact, in the low-γ regime the A_1 - B_1 degeneracy makes the system intrinsically unstable with respect to any perturbation that, breaking i' symmetry, allows the mixing of A_1 and B_1 states, yielding site-CDW condensation. Analogously, in the $\gamma > \gamma^*$ regime the system tends to break i symmetry, giving rise to bond-CDW condensation. We also notice that, for large U, the A_1 - A_2 (A_1 - B_1) gap abruptly departs from zero as γ decreases (increases) from γ^*. This fact suggests that large electron-phonon interactions are required, even in the γ^* neighborhood, to have the coexistence of bond- and site-CDW instabilities in highly correlated systems. On the contrary, for small U, the gaps vary quite smoothly around γ^*, and the coexistence of bond- and site-CDW should be observable, even for weak electron-phonon coupling.

The A_1 - A_2 and A_1 - B_1 energy gaps for the (t, U, X) model are shown in Fig. 2. In agreement with the previously proposed picture, the X term added to SHM is not able to induce

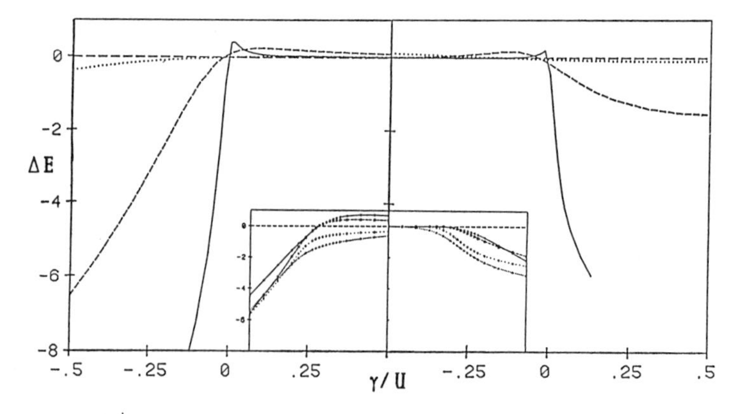

Fig. 1. Energy gaps ($\sqrt{2}|\,t\,|$ units) between lowest singlets in A_1 and A_2 subspaces (left) and in A_1 and B_1 subspaces (right). Dotted, dashed and full lines refer to U = 1, 4, and 12, respectively. Inset: the same gaps for U = 4 finite-size rings. Full, dotted, dashed and dotted (long period) lines refer to N = 4, 6, 8, 10, respectively.

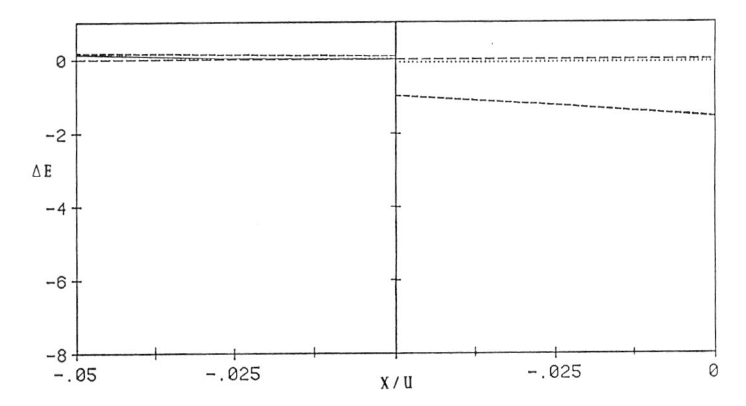

Fig. 2. The same energy gaps as in Fig. 1, but for the (t, U, X) infinite stack.

any dramatic variation on the instabilities of the system. By increasing X from zero, the A_1 - A_2 gap remains zero, whereas the A_1 - B_1 one remains finite: in the (t, U, X) model the system is unstable with respect to bond-CDW, as it is in the absence of X.

The instabilities of the system can also be investigated by looking at correlation functions. We evaluate those relevant to site-CDW [12] and SDW [13], defined as:

$$f_{site\text{-}CDW} = < \Sigma_i\, n_i\, (n_i - n_{i+1}) > / \, 2\, N, \qquad f_{SDW} = -\, 8 < \Sigma_i\, \mathbf{s}_i \cdot \mathbf{s}_{i+1} > / \, N, \tag{5}$$

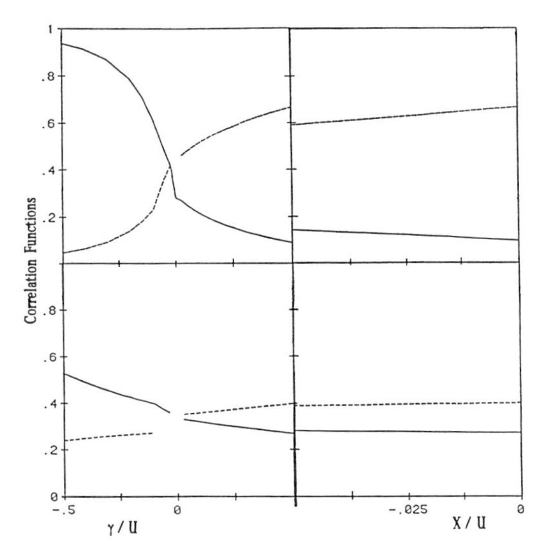

Fig. 3. $f_{site\text{-}CDW}$ (full line) and f_{SDW} (dashed line) for EHM (left) and (t, U, X) (right) infinite stack. Upper panels: U = 4; lower panels: U = 1.

where brackets indicate integration over the ground state, and the dot means the scalar product of the two spin vectors (**s**). In Fig. 3 we report the correlation functions obtained for the infinite stack, described by EHM (left panels) and the (t, U, X) model (right panels), with U = 4 and 1. In EHM, by increasing γ, the system switches, from a region of dominant site-CDW fluctuations, to a region of dominant SDW fluctuations. On the other hand, in the (t, U, X) model the SDW instability always dominates over the site-CDW instability. In agreement with our previous discussion, it turns out that, for both EHM and the (t, U, X) model, the SDW instability coexists with the bond-CDW instability. Finally, the S-shaped curves obtained for EHM with U = 4 strongly suggest a first-order phase change from site-CDW to SDW (bond CDW) regime, whereas the smoother curves reported for U = 1 suggest the occurrence of a continuous transition [14].

CONCLUSIONS

The main results of this paper can be summarized as follows. If one wants to account for e-e interactions in a 1D system, whose one-electron hamiltonian is conveniently described in terms of a simple tight-binding model, one certainly has to introduce the interaction between electrons on the same site, the usual Hubbard U. Moreover, in the case of poorly screened, long range e-e interactions, intersite e-e repulsion terms (V_{ij}) must also be considered, yielding the EHM (also known as Pariser-Parr-Pople in the chemical literature). In the opposite limit of a short range potential, if the overlap between nearest neighbor sites is not very small, SHM is adequate, otherwise an effective attraction between site and bond charge (X term) has to be accounted for.

For half-filled systems the addition of e-e intersite interactions to SHM can have dramatic effects on the ground state of the system, as it can drive it from a bond-CDW (or SDW: the two instabilities always coexist for half-filling) to a site-CDW regime. On the contrary, adding the X term to SHM just yields a smooth variation of the ground state properties: the system is, and stays, in the bond-CDW (SDW) regime. It will now be interesting to investigate how this picture varies with band filling, and, in particular, whether it is affected by small departures from half-filling, in correspondence with doping of the system.

ACKNOWLEDGMENTS

We thank Z. G. Soos and J. Voit for useful discussions and correspondence. This work has been supported by the Ministero della Pubblica Istruzione and by the National Research Council (CNR) of Italy through its Centro Studi Stati Molecolari Radicalici ed Eccitati.

REFERENCES

[1] W.-P. Su, J. R. Schrieffer, and A. J. Heeger, Phys. Rev. Lett. 42: 1698 (1979) and Phys. Rev. B 22: 2099 (1979).
[2] D. K. Campbell, D. Baeriswyl, and S. Mazumdar, Synth. Metals 17: 197 (1987) and references therein.
[3] S. Mazumdar and S. N. Dixit, Phys. Rev. B 34:3683 (1986) and references therein.
[4] S. Kivelson, W.-P. Su, J. R. Schrieffer, and A. Heeger, Phys. Rev. Lett. 58: 1899 (1987); C. Wu, X. Sun, and K. Nasu, Phys. Rev. Lett. 59: 72 (1988).
[5] A. Painelli and A. Girlando, Solid State Commun. 66: 273 (1988).
[6] A. Painelli and A. Girlando, Synth. Metals 27: 15 (1988) and Phys. Rev. B 39: 2830 (1989).
[7] A. Painelli and A. Girlando in: "Interacting Electrons in Reduced Dimension,", ed. by D. Baeriswyl and D. K. Campbell, NATO ASI B (Plenum, New York, in press).
[8] S. Ramasesha and Z. G. Soos, J. Chem. Phys. 80: 3278 (1984) and Int. J. Quantum Chemistry 25: 1003 (1984).
[9] E. H. Lieb and D. Mattis, Phys. Rev. 125: 164 (1962).
[10] A. Painelli and A. Girlando, Synth. Metals 29: F181(1989) and in "Interacting Electrons in Reduced Dimension,", ed. by D. Baeriswyl and D. K Campbell, NATO ASI B (Plenum, New York, in press).
[11] A. Painelli and A. Girlando, in preparation.
[12] L. Milans del Bosch and L. M. Falicov, Phys. Rev. B 37: 6073 (1988).
[13] S. Jagannathan and Z. G. Soos, J. Chem. Phys. 87: 4609 (1983).
[14] V. Waas, H. Buettner, and J. Voit, preprint.

THEORETICAL CHARACTERIZATION OF THE ELECTRONIC STRUCTURE OF POLY(HETEROAROMATIC VINYLENES)

E. Ortí, M. C. Piqueras, R. Crespo, and F.Tomás

Departament de Química Física
Facultat de Ciències Químiques
Universitat de València
46100-Burjassot (Valencia) Spain

INTRODUCTION

Since the pioneering work of the mid-seventies, the number of organic π-conjugated polymers,that can be made highly conducting upon appropriate chemical treatment,has grown tremendously. Among the most studied of these compounds are those based on five-membered heteroaromatic rings, such as polythiophene (PT) and polypyrrole (PPy). The attractiveness of these polymers,relative to polyacetylene (PA) and poly(p-phenylene) (PPP), is the high chemical and thermal stability they show.[1]

In addition to environmental stability, convenient processibility of conducting polymers is desired. This processibility has recently been achieved for polythiophene in two different ways: (i) by incorporating long-chain alkyl substituents on the thiophene rings to yield soluble poly(3–alkylthiophenes)[2] and (ii) by introducing vinylene units between adjacent thiophene rings to obtain poly(2,5-thienylene vinylene) (PTV). PTV was firstly synthesized by Kossmehl,[3] but the procedure used formed intractable, low molecular weight powders with low conductivities. More recently, Jen et al.[4] and Yamada et al.[5] have reported the synthesis of flexible, high molecular weight PTV films via a water-soluble precursor polymer. The technique used by these authors is similar to that previously developed for the synthesis of PA[6], PPP[7], and poly(p-phenylene vinylene) (PPV).[8] In addition to the desired processibility and environmental stability, conductivities as high as 60 S/cm have been measured for PTV films on exposure to iodine vapors at room temperature.[4,5,9] Conductivity values up to 300 S/cm are expected for doped films of PTV.[9]

Lower-Dimensional Systems and Molecular Electronics
Edited by R.M. Metzger *et al.*, Plenum Press, New York, 1991

These findings prompted our theoretical work on PTV. This polymer can be regarded as an alternating copolymer of thiophene and acetylene, and our main goal is to study how the introduction of the vinyl groups affects the electronic properties of the parent polymers. We have extended our study to poly(2,5-furylene vinylene) (PFV) and poly(2,5-pyrrylene vinylene) (PPyV) (see Fig. 1), in order to compare with polyfuran (PF) and polypyrrole. The synthesis of PFV has recently been reported, and conductivities as high as 36 S/cm have been measured on acceptor doping for this polymer.[10] To our knowledge, no synthesis is found in the literature for PPyV.

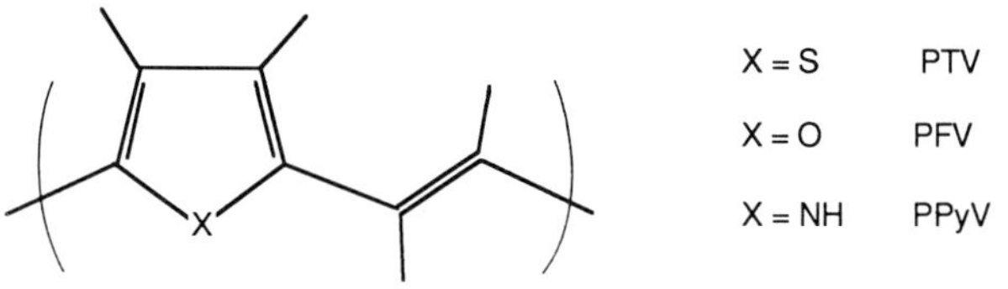

X = S PTV

X = O PFV

X = NH PPyV

Fig. 1. Structure of poly(2,5-thienylene vinylene) (PTV), poly(2,5-furylene vinylene) (PFV), and poly(2,5-pyrrylene vinylene) (PPyV).

Band structure calculations have been performed with the valence effective Hamiltonian (VEH) nonempirical pseudopotential technique.[11] The VEH method yields one-electron energies of ab initio double-zeta quality, and has been demonstrated to provide accurate estimates of essential electronic properties such as ionization potentials (IP), bandwidths (BW), bandgaps (E_g), and electron affinities (EA) in the context of conducting polymers.[12] All the calculations have been carried out using the VEH parameters previously reported for sulfur,[13] oxygen,[14] and nitrogen[15] atoms and those recently obtained for carbon and hydrogen atoms.[16]

RESULTS AND DISCUSSION

Since no detailed structural study is available for PTV, PFV, and PPyV polymers, we have used ab initio SCF-LCAO-MO methods to perform reliable geometry optimizations on the monomers. The 3-21G fully-optimized geometries calculated for the trans conformers of 2–vinylthiophene (2-VT), 2-vinylfuran (2-VF), and 2-vinylpyrrole (2-VPy) are shown in Fig. 2. These geometries correlate very well with the experimental structures reported for thiophene,[17] furan,[18] and pyrrole[19] rings. Significant differences between theory and experiment are only detected for 2-VT, and are mainly due to the overestimation of the experimental carbon-sulfur bond length (1.72 Å)[17] by the theoretical calculations (*ca* 1.80 Å). Bond length values in Fig. 2 show a clear alternation of single and double bonds along the carbon backbone. A bond length of 1.45 Å is predicted for the single C-C bond joining the vinyl group to the ring, in very good agreement with the X-ray bond lengths reported for the closely related 1,4–bis(2–thienylvinyl)benzene molecule.[20] A bond length of 1.45 Å was also employed for the C-C inter-ring distance in PT,[21] PF,[14] and PPy.[22]

The VEH band structure calculations have been carried out using the geometries depicted in Fig. 2 as unit cells, and assuming a fully coplanar conformation, with all the vinyl groups in trans position with respect to adjacent rings (see Fig. 3a). This assumption has been taken on the basis of the accurate ab initio calculations performed to study the rotation potential of the vinyl group in 2-VT, 2-VF, and 2-VPy.[23] For all of these monomers, the trans rotamer is predicted as the most stable conformation. The VEH band structures calculated for the all-cis conformations displayed in Fig 3b show no significant difference with respect to those obtained for trans conformations and they will not be discussed here.

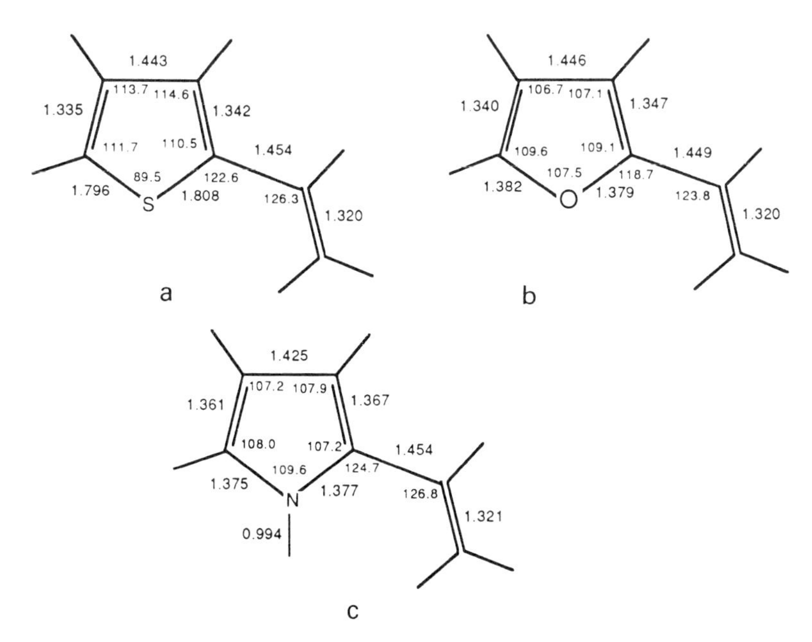

Fig. 2. 3-21G fully-optimized structures for trans conformers of 2-vinylthiophene (a), 2–vinylfuran (b), and 2-vinylpyrrole (c). Bond lengths are given in Å and angles in degrees. Carbon-hydrogen bond lengths have values around 1.07 Å and are not indicated.

The VEH band structure computed for PTV is presented in Fig. 4, where all the crossings among π bands and among σ bands are avoided, due to the low symmetry of the unit cell. The HOMO (highest occupied molecular orbital) and LUMO (lowest unoccupied molecular orbital) bands are, as expected, π bands, and show the same atomic orbital composition patterns as those reported for PT.[24] The HOMO band of PTV corresponds to wave functions delocalized over the carbon backbone, with essentially no contribution from

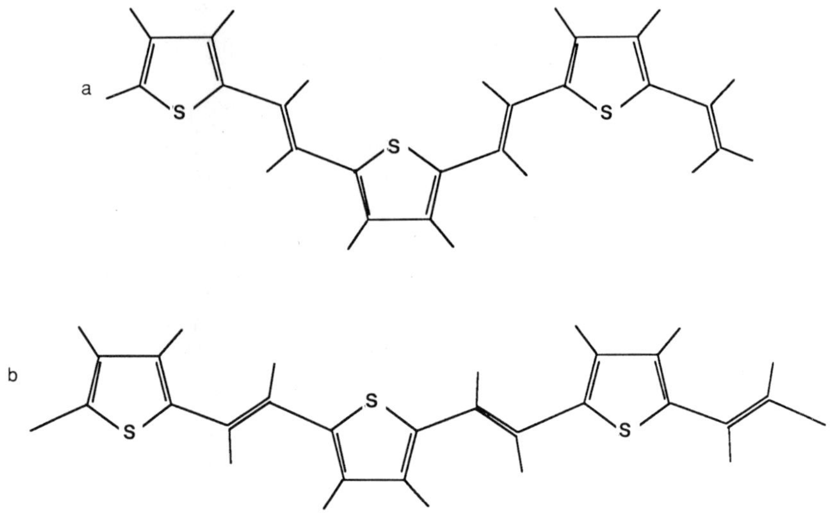

Fig. 3.. Conformation of a trimer of PTV. Vinyl groups are in trans (a) and in cis (b) positions with respect to adjacent thiophene rings.

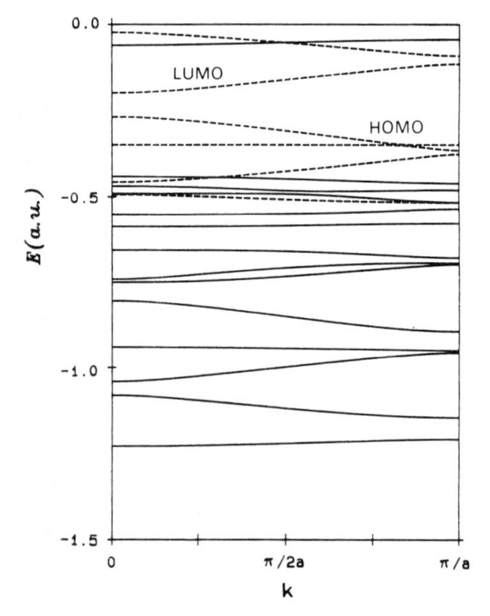

Fig. 4. VEH band structure calculated for PTV. Energies are in atomic units. Unoccupied bands with energies larger than 0.0 a.u. are not shown. Solid lines indicate σ bands. Dashed lines indicate π bands. HOMO and LUMO denote the highest occupied and lowest unoccupied molecular orbital bands, respectively.

the sulfur atoms, and has a width of 2.18 eV. This bandwidth is of the order of that calculated for PT (2.1-2.6 eV)[21,25] and much smaller than that found for PA (6.5 eV).[26] The LUMO band of PTV shows a width of 2.30 eV.

The ionization potential, after substracting 2.3 eV to take into account the polarization energy of the solid, is calculated to be 5.04 eV for PTV. This value is intermediate between those of PT (5.4 eV)[25] and PA (4.7-4.8 eV).[26] The reduction of 0.4 eV, caused in the ionization potential of PT by the incorporation of vinyl units, is in excellent agreement with the decrease of 0.4 V, measured for the oxidation potential on going from PT to PTV.[9] The

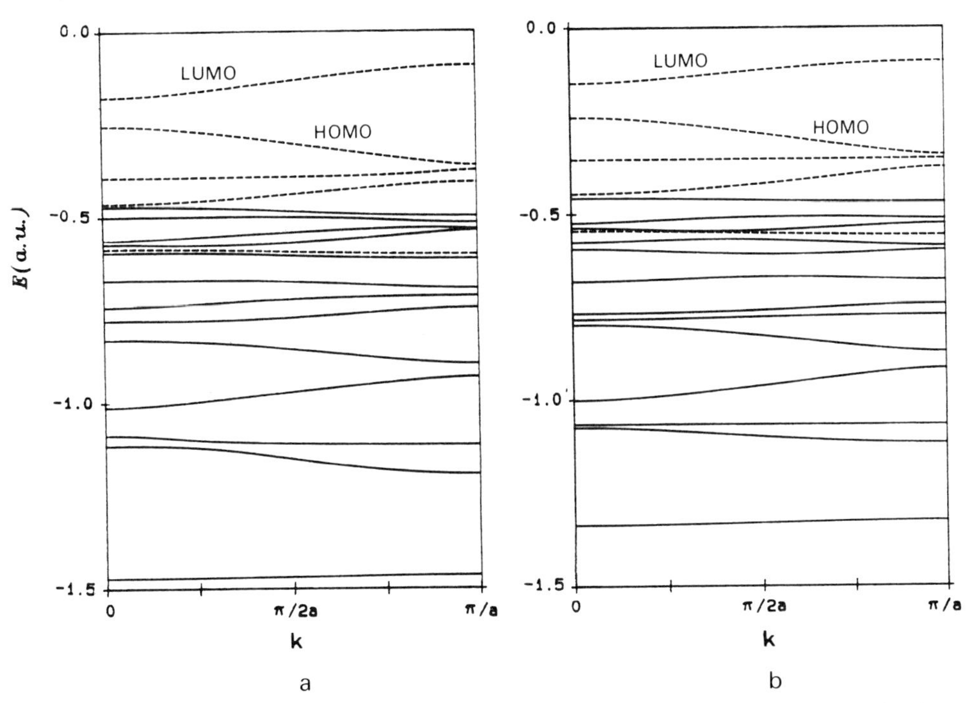

Fig. 5. VEH band structures calculated for PFV (a) and PPyV (b). See Fig 4 for further details.

presence of vinyl groups also determines a narrowing of the energy gap separating the HOMO and LUMO bands. A bandgap of 1.90 eV is computed for PTV, which is 0.30 eV smaller than that reported for PT (2.2 eV).[25] This bandgap correlates very well with the optical bandgap (1.7-1.8 eV) estimated from the onset of the optical absorption spectra.[5,9] The narrowing of 0.30 eV is also in very good agreement with experimental data, since an optical bandgap of about 2.0 eV (0.2-0.3 eV larger than that observed for PTV) is reported for PT.[27] It is to be noted that the differences between the results presented here, and those previously obtained for PTV[21], are due to the use of different input geometries and VEH parameterizations.

Table I. VEH results for the widths of the HOMO band (BW), ionization potentials (IP), bandgaps (E_g), and electron affinities (EA) of poly(heteroaromatic vinylenes) (PTV, PFV, and PPyV) and polyheterocycles (PT, PF, and PPy). All values are in eV.

Polymer	BW	IP	E_g	EA
PTV	2.18	5.04	1.90	3.14
PFV	2.92	4.64	2.15	2.49
PPyV	2.81	4.20	2.48	1.72
PT[a]	2.1	5.4	2.2	3.2
PF[a]	3.7	4.9	3.1	1.8
PPy[a]	3.8	3.9	3.6	0.3

[a] Data from Refs.14, 22, and 25.

The VEH band structures calculated for PFV and PPyV are very similar to that of PTV, and are displayed in Fig. 5. The most relevant VEH electronic parameters calculated for PFV and PPyV are collected in Table I. For the sake of comparison, the VEH parameters discussed above for PTV, and those previously reported for PT[21,25] PF,[14] and PPy[22], are included in Table I. As it was previously observed for PF and PPy, the HOMO bands show similar widths for PFV(2.92 eV) and PPyV (2.81 eV), but now are narrower by about 1 eV.

The introduction of the vinyl groups induces a different effect on the ionization potentials of PF and PPy. While a slight reduction of the IP is calculated when passing from PF (4.9 eV) to PFV (4.64 eV) an increase of 0.30 eV is obtained on going from PPy (3.9 eV) to PPyV (4.20 eV). This different behaviour is understood as a consequence of the intermediate value the IP of polyacetylene (4.7 eV)[26] has with respect to the IPs of PF (4.9 eV) and PPy (3.9 eV). The decrease of 0.4 eV calculated in ionization potential, on going from PTV (5.04 eV) to PFV (4.64 eV), is in excellent agreement with the difference of 0.2-0.4 V observed between the oxidation potentials of PTV (3.36 to 3.54 V vs Li/Li^+, depending on solvent)[9] and PFV (3.15 eV vs Na/Na^+).[10] Both for PFV and PPyV, the presence of the vinyl groups causes a narrowing of the band gap with respect to PF and PPy. The band gap calculated for PFV (2.15 eV) overestimates by 0.4 eV the optical band gap (1.76 eV) inferred from the absorption spectra.[10] The band gap obtained for PPyV (2.48 eV) is *ca* 1.1 eV smaller than that calculated for PPy (3.6 eV).[22] This fact determines the high electron affinity predicted for PPyV (1.72 eV) compared with that reported for PPy (0.3 eV).[22] No experimental data are available for PPyV.

As a main conclusion of this work, a summary of the influence of the vinyl groups on the electronic properties of the parent polymers is presented in Fig. 6. The same trends are observed, when changing the heteroatom, for the ionization potential and the energy bandgap of polyheterocycles and poly(heteroaromatic vinylenes). A continuous decrease of IP and increase of E_g are obtained on going from sulfur to oxygen-containing polymers and from

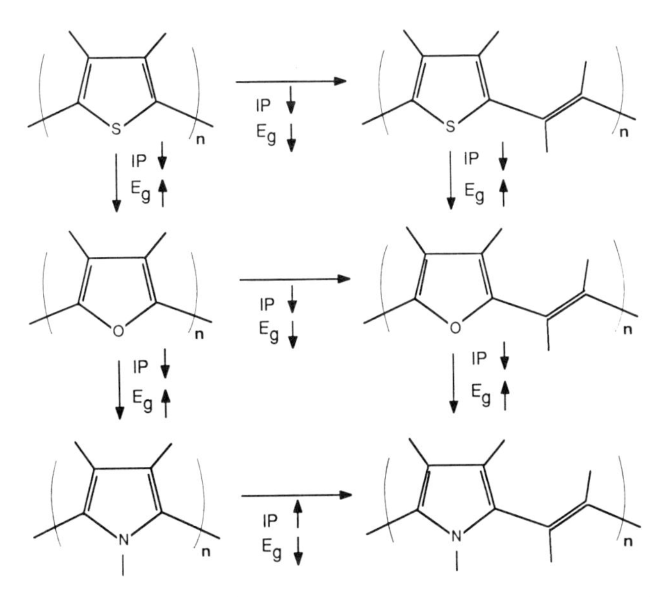

Fig. 6. Evolution of the ionization potentials (IP) and energy bandgaps (E_g) for polyheterocycles and poly(heteroaromatic vinylenes).

oxygen to nitrogen-containing polymers. Most importantly, the incorporation of the vinyl groups, in all cases, produces a reduction of the ionization potential, but for PPy, and a decrease of the band gap

ACKNOWLEDGEMENTS

We thank the CIUV (Centro de Informática de la Universidad de Valencia) for use of computing facilities. One of us (RC) is grateful to the Ministerio de Educación y Ciencia for a doctoral grant. This work has been partially supported by the Project No 683.2/89 of the University of Valencia and the CAYCIT Project No 714/84.

REFERENCES

1. R. J. Waltman, J. Bargon, and A. F. Diaz, J. Phys. Chem. **87**, 1459 (1983); G. Tourillon and F. Garnier, J. Electrochem. Soc. **130**, 2042 (1983).
2. S. D. D. V. Rughooputh, M. Nowak, S. Hotta, A. J. Heeger, and F. Wudl, Synth. Met. **21**, 41 (1987); S. Hotta, S. D. D. V. Rughooputh, A. J. Heeger, and F. Wudl, Macromol. **20**, 212 (1987).
3. For a review see G. A. Kossmehl in "Handbook of Conducting Polymers", edited by T. A. Skotheim (Marcel Dekker, New York, 1986), Vol 1, Chap 10, p.351.
4. K.-Y. Jen, M. Maxfield, L. W. Shacklette, and R L. Elsenbaumer, J. Chem. Soc., Chem. Commun., 309 (1987).
5. S. Yamada, S. Tokito, T. Tsutsui, and S. Saito, J. Chem. Soc., Chem. Commun., 1448 (1987).
6. J. H. Edwards and W. J. Feast, Polym. Commun. **21**, 595 (1980).

7. D. G. H. Ballard, A. Courtis, I. M. Shirley, and S. C. Taylor, J. Chem. Soc., Chem. Commun., 954 (1983).
8. I. Murase, T. Ohniski, T. Noguchi, and M. Hirooka, Polym. Commun. **25**, 327 (1984).
9. K.-Y. Jen, R. Jow, L.W. Shacklette, M. Maxfield, H Eckhardt, and R L. Elsenbaumer, Mol. Cryst. Liq. Cryst. **160**, 69 (1988).
10. K.-Y. Jen, T. R. Jow, and R L. Elsenbaumer, J. Chem. Soc., Chem. Commun., 1113 (1987).
11. G. Nicolas and Ph. Durand, J. Chem. Phys. **70**, 2020 (1979); **72**, 453 (1980); J. M. André, L. A. Burke, J. Delhalle, G. Nicolas, and Ph. Durand, Int. J. Quantum Chem. Symp. **13**, 283 (1979).
12. For an extense review see J. L. Brédas in "Handbook of Conducting Polymers", edited by T. A. Skotheim (Marcel Dekker, New York, 1986), Vol 2, Chap 25, p.859.
13. J. L. Brédas, R. R. Chance, R. Silbey, G. Nicolas, and Ph. Durand, J. Chem. Phys. **77**, 371 (1982).
14. B. Thémans, J. M. André, and J. L. Brédas, Mol. Cryst. Liq. Cryst. **118**, 121 (1985).
15. J. L. Brédas, B. Thémans, and J. M. André, J. Chem. Phys. **78**, 6137 (1983).
16. B. Thémans, J. M. André, and J. L. Brédas, to be published.
17. B. Bak, D. Christensen, L. Hansen-Nygaard, and J. Rastrup-Andersen, J. Mol. Spectrosc. **7**, 58 (1961); P. B. Liescheski and D. W. H. Rankin, J. Mol. Struct. **178**, 227 (1988).
18. F. Mata and M. C. Marín, J. Mol. Struct. **48**, 157 (1978).
19. L. Nygaard, J. J. Nielsen, J. Kirchheiner, G. Maltesen, J. Rastrup-Andersen, and G. O. Sorensen, J. Mol. Struct. **3**, 491 (1969).
20. D. Zobel, Acta Crystallogr. B **32**, 2838 (1976).
21. J. L. Brédas, R. L. Elsenbaumer, R. R. Chance, and R. Silbey, J. Chem. Phys. **78**, 5656 (1983).
22. J. L. Brédas, B. Thémans, and J. M. André, J. Chem. Phys. **78**, 6137 (1983).
23. M. C. Piqueras, R. Crespo, and E. Ortí, to be published.
24. J. L. Brédas, A. J. Heeger, and F. Wudl, J. Chem. Phys. **85**, 4673 (1986).
25. J. L. Brédas, B. Thémans, J. G. Fripiat, J. M. André, and R. R. Chance, Phys. Rev. B **29**, 6761 (1984).
26. J. L. Brédas, R. R. Chance, R. H. Baughman, and R. Silbey, J. Chem. Phys. **76**, 3673 (1982).
27. T. C. Chung, J. H. Kaufman, A. J. Heeger, and F. Wudl, Phys. Rev. B **30**, 702 (1984); M. Kobayashi, J. Chen, T. C. Chung, F. Moraes, A. J. Heeger, and F. Wudl, Synth. Met. **9**, 77 (1984).

THEORETICAL DESIGN OF ORGANIC METALS BASED ON THE PHTHALOCYANINE MACROCYCLE

E. Ortí, M. C. Piqueras, and R. Crespo

Departament de Química Física
Facultat de Ciències Químiques
Universitat de València
46100-Burjassot (Valencia) Spain

INTRODUCTION

Phthalocyanine molecular crystals and cofacially linked polymers are well documented as low-dimensional materials that may attain high electrical conductivities. Air-stable conductivities on the order of 1 to 1000 S/cm after partial oxidation by iodine have been reported.[1-3] These conductivity studies indicate that the electrical conductivity has very little dependence on the identity of the atom complexed in the cavity, but is strongly dependent on the orientation and spacing of the phthalocyanine rings. A columnar stacking with minimum spacing leads to a maximum interaction between π-molecular orbitals on adjacent rings and promotes the highest conductivity. More effective π-interactions and, therefore, better conductivity properties could then be expected when the conjugated system of the phthalocyanine molecule is extended. Several synthetic efforts have been recently dedicated to crystals and polymers derived from large tetraaza-macrocycles.[4,5]

The main purpose of this contribution is to analyze how the molecular electronic properties more closely related with electrical conductivity, i.e., ionization potentials, redox potentials, and optical transitions, are affected by the extension of the aromatic structure of the macrocycle. The extension of the macrocycle is carried out by succesive annellation of benzene rings and all the molecular systems depicted in Fig. 1 are studied in this work. Note that phthalocyanine (H_2Pc) results from the condensation of four benzene rings to the parent tetraazaporphyrin (H_2TAP) molecule. In the same way, 1,2-naphthalocyanine (1,2-H_2Nc) and 9,10-phenanthrenocyanine (H_2Phc) can be regarded as derivatives of phthalocyanine with angular annellation of the benzene rings, while 2,3-naphthalocyanine (2,3-H_2Nc) presents a linear annellation of the benzene rings.

Lower-Dimensional Systems and Molecular Electronics
Edited by R.M. Metzger *et al.*, Plenum Press, New York, 1991

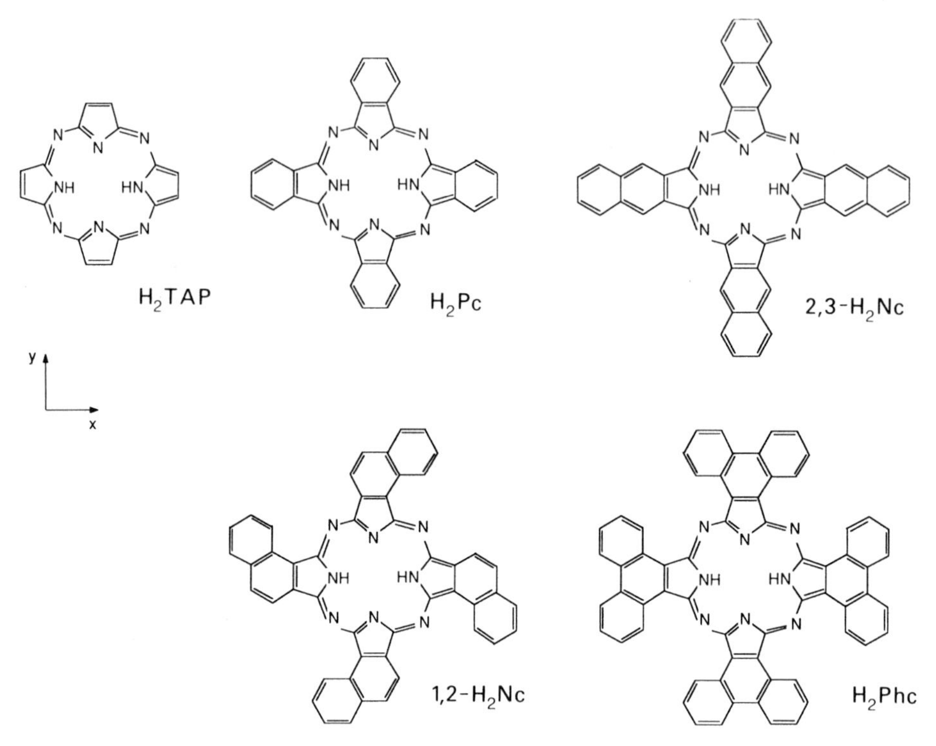

Fig. 1. Structure of the metal-free macrocycles tetraazaporphyrin (H_2TAP), phthalocyanine (H_2Pc), 2,3-naphthalocyanine (2,3-H_2Nc), 1,2-naphthalocyanine (1,2-H_2Nc), and 9,10-phenanthrenocyanine (H_2Phc).

The study of the electronic structures of the molecular systems sketched in Fig.1 have been carried out using the nonempirical valence effective Hamiltonian (VEH) quantum-chemical technique.[6] The VEH method is especially useful to deal with large molecular systems,since it yields one-electron energies of ab initio double-zeta quality at a reasonable computer cost. The adequacy of the VEH method to study phthalocyanine-type macrocycles has been previously shown for H_2Pc.[7] All the calculations have been performed using the VEH parameters previously reported for the nitrogen atoms[8] and those recently obtained for the carbon and hydrogen atoms.[9] The neutron diffraction structure reported for H_2Pc[10],and averaged to a D_{2h} symmetry,has been employed as the input geometry for H_2Pc,and to build up the tetraazaporphyrin ring in H_2TAP, 2,3-H_2Nc, and H_2Phc. The atomic coordinates for 2,3-H_2Nc and H_2Phc have been obtained by fusing the experimental structures reported for naphthalene[11] and phenanthrene[12] to the tetraazaporphyrin ring. The input geometry used for 1,2-H_2Nc has been taken from X–ray diffraction data.[13]

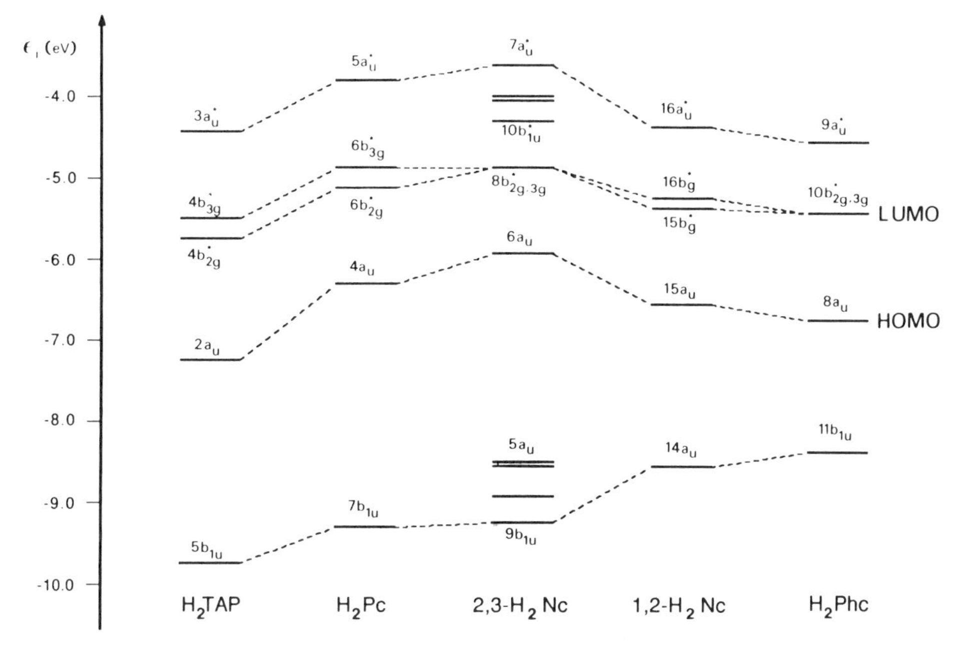

Fig. 2. Schematic display of the VEH one-electron energy level distributions obtained for the molecules sketched in Fig.1. All the orbitals are of π-type and they have been correlated on the basis of their atomic orbital composition.

RESULTS AND DISCUSSION

The VEH one-electron energy level distributions calculated for the molecular systems depicted in Fig.1 are schematically displayed in Fig.2. Only the highest occupied and lowest unoccupied molecular orbitals are included in Fig.2,since they determine the most interesting electronic properties. All of these orbitals are of π-nature, and they have been classified according to a D_{2h} symmetry, but for 1,2-H_2Nc, for which a C_{2h} geometry has been assumed.

Fig. 2 indicates that the highest occupied molecular orbital (HOMO) corresponds, in all cases, to a level of a_u symmetry. This level shows the same atomic orbital patterns for all the molecules studied in this work,with no contribution from the nitrogen atoms. An energy destabilization is observed for the HOMO level when passing from H_2TAP to H_2Pc. The $4a_u$ HOMO of H_2Pc lies at -6.32 eV, about 1 eV above the $2a_u$ HOMO of H_2TAP located at -7.27 eV. These values are in very good agreement with the gas-phase first ionization potentials reported for H_2Pc (6.41 eV)[14], and the metal-free porphine (H_2P) molecule (7.1 eV).[15] Since the only difference between H_2TAP and H_2P is the four aza-nitrogen atoms linking the pyrrole units, and they do not contribute to the HOMO level, similar ionization potentials can be expected for both molecules.

An additional destabilization of the HOMO level is calculated when four benzene rings are fused to the phthalocyanine ring, to obtain the 2,3-H_2Nc molecule. The $6a_u$ HOMO is now located at -5.93 eV, 0.39 eV above the $4a_u$ HOMO of H_2Pc. On the contrary, the condensation of benzene rings to yield 1,2–H_2Nc and H_2Phc gives rise to a slight stabilization of the HOMO level with respect to H_2Pc.

Since oxidation involves removing an electron from the HOMO, it is reasonable to expect molecules with lower-lying HOMOs to have larger oxidation potentials. To this effect, the VEH results suggest that the extension of the conjugated system by annellation of benzene rings following the axis of pyrrole units (linear annellation) produces a continuous destabilization of the HOMO and, as a consequence, a decreasing of the oxidation potential of the macrocycle can be expected on going from H_2TAP to H_2Pc and from H_2Pc to 2,3-H_2Nc. This trend is supported by the oxidation potentials of 1.00 and 0.58 V measured in CH_2Cl vs SCE (saturated calomel electrode) for $SiPc(OR)_2$ and $SiNc(OR)_2$, respectively.[16] The difference between these potentials (0.42 V) is in excellent agreement with the VEH energy difference (0.39 eV) between the $4a_u$ HOMO of H_2Pc and the $6a_u$ HOMO of 2,3-H_2Nc. On the other hand, the VEH results predict that angular annellation of benzene rings, as for 1,2-H_2Nc and H_2Phc, slightly stabilizes the HOMO level with respect to H_2Pc and similar oxidation potentials can be expected. Experimental redox data are only available for 1,2–FeNc which shows an oxidation potential of 1.01 V, very similar to that reported for FePc (1.10 V).[5]

Turning to the optical spectrum, the energies and oscillator strengths calculated for the lowest-energy electronic transitions are collected in Table I. These transitions correspond to the promotion of an electron from the a_u HOMO to the lowest unoccupied molecular orbital (LUMO), and can be clearly correlated with the intense absorption band experimentally observed in the visible (Q band). Due to the presence of the central two hydrogen atoms, the LUMO is split in two π*-molecular orbitals of b_{2g} and b_{3g} symmetries (b_g for 1,2-H_2Nc). The quasi-degeneracy of these two orbitals results from the almost identical geometry of the carbon-nitrogen backbone along x and y axes (see Fig. 1). This degeneracy is more pronounced for 2,3-H_2Nc and H_2Phc, since identical aromatic units have been fused to the tetraazaporphyrin ring along x and y axes, as discussed above.

The transition energies in Table I have been obtained after a systematic shift of 0.59 eV to higher energies of the unoccupied molecular orbitals in order to obtain the best correlation with experimental absorption data. For H_2TAP, the $2a_u \rightarrow 4b_{2g}$* and $2a_u \rightarrow 4b_{3g}$* electronic transitions have energy values of 2.12 eV (586 nm) and 2.30 eV (539 nm), respectively. These values are in very good agreement with the sharp absorption peaks recorded at 617 nm (Q_x band) and 545 nm (Q_y band) for H_2TAP in chlorobenzene solution.[17] Both the splitting of the Q band (0.25 eV) and the relative intensity of the constituent peaks (Q_x more intense than Q_y) are very well reproduced by the theoretical calculations. By general convention, the lowest-energy Q band is referred to as Q_X, without implying any absolute orientation. The important fact is that, when adopting as x axis the axis defined by the central two hydrogens (see Fig. 1), the lowest energy Q_x band is calculated to be polarized along the y axis (see Table I), i.e., perpendicular to the H-H axis.

Table I. VEH lowest energy electronic transitions. Energies are in eV (nm in parentheses).

	Transition	Energy	Oscillator Strength	Polarization[a]
H_2TAP	$2a_u \rightarrow 4b_{2g}^*$	2.12 (586)	4.3	y
	$2a_u \rightarrow 4b_{3g}^*$	2.30 (539)	3.9	x
H_2Pc	$4a_u \rightarrow 6b_{2g}^*$	1.81 (685)	5.1	y
	$4a_u \rightarrow 6b_{3g}^*$	1.98 (625)	4.7	x
2,3-H_2Nc	$6a_u \rightarrow 8b_{3g}^*, 8b_{2g}^*$	1.64 (757)	5.2	x,y
1,2-H_2Nc	$15a_u \rightarrow 15b_g^*$	1.84 (675)	4.8 (1.5)[b]	x,y
	$15a_u \rightarrow 16b_g^*$	1.97 (629)	1.5 (4.6)	x,y
H_2Phc	$8a_u \rightarrow 10b_{3g}^*, 10b_{2g}^*$	1.88 (661)	4.9	x,y

[a] Axes refer to those presented in Fig. 1.
[b] Numbers in parentheses refer to oscillator strengths along y axis.

As can be seen from Table I, a bathochromic shift of the long-wave transitions is observed in going from H_2TAP to H_2Pc. Now, the $4a_u \rightarrow 6b_{2g}^*$ and $4a_u \rightarrow 6b_{3g}^*$ electronic transitions appear at 1.81 eV (685 nm) and 1.98 eV (625 nm). The bathochromic shift is consequence of the narrowing of the HOMO-LUMO energy gap in H_2Pc, compared with that calculated for H_2TAP (see Fig. 2) and it is in perfect agreement with the longer wave lengths experimentally observed for H_2Pc. Absorption Q_x and Q_y bands are measured at 686 and 623 nm, respectively, for phthalocyanine vapors.[18]

The condensation of additional benzene rings leads to different and interesting results, depending on the way the rings are fused. Linear annellation to obtain 2,3-H_2Nc produces an additional narrowing of the HOMO-LUMO energy gap (see Fig.2) and the consequent bathochromic shift of the lowest-energy electronic transition (see Table I). This transition ($6a_u \rightarrow 8b_{3g}^*, 8b_{2g}^*$) is now located at 1.64 eV (757 nm), in excellent agreement with the absorption Q_x band measured at 750 nm for a derivative of 2,3-FeNc in chloroform.[13] In contrast to 2,3-H_2Nc, angular annellation, to obtain 1,2-H_2Nc and H_2Phc, produces no significant effect on the HOMO-LUMO energy-gap, and these molecules retain the main absorption characteristics of H_2Pc. The lowest-energy electronic transitions occur at energy values of 1.84 eV (675 nm) and 1.88 eV (661 nm) for 1,2-H_2Nc ($15a_u \rightarrow 15b_g^*$) and H_2Phc ($8a_u \rightarrow 10b_{3g}^*, 10b_{2g}^*$), respectively. These values are almost equal to that calculated for H_2Pc (1.81 eV) and are in perfect accord with those reported for octahedrally complexed iron derivatives of 1,2-H_2Nc (653 nm) and H_2Phc (657 nm) in chloroform.[13] Note that the same derivative of H_2Pc presents the first absorption peak at 658 nm.[13]

As a main result, it can be concluded that linear annellation leads to a more effective conjugation in the macrocyclic system and causes:
(i) a continuous destabilization of the HOMO level and, as a consequence, a decreasing of the oxidation potential of the macrocycle,
(ii) a continuous narrowing of the HOMO-LUMO energy gap and, as a consequence, a bathochromic shift of the long-wave electronic transition

along the series H_2TAP, H_2Pc, and 2,3-H_2Nc. On the other hand, angular annellation produces no important effect on the HOMO stability and on the HOMO-LUMO energy gap with respect to H_2Pc.

Thus, the VEH results suggest that, due to their lower oxidation potentials and narrower energy gaps, better conductivity properties are to be expected for linearly annellated than for angular annellated phthalocyanines. Indeed, this trend explains the high intrinsic electrical conductivity reported for 2,3-FeNc (4×10^{-5} S/cm),[19] compared with the similar conductivities measured for FePc (4×10^{-9} S/cm),[20] 1,2-FeNc (4×10^{-9} S/cm),[13] and FePhc (8×10^{-9} S/cm)[13] compounds. Preliminary VEH calculations on more extended linearly annellated phthalocyanines suggest the possibility of obtaining crystals and polymers with very small band gaps.

ACKNOWLEDGEMENTS

We thank the CIUV (Centro de Informática de la Universidad de Valencia) for use of computing facilities. One of us (RC) is grateful to the Ministerio de Educación y Ciencia for a doctoral grant. This work has been partially supported by the Project No 683.2/89 of the University of Valencia and the CAYCIT Project No 714/84.

REFERENCES

1. T. J. Marks, Science **227**, 881 (1985).
2. B. M. Hoffman and J. A. Ibers, Acc. Chem. Res. **16**, 15 (1983); S. M. Palmer, J. L. Stanton, J. Martinsen, M. Y. Ogawa, W. B. Hener, S. E. Van Wallendael, B. M. Hoffman, and J. A. Ibers, Mol. Cryst. Liq. Cryst. **125**, 1 (1985).
3. M. Hanack, A. Datz, R. Fay, K. Fischer, U. Keppeler, J. Koch, J. Metz, M. Mezger, O. Schneider, and H. J. Schulze in "Handbook of Conducting Polymers", edited by T. A. Skotheim, (Marcel Dekker, New York, 1986), Vol 1, Chap. 5, p. 133.
4. A. W. Snow and T. R. Price, Synth. Met. **9**, 329 (1984).
5. M. Hanack, A. Lange, M. Rein, R. Behnisch, G. Renz, and A. Leverenz, Synth. Met. **29**, F1 (1989).
6. G. Nicolas and Ph. Durand, J. Chem. Phys. **70**, 2020 (1979); **72**, 453 (1980); J. M. André, L. A. Burke, J. Delhalle, G. Nicolas, and Ph. Durand, Int. J. Quantum Chem. Symp. **13**, 283 (1979).
7. E. Ortí and J. L. Brédas, J. Chem. Phys. **89**, 1009 (1988).
8. J. L. Brédas, B. Thémans, and J. M. André, J. Chem. Phys. **78**, 6137 (1983).
9. B. Thémans, J. M. André, and J. L. Brédas, to be published.
10. B. F. Hoskins, S. A. Mason, and J. C. B. White, J. Chem. Soc., Chem. Commun., 554 (1969).
11. O. Bastiansen and P. N. Skancke, Adv. Chem. Phys. **3**, 323 (1961).
12. J. Kao, J. Am. Chem. Soc. **109**, 3817 (1987).
13. M. Hanack and G. Renz, personal communication.
14. J. Berkowitz, J. Chem. Phys. **70**, 2819 (1979).
15. P. Dupuis, R. Roberge, and C. Sandorfy, Chem. Phys. Lett. **75**, 434 (1980).
16 B. L. Wheeler, G. Nagasubramanian, A. J. Bard, L. A. Schechtman, D. R. Dininny, and M. E. Kenney, J. Am. Chem. Soc. **106**, 7404 (1984).
17 R. P. Linstead and M. Whalley, J. Chem. Soc., 4839 (1952).
18 L. Edwards and M. Gouterman, J. Mol. Spectrosc. **33**, 292 (1970).
19. S. Deger and M. Hanack, Synth. Met. **13**, 319 (1986).
20. P. Day, G. Soregg, and R. J. P. Williams, Nature **197**, 589 (1963).

ELECTRICALLY CONDUCTIVE PHTHALOCYANINE ASSEMBLIES. STRUCTURAL AND NON-INTEGER OXIDATION NUMBER CONSIDERATIONS

Francisco Torrens, Enrique Ortí, and José Sánchez-Marin

Department de Quimica Fisica, Facultat de Quimica
Universitat de València
Dr. Moliner 50
46100 Burjassot, València, SPAIN

INTRODUCTION

Aggregation is a well-known phenomenon in phthalocyanine chemistry. Interactions can occur between adjacent phthalocyanine rings, both in organic and aqueous phases, resulting in coupling between the electronic states of two, or more, phthalocyanine units [1].

Phthalocyanine compounds, metallophthalocyanines, are known to be electrical semiconductors. When metallophthalocyanines or metal-free phthalocyanine are partially oxidized with a halogen (typically I_2), they become conducting, while simultaneously adopting a face-to-face stacking [2].

The structure of the phthalocyanine dimer was determined by theoretical calculations, utilizing Fraga's atomic pair potential function ($R^{-1-4-6-12}$ type). A face-to-face, slipped structure is reported for the dimer. Structural and oxidation state effects on the association energy are discussed. A growth mechanism of phthalocyanine clusters and β-crystalline phthalocyanine is proposed.

PAIR-POTENTIAL FUNCTION AND ITS IMPLEMENTATION

The intermolecular interaction between two phthalocyanine molecules was studied theoretically by using the atom-to-atom pair potential proposed by S. Fraga [3,4]:

$$\begin{aligned} E_{ij} = \; & 1389.4168\ q_i q_j/R_{ij} \\ - \; & 694.70838\ (f_i \alpha_i q_j + f_j \alpha_j q_i)/R_{ij}^4 \\ - \; & 1516.0732\ f_i \alpha_i f_j \alpha_j/[(f_i \alpha_i/n_i)^{1/2} + (f_j \alpha_j/n_j)^{1/2}]R_{ij}^6 \\ + \; & 4.184\ c_i c_j/R_{ij}^{12} \end{aligned} \qquad (1)$$

where q (atomic net charge), f (optimized fitting factor), α (atomic polarizability), n (effective number of electrons in atom), and c (overlap coefficient) are assigned to each (a, b) class of (i, j) atoms belonging to the (A, B) molecules using Clementi's [5] classification.

An empirical correction to the dispersion energy term is included by using a second R^{-6} term [6], that is added to the original potential [7-9]. The R^{-4} and the two R^{-6} terms are damped, by using a formula by Douketis *et al.* [10]:

$$E^{DAMP}(R) = E(R)\, f(R)\, g(R, n) \quad (2)$$

Here, the f function corrects for the exchange effects:

$$f(R) = 1 - R^{1.68} \exp(-0.78\, R)$$

and the g function corrects for the overlap effects:

$$g(R, n) = [1 - \exp(-2.1\, R / n - 0.109\, R^2/n^{1/2})]^n$$

The values of n are n = 4 and 6 for the R^{-4} and R^{-6} terms, respectively.

Two different procedures of renormalization of the molecular electrostatic charge were implemented: (i) local renormalization of the electrostatic charge in a molecular fragment; (ii) global renormalization of the electrostatic charge through the overall molecule. A variable metric algorithm was implemented, in order to optimize various starting geometries.

RESULTS

Stacking face-to-face, slipped (see Fig. 1) and face-to-face (see Fig. 2) configurations of the phthalocyanine dimer were revealed as the most important orientations in the crystal structures of β-phthalocyanine [11] and halogen-doped salts [12]. The association energies and distances of these minima are reported in Table 1.

In order to test the effect of the oxidation state of the phthalocyanine molecules, the structures of the dimer were optimized with various oxidation numbers. The results are given in Table 2.

The growth mechanism of the phthalocyanine clusters was simulated by building molecular stackings, with a number of units from 1 to 10 in the β–crystal structure (see Fig. 3). The geometrical arrangement between the stackings was optimized. The results are shown in Table 3. For stackings containing one molecule, the optimized structure is face-to-face, slipped (see b minimum in Fig. 3a). Two stackings, formed by two units, present a perpendicular arrangement (see d minimum in Fig. 3b). For stackings with a size greater than 2, the side-by-side organization of the β-phthalocyanine crystal [11] appears (see Fig. 3c, 3d).

DISCUSSION

Two basic stacking structures are important for the description of the phthalocyanine dimer (see Table 1): face-to-face slipped (a-c minima, Fig. 1) and face-to-face (d-f minima, Fig. 2). The stabilities of the slipped a-c dimers are much greater than those of the face-to-face dimers. The horizontal slipping effects reduce the vertical distance between the molecular planes by 0.2 Å. Changes in the rotational angle yield a small effect on the association energies.

The effect of the oxidation state is very significant. When the molecular charges of the molecules are increased, the association energies increase greatly (see Table 2). For the less stable face-to-face structures, the association energies increase the most, and the structures with the most stable minimum show the smallest increases with oxidation.

A possible growth mechanism of phthalocyanine clusters and β-crystalline phthalocyanine is shown in Fig. 3. Small stackings show neither horizontal nor vertical minima, indicating no growth in these directions. The optimized structures are the b and d minima reported in Table 3 (see Figs. 3a and 3b). These structures are "a priori" difficult to imagine. However, horizontal side-by-side stackings are not unstable (see the a and c structures in Table 3). Nevertheless, stackings consisting of more than two units show horizontal minima (see Figs. 3c and 3d). A horizontal growth mechanism is proposed for stackings of three or more units.

CONCLUSION

Two clear advantages appear, when oxidation state effects are used for describing the phthalocyanine dimer: (i) The association energy is increased. (ii) the interaction energy is less sensitive to the effect of the geometrical parameters (distance between the molecular planes and rotation angle).

A possible growth mechanism of phthalocyanine clusters and β-crystalline phthalocyanine is reported. A horizontal growth is proposed for stacking containing more than two molecules.

Table 1. Energies and geometrical parameters of the phthalocyanine dimer.

minimum	-E (kJ/mole)	horizontal slipping (Å)	vertical stacking (Å)	rotation angle (°)
a	236.0	7.8	3.2	0
b	233.3	7.9	3.2	93
c	230.6	8.0	3.1	0
d	5.1	0.0	3.4	0
e	6.7	0.0	3.4	10
f	3.9	0.0	3.4	90

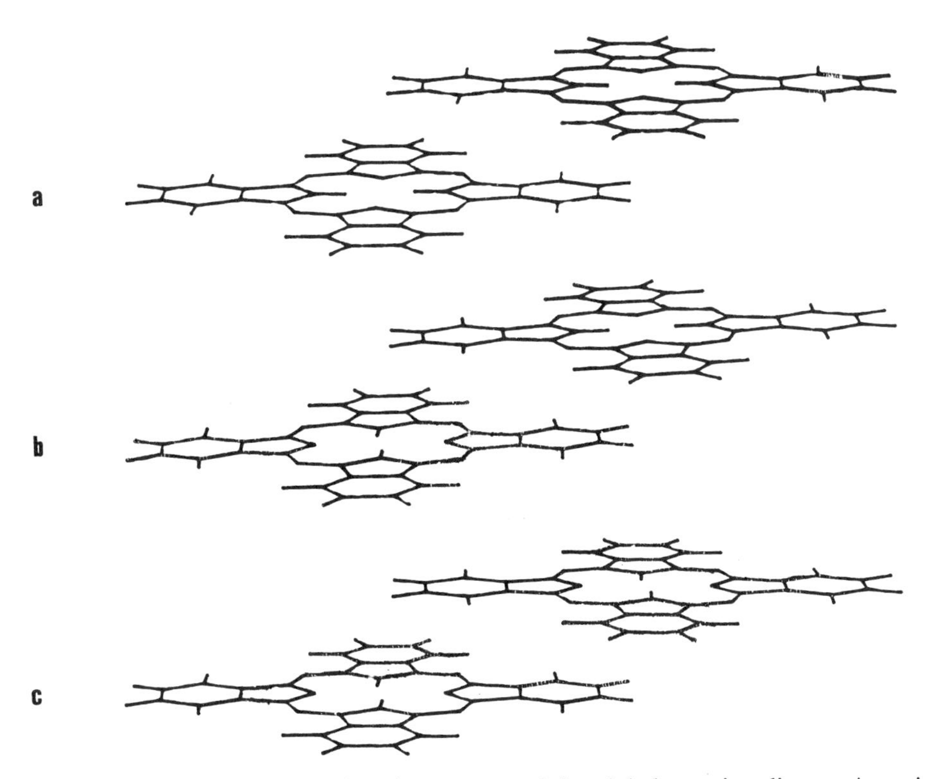

Fig. 1. Stacking face-to-face, slipped structures of the phthalocyanine dimer. A conical projection on the ZY plane is shown. The Z and Y axes denote the vertical and horizontal directions, respectively. The X axis lies in the viewing direction. Structures a (b, c) correspond to the a (b, c) minima in Table 1.

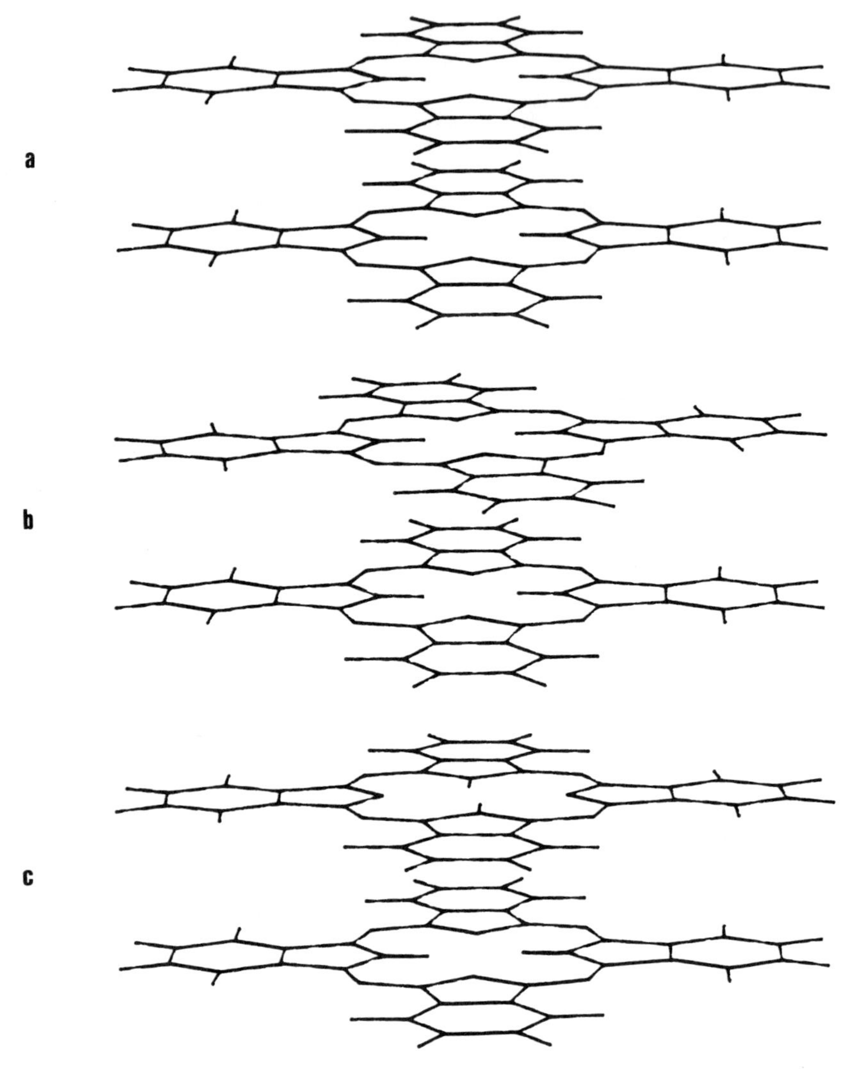

Fig 2. Face-to face structures of the phthalocyanine dimer. A conical projection on the ZY plane is shown. Structures a (b, c) correspond to the d (e, f) minima in Table 1.

Table 2. Association energies (-E, kJ/mole) for various oxidation states of the phthalocyanine dimer.

	oxidation states of molecules A;B		
minimum	0;0	1/3;1/3	0;1
a	236.0	243.6	276.1
b	233.3	240.2	273.6
c	230.6	236.9	268.4
d	5.1	55.1	---*
e	6.7	57.3	117.5
f	3.9	54.1	116.2

* The trial geometry of the d minimum optimizes into the final geometry of the e minimum.

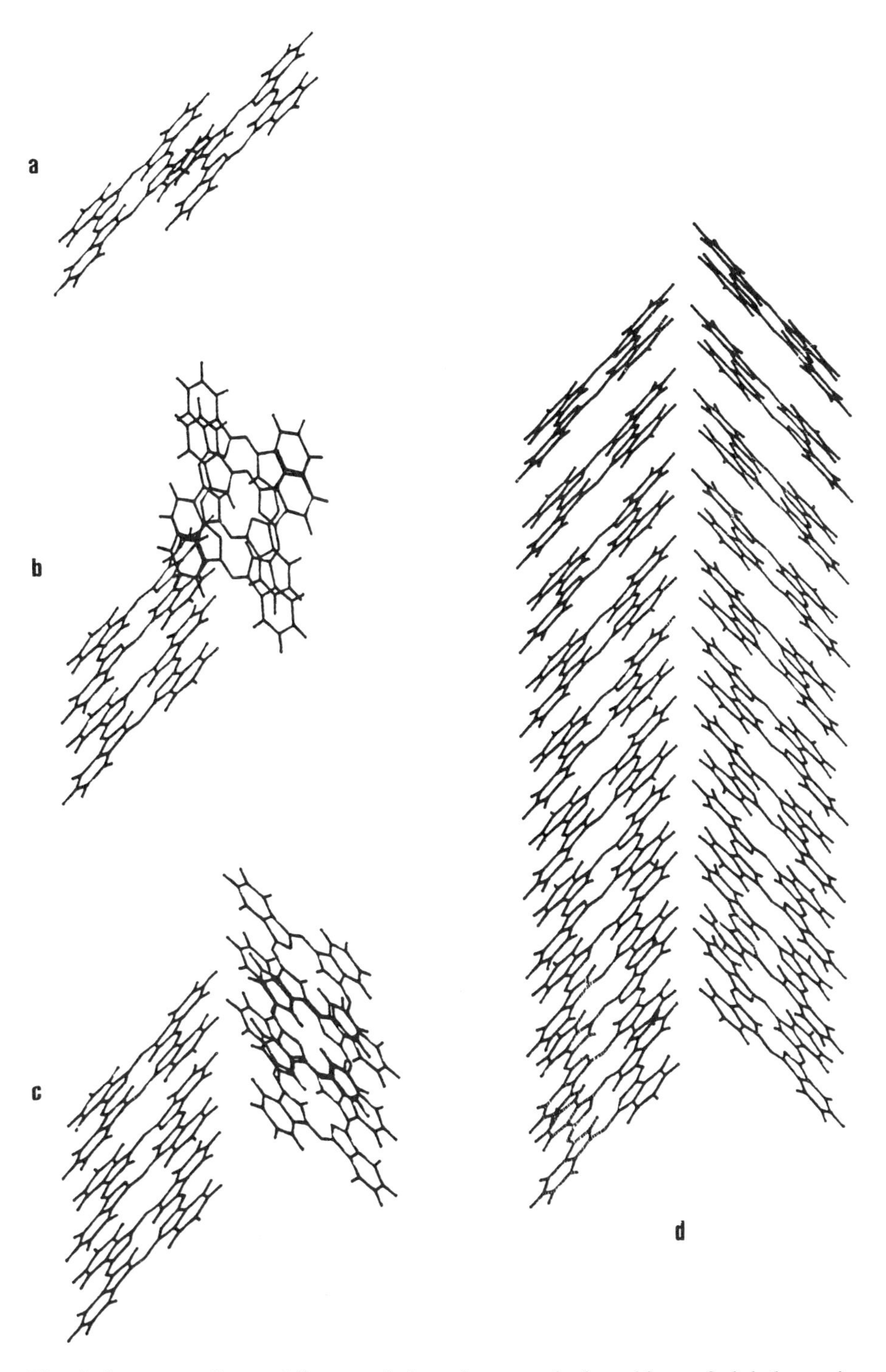

Fig. 3. Structure of assemblies, consisting of two vertical stackings of phthalocyanine molecules. A conical projection on the ZY plane is shown. Structures a (b, c, d) correspond to the b (d, e, 1) minima in Table 3. The structures of the 4:4 to the 9:9 assemblies are similar to those of the 3:3 and 10:10 clusters.

Table 3. Association energies (-E, kJ/mole) for two vertical stackings of phthalocyanine molecules. The number of molecules in each stacking is varied from 1 to 10 units.

				geometrical parameters (Å)#		
minimum	type¶	units	-E	x	y	z
a	SS	1:1	(12.3)¤	-5.10	9.83	2.44
b*	FFS	1:1	233.3	-5.56	6.19	1.87
c	SS	2:2	(64.1)¤	-5.10	9.83	2.44
d	PERP	2:2	192.3	-9.88	5.84	5.56
e	SS	3:3	175.1	-7.28	9.81	4.22
f	SS	4:4	251.2	-6.54	10.00	3.41
g	SS	5:5	330.9	-5.71	9.94	2.87
h	SS	6:6	419.9	-5.30	9.86	2.60
i	SS	7:7	507.9	-5.22	9.83	2.53
j	SS	8:8	599.2	-5.17	9.83	2.48
k	SS	9:9	687.7	-5.12	9.83	2.46
l	SS	10:10	779.6	-5.10	9.83	2.44
m	CRYS§	:	-	-3.26	9.38	2.37

¶ The types of minima are: SS (Side-by-Side), FFS (Face-to-Face, Slipped), PERP (Perpendicular), and CRYS (β-crystal).

The geometrical parameters refer to translations along the axes: x (viewing direction), y (horizontal side-by-side direction), and z (vertical stacking direction).

* The b minimum in this Table corresponds to the b minimum in Table 1.

§ Data taken from Ref. [11].

¤ Data in parentheses indicate geometries that are not actual minima. Trial geometries a and c are stable, but optimize into the b and d minima, respectively.

ACKNOWLEDGEMENT

One of us (F. T.) is grateful to the Conselleria de Cultura, Educació i Ciència de la Generalitat Valenciana for a Grant. This work has been partially supported by the project 714/84 of the C.A.I.C.Y.T. and the project 683.2/89 of the Universitat de València. This research has used the resources of the Servei d'Informàtica de la Universitat de València.

REFERENCES

[1] E. S. Dodsworth, A. B. P. Lever, P. Seymour, and C. C. Leznoff, J. Phys. Chem. 89: 5698 (1985).
[2] E. Canadell and S. Alvarez, Inorg. Chem. 23: 573 (1984).
[3] S. Fraga, J. Comput. Chem. 3: 329 (1982).
[4] S. Fraga, Comput. Phys. Commun. 29: 351 (1983).
[5] E. Clementi, "Computational Aspects for Large Chemical Systems," (Springer, Berlin, 1980).
[6] J. C. Slater and J. G. Kirkwood, Phys. Rev. 37: 682 (1931).
[7] F. Torrens, J. Sánchez-Marín, E. Ortí, and I. Nebot-Gil, J. Chem. Soc. Perkin Trans. II 943 (1987).
[8] F. Torrens, A. M. Sánchez-de-Merás, and J. Sánchez-Marín, J. Mol. Struct. (Theochem) 166: 135 (1988).
[9] F. Torrens, J. Sánchez-Marín, and F. Tomás, J. Phys. Chem. (submitted).
[10] C. Douketis, G. Scoles, S. Marchetti, M. Zen, and A. J. Thakkar, J. Chem. Phys. 76: 3057 (1982).
[11] R. Mason, G. A. Williams, and P. E. Fielding, J. Chem. Soc. Dalton Trans. 676 (1979).
[12] T. J. Marks, Science 227: 881 (1985).

VALENCE-BOND TREATMENT OF 3/4 FILLED DIMERIZED CHAINS: EXTENDED HUBBARD RESULTS

A. Fritsch and L. Ducasse

Laboratoire de physicochimie théorique
Université de Bordeaux I
F-33405 Talence, FRANCE

INTRODUCTION

Some organic salts, based on derivatives of TTF and centrosymmetrical anions [$(DIMET)_2SbF_6$, β-$(ET)_2ICl_2$, $(ET)_2AuCl_2$], show a very large degree of electronic localization, even at room temperature [1]. It is clear from the structural data (Fig. 1) that in these salts, the organic stacks are much more dimerized than in type-I salts, such as Bechgaard salts [2]. The physical properties also give a very localized picture of these compounds, which are already semiconductors at room temperature, and exhibit spin chain magnetic properties (Fig. 2). The monoelectronic description of the electronic properties is not relevant in these systems, and we focus in this paper on the influence of the electronic correlation on the localization process and the magnetic properties.

We applied the Diagrammatic Valence-Bond (DVB) method [3,4] to an extended Hubbard Hamiltonian [5], including second-neighbor repulsion. The DVB method allows us to go beyond mean-field theory, since it performs a complete configuration interaction. Unfortunately, the size of the configuration basis grows so rapidly, with the number of site orbitals, that we are compelled to restrict ourselves to finite clusters of N = 8 or 12 sites. This is a serious drawback of the method, but the advantage is that we can solve the model Hamiltonian exactly, and obtain an accurate wavefunction, which could be a good indication of that of the infinite system, since we are dealing with local interactions.

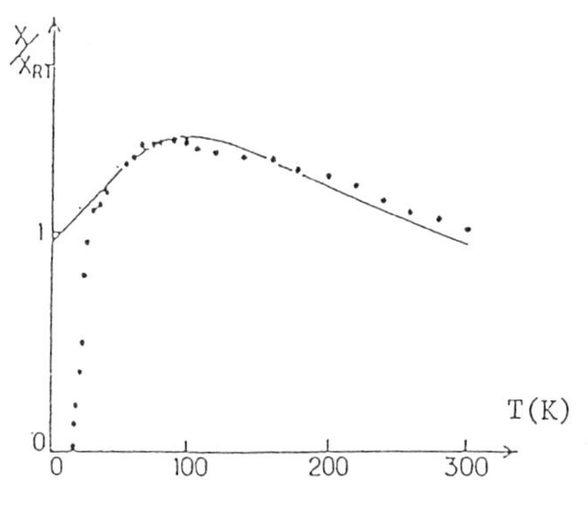

Fig. 1. ESR susceptibility of $(DIMET)_2SbF_6$ with spin-1/2 Heisenberg fit, from Ref. [1].

Table 1. Bond energies of the 12-sites cluster

	E_{S1}/meV	E_{S2}/meV
V'=0	79	14
V'=$V_{S2}/2$	84	16
Hückel	78	37

Lower-Dimensional Systems and Molecular Electronics
Edited by R.M. Metzger *et al.*, Plenum Press, New York, 1991

MODELLING THE SYSTEM

Our interest lies in the localization, which is already established at room temperature. In such anisotropic systems, the localization belongs to 1-D physics. We then choose to use a chain of equivalent sites, as there is no experimental evidence of independent molecules in these systems. As in the usual band calculations [6], we replace each molecule by its highest occupied molecular orbital (HOMO), providing 3 conduction electrons for two sites.

A complete set of electronic configurations is built up upon these orbitals, following the Rumer procedure [3], along with a matrix representation of the Hamiltonian. The use of only one orbital per site allows us to draw simple diagrams to symbolize each configuration [3].

The model Hamiltonian is of the extended Hubbard type:

$$H = H_t + H_U + H_V$$

Here:

$H_t = -\Sigma_{pq} t_{pq} a_{p\sigma}^+ a_{q\sigma}$ is the transfer Hamiltonian, restricted to first-neighbor interactions; the operator product $a_{p\sigma}^+ a_{q\sigma}$ annihilates an electron (with S_z projection of the spin σ) on site q and creates it on site p; t_{pq} is the transfer integral;

$H_U = 1/2\ \Sigma_p U n_p (n_p - 1)$ is the intrasite correlation Hamiltonian; U is the intrasite coulomb potential, and n_p is the number operator on site p;

$H_V = 1/2\ \Sigma V_{pq} (n_p - z_p)(n_q - z_q)$ is the intersite correlation Hamiltonian; V_{pq} is the intersite coulomb potential, relative to sites p and q; z_p and z_q are the core charges of sites p and q ($z_p = z_q = 2$).

H_U and H_V are diagonal in the DVB basis, and give the electrostatic contribution of each configuration. We assume cyclic boundary conditions, by including t and V interaction terms between the boundaries. The diagrams, consequently, show a ring geometry (see Fig. 2).

PARAMETER CALCULATION

The transfer integrals are, as usual, calculated using the dimer splitting approximation [6], and the extended Hückel [7] or CNDO/2 [8] methods; the latter includes corrections due to the Fock potential [9]. We also tried to evaluate the two-electron terms U and V_{pq}, using CNDO/2 or MINDO/3 methods.

The intrasite potential is $U = \langle\Psi_p \Psi_p \mid 1/r \mid \Psi_p \Psi_p\rangle$, calculated from the HOMO of the monomer p. The intersite potential between sites p and q is $V_{pq} = \langle\Psi_p \Psi_q \mid 1/r \mid \Psi_p \Psi_q\rangle$, obtained from the HOMO of the monocation dimer (p,q).

The resulting values of the two-electron terms are quite large: around 6 eV and 3 eV for U and V_{pp+1}, respectively, for both methods and for various molecules. Such large values imply that a correct calculation should be based on a cluster with a size larger than the range of the intersite potential V_{pq} (the range being determined when $V_{pq} < t$). Moreover, the "perfect" cluster should then be three-dimensional. The present computational capacities are still not sufficient to handle such calculations. It seems more promising to use semi-phenomenological parameters, assuming that the most important feature of the correlation, when one is looking at localization effects, is the effective electrostatic cost of the hopping of one electron from one site to another along the chains.

Such phenomenological parameters are chosen roughly equal to the ones deduced from experiment [10], except that we take into account the dimerization of V_{pp+1} ($V_{2p+1,2p+2} = V_{S1}$ and $V_{2p,2p+1} = V_{S2}$) and the second-neighbor potential V'. We believe that V' is quite relevant in 3/4-filled bands, since the range of the effective potential is not restrained to the first p+n+neighbors: with $t_{S1} = 0.15$ eV and V_{S1} and V_{S2} around 0.4 eV, V' should be comparable to t_{S1}. Nevertheless, there are no reliable experimental determinations of the parameters, since they rely upon fits to simpler models, and the parameters obtained are model-dependent.

The inclusion of V' emphasizes the competition between two types of perfectly ordered configurations, called A and B. A shows simple alternation of singly and doubly occupied sites, while B introduces alternation of neutral and doubly charged dimers. The mixing of A and B arises from successive one-electron transfers, and involves "excited" configurations (C in Fig. 2), pointing to the fact that the role of such charge fluctuations is fundamental to the electronic behavior. The degree of mixing of A and B depends on the relative electrostatic energies associated with these two types of configurations.

DESCRIPTION OF THE ELECTRONIC SYSTEM

The localization/delocalization of the ground state is estimated through the contribution of the transfer Hamiltonian to the total energy.

$$<\Psi_0 | H_t | \Psi_o> = - N (p_{S1} t_{S1} + p_{S2} t_{S2}) = N/2 (E_{S1} + E_{S2}),$$

where p_{S1}, p_{S2} are the bond orders of bonds S1 and S2, and E_{S1} , E_{S2} are the contributions of the S1 and S2 bonds to the ground state energy.

The ratio E_{S1}/E_{S2} gives the dimerization of the bond energies, which can be related to the charge density between the corresponding sites. These values must be compared to the one obtained with an uncorrelated Hamiltonian (Band or Hückel Model).

The charge correlation function C_{nm} for sites p and q, defined as the probability of having n electrons on site p and m electrons on site q, is obtained by calculating the weight in the wavefunction norm of such a distribution. This is a useful tool to investigate correlation effects. Since the wavefunction is expressed as a linear combination of configurations, the calculation reduces to summing the overlaps between the configurations having the desired electronic distribution.

Calculations on the singlet, triplet and quintuplet states allow us to evaluate the spin susceptibility [3] of a cluster of N = 8 sites. Since the size of the system is finite, there is a gap in the spin excitation spectrum. The resulting temperature-dependent susceptibility curve cannot reproduce the macroscopic curve near T = 0 K, but at least we may compare the absolute value of the susceptibility at 300 K, and the temperature of the maximum of the susceptibility, to the experimental values, as this has already been done for Heisenberg phenomenological Hamiltonians [11].

RESULTS

We focus here on two cases, which are expected to show qualitative differences with the type of correlation of the charge.

The first case assumes values of the parameters comparable to the ones proposed in the literature for $(DIMET)_2SbF_6$ [10]; say U = 1.3 eV, V_{S1} = 0.5 eV, V_{S2} = 0.3 eV, t_{S1} = 0.17 eV, and t_{S2} = 0.10 eV. We are here in the intermediate case of correlation, but with a ratio U/t sufficiently large to avoid hopping of the carriers, with creation of additional holes in the organic stacks. The major configurations, involved in the ground state, will then be A, and the low-lying excited configurations obtained from A by one-electron transfers.

The most striking result (Table 1) is that the S2 bond energy is very much weakened by the correlation, as we obtain a value 5.6 times smaller than the S1 bond energy (the ratio being only 2.1 for the Hückel calculation). The DVB process, thus, strongly enhances the bond dimerization. It should also be noticed that the motion of the carriers through the S1 dimer doesn't seem much affected by the intersite potential V_{S2}; the ratio E_{S1} / Hückel bond energy is close to 1.

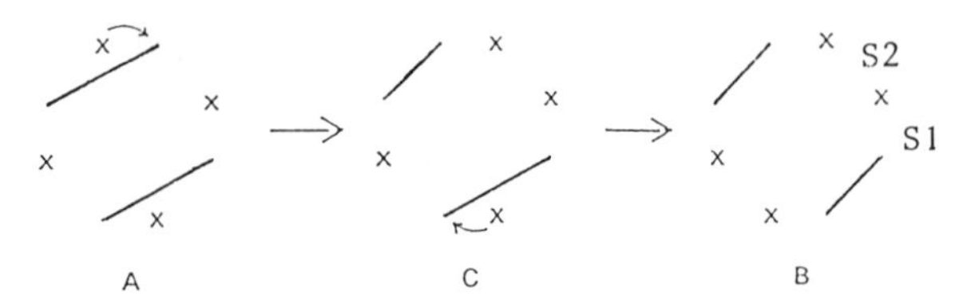

Fig 2. Diagrammatic representation of A and B mixing. X = doubly occupied site, lines connect singly occupied sites. S1 is the intradimer interaction, S2 the interdimer interaction.

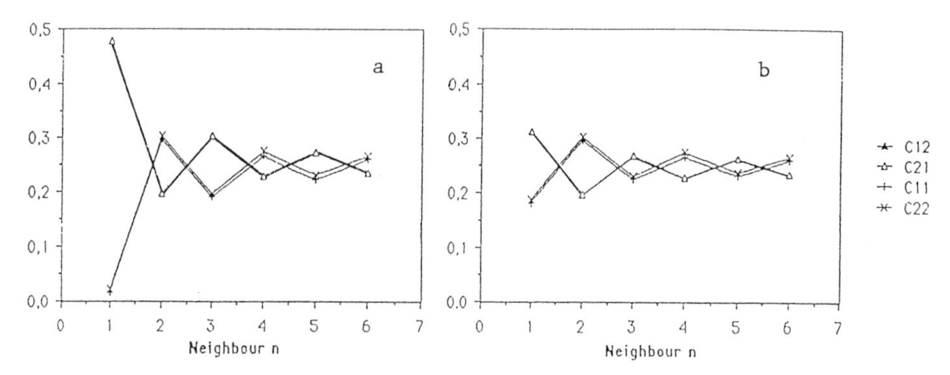

Fig 3. Charge correlation functions for N = 12 sites and V' = 0 (a) for sites 2p+1 and 2p+n+1, (b) for sites 2p and 2p+n.

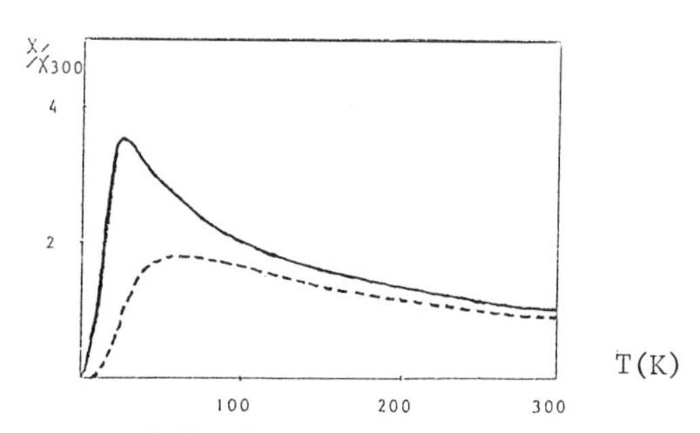

Fig 4. Theoretical spin susceptibility for N = 8 sites when V' = 0 (——) and $V' = V_{S2}/2$ (---).

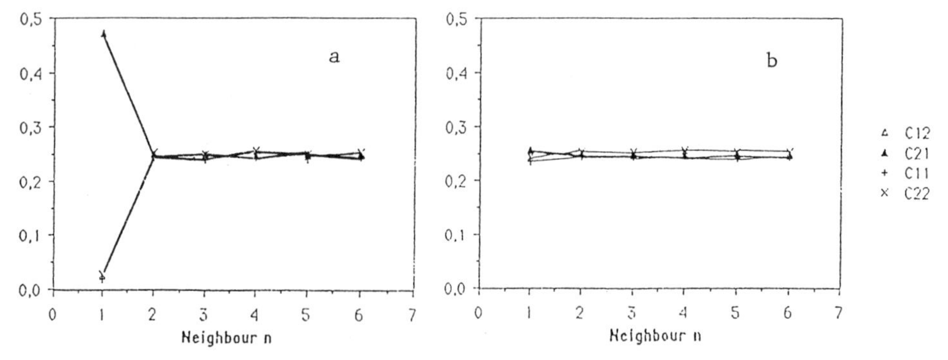

Fig 5. Charge correlation functions for N = 12 sites and $V' = V_{S2}/2$, (a) for sites 2p+1 and 2p+n+1, (b) for sites 2p and 2p+n.

Charge correlation functions are given in Fig. 3. There is a very strong correlation of the charge within the S1 dimer, which acts so as to assure 3 electrons on the two molecules (C12 + C21 = 0.96), and prevents the creation of a supplementary hole or electron, since the probability of having 2 or 4 carriers (C11 and C22) is very small, avoiding the delocalization of the electrons from one dimer to another. Since U is very large, compared to the other parameters, probabilities involving empty sites are negligible. The correlation range extends outside the S1 dimer, first by correlating the charges through the S2 interdimer interaction, but also by coupling successive dimers, and favoring situations with the same number of electrons on sites n and n+2. However, the intensity of the charge correlation decreases, as we consider more distant sites, and the four main situations become equally probable.

The temperature-dependent reduced spin susceptibility is shown in Fig. 4, and resembles that of a finite spin chain, emphasizing the influence of electron correlation, and especially of the intersite potential dimerization, on the magnetic properties. There is a sharp maximum at 24 K, contrasting with the experimental curves, which show very smooth maxima at higher temperature (around 60 K for $(DIMET)_2SbF_6$); nevertheless, the absolute value at room temperature is in the right order of magnitude, with 5×10^{-4} μemcgs/mol of dimer against 7×10^{-4} for $(DIMET)_2SbF_6$.

The second case accounts for a more continuous evolution of the intersite potential, as we consider more distant sites. Inclusion of the second-neighbor potential V' allows the relative electrostatic energy of the B configurations to be lowered, and the mixing with A to be improved. We focus here on an "extreme" case where $V' = V_{S2}/2$; the other parameters are the same as in the first case. This value of V' implies that, e.g. in the 8-sites cluster A, C and B are energetically equivalent; electrons are then free to cross S1, but there still remains a smaller barrier $V_{S1} - 2V' = 0.2$ eV for crossing the interdimer space (in the case V'=0 the barriers were, respectively, 0.3 and 0.5 eV). Meanwhile, even with such low values of the barriers, Table I shows that there is no qualitative difference in the bond energy dimerization between the cases V'=0 and $V'=V_{S2}/2$. On the other hand, the charge correlation functions are quite different (Fig. 5). The S1 intradimer correlation is still very large, compared with the S2 interdimer interaction, which shows an almost uncorrelated situation, as all the functions are intersecting. The consequences are seen on the long-range charge correlation, which drops quickly, and the different curves superimpose. We thus can see the system as a collection of uncorrelated dimers, with three delocalized electrons on each, but with almost no possibility of charge transfer between them.

The temperature-dependent reduced spin susceptibility (Fig. 4) is smoother, and presents now a very flat maximum at 60 K; thus, the behavior near the maximum is rather dependent on the V' parameter, and indicates the original organization of the electronic system. The room temperature value is 4.7 μemcgs/mol, slightly smaller than in the former case, but still in the right range. We recall that the parameters used in the calculation do not result from a fit of the susceptibility, and may be improved.

CONCLUSION

Using a Valence-Bond method on an extended Hubbard model, we have reproduced finite-spin-chain behavior using dimerized first-neighbor parameters. The inclusion of the second-neighbor repulsion V' showed that we could also reproduce the characteristic temperature dependence of the magnetic susceptibility near its maximum. Athough the DVB method cannot compete with Heisenberg models in extensions to infinite chains, it may give better results on finite rings, and, overall, shed some light on the origin of the magnetic properties.

We believe that V' is an important feature in these systems, since it may lead to a better simulation of the magnetic properties and, above all, because of its role in the charge correlation along the chains. Neglecting V' assures a correlation of the charge, over a finite range longer than the unit cell, while values of V' close to $V_{S2}/2$ decouple successive dimers.

The localization of the ground state is understood in terms of bond energy dimerization. It is strongly enhanced by the electronic correlation, even with relatively small values for V_{S1}

and V_{S2}, and thus very different from the Mott localization observed in regular systems [9], which require very large correlation effects. Further work is in progress on quasi-regular systems, in order to investigate the localization process in the Bechgaard salts $(TMTTF)_2X$ [12].

We thank Z. G. Soos for providing us his DVB program.

REFERENCES

[1] R. Laversanne, J. Amiell, C. Coulson, E. Dupart, P. Delhaès, J. P. Morand, and C. Manigand, Solid State Commun. 58: 765 (1986).
[2] K. Bechgaard, C. S. Jacobsen, K. Mortensen, H. J. Pedersen, and N. Thorup, Solid State Commun. 33: 1119 (1980).
[3] S. Mazumdar and Z. G. Soos, Synth. Metals 1: 77 (1980).
[4] S. Ramasesha and Z. G. Soos, Int. J. Quantum Chem. 25: 1003 (1984).
[5] J. Hubbard, Phys. Rev. B 17: 494 (1978).
[6] L. Ducasse, A. Abderraba, J. Hoarau, M. Pesquer, B. Gallois, and J. Gaultier, J. Phys. C: Solid State Phys. 19: 3805 (1986).
[7] R. Hoffmann, J. Chem. Phys. 39: 1397 (1963).
[8] J. A. Pople and D. L. Beveridge, "Approximate Molecular Orbital Theory," (McGraw-Hill, New York, 1970).
[9] A. Fritsch, Ph.D. thesis, Univ. of Bordeaux (1989).
[10] P. Delhaès and C. Garrigou-Lagrange, Phase Transitions 13: 27 (1986).
[11] J. C. Bonner and M. E. Fischer, Phys. Rev. 135: A640 (1964).
[12] R. Laversanne, C. Coulon, B. Gallois, J. P. Pouget, and R. Moret, J. Physique Lett. 45: L399 (1984).

POSSIBLE NICHES FOR LANGMUIR BLODGETT FILMS

G. G. Roberts

Department of Engineering Science
University of Oxford
Parks Road, Oxford, OX1 3PJ, England

1 INTRODUCTION

During the past decade many papers have been published in the field of Langmuir-Blodgett (LB) films[1]. Considerable progress has been made in improving the deposition system (Langmuir-trough) and associated equipment for monitoring the quality of a floating monolayer. There has also been much innovation in the design of suitable molecules. Consequently, there is now a reasonable body of information for a range of different materials deposited in LB film form. Moreover, virtually every relevant analytical technique has now been used to show that well-engineered molecules can form high-quality ordered films on a variety of substrates.

To date there has been little or no success in commercializing monomolecular assemblies of this type. However, certain areas of investigation have exhibited promising results, which have prompted investment in applied research and development. These are described in this paper. They have been grouped into three sections. The first two deal with passive and active applications of LB films. The final section concentrates on the strong claims for pyroelectric LB films to be used in thermal imaging systems.

2 PASSIVE APPLICATIONS OF LB FILMS

All the early proposed applications of LB films, including those in the areas patented by Langmuir and Blodgett[2-4] focussed on their passive use. Examples included coatings for low reflectance glass, step thickness gauges and mechanical filters. During the past few years, more interesting possibilities have emerged, especially those involving the deposition of a low number of monolayers.

Lower-Dimensional Systems and Molecular Electronics
Edited by R.M. Metzger *et al.*, Plenum Press, New York, 1991

2.1 Electron Beam Lithography

In integrated circuit technology there is a continuing demand for faster response and increased memory storage capacity. This has led to refinements of the microlithographic techniques for producing even smaller circuit elements,and this has necessitated a drift away from conventional photo-lithography to technologies involving X-rays and electron or ion beams. The use of scanned electron beams are an attractive possibility, owing to the excellent depth of focus and the high registration accuracy that may be achieved. However, using this method, because of scattering, high resolution can only be achieved with very thin resists. Below thicknesses of about 1μm, conventional spin coated polymer films, such as polymethylmethacrylate (PMMA), display unacceptably high pin-hole densities and variations of thickness. Following the early work of Barraud and his co-workers,[5] there are now good examples of both negative and positive resists. The most exciting results have been reported by Pease[6] using multilayer assemblies based on substituted PMMA molecules. The observed defect densities are extremely low, and appear to offer a realistic chance of being utilized in commercial microstructure fabrication, particularly for specialist applications requiring nanolithography.

2.2 Lubrication of Magnetic and Optical Storage Media

Lord Raleigh was the first to demonstrate that even a single monolayer deposited onto glass or porcelain was sufficient to drastically reduce static frictional effects. The mechanisms influencing the frictional properties and responsible for maintaining the monolayer's integrity during sliding have been discussed by Briscoe and Evans[7]

The magnetic recording industry requires a good permeability barrier layer, which both protects the oxidized magnetic media, and acts as a lubricant to avoid wear and tear during transit of the tape across the magnetic head. For use in high-density recording systems it is essential that the spacing between the magnetic layer and the recording head be very thin, typically less than 10nm. Two reports have appeared describing the use of LB films. Figure 1 illustrates the significant effect on the frictional coefficient of coating an evaporated cobalt tape with an LB film of barium stearate. Further experiments to confirm the beneficial effects of the LB film coating were obtained by analyzing the playback signal of a video recorder. These initial investigations have confirmed the strong relationship between the adhesive properties of monolayers and their effectiveness as lubricants. Optical discs could also benefit from specialized coatings of this kind. In principle there is no difficulty in envisaging a special production unit for coating LB films onto rapidly moving belts of magnetic tape. However, to be successful commercially, the film deposition process would have to be extremely cheap.

2.3 Enhanced Device Processing

There may be a useful niche for multilayer LB films as passive layers in electronic devices. For example, even a single monolayer is sufficient to increase the breakdown strength of a leaky 'oxide' film[10]. In order to illustrate the effectiveness

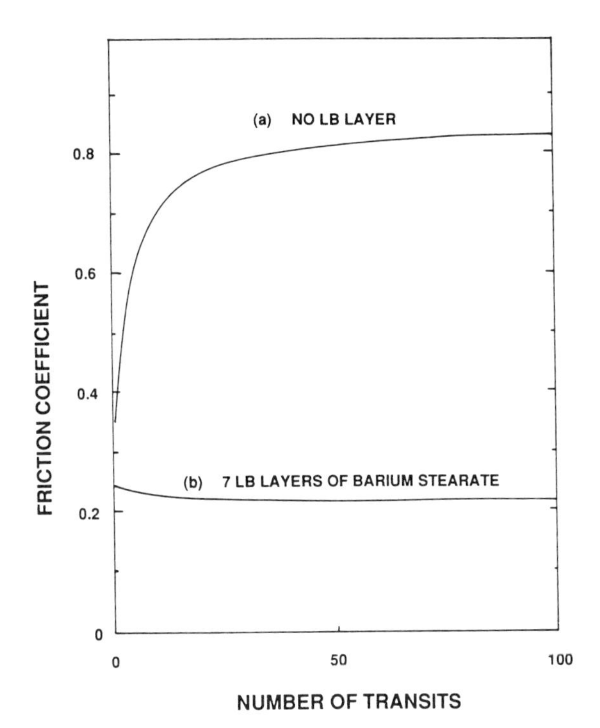

Fig.1. The influence of 7 layers of barium stearate (curve b) in reducing the frictional coefficient between a cobalt magnetic tape and a recording head. Curve (a) is for an unprotected metal surface. (See Reference 8)

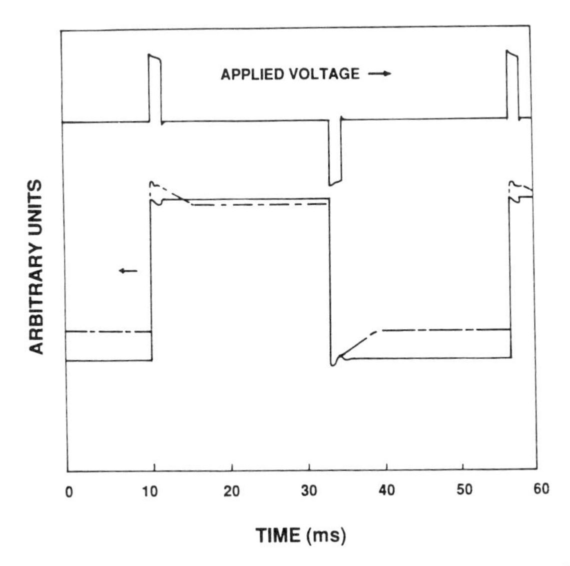

Fig.2. The electro-optic response curve of a ferroelectric liquid crystal display to a voltage pulse of approximately ±20V. The solid and dashed curves are for polyimide alignment layers produced using LB film and rubbing techniques, respectively. (See Reference 11)

of LB films for improving device performance two contrasting but important situations of practical importance are now discussed.

2.3.1 Liquid Crystal Alignment. Exciting research work is currently in progress on ferroelectric liquid crystal displays. The advantage of these molecules over the conventional smectic or nematic types is that they have an intrinsic polarization and can be switched much faster. Thus, when a pulsed voltage is applied, the liquid crystal molecules are switched and locked into a specific state, which can be reversed when the voltage is changed in sign. When used in television screen applications, there is then no need to address them with a thin film transistor array. Because of their memory capability, all that is required is flat plates of glass coated with a relatively straightforward array of conductors. A key factor is the alignment of the liquid crystal molecules on the substrate. Rubbing an evaporated or spin-coated polymer film is one way of achieving the required high pre-tilt angle. However, LB films, because of their controlled architecture and large area uniformity, are well worth investigating for this purpose. A good example of what can be achieved is shown in figure 2, which compares the electro-optic response curve for liquid crystal cells aligned with polyimide films prepared using the Langmuir trough and conventional means. The almost perfect bistability characteristics obtained using LB films were reported to be very reproducible.

2.3.2 Surface Acoustic Wave Devices. The majority of surface acoustic wave (SAW) devices are based on quartz or lithium niobate, and are fitted with complex patterns of interdigitated electrodes. For more specialized applications the substrate also contains other two-dimensional structures, which influence the movement of the sound wave along the surface. The accuracy and usefulness of a device depends on the precision with which the patterns have been defined using lithography. Sometimes, it is necessary to fine-tune the characteristics after the device has been fabricated and tested, in order to meet the exacting specifications demanded. This costly procedure normally requires additional expensive photolithographic stages, but can be accomplished more easily by depositing specified thicknesses of monolayers on different regions of the substrate. Figure 3 shows a compression pulse for an in-line reflective array compressor device before and after deposition of corrective coatings of an LB film. The phase correction process has the effect of reducing the signal strength of the sidelobes by several dB below that of the main pulse.

2.3.3 LB Films on Semiconductors Frequently, a single monolayer deposited onto a semiconductor can strongly influence its surface properties. This was first demonstrated[13] for LB films coated on cadmium telluride single crystals. The physics occurring at the inorganic/organic interface is not at all well understood, but the most dramatic effect is the observed increase in the effective Schottky barrier height. This result has been capitalized upon in photovoltaic cells[14,15] and in electroluminescent diodes[16]. In both cases the efficiency of the devices was shown to be optimum when they were coated with a definite number of monolayers. Figure 4 confirms this result in the case of gallium phosphide.

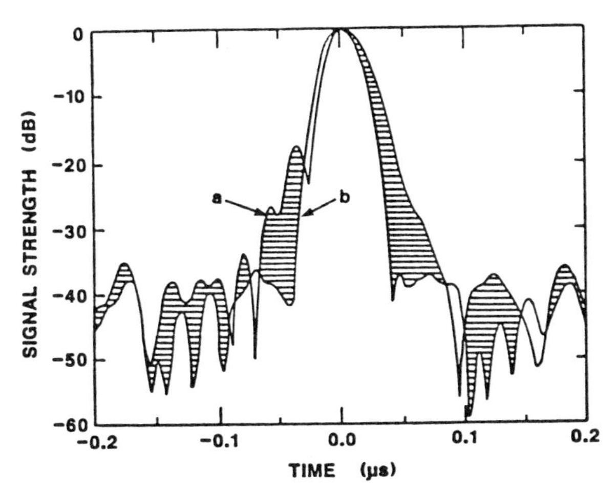

Fig.3. The resonance curve for a surface acoustic wave device. Curve (a) is for the imperfect device; Curve (b) shows the dramatic improvement in terms of sharpening the resonance main peak when the device is coated with an appropriate number of LB monolayers. (See Reference 12)

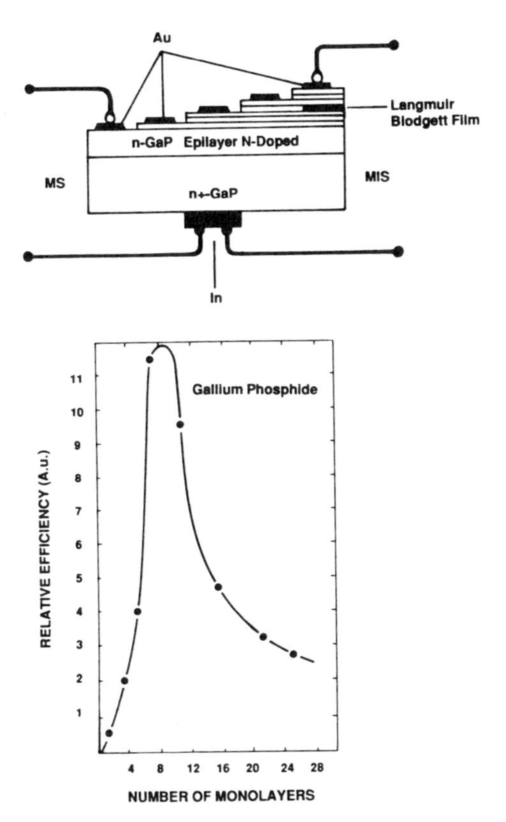

Fig.4. The electroluminescent efficiency versus number of ωTA monolayers for the metal-LB film-semiconductor structure shown in the upper section of the diagram.

The first transistor incorporating LB multilayers as the insulator was reported over a decade ago[17]. The subject is still the focus of attention[18], especially where group III - V semiconductors are concerned. Interesting data have been reported recently by Chan et al[19] who used a single monolayer of cadmium arachidate to modify the surface of the alloy $In_{0.53}Ga_{0.47}As$. The resulting barrier height, after treatment in an oxygen plasma, increased by about 0.32eV, sufficient to produce a modulation-doped field effect transistor with excellent dc and microwave performance (cut-off frequency = 19 GHz, transconductance = 170 mS mm^{-1}).

Because of the high investment in integrated circuit processing technologies based on conventional inorganic materials, especially silicon, it seems unlikely that organic films will play a major role in microelectronic signal processing devices. However, it may well be that a semiconductor structure, incorporating an organic LB film at its centre, may be useful as a sensor.

3 ACTIVE APPLICATIONS OF LB FILMS

There are several fields where LB films perform an active function; the areas highlighted in this section are chemical sensors, optoelectronics and information storage. The promising application utilizing their pyroelectric properties is discussed in the following section. In the longer term, LB films may play an important role in the processing and transmission of information. However, switching or rectification in organic monolayers has not yet been observed, despite intensive and creative attempts to design suitable molecules[20].

3.1 Chemical Sensors

There is a great demand for improved analytical techniques in industry. The tremendous scope available with organic molecular systems to produce carefully designed, optimized and engineered materials, that are specific (ability to reorganise), and selective (ability to discriminate), should lead to important commercial devices. The whole field is, however, fraught with difficulties for, unlike other application areas, it is not possible to completely encapsulate the fragile organic layer. Consequently, more emphasis needs to be placed on mechanical stability requirements. The first layer in an LB film assembly normally adheres very firmly to the substrate; the weakness is more likely to occur due to lack of cohesion between adjacent molecules. For this reason, it seems sensible to explore techniques which require the deposition of a single monolayer to sense the gas or fluid.

From a device perspective, the operation of a chemical sensor can be divided into three parts. By far the most challenging is designing a lock-key mechanism governed by a specific interaction between a molecule and a sensing surface. This has been achieved in a few isolated cases, but most reports of gas sensing are disappointing, in that discrimination is often inferred via indirect effects, such as differing temperature responses or recovery times etc. Another method involves the use of an array of different sensing materials, each responding slightly differently to the gas to be sensed. Naturally, the binding process must result in an detectable systematic change in a physical parameter. The final stage requires a transducer, to convert the change into an observable signal.

There are many types of device that can be used in conjunction with a suitable sensing element. The most common rely on the measurement of the dc conductivity across the surface of a material fitted with appropriately positioned planar electrodes. It has also been fashionable to employ a transistor-type structure. However, there are more convenient transducers that can be used in conjunction with LB films. These involve optical and acoustical methods of detecting the minute changes introduced by the presence of a gas or fluid.

3.1.1. Surface Plasmon Resonance Optical fibres and optical waveguides can be used in a variety of sensors, which exploit a critical resonance or threshold effect. The most elegant technique involves quanta called surface plasmons. Surface plasma waves are electromagnetic waves at the interface of a metal and dielectric. Any changes in the properties of the dielectric layer in the immediate vicinity of the interface influences the excitation of the surface plasmons. Figure 5 shows schematically a photon reflection technique for generating plasmons. At the resonance condition, there is considerable reduction in light throughput because of plasmon absorption. The thickness of the metal layer (normally silver) is very important in achieving a sharp resonance. If an organic layer, e.g. a monolayer of a chemically specified material, is deposited onto the metal surface, the resonance position is shifted to a higher angle of incidence, and also some broadening occurs. When the film is exposed to the vapour, a further shift occurs. This can be noted either by mapping out the overall curve or, more conveniently, by setting the angle of incidence to be at the steepest part of the resonance curve, and monitoring the intensity change. A good example of the use of SPR to monitor acceptor-type gases has been reported by Lloyd et al[21].

There are excellent prospects for exploiting the effect in medical diagnostics[22]. Some interest has been shown in using LB films as a matrix for the immobilization of a recognition molecule which is suitable for antibody-antigen type reactions. Good reviews for this type have been given by Moriizumi[23] and Reichert et al[24].

3.1.2. Surface Acoustic Wave Sensors A surface acoustic wave (SAW) structure can serve as a sophisticated acoustoelectric microgravimetric sensor. In such devices, as has been described earlier, input and output interdigitated electrodes perform the conversion between electric and acoustic energy. The piezoelectric single crystal (quartz or lithium niobate) surface region between the electrodes serves as a propagation path for acoustic waves, and thus forms a delay line. A selective coating is placed in the propagation path of one of two identical SAW oscillators positioned alongside each other, thus affecting the delay time. The change in frequency between the two channels is then directly attributed to the sensor layer, and extraneous effects, such as temperature changes, are eliminated, or very much reduced.

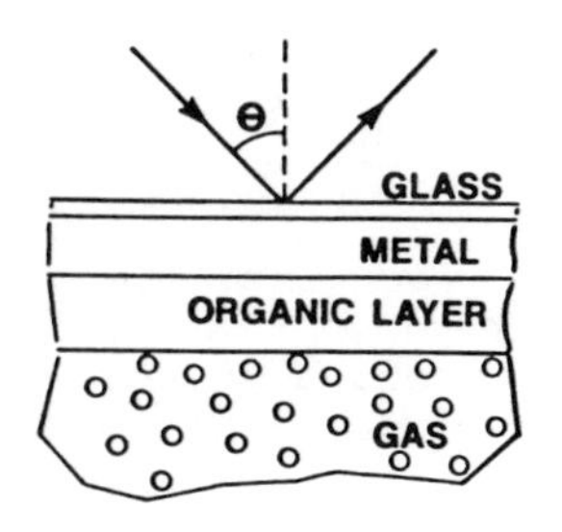

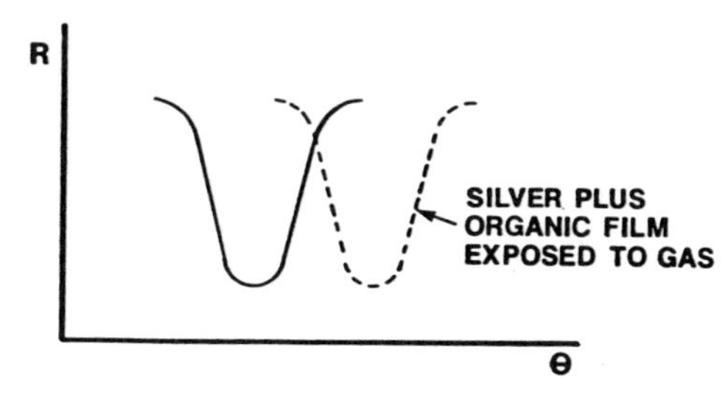

Fig.5. Schematic representation of Surface Plasmon Resonance (SPR) method for detecting gases or fluids. The minimum in the reflectivity occurs due to plasmon absorption.

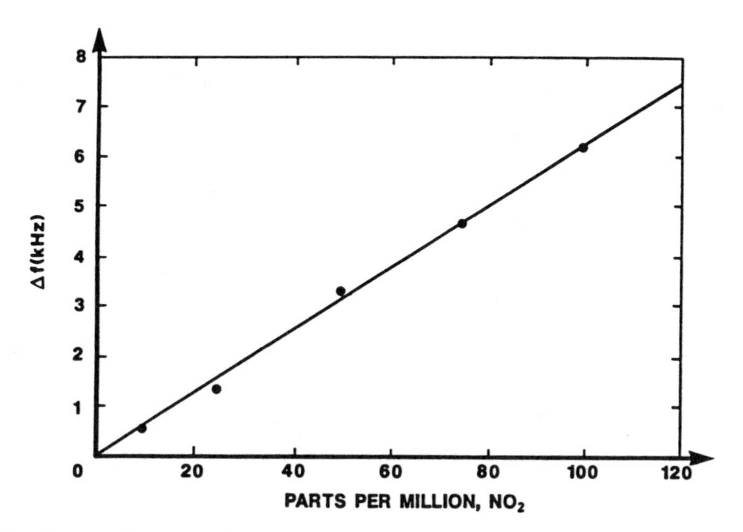

Fig.6. Change in resonant frequency of 98.6 MHz SAW oscillator exposed to NO_2. The chemically sensitive coating is pyridinium TCNQ.

Figure 6 shows data for a substituted pyridinium TCNQ molecule used to detect minute quantities of NO_2[25] Such sensors have no inherent specificity, and, therefore, as for optical sensors, there is a requirement for specially tailored materials to bind specific molecules, or for multiplexed sensing techniques, to alleviate problems associated with response selectivity. Sensitivity presents no problem, for mass changes as low as a femtogram per square centimetre can be detected.

3.2 Optoelectronics

The incorporation of thin-film layer organic materials in waveguides has been of particular interest to those working in the field of non-linear optics. These seek to exploit the high non-linear coefficients observed in organic non-centrosymmetric molecules. Several review articles[26,27] emphasize the strong possibility that organic materials will soon replace inorganic compounds for certain applications, provided that stability problems can be overcome. An added attraction of using a thin organic film geometry is that it might then be possible to integrate non-linear interactions, linear filtering and transmission functions into one precision monolithic structure.

The microscopic polarization of a solid comprising many individual molecules may be expressed as

$$P = P_o + \chi E + \chi_2 E^2 + \chi_3 E^3 + ..$$

where P_o is a constant, E is the electric field and χ_n is the n'th order susceptibility tensor. It is the coefficients χ_2 and χ_3 that are responsible for second and third order harmonic generation effects, respectively. In order to observe second order effects a non-centrosymmetric geometry is essential. Anisotropic features, such as parallel conjugated chains, are useful for third order effects. The value of χ_2 for lithium niobate is approximately 1.4 $\times 10^{-8}$ esu. By comparison, the organic material methyl (2, 4 - dinitrophenyl) aminopropanoate has a χ_2 value nearly 3 times larger, and is phase-matchable over its entire transparency range. For second harmonic generation applications a figure of merit (χ_2^2/n^3) is often used, in which case this compound is fifteen times better than the best inorganic material.

It is generally agreed that LB films are likely to be of most use in areas capitalizing on their large non-centrosymmetry. There are many reports of second harmonic generation in different types of LB film. It is difficult to make comparative assessments of their χ_2 values because the magnitudes of these coefficients are a strong function of the degree of order, level of aggregation, degree of polymerization etc in the film. Many of the LB films studied in this context rely on the deposition of alternate layers of two different materials. For example, Neal et al[28] reported the non-linear behaviour of a hemicyanine/nitrostilbene organic superlattice; their result, when translated into macroscopic terms, gives a χ_2 value approximately 50 times that of lithium niobate.

In order to meet the stringent thermal, mechanical and chemical stability requirements demanded in the field of optoelectronics, Richardson et al[29] have developed a novel group of non-centrosymmetric compounds based on organotransition metal complexes. Figure 7 shows two examples of specially engineered molecules which display interesting SHG data. Figure 8 shows the expected square-root dependence of the second harmonic intensity with number of deposited layers.

For many applications in optics, film thicknesses approaching the wavelength of radiation are required. This may well exceed the range of convenience with LB films. Therefore, it is important to design optical devices that can capitalize on their ultra-thinness and uniformity. Another important requirement is to have low scattering losses. These problems have not yet been resolved, and are likely to hinder the introduction of LB films in this important field.

3.3 Optical Information Storage

A variety of organic-polymeric based materials have been studied for optical recording. A comprehensive review has been given by Pearson[30]. In many cases the light absorption function is provided by a dye with the polymer serving as the binder. Apart from the benefits associated with uniformity of thickness there is little to suggest that LB films should be used in an ablative, write-once situation. However, the first result of an LB film system suitable for the purpose has just been reported[31] using a specially substituted near-infra-red absorbing dye based on 1,4-naphthoquinone. Ablative data marks with good form and contrast were recorded using 830nm GaAlAs diode laser light of pulse energy 3.6nJ. Substrate conduction was minimized using thermal buffer layers of fatty acid.

In ablative recording there is material displacement to create some form of pit. However, it is also possible to obtain optical contrast by introducing some optically detectable transformation in the film, for example, a phase change or aggregation - deaggregation change. The performance criteria required for an erasable medium are similar, but there is an additional need for the medium to undergo rapid, laser-induced reversibility from its marked state to its original state, many times over. Practical devices require this repeatable step to occur in the micro to nanosecond time scale, and be reproducible for typically one million erase-record cycles. Mey[32] has described an organic thin-film system which consists of tiny dye/polymer complex particles dispersed in a continuous phase of the same polymer. Exciting this aggregated film with a laser pulse results in disaggregation, with an attendant shift in the absorption spectrum. Reaggregation of the dye/polymer complex is achieved using a thermal process. This approach is essentially that used by Ishimoto et al[33] using LB films. They capitalized on the narrow absorption bands of J aggregates of a cyanine dye formed in a mixture of dioctadecyldimethyl ammonium chloride and methyl stearate. The original state was recovered by placing the film in an environment of 100% humidity at 30°C.

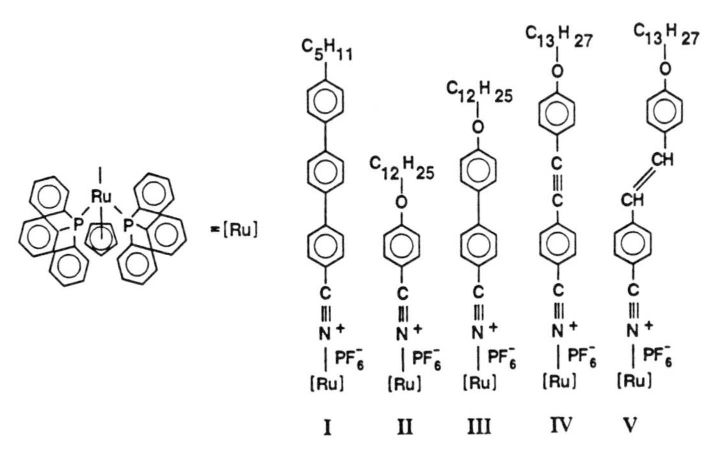

Fig.7. A selection of organoruthenium complexes which display second harmonic generation and pyroelectric effects.

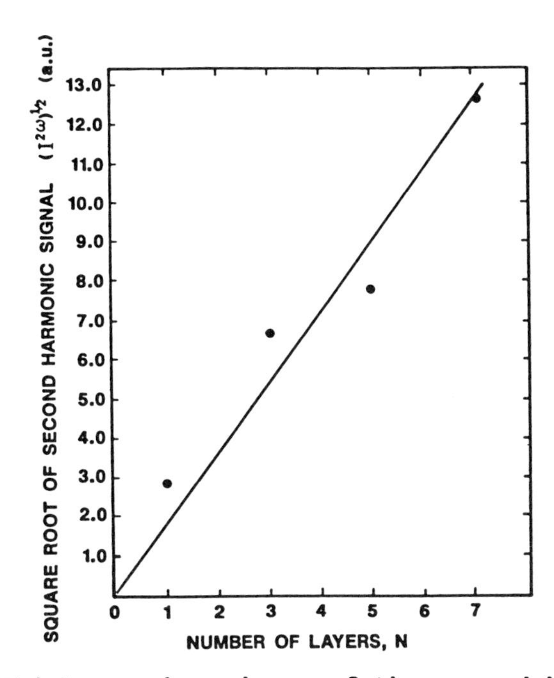

Fig.8. Thickness dependence of the second harmonic generation signal in a Z type film of compound 1 shown in figure 7

Should photochemical hole burning[34] ever be considered seriously for optical information storage, then LB films would be strong contenders in which to store information in the frequency domain. However, until the requirement to operate at liquid helium temperatures is lifted, there seems little chance of commercializing this technology.

4 PYROELECTRIC LB FILMS

The previous sections have outlined several possible niches for LB films. In all cases, however, monomolecular assemblies face severe competition from inorganic materials and organic films produced using alternative techniques. For commercial applications, the LB layers will need to play an essential role before a company is prepared to invest in new technology. That is, one must be seen to be capitalizing on their special features, such as the degree of control over molecular architecture, their ultra-thinness, and the selective way in which they might react with the environment. These considerations, taken together with the requirements of mechanical and thermal stability, will certainly restrict the areas of applicability to a relatively small number. One of the strong contenders to be the first field of practical importance for LB films is that of thermal imaging.

A traditional night-vision device is based on the narrow band-gap semiconductor cadmium mercury telluride. Reliable and high-resolution images are possible, using this quantum detector approach. However, a major restriction on their usefulness is that they must be operated using liquid refrigerants. Pyroelectric detectors, on the other hand, do not require cooling, and are relatively rugged and inexpensive. The pyroelectric effect arises due to the temperature dependence of the spontaneous polarization of a solid. Devices based on this principle respond to a rate of change of temperature, rather than to changes of temperature. Therefore, by modulating the incoming infrared radiation, they are capable of operation at ambient temperature.

The pyroelectric coefficient p, is a useful parameter with which to compare different materials. If the thin film acts as a dielectric in a capacitor, and an external resistance is connected between the electrodes, a pyroelectric current, I, flows in the circuit; this can be expressed as $I = pA(dT/dt)$ where dT/dt is the rate of change of temperature, and A is the cross-sectional area of the device. In a thermal imager many considerations, other than a high value of p, must be borne in mind, when designing a pyroelectric detector capable of resolving a temperature difference in the scene temperature of 0.1K. For example, the figure of merit for a thermal imaging device requires the pyroelectric materials to have low values of permittivity.

Table 1 illustrates typical data for a number of different materials. They are only approximate, for the values are dependent on preparation conditions, and proximity to Curie point etc. The single-crystal materials are not suitable, in that they are only available in bulk form. The ceramic materials have a high value of p, but this is offset by their large values of relative permittivity. The polymeric materials are of some

TABLE 1
Comparison of Pyroelectric and Dielectric properties of a range of different materials

MATERIAL	p ($\mu Cm^{-2}K^{-1}$)	ε_r	p/ε_r ($\mu Cm^{-2}K^{-1}$)
SINGLE CRYSTAL Li Ta O_3	190	46	4.1
TGS	300	50	6.0
CERAMIC $Sr_{0.6}Ba_{0.4}Nb_2O_6$	850	607	1.4
PZFNTU *	380	290	1.3
POLYMER PVF_2	30	10	3.0
VF_2/VF_3	10	11	0.9
LIQUID CRYSTAL Biphenyl/Ester +	4.0	3.6	1.1
Polysiloxane +	3.7	3.1	1.2

* PZFNTU : $PbZrO_3$ - $PbTiO_3$ - PbFe $Nb_{1/2}$ $(O_3)_{1/2}$ + UO_3

+ Properties measured 5° from Curie point

TABLE 2
Pyroelectric and Dielectric properties of a range of LB Film materials.

MATERIAL	p ($\mu Cm^{-2}K^{-1}$)	ε_r	p/ε_r ($\mu Cm^{-2}K^{-1}$)	Reference
Azo Compound X-Type, LB Film	3.0	2.6	1.2	36
Liquid Crystal Z-Type, LB Film	1.0	5.8	0.2	37
ωTA-Docosylamine LB Film Superlattice	1.9	2.7	0.7	38 & 39
ωTA-Ruthenium LB Film Superlattice	2.7	4.0	0.7	35

interest; they fall into the two categories, of plastic polymer membranes, such as polyvinylidene fluoride, and ferroelectric liquid crystal polymers.

Table 2 shows equivalent data for the few LB film pyroelectric materials reported to date. It may be seen that the values of (p/ε) are as good, if not better than, those reported for other compounds. Using molecular engineering principles, there is every prospect of improving substantially on these figures.

There are several other reasons why LB films are likely to be used in preference to other pyroelectric materials. The figure of merit referred to above, (p/ε), is only a measure of the signal strength, and does not take into account the noise sources in a complete infrared detection system. If this is done, one discovers that it is essential to have low dielectric loss in the film over an extended frequency range. All ferroelectric liquid crystals are ruled out of contention on this basis, because they have high tan δ values in the range 1Hz to 10kHz. Moreover, it is not possible to prepare very thin films in structures with low thermal mass and low thermal conductivity.

Like the ferroelectric liquid crystal polymers, the co-polymers of the PVDF type require poling. By contrast, the polarization of an LB film is solidly built into the structure, and is only removed, if fields in excess of the dielectric breakdown field are applied. The main practical disadvantage associated with VF_2/VF_3 co-polymers is their microphonic activity. Fortunately, the LB films studied to date, with the exception of the azobenzene compounds, have piezoelectric coefficients low enough to be undetectable.

The final point to be mentioned in favour of LB films relates to the optimum thickness of a pyroelectric detector. When due consideration is given to this factor, ceramics show up poorly by comparison. In order to capitalize on the advantages of thin organic films as pyroelectric detectors, a suitable supporting structure must be designed. This must provide a means for extracting the electrical signal, whilst, at the same time, providing good thermal isolation of the individual sensors of the array. This can be achieved by combining the planar techniques of microelectronics with silicon micromachining; then the same silicon substrate can be used for both the supporting structure and for the electronic circuitry. A useful indicator of performance is the voltage responsivity, V_r, which is defined as the amplitude of the voltage response, divided by the total modulated power incident on the detector. The optimum thickness, where V_r is a maximum, is strongly dependent on G, the thermal conductance per unit area of the detector array. Figure 9 shows that, if the value of G is reduced to $100Wm^{-2}K^{-1}$, then the optimum thickness is approximately 100nm, a value corresponding to, typically, 30 monolayers.

We thus conclude that there is every prospect of seeing carefully optimized LB films on carefully engineered, low thermal conductance substrates being used commercially in thermal imaging applications. Clearly, a great deal more needs to be done to understand the basic physical processes involved, and thereby design more appropriate molecules, displaying enhanced pyroelectric coefficients. At the present time there is some uncertainty in deciding which mechanism contributes to the

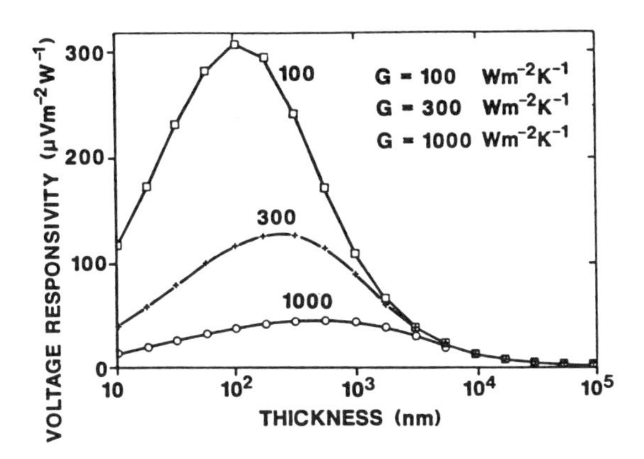

Fig.9. Theoretical modelling results for an acid-amine LB film deposited onto substrates of different thermal conductance. Standard values, e.g. a modulating frequency of 25Hz, have been assumed for the other parameters used in the calculation.

temperature dependence of the spontaneous polarization in LB films. Colbrook and Roberts[35] have discussed the relative likelihood of ionic interactions between molecular headgroups or the realignment of polarized molecules. Because it is essential to have a net polarization in the structure, there is considerable advantage, as in the case of second harmonic generation, to assemble organic superlattices.

5 SUMMARY

This paper has highlighted areas which, in the medium term, might provide an exploitation route for LB films. Those most likely to succeed will probably combine at least two of the main attributes of monomolecular assemblies, e.g. their architecture and their ultra-thinness. It also seems probable that the successful application area will rely on an active property of the film, possibly a collective phenomenon, where cooperative properties are involved, that are different from those of the individual molecular components. In the longer term, the future of LB films lies in designing structures that perform novel functions, rather than in the inorganic replacement mode technologies discussed in this article. There is a need to sustain the present level of research investment, because good science will emerge from such studies. It is worth remembering Langmuir's own words "experiments started perhaps just for the fun of it, or to satisfy scientific curiosity, often lead to unexpected useful results in fields that could not have been predicted".

ACKNOWLEDGEMENTS

It is a pleasure to acknowledge the considerable satisfaction and benefit gained from the collaboration with my research students at Oxford University, specifically, Richard Colbrook, Simon Cooke, David Heard, Tim Richardson and Mark Poulter.

REFERENCES

(1) Roberts, G.G., Advances in Physics 34 475 (1985)

(2) Blodgett, K.B., US Patent 2220860 (1940)

(3) Langmuir, I., US Patent 2232539 (1941)

(4) Blodgett, K.B., US Patent 2587282 (1952)

(5) Barraud, A., Rosilio, C. and Ruaudel-Teixier, A., J. Colloid and Interface Sci. 62 509 (1977)

(6) Pease, F., Engineering Foundation Conference on Molecular Electronics, Hawaii (1989)

(7) Briscoe, B.J. and Evans, D.C.B., Proc. Roy. Soc. Lond. A380 (1982)

(8) Seto,J. Nagai, T., Ishimoto, C and Watanabe, H., Thin Solid Films 134 101 (1985)

(9) Suzuki, M., Saotome, Y. and Yanagisawa, M., Thin Solid Films 160 453 (1988)

(10) Holcroft, B., D.Phil Thesis, University of Oxford (1988)

(11) Nishikata, Y., Kakimoto, M., Morikawa, A. and Imai, Y., Thin Solid Films 160 15 (1988)

(12) Huang, F., UK Patent Application 8729310 (1987)

(13) Dharmadasa, I.M., Roberts, G.G. and Petty, M.C., Electronics Letts. 16 201 (1980)

(14) Roberts, G.G., Petty, M.C. and Dharmadasa, I.M., Proc. IEE Pt 1 128 197 (1981)

(15) Tredgold, R.H. and El-Badawy, Z.I., J. Phys. D. 18 103 (1985)

(16) Batey, J., Roberts, G.G. and Petty, M.C., Thin Solid Films 99 283 (1983)

(17) Roberts, G.G., Pande, K.P. and Barlow, W.A., Proc. IEE Pt 1 2 169 (1978)

(18) Fowler, M.T., Suzuki, M., Engel, A.K., Asano, K. and Itoh, T., J. Appl. Phys. 62 3427 (1987)

(19) Chan, W.K., Cox, H.M., Abeles, J.H. and Kelty, S.P., Electronics Letts (to be published)

(20) Metzger, R.M., Panetta, C.A., Heimer, N.E., Bhatti, A.M., Torres, E., Blackburn, G.F., Tripathy, S. and Samuelson, L.A., J. Mol. Electronics 2 119 (1986)

(21) Lloyd, J.P., Pearson, C. and Petty, M.C.,Thin Solid Films 160 431 (1988)

(22) Nylander, C., Liedberg, B. and Lind, T., Sensors and Actuators 3 79 (1982)

(23) Moriizumi, T., Thin Solid Films 160 413 (1988)

(24) Reichert, W.M., Bruckner, C.J. and Joseph, T., Thin Solid Films 152 345 (1987)

(25) Roberts, G.G., Holcroft, B., Barraud, A., and Richard, J., Thin Solid Films 160 53 (1988)

(26) Williams, D., Angew Chem. Inst. Ed. 23 690 (1984)

(27) Zyss, J., J. Molec. Electronics 1 25 (1985)

(28) Neal,D.B., Petty, M.C., Roberts, G.G., Ahmad, M.M., Feast, W.J., Girling, I.R., Cade, N.A., Kolinsky, P.V. and Peterson, I.R., Electronics Letts. 22 460 (1986)

(29) Richardson, T., Roberts, G.G., Polywka, M.E.C. and Davies, S.G., Thin Solid Films 160 231 (1988)

(30) Pearson, J.M., Proc. Int. Conf. on Polymers in Electronics, American Chemical Society (1988)

(31) Heard, D., Roberts, G.G., Goringe, M.J. and Griffiths, J., Thin Solid Films (to be published)

(32) Mey, W., US Patent 4513071 (1985)

(33) Ishimoto, C., Tomimoro, H. and Seto, J., Appl. Phys. Letts 49 1677 (1986)

(33) Ishimotor, C., Tomimoro, H. and Seto, J., Appl. Phys. Letts 49 1677 (1986)

(34) Moerner, W.E., J. Molec. Electronics 1 55 (1985)

(35) Colbrook, R., and Roberts, G.G., Thin Solid Films (to be published)

(36) Blinov, L.M., Mikhnev, L.V., Sokoluva, E.B. and Yudin, S.G., Sov. Tech. Phys. Letts 9 640 (1983)

(37) Sukuhara, T., Nakahara, H. and Fukuda, K., Thin Solid Films 159 345 (1988)

(38) Christie, P., Roberts, G.G. and Petty, M.C., Appl. Phys. Letts 48 1101 (1986)

(39) Jones, C.A., Petty, M.C. and Roberts, G.G., Thin Solid Films 160 177 (1988)

INVESTIGATIONS OF THE MICROSTRUCTURE OF LIPID INTERFACE FILMS

Mathias Lösche

Institut für Physikalische Chemie
Johannes-Gutenberg Universität Mainz
D6500 Mainz, West Germany

ABSTRACT

The phase behaviour of phospholipid monolayers at electrolyte/gas interfaces is studied by fluorescence microscopy. At the LE/LC phase transition, phase separation leads to a Wigner-type lattice structure. The observations are quantified using digital image processing. The results show that the phase transition comprises three different regimes.

INTRODUCTION

Langmuir-Blodgett (LB) multilayers bear a large amount of anticipation as (quasi-) two-dimensional structured systems, with a great potential for the design of biosensors and molecular electronic devices.[1] Naturally, the main objective, when developing such a novel device, is the molecular characterization and optimization of the working horses, i.e. of the functional entities that either comprise the LB multilayer, or that are kept in place by amphiphilic matrix molecules within the structure. However, a task of equal importance is the structural characterization of the LB multilayer *itself*, both on a microscopic and a macroscopic scale. It is evident that the structure one finds in the multilayer will depend on structure formation on the Langmuir trough, since it is assembled on a water surface and subsequently transferred to the solid support. Therefore, one has to monitor the system from its infancy (on the water surface) continuously to its old age, including the rearrangement it experiences by the transfer procedure.

Despite a rich history of systematic investigations of monomolecular interface films since the days of Langmuir,[2] only recently have methods been devel-

Lower-Dimensional Systems and Molecular Electronics
Edited by R.M. Metzger *et al.*, Plenum Press, New York, 1991

oped to investigate the microstructure of monolayers on the water interface; these techniques include fluorescence microscopy[3-6] and synchrotron X-ray diffraction.[6-8] This contribution focuses on the former technique, and reports the assessment of the liquid expanded (LE) to liquid condensed (LC) phase transition (which occurs on continuously compressing a phospholipid monomolecular interface film) by observation of the spatial distribution of fluorescent tracers and digital processing of the images. Although, at present, not all the observations rendered are quantitatively understood in detail, emphasis here is placed on displaying the potential of the method, and the conclusion is drawn that it is essential to monitor the microstructure of these systems as early as on the water surface, in order to be able to control the structure of any completed device.

PHASE TRANSITIONS IN PHOSPHOLIPID INTERFACE FILMS

Isotherms (interface pressure π [mN/m] vs. area per molecule A [$Å^2$]) are usually measured by continuously compressing the film, which has been prepared on top of an electrolytic subphase. In this experiment, the interface pressure $\pi(A) = \gamma_0 - \gamma(A)$ [i.e. the difference of surface tensions without (γ_0) and with an interface film] is determined by monitoring the surface tension $\gamma(A)$. For the phospholipid L-α-dimyristoyl phosphatidic acid (DMPA) a set of data at pH = 6 and various temperatures of the subphase was measured by Albrecht et al.,[9] and is reproduced in Fig.1.

The schematic, Fig.2, analyzes the phase transitions that occur for a typical isotherm in Fig.1 (e.g. at T = 30 °C).[10-11] At high average molecular area, the film is in a two-dimensional *gaseous* state (G), where individual lipid molecules do not interact. On compression, a first-order phase transition (LE/G) to a *liquid expanded* state LE occurs. Both of these regimes are accompanied by magnitudes of π too small to be observed on conventional film balances, and are not observed in Fig.1. After termination of the LE/G transition at A_{fl}, the surface film is in the homogeneous state LE until, at a point (π_c,A_c), another discontinuity of the

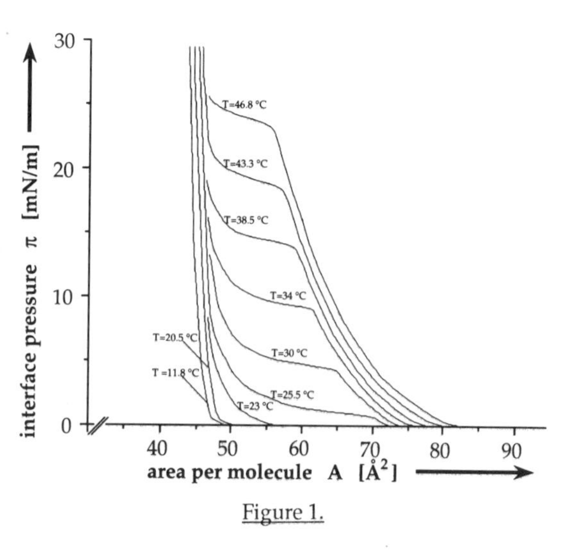

Figure 1.

Isotherms of a DMPA monolayer on pure water (pH = 6) at various temperatures (from Albrecht et al.[9])

compressibility $\kappa = -\frac{1}{A}\left(\frac{\partial A}{\partial \pi}\right)$ is observed. Here, the system enters a phase transition region designated LC/LE. This phase transition is one of the most distinguished features of phospholipid monolayer isotherms, and its nature has been disputed for years.[11-13] Although it shows features of a first-order phase transition, the compressibility κ remains finite, and, at its high-density terminus, κ decreases continuously, rather than showing the discontinuity expected (see dotted trace in Fig.2). At high density, the surface film is in a state thought of as a homogeneous *liquid condensed* or *gel-like* phase, LC, and finally, at (π_S, A_S), a second-order phase transition to a *solid condensed* state SC occurs. Depending on the subphase temperature, not all of the phases shown in the schematic, Fig.2, are actually observed in the measurement, Fig.1. Rather, at low temperatures (T < 20 °C) a transition from the gaseous to the condensed state (analogous to a three-dimensional sublimation) occurs, whereas at high temperatures (T > 50 °C) the surface film remains in the expanded state, and no condensation occurs at all.

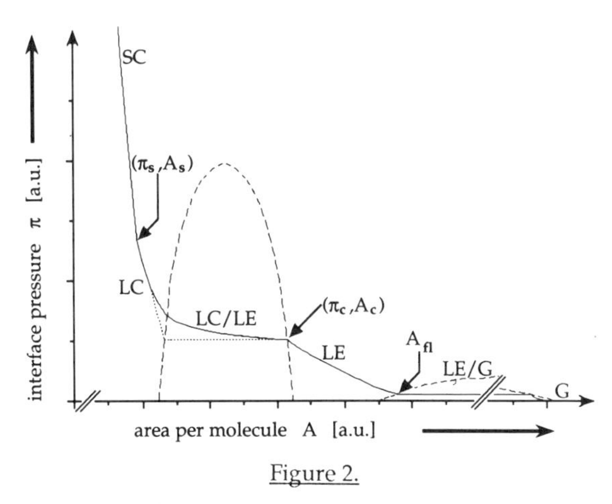

Figure 2.

Schematic of the phase behaviour of a phospholipid monolayer along an isotherm. The nomenclature is according to Cadenhead et al.[11] In the regions of the phase transition LC/LE and the pure LC phase, ideal behaviour is shown as a dotted line.

THE FLUORESCENCE MICROSCOPE AND IMAGE ANALYSIS

In order to investigate their microscopic structure, fluorescence microscopic methods have been developed, to directly observe phospholipid monolayers on the film balance.[14,3-4] The fluorescence microscope employed in this investigation has been described in detail previously.[15] A simplified block diagram is given in Fig.3. Briefly, a 40x water immersion objective (Zeiss) is built into the bottom of the trough of a microcomputer-controlled film balance. A fluorescently labeled surface film is prepared on an aqueous subphase under an Argon atmosphere, to avoid bleaching of the tracer molecules. The inverted microscope (resolution: 2 μm) is home-built from commercial parts (Zeiss). The interface film is imaged onto the cathode of a SIT video camera (Hamamatsu). A typical chromophore density required for the images shown below was $4 \cdot 10^{11}/cm^2$ (nitrobenzoxadiazol fluorophores). Images were stored on tape (U-matic, Sony),

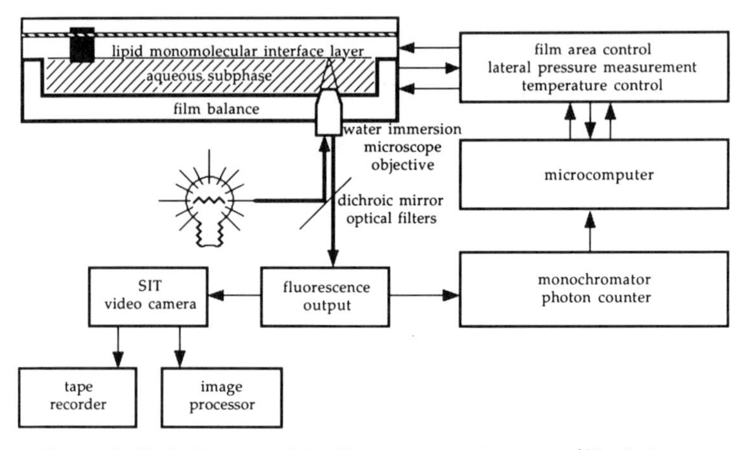

Figure 3. Block diagram of the fluorescence microscope/film balance

or digitally processed. Alternatively, the emission spectrum can be analyzed in a fluorescence spectrometer.

Phase-separated areas, in the LC/LE coexistence regime of the surface film under investigation, are visible, due to rejection of tracer molecules from the more condensed phase LC. Hence, the more expanded phase LE fluoresces brighter than LC. Images were processed in a home-built analysis system by standard digital image filtering procedures,[10,16] yielding the boundaries between different coexisting phase areas, from which proportions of molecules in either phase state, nucleation densities, average shapes, and distances between condensed phase areas were computed.

MATERIALS

The DMPA (Sigma, München, West Germany) was chromatographically pure.[16] As a fluorescent tracer, dipalmitoyl-nitrobenzoxadiazol-phosphatidyl-ethanolamine (DP-NBD-PE, Avanti, Birmingham, AL) was used in a concentration of 0.25 mol%. The lipid mixture was spread from a 3:1 chloroform/ methanol solution.

The aqueous subphase was prepared from ultrapure water ($R > 18$ MΩcm), and comprised 1 μM ethylenediamine tetraacetic acid (EDTA, tetrasodium, Sigma, practical grade), 2 μM $CaCl_2$ (purity > 99.5 %, Merck, Darmstadt, West Germany), 1 mM NaOH, 10 μM Na_2HPO_4 (both: Merck, p.a.), and 100 mM NaCl (Fluka, Buchs, Switzerland, divalent ion content $< 5*10^{-6}$). Its pH value was 11.3.

RESULTS

Fig.4 shows the isotherm of a DMPA monolayer at pH = 11.3 of the subphase; T = 10.5 °C. It contains 0.25 mol% DP-NBD-PE, and was measured during microscopic observation of the film. The original data show small irregularities at points where the continuous compression [compression rate: 0.8 Å^2/(molecule*min)] was stopped for focusing or taking photographs from the TV screen.[16] The isotherm data shown here are smoothed. Since the DMPA head group at pH = 11.3 is twice deprotonated (as opposed to incomplete deprotonation at pH = 6),[17] the surface film is more expanded (compare to Fig.1), due to strong mutual electrostatic repulsion. Indicated are the points A_{fl} = 135 Å^2, (π_c = 8.5 mN/m, A_c = 80 Å^2) and (π_s = 48 mN/m, A_s = 41.5 Å^2), as well as 8 points along the trace, where the microfluorescence photographs shown in Fig.5 were taken. The impact of the tracer molecules on the isotherm has been investigated earlier[10,18] and is negligible at the concentration employed here.

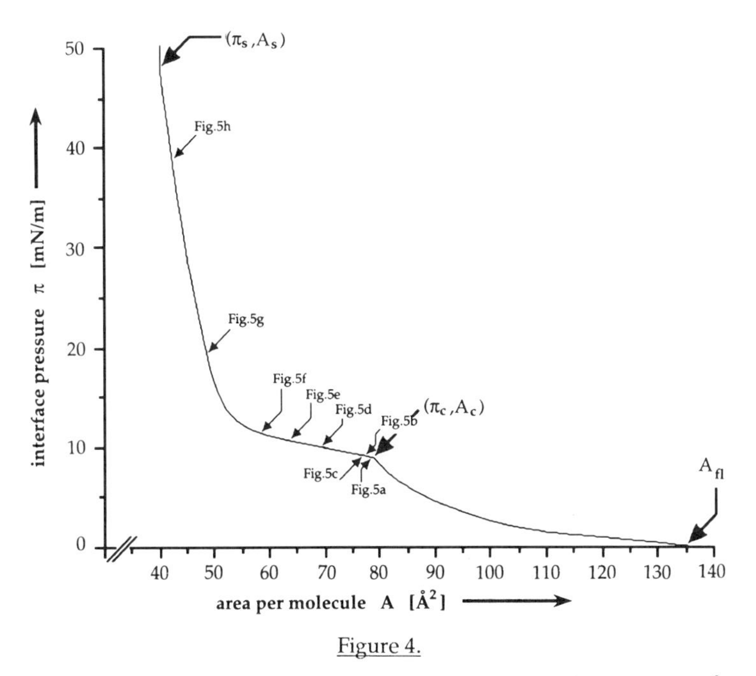

Figure 4.

Isotherm of a DMPA monolayer at pH = 11.3; 100 mM NaCl; T = 10.5 °C. Marked are distinct points indicating phase transitions as in the schematic, Fig.2, and the positions, where the microfluorescence photographs of Fig.5 have been obtained.

Fluorescence microscopic observation, on compression of the monomolecular film of Fig.4, gave the following results: Between A_{fl} and A_c, the film showed homogeneous fluorescence within the resolution of the microscope.

The first indication of the nucleation of a new phase occurred immediately beyond (π_c,A_c), as indicated in Fig.4, and shown in Fig.5a. The domains observed were compact and had a monodisperse size distribution. Obviously, there is a minimum distance between them, as no closely neighbouring dark spots (compared to their diameter) are observed. However, distances between nearest neighbours differ largely.

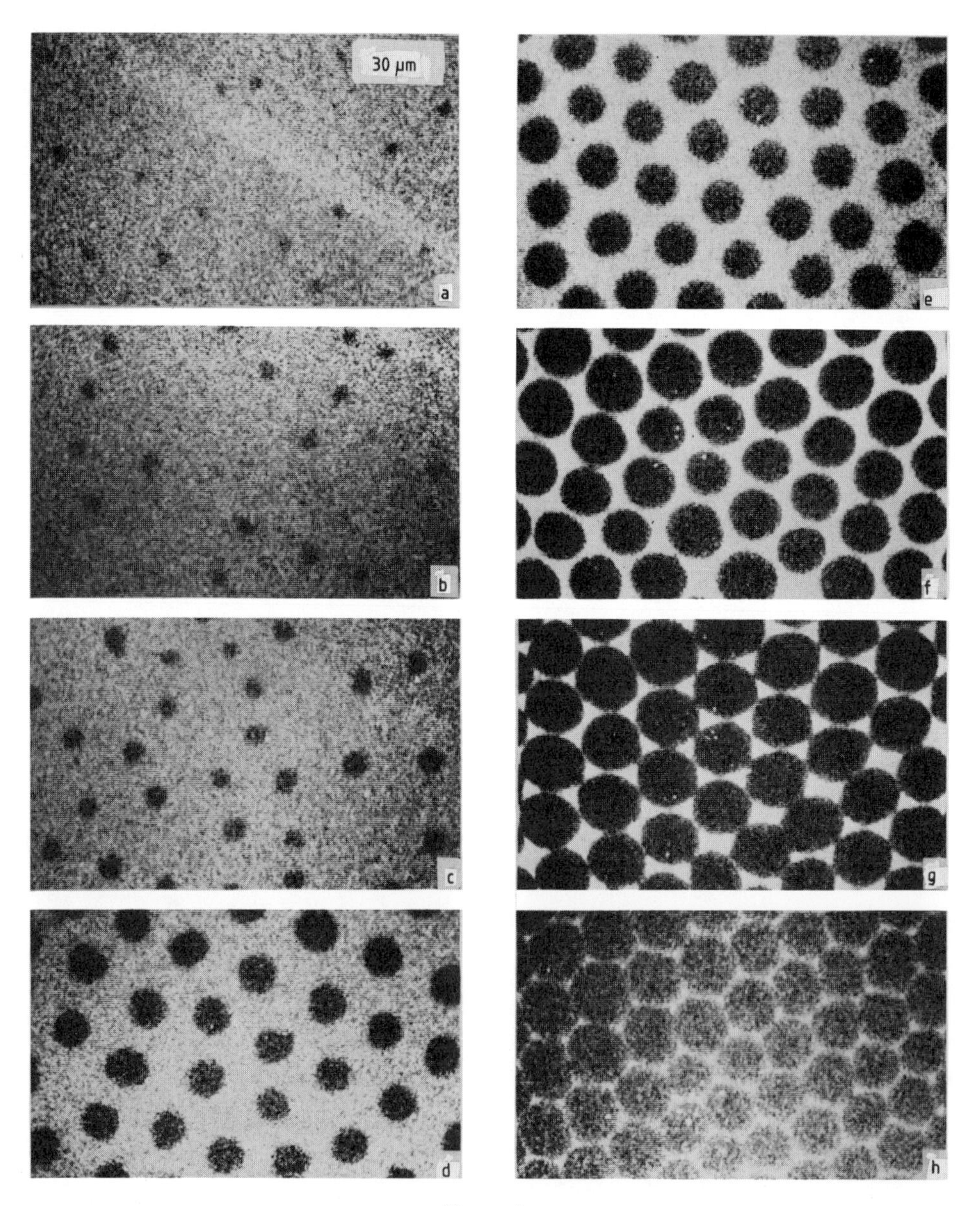

Figure 5.

Microfluorescence photographs obtained from a DMPA monolayer at pH = 11.3; 2 μM $CaCl_2$; 1 μM Na_4EDTA; 100 mM NaCl; T = 10.5 °C. The positions in the isotherm where the frames have been taken are indicated in Fig.4. The bar in Fig.5a is 30 μm in length.

From the series of photographs, Fig.5, taken along the isotherm, three different regimes can be distinguished:

(a) In the initial regime (Figs.5a-c), nucleation of macroscopic domains occurs. Nearest–neighbour distances differ, but become progressively more uniform, as a regular pattern of the domains is formed on compression.

(b) Subsequently, with a hexagonal lattice of the domains established, a growth regime of the new LC phase is observed (Figs.5d-f).

(c) Finally (Figs.5g,h), the observations suggest a regime of strong mutual interaction between the condensed phase domains. It falls in a region of the isotherm where a homogeneous phase LC is expected, after the termination of the LC/LE phase transition. From the fluorescence microscopic assessment, however, it is not clear if the inhomogeneous chromophore distribution now truly reflects a coexistence of different lipid phases, because here, film areas might solidify, with significant amounts of tracer molecules incorporated. It can only be stated that the region, where the isotherm begins to deviate from a linear π-A relation, coincides with the observation of close spatial contact between the domains.

If, in a similar experiment, the compression was stopped at any position within regimes (b) or (c) of the phase transition, no dynamic rearrangement was observed. In particular, within an observation time of hours, there was no indication of the growth of some domains at the expense of others. The size distribution remained monodisperse, indicating the existence of a metastable state.

These observations have been quantified by processing 30 images which yielded the results summarized in Fig.6. The broken lines are guides for the eye. In Fig.6a the portion of lipid molecules in the gel state,[16] Φ, (in %) is plotted versus A. The relation is linear in a region between the onset of the phase transition at A = 80 Å^2 down to A < 55 Å^2, where it starts leveling off. Fig.6b shows the number of domains, normalized to 10^{10} lipid molecules, N, in the surface film. The scatter in the data reflects the number of domains ($n \sim 50$) in the field of view on a typical photograph (cf. Fig.5). Clearly, different regimes can be distinguished. From the onset of the transition, N increases roughly linearly (although the small data base in this region does not allow for a conclusive quantitative statement). This regime is terminated at $A \sim 75$ Å^2; thereafter N remains constant to A < 55 Å^2. Subsequently, N decreases slightly, though not significantly. Fig.6c quantifies the shape of the gel domains by means of a parameter $\varphi = \dfrac{\bar{u}}{2\sqrt{\bar{f}\,\pi}}$, where $\bar{u}$ and $\bar{f}$ are average values of perimeter u and area f, respectively.[16] For a monodisperse population of circular disks, $\varphi = 1$, and $\varphi > 1$ quantifies deviations from circular shape. In the nucleation regime, 80 Å$^2 > A >$ 75 Å^2, φ has large values, that decrease rapidly to $\varphi \sim 1.07$. At the terminus of the phase transition, the shape parameter again sharply increases to $\varphi > 1.10$, reflecting the deformation of domains due to the mutual interaction observed in Figs.5g,h. Finally, Fig.6d displays the mean deviation of nearest-neighbour distances of the domains from the average value,[16]

$$\Delta D = \frac{1}{\sqrt{\nu - 1}} \sqrt{\sum_{i \neq j} \left\{ (d_{ij}^{(1)} - \bar{d}^{(1)}) \right\}^2} ,$$

where $d_{ij}^{(1)}$ is the distance between centers of gravity of nearest-neighbour domains i and j, and ν is the number of terms in the sum. Again, a significant decrease is observed, until, at the terminus of the phase transition, ΔD increases, due to the distortion of the lattice.

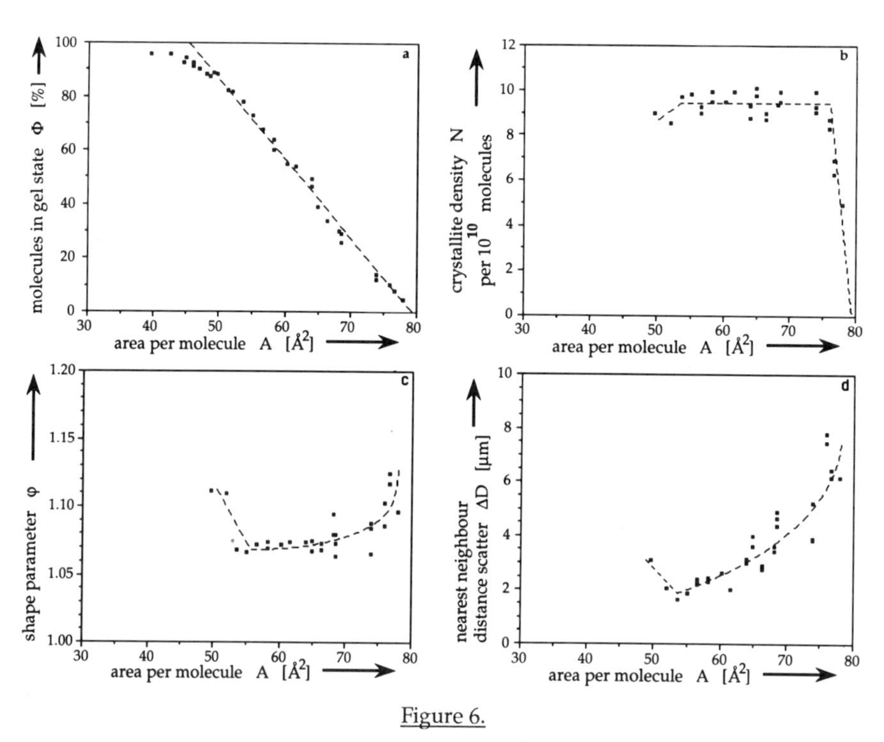

Figure 6.

Results of the image analysis of 30 frames similar to those in Fig.5 along different positions of the isotherm during the LE/LC phase transition. The broken lines are guides for the eye.

Fig.6a. proportion of the molecules, Φ, [%] in the gel state LC.

Fig.6b. Number density of phase separated areas, N, per 10^{10} molecules in the monolayer.

Fig.6c. shape parameter φ as described in the text.

Fig.6d. mean deviation of nearest neighbour distances, ΔD, as described in the text.

DISCUSSION

Phase-separated structures in lipid monolayers, such as the ones described here, have been shown to have their origin in mutual electrostatic interaction between the high-density domains.[19-22] The spatial arrangement of the domain lattice has been shown to be readily manipulated in inhomogeneous electric fields[19], as a consequence of the excess dipole density in the gel phase. It was therefore argued that the observed structure has the properties of a Wigner lattice. This report demonstrates how fluorescence microscopy, in combination with digital image analysis, serves to quantify the properties of the underlying phase transition. The results show that the Wigner lattice observed has its parentage in the LC/LE phase transition of the interface film. Within the transition itself, three regimes can be distinguished: the nucleation; the growth period; and the termination. From the assumption that, during the phase transition, an individual lipid molecule condenses from a configuration, where it occupies an area A_{fluid} (= A_c) in the film, to another one with an area A_{gel},[16] the linear relation displayed in Fig.6a is understood, and A_{gel} = 45.5 Å^2 is quantified from the extrapolation of the linear part of this plot to a fully condensed film (100% of the molecules in the high density configuration).

The deviation of this relation from ideal behaviour, near the high-density terminus of the phase transition, may be due to strong electrostatic or steric interactions between domains, may be an artifact due to dye accumulation between gel domains, or may reflect an influence of the phase transition at (π_s,A_s). The difference between A_{gel} = 45.5 Å^2 and A_s = 41.5 Å^2 shows the different identities of the monolayer states LC and SC, a result which was clearly demonstrated by synchrotron X-ray diffraction[8] where a decoupling of orientational and positional order parameters was discussed.

As is derived from Fig.4, at A_{gel} = 45.5 Å^2, the corresponding value of the lateral pressure is, π = 30 mN/m. This is surprisingly high, as the onset of the phase transition is at π_c = 8.5 mN/m, and demonstrates the presence of significant contributions from interactions between domains (most probably electrostatic and/or steric ones) to the energy of the system. The origin of the rounding of the shoulder in the isotherm, at the high-density terminus of the phase transition (compare solid and dotted lines in the schematic, Fig.2), is very poorly understood. The technique presented here should enable identification and quantitatation of the interactions, that lead to the observed deviation from ideal behaviour.

Finally, the question remains to be answered: how can we quantitatively understand the formation of a monodisperse lattice of gel-phase domains?

Clearly both the electrostatic interaction[17,22] and the line tension between gel and fluid phases[23-24] have been shown to dramatically affect size and morphological shapes of the domains formed. In the extreme case, arising from diffusion limited aggregation effects, even dendritic growth on pressure–jump experiments has been reported.[25]

At present, it is difficult to quantitatively interpret all of these observations,[22] because they represent manifestations of equilibrium and non-equilibrium features.[26] However, fluorescence microscopy on monomolecular lipid interface layers has provided access to a whole new microcosm of structure formation, and is a promising technique for the observation and manipulation of these systems at a microscopic level.

ACKNOWLEDGMENTS

I am indebted to Prof. H. Möhwald for his encouragement in the course of this project. This work was done at the TU München/West Germany and supported by the Deutsche Forschungsgemeinschaft.

REFERENCES

1. G.C. Roberts, An applied science perspective of Langmuir-Blodgett films, Adv. Phys. 34:475 (1985).

2. I. Langmuir, The constitution and fundamental properties of solids and liquids. II. Liquids, J. Am. Chem. Soc. 39:1848 (1917)

3. M. Lösche, E. Sackmann, and H. Möhwald, A Fluorescence Microscopic Study Concerning the Phase Diagram of Phospholipids, Ber. Bunsenges. Phys. Chem. 87:848 (1983).

4. R. Peters and K. Beck, Translational diffusion in phospholipid monolayers measured by fluorescence microphotolysis, Proc. Natl. Acad. Sci. USA 80:7183 (1983).

5. R.M. Weis and H.M. McConnell, Two-dimensional chiral crystals of phospholipid, Nature 310:47 (1984).

6. H. Möhwald, Microstructure of Organic Mono- and Multilayers, in: "The Physics and Fabrication of Microstructures", M.J. Kelly and C. Weisbuch, eds., Springer Series in Physics, Vol. 15, Berlin (1986).

7. K. Kjær, J. Als-Nielsen, C.A. Helm, L.A. Laxhuber, and H. Möhwald, Ordering in Lipid Monolayers Studied by Synchrotron X-Ray Diffraction and Fluorescence Microscopy, Phys. Rev. Lett. 58:2224 (1987).

8. C.A. Helm, H. Möhwald, K. Kjær, and J. Als-Nielsen, Phospholipid monolayers between fluid and solid states, Biophys. J. 52:381 (1987).

9. O. Albrecht, H. Gruler, and E. Sackmann, Polymorphism of phospholipid monolayers, J. Physique (Paris) 39:301 (1978).

10. M. Lösche, Das Phasenverhalten eines quasi-zweidimensionalen Dielektrikums an der Elektrolyt/Gas-Grenzfläche, Ph.D. thesis, München (1986).

11. D.A. Cadenhead, F. Müller-Landau, and B.M.J. Kellner, Phase transitions in insoluble one and two-component films at the air/water interface, in: "Ordering in Two Dimensions", S.K. Sinha, ed., Elsevier North Holland, Amsterdam (1980).

12. J.F. Nagle, Theory of Lipid Monolayer and Bilayer Phase Transitions: Effect of Headgroup Interactions, J. Membr. Biol. 27:233 (1976).

13. J.F. Baret, Phase transitions in two-dimensional amphiphilic systems, Progr. Surf. Membr. Sci. 14:291 (1981).

14. V. v.Tscharner and H.M. McConnell, An alternative view of the phospholipid phase behaviour at the air-water interface, Biophys. J. 36:409 (1981).

15. M. Lösche and H. Möhwald, Fluorescence microscope to observe dynamical processes in monomolecular layers at the air/water interface, Rev. Sci. Instrum. 55:1968 (1984).

16. M. Lösche, H.-P. Duwe, and H. Möhwald, Quantitative analysis of surface textures in phospholipid monolayer phase transitions, J. Coll. Interf. Sci. 126:432 (1988)

17. C.A. Helm, L. Laxhuber, M. Lösche, and H. Möhwald, Electrostatic interactions in phospholipid membranes I: Influence of monovalent cations, Coll. Polym. Sci. 264:46 (1986).

18. A. Miller and H. Möhwald, Diffusion limited growth of crystalline domains in phospholipid monolayers, J. Chem. Phys. 86:4258 (1987).

19. A. Miller and H. Möhwald, Collecting Two-Dimensional Phospholipid Crystals in Inhomogeneous Electric Fields, Europhys. Lett. 2:67 (1986).

20. A. Miller, C.A. Helm, and H. Möhwald, The colloidal nature of phospholipid monolayers, J. Physique (Paris) 48:693 (1987).

21. D. Andelman, F. Brochard and J.-F. Joanny, Phase transitions in Langmuir monolayers of polar molecules, J. Chem. Phys. 86:3673 (1987).

22. M. Lösche and H. Möhwald, Electrostatic interactions in phospholipid membranes II: Influence of divalent ions on monolayer structure, J. Coll. Interf. Sci., in press

23. H.M. McConnell, D. Keller, and H. Gaub, Thermodynamic Models for the Shapes of Monolayer Phospholipid Crystals, J. Phys. Chem. 90:1717 (1986).

24. W.M. Heckl, M. Lösche, D.A. Cadenhead, and H. Möhwald, "Electrostatically induced growth of spiral lipid domains in the presence of cholesterol", Eur. Biophys. J. 14:11 (1986).

25. A. Miller, W. Knoll, and H. Möhwald, Fractal Growth of Crystalline Phospholipid Domains in Monomolecular Layers, Phys. Rev. Lett. 56:2633 (1986).

26. C.A. Helm and H. Möhwald, Equilibrium and non-equilibrium features determining superlattices in phospholipid monolayers, J. Phys. Chem. 92:1262.

CHARGE-TRANSFER CONDUCTING LANGMUIR-BLODGETT FILMS

Michel Vandevyver

DESCICP-DLPC, Service de Chimie Moléculaire
Institut de Recherche et de Développement Industriel
Commissariat à l'Energie Atomique, BP 121
91191 Gif-sur-Yvette, FRANCE

INTRODUCTION

Advantages of the Langmuir-Blodgett (LB) method in building conducting films are pointed out, together with the main results obtained in this field by the use of charge-transfer (CT) systems.

The LB method has proved to be a very convenient tool for building organic films of molecular thickness. Starting with amphiphilic molecules, the LB sequence consists in the spreading of a stable monolayer at the air-water interface, followed by successive depositions of the "floating" monolayer onto a solid substrate. Due to the necessary amphiphilic character of the molecules, the resulting film is basically a lamellar one. It is made of the superimposition of hydrophilic sheets (a few Ångströms thick) separated by aliphatic ones (a few tens of Ångströms thick). Several amphiphilic molecules, possessing aromatic head groups, are quite suitable for this purpose [1] and, in some cases, aromatic sheets exhibit a weak dc conductivity: an ohmic behavior was observed in LB films of lightly substituted anthracene [2], metal-free phthalocyanine [3] porphyrins [4] and copper phthalocyanine [5]. As expected from their lamellar structure, those films exhibit a very high anisotropy in conductivity. As an example, the in-plane conductivity was found to be greater, by eight orders of magnitude, than that for current flow perpendicular to the support [2]. However, the value of the in-plane conductivity remains very low (< 10^{-6} Siemens cm^{-1}). Later, Ruaudel *et al.* have shown that the use of molecular associations involving CT systems, more precisely TCNQ-containing LB films, can give rise to a considerable increase in the lateral conductivity (about five orders of magnitude) [6]. The present paper is devoted to the main results, together with the most recent developments in this field.

PYRIDINIUM.TCNQ.IODINE CHARGE-TRANSFER SYSTEMS

In a bulk CT conducting material, donors and acceptor molecules (or both) crystallize into segregated stacks, making possible, in this way, the emergence of a band structure. Under some special conditions, a high electronic conductivity is observed. For a TCNQ-containing material, the conductivity along the stacking direction is much higher than that for the two other directions, and the conductivity is basically one-dimensional (1D), with the molecular planes of TCNQ perpendicular to the electron flow. On the other hand, a stringent condition to obtain a conducting behavior is related to the net charge of the associated molecules: in the ground state, electron transfer between donors and acceptors should not be complete, to allow the formation of a mixed-valence state. Obviously, these necessary conditions should be retained in the final LB structures.

In the paper of Ruaudel *et al.* [6] only the stacks of TCNQ carry the electron flow. In fact, the hydrophilic planes consist of a ternary association of TCNQ conducting stacks on one hand, and iodine and pyridinium molecules (inactive with respect to the conducting properties), on the other hand. The main problem was to bring TCNQ molecules into LB films in such a way, that a free crystallization could take place. The trick was to use a semi-amphiphilic TCNQ salt, instead of a regular amphiphilic molecule classically used in the LB method (Fig. 1a). Beside the TCNQ anion, the amphiphilic cation consists of a pyridinium ring associated to a long hydrophobic chain. In that case, both $TCNQ^-$ and $pyridinium^+$ are hydrophilic. In such a molecular association, the amphiphilic cation could be expected to be layered as regular LB films, and the $TCNQ^-$ moiety dragged into the hydrophilic planes, just through Coulombic attraction between cations and anions. $TCNQ^-$ anions, free of any substituent and submitted only to long-range Coulombic forces, were expected to crystallize quite easily. In fact, conducting films are obtained in two successive steps:

(i) Building up successive layers of N-docosyl $pyridinium^+$ (NDP^+) $TCNQ^-$ (Fig. 1a). The pristine film obtained in that way is an insulator.

(ii) In a second step, the pristine film is exposed to iodine vapor, and, under some special conditions, conducting properties are obtained [7].

The compressibility of the NDP^+TCNQ^- monolayer at the air-water interface has been carefully investigated. The key point is the existence of a plateau in the compression isotherm, appearing between 16 and 10 mN/m, depending on the molecular compression speed [8]. Direction observation of the film, either on the subphase, or transferred onto a solid support, reveals, unambiguously, that the plateau regime corresponds to an irreversible 2D-to-3D phase transition, and appears to be a crystallization phenomenon. It was shown to be the consequence of an instability, arising from the molecular arrangement in the film compressed at the air-water interface beyond its elastic limit. The result is the formation of bilayer-thick, large-size monocrystals, superimposed on the top of the floating monolayer, and exhibiting very well-defined angles, together with very sharp edges (Fig. 2a) [8].

Built-up multilayers can be obtained only for transfer achieved at a surface far beyond the plateau regime, generally around 35 mN/m. Hence the resulting films are just a disordered superposition of broken monocrystals, with a random orientation in the plane of the support, as shown in Fig. 2b. As a consequence, the normal to the support behaves as an axis of full rotation. Such a macroscopic configuration is easy to investigate. This was done by the use of two different resonance methods, namely, linear dichroism and electron spin resonance, with the following result: $TCNQ^-$ radical anions lie completely flat on the substrate in the dimeric form [9]. Obviously, such a structure is not compatible with an in-plane conductivity, for which TCNQ molecular planes should be perpendicular to the substrate. In addition, the complete charge transfer between $pyridinium^+$ and $TCNQ^-$ prevents any kind of electron conductivity in the film and, as expected, the electrical properties of the pristine film are those of an insulator.

A conducting behavior is obtained after a controlled exposure of the pristine film to iodine vapor. Advantage is taken of the extreme thinness of the film, to diffuse iodine homogeneously into the structure: $TCNQ^-$ is partially oxidized, and iodine is incorporated into the film as I_3^- [7]. However, the most striking feature is that the chemical reaction with iodine also triggers a huge molecular reorganization of TCNQ species, making possible the emergence of an in-plane conductivity. Starting with the TCNQ molecular axes parallel to the substrate, the pristine film progressively turns towards a conducting state, for which the TCNQs stand on edge, with their molecular planes perpendicular to the substrate. At the same time, a considerable modification in optical spectra, especially in the infrared spectrum, is observed (Figs. 3a and 3b). The room-temperature dc conductivity is *ca.* 0.1 Siemens cm^{-1}.

At the same time, a huge increase in the molecular order, along the normal to the substrate, can be checked by the low-angle X-ray method. The corresponding X-Ray pattern is very informative. It allows the determination of the electron profile, and leads to a precise molecular model. It turns out that molecular planes of TCNQ are perpendicular to the support, with their long axes perpendicular to it. An accurate localization of I_3^- columns is obtained, together with the unambiguous conclusion that the aliphatic chains are interdigitated [10].

This latter finding can easily explain the relatively long-range molecular order along the normal to the support for the conducting films. The electrical conductivity is typical of a

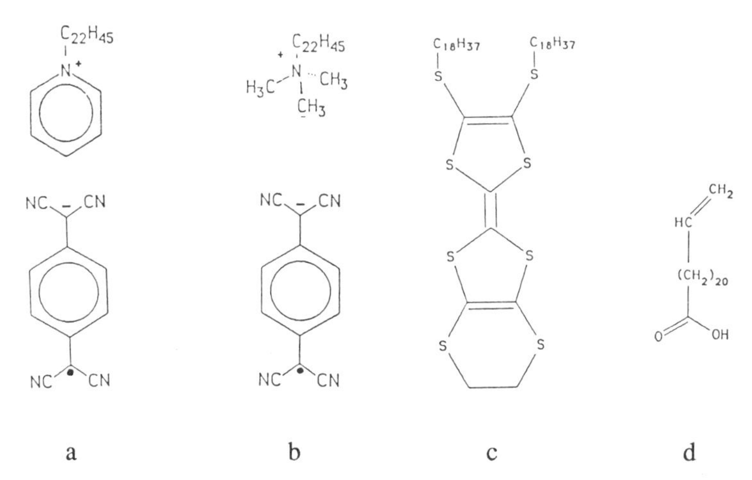

Fig. 1 (a) N-docosyl pyridinium$^+$TCNQ$^-$ semi-amphiphilic salt (NDP$^+$ TCNQ$^+$);
(b) N docosyl trimethyl ammonium TCNQ$^-$ semi-amphiphilic salt (NDTA$^+$TCNQ$^-$);
(c) EthylenedithiodialkylthioTTF (EDT-TTF);
(d) ω-tricosenoic acid.

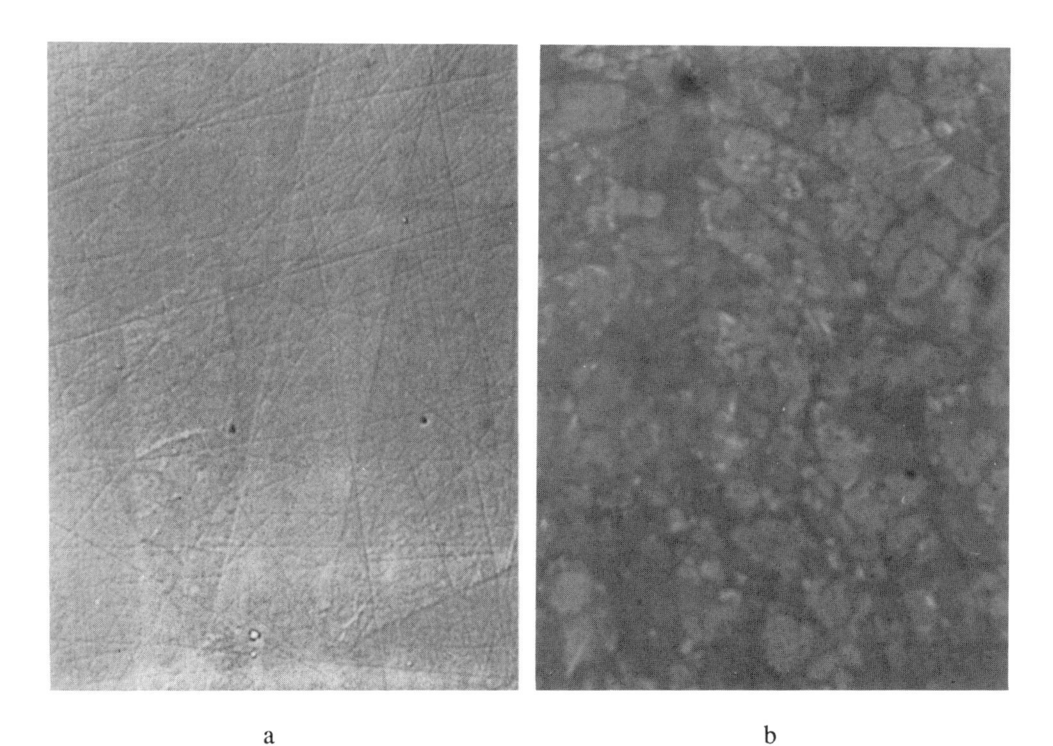

Fig. 2 (a) Optical microscopic image of bilayer-thick monocrystals of NDP$^+$TCNQ$^-$, transferred near the end of the plateaus regime (16 mN/m), under the conditions given in Ref.[8], and using the horizontal dipping method [13]. Nomarski reflection. Magnification: 1000X (unwanted straight lines are due to the polishing of the CaF_2 substrate).

(b) Optical microscopic image of a 10-layer thick film of NDP$^+$TCNQ$^-$, transferred at 35 mN/m by the vertical dipping method. Nomarski reflexion. Magnification: 1000X (see text).

narrow-gap semiconductor, with a temperature-activated behavior (0.15 eV and 0.08 eV for dc and microwave measurements, respectively [7,11]) and a carrier mobility *ca.* 1 $cm^2 V^{-1} s^{-1}$. [12]. In addition, due to the random superposition of 1D conducting microscopic platelets, the macroscopic conductivity is found to be isotropic in the plane of the substrate. This property proves the efficiency of electrical contacts between overlapping platelets of different orientation belonging to adjacent layers. Films are very stable (several months) in the room atmosphere, and even in a high vacuum. However, when the films are extremely thin, (roughly up to three or four monolayers, and especially for a single monolayer), abnormally weak, or even vanishing, conductivity is observed. The origin of this behavior could be found in short-range interactions between the substrate and the inner conducting sheets, or, more likely, in the lack of overlapping between independent conducting platelets.

The same general behavior, involving the same molecular reorganization upon exposure, was also observed for the sulfonium and phosphonium salts [13] or the N-docosyl-trimethylammonium$^+$ $TCNQ^-$ salt ($NDTA^+$ $TCNQ^-$) (see Figs. 1b and 3c). More recently, the role of the cation in such a process was investigated: both the size and the aromaticity of the cation head group play a role. However, a molecular structure of the pristine film similar to that of NDP^+ $TCNQ^-$ (crystalline structure of $TCNQ^-$ dimers with molecular planes parallel to the support) seems to be a key condition in obtaining a conducting film. A more complete discussion will be given elsewhere [14].

DIRECT BUILDING OF CONDUCTING FILMS

Conducting films can be obtained directly, without any chemical treatment, when starting with N-alkylpyridinium$^+$ $(TCNQ)_n^-$, with n between 1.4 and 2. Such films exhibit a dc conductivity comparable to the one of iodinated NDP^+ $TCNQ^-$. Again, their conductivity drastically depends on the number of superimposed monolayers: films that are less than ten layers thick are generally found to be non-conducting [5]. Such a behavior was explained through the discontinuous crystalline nature of the films: the overlapping of conducting crystals is obtained only for a certain number of superimposed layers. An important property is the sharp decrease in the conductivity σ upon electron irradiation (accelerating voltage: 3 kV). More precisely, a decrease in conductivity, by fifty percent, is obtained for an electron dose $D = D_{50} = 5 \times 10^{-2}$ coulomb m^{-2}. The corresponding value for the "contrast," $\gamma \equiv d(\sigma/\sigma_o)/d(\log D)$, for $D = D_{50}$ is 2.3. This finding opens an entirely new field in microlithography, and more especially in the fabrication of patterned conducting microstructures [16]. It is worth noticing that the conductivity can be increased (and optimized), after a controlled iodine vapor exposure. As for the case of NDP^+TCNQ^-, the iodination process leads to very stable films. However, for both the resistivity is very sensitive to a reducing or oxidizing atmosphere. This makes possible the building of gas sensors. In this respect, very encouraging results have been obtained in detecting phosphine, for example, for which the detection limit is close to 0.5 ppm [17].

An alternative strategy, successfully proposed by a Japanese team [18,19], consists in the direct spreading, at the air-water interface, of a solution containing both "active" donors and acceptors, namely TMTTF and octadecyl TCNQ. For films more than eight monolayers thick, they obtain a conductivity of *ca.* 10^{-1} Siemens cm^{-1}. The lack of results in Refs. [18,19] for films thinner than eight layers, suggests than the thinner films were insulators. On the other hand, an accident is clearly visible around 18 mN/m on the compression isotherm curve of Ref. [18], suggesting some structural change during the compression of the film. Such an experiment was recently reproduced in our laboratory [20]. For a relatively high molecular compression speed, the compression isotherm is similar to the one of Ref. [18]. In contrast, compressing the film at very low speed clearly reveals a strong instability, for a surface pressure below 15-20 mN/m. In fact, when transferred onto a solid support, the film appears as a set of overlapping, more of less fibrous, entities. Mixing TMTTF, octadecyl TCNQ and ω-tricosenoic acid in the spreading solution (molecular ratio 1:1:1) significantly improves the stability of the film, together with the dc conductivity (Fig. 4). At the same time, the transfer efficiency is slightly better. TMTTF, octadecyl TCNQ and ω-tricosenoic acid-containing films were recently used as antistatic sheets, instead of evaporated gold electrodes, in the preparation of radiation standard sources, to reduce the self absorption of particles by the support [21].

INCREASING THE DIMENSIONALITY OF THE CONDUCTIVITY

Another possibility is the use of substituted TTF derivatives: ethylenedithio-dialkylthio-TTF (EDT-TTF, Fig. 1c). This is a qualitative step in building conducting films, because new specific properties can be expected from the no longer 1D character of the conductivity, especially the dependence on lattice concentration defects. An example of such a strategy is given in Refs. [21,23]. The neutral molecule (EDT-TTF) is entirely hydrophobic, and yields completely unstable films at any surface pressure value. In contrast, the complex obtained by mixing with $TCNQF_4^0$, in the molecular ratio 1:1, yields very stable monolayers, which can be readily transferred as thick films, with only a very small amount of defects. At this step, the

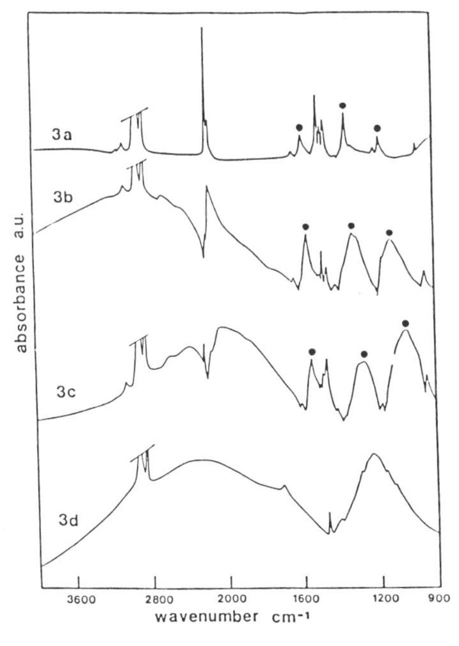

Fig. 3. Infrared spectra:

a) of a pristine LB film of NDP^+TCNQ^- (Fig. 1a). The totally symmetrical (a_g) activated modes (ν_3, ν_4, ν_5 at 1585, 1356, 1182 cm^{-1} respectively, ●) prove the $TCNQ^-$ dimerization [9].
b) of dc conducting LB film of NDP^+TCNQ^- (after a controlled exposure to I_2 vapor). Note the broad C.T. band around 2800 cm^{-1}, and the Fano-like structure around the C≡N stretching modes (2200 cm^{-1}). The forbidden activated modes (●) are broadened and red-shifted to 1563, 1329 and 1128 cm^{-1}, respectively.
c) of LB films of the $NDTA^+TCNQ^-$ conducting salt (Fig. 1b) after a controlled I_2 exposure. The a_g activated modes (ν_3, ν_4, ν_5, ●) are at 1544, 1273 and 1053 cm^{-1} respectively; the broad CT band is around 2000 cm^{-1}: note in the antiresonant dip a narrow line located at 2225 cm^{-1}, attributed to the b_{1u} ν_{19} IR-active C≡N stretch mode of neutral TCNQ [27]. The films are probably the juxtaposition of conducting plates, separated by insulating neutral TCNQs.
d) of a dc conducting LB film of a 1:1 mixture of the EDT-TTF (Fig. 1c) with ω-tricosenoic acid (Fig. 1d), after a controlled expoure to I_2 vapor [24].

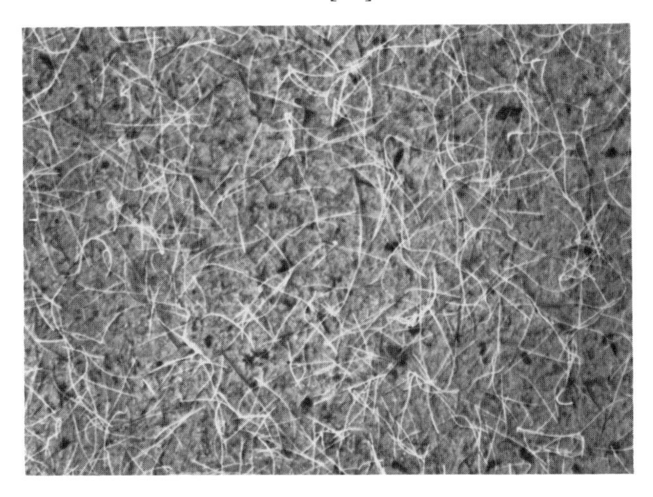

Fig. 4. Scanning electron microscopic image of a conducting 5-layer thick film of (TMTTF: octadecylTCNQ: ω-tricosenoic acid, molecular ratio 1:1:1). Magnification: 1000X. See text.

electronic and spectroscopic properties of the films are those of an insulator. In addition, inspection of the IR spectrum reveals that $TCNQF_4$ is incorporated in the film as $TCNQF_4^-$ monomers. Upon iodine exposure, followed by a moderate heating, the film acquires a lateral dc conductivity *ca.* 5 x 10^{-2} Siemens cm^{-1} for the conducting phase. At the same time, the infrared spectrum is typical of an electron conducting material. In addition, the IR lines of $TCNQF_4^-$ are not modified, indicating that the conductivity is only on the cation stacks, without any contribution of the $TCNQF_4^-$ radical anions. An even better result can be obtained by mixing EDT-TTF with ω-tricosenoic acid in 1:1 molar ratio (Fig. 1d) [24]. Reasonably good-quality films are obtained in this way, and, after iodine exposure, the room-temperature dc lateral conductivity is close to 2 x 10^{-1} Siemens cm^{-1}, while the dc activation energy is only 80 meV; its IR spectrum is typical for a conductor, and is shown in Fig. 3d. Preliminary experiments, in which ω-tricosenoic acid is replaced by stearic acid, strongly suggest that both the lateral resistivity and the activation energy can be reduced.

With the aim of increasing the dimensionality of the conductivity, Nakamura and co-workers pointed out the special properties of alkylammonium-metal $(dmit)_2$ [28]. By mixing the alkylammonium gold$(dmit)_2$ with eicosanoic acid, then oxidizing carefully, either with bromine or electrochemically, they obtained a dc lateral conductivity (for 20-layer samples) in the range of 20-30 Siemens cm^{-1}. Such a value is far beyond the values of conductivity for LB films published to date. Moreover, the temperature dependence of the conductivity reveals a metal-like behavior, down to about 200 K. Despite the lack of structural information on these films, and the lack of information for films less than 20 layers thick, the results of Nakamura and coworkers are very encouraging, and stress the importance of increasing the dimensionality of conductivity.

Another strategy is to take advantage of the solid-state polymerization in LB structures. Several studies have proved that polymeric conducting LB films can be obtained in this way. As an example, substituted pyrrole can be electropolymerized, leading to conducting films [25]. The lamellar structure is retained, and, for 200 monolayer-thick films, the in-plane conductivity is estimated to be 10^{-1} Siemens cm^{-1}, while the anisotropy ratio (in-plane conductivity / pependicular conductivity) is found as high as 10^{10} [26]. A detailed compilation of results in this field is not in the scope of the present paper. Nevertheless, such a process is most interesting, because it takes advantage of the initial molecular order existing in LB films, to facilitate polymerization, through a close control of the chemical environment.

CONCLUSION

Several ways of using the LB method have been proved to yield organic ultra-thin conducting films. Very promising results have been published, the main features of which can be summarized as follows:

(i) Charge transfer systems, as well as conducting polymers, can give rise to films possessing high lateral conductivity, and a relatively high measured anisotropy (in-plane / out-of-plane conductivity ratio).

(ii) Except for the results reported by Nakamura *et al.* [28], the best values for the lateral conductivity lie in the range 0.1 - 1 Siemens cm^{-1}.

(iii) Considering the possibilities offered by the LB method, spectacular progress should be expected, both by synthesizing new molecules, and by improving the building conditions of films.

However, some puzzling problems remain. As an example, no answer has been given, so far, to the following question: why are films, that consist of a very few monolayers, non-conducting? This finding could be, presumably, explained through some special texture of the film. More generally, in the special case of conducting films, the interpretation of results is strongly dependent on the film texture, which often consists of several phases, even containing macroscopic defects. Hence, significant progress could be expected by continuously increasing the quality of the films, followed by a careful control of their structural and morphological properties.

REFERENCES

[1] See, for example, H. Kuhn, Thin Solid Films 99: 1 (1983) and references therein.

[2] G. G. Roberts, T. M. McGinnity, W. A. Barlow, and P. S. Vincett, Thin Solid Films 68: 223 (1980).

[3] S. Baker, M. C. Petty, G. G. Roberts,and M. V. Twigg, Thin Solid Films, 99: 53 (1983).

[4] R. Jones, R. H. Tredgold, A. Hoorfar, and P. Hodge, Thin Solid Films 113: 115 (1984).

[5]. R. A. Hann, S. K. Gupta, J. R. Fryer, and B. L. Eyres, Thin Solid Films 134: 35 (1985).

[6] A. Ruaudel-Teixier, M. Vandevyver, and A. Barraud, Mol. Cryst. Liq. Cryst. 120: 319 (1985).

[7] J. Richard, M. Vandevyver, P. Lesieur, A. Barraud, and K. Holczer, J. Phys. D. Appl. Phys. 19: 2421 (1986).

[8] A. Barraud, M. Florsheimer, H. Möhwald, J. Richard, A. Ruaudel-Teixier, and M. Vandevyer, J. Coll. Interf. Sci. 121: 491 (1988).

[9] J. Richard, M. Vandevyver, P. Lesieur, A. Ruaudel-Teixier, A. Barraud, R. Bozio, and C. Pecile, J. Chem. Phys. 86: 2428 (1987).

[10] B. Belbeoch, M. Roulliay, and M. Tournarie, Thin Solid Films 134: 89 (1985).

[11] J. Richard, Thèse, Université de Paris VI, December 20, 1987.

[12] J. Richard, M. Vandevyver, and A. Barraud, Thin Solid Films 161: L738 (1988).

[13] M. Vandevyver, J. Richard, A. Barraud, A. Ruaudel-Teixier, M. Lequan, and R. M. Lequan, J. Chem. Phys. 87: 6754 (1987).

[14] M. Vandevyver, A. Ruaudel-Teixier, S. Palacin, A. Barraud, J. P. Bourgoin, R. Bozio, C. Pecile, and M. Meneghetti, to be published in Mol. Cryst. Liq. Cryst.

[15] M. Vandevyver, A. Barraud, P. Lesieur, J. Richard, and A. Ruaudel-Teixier, J. Chim. Phys. 83: 599 (1986).

[16] A. Barraud, J. Richard, and M. Vandevyver, to be published.

[17] L. Henrion, G. Derost, A. Ruaudel-Teixier, and A. Barraud, Sens. Actuators, 17: 493 (1989).

[18] Y. Kawabata, T. Nakamura, M. Matsumoto, M. Tanaka, T. Sekiguchi, H. Komizu, and E. Manda, Synth. Metals 19: 663 (1987).

[19] M. Matsumoto, T. Nakamura, E. Manda, Y. Kawabata, K. Ikegami, S. I. Kuroda, M. Sugi, and G. Saito, Thin Solid Films, in press.

[20] A. Ruaudel-Teixier and M. Vandevyver, unpublished results.

[21] P. Blanchis, J. Page, J. Bouchard, M. Vandevyver, and A. Ruaudel-Teixier, to be published.

[22] J. Richard, M. Vandevyver, A. Barraud, J. P. Morand, R. Lapouyade, J. F. Jacquinot, and M. Roulliay, J. Chem. Soc. Chem. Commun. 755 (1988).

[23] J. P. Morand, R. Lapouyade, P. Delhaès, M. Vandevyver, J. Richard, and A. Barraud, Synth. Metals 27: B569 (1988).

[24] J. Richard, M. Vandevyver, A. Barraud, J. P. Morand, and P. Delhaès, J. Coll. Interf. Sci. 129: 254 (1989).

[25] T. Iyoda, M. Ando, T. Kaneko, A. Ohtani, T. Shimidzu, and K. Honda, Tetrahedron Lett. 24: 5633 (1986).

[26] T. Shimidzu, T. Iyoda, M. Ando, A. Ohtani, T. Kaneko, and K. Honda, Thin Solid Films 160: 67 (1988).

[27] R. Bozio, I. Zanon, A. Girlando, and C. Pecile, J. Chem. Soc. Faraday Trans. II 74: 235 (1978).

[28] T. Nakamura *et al.*, elsewhere in this volume.

SOLID-STATE MOLECULAR ENGINEERING IN LANGMUIR-BLODGETT FILMS

A. Ruaudel-Teixier

Service de Chimie Moléculaire
Département d'Etude des Lasers et de Physico-Chimie
Division d'Etudes de Séparation Isotopique et de Chimie Physique
Institut de Recherche Technologique et de Développement Industriel
Commissariat à l'Energie Atomique, BP 121
91191 Gif sur Yvette Cedex, FRANCE

INTRODUCTION

The Langmuir-Blodgett (LB) films are solid organized media, where the studies of the chemical reactivity lead to a better comprehension of molecular interactions in the solid state, and to a development of a true molecular engineering. The LB technique allows the building of different structures at the molecular level, such as mixed layers and alternating layers. These molecular assemblies appear as sets of molecules, which can interlock, to give rise to specific physico-chemical properties.

Solid-state molecular engineering is a challenge to obtain super-molecular systems, which can propagate, or modulate, signals or information from moving particles, such as photons, electrons, protons or ions, as is seen in biological systems. General concepts for the design of molecules, tailored for this solid-state architecture, will be briefly described here, as well as their work-up by the LB technique, and their chemical reactivity. Two examples of the first attempts of such molecular engineering will be given, namely the dioxygen trap, and an approach towards magnetic memories in cooperative reversible-spin transition complex multilayers.

Molecular engineering consists in the design of a complex system of molecules, which interlock and interact, like the parts of a machine, as exhibited in biological mechanisms [1]. Organized molecules in the solid state are easily achieved by the Langmuir-Blodgett technique, a convenient tool to superimpose or to alternate, arrays of molecules, tailored for a specific activity [2]. These geometrical arrangements are greatly enhanced by selecting "smart" molecules: this can lead to interactions between sets of molecules, or to interlockings between assemblies. This is one of the challenges today for solid-state preparative chemistry, where the potential interest is to develop micro devices in the field of advanced technologies (integrated optics, molecular electronics, high storage capacity memories, molecular recognition). A desired architecture is, nevertheless, a difficult enterprise, since it requires multifunctional molecules (hence difficult organic syntheses), and the testing of the planned supermolecular arrangement (i.e. physical measurements at the molecular level). In this paper only the preparative point of view will be discussed.

THE DESIGN OF TAILORED MOLECULES

Structural requirements

Amphiphilicity. At the air-water interface, a closed-packed molecular plane can be obtained under a lateral pressure, if lateral interactions exist between the individual molecules.

Lower-Dimensional Systems and Molecular Electronics
Edited by R.M. Metzger *et al.*, Plenum Press, New York, 1991

Van der Waals forces between long aliphatic chains (≥ 18 C) are generally chosen for the lattice cohesion, i.e. the structure of the molecular assembly. A hydrophilic moiety is also required, to interact with the aqueous subphase: this polar head is generally afforded by a chemical functionality containing O, N, H (S alone is rather hydrophobic).

Other forces, like π-interactions, are also of value. Benzene, or fused rings, contribute to stronger lateral interactions, and, by these means, the aliphatic chains can be shortened, or even eliminated in some cases. For such "in situ" chemical reactions, the drawbacks are that the syntheses are more difficult, and that the aromatic rings are more reactive than the inert aliphatic chains.

In-plane cross-linking. The monolayer is ordered, in the direction perpendicular to the air-water interface, by the sequential LB deposition, but no in-plane ordering occurs, unless side groups in the polar part can interact, and link the molecules that are in the same plane. The problem arises in the case of molecules mixed in the same layer. If no care is taken, a heterogeneous solution is obtained in the solid state: patches of molecules A are randomly distributed among patches of molecules B. But if A and B are designed so as to make a molecular recognition between each other (ligand + substrate = true complex) or even only an attractive interaction (J aggregates [3, 4], H aggregates, π complexes, etc.) then a better in-plane order is obtained, and the assembly could be considered as a supermolecule, and hence could exhibit new specific properties from this long-range ordering (conductivity, cooperativity, photoconductivity, etc).

Plane-to-plane connection. In LB films, the polar planes are separated from each other by the aliphatic moiety (e.g. 30 to 60 Å), as in biological membranes. But, in biological bilayers, channels allow ionic transfers for signal propagation or chemical reactions. It could be of value, in solid-state organizates, to provide paths for electrons, or for the transfer of other mobile species. If a channel, or a molecular wire, must be grafted along the axis perpendicular to the polar plane, then the basic molecule should be designed to support this possibility.

Reactive site. The requirements just discussed above will give a higher in-plane ordering, but no new specificity. To afford a novel molecular machinery with specificity, a reactive site must be included in the design of the molecule (Fig. 1).

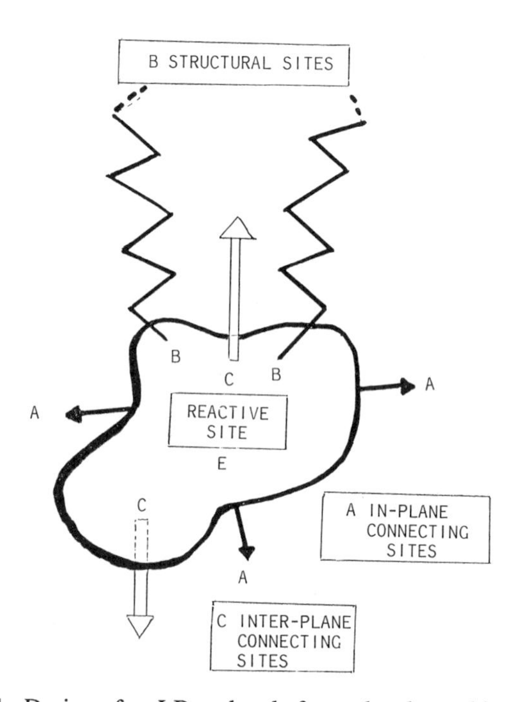

Fig 1. Design of an LB molecule for molecular architecture.

Finally, if all the above requirements must be fulfilled, the molecule will become more and more sophisticated and difficult to synthesize; however, biological molecules have already shown the way! It is beyond all doubt that synthetic organic chemistry will be the price to be paid for intelligent molecular machineries, but the work-up of molecules by the LB technique can relieve us, to some extent, from these difficulties.

Work-up with the LB technique

Interlocking sets of molecules, or assemblies, requires chemical planning, as previously seen, i.d. the choice and the design of the sets of molecules. The LB technique affords a great help, as earlier described by Kuhn [5] and more recently by Palacin [6].

Mixed LB monolayers (a) For a given set of molecules mixed in the spreading solvent, the supermolecule may occur during the compression of the monolayer, when attractive chemical interactions (complex formation or reaction product formation) and attractive van der Waals forces, work together. The resulting film is homogeneous, made of the supermolecules formed by the interlocking of the components. With π-electron-rich molecules J, S, or H aggregates can be obtained, producing a better lateral ordering. Host and guest reactions may also be performed at this stage [6].

(b) The supermolecular building can occur by interlocking the molecules of the fully compressed film (solid state) with dissolved species in the water bath [7]. The reactivity isgoverned by local structure, electric field of the dipoles, or reaction equilibria. By these means, non-amphiphilic molecules can be transferred into LB films.

Alternate LB layers. The LB technique provides the possibility for alternate layers of different sets of molecules, which are forced to come in contact in the polar plane. The only requirement is that the attraction between the polar head is greater than the attraction with water. This powerful procedure allows a great many combinations for solid-state architecture (non centro-symmetric crystals, hetero dimers, solid-state synthesis, molecular recognition).

Work-up with chemical reactivity.

Even when supermolecular arrangements have been achieved, increased specificity may

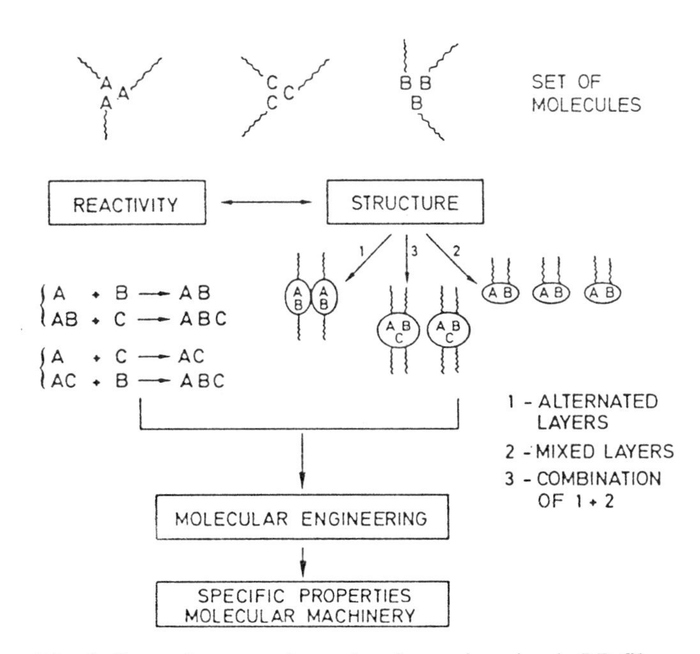

Fig. 2. General concept for molecular engineering in LB films.

be introduced by "in situ" reactions in the solid state. Diffused gaseous or aqueous species can undergo reactions with the reactive matrix [8]. Structural changes are often produced, and are closely related to reactivity, but the LB structure is generally preserved. Moreover, synthesis under matrix control can be performed, e.g. by inserted inorganic compound synthesis [9]. Physical treatments, such as irradiation (γ rays [10] UV light [11, 14] or electron beams [12-14]), lead to supermolecular networks, by relying on the chemical bond possibilities introduced in the molecular design. This concept is illustrated in Fig. 2.

SUPERMOLECULAR ENGINEERING

However, today, at the very early stage of this solid-state architecture, some examples have already been described, where rather simple molecules are manipulated by the help of the LB technique, combined with *in situ* chemistry.

H. Kuhn and co-workers pioneered this work, having in mind the tremendous efficiencies of biological machineries (the synthetic replicates of which have never been envisaged, because of their high complexities). But devices simulating partial properties of those machineries have been described by Fromhertz [15] and by Kuhn: energy storage simulation as in the reaction center in photosynthetic bacteria [16], electron and energy transfer in surface layers [1], photoconduction with conducting molecular components [17], reactions at organized interfaces [7]. Two other recent examples will be discussed here: the "dioxygen trap" [18,19] and the spin-transition complex [20].

The "dioxygen trap"

An attempt to realize a solid-state architecture acting as a molecular machinery was made by the Saclay team. The challenge was to bind oxygen selectively and reversibly, as does hemoglobin. The mechanism involves, in biological systems, the hemoglobin supermolecule (Fig. 3): iron-II porphyrin, axially coordinated by one imidazole molecule (residue from a proximate histidine), which binds dioxygen on the remaining axial position. This structure has to be persist, to ensure the activity toward oxygen.

The interest of this example lies in the fact that the supermolecular engineering has been worked up in solution [19] and also in the solid state [18]. In this way, we can compare the two procedures.

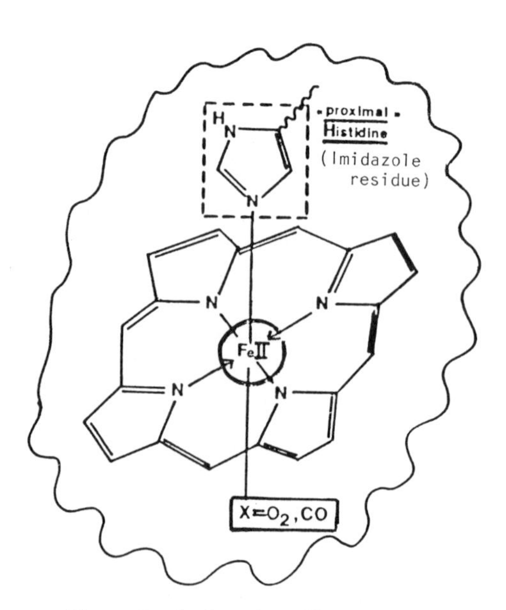

Figure 3. Active site of hemoglobin.

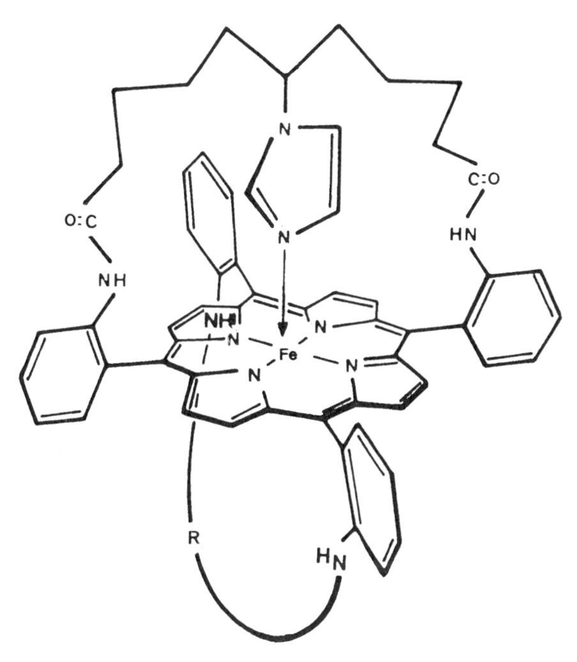

Fig. 4. Design of the "dioxygen trap" in solution [18].

For solution experiments, the molecular engineering requires a highly complex synthesis, to avoid undesirable side-reactions, and preserve the compulsory structural organizate:

– One "basket handle" is necessary on the porphyrin, to prevent dimerization of the macrocycle in solution.

– A second "basket handle", with one hanging base (imidazole) is also necessary, to avoid the bis complex formation, which would leave no position to bind dioxygen. This work was sorted out by the team of Momenteau [18] and the supermolecules (Fig. 4) were synthesized.

In the solid state, by making use of both the LB technique and the *"in situ"* chemical reactivity, the same structure was obtained in an easier way [19] by the Saclay group.

Two amphiphilic molecules were synthesized: the porphyrin P = Co (II) 5, 10, 15, 20-tetra-(4-oxy-(2'-docosanoic acid)phenyl)porphyrin, and the imidazole Im = N-octadecylimidazole. These two molecules interlock, to produce the planned architecture with the same activity toward dioxygen. The geometry is controlled by the combination of mixed and alternated layers, obtained by the LB technique, and by the chemical reactivity, which allows the long-chain imidazole to insert in the macro ring, and stand above the Co atom. The mixed monolayers of the set imidazole-porphyrin are homogeneous, (area per mole = area of P alone, and not area of P + area Im) because an efficient chemical interaction is involved between P and Im. To avoid dimerization, alternate layers were built (with the help of an automatic trough) of either behenic acid interspersed with mixed layers of P + Im, or else layers of P alternated with layers of Im, as shown in Fig. 5.

The net result is that the molecular machinery works towards selection of oxygen from air as does hemoglobin. From the preparative point of view, from both experiments, in solution and in the solid state, it appears that the LB technique affords structural advantages [19], which avoid the difficult syntheses for solution experiments [18].

This example has been described to give the proof that this way of interlocking molecules is easy, and leads to an efficient machinery. This example is only a mimic of a biological

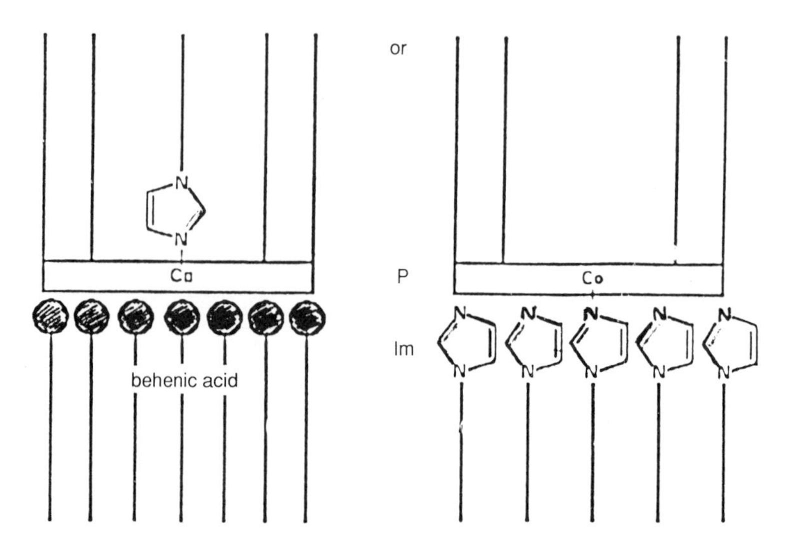

Fig. 5. LB film architecture for the "dioxygen trap". Here P = Co (II) 5,10,15,20-tetra-(4-oxy-(2'-docosanoic acid)phenyl)porphyrin, Im = N-octadecylimidazole.

function, but it can encourage the search for other architectures, tailored for advanced technologies such as "molecular electronics".

The spin-transition complex

In this experiment, the goal was to produce a magnetic reactive center in an LB film, exhibiting, as in a 3-D crystal, the sharp transition between a high-spin state and a low-spin state, proceeding from cooperative phenomena [20]. (A cooperativity exists, when an assembly of individual molecules works as a single supermolecule).

The transition of spin is induced by an external perturbation (temperature, pressure, etc.). The molecule designed for the LB experiment has the polar head of a well-studied spin-

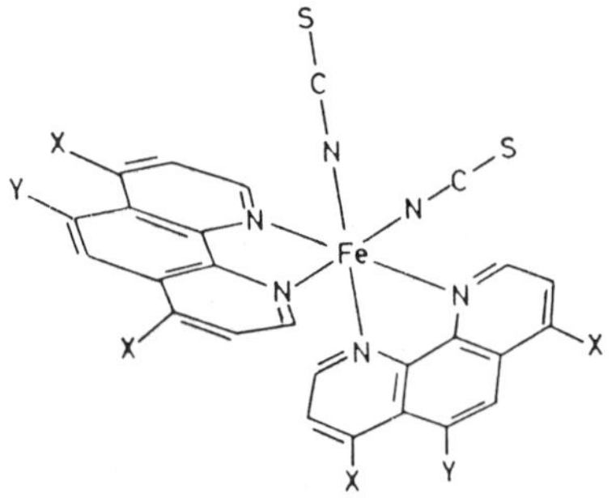

(1) X = Y = H

(2) X = $C_{17}H_{35}$, Y = $C_{18}H_{37}O$

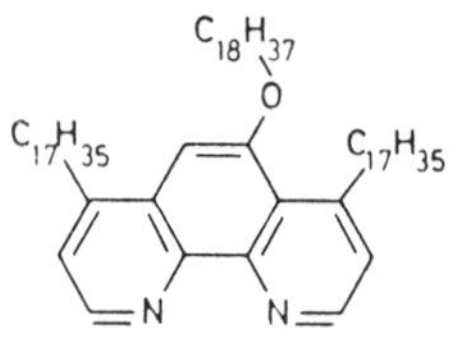

(3)

Fig. 6. Design of the LB film spin-transition complex, $[(OP_3)_2Fe^{II}(NCS)_2]$, where OP = substituted ortho-phenanthrolines (**1**, **2**, **3**).

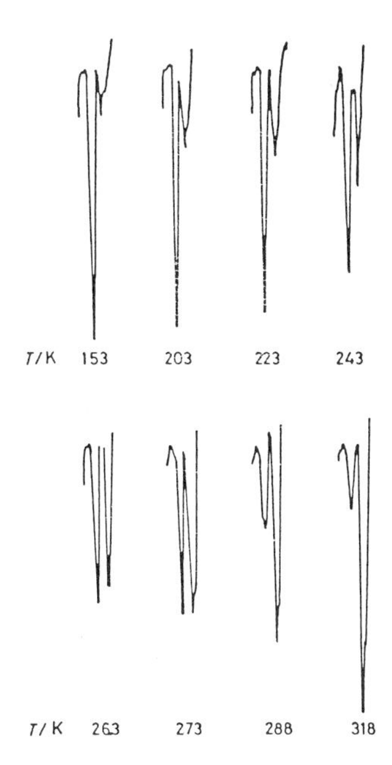

Fig. 7. Infrared absorption versus temperature.

transition complex [$(phen)_2Fe(NCS)_2$] and the hydrophobic part was constituted by 3 aliphatic chains grafted at specific positions on the ligands (Fig. 6).

This amphiphilic molecular complex forms monolayer films at the air water interface, and is transferred, as X-type LB multilayers, onto CaF_2 substrates (d = 28Å). The spin transition is induced by temperature changes from 77 K to 300 K and vice versa. The high-spin state is characterized in the infrared spectrum by a doublet (ν NCS: 2065-2075 cm^{-1}) and the low-spin state by another doublet (ν NCS: 2108-2118 cm^{-1}). Through the transition, the shift is not continuous, but the two doublets coexist, one decreasing while the other increases (Fig. 7).

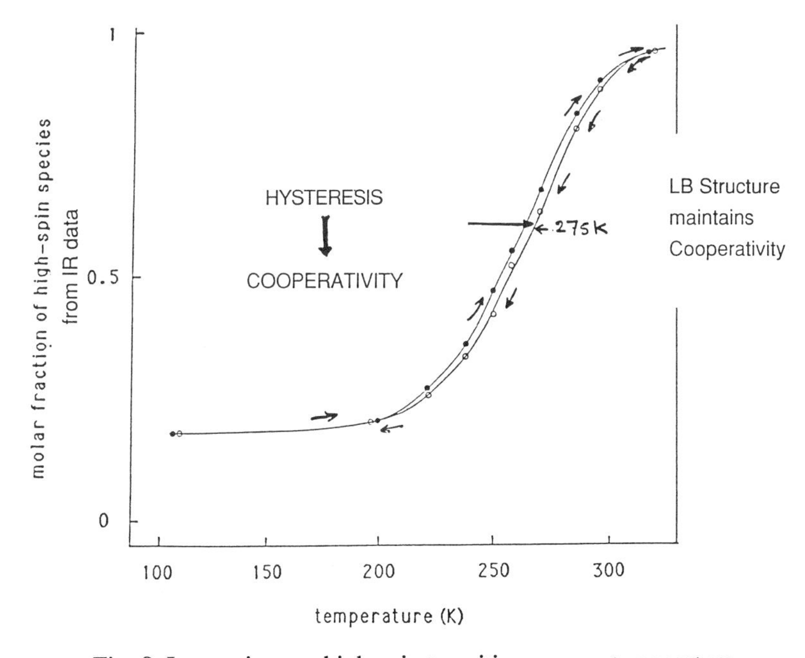

Fig. 8. Low-spin <--> high-spin transition versus temperature.

The temperature dependence of the molar fraction of high-spin species (relative intensity of the two doublets) is plotted in Fig. 8, for both cooling and warming modes. The transition is far from being as abrupt ($\Delta T = 75$ K) as in the correlated 3-D complex ($\Delta T = 2$ K), but is definitely less smooth, than that given by a simple Boltzmann distribution law. The slope is also certainly smoothed by the presence of a high rate of a non-magnetic complex [21], arising from a solvent effect in the spreading solution.

However, the observed hysteresis supports a cooperativity effect. The hydrophobic lattice orients the magnetic polar heads, so that the neighboring molecules interact, and behave as supermolecular domains. It has been demonstrated that the spin transition is accompanied by a variation of the molecular volume in 3-D crystal experiments, and a mechanical coupling occurs. This coupling is clearly maintained in the LB films. Experiments towards purification and mechanism studies are in progress. But this supermolecular architecture opens the way for passive information storage at the molecular level.

CONCLUSION

Other pioneering experiments have been performed by Japanese teams [22], or Kuhn's team, and encourage us to follow this strategy: cross-linking the reactivity of "smart molecules" with the LB technique. If organic chemists can take on the challenge of thinking about solid-state molecular engineering, then clever machineries will emerge, and evolve, and "molecular electronics" will arise.

REFERENCES

[1] H. Kuhn, Pure Appl. Chem. 53: 2105 (1981).
[2] M. Sugi, Thin Solid Films 152: 305 (1987).
[3] D. Möbius and H. Kuhn, J. Appl. Phys. 64: 5138 (1988).
[4] Y. Yonezawa, D. Möbius, and H. Kuhn, Nippon Schashin Gakkaishi 48(4): 250 (1985).
[5] H. Kuhn, Thin Solid Films 99: 1 (1983).
[6] S. Palacin, in Fourth Intern. Conf. on LB Films, Tsukuba, 1989, Thin Solid Films, in press.
[7] H. Kuhn in "Modern Trends in Colloid Science in Chemistry and Biology, an International Symposium in Colloid and Surface Science", edited by H. F. Eicke, (Birkhausen, Basel, Switzerland,1984) p. 97.
[8] A. Ruaudel-Teixier, A. Barraud, and C. Rosilio, Thin Solid Films 68: 7 (1980).
[9] C. Zylberajch, A. Ruaudel-Teixier, and A. Barraud, Synth. Metals 27: B609 (1988).
[10] A. Cemel, T. Fort, and J. B. Lando, J. Polym. Sci. A10: 2061 (1972).
[11] B. Tiecke, G. Wegner, D. Naegele, and H. Ringsdorf, Angew. Chem. Intl. Ed. Engl. 15: 764 (1976).
[12] A. Barraud, C. Rosilio, and A. Ruaudel-Teixier, J. Colloid Interf. Sci. 62: 509 (19777).
[13] T. Iyoda, M. Ando, T. Kaneko, A. Ohtani, T. Shimadzu, and K. Honda, Tetrahedron Lett. 27 (1986).
[14] A. Barraud, A. Ruaudel-Teixier, and C. Rosilio, Polymer Preprints, ACS Div. Polym. Chem. 19(2): 179 (1978).
[15] P. Fromhertz, Chem. Phys. Lett. 77: 460 (1981).
[16] H. Kuhn, in "Advances in Chemical Reaction Dynamics", NATO ASI Series C184: 525 (1986).
[17] E. E. Polymeropoulos, D. Möbius, and H. Kuhn, J. Chem. Phys. 68: 3918 (1978).
[18] M. Momenteau, B. Loock, D. Lavalette, C. Tetreau, and J. Mispelter, J. Chem. Soc. Chem. Commun. 962 (1983).
[19] C. Lecomte, C. Baudin, F. Berleur, A. Ruaudel-Teixier, A. Barraud, and M. Momenteau, Thin Solid Films 133: 103 (1985).
[20] P.Coronel, A. Barraud, R. Claude, O. Kahn, A. Ruaudel-Teixier, and J. Zarembovitch, J. Chem. Soc. Chem. Commun. 193 (1989).
[21] P. Coronel, A. Ruaudel-Teixier, A. Barraud, and O. Kahn, Thin Solid Films 160: 107 (1988).
[22] K. Fukuda,Y. Shibasaki, and H. Narahara, Thin Solid Films 133: 39 (1985).

DEVELOPMENT OF NOVEL CONDUCTIVE LANGMUIR-BLODGETT FILMS: METALLIC PROPERTIES AND PHOTOCHEMICAL SWITCHING PHENOMENA

Takayoshi Nakamura, Hiroaki Tachibana, Mutsuyoshi Matsumoto, Motoo Tanaka and Yasujiro Kawabata

National Chemical Laboratory for Industry
Tsukuba, Ibaraki 305, JAPAN

INTRODUCTION

Recently, there has been a great interest focused on the fabrication of Langmuir-Blodgett (LB) films from electrically active materials. Following the initial report of Ruaudel-Teixier *et al.*. [1], a great deal of progress has been made towards the development of conductive LB films. To date, more than ten different types of conductive LB films have been reported (Table I) [2-25]. These films are, however, made mostly of long-chain derivatives of low-dimensional organic metals: radical salts, charge-transfer complexes, or conductive polymers. Due to the one-dimensional nature of the conducting stacks, these materials are very sensitive to defects and disorder, and are intrinsically unstable. To suppress this instability, and accomplish high, metallic conductivity, one of the most important tasks in the next stage is to prepare conductive LB films of higher dimensionality [26]. The external control of the conductivity has also become a matter of general interest, from the viewpoint of fabricating molecular switching devices using LB films.

In this paper we describe two newly developed conductive LB films: an LB film with metallic conductivity using transition metal complexes [23], and a photochemical switching LB film using a TCNQ radical salt, containing an azobenzene unit in the hydrophobic part of the amphiphilic molecule [13].

CONDUCTIVE LB FILMS OF METAL-dmit COMLEXES

A mixed-valence complex of metal(dmit)$_2$ (H_2dmit = 4,5-dimercapto-1,3-dithiolo-2-thione) system is one of the promising candidates for the preparation of conductive LB films with higher dimensionality [27]. This sulfur-rich 1,2-dithiolene complex has provided three molecular superconductors, TTF[Ni(dmit)$_2$]$_2$ [28], TTF[Pd(dmit)$_2$]$_2$ [29], and Me_4N[Ni(dmit)$_2$]$_2$ [30]. By the introduction of the alkylammonium group as a counter cation, the complexes become amphiphilic, and suitable for the construction of LB films.

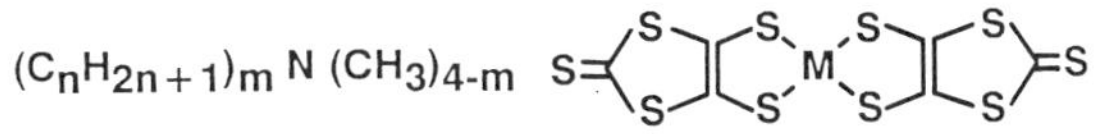

$(C_nH_{2n+1})_m N (CH_3)_{4-m}$ M(dmit)$_2$

mCn - M

Fig. 1 Molecular structure of the alkylammonium-metal(dmit)$_2$.

Lower-Dimensional Systems and Molecular Electronics
Edited by R.M. Metzger *et al.*, Plenum Press, New York, 1991

Table 1. Classification of conductive LB films.

Materials	Secondary Treatment	Conductivity (Bulk) S/cm	Remarks	Ref.
1. Anion Radical Salts				
$C22py^+ TCNQ^-$	I_2 doping	0.01	$E_a = 0.15$ eV, semicond.	[1]
$C18S^+ TCNQ^-$	I_2 doping	?		[7]
$C18P^+ TCNQ^-$	I_2 doping	?		[7]
$C22py^+ (TCNQ)_2^-$	none	0.01	$E_a = 0.3$ eV, semicond.	[2]
$C18py^+ TCNQ^-$	none	0.01	$E_a = 0.13$ eV	[14]
$C18py^+ (TCNQ)_2^-$	none	0.03		[15]
APT(8-12)	none	0.01		[13]
2. Cation Radical Salts				
C16TTF	I_2 doping	0.01	$E_a= 0.19$ eV	[5]
TTF-C17	I_2 doping	?		[8]
BEDT-TTF-C18	ICl doping	?		[16]
C8TTT	I_2 doping	0.01		[17]
3. Charge-Transfer Complexes				
(TMTTF)(C18TCNQ)	none	0.1	$E_a = 0.08$ eV, metal ?	[3]
(TMTTF)(C14TCNQ)	none	1	$E_a = 0.06$ eV, metal ?	[4]
(C17DMTTF)(TCNQ) + (TMTTF)(C16TCNQ)	none	0.1		[18]
(C16TET-TTF)($TCNQF_4$)	I_2 doping	0.25		[19]
(C18TET-TTF)($TCNQF_4$)	I_2 doping	0.01	$E_a = 0.08$ eV	[12]
(C16TDMT-TTF)($TCNQF_4$)	I_2 doping	10^{-3}		[19]
(TTF)(TCNQ)	none	5.5		[20]
4. Transition Metal Complexes				
C18-$Ni(pdt)_2$	none	0.01		[21]
2C10-$Ni(dmit)_2$	Br_2 doping	1		[22]
3C10-$Au(dmit)_2$	electrooxidation	30	metallic	[23]
5. Polymers				
C18Mepy	electrochemical polymerization	0.1		[6]
poly-2C4Py + $C_7F_{15}COOH$	I_2 doping	10^{-6}		[24]
C18Py	pyrrole+$FeCl_3$	0.01		[25]
P-BT	$NOPF_6$ doping	5		[9]
poly-(C18OAn)	I_2 doping	10^{-6}		[24]
PPV	SO_3 doping	0.5		[10]

Synthesis of complexes and preparation of LB films

Amphiphilic dmit complexes with various ammonium cations and central metal cations (mCn-M, Fig 1) were synthesized, according to procedures described previously [22]. 1:1 mixtures with eicosanoic acid formed monolayers on pure water. The monolayers were deposited on hydrophobized solid substrates by the horizontal lifting method. Gold electrodes (gap distance: 0.1 mm) were formed, by vacuum deposition, before the transfer of monolayers. The conductivity of the films was measured for a 20-layer sample by dc 2-probe or 4-probe method after oxidation. The oxidation of the LB films was performed by two different methods: oxidation by bromine gas and electro-oxidation. The latter was achieved by a constant-current method in 0.1 mol/L $LiClO_4$ aqueous solution.

Conductivity of the LB films

Table 2 summarizes the bulk conductivities of the oxidized LB films at room temperature (assuming a monolayer thickness of 3 nm, the thickness of the matrix eicosanoic acid). The conductivities were measured by the dc 2-probe method, using deposited gold electrodes,

Table 2. Conductivity of the LB films of metal(dmit)$_2$ complexes

Materials	Mixing Ratio[a]	Bulk Conductivity (S/cm)	
		Bromine Oxidation	Electrochemical Oxidation
1C18-Ni(dmit)$_2$	1:1	0.11	0.9
2C10-Ni(dmit)$_2$	1:1	1.0	1.4
2C12-Ni(dmit)$_2$	1:1	0.03[d]	
2C14-Ni(dmit)$_2$	1:1	0.09[d]	
2C16-Ni(dmit)$_2$	1:1[b]	0.05[d]	
2C18-Ni(dmit)$_2$	1:1[b]	0.009[d]	
2C22-Ni(dmit)$_2$	pure[b]	0.002[d]	
2C10-Au(dmit)$_2$	1:1	0.12[d]	1.4
2C14-Au(dmit)$_2$	1:1	0.15[d]	
2C18-Au(dmit)$_2$	1:1	0.005[d]	
2C22-Au(dmit)$_2$	1:1	----[e]	
3C10-Ni(dmit)$_2$	1:1	1.5	1.4
3C14-Ni(dmit)$_2$	1:1	1.3	0.87
3C10-Au(dmit)$_2$	1:1	15	33
3C14-Au(dmit)$_2$	1:1	5.4	19
3C16-Au(dmit)$_2$	1:1	2.6	0.46
3C18-Au(dmit)$_2$	1:1	1.4	0.12
4C10-Ni(dmit)$_2$	1:1	1.6	0.012
C14py-Ni(dmit)$_2$	1:1	1.5	1.2
C22py-Ni(dmit)$_2$	1:1	0.23	0.32
(3C10)$_2$-Pd(dmit)$_2$[c]	pure	0.3[f]	1.0
(2C10)$_2$-Pt(dmit)$_2$[c]	pure	0.001[f]	

a) Molar mixing ratio with eicosanoic acid. b) Transferred by the vertical dipping method. c) 2:1 salt. d) Silver paste was used for the electrodes. e) Below the limit of detection. f) Oxidized by iodine vapor.

unless otherwise noted. Fairly conductive LB films were obtained for the 3Cn-Au system; among these, the LB films of 3C10-Au and 3C14-Au exhibit a conductivity of 20 to 30 S/cm after electrooxidation, which, to the best of our knowledge, is the highest recorded value for conductive LB films.

Transport properties of the LB film of tridecylmethylammonium-Au(dmit)$_2$

Figure 2 shows the temperature dependence of the electrical conductivity of a 3C10-Au LB film, after electrooxidation, in the range of 280 - 10 K, measured by the dc 4-probe method. The film showed metallic behavior down to around 200 K, with weak temperature dependence, then the conductivity decreased with the decrease in temperature. The conductive LB films reported so far exhibit activated-type conductivity. Due to the polycrystalline nature of the film, the electronic transport properties are not defined precisely. Several studies, however, suggest that they are either narrow-gap semiconductors [31,32] or metallic only in the individual microscopic domains [33,34]. In contrast to those films, the LB film of 3C10-Au remains macroscopically metallic, in the range of around room temperature to 200 K, suggesting the predominance of a scattering mechanism over thermal activation. Probably, short S···S contacts (shorter than the van der Waals radii between dmit molecules in adjacent stacks) may occur in the LB film of 3C10-Au after oxidation; this would reduce the unfavorable effects of the defects in the arrangement of the complexes, and result in the observed macroscopically metallic conductivity of the LB film.

Below 200 K, the curve follows the Arrhenius equation, with an activation energy of 0.002 eV in the range of 200 - 50 K. The value of the activation energy is smaller, than that of the LB films reported so far, by about one to two orders of magnitude. There is a distinct deviation from the Arrhenius plot below 50 K, which can be explained by variable-range hopping. The film still exhibits a conductivity of ca. 10 S/cm at 10 K, suggesting that a metal-insulator transition does not occur down to 10 K.

PHOTOCHEMICAL SWITCHING PHENOMENA IN LB FILM

One of the promising ideas for controlling the conductivity of the LB film externally is to introduce a switching unit into the amphiphilic part of the conductive LB film. The switching unit is triggered by an external stimulus, which induces the change in structure and conductivity of the functional moiety located in the hydrophilic part [13].

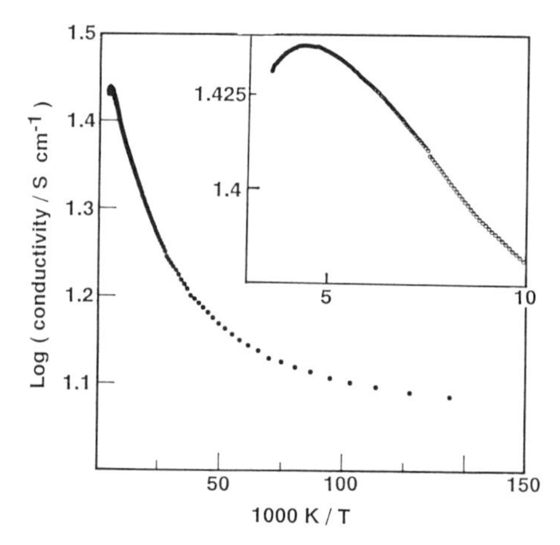

Fig. 2 Temperature-dependent electrical conductivity of the LB film of a 1:1 mixture of 3C10-Au and eicosanoic acid after electro-oxidation. The measurement was carried out by the dc 4-probe method in vacuum. The cooling rate was 0.5 K/min.

Molecular design of photochemical switching LB film

We designed an amphiphilic molecule, in which three functional parts are involved: azobenzene as the switching unit, an alkyl chain as the transmission unit, and a TCNQ charge transfer salt as the working unit (APT(8-12), Fig. 3). The molecule is assembled in an ordered array by a horizontal lifting method on a solid substrate. In this arrangement, the *cis-trans* photoisomerization of azobenzene induces the reversible change in the lateral conductivity of the LB film.

Cis-trans isomerization of azobenzene in LB film

The spectral change in the LB film of APT(8-12), upon photo-irradiation, is shown in Fig. 4. Reversible *trans*- to-*cis* and *cis*- to-*trans* isomerization occurred with irradiation of 365 and 436-nm light, respectively. The absorption maximum at 365 nm, due to the *trans*-azobenzene, was the same as that in solution, indicating that the azobenzene in the LB film is in

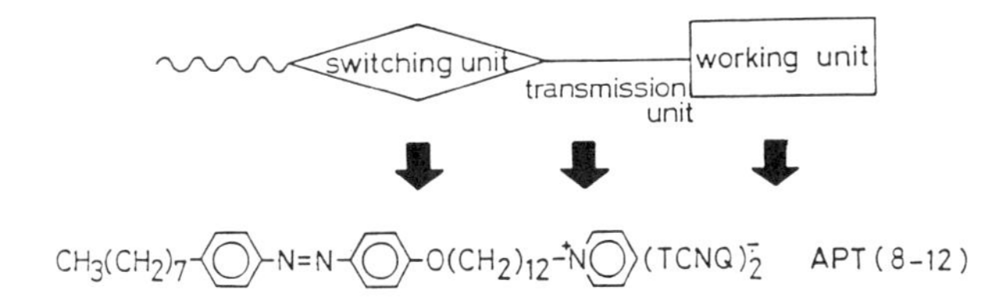

Fig. 3. Molecular design of photochemical switching LB film.

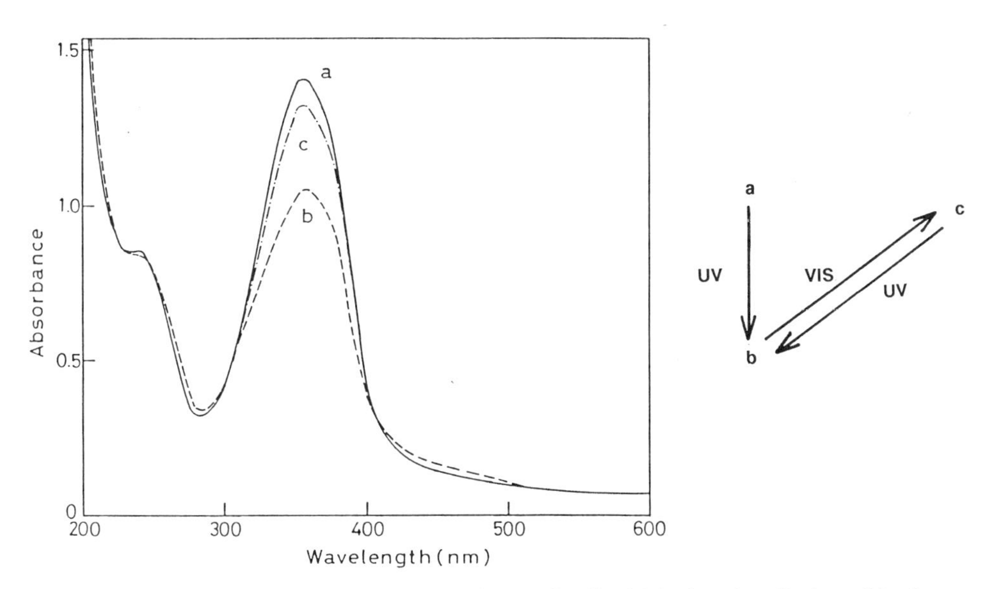

Fig.4. Spectral change in the LB film of APT(8-12): (a) before irradiation, (b) after irradiation with 365-nm light, (c) after irradiation with 436-nm light.

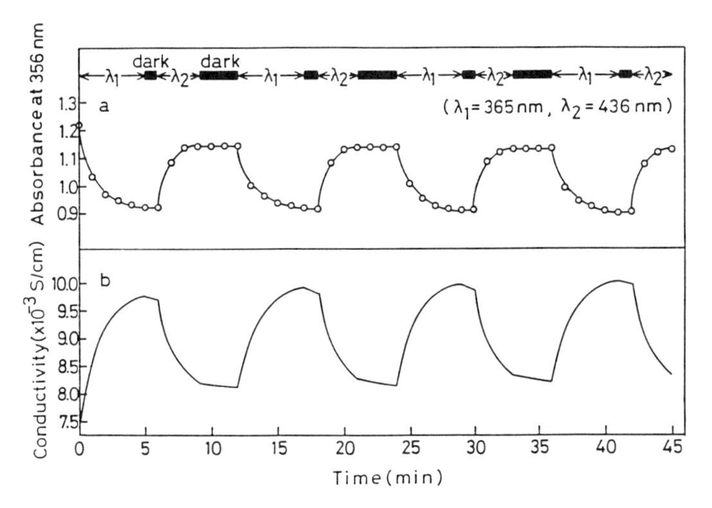

Fig.5. Change in (a) the absorbance at 356 nm due to *trans*- azobenzene and (b) the conductivity of the LB film of APT(8-12) on alternative irradiation with 365 and 436-nm light.

the monomeric state. Hence, the intermolecular interaction between azobenzene parts is rather weak, which allows the *cis-trans* isomerization of azobenzene.

Photoresponsive switching in LB film

The change in conductivity of the LB film, and absorbance at 356 nm, due to *trans*-azobenzene, are shown in Fig. 5, as a function of alternate irradiation time with UV and visible light. The conductivity of the LB film increased by ca. 25% upon the irradiation with 365-nm light. Then, the conductivity reverted with *cis* to *trans* isomerization, upon irradiation with 436-nm light. The change in conductivity occurs almost concomitantly with the isomerization of azobenzene. This reversible cycle can be repeated more than a hundred times. The half-life of the thermal recovery of the conductivity was several hours, which was the same as that of the thermal isomerization from *cis* to *trans* azobenzene.

CONCLUSION

In this work, highly conductive LB films of metal-$(dmit)_2$ complexes and photochemical switching phenomena in the LB film are described. The LB films of tridecylmethylammonium-$Au(dmit)_2$ showed a conductivity of 30 S/cm after electrochemical oxidation. Moreover, the film exhibited a metallic temperature dependence of the conductivity around room temperature. The conductivity of APT(8-12) was reversibly controlled by the photo-induced conformational change in azobenzene. This photo-responsive switching phenomenon can be repeated more than a hundred times.

These systems may afford promising candidates for superconducting LB films and for molecular switching devices.

REFERENCES

[1] A. Ruaudel-Teixier, M. Vandevyver, and A. Barraud, Mol. Cryst. Liq. Cryst. 120: 319 (1985).
[2] M. Matsumoto, T. Nakamura, F. Takei, M. Tanaka, T. Sekiguchi, M. Mizuno, E. Manda, and Y. Kawabata, Synth. Met. 19: 675 (1987).
[3] Y. Kawabata, T. Nakamura, M. Matsumoto, M. Tanaka, T. Sekiguchi, H. Komizu, E. Manda, and G. Saito, Synth. Met. 19: 663 (1987).
[4] M. Matsumoto, T. Nakamura, E. Manda, Y. Kawabata, K. Ikegami, S. Kuroda, M. Sugi, and G. Saito, Thin Solid Films 160: 61 (1988).
[5] A. S. Dhindsa, M. R. Bryce, J. P. Lloyd, and M. C. Petty, Thin Solid Films, 165: L97 (1988); Synth. Met 27: B563 (1988).
[6] T. Iyoda, M. Ando, T. Kaneko, A. Ohtani, T. Shimidzu, and K. Honda, Tetrahedron Lett. 27: 5633 (1986).
[7] M. Vandevyver, J. Richard, A. Barraud, A. Ruaudel-Teixier, M. Lequan, and R. M. Lequan, J. Chem. Phys. 87: 6754 (1987).
[8] F. Bertho, D. Talham, A. Robert, P. Batail, S. Megtert, and P. Robin, Mol. Cryst. Liq. Cryst. 156: 339 (1988).
[9] I. Watanabe, K. Hong, and M. F. Rubner, Synth. Met., 28: C473 (1989); J. Chem. Soc. Chem. Commun. 123 (1989).
[10] Y. Nishikata, M. Kakimoto, and Y. Imai, J. Chem. Soc. Chem. Commun. 1040 (1988).
[11] A. Otsuka, G. Saito, T. Nakamura, M. Matsumoto, Y. Kawabata, K. Honda, M. Goto, and M. Kurahashi, Synth. Met. 27: B575 (1988).
[12] J. Richard, M. Vandevyver, A. Barraud, J. P. Morand, R. Lapouyade, P. Delhaès, J. F. Jacquinot, and M. Roulliay, J. Chem. Soc. Chem. Commun. 754 (1988).
[13] H. Tachibana, T. Nakamura, M. Matsumoto, H. Komizu, E. Manda, H. Niino, A. Yabe, and Y. Kawabata, J. Am. Chem. Soc. 111: 3080 (1989).
[14] A. S. Dhindsa, M. R. Bryce, J. P. Lloyd, and M. C. Petty, Synth. Met. 22: 185 (1987).
[15] P. S. Sotnikov, T. S. Berzina, O. Ya. Neiland, and V. I. Troitsky, Thin Solid Films, in press.

[16] C. Lalanne, P. Delhaès, E. Dupart, Ch. Garrigou-Lagrange, J. Amiell, J. P. Morand, and B. Desbat, Thin Solid Films, in press.
[17] B. Tieke and A. Wegmann, Thin Solid Films, in press.
[18] V. I. Troitsky, T. S. Berzina, O. Ya. Neiland, and P. S. Sotnikov, Thin Solid Films, in press.
[19] D. Zhu, X. Wu, Y. Liu, X. Wang, Y. Hua, X. Pang, and D. Jiang, Thin Solid Films, in press.
[20] M. Fujiki and H. Tabei, Synth. Met. 18: 815 (1987).
[21] M. Watanabe, H. Kamikita, K. Sanui, and N. Ogata, Polym. Prep. Jpn. 36: 3242 (1987).
[22] T. Nakamura, H. Tanaka, M. Matsumoto, H. Tachibana, E. Manda, and Y. Kawabata, Chem.Lett. 1667 (1988); Synth. Met. 27: B601 (1988).
[23] T. Nakamura, K. Kojima, M. Matsumoto, H. Tachibana, M. Tanaka, E. Manda and Y. Kawabata, Chem. Lett. 367 (1989).
[24] M. Ando, Y. Watanabe, T. Iyoda, K. Honda, and T. Shimidzu, Thin Solid Films, in press.
[25] K. Hong and M. F. Rubner, Thin Solid Films 160: 187 (1988).
[26] G. Saito, Pure Appl. Chem. 59: 999 (1987).
[27] G. Steimecke, H. J. Sieler, R. Kirmse, and E. Hoyer, Phosphorus and Sulfur 7: 49 (1979).
[28] M. Bousseau, L. Valade, J. P. Legros, P. Cassoux, M. Garbauskas, and L. V. Interrante, J. Am. Chem. Soc. 108: 1908 (1986).
[29] L. Brassard, H. Hurdequint, M. Ribault, L. Valade, J. P. Legros, and P. Cassoux, Synth. Met. 27: B157 (1988).
[30] A. Kobayashi, H. Kim, Y. Sasaki, R. Kato, H. Kobayashi, S. Moriyama, Y. Nishio, K. Kajita, and W. Sasaki, Chem. Lett. 1819 (1987).
[31] J. Richard M. Vandevyver, P. Lesieur, A. Barraud and K. Holczer, J. Phys. D: Appl. Phys. 19: 2421 (1986).
[32] K. Ikegami, S. Kuroda, M. Saito, K. Saito, M. Sugi, T. Nakamura, M. Matsumoto and Y. Kawabata, Phys. Rev. B35: 3667 (1987); Thin Solid Films 160: 139 (1988) .
[33] K. Ikegami, S. Kuroda, M. Sugi, M. Saito, S. Iizima, T. Nakamura, M. Matsumoto, Y. Kawabata and G. Saito, Synth. Met. 19: 669 (1987).
[34] K. Ikegami, S. Kuroda, K. Saito, M. Saito, M. Sugi, T. Nakamura, M. Matsumoto, Y. Kawabata and G. Saito, Synth. Met. 27: B587 (1988).

THE QUEST FOR HIGHLY CONDUCTING LB FILMS

K. Lerstrup, J. Larsen, P. Frederiksen, and K. Bechgaard

Chemistry Department
University of Copenhagen, Universitetsparken 5
DK-2100 Copenhagen Ø, DENMARK

INTRODUCTION

Alongside our ongoing research in the field of conducting organic materials, we have investigated the possibilities of attaching long-chain fatty acids to various donor molecules, in order to prepare conducting Langmuir-Blodgett (LB) films.

The LB film technique provides us with a quite powerful tool for the preparation of new materials with an inherently high degree of organization in a truly two-dimensional fashion. The idea of combining LB films with conducting organic materials, in order to obtain highly conducting thin films, is not new [1], but its realization as a routine bench-top technique has yet to come.

In the following, we wish to describe - not the preparation of conducting LB films - but the synthesis of a number of candidate compounds for this task. Since long-chain carboxylic acids seem to be superior for LB film formation, we have concentrated our synthetic efforts on the preparation of fatty acids with a π-donor molecule attached to the terminal carbon. The film-forming properties of this series of terminally π-donor-substituted long-chain fatty acids is, at present, only partly investigated, but preliminary results and guidelines for synthetic approach will be given, together with important design criteria for what we, at present, consider as good film-forming properties.

PACKING CRITERIA

It is commonly acknowledged that a primary requirement for high metallic conduction in organic materials is a good π-π contact between the donor molecules. For donors of the tetrathiafulvalene (TTF) type, the inter-planar distances are, typically, around 3.5 Å, in the usual ring-over-bond packing mode. As the diameter of a carbon chain in an LB film is roughly 5 Å, it is obvious, from a simple model study, that the concept of a highly conducting LB film, based on TTF-type π-donors, will not work very well (Fig. 1a). Close contact could be obtained by tilting the donor end of the molecules, but then the "slip" distance becomes too long, resulting in a very small area of overlap (Fig. 1b). One way to overcome this problem would be to use larger (longer) donor molecules, e.g. pyrene or perylene-type donors (Fig.1c).

The experimental verification of these very simple models have yet to come. Structure determination of LB films is difficult, due to the very small volume accessible, and the tendency to form domains.

It has been shown that the long alkyl chains in LB films order in a regular manner. On the other hand, no definite information on the orientation of the attached groups is available. Due to the flat shape of the attached donors, strong π-π overlap can only be established in one direction. In a simple picture of the donor layer, one would thus have to consider two ordering

Lower-Dimensional Systems and Molecular Electronics
Edited by R.M. Metzger *et al.*, Plenum Press, New York, 1991

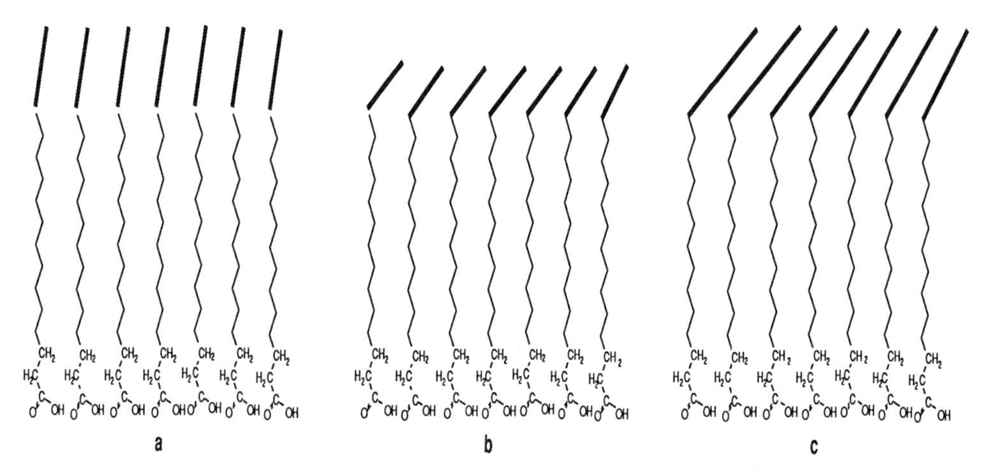

Fig. 1. (a) Short, flat aromatic planes (seen end-on), when bonded to 5 Å-wide alkane chains, pack poorly if not tilted; (b) the same aromatic planes, if tilted and side-slipped, will achieve better packing, at the expense of good overlap. (c) Longer aromatic planes, e.g. perylene, can tilt and also overlap better, with best packing.

vectors: parallel and transverse to the donor plane. As a result, the two-dimensional (2D) layer of donors would be expected to exhibit some degree of one-dimensional properties.

Thus, the tendency of ordering, which is important for the formation of highly conducting LB films, will be very dependent on the size and shape of the donors employed.

SYNTHETIC SCHEMES

The investigation of these effects has been undertaken through the synthesis of a number of fatty-acid-sunstituted π-donors. Preliminary results, early in the study, showed that the films prepared from TTF and dithiapyrene-fatty acids did exhibit electrical conductivity, and that the conductivity did increase, when the film was doped with iodine vapor. We also found that the film-forming properties, as well as the conductivity, was better for the larger dithiapyrene-derived fatty acid. This encouraged us to carry out the synthetic work outlined below.

In all cases, the donors were prepared separately, functionalized, and, as the final step in the synthesis, linked to the fatty acid. Two basic coupling reactions were used. Various TTF's were converted to their corresponding mono-carboxaldehydes [2], and reacted with a Wittig reagent, prepared from the desired ω-phosphonium-C_{11} or -C_{16} carboxylic acid. A similar approach was used in the derivatization of some of the perylene-type donors. The triphenylphosphonium bromides were prepared in the usual manner, by heating ω-bromo-

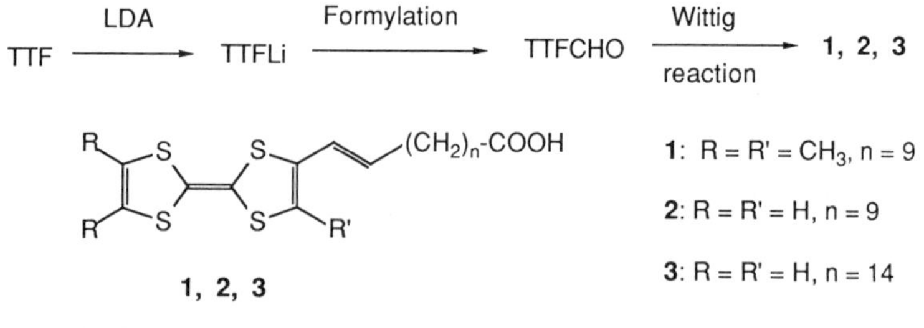

TTF —LDA→ TTFLi —Formylation→ TTFCHO —Wittig reaction→ **1, 2, 3**

$(CH_2)_n$-COOH

1: R = R' = CH_3, n = 9

2: R = R' = H, n = 9

3: R = R' = H, n = 14

1, 2, 3

Fig. 2. Vinyl link of fatty acid to TTF: Compounds **1**, **2**, and **3**.

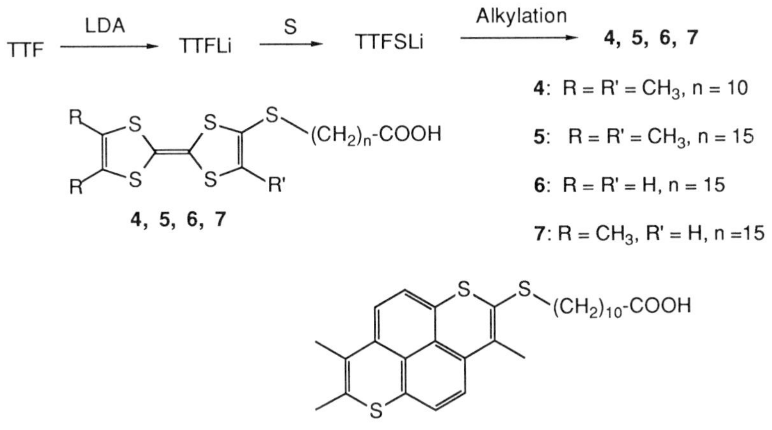

TTF $\xrightarrow{\text{LDA}}$ TTFLi $\xrightarrow{\text{S}}$ TTFSLi $\xrightarrow{\text{Alkylation}}$ **4, 5, 6, 7**

4: R = R' = CH_3, n = 10

5: R = R' = CH_3, n = 15

6: R = R' = H, n = 15

7: R = CH_3, R' = H, n =15

$(CH_2)_n$-COOH

4, 5, 6, 7

$(CH_2)_{10}$-COOH

8

Fig. 3. Sulfur link of fatty acid to TTF and dithiapyrene: compounds **4** through **8**.

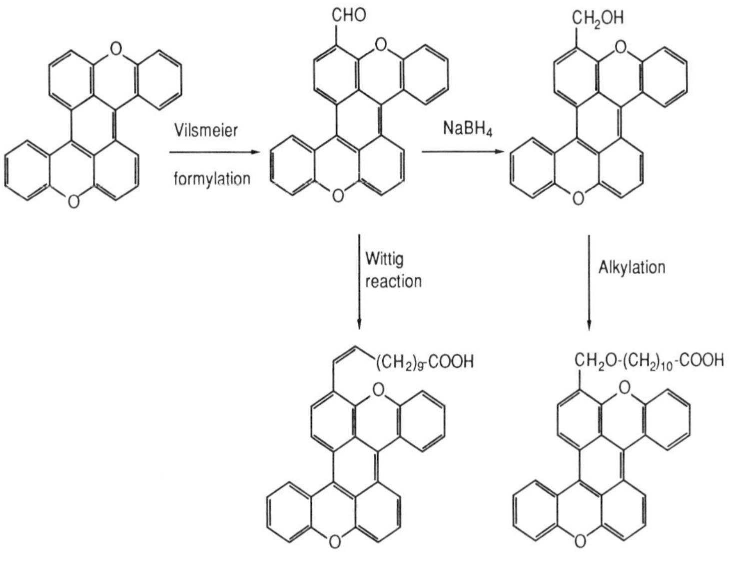

Vilsmeier formylation

CHO

$NaBH_4$

CH_2OH

Wittig reaction

Alkylation

$(CH_2)_9$-COOH

CH_2O-$(CH_2)_{10}$-COOH

9

10

Fig. 4. Vinyl or oxygen link of fatty acid to corbien: compounds **9** and **10**.

undecanoic or -hexadecanoic acid with an equivalent amount of triphenylphosphine in a high-boiling solvent, like mesitylene or decalin (Fig. 2, compounds **1**, **2**, **3**).

Alternatively, the TTF or the dithiapyrene was lithiated, treated with elemental sulfur, and allowed to react with the potassium salts of ω-bromo- or ω-iodo-undecanoic acid or -hexadecanoic acid (Fig. 3, compounds **4**, **5**, **6**, **7**, and **8**).

The vinyl link and also an oxygen link were made to corbien (Fig. 4, compounds **9** and **10**).

EXPERIMENTAL

Vinyl link

Hexadecanoic acid-ω-triphenylphosphonium bromide (5 mmol) was dissolved in tetrahydrofuran (THF) containing 10% HMPTA, and treated with two equivalents of potassium tert-butoxide at -10°C. One equivalent of TTF-carboxaldehyde, dissolved in THF, was added, and the mixture was stirred for 0.5 hour; diethyl ether was added, and the product filtered off, and purified through acidic water-ether extraction, column chromatography (silica/ethyl acetate), followed by recrystallization from methanol, to provide ca. 500 mg of **3**.

The C_{11}-analog (**2**) was prepared similarly, and the product was isolated through recrystallization from toluene-hexane. Crystals of X-ray quality were obtained in this way, and the crystal structure was reported [3]. In a similar manner, **9** was prepared in ca. 25% yield.

Sulfur link

Trimethyl-TTF was dissolved in dry THF, and treated with a slight excess of LDA (lithium diisopropylamide) for one hour at -78°C. One equivalent of sulfur was added, and the mixture stired for 2 hours, while the temperature was allowed to rise to -10°C. The mixture was buffered by the addition of one mL of tert-butylalcohol, and a THF-DMF slurry of the potassium salt of ω-bromohexadecanoic acid was added, and the mixture stirred for 20 hours. Water was added, the mixture was acidified with HCl, and extracted several times with diethyl ether. The product was purified through flash column chromatography (silica/dichloromethane-acetone gradient) and recrystallization from toluene-hexane. Yields of **5** were typically 30-40%. Compounds **4**, **6**, **7**, and **8** were prepared in a similar manner.

Oxygen link

In one instance, the corbien-carboxaldehyde was reduced with sodium borohydride to the corresponding alcohol, and O-alkylated with ω-undecanoic acid in THF, to produce compound **10** in moderate yield.

REFERENCES

[1] G. G. Roberts, M. McGinnity, W. A. Barlow, and P. S. Vincett, Solid State Commun. 32: 636 (1979).
[2] D. C. Green, J. Org. Chem. 44: 1476 (1979).
[3] G. Rindorf, N. Thorup, K. Lerstrup, and K. Bechgaard, Synth. Metals 30: 391 (1989).

LANGMUIR-BLODGETT FILMS FROM
DONOR-ACCEPTOR SUBSTITUTED POLYENES

S. Hagen, H. Schier, S. Roth, M. Hanack*

Max-Planck-Institut für Festkörperforschung
D-7000 Stuttgart 80, Heisenbergstr. 1, FRG
* Institut f. Organische Chemie der Universität
D-7400 Tübingen, Auf der Morgenstelle, FRG

INTRODUCTION

LB-films play a major role as model systems in the development of future devices, especially in fields related to nonlinear optics and molecular electronics. The LB-technique provides a powerful method to assemble molecules with specific electronic functions.

Polyene chains are often discussed as possible elements in molecular electronics[1]. They are regarded as molecular wires, which connect other molecular switching elements. If this molecular wire is modified by attaching donor and acceptor groups to the ends of the polyene chain, optical excitation of the donor group at a selected photon energy is accompanied by a transfer of charge from the donor group to the acceptor group via the polyene chain. In a chemical view this is expressed by shifting the positions of the double-bonds by one C-atom. This optically-induced change of the molecular dipole moment can also be discussed in terms of solitons[2].

The Langmuir-Blodgett-technique (LB)[3] is a method to order molecules on the interface water-air and to transfer these ordered films to solid substrates. Amphiphilic molecules, consisting of a hydrophilic headgroup and a hydrophobic 'tail' are spread on a clean water surface. The molecules lying on the surface form a two-dimensional gas. On reducing the covered surface, the mean area per molecule [$Å^2$/mol] is reduced until the interaction of the molecules induces a phase transition into a liquid-analogous state. This phase transition is characterized by a change of slope in the surface-pressure/area (σ/F)-diagram. Further reduction of the phase induces a second transition into 2D-solid state. This face is comparable to a smectic-b phase, all molecules standing perpendicular to the water surface. In this solid-analogous state the ordered film can be transferred to a solid substrate. The film is held under constant pressure and the

Lower-Dimensional Systems and Molecular Electronics
Edited by R.M. Metzger *et al.*, Plenum Press, New York, 1991

substrate is dipped into the water, thereby transfering the molecules from the water surface to the substrate.

We have prepared ultra-thin films with the Langmuir-Blodgett technique from donor-acceptor substituted polyenes. To improve the formation of such well-ordered structures we have synthesized polyenes to which long aliphatic chains were attached[4].

CHEMICAL PREPARATION

The synthesis of 7-N,N-dioctylaminophenyl-7,8'-diapo-carotene-8'-al (7) is shown in Fig. 1.

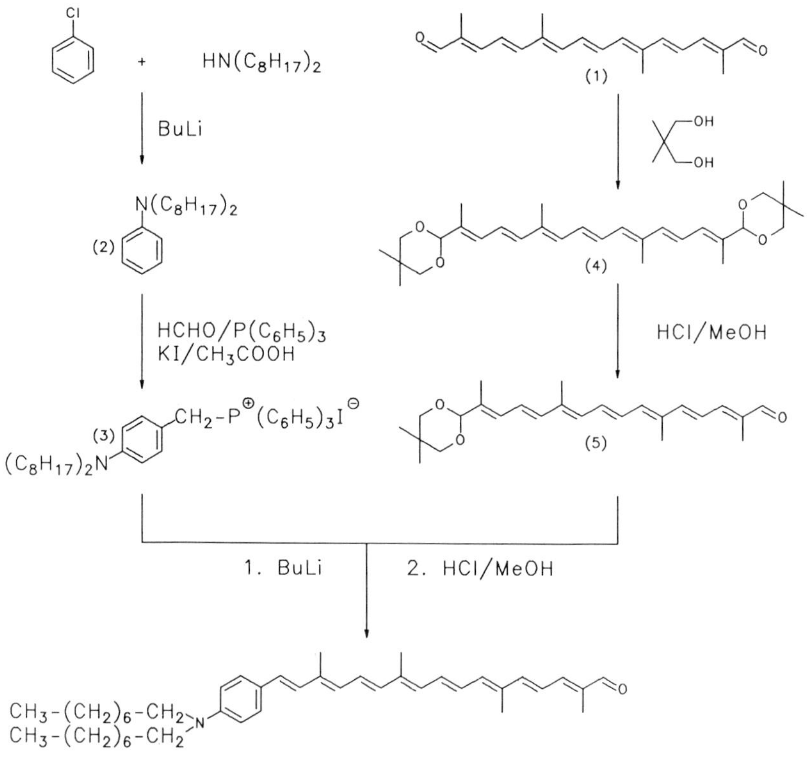

Fig. 1

For inserting the long aliphatic chains into the donor group, chlorobenzene was treated with di-n-octylamine and n-BuLi in a nucleophilic aromatic substitution reaction. The resulting N,N-dioctylaniline (2) was converted into the phosphonium salt (3) in a reaction with HCHO, $P(C_6H_5)_3$, CH_3COOH,

KI, in $CHCl_3$ at room temperature and occasionally stirring for three weeks[5]. The donor was linked to the crocetindialdehyde (1), which is commercially available, by the conversion of one carbonyl group into a new C=C double bond in a Wittig reaction[6]. To avoid the formation of the undesired disubstituted product, we protected one of the carbonyl groups of the dialdehyde. In order to do this, crocetindialdehyde was converted into its diacetal (4) with 2,2-dimethyl-1,3-propanediol. Partial hydrolysis of the diacetal led to the monoacetal (5), which was purified by flash chromatography[7]. In the Wittig reaction a solution of the monoacetal in THF was slowly added to the dark-red solution of the donor-triphenylphosphorane, which was generated in a reaction between the phosphonium salt (3) and n-BuLi in THF at 0°C. After four hours of stirring (control by TLC) at room temperature and final hydrolysis of the acetal group, we applied purification by flash chromatography on silicagel with a mixture of CH_2Cl_2 and ethylacetate 60:1 and obtained monosubstituted polyene with a high yield of 75 %.

NMR-SPECTROSCOPY

The NMR-spectra were taken on a Bruker WM 250 NMR-spectrometer. $CDCl_3$ was used as solvent, TMS as internal standard. The ^1H and ^{13}C-NMR-spectra are shown in Fig. 2 and Fig. 3. The values of the chemical shifts are indicated in the inset of the figures. The interpretation of the spectra was done by comparison with already known NMR data of similar compounds[8].

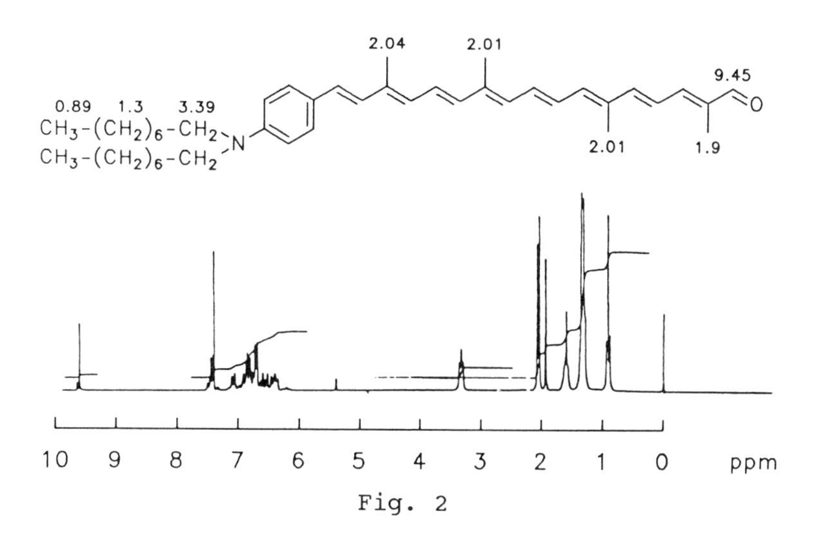

Fig. 2

^1H-NMR-SPECTRUM (250 MHz)

The signal of the methylene-groups, neighbouring the tertiary nitrogen, appears at 3.39 ppm as a broad triplet. The methylene and methyl groups of the aliphatic chain give signals at 1.3 and 0.89 ppm. The signal of aldehyde proton appears at 9.45 ppm. From these two signals it is possible to estimate the ratio of Z and E isomers. On the supposition that the all trans-form is preferred one can determine a ratio of 20:1.

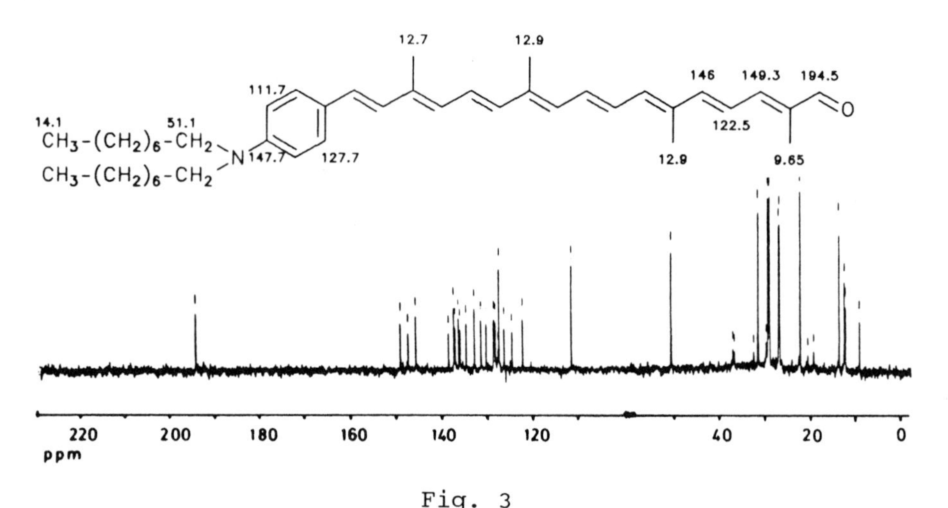

Fig. 3

^{13}C-NMR-SPECTRA (62.89 MHz)

The carbonyl-group polarizes the polyene-chain. Due to this effect, the 9'-C-atom (α-position) is strongly shielded, which causes a shift towards higher field. This effect decreases over the 11'- and 13'-position. On the other hand, the electron density on the 10'-C-atom (β-position) is reduced, which leads to a shift towards lower field. From this influence onto the polyene chain, it is possible to identify the signals in the region of the carbonyl group. The electron density of the C-atom in the aromatic ring, neighbouring the tertiary nitrogen, is reduced, due to the difference in electronegativity, which leads to a shift towards lower field. The meta-position has a lower electron density than the ortho-position, and is therefore more deshielded. The position of the six quarternary C-atoms can be identified by comparing the normal ^{13}C-spectrum with a DEPT-spectrum. DEPT[9] (distortionless enhancement by polarization transfer) is a special pulse sequence, which can be used to distinguish between methine-, methylene- and methyl-signals depending on the width of the Θ-pulse (tip angle). Primary and tertiary C-atoms will give a positive signal, methylene C-atoms a nega-

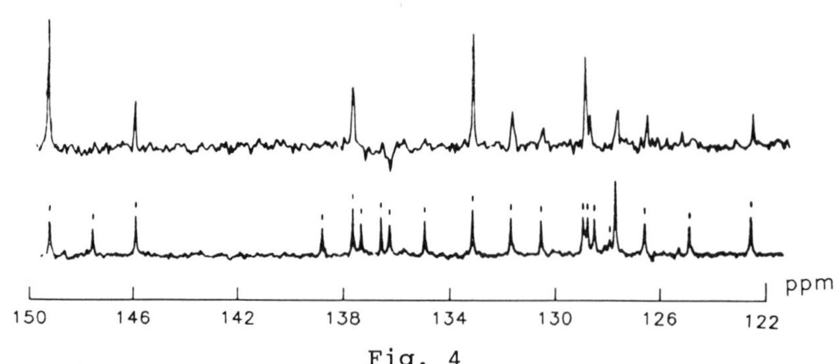

Fig. 4

tive signal. Quarternary carbon signals will be missing from the polarization transfer spectra. The DEPT-spectrum of (7) is shown in Fig. 4. In the olefinic region eight signals are missing, therefore it is not possible to find an exact identification of the quarternary carbons.

PREPARATION OF LB-FILMS

The LB-Layers were prepared with a commercial Lauda film balance, installed in a class 1000 clean room under yellow light. Merck Uvasol quality chloroform and ethanol were used. Water was purified in a Milli-Q-System (Millipore) to 18 MΩ·cm. Commercial Bixin ($C_{25}H_{30}O_4$, Fa. Roth) was purified by chromatography on ODS-phase (MeOH/Acetone=95:5). Two different substrates, i.e. Au and Si were used. Glass slides (75·10·1 mm) were coated with layers of 200 Å Cr and 2000 Å Au by thermal evaporation. The gold substrates were coated immediately after vacuum deposition of the metals. Si substrates (FZ (111), p(B), 1000 X·cm) were etched in 30 % HF and rinsed in Mili-Q water prior to use. For Bixin, the subphase contained $5 \cdot 10^{-4}$ mol/l $CdCl_2$ and was buffered with KOH to pH 7.5. The subphase temperature was 19 °C. (7) was spread on a subphase of pure water. Bixin was spread from solution containing $2 \cdot 10^{-3}$ mol/l Bixin in $CHCl_3$/MeOH=10:1 and rested for 5 min before compression. The monolayer was compressed at a constant rate of 1.14 cm^2/s up to the transfer pressure of 32 dyn/cm, corresponding to a mean area of 22 $Å^2$/molecule. We transferred the films at a dipping speed of 0.83 mm/s for all cycles with a transfer ratio of 0.98. We spread (7) from a $8.9 \cdot 10^{-4}$ M solution in $CHCl_3$ and rested 5 min before compression. We compressed the film at a constant rate of 0.57 cm^2/s up to the transfer pressure of 13 dyn/cm with a mean area of 56 $Å^2$/molecule. Transfer was done at a speed of 0.415 mm/s and yielded a ratio of 0.8. The (σ/F)-diagrams are shown in Fig. 5.

RESULTS

In order to study highly ordered structures of donor-acceptor substituted polyenes, we synthesized (7) in all-trans configuration. The polyene (7), substituted with two aliphatic chains, showed a σ_{max} of 18 dyn/cm. At higher surface pressure the films seem to be stable but consist of, at least partly, folded 'floes' (dashed part of curve in Fig. 5). At the transfer pressure of 13 dyn/cm the film was very insensitive to mechanical vibrations. The transferred film was characterized by IR and UV spectroscopy.

The compressed solid-analogous Langmuir film of Bixin is extremely sensitive to mechanical perturbations. At a surface pressure of 32 dyn/cm the film collapses under the influence of surface waves. We managed this problem by reducing the water level in the trough. The ratio of transferred film area for up- to downstroke motion was almost 1. By comparison of the IR spectra of a multilayer sample on Si (20 LB) to a KBr pellet we could not yet determine if the molecules are stacked in head-to-tail or head-to-head orientation.

From the IR-spetra we deduce that it is possible to get highly ordered structures from donor-acceptor substituted polyenes with the LB-technique. The main problem is still the high sensitivity against mechanical perturbations during the film transfer and therefore up to now most work on polyenes was done by embedding the active molecules in an inert matrix[10,11]. In this work we could show that it is possible to transfer LB-films from pure polyenes. The handling of polyenes substituted with aliphatic sidechains is more convenient, but the maximum surface pressure and perhaps the ordering is lower than for unsubstituted polyenes.

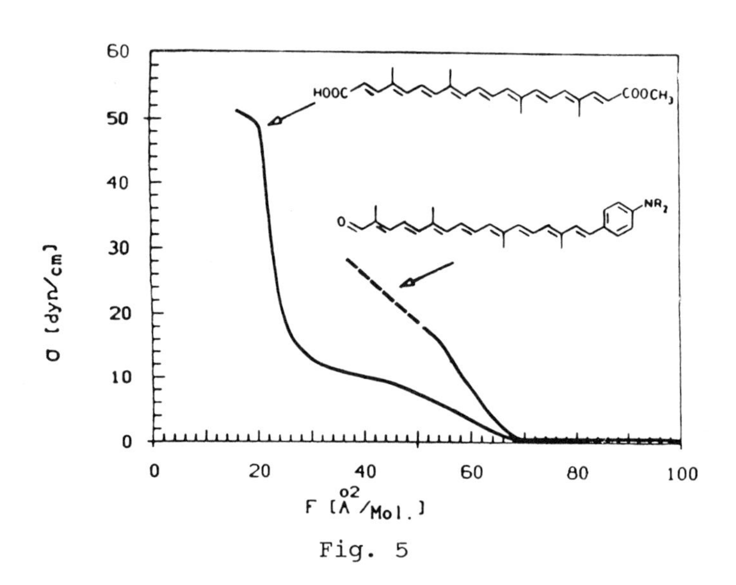

Fig. 5

REFERENCES

1. F.L. Carter (Ed.), "Molecular Electronic Devices", New York and Basel, 1982
2. S. Roth, Physica Scripta, in print
3. H. Kuhn, D. Möbius, Angew. Chem. 83 (1971) 672
4. S. Hagen, H. Schier, S. Roth, M. Hanack, in: "Electronic Properties of Conjugated Polymers (Kirchberg III)", H. Kuzmany, M. Mehring, S. Roth (Eds.), Springer Series in Solid State Sciences, Heidelberg, in print
5. H. Bredereck, Chem. Ber. 106 (1973) 3732
6. W.S. Wadsworth, Jr., Org. React. 25 (1978) 73
7. K. Bernhard, G. Englert, H. Mayer, R. Müller, A. Rüttimann, N. Vecchi, E. Widmer, R. Zell, Helv. Chim. Acta 64 (1981), 2964
8. G. Englert, P. Appl. Chem. 57 (1985) 801
9. J.K.M. Sanders, B.K. Hunter, "Modern NMR-Spectroscopy", Oxford University Press, 1987
10. G. Duda, G. Wegner, Makromol. Chem., Rapid Commun. 9 (1988), 495
11. R.M. Metzger, R.R. Schumaker, M.P. Cava, R.K. Laidlaw, C.A. Panetta, E. Torres, Langmuir 4 (1988), 298

LANGMUIR-BLODGETT FILMS OF A PYRROLE AND FERROCENE MIXED SURFACTANT SYSTEM

L. Samuelson*, A.K.M. Rahman*, S. Clough*, S. Tripathy*, P.D. Hale+, T. Inagaki+, T.A. Skotheim+, Y. Okamoto^

*Dept. of Chemistry, University of Lowell, Lowell, MA, 01854, U.S.A.
+Division of Materials Science, Department of Applied Science, Brookhaven National Laboratory, Upton, NY, 11973, U.S.A.
^Dept. of Chemistry, Polytechnic University, Brooklyn, NY, 11201, U.S.A.

ABSTRACT

The Langmuir-Blodgett technique was used to study the molecular organization of a mixed 3-hexadecyl pyrrole (3HDP) and ferrocene-derivatized pyrrole (Fc-Py) surfactant system. It has been determined that stable monolayer films of the mixed system could be formed at the air-water interface. The growth and assembly process led to polypyrrole 2-D lattices with heretofore unsurpassed order. In fact, the process of template polymerization, it appears, leads to a new crystal phase for the polypyrrole component of the thin film structure. Various monolayer and multilayer films were prepared on platinum-coated substrates for surface spectroscopic characterization. Near Edge X-Ray Absorption Fine Structure (NEXAFS) studies revealed that highly ordered multilayer structures are being formed. Electrochemical studies have been initiated to determine the feasibility of these films in molecular electronic device applications.

INTRODUCTION

Langmuir-Blodgett films of organometallic complexes and conducting polymers are of particular interest for molecular electronic device applications. Here, one may define and develop unusual molecular architectures and assemblies having significant control over the molecular arrangement and resultant electronic structure of the assembly. One relevant problem to address, at this stage, is the development of methodologies which will effectively arrange and couple such molecular organizates to electronic circuitry, to elicit specific desired chemical, electrical, optical and/or magnetic properties. These molecular architectures present an opportunity to investigate the structures and structure-function relationships in two-dimensional conductive and redox active films. In addition, if these assemblies can be effectively coupled to appropriate solid supports including sensing devices, they may find application in such molecular electronic devices as ultrathin electrodes and interconnects in multilayer superlattices, optical waveguides and switches, and chemical and biochemical sensors.

In our laboratory, we have been developing a class of ordered electroactive materials from first principle. The desire is to build, in a complex hierarchy, a material with components that either dictate their own assembly, or can be organized using appropriately tailored techniques. Using the Langmuir-Blodgett technique, mono- and multilayer structures of 3-hexadecyl pyrrole and ferrocene-derivatized surfactant systems and mixtures thereof, have been prepared. These systems have been observed to self-assemble into such a unique structural hierarchy, which possesses novel functional behavior. The pyrrole moiety was chosen for its well-established electronic and optical properties when polymerized. Ferrocene, it is theorized, if properly oriented into a Langmuir-Blodgett monolayer film in conjunction with polypyrrole, may show a layered array of a transition metal, and a redox couple system. The ferrocene group may thus provide the possibility of charge coupling between neutral ferrocene and oxidized ferricenium, which could be controlled electrochemically or photochemically. The array of transition metals, in addition, may serve as a two-dimensional model magnetic system.

The knowledge gained from the ability to dictate these materials into well-ordered organizations, with controlled molecular architectures, should allow for the full utilization of their unique electroactive properties. As a result, materials may be prepared with widely varying electrical, optical, magnetic, and structural properties, and, if fabricated onto suitable substrates, lead to the development of novel molecular electronic devices.

EXPERIMENTAL

The surfactant materials used in this investigation, 3-hexadecyl pyrrole (3HDP) and 3-(13-ferrocenyl,13-hydroxytridecyl) pyrrole (Fc-Py) were prepared using techniques reported elsewhere [1]. Mixed monolayers of 3HDP and pyrrole, in a 1 to 500 ratio, were prepared at the air-water interface, using a commercial Lauda film balance. The monolayers were spread from chloroform solutions (concentration of 3HDP ca. 1-2 mg/ml) onto the subphase, which contained an oxidizing solution of 0.01M $FeCl_3$. A Millipore Milli-Q purification system was used for subphase preparation, and a constant temperature bath was used to control the subphase temperature. The mixed spreading solutions were dispersed at the air-water interface, and then slowly compressed at speeds of about 5 $Å^2$ mol^{-1} min^{-1} to surface pressures of 10-15 mN/m prior to deposition. Monolayers were transferred onto electron microscope grids for transmission electron microscopy and electron diffraction, using both the horizontal and vertical dipping techniques. Multilayer assemblies were prepared onto platinum-coated substrates using the vertical dipping technique for Near Edge X-Ray Absorption Fine Structure Spectroscopy (NEXAFS).

Monolayers of a Fc-Py surfactant system and mixed monolayers with 3HDP of various ratios were prepared under similar conditions. Pressure-area isotherms were obtained for pure Fc-Py, 1:1, 1:2, 1:5, 1:10, and 1:20 ratios of Fc-Py to 3HDP. Monolayer and multilayer assemblies of the mixed systems were subsequently transferred onto platinum-coated substrates for NEXAFS and electrochemical studies. Electrochemistry (cyclic voltammetry) of these films was performed using a BAS CV-27 voltammograph. A conventional electrochemical cell was used, containing .2M tetraethylammonium tetrafluoroborate ($TEABF_4$) in acetonitrile, platinum or indium tin oxide (ITO) with a polymerized LB film as the working electrode, a clean platinum coated slide as the counter electrode, and saturated calomel as the reference. All solutions were deaerated by bubbling nitrogen through the solutions for at least 10 minutes, and then blanketing the solution during measurements.

RESULTS AND DISCUSSION

500:1 Pyrrole to 3HDP

3-hexadecyl pyrrole (3HDP) was selected as a surfactant polypyrrole analog, in an attempt to prepare monolayer films at the air-water interface [2], and subsequently polymerize the pyrrole moiety into a highly ordered, two-dimensional, ultrathin, polypyrrole conductive film. Figure 1 shows the pressure-area isotherm of 3HDP at 20°C on a pure water subphase. As indicated, 3HDP is a well-behaved system, with a relatively steep isotherm and area per molecule, at maximum surface pressure, of 22Å^2. These results suggest that the molecules pack optimally into a vertical orientation, as is typical of an ideal Langmuir-Blodgett system. In addition, the monolayer films are extremely stable, as they may be annealed at constant pressure for long periods of time, and allow facile deposition onto various substrates with good transfer ratios. This behavior demonstrates that the pyrrole ring of the 3HDP is sufficiently hydrophilic to provide deposition characteristics which are similar to those of a classic surfactive molecule.

It was previously reported [3], that introduction of $FeCl_3$ into the subphase of the 3HDP monolayer, did not lead to polymerization to a sufficient extent through the pyrrole group. Instead, a premature buckling of the film was observed, simultaneously with an area expansion, which was found to be dependent on the concentration of $FeCl_3$ in the subphase. In an attempt to explain this behavior, one must consider the lowest-energy conformation of the polypyrrole chain, which is a planar extended chain, where the adjacent pyrrole rings point in the opposite directions to each other [4]. If all the pyrrole rings in polypyrrole were to point in the same direction, a helical structure would be generated. In the 3HDP monolayer, with anticipated vertical arrangement of the molecules, the pyrrole rings are all expected to point in the same direction, and it is possible to argue why polymerization through the pyrrole ring, in such a monolayer structure, may not occur.

If pyrrole molecules could be inserted between adjacent 3HDP molecules, so as to permit alternate pyrrole groups to point in opposite directions, kinetic restriction may be alleviated. Unfortunately, pyrrole is extensively soluble in water, and a 1:1 ratio of pyrrole to 3HDP, when deposited at the air-water interface, does not ensure this preferred arrangement either. Therefore, a much higher ratio, 500:1, of pyrrole to 3HDP, spread on the subphase containing 0.1M$FeCl_3$, leads to extensive polymerization, and a conductive film is obtained [5]. Alternatively, electrochemical polymerization, in mixed multilayers of n-octadecane with surfactive pyrrole derivatives, has been carried out by Shimidzu et al. [6]. It is possible, that the pyrrole groups, from adjacent monolayers of a Y-type bilayer, in this case, provide the requisite configuration for an adequate polymerization condition.

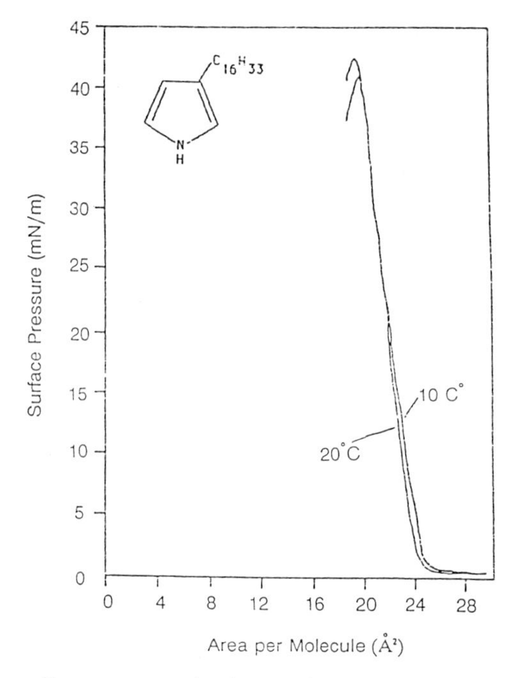

Figure 1. Pressure-area isotherm of 3HDP on a water subphase.

Monolayers of 500:1 pyrrole to 3HDP were transferred onto electron microscope grids for transmission electron microscopy and electron diffraction. Electron micrographs were taken, before and after washing with HCl to remove $FeCl_3$. The micrographs, from the monolayers before washing with HCl, indicate the possibility of inclusion of excess $FeCl_3$ from the subphase. A relatively thick texture also indicates the possibility of polymerization of the excess pyrrole in the subphase, once polymerization is initiated at the air-water interface. Thus, we propose that pyrrole monomers may organize themselves at the air-water interface onto the template of a 1:1 3HDP to pyrrole layer.

Electron diffraction from the monolayers was done, before and after washing with HCl. The "d" spacings of the sample, before and after washing, agree within 2-4 %

experimental error, thus suggesting that the diffraction pattern is, indeed, resultant of the polymer, and not of $FeCl_3$ inclusions. Further evidence was obtained by allowing a $FeCl_3$ aqueous solution to evaporate onto a carbon coated grid, upon which subsequent diffraction revealed completely different diffraction patterns. The diffraction patterns from 500:1 pyrrole to 3HDP were obtained, and the corresponding experimental "d" spacings are shown in Table 1. These "d" spacings do not conform to the hexagonal packing of the alkyl side chains, nor do they agree with those reported by Geiss et al. [7] and Buckley et al. [8] for electrochemically polymerized polypyrrole. This well-defined diffraction pattern is also not observed from a monolayer of 3HDP.

The "d" spacings for this polymer were also calculated, assuming the lattice to be monoclinic. These calculated values are listed in Table 1 for comparison, and are found to be in excellent agreement with the experimental values for a monoclinic lattice, with a = 3.28 Å, b = 7.0 Å, and $\gamma = 50^0$. These values are also close to those obtained by Orchard et al [9] for a possible crystalline lattice of polypyrrole.

TABLE 1

Experimental "d" spacings (Å)	Calculated "d" spacings (Å)
5.33	d_{010} = 5.36
2.57	d_{100} = 2.51
1.60	d_{220} = 1.60
1.35	d_{040} = 1.34
1.03	d_{330} = 1.06
0.90	d_{310} = 0.92
0.79	d_{440} = 0.80

We conclude that the diffraction pattern obtained is from the polypyrrole produced at the air-water interface. Further, that this monolayer is, in fact, dictated into a new crystalline organization, as growth and polymerization is initiated from a 1:1 3HDP to pyrrole two-dimensional template, in which the horizontal lattice is comprised of electroactive polypyrrole. Figure 2 illustrates a schematic of the proposed two-dimensional template polymerization. A continuation of this investigation next involved determining the effect of incorporation of a bulky ferrocene moiety into the monolayer organization.

Fc-Py and 3HDP Mixed Monolayer System

The ferrocene-derivatized pyrrole (Fc-Py) molecule consists of a pyrrole group at one end of a 12 carbon aliphatic chain, and a hydroxyl ferrocene group at the opposite end.

It was hoped that this molecule, which is structurally similar to the well-behaved 3HDP molecule, would also form high-quality monolayer films. However, this proved not to be the case. As shown in Figure 3a, a pressure-area isotherm of the Fc-Py molecule on a pure water subphase, a stable monolayer is not formed. The long plateau region and very low surface pressure are indication of a lack of vertical packing and regular conformation. This is most likely due to the fact that both ends of the molecule possess hydrophilic character, and result in a poorly ordered film.

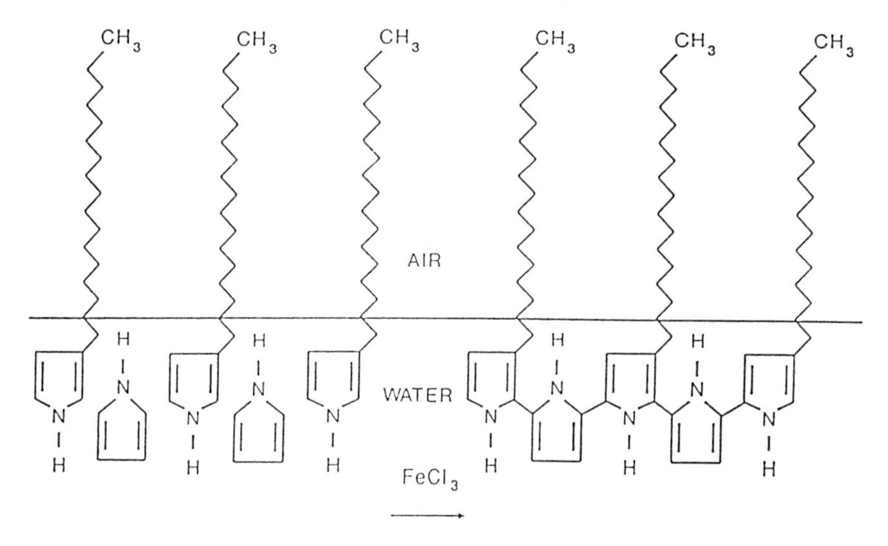

Figure 2. Schematic of the proposed two-dimensional template polymerization of 500:1 pyrrole to 3HDP at the air-water interface.

To optimize the vertical orientation of this molecule, and yet maintain the pyrrole moiety in the film, mixed monolayer systems were prepared, using the well-behaved 3HDP molecule. Solutions of Fc-Py and 3HDP, in ratios of 1:0, 1:1, 1:2, 1:5, 1:10, and 1:20 of Fc-Py to 3HDP, were prepared, and their monolayer formation was characterized via pressure-area isotherms. As shown in Figures 3b, 3c, and 3d (1:2, 1:5, 1:20 respectively), addition of 3HDP, to prepare a mixed system, has a pronounced effect on the behavior of monolayer formation. It was observed that, with increasing amounts of 3HDP to the mixture, multiple transition regions developed, and a corresponding decrease in length of the plateau region occurred. Also, a substantially steeper isotherm appears at smaller areas per molecule. These isotherms were found to be very reproducible, and indicate that the 3HDP is optimizing the vertical orientation of the Fc-Py.

For further structural characterization, mono- and multilayer films of the various mixtures were transferred onto hydrophobic and hydrophilic (oxidized) platinum substrates. It was determined that higher surface pressures, (about 20mN/m) and higher ratio mixtures of 3HDP resulted in the most stable monolayers, which could be transferred facilely onto the platinum substrates. Near Edge X-Ray Absorption Fine Structure (NEXAFS) techniques, using synchrotron radiation, were used, to study the ordering of these Langmuir-Blodgett films at the molecular level. Details of these measurements for the 500:1 (pyrrole to 3HDP) system have been reported elsewhere [10]. Figure 4 shows the NEXAFS spectra of a 4 layer film of 1:5, Fc-Py to 3HDP on a platinum substrate.

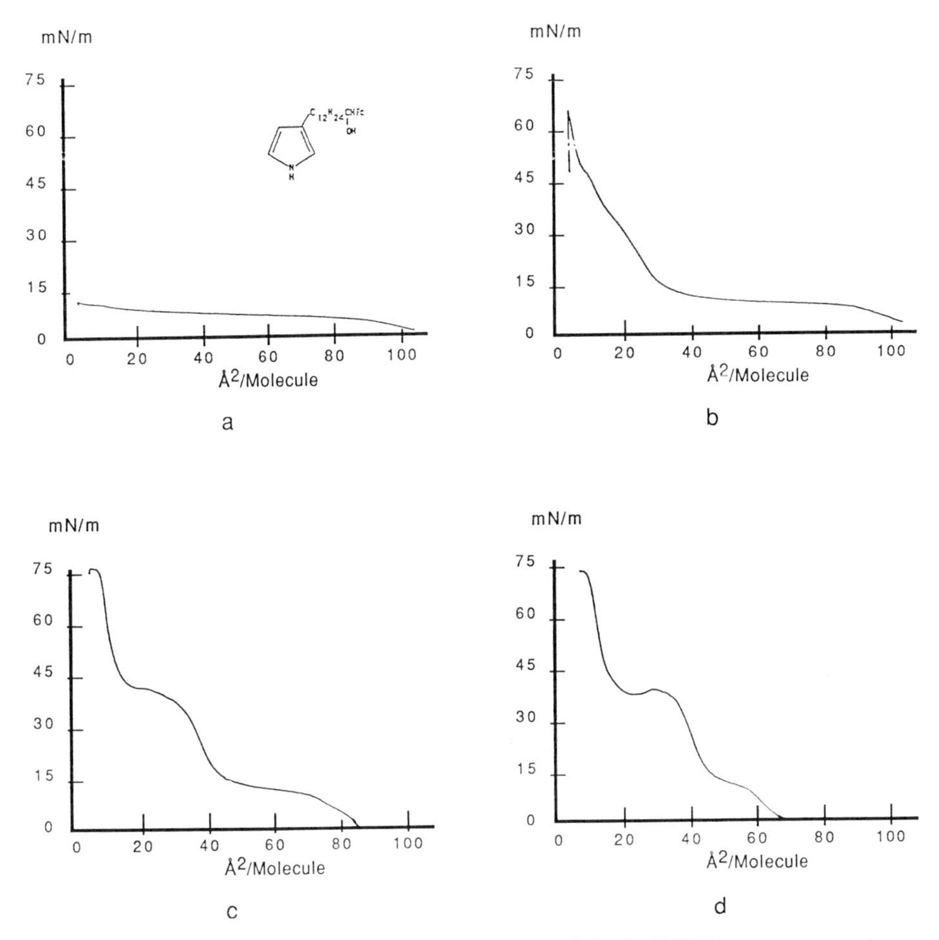

Figure 3. Pressure-area isotherms of mixtures of Fc-Py:3HDP on a pure water subphase (a) 1:0 (b) 1:2 (c) 1:5 (d) 1:20.

As observed, there is a marked dependence of the spectral features in the near-edge region on polarization. The carbon spectrum is predominately characteristic of the hydrocarbon chains in the multilayer, since they contain the majority of carbon atoms. Thus, the strong polarization dependence of the spectrum suggests that the chains are highly ordered. The sharp peak, observed at approximately 290 eV, corresponds to excitation to an empty orbital, associated with the C-H bonds on the chain. This orbital is perpendicular to the chain direction, and its associated feature is most pronounced with the incident electric field vector parallel to the substrate. The broad peak observed at approximately 295 eV, is assigned to the C-C bond which, on the average, is along the chain direction. This feature is most pronounced with the electric field vector normal to the substrate. Therefore, it may be concluded that the chains are highly ordered, and oriented at an angle which approaches the normal to the substrate.

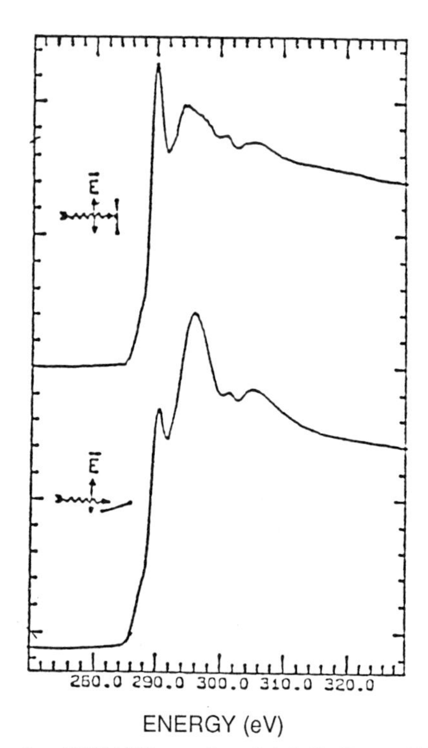

Figure 4. Carbon K-edge NEXAFS spectra of 1:5 Fc-Py to 3HDP 4 layer LB Film on platinum

Figure 5 shows a schematic of the proposed configuration of the two-dimensional polypyrrole lattice, with mixed alkyl and ferrocene derivitized alkyl units extended into the third dimension. It is not confirmed from our studies whether the ferrocene units are clustered, or are molecularly dispersed in the Py-Fc-3HDP system. The exact orientation of the Py-Fc molecule in the mixed monolayer system, at different phases of monolayer organization, is currently being investigated. However, it does appear that a unique structural hierarchy is maintained, when ferrocene moieties are incorporated into the films.

Biosensor Investigation

Ferrocene may be employed in biosensors, as a non-physiological redox couple, to shuttle electrons between the biological component and the electrode. The high reversibility, and low redox potentials, make it more desirable than the more common natural acceptors for the oxidases, such as oxygen, for facilitating electron transfer. Sensors based on the ferrocene/ferricinium redox couple have been previously reported [11,12,13].

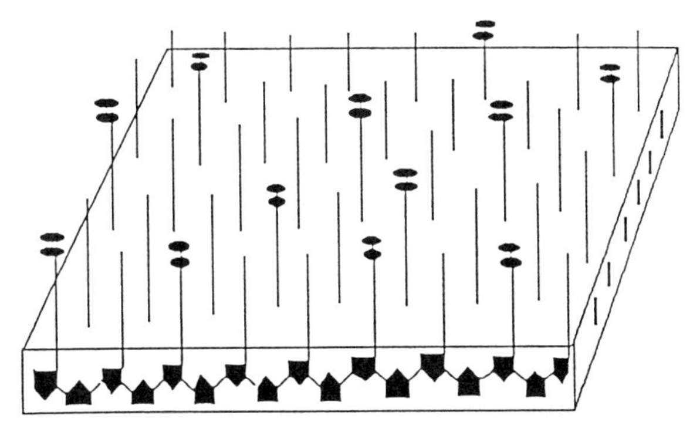

Figure 5. Schematic of the two-dimensional polypyrrole lattice with mixed alkyl and ferrocene derivitized unit extended into the third dimension.

This study describes initial electrochemical experiments, aimed at the development of an amperometric glucose biosensor, using ferrocene as the electron mediator, incorporated into an oriented polypyrrole LB film. A variety of monolayer samples were prepared for cyclic voltammetry (CV) studies to determine if the ferrocene was detectable electrochemically in such a configuration. This was an important criterion, as

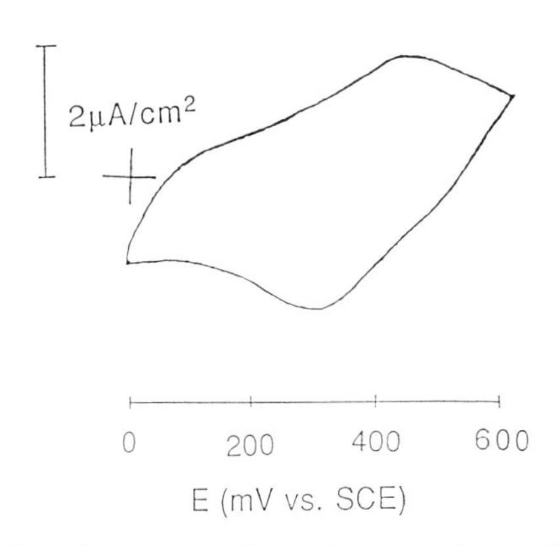

Figure 6. Cyclic voltammetry of a polymerized monolayer of 2500:5:1 (pyrrole to 3HDP to Fc-Py) on an ITO substrate

the ferrocene moieties must be accessible to the biological complexes for electron mediation to occur. Polymerized monolayer films of the mixed 1:5 system, with excess pyrrole monomer, were prepared on a ferric chloride subphase, to enhance the film ruggedness. The films were subsequently transferred to platinum and ITO solid supports for electrochemical studies. Cyclic voltammetry was carried out by scanning from 0 to 600 mV.

All samples which contained the ferrocene moiety showed the reversible behavior of the ferrocene/ferricinium redox couple (Figure 6). This is significant, as there is relatively little ferrocene present in these samples (approximately 1 in every 10 molecules for films which are only 1 molecule in thickness) and suggests that the ferrocene moiety may be accessible for electron mediation. The change in the cyclic voltammogram, upon addition of glucose oxidase and glucose, is a signature of electron mediation from the biological complex through the ferrocene groups to the electrode. Investigations of these signatures, as a function of monolayer organization, are currently underway.

CONCLUSION

The 3HDP/pyrrole mixed system has been shown to form large-area thin films in a structural organization, which is substantially different from bulk polymerized films. This organization is a direct result of confinement and controlled reaction in a two-dimensional template at the air-water interface. An extension of these studies has included a mixed ferrocene-derivatized pyrrole and 3HDP system, in an attempt to combine the electroactive polypyrrole planar lattice with the unique electroactive and magnetic properties of ferrocene. It has been determined, that this system is, in fact, able to self-assemble into a unique structural hierarchy. Ferrocene is clearly detected electrochemically in the polymerized monolayer films, and suggests that these groups may be accessible for electron mediation processes. The anticipated electroactive properties of these films, in conjunction with their anisotropy and complex architecture in multilayer lattices, are expected to yield rich dividends in molecular electronic device applications, and are currently under investigation.

REFERENCES

1 T. Inagaki, T.A. Skotheim, H.S. Lee, Y. Okamoto, L. Samuelson, and S. Tripathy, Synthetic Metals, 28, (1989), C245.

2 X.Q. Yang, J. Chen, P.D. Hale, T. Inagaki, T.A. Skotheim, Y. Okamoto, L. Samuelson, S. Tripathy, K. Hong, M.F. Rubner, and M.L. denBoer, Synthetic Metals, 28, (1989), C251.

3 A.K. Rahman, L. Samuelson, D. Minehan, S. Clough, S. Tripathy, T. Inagaki, W.Q. Yang, T.A. Skotheim, and Y. Okamoto, Synthetic Metals, 28, (1989), C237.

4 B.J. Orchard, B. Freidenreich and S.K. Tripathy, Polymer, 27, (1986), 1533.

5 K. Hong and M.F. Rubner, Thin Solid Films, 160, (1988), 187.

6 T. Shimidzu, T. Iyoda, M. Ando, A. Ohtani, T. Kaneko, K. Honda, Proc. of 3rd International Conf. on Langmuir-Blodgett Films, Gottingen, July 26-31, 1987.

7 R.H. Geiss, G.B. Street, W. Volksen and J. Economy, IBM J. Res. Develop., 27 (1983), 321.

8 L.J. Buckley, D.K. Roylance and G.E. Wnek, J. Polym. Sci. Polym. Chem. Ed., 25 (1987), 2179.

9 B.J. Orchard, B. Freidenreich and S.K. Tripathy, Polymer, 27, (1986), 1533.

10 T.A. Skotheim, X.Q. Yang, J. Chen, P.D. Hale, T. Inagaki, L. Samuelson, S. Tripathy, K. Hong, M.F. Rubner, M.L. denBoer, and Y. Okamoto, Synthetic Metals, 28, (1989), C229.

11 A.E.G. Cass, G. Davis, G.D. Francis, H.A.O. Hill, W.J. Aston, I.J. Higgins, E.V. Plotkin, L.D.L. Scott and A.P.F. Turner, Anal. Chem., 56,(1984), 667.

12 J.M. Dicks, W.J. Aston, G. Davis and A.P.F. Turner, Anal. Chim. Acta., 182, (1986), 103.

13 M.A. Lange and J.Q. Chambers, Anal. Chim. Acta., 175 ,(1985), 89.

MEASUREMENT TECHNIQUES IN PYROELECTRIC LANGMUIR-BLODGETT FILMS

M.W. Poulter, R. Colbrook and G.G. Roberts

Department of Engineering Science
University of Oxford
Parks Road
Oxford, OX1 3PJ
UK.

Abstract

The importance of the pyroelectric effect is considered, with particular reference to the application of thermal imaging. A range of pyroelectric materials is reviewed, and the suitability of Langmuir-Blodgett (LB) films to this application is highlighted.

Experimental techniques are discussed for the characterisation of potentially useful thin film materials, including measurement of pyroelectric coefficient and dielectric data (permittivity and dielectric loss). It is noted that, when considering a complete thermal imaging system, it is not sufficient to consider material parameters in isolation, and that the combined features of LB films render them particularly suitable to high system performance.

Introduction

The pyroelectric effect is well documented, and has been reported for a wide range of inorganic and organic materials [1]. One of the most promising applications of pyroelectricity is in the area of infra-red or thermal imaging. Thermal information from a scene is derived by measuring the pyroelectric currents from elements in an array onto which the scene has been projected. In order to optimise the performance of thermal imaging devices, it is necessary to be able to characterise pyroelectric materials, both in terms of their intrinsic properties, and their performance in model systems.

This work introduces experimental techniques for materials characterisation, the importance of such studies for the application of thermal imaging, and the suitability of Langmuir-Blodgett (LB) films in this application.

The Pyroelectric Effect

A pyroelectric material is one which possesses a temperature-dependent, spontaneous electric polarisation. If such a material experiences a change in temperature, a charge is developed across the material in response to the change in electric dipole, δP; if an electrical circuit is completed, then a short-circuit current results which is termed the pyroelectric current (figure 1). This can be expressed as:

Lower-Dimensional Systems and Molecular Electronics
Edited by R.M. Metzger *et al.*, Plenum Press, New York, 1991

$$I_p = pA\frac{dT}{dt} \qquad (1)$$

where dT/dt is the rate of change of temperature, A is the electrode area of the device and p is the pyroelectric coefficient. By considering a parallel plate capacitor structure with the active material as the dielectric, the temperature dependence of the induced voltage can be shown to be:

$$\frac{dV}{dT} = \frac{pd}{\varepsilon_0 \varepsilon_r} \qquad (2)$$

where d is the material thickness and ε_r the relative permittivity. Thus, when considering the voltage response of a thermal imaging element, the ratio p/ε_r is an important figure of merit.

Various workers have developed models relating the thermal and electrical properties of a pyroelectric element [2] yielding an expression for voltage responsivity (amplitude of signal per unit input power):

$$R_v \propto \frac{p\omega}{[(G^2 + \omega^2 H^2)(R^{-2} + \omega^2 C^2)]^{\frac{1}{2}}} \qquad (3)$$

where ω is the modulation frequency of incident radiation, H is the substrate heat capacity, G is the thermal conductance, and C and R are the electrical capacitance and resistance. Maximisation of the voltage responsivity is strongly dependent on the thermal characteristics of the element which, in turn, are strongly dependent on the material thickness. This thickness dependence gives rise to an optimum thickness for a particular material system.

Pyroelectric Materials

Ceramic and single crystal inorganic pyroelectrics exhibit very high values of pyroelectric coefficient. However, the disadvantages of such materials are correspondingly high values of ε_r and the processing problems associated with producing thin films approaching optimum thickness.

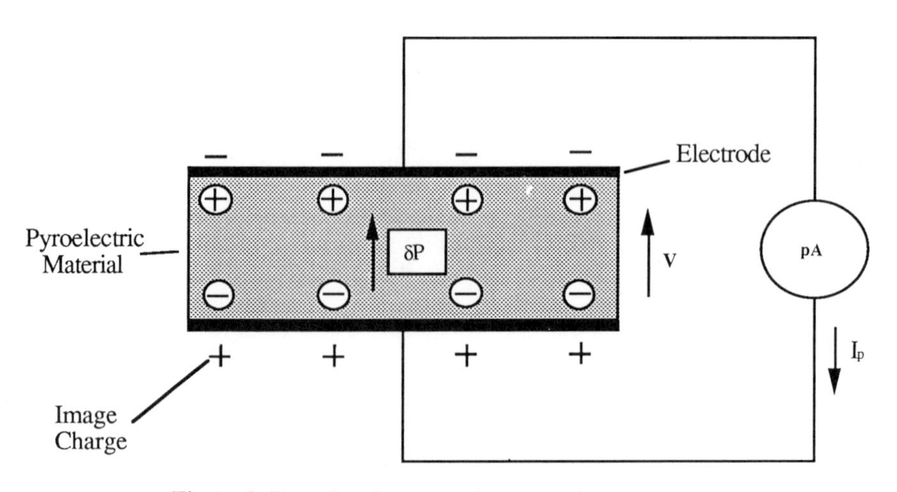

Figure 1. Pyroelectric current in a capacitor structure.

Spun-cast polymers, such as polyvinylidenefluoride and its copolymers, show good values of pyroelectric coefficient, and favourable values of the figure of merit p/ε_r. The major disadvantages of such materials are their large piezoelectric activity, and the necessity for electrical poling. In practical devices, piezoelectricity is a problem, because it can cause problems associated with microphony.

The phase change in ferroelectric liquid crystals can give rise to the pyroelectric effect [3]. Values of p and p/ε_r are also respectable. However, all the materials reported to date have a large dielectric loss. In addition, it is not possible to obtain optimum thickness structures, whilst still maintaining acceptable thermal mass and conductivity.

The Langmuir-Blodgett technique [4] allows the transfer of highly ordered organic monolayers from the surface of water onto a suitable substrate. Repeated deposition of monolayers makes it possible to build up highly ordered molecular superlattices. Considerable interest has been shown in a variety of LB film forming materials which have been found to be pyroelectric [5,6]. By judicious use of molecular engineering, molecules can be tailored to enhance their pyroelectric properties, thus offering great potential for improvement of system performance.

Although pyroelectric coefficients for LB films tend to be smaller than for most other materials, the values of p/ε_r obtained compare favourably. Also, in contrast, dielectric loss appears to be low over a wide frequency range. Due to the unique deposition method involving individual monolayers, it is possible to obtain films with the exact optimum thickness. To date, it has been found that LB films are not significantly piezoelectric.

Measurement of Pyroelectric Coefficient

Values of pyroelectric coefficient are calculated from equation (1), by measuring the pyroelectric current for a known controlled temperature profile. A computer-controlled system has been developed, to force the temperature of a sample under test, and log the corresponding pyroelectric current. This system is shown schematically in figure 2. Samples are mounted in an earthed vacuum chamber. Evacuation of the sample chamber removes a significant proportion of absorbed water, which is otherwise responsible for spurious, thermally induced currents. The temperature of the sample is controlled, in both heating and cooling, by a Peltier thermoelectric heat pump.

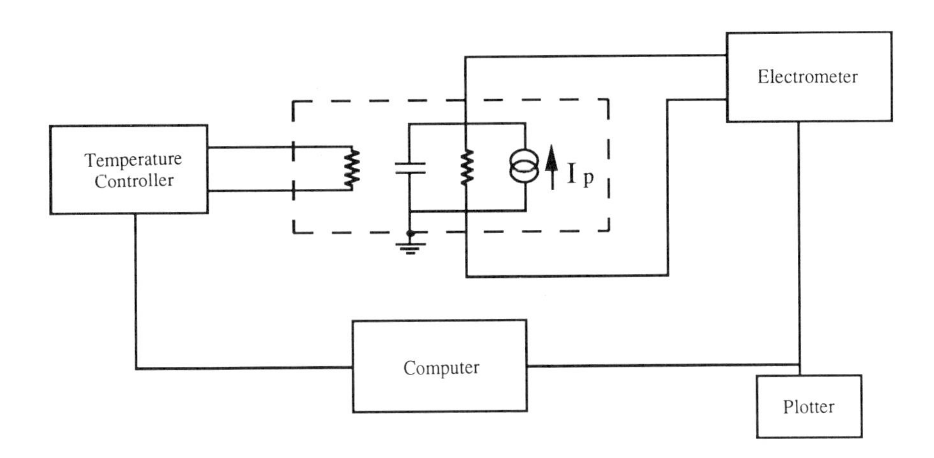

Figure 2. Schematic of pyroelectric test rig

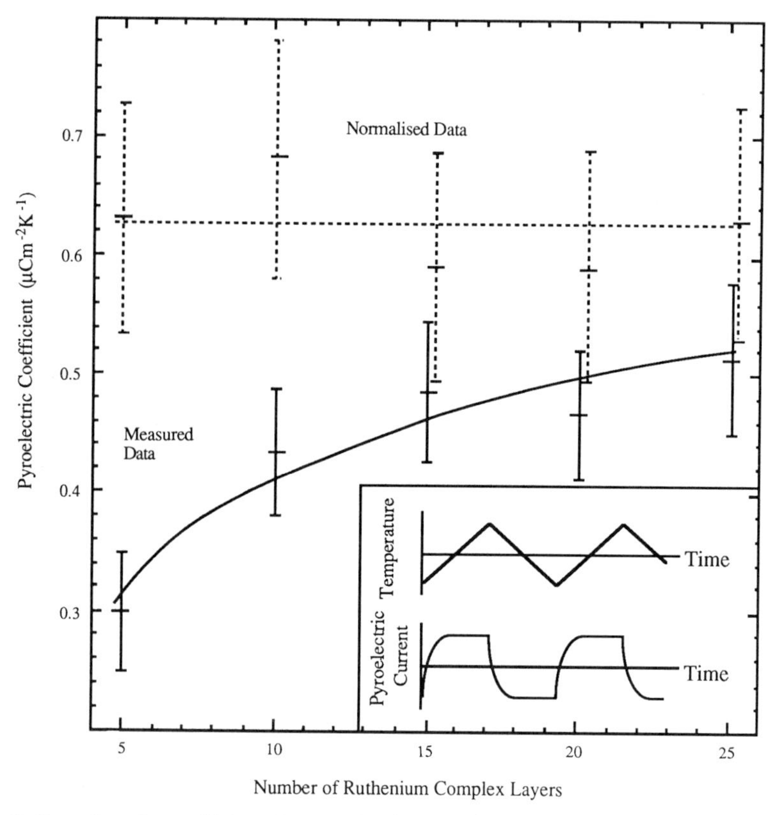

Figure 3. Pyroelectric coefficient for a ruthenium complex as a function of the number of active layers

Langmuir-Blodgett films are deposited onto glass microscope slides, onto which an aluminium back electrode has been evaporated. The pyroelectric LB films studied consist of alternate layers of two materials, which produce a superlattice with a net polarisation. Buffer layers of the fatty acid, behenic acid, are included, both before and after deposition of the active layers. This ensures that the active layers are free from any interactions with the aluminium back and top electrodes. The capping layers also provide protection during top electrode evaporation.

The results, for an alternate layer structure of a ruthenium complex and behenic acid, are shown in figure 3. The structure of the ruthenium complex is shown schematically in figure 4 [7]. When normalisation is performed, to take account of the dilution of dipole moment due to the buffer layers, it is found that the pyroelectric coefficient is independent of the number of active layers. This is the result to be expected, if the effect observed is true pyroelectricity, since the pyroelectric coefficient is dependent only on the number of dipoles per unit volume. Work is still in progress to attempt to isolate the exact mechanism responsible. These results compare favourably with the largest pyroelectric coefficients found in acid/amine LB films, typically $1.3\mu Cm^{-2}K^{-1}$ [7].

Measurement of Dielectric Parameters

The importance of low relative permittivity of pyroelectric materials to be used in thermal imaging devices has been stressed. However, the figure of merit p/ε_r only measures signal strength. Noise sources, in a practical system must also be considered, and, consequently, potential materials must also possess low dielectric loss. In cases where dielectric loss becomes a significant noise source for a pyroelectric element, a more applicable figure of merit is:

$$\frac{p}{\sqrt{\varepsilon_r \tan\delta}} \tag{4}$$

where tan δ is the dielectric loss tangent.

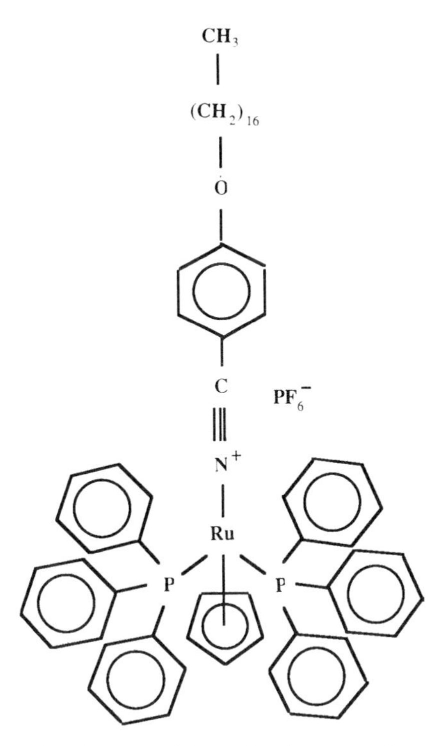

Figure 4. Schematic diagram of the ruthenium complex material studied.

A computer automated system, shown schematically in figure 5, has been developed to yield results for permittivity and dielectric loss over a range of temperatures. A real dielectric is considered to have an admittance, $Y=j\omega C^*$. In practice, the measured admittance is found to have a conductive as well as a capacitive component, $Y = G + j\omega C$. In order to account for this, the capacitance of the dielectric, C^*, is considered to be characterised by a complex relative permittivity, ε_r^*:

$$C^* = \frac{\varepsilon_0 \varepsilon_r^* A}{d} = \frac{\varepsilon_0 (\varepsilon_r' - j\varepsilon_r'')A}{d} \tag{5}$$

Using this model, the measured components of admittance (conductance G, and capacitance C) are $G = \omega\varepsilon_r'' C_0$ and $C = \varepsilon_r' C_0$, where C_0 is the capacitance of the structure if the dielectric were replaced by free-space. By further manipulation, it is found that tan δ is given by:

$$\tan\delta = \frac{\varepsilon_r''}{\varepsilon_r'} = \frac{G}{\omega C} \tag{6}$$

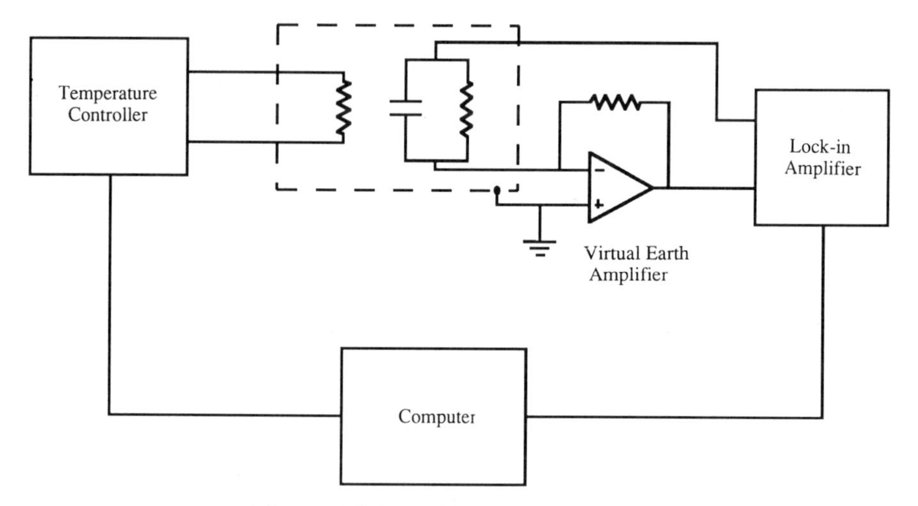

Figure 5. Schematic of dielectric test rig.

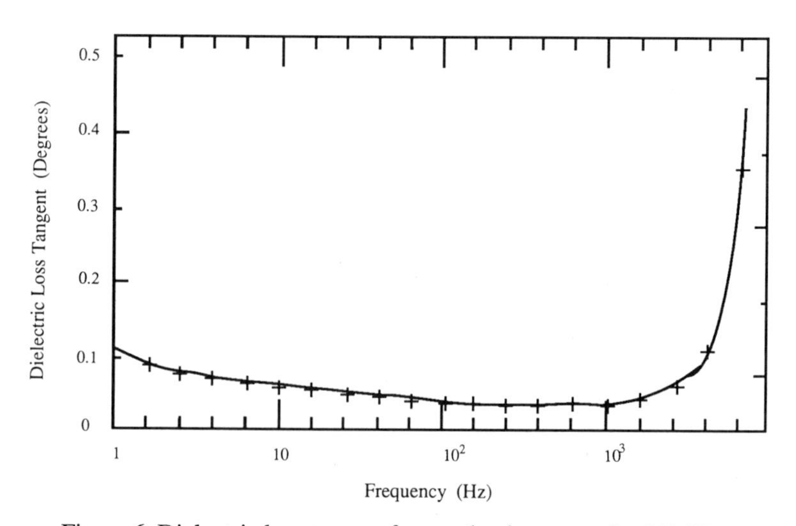

Figure 6. Dielectric loss tangent for a ruthenium complex LB film.

The dielectric loss tangent, as a function of frequency, is shown in figure 6 for the same ruthenium complex. For a practical device, the frequency range of interest lies between 1 and 100Hz. For this LB film material, an acceptably low figure is obtained over this range; at 25Hz, a typical frame scan rate, the dielectric loss is 0.05 degrees.

Summary

The development of pyroelectric thermal imaging devices is dependent upon the screening of suitable materials. The materials, which are most likely to find application, will have large values of pyroelectric coefficient, and small permittivity and dielectric loss. The ability to produce thin films of optimum thickness will also be significant.

Langmuir-Blodgett films show great potential, due to a combination of features including good figure of merit p/ε_r, low dielectric loss and low piezoelectric response, coupled with the capability of the LB technique to produce highly polar thin films of precisely controlled thickness. It is hoped that molecular engineering will be able to further develop LB film forming materials, with properties tuned to the needs of thermal imaging devices.

Acknowledgement

The authors are grateful for the support of Mr T. Richardson and the other members of the Oxford University Molecular Electronics group. This work has been sponsored by Thorn EMI plc.

References

[1] Whatmore R.W., Rep. Prog. Phys. 49 (1986) 1335.

[2] Putley E.H., Semiconductors and Semimetals, ed. Willardson and Beer, 5 (1970) 259.

[3] Glass A.M., Patel J.S., Goodby J.W., Olson D.H. & Geary J.W., J. Appl. Phys. 60 (1986) 2778.

[4] Roberts G.G., Adv. Phys. 34 (1985) 475.

[5] Christie P., Petty M.C. & Roberts G.G., Appl.Phys. Lett. 48 (1986) 1101.

[6] Colbrook R., Holcroft B., Roberts G.G., Polywka M.E.C. & Davies S.G., Proceedings of the 1st European Conference on Applications of Polar Dielectrics, Zurich, 1988, to be published in Ferroelectrics.

[7] Colbrook R. & Roberts G.G., 4th International Conference on Langmuir-Blodgett Films, April 1989, Tsukuba (Japan), To be published in Thin Solid Films.

EFFECT OF LIGHT INTENSITY AND TEMPERATURE ON THE PHOTOVOLTAIC PARAMETERS OF CHLOROPHYLL *b* LANGMUIR-BLODGETT FILMS

André Désormeaux and Roger M. Leblanc

Centre de recherche en photobiophysique, Université du Québec à Trois-Rivières, C.P. 500, Trois-Rivières (Québec) G9A 5H7, Canada

ABSTRACT

The photovoltaic characteristics of Al/44 monolayers of Chl *b*/Ag sandwich cells are reported. Open-circuit photovoltages as high as 1.1 V were observed for most photovoltaic cells. The variation of the short-circuit photocurrent with light intensity is shown to be strongly dependent upon the wavelength of illumination. The low fill factor values of the Chl *b* cells (0.15-0.19) are related to the large internal resistance of the devices, which is mainly caused by the insulating layer present at the aluminium/pigment junction. The short-circuit photocurrent is found to increase exponentially with temperature. An activation energy of 0.19 eV was determined at the maximum absorption in the red region.

INTRODUCTION

Organic photovoltaic devices have been intensively investigated with a view to produce efficient solar cells. Among the various organic pigments used, chlorophyll *a* molecules have been largely studied in order to characterize the primary steps of photosynthesis[1-4]. The Langmuir-Blodgett films[5] constitute an important model system for such studies since it produces a well-ordered environment. However, the photovoltaic cells made with chlorophyll pigments exhibit low power conversion efficiency (0.02-0.20%) compared to the efficient energy transduction mechanisms observed in higher plants.

Various aspects may explain the low power conversion efficiency of such organic devices. For example, some parameters which are intrinsic to the pigment, such as the molecular structure which may confer poor semiconducting properties to the pigment, or a bad morphological state of the pigment in the photovoltaic cells, so that the energy transduction mechanisms within the cell are ineffective. On the other hand, the low energy conversion efficiency can also be due to some characteristics extrinsic to the pigment, such as the choice of the metal electrodes, the inappropriate experimental conditions or/and the presence of an interfacial layer of high resistance at the metal/pigment junction. Recent studies on the influence of molecular structure of pigments on the photovoltaic properties of sandwich cells of chlorophyll *a*, *b* and of a zinc porphyrin derivative suggest that the poor photovoltaic characteristics

Lower-Dimensional Systems and Molecular Electronics
Edited by R.M. Metzger *et al.*, Plenum Press, New York, 1991

of these cells may be due to the large internal resistance of the devices (30 to over 1000 MΩ cm^2)[6]. These high values of resistance were mainly attributed to an insulating layer of aluminium oxide and of a single monolayer of fatty acid present at the rectifying junction.

The present report is concerned with the photovoltaic properties of sandwich cells of Langmuir-Blodgett films of chlorophyll b (Chl b). The molecules of Chl b are known to be important accessory pigments for the transfer of light energy to the reaction centers in green plants[7]. The objective of this study is to examine the influence of light intensity and of temperature on the photovoltaic parameters of these organic devices.

EXPERIMENTAL

Chl b was purchased from Sigma (St. Louis, MO) and was used without further purification. The photovoltaic cells were prepared by interposing the Langmuir-Blodgett films of Chl b between two dissimilar metal electrodes, i.e. aluminium and silver, with work functions (ϕ), such that $\phi_{Al} < \phi_{pigment} < \phi_{Ag}$. Both metal electrodes were evaporated under vacuum according to the method previously described[3]. The active area of the cell is 0.45 cm^2. Before the deposition of Chl b monolayer, the top aluminium electrode was covered with one monolayer of cadmium arachidate transferred at a surface pressure of 30 mN m^{-1}. The sandwich cells contained 44 monolayers of Chl b transferred on top of the arachidate layer at a surface pressure of 20 mN m^{-1}.

The photovoltaic cells were illuminated in a Faraday cage with monochromatic light ranging from 0.1 to 24.0 μW cm^{-2}. The temperature of the sample (248-318 K) was controlled by using a thermostated bath (Neslab Instruments). Currents and voltages developed by the cells were measured with a Keithley 616 electrometer.

RESULTS AND DISCUSSION

Fig. 1 shows typical light intensity dependence of the open-circuit photovoltage (V_{oc}) and of the short-circuit photocurrent (I_{sc}) for the illumination of a Chl b cell through the aluminium electrode at the maximum absorption in the red region. One can observe that the initial large increase of V_{oc} at low incident light intensity is followed by a slight variation at higher light intensity. Maximum V_{oc} values as high as 1.1 V were generally observed for most photovoltaic cells. However, large variations in the dark open-circuit voltage (0.015-0.580 V) have been observed from one cell to another. As pointed out before, our cell devices possess an insulating layer of aluminium oxide and of a monolayer of cadmium arachidate. This insulating layer can strongly alter the electrical and photovoltaic parameters of the overall system. This interfacial layer tends to increase the potential barrier of the cells to produce a higher open-circuit voltage. It was noticed that the presence of a single insulating monolayer of stearic acid increases the open-circuit photovoltage developed by the n^+-GaP/Au diodes from 400 to 900 mV without significantly altering the short-circuit photocurrent[8]. Consequently, the variations in the open-circuit voltages observed in our cells can be related to the quality of the rectification properties of the aluminium/pigment contact.

The short-circuit photocurrent shows a much stronger light intensity dependence. The values of photocurrent follow the relationship $I_{sc} \propto I_{inc}^{\gamma}$, where γ, the light exponent, ranges from 0.74 to 0.82 for all

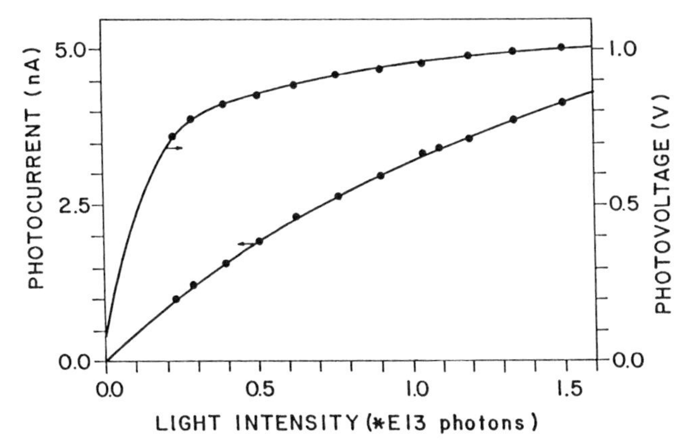

Fig. 1 Dependence of the short-circuit photocurrent and of the open-circuit photovoltage on light intensity for a sandwich cell of Chl b illuminated at 656 nm.

photovoltaic cells at the considered wavelength. For most organic devices, values of light exponent from 0.3 to 1.0 are generally observed[9-11]. Such a power-law dependence is often attributed to some exponentially distributed traps between the valence and conduction bands[12]. However, the interfacial layer at the aluminium/pigment junction can also strongly affect the light intensity dependence of the short-circuit photocurrent. Much lower light exponents have been observed in photovoltaic cells of phthalocyanines using aluminium contacts[13]. This characteristic has been analyzed in terms of coupled diffusion and second-order decay of the photogenerated carriers produced near the rectifying junction.

The light exponent values of the sandwich cells of Chl b depend of the wavelength of illumination. To illustrate this effect, Fig. 2 shows the dependence of the short-circuit photocurrent on the incident light intensity at four different wavelengths. All photovoltaic cells invariably exhibit lower light exponent values with decreasing energy

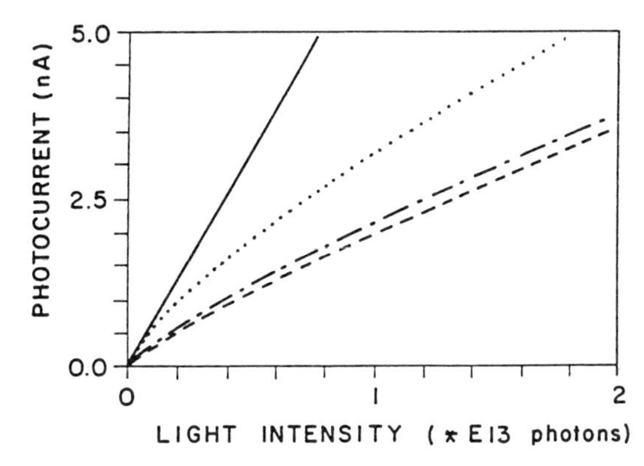

Fig. 2. Light intensity dependence of the short-circuit photocurrent for a sandwich cell of Chl b illuminated at 468 nm (——), 560 nm (---), 610 nm (-•-) and 656 nm (•••).

of wavelength of illumination. Average light exponent values of 1.00, 0.85, 0.83 and 0.77 have been observed for the 468, 560, 610 and 656 nm, respectively. However, it was demonstrated that the low pigment thickness used in our photovoltaic cells (≃ 616 Å) allows strong multiple internal reflections of light between the two semitransparent metal electrodes which tend to increase the values of the short-circuit photocurrent[6]. Furthermore, the number of light reflections within the cell strongly depends upon the optical absorption coefficient of the pigment. Since the light intensity dependence of the photocurrent presented in Fig. 2 takes into consideration the incident number of photons reaching the pigment, the values of photocurrent are more or less overestimated according to the optical absorption coefficient of the Chl b multilayer at each wavelength.

Fig. 3 shows typical photocurrent-photovoltage characteristics of a sandwich cell of Chl b illuminated through the aluminium electrode at various wavelengths. The fill factor values, defined as the ratio of the maximum power output to the product of the open-circuit photovoltage and the short-circuit photocurrent, vary between 0.15 and 0.19 for all photovoltaic cells at all wavelengths. These values are very low compared to some inorganic solar cells. The low fill factor values for the Chl b cells are mainly attributed to the high internal resistance of our cell devices (30 to over 1000 MΩ cm^2). It has recently been shown, from capacitance measurements of unpigmented cells, that the high resistance of our photovoltaic cells is mainly caused by the insulating layer present at the aluminium/pigment contact[14]. A maximum power output from 0.3 to 1.1 nW cm^{-2} was developed by the Chl b cells when the pigment was illuminated at 656 nm with a light intensity ranging from 4.8 to 10.0 μW cm^{-2}. A corresponding power conversion efficiency ranging from 0.003 to 0.030% was observed for all photovoltaic cells at this light intensity level. Furthermore, the power conversion efficiency of the cells decreased with increasing light intensity. This is attributed to the sublinear light intensity dependence of the short-circuit photocurrent.

We have next studied the effect of temperature on the photovoltaic properties of the sandwich cells of Chl b in the temperature range of 248-318 K. Fig. 4 shows the typical temperature dependence of the short-circuit photocurrent of a Chl b cell illuminated at 656 nm with a light

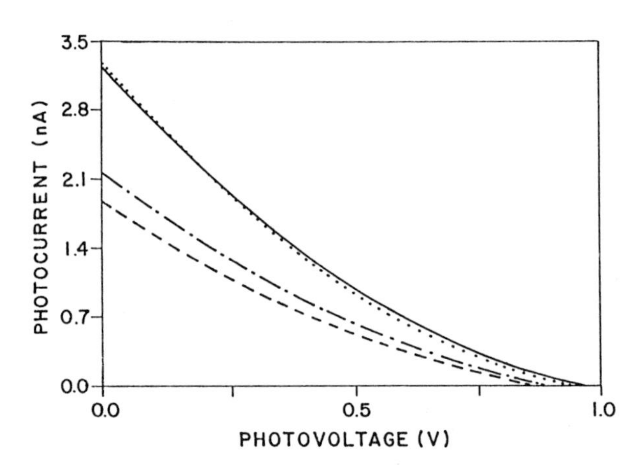

Fig. 3. Photocurrent-photovoltage characteristics of a sandwich cell of Chl b illuminated at 468 nm (——), 560 nm (---), 610 nm (-•-) and 656 nm (•••).

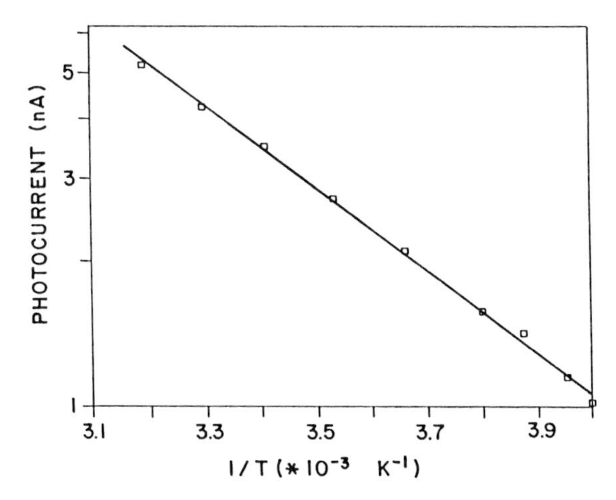

Fig. 4. Semilogarithmic plot of the short-circuit photocurrent vs 1/T for a sandwich cell of Chl b illuminated at 656 nm (light intensity = 6.7 μW cm^{-2}).

intensity of 6.7 μW cm^{-2}. The short-circuit photocurrent is found to increase exponentially with temperature. The activation energy, calculated from the slope of the Arrhenius plot, was found to be 0.19 eV. This value of activation energy is the same as observed for in-plane conduction of copper tetra-tert-butyl phthalocyanine Langmuir-Blodgett films[15]. Furthermore, this activation energy value agrees very well with the value of 0.14 eV observed for photoconduction induced by red photons in a SnO_2/CdS/x-H_2Pc/Au cell[16]. Table 1 summarizes the effect of temperature on various photovoltaic parameters of the Chl b devices. One can notice a large increase of the dark current and dark voltage by increasing the temperature. The origin of these dark characteristics is often related to an electrochemical reaction involving the interfacial oxide layer at the aluminium/pigment junction[2,17]. Furthermore, both short-circuit photocurrent and open-circuit photovoltage increase with temperature resulting in a large increase in the power conversion effi-

Table 1. Effect of temperature on the photovoltaic parameters of an Al/44 monolayers of Chl b/Ag cell illuminated at 656 nm (light intensity = 11.5 μW cm^{-2}).

	248 K	293 K	318 K
Dark current (nA cm^{-2})	0.015	0.120	0.600
Dark voltage (V)	0.02	0.15	0.35
Short-circuit photocurrent (nA cm^{-2})	2.2	12.6	16.1
Open-circuit photovoltage (V)	0.49	0.86	1.06
Internal resistance (MΩ cm^{-2})	200.0	57.0	54.5
Cell power (nW cm^{-2})	0.22	2.18	3.00
Power conversion efficiency (%)	0.002	0.019	0.026
Light exponent	0.61	0.71	0.87

ciency of the sandwich cells of Chl b. Moreover, the internal resistance of the photovoltaic cells tends to decrease with increasing the temperature. This can be explained by the fact that the increase of the short-circuit photocurrent at higher temperature tends to reduce the resistance offered by the insulating layer at the aluminium/pigment contact. Finally, large variations are observed in the light exponent according to the temperature of the photovoltaic devices. This characteristic can be related to some variations in the electrical properties of the interfacial layer with temperature.

REFERENCES

1. C. W. Tang, and A. C. Albrecht, Photovoltaic effects of metal-chlorophyll a-metal sandwich cells, J. Chem. Phys. 62:2139 (1975).
2. R. Jones, R. H. Tredgold, and J. E. O'Mullane, Photoconductivity and photovoltaic effects in Langmuir-Blodgett films of chlorophyll a, Photochem. Photobiol. 32:223 (1980).
3. M. F. Lawrence, J.-P. Dodelet, and L. H. Dao, Improvement of Al/Al_2O_3/multilayer array of chlorophyll a/Ag photovoltaic cell characteristics by treatment of the aluminium electrode, J. Phys. Chem. 88:950 (1984).
4. A. Diarra, S. Hotchandani, J.-J. Max, and R. M. Leblanc, Photovoltaic properties of mixed monolayers of chlorophyll a and carotenoid canthaxanthin, J. Chem. Soc., Faraday Trans 2 82:2217 (1986).
5. G. L. Gaines, Jr. "Insoluble monolayers at liquid-gas interfaces", Interscience, New York (1966).
6. A. Désormeaux, J.-J. Max, and R. M. Leblanc, Photovoltaic properties of Al/Langmuir-Blodgett films/Ag sandwich cells incorporating either chlorophyll a, chlorophyll b or zinc porphyrin derivative, J. Phys. Chem. submitted (1989).
7. Govindjee, "Bioenergetics of photosynthesis", Academic Press, New York (1975).
8. R. H. Tredgold, and R. Jones, Schottky-barrier diodes incorporating Langmuir-film interfacial monolayers, IEE Proc. 128:202 (1981).
9. J.-P. Dodelet, H.-P. Pommier, and M. Ringuet, Characteristics and behavior of electrodeposited surfactant phthalocyanine photovoltaic cells, J. Appl. Phys. 53:4270 (1982).
10. J. A. Bardwell, and J. R. Bolton, Monolayer studies of 5-(4-carboxyphenyl)-10,15,20-tritolyl-porphyrin. II. Photovoltaic study of multilayer sandwich cells, Photochem. Photobiol. 40:319(1984).
11. H. Meier, W. Albrecht, D. Wöhrle, and A. Jahn, Correlation of chemical structure to photoconductivity: octacyano- and octamethoxy-substituted zinc phthalocyanine, J. Phys. Chem. 90:6349 (1986).
12. H. Meier, "Organic semiconductors, Monographs in modern chemistry", H. F. Ebel (Ed.), Verlag-Chemie, Weinheim (1974).
13. F.-R. Fan, and L. R. Faulkner, Photovoltaic effects of metal free and zinc phthalocyanines. II. Properties of illuminated thin-film cells, J. Chem. Phys. 69:3341 (1978).
14. J.-J. Max, S. Hotchandani, and R. M. Leblanc, Influence of the insulating layer on the properties of Al/Chl a/Ag sandwich cells. II. Capacitance characteristics, J. Appl. Phys. submitted (1989).
15. R. A. Hann, S. K. Gupta, J. R. Fryer and B. L. Eyres, Electrical and structural studies on copper tetra-tert-butyl phthalocyanine Langmuir-Blodgett films, Thin Solid Films 134:35 (1985).
16. R. O. Loutfy, Photoconduction of phthalocyanines in the near-infrared, J. Phys. Chem. 86:3302 (1982).
17. R. O. Loutfy, and J. H. Sharp, Photovoltaic properties of metal-free phthalocyanines. I. Al/H_2Pc Schottky barrier solar cells, J. Chem. Phys. 71:1211 (1979).

PHOTONICS AND NONLINEAR OPTICS - MATERIALS AND DEVICES

Paras N. Prasad

Photonics Research Laboratory
Department of Chemistry
State University of New York at Buffalo
Buffalo, New York 14214

PHOTONICS AND NONLINEAR OPTICS

(a) The Technology of Photonics

The area of photonics is a new frontier of science and technology, which is capturing the imagination of scientists and engineers world wide, because of its potential applications to many areas of present and future technologies. Photonics is the analog of electronics, where photons, instead of electrons, are used to carry information. Applications range from fiber-optics communication and link, to optical computing, and other optical processing of information and images[1,2]. In addition, the technology of photonics can also be used for new applications not provided by electronics. One example is sensor protection (or even eye protection) against laser threats. To emphasize the importance of photonics, other terms such as light-wave technology and optical circuitry have also been used.

Nonlinear optical processes provide important functions of frequency conversion (like frequency-doubling, to increase the density of data storage), light control by electric field, and even all-optical processing such as light control by light. Utilizing these manifestations of optical nonlinearities, one can build devices, such as frequency converters, light modulators, all-optical switches, optical logic, optical memory storage, and optical limiters. The technology of photonics, with devices using nonlinear optical processes, offer many advantages; among them the most important ones are: (i) gain in speed of signal processing, (ii) three-dimensional connectivity (optical neural network) to produce smaller integrated optical chips, (iii) no electrical and magnetic interference, and (iv) compatibility with fiber-optics link.

(b) Nonlinear Optical Effects

Nonlinear optical processes occur when a medium is subject to an intense light pulse, such as that from a high peak power laser. The strong oscillating electric field of the laser creates a nonlinear polarization in the medium, yielding nonlinear response.

Lower-Dimensional Systems and Molecular Electronics
Edited by R.M. Metzger *et al.*, Plenum Press, New York, 1991

Nonlinear optical processes can, conveniently, be viewed as dielectric phenomena. For a linear dielectric, application of an electric field polarizes the medium, to produce a polarization P, which is linearly proportional to the applied field, the proportionality constant $\chi^{(1)}$ being the linear susceptibility. The response is also described by a dielectric constant ε. If the electric field is due to an optical field, then the response is also described by a refractive index n. For a plane-wave propagation

$$E = E_{\omega}(z,t) = E_o \cos(kz-\omega t) \text{ or } \frac{1}{2} [E_o e^{i(\omega t-kz)}+c.c] \quad (1)$$

The refractive index n, the wave vector $k = \frac{n\omega}{c}$, and the phase velocity $v = \frac{c}{n}$ are all independent of the field strength E.

For a nonlinear dielectric one has the polarization[3]

$$P = \chi^{(1)} \cdot E + \chi^{(2)} : E\,E + \chi^{(3)} \vdots E\,E\,E + \ldots$$

$$= \chi_{eff} \cdot E \quad (2)$$

χ_{eff}, the effective susceptibility, now depends on the field strength E. Therefore n, k and v are all dependent on E at optical frequencies. $\chi^{(2)}$ and $\chi^{(3)}$ are the second and third-order nonlinear optical susceptibilities. Two important manifestations of optical nonlinearities are harmonic generation and refractive index modulation by electric and optical fields. Their origin can conveniently be explained by considering a plane wave propagation through the nonlinear medium. The polarization is then given by

$$P = \chi^{(1)} E_o \cos\alpha + \chi^{(2)} E_o^2 \cos^2\alpha + \chi^{(3)} E_o^3 \cos^3\alpha$$

$$= \chi^{(1)} E_o \cos\alpha + \frac{1}{2}\chi^{(2)} E_o^2 (\cos 2\alpha + 1) + \chi^{(3)} E_o^3 [\tfrac{3}{4}\cos\alpha + \tfrac{1}{4}\cos 3\alpha] \quad (3)$$

where $\alpha = (kz-\omega t)$. Equation 3 shows that, due to nonlinear optical effects, higher-frequency components (2α and 3α terms) are generated, which describe higher harmonic generation; examples being second-harmonic generation, due to $\chi^{(2)}$, and third-harmonic generation, due to $\chi^{(3)}$. In addition, $\chi^{(3)}$ leads to a term with $\cos\alpha$, which describes the intensity dependence of refractive index. The dependence of refractive index on the electric field actually consists of two terms: (i) one derived from $\chi^{(2)}$, which is linearly dependent on E, and describes the electro-optic effect, also known as Pockels effect, in which the application of an electric field modulates the refractive index and, (ii) that derived from $\chi^{(3)}$, which is quadratically dependent on E, and hence linearly dependent on I; it describes the optical Kerr effect. It is the latter which provides a mechanism for light control by light, because an intense beam can be used to change the refractive index of the medium, and influence either its own propagation, or propagation of another beam of different or same frequency.

The intensity-dependence of refractive index can also be used to generate a phase conjugate signal, in which the phase of a carrier wave is reversed to correct for any phase distortions.

NONLINEAR OPTICAL MATERIALS

(i) General Classification

Nonlinear optical materials can be classified in two different categories: (i) molecular materials which consist of chemically bonded molecular units interacting in the bulk through weak van der Waals interactions. For these classes of materials, the optical nonlinearity is primarily derived from the molecular structure, and one can define microscopic nonlinear coefficients β and γ which are the molecular equivalents of the bulk susceptibilities $\chi^{(2)}$ and $\chi^{(3)}$. Examples of molecular materials are organic crystals and polymers[4]. (ii) Covalent and ionic bulk materials where the optical nonlinearity is a bulk effect. The examples of this class of materials are most inorganic systems, multiple quantum well semiconductors, and inorganic photorefractive crystals[5]. The molecular materials are emerging as an important class of materials, because they provide flexibility to tailor their molecular structure, and use molecular engineering to maximize the optical nonlinearity[4].

(ii) Second-Order Molecular Materials

Although our theoretical understanding of relationship between the structure and the microscopic optical nonlinearity is rather limited, some features of structural requirements for nonlinear optical molecular materials have been identified[4]. Since $\chi^{(2)}$ and the corresponding microscopic coefficient β are third-rank tensors, in order for them to be nonvanishing, the molecular structure must be noncentrosymmetric. An asymmetric charge distribution in a molecular structure

D—A

is found to enhance β. In the above structure, D and A represent the electron donating and accepting groups, respectively, which are separated by a π structure (aromatic ring in the above example). Even with a nonzero β structure, one needs to have a noncentrosymmetric bulk structure to achieve $\chi^{(2)} \neq 0$. To recognize inadequacies in our understanding of optical nonlinearities, it is to be noted that we are investigating several systems in our laboratory, which do not conform to the above structure, but exhibit large $\chi^{(2)}$. Examples are CHI_3: sulfur, SbI_3: sulfur and CHI_3: hexamethylenetetramine complexes, which do not have π-electrons, or typical electron donor and acceptor groups.

The $\chi^{(2)}$ organic systems do not necessarily require polymeric structures. Consequently, an important group of organic materials for $\chi^{(2)}$ processes are organic crystals of relatively small molecules. It should be pointed out that one does not need crystallinity in order to have non-vanishing $\chi^{(2)}$. Simply, a noncentrosymmetric ordering of dipoles in an otherwise noncrystalline medium (such as a glass, an amorphous polymer or a Langmuir-Blodgett film) is sufficient to produce non-vanishing $\chi^{(2)}$, provided $\beta \neq 0$. In Langmuir-Blodgett multilayer films such a noncentrosymmetric ordering can be generated by so called X or Z type depositions, or by using an alternate layer structure. In a glassy medium or a polymer, a noncentrosymmetric ordering of dipoles can be created by electric field poling of the medium in a more fluid state. Among the polymeric structures, side-chain liquid crystalline polymers are becoming recognized as an important group of $\chi^{(2)}$ materials, where a nonlinear optically active group (with $\beta \neq 0$) is attached to the polymer backbone through flexible spacers[4]. Such polymers are easier to pole, and also offer considerable flexibility for molecular engineering. In addition, we have found that some liquid crystalline polymers can also be deposited as Langmuir-Blodgett films, which may be useful for some device structures[6].

(iii) Third-order Materials

Structural requirements for $\chi^{(3)}$ materials are different from that for $\chi^{(2)}$ materials[4]. Organic structures with extensive π conjugation exhibit large γ (molecular analog of $\chi^{(3)}$). These structures do not have to be noncentrosymmetric, because γ is a fourth-rank tensor. Conjugated polymers provide a molecular frame for extensive conjugation and have, therefore, emerged as the dominant group of $\chi^{(3)}$ organic materials[4]. Examples of conjugated polymers are polydiacetylene, poly-p-phenylenevinylene and polythiophene. Our work has revealed that all conjugated polymers do not show the same effective delocalization in relation to optical nonlinearity. Therefore, in addition to being a conjugated polymeric structure, it must provide effective delocalization. In a sequentially built structure, we have found that the π-delocalization effect on γ is more effective for the thiophene oligomers, than it is for the benzene oligomers[7]. The largest component of the γ-tensor is along the conjugation direction. Therefore, even though no order in the bulk is required for $\chi^{(3)} \neq 0$, a unioriented bulk, in which all conjugated polymeric chains align in the same direction, will have a large $\chi^{(3)}$ value, compared to that in a truly amorphous form of the same polymer. Our study of $\chi^{(3)}$ measurements in stretch-oriented polymers, as discussed below, confirms this prediction. Finally, to increase the number density, these polymeric chains must pack closely so that $\chi^{(3)}$ is maximized.

From the above discussion, it is apparent that amorphous polymeric or glassy structures are more suitable as media for third-order nonlinear optical processes. Interest recently has shifted considerably to the third-order processes because they provide mechanisms for all-optical switching and all-optical signal processing.

It is to be noted that a structural requirement for conducting polymers is also extensive π-conjugation. Therefore, conducting polymers are also important as nonlinear optical polymers. Table I compares some of the similarities and differences for conducting and nonlinear optical polymers.

Our research program focuses heavily on the study of third-order nonlinear optical processes. We measure the third-order optical nonlinearity by a variety of methods. The method most frequently used in our laboratory is that of time-resolved degenerate four-wave mixing (DFWM).[8,9] We have used subpicosecond pulses to determine both the magnitude and the time-response of the nonlinearity. The latter is important in the case of resonant processes where excited-state dynamics plays an important role. Table II lists the $\chi^{(3)}$ values for some of the conjugated organic structures investigated in our laboratory. In the case of poly-p-phenylenevinylene and its 2,5-dimethoxy derivatives, we have measured the $\chi^{(3)}$ anisotropy in various uniaxially-stretched films.[10,11] The polar plot of the $\chi^{(3)}$ value conforms to its fourth-rank tensor behavior.[10]

NONLINEAR OPTICAL DEVICES

Examples of second-order devices being currently pursued are second harmonic generators, electro-optic spatial light modulators, and electro-optic switches. However, the biggest pay-off is for $\chi^{(3)}$ devices which can be used for all optical communications and signal processing. Organic polymeric structures provide the largest non-resonant $\chi^{(3)}$ and, therefore, are the best candidates for guided-wave structures using fibers and channeled waveguides because these structures would require optical transparency for long propagation distances. Furthermore, organic polymers

TABLE I

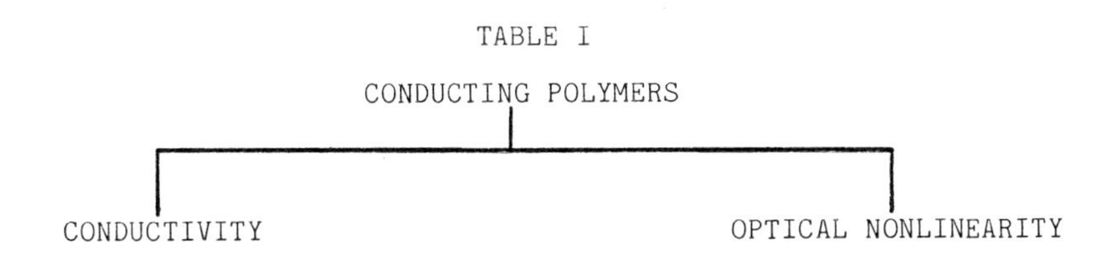

SIMILARITIES

Extended conjugation and effective π-electron delocalization

Large anisotropy - the largest value along the direction of conjugation

In an oriented polymer - largest value along the draw (chain orientation) direction

Novel excitations, such as solitons (in polymers with degenerate ground state) and bipolarons (with non-degenerate ground state) play important role

DIFFERENCES

CONDUCTIVITY	OPTICAL NONLINEARITY
Bulk effect	Primarily microscopic
Doping increases conductivity by many orders of magnitude	Doping with electron acceptors decreases $\chi^{(3)}$ drastically
Carrier transport important	Carrier transport not important
Important temperature effect - phonon interactions	Not as important, at least for non-resonant processes

DEVICE MATERIAL REQUIREMENTS ARE DIFFERENT

TABLE II

$\chi^{(3)}$ VALUES OF VARIOUS POLYMERS

Systems	Repeat Structures	$\chi_{1111}^{(3)}$ (in esu) at λ = 603 nm
LARC-TPI	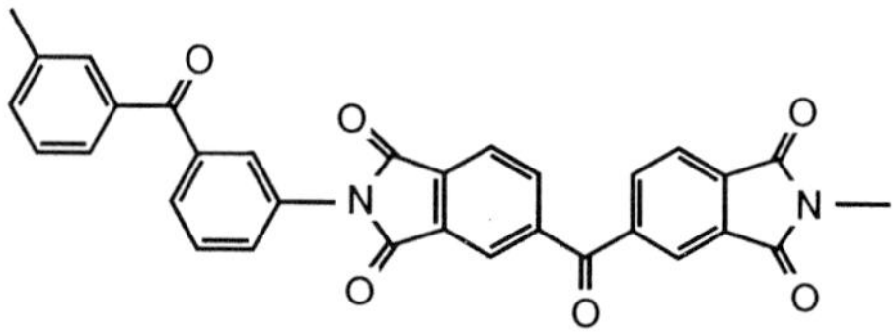	~ 2 x 10^{-12}
PBT	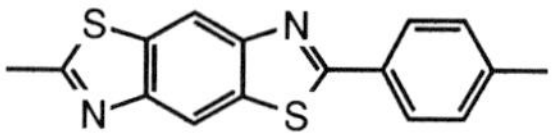	~ 10^{-11}
PBO	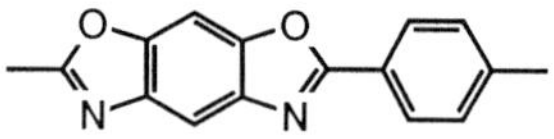	~ 10^{-11}
PPV	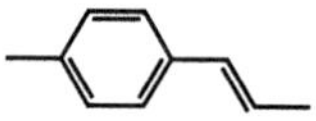	~ 4 x 10^{-10}
poly-PHENYLACETYLENE	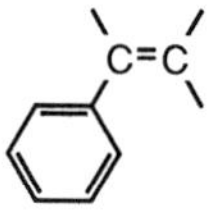 C=C	~ 5 x 10^{-11}
poly-n-BCMU red	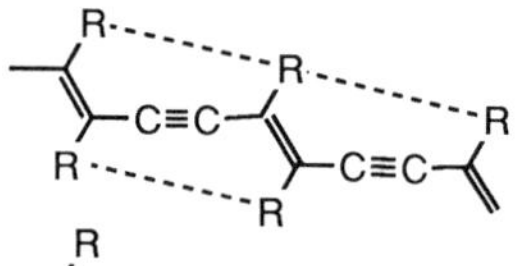 R, R, R, R, R; C≡C, C≡C	~ 3 x 10^{-10}
poly-n-BCMU yellow	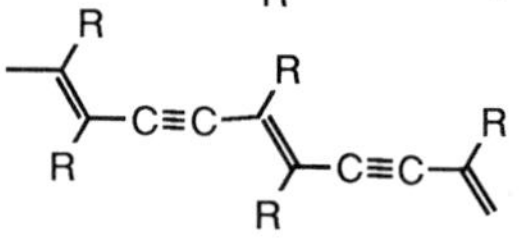 R, R, R, R, R; C≡C, C≡C	~ 2.5 x 10^{-11}
where:	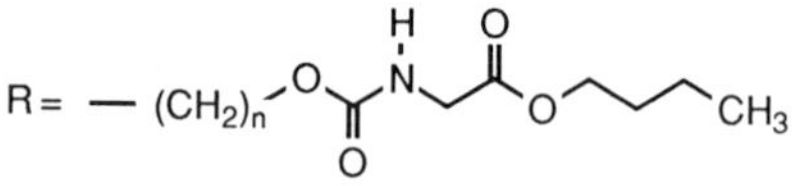 R = — $(CH_2)_n$–O–C(=O)–NH–CH_2–C(=O)–O–$CH_2CH_2CH_2$–CH_3	
poly-THIOPHENE	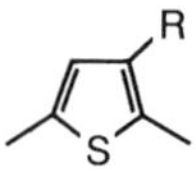 R; S	~ 5 x 10^{-11}

can easily conform to device structures, by being able to be shaped into fibers and films.

Optical signal processing will utilize the phenomenon of optical switching and optical bistability. The principle of optical switching is based on the intensity-dependent refractive index derived from $\chi^{(3)}$. This intensity-dependent refractive index leads to intensity-dependent phase shift in the nonlinear medium and changes the coupling angle to the waveguide, or the resonance parameters of a cavity. We have investigated these processes in different geometries: (i) Fabry-Perot etalon and (ii) planar waveguide.

In a Fabry-Perot etalon geometry we have investigated both the absorptive and dispersive bistability behaviors. A bistable behavior implies that, for a certain input range, the system exhibits two output states.[12] By increasing the input intensity, switching from a low-output state to a high-output state occurs at one intensity level. However, in the reverse cycle, when the input intensity is decreased, switching from the high to low state occurs at a different input level. Therefore, by a proper choice of the input hold level, one can use the optical bistability behavior for latching or memory operation. In absorptive bistability, one uses the saturable absorption behavior of the nonlinear medium. The transmission of the Fabry-Perot cavity is initially low, due to absorption. As the input intensity increases, at certain level the medium bleaches, due to saturation of absorption. The cavity then switches from the low to high-output state. In the reverse cycle, as the input is lowered, the local field within the cavity remains high, to keep the cavity in the high output level, well below the initial input switching level, thus generating a hysteresis.[8]

In dispersive bistability in a Fabry-Perot etalon geometry, one takes advantage of the intensity-dependent refractive index. The Fabry-Perot etalon is detuned, so that the transmission is low. Now, by increasing the intensity of the input signal, the refractive index of the nonlinear medium is changed to bring the cavity in resonance condition, at which point the device switches to the high-output level. Again in the reverse cycle, due to the high field inside the cavity, a hysteresis results. We have observed dispersive bistability in a film of a polyimide.

We have also conducted studies of optical switching in a polymer waveguide. We have reported on optical switching in a poly-4-BCMU quasi-waveguide.[13] A quasi-waveguide is formed when the wave is guided at only one interface but the other interface is leaky, because the refractive index of the film (the guiding medium) is less than that of the cladding medium (in our case the prism substrate). The physics of a quasi-waveguide process is similar to that for an interferometer, as described above in the case of a Fabry-Perot device. Recently, we have also successfully observed nonlinear optical processes in an optical waveguide of polyamic acid, in which a propagation distance of over 5 cm was achieved.[14] The process of intensity-dependent phase shift (coupling efficiency changes, due to change of refractive index, as the input intensity changes) has been observed using nanosecond, 80 picosecond and 400 femtosecond pulses. The observed corresponding limiter action (ie. leveling off of the output power at certain input power levels) has been observed. Our analysis shows that the observed switching with the femtosecond and picosecond pulses is due to electronic nonlinearity, but that observed with nanosecond is mostly dominated by thermal nonlinearity.

CURRENT STATUS

For second-order nonlinear optical materials, organic systems have demonstrated orders of magnitude larger non-resonant $\chi^{(2)}$ compared to that for inorganics, and are extremely promising. This area has been the focus of past research. The limitations for commercial development of devices have been materials requirements, which the organics have not adequately met. For second harmonic generation, the problems have been to produce defect-free, high optical and surface quality crystals. Meeting the phase-matching requirement is not a problem, but again good quality materials are needed. For electro-optic applications, promising materials, at the present time, are electrically-poled guest-host polymeric systems and side-chain polymers. However, questions of long-term stability and relaxation effects in electrically poled systems have to be answered.

The area of third-order nonlinear optical effects in organic systems has been a focus of attention only recently. Consequently, on a relative basis, this area is not as well explored. Here the limitation to device production is due to both the low magnitude of $\chi^{(3)}$, and the lack of material quality required for device structures.

Although we have begun to build some microscopic understanding of third-order optical nonlinearity, it is still in its infancy. We need to considerably improve upon our understanding of structure-property relationships, in order to project structural requirements for enhanced third-order optical nonlinearity. Currently achieved $\chi^{(3)}$ values in organic systems are the highest for non-resonant cases but still need to be increased to be compatible with the diode laser technology. Electronic resonance enhancement does not seem to increase the $\chi^{(3)}$ value as dramatically as is observed in inorganic multiple quantum well semiconductors.

In order to improve our understanding of the structure-property relation, we must study the role of substituents in addition to that of effective conjugation. For this objective, studies utilizing theoretical calculations and experimental measurements of γ on sequentially built and systematically derivatized structures are necessary. We have taken this approach.

THEORETICAL AND EXPERIMENTAL STUDIES OF SEQUENTIALLY BUILT AND SYSTEMATICALLY DERIVATIZED COMPOUNDS

To fulfill the need for understanding what structures will allow enhancement of optical nonlinearity, we have coupled ab-initio theoretical calculations of optical nonlinearity with synthesis of sequentially built and systematically derivatized model compounds, and the measurement of their optical nonlinearities. Now I would like to discuss very briefly our efforts to compare microscopic optical nonlinearities. An expression, similar to the expansion of the bulk polarization as a function of the applied field, can be written for the induced dipole moment. Naturally, the nonlinear term γ, for example, is the third derivative of the induced dipole moment with respect to the applied field. Also, using the Stark energy analysis, one can write the nonlinear terms β (and γ) as a sum over all excited states terms involving transition-dipoles and permanent dipoles, similar to what one does for polarizability. Consequently, the two theoretical approaches are: (i) the derivative method and (ii) the sum-over-states method. We have used the derivative method at the ab-initio level[15]. We correlate the predictions of these calculations with measurements on systematically derivatized and sequentially built model compounds. Some conclusions of our theoretical computations are as follows:

1. The choice of basis functions, particularly the inclusion of diffuse polarization functions, is important in the calculation of β and γ.
2. π-electron contribution to optical nonlinearity dominates.
3. π-electron delocalization (conjugation effect) increases γ significantly, exhibiting a N^3 to N^4 power dependence with respect to the number of repeat units N in a sequentially built oligomeric structure.
4. For nonresonant cases γ is positive.

One group of sequentially built structure is the thiophene oligomers and related derivatized structures, which we have synthesized and experimentally investigated. The agreement with theoretical predictions appears to be satisfactory[16].

FUTURE DIRECTIONS OF RESEARCH

To conclude, a discussion of future direction of research is appropriate. In our opinion, the following areas need to be developed: (i) Improvement of our microscopic understanding of optical nonlinearity. For this purpose, theoretical and experimental studies of sequentially built and systematically derivatized structures are needed. (ii) Increase of data base. Even though more studies are being reported on the measurement of third-order optical nonlinearities, still the data base is limited. We need input from syntheses of novel structures, and measurements of optical nonlinearities on them. (iii) Finally, materials research. This is a very important direction of future research. Work in this area will focus on developing new composites, designing processing schemes to improve optical quality, properly characterizing the ultrastructures of these materials, and fabricating various device structures.

In our photonics research laboratory at SUNY at Buffalo, we have developed a comprehensive research program, which covers all the relevant issues, including theoretical modeling, synthesis of new structures, materials processing to produce high optical quality waveguides, study of nonlinear optical processes, using state-of-the-art ultrashort laser pulses, and study of device processes. A multidisciplinary approach like ours, with an interactive feedback and optimization at each stage, is needed to bring this new technology to its maturity in a near future.

The message I would like to leave is that it is a highly active area, with exciting opportunities for chemists, physicists, engineers and materials scientists. One can expect significant developments to occur in the near future.

ACKNOWLEDGMENT

This work was supported by the Defense Initiative Organization, the office of Innovative Science and Technology; the Air Force Office of Scientific Research, Directorate of Chemical Sciences; and the Air Force Wright Aeronautical Laboratory, through contract numbers F4962087C0097 and F4962087C0042.

We wish to acknowledge valuable collaborations with Professor F. E. Karasz of University of Massachusetts, Professor G. I. Stegeman of University of Arizona and Dr. B. Reinhardt of Air Force Wright Aeronautical Laboratory, Polymer Branch. My thanks are to my research group members, Dr. J. Swiatkiewicz, Dr. R. Burzynski, Dr. M. Samoc, Dr. A. Samoc, Dr. E. Perrin, Dr. K. Wijekoon, Dr. B. P. Singh, Ms. P. Chopra, Mr. Y. Pang, Mr. M. T. Zhao, and Mr. J. C. Wung, whose research efforts have contributed to this paper.

REFERENCES

1. B. Clymer and S. A. Collins, Jr. Opt. Eng. 24:74 (1985).
2. G. I. Stegeman, Thin Solid Films 152:231 (1987).
3. Y. R. Shen "The Principles of Nonlinear Optics" Wiley & Sons, New York (1984).
4. "Nonlinear Optical and Electroactive Polymers" Eds. P. N. Prasad and D. R. Ulrich, Plenum Press, New York (1988).
5. "Optical Nonlinearities and Instabilities in Semiconductors" Ed. H. Haug, Academic Press, London (1988).
6. M. M. Carpenter, P. N. Prasad, and A. C. Griffin, Thin Solid Films 161:351 (1988).
7. M. T. Zhao, M. Samoc, B. P. Singh, and P. N. Prasad, J. Phys. Chem. (in press).
8. D. N. Rao, J. Swiatkiewicz, P. Chopra, S. K. Ghoshal, and P. N. Prasad, Appl. Phys. Lett. 48:1187 (1986).
9. P. N. Prasad in "Nonlinear Optical Effects in Organic Polymers" eds. J. Messier, F. Kajzar, P. Prasad, and D. Ulrich, NATO ASF Series E, Vol. 162 (1989) p. 351.
10. B. P. Singh, P. N. Prasad, and F. E. Karasz, Polymer 29:1940 (1988).
11. J. Swiatkiewicz, P. N. Prasad, F. E. Karasz, and M. Druy, unpublished results.
12. H. M. Gibbs "Optical Bistability: Controlling Light with Light" Academic Press, New York (1985).
13. B. P. Singh and P. N. Prasad, J. Opt. Soc. Am. B5:453 (1988).
14. R. Burzynski, B. P. Singh, P. N. Prasad, R. Zanoni, and G. I. Stegeman, Appl. Phys. Lett. 53:2011 (1988).
15. P. Chopra, L. Carlacci, H. F. King, and P. N. Prasad, J. Phys. Chem. (in press).
16. M. T. Zhao, B. P. Singh, and P. N. Prasad, J. Chem. Phys. 89:5535 (1988).

LANGMUIR-BLODGETT FILMS FOR NONLINEAR OPTICS

Paras N. Prasad

Photonics Research Laboratory
Department of Chemistry
State University of New York at Buffalo
Buffalo, N.Y. 14214

INTRODUCTION

The Langmuir-Blodgett (L-B) technique provides a useful method to prepare novel molecular assemblies with unique structure, in order to probe structure-property relationship for nonlinear optics, and to explore new device configurations. One can take advantage of the monolayer thickness control to fabricate layered structures, leading to the possibility of molecular engineering of alternate-layer heterostructures and superlattice structures. Furthermore, as our work shows, in certain favorable cases, one can also control the microscopic structure in the L-B films, by appropriately selecting the film-forming and transferring conditions. This prospect leads to opportunities for studying the effects of these structural variations on the nonlinear optical properties and, therefore, to enrich our microscopic understanding of these processes.

In order to fully use the potential of various structures prepared by this technique, an important step is a careful characterization of L-B films. In our laboratory, a number of spectroscopic and surface-sensitive techniques are used. Both second and third-order nonlinear optical processes have been observed using L-B films. The organization of this paper is as follows. First, some of the techniques used for the characterization of L-B films are discussed. Then, some interesting examples of control of order and conformation in the L-B films are presented. This is followed by a subsection, which presents results of the study of both second and third-order nonlinear optical processes in L-B films. Finally, possible applications of L-B films in nonlinear optical devices are discussed.

CHARACTERIZATION OF L-B FILMS

In order to make meaningful physical measurements on monolayer and multilayer L-B films, it is essential that the films be characterized to determine the molecular structure, unformity of the film and layer-to-layer structural correlation. To probe the molecular structure, our research group has utilized a variety of spectroscopic techniques such as u.v.-visible spectroscopy, ATR-IR, wave-guide Raman, surface-plasmon wave Raman spectroscopy and inelastic electron tunneling. The spectroscopic techniques

Lower-Dimensional Systems and Molecular Electronics
Edited by R.M. Metzger *et al.*, Plenum Press, New York, 1991

have successfully revealed the molecular nature of a monolayer-to-bilayer transition, observed in the case of a class of soluble poly-diacetylenes (poly-BCMU). When a poly-4-BCMU monolayer film is compressed on a water subphase, it undergoes a monolayer-to-bilayer transition.[1,2] The u.v.-visible spectra change revealing that the monolayer is in the yellow form, known as the coiled-polymer conformation (λ_{max} = 460 nm), and the bilayer is in the red form, labelled as the extended-rod polymer conformation (λ_{max} = 530 nm). Waveguide Raman studies have been performed, using the L-B film as a cladding medium on the top of a waveguide.[3] The evanescent field from the waveguide interacts with the L-B film, to give rise to Raman scattering. In the case of surface-plasmon enhanced Raman scattering, the L-B film is coated on the top of a silver metal film of thickness ~400 Å. Light is coupled into the metal film as a surface plasmon, by the prism coupling technique. Again, through evanescent-field coupling, Raman scattering from the L-B film is observed.[4] Both the waveguide Raman and the surface plasmon Raman studies of the L-B films, of poly-4-BCMU have revealed that the yellow (coil conformation) film has its -C=C- vibrational frequency higher in value than that observed with the bilayer red films transferred past the phase transition.[4] Hence, it clearly established that the red form of the polymer has a larger effective π-conjugation than the yellow form.

The uniformity of a monolayer film over microscopic dimensions can conveniently be investigated by using the surface plasmon technique. By using the surface plasmon technique, one can measure the thickness of a film, provided the refractive index of the material can be approximated by using other techniques. Our group has successfully used this method to measure film thicknesses of monolayer and multilayer L-B films. Then, by spatially scanning the probing region of the film, one can investigate the uniformity of a film. In the case of a side-chain liquid crystalline polymer, the use of this method clearly demonstrated that films transferred below a surface pressure 5 dynes/cm were not uniform, while those transferred at higher surface pressure were more condensed and uniform.[5]

Structures of L-B films and the dynamics of electron transport within them can be conveniently studied by inelastic electron tunneling spectroscopy (IETS).[6] Basically, in this method, when a bias voltage is applied across a metal/insulator(barrier)/metal sandwich structure, inelastic tunneling occurs via the transfer of energy from the electrons to the vibrational modes of the sample molecules. After undergoing inelastic energy loss, the electrons continue across the junction. This electron flow (or current) is monitored as a function of the bias voltage. When the second derivative (d^2I/dV^2) is plotted against the bias voltage V (which consists of a ramp d.c. voltage plus a small modulation voltage), it gives the vibrational spectrum. For increased resolution, the junction is kept at a temperature of 4 K. Since there is no IETS spectrometer commercially available, our spectrometer is home-built and was successfully used to obtain vibrational spectra of several polymeric films.

This technique has several advantages unique to it, when compared with Raman and IR vibrational spectroscopic techniques. Firstly, IETS has a submonolayer sensitivity. Therefore, it is much more sensitive than either IR or Raman spectroscopy. Secondly, since there are no optical selection rules, both Raman and IR vibrational modes appear in the IETS spectra. Thirdly, the intensity of each vibrational band is proportional to the inelastic scattering of the electron by a particular vibrational state. In spite of these advantages, the application of this technique for the study of L-B films appears to have had only limited success. The problem seems to stem from the lack of a convenient way of making junctions of appropriate resitivity, using an LB film as an insulating barrier. Furthermore, problems also arise, due to damage of the L-B film during the process of depositing the top electrode. Pinholes in the film also give rise to problems. We have some success with the polymeric films such as that of poly-4-BCMU.[6] This method will be more suitable for the L-B films of conducting polymers, because those films are more rigid, and also would not have very high resistivity.

In order to study the structural correlation from layer to layer, we have successfully used a combination of the quartz-crystal microbalance method and u.v.-visible spectroscopy.[7] A quartz-crystal microbalance, utilizing the change in the resonance frequency upon film deposition, provides a convenient method of determining the mass of the film. Therefore, the uniformity of mass transfer can be determined from the layer dependence. In the case of a uniform mass transfer, a linear relation will be obtained between the number of layers and the change in the frequency. The u.v.-visible spectra, in the linear absorption regime, would also yield a linear dependence of the absorbance as a function of the number of layers, provided the structural correlation from layer to layer is maintained. The results of the quartz-crystal microbalance and the u.v.-visible spectral studies for L-B films of a macrocycle, tetrakis-butyl-phenoxy-phthalocyanine show a linear relation in each case, indicating a uniform transfer from layer to layer with the same molecular structure.[7,8] In contrast, the results on the L-B films of poly-(3-dodecyl-thiophene) reveal a uniform mass transfer (linear relation between the frequency change and the number of layers), but the structural correlation from layer-to-layer changes around the tenth layer, as evidenced by the change of the slope of the absorbance curve, as a function of the number deposited layers.[7]

CONTROL OF ORDER AND CONFORMATION

Studies, conducted by our research group, have revealed that a group of soluble polydiacetylenes abbreviated as poly-n-BCMU can be manipulated in the monolayer film to control the polymer conformation, order and π-conjugation. The coefficient, n, simply represents the number of methylene units in the side chain. We have investigated poly-3-BCMU which undergoes a monolayer yellow form-to-bilayer blue form transition, while poly-4-BCMU discussed above shows a monolayer yellow form to a bilayer red form transition. The blue form is the most conjugated structure. π-conjugation strongly influences the optical nonlinearity. Therefore, by controlling the effective conjugation by the L-B technique, one can judiciously tune the optical nonlinearity. In addition to the pressure-dependent monolayer-to-bilayer transitions, effective π-conjugation can be changed, by virtue of the fact that the monolayers of monomers can be polymerized by u.v. light.

The 9-BCMU multilayer films show more subtle differences in the degree of order and conjugation, when prepared differently.[9] The solution-cast film, when polymerized by u.v. light, shows the least order (λ_{max} = 635 nm) for a blue form. When a 9-BCMU monomer monolayer film is polymerized at the air-water interface, and then transferred to a substrate, it shows a greater degree of order and conjugation (λ_{max} = 645 nm). The most ordered polymerized multilayer film is obtained when the monomer is first transferred as a multilayer, and then polymerized (λ_{max} = 655 nm).

NONLINEAR OPTICAL STUDIES

Our group has been involved in the study of both second and third-order nonlinear optical processes in Langmuir-Blodgett films.[7,8,10-12] For the second-order nonlinear optical studies, we have used side-chain liquid crystalline polymers, with side groups that have nonlinear optical properties. These polymers form condensed rigid monolayer films, which can be transferred by the horizontal lifting technique. A second harmonic generation study has been conducted on these films. Furthermore, electrooptic modulation experiments have been performed on monolayer L-B films in a surface plasmon geometry.

Although third-order nonlinear optical effects are much weaker, several studies of third-order processes in L-B films have successfully been performed.[7,8,10-12] A third harmonic generation study was conducted on poly-4-BCMU as a function of film compression.[10,11] The monolayer-to-bilayer transition of this film is clearly revealed by a rapid increase in third-harmonic generation. This result is in accordance with the strong dependence of $\chi^{(3)}$ on the effective π-conjugation length.

The value of a resonantly enhanced $\chi^{(3)}$ is often sufficiently large to obtain the degenerate four-wave mixing (DFWM) signal from even a monolayer film. DFWM studies have been performed on L-B films of both poly (3-dodecyl thiophene) and various substituted phthalocyanines.[7,8,12] An interesting system is Si-phthalocyanine, $SiPc(OSiMePhOH)_2$. This compound, in addition to having the desirable properties of other phthalocyanines, is uncharacteristically soluble in common solvents, and produces good L-B films.

The degenerate four-wave mixing experiments on Si-phthalocyanine Langmuir-Blodgett films were performed using approximately 400 fs pulses at 602 nm. The working wavelength coincides with an absorption band. By delaying one of the beams, one can study the temporal profile of the transient grating formed in the LB films. It has been found that the decay of the four-wave mixing signal is power-dependent, namely, the signals decay faster for higher input power. The decays are also clearly nonexponential. This behavior can be rationalized, assuming that the relaxation of the excited-state grating occurs via at least two channels, one of them being the first-order decay and the other a bimolecular exciton-exciton annihilation process.[12] The effective value of $\chi^{(3)}$ obtained from the measurements using 400 fs pulses is about 10^{-9} esu.[12] For the poly (3-dodecyl thiophene) film, the decay of resonant $\chi^{(3)}$ is within the resolution of our optical pulses. The effective $\chi^{(3)}$ value is $\sim 1.1 \times 10^{-9}$ esu.[7]

APPLICATIONS IN NONLINEAR OPTICS

Possible device structures incorporating L-B films can be classified in two categories: (i) devices which could utilize monolayers or only a small number of layers, (ii) devices which would require multilayered assemblies of dimensions comparable to optical wavelengths. The latter case would require a deposition of thousands of layers. Examples of devices of the first category are surface-plasmon electrooptic modulators and optical sensors. The L-B films are promising, as they provide a method of grafting novel components or heterostructure molecular assemblies with a monolayer resolution. Also, if the number of layers deposited is not very large, structural correlation can be maintained to get a uniform and homogeneous film. Another example of a possible device in this category would be a guided-wave device, in which the L-B films can be used as a cladding medium. In this case the nonlinear effect is not as much manifested because the interaction is provided only by the evanescent field, which decays exponentially in the cladding medium.

The examples of the second category of devices will be guided-wave optics, in which successive deposition of L-B films can be used to fabricate a waveguide or a composite waveguide. In principle, this method offers tremendous flexibility to fabricate many unusual waveguide structures. However, it is very difficult to maintain structural correlation and uniform layer-to-layer deposition for thousands of layers. Our experience has been that domain structures develop, and the resulting films are highly lossy. At the present, it appears doubtful that the L-B film technique would produce high optical quality waveguides. For this particular application, it may be advantageous to go to self-assembling systems, where chemical bonding to the substrate and interlayer chemical interaction helps maintain layer-to-layer structural correlation.

ACKNOWLEDGEMENT

This research was sponsored by the Office of Innovative Science and Technology, Defense Initiative Organization, and the Directorate of Chemical Sciences, Air Force Office of Scientific Research, under contract numbers: F4962087C0097 and F4962087C0042. Partial support from the Health Instrument and Devices Institute (HIDI) at SUNY at Buffalo is also acknowledged.

REFERENCES

1. J. Biegajski, R. Burzynski, D. A. Cadenhead and P. N. Prasad, Macromolecules 19:2457 (1986).
2. J. E. Biegajski, D. A. Cadenhead and P. N. Prasad, Langmuir 4:689 (1988).
3. R. Burzynski, P. N. Prasad, J. Biegajski and D. A. Cadenhead, Macromolecules 19:1059 (1986).
4. X. Huang, R. Burzynski and P. N. Prasad, Langmuir 5:325 (1989).
5. M. M. Carpenter, P. N. Prasad and A. C. Griffin, Thin Solid Films 161:315 (1988).
6. R. Burzynski, X. Huang and P. N. Prasad, Polymer Preprints 28:213 (1987).
7. P. Logsdon, J. Pfleger, and P. N. Prasad, Synthetic Metals 26:369 (1988).
8. P. N. Prasad, M. K. Casstevens, J. Pfleger and P. Logsdon, Symposium on Multifunctional Materials, SPIE Proceedings, Vol 878 (1988) p. 106.
9. J. E. Biegajski, D. A. Cadenhead and P. N. Prasad, Unpublished Results.
10. G. Berkovic, Y. R. Shen and P. N. Prasad, J. Chem. Phys. 87:1897 (1987).
11. G. Berkovic, R. Superfine, P. Guyot-Sinnoset, Y. R. Shen and P. N. Prasad, J. Opt. Soc. Am. B5:668 (1988).
12. P. N. Prasad, M. Casstevens and M. Samoc in Proceedings of Symposium on "Photochemistry of Thin Films" SPIE Meeting, January 1989, Los Angeles.

ALL-OPTICAL SWITCHING USING OPTICAL FIBERS AND NONLINEAR ORGANIC LIQUIDS

L. Domash[^], P. Levin[^], J. Ahn[*], J. Kumar[°] and S. Tripathy[*]

[^]Foster-Miller, Inc., Waltham, MA, 02254 USA
[*]Dept. of Chemistry, Univ. of Lowell, Lowell, MA, 01854 USA
[°]Dept. of Physics, Univ. of Lowell, Lowell, MA, 01854 USA

ABSTRACT

Nonlinear optical organic liquids, based on 2-methyl-4-nitroaniline (MNA)-type molecules in solution, are potentially practical as third-order materials, featuring a large and fast intensity-dependent index of refraction (n_2), low optical loss, insusceptibility to laser damage, and a freely adjustable refractive index. The latter property makes them compatible with glass substrate devices, and permits ease of fabrication for demonstration devices. Concepts and preliminary data are presented for a high-speed all-optical switch, based on glass waveguides suspended in a nonlinear liquid cladding of index-matched organic solutions.

1. Introduction

All-optical waveguide devices, based on ultrafast third-order nonlinear optical properties, are desired for a variety of functions, including fiber optic switching, digital logic, optical limiting and optical computing systems. Materials potentially useful in these applications must possess (a) high third-order nonlinearity, (b) ultrafast response time and (c) excellent additional requisite physical properties. These properties include superior transparency, freedom from scattering centers, uniformity, optically flat surfaces, chemical and environmental stability, and processibility. Currently, a number of inorganic optical materials satisfy these criteria, but no single material has emerged that may be termed as an optical "silicon." Further enhancement of nonlinear properties is still desirable for better performance, and, in addition, increased flexibility in tailoring the requisite physical properties is crucial. Organic materials are often cited as having the best long-range promise for such devices (1), which may exceed efficiency of inorganic materials in many aspects, such as faster response time with a low dielectric constant, and higher third-order nonlinearity in the materials' transparent regime(2,3). In addition, great flexibility in molecular design and modification, for the enhancement of

Lower-Dimensional Systems and Molecular Electronics
Edited by R.M. Metzger *et al.*, Plenum Press, New York, 1991

properties is easily afforded. In spite of intensive research using organic materials, relatively few working devices have been demonstrated to date,due to a variety of problems in selecting, processing and fabricating such materials into appropriate device configurations.

Among organic nonlinear optical (NLO) solids available today, conjugated macromolecular systems,such as the polydiacetylene(PDA) family and guest-host systems (in which an NLO material, e.g. MNA, is composited with a highly transparent and chemically and dimensionally stable polymer, e.g. polymethylmethacrylate), seem to be good candidates for the application. However, when these materials are in a solid form, it is not easy to fabricate them into desirable configurations, and, at the same time, match the refractive indices with associated optical elements, as may be dictated in a device configuration.

We discuss concepts for a class of switching devices,intended to be easy to fabricate and test, based on combining nonlinear organic solutions with glass fiber waveguides. In constructing practical nonlinear waveguide devices for the application discussed above, many characteristics,other than large, high speed n_2 effects, become important. In particular, low loss (the sum of absorption and scattering) is significant. The effective nonlinear response for many device configurations is not the intensity-induced index modulation $n_2 \cdot I$ but the figure of merit, $n_2 \cdot I/\alpha\lambda$, where I is the intensity of the incident beam, α is the loss per cm and λ is the wavelength of the incident beam.

A more mundane, and often overlooked, factor is that, for maximum flexibility in device design, it is desirable that a third-order material should have a linear refractive index,compatible with the other materials typically used in fiber optics and waveguide structures. For example, some high-performance organic films, or crystals, have indices in the 1.6-1.9 range, making it difficult to integrate them with glass structures of index $\approx$ 1.45. In addition, laser damage threshold of the nonlinear material under consideration must be high enough to withstand the fields required to generate significant nonlinear effects. Finally, the usefulness of a material is severely limited, unless a fabrication technology exists to produce high-precision, low-loss waveguide forms. Organic materials exist,which satisfy each of these requirements, but no one material, so far, satisfies them all at once. In an attempt to develop approaches to maximize utilization of organic NLO materials,and minimize processing and fabrication complexities, we employed NLO materials in solution form.

Solution form of NLO materials provides at least two advantages for research and feasibility demonstrations. First, adjusting the linear refractive index to match a given fiber cladding will be significantly easier, than in the solid state. Second, the transparency problem will be drastically reduced,since the scattering and inhomogeneity prevailing in most solid films will be avoided. As a tradeoff for these advantages, solutions raise problems,such as concentration limits and temperature dependence of the refractive index. Therefore, several solvents must be evaluated to identify the range of refractive index available, and temperature must be controlled as accurately as possible, to minimize the temperature-dependent changes of refractive index.

With these factors in mind, it is instructive to note that Friberg et. al. (4) were able to demonstrate all-optical switching in a simple dual-core optical fiber, operating as a Jensen coupler (5). In their early experiments, the "nonlinear" material was ordinary silica, whose n_2 is only 1/10,000 that of a prototypical high performance polymer, such as PDA, in the material's transparent regime (6). Because of the extremely high transparency of silica, however, combined with the existing technology to form long low-loss fibers, this small nonlinearity was available over a long optical path length (many cms) in a "pipelined" switching configuration capable of subpicosecond speeds. Switching power threshold was on the order of 1 kW. Using the organic NLO solutions of the present research, larger n_2 effects may yield switching times an order of magnitude faster, with optical switching power on the order of 100mW.

2. Dual-Waveguide Switching Device Using Liquid Cladding

In view of the difficulty of fabricating high-precision nonlinear thin-film waveguides, it is clearly a practical advantage to work with devices based on glass fibers as an existing high-quality waveguiding structure, provided more strongly nonlinear materials can also be introduced. Clark, Andonovic and Culshaw (7) modeled dual waveguide devices, in which the nonlinear material in a dual-core optical switch was located contiguous with the cladding, as shown schematically in Figure 1. A directional coupler is first fabricated which, under low power conditions, transfers 100% of the optical power from fiber 1 to fiber 2. Under the control of a separate optical pump beam, or else a simple increase in power of the signal itself, the nonlinear core or cladding index $n = n_o + n_2 \cdot I$ (where n_o is the linear refractive index) is modulated sufficiently to alter the coupling ratio between the two fibers. If the coupler length and power threshold are set correctly, this can cause the signal to exit 100% from fiber 1, instead of transferring to fiber 2, effecting an optically controlled switch. This class of devices is denoted nonlinear coherent couplers (NLCC). Figure 2 suggests that a simple NLCC can, in principle, be made by immersing a rather ordinary bidirectional coupler with etched cladding in an index-controlled bath of a nonlinear liquid. Although it is inherently less efficient to locate the active nonlinear material in the cladding than the core, this can be compensated by large n_2 values, long path lengths, and very accurate index control of the nonlinear cladding element. The goal, in developing such a switch, is subpicosecond speed,

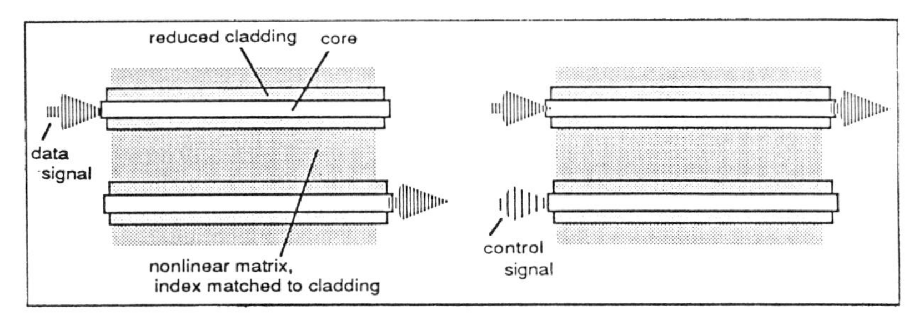

Figure 1. The Nonlinear Coherent Coupler, NLCC.

combined with a power threshold for switching on the order available from diode lasers, 10-100mW.

Highly transparent third-order materials, whose linear index can be precisely adjusted to match glass cladding values, are essentially limited to glasses, doped glasses or liquids. The latter have the advantage of potentially much larger nonresonant nonlinearities. Although liquid CS_2 is a standard for third-order nonlinear optics, relatively little effort has been devoted to developing new, higher performance liquids as practical device materials. Using high concentrations of NLO solutions, it is possible to precisely match the index to a range of desired values. In high concentrations, n_2 values comparable to the best solid organic materials may be possible, along with the low loss and self-healing characteristic of high-purity liquids. A number of approaches, such as gelation, may be adopted to "fix" the solution for transition to a solid device, if desired.

3. Device Model

The equations describing power transfer within a NLCC were derived by Jensen (5). The NLCC is defined as two parallel waveguides spaced closely enough for evanescent wave overlap, and separated by a nonlinear medium. For a coupler with active region equal to the characteristic coupling length, L_c, the coupler will operate in the "crossed state" at low powers (i.e. all of the power launched into one waveguide will exit the second waveguide). As the input power to waveguide 1 is increased, the power out of waveguide 1 is described by

$$P_{out}(1) = P_{in}(1)\ [1 + CN(\pi|(P_{in}(1)/P_c)^2)]/2$$

where CN is a Jacobi elliptic function and

$$P_c = A\ \lambda/L_c\ n_2$$

where n_2 is the nonlinear refractive index of the medium constituting and surrounding the guides, and A is the cross-sectional area of each guide. The nonlinear dynamics

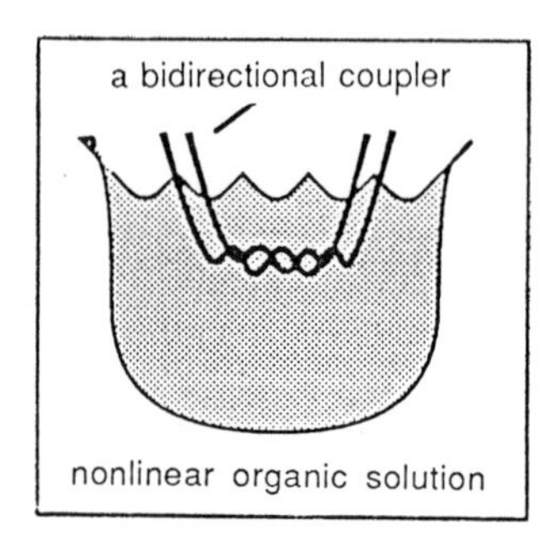

Figure 2. A Simple NLCC with Glass Fibers in a Nonlinear Liquid Bath.

responds to both core and cladding n_2. A goal of this investigation is enhancement of the effective nonlinear index of the medium surrounding the coupler guides,by using organic NLO solutions.

Switching of the NLCC from the crossed to the parallel state is shown as a function of input power in Figure 3. The effect of a factor of two increase in the nonlinear index of the medium,constituting and surrounding the coupler guides,is shown by the two curves; the coupler with the higher nonlinearity shows 100% switching at a lower power. Since some nonlinearity is always present in the core, and the higher field there emphasizes this contribution over that of the cladding, the effect of a nonlinear liquid bath is, essentially,to enhance the performance (i.e. reduce the power threshold for switching) which is already present,due to the small nonlinearity of the glass core.

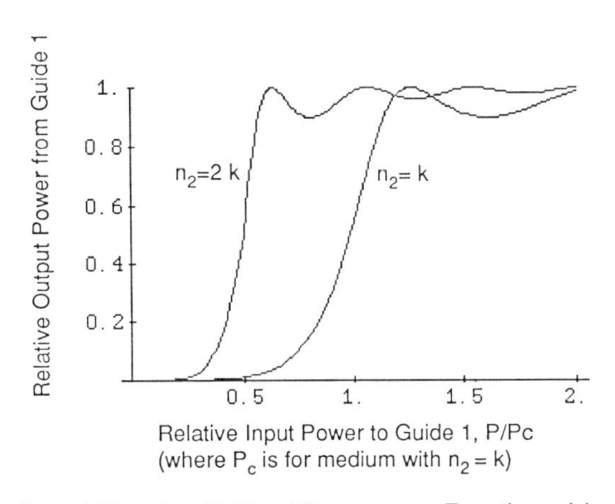

Figure 3. Signal Fraction Exiting Fiber 1 as a Function of Input Power.

4. Organic NLO Materials and their Solvent Systems

For the preliminary demonstration of all-optical switching using NLO solutions, commercially available 2-methyl-4-nitroaniline (MNA) was selected, which is primarily known as a second order material (2). MNA also possesses considerable third order nonlinear optical properties, whose high speed nonresonant nonlinear index of refraction (n_2 = 25.0 x 10^{-11}esu) is 2-3 times greater than that of Si or p-nitroaniline (3). Exceptionally high solubility of MNA in polar solvents allows a wide range of refractive indices to be available. MNA also provides a wide transparent regime (from mid visible to near infrared), in which optical losses are minimum (Figure 4).

Suitable solvent systems for MNA were identified for the solution-based all-optical switching device. Dimethylformamide (DMF), p-dioxane, propylene carbonate (PC) and tetrahydrofuran (THF) provided high solubility with a wide range of index choices. Solubility of MNA in each solvent (HPLC grade) at room temperature was measured for the determination of the range of refractive index available (Table 1).

Table 1. Solubility of MNA in Different Solvents at Room Temperature.

SOLVENT	SOLUBILITY (grams/liter)
DMF	527
p-DIOXANE	261
PC	154
THF	349

Table 2. Temperature Dependence of Refractive Indices of MNA Saturated Solutions.

SOLVENT	REFRACTIVE INDEX	
	15°C	25°C
DMF	1.5660	1.5613
p-DIOXANE	1.4766	1.4725
PC	1.4580	1.4546
THF	1.4854	1.4817

Variation of refractive index of the MNA solution in various solvents at 20°C as a function of concentration was obtained using an Abbe refractometer. Refractive indices available, as shown in Figure 5, range from 1.4 to 1.5 with the solvents used, from which one can easily match the refractive index of silica optical fibers. Temperature dependence of refractive index in MNA saturated solutions, measured at 15°C and 25°C, is tabulated in Table 2. Fluctuation of refractive index, as a result of the temperature variation, turned out to be approximately 0.0004/°C, so that it is essential to control the solution temperature to better than ±0.1°C in this experiment, in order for only electronic nonlinear refractive index change to be effective.

One can tailor the refractive index to the desired value, by varying one or more of the following factors; (a) concentration of the NLO solution, (b) temperature of the solution and (c) solvent type. It is true, that the higher the number density of the NLO molecules in solution, the larger the nonlinear effects are. In practice, however, not only nonlinearity but also optical loss, solvent volatility, stability of polymeric components in the device assembly, such as epoxy encapsulant and so on, must be taken into consideration. Consequently, non-volatile PC was selected as the best choice as a test solvent for the demonstration, in which epoxy encapsulant was found to be very stable for long periods of time.

A UV-VIS transmission spectrum of MNA was obtained to determine the transparent window available for the probe and pump beams. As shown in Figure 4, in MNA solution, excellent transparency is seen for wavelength of 450nm through near infrared, with an absorption peak at 380nm. Output from a HeNe laser (633nm) for probe beam and from

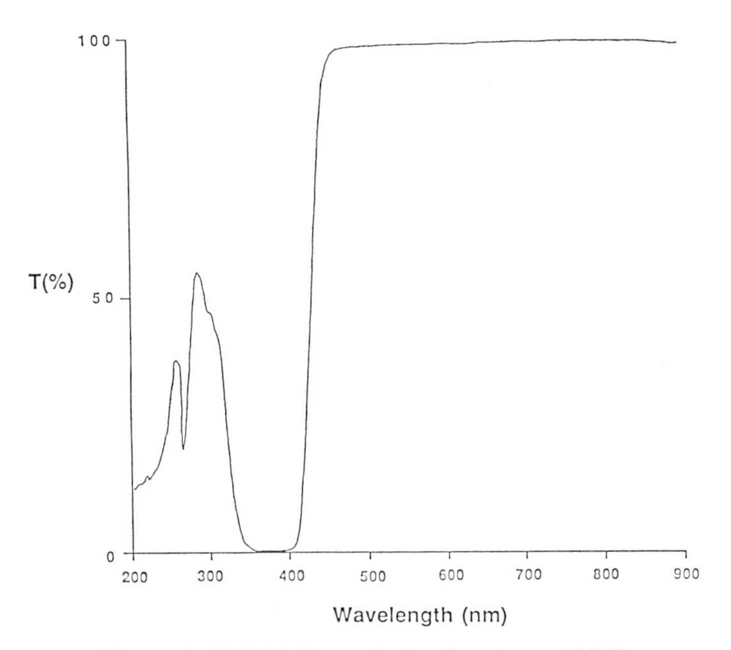

Figure 4. UV-VIS Transmission Spectrum of MNA.

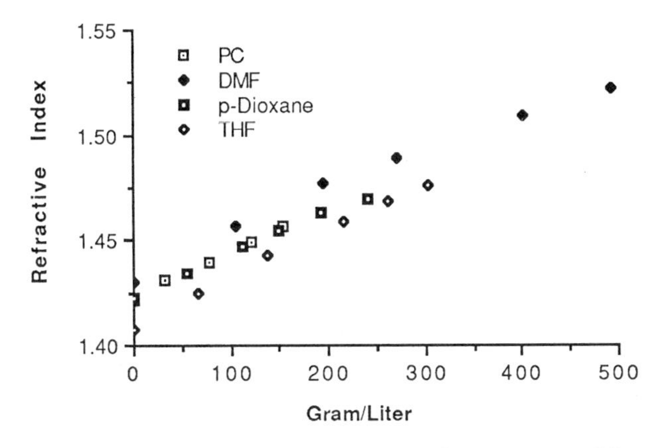

Figure 5. Refractive Indices of MNA Solutions at 20°C.

an Argon-ion laser (515nm) for pump beam were chosen in this research. All-optical switching demonstration,using optical fiber waveguides with organic nonlinear solutions, is in progress.

A large number of organic and polymeric molecular systems have been identified with significantly larger third-order nonlinearities,compared to MNA. While polymeric solutions with high concentrations may not be possible to prepare, gels and oligomers may be prepared with large nonlinear coefficients and high number density of the active molecules. A number of organic NLO molecular systems have been designed and synthesized in our laboratory (8), which are expected to possess significantly larger third order nonlinearities. In addition, several of these molecules incorporated diacetylene and acetylene units. Polymerization in these systems may lead to soluble materials with large nonlinear coefficients. Eventually, these polymeric NLO materials are expected to improve the device performance,by at least an order of magnitude,leading to a switching power threshold of less than 100mW.

5. Preliminary Experiments

As a preliminary test, we have constructed a device as shown in Figure 2, developed a high precision laboratory method of matching liquid indices to glass cladding, and demonstrated modulation of the probe beam by a pump beam,at low speed,using thermal effects. This demonstration, which has been carried out with thermal changes in the index of refraction of the surrounding liquids, gives some indication of the dynamics to be expected from electronic nonlinearities of the NLO solutions in subsequent experiments.

A cross-section of the experimental apparatus is shown in Figure 6. The basic coupler is of the fused type,and is fabricated from two optical fibers with single mode cutoff at about 590 nm. The fused or "active" region of the coupler is about 1 cm long, and has an estimated minimum diameter of about 40 microns. At both ends of the active region, at the points of bifurcation, an epoxy encapsulant is used to anchor and seal the coupler. As shown, after encapsulation, only the active region remains available for

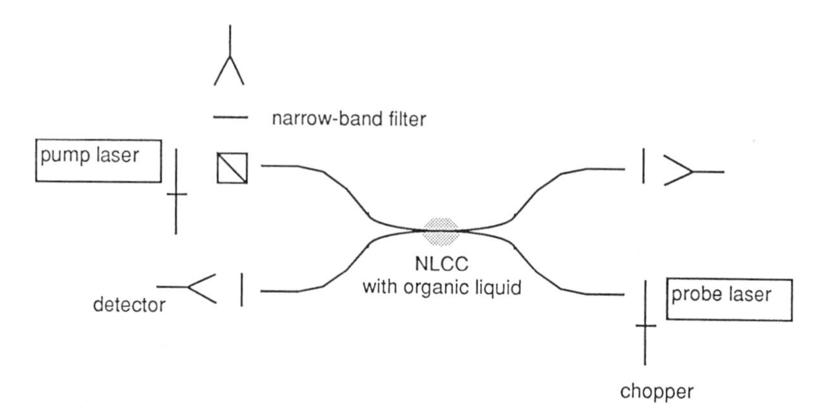

Figure 6. Preliminary Experiment.

exposure to the test liquids. The coupler sits within a stainless steel tube, with central section cut out. The tube passes diametrically through a liquid reservoir,and serves as a liquid-tight bulkhead,through which the input and output fiber ends of the coupler pass. As some of the test solvents are highly volatile, the reservoir is fitted with a removable cap,to prevent evaporation. The liquids are tested by filling the reservoir,until the active region is submersed. The temperature of the chamber may be controlled by heating or cooling through the walls of the reservoir; close temperature control is important to stabilize the splitting ratio. If necessary, the liquid in the reservoir may be stirred with a magnetic bar.

The apparatus may be used with a single laser input beam of adjustable intensity, which serves as both the pump beam and the probe beam, or else two lasers,operating at different wavelengths,may be used to separate the probe/signal and pump/control functions. In our initial tests,it was convenient to use separate lasers for the probe and pump beams; the arrangement for launch and detection of transmitted optical signals is shown in Figure 6. The probe beam input to the coupler was stripped of cladding modes by use of index-matching fluid,applied over the first 3 to 5 cm of uncoated fiber tip.

With separate pump and probe lasers,it is easy to maximize the amount of pump beam power,which enters into the organic liquid under test. While the probe beam is restricted to single-mode operation,for proper functioning of the switch, the pump beam is not similarly restricted. Excitation of cladding modes by the pump beam assures that substantial pump power will enter the nonlinear liquid. To further enhance interaction with the liquid, the wavelength of the pump beam may be chosen to lie just below cut-off, where the second mode is loosely bound, or far above cut-off,where the fundamental mode is loosely bound. Balancing these advantages, however, is the added complexity that a dual laser arrangement poses. The pump and probe channels must be separated at the detectors by use of, for example, narrowband filters, and/or lock-in detection, using modulation frequencies for the two lasers which are not related harmonically.

As a preliminary test of the device apparatus, an argon ion laser (Coherent Innova 90-6),producing 0.5-4 W at 515 nm,was used as the pump beam in Figure 6. The distance from the pump launch end to the coupler active region was less than 1 meter. A comparison of coupler pump beam throughput (sum of both output arms),before and after the application of index matching fluid to the active region,revealed that 15 %,or more,of the launched power could be made available for excitation of the organic fluid.

Optical modulation of the HeNe probe beam was demonstrated under control of the argon pump beam, by means of thermal modulation of the active region of the coupler, as shown in Figure 7. This effect was quite slow,as expected, on the order of 10 ms, but served to demonstrate principles similar to those which will operate in subpicosecond experiments,using purely electronic index modulations. Addition of an index-matching fluid to the active region reduced the amplitude of the probe beam modulation, possibly because the liquid stabilized the active region temperature through convective cooling. The frequency response of the probe beam modulation rolled off at only a few kHz, characteristic of a cooling-rate limited system. Note that it is difficult to distinguish, experimentally, between the contributions of the core and cladding nonlinearities, whether thermally or electronically induced.

The experimental apparatus is now being readied for high speed all-optical switching experiments, using a sub-nanosecond pulsed laser pump, in conjunction with the organic NLO liquids as described above. To index-match such liquid baths to the cladding, or other selected glass references accurately, a scattering technique was developed using a HeNe laser beam directed transversely onto the etched fiber/liquid interface in a test cell. An index match is indicated by titrating the liquid index, until the scattered light is minimized. By careful collection of scattered light, and use of a micropipette, to adjust the solvent mixture, it was possible to match the index to better than 0.0001. The liquids used were mixtures of organic solvents 1,2 dimethoxymethane ($n \approx 1.42$) and acetylene tetrabromide ($n \approx 1.48$), selected to bracket silica ($n = 1.4570$ at 633 nm). Using improved sensitivity, it is expected that 0.00001 index control can be achieved, which is sufficiently precise to set the propagation parameters of single-mode waveguides.

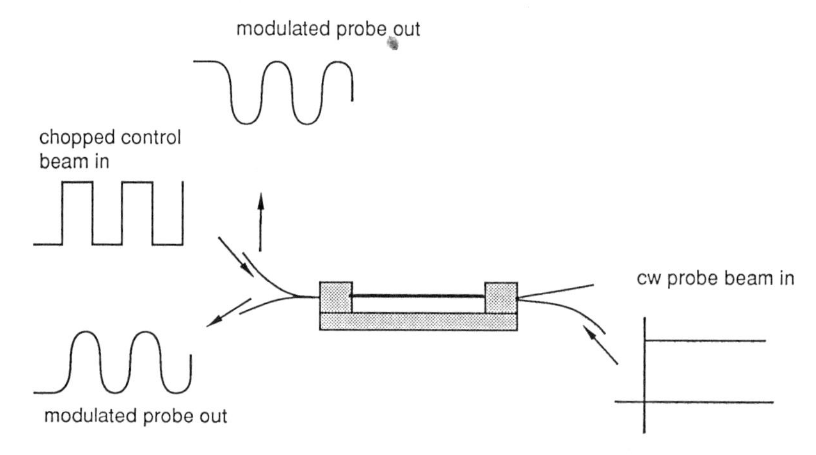

Figure 7. Optical Coupling due to Thermal Index Modulation.

6. Concluding Remarks

To demonstrate all-optical switching using organic NLO solutions, MNA was selected as a test NLO material, and its suitable solvent systems were identified. The range of refractive index available was also obtained from the MNA solutions in various solvents and at different temperatures. The transparent regime in MNA solution was determined for the probe and pump beams, employing UV-VIS spectroscopy. A test apparatus has been designed and assembled, which enables optical interaction between a fiber-optic single-mode coupler and NLO liquids. Thermal modulation of an input signal has been demonstrated with this apparatus. Subpicosecond all-optical switching, using organic NLO solution (e.g. MNA/PC), including new NLO materials synthesized in our laboratory, is now under investigation for enhanced switching efficiency (faster switching speed, smaller switching power threshold and shorter interaction length).

ACKNOWLEDGEMENTS

This University of Lowell/Foster-Miller collaboration is supported by the Massachusetts Center of Excellence Corporation in Polymer Science, a state agency.

REFERENCES

1. G. Stegeman, C. Seaton, J. Appl. Phys., 58, R57 (1985).
2. A. Yariv, "Quantum Electronics", 2nd Ed., John Wiley & Sons, Inc., New York, 1975.
3. T. Chang, Opt. Eng., 20, 220(1981).
4. S. Friberg, Y. Silverberg, P. Smith, Appl. Phys. Lett., 51, 1135(1987).
5. S. Jensen, IEEE JQE-18, 1580(1982).
6. D. Williams, "Nonlinear Optical Properties of Organic and Polymeric Materials", ACS, Washington DC, 1983
7. D. Clark, I. Andonovic, B. Culshaw, Opt. Lett., 11, 540(1986).
8. R. Kumar, J. Kumar, S. Kumar, A. Blumstein, S. Tripathy, Proc. IUPAC MACRO 88, 32nd Int'l Symp. Macromol., Kyoto, Japan, Blackwell Sci. Publ., 1988.

OPTICAL STUDIES OF AMPHIPHILIC MOLECULES WITH INTERESTING ELECTRO-OPTICAL AND NON-LINEAR OPTICAL PROPERTIES

Mathias Lösche and Gero Decher

Institut für Physikalische Chemie
Johannes-Gutenberg Universität Mainz
D6500 Mainz, West Germany

ABSTRACT

Structural control is a major issue in both life science, investigating the function of the biological machinery, and in materials science, aiming at the design of novel devices. In part one, recent electro-optical investigations of the primary event of photosynthesis on purified protein preparations are described. Part two focuses on structural studies of monolayers at an air/water interface, and of Langmuir-Blodgett multilayers from a new molecule designed for non-linear optical applications.

INTRODUCTION

Molecularly thin organic films have attracted much interest as proposed two-dimensional structured systems, for applications in a wide spread of fields, ranging from nonlinear optics and sensorics to microelectronics. One crucial task, for designing organic thin film structures is the engineering of suitable molecules, and/or the utilization of moieties prefabricated by nature, for use as functional entities. In any case, the first aim is to understand function, structure, and structure/function relationship of these entities. Work in our group is focused mainly on experimental investigation of such problems, and the objective of this paper is to report on recent progress which has been achieved with two different systems:

- the electro-optical characterization of proteins from the photosynthetic apparatus and
- the structural investigation of Langmuir-Blodgett (LB) films built from a new type of amphiphilic molecules, that show strong second-harmonic generation.

Lower-Dimensional Systems and Molecular Electronics
Edited by R.M. Metzger *et al.*, Plenum Press, New York, 1991

I. THE CHARGE SEPARATION PROCESS IN BACTERIAL REACTION CENTERS

The bacterial reaction center (RC) has been the subject of intense research efforts during the past 25 years. The reasons are that it is an evolutional precursor to the much more elaborate systems of green algae and plants, and that the initial charge separation is located on one protein which, for some species of purple bacteria, can easily be purified in significant amounts from fresh cells. In fact, for two species, *Rhodopseudomonas viridis* (*Rp. viridis*)[1] and *Rhodobacter sphaeroides* (*Rb. sphaeroides*),[2-7] this protein has been crystallized, and the three-dimensional structure has been solved.

Despite the fact that spectroscopic investigations have revealed many details of the process of the primary charge separation, one is far from a full understanding of the events that lead to one of the most important electrochemical reactions that exist in nature. Traditional techniques of optical and electronic spectroscopy have lead to the identification of the participating prosthetic groups[8] and gave first indications about their mutual geometric arrangement.[9,10] Time-resolved spectroscopy revealed a view of the migration of the excitation between the chromophores.[11] It showed extraordinarily fast reaction constants (3 psec for the transfer of the electron from the primary donor, P, to the first intermediate acceptor, I, across a distance of roughly 15 Å!). Results from photochemical hole burning studies suggested that the life-time of the optical excitation P^* is only about 10 fsec.[12]

The determination of the three-dimensional structure naturally solved most of the structural questions. The pigments (6 porphyrins and 2 quinones) are organized in two branches, A and B, within the protein in almost perfect C_2 symmetry, see Fig.1. The donor P is a bacteriochlorophyll (BChl) dimer, D_A and D_B (i.e. BChl moieties closely related to the A or B branch, respectively), and interacts with the two quinone (Q) acceptors, via a BChl monomer, B_A, and a bacteriopheophytin (BPh) monomer, Φ_A. Φ_A is the intermediate acceptor I and transfers the electron to Q_A, the primary, and via a high-spin iron to Q_B, the secondary acceptor, which communicates with a quinone pool in the lipid membrane. P^+, in turn, is re-reduced from an associated heme system on a close-by cytochrome protein. Cytochrome and the quinone pool are not depicted in Fig.1.

Questions that are open to date include:[13]

> How does the extraordinarily high quantum yield for the charge separation come about? Why are the rate constants for the forward reactions at least two orders of magnitude larger, than the rates of the back and the waste reactions (c.f. fig.1)?

- What is the identity of the excited state P^*? Where does the electron go first? Are there charge–transfer states involved? Which mechanism is responsible for the very efficient charge transfer within the protein?
- What is the role of the BChl monomer, B_A, which is located half-way between P and I? The excitation has not been found on this structurally intermediate molecule, even on the fastest time scales.
- How does the strong directionality of the charge transfer along the A branch come about, given the structural symmetry of the system?

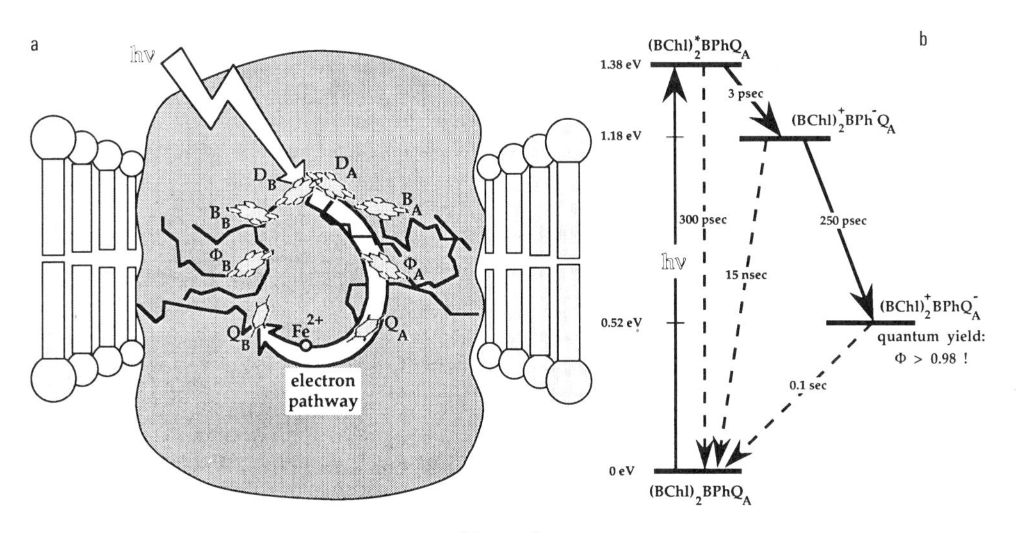

Figure 1.

(a) Schematic of the RC from *Rb. sphaeroides* with the prosthetic groups within the lipid membrane. The outline of the protein backbone is shown in dark grey. Also depicted is the electron pathway after light excitation.
(b) Energy relations and kinetics of the electron transfer reactions in *Rb. sphaeroides* RCs at room temperature (from Ref. 14)

STARK SPECTROSCOPIC INVESTIGATIONS OF REACTION CENTERS

Electro-optical studies have been undertaken, to contribute to the solution of some of these questions.[15-20] Stark spectroscopy is sensitive to changes in the electronic structure of chromophores during their optical excitation, and hence is especially suited to tackle the second set of the questions raised above.

METHODS

The theory of electro-optical excitations of chromophores, immobilized in thin film samples, has been worked out by Reich and Schmidt.[21] An expansion of the optical absorption, A, into the energy, $\varepsilon = hc\tilde{\nu}$, leads to a difference spectrum

$$\Delta A = \frac{\Delta\mu * <\cos\vartheta>}{hc} * \tilde{\nu} * \frac{\partial}{\partial\tilde{\nu}}\left(\frac{A}{\tilde{\nu}}\right) * F + \frac{1}{2}\frac{\Delta\alpha}{hc} * \tilde{\nu} * \frac{\partial}{\partial\tilde{\nu}}\left(\frac{A}{\tilde{\nu}}\right) * F^2$$

$$+\frac{1}{2}\frac{\Delta\mu^2 * <\cos^2\vartheta>}{h^2c^2} * \tilde{\nu} * \frac{\partial^2}{\partial\tilde{\nu}^2}\left(\frac{A}{\tilde{\nu}}\right) * F^2 \tag{1}$$

(h: Planck's constant; c: velocity of light; $\tilde{\nu}$: wavenumber; ϑ: angle between the static dipole moment change on excitation, $\Delta\vec{\mu}$, and the applied electric field, $\vec{F}$; brackets denote macroscopically averaged values).

If the static dipole moment changes, $\Delta\vec{\mu}$, dominate the changes in polarizability, $\Delta\alpha$, and, in the case of random orientation of the transition moments, $\vec{\mu}_{tr}$, in the macroscopic sample, the observed absorbance change, ΔA, due to the applied electric field is,[16,21]

$$\Delta A = \frac{(\Delta\mu)^2 * F^2}{10\, h^2c^2} * \tilde{\nu} * \frac{\partial^2(A/\tilde{\nu})}{\partial\tilde{\nu}^2} * (3\cos^2\delta\cos^2\chi + 2 - \cos^2\delta - \cos^2\chi) \tag{2}$$

(χ: angle between the applied electric field, $\vec{F}$, and the electric field vector, $\vec{E}$, of the probing light; δ: angle between dipole moment change, $\Delta\vec{\mu}$, and transition moment, $\vec{\mu}_{tr}$.).

Purified RC proteins have been incorporated into thin polymer (poly vinyl alcohol; PVA) films, which were glued in place between transparent glass electrodes, as described earlier.[16] The sample sandwich was mounted inside a liquid nitrogen dewar. Phase-sensitive detection of the second harmonic light modulation $I^{2\omega}_{rms}$, with respect to a f = 1 kHz field oscillation at $\chi = 90°$, revealed the magnitude of the absorbance change, $\Delta A(\chi{=}90°)$. δ was determined independently, from the slope s of the linear relation between $\frac{\Delta A(\chi)}{\Delta A(\chi=90°)}$ and $\cos^2\chi$, c.f. Eq. (1), as[16]

$$\delta = \cos^{-1}\left(\sqrt{\frac{2s+1}{s+3}}\right). \tag{3}$$

Finally, $\Delta\mu$ was determined as

$$\Delta\mu_{app} = \frac{hc}{F_{rms}} * \left[\frac{10}{1+\sin^2\delta} * \frac{\sqrt{2}\,\Delta A(\chi=90°)}{\tilde{\nu}\,\dfrac{\partial^2(A/\tilde{\nu})}{\partial\tilde{\nu}^{\,2}}} \right]. \tag{4}$$

Here, the notation $\Delta\mu_{app}$ indicates that the local field, at the site of the chromophore (i.e. inside the protein), is unknown, and is augmented, with respect to the applied field, by dielectric enhancement. The *apparent* dipole moment changes are therefore overestimated, by a factor of 1.3 - 1.7 at 77 K with respect to the *intrinsic* ones.[17]

RESULTS

Figure 2 shows the Q_y band absorption (top panel), second derivative (center), and Stark spectra (bottom panel) of RCs from *Rb. sphaeroides* (left; $F_{rms} = 7.14{*}10^4$ V/cm) and *Rp. viridis* (right; $F_{rms} = 4.5{*}10^4$ V/cm) in PVA matrices. In either case, the lowest energy spectral regimes correspond to the excitations of the primary donors, whereas the features above 12,000 cm^{-1} are associated with the accessory chromophores. Both samples show a striking similarity between Stark and second derivative spectra, indicating the dominance of the static over the induced dipole moment changes. In fact, according to eq.(1), changes in polarizability on excitation are expected to contribute a first derivative component to the Stark spectra. In view of the lack of such features in the spectra, we assumed it a good approximation to neglect polarizability contributions, when quantifying our results. The second immediate conclusion from Fig. 2 is that the special pairs, in either case, have a much larger proportionality factor between Stark and second derivative spectral bands, than the accessory pigments. Accordingly, the dipole moment change, $\Delta\mu$, is much greater in these cases.

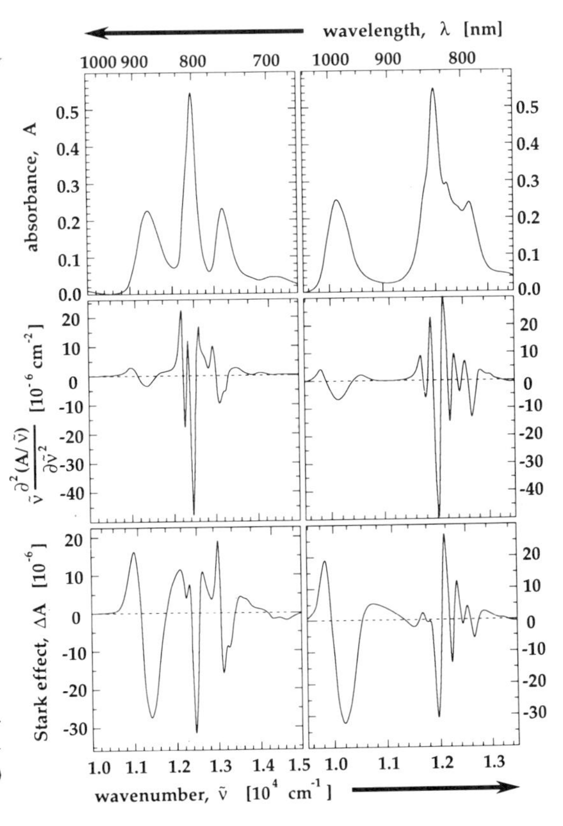

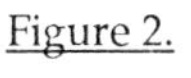

Figure 2.

Absorption (top), second derivative (center) and Stark (bottom) spectra of the Q_y region of RCs from *Rb. sphaeroides* (left) and from *Rp. viridis* (right) at T = 77 K.

For the case of *Rb. sphaeroides*, Fig.3 exemplifies the determination of the angles δ, for three large bands at λ = 877 nm (low energy band of the excitonically coupled special pair D_A and D_B); λ = 802 nm (*mainly* due to the BChl monomer B_A); and λ = 761 nm (*mainly* due to the BPh molecule Φ_A). The values, derived from these data, are $\delta = 38 \pm 1°$; $\delta = 23 \pm 2°$; $\delta = 8 \pm 4°$ (λ = 877 nm; 802 nm; 761 nm). Utilizing eqs. (2)-(4), Table 1 summarizes the information revealed from the data shown in Fig. 2.

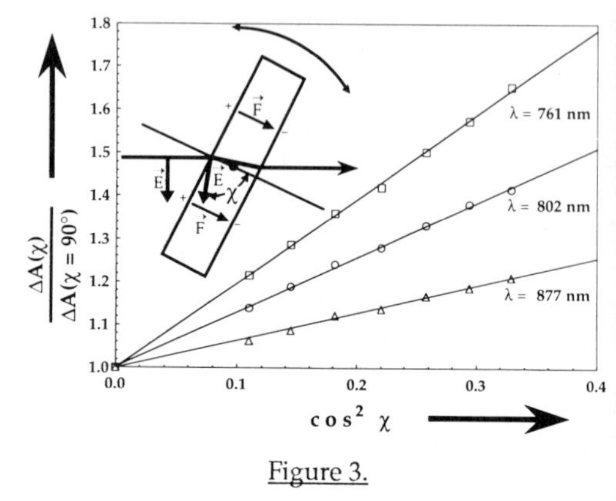

Figure 3.

Dependence of the magnitude of the Stark effect on the angle of incidence of the probing light on the sample for 3 prominent Q_y bands in *Rb. sphaeroides* RCs at T = 77 K.

Table 1.

Magnitude and direction of the apparent static dipole moment changes, $\Delta\mu_{app}$, in chromophores of different bacterial reaction centers at T = 77 K.

bacterial species	λ [nm]	δ [deg]	$\Delta\mu_{app}$ [Debye]
Rb. sphaeroides	877	38 ± 1	6.5 ± 0.2
	817	n.d.	≈ 1
	802	23 ± 2	2.1 ± 0.1
	761	8 ± 4	3.5 ± 0.2
	754	≈ 5	≈3.5
Rp. viridis	989	40 ± 2	8.2 ± 0.8
	849	n.d.	n.d.
	833	≈ 50	1.8 ± 0.6
	818	14 ± 6	3.4 ± 0.3
	803	n.d.	n.d.
	787	0 ± 5	2.7 ± 0.3

DISCUSSION AND FUTURE PERSPECTIVES

The important result of the Stark spectroscopic investigations is that the charge redistribution on light excitation has a much larger impact on the primary donor within the protein, than it has on the accessory pigments. This finding, of course, does not come unexpected, and it has been argued[17] that the large dipole moment change is due to mixing of charge transfer states into the dimer excitation. However, in order to elucidate the mechanism of charge separation, the charge redistribution within the protein has to be investigated in even more detail. It is clear, from the magnitude of the effect (a dipole moment change of $\Delta\mu = 8$ Debye corresponds to the transfer of an electron across 1.6 Å only!), that the electron is *not* transferred directly to the intermediate acceptor *I*. For the use as a critical test, to decide which of several theoretical models is the appropriate one, the precision of the results is not sufficient at present. The reasons are *not* precision of the measurement, but systematic errors, most notably the problem of the local field and, in the second place, the determination of the sample thickness, non-uniformity of the sample thickness, and contributions from the induced dipole moment changes.

Eq.(1) shows that the full power of Stark spectroscopy is exploited only if the method is used with oriented samples. One obvious way to accomplish this aim is to incorporate the protein into LB structures. From a materials science view, this comprises the first step toward the design of a composite system, incorporating a functional entity into a lipid matrix. More important, this will facilitate an independent measurement of static and induced dipole moment changes. The directionality of the charge translocation can be determined in oriented samples, whereas, at present, it is only evaluated as a cone angle, δ, around the axis defined by the transition moment.

Recently, Blinov and coworkers[22] have shown that Stark spectroscopy provides an elegant means to measure local field effects in LB films. Even if this does *not* solve the problem of assessment of the local field *within the protein*, it does provide a prospect to investigate the influence of the sample preparation and, by comparison of lipid matrix local field effects to electrostatic manipulations within the protein (e.g. trapping of light-induced or chemically produced charge-separated states; protein manipulations by site-directed mutagenesis), can bring a solution to the original problem within reach.

Finally, once the magnitude of the charge redistribution is *known*, this information can be utilized to, in turn, characterize the orientation of the molecules within the thin-film structure. It has been worked out[23] that the high-order harmonic Stark spectra of a chromophore depend on the corresponding moments of the orientation distribution function, $f(\vartheta)$, which can be reconstructed, taking into account the appropriate geometrical symmetry of the sample. In future work, we will try to employ this property to spectroscopically characterize the structure of transferred LB films, since the assessment of structural control is the major issue in the work of our group, as will be reported on in more detail in the second contribution from our laboratory to this conference.

II. STRUCTURAL INVESTIGATIONS OF MONOLAYERS AND LB MULTILAYERS WITH NON-LINEAR OPTICAL PROPERTIES

Non-linear optical thin-film devices are expected to play an important role in future technologies, such as optical computing and other applications of opto-electronics. In optical storage systems, second harmonic generation (SHG) can be used to double the frequency of a laser beam, and thus enhance the storage capacity of the medium by a factor of four. Current devices are made from inorganic single crystals, and therefore too costly for mass production. It is expected that the performance of organic materials may reach or even surpass[24] that of the best inorganics, such as $LiNbO_3$.

One major advantage of organic materials is that their molecular structure can be designed to an optimum performance. On the other hand, the preparation of crystalline non-centrosymmetric structures is still a matter of trial and error. An excellent method for the engineering of tailor-made supramolecular assemblies is the Langmuir-Blodgett technique.[25] It offers the possibility to prepare ultrathin films, which can directly be used for integrated optics. The suitability to fulfill both basic prerequisites for non-linear optics, the design of molecular structure and proper molecular arrangement, has stimulated considerable research in the area.

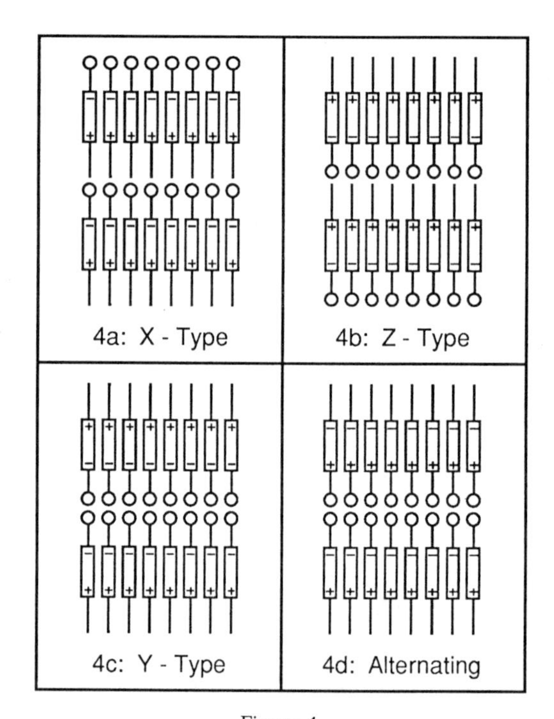

Figure 4.

Schematic of the arrangement of polar molecules in Langmuir-Blodgett multilayers.

Figs.4a and 4b depict a general scheme of how polar multilayers of one single component can be organized in a head-to-tail arrangement of the molecules (Z-type[26,27] and X-type multilayers or $[\vec{A}\,\vec{A}\,..]$–film structures). Due to the close contact of hydrophilic head groups and aliphatic tails, however, these structures are not easy to prepare, and rather unstable. Nevertheless, the first report on a quadratic enhancement of second-harmonic generation (SHG) with increasing numbers of layers[26] was from Z-type LB films. Fig.4c shows the usual head-to-head and tail-to-tail one component multilayer (Y-type or $[\overrightarrow{A}\overleftarrow{A}\ldots]$–film structure), which is centrosymmetric, and thus not suitable for SHG. Fig.4d displays an alternating multilayer (Y-type or $[\overrightarrow{A}\overrightarrow{B}\ldots]$–film structure), prepared from two components with opposing dipole moments. LB films of this type exhibit strong SHG,[28,29] but are difficult to prepare with sufficient thickness for optical waveguiding, since dipole-dipole repulsion increases with rising numbers of layers. An additional drawback is the requirement of a dedicated two-compartment LB trough for the alternating film deposition. Recently, a new type of alternating multilayer (Y-type or $[\overrightarrow{A}\overrightarrow{B}CC\ldots]$–film structure) was introduced,[30] where the active layers $[\overrightarrow{A}\overrightarrow{B}]$ are separated by a spacer bilayer [CC], which acts as an insulator for the local electric field. Again, a disadvantage here is that the multilayers are composed of three components, and thus are extremely troublesome to deposit.

MONOLAYERS AND NON-CENTROSYMMETRIC Y-TYPE MULTILAYERS OF DCANP: STRUCTURE AND OPTICAL PROPERTIES

In order to use the advantages of Y-type multilayers of one single material one must employ an alternative concept for the arrangement of the molecules in order to obtain non-centrosymmetric LB films. We have recently introduced 2-docosylamino-5-nitropyridine (DCANP, Fig.5a), a substance that can easily be deposited onto hydrophobic substrates, yielding LB films of at least up to 1.2 µm thickness.[31]

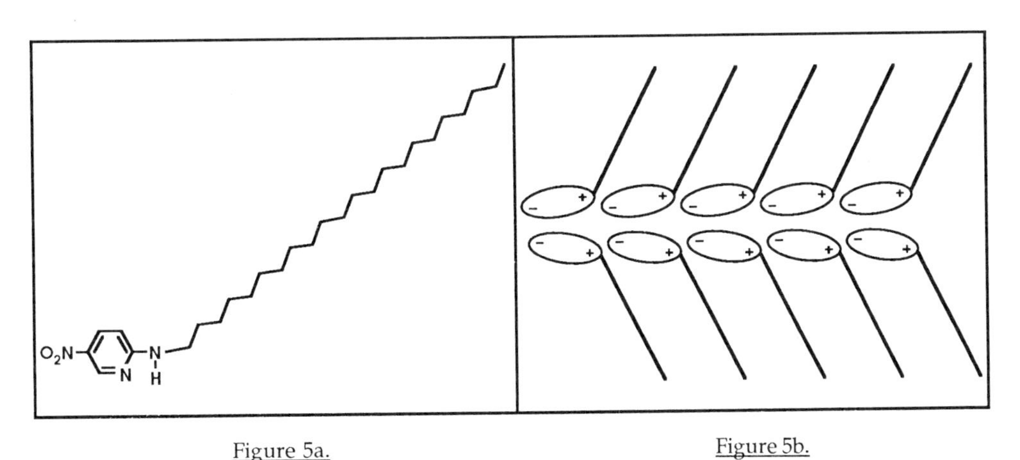

Figure 5a.

Structure of the DCANP molecule.

Figure 5b.

Proposed arrangement of DCANP molecules, forming non-centrosymmetric Y-type LB films with a herringbone structure.

Monolayer transfer, for both up- and downstroke, was 100 % within experimental error, and a long spacing of 4.42 nm was determined by X-ray diffraction. Upon irradiation at $\lambda_{fundamental}$ = 1064 nm, an LB film of 1.2 μm thickness, corresponding to 540 monolayers, had a peak conversion efficiency $(I_{2\omega}/I_{\omega}) = 2 * 10^{-7}$ at approximately 220 MW cm^{-2} input power. The green light (λ_{SH} = 532 nm) is clearly visible to the naked eye. The emitted light is polarized parallel to the dipping direction over the whole film area, thus indicating a homogeneous non-centrosymmetric film structure. The combination of transfer, X-ray and SHG data suggests a Y-type herringbone multilayer structure, as shown in Fig.5b. Here, the non-centrosymmetry arises from a preferential alignment of the dipolar chromophores parallel to the layer plane. Dipole-dipole repulsion is reduced by the aliphatic chains separating the polar layers, thus enabling deposition of films thick enough for waveguiding purposes.

Fig.6a shows that the optical density, at the maximum optical absorbance wavelength λ_{max} = 376 nm, increases linearly with film thickness, up to approximately 0.4 μm. The linearity demonstrates that the film structure is independent of the number of deposited layers. When film thickness is further increased, the absorbance becomes so large, that the Lambert-Beer law is no longer satisfied. In Fig.6b the square root of the relative second-harmonic intensity is plotted versus the number of bilayers.

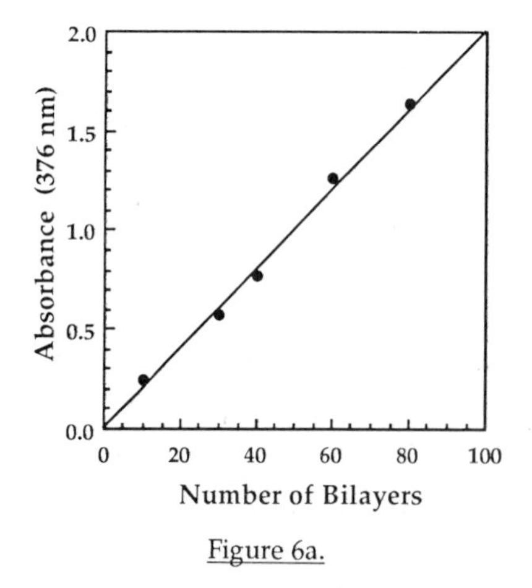

Figure 6a.

Optical density of DCANP multilayers as a function of film thickness (λ_{max} = 376 nm).

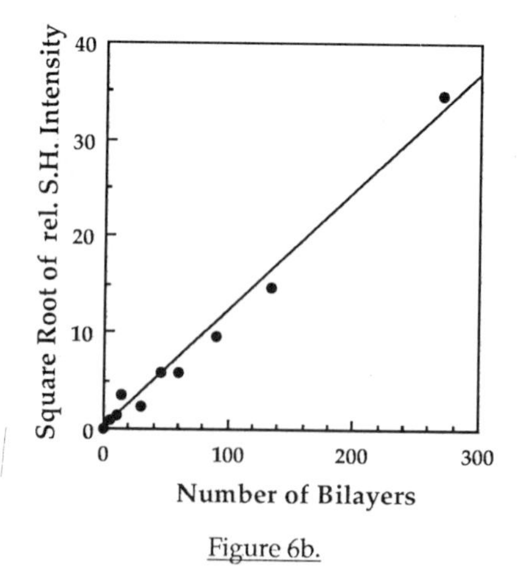

Figure 6b.

Square Root of the relative second harmonic intensity at λ = 532 nm as a function of film thickness.

As required by theory[32] a linear increase is observed, which confirms both the structural homogeneity of the LB film, up to 270 bilayers, and the absence of phase mismatch.

$$I_{2\omega} = \frac{2\omega^2 d_{eff}^2 l^2}{c^3 \varepsilon_o n_\omega^2 n_{2\omega}} I_\omega^2 \ \mathrm{sinc}^2\left(\frac{\Delta k\, l}{2}\right) \tag{5}$$

The SH intensity is given by Eq.(5), where sinc(x) denotes $\frac{\sin(x)}{x}$, l is the sample thickness, d_{eff} is the effective nonlinear optical susceptibility, n is the refractive index, c is the speed of light, and $\Delta k = k_2 - k_1$ is the phase mismatch between the fundamental and the SH waves, with the wavevectors k_1 and k_2, respectively. The nonlinear optical susceptibility perpendicular to the layer plane, d_{311}, was measured against a quartz reference, and a value of d_{311} = 6.8 ± 1.2 pm/V, for polarization parallel to the dipping direction has been obtained.[33] d_{eff} for LB films of DCANP in transmission is approximately 25% of the value for $LiNbO_3$.

ON THE ORIGIN OF THE OPTICAL ANISOTROPY IN MULTILAYERS

LB multilayers of DCANP are highly anisotropic with respect to the dipping direction. Preferential orientation of the aminonitropyridine chromophore was established both by SHG measurements[31] and by UV/Vis spectroscopy of transferred multilayers.[33] Since this orientational anisotropy is essential for nonlinear optical purposes, its origin must be assessed, in order to achieve the necessary structural control. To date, there is only little understanding of the processes leading to the in-plane anisotropy of LB films.

It has been postulated[34] that the orientation of molecular aggregates on the solid substrate can be established by viscous flow[35] within the monolayer on the water subphase. In this model, the flow is induced by the dipping process, in which the substrate acts as a sink. In order to test if this mechanism is relevant in our situation, optical anisotropy measurements of DCANP monolayers on the water surface have been carried out with a set-up described earlier.[36] Within experimental error, no anisotropy with respect to the directions parallel and perpendicular to the line connecting the observation spot and the pickup (drain) region has been detected, independent of the number of dipping cycles. Since, after transfer, a dichroic ratio of 1.6 has been determined, we therefore conclude that the anisotropy is induced, rather, during the transfer process itself, and may be due to flow of water and/or friction occuring in the meniscus region between the monolayer and the solid substrate.

SUMMARY AND CONCLUSIONS

Non-centrosymmetric Y-type LB films of DCANP can easily be prepared in a thickness more than sufficient for optical waveguiding. Their non-linear optical susceptibility in transmission is approximately 25% of that of $LiNbO_3$ and there is no phase mismatch, up to a thickness of 270 bilayers. Multilayers show a strong optical and non-linear optical anisotropy with respect to the dipping direction, which is a consequence of the transfer process.

We have demonstrated how electro-optical and non-linear optical techniques can be utilized to characterize the structure/function relationship in biologically as well as in technically important supramolecular systems. In both cases, interaction of the molecules under investigation with a liquid interface can be used to bring about system orientation, which is utilized to enhance the performance of the spectroscopic experiment, or to induce the desired macroscopic properties.

ACKNOWLEDGMENTS

Much of the work reported on has been done in collaboration with other research groups. M.L. is indepted to Professors G. Feher and M. Okamura of the University of California at San Diego for the support he received as a postdoctoral fellow there. Special thanks are due to E. Abresch for the purification of the proteins. Financial support by the Deutsche Forschungsgemeinschaft (Lo 352-1) is gratefully acknowledged. G. D. is grateful to Professor P. Günter and Ch. Bosshard from the Federal Institute of Technology at Zürich and to Dr. B. Tieke from CIBA-GEIGY AG in Fribourg, all in Switzerland, for their support during his postdoctoral work.

REFERENCES

1. J. Deisenhofer, O. Epp, K. Miki, R. Huber, and H. Michel, X-ray Structure Analysis of a Membrane Protein Complex; Electron Density Map at 3 Å Resolution and a Model of the Chromophores of the Photosynthetic Reaction Center from *Rhodopseudomonas viridis*, J. Mol. Biol. 180:385 (1984).

2. J.P. Allen, G. Feher, T.O. Yeates, H. Komiya, and D.C. Rees, Structure of the reaction center from *Rhodobacter sphaeroides* R-26: The cofactors, Proc. Natl. Acad. Sci. USA 84:5730 (1987).

3. J.P. Allen, G. Feher, T.O. Yeates, H. Komiya, and D.C. Rees, Structure of the reaction center from *Rhodobacter sphaeroides* R-26: The protein subunits, Proc. Natl. Acad. Sci. USA 84:6162 (1987).

4. T.O. Yeates, H. Komiya, D.C. Rees, J.P. Allen, and G. Feher, Structure of the reaction center from *Rhodobacter sphaeroides* R-26: Membrane-protein interactions, Proc. Natl. Acad. Sci. USA 84:6438 (1987).

5. T.O. Yeates, H. Komiya, A. Chirino, D.C. Rees, J.P. Allen, and G. Feher, Structure of the reaction center from *Rhodobacter sphaeroides* R-26 and 2.4.1.: Protein-cofactor (bacteriochlorophyll, bacteriopheophytin, and carotenoid) interactions, Proc. Natl. Acad. Sci. USA 85:7993 (1988).

6. J.P. Allen, G. Feher, T.O. Yeates, H. Komiya, and D.C. Rees, Structure of the reaction center from *Rhodobacter sphaeroides* R-26: Protein-cofactor (quinones and Fe^{2+}) interactions, Proc. Natl. Acad. Sci. USA 85:8487 (1988).

7. H. Komiya, T.O. Yeates, D.C. Rees, J.P. Allen, and G. Feher, Structure of the reaction center from *Rhodobacter sphaeroides* R-26 and 2.4.1.: Symmetry relations and sequence comparisons between different species, Proc. Natl. Acad. Sci. USA 85:9012 (1988).

8. G. Feher, Some Chemical and Physical Properties of a Bacterial Reaction Center Particle and its Primary Photochemical Reactants, Photochem. Photobiol. 14:373 (1971).

9. J.R. Norris, R.A. Uphaus, H.L. Crespi, and J.J. Katz, Electron Spin Resonance of Chlorophyll and the Origin of Signal I in Photosynthesis, Proc. Natl. Acad. Sci. USA 68:625 (1971).

10. P.L. Dutton, K.M. Petty, H.S. Bonner, and S.D. Morse, Cytochrome c_2 and Reaction Center of *Rhodopseudomonas spheroides* Ga. Membranes. Extinction Coefficients, Content, Half-Reduction Potentials, Kinetics and Electric Field Alterations, Biochim. Biophys. Acta 387:536 (1975).

11. W.W. Parson, Photosynthetic Bacterial Reaction Centers: Interactions among Bacteriochlorophylls and Bacteriopheophytins, Annu. Rev. Biophys. Bioenerg. 11:57 (1982).

12. S.R. Meech, A.J. Hoff, and D.A. Wiersma, Role of charge-transfer states in bacterial photosynthesis, Proc. Natl. Acad. Sci. USA 83:9464 (1986).

13. J. Jortner and M.E. Michel-Beyerle, Some Aspects of Energy Transfer in Antennas and Electron Transfer in Reaction Centers of Photosynthetic Bacteria, in "Antennas and Reaction Centers of Photosynthetic Bacteria", M.E. Michel-Beyerle, ed., Springer-Verlag, Berlin (1985).

14. P.L. Dutton, G. Alegria, and M.R. Gunner, The Possible Existence of a Charge Transfer State which Preceeds the Formation of $(BChl)_2^{+}BPh^{-}$ in *Rhodobacter sphaeroides* Reaction Centers, in "The Photosynthetic Bacterial Reaction Center", J. Breton and A. Vermeglio, eds., Plenum Publishing Corp., San Diego (1988).

15. D.L. deLeeuw, M.M. Malley, G. Butterman, M.Y. Okamura, and G. Feher, The Stark Effect in Reaction Centers from R. Sphaeroides, Biophys. J. 37:111a (1982).

16. M. Lösche, G. Feher, and M.Y. Okamura, The Stark effect in reaction centers from *Rhodobacter sphaeroides* R-26 and *Rhodopseudomonas viridis*, Proc. Natl. Acad. Sci. USA 84:7537 (1987).

17. M. Lösche, G. Feher, and M.Y. Okamura, The Stark Effect in Reaction Centers from *Rhodobacter Sphaeroides* R-26, *Rhodopseudomonas Viridis* and the D_1D_2 Complex of Photosystem II from Spinach, in "The Photosynthetic Bacterial Reaction Center", J. Breton and A. Vermeglio, eds., Plenum Publishing Corp., San Diego (1988).

18. D.J. Lockhart and S.G. Boxer, Magnitude and Direction of the Change in Dipole Moment Associated with Excitation of the Primary Electron Donor in *Rhodopseudomonas sphaeroides* Reaction Centers, Biochemistry 26:664 (1987).

19. D.J. Lockhart and S.G. Boxer, Stark effect spectroscopy of *Rhodobacter sphaeroides* and *Rhodopseudomonas viridis* reaction centers, Proc. Natl. Acad. Sci. USA 85:107 (1988).

20. H.P. Braun, M.E. Michel-Beyerle, J. Breton, S. Buchanan, and H. Michel, Electric field effect on absorption spectra of reaction centers of *Rb. sphaeroides* and *Rps. viridis*, FEBS Lett. 221:221 (1987).

21. R. Reich and S. Schmidt, Über den Einfluß elektrischer Felder auf das Absorptionsspektrum von Farbstoffmolekülen in Lipidschichten. I. Theorie, Ber. Bunsenges. Phys. Chem. 76:589 (1972).

22. L.M. Blinov, S.P. Palto, and S.G. Yudin, Stark Spectroscopic Technique for Probing the Local Field in Dielectric Langmuir-Blodgett Films, J. Molec. Electronics 5:45 (1989).

23. N.V. Dubinin, S.G. Yudin, and L.M. Blinov, Quadratic Stark effect and orientation of molecules in polar Langmuir multilayers, Opt. Spektrosk. 59:92 (1985).

24. D.S. Chemla and J. Zyss, Nonlinear Optical Properties of Organic Molecules and Crystals, Academic Press, New York (1987), Vols. 1 and 2.

25. H. Kuhn, D. Möbius, and H. Bücher, "Spectroscopy of Monolayer Assemblies", in: Physical Methods of Chemistry, Vol. 1, Part IIIB, A. Weissberger and P. Rossiter, eds., Wiley Interscience, New York (1972).

26. O.A. Aktsipetrov, N.N. Akhmediev, I.M. Baranova, E.D. Mishina, and V.R. Novak, Multilayer Langmuir films for optoelectronics; nonlinear-optics method for studying them, Sov. Tech. Phys. Lett. 11:249 (1985).

27. L.M. Hayden, S.T. Kowel, and M.P. Srinivasan, Enhanced second harmonic generation from multilayered Langmuir/Blodgett films of dye, Opt. Commun. 61:351 (1987).

28. I.R. Girling, P.V. Kolinsky, N.A. Cade, J.D. Earls, and I.R. Peterson, Second Harmonic Generation From LB Superlattices Containing Two Active Components, Opt. Commun. 55:289 (1985).

29. I. Ledoux, D. Josse, P. Fremaux, J.P. Piel, G. Post, J. Zyss, T. McLean, R.A. Hann, P.F. Gordon and S. Allen, Second harmonic generation in alternate non-linear Langmuir-Blodgett films, Thin Solid Films 160:217 (1988).

30. B.L. Anderson, R.C. Hall, B.G. Higgins, G. Lindsay, P. Stroeve, and S.T. Kowel, Quadratically enhanced second harmonic generation in polymer-dye Langmuir-Blodgett films: a new bilayer architecture, Synth. Met. 28:D683 (1989).

31. G. Decher, B. Tieke, C. Bosshard, and P. Günter, Optical Second-harmonic Generation in Langmuir-Blodgett Films of 2-Docosylamino-5-nitropyridine, J. Chem. Soc. Chem. Commun. 933 (1988).

32. See e. g., A. Yariv, Quantum Electronics, Wiley Interscience, New York (1975)

33. G. Decher, B. Tieke, C. Bosshard, and P. Günter, Optical Second-Harmonic Generation in Langmuir-Blodgett Films of Novel Donor-Acceptor Substituted Pyridine and Benzene Derivatives, Ferroelectrics, in press.

34. N. Minari, K. Ikegami, S. Kuroda, K. Saito, M. Saito, and M. Sugi, Origin of the In-Plane Anisotropy in Langmuir-Blodegett Films, J. Phys. Soc. Japan 58:222 (1989).

35. M.F. Daniel and J.T.T. Hart, Effect of Surface Flow on the Morphology of Langmuir-Blodgett Films, J. Mol. Electronics 1:97 (1985).

36. G. Decher, F. Klinkhammer, I.R. Peterson, and R. Steitz, Structural Investigations of Langmuir-Blodgett Films of 2-Docosylamino-5-nitropyridine, a New Type of Non-Centrosymmetric Multilayer for use in Non-Linear Optics, Thin Solid Films, in press.

THE USE OF THE SURFACE PLASMON RESONANCE TECHNIQUE IN NON-LINEAR OPTICS

S. J. Cooke and G. G. Roberts

Department of Engineering Science
University of Oxford
Parks Road
Oxford, OXI 3PJ, UNITED KINGDOM

INTRODUCTION

The Surface Plasmon Resonance (SPR) technique is presented as a sensitive system of measurement, and the underlying physical mechanism is described. The method is ideal for determining the linear and non-linear optical properties of dielectric media in thin film form, specifically those prepared by the Langmuir-Blodgett (LB) method. The LB method of film deposition and its inherent advantages as a process for non-linear optical device fabrication are discussed briefly.

The application of the SPR technique to the measurement of the electro-optic coefficient is described, as an extension to the measurement of film thickness and permittivity. Preliminary results are presented of the measurement of linear optical parameters, showing that for multilayers of ruthenium (η^5-cyclopentadienyl) - (bistriphenylphosphine) - 4 - cyano - 4" -*n* -pentyl-*p* -terphenyl hexafluorophosphate (RuCTP) [1], the first monolayer is influenced by the substrate, while subsequent layers are mutually identical.

The Surface Plasmon Resonance technique has proved to be very versatile, finding a number of applications where small changes in the properties of a material characterise a physical change in the vicinity of the measurement [2]. As a technique for investigating the linear and non-linear bulk optical properties of novel materials in thin film form, the SPR method is both accurate and sensitive [3]. For non-linear organic materials, particularly in LB film form, information on film thickness, refractive index and electro-optic coefficient may be straightforwardly obtained by computer analysis, even for a molecular monolayer with thickness of the order of only 3 nm or less.

THE SPR METHOD

A film of metal evaporated onto a glass microscope slide dramatically increases the internal reflectance at angles below the critical angle for total internal reflectance, θ_c, and causes some loss in the total internal reflection (TIR) region. For monochromatic light incident on a metal film roughly 50 nm thick, polarized in the plane of incidence (p-polarized), a resonant absorption is observed just above θ_c. For metals with loss loss at optical frequencies, particularly silver, the reflectance at the resonance angle may fall to practically zero, with a resonance width typically less than 0.5°. Fig. 1 shows such a resonance in the reflectance curve for a film of silver.

Lower-Dimensional Systems and Molecular Electronics
Edited by R.M. Metzger *et al.*, Plenum Press, New York, 1991

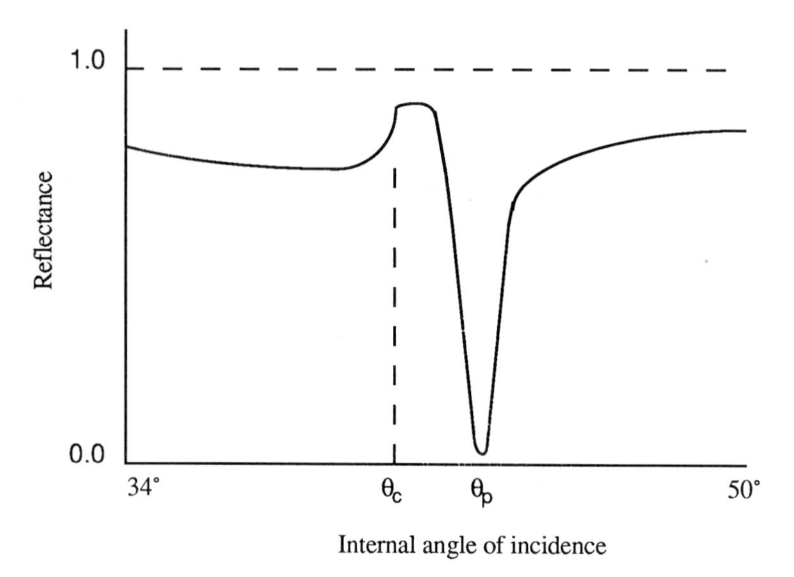

Fig. 1. Reflectance of a bare silver film

The absorption is caused by the interaction of the electric field of the incident light with charges (delocalized electrons) in the metal, near the metal-air interface. A dispersion relation governing the propagation of waves of charge density at the metal surface (surface plasmons) is depicted graphically in Fig. 2 [4]. For a given frequency of incident light, the component of the wavevector in the plane of the surface may be matched to the allowed wavevector of a plasmon at that frequency, by varying the incidence angle. When the dispersion conditions are matched, resonant energy transfer takes place, and the reflected beam is strongly attenuated.

For the analysis of the optical properties of thin films of dielectric material, this effect provides an ideal mechanism. Since the surface plasmon is localized at the metal-air interface,

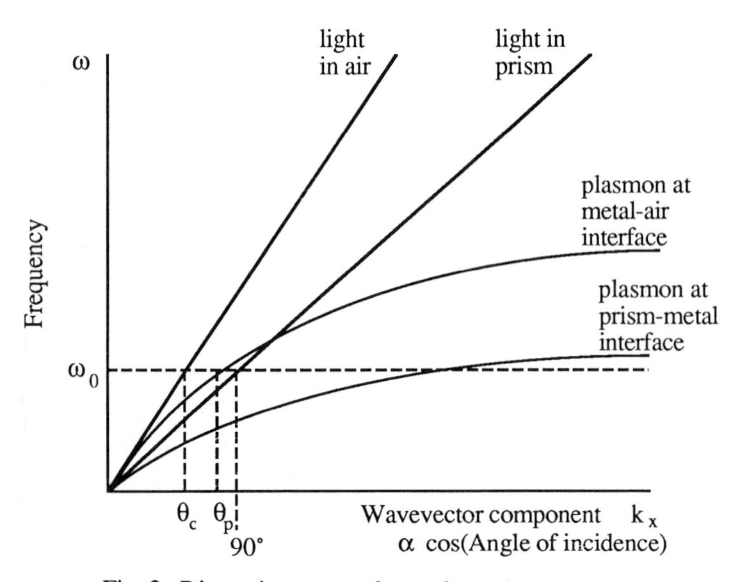

Fig. 2. Dispersion curves for surface plasmon wave

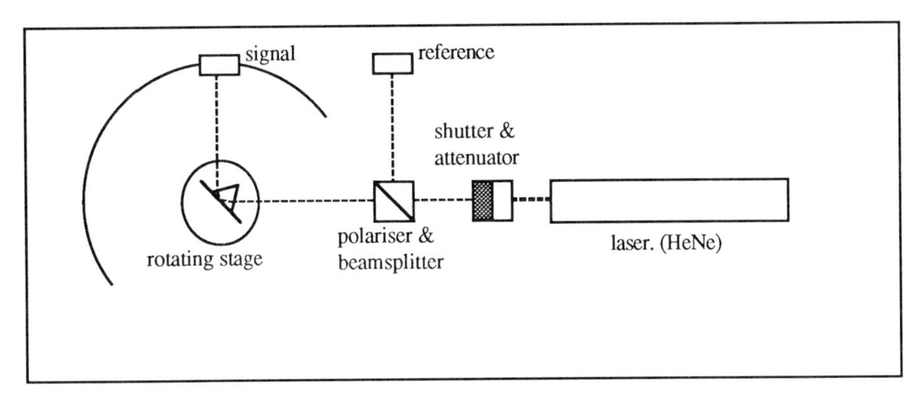

Fig. 3. Schematic diagram of the SPR Reflectometer

a film of dielectric material in this region will strongly influence the dispersion curve, hence shifting the optimum angle for resonant coupling. Shifts of the order of 1° are observed for a film thickness of only 10 nm.

A reflectometer is used to measure the internal reflectance at the silver film, using a prism to couple in the light as shown in Figs. 3 and 4. Typically, the angular resolution of the measurement is better than 0.01°, and the intensity resolution better than 0.1%. The light, from a low-power laser, is multiply reflected coherently within the glass-metal-dielectric-air boundary system, creating a region of enhanced electric field. This may be modeled by repeated combination of Fresnel's equation for reflection at a dielectric interface, applied at each boundary. Adjusting the film parameters in these equations allows theoretical data to be fitted to the experimentally determined reflectance curve. In this way, the thickness and complex permittivity of both the silver and dielectric films may be deduced.

Langmuir-Blodgett Films

The LB technique produces layered films, building up a superlattice structure from monomolecular layers of aligned organic molecules [5]. A film of amphiphilic molecules, spread on the surface of water, is compressed to form a densely packed, ordered monolayer of upright, aligned molecules. By passing a substrate through the water surface, this layer may be transferred to the substrate surface. Repeated deposition, using either a single material or two materials alternately, produces multilayers, giving fine control over film thickness, while incorporating active molecular units in a controlled fashion.

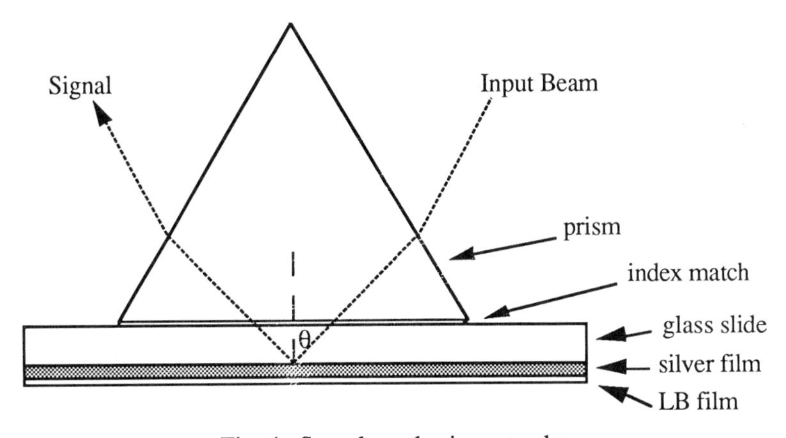

Fig. 4. Sample and prism coupler

LB films are ideally suited to analysis by the SPR technique. They may easily be deposited onto a silver-coated microscope slide, and produce a uniform reproducible dielectric layer with a thickness of roughly 2-3 nm per layer, depending on molecular size and orientation. For such films, the linear optical properties (refractive index and absorption coefficient) and the thickness may be obtained straightforwardly as described.

Non-linear Optics in Langmuir-Blodgett Films

A number of exploitable effects exist, due to the non-linear response of certain dielectric materials to applied electric and optical fields. An applied field, **E**, gives rise to a polarization field, **P**, within any dielectric medium. In a linear material, the relationship between **P** and **E** may be characterized by a single (first-order, second-rank) susceptibility tensor

$$\mathbf{P} = \varepsilon_0\, \chi : \mathbf{E}$$

For non-linear materials, the polarization is not linear with applied field, and terms must be included proportional to higher powers of the field [6]. In tensor notation,

$$\mathbf{P} = \varepsilon_0\, \chi^{(1)}:\ \mathbf{E} + \chi^{(2)}:\ \mathbf{E}\,\mathbf{E} + \chi^{(3)}:\mathbf{E}\,\mathbf{E}\,\mathbf{E} + \ldots$$

where $\chi^{(2)}$ (the second-order non-linear susceptibility) is a third-rank tensor and $\chi^{(3)}$ is a fourth-rank tensor. Higher orders may be considered, but most applications are due to second or third-order effects. These effects include the electro-optic (Pockels) effect, second and third harmonic generation, parametric oscillation and sum or difference frequency generation.

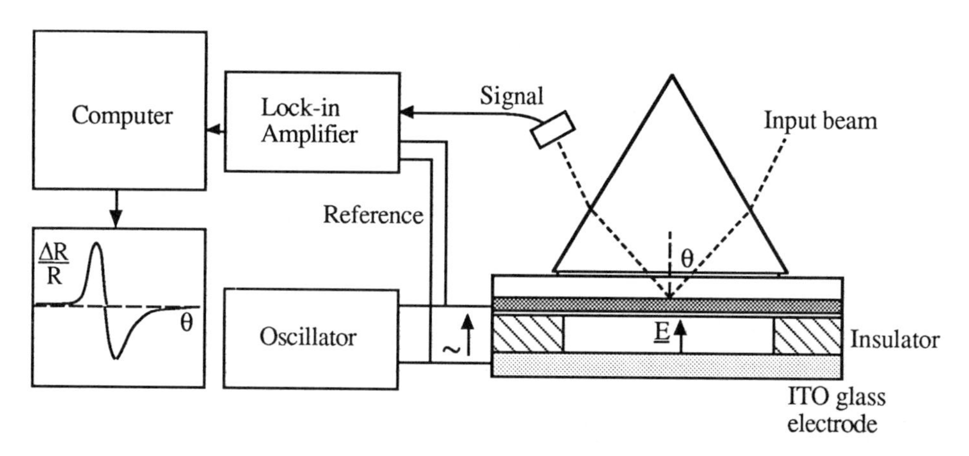

Figure 5. Electro-optical measurement system.

LB films possess many features advantageous to applications based on non-linear optical processes. Large non-linear coefficients may be achieved by the design of molecules containing units, which exhibit a highly non-linear response in the presence of an optical electromagnetic field. When incorporated into a superlattice structure, the alignment of the molecules ensures a cooperative response, while interaction between molecules may enhance the effect [7]. Both the active molecular moieties and intermolecular interaction may be optimized, to produce non-linearity far greater, than has been achieved in either organic or inorganic crystalline materials.

Practical device fabrication is an important consideration. The planar nature of LB films and their precisely controllable thickness and composition make them compatible with existing electro-optic technology, and provide the opportunity to construct novel active optical structures.

SPR and Electro-optic Measurement

Surface Plasmon Resonance may be used to measure the electro-optic effect, using the sample configuration shown in Fig. 5 [8].

In the presence of the applied field, the refractive index of the LB film will be slightly modified, producing an imperceptible shift in the resonance. By applying a sinusoidal voltage and monitoring the in-phase reflectance variation (roughly 1 part in 10^5), the derivative of the reflectance curve may be measured. Once again, fitting to theoretical values allows the magnitude of the permittivity change to be deduced, and enables the electro-optic coefficient to be calculated.

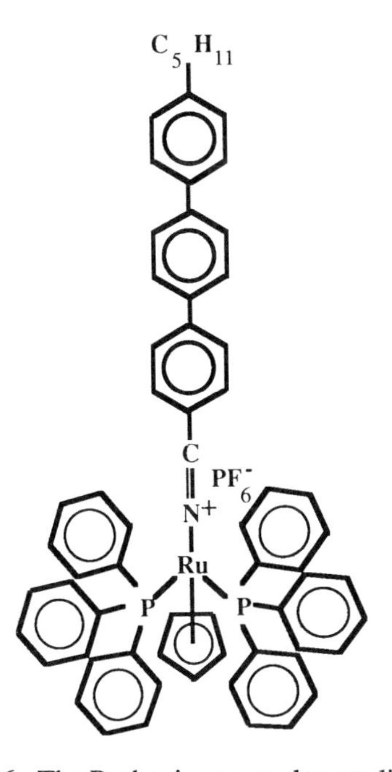

Fig. 6. The Ruthenium complex studied.

RESULTS

Measurement of resonance shifts for 1, 2 and 3 layers of RuCTP (illustrated in Fig. 6) indicates an angular shift varying linearly with the number of layers deposited (Fig. 7). To a first-order approximation, theory predicts a shift proportional to the number of layers deposited. For the measured data, however, the shift due to the first layer is different, and the graph, though linear for subsequent layers, does not pass through the origin. This suggests an interaction between the initial monolayer and the silver substrate. Such an anomaly is frequently found in studies of LB films, for example in capacitance measurements.

Preliminary analysis indicates a layer thickness of 3.09 nm and a complex permittivity of (2.34 + 0.25 i), corresponding to a refractive index, n, of 1.53 and absorption coefficient, k, of 0.08. These values are in agreement with figures produced by ellipsometric analysis of multilayer films (7 to 29 layers) of the same material [9]. The film thickness also coincides with the molecular length of 3.0 nm, measured from a precision space-filled model, suggesting vertical alignment of the molecules within the layers.

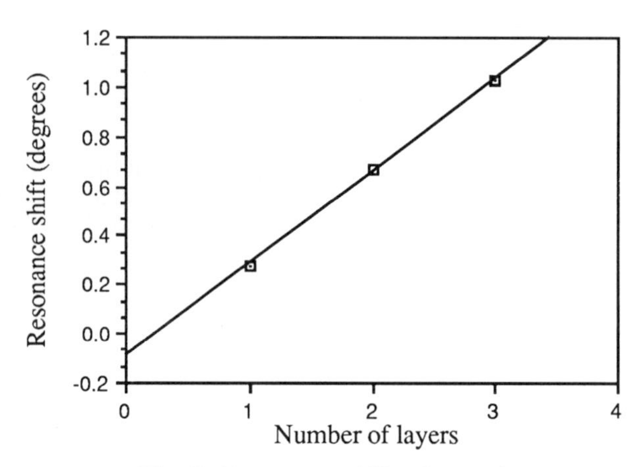

Fig. 7. Resonance shifts observed.

CONCLUSIONS

The SPR technique has proved capable of analyzing the optical properties of a material, by observing as little as two molecular monolayers. Full analysis promises to give accurate thickness and complex permittivity data, providing information on molecular orientation and packing. The sensitivity to a single monolayer will allow electro-optic measurements to be carried out on novel materials, without requiring that a large superlattice be fabricated.

Further research will investigate the nature of interactions within a monolayer and between layers in superlattices, in order to understand the mechanisms involved in producing large second-order optical non-linearity.

REFERENCES

[1] T. Richardson, G. G. Roberts, M. E. C. Polywka, and S. G. Davies, Preparation and characterization of organotransition metal Langmuir-Blodgett films, Thin Solid Films 160: 231-239 (1988).

[2] C. Nylander, B. Liedberg, and T. Lind, Gas detection by means of Surface Plasmon Resonance, Sensors & Actuators 3: 79 (1982-3).

[3] K. R. Welford, J. R. Sambles, and M. G. Clark, Guided modes and surface plasmon-polaritons observed with a nematic liquid crystal using attenuated total reflection, Liquid Crystals 2: 91 (1986).

[4] H. Raether, Surface Plasma Oscillations and Their Applications, in: "Physics of Thin Films, Vol. 9", G. Hass, M. H. Francombe, and R. W. Hoffman, eds., Academic Press, New York (1977).

[5] G. C. Roberts, An applied science perspective of Langmuir-Blodgett films, Advances in Physics 34: 475 (1985).

[6] A. Yariv, "Optical Electronics", Holt-Saunders, Japan (1985).

[7] D. B. Neal, M. C. Petty, G. G. Roberts, M. M. Ahmad, W. J. Feast, I. R. Girling, N. A. Cade, P. V. Kolinsky, and I. R. Peterson, Second harmonic generation from LB superlattices containing two active compoents, Electronics Letters 22: 460 (1986).

[8] G. H. Cross, I. R. Peterson, I. R. Girling, N. A. Cade, M. J. Goodwin, N. Carr, R. S. Sethi, R. Marsden, G. W. Gray, D. Lacey, A. M. McRoberts, R. M. Scrowston, and K. J. Toyne, A comparison of second harmonic generation and electro-optic studies of Langmuir-Blodgett monolayers of new hemicyanine dyes, Thin Solid Films 156: 39 (1988).

[9] Private communication with T. Richardson.

REVIEW OF THE ORGANIC RECTIFIER PROJECT: LANGMUIR-BLODGETT FILMS OF DONOR- SIGMA-ACCEPTOR MOLECULES

Robert M. Metzger

Department of Chemistry
University of Alabama
Tuscaloosa, AL 35487-0336, USA

Charles A. Panetta

Department of Chemistry
University of Mississippi
University, MS 38677, USA

INTRODUCTION

The Organic Rectifier Project (ORP) endeavours to verify the tantalizing proposal by Ari Aviram, Mark A. Ratner and coworkers [1-3] that a single organic molecule of the type D-σ–A could be a rectifier of electrical current. The molecule would do so because the D end is a good organic one-electron donor, σ is a covalent saturated ("sigma") bridge, and A is a good organic one-electron acceptor. The progress of the ORP [4-24] has been chronicled often [5,8,10,16,19-21,23] but the rectification by a single organic molecule, or by an organized Langmuir-Blodgett (LB) [25-31] monolayer of such molecules, has not yet been demonstrated. The driving force for the ORP, which may be one of the key experiments in molecular electronics, is that the working thickness promised by such a D-σ-A device is of the order of a few molecular lengths, i.e. about 5 nm: such a small size is predicted to be unattainable even by the rosiest forecasts for silicon or gallium arsenide technology.

A recent report of rectification observed in an internally H-bonded and σ–bridged ortho-quinone-catechol molecule by a modified scanning tunneling microscope (STM) [32], as well as preliminary STM work on one of the ORP LB films [20,21] have both turned out to be in error [20,21,23].

Of course, electrical rectification has been seen in LB multilayer sandwiches [33,34] $D\text{-}CA_n \mid CA_m \mid A\text{-}CA_p$ of n monolayers of cadmium arachidate doped by donor dyes ($D\text{-}CA_n$), followed by m undoped monolayers CA_m, followed by p monolayers of cadmium arachidate doped by electron acceptor dyes ($A\text{-}CA_p$) (where $n+m+p \geq 7$). Photodiode behavior by an LB film of pyrene-σ-viologen-σ-ferrocene has been seen [35,36]. The interest in detecting rectification by a single D-σ-A molecule (by STM), or by an LB film (by either STM or by conventional methods) remains.

This review first discusses the Aviram-Ratner Ansatz, then presents the criteria required for a successful device, then reviews the formation of LB films with molecules synthesized by the ORP; a summary is given of the cyclic voltammograms measured for these molecules, of solved crystal structures for related molecules, and of theoretical studies; a preliminary account is given of FTIR spectra of these LB films. Finally, the efforts to detect rectification are reviewed.

Lower-Dimensional Systems and Molecular Electronics
Edited by R.M. Metzger *et al.*, Plenum Press, New York, 1991

THE ORGANIC RECTIFIER ANSATZ

The Ansatz of Aviram and Ratner starts with the discovery, reviewed elsewhere in these Proceedings, of highly conducting lower-dimensional organic charge-transfer systems based on good one-electron donors (D) such as tetrathiafulvalene (TTF, **1**) and good organic one-electron acceptors (A) such as 7,7,8,8-tetracyanoquinodimethan (TCNQ, **2**). Good donors (i.e. molecules with relatively low gas-phase first ionization potentials I_D) are poor acceptors (have low electron affinity A_D), and good acceptors (i.e. molecules with a relatively high first electron affinity A_A); are poor donors (have high I_A) thus the gas phase energy required for reaction 3 (both components at infinite separation) is about 4 eV, while reaction 4 would need over 9 eV:

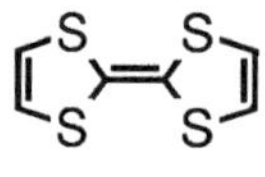

NC CN / NC CN — **1** TTF — **2** TCNQ — **5** — M_1 | D-σ-A / D-σ-A / D-σ-A | M_2 — **6**

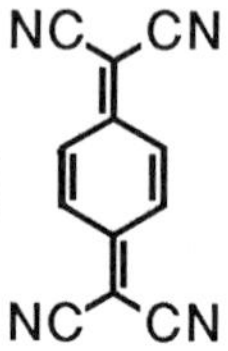

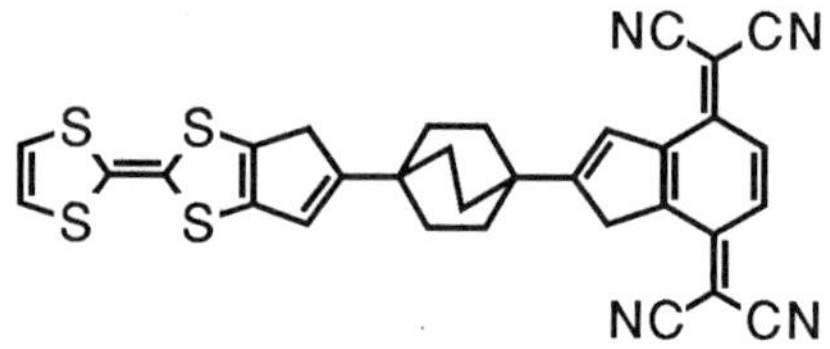

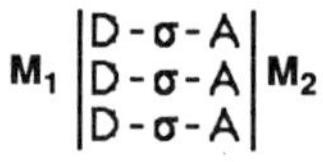

$$\mathrm{TTF(g) + TCNQ(g) \dashrightarrow TTF^{+}(g) + TCNQ^{-}(g) \quad I_D - A_A = 4.0\ eV} \qquad (3)$$

$$\mathrm{TTF(g) + TCNQ(g) \dashrightarrow TTF^{-}(g) + TCNQ^{+}(g) \quad I_A - A_D = 9.6\ eV\ (est)} \qquad (4)$$

Thus, if one makes a D-σ-A molecule like **5**, and assembles it somehow between two metal electrodes M_1 and M_2, as in **6**, then the direction of easy electron flow is from M_2 to M_1 because it utilizes the zwitterionic state D^+-σ-A^- (while the electron flow from M_1 to M_2 would be inefficient, because the barrier to forming the zwitterion D^--σ–A^+ would be several eV higher). Using terms popularized by Hoffman [37], the Aviram-Ratner device can work if the tunneling of electrons from A to D is through the bond system, and will fail if the tunneling between the metal electrodes M_1 and M_2 is predominantly through space. Molecule **5** was never synthesized, and the idea languished until the ORP started in earnest.

DECISION: HOW TO ASSEMBLE D-σ-A DEVICES

To address a single molecule electrically requires a "molecular wire" (e.g. a polyacetylene strand) or a "molecular antenna" (e.g. the conjugated portion of β-carotene), neither of which can be easily connected to an external potential source, at present. Until the recent advent of the STM, one could not connect an external circuit to a single molecule. Therefore, when the ORP was initiated, one had to content oneself with assemblies of molecules. The three techniques that showed promise were (i) the LB technique [25-31], discussed elsewhere in these Proceedings, and the technique of covalently bonding molecules to electrode surfaces, either by (ii) silanizing a hydroxyl-coated electrode, then attaching molecules covalently [38], or (iii) by silanizing the molecule, and attaching it directly by spin-coating by the oleophobic method [39] to a hydroxyl-coated electrode [40]. Good monolayer coverage is claimed for the latter method [40], but not for the former method [38].

The ORP is committed to use the LB technique. With it, of course, one has to be sure that the films are defect-free in the lateral dimension to within the area probed electrically. It is well known, alas, that monolayers do have microscopic and macroscopic defects [41].

ELECTRONIC AND SYNTHETIC CRITERIA

There are several interlocking criteria that must be satisfied for the rational synthesis of suitable D-σ-A systems:

(i) I_D for the donor end must be as small as possible, and as close as possible to the work function ϕ of the metal layer M_1. Typical values are listed in Table 1.

(ii) A_A for the acceptor end (Table 1) must be as large as possible, and as close as possible to the work function ϕ of the metal layer M_2. It is clear from Table 1 that requirements (i) and (ii) can be satisfied only approximately.

Table 1. Experimental ionization potentials I_D for selected good donors D, experimental electron affinities A_A for selected good acceptors A, and work functions ϕ for selected metals M_1, M_2 (from Ref.[19])

Donor D	I_D/eV	Acceptor A	A_A/eV	Metal	ϕ/eV
TMPD (**7**)	6.25	DDQ (**9**)	3.13	Al	3.74
TTF (**1**)	6.83	TCNQ (**2**)	2.8	Au	4.58
Pyrene (**8**)	7.41	Chloranil (**10**)	2.76	Pt	5.29

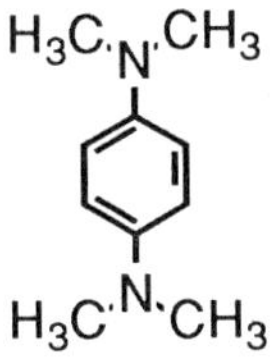

7 TMPD   **8** Py   **9** DDQ   **10** Chl

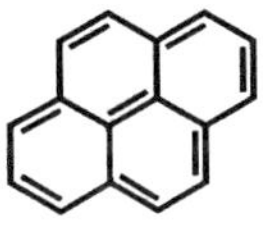

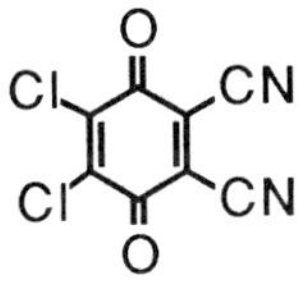

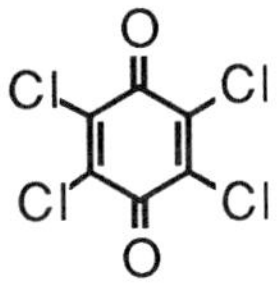

(iii) It is synthetically difficult to convert a weak donor into a strong donor, and a weak acceptor into a strong acceptor in situ on a molecule in which the bridge has already been built. Therefore, instead, one must use mono-substituted strong donors, and mono-substituted strong acceptors, which can be coupled efficiently by some coupling reaction which avoids the usual formation of ionic charge-transfer complexes. Such a coupling reaction is the urethane, or carbamate coupling reaction, pioneered for a $(-\sigma-\text{TTF-}\ \sigma\text{-TCNQ-})_x$ copolymer by Hertler [42] and adapted to monofunctional derivatives by Baghdadchi [43].

(iv) The molecules must pack well into self-assembling monolayers. If the designed molecule does not form Pockels-Langmuir (PL) [15, 24] self-assembling monolayers at the air-water interface, then either long aliphatic "greasy tails" must be added to form a hydrophobic tail, or an ionic, hydrophilic "head" should be added. The molecules should be reasonably flat, so as to form compact films, yet flexible enough so as transfer well (by the vertical dipping method) as LB films.

(v) The synthesis of the acceptor should be facile. The acceptor used in the early work of the ORP was BHTCNQ, **11**, whose synthesis [42, 43] was very inefficient. A better monofunctionalized TCNQ acceptor made in high yield was HETCNQ, **12** [17].

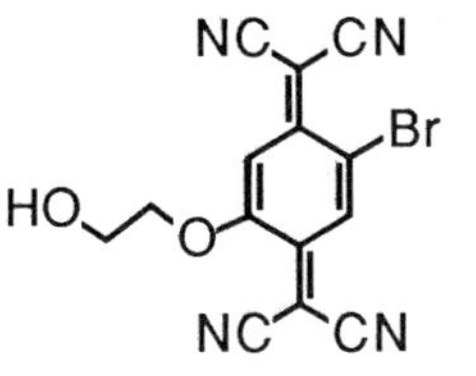

11 BHTCNQ   **12** HETCNQ

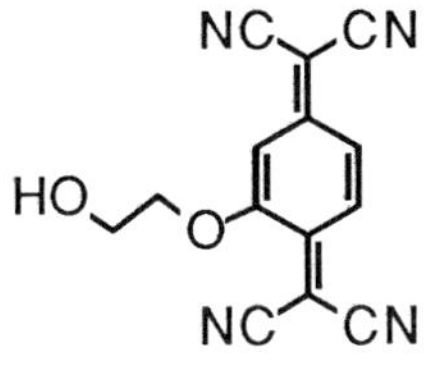

(vi) The electron transfer through the D-σ-A molecule, and through its hydrophobic or hydrophilic tails, must be fast: a molecular device that is small but slow is not predicted to be useful. That electron transfer is fast through properly designed molecules, e.g. the photosynthetic reaction center, is well known. Miller and co-workers [44,45] proved the Marcus "inverted region" theory [46] of electron transfer rates, and showed conclusively that the through-bond electron transfer rate first increases dramatically with I_D-A_A, then reaches a maximum, then the electron transfer rates starts to decrease with further increases of I_D-A_A, where the reorganization of the donor into the geometry of the product D^+, and of the acceptor into the product A^- is so great, that the electron transfer rate is significantly reduced by the increasing Franck-Condon factor. Thus we must balance the requirements of Table 1 with those of the Franck-Condon factor.

(vii) The device **6** (which we will indicate as M_1 | D-σ-A | M_2) will have a finite tolerance for high voltages or heating.

LANGMUIR-BLODGETT FILMS

We present in Table 2 a "zoo" of molecules (**13-22**) prepared by the ORP, which we have shown to form PL monolayers at the air-water interface, and which transfer well onto Al or glass or other slides as LB monolayers. TTF-C-BHTCNQ, **13**, was difficult to purify; the "neutral" form seemed to deposit "pancake-style" onto the water, and synthetic difficulties forced its abandonment. The strongest films (highest collapse pressure, most vertical pressure-area isotherm) were obtained with BDDAP-C-BHTCNQ, **15** (Fig. 1). The acceptor HMTCAQ used in **17** and **18** was easy to prepare [9], but is well known to be a weak two-electron acceptor, with a highly non-planar geometry [22].

Molecules **19-22** are model systems for a related project, which aims to incorporate D-σ-A systems into LB-film-forming diacetylenes, which may be polymerized in situ on the film balance, for the purpose of preparing new systems with promise as non-linear optical devices. Interestingly, **19-22** form Z-type multilayers on a glass substrate (the film subphase is held at 5° C but the slide is at room temperature). An attempt was made to see whether any second harmonic signals could be detected from Z multilayers of **21**, but the result was negative [47].

Table 2. Pressure-Area isotherm data for Pockels-Langmuir films. Π_c and A_c are the pressure and molecular area, respectively, at the collapse point.

Molecule	No.	Type	T K	Π_c mN/m	A_c Å^2	Ref.
TTF-C-BHTCNQ	**13**	strong D strong A	292	12.7	134±50	[5]
DDOP-C-BHTCNQ	**14**	weak D strong A	292	20.2	50	[7]
BDDAP-C-BHTCNQ	**15**	medium D strong A	293	47.3	57	[15]
Py-C-BHTCNQ	**16**	medium D strong A	283	28.2	53	[7]
BDDAP-C-HMTCAQ	**17**	medium D weak A	293	22.3	58	[15]
BHAP-C-HMTCAQ	**18**	medium D weak A	293	35.8	42	[16]
DDOP-C-ENP	**19** a	weak D weak A	278	23.7	38	[24]
TDDOP-C-ENP	**20** a	weak D weak A	278	34.0	76	[24]
TDDOP-C-HETCNQ	**21** a	weak D strong A	283	47.5	54	[24]
MTDAP-C-ENP	**22** a	weak D weak A	278	16.5	63	[23]

(a) Z-type LB multilayers on glass (substrate at 22°C, Pockels-Langmuir film on water subphase at 5°C).

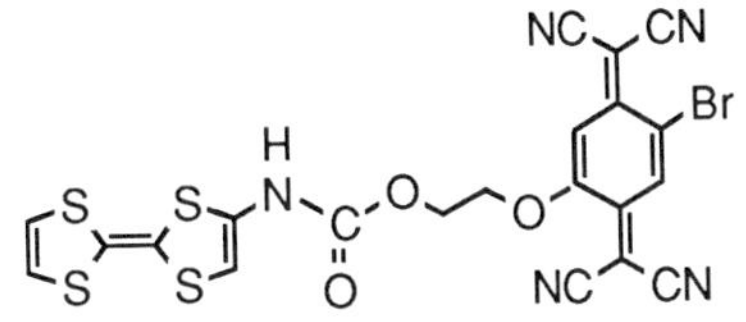

13
TTF-C-BHTCNQ

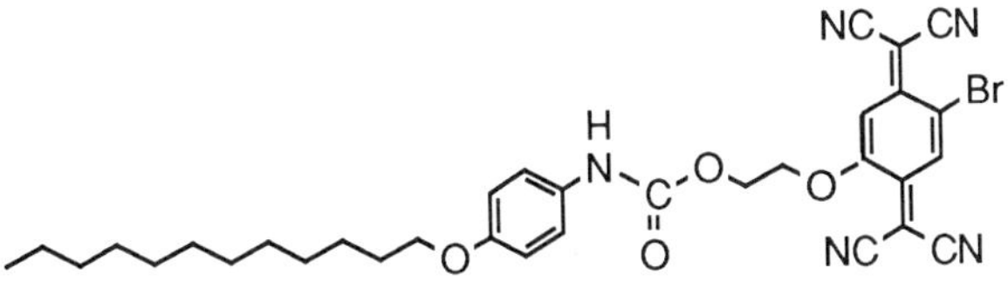

14
DDOP-C-BHTCNQ

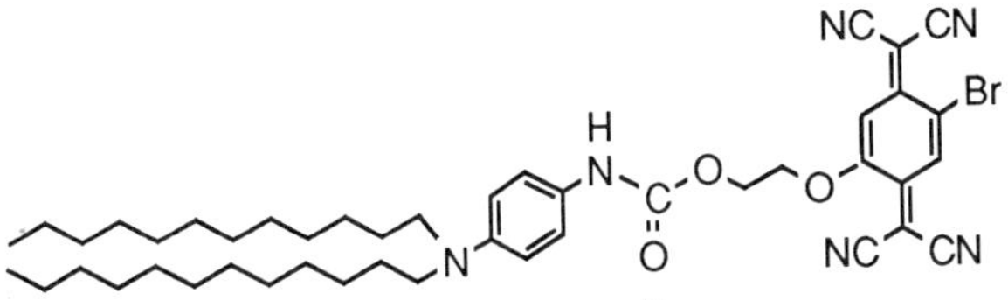

15
BDDAP-C-BHTCNQ

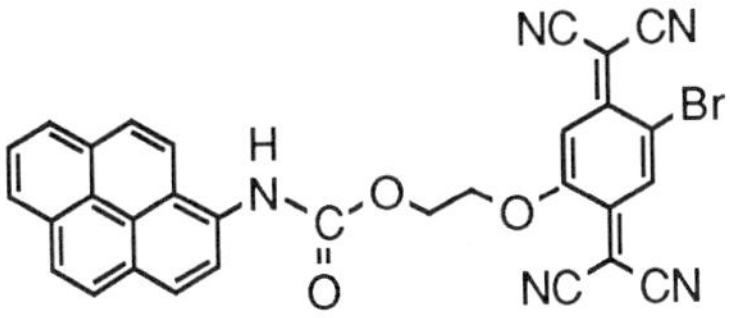

16
Py-C-BHTCNQ

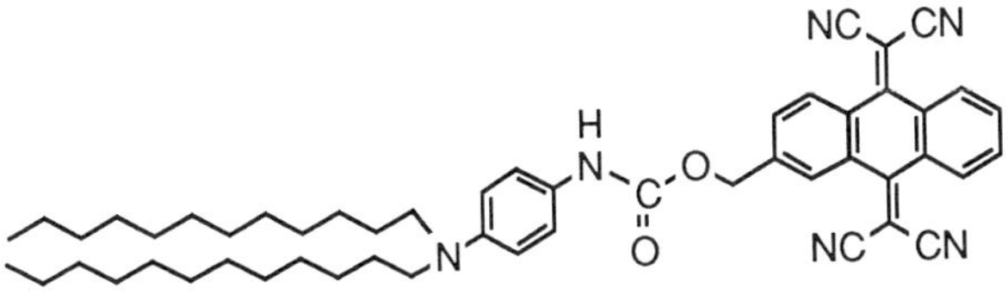

17
BDDAP-C-HMTCAQ

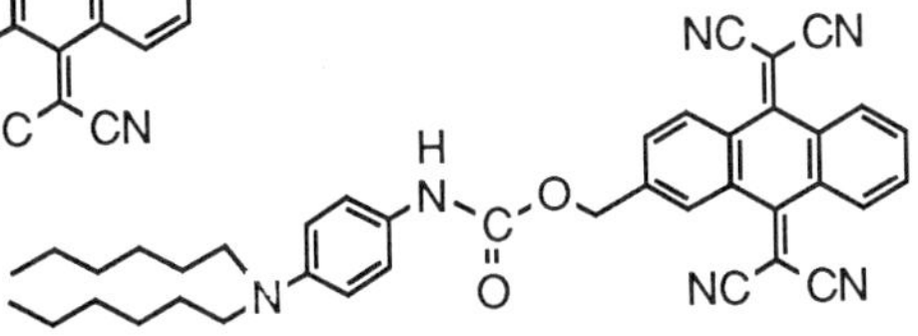

18
BHAP-C-HMTCAQ

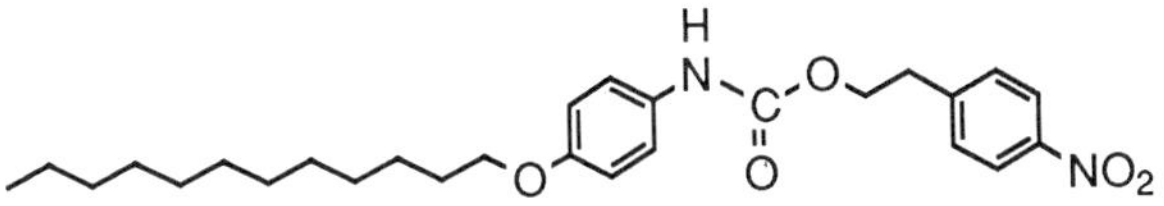

19
DDOP-C-ENP

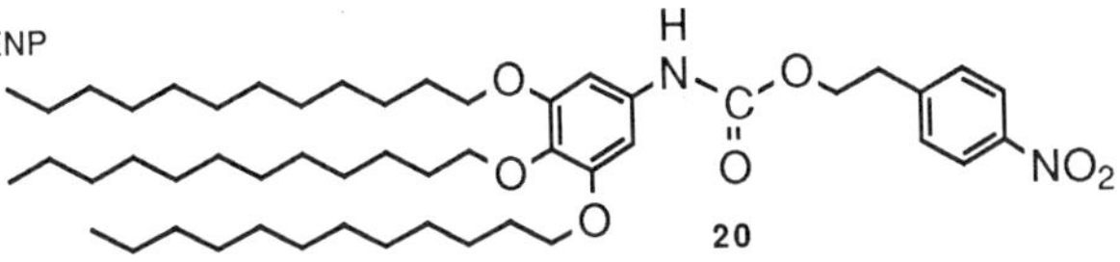

20
TDDOP-C-ENP

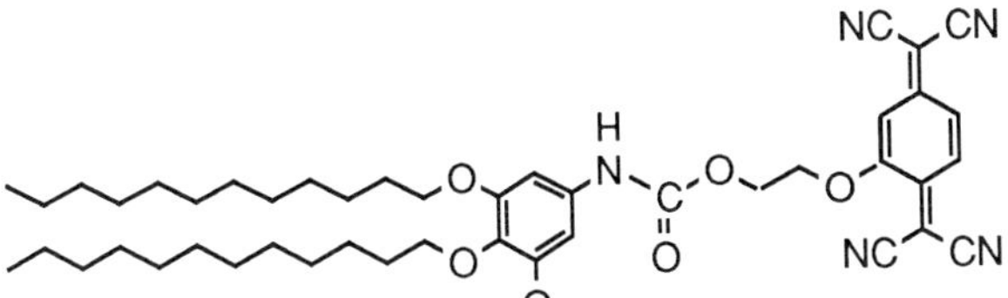

21
TDDOP-C-HETCNQ

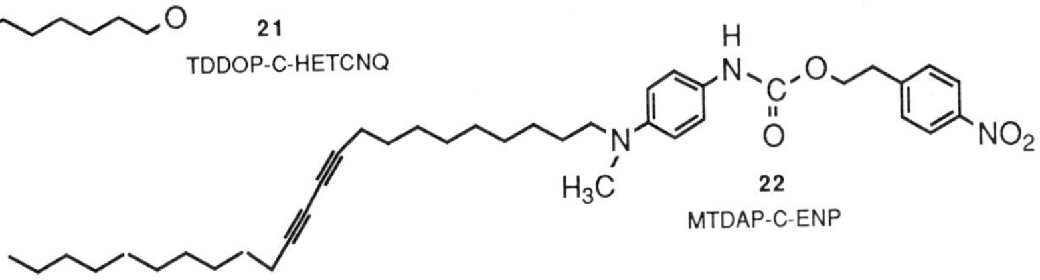

22
MTDAP-C-ENP

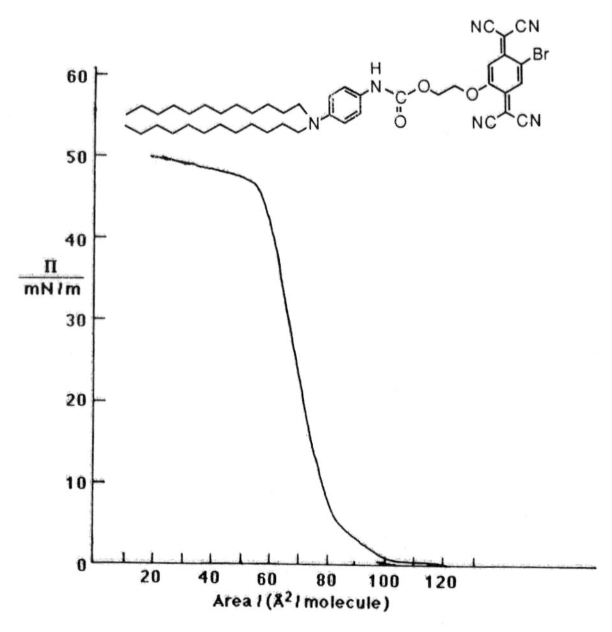

Fig. 1. Pressure-area isotherm of BDDAP-C-BHTCNQ, **15**, at 293 K [15].

CYCLIC VOLTAMMETRY

We have previously characterized donor DMAPCMe, **23** , and acceptors **2**, **11**, **12**, and HMTCAQ, **24**, and D-σ-A molecules **14**, **19**, **20** and **21** by cyclic voltammetry (CV). CV data have also been obtained recently for BHAP-C-HMTCAQ, **18**, MTDAP-C-ENP, **22**, Py-C-HETCNQ, **25** (a minor modification of **16**; Fig. 2) and for BDDAP-C-HETCNQ, **26** (a minor modification of **15**, Fig. 3). The results are summarized in Table 3.

23
DMAPCMe

24
HMTCAQ

25
Py-C-HETCNQ

26
BDDAP-C-HETCNQ

Table 3. Solution cyclic voltammetric potentials for donor **23**, acceptors **2, 11, 12,** and **24**, and D-σ-A molecules **14, 18-22, 25** and **26**. All data were obtained at a Pt electrode, and are given in Volts vs SCE.

Molecule	No.	Oxid.(1) D ---> D+		Oxid.(2) D+ --->D++		Red.(1) A--->A⁻		Red. (2) A⁻ --->A⁻ ⁻		Ref.
		E_p	$E_{1/2}$	E_p	$E_{1/2}$	E_p	$E_{1/2}$	E_p	$E_{1/2}$	
Donor:										
DMAP-C-Me	**23** a	0.58	0.55	-	-	-	-	-	-	[12]
Acceptors:										
TCNQ	**2** b	-	-	-	-	-	0.19	-	-0.35	[48]
TCNQ	**2** c	-	-	-	-	0.11	0.13	-0.46	-0.43	[24]
BHTCNQ	**11** b	-	-	-	-	-	0.305	-	-0.170	[7]
HETCNQ	**12** b	-	-	-	-	-	0.107	-	-0.398	[17]
HMTCAQ	**24** b	-	-	-	-	-	--	-0.372	-0.333	[9]
D-σ-A:										
DDOP-C-ENP	**19** c	1.42	1.39	-	-	-1.16	-1.13	-	-	[24]
TDDOP-C-ENP	**20** c	1.17i	-	-	-	-1.12	-1.15	-	-	[24]
MTDAP-C-ENP	**22** c	0.57	0.54	-	-	-1.09	-1.06	-	-	[49]
BHAP-C-HMTCAQ	**18** c	0.63	0.60	-	-	-	-	-0.39	-0.36	[49]
DDOP-C-BHTCNQ	**14** d	-	1.21	-	-	-	0.25	-	-0.07	[7]
TDDOP-C-HETCNQ	**21** c	1.02i	-	-	-	0.10	0.07	-0.50	-0.47	[24]
BDDAP-C-HETCNQ	**25** e	0.66i	-	1.10	-	0.02	-	-0.49i	-	[49]
Py-C-HETCNQ	**26** e	1.04	1.01	1.18	1.15	0.11	0.08	-0.32	-0.35	[49]

(a) Solvent: CH_3CN. Reference electrode: SCE. A peak at 0.37 V (return scan) grows with successive cycles, and is indicative of dimer or polymer formation [12].
(b) Solvent: CH_3CN. Reference electrode: SCE.
(c) Solvent: CH_2ClCH_2Cl. Reference electrode: Ag|AgCl. An offset correction of 0.15 V has been applied to convert the values to V vs. SCE.
(d) Solvent: CH_3CN. Reference electrode: Ag|AgNO3. An offset correction of 0.320 V has been applied to convert the values to V vs. SCE.
(e) Solvent: CH_2ClCH_2Cl. Reference electrode: Ag|AgCl. An offset correction of 0.19 V has been applied to convert the values to V vs. SCE.

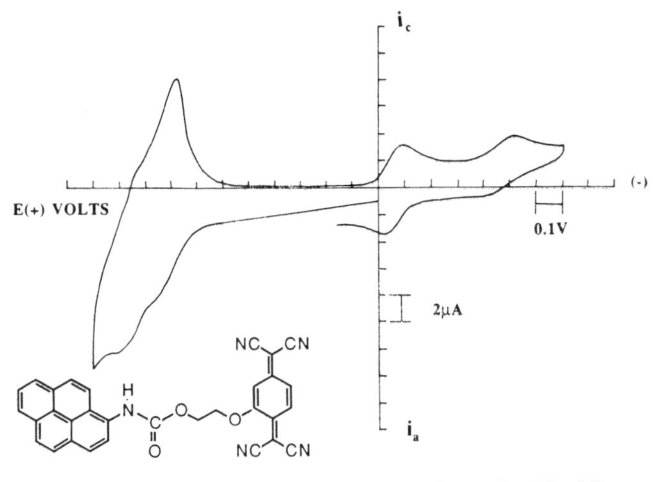

Fig. 2. Cyclic voltammogram of Py-C-HECTNQ, **25**.

These data were collected to confirm that the carbamate linkage does preserve the oxidation (reduction) potentials of the D(A) ends of the D-σ-A molecules. Roughly speaking, these are indeed preserved, i.e. the CV's are reasonably close to being superpositions of the

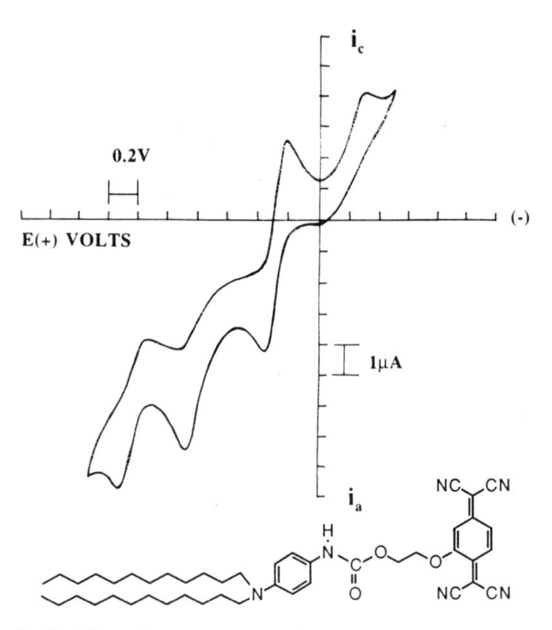

Fig. 3. Cyclic voltammogram of BDDAP-C-HETCNQ, **26**.

CV's of the donors and the acceptors. One has to be careful with the data of Table 3, in that different solvents have been used, and conversions from different reference electrodes have been made. Also, vastly differing molecular sizes can affect the location of peak potentials.

The oxidation potentials of the D of DMAP-C-Me, BHAP-C-HMTCAQ, and BDDAP-C-HETCNQ (E_p = 0.58, 0.63, 0.66 V) are fairly close to each other, since they all involve the same N-substituted p-phenylenediamine moiety. The oxidation potentials for DDOP-C-ENP and DDOP-C-BHTCNQ ($E_{1/2}$ = 1.39, 1.21 V) should be identical, but aren't. As more dodecoxy groups are added to the aminobenzene moiety, in the sequence DDOP, BDDOP, TDDOP, the oxidation potentials decrease slightly, as expected ($E_{1/2}$ = 1.21 V for DDOP-C-BHTCNQ, and 1.39 V for DDOP-C-ENP, versus E_p = 1.17 V for TDDOP-C-ENP, and 1.02 V for TDDOP-C-HETCNQ). The "best" donor (lowest value of oxidation potential, which should correlate with the lowest value of the ionization energy I_D) is seen to be the BDDAP end of BDDAP-C-HECTNQ. It is quite suprising how low $E_{1/2}$ is for MTDAP-C-ENP, **22**.

The lowest (most positive) first reduction potential (which should correlate with the highest electron affinity A_A) is for BHTCNQ ($E_{1/2}$= 0.305 V), followed by TCNQ (0.13 V), and by HETCNQ (0.107 V). The single reduction peak for HMTCAQ ($E_{1/2}$= -0.333 V) corresponds to a two-electron process [50] and is fairly close to the second reduction potentials of the other, "better" acceptors BHTCNQ (-0.170 V), TCNQ (-0.43 V), and HETCNQ (-0.398 V). The very weak donor ENP has reduction potentials exceeding -1.0 V. The reduction potential for BHAP-C-HMTCAQ (-0.36 V) corresponds closely to that of HMTCAQ (-0.333 V); the reduction potential of DDOP-C-BHTCNQ (0.25 V) is reasonably close to that of BHTCNQ (0.305 V); the reduction potentials of Py-C-HETCNQ (0.02 V), TDDOP-C-HETCNQ (0.07 V), and of BDDAP-C-HETCNQ (0.02 V) are very close to that of HETCNQ (0.107 V).

CRYSTAL STRUCTURES OF DONOR, ACCEPTORS AND MODEL D-σ-A MOLECULES

A certain number of crystal structures have been solved: of the donor DMAP-C-Me, **23** [11] (Fig.4) the acceptor BHTCNQ, **11** [13] (Fig.5), and of the methyl ester (AETCNQ, **27**) of the acceptor HETCNQ, **12** [14] (Fig.6). The difference in conformation between AETCNQ and BHTCNQ must be due to crystal packing forces, rather than to intramolecular effects. Of course, molecules that form PL and LB films will not usually crystallize, because of the usual aliphatic "tails" added to them to help them self-assemble as PL and LB films. However, we

27
AETCNQ

28
Ph-C-BHTCNQ

29
BMAP-C-HMTCAQ

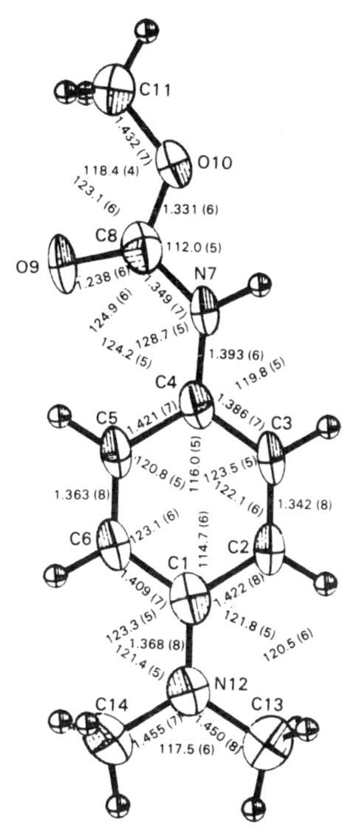

Fig. 4. ORTEP-II plot of the crystalline structure of DMAP-C-Me, **23** [11]. Space group Pbca (#61), a = 13.926, b= 9.999, c= 14.854 Å, Z =8, R =5.9% for 786 reflections.

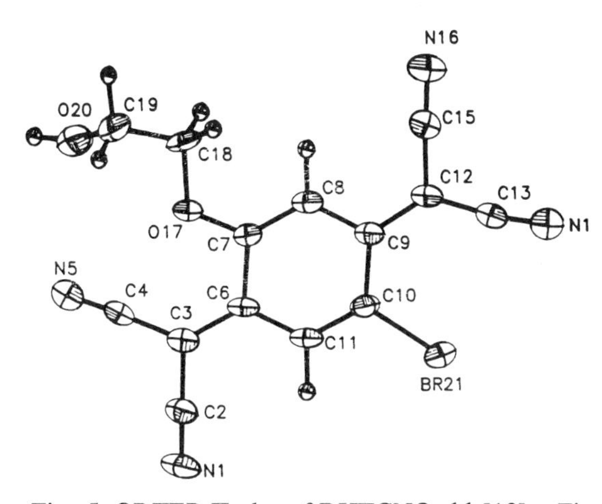

Fig. 5. ORTEP-II plot of BHTCNQ, **11** [13]. Space group $P2_1/n$ (#14) , a = 9.258, b= 13.618, c= 10.947 Å,β = 92.14°, Z = 4, R =3.9% for 1395 reflections

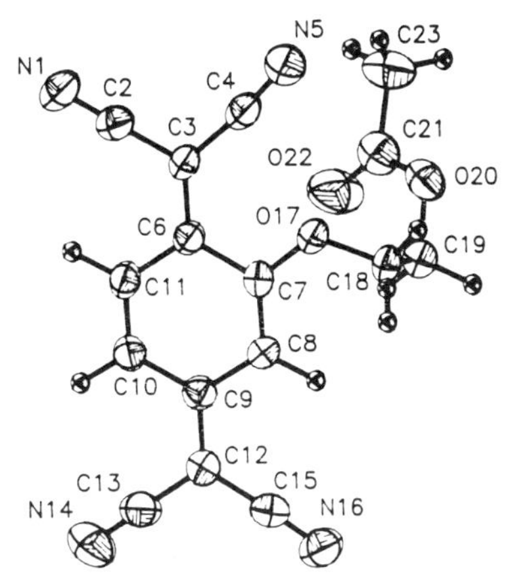

Fig.6. ORTEP-II plot of AETCNQ, **27** [14] Space group P1 bar (#2), a = 7.165, b= 9.058, c= 13.244 Å, α = 70.06°, β = 87.14°, γ = 68.22°, Z = 2, R = 3.4% for 1143 reflections

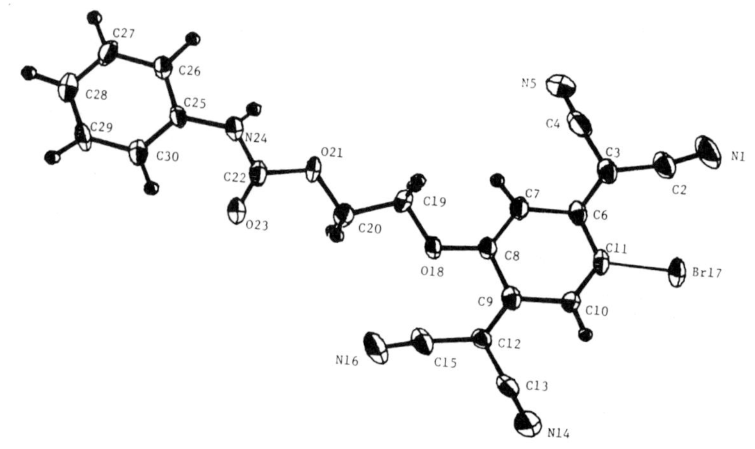

Fig. 7. ORTEP-II plot of the crystalline structure of Ph-C-BHTCNQ, **28** [6].Space group $P2_1/n$ (#14), a = 8.310, b= 9.278, c= 25.383 Å, β = 96.15°, Z = 4, R =7.9 % for 2229 reflections

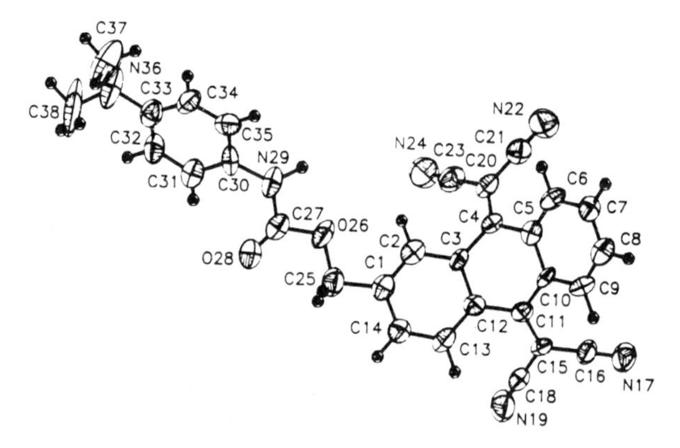

Fig. 8. ORTEP-II plot of the crystalline structure of BMAP-C-HMTCAQ, **29** [22]. Space group P1 bar (#2), a = 8.748, b= 10.989, c= 13.541Å, α = 90.67° β = 99.15°, γ = 98.62°, Z = 2, R =12.7 % for 1839 reflections.

have solved the structures of two D-σ-A molecules which do not form PL or LB films: Ph-C-BHTCNQ, **28** [6] (Fig. 7), and BMAP-C-HMTCAQ, **29** [22] (Fig. 8). Both structures show that the carbamate linkage is extended, rather than bent; in Ph-C-BHTCNQ, the dihedral angle between the phenyl ring and the six-membered central ring of BHTCNQ is only 8° [6].This gives hope that in LB films of related D-σ-A molecules the carbamate linkage will also be extended.

MOLECULAR ORBITAL CALCULATIONS

As reported previously [20] semi-empirical molecular orbital (MO) calculations, using the MNDO algorithm in program MOPAC with full geometry optimization, have been performed on D-σ-A molecules, to predict their geometry, and also their HOMO and LUMO energies. The structures, as drawn by program ORTEP-II using typical van der Waals atomic radii (1.2 Å for H, 1.7 Å for C, 1.5 Å for O, 1.35 Å for F, and 1.65 Å for S) are shown in Figs. 9-13 (the van der Waals shape is viewed from the A end of the molecule). Also given are the HOMO ($-I_D$) and LUMO ($-A_A$) energies, as well as the differences I_D-A_A and I_A-A_D discussed in Eqs. 3 and 4 above.

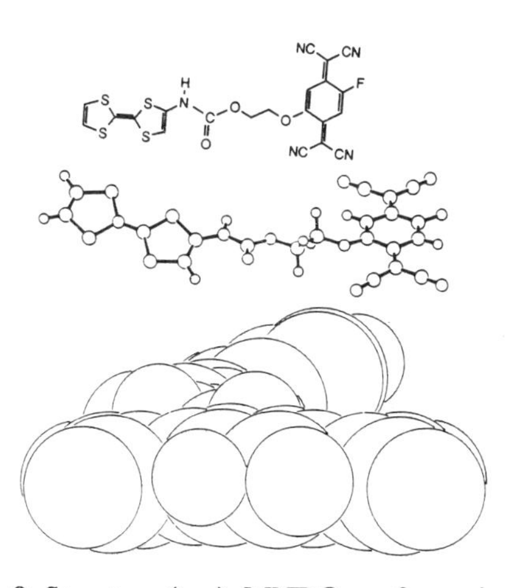

Fig. 9. Structure (top), MNDO conformation (middle), and MNDO van der Waals shape, viewed from the acceptor end (bottom) for TTF-C-FHTCNQ, a variant of TTF-C-BHTCNQ, **13**; I_D=8.197 eV, A_A = 2.975 eV, I_D-A_A = 5.222, I_A-A_D = 9.447 eV [20]

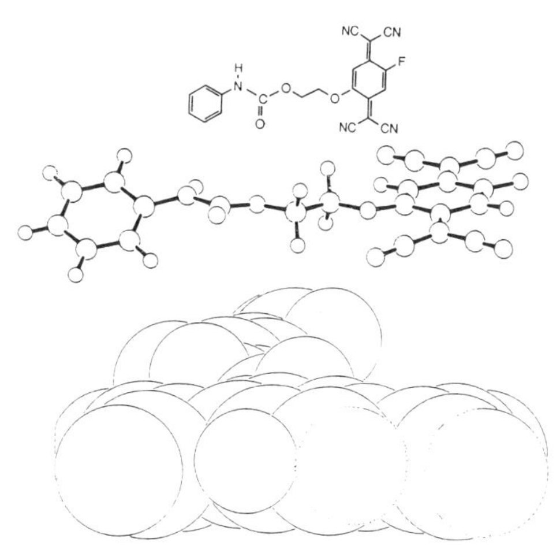

Fig. 10. Structure (top), MNDO conformation (middle), and MNDO van der Waals shape, viewed from the acceptor end (bottom), for Ph-C-FHTCNQ, **28**, a variant of Ph-C-BHTCNQ; I_D=9.141 eV, A_A = 2.941 eV, I_D-A_A = 6.200 eV, I_A-A_D = 9.493 eV [20]

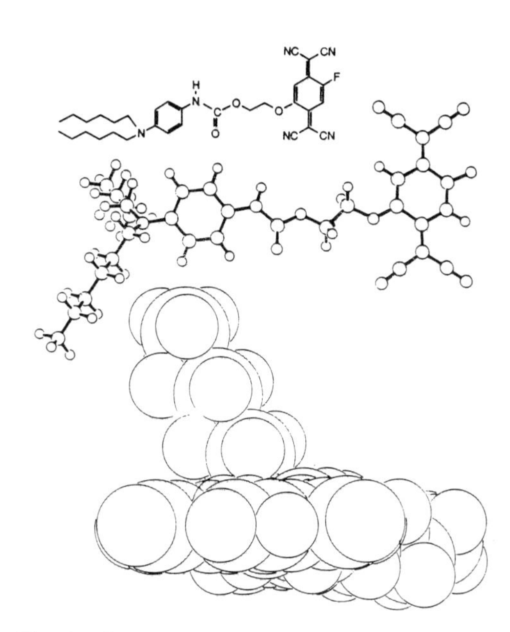

Fig. 11. Structure (top), MNDO conformation (middle), and MNDO van der Waals shape, viewed from the acceptor end (bottom) for BHAP-C-FHTCNQ, a variant of BDDAP-C-BHTCNQ, **15**; I_D=8.767 eV, A_A = 2.936 eV, I_D-A_A = 5.831, I_A-A_D = 9.461 eV [20]

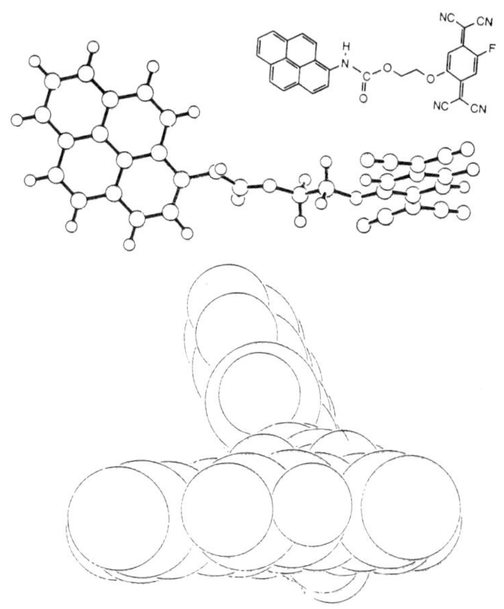

Fig. 12. Structure (top), MNDO conformation (middle), and MNDO van der Waals shape, viewed from the acceptor end (bottom), for Py-C FHTCNQ, a variant of Py-C-BHTCNQ; **16**; I_D=8.192 eV, A_A = 2.957 eV, I_D-A_A = 5.235 eV, I_A-A_D = 8.347 eV [20]

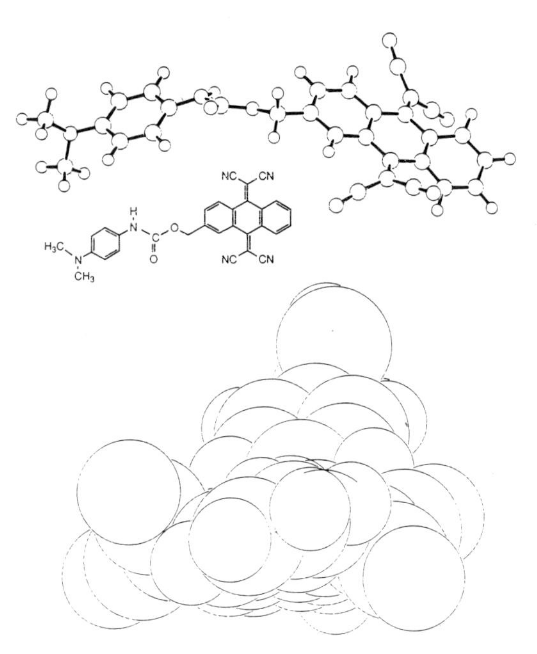

Fig. 13. MNDO conformation (top), structure (middle), and MNDO van der Waals shape, viewed from the acceptor end(bottom) of BMAP-C-HMTCAQ, **29**; I_D=9.052 eV, A_A = 2.627 eV, I_D-A_A = 6.425, I_A-A_D = 9.220 eV [20].

Overall, the MNDO structures are extended, as expected, but there are some small surprises: (i) while the data of Table 2 suggest that in a PL film Py-C-BHTCNQ is a relatively flat molecule, Fig. 12 introduces a twist in the carbamate linkage; (ii) while the data of Table 2 suggest a well-packed BDDAP-C-BHTCNQ molecule, Fig. 11 shows a conformer in which the bis-hexyl "tails" are not well aligned with each other; (iii) while the crystal structure of BMAP-C-HMTCAQ [22] shows that both dicyanomethylene substituents on the anthracene ring deviate from the plane in the same direction, MNDO decided to give them a "corkscrew" twist in Fig. 13. The MNDO ionization energies I_D seem 1 to 2 eV high, as is to be expected from a Koopmans' theorem calculation; the size of the difference (I_D-A_A) - (I_A-A_D) is as large as expected, but some of the I_D values are not as low as hoped. Further calculations are planned.

FOURIER TRANSFORM INFRARED SPECTRA OF D-σ-A MONOLAYERS

Grazing-angle Fourier transform infrared (FTIR) spectra of monolayers of BDDAP-C-HMTCAQ, BDDAP-C-BHTCNQ, and TDDOP-C-HETCNQ, have been measured, and reported previously [19-21, 23]. The C-H stretch bands are well resolved, even for a single monolayer, and a broad structure at about 3500 cm^{-1} is seen for "fresh" samples, but disappears for samples older than about 60 days; this may be water trapped between the LB film layer and the aluminum layer, but the identification is not certain. Further studies are planned.

RECTIFICATION ATTEMPTS

We discuss here three experiments [7,19-21] to detect rectification. In the first experiment, a 2 mm diameter droplet of Hg was used to probe the conductivity across a single monolayer LB film deposited on either Pt or conducting tin oxide glass: the sandwiches:
(i) Pt | DDOP-C-BHTCNQ | Hg,
(ii) Pt | Py-C-BHTCNQ | Hg, and

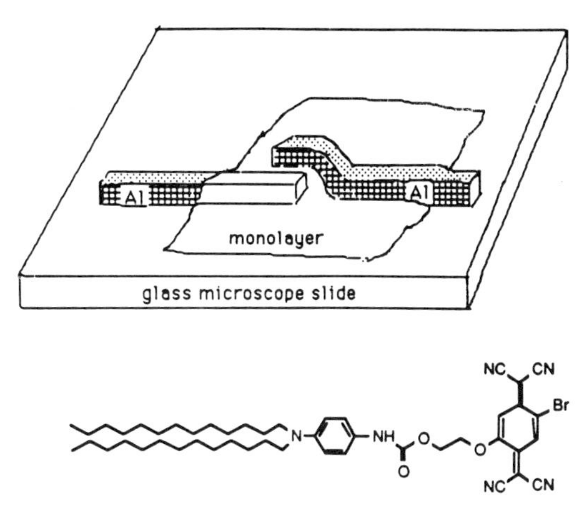

Fig. 14. Arrangement of one Al finger, covered by a monolayer of BDDAP-C-BHTCNQ, and by a second finger of Al with an area of overlap 0.5 mm x 0.5 mm.

(iii) conducting SnO_2 glass I DDOP-C-BHTCNQ I Hg
were thus tested [7]: in all cases the background conductivity of the solid support was measured, presumably because of microscopic pinholes in the LB film.

The second experiment [19-21], tried to avoid defects, statistically, in a domain of the order of about 0.5 mm x 0.5 mm, by searching through enough samples. The left-hand half of fifteen glass microscope slides was coated (using a mask) with five parallel fingers of Al at least 500 nm thick, 3.5 mm long, and 1.6 mm wide. Next, the fifteen slides were covered with a single LB monolayer of BDDAP-C-BHTCNQ, **15**, at room temperature. Finally, the slides were coated again with five fingers of Al, but this time on the right hand side of the slide, so that the vertical overlap Al I BDDAP-C-BHTCNQ I Al would be in an area of only about 0.5 mm x 0.5 mm. This is shown schematically (for only one of the fingers) in Fig. 14. Of the 75 junctions thus prepared, many were open circuits; the rest were short circuits. Thus, a defect-free domain of BDDAP-C-BHTCNQ had not been found.

It had been reported by Aviram, Joachim, and Pomerantz that "rectification" had been observed in a modified home-built scanning tunneling microscope (STM) for the molecule **30**, which had been originally designed as internal hydrogen atom transfer switch (and not as a rectifier). First, **30** was chemisorbed onto a Au surface (formed by shadowing mica first with Ag and then with Au); then, the molecule **30** was located in a standard STM experiment, using an atomically sharp W "nanotip" over the Au surface; next, the piezoelectric scanning feedback loop was turned off long enough to put a fast voltage ramp across the device: the junction W I **30** I Au was found to be a rectifier of electrical current when W was at -0.20 V versus Au [32].

$(H_3C)_3C$ OH ········ O $C(CH_3)_3$
HOOC S OH ········ O
H_3C CH_3 CH_3 CH_3

30

In the third experiment, the same effect was obtained by Dr. Pomerantz, when an LB monolayer of BDDAP-C-BHTCNQ, **15**, was deposited on the Au I Ag I mica substrate (deposition occurred only during withdrawal from the water subphase, i.e. as if the hydrophilic part of BDDAP-C-BHTCNQ, presumably the TCNQ end, would attach itself to the "nominally hydrophilic" (normally hydrophobic if fresh) Au layer). In this case, the couple W | BDDAP-C-BHTCNQ | Au passed large amounts of current if the W tip was at -0.15 V versus Au [20, 21] (i.e. electrons pass easily through the molecule BDDAP-C-BHTCNQ in the direction BHTCNQ to BDDAP, and not vice versa) [20,21].

However, later control experiments by Dr. Pomerantz showed that this "rectification effect", with very large currents, could occur in the absence of any molecule, if the tip was -1.5 V versus Au. The magnitude of the current, and the strange direction of the current through BDDAP-C-BHTCNQ, both suggest that some unexplained effect, but not necessarily rectification, had been observed, for both **15** and **30**. Disclaimers for the preliminary results have been issued [20, 21, 23]. A Digital Instruments Nanoscope II STM is now on order, and the conductivity of single D-σ-A molecules, using many different metal substrates and nanotips, will soon be measured.

CONCLUSION

We have shown that the synthesis and the Langmuir-Blodgett film formation of promising D-σ-A molecules is now well under control. Simple tests of rectification having failed, we still hope to perform a realistic test of the asymmetric electrical conductivity through single D-σ-A molecules by STM in the not too distant future.

ACKNOWLEDGEMENT

The support of the National Science Foundation (Grant DMR-88-01924) is gratefully acknowledged.

REFERENCES

[1] A. Aviram, M. J. Freiser, P. E. Seiden, and W. R. Young, U.S.Patent US-3,953, 874 (27 April 1976).

[2] A. Aviram and M. A. Ratner, Chem. Phys. Lett. 29: 277-283 (1974).

[3] A. Aviram, P. E. Seiden, and M. A. Ratner, in "Molecular Electronic Devices", F. L. Carter, ed. (Dekker, New York, 1982) page 5.

[4] R. M. Metzger and C. A. Panetta, J. Phys. (Les Ulis, Fr.) Colloque 44: C3-1605 (1983).

[5] R. M. Metzger and C. A. Panetta, in "Molecular Electronic Devices, Vol. II", F. L. Carter, ed. (Dekker, New York, 1987) page 1.

[6] C. A. Panetta, J. Baghdadchi, and R. M. Metzger, Mol. Cryst. Liq. Cryst. 107: 103 (1984).

[7] R. M. Metzger, C. A. Panetta, N. E. Heimer, A. M. Bhatti, E. Torres, G. F. Blackburn, S. K. Tripathy, and L. A. Samuelson, J. Mol. Electronics, 2: 119 (1986).

[8] R. M. Metzger, C. A. Panetta, Y. Miura, and E. Torres, Synth. Met. 18: 797 (1987).

[9] E. Torres, C. A. Panetta, and R. M. Metzger, J. Org. Chem. 52: 2944 (1987).

[10] R. M. Metzger and C. A. Panetta, in "Proc. of the Eighth Winter Conference on Low-Temperature Physics, Cuernavaca, Mexico", 81 (1987).

[11] R. K. Laidlaw, Y. Miura, C. A. Panetta, and R. M. Metzger, Acta Cryst. C44: 2009 (1988).

[12] R. K. Laidlaw, Y. Miura, J. L. Grant, L. Cooray, M. Clark, L. D. Kispert, and R. M. Metzger, J. Chem. Phys. 87: 4967 (1987).

[13] R. K. Laidlaw, J. Baghdadchi, C. A. Panetta, Y. Miura, E. Torres, and R. M. Metzger, Acta Cryst. B44: 645 (1988).

[14] Y. Miura, R. K. Laidlaw, C. A. Panetta, and R. M. Metzger, Acta Cryst. C44: 2007 (1988).

[15] R. M. Metzger, R. R. Schumaker, M. P. Cava, R. K. Laidlaw, C. A. Panetta, and E. Torres, Langmuir 4: 298 (1988).

[16] R. M. Metzger and C. A. Panetta in "Organic and Inorganic Lower-Dimensional

Materials", ed. by P. Delhaès and M. Drillon, NATO ASI Series, B168: 271 (Plenum, New York, 1988).
[17] Y. Miura, E. Torres, C. A. Panetta, and R. M. Metzger, J. Org. Chem. 53: 439 (1988).
[18] Y. Miura, C. A. Panetta, and R. M. Metzger, J. Liquid Chrom. 11: 245 (1988).
[19] R. M. Metzger and C. A. Panetta, J. Mol. Electronics 5: 1 (1989).
[20] R. M. Metzger and C. A. Panetta, J. Chim. Phys. 85: 1125 (1988).
[21] R. M. Metzger and C. A. Panetta, Synth. Met. 28: C807 (1989).
[22] R. M. Metzger, R. K. Laidlaw, E. Torres, and C. A. Panetta, J. Cryst.Spectr.Res. 19: 475 (1989).
[23] R. M. Metzger and C. A. Panetta in "Molecular Electronics - Science and Technology", ed. by A. Aviram and A. Bross (New York Engineering Foundation), in press.
[24] R. M. Metzger, D. C. Wiser, R. K. Laidlaw, M. A. Takassi, D. L. Mattern, and C. A. Panetta, Langmuir, accepted and in press.
[25] See e.g. G. L. Gaines, Jr. "Insoluble Monolayers at Liquid - Gas Interfaces" (Interscience, New York, 1966).
[26] K. B. Blodgett, J. Am. Chem. Soc. 57: 1007 (1935).
[27] K. B. Blodgett and I. Langmuir, Phys. Rev. 51: 964 (1937).
[28] H. Kuhn, D. Möbius, and H. Bücher in "Techniques of Chemistry, Vol. I - Physical Methods of Chemistry - Part V - Determination of Thermodynamic and Surface Properties" ed. by A. Weissberger and B. W. Rossiter (Wiley - Interscience, New York, 1972) page 577.
[29] H. Kuhn, Pure Appl. Chem. 51: 341 (1979).
[30] H. Kuhn, Pure Appl. Chem. 53: 2105 (1981).
[31] See e. g. Thin Solid Films Vols. 68 (1980), 99 (1983), 132-134 (1985), 160 (1987).
[32] A. Aviram, C. Joachim, and M. Pomerantz, Chem. Phys. Lett. 146: 490 (1988).
[33] E. E. Polymeropoulos, D. Möbius, and H. Kuhn, Thin Solid Films 68: 173 (1980).
[34] M. Sugi, K. Sakai, M. Saito, Y. Kawabata, and S. Iizima, Thin Solid Films 132: 69 (1985).
[35] M. Fujihira, K. Nishiyama, and H. Yamada, Thin Solid Films 132: 77 (1985).
[36] M. Fujihira and Y. Yamada, Thin Solid Films 160: 125 (1988).
[37] R. Hoffman, Acc. Chem. Res. 4: 1 (1971).
[38] R. W. Murray, Acc. Chem. Res. 13: 135 (1980).
[39] W. C. Bigelow, D. L. Pickett, and W. A. Zisman, J. Colloid Sci. 1: 513 (1946).
[40] R. Maoz, L. Netzer, J. Gun, and J. Sagiv, J. Chim. Phys. 85: 1059 (1988).
[41] I. R. Peterson, J. Chim. Phys. 85: 997 (1988).
[42] W. R. Hertler, J. Org. Chem. 41: 1412 (1976).
[43] J. Baghdadchi, Ph. D. dissertation, Univ. of Mississippi, Dec. 1982.
[44] L. T. Calcaterra, G. L. Closs, and J. R. Miller, J. Am. Chem. Soc. 105: 670 (1983).
[45] J. R. Miller, L. T. Calcaterra, and G. L. Closs, J. Am. Chem. Soc. 106: 3047 (1984).
[46] R. A. Marcus, Disc. Faraday Soc. 29: 21 (1960).
[47] A. C. Cephalas, private communication.
[48] J. R. Anderson and O. Jorgensen, J. Chem. Soc. Perkin Trans. I, 3095 (1979).
[49] M. A. Takassi, Ph. D. Dissertation, University of Mississippi, Aug. 1989
[50] A. M. Kini, D. O. Cowan, F. Gerson, and R. Mockel, J. Am. Chem. Soc. 107: 556 (1985).

SOLID-STATE MICROELECTROCHEMICAL DEVICES: TRANSISTOR AND DIODE DEVICES EMPLOYING A SOLID POLYMER ELECTROLYTE

Daniel R. Talham, Richard M. Crooks,
Vince Cammarata, Nicholas Leventis,
Martin O. Schloh and Mark S. Wrighton*

Department of Chemistry
Massachusetts Institute of Technology
Cambridge, Massachusetts 02139 USA

INTRODUCTION

Not long ago, this group first described microelectrochemical devices, which are based on microfabricated arrays of electrodes, connected by electroactive materials.[1] Because the active components of these devices are chemical in nature, many of these devices are chemically sensitive, and comprise a potentially useful class of chemical sensors. Devices showing sensitivity to pH, O_2, H_2, Li^+, and Na^+ have been demonstrated.[2,3] These devices are, typically, operated in fluid solution electrolytes. If this class of devices is to be useful as gas sensors, systems which are not dependent on liquid electrolytes need to be developed. We have recently reported solid state microelectrochemical transistors, which replace conventional liquid electrolytes with polymer electrolytes based on polyethyleneoxide (PEO) and polyvinylalcohol (PVA).[4] In this report, we discuss additional progress toward solid-state devices by employing a new polymer ion conductor based on the polyphosphazene comb-polymer, MEEP[5] (shown below). By taking advantage of polymer ion conductors, we have developed microelectrochemical devices, where all of the components of the device are confined to a chip.

Poly[bis(2-(2-methoxyethoxy)ethoxy)phosphazene]

MEEP/$LiCF_3SO_3$ (4:1)

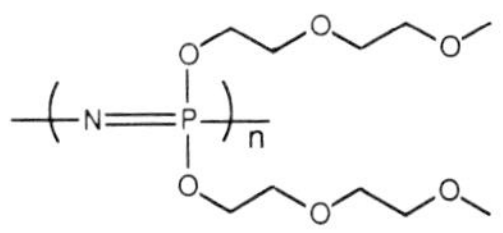

O, O, O, N, P, n, O, O, O

*Author to whom correspondence should be addressed.

Lower-Dimensional Systems and Molecular Electronics
Edited by R.M. Metzger *et al.*, Plenum Press, New York, 1991

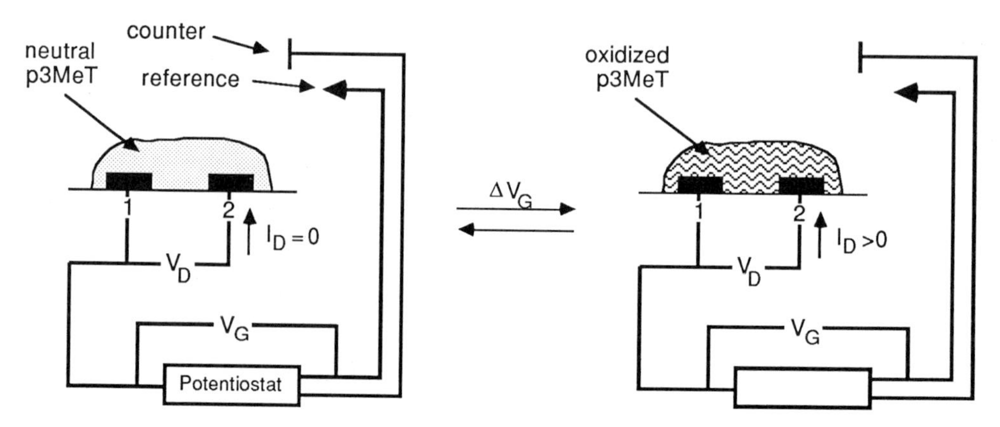

Figure 1. A conducting polymer-based microelectrochemical transistor. P3MeT connects two wires of a microfabricated array. Electrodes 1 and 2 are source and drain, respectively. At left, V_G is such that the polymer is neutral and the device is "off", $I_D = 0$. Switching V_G to an oxidizing potential (right) turns the device "on", $I_D > 0$.

MICROELECTROCHEMICAL TRANSISTORS

Figure 1 shows a cross-sectional view of a conducting polymer-based microelectrochemical transistor. Two electrical contacts, the source and drain, are connected by an electroactive material, whose resistance can be varied as a function of its state of charge. In the example in Figure 1, the electroactive material is the conducting polymer poly(3-methylthiophene), p3MeT, which in its neutral state is insulating, but when oxidized becomes conducting. A small bias, V_D, is applied between the source and the drain, resulting in a drain current, I_D, depending on the state of charge of the conducting polymer. For a given V_D, I_D can be varied by changing the gate potential, V_G, of the system, which, in these devices, is electrochemical potential. The faradaic current required to switch the device is the gate current, I_G. The device output is usually represented as a plot of I_D *vs.* V_G for a constant V_D. This representation provides a measure of the relative conductivity of the conducting polymer, as a function of electrochemical potential, for small values of V_D.

In our devices, the source and drain are microfabricated wires, typically ~70 µm long x ~2.4 µm wide x ~0.1 µm high.[6] An array of eight gold or platinum microelectrodes, with interelectrode spacing of 1.4 µm, is fabricated on a 3 mm x 3 mm chip.[1,6] The switching speeds of these devices are enhanced by small electrode spacings.[6] In addition, reduced spacing between electrodes and small electrode area permit studies in solid electrolyte systems, where the resistances are, typically, much higher than in liquid electrolyte solutions.

Polymer Electrolytes

Preparation of solid-state electrochemical devices require that conventional fluid solution electrolytes be replaced with a solid ion conductor. Solid electrolytes have been widely studied, primarily for the development of high energy-density batteries.[7] Classes of solid state electrolytes include classical solids, such as the β-aluminas, polyelectrolytes such as Nafion, gel electrolytes and polymer electrolytes.[7] For the purpose of developing solid-state electrochemical devices, polymer electrolytes are promising, because they are easily confined to microelectrochemical arrays, and are gas permeable.

For application to the surface of the device, the polymer and electrolyte are codissolved in a spreading solvent. After evaporation of the solvent, the polymer electrolyte remains as a thin film. The concentration of the salt in the polymer is expressed as a ratio of the number of polymer repeat units per unit of salt. In these studies, the MEEP/$LiCF_3SO_3$ ratio is 4:1 or 5:1.[5] The ionic conductivity of these polymer electrolytes is thought to be due to ion hopping between Lewis base sites along and between chains. This process is facilitated by small amounts of a coordinating solvent.[8] The devices described here are operated in the presence of some solvent vapor. Usually, 20 μl of THF per 50 ml N_2 atmosphere over the device is added as a plasticizer.

Transistor Devices

A schematic of a solid-state p3MeT transistor, employing the MEEP/$LiCF_3SO_3$ electrolyte is shown in Figure 2. Ag, plated onto one microelectrode, serves as a reference electrode, and a small amount of Ag epoxy, close to the array, serves as the counter electrode. Alternatively, Ag epoxy can be used as both reference and counter electrode. Monomeric 3-methylthiophene is polymerized electrochemically onto electrodes 2-4 in CH_3CN/0.1 M $(n\text{-}Bu_4N)ClO_4$.[9] The polymer is confined to these electrodes, by holding the other electrodes in the array at a reducing potential, to discourage polymer growth. The device is characterized before and after the addition of the MEEP electrolyte. This permits comparison of the device characteristics in solution and in the polymer electrolyte. Cyclic voltammetry at each of the derivatized electrodes in CH_3CN/0.1 M $LiCF_3SO_3$ is shown on the left side of Figure 2. The magnitude of the current is the same, whether scanning the derivatized electrodes together, or individually, showing that the p3MeT connects the electrodes. Cyclic voltammetry of the same array, now employing MEEP/$LiCF_3SO_3$ electrolyte, is shown on the right side of Figure 2. The shape of the curves is essentially the same as in solution electrolyte. The currents are smaller, due to the slower scan rates and reduced counterion mobility in the polymer.

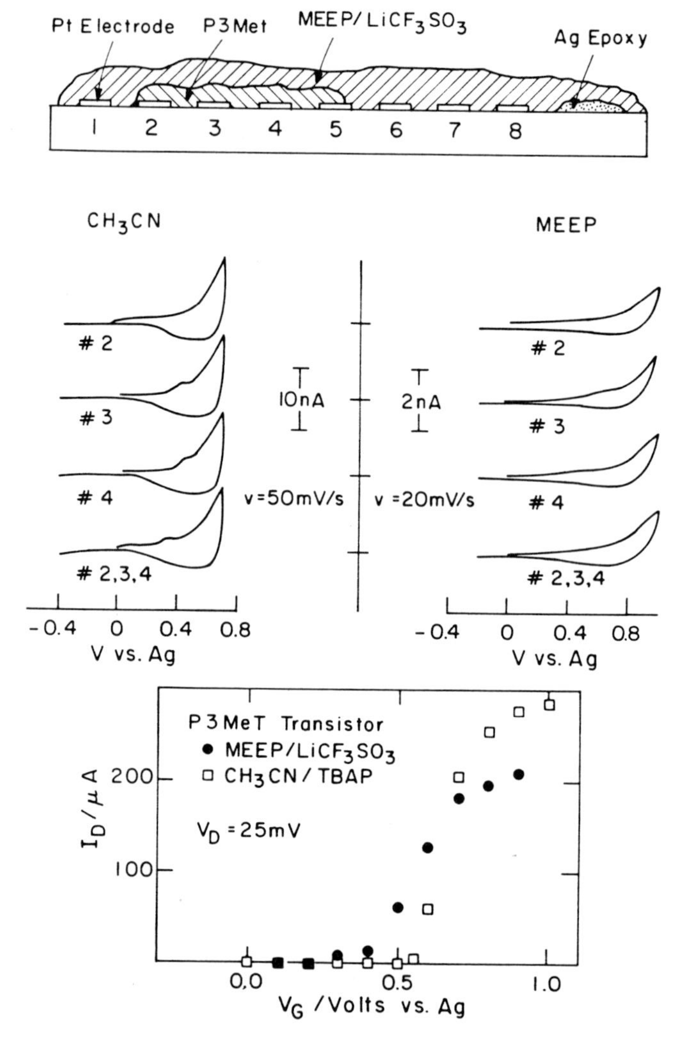

Figure 2. **Top.** Schematic of a p3MeT-based solid-state microelectrochemical device. **Center.** Cyclic voltammetry at the p3MeT derivatized electrodes. At left, the device is characterized in the solution electrolyte CH_3CN/0.1 M $LiCF_3SO_3$ before the application of MEEP. At right, the same device is characterized under MEEP/$LiCF_3SO_3$ (5:1). **Bottom.** Comparison of the steady-state I_D *vs.* V_G of the p3MeT device in fluid solution electrolyte and under MEEP/$LiCF_3SO_3$. Electrodes 3 and 4 are source and drain respectively (see Figure 1).

The steady-state I_D *vs.* V_G characteristics of the p3MeT device, in both solution and polymer electrolytes, are compared in Figure 2. At gate potentials, where the polymer is neutral, the device is "off", but it turns "on", when V_G is moved positive to an oxidizing potential. There are two key points. First, the drain current of the device in the "on" state is nearly the same under the polymer electrolyte, as it is in the solution electrolyte. Second, the transistor device amplifies small electrochemical signals, an important feature of all transistors. This is particularly important in solid electrolytes, where diffusion coefficients are small. When the device is turned on, the leakage current through the gate is very small (~1 nA) when the I_D is ~200 μA. However, the gain is restricted to very low frequency, because electrochemical switching is very slow.[4]

WO_3 is an example of another class of electroactive material, metal oxides, which has been used to construct microelectrochemical devices.[10] WO_3 is a wide-band-gap semiconductor, with high resistance in its neutral state.[11] Upon reduction, WO_3 intercalates cations such as H^+, Li^+, and Na^+, and becomes conducting. WO_3-based transistors, showing sensitivity to pH and to Li^+ concentration, have been demonstrated in solution electrolytes.[3] A schematic of a MEEP/WO_3 device is shown in Figure 3. WO_3 is confined to the required electrodes, using standard photolithographic techniques.[10] The transistor characteristics of this device are also presented in Figure 3. At potentials where WO_3 is neutral, the device is "off", however at negative potentials, WO_3 is reduced, and the device turns "on".

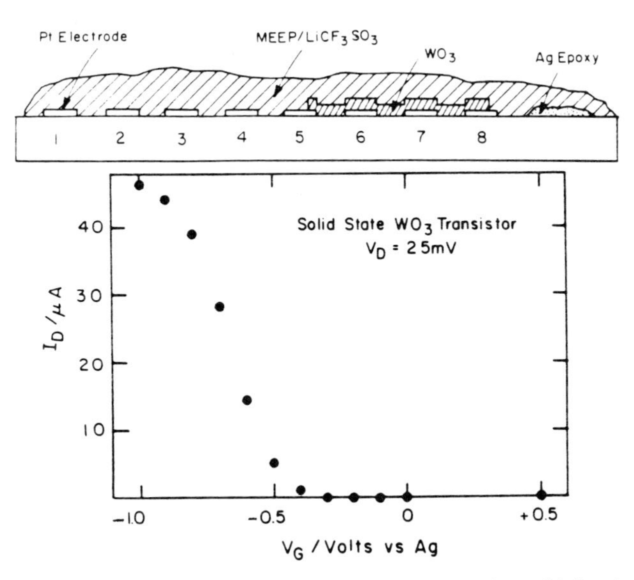

Figure 3. **Top.** Schematic of a WO_3-based solid-state microelectrochemical device. **Bottom.** Steady-state I_D *vs.* V_G of the device diagrammed at top. Electrodes 6 and 7 are source and drain.

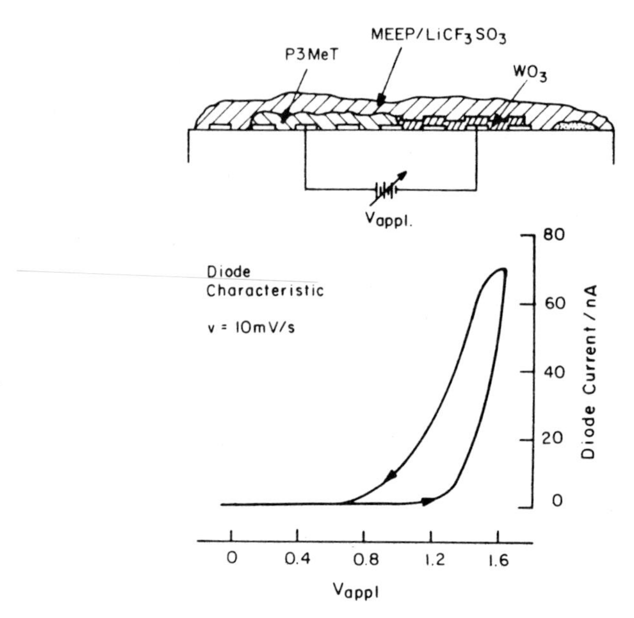

Figure 4. Top. Schematic of a solid-state microelectrochemical diode based on p3MeT and WO_3 under the polymer electrolyte MEEP/$LiCF_3SO_3$. **Bottom.** Diode characteristics of the two-terminal device at top. The device turns on when the applied voltage is equal to the difference in redox potentials of the two materials, and the bias is such that the p3MeT is oxidized and the WO_3 is reduced. If the bias is reversed (not shown), no current flows.

Diode Devices

By confining both p3MeT and WO_3 to the same array, we form the basis of a two-terminal microelectrochemical diode. The underlying principle is that one material, the p3MeT, will only be conducting when it is oxidized, and the other, the WO_3, will only be conducting when reduced. A schematic of the device and the diode characteristic are shown in Figure 4. Current will only flow in the two-terminal device, if the magnitude of the applied voltage is equal to the difference in redox potential of the two materials, and the bias is such, that the conducting polymer is oxidized and the metal oxide is reduced. No current flows when it is reverse-biased.

Devices Based on Redox Conduction

Conventional redox polymers can also form the basis of electrochemical transistors.[12] Conventional redox polymers have lower maximum conductivity, and yield devices having lower values of I_D, than conducting polymers or metal oxides. Conventional redox polymers offer an important design advantage, however. Nearly any stable redox active material can be incorporated into a polymeric system to form a conventional redox polymer. This allows the fabrication of devices with a wide range of chemical sensitivities.

Examples of conventional redox polymer-based devices include those based on viologen, ferrocene, and quinone-based redox polymers.[12] Figure 5 shows an example from a new class of solid-state microelectrochemical transistors, which are based on redox-active molecules dissolved in the polymer. The redox-active material is N,N,N',N'-tetramethyl-*p*-phenylenediamine (TMPD) which is sublimed into the MEEP/$LiCF_3SO_3$ film. Here, the MEEP/$LiCF_3SO_3$ acts as both polymer host and electrolyte. The transistor characteristic of this device is also shown in Figure 5. Below 0.0 V *vs.* Ag, the device is off, $I_D = 0$, since all the TMPD is neutral. As TMPD is oxidized, the device turns on, with a maximum I_D near $E_{1/2}$ of $TMPD^{+1/0}$. We have determined the diffusion coefficient for charge transport, D_{ct}, of $TMPD^+$ in MEEP to be 2×10^{-8} cm^2/s. It is not clear if the charge transport mechanism is *via* self-exchange or physical diffusion, but most likely involves both mechanisms. This value of D_{ct} is comparable to those of conventional redox polymers, where the redox centers are covalently bonded to the polymer backbone, and the sole mechanism of charge transfer is *via* self-exchange.[13] The ability to fabricate a solid-state microelectrochemical transistor, where the active molecules are not required to be covalently attached to a polymer backbone, suggests a wide range of new devices. Of particular interest is the prospect of including chemically sensitive molecules into this class of device, for potential use as gas sensors.

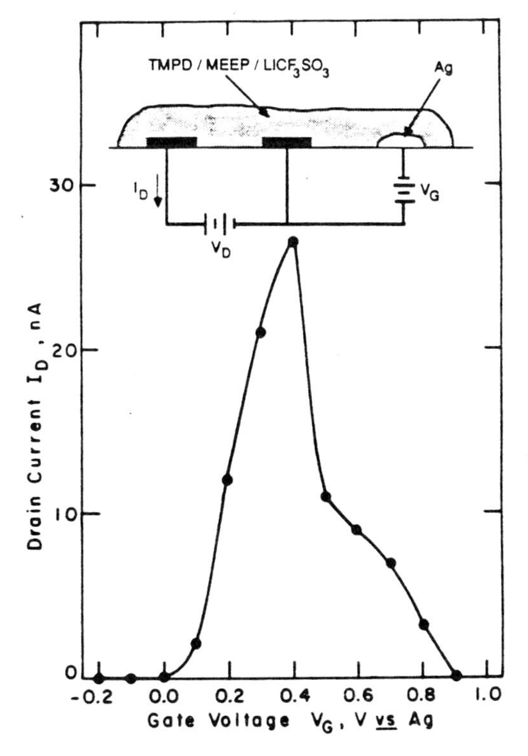

Figure 5. Schematic and transistor characteristic of a new class of microelectrochemical device, which is based on a redox-active material dissolved in a polymer ion conductor. Here, TMPD is sublimed into and saturates the MEEP/$LiCF_3SO_3$ film. The drain voltage, V_D, is 25 mV.

AKNOWLEDGEMENTS

We thank the Lockheed Missle and Space Corporation, the Office of Naval Research, and the Defense Advanced Research Projects Agency each for partial support of this research. We also thank Professor H. R. Allcock of the Pennsylvania State University for providing us with a sample of MEEP.

REFERENCES

1. H. S. White, G. Kittlesen and M. S. Wrighton, *J. Am. Chem. Soc. 106*:5375 (1984); H. S. White, G. P. Kittlesen and M. S. Wrighton, *J. Am. Chem. Soc. 106*:7389 (1984); E. W. Paul, A. J. Ricco and M. S. Wrighton, *J. Phys. Chem. 81*:1441 (1985).
2. J. W. Thackeray and M. S. Wrighton, *J. Phys. Chem. 90*:6674 (1986); M. S. Wrighton, J. W. Thackeray, M. J. Natan, D. K. Smith, G. A. Lane and D. Belanger, *Phil. Trans. Roy. Soc. Lond. B316*:13 (1987).
3. N. Leventis, M. O. Schloh, J. J. Hickman and M. S. Wrighton, (to be published).
4. S. Chao and M. S. Wrighton, *J. Am. Chem. Soc. 109*:2197 (1987); S. Chao and M. S. Wrighton, *J. Am. Chem. Soc. 109*:6627 (1987); N. Leventis, M. J. Natan, M. O. Schloh, J. J. Hickman and M. S. Wrighton, (to be published).
5. H. R. Allcock, P. E. Austin, T. X. Neenan, J. T. Sisko, P. M. Blonsky and D. F. Shriver, *Macromolecules 19*:1508 (1986); P. E. Austin, G. H. Riding and H. R. Allcock, *Macromolecules 16*:719 (1983); P. M. Blonsky, D. F. Shriver, P. E. Austin and H. R. Allcock, *J. Am. Chem. Soc. 106*:6854 (1984); P. M. Blonsky, D. F. Shriver, P. E. Austin and H. R. Allcock, *Solid State Ionics 18*:258 (1986).
6. M. S. Wrighton, S. Chao, O. M. Chyan, E. T. Jones, N. Leventis, E. A. Lofton, M. O. Schloh and C. F. Shu, *in* "Chemically Modified Electrode Surfaces in Science and Industry," D. E. Leyden and W. T. Collins, eds., Gordon & Breach: New York (1987), p337.
7. C. A. Vincent, *Prog. Solid St. Chem. 17*:145 (1987); M. A. Ratner and D. F. Shriver, *Chem. Rev. 88*:109(1988).
8. L. Geng, M. L. Longmire, R. A. Reed, J. F. Parker, C. J. Barbour and R. W. Murray, *Chem. Mater. 1*:58 (1989); L. Geng, R. A. Reed, M.-H. Kim, T. T. Wooster, B. N. Oliver, J. Egekeze, R. T. Kennedy, J. W. Jorgenson, J. F. Parcher and R. W. Murray, *J. Am. Chem. Soc. 111*:1614 (1989).
9. J. W. Thackeray, H. S. White and M. S. Wrighton, *J. Phys. Chem. 89*:5133 (1985).
10. M. O. Schloh, N. Leventis and M. S. Wrighton, *J. Appl. Phys. 66*:965 (1989).
11. W. C. Dautremont-Smith, *Displays 3*:67 (1982).
12. G. P. Kittlesen, H. S. White and M. S. Wrighton, *J. Am. Chem. Soc. 107*:7373 (1985); D. K. Smith, G. A. Lane and M. S. Wrighton, *J. Am. Chem. Soc. 108*:3522 (1986).
13. P. G. Pickup and R. W. Murray, *J. Am. Chem. Soc. 105*:4510 (1983).

POLYMER FIELD-EFFECT TRANSISTORS FOR TRANSPORT PROPERTY STUDIES

J. Paloheimo (a), E. Punkka (a), H. Stubb (a), and
P. Kuivalainen (b)

(a) Semiconductor Laboratory, Technical Research Centre of Finland, Otakaari 7 B, SF-02150 Espoo, Finland

(b) Electron Physics Laboratory, Helsinki University of Technology, Otakaari 7 A, SF-02150 Espoo, Finland

ABSTRACT

Thin film field-effect transistors have been prepared of poly(3-alkylthiophenes) by using spin-coating techniques. The devices are used in the determination of the charge carrier mobility μ, dc conductivity σ, and the carrier concentration p_0. Poly(3-hexylthiophene) is characterized in a wide temperature range T = 130 - 430 K, and possible transport mechanisms are discussed.

I INTRODUCTION

Rather little work has been reported on the use of conducting polymers in active semiconductor devices. A problem in organic pn-junctions is the interdiffusion of dopants destroying the abrupt interface. In Schottky junctions, made of (the more stable) p-type polymers, metals with low work functions must be used. These metals, such as aluminium (work function 4.3 eV), are typically reactive and oxidize easily.

Polymer thin-film field-effect transistors (POLYFETs) have some advantages over the pn- or Schottky-junction devices. In POLYFETs nearly undoped polymers must be used, thus increasing stability and avoiding dopant diffusion. Considering FET applications, the main drawback is the low charge carrier mobility.

In conducting polymers the extra charge does not go into extended states, but is self-trapped. If the polymer has a non-degenerate ground state, the charge is mainly stored in polarons or bipolarons, the latter being energetically more favourable. These are normally not free to move along the polymer chains, but bound to ionized dopant molecules by strong coulombic attraction. The co-existence of an important number of free polarons or bipolarons is possible only at high temperatures or in inversion or strong accumulation in the POLYFET channel, but even then their free motion is hindered by disorder and defects, such as chain ends. Thus, some kind of hopping mechanism dominates the charge transport, although the details are unknown to some extent.

Lower-Dimensional Systems and Molecular Electronics
Edited by R.M. Metzger *et al.*, Plenum Press, New York, 1991

The conductivity for a p-type material can be written in the form

$$\sigma = |e|\mu p_0 \tag{1}$$

where p_0 is the concentration of charge carriers arising from (often unintentional) acceptors, and μ is the mobility of the carriers. The number of active acceptors N_a may slightly exceed p_0, because some donated holes may reside in deep defect or donor states, that are not involved in the transport. Here we assume that $p_0 \approx N_a$.

The conductivity of conducting polymers increases overlinearly with increasing doping N_a. This implies that also μ must be N_a-dependent, and the dopant ions have an important role in the conduction mechanism, acting as hopping centers, or assisting the carriers to move between chains.

In this work the POLYFET structure is utilized in the study of charge transport in poly(3-hexylthiophene) (PHT) in a wide temperature range T = 130 - 430 K. Poly(3-butylthiophene) (PBT), poly(3-octylthiophene) (POT), and poly(3-decylthiophene) (PDT) have so far been characterized only at room temperature.

II RESULTS AND DISCUSSION

The POLYFETs studied in this work were processed on a silicon wafer (gate) covered with thermally grown 3000 Å thick SiO_2 (gate oxide). Unintentionally doped PAT thin film (PHT 700 Å) was spun from chloroform solution on the oxide and the gold (work function > 5 eV) drain and source electrodes. The length and width of the channel were 5 μm and 8 cm, respectively. The detailed PHT-FET-structure and its fabrication are described elsewhere[1]. The work function of PHT is about 4.7 eV, evaluated from small-signal capacitance measurements of an In-PHT-junction[2]. This value ensures an ohmic contact with the gold electrodes.

After the room temperature processing of the polymer films, our POLYFETs were normally-on devices, i.e. positive (negative) gate voltages V_G decrease (increase) the drain current I_D[1]. This indicates that the polymer is a p-type

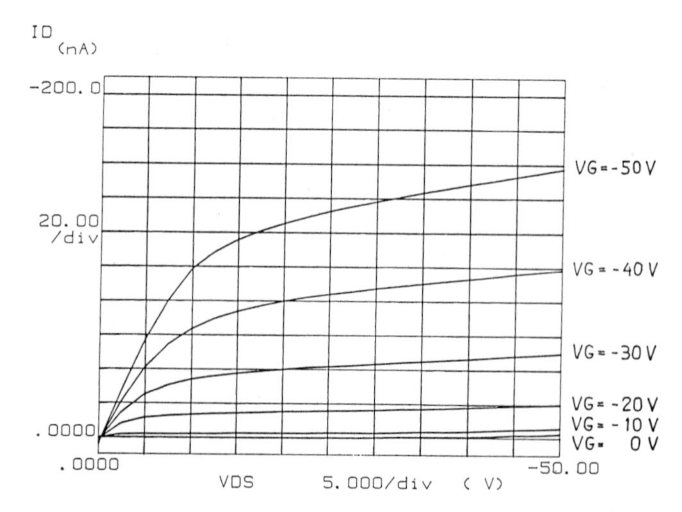

Fig. 1. I-V characteristics of a PHT-FET at room temperature after heat treatment at 430 K.

semiconductor. A heat treatment of the PHT-FET at 430 K changes the I-V characteristics significantly[1]. Due to the increase in resistivity, the FET becomes a normally-off device, as shown in Fig. 1. The modulation ratio of I_D ($I_{D,on}/I_D(V_G = 0\ V)$) in the linear region increases to more than 10^3. No clear sign of inversion mode operation could be found with large positive gate voltages $V_G \leq +50$ V in this heat-treated PHT-FET. $|I_D(V_G = +50\ V)|$ was at least two orders of magnitude less than $|I_D(V_G = -50\ V)|$. Also, the as-prepared PAT-FETs did not operate in inversion mode at $V_G \leq +100$ V. Probably the gold electrodes, having high work function, do not form ohmic contacts with an inversion layer, as they do with an accumulation layer and the bulk. Also, the mobility of negative charge carriers may be very small, since there are few positive donor dopant ions which could act as hopping centers, or assist charge to move from one chain to another. Therefore the existence of an inversion layer does not necessarily have a notable effect on I_D.

When the thickness of the oxide film d_{ox} is much larger than the thickness of the accumulation or depletion layer, the total amount of extra charge per unit surface area between the drain and source depends linearly on V_G for $V_{DS} \approx 0$ V

$$|e| \cdot (\langle p \rangle - p_0) \cdot d = C_{ox} \cdot (V_{FB} - V_G) \tag{2}$$

Here C_{ox} is the oxide capacitance per unit area, V_{FB} the flat-band voltage, p the positive charge carrier concentration, and $\langle\ \rangle$ denotes an average over the polymer film thickness d. We have assumed that p_0 is constant in the film.

The drain conductance is

$$g_D = \frac{\partial I_D}{\partial V_{DS}}\bigg|_{V_{DS} \to 0} = |e| \cdot \frac{Wd}{L} \cdot \langle \mu(p) \cdot p \rangle = -\frac{W\mu_p C_{ox}}{L} \cdot (V_G - V_T) \tag{3}$$

where V_{DS} is the voltage between the drain and source, W (L) the width (length) of the transistor, V_T the threshold voltage

$$V_T = \frac{|e|\ dp_0}{C_{ox}} + V_{FB} \tag{4}$$

and μ_p the average mobility

$$\mu_p = \frac{\langle \mu(p) \cdot p \rangle}{\langle p \rangle} \tag{5}$$

For $V_G = V_{FB}$

$$\sigma = \frac{L}{Wd} \cdot g_D \tag{6}$$

In thin-film transistors the drain current in the linear region ($|V_{DS}| < |V_G - V_T|$) can be expressed as[3]

$$I_D = -\frac{W\mu'_p C_{ox}}{L} \cdot \left[(V_G - V_T) \cdot V_{DS} - \frac{V_{DS}^2}{2}\right] \tag{7}$$

Here μ'_p equals μ_p only when $|V_{DS}| \ll |V_G - V_T|$ if the mobility μ is a function of p or the electric field E.

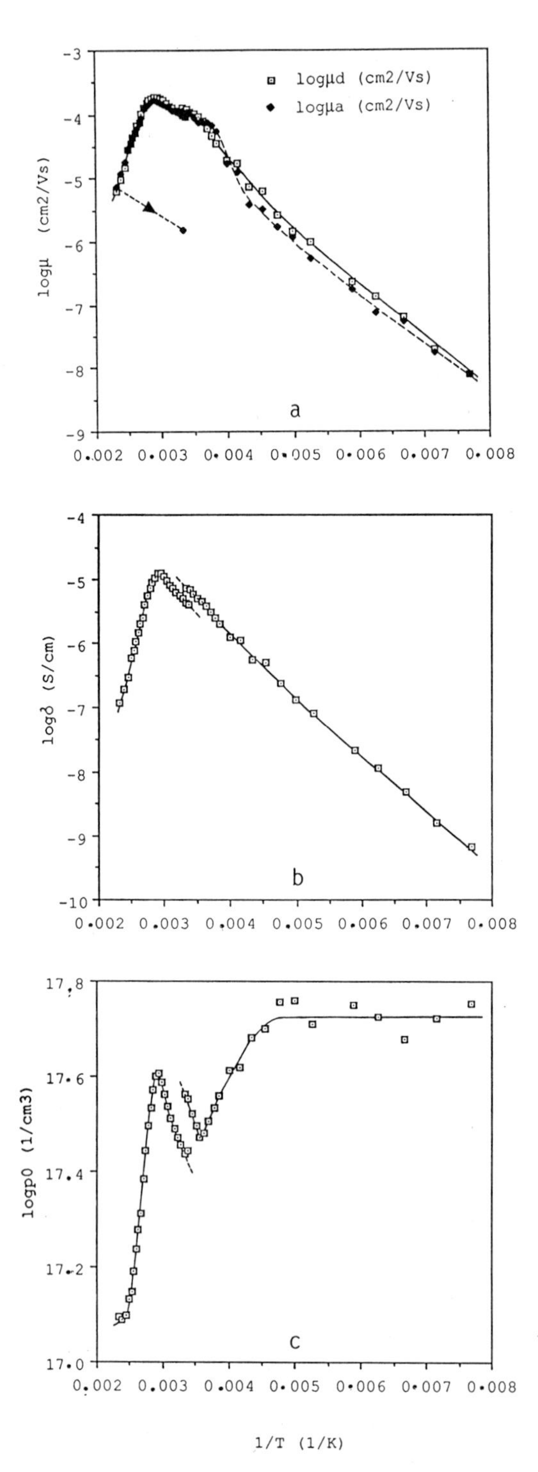

Fig. 2. (a) Mobilities μ_d and μ_a, (b) dc conductivity σ, and (c) charge carrier concentration p_0 vs temperature.

By measuring g_D for two different values of V_G, we can determine μ_p (Eq. 3). In the case $V_G \geq V_{FB}$ (depletion), the resulting mobility μ_d can be identified to be the bulk mobility $\mu(p_0)$ under the depletion approximation. When $V_G \leq V_{FB}$ (accumulation), the calculated mobility μ_a tells us qualitatively how strongly dependent $\mu(p)$ is on p. The flat-band voltage V_{FB} is very near 0 V, because the work functions of PATs and silicon (gate) are about the same. Notice that the accumulation layer can be very thin, and therefore it is possible that, at the polymer-SiO_2-interface, $p \gg p_0$. New effects may also appear if the hopping distance, or the length of polarons or bipolarons, become comparable to the width of the accumulation layer in strong accumulation (inversion) and/or at low temperatures.

The room temperature mobilities μ_d (V_G = 0 V, +10 V) and μ_a (V_G = -10 V, 0 V), estimated from Eq. 3, are about the same, i.e. $\mu_d \approx \mu_a \approx 2 \cdot 10^{-4}$, 10^{-4}, $3 \cdot 10^{-6}$, and $6 \cdot 10^{-7}$ $\frac{cm^2}{Vs}$ in as-prepared PBT, PHT, POT, and PDT, respectively. These values are typical for very lightly doped polymers[1,4,5,6,7]. We have not found any sign of metallic mobilities, even in very strong accumulation (V_G = -100 V). It seems that μ is nearly independent of p. In PHT $\sigma \approx 6 \cdot 10^{-6}$ $\frac{S}{cm}$ (Eq. 6), and $p_0 \approx 3 \cdot 10^{17}$ $\frac{1}{cm^3}$ (Eq. 1) the latter being near the values of p_0 in other PATs, too.

We have determined the dependence of μ_d, μ_a, σ, and p_0 on the temperature in PHT. Fig. 2 shows that the strongly T-dependent σ is mainly due to $\mu(T)$. When the temperature of a PHT-FET increases over 340 K, which is about the glass transition temperature for PHT, the mobility and conductivity decrease, Fig. 2. This has been interpreted to arise from some structural changes, such as conformons[1,7], and/or the outdiffusion of acceptor dopants[1], Fig. 2 (c). The conformons cause further disorder and localization of the states. The hopping rates decrease rapidly with decreasing wave function decay lengths.

At low temperatures (< 250 K), where the mobility is low, we have found some nonlinear effects, as shown in Fig. 3. From the temperature T and the electric field E_0, where the non-ohmic behaviour starts, we can give an estimation of the hopping distance $R \approx kT/|e|E_0 \approx 60$ Å. This is about the same as

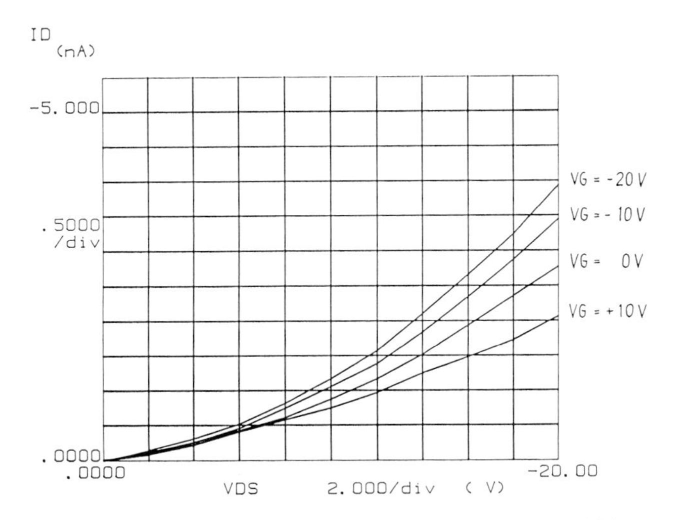

Fig. 3. I-V characteristics of a PHT-FET at 130 K.

the typical separation between acceptor ions, calculated assuming $N_a \approx p_0$. At room temperature, and above, there is no sign of nonlinearities in PHT, and the hopping distances must be smaller. In PBT there are non-ohmic effects, even at room temperature.

After a heat treatment of the PHT-FET at 430 K, the room-temperature mobility μ_a is about $10^{-6} \frac{cm^2}{Vs}$. The negative sign of V_T, in this case, can be explained by some deep defect states. A preliminary study shows that, if PHT powder is heat treated before solvation and spinning, this has qualitatively same effects on the room-temperature characteristics: μ and σ decrease.

III CONCLUSIONS

We have demonstrated that a POLYFET structure is a powerful tool in the characterization of the transport properties of lightly doped conducting polymers. The strongly temperature-dependent mobility and non-ohmic behaviour at low temperatures in lightly doped PHT imply that a phonon-assisted hopping between dopant ions acting as hopping centers probably dominates the charge transport when $T \lesssim 200$ K. At higher temperatures ($T \gtrsim 300$ K) some free polarons or bipolarons may move along chains, but the limiting process is hopping between chains. At temperatures above the glass transition temperature (65 °C), the mobility and conductivity decrease, because of some structural changes, and/or the diffusion of acceptor molecules out of the film.

In the future, POLYFETs may have some applications as sensors, switching elements in displays etc. However, nearly all FET applications are limited by the very low mobility ($\mu \approx 10^{-6} \ldots 10^{-3} \frac{cm^2}{Vs}$ at room temperature), and maximum operating frequency f_{max}. These problems may be reduced by smaller FET dimensions ($f_{max} \sim \frac{\mu}{L^2}$), other transistor structures (e.g. MESFETs), more suitable materials, higher order (e.g. Langmuir-Blodgett films) or the possible increase of μ at high operating frequencies. Very important is the optimization of the dopant concentration and film thickness, to obtain large μ but low $g_D(V_G = 0)$. If one were able to dope a polymer with nearly equal amounts of acceptors and donors, the material would become compensated i.e. the carrier concentration $|N_a - N_d| \approx 0$ (and therefore $\sigma \approx 0$, $g_D(V_G = 0) \approx 0$), but the mobility for both positive and negative charge carriers, in an accumulation or inversion layer, could be very high, because of the high concentration of hopping centers.

ACKNOWLEDGEMENTS

This work was part of our conducting polymers research sponsored by the Finnish Academy, the Technology Development Centre, Finland (TEKES), and the Nordic Fund for Industrial and Technological Development. The polyalkylthiophenes used in our work were synthesized by J. - E. Österholm and coworkers at Neste Ltd. P. Yli-Lahti assisted in the sample preparation.

REFERENCES

1. J. Paloheimo, E. Punkka, P. Kuivalainen, H. Stubb, and P. Yli-Lahti, New dimensions for semiconductor devices: polymeric field-effect transistors, Acta Polytechnica Scandinavica, El. Eng. Series 64:178 (1989).
2. H. Tomozawa, D. Braun, S. Phillips, and A. J. Heeger, Metal-polymer Schottky barriers on cast films of soluble poly(3-alkylthiophenes), Synth. Met., 22:63 (1987).

3. S. M. Sze, "Physics of Semiconductor Devices", John Wiley & Sons, 1969.
4. A. Tsumura, H. Koezuka and T. Ando, Polythiophene field-effect transistor: its characteristics and operation mechanism, Synth. Met., 25:11 (1989).
5. N. Oyama, F. Yoshimura, T. Ohsaka, H. Koezuka, and T. Ando, Characteristics of a field-effect transistor fabricated with electropolymerized thin film, Jpn. J. Appl. Phys., 27:L448 (1988).
6. J. H. Burroughes, C. A. Jones, and R. H. Friend, New semiconductor device physics in polymer diodes and transistors, Nature, 335:137 (1988).
7. A. Assadi, C. Svensson, M. Willander and O. Inganäs, Field-effect mobility of poly(3-hexylthiophene), Appl. Phys. Lett., 53:195 (1988).

MOLECULAR LINES

Larry L. Miller

Department of Chemistry
University of Minnesota
Minneapolis, MN 55455, USA

INTRODUCTION

Molecular lines are stiff, linear molecules of precisely known dimensions. They represent a starting point for the design of complex molecular surface structures of interest, for what can be broadly classified as molecular electronics. We envisage the possibility of an alternative technology, in which molecular materials replace semiconductor materials, and where synthesis replaces etching and lithography. Molecular circuitry would have the advantage of extremely small scale (nanometers) and very high resolution (Ångstroms). Molecular structures also have the potential advantage of a more variably detailed and "tuned" structure. Such structures might act usefully by mechanisms quite different from those of microelectronics. Thus, one can imagine optical, energy storage and chemical communication events, which are entirely different from what is achieved using semiconductor materials. Obviously, it is a grand idea, and there is a wealth of good science to do. The problem is where to begin to work in it.

Our strategy involves chemically coupling molecular components together to produce a "structure". The simplest of these structures in microelectronics is a line, and we have set out to: (1) produce molecular lines with nanometer-scale lengths, (2) image such lines on surfaces, (3) develop physical properties of interest to electronics and optics, and materials science in general. In this paper are included some specific examples of synthesis [1], imaging [2], Langmuir-Blodgett (LB) film formation [3], and solid-state conductivity [4]. Results on optical effects (non-linear optics and near-infrared spectra) [5] are not included.

SYNTHESIS

Our synthesis tactics involve the development of rigid aromatic units, which can be joined together by linear imide linkages. We have developed several syntheses of polyacenequinones, with the idea that these will provide building blocks that are stable, but whose structure can be varied via quinone chemistry. As an example of a successful synthesis of nanometer-scale lines from the simple quinone precursor **1**, consider the last steps of the 7.5 nm-long molecule **2**. Note that, although there are imide linkages in this line, it is prepared by Diels-Alder reactions. Second, note that this compound is soluble in chloroform. This allowed purification and complete spectroscopic identification. Solubility is a major concern for these long flat molecules, and in this case the synthesis was specifcally designed to provide solubilizing Ar = t-butylphenyl groups along the line.

A second approach to molecular lines has been developed in our laboratory [6]. It owes its success to the discovery that oligomeric imides can be made soluble in strongly hydrogen-bond donating solvent mixtures. In this synthesis, repetitive imide-forming reactions lead from monomer to trimers to the nonamer **3**. It was characterized by IR and NMR (end-group analysis).

1

a) 1

b) H_2SO_4

2 (7.5 nm)

3 (7.84 nm)

Since the components in each series can be varied in their detailed structure, and mixed between the two series, it seems that a valuable approach has been devised to prepare nanometer-scale lines. Although these lines can flex and rotate internally about single bonds, they are as rigidly linear as any molecules we have imagined.

IMAGING

For surface structure determination, the usual techniques of organic chemistry are mostly inapplicable. Beyond that, it is necessary, for the development of this strategy, that the surface structures be imaged. We have investigated two approaches to this problem. One utilizes scanning transmission electron microscopy (STEM), at a resolution of 2 Å [2]. Since these organic lines are invisible to this method (they do not scatter electrons very well), we took the approach of attaching electron-dense metal clusters to each end of the molecular lines, as in the tetrairidiumundecacarbonyl-labelled compound. We expected to see pairs of spots in the micrograph, with each pair separated by distances predicted from molecular models. Appropriate pairs of 6 Å diameter spots were detected, and pairing was shown to be statistically significant. This method allowed the first microscopic measurement of the length of **individual** small organic molecules. It is not a very attractive approach to our problem, however, because the cluster attachment is a complication on an already complex synthesis, and because only spots - not full images - are obtained.

Scanning tunneling microscopy (STM) promises to revolutionize imaging of molecules on surfaces. Recent reports have provided atomic resolution pictures of a porphyrin, and of benzene co-adsorbed with CO on rhodium. Images of DNA, showing the expected molecular shape features, have also been obtained. We are now using an STM, and, although we have been unable to image molecules like **2**, we have images of long "lines" of polypyrrole [7]. This is actually a polymer of polypyrrole tosylate grown on highly ordered pyrolytic graphite [8]. The possibilities for STM seem enormous. It, and related microscopic methods, should provide good methods for imaging surface molecular electronic structures.

LANGMUIR-BLODGETT FILMS

Attention has been given to the use of organized LB films for electronics. LB films are of specific interest here, as a way to organize molecular lines on a surface. The basic materials for LB films are amphiphilic surfactants, which have a long aliphatic hydrocarbon chain and a polar head group. Indeed, the field has been curiously narrow in terms of the structures realized.

Using a Langmuir trough, molecular lines, with structures like **2**, were applied to the water surface [3]. Isotherms were recorded, which demonstrated that organized layers resulted. The area/molecule in these layers corresponds to the largest cross-sectional area that is found along the line, i.e. from aryl-to-aryl across the molecule. Using lines of different length shows that the area is independent of the length. Thus, we suggest that these lines are oriented approximately perpendicular to the water surface, like soap molecules.

It is of considerable interest that these lines organize even when they are not amphiphilic in the usual sense. The organizing factor may be the preferred contact of the molecular surfaces, instead of the contact of the π-molecular surface with water. Molecular length is also a factor, since several shorter quinones do not form organized layers; it appears that about five rings are required (Table 1).

Table 1. Four quinones, of which two (**4**, **5**) form LB monolayers, and two (**6**, **7**) do not (from Ref. [3]).

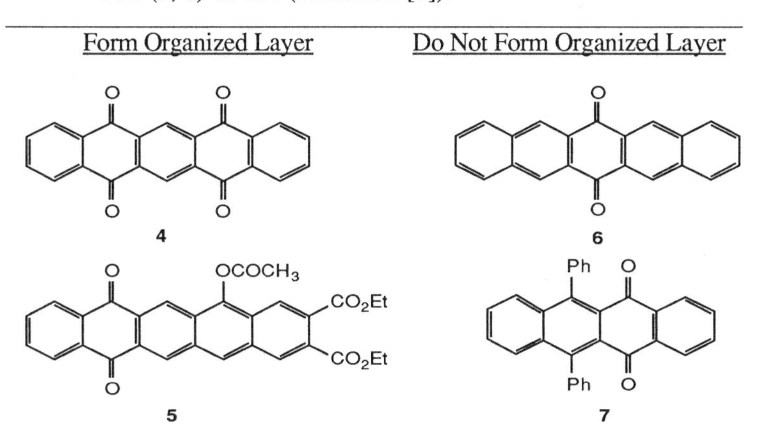

SOLID-STATE PROPERTIES

Studies of molecular lines in the solid state are prompted by considerations of conductivity, magnetism, and optical effects. Clearly, the molecular length provides an unusual organizing factor in the long dimension, and the propensity to organize (LB studies) suggests π-stacking. Our studies of solids are very preliminary, but have provided promising results in terms of conductivity from rather small molecules [4].

Pentacene diquinone (**4**) was reduced electrochemically with one Faraday/mole, and the semiquinone salt was precipitated. Four-point-probe measurements were made on pressed pellets at room temperature. The lithium salt and tetramethylammonium salt had $\sigma = 1$ [4] and $\sigma = 0.02$ S cm^{-1}, respectively. Such high conductivities are not measured on mono-semiquinones. For example, naphthacene quinone has $\sigma < 10^{-6}$ S cm^{-1}.

The $(CH_3)_4N^+$ $4^{\bullet -}$ salt has been crystallized [9]. X-ray diffraction shows a 1-D structure of stacked semiquinone units. Each anion is flat and symmetrical, and the spacing between anions in the stack is constant (3.31 Å). Each $(CH_3)_4N^+$ is held between the four oxygens of two anions in separate stacks. These stacks are at about 90° to each other. It seems likely that the conductivity is along the stacks, so that the molecular length plays a minor role. It is important, however, for conductance, that there are two quinone units in one molecule. The presence of two quinones and one electron may impart some mixed-valence character to the species in the solid.

CONCLUSION

The properties of molecular lines make them scientific objects of desire. Ongoing work will explore more complex shapes and other physical phenomena, and continues to develop surface synthesis and imaging methods.

ACKNOWLEDGEMENT

This work has been supported by NSF and ONR.

REFERENCES

[1] L. L. Miller and P. W. Kenny, J. Chem. Soc. Chem. Comm. 84 (1988); W. Christopfel and L. L. Miller, Tetrahedron 43: 3681 (1987); T. Chiba, P. W. Ke, and L. L. Miller, J. Org. Chem. 52: 4327 (1987); W. C. Christopfel and L. L. Miller, J. Org. Chem. 51: 4169 (1986); A. D. Thomas and L. L. Miller, J. Org. Chem. 51: 4160 (1986).

[2] F. Furuya, L. L. Miller, P. W. Kenny, W. C. Christopfel, and J. H. Hainfeld, J. Am. Chem. Soc. 110: 641 (1988).

[3] P. W. Kenny, L. L. Miller, S. F. Rak, T. H. Jozefiak, W. C. Christopfel, J. H. Kim, and R. A. Uphaus, J. Am. Chem. Soc. 110: 4445 (1988).

[4] L. L. Miller and T. H. Jozefiak, Synth. Metals 27: B431 (1988).

[5] T. H. Jozefiak, J. E. Almlöf, M. W. Feyereisen, and L. L. Miller, J. Am. Chem. Soc. 111: 4105 (1989); P. W. Kenny, T. H. Jozefiak, and L. L. Miller, J. Org. Chem. 53: 5007 (1988); T. H. Jozefiak and L. L. Miller, J. Am. Chem. Soc. 109: 6560 (1987).

[6] A similar idea has been developed by P. Kasaynski and J. Michl, J. Am. Chem. Soc. 110: 5225 (1988).

[7] Unpublished work of N. Phillips.

[8] R. Yang, K. M. Dalsin, D. F. Evans, L. Christensen, and W. A. Hendrickson, J. Phys. Chem. 93: 511 (1989).

[9] Unpublished work of M. Forkner.

MULTIFREQUENCY PHOTOCHROMIC MEMORY MATERIALS

G. J. Ashwell*, M. Szablewski and A. P. Kuczynski

Centre for Molecular Electronics
Cranfield Institute of Technology
Cranfield MK43 0AL, UK

There is renewed interest in organic materials for optical data storage and, recently, a multifrequency memory, comprising a Langmuir-Blodgett film with different photochromic layers, was proposed by Ando et al.[1-4]. To store several bits per pixel requires layers, with non-overlapping photochromic bands, which may be independently switched. The J-aggregated spiropyrans show some promise[4] but their frequency range is restricted. It is necessary to design alternative materials with photochromic bands in the visible and near IR. An interesting series of photochromics, with tunable absorption bands, has been studied at Cranfield[5-7]. The materials are zwitterionic[8] and have the general formula $D^+\text{-CH=C(CN)-C}_6H_4\text{-C(CN)}_2^-$ where D^+ is a pyridinium or quinolinium group linked at the alpha or gamma position. They may be switched at visible wavelengths and the photochromism has been attributed[5,6] to electron transfer between a coloured zwitterionic form and a neutral colourless form.

$$D^+\text{-CH=C(CN)-C}_6H_4\text{-C(CN)}_2^- \longrightarrow \text{D=CH-C(CN)=C}_6H_4\text{=C(CN)}_2$$

Zwitterions were synthesised from the reaction of a picolinium, quinaldinium or lepidinium halide with either neutral TCNQ and piperidine or Li^+TCNQ^- using procedures described previously[5-7]. Within this class, three suitable Langmuir-Blodgett (LB) film forming materials, with photochromic bands at 495, 565 and 614 nm and half widths at half maximum (HWHM) of 27, 22 and 37 nm respectively, have been identified.

PYRIDINIUM AND QUINOLINIUM ZWITTERIONS

Z-β-(N-alkyl-4-pyridinium)-α-cyano-4-styryldicyanomethanide is photochromic (see Figure 1) and bleaches when irradiated at wavelengths which overlap the charge transfer band. Bleached solutions recolour in the dark whereas LB films do not revert to their pre-switched form and may be used in optical memory devices. An unusual feature of the LB films is the very sharp charge transfer band at 495 nm with a HWHM of 27 nm (Figure 1)

*To whom correspondence should be addressed.

Lower-Dimensional Systems and Molecular Electronics
Edited by R.M. Metzger *et al.*, Plenum Press, New York, 1991

whereas, in solution, the corresponding band is broad, the HWHM of the pyridinium zwitterion being 76 nm in acetonitrile. The narrowing is attributed to isolation of the intermolecular and intramolecular charge transfer bands in the aligned films (see below for the quinolinium zwitterions). Both transition types have been observed from single crystal reflectance studies on an alpha-bridged analogue[9] and probably both contribute to the broadening of the solution charge transfer band.

The quinolinium analogue, Z-β-(N-alkyl-4-quinolinium)-α-cyano-4-styryldicyanomethanide, is also photochromic. In this case the LB film spectra are dependent upon the alkyl chain length with λ_{max} = 614 ± 4 nm and HWHM = 37 ± 2 nm for the hexyl to tetradecyl homologues and λ_{max} = 565 ± 4 nm and HWHM = 22 ± 1 nm for the pentadecyl to eicosyl homologues (Table 1). A change of alignment occurs between tetradecyl and pentadecyl, the areas at 25 mN m^{-1} being 28 to 34 $Å^2$ $molecule^{-1}$ for the short tail zwitterions and 40 to 50 $Å^2$ $molecule^{-1}$ for those with long tails[7].

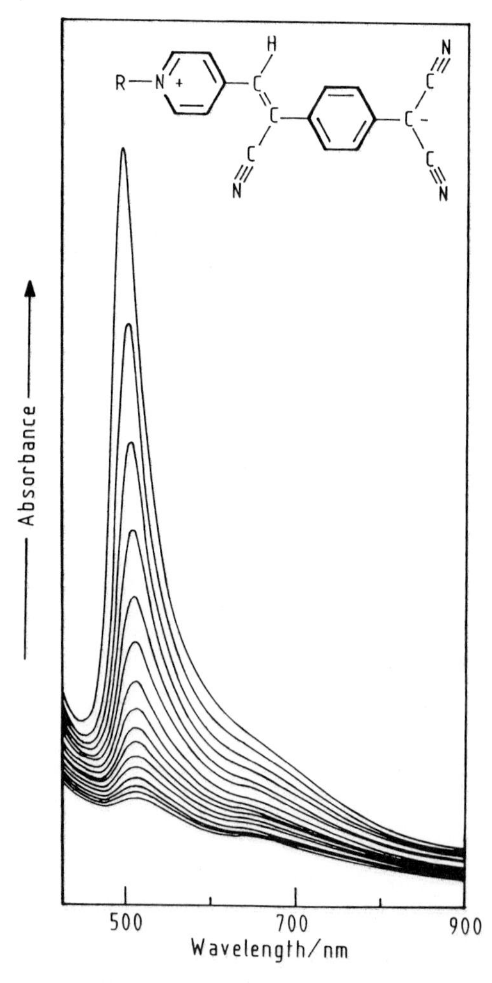

Fig. 1. LB film spectra of the N-octyl-4-pyridinium zwitterion showing the progressive photochromic bleaching of the charge transfer band.

Table 1. Charge transfer bands of D^+-CH=C(CN)-A^- where D^+ is an N-alkyl-4-pyridinium or quinolinium cation and A^- is p-phenyldicyanomethanide.

Amphiphilic cation (D^+)	LB film		
	λ_{max}/nm	HWHM/nm	Abs/layer^{-1}
Pyridinium zwitterions[6]			
$C_{16}H_{33}$-Py	495	27	0.020
C_8H_{17}-Py	495	27	variable
Quinolinium zwitterions[7]			
$C_{20}H_{41}$-Qn	568	21	0.020
$C_{18}H_{37}$-Qn	565	23	0.020
$C_{16}H_{33}$-Qn	565	22	0.020
$C_{15}H_{31}$-Qn	561	22	0.016
$C_{14}H_{29}$-Qn	614	37	0.020
$C_{13}H_{27}$-Qn	615	36	0.022
$C_{12}H_{25}$-Qn	613	37	0.018
$C_{11}H_{23}$-Qn	616	38	0.018
$C_{10}H_{21}$-Qn	616	39	0.017
C_9H_{19}-Qn	617	35	0.018
C_8H_{17}-Qn	616	37	0.020
C_7H_{15}-Qn	614	39	0.017
C_6H_{13}-Qn	610	37	0.014

For comparison, the end-on van der Waals area[10] of the $-C(CN)_2$ swallowtail is 24 $Å^2$ whereas the cross-sectional area[11] of the quinolinium group is 30 $Å^2$. Thus, the alignment of the short tail homologues is nearly perpendicular to the substrate. It follows that the transition at about 614 nm corresponds to an intermolecular charge transfer band because the intramolecular transition moment and electric vector are orthogonal in this case. In contrast, larger areas of 40 to 50 $Å^2$ molecule^{-1} imply that the long tail homologues tilt towards the plane of the substrate and, thus, the sharper band at about 565 nm probably corresponds to an intramolecular transition.

The charge transfer spectra of the pyridinium zwitterions are less sensitive to the alkyl chain length with the octyl and hexadecyl homologues showing similar properties (see Table 1). The charge transfer bands at 495 nm are sharp but show a broad low intensity shoulder on the high wavelength side (Figure 2). Poor quality films also reveal a small peak with its maximum at 634 nm. The peak and shoulder are both attributed to a broad intermolecular charge transfer band whereas the main absorption at 495 nm is attributed to an intramolecular transition. Both bands bleach when irradiated.

The spectra of the pyridinium and quinolinium analogues are superimposed in Figure 2 to show three photochromic bands with exploitable maxima at 495, 565 and 614 nm. The absorption maximum at 495 nm of the pyridinium zwitterion corresponds to minima in the spectra of the other materials whereas the maxima at 565 and 614 nm of the quinolinium zwitterions only partially overlap the adjacent charge transfer bands. The fact that there

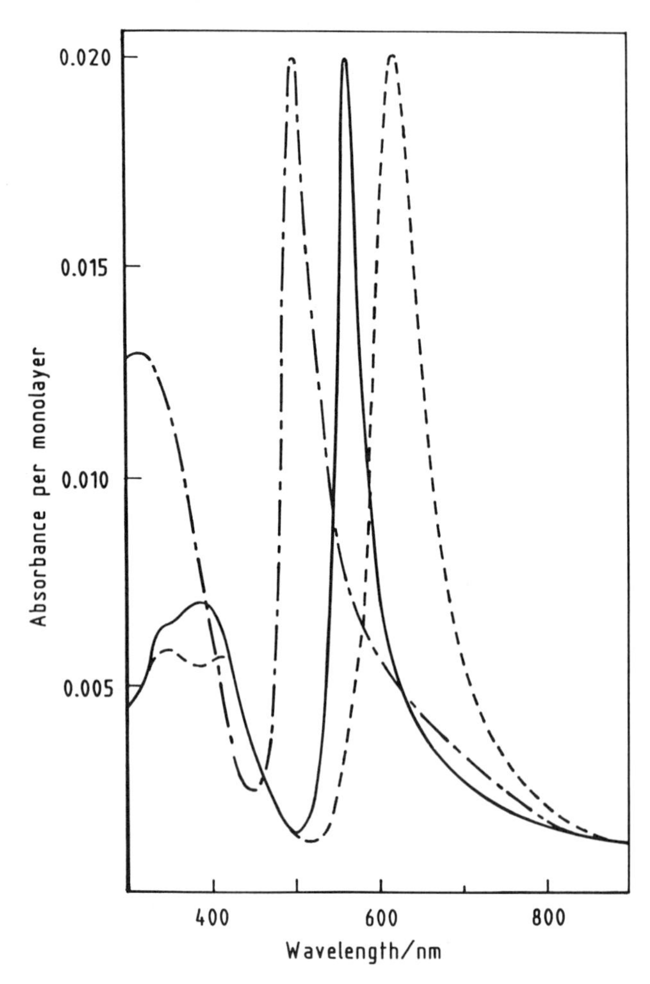

Fig. 2. LB film spectra of N-hexadecyl-4-pyridinium (λ_{max} = 495 nm), N-hexadecyl-4-quinolinium (565 nm) and N-octyl-4-quinolinium (614 nm) zwitterions.

Table 2. Photochromic charge transfer bands of D^+-CH=C(CN)-A^- where D^+ is N-hexadecyl-4-quinolinium and A^- is a substituted p-phenyldicyanomethanide group.

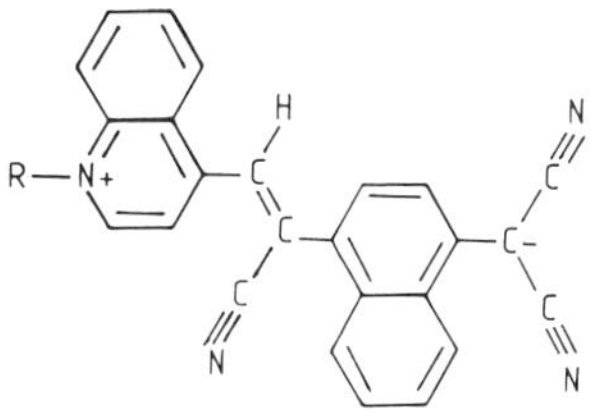

I

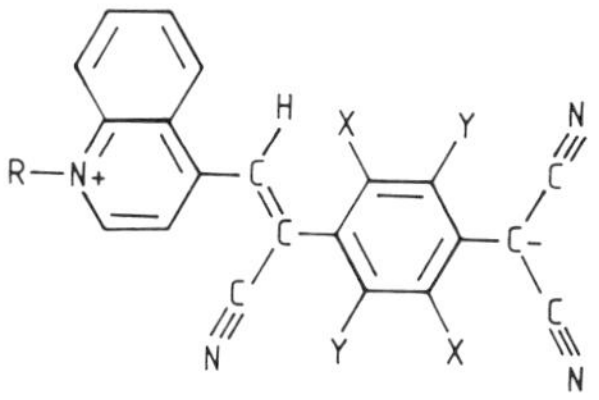

II

Molecular structure	Anion end (A^-)	LB film λ_{max}/nm	$E_{1/2}$/V*
(I)	$-C_{10}H_6-C(CN)_2^-$	623	-0.09
(II)	$-C_6H_4-C(CN)_2^-$	565	+0.17
(II)	$-C_6H_2Cl_2-C(CN)_2^-$	545	+0.41
(II)	$-C_6H_2Br_2-C(CN)_2^-$	545	+0.41
(II)	$-C_6F_4-C(CN)_2^-$	480	+0.53

*Half wave reduction potentials versus SCE of TCNQ[12] and its derivatives[12,13].

is little direct overlap at wavelengths which correspond to the maxima implies that the bands may be independently switched. To exploit these materials it is necessary to increase the number and to shift the absorption towards the infrared.

TUNING THE CHARGE TRANSFER BAND

The frequency of the intramolecular charge transfer band may be tuned by substitution to optimise its position relative to the neighbouring bands. The energy of the transition is crudely given by:

$$E_{CT} = (I_D - E_A) - C$$

where I_D and E_A are the ionisation energy and electron affinity of opposite ends of the zwitterion and C is the Coulomb energy. The effect of substitution is evident from the red shift of the intramolecular back charge transfer band, from 495 nm for the pyridinium zwitterion to 565 nm for the quinolinium zwitterion, the quinolinium heterocycle being a stronger electron acceptor than pyridinium. Also, for materials in Table 2, substituted only at the anion end, the intramolecular charge separation and donor ionisation energy are assumed to be constant and, thus, the energy of back charge transfer is mainly dependent upon the electron affinities of the $=C_{10}H_6=C(CN)_2$ and $=C_6X_2Y_2=C(CN)_2$ acceptor groups of the neutral forms of (**I**) and (**II**) above. The values are unavailable but should follow the same general trend as the half wave reduction potentials of the TCNQ derivatives[12,13] from which the zwitterions were prepared. The notable effect of electron withdrawing groups on the position of the charge transfer band is shown by the data in Table 2. With decreasing reduction potential, the absorption maximum is progressively blue shifted from 623 nm for the benz analogue to 480 nm for the tetrafluoro analogue. Thus, the photochromic band may be tuned by substitution.

ACKNOWLEDGEMENTS

One of us (GJA) acknowledges the SERC for support of this work (GR/E/88974 and GR/F/45813) and for providing studentships to MS and APK. We are grateful to Dr Martin Bryce for supplying the 2,3-benz and 2,5-dihalo TCNQ derivatives from which the substituted zwitterions were synthesised.

REFERENCES

1. J. Miyazaki, E. Ando, K. Yoshino and K. Morimoto, Eur. Patent Appl. EP 193931 A2 (1986); US Patent 4737427A (1988).
2. E. Ando, J. Miyazaki, K. Morimoto, H. Nakahara and K. Fukuda, Int. Symp. on Future Electronic Devices - Bioelectronic and Molecular Electronic Devices, Tokyo, 47 (1985).
3. E. Ando, J. Miyazaki, K. Morimoto, H. Nakahara and K. Fukuda, Thin Solid Films 133:21 (1985).
4. E. Ando, J. Hibino, T. Hashida and K. Morimoto, Thin Solid Films 160:279 (1988).
5. G. J. Ashwell, UK Patent Appl. 8907311.8, Filed March 31 (1989).
6. G. J. Ashwell, Thin Solid Films, in the press (1989).
7. G. J. Ashwell, H. Block, A. P. Kuczynski, I. M. Sandy, M. Szablewski, M. R. Bryce, A. M. Grainger, M. Hasan, N. A. Bell, R. A. Broughton, A. T. Jones and R. C. Thorpe, Thin Solid Films, in the press (1989).
8. R. M. Metzger, N. E. Heimer and G. J. Ashwell, Mol. Cryst. Liq. Cryst. 107:133 (1984).
9. S. Akhtar, J. Tanaka, R. M. Metzger and G. J. Ashwell, Mol. Cryst. Liq. Cryst. 139:353 (1986).
10. Calculated from the molecular dimensions of the zwitterion in ref. 8.
11. Calculated from the molecular dimensions of the quinolinium cation in G. J. Ashwell, D. D. Eley, S. C. Wallwork, M. R. Willis, G. F. Peachey and D. B. Wilkos, Acta Cryst. B33:843 (1977).
12. R. C. Wheland and J. L. Gillson, J. Amer. Chem. Soc. 98:3916 (1976).
13. S. Chatterjee, J. Chem. Soc. B 1170 (1967).

FUNCTIONALIZED CONDUCTING POLYMERS: TOWARD MOLECULAR DEVICES

Takeo Shimidzu

Division of Molecular Engineering
Graduate School of Engineering
Kyoto University
Kyoto 606, JAPAN

INTRODUCTION

There are many functional molecules. For their exploitation, conducting polymers are considered to be suitable matrices, because their conductive properties may be used to prove the electronic structure and state, which relate directly to the function of these molecules. To construct a molecular device, the fabrication and the materialization of those functional molecules at the molecular level is one of the most important problems. The present study demonstrates three approaches toward molecular device construction:

(1) materialization of functional molecules by their incorporation in conducting polymers
(2) molecular level fabrication of conducting polymers
(3) integration of functional molecules to be incorporated, as shown in the following scheme (Fig. 1).

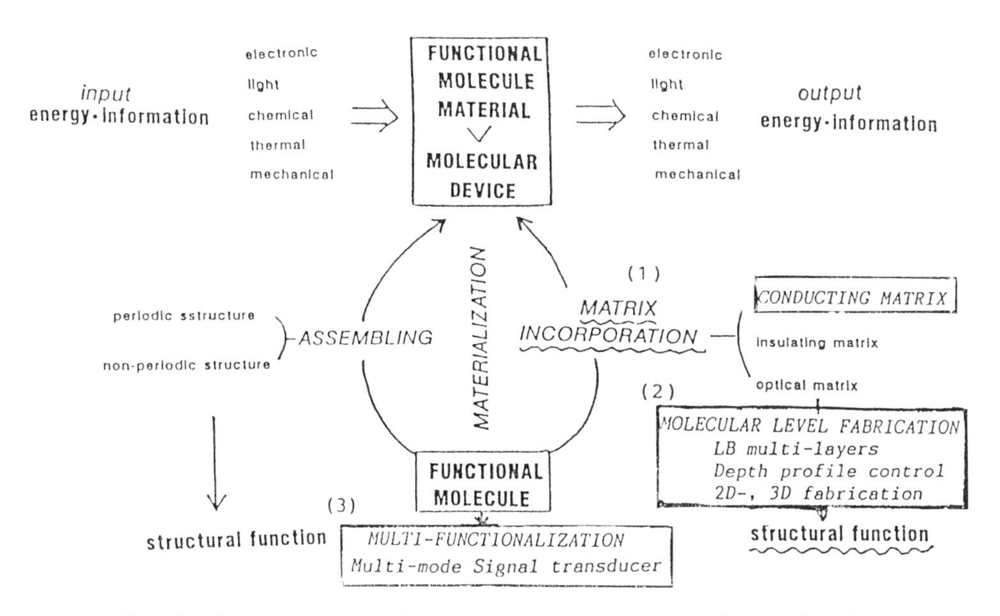

Fig. 1. Construction of molecular devices from functionalized molecules.

Lower-Dimensional Systems and Molecular Electronics
Edited by R.M. Metzger *et al.*, Plenum Press, New York, 1991

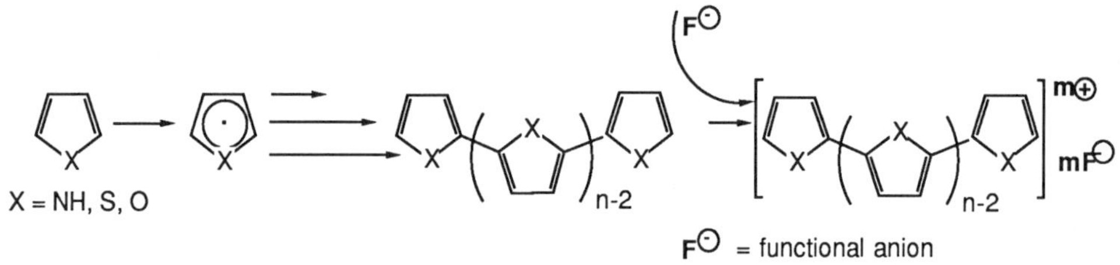

X = NH, S, O

$F^{\ominus}$ = functional anion

Fig. 2. Electrolytic polymerization of conducting polymers using a functional molecular anion.

MATERIALIZATION OF FUNCTIONAL MOLECULES BY THEIR INCORPORATION IN CONDUCTING POLYMERS [1-5].

The materialization of a functional molecule, by incorporation into a conducting polymer matrix, is achieved in the electrolytic polymerization of pyrrole, thiophene, furan, aniline, etc., in the presence of negatively charged functional molecules. The incorporation of the functional molecules is driven electrostatically, by the positive charges of the partially oxidized conducting polymer matrices, through a doping process, as shown in Fig. 2.

Electrolytic polymerization was easily achieved with a monomer solution, in the presence of negatively charged functional molecules. A key concept of this preparative approach is that the resulting conducting polymer has a lower redox potential (doping-undoping), than its monomer, therefore anodic doping occurs simultaneously with electropolymerization. An electrolyte anion, or functional anionic molecule, is incorporated into the partially oxidized conducting polymer matrix, so as to conserve its electroneutrality.

The resulting functionalized conducting polymers display their native functions; the functionalities and the conducting properties are not much reduced. Table 1 shows some examples.

Chemical polymerization methods are also available for preparation of functionalized conducting polymers. Several techniques are shown in Table 2. In these cases, all kinds of functional molecules (negatively charged, neutral, positively charged) can be incorporated.

Utilizing these functionalized conducting polymers, a solar cell, an electrochromic device, a chemical luminescent material, a sensor, a polymer battery, etc. can be constructed, in which the functionality is attributed to the incorporated functional molecule.

Table 1. Negatively charged functional molecules suitable for incorporation in conducting polymer matrices.

Group	Examples[a]	Function[b]
Organic molecule	AQS, rose bengal, indigo carmine	A, B
Metal complex	$Lu(PTS)_2$, $Fe(BPS)_n(BP)_{3-n}$, $Ru(BPS)_n(BP)_{3-n}$	A, B, C
	M(TPPS) (M = H_2, Zn, Pd)	B
	Fe(PTS) [5], M(TPS) (M = Fe, Co, Mn) [6]	E
Inorganic metal ion	RuO_4^-[4], $PtCl_4^{--}$, $AuCl_4^{--}$, MnO_4^{--}, phosphotungstate	A, D, E
Polyelectrolyte	PVSK, PSSNa, Nafion®	F, G
Biomaterial	enzyme, nucleotide material	E, H

a AQS = Anthraquinone-2-sulfonic acid; M(TPS) = metal tetrasulfophthalocyanine; BPS = bathephenanthroline disulfonic acid; M(TPPS) = metal tetra(4-sulfophenyl)porphyrin: PVSK = potassium poly(vinylsulfate); PSSNa = sodium poly(styrene sulfonate) Nafion® = Salt of sulfonated and highly fluorinated polymer (trademark of Du Pont).

b A = electrochromism; B = photoelectric conversion; C = electrogenerated chemiluminescence; D = highly dispersed noble metal; E = catalyst; F = high mechanical strength conducting polymer; G = charge-controllable transport membrane; H = sensor.

Table 2. Methods for preparation of conducting polymer composites by chemical polymerization.

Chemical Polymerization	SEPTAL	DIPPING	BLENDING
	O M S	S S S O or M M or O	M or O +S O or M

Monomer (M)	pyrrole, thiophene, furan, aniline, indole, their derivatives
Support (S)	ion-exchange membrane, natural and synthetic membrane, porous ceramic
Oxidant (O)	Lewis acid, halogen, peroxide, metal acid
Solvent	water, organic solvent, gas-phase reaction

MOLECULAR-LEVEL FABRICATION OF CONDUCTING POLYMERS

Molecular-level fabrication of a conducting polymer matrix is an important hurdle to constructing molecular devices. Three approaches have been demonstrated: (i) electrochemical polymerization of Langmuir-Blodgett (LB) multilayers, (ii) photo-patterning of conducting polymers [9], and (iii) depth profile control of conducting polymer multilayers. These are discussed in turn.

Electrochemical Polymerization of LB multilayers of amphiphile pyrrole derivatives [6-8]

A mixed monolayer of an amphiphilic pyrrole, CPy or EPy (Fig. 3) and octadecane was very stable over a neutral aqueous subphase. Distinct built-up points of surface pressure and sharp π-A curves were observed (the collapse pressure exceeded 50 mN/m). More than 600 layers of the mixed monolayers (30 mN/m) could be transferred as Y-type LB films onto a silanized ITO-deposited glass or ITO-coated polyester film. The oxidative electrolysis of 200 layers of the mixed EPy-octadecane (2:1) on ITO-deposited substrate, in acetonitrile containing $LiClO_4$, resulted in a color change from transparent to reddish brown. Both X-ray diffraction analysis and the TEM image of the cross-section showed a fine multilayered structure. This clear pattern of the layered structure shows that one can regulate, at the molecular level, the structure of a macroscopic (μm thick) conducting polymer material (Fig. 4). The dark region is considered to be the polypyrrole moiety, while the light region is the alkyl chain region. The TEM bilayer spacing is d = 55-62 Å, which almost the same as obtained from the X-ray diffraction method.

This electropolymerized LB multilayer had a highly anisotropic dc-conductivity [by ca. 10 orders of magnitude ($\sigma_{\parallel} = 10^{-1}$ S/cm, $\sigma_{\perp} = 10^{-11}$ S/cm)], as was suggested from its layered structure.

Photo-patterning of conducting polymers [9]

The fabrication of a conducting polymer pattern is considered to be one of the most important subjects in the field of molecular electronic devices. A novel photo-sensitized polymerization of pyrrole was investigated, in order to form fine conducting polymer patterns

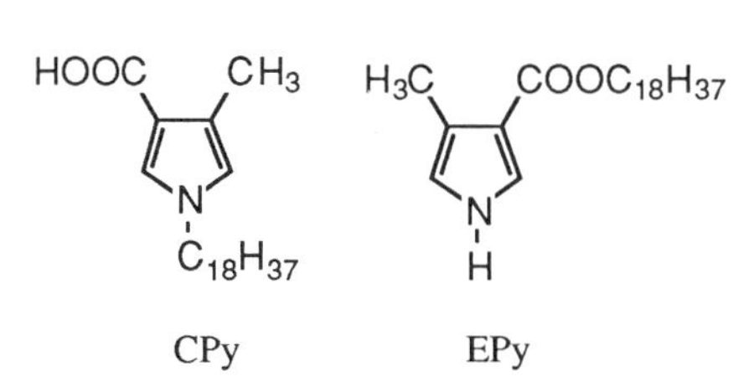

Fig. 3. Amphiphilic pyrroles.

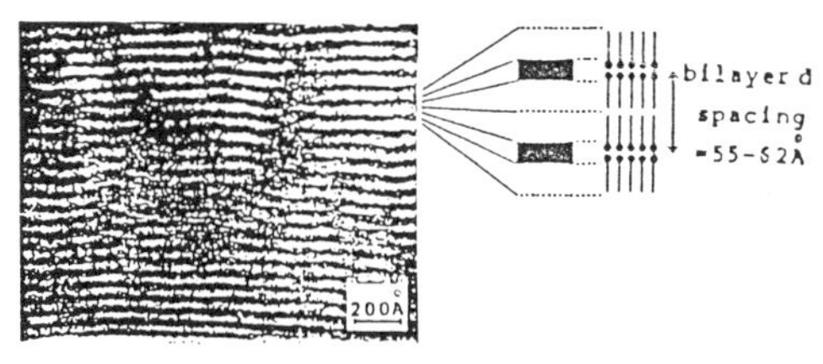

Fig. 4. Transmission electron micrograph (TEM) of the cross section of an electropolymerized EPy-C_{18} LB film.

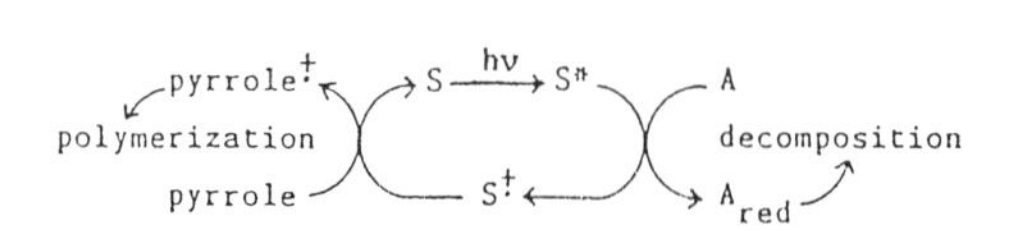

Fig. 5. Mechanism of photosensitized polymerization of pyrrole.
$S = Ru(bpy)_3^{++}$
$A = Co(NH_3)_5Cl^{++}$.

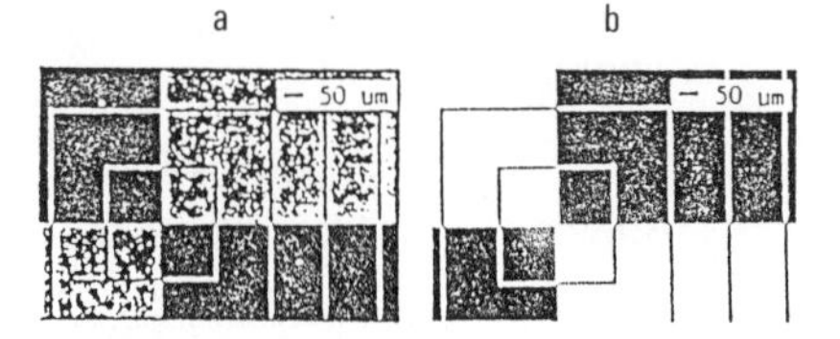

Fig. 6. Optical microscope photographs of: (A) polypyrrole synthesized on Nafion film by photosensitized polymerization; (B) photo mask used for photoirradiation (width of line = 10 μm).

on insulating materials. Here, both a photosensitizer (which can oxidize pyrrole) and a sacrificial oxidant (which does not directly oxidize pyrrole) are required for polymerization of pyrrole (Fig. 5). It was proved that the polymerization proceeded by the oxidation of pyrrole by the photo-oxidized sensitizer, through an oxidative electron transfer process. A fine conducting pattern, with ca. 10 μm width of polypyrrole, was formed on organic membranes such as Nafion, which absorbed the cationic sensitizer by this method (Fig. 6).

Depth profile control of different conducting polymer multilayers at the nm level

Utilizing an electrocopolymerization technique, various conducting copolymer layered structures were constructed. The copolymer composition, and the thickness of the polymerized copolymer, are controlled easily in the electrolytic polymerization process (Fig. 7). Fig. 8 is an example of such depth profile-controlled multilayers, consisting of polypyrrole and copoly(pyrrole/3-methylthiophene). The present method will allow nm-level thickness control.

MULTI-MODE CHEMICAL TRANSDUCER

In order to design a molecule for the coming chemical transducer, one will be forced, not only to improve the intrinsic functions, but also to endow highly-integrated transformation modes to one molecule. Here, a new class of chemical transducers, which have multiple transformation modes, by independent stimulations, and whose response by one stimulation can be regulated by the other stimulations, is demonstrated. The basic requirements are as follows: (i) the molecule has multiple regions, i.e., functional groups, responsive independently to stimulations; (ii) these regions have appropriate interaction between them, e.g., through a conjugated system. Thus, a molecule with n responsive regions would have 2n sites, corresponding to n transformation processes by independent stimulation. Also, all the transformations would be, more or less, affected by the states of the other n-1 regions. Such independent single transformations are termed "conjugated functions."

Based on the above concept, the following multi-mode chemical transducer molecule was synthesized (Fig. 9), and its function was demonstrated. This molecule showed both electro-

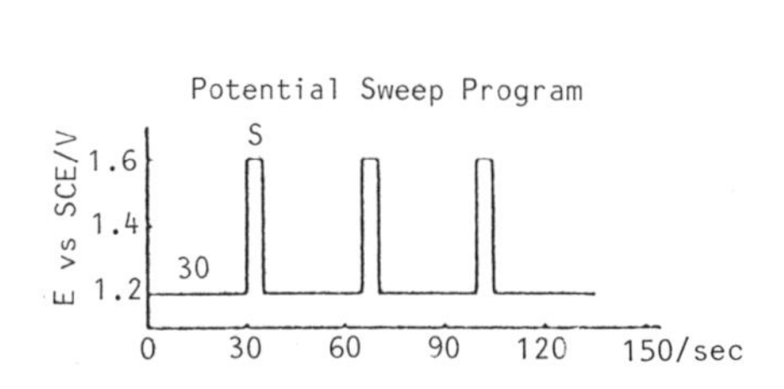

Fig. 7. Potential sweep program in electro-copolymerization.

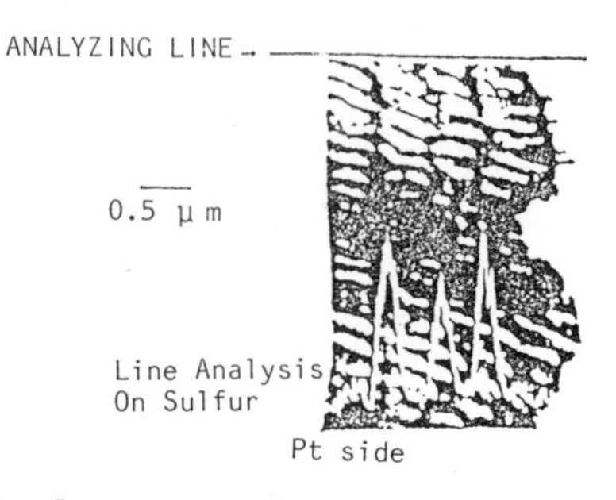

Fig. 8. Depth profile analysis by X-ray microanalytic method.

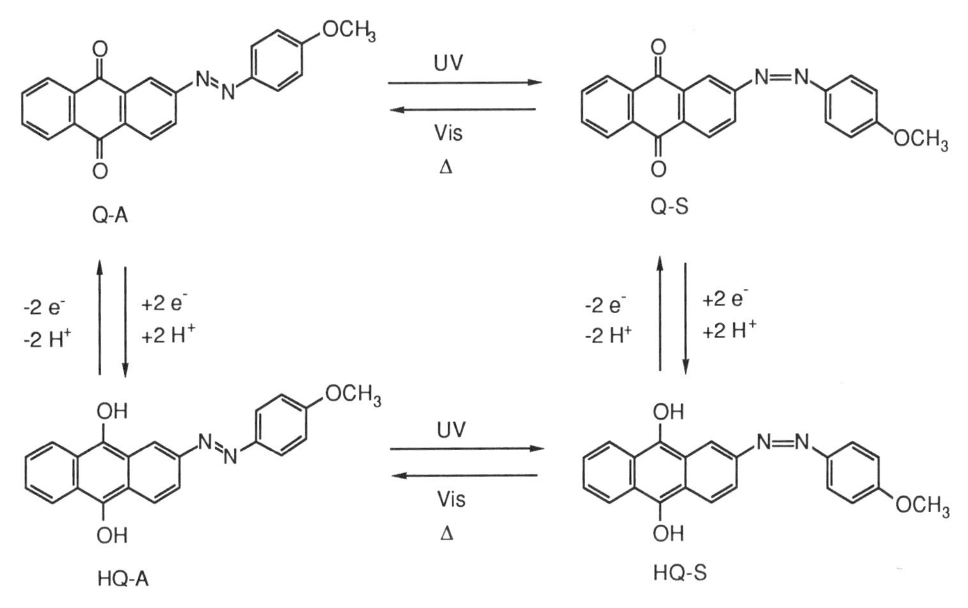

Fig. 9. Multimode (electrochromic and photochromic) chemical transducer molecule.

chromism, due to its quinone-hydroquinone region, and photo-chromism, due to its azo region, as expected. These observations illustrated that this molecule had 4 ($=2^2$) separate states, corresponding to both redox states of the quinone-hydroquinone region, and geometric isomers of the azo region. Accordingly, this molecule does play a significant role as a new chemical transducer in two modes, electrochromism and photochromism.

The present conjugated function, characteristic of a multi-mode chemical transducer, may be applicable for dual-mode memory: One is a "deep" memory mode (hydroquinone-type photochromism, hard to transform) and the other is a "shallow" memory mode (quinone-type photochromism, easy to transform).

REFERENCES

[1] T. Shimidzu, Reactive Polymers 6: 221 (1987).
[2] T. Iyoda, A. Ohtani, T. Shimidzu, and K. Honda, Synth. Metals 18: 725 (1987).
[3] T. Shimidzu, A. Ohtani, T. Iyoda, and K. Honda, J. Chem. Soc. Chem. Commun. 1415 (1986).
[4] T. Shimidzu, T. Iyoda, A. Ohtani, M. Ando and K. Honda, in "Molecular Electronic Devices", Vol. III, ed. by F. L. Carter, R. E. Siatkowski, and H. Wohltjen (North-Holland, Amsterdam, 1988) p. 599.
[5] T. Shimidzu, A. Ohtani, T. Iyoda and K. Honda, J. Chem. Soc. Chem. Commun. 327 (1987).
[6] T. Iyoda, M. Ando, T. Kaneko, A. Ohtani, T. Shimidzu, and K. Honda, Tetrahedron Lett. 27: 5633 (1986).
[7] T. Iyoda, M. Ando, T. Kaneko, A. Ohtani, T. Shimidzu, and K. Honda, Langmuir 3: 1169 (1987).
[8] T. Shimidzu, T. Iyoda, M. Ando, A. Ohtani, T. Kaneko, and K. Honda, Thin Solid Films 160: 67 (1988).
[9] H. Segawa, T. Shimidzu, and K. Honda, J. Chem. Soc. Chem. Commun. 132 (1989).

PROSPECTS FOR TRULY UNIMOLECULAR DEVICES

Robert M. Metzger

Department of Chemistry
University of Alabama
Tuscaloosa, AL 35487-0336, USA

INTRODUCTION

This review discusses some of the issues that have occupied the attention of several recent conferences on "Molecular Electronics" (ME). The term ME has been popularized by the late Forrest L. Carter, who chaired three international workshops on ME devices in Washington in 1981 [1], 1983 [2], and 1986 [3]. There was also a conference in Varna, Bulgaria in 1986 [4]. A recent conference on ME elucidated the present challenges [5]. By now several conferences world-wide have used ME in all or part of their title, so the term has gained respectability. There is a Journal of ME. Several Centers for ME have appeared (at Carnegie-Mellon University in the US, at Cranfield Institute of Technology in the UK, etc.). A fair assessment of the present status has recently appeared in a publication for the "lay audience" [6].

In the period 1983-1984, some visionary (but unscientific) ideas about self-assembling biological "computers" ("biochips") grabbed undeserved and uncritical world-wide media attention; the critical reaction to such exaggerations almost drowned, in the ocean of righteous disbelief, the infant field of ME! A very sobering note outlined what technical accomplishments were still needed [7]. Since then, more conservative chemists, physicists, and materials scientists have broadened the definition of ME, re-labelled some of their research areas as ME, and thus have given ME its present nascent respectability.

Two definitions of ME exist: in a broader sense, *(sensu lato)* ME includes the study of all molecule-based electronic processes in lower-dimensional systems:

(A1) charge-transfer salts (organic conductors and superconductors) [8];
(A2) conducting low-dimensional polymers [8, 9];
(A3) intercalated graphites ($C8K$, etc.) [10];
(A4) intercalated 2-D chalcogenides ($TaS2$, etc) [11];
(A5) inorganic linear-chain compounds [12];
(A6) Langmuir-Blodgett (LB) multilayers [13].

These research areas have been studied for almost two decades, and relabelling them as ME may be a convenient short-hand, an attention-getting method for funding research proposals, or just a current fad.

In a more narrow sense, *(sensu stricto)* the "true" ME devices exploit the electronic properties of single molecules or clusters.These have been called "zero-dimensional devices". As will be shown below, these form a long list of proposals and ideas, one experimental accomplishment (*), and, as of this writing, no marketable devices:

(B1) Aviram and Ratner's rectifier [14];
(B2) Aviram and Ratner's hydrogen atom switch [15, 16];
(B3) Mitani's proton transfer system [17];
(B4) Carter's soliton switch [18];
(B5) * Fujihira's LB monolayer photodiode [19].

Table 1. Approximate Chronology of Molecular Electronics.

Date	Event	Ref.
1960	Feynman proposes that "there is plenty of room at the bottom"	[26]
1960	Du Pont group synthesizes TCNQ	[27]
1960's	Optical computing discussed	[28]
1964	Little proposes room-temperature superconductivity in organic polymer	[29]
1965	Melby finds high conductivity in NMP TCNQ salt	[30]
1966	von Neumann's cellular automata ideas published	[31]
1973	Johns Hopkins group finds first organic metallic conductor: TTF TCNQ	[32]
1978	Penn group finds first highly conducting polymer: doped polyacetylene	[33]
1981	Bechgaard-Jérome group finds first organic superconductor: $(TMTSeF)_2PF_6$	[34]
1983	IBM group finds superconductivity in $(BEDT\text{-}TTF)_2ReO_4$	[35]
1981	First International Workshop on Molecular Electronic Devices	[1]
1983	Second International Workshop on Molecular Electronic Devices	[2]
1986	Third International Workshop on Molecular Electronic Devices	[3]
1986	Molecular Electronics: Fourth Intern. School on Condensed Matter	[4]
1989	First International Conf. on Molecular Electronics- Science and Engineering	[5]

As "passive" molecular connectors, which may, one day, be useful to interrogate true ME devices, one can consider a few ideas and one accomplishment (*):

(C1) molecular wires and antennas, e.g. carotenes and other polyenes [21];
(C2) Aviram's linked thiophene "connects" [22];
(C3)* Miller's "molecular spacers" [23];
(C4) molecular inclusion compounds, e.g. cyclodextrins [24] and calixarenes [25].

These ideas and potential devices, which represent the best present long-range hope for ME, should exhibit the inherent speed of intramolecular electron transfer (>1 GHz). In my opinion, ME can compete with present-day semiconductor electronics only if ME can couple the advantages of molecular size and fast intramolecular electron transfer: there is little use for a small but slow device.

This brief review will concentrate on the results of other researchers. To all those, whose research I may unwittingly omit, slight, or misrepresent, my sincere apologies....

HISTORICAL REVIEW

In Table 1 are given some dates which help to chronicle the growth of Molecular Electronics. In the last thirty years we have learned how organic molecular crystals and polymers can exhibit many of the properties of their inorganic counterparts: high electrical conductivity, dopability, and even superconductivity (which in 9 years after the initial discovery [34] has reached 11 K [36] for organic crystals; in the 87 years since the discovery of superconductivity in Hg, the inorganic ceramic cuprate superconductors have reached a critical temperature of 125 K [37]). Organic systems can be tailored, through laborious chemical synthesis ("designer molecules"), to exhibit the desired molecular or bulk properties, thanks to the molecular orbitals of the individual molecules, or to the intermolecular overlap and exchange of energy or of electrons. Organic systems can exhibit lower-dimensional effects more easily than the inorganic systems, but are more susceptible to thermal degradation (but some of them, e.g. the ladder polymers, can withstand very high temperatures).

First, we will first discuss the "bulk" molecular devices which have been popular, even though they are molecular devices only in the larger sense (i.e. because of the band structure of the solids, or because of phase change properties, or other bulk effects). Second, a small detour is made to discuss electron transfer within the primary reaction center in photosynthesis, and the controlling factors in intramolecular electron transfer. Third, we will concentrate on the truly unimolecular devices. Fourth and last, a discussion on molecular "wires" and connectors is given.

MOLECULAR DEVICES THAT DEPEND ON BULK PROPERTIES

Bulk Organic Rectifiers

It was of academic interest, ever since the discovery of the transistor, to see whether organic molecules could function as bulk pn rectifiers (diodes), or as npn transistors. This would occur if a film or crystal of an organic electron donor were brought in contact with that of an organic electron acceptor. This was indeed verified in the 1960's[38].

Multilayer LB Organic Rectifiers

Kuhn and co-workers [39] showed that one can obtain a pn (or DA) rectifier in a LB multilayer sandwich Al | $(CA\text{-}D)_q$|$(CA)_r$ |$(CA\text{-}A)_s$|Al, where Al denotes the bulk Al contacts, $(CA\text{-}D)_q$ denotes the electron donor system D [q LB monolayers of cadmium arachidate (CA) doped in the ratio 5:1 with suitable organic π electron donors D], $(CA)_r$ denotes a spacer layer of r undoped monolayers of CA, and $(CA\text{-}A)_s$ denotes the electron acceptor system A [s LB monolayers of CA doped with suitable organic π electron acceptors A]. This work was confirmed by Sugi and co-workers [40], who observed rectification properties in LB multilayers, provided that $q \geq 3$, $r \geq 1$, and $s \geq 3$.

A Phase-Change Organic Switch

A fast switch was discovered by Potember in CuTCNQ [41]. This was due to the bulk thermodynamic metastability (in crystals or amorphous powders) of the violet, low-conductivity ionic state (IS) Cu^+ $TCNQ^{-}$(c), relative to the yellow, low-conductivity, neutral state (NS) Cu(0) TCNQ(0)(c), with, presumably, an intermediate, mixed-valent, higher-conductivity state (CS); one could switch between the two states IS <--> CS, either with an applied voltage over a certain threshold value, or by a moderate laser beam, while heat will restore the IS. This effect is found also in AgTCNQ, and in a few related systems [42]. The switching rate is fast, and could be used for optical data storage [43].

A Molecule-based Transistor Using Conducting Polymers

Wrighton and co-workers developed a "molecule-based transistor" which uses conducting polymers: either chemically doped polyaniline layers deposited atop Au interdigitated electrodes [44] or on 50-100 nm "gate" polyaniline polymer between two Au electrodes shadowed with SiO2: this device still has a rather slow switching rate (10 kHz) and a gain of almost 1,000 [45]. The related work of Stubb and co-workers is discussed elsewhere in this volume.

Phase-Change Optical Memories

There are three established technologies for mass information storage [46-48]:

(i) ferrite, iron oxide and chromium oxide-based magnetic memories, disks and tapes (MM): these are erasable, or write many, read many (WMRM) media

(ii) Si and GaAs-based semiconductor computer memories (SM): these can be considered as very fast write many, read many (FWMRM) media;

(iii) Al pits on polycarbonate-based laser-readable compact disk (CD) technology; this is a permanent, or write once, read many (WORM) medium.

The limitations in storage density, access time, and cost of the above three techniques have spurred extensive research in alternative storage strategies. For instance, CD magneto-optical recording provides WMRM capabilities, but no increase in storage density. Phase-change methods of storing WORM data on CD-type surfaces can use inorganic (Te) or organic (phthalocyanine) substrates [49]. The need to find WMRM optically addressable media is acutely felt, and many organic systems are being considered. Organic phase-change systems will have no real size or speed advantage over inorganic systems, but may show a cost advantage.

PHOTOSYNTHESIS AND INTRAMOLECULAR ELECTRON TRANSFER

Much effort has been expended in understanding the nature of the charge-separation process in photosynthesis [50]. Within the chloroplasts of plants, the chlorophyll or electron donor molecule D is at some fixed distance from an electron acceptor or quinone molecule A (let us denote this as D...A); the absorption of a visible light photon by D is followed by the extremely rapid formation of a radical ion pair D+...A- (charge-transfer reaction CTR),with extremely high quantum efficiency and no appreciable back charge transfer reaction (BCTR):

$$\underset{\text{CTR}}{\text{D..A -----------> }} \text{D}^{+}\text{....A}^{-} \underset{\text{No BCTR}}{\text{ ------XXXXX----> }} \text{D...A}$$

The detailed physical mechanism of this process has been a long-standing puzzle, whose resolution has been aided in part by the structure elucidation of the photosynthetic reaction center in purple bacteria [51]. For photosystem II, D = dimer of chlorphyll P-680 is excited by light, and an exothermic (0.04 to 0.08 eV) and rapid (less than 1 ns) electron transfer occurs to an intermediate acceptor A = pheophytin (forming the radical pair $P680^{+}..Pheo^{-}$). The back reaction (which would otherwise occur within 2-4 ns) is prevented by a faster (<200 ps), very efficient, and very exothermic (0.5 eV) electron transfer, in one or several steps, from $Pheo^{-}$ to the ultimate acceptor A = a complex of plastoquinone with Fe.

The open question is whether artificial D...A, or D-σ-A, or D-π-A molecules could be synthesized, which could mediate either (i) photon-induced controlled electron transfer reactions or (ii) photon-induced water splitting.

Understanding Intramolecular Electron Transfer

Taube pioneered the understanding of intramolecular electron transfer in solution involving binuclear transition metal complexes, both D-σ-A and D-π-A [52]. More recently, efforts have been made to control the electron transfer in solution between D-σ = Creutz-Taube ion $[Ru(bpy)_3]^{++}$ (bpy is 2,2'-bipyridine) and A = methylviologen [53] It was shown by Miller, Closs and co-workers [54, 55] that the intramolecular electron transfer rate through a molecule D-σ-A (i) increases with an increasing I_D-A_A (where I_D is the ionization potential of the donor moiety D, A_A is the electron affinity of the acceptor moiety A), but (ii) decreases again, if the Franck-Condon reorganization is large (because the geometries of D^+ and A^- are quite different from the geometry of D and A, respectively) (Marcus "inverted region" [56]).

Potential Artificial Photosynthetic Molecules D-σ-A

Efforts have been made to make artificial photosynthetic molecules, of the type D-σ-A, where D is an electron donor, such as a porphyrin, phthalocyanine, or related molecule, σ is a covalent "sigma" bridge, i.e. a saturated linkage, such as one or several [2.2.2]-bicyclooctane rings, and A is the electron acceptor (usually p-benzoquinone). The groups of Mauzerall (Rockefeller) [57], Weedon and Bolton (Western Ontario) [58], Dervan and Hopfield (Caltech) [59], Mataga (Osaka) [60], Gust (Arizona State) [61], Staab (Heidelberg) [62], and Verhoeven (Amsterdam) [63] have prepared several interesting D-σ-A molecules. The results sofar have shown rapid CRT, but also a very rapid recombination of the charge-separated state by BCTR. It is hoped by some researchers that, if one organizes such D-σ-A molecules in LB multilayers or other assemblies, the BCTR may be suppressed.

Potential Artificial Photosynthetic Molecules D-π-A

Nature produces an abundant supply of carotenes as sacrificial anti-oxidants in living organisms; these carotenes have discussed either as "molecular antennas" [64] or as "molecular wires" [21]. Carotenes have been used as the "pi" bridge π between D and A, to yield D-π-A structures (D = Zn porphyrin, π = carotene, A = anthraquinone). It was seen, however, that both CTR and BCTR occurred at comparable rates, i.e. within a few ps of the laser excitation [64].

TRULY UNIMOLECULAR, OR ZERO-DIMENSIONAL DEVICES

Proposed D-σ-A Rectifier

Aviram and Ratner's proposed D-σ-A unimolecular rectifier [14] and the efforts to prepare a practical device are discussed elsewhere in this volume [65].

Proposed H Atom and Proton Transfer Switches

Aviram also proposed an intramolecular hydrogen atom transfer switch, based on H bonding in ortho-quinone-cathecol systems [15,16]. It has been confirmed that there is interamolecular H atom transfer in such a system [64]. Mitani is working toward a proton transfer switch [17].

Proposed Soliton Switch

Carter [18] proposed that soliton motion in polyacetylenes could be harnessed in various switches, gates, and logic circuits. However, apart from the difficulties of synthesizing or addressing such molecules, soliton switches would be relatively slow.

Fujihira's LB Photodiode

Fujihira and co-workers have demonstrated that a single LB monolayer can function as a photodiode [19]: this is probably the first truly unimolecular device. They synthesized a D-σ_1-A-σ_2-S molecule, where D = electron donor = ferrocene, $\sigma_1 = (CH_2)_{11}$ chain, A= ultimate electron acceptor = viologen , $\sigma_2 = (CH_2)_6$ chain, S = sensitizer = pyrene. This molecule was transferred as an LB monolayer onto a semitransparent Au electrode (with the viologen, or A, part of molecule closest to Au); this electrode was used as a window of an electrochemical cell, which also contained a 0.1 M KCl solution and a Pt counter electrode. Under bias, an electron was transferred from solution to the ferrocene end of the LB film, and then to the ground state of the pyrene molecule. Light at 330 nm excited the pyrene radical cation from the ground state to an excited state, from which the electron was transferred to the viologen, thus completing the circuit. A photocurrent of 2 nA at 0.0 V vs SCE was observed only when the light was turned on [19].

Proposed Keto-Enol Switch

N-salicylideneaniline is photochromic: the enol configuration can be converted to the trans-keto conformation by using light of frequency ν_1, and converted back by either heat, or by light of frequency ν_2: here the tautomerism causes an intramolecular rotation [67].

Conductivity Change in LB films Due to Light-Induced Conformational Change

As discussed elsewhere in this volume, Kawabata and co-workers showed that a light-induced cis-trans conformational change in azobenzene could affect the electrical conductivity of an LB film [68].

Potential Conformational Change Systems

There are many ideas for using organic molecules which undergo conformational changes in optical storage devices. Birge [69] has studied the primary step in the photocycle of bacteriorhodopsin (the light-harvesting protein of *Halobacterium halobium*): here the bacteriorhodopsin containing all-trans retinal (bR568) can be converted (at 77 K), with two-photon illumination at 540 nm and 635 nm, into K610, or bathorhodopsin; at lower laser powers, the same two-photon illumination scheme can interrogate which molecules have converted to K610, and which remained in the bR568 state. The K610 state may consist of a cis-retinal segment. The harnessing of such cis-trans isomerizations (also involved in the physiology of the visual pigments) may, some day, yield interesting molecular devices.

DEVICE ASSEMBLY IDEAS

Macroscopic Connections - LB Flms and Polydiacetylenes

For the proposed unimolecular devices, the problem of assembly and device interrogation or control is a very serious one. If one cannot talk to a single molecule, one can perhaps talk to a monolayer of identical molecules transferred to a suitable (metal) substrate by the Langmuir-Blodgett film method [70]; then, many identical molecules can be addressed electrically, provided that electrical short circuits through defects do not occur. However, LB monolayers are only weakly physisorbed to surfaces, and can desorb with time. This difficulty can be circumvented, if a photopolymerizable diacetylene [71] group is included in the monolayer-forming molecule, and if the molecular geometry allows a topotactic polymerization: this can result in a very robust, yet electroactive polymer.

Macroscopic Connections: the Silyl Bridge

An alternative strategy, devised by Murray [72], consists of derivatizing an oxide-coated metal electrode surface with trichlorosilyl groups, then reacting this surface with the desired molecule, which has a terminal alcohol group: this, however, does not usually give full monolayer coverage on the metal. The reverse strategy, of silanizing the molecule and attaching it to an oxide and hydroxyl-group-covered metal surface, claims full monolayer coverage [73].

Proposed Macroscopic Connections: Cyclodextrins and Calixarenes

The existence of cavities of precisely controlled size within cyclodextrins and calixarenes allows the inclusion of small electroactive molecules within these cavities, and yet provide well-formed macroscopic crystals with the desired inclusions precisely oriented within them [24,25].

HOW TO CONNECT TO A UNIMOLECULAR ELECTRONIC DEVICE

Proposed Molecular Wires

The carotenes and other conjugated linear polyenes have been touted as "molecular wires" or as "molecular antennas" [21], and certainly will provide fast electronic access to single molecules. These polyenes are, unfortunately, very susceptible to air oxidation (as is the simplest conducting polymer, doped polyacetylene).

Proposed Spiro-Linked Thiophene Intercircuit Connections

Aviram has proposed [22] a spiro-linked polythiophene system as a possible interconnection between conducting "molecular wires' based on n or p-doped polythiophene.

Molecular Spacers

Miller has shown how simple Diels-Alder condensation can yield insulating but sturdy, LB film-forming oligomers of controlled length [23].

Scanning Tunneling Microscopy

The recent advent of affordable scanning tunneling microscopy (STM) [74] allows us the luxury of addressing electronically a single molecule; the use of such microscopes will certainly yield new ways for connecting to a single molecule or cluster.

CONCLUSION

This short review should whet the appetite for new ME devices However, molecular designers are cautioned: "graphite and cellulose chemistry" , or "chalk and chalkboard syntheses" do not easily become real molecules, or real devices. Much time and plenty of grief will, as usual, pass between idea and realization. But the challenge of ultimate miniaturization is upon us.

ACKNOWLEDGEMENT

This work was supported in part by the Natioonal Science Foundation, Grant DMR-88-01924.

REFERENCES

[1] F. L. Carter, Ed., "Molecular Electronic Devices", (Dekker, New York, 1982).
[2] F. L. Carter, Ed., "Molecular Electronic Devices II" (Dekker, New York, 1987).
[3] F. L. Carter, R. E. Siatkowski, and H. Wohltjen, Eds., "Molecular Electronic Devices, Proceedings of the 3rd International Symposium" (North-Holland, Amsterdam, 1988).
[4] M. Borissov, Ed., "Molecular Electronics - IVth International School on Condensed Matter" (World Scientific, Singapore, 1987).
[5] A. Aviram and A. Bross, Eds., "Molecular Electronics - Science and Technology" (New York Engineering Foundation, in press).
[6] M. La Brecque, "Circuits and devices a molecule wide" and "Devices that assemble themselves", Mosaic (National Science Foundation) 20(1): 16, 28 (1989).
[7] R. C. Haddon and A. C. Lamola, Proc. Natl. Acad. Sci. U.S. 82: 1874 (1985).
[8] For the proceedings of a recent conference, see Synth. Metals 27-29 (1988-1989).
[9] T. A. Skotheim, Ed., "Handbook of Conducting Polymers", Vol. I and II (Dekker, New York, 1986).
[10] D. Davidov and H. Selig, Eds. "Graphite Intercalation Compounds" Synth. Metals 23 (1988)
[11] J. V. Acrivos, NATO ASI Ser. C130: 479 (1984).
[12] Relevant articles in "Extended Linear-Chain Compounds" Vols. I, II, and III, ed. by J. S. Miller, (Plenum, New York, 1982, 1982, 1983).
[13] For proceedings of recent conferences, see Thin Solid Films Vols. 68 (1980), 99 (1983),132-134 (1985), and 159-160 (1987)
[14] A. Aviram and M. A. Ratner, Chem. Phys. Lett. 29: 277 (1974).
[15] A. Aviram and P. E. Seiden, U. S. Patent 3,833,894 (1974).
[16] A. Aviram, P. E. Seiden, and M. A. Ratner, in "Molecular Electronic Devices", ed. by F. L. Carter, (Dekker, New York, 1982) p. 5.
[17] T. Inabe, N. Hoshino, T. Mitani, and Y. Maruyama, Bull. Chem. Soc. Jpn. 62: 2245 (1989).
[18] F. L. Carter in "Molecular Electronic Devices", ed. by F. L. Carter, (Dekker, New York, 1982) p. 51.
[19] M. Fujihira, K. Nishiyama, and H. Yamada, Thin Solid Films 132: 77 (1985).
[21] J.-M. Lehn, Angew. Chem. Intl. Ed. Engl. 27: 89 (1988).
[22] A. Aviram, J. Am. Chem. Soc. 110: 5687 (1988).
[23] P. W. Kenny, L. L. Miller, S. F. Rak, T. R. Jozefiak, W. C. Christopfel, J. H. Kim, and R. A. Uphaus, J. Am. Chem. Soc. 110: 4445 (1988).
[24] J. Szejtli, "Cyclodextrin Technology" (Kluwer, Dordrecht, 1988).
[25] J. Vicens and V. Böhmer, Eds., "Calixarenes, a Versatile Class of Macrocyclic Compounds" (Kluwer, Dordrecht, Holland, 1990).
[26] R. P. Feynman, "There is plenty of room at the bottom", Engr. and Sci. (Feb.22, 1960) p.36.
[27] D. S. Acker, R. J. Harder, W. R. Hertler, W. Mahler, L. R. Melby, R. E. Benson, and W. E. Mochel, J. Am. Chem. Soc. 82: 6408 (1960).
[28] T. Coffey, unpublished remarks at II International Workshop on Molecular Electronic Devices, 1983.
[29] W. A. Little, Phys. Rev. 134: 1416 (1964).
[30] L. R. Melby, Can. J. Chem. 43: 1448 (1965).
[31] J. von Neumann, "The Theory of Self-Reproducing Automata" (Univ. of Illinois Press, Urbana, IL, 1966).
[32] J. Ferraris, D. O. Cowan, V. Walatka, and J. H. Perlstein, J. Am. Chem. Soc. 95: 948 (1973).
[33] H. Shirakawa, E. J. Louis, A. G. MacDiarmid, C. K. Chiang, and A. J. Heeger, Chem. Commun. 578 (1978).
[34] D. Jérome, A. Mazaud, M. Ribault, and K. Bechgaard, J. Physique (les Ulis) 41: L95 (1980).
[35] S. S. P. Parkin,E. M. Engler, R. R. Schumaker, R. Lagier, V. Y. Lee, J. C. Scott, and R. L. Greene, Phys. Rev. Lett. 50: 270 (1983).
[36] H. Urayama, H. Yamochi, G. Saito, K. Nozawa, T. Sugano, M. Kinoshita, S. Sato, K. Oshima, A. Kawamoto, and J. Tanaka, Chem. Lett. 55 (1988).
[37] S. S. P. Parkin, V. Y. Lee, A. I. Nazzal, T. C. Huang, G. Gorman, R. Savoy, and R. Beyers, Phys. Rev. Lett. 60: 2539 (1988).
[38] J. E. Meinhard, Appl. Phys. Lett. 35: 3059 (1964).

[39] E E. Polymeropoulos, D. Möbius, and H. Kuhn, Thin Solid Films 68: 173 (1980).
[40] M. Sugi, K. Sakai, M. Saito, Y. Kawabata, and S. Iizima, Thin Solid Films 132: 69 (1985).
[41] R. S. Potember, T. O. Poehler, and D. O. Cowan, Appl. Phys. Lett. 34: 405 (1979).
[42] R. S. Potember, R. C. Hoffmann, H. S. Hu, J. E. Cocchiaro, C. A. Viands, R. A. Murphy, and T. O. Poehler, Polymer 28: 574 (1987).
[43] R. S. Potember, R. C. Hoffman, R. C. Benson, and T. O. Poehler, J. Phys. (Les Ulis) 44: C3-1597 (1983).
[44] H. S. White, G. P. Kittleson, and M. S. Wrighton, J. Am. Chem. Soc. 106: 5375 (1984).
[45] E. T. Turner Jones, O. M. Chyan, and M. S. Wrighton, J. Am. Chem. Soc. 109: 5526 (1987).
[46] H. Ryu, K. Akagane, and H. Kori, Chem. Economy and Engin. Rev. 18(12): 26 (1986).
[47] J. Slonim, D. Mole, and M. Bauer, Library Hi-Tech 3(4): 27 (1985).
[48] P. S. Kalanaraman, J. E. Kuder, and R. S. Jones, in "Functional Polymers" ed. by D. E. Bergbreiter and C. R. Martin, (Plenum, New York, 1989) p. 173.
[49] D. E. Nikles, C. E. Forbes, H. A. Goldberg, R. E. Johnson, R. S. Kohn, and F. J. Onorato, Proc. Soc. Photo-optical Instrum. Engineers 107B: 43 (1989).
[50] D. E. Budil, P. Gast, C. H. Chang, M. Schiffer, and J. R. Norris, Ann. Rev. Phys. Chem. 38: 561 (1987).
[51] J. Deisendorfer and H. Michel, Angew. Chem. Intl. Ed. Engl. 28: 829 (1989); R. Huber, ibid. 848 (1989).
[52] H. Taube, Pure Appl. Chem. 44, 25 (1975).
[53] T. J. Meyer in "Photochemical Energy Conversion", J. R. Norris, Jr. and D. Meisel, Eds. (Elsevier, New York, 1989)p. 75
[54] L. T. Calcaterra, G. L. Closs, and J. R. Miller, J. Am. Chem. Soc. 105: 670 (1983).
[55] J. R. Miller, L. T. Calcaterra, and G. L. Closs, J. Am. Chem. Soc. 106: 3047 (1984).
[56] R. A. Marcus, Disc. Faraday Soc. 29: 21 (1960).
[57] J. Lindsay, D. Mauzerall,and H. Linschitz, J. Am. Chem. Soc.105: 6528 (1983).
[58] A. R. McIntosh, J. R. Bolton, J. S. Connolly, K. L. Marsh, D. R. Cook, T.-F. Ho, and A. C.Weedon, J. Phys. Chem. 90: 5640 (1986).
[59] A. D. Joran, B. A. Leland, G. G. Geller, J. J. Hopfield, and P. B. Dervan, J. Am. Chem. Soc.106: 6090 (1984).
[60] S. Nishitani, N. Kurata, Y. Sakata, S. Misumi, A. Karen, T. Okada, and N. Mataga, J. Am. Chem. Soc. 105: 7771 (1983).
[61] T. A. Moore, D. Gust, P. Mathis, J.-C.Mialocq, C. Chachaty, R. V. Bensasson, E. J. Land, D. Doizi, P. A. Liddell, W. R. Lehman, G. A. Nemweth, and A. L. Moore, Nature 307: 630 (1984).
[62] C. Krieger, J. Weiser, and H. A. Staab, Tetrahedron Lett. 26: 6050 (1985).
[63] N. S. Hush, M. N. Paddon-Row, E. Cotsaris, H. Oevering, J. W. Verhoeven, and M. Heppener, Chem. Phys. Lett. 117: 8 (1985).
[64] M. R. Wasielewski, D. G. Johnson, W. A. Svec, K. M. Kersey, D. E. Cragg, and D. W. Minsek in "Photochemical Energy Conversion", J. R. Norris, Jr. and D. Meisel, Eds. (Elsevier, New York, 1989)p. 135
[65] R. M. Metzger and C. A. Panetta, this volume.
[66] A. Aviram, C. Joachim, and M. Pomerantz, Chem. Phys. Lett. 146: 490 (1988).
[67] H. Sixl and D. Higelin, in "Molecular Electronic Devices II" , ed. by F. L. Carter, (Dekker, New York, 1987), p. 27.
[68] H. Tachibana, T. Nakamura, M. Matsumoto, H. Komizu, E. Manda, H. Niino, A. Yabe, and Y. Kawabata, J. Am. Chem. Soc. 111: 3080 (1989).
[69] R. R. Birge and T. M. Cooper, Biophys. J. 42: 61 (1983).
[70] H. Kuhn, D. Möbius, and H. Bücher in "Techniques of Chemistry, Vol. I - Physical Methods of Chemistry - Part V - Determination of Thermodynamic and Surface Properties", ed. by A. Weissberger and B. W. Rossiter (Wiley-Interscience, New York, 1972) page 577.
[71] E. G. Wilson, Mol. Cryst. Liq. Cryst. 121: 271 (1985).
[72] R. W. Murray, Acc. Chem Res. 13: 135 (1980).
[73] R. Maoz, L. Netzer, J. Gun, and J. Sagiv, J. Chim. Phys. 85: 1059 (1988).
[74] G. Binnig, H. Rohrer, Ch. Gerber, and E. Weibel, Phys. Rev. Lett. 49: 57 (1982).

PARTICIPANTS

Mr. Bernhard **Alberti**, Physical Chemistry Inst., University of Mainz, 11 Jakob Welder Weg, D-6500 Mainz, German Federal Republic
Dr. Geoffrey J. **Ashwell**, Center for Molecular Electronics, School of Industrial Science, Cranfield Institute of Technology, Cranfield, Bedford MK43 0AL, United Kingdom
Miss Aleka **Bakou**, Dept. of Chemistry, University of Crete, P. O. Box 1470, Iraklion, Crete, Greece
Dr. Anna **Berlin**, CNR Center for Synthesis and Stereochemistry of Special Organic Systems, University of Milano, 19 Via C. Golgi, I-20133 Milano, Italy
Prof. Renato **Bozio**, Dept. of Physical Chemistry, University of Padova, 2 Via Loredan, I-35131 Padova, Italy
Dr. Daniel **Chasseau**, CNRS Crystallography Lab., University of Bordeaux I, 351 Cours de la Liberation, F-33405 Talence, France
Prof. Robin J. H. **Clark**, Christopher Ingold Labs., University College London, 20 Gordon Street, London WC1H 0AJ, United Kingdom
Mr. Simon **Cooke**, Dept. of Engineering Science, Oxford University, Parks Road, Oxford OX1 3PJ, United Kingdom
Prof. Athanassios G. **Coutsolelos**, Dept. of Chemistry, University of Crete, P. O. Box 1470, Iraklion, Crete, Greece
Prof. Dwaine O. **Cowan**, Dept. of Chemistry, The Johns Hopkins University, 3400 Charles Street, Baltimore, MD 21218, USA
Mr. Raúl **Crespo**, Dept. of Physical Chemistry, Faculty of Chemical Sciences, University of Valencia, 50 Doctor Moliner, Burjassot, Valencia, Spain
Mr. Vasco Pires Silva da **Gama**, Dept. of Chemistry, ICEN, National Laboratory for Engineering and Industrial Technology, P-2686 Sacavem Codex, Portugal
Prof. Peter **Day**, Director, Institut Laue-Langevin, Boite Postale 156 X, F-38042 Grenoble Cedex, France
Dr. Pierre **Delhaès**, CNRS Center "Paul Pascal", University of Bordeaux I, Domaine Universitaire, F-33405 Talence Cedex, France
Mr. André **Désormeaux**, Photobiophysics Research Center, University of Quebec at Trois Rivières, C. P. 500, Trois Rivières, Quebec G9A 5H7, Canada
Mr. Stephen **Edge**, Dept. of Chemistry, University College of North Wales, Bangor, Gwynedd, United Kingdom
Prof. John E. **Eldridge**, Dept. of Physics, University of British Columbia, Vancouver, B. C. V6T ZA6, Canada
Mrs. Patricia **Enzel**, Dept. of Chemistry, University of New Mexico, Albuquerque, NM 87131, USA
Prof. Arthur J. **Epstein**, Dept. of Physics, Ohio State University,Columbus, OH 43210, USA
Mr. John A. **Fortkort**, Dept. of Chemistry, The Johns Hopkins University, 3400 N Charles Street, Baltimore, MD 21218, USA
Mr. Alain **Fritsch**, CNRS Theworetical Physical Chemistry Lab., University of Bordeaux I, 351 Cours de la Libération, F-33405 Talence Cedex, France
Dr. Vassilis **Gionis**, National Hellenic Research Foundation, 48 Vassiliou Konstantinou, GR-11635 Athens, Greece
Prof. Alberto **Girlando**, Dept. of Chemistry, University of Parma, Parma, Italy
Dr. Thierry **Granier**, CNRS Crystallography Lab., University of Bordeaux I, 351, Cours de la Libération, F-33405 Talence Cedex, France

Mr. Jean **Guay**, National Scientific Research Inst. - Energy, University of Quebec, 1650 Montée Ste-Julie, Varennes, Quebec J0L 2PO, Canada

Mr. Göran **Gustafsson**, Inst. for Physics and Measurement Technology, University of Linköping, S-58183 Linköping, Sweden

Mr. Sigurd **Hagen**, Max Planck Inst. for Solid State Physics, 1 Heisenbergstrasse, D-7000 Stuttgart 80, German Federal Republic

Prof. Alan J. **Heeger**, Inst. for Polymers and Organic Solids, University of California, Santa Barbara, Santa Barbara, CA 93106, USA

Mr. Claus-Peter **Heidmann**, Walther Meissner Inst., Technical University of München, 8 Walther Meissnerstrasse, D-8046 Garching, German Federal Republic

Mr. Matthias **Hoch**, Inorganic Chemistry Inst., University of Heidelberg, 270 Im Neuenheimer Feld, D-6900 Heidelberg 1, German Federal Republic

Dr. András **Jánossi**, Central Research Inst. for Physics, Hungarian Academy of Sciences, P. O. Box 49, H-1525 Budapest, Hungary

Prof. Denis **Jérome**, CNRS Lab. of Solid State Physics,University of Paris Sud, Batiment 510, Centre Universitaire, F-31405 Orsay Cedex, France

Prof. M. Thomas **Jones**, Dept. of Chemistry, University of Missouri at Saint Louis, Saint Louis, MO 63121, USA (present address: Dept. of Chemistry, University of Houston, Houston, TX 77204, USA).

Miss Venetziana C. **Kakoussis**, Theoretical and Physical Chemistry Inst., National Hellenic Research Foundation, 48 Vassiliou Konstantinou, GR-11635 Athens, Greece

Dr. E. I. **Kamitsos**, Theoretical and Physical Chemistry Inst., National Hellenic Research Foundation, 48 Vassiliou Konstantinou, GR-11635 Athens, Greece

Prof. Mercouri G. **Kanatzidis**, Dept. of Chemistry, Michigan State University, East Lansing, MI 48824, USA

Prof. Heimo J. **Keller**, Inorganic Chemistry Inst., University of Heidelberg, 270 Im Neuenheimer Feld, D-6900 Heidelberg, German Federal Republic

Dr. Koichi **Kikuchi**, Dept. of Chemistry, Faculty of Science, Tokyo Metropolitan University, Fukazawa, Satagaya-ku, Tokyo 158, Japan

Miss Hilde **Krikor**, National University Center Antwerp, University of Antwerp, 1 Middelheimlaan, B-2020 Antwerp, Belgium

Prof. Susumu **Kurita**, Lab. of Applied Physics, Fac. of Engineering, Yokohama National University, Hodogaya, Yokohama 240, Japan

Dr. Mohammed **Kurmoo**, Inorganic Chemistry Lab., Oxford University, South Parks Road, Oxford OX1 3QR, United Kingdom

Dr. Guglielmo **Lanzani**, Dept. of Physics "Alessandro Volta", University of Pavia, 6 via Bassi, Pavia, Italy

Mr. Jan **Larsen**, Dept. of General and Organic Chemistry, H. C. Ørsted Inst.,University of Copenhagen,5 Universitetsparken, DK-2100 Copenhagen Ø, Denmark

Prof. Serge **Lefrant**, Lab. of Crystalline Physics, University of Nantes, 2 rue de la Houssinière, F-44072 Nantes Cedex 03, France

Dr. Mathias **Lösche**, Inst. of Physical Chemistry, University of Mainz, 11 Jakob Welder Weg, D-6500 Mainz, German Federal Republic

Mr. Li **Lu**, Inst. of Physics, Chinese Academy of Sciences, Beijing 100080, People's Republic of China

Dr. Silvia **Luzzati**, CNR Institute of Macromolecular Chemistry, Universiy of Milano, 15 via E. Bassini, I-20133 Milano, Italy

Prof. Alan G. **MacDiarmid**, Dept. of Chemistry, University of Pennsylvania, Philadelphia, PA 19104, USA

Mr. Ian R. **Marsden**, Cavendish Lab., Cambridge University, Madingley Road, Cambridge CB3 0HE, United Kingdom

Dr. Elisabeth A. **Marseglia**, Cavendish Lab., Madingley Road, Cambridge CB3 0HE, United Kingdom

Mr. Josef **Martens**, Wolfron College, Cambridge University, Cambridge CB3 9BB, United Kingdom

Mr. Manuel José **Matos**, Dept. of Chemistry, ICEN, National Laboratory for Engineering and Industrial Technology, P-2686 Sacavem Codex, Portugal

Dr. Irene M. **Mavridis**, Inst. of Physical Chemistry, National Physical Sciences Res. Center "Democritos", P. O. Box 60228, Aghia Paraskevi, GR-15310 Athens, Greece

Prof. Robert M. **Metzger**, Dept. of Chemistry, University of Alabama, Tuscaloosa, AL 35487-0336, USA

Prof. Larry L. **Miller**, Dept. of Chemistry, University of Minnesota, 207 Pleasant Street S.E., Minneapolis, MN 55455-0431, USA
Mr. Andrew **Millington**, Dept. of Chemistry and Applied Chemistry, The University of Salford, Salford M5 4WT, United Kingdom
Mr. Jorge Manuel F. **Morgado**, Dept. of Chemical Engineering, Advanced Technical Institute, Av. Rovisco Pais, P-1096 Lisboa Codex, Portugal
Mr. G. A. **Mousdis**, Theoretical and Physical Chemistry Inst., National Hellenic Research Foundation, 48 Vassiliou Konstantinou, GR-11635 Athens, Greece
Mr. Harald **Müller**, Walther Meissner Inst., Technical University of München, 8 Walther Meissnerstrasse, D-8046 Garching, German Federal Republic
Dr. Takayoshi **Nakamura**, National Chemical Laboratory for Industry, Higashi Tsukuba, Ibaraki-gun 305, Japan
Mr. Helmut **Neugebauer**, Inst. of Physical Chemistry, University of Wien, 42 Währingerstrasse, A-1090 Wien, Austria
Mr. Maciej E. **Orczyk**, Inst. of Organic and Physical Chemistry,Technical University of Wroclaw, 27 Wybrzeze Wyspianskiego, PL-50370 Wroclaw, Poland
Mr. Lahcène **Ouahab**, Lab. of Inorganic Molecular and Solid State Chemistry,URA 254 CNRS, University of Rennes I, F-35042 Rennes Cedex, France
Prof. Giorgio A. **Pagani**, Dept. of Organic and Industrial Chemistry, University of Milano, Via C. Golgi 19, I-20133 Milano, Italy
Dr. Anna **Painelli**, Dept. of Physical Chemistry, University of Padova, 2 via Loredan, I-35131 Padova, Italy
Dr. Karl **Pampus**, Dept. of Applied Physics, Hoechst Aktiengesellschaft, Postfach 800 320, D-6230 Frankfurt am Main 80, German Federal Republic
Dr. George C. **Papavassiliou**, Theoretical and Physical Chemistry Inst., National Hellenic Research Foundation, 48 Vassiliou Konstantinou, GR-11635 Athens, Greece
Dr. Danilo **Pedron**, Dept. of Physical Chemistry, University of Padova, 2 via Loredan, I-35131 Padova, Italy
Dr. François **Pesty**, CNRS Lab. of Solid State Physics, University of Paris Sud, Batiment 510, Centre Universitaire, F-31405 Orsay Cedex, France
Miss Marí Carmen **Piqueras**, Dept. of Physical Chemistry, Faculty of Chemical Sciences, University of Valencia, 50 Doctor Moliner, Burjassot, Valencia, Spain
Dr. William **Porzio**, CNR Inst. of Macromolecular Chemistry, 15 Via E. Bassini, I-20133 Milano, Italy
Mr. Mark **Poulter**, Dept. of Engineering Science, Oxford University, Parks Road, Oxford OX1 3PJ, United Kingdom
Prof. Paras N. **Prasad**, Dept. of Chemistry, State University of New York at Buffalo, Buffalo, NY 14214, USA
Mr. Jürgen **Reichenbach**, Max Planck Inst. for Solid State Research, 1 Heisenbergstrasse, D-7000 Stuttgart 80, German Federal Republic
Mr. Andrew **Rohl**, Inorganic Chemistry Lab., Oxford University, South Parks Road, Oxford OX1 3QR, United Kingdom
Prof. Gareth G. **Roberts**, Central Research Laboratories, Thorn-EMI plc, Dawley Road, Hayes, Middlesex UB3 1HH, United Kingdom (present address: Vice-Chancellor's Office, University of Sheffield, Western Bank, Sheffield S10 2TN, United Kingdom)
Dr. Annie **Ruaudel-Teixier**, Molecular Chemistry Service, Technological Research and Industrial Development Inst., Atomic Energy Commission, Saclay Nuclear Studies Center, F-91191 Gif-sur-Yvette Cedex, France
Mr. Olivier **Sagnes**, LCR 071, Thomson CSF, Domaine de Courbeville, BP 10, F-91401 Orsay Cedex, France
Prof. Gunzi **Saito**, Inst. of Solid State Physics, University of Tokyo, 7-22-1 Roppongi, Minato-ku, Tokyo 106, Japan (present address: Dept. of Chemistry, Kyoto University, Kyoto, Japan)
Mr. Ifor D. W. **Samuel**, Cavendish Lab., Cambridge University, Madingley Road, Cambridge CB3 0HE, United Kingdom
Miss Lynne **Samuelson**, Dept. of Chemistry, University of Lowell, Lowell, MA 01854, USA
Ms. Nuria **Santaló Bel**, Research Center, University of Barcelona, 18-26 calle Jordi Girona, E-08034 Barcelona, Spain
Mr. John **Schlueter**, Dept. of Chemistry, Northwestern University, 2145 Sheridan Road, Evanston, IL 60208, USA

Prof. Dieter **Schweitzer**, Third Physics Inst., University of Stuttgart, 57 Pfaffenwaldring, D-7000 Stuttgart 80, German Federal Republic
Prof. Takeo **Shimidzu**, Div. of Molecular Engineering, Kyoto University, Kyoto 606, Japan
Miss June H. **Smith**, Dept. of Chemistry, University of Aberdeen, Old Aberdeen, Aberdeen, Scotland AB9 2UE, United Kingdom
Mr. Ulrich **Sohling**, Inst. of Physical Chemistry, University of Mainz, 11 Jakob Welder Weg, D-6500 Mainz, German Federal Republic
Dr. Henrik **Stubb**, Semiconductor Lab., Technical Research Centre of Finland, SF-02150 Espoo, Finland
Mr. Thomas E. **Sutto**, Dept. of Chemistry, University of Virginia, Charlottesville, VA 22901, USA
Dr. Daniel R. **Talham**, Dept. of Chemistry, Room 6-427, Massachusetts Institute of Technology, Cambridge, MA 02139, USA (present address: Dept. of Chemistry, University of Florida, Gainesville, FL 32611, USA)
Dr. A. **Terzis**, Inst. of Materials Science, National Research Center "Democritos", Aghia Paraskevi, GR-15310 Athens, Greece
Dr. Silvia **Tomic**, Inst. of Physics, University of Zagreb, P. O. Box 304, YU-41001 Zagreb, Yugoslavia
Mr. Francisco **Torrens**, Dept. of Physical Chemistry, University of Valencia, Faculty of Chemical Sciences, 50 Doctor Moliner, E-46100 Burjassot, Valencia, Spain
Prof. Sukant K. **Tripathy**, Dept. of Chemistry, University of Lowell, Lowell, MA 01854, USA
Dr. Michel **Vandevyver**, Molecular Chemistry Service, Technological Research and Industrial Development Inst., Atomic Energy Commission, Saclay Nuclear Studies Center, F-91191 Gif-sur-Yvette Cedex, France
Mr. F. J. **Vergeldt**, Jr., 23 Beurydingsstraat, NL-6701 AA Wageningen, The Netherlands
Dr. Jack M. **Williams**, Chemistry Div., Argonne National Laboratory, Argonne, IL 60439, USA
Prof. Attila **Yildiz**, Dept. of Chemistry, Faculty of Engineering, Hacettepe University, 06532 Beytepe, Ankara, Turkey
Dr. Roberto **Zamboni**, CNR Molecular Spectroscopy Inst., 1 Via de' Castagnoli, I-40126 Bologna, Italy
Prof. Giuseppe **Zerbi**, Dept. of Industrial Chemistry and Chemical Engineering, Politecnico di Milano, Piazza Leonardo da Vinci, I-20133 Milano, Italy

AUTHOR INDEX

Ahn, J., 579
Alcácer, Luis, 211
Almeida, M., 205
Andres, K., 227, 233
Arca, Emin, 435
Arca, Meral, 435
Ashwell, Geoffrey J., 647
Averill, Bruce A., 285, 289

Bechgaard, Klaus, 527
Bein, Thomas, 421
Bencharif, M., 185
Beno, Mark A., 91
Berlin, Anna, 359, 363
Biberacher, W., 233
Bleier, Hartmut, 415
Bolognesi, A., 411
Borghesi, A., 387
Bozio, Renato, 23, 129
Bradley, D. D. C., 393
Brückner, S., 411
Buisson, J. P., 375
Burn, P., 393

Callaerts, Robert, 381
Cammarata, Vince, 627
Cao, Y., 317
Carlson, K. Douglas, 91
Catellani, Marinella, 411
Chen, Ben-Ming, 175
Chryssikos, G. D., 197
Clark, Robin J. H., 263
Clough, S., 537
Colaneri, N. F., 393
Colbrook, R., 549
Cooke, Simon J., 605
Coulon, Claude, 163
Coutsolelos, Athanassios G., 217
Cowan, Dwaine O., 1
Crespo, Raul, 447, 455
Crooks, Richard M., 627

Day, Peter, 115, 169
Decher, Gero, 591
DeGroot, Donald C., 427
Deilacher, F., 175
Delhaès, Pierre, 43, 185
Dellepiane, G., 387
Désormeaux, André, 557
Destri, Silvia, 411
Domash, L., 579
Duan, Hong-Min, 257
Ducasse, Laurent, 163, 467

Enzel, Patricia, 421
Epstein, Arthur J., 303, 335

Feis, A., 23
Ferraro, John R., 91
Fortkort , John A., 1
Frederiksen, P., 527
Friend, Richard H., 393
Fritsch, Alain, 163, 467
Fuchs, H., 227

Gallois, Bernard, 163
Gama, Vasco Pires Silva da, 205
Garoche, Pierre, 245
Garrigou-Lagrange, C., 185
Gärtner, S., 97, 175
Geiser, Urs, 91
Gionis, Vassilis, 197
Girlando, Alberto, 441
Gogu, E., 97
Grandjean, D., 185
Granier, Thierry, 163
Grimm, H., 97
Gustafsson, Göran, 369
Güven, Ogun, 435

Hagen, Sigurd, 531
Hale, P. D., 537
Hanack, Michael, 531
Heeger, Alan J., 293, 317
Heidmann, Claus- Peter, 227, 233
Heinen, I., 97
Henriques, Rui T., 205, 211
Hoch, Matthias, 175
Honda, Yoshiaki, 223

Ikemoto, Iaso, 223
Inagaki, T., 537
Inganäs, Olle, 369
Ishikawa, Yoshimitsu, 223

Jánossy, András, 277
Jérome, Denis, 85
Jones, M. Thomas, 191
Joss, W., 233

Kahlich, S., 97, 175
Kakoussis, Venetziana C., 181
Kamitsos, E. I., 197
Kanatzidis, Mercouri G., 427
Kannewurf, Carl R., 427
Kawabata, Yasujiro, 519
Keller, Heimo J., 97, 175, 239
Kikuchi, Koichi, 223
Kini, Aravinda M., 91
Kobayashi, Keiji, 223
Krikor, Hilde, 381
Kuczynski, A. P., 647
Kuivalainen, P., 635
Kumar, J., 579
Kurita, Susumu, 271
Kurmoo, Mohammed, 169
Kutz, T., 97

Lanzani, Guglielmo, 387
Larsen, Jan, 527
Leblanc, Roger M., 557
Lefrant, Serge, 375
Lerstrup, Knud, 527
Lerf, A., 227
Leventis, Nicholas, 627
Levin, P., 579
Lin, Shu-yuan, 257
Lösche, Mathias, 491, 591
Lu, Li, 257

Ma, Bei-Hai, 257
MacDiarmid, Alan G., 303, 335
Marcy, Henry O., 427
Marseglia, Elisabeth A., 393
Martens, Josef, 393
Martina, Stefano, 359
Maruo, Toshio, 191
Matos, Manuel José, 211
Matsumoto, Mutsuyoshi, 519
Meneghetti, Moreno, 129
Mertens, Robert, 381
Metzger, Robert M., 1, 611, 659
Mevellec, J. Y., 375
Miller, Larry L., 643
Mousdis, G. C., 181
Müller, Harald, 227, 233
Murata, Keizo, 223

Nagels, Piet, 381
Nakamura, Takayoshi, 519
Neugebauer, Helmut, 401

Okamoto, Yoshiyuki, 537
Orczyk, Maciej, 407
Ortí, Enrique, 447, 455, 461
Ouahab, Lahcène, 185

Pagani, Giorgio A., 359
Painelli, Anna, 441
Paloheimo, J., 635
Panetta, Charles A., 611
Papavassiliou, George C., 143, 181
Pecile, Cesare, 23, 129
Pedron, Danilo, 23, 129
Pesty, François, 245
Piaggi, A., 387
Piqueras, Marí Carmen, 447, 455
Porzio, William, 411
Poulter, Mark W., 549
Prasad, Paras N., 563, 573
Punkka, E., 635

Rahman, A. K. M., 537
Reichenbach, Jürgen, 415
Remaut, Germain, 381
Renner, G., 97
Roberts, Gareth G., 473, 549, 605
Roth, Siegmar, 415, 531
Ruaudel-Teixier, Annie, 511
Rühe, Jürgen, 363

Saito, Gunzi, 67
Saito, Kazuya, 223
Samuelson, Lynne, 537
Sánchez-Marin, José, 461
Sariciftci, N. S., 401
Schiavon, Gilberto, 359
Schier, H., 531
Schilling, J. S., 227
Schloh, Martin O., 627
Schultz, Arthur J., 91
Schweitzer, Dieter, 97, 175, 239
Shen, Y. Q., 415
Shimidzu, Takeo, 653
Sieburger, R., 227
Singh, Megh, 191
Skotheim, Terje A., 537
Smith, P., 317
Stubb, Henrik, 635
Sutto, Thomas E., 285, 289
Sworakowski, Juliusz, 407
Szablewski, Marek, 647

Tachibana, Hiroaki, 519
Talham, Daniel R., 169, 627
Taliani, Carlo, 411
Tanaka, Motoo, 519
Tomás, F., 447
Tomić, Silvia, 251
Torrens, Francisco, 461
Triki, S., 185
Tripathy, Sukant K., 537, 579

Vandevyver, Michel, 503

Wang, Hau-H., 91
Wang, Xiao-Hong, 257
Ward, D. L., 217

Wegner, Gerhard, 363
Whangbo, Myung-Hwan, 91
Williams, Jack M., 91
Wrighton, Mark S., 627
Wu, Chun-Guey, 427
Wu, Pei-Ji, 175

Yao, You-Xin, 257
Yildiz, Attila, 435

Zamboni, Roberto, 97, 239, 411
Zanon, I., 23
Zeraoui, S., 375
Zerbi, Giuseppe, 345
Zhang, Dian-Lin, 257
Zhu, Dao-Ben, 257
Zotti, Gianni, 359

SUBJECT INDEX

A_A *see* electron affinity (of acceptor)
ab initio molecular orbital calculations
---of hyperpolarizabilities, 570
---of molecule with 12 double bonds, 348
---for 2-vinylthiophene (321-G), 448
---for 2-vinylfuran (321-G), 448
---for 2-vinylpyrrole (321-G), 448
acceptors *see* electron acceptors
acoustic phonon, 132
activation energy for hopping, 366
activation energy for conductivity, 9, 366
adiabatic approximation, 133
AETCNQ (2-acetoxy-7,7,8,8-tetracyanoquinodimethane)
---crystal structure, 619
$Ag(DM\text{-}DCNQI)_2$
---infrared spectrum: peaks at 4,000 and 3500 cm^{-1}; strong triplet at 2160, 2130, and 2105 cm^{-1}, 197
---Raman spectrum: peaks at 2170 cm^{-1} (C≡N stretch), 1618 (C=C stretch), 1460 cm^{-1} (C≡N stretch), 203
---synthesis, 197
---visible-ultraviolet spectrum, 197
Ag epoxy, 629
aggregates: J, S, and H, 512, 513
Ag(TCNQ)
---organic phase-change switch, 661
alkylammonium metal$(dmit)_2$ [alkyl=$R_m(CH_3)_{4-m}$, R=$C_{10}H_{21}$, $C_{12}H_{25}$, $C_{14}H_{29}$, $C_{16}H_{33}$, $C_{18}H_{37}$, $C_{22}H_{45}$, m=1, 2,3, 4, metal=Ni, Au, Pd, Pt] oxidized with bromine or electrochemically, and mixed 1:1 with eicosanoic acid: semiconductive LB films, 520-522
N-alkyl-4-pyridinium-tricyanoquinodimethanide (LB films), where alkyl= hexadecyl, octyl, 649
N-alkyl-4-pyridinium TCNQ (LB films) used to detect NO_2, 480, 481
N-alkyl-4-quinolinium-tricyanoquinodimethanide (LB films), where alkyl= decyl, dodecyl, eicosyl, heptyl, hexadecyl, hexyl, nonyl, octadecyl, octyl, pentadecyl, terdecyl, tetradecyl,undecyl 649
alternating Langmuir-Blodgett films, 598
β-alumina, 629
aluminum pits in compact discs, 661
amphiphilic molecule, 607
Anderson localization, 58, 156
[13]aneN_4 *see* 1,4,7,10-tetraazacyclotridecane
[14]aneN_4 *see* 1,4,8,11-tetraazacyclotetradecane
[15]aneN_4 *see* 1,4,8,12-tetraazacyclopentadecane
anellated dipyrrolyl polymers, 359
anion-radical salts
---counter-cations: metals, quaternary ammoniums, phosphoniums, sulfoniums, 45
---radical anions: DCNQI, dithiolates, quinones, TCNQ, 45
anthracene
---structure classification, 8
anthracene-TCNQ
---structure classification, 8
anthraquinone-2-sulfonic acid, for anionic dopant in conducting polymers (for electrochromism, photoelectric conversion), 654
anthraquinone-2-(diazo-4'-anisole), as transducer, 657
---electrochromism and photochromism, 657
---"shallow memory" as hydroquinone, "deep memory" in quinone form, 657
antiferromagnet, low-dimensional
--- α-$(BEDT\text{-}TTF)_2I_3$, 126
antiferromagnetic resonance, 86
antiferromagnetic state *see* SDW state
APT(8-12) *see* (N-[p-(p-octylphenylazo)-phenyloxy]-dodecylpyridinium) $(TCNQ)_2$
arenes, 45
Aviram-Ratner organic rectifier, 611, 659, 662
Aviram-Ratner hydrogen atom switch, 623, 659, 663
Aviram spiro-connected trithienyls, 660, 664

Bacterial reaction center, 592
---in *Rhodopseudomonas viridis* : protein crystal structure solved, 592
---in *Rhodobacter sphaeroides* : protein crystal structure solved, 592
---photochemical hole burning: lifetime of excited state of P is 10 fsec, 592
---structure: branches A, B with almost 2-fold symmetry, with 6 porphyrins and 2 quinones. Primary donor P = 2 bacteriochlorophyll molecules, D_A and D_B (excited state P*); these interact, via another BChl monomer B_A (role unknown) and a bacteriopheophytin monomer Φ_A (intermeditate acceptor I) with 2 quinone (Q) acceptors Q_A (primary acceptor) and Q_B (secondary acceptor). Quantum yield > 0.98, back transfer and waste reaction rates > 100 times slower than forward rate, 592
---Stark effect: changes in apparent static electric dipole moment (magnitude and direction), 596
---Stark effect largest on primary donor P, 596
---Stark effect best in oriented samples, 597
---time-resolved spectroscopy: 3 psec for electron transfer from primary donor P to first acceptor I (15 Å away), 592
bacteria, purple
---*Rhodopseudomonas viridis* , 592
---*Rhodobacter spaeroides* , 592
bacteriochlorophyll (BChl), 592
bacteriopheophytin, 592
bacteriorhodopsin, 663
Ballard synthesis of poly(*para* -phenylene), 381
band
---charge-transfer, 649
---conditions for organic metals, 5
---conductivity, 3
---decrease bandgap, increase bandwidth, 5
---filling, 50
---gap, 3
---inverted, 2
---narrow-band metals, 3
---narrow-band solids, 2
---one-dimensional, 129
---origin, 2
---partially filled, 2, 3
---quarter-filled, 129
---quarter-filled band, tetramer model, 137
---structure for 4 electrons on 4 sites, 9
---structure: extended Hückel, 49, 72
---structure: extended zone, 337, 338
---structure: valence effective hamiltonian, 327, 448-453
---theory, 1
---valence, 3
---very narrow, 49
---width (4t), 1, 131
$BaPb_{1-x}Bi_xO_3$, a superconductor, 116
Bardeen-Cooper-Schrieffer (BCS) theory, 75, 93, 247, 251
barium stearate, LB films, as lubricants for magnetic tape, 474
bathorhodopsin, 663
battery, polymer, 654
BChl *see* bateriochlorophyll
BCS see Bardeen-Cooper-Schrieffer
BDDAP-C-BHTCNQ (bis(dodecyl)-4-amino-phenyl-carbamate of 2-bromo, 5-hydroxyethoxy-7,7,8,8-tetracyano-quinodimethane
---infrared spectra of multilayers, 622
---LB films, 614, 615
---rectification attempts, 623
---scanning tunneling microscopy,623, 624
---structure, 615
BDDAP-C-HETCNQ (bis(dodecyl)-4-amino-phenyl-carbamate of 2-hydroxyethoxy-7,7,8,8-tetracyanoquinodimethane
---oxidation and reduction potentials, 617
---structure, 616
BDDAP-C-HMTCAQ (bis-(4-dodecyl)-amino-phenyl-carbamate of hydroxy-methyl-7,7,8,8-tetracyanoanthraquino-dimethane
---infrared spectra of multilayers, 622
---LB films, 614
---structure, 615
Bechgaard salts *see* $(TMTSF)_2X$ salts, 16, 129, 141, 249
Bednorz-Müller high-temperature superconductor *see* $La_{2-x}M^{II}_xCuO_{4-y}$
BEDO-TTF *see* bis(ethylenedioxo)tetrathiafulvalene
$(BEDO\text{-}TTF)_2X$
---metallic to very low temperatures, 145
BEDS-TTF *see* bis(ethylenediselenolo)tetrathiafulvalene
BEDT-TSF *see* bis(ethylenedithiolo)tetraselenafulvalene
BEDT-TTF *see* bis(ethylenedithiolo)tetrathiafulvalene
$(BEDT\text{-}TTF)_3Cl_2.2H_2O$
---superconducts, T_c=5.1 K at 14 kbar, 116
α'-$(BEDT\text{-}TTF)_2Ag(CN)_2$
---EPR: 21-34 Gauss LW, g=2.003-2.009, 170
---EPR linewidth decreases with T, 172
---excess susceptibility (9.0-9.2) x 10^{-4} emu/mol, 171
---semiconductor, 170
---susceptibility: Bonner-Fischer/spin-Peierls, 170
---structure: stacks of strongly dimerized pairs; molecules rotated by 30°, 170
κ-$(BEDT\text{-}TTF)_2Ag(CN)_2.H_2O$
---EPR: 54-84 Gauss LW, g=2.003-2.010, 170
---EPR linewidth decreases with T, 170
---excess susceptibility=4.6 x 10^{-4} emu/mol, 171

---metal, conductivity 20 S/cm at 300 K, 170
---susceptibility: Pauli, 170
---structure: face-to-face dimerized pairs, each rotated by 90°, 170
(BEDT-TTF)$Ag_{1.5}(SCN)_2$
---conductivity (300 K)= 1 S/cm, 231
---ESR: linewidth (330 K)=36-40 G, 228
---semiconductor, gap=0.185 eV, 231
---space group Pbca, Z=4, V=1956.4 Å^3, 228
(BEDT-TTF)$Ag_{1.6}(SCN)_2$, 228
α'-(BEDT-TTF)$_2AuBr_2$, 170
β-(BEDT-TTF)$_2AuBr_2$
---conductivity anisotropy, 122
---crystal structure, 122
---interstack transfer integral ratio small, 122
β''-(BEDT-TTF)$_2AuBr_2$, 170
---EPR: 35-63 Gauss LW, g=2.002-2.013, 170
---EPR linewidth increases with T, 172
---EPR linewidth changes at 30 K, 172
---metal, conductivity (300 K)=500 S/cm, 170
---microwave conductivity, 172
---susceptibility: linear T, 170
---structure: stacks at about 60° to molecular plane; in transverse direction, coplanar within 5°, 170
β-(BEDT-TTF)$_2AuCl_2$
---anion length=8.1 Å, 92
---antiferromagnetic state below 27 K, 57
---semiconductor, 55
---more strongly dimerized than the (TMTSF)$_2$X salts, 61, 467
---Hubbard-Peierls gap, 54, 55
---Heisenberg paramagnet, 55
---longitudinal conduction is diffusional, 55
---metal-to-insulator transition > 32 K, 92
---unit cell volume, 92
β'-(BEDT-TTF)$_2AuCl_2$
---antiferromagnetic resonance, 172
---EPR: 5-8 Gauss LW, g=2.003-2.010, 170
---EPR linewidth increases with T, 170
---EPR linewidth changes at 30 K, 172
---SDW onset at 30 K, 172
---semiconductor, 170
---susceptibility: Bonner-Fischer/SDW, 170
---structure: stack of strongly dimerized face-to-face pairs, with steplike side-by-side arrangement, 170
β''-(BEDT-TTF)$_2$AuIBr, 170
β-(BEDT-TTF)$_2AuI_2$
---anion length 9.4 Å, 92
---eclipsed ethylene groups, 94
---EPR: 14-20 Gauss LW, g=2.003-2.011, 170
---EPR linewidth increases with T, 170
---excess susceptibility (3.4-4.6) x 10^{-4} emu/mol, 172
---microwave conductivity, 172
---resistivity, 119
---structure: stack of strongly dimerized face-to-face pairs, with steplike side-by-side arrangement, 170
---superconductor, T_c=3.4-5 K, 18, 67, 85, 91, 92, 122, 145, 170
---susceptibility: Pauli, 170
---unit cell volume, 92
(BEDT-TTF)Br (powder)
---Raman spectrum: weak band at 246 cm^{-1}, 241
(BEDT-TTF)$_2Cl.H_2O$
---superconducts, 67, 227
β-(BEDT-TTF)$_2$ClIBr
---anion length 9.0 Å, 92
---unit cell volume, 92
α'-(BEDT-TTF)$_2CuCl_2$, 170
(BEDT-TTF)$_3CuCl_4.H_2O$
---not superconductive, 126
κ-(BEDT-TTF)$_2Cu(NCS)_2$
---anions: 2 linear NCS: one forms chain, the other is a pendant group, 72
---conducting sheet of cation dimers in **bc** plane, 70
---conductivity (300 K)=10-40 S/cm, 75
---conductivity anisotropy 0.0016:1:1.2, 75
---conductivity ≈ isotropic in **bc** plane, 240
---conductivity minimum around 100 K, 124
---conductivity: temperature dependence, 75, 76, 119
---critical current density 1060 A/cm^2, 79
---crystal structure at 298 K and 104 K, 68, 70-72, 99
---crystal structure below 50 K not known, 98
---crystals black, shiny, distorted hexagonal, 69
---cylindrical Fermi surface, 233, 235
---cyclotron mass=3.2 electron masses, 236
---density of states at Fermi energy= 7.1-7.5 /eV, 79
---Dingle temperature=0.7 K, 236
---effective mass=3.5 m_e at E_F, 77
---electrocrystallization, 233
---energy gap (tunneling) $2\Delta_o$=4 meV, close to BCS value, 67, 79
---EPR: no Cu(II) signal (300 K to 4 K); broad signal (61 G at 300 K) g=2.0078-2.0070, plus additional sharp (defect?) signal (10-20 G, g=2.0075); Elliott formula not obeyed, 67, 79, 170, 172
---ESCA: Cu valency=+1, 79
---Fermi energy=0.13 eV, 79
---Fermi surfaces (extended Hückel MO): one closed cylindical surface, one open modulated 1-D sheet, 67, 72, 73
---Ginzburg-Landau coherence length

anisotropic: long (182 Å) in **bc** plane, short (9.6 Å < **a** * along **a***), 77
---inverse isotope effect (T_c rises upon deuteration) may be due to increased (ethylene-anion?) electron-phonon coupling, 67, 75
---ionicity=0.5 (uniform), 48
---layered 2-D superconductor, 98
---magnetic susceptibility (ac, dc): Pauli-like (10-300 K: metal), 67, 77, 98
---magnetoresistance, 233
---Meissner effect: ≈100% flux exclusion, 98
---metallic 10.5 K-100 K (1 bar), 121, 124
---metallic below 300 K (5 kbar), 124
---muon spin resonance, 81
---NMR: $1/T_1$ enhanced well below T_c, 67, 79
---no stacking direction, 233
---optically active: specific rotatory power =230° at 298 K and 632.8 nm, 72
---orthogonal dimers, 48
---phase transitions around 100 and 50 K, 98
---polycrystalline pressed pellet: no superconductivity, 104
---quasi-two-dimensional metal, 98
---resonance Raman spectrum (polarized): bands at 1470 cm^{-1} (intense: C=C stretch in central ring, coupled to 20,000 cm^{-1} electronic transition), 1492 cm^{-1} (weak: ring C=C stretch), 1450 cm^{-1}, 1506 cm^{-1} (weak: other C=C vibrations; latter may be antisymmetric b_{1u}), 774 cm^{-1} (very weak, b_{1u} C-S stretch), 1037 cm^{-1} (C-C-H bend: symm.), 246 cm^{-1} (intense: b_{2u}, Herzberg-Teller mixing or symm. Cu-NCS vibration), 166 cm^{-1} (symm. BEDT-TTF major axis skeletal vibration), 97 cm^{-1} (lattice mode), 239-242
---scanning tunneling microscopy, 81
---semiconductor 100 - 300 K (1 bar), 124
---Shubnikov-de Haas oscillations (frequency 597 T, orbit 6.37 x 10^{14} cm^{-2}, area of closed cylindrical Fermi surface, proof of 2D superconductivity), 67, 77, 233
---space group $P2_1$ (acentric), 71, 72
---structure classification, 8
---superconducting critical fields, 67, 77, 233
---superconducting T_c decreases 1.3 K per kbar, 75
---superconductor, T_c=10.2-10.4 K, 15, 57, 75, 85, 97, 98, 118, 227, 233, 240
---synthesis, 69, 70
---thermally stable to 190°C, decomposes around 230°C, 70
---thermopower is anisotropic (positive along **c**, negative along **b**) and vanishes at 10 K, 67, 77
---transition to superconductivity sharp, 98
---transverse magnetoresistance, 233
---two-dimensional dimeric tiling, 15
---two types of superconducting crystals, A (T_c(onset)=9.4 K) and B (T_c(onset)= 10.3 K) differ in anion sources, solvents used, ESR linewidth (A: 72 G); A may have SCN^- ions, 230, 233
---unit cells at 298 K and 104 K, 71
---upper critical field (20.2 T at 1.17 K) exceeds Chandrasekhar-Clogston limit, 233
---upper critical magnetic field H_{c2} exceeds the Pauli value, 67, 77

κ-(BEDT-TTF-d_8)$_2$Cu(NCS)$_2$
---superconductor, T_c=10.8-11.0 or 11.1 K (one deuterated sample: 12.8 K), 18, 67, 75, 76, 98

(BEDT-TTF)$_2$FeCl$_4$
---not superconductive, 126

(BEDT-TTF)$_4$(Hg$_{2.89}$Br$_8$)
---superconductor, 67

(BEDT-TTF)$_2$(Hg$_{2.78}$Cl$_8$)
---superconductor, 67

(BEDT-TTF)$_4$(Hg$_3$Br$_8$)
---superconducts, 227

(BEDT-TTF)$_4$(Hg$_3$Cl$_8$)
---superconducts, 227

(BEDT-TTF)$_3$Hg(SCN)$_3$
---conductivity (300 K)=23 S/cm, 70
---metal-to-insulator transition ≈ 180 K, 70

(BEDT-TTF)$_4$Hg$_{1.4-1.5}$(SCN)$_4$
---black fragile crystals from 1:1 chlorobenzene:acetonitrile, 229
---semiconductor gap=0.084 eV, 231

(BEDT-TTF)$_4$Hg(SCN)$_4$(1,1,2-trichloroethane)
---black fragile crystals from trichloroethane, 228

α-(BEDT-TTF)$_2$I$_3$ (not superconducting)
---conductivity (300 K)=60-250 S/cm, 261
---difference between α and β structures: rotation of BEDT-TTF by 35° around long axis, 99
---electrochemical synthesis, 257
---ESR: polycrystalline powder: metal-to-insulator transition can be seen, 106
---if tempered (heated above 70°C for a few days), becomes superconducting (α_t or β" phase, similar to β_H) T_c=8 K, 67, 69, 97-99, 124
---if iodine-doped, superconducts, 67
---low-dimensional antiferromagnet, 124
---metal-insulator transition at 135 K, 100, 257
---Mott-Hubbard insulator, 124
---polycrystalline pressed pellet: no super-

conductivity; broad transition to semiconductor (conductivity=0.5 S/cm at 4 K), 104, 106
---resonance Raman spectrum at 1.3 K: peak at 110 cm^{-1}, shoulder at 120 cm^{-1} (partially non-linear I_3^- ion), 106, 107, 239, 243
---transfer integral (thermopower)=0.2 eV, 261
---unit cell, 258
α_t-$(BEDT\text{-}TTF)_2I_3$
---α phase, when heated above 70°C for a few days, becomes stable "tempered" superconducting α_t (or β") phase, ($\approx \beta_H$), T_c= 8 K, 75, 97, 99, 124
---crystal structure unknown (large mosaic 10 μm), 99
---EPR: polycrystalline pellet: pressure during pellet formation forces transition to semiconductor (undone if annealed), 106
---ethylene groups may be eclipsed or ordered, 99
---Ginzburg-Landau coherence lengths =120, 20 Å, 102
---"glassy" resisitivity, 100
---London penetration depths=4300, 750 Å, 102
---Meissner effect: 50% flux exclusion, 101
---Meissner effect for polycrystalline annealed sample: onset at 7 K, 50% flux exclusion at 2 K, 104
---metal-to-insulator transition at 135 K seen for partially tempered sample, 100
---NMR coherence length 6-10Å, 102
---optical conductivity, 241
---phase transition at 60 K, 100
---polycrystalline pressed pellet: metallic, not superconducting, 300 K to 1.3 K, 104
---polycrystalline pressed pellet, annealed: broad superconductivity, onset at 9 K, midpoint at 5.5 K, zero resistance at 2.2 K, 97, 104
---polycrystalline pressed pellet, annealed, surface polished with razor blade: no superconductivity; Raman spectrum: 110 and 120 cm^{-1} peaks (disordered or non-linear I_3^-, 243, 244
---resonance Raman spectrum for polycrystalline sample at 1.3 K: 120 cm^{-1} (linear I_3^- symm. stretch): disordering of I_3^- caused by pressure, 106, 107, 239
---superconductivity (single crystal): onset at 9 K, midpoint at 8 K, zero resistance at 6 K, 97, 99, 104
---stable superconductivity, T_c=8 K, 97, 99, 101
---upper critical magnetic field H_{c2} higher in polycrystalline annealed sample than in single crystal, 105
β-$(BEDT\text{-}TTF)_2I_3$ (also called β_L phase)
---anion length=10.1 Å, 92
---conductivity (300 K)=20-70 S/cm, 16, 261
---conductivity (pressed pellet, 300 K)= 5-10 S/cm, 108
---conductivity max. (broad, 4.2K)= 15,000 S/cm, 16
---conductivity minimum (pressed pellet) at 220 K with hysteresis, 109
---difference between α and β crystal structures: rotation of BEDT-TTF by 35° around long axis, 99
---eclipsed ethylene groups, 94
---electrochemical synthesis, 257
---EPR, 170, 171
---incommensurately modulated superstructure below 200 K, 18
---incommensurate superstructure between 175 K and 1.5 K, suppressed by moderate pressure (yields β_H) or by annealing α-phase for 20-40 hr at 110 K, 67, 69, 97
---ionicity=0.50, 16
---phase diagram, pressure-temperature, 68
---polycrystalline pressed pellet: superconductor, onset T_c=9 K, midpoint 7.5 K, zero resistance at 3.2 K, Meissner effect 50% (broad onset), 104, 108
---resonance Raman spectrum of I_3^-, 107
---superconducts, T_c=1.5-2.5 K, 18, 67, 91, 92, 122, 145, 227
---superconducts, T_c=1.3 K (β_L or β), 97, 257
---structure classification, 8
---transfer integral=0.12 eV (thermopower), 261
---unit cell, 258
---unit cell volume, 92
β_H-$(BEDT\text{-}TTF)_2I_3$
---also called β^* phase, 92
---formed at high pressure, stabilized at 1 bar if transition to an incommensurate modulation is avoided, 67, 88, 97, 257
---formed by annealing β-salt at 110 K for 25-40 hr, 69
---competition of superconductivity versus antiferromagnetism is decreased, 88
---ethylene groups staggered, 94, 97
---metastable superconductor, T_c(0.5 kbar) =7.5 K (at 8 K and 1 bar, after pressurization to 1 kbar at 300 K, cooling to 125 K, and releasing pressure to 1 bar), 97, 98
---magnetoresistance (Shubnikov-de Haas) oscillations confirm cylindrical Fermi surface, 85, 88
---phase diagram, pressure-temperature, 68
---superconducts, T_c (1.8 kbar)=8.1 K, 18, 57, 67, 88, 91, 122, 257
β_d'-$(BEDT\text{-}TTF)_2I_3$
---conductivity decreases from 110 K to 60

K, but increases from 60 Kto 20 K, 258, 259
---conductivity (300 K)= 100-1000 S/cm, 261
---crystal structure, 257, 258
---energy gap smeared at 300 K by disorder in I_3^- chain, 258
---energy gap=0.07 eV, 259
---ESR susceptibility, 249, 260
---hexagonal-shaped crystals, 257
---iodine positions partially occupied, 258
---metal-semiconductor transition at 140 K, 259
---Raman spectrum: I_3^- orders at 140 K, 258, 259
---synthesized by diffusion method, 257
---thermopower, 259
---transfer integral (thermopower)=0.12 eV, 261
---unit cell, 258
γ-(BEDT-TTF)$_2$I$_3$
---superconductor, 67
δ-(BEDT-TTF)$_2$I$_3$
---structure classification, 8
θ-(BEDT-TTF)$_2$I$_3$
---conductivity (300 K)=30-100 S/cm, 261
---electrochemical synthesis, 257
---superconducts, T_c=3.6 K, 67, 257
---unit cell, 258
κ-(BEDT-TTF)$_2$I$_3$, 170
---conductivity (300 K)=40-150 S/cm, 261
---electrochemical synthesis, 257
---space group P2$_1$/c (centrosymmetric), 72
---superconducts, T_c=3.6 K, 67, 157, 257
---unit cell, 258
λ_d-(BEDT-TTF)$_2$I$_3$
---conductivity (300 K)=500-5000 S/cm, 261
---conductivity constant 300-230 K, then decreases to 4 K, is sensitive to thermal cycling, may be limited by disorder, 260
---plate-shaped crystal, 258
---synthesized by diffusion method, 257
---thermopower, 260
---transfer integral (thermopower)=0.6 eV, 261
---unit cell, 258
β-(BEDT-TTF)$_2$IBr$_2$
---anion length 9.4 Å, 92
---eclipsed ethylene groups, 94
---EPR, 170
---superconducts, T_c=2.5-2.8 K, 18, 67, 91, 92, 122, 145, 227
---non-superconducting phase with disordered anions exists, 69
---unit cell volume, 92
β-(BEDT-TTF)$_2$I$_2$Br
---anion length 9.7 Å, 92
---does not superconduct, 92
---unit cell volume, 92
β-(BEDT-TTF)$_2$ICl$_2$
---anion length 8.7 Å, 92
---more strongly dimerized than (TMTSF)$_2$X salts, 467
---metal-to-insulator transition 22.1 K, 92
---unit cell volume, 92
β'-(BEDT-TTF)$_2$ICl$_2$, 170
β''-(BEDT-TTF)$_2$ICl$_2$, 170
(BEDT-TTF)$_2$KHg(SCN)$_4$
---conductivity 200-100 S/cm at 300 K, 82
---conductivity metallic to 1.8 K, 82
---crystal structure at 298 K and 104 K, 72, 74, 75
---extended Hückel band structure, 82
---Fermi surface: closed 2D and open 1-D regions (extended Hückel basis), 72, 74
---magnetoresistance saturates at 10 T, 82
---organic metal, 67
---Shubnikov-de-Haas oscillations, 82
---space group P1, 73
---synthesis, 70
---three crystallographically independent BEDT-TTF species (charge ≈ 0.5 each) stacked like in α-I$_3$ salt, 72
---triple sheet of KHg(SCN)$_4$with chain of SCN and NCS linked to K and Hg, and two pendant NCS or SCN groups, 74, 75
---unit cell at 298 K and 104 K, 73
(BEDT-TTF)$_2$NH$_4$Hg(SCN)$_4$
---synthesis, 70
(BEDT-TTF)$_4$Ni (CN)$_4$
---conductivity (300 K)=2.04 S/cm, 186
---crystal structure, 188
---metal-insulator transition around 200 K, 188
---Peierls-like phase transition at 200 K, 188
---temperature-dependent IR modes: electron-molecular vibration coupling, 188
---tetramer of BEDT-TTF slipped stacks, 188
---two optical charge-transfer bands, 188
---unit cell and space group, 186
(BEDT-TTF)$_x$Ni(dcit)$_2$
---black microcrystals, 183
δ-(BEDT-TTF)$_2$NO$_3$, 45
(BEDT-TTF)$_4$Pt (CN)$_4$
---conductivity (300 K)=12 S/cm, 186
---crystal structure, 187, 188
---Peierls-like phase transition at 200 K, 188
---temperature-dependent IR modes: electron-molecular vibration coupling, 188
---tetramer of BEDT-TTF slipped stacks, 188
---two different unit cells and space group, 185, 186
---two optical charge-transfer bands, 188

$(BEDT\text{-}TTF)_2ReO_4$
---superconducts, T_c=1.5 K, p>5 kbar, 18, 67
$(BEDT\text{-}TTF)_3(PW_{12}O_{40})$
---fully ionized BEDT-TTF trimers, 189
---insulator at 300 K, 186, 189
---one optical band at 11,000 cm^{-1} (no CT), 189
---unit cell and space group, 186
(BEDT-TTF) X salts, 18
$(BEDT\text{-}TTF)_2X$ salts
---α, α_t, α', β, β_H, β', β'', β_d', γ, δ, ε, ζ, θ, κ, λ, λ_d, etc. phases, 67, 97, 124, 169, 257
---α: Mott-Hubbard insulator, 124
---α' : X = $Ag(CN)_2$, $Au(CN)_2$, $AuBr_2$, $CuCl_2$, 170
---α' : stacks of strongly dimerized pairs; molecules rotated by 30°, 170
---α': open 1- D Fermi surface, 170
---β: almost two-dimensional sheets, 122
---β: critical temperature T_c correlates with lattice "softness" (inversely with non-bonded $=CH_2^-$..halide distance), 94
---β: face-to-face stacks with steplike side-by-side arrangement, 85, 170
---β: linear anion length and lattice pressure:"negative pressure" should raise T_c above 8 K, 69
---β: linear anion length correlates with superconducting critical temperature T_c: T_c up to 40 K may be possible with longer anions [$Cu(NCS)_2^-$, $Ag(NCS)_2^-$, $Cu(NCSe)_2^-$, $Au(NCS)_2^-$, $Ag(NCSe)_2^-$, $Au(NCSe)_2^-$, $Cu(NCTe)_2^-$, $Ag(NCTe)_2^-$, $Au(NCTe)_2^-$], 93
---β: low-dimensional antiferromagnet, 126
---β: Pauli paramagnetism in normal state, 126
---β: 2-D closed Fermi surface (50% of Brillouin zone), 170
---β: unit cell volume correlates with T_c, 92
---β: X = AuI_2, IBr_2, I_3 superconductors, 92, 170
---β': X = $AuCl_2$, ICl_2, 170
---β': stack of strongly dimerized face-to-face pairs, with steplike side-by-side arrangement, 170
---β': open 1-D Fermi surface, 170
---β'': X= $AuBr_2$, ICl_2, AuIBr, 170
---β'': stacks at about 60° to molecular plane; in transverse direction, coplanar within 5°, 170
---β'': both open 1-D Fermi surface and closed pockets (20% of Brillouin zone), 170
---κ: X = $Ag(CN)_2.H_2O$, $Cu(NCS)_2$, I_3, 170
---κ: face-to-face dimerized pairs, each rotated by 90°, 68, 70, 170
---κ: centrosymmetric salt (X = I_3, space group $P2_1$), acentric, optically active salt (X = $Cu(NCS)_2$, space group $P2_1$), 72
---conductivity at 300 K scales with transfer integral (thermopower), 261
---critical temperature highest for lowest room-temperature conductivity, 18
---correlation: T_c versus effective volume, 82
---estimated Hubbard U(solid)=1.2 eV, 58
---estimated Hubbard V(solid)=0.4 eV, 58
---Fermi energy may be close to a van Hove singularity in density of states, 89
---Fermi surface isotropic in 2 dimensions, 18
---four anion types:
(1) tetrahedral anions (ReO_4^-), (2) linear trihalides (I_3^-, IBr_2^-) or metalhalides (AuI_2-), (3) clusters ($Cl^-.H_2O$) or polymers ($Hg_{2.78}Cl_8$, $Hg_{2.89}Br_8$) (4) boomerang-shaped anions in zig-zag 1-D polymer ($Cu(NCS)_2$), 67, 72
---κ: isolated orthogonal dimers, 122
---kinetic vs. thermodynamic stability, 7
---"lattice pressure" for X=linear anions, 69
---metallic compounds, 53
---multiple phases due to conformational freedom of ethylene groups, cooperative π-π overlap and S...S interactions, 69
---ordering of non-centrosymmetrical X, 52
---Pauli-type paramagnetism, 53
---Robin-Day class IIIB, 53
---superconductivity in β and κ forms, 122
---two-dimensional network of S..S, 18
$(BEDT\text{-}TTF)_3X_2$ salts, 18
behenic acid, Langmuir-Blodgett films, 515, 552
berthollitic stoichiometry, 7, 49
β-alumina, 629
BHAP-C-HMTCAQ (bis-(hexyl)-amino-phenyl-carbamate of hydroxymethyl-7,7,8,8-tetracyanoanthraquinodimethane
---LB films, 614
---oxidation and reduction potentials, 617
---structure, 615
BHTCNQ *see* 2-bromo-5-hydroxyethoxy-7,7,8,8-tetracyanoquinodimethane
bidentate ligands, 264
---cyclohexanediamine (chxn), 265
---1,2-diaminoethane (en), 264
---1,2-diaminopropane (pn), 264
---1,3-diaminopropane (tn), 264
---2,3-diaminobutane (btn), 266
---diphosphite (pop), 263, 267
---methylenebisphophinate (pcp), 267
biochips, 659
biological machinery, 514
bipolarons
---indistinguishable from polarons or

solitons, in IR spectra of doped poly(acetylene), 356
---in conducting polymers, 411, 415
---in g-ology models, 59
---in cuprate superconductors, 59
---in ligand-bridged mixed-valent systems, 59
---in poly(aniline) at low protonation levels, 340
---in poly(pyrrole), 425
---in poly(pyrrole) encapsulated in zeolites, 425
---in poly(3-methylthiophene), 425
---in poly(3-methylthiophene) encapsulated in zeolites, 425
---mixed-valent, for future organic conductors or superconductors, 59
---spinless, in conducting form of poly(aniline) with short-range order, 339
2,2'-bipyrrolyl, 359
[bis(decyl)bis(methyl)ammonium] Au(dmit)$_2$
---mixed 1:1 with eicosanoic acid and oxidized with Br_2, LB film, conductivity (330 K)=0.12 S/cm, 521
---mixed 1:1 with eicosanoic acid and electro-oxidized, LB film, conductivity (330 K)=1.4 S/cm, 521
[bis(decyl)bis(methyl)ammonium] Ni(dmit)$_2$
---mixed 1:1 with eicosanoic acid and oxidized with Br_2, LB film, conductivity(300 K) =1 S/cm, 521
---mixed 1:1 with eicosanoic acid and oxidized with Br_2, LB film, conductivity (300 K)=1.4 S/cm, 521
[bis(decyl)bis(methyl)ammonium]$_2$ Pd (dmit)$_2$
---oxidized with I_2, LB film, conductivity (300 K)=0.3 S/cm, 521
---electro-oxidized, LB film, conductivity (300 K)=1.0 S/cm, 521
[bis(decyl)bis(methyl)ammonium]$_2$ Pt (dmit)$_2$
---oxidized with I_2, LB film, conductivity(300 K)=0.001 S/cm, 521
(bis(2',3'-dimethylethylenediselenolo)-diselena)dithiafulvalene, 150
bis(2',3'-dimethylethylenediselenolo)-TSF, 150
bis(2',3'dimethylethylenediselenolo)-TTF, 150
bis(2',3'-dimethylethylenedithiolo)-dithiadiselenafulvalene, 150
bis(2',3'-dimethylethylenedithiolo)-TSF, 150
bis(2',3'-dimethylethylenedithiolo)-TTF, 150
[bis(docosyl)bis(methyl)ammonium] Au(dmit)$_2$ mixed 1:1 with eicosanoic acid and oxidized with Br_2, LB film, low conductivity, 521
[bis(docosyl)bis(methyl)ammonium] Ni (dmit)$_2$ oxidized with Br_2, LB film, conductivity(300 K)=0.002 S/cm, 521
[bis(dodecyl)bis(methyl)ammonium] Ni (dmit)$_2$
---mixed 1:1 with eicosanoic acid and oxidized with Br_2, LB film, conductivity (300 K)=0.03 S/cm, 521
(bis(ethylenediselenolo)-diselena)dithiafulvalene, 150
bis(ethylenediselenolo)-tetraselenafulvalene, 150
bis(ethylenediselenolo)tetrathiafulvalene (BEDS-TTF), 47
bis(ethylenediselenolo)-TTF, 150
(bis(ethylenedithiolo)-dithia)diselenafulvalene, 150
bis(ethylenedithiolo)tetraselenafulvalene (BEDT-TSF), 47, 150
bis(ethylenedithiolo)tetrathiafulvalene (BEDT-TTF or ET)
---*see also* related electron donors: BEDS-TTF, BEDT-TSF, DIET, TMET, etc.
---chloride salt (3:2), 116
---copper isothiocyanate salt (2:1), 8, 15
---gold bromide salt (2:1), 119, 122
---gold iodide salt (2:1), 122
---good donor, 5
---iodide salts (2:1), 8, 45
---nitrate salt, 45
---perchlorate salt (2:1), 119
---perrhenate salt (2:1), 18
---polymorphism in salts, 18, 45, 124
---small non-planarities, 5
---structure, 6, 47, 91, 150
bis(ethylene)propylenetetramine (2,3,2-tet), 266
[bis(heptadecyl)dimethylTTF] (TCNQ) and (tetramethylTTF)(hexadecyl-TCNQ), LB film semiconductor, 520
bis(hexadecyl)bis(methyl)ammonium nickel (dmit)$_2$
---mixed 1:1 with eicosanoic acid and oxidized with Br_2, LB film, conductivity (300 K)=0.05 S/cm, 521
[bis(hexadecylthio)(ethylenedithiolo)TTF] (TCNQF$_4$)I$_x$, LB film semiconductor, 520
[bis(hexadecylthio)(ethylenedithiolo)TTF] (TCNQF$_4$)I$_x$, LB film semiconductor, 520
(bis(iso-propylenediselenolo)-diselena)dithiafulvalene, 150
bis(iso-propylenediselenolo)-TSF, 150
bis(iso-propylenediselenolo)-TTF, 150
bis(iso-propylenedithiolo)-dithiadiselenafulvalene, 150
bis(iso-propylenedithiolo)-tetraselenafulvalene, 150
bis-(isopropylenedithiolo)-tetrathiafulvalene (DIET) (*cis* and *trans*), 150

---crystal structure (disordered *trans* isomer): central C=C bond distance =1.4 Å, 177
---synthesis, 177, 178
4,5-bis(methylacetate)diseleno-1,3-dithiole-2-thione
---synthesis, 181
4,5-bis(methylcarboxy)-1,3-dithiole-2-thione, 146, 147
4,5-bis(methylcarboxy)-1,3-diselenole-2-selenone, 146, 147
bis(methylenediselenolo)-diselena)dithiafulvalene, 150
bis(methylenediselenolo)-TSF, 150
bis(methylenediselenolo)-TTF, 150
(bis(methylenedithiolo)-dithia)diselenafulvalene, 150
bis(methylenedithiolo)-TSF, 150
bis(methylenedithiolo)-TTF (BMDT-TTF), 150
(bis(methylseleno)-diselena)-(bis(methylthio)-dithia)fulvalene, 150
(bis(methylseleno)diselena)-dithiafulvalene, 150
bis(methylseleno)-TSF, 150
bis(methylseleno)-TTF, 150
4,5-bis(methyl-selenoacetoxy)-1,3-dithiole-2-thione, 146, 147
4,5-bis(methyl-thioacetoxy)-1,3-dithiole-2-thione, 146, 147
bis(methylthio)-bis(methylseleno)-TTF, 150
(bis(methylthio)-diselena)-(bis(methylseleno)-diselena)fulvalene, 150
(bis(methylthio)-diselena)-(bis(methylseleno)-dithia)fulvalene, 150
(bis(methylthio)-diselena)dithiafulvalene, 150
(bis(methylthio)-dithia)-(bis(methylseleno)-diselena)fulvalene, 150
bis(methylthio)-TSF, 150
bis(methylthio)-TTF (BMT-TTF), 150
bismuth, a semimetal, 5
bis-[bis(heptadecyl)-octadecyl-oxy-ortho-phenanthroline]Fe(II)$(NCS)_2$
---LB films (X-type) on CaF_2 516-518
---high-spin to low-spin transition has hysteresis, 517, 518
bis(octadecyl)bis(methyl)ammonium gold $(dmit)_2$ mixed 1:1 with eicosanoic acid and oxidized with Br_2, LB film, conductivity(300 K)=0.005 S/cm, 521
bis(octadecyl)bis(methyl)ammonium nickel $(dmit)_2$ mixed 1:1 with eicosanoic acid and oxidized with Br_2, LB film, conductivity(300 K)=0.009 S/cm, 521
[bis(octadecylthio)bis(methyl)TTF] $(TCNQF_4)I_x$, LB film semiconductor, 520
(bis(propylenediselenolo)-diselena)dithiafulvalene, 150
bis(propylenediselenolo)-TSF, 150
bis(propylenediselenolo)-TTF, 150
(bis(propylenedithiolo)-dithia)diselenafulvalene, 150
bis(propylenedithiolo)-TSF, 150
bis(propylenedithiolo)-TTF (BPDT-TTF), 150
bis(propylene)ethylenetetramine (3,2,3-tet), 266
bis(pyrrolo)benzenes, 359
---electropolymerization, 361
---oxidation potential, 361
---polymer, 361
4,5-bis(selenoalkyl)-1,3-diselenole-2-selenone, 147
4,5-bis(selenoalkyl)-1,3-dithiole-2-thione, 147
$Bi_2Sr_2Ca_{n-1}Cu_nO_{2n+4}$ superconductors, 123
---layers M_2O_2 with NaCl structure, 123
---1 to 4 layers of CuO_5 square pyramids, 123
---T_c=80-110 K, 123
bis(tetradecyl)bis(methyl)ammonium $Au(dmit)_2$ mixed 1:1 with eicosanoic acid and oxidized with Br_2, LB film, conductivity=0.09 S/cm at 300 K, 521
bis(tetradecyl)bis(methyl)ammonium $Ni(dmit)_2$ mixed 1:1 with eicosanoic acid and oxidized with Br_2, LB film, conductivity=0.15 S/cm at 300 K, 521
4,5-bis(thioalkyl)-1,3-diselenole-2-selenone, 147
4,5-bis(thioalkyl)-1,3-dithiole-2-thione, 147
(bis(vinylenediselenolo)-diselena)dithiafulvalene, 150
bis(vinylenediselenolo)-TSF, 150
bis(vinylenediselenolo)-TTF, 150
(bis(vinylenedithiolo)-dithia)diselenafulvalene, 150
bis(vinylenedithiolo)-TSF, 150
bis(vinylenedithiolo)-TTF (BVDT-TTF), 150
bixin, $C_{25}H_{30}O_4$
---area per molecule=22 $Å^2$, 535
---collapse pressure=48 mN/m, 535
---compressed Langmuir film collapses, 535
---LB films, transfer ratio=0.98, 534
blue bronzes $A_{0.3}MoO_3$ (A= K, Rb, Tl), 277
BMAP-C-HMTCAQ (bismethylaminophenyl-carbamate of hydroxymethyltetracyanoanthraquinodimethane
---crystal structure, 620
BMDT-TTF *see* bis(methylenedithiolo)-TTF
$(BMDT\text{-}TTF)_2Ni(dcit)_2$
---metallic, metal-to-insulator transition at ≈ 125 K, 156
---red-black chips, 183
bond-order waves, 132

Bonner-Fischer one-dimensional antiferromagnet, 169, 170
Born-Oppenheimer states, 32
Bose condensation, 59
Bourbonnais-Caron model, 57, 163
BPDT-TTF *see* bis(proylenedithiolo)-TTF
$(BPDT\text{-}TTF)_xNi(dcit)_2$
---black microcrystals, 183
Bragg scattering
---at 2 k_F in metals, 9
---at 2 k_F in Peierls semiconductors, 9
Bragg scattering
---at 4 k_F in Mott-Hubbard semiconductors, 9
Breslow proposal for organic ferromagnet, 60
Brillouin zone, 130, 348
---partial filling, 11
---in one dimension: planar, 50, 52
---in two dimensions: elliptical, 52
para -bromanil
---structure, 46
2-bromo-5-hydroxyethoxy-7,7,8,8-tetra-cyanoquinodimethane (BHTCNQ)
---crystal structure, 619
---reduction potentials, 617
---structure, 615
bronzes *see* blue bronzes *and* tungsten bronzes
btn *see* 2,3-diaminobutane
butadiene
---bond length (from IR frequency) d_{C-H}=1.088, 352
---effective conjugation force constant, from MNDO, 355
---infrared intensity parameters, 352
N-butylphenazinium (NBP), 8
$(n\text{-}Bu_4N)ClO_4$, electrolyte, 629
BVDT-TTF *see* bis(vinylenedithiolo)-TTF
$(BVDT\text{-}TTF)_xNi(dcit)_2$
---black microcrystals, 183

Cadmium arachidate, 558, 611
---bilayer spacing (LB film)=55.13 Å, 414
---mean coherence length 7 bilayers, 414
---monolayer cover on $In_{0.53}Ga_{0.47}As$: modulation-doped FET (cut-off frequency 19 GHz, transconductance 0.170 S/mm), 478
cadmium mercury telluride *see* mercury cadmium telluride
cadmium telluride
---Schottky barrier increased by surface coating of LB multilayers, 476
calcium fluoride, 517
calixarenes, 660, 664
carbon disulfide *see* CS_2
Caron-Bourbonnais model, 57, 87, 163
carotene derivatives, 532
---as molecular wires, 660
---bixin, 535
---β-carotene: effective conjugation force constant from MNDO, 354
---crocetindialdehyde, 533
---7-N,N-dioctylaminophenyl-7,8'-diapo-carotene-8'-al, 532
Carter's soliton switch, 659
cation-radical salts
---counter anions: MX_6^-, MO_4^-,MX_6^{--}, clusters, 45
---radical cations: arenes, Por, Pc,TTF, TTT, 45
CDW *see* charge-density wave
chain-stretching mode
---in platinum chain complexes, 264
---in polyacetylene, 266
Chandrasekhar-Clogston critical field limit, 233
charge-density wave (CDW)
---basic interaction is electron-phonon (*versus* electron-electron for SDW) 251
---bond CDW and site CDW, 443, 446
---bond CDW coexists with SDW in half-filled band, 446
---caused by electron-phonon coupling, 52
---commensurate (Peierls-Fröhlich), 52, 59, 132
---commensurate in $TEA(TCNQ)_2$, 118
---depinning, 282
---distortion, 9
---ground state exists below a temperature T_c (related to the single-particle gap by the BCS expression), 251
---incommensurate (Fröhlich), 277
---incommensurate in TTF TCNQ, 12, 118
---in TTF TCNQ, 12
---phase vortices, 282
---pinned, 9
---site-CDW and bond-CDW, 443, 446
---sliding, in blue bronzes $A_{0.3}MO_3$ (A=K, Rb, Tl), 277-283
---sliding, in $NbSe_3$, 277
---sliding, in TaS_3, 277
---sliding, in TTF TCNQ, 277
---sliding, confirmed by X-ray diffraction, 277
---sliding, confirmed by nuclear magnetic resonance, 277
---sliding, in propagating voltage pulses, 282
---sliding, seen in incommensurate systems, 278
charge-transfer (CT)
---band, 3, 43, 44, 649
---band affected by intrasite (U) and intersite (V) Coulomb repulsion, 54
---complexes (Mulliken) (CTC), 3, 43
---conductors, 129, 131
---crystals, 23, 129
---crystals, two-stack, 7
---frequency, 24
---integral (Mulliken), 1
---oscillator strength, 25
---salts, 3
---salts, divided as CT complexes (CTC)

and ion radical salts (RIS), 43
---theory (Mulliken), 3
---transition, 138
chemical transducer, 657
chemiluminescence, 654
chiral electron donors, 175
Chl *see* chloranil
para -chloranil (Chl)
---electron acceptor, 45
---electron affinity 2.76 eV, 613
---structure, 46, 613
---TTF complex, 8
chloro-gallio-octamethylporphyrin [Ga(OMP)F], 218
chloro-gallio-octaethylporphyrin [Ga(OEP)Cl], 219
chloro-gallio-tetraphenylporphyrin [Ga(TPP)Cl], crystal structure, 219
chlorophyll
---photovoltaic cells: low power conversion efficiency (0.02 - 0.20 %), 557
---photovoltaic cells: large internal resistance, 558
---and photosynthesis, 592, 661
chlorophyll **a**, 557
chlorophyll **b** (chl **b**), 557
---accessory pigments in photopigments of green plants, 558
---photovoltaic sandwich cells: Al | 1 LB monolayer cadmium arachidate | 44 LB monolayers of chl **b** | Ag: large internal resistance, power conversion efficiency 0.002-0.026%, gap 0.19 eV, 558
chlorophyll P-680, 662
chloroplasts, 661
chromatography
---column, 147
---flash, 533
---liquid, 152
---thin-layer (TLC), 533
chromium oxide, 661
chromophores, electro-optical excitations, 594
$(CH)_x$ *see* poly(acetylene)
chxn *see* cyclohexanediamine
cluster anions, 185
CNDO/2 molecular orbital method, 468
coatings for glasses, LB films, 473
cobalt tetrasulfophthalocyanine, for anionic dopant in conducting polymers (for catalysis), 654
cohesive energy
---calculations do not yet predict crystal structure, 49
---covalent bonding effects, 49
---Madelung energy component, 10, 49
Cole-Cole analysis, 340
collective ground state
---antiferromagnet, spin density wave, and superconductor, 52, 53, 57, 118
---charge density wave (insulator), 52, 118
---singlet superconductor, 52
---triplet superconductor, 52, 53
---superconductor or antiferromagnet possible if Hubbard U≈2V≈4t, 59
complex, charge-transfer, 3
complex permittivity, 607
complex stoichiometry, 7
conductance, thermal, 550
conducting Langmuir-Blodgett films *see* Langmuir-Blodgett films
conducting liquid crystals, 61
conducting polymers, 129
---dopant anion types: organic molecules, metal complexes, inorganic metal ions, polyelectrolytes, biomaterials, 654
---as self-organizing systems, 293
---carrier transport important for conductivity, not for optical non-linearity, 567
---charge-transport theories: soliton, polaron-bipolaron, 411
---comparison for conductivity and for optical non-linearity, 567
---conditions for metallic regime, 297
---conductivity is bulk effect, 567
---conductivity limited by imperfections, not by scattering from dopants, 293
---dirty (conductivity < 1 S/cm), 294
---dopant potential, with fair local order, will not contribute to back-scattering, 300
---doping increases conductivity greatly, 567
---doping with electron acceptors decreases third-order non-linearity, 567
---excitations (solitons, polarons) important for conductivity and for optical non-linearity, 567
---large anisotropy of conductivity and of optical non-linearity, 567
---metals (conductivity > 300 S/cm), 294
---need for processing improvements, 293
---optical non-linearity is mostly microscopic effect, 567
---phonon frequencies higher than in conventional metals, 300
---π-electron conjugation, delocalization important for conductivity and for optical non-linearity, 567
---seek high molecular weight and high degree of chain perfection, 300
---temperature affects conductivity, but not optical non-linearity, 567
conduction band, 2
conductivity, electrical (*also* resistivity), 14, 55, 56, 76, 81, 98, 100, 101, 102, 104-106, 108, 110, 112, 113, 119, 121, 125, 157, 163, 178, 179, 206, 208, 212, 224, 229, 231, 232, 254, 255, 259, 260, 366, 431, 522
---anisotropy, 14
---anisotropy affected by transverse transfer integral, 54
---infra-red signature, 133, 134

---≈300-500 S/cm in organic metals, at border between hopping and wavelike scattering, 54
---≈ 30 S/cm in conducting polymers, at border between hopping and wavelike scattering, 294
---maximum, 16, 132
---of a solid, 2, 294
---of (BEDT-TTF)$_2$Ag$_x$(SCN)$_2$, 229
---of β-(BEDT-TTF)$_2$AuCl$_2$, 55
---of (BEDT-TTF)$_2$AuI$_2$, 119
---of κ-(BEDT-TTF)$_2$Cu(NCS)$_2$, 18, 76, 98, 119, 125, 157, 229
---of κ-(BEDT-TTF-d$_8$)$_2$Cu(NCS)$_2$, 76
---of (BEDT-TTF)$_2$Hg$_x$(SCN)$_4$, 229
---of α-(BEDT-TTF)$_2$I$_3$, 106, 261
---of α$_t$-(BEDT-TTF)$_2$I$_3$, 100, 102, 104, 105
---of β-(BEDT-TTF)$_2$I$_3$, 16, 261
---of β$_d$'-(BEDT-TTF)$_2$I$_3$, 257, 261
---of β-(BEDT-TTF)$_2$I$_3$ pressed pellet, 108, 110
---of θ-(BEDT-TTF)$_2$I$_3$, 261
---of κ-(BEDT-TTF)$_2$I$_3$, 157, 261
---of λ$_d$-(BEDT-TTF)$_2$I$_3$, 257, 261
---of (BEDT-TTF)$_2$IBr$_2$, 18
---of (BEDT-TTF)$_2$[KHg(SCN)$_4$], 81
---of (BEDT-TTF)$_2$ReO$_4$, 18
---of (DIET)$_2$(BF$_4$)$_3$, 178
---of (DIET)$_x$(NO$_3$)$_y$, 179
---of (DIMET)$_2$MF$_6$, 55
---of (DMET)$_2$AuBr$_2$, 157, 224
---of (DMET)$_2$Au(CN)$_2$, 224
---of (DMET)$_2$BF$_4$, 224
---of (DMET)$_2$I$_3$, 224
---of (DMET)$_2$PF$_6$, 224
---of (DMtTSF)$_2$X, 55
---of (DMtTTF)$_2$X, 55
---of (Li)(TCNQ), 11
---of (HMTSF)(TCNQ), 16, 121
---of KCP, 263
--of K$_4$[Pt$_2$(H$_2$P$_2$O$_5$)$_4$Br].3H$_2$O, 263
---of (MDT-TTF)$_2$AuI$_2$, 157
---of (NMP)(TCNQ), 11
---of α-(Per)$_2$Au(mnt)$_2$, 206
---of α-(Per)$_2$Co(mnt)$_2$, 206
---of β-(Per)$_2$Co(mnt)$_2$, 206
---of (Per)$_2$Co(mnt)$_2$.0.5CH$_2$Cl$_2$, 206
---of β-(Per)$_2$Cu(mnt)$_2$, 206
---of α-(Per)$_2$Fe(mnt)$_2$, 206
---of α-(Per)$_2$Ni(mnt)$_2$, 206
---of α-(Per)$_2$Pd(mnt)$_2$, 206
---of α-(Per)$_2$Pt (mnt)$_2$, 206
---of poly(aniline) intercalated into V$_2$O$_5$, 431
---of poly(2,2'-bithiophene) intercalated into V$_2$O$_5$ xerogel, 431
---of poly(Cu(ethylene-tetrathiolate)), 113
---of poly(Cu(TTF-tetrathiolate)), 113
---of poly(3,4-cycloalkylthiophenes) doped with FeCl$_4^-$ and PF$_6^-$, 366
---of poly(N-methylpyrrole) intercalated into V$_2$O$_5$ xerogel, 431
---of poly(Ni(ethylene-tetrathiolate)), 112
---of poly(Ni(TTF-tetrathiolate)), 112
---of poly(*para* -phenylene), 386
---of (quinolinium)(TCNQ)$_2$, 11
---of Reihlen's green, 263
---of (TEA)(TCNQ), 119
---of (TMTSF)$_2$ClO$_4$, 18, 119
---of (TMTSF)$_2$NO$_3$, 254, 255
---of (TMTSF)$_2$PF$_6$, 18, 119, 254
---of (TMTSF)$_2$X, 55
---of (TMTSF)(TCNQ), 16, 121
---of (TMTTF)(TCNQ), 16
---of (TMTTF)$_2$AsF$_6$, 163
---of (TMTTF)$_2$PF$_6$, 163
---of (TMTTF)$_2$SbF$_6$, 163
---of (TMTTF)$_2$X, 55
---of (triphenylmethylphosphonium) (TCNQ), 11
---of (triphenylmethylphosphonium) (TCNQ)$_2$, 11
---of trisdecylmethylammonium Au(dmit)$_2$: eicosanoic acid (1:1), LB film, 521
---of (TST)$_2$I, 55
---of (TTF)(TCNQ), 16, 119
---of V$_2$O$_5$, 431
---of V$_2$O$_5$, sodium-doped, 431
---temperature dependence, 3
---van der Pauw's method, 218
conductivity, optical
---of Ag(2,5-DM-DCNQI)$_2$, 201
---of κ-(BEDT-TTF)$_2$Cu(NCS)$_2$, 241
---of Cu(2,5-DM-DCNQI)$_2$, 201
---of Na(2,5-DM-DCNQI)$_2$, 201
---of (TMTSF)$_2$ClO$_4$, 136
---of (TMTSF)$_2$PF$_6$, 134, 136
---of (TMTSF)$_2$ReO$_4$, 138
---of (TMTTF)$_2$Br, 136
---of (TMTTF)$_2$PF$_6$, 136
conductivity, thermal, 550
conformon, 639
Co(NH$_3$)$_5$Cl^{++} sacrificial oxidant in polymerization of pyrrole, 656
Cooper pairs, 52
cooperative ground state *see* collective ground state
corbien-methoxy-undecanoic acid, 529, 530
correlation energy, on-site *see* Coulomb on-site repulsion energy
Co(II)-5,10,15,20-tetra[4-oxy-(2'docosanoic acid)]-phenylporphyrin and N-octadecylimidazole: alternate-layer LB film shows dioxygen selectivity like hemoglobin, 515
Coulomb
---interionic attraction energy, 3
---on-site repulsion energy (U), 5, 15, 131
---on-site repulsion energy (U): affects charge-transfer absorption band and paramagnetic susceptibility, 54
---nearest-neighbor repulsion (V), 51, 131
---nearest-neighbor repulsion (V) affects charge-transfer absorption band and

paramagnetic susceptibility, 54
---transverse energy: affects diffuse scattering, transition temperature, 54
creation operator, 3
Creutz-Taube ion, 662
criteria for organic metals
---acceptor electron affinity not too high, 12
---cation and/or anion are divalent, 5
---compatible sizes of components, 10
---donor ionization potential not too low, 12
---fractional charge transfer, 11
---interchain coupling to suppress CDW and SDW, 13
---large molecular polarizability, 5
---low disorder, 15
---mixed valence, 11
---no Peierls distortion, 13
---planar molecules, delocalized π orbitals, 5
---partially filled electronic band, 44
---redox potential difference range, 12
---Saito-Ferraris, 12
---segregated stacks of radicals, 11
---smaller bandgaps, larger bandwidths, 4, 5
---stable free radicals, 3, 4
---Torrance, 12
---uneven charge and spin distribution, 12
criteria for organic rectifiers (M_1|D-σ-A |M_2)
---assembly onto M_1 electrode by Langmuir-Blodgett (self-assembly) or by covalent binding methods, 612
---D^+-σ-A^- state must be much closer to D-σ-A ground state than D^--σ-A^+ state, 612
---extended geometry, 620-622
---facile syntheses of D and A, 613
---finite tolerance for voltages or heat, 614
---high electron affinity (acceptor end A), 612
---low ionization potential (donor end D), 612
---one-molecule thickness, 611
---reaction for building bridge (σ) fast or favored so no salt is formed, 613
---through-bond electron transfer rather than through-space tunneling, 613
---work functions for metal electrodes M_1 and M_2 compatible with I_D and A_A, 613
criteria for organic superconductivity
---decrease disorder, 110
---high symmetry for crystal or repeat unit (molecule can be unsymmetrical), 110
---increase dimensionality from 1 to 2, 18, 120, 145
---infinite regular stacks are not necessary, 158
---in polymers: high symmetry is difficult, but possible if coherence length = 10 Å, 97, 110
---metallic at higher temperature, 16
---residual conductivity > 6,000 S/cm (to exceed Anderson localization), 156
---suppress CDW and CDW states, 13
---room-temperature conductivity modest, 18, 57
---unsymmetrical molecules are also good, 158
criteria for second-order non-linear optical materials
---bulk needs non-zero second-order electric suceptibility $\chi^{(2)}$, 565
---bulk must be non-centrosymmetric, 565
---molecules need large second-order hyperpolarizability β, 565
---good crystal quality, 570
---long-term stability in poled polymers for electro-optic applications, 570
criteria for third-order non-linear optical materials
---bulk needs non-zero third-order electric suceptibility $\chi^{(3)}$, 565
---bulk needs no symmetry: organic polymers OK, 565
---high π-electron conjugation, delocalization, and orientation, 567, 576
---need for microscopic understanding of γ, 570
---organic conducting polymers worse if doped, 567
---molecules need large third-order hyper-polarizability γ, 565
---present $\chi^{(3)}$ for organics still too low, 570
---electronic resonance enhancement of $\chi^{(3)}$ smaller for organics than for inorganic multiple quantum well structures, 570
critical
---angle for total internal reflection, 605
---magnetic field of type-II superconductor, H_{c2}, 67
---temperature (T_c), for superconductivity, defined as mid-point of resistance drop, 67
---temperature (T_c), metal-to-supercon-ductor transition, 8, 16, 18
crocetin dialdehyde, 533
crystal electrostatic binding energy *see* Madelung energy
crystal potential, static, 129
crystal structure
---of AETCNQ, 619
---of $(BEDT\text{-}TTF)_2AuBr_2$, 122
---of κ-$(BEDT\text{-}TTF)_2Cu(NCS)_2$, 68, 71, 98, 123
---of α-$(BEDT\text{-}TTF)_2I_3$, 99
---of β-$(BEDT\text{-}TTF)_2I_3$, 99
---of β_d'-$(BEDT\text{-}TTF)_2I_3$, 257
---of $(BEDT\text{-}TTF)_2[KHg(SCN)_4]$, 73, 74
---of $(BEDT\text{-}TTF)_4Pt(CN)_4$, 186, 187

---of $(BEDT-TTF)_3PW_{12}O_{40}$, 186, 187
---of β-$(BEDT-TTF)_2X$, 91
---of BHTCNQ, 619
---of $(DIET)_2(FSO_3)_3$, 178
---of DMAP-C-HMTCAQ, 620
---of DMAP-C-Me, 619
---of $(DMET)_2Au(CN)_2$, 225
---of $(DMET)_2AuBr_2$, 225
---of (HMTSF)(TCNQ), 121
---of $K_4[Pt_2(H_2P_2O_5)_4Cl].3H_2O$, 268
---of Ph-C-BHTCNQ, 620
---of poly(aniline) phases, 339
---of tetrabutylammonium pentacenedi-quinone, 646
---of Tl(TPP)F (X-ray), 219, 220
---of $(TMTSF)_2ClO_4$, 17, 120
---of $(TMTSF)_3W_6O_{19}.2DMF$, 186, 187
---of $(TMTTF)_2SbF_6$ at 135 K (X-ray), 164-167
---of (TTF)(TCNQ), 13
---of $(TTF)_5Pt(CN)_4.2CH_3CN$, 186, 187
---of $(TTF)_3W_6O_{19}$, 186, 187
---of $(TEA)_2(TTF)_6PW_{12}O_{40}$, 186, 187
crystallization, kinetics of, 7
CS_2 (a standard for third-order NLO), 582
Cs_2TCNQ_3
---charge-transfer excitations, 53
---discrete valences in, 11, 52
---low conductivity, 11
---paramagnetic insulator or large-gap semiconductor, 53
---Robin-Day class II, 53
---structure classification, 8
CT *see* charge transfer
CTC *see* charge transfer complexes
Cu $(DM-DCNQI)_2$
---absorption peaks at 10,500 cm^{-1} (CT) 27,000 and 40,000 cm^{-1} (DCNQI), 200
---antiferromagnetic below 5 K, 58
---conductivity maximum (3.5 K, broad)= 500,000 S/cm, 16, 197
---conductivity (300 K)=800 S/cm, 16
---Cu average oxidation number=1.3, 203
---ESR spectra, 197
---infrared spectrum: broad absorption; weak CN around 2100 cm^{-1}; strong peak at 140 cm^{-1}, 198
---magnetic susceptibility, 197
---metallic to 1.3 K, 58, 197
---optical reflectance, 197
---pressure effects, 197
---Raman spectrum: peaks at 2140 cm^{-1} (C≡N stretch), 1615 cm^{-1} (C=C stretch), 1420 cm^{-1} (C≡N stretch), 202
---visible-ultraviolet spectrum, 197
---X-ray photoelectron spectra, 197
Cu(Pc)I (copper phthalocyanine iodide)
---conductivity maximum (>>30 K, broad)=7,000 S/cm, 16
---conductivity (300 K)=800 S/cm, 16
---indirect exchange coupling of the Cu^{++} mediated by π conduction electrons, 61
---ionicity=0.33, 16
---spin glass state below 2 K, 61
Curie paramagnetism
---in $(perylene)_2Au(i\text{-}mnt)_2$, 213
---in poly(aniline), 341
Cu(TCNQ)
---absorption peak at 10,000 cm^{-1} (CT), 200
---phase-change switch, 661
Cu(tetra-*tert* -butylphthalocyanine)
---LB films, in-plane conduction, 561
Cu-TTF-TT [copper (I)/(III) tetrathiafulvalene-tetrathiolate coordination polymer], 112
---conductivity (300 K)=150 S/cm, (4.2 K)=40 S/cm, 113
---Cu polymer less stable than Ni polymer, 111
---thermopower: metallic, 111
Cu-TT [copper (I)/(III) ethylene-tetrathiolate coordination polymer], 111, 112
---conductivity (300 K)=2 S/cm, (~20 K)= 10^{-3} S/cm, 113
---Cu polymer less stable than Ni polymer, 111
---thermopower: metallic, 111
5-cyanoisothiazolo-bis(methylseleno)-TSF, 150
5-cyanoisothiazolo-bis(methylseleno)-TTF, 150
5-cyanoisothiazolo-bis(methyl)-TSF, 150
5-cyanoisothiazolo-bis(methyl)-TTF, 150
5-cyanoisothiazolo-bis(methylthio)-TSF, 150
5-cyanoisothiazolo-bis(methylthio)-TTF, 150
5-cyanoisothiazolo-(dimethylethylene-diselenolo)-TSF, 150
5-cyanoisothiazolo-(dimethylethylenc-diselenolo)-TTF, 150
5-cyanoisothiazolo-(dimethylethylene-dithiolo)-TSF, 150
5-cyanoisothiazolo-(dimethylethyle-nedithiolo)-TTF, 150
(5-cyanoisothiazolo-diselena)-(bis(methyl-thio)-dithia)fulvalene, 150
(5-cyanoisothiazolo-diselena)-(dimethyl-ethylene-dithiolo)-dithia)fulvalene, 150
(5-cyanoisothiazolo-diselena)-dithiafulva-lene, 150
(5-cyanoisothiazolo-diselena)-(ethylene-dithiolo)-dithia)fulvalene, 150
(5-cyanoisothiazolo-diselena)-(iso-propy-lenedithiolo)-dithia)fulvalene, 150
(5-cyanoisothiazolo-diselena)-(methylene-dithiolo)-dithia)fulvalene, 150
(5-cyanoisothiazolo-diselena)-(propylene-dithiolo)-dithia)fulvalene, 150
(5-cyanoisothiazolo-dithia)-(bis(methyl-seleno)-diselena)fulvalene, 150

(5-cyanoisothiazolo-dithia)-(bis(methyl)-diselena)fulvalene, 150
(5-cyanoisothiazolo-dithia)-(2,3-dimethylethylene-diselenolo-diselena)fulvalene, 150
(5-cyanoisothiazolo-dithia)-diselenafulvalene, 150
(5-cyanoisothiazolo-dithia)-(ethylene-diselenolo-diselena)fulvalene, 150
(5-cyanoisothiazolo-dithia)-(iso-propylenediselenolo-diselena)fulvalene, 150
(5-cyanoisothiazolo-dithia)-(methylene-diselenolo-diselena)fulvalene, 150
(5-cyanoisothiazolo-dithia)-(propylene-diselenolo-diselena)fulvalene, 150
1'-cyano-4,5-isothiazolo-1,3-dithiole-2-thione, 147
5-cyanoisothiazolo-(ethylenediselenolo)-TSF, 150
5-cyanoisothiazolo-(ethylenediselenolo)-TTF, 150
5-cyanoisothiazolo-(ethylenedithiolo)-TSF, 150
5-cyanoisothiazolo-(ethylenedithiolo)-TTF, 150
5-cyanoisothiazolo-(iso-propylenediselenolo)-TSF, 150
5-cyanoisothiazolo-(iso-propylenediselenolo)-TTF, 150
5-cyanoisothiazolo-(iso-propylenedithiolo)-TSF, 150
5-cyanoisothiazolo-(iso-propylenedithio lo)-TTF, 150
5-cyanoisothiazolo-(methylenediselenolo)-TSF, 150
5-cyanoisothiazolo-(methylenediselenolo)-TTF, 150
5-cyanoisothiazolo-(methylenedithiolo)-TSF, 150
5-cyanoisothiazolo-(methylenedithiolo)-TTF, 150
5-cyanoisothiazolo-(propylenediselenolo)-TSF, 150
5-cyanoisothiazolo-(propylenediselenolo)-TTF, 150
5-cyanoisothiazolo-(propylenedithiolo)-TSF, 150
5-cyanoisothiazolo-(propylenedithiolo)-TTF, 150
5-cyanoisothiazolo-TSF, 150
5-cyanoisothiazolo-TTF, 150
5-cyanoisothiazolo-(vinylenediselenolo)-TSF, 150
5-cyanoisothiazolo-(vinylenedithiolo)-TTF, 150
cyclam *see* 1,4,8,11-tetraazacyclotetradecane
cyclic voltammogram
---of BDDAP-C-HETCNQ, 618
---of benzoquinone on γ-irradiated poly-(pyrrole), 438
---of dipyrrolopyrazine, 360
---of ferrocene on γ-irradiated poly-(pyrrole), 438
---of poly(cyclohexa[c]thiophene), 364
---of poly(3-methylthiophene), 630, 632
---of poly(pyrrole + 3-hexadecylpyrrole + 3-(13-ferrocenyl,13-hydroxytridecyl)-pyrrole), 545
---of Py-C-HETCNQ, 617
---of (TTF)Br_x powder, 33
cyclodextrins, 660, 664
cyclohepta[c]thiophene, 364
---conductivity of polymer, 366
---oxidation potential, 365
---peak potentials of polymer, 365
cyclohexa[c]thiophene, 364
---conductivity of polmer, 366
---oxidation potential, 365
---peak potentials of polymer, 365
cyclopenta[c]thiophene, 364
---conductivity of polymer, 366
---oxidation potential, 365
---peak potentials of polymer, 365
cyclohexanediamine (chxn), 265
cyclotron
---energy, period, 248
---mass, 236

DBTTF *see* dibenzotetrathiafulvalene
DBTTF-TCNQ
---structure classification, 8
DBTTF-$TCNQF_2$
---structure classification, 8
dcit *see* dithiolates, ---4,5-dithiolato-5-cyano-isothiazole
DCNQI *see* dicyanoquinonediimine
DCNQI anion radical salts, 58
---$Cu(DM\text{-}DCNQI)_2$ metal to 1.3 K, 58, 197
---$Cu(DM\text{-}DCNQI)_2$ antiferromagnetic below 5 K, 58
DDOP-C-BHTCNQ (4-dodecyl-oxy-phenyl-carbamate of 2-bromo, 5-hydroxyethoxy-7,7,8,8-tetracyano-quinodimethane)
---LB films, 614
---oxidation and reduction potentials, 617
---rectification attempts, 622, 623
---structure, 615
DDOP-C-ENP (dodecyloxy-amino-phenyl-carbamate of ethoxy-para-nitrobenzene)
---LB films, 614
---oxidation and reduction potentials, 617
---structure, 615
---Z-type multilayers, 614
DDQ *see* 2,3-dichloro-5,6,-dicyano-1,4-benzoquinone
Debye frequency, 75
Debye-Scherrer powder diffraction lines, 311, 397
decamethylferrocene-tetracyanoethylene
---ferromagnetic below 4.8 K, 60

degenerate Fermi gas, 294
degenerate four-wave mixing, 576
---of poly(3-dodecylthiophene) LB films, 576
---of $Si(Pc)(OSiMePhOH)_2$ LB films, 576
degree of charge transfer *see* ionicity
delocalized charges, 46
delocalized electron states, 129
density of states, 2, 49, 50
depletion layer, 637
derivatized metal electrode, 612, 664
destruction operator, 3
device fabrication, 608
diacetylenes, 407, 614, 663, 664
---solid-state polymerization, 407
---MTDAP-C-ENP, 614, 615, 617
---pFBS [bis-(methylene-*para* -fluorobenzenesulfonate)-diacetylene], 407
---pTS [bis-(methylene-*para* -toluenesulfonate)-diacetylene], 407
---poly-n-BCMU, 568, 574, 575
diamagnetism
---and structure classification, 8
---perfect, in superconductors, 115
2,3-diaminobutane (btn), 266
1,2-diaminoethane (en), 264
1,2-diaminopropane (pn), 264
1,3-diaminopropane (tn), 264
dibenzotetrathiafulvalene (DBTTF), 8
2,3-dichloro-5,6,-dicyano-1,4-benzoquinone (DDQ)
---electron affinity 3.13 eV, 613
---structure, 613
dicyanoquinonediimine (DCNQI)
---good acceptor, 6
---derivative *see* 2,5-DM-DCNQI
---structure, 6, 50
dielectric constant and non-linear optics, 564
dielectric function, 24, 129
---Lorentz model, 24
---of poly(diacetylenes), 407
dielectric, 550
---interface, 607
---loss tangent, 554
Diels Alder adduct polyacenequinones, 643, 644
dien *see* diethylenetriamine
DIET *see* bis-(isopropylenedithiolene)-tetrathiafulvalene
DIET salts:
---with AuI_2^-, BF_4^-, Br_3^-, ClO_4^-, FSO_3^-, HSO_4^-, I_3^-, IO_4^-, NO_3^-, PF_6^-, ReO_4^-, $SO_3CF_3^-$, 177
$(DIET)_2 (BF_4)_3$
---broad metal-to-insulator phase transition around 50 K, 178
---central C=C bond distance=1.292 Å, 177
---charge on cation=+1.5, 177
---chiral donor, 176
---conductivity (10-300 K), 178
---isostructural to FSO_3^- salt, 177
---racemic salt, 179
---thermopower, 179
$(DIET)_2 (FSO_3)_3$
---central C=C bond length=1.322 Å, 177
---charge on cation=+1.5, 177
---chiral donor, 176
---crystal structure, 177, 178
---racemic salt, 179
$(DIET)_x (NO_3)_y$
---conductivity (10-300 K), 179
diethyl alkoxypentacene diquinone, LB films, 645
diethylenetriamine (dien), 264
difference frequency generation, 608
2,5-difluoro-7,7,8,8-tetracyanoquinodimethane ($TCNQF_2$)
---TTF complex, 8
diffuse X-ray scattering (X-ray, neutron), 9, 57
---affected by transverse electrostatic interactions, 54
---4 k_F charge density wave (in Mott-Hubbard semiconductors), 9
---4 k_F in TTF TCNQ, 12
---spots (2 k_F), 9
---streaks (2 k_F), 9
---2 k_F charge density wave (in Peierls semiconductors), 9
---2 k_F in Krogmann's salt, 57
---2 k_F in TMTSF TCNQ, 12
---2 k_F in TTF TCNQ, 12, 57
---2 k_F spin density wave (in spin-Peierls semiconductors), 9
dihydroxycyclohexa-1,3-diene, 382
dimensionality problem and Peierls instability, 13
dimer charge oscillations, 30, 32
dimeric lattice distortion, 131
dimerization gap, 130, 131
dimerized half-filled systems, 24
---donor-acceptor crystals, 34
---ion-radical salts, 34
dimerized mixed stack, 35
---symmetric mode Raman, IR perturbed, 36
dimer model, 129
---symmetric, 27
---unsymmetric (donor-acceptor), 33
dimerized segregated stack, 35
---symmetric mode Raman unperturbed, 36
dimer splitting, 49
DIMET *see* dimethyl(ethylenedithiolo)-tetrathiafulvalene
$(DIMET)_2SbF_6$
---deformed stacks (1 site/cation), 48
---Heisenberg paramagnet, 54, 55
---Hubbard-Peierls gap, 54, 55
---ionicity=0.5 (uniform), 48
---longitudinal conductivity diffusional, 55
---max. spin susceptibility at 60 K, 471
---more strongly dimerized than

$(TMTSF)_2X$ salts, 469
---nearest-neighbor Coulomb repulsion V=0.3, 0.5 eV, 471
---on-site Coulomb repulsion U=1.3 eV, 471
---off-site transfer integrals t=0.10, 0.17 eV, 471
---paramagnetism fits Heisenberg model, 467
---semiconductor, 48, 55
---strongly dimerized, 54
$(DIMET)_2X$
---estimated Hubbard U(solid)=1.5 eV, 58
---estimated Hubbard V(solid)=0.3 eV, 58
2,5-dimethyl-dicyanoquinodiimine (DM-DCNQI)
---Ag salt (1:2), 197
---Cu salt (1:2), 16, 58, 197
---Na salt (1:2), 197
---Raman spectrum: peaks at 2170 cm^{-1} (C-N stretch), 1635 (C=C stretch), 1502 cm^{-1} (C-N stretch), 202
---synthesis, 198
dimethyl(diselena)ethylenedithiolo(dithia)-fulvalene (DMET)
---electron donor, 47
---structure, 47, 223
(dimethylethylenediselenolo-diselena)-(dimethylethylenedithiolo-dithia)fulvalene, 150
(dimethylethylenediselenolo-diselena)dithiafulvalene, 150
dimethylethylenediselenolo-TSF, 150
dimethylethylenediselenolo-TTF, 150
dimethylethylenedithiolo-bis(methylthio)TTF (DMEDT-BMT-TTF), 182
dimethylethylenedithiolo-dimethylethylenediselenolo-TTF, 150
(dimethylethylenedithiolo)diselena)-(dimethylethylenediselenolo)diselena)-fulvalene, 150
(dimethylethylenedithiolo-diselena)-(dimethylethylenediselenolo-dithia)fulvalene, 150
(dimethylethylenedithiolo-diselena)-dithiafulvalene, 150
(dimethylethylenedithiolo-dithia)-(dimethylethylenediselenolo-diselena)fulvalene, 150
dimethyl(ethylenedithiolo)-tetraselenafulvalene, 150
dimethyl(ethylenedithiolo)-tetrathiafulvalene (DIMET)
---electron donor, 47
---structure, 47, 150
N,N'-dimethylphenazine TCNQ, 37
---polarized infrared spectrum, 38
---stack dimerization, 37
dimethylpyrazino-bis(methylseleno)-TSF, 150
dimethylpyrazino-bis(methylseleno)-TTF, 150
dimethylpyrazino-bis(methyl)-TSF, 150
dimethylpyrazino-bis(methyl)-TTF, 150
dimethylpyrazino-bis(methylthio)-TSF, 150
dimethylpyrazino-bis(methylthio)-TTF, 150
dimethylpyrazino-(dimethylethylenediselenolo)-TSF, 150
dimethylpyrazino-(dimethylethylenediselenolo)-TTF, 150
dimethylpyrazino-(dimethylethylenedithiolo)-TSF, 150
dimethylpyrazino-(dimethylethylenedithiolo)-TTF, 150
(dimethylpyrazino-diselena)-(bis(methylthio)-dithia)fulvalene, 150
(dimethylpyrazino-diselena)-(dimethylethylene-dithiolo)-dithia)fulvalene, 150
(dimethylpyrazino-diselena)-dithiafulvalene, 150
(dimethylpyrazino-diselena)-(ethylenedithiolo)-dithia)fulvalene, 150
(dimethylpyrazino-diselena)-(iso-propylenedithiolo)-dithia)fulvalene, 150
(dimethylpyrazino-diselena)-(methylenedithiolo)-dithia)fulvalene, 150
(dimethylpyrazino-diselena)-(propylenedithiolo)-dithia)fulvalene, 150
(dimethylpyrazino-dithia)-(bis(methyl)-diselena)fulvalene, 150
(dimethylpyrazino-dithia)-(bis(methylseleno)-diselena)fulvalene, 150
(dimethylpyrazino-dithia)-(dimethylethylene-diselenolo)-diselena)fulvalene, 150
---2:1 salt with I_3^-: metallic, metal-to-insulator transition at ≈ 120 K, 155
(dimethylpyrazino-dithia)-diselenafulvalene, 150
(dimethylpyrazino-dithia)-(ethylenediselenolo)-diselena)fulvalene, 150
(dimethylpyrazino-dithia)-(iso-propylenediselenolo)-diselena)fulvalene, 150
(dimethylpyrazino-dithia)-(methylenediselenolo)-diselena)fulvalene, 150
(dimethylpyrazino-dithia)-(propylenediselenolo)-diselena)fulvalene, 150
1',2'-dimethyl-4,5-pyrazino-1,3-dithiole-2-thione, 147
dimethylpyrazino-(ethylenediselenolo)-TSF, 150
dimethylpyrazino-(ethylenediselenolo)-TTF, 150
---2:1 TCNQ complex: semiconductor, 155
dimethylpyrazino-(ethylenedithiolo)-TSF, 150
dimethylpyrazino-(ethylenedithiolo)-TTF, 150
dimethylpyrazino-(isopropylenediselenolo)-TSF, 150
dimethylpyrazino-(iso-propylenediselenolo)-TTF, 150

dimethylpyrazino-(iso-propylenedithiolo)-TSF, 150
dimethylpyrazino-(iso-propylenedithiolo)-TTF, 150
dimethylpyrazino-(methylenediselenolo)-TSF, 150
dimethylpyrazino-(methylenediselenolo)-TTF, 150
dimethylpyrazino-(methylenedithiolo)-TSF, 150
dimethylpyrazino-(methylenedithiolo)-TTF, 150
dimethylpyrazino-(propylenediselenolo)-TSF, 150
dimethylpyrazino-(propylenediselenolo)-TTF, 150
dimethylpyrazino-(propylenedithiolo)-TSF, 150
dimethylpyrazino-(propylenedithiolo)-TTF, 150
dimethylpyrazino-TSF, 150
dimethylpyrazino-TTF, 150
dimethylpyrazino-(vinylenediselenolo)-TSF, 150
dimethylpyrazino-(vinylenedithiolo)-TTF, 150
dimethyltetracyanoquinodimethane (DMTCNQ)
---HMTTeF complex (1:1), 16
---structure, 6
---TMTSF complex (1:1), 16
dimethyl-tetrathiafulvalene-1-thia-heptadecanoic acid, 529, 530
dimethyltetrathiotetracene (DMTTT), 191, 192
---lower oxidation potential than TTT, 192
---optical absorptions at 470, 635, 693 nm, 192
---radical cation; ESR : 11 lines, g=2.00778, splitting 0.63 Gauss, 193
---radical cation: absorbs at 463, 503, and 591 nm, 189
---synthesis, 192
di(methylthio)dithiole $[Ni(dmit)_2]_2$
---semiconductor, 155
dimethyltrimethylenetetraselenafulvalene (DMtTSF), 55
dimethyltrimethylenetetrathiafulvalene (DMtTTF), 55
L-α-dimyristoyl phosphatidic acid
---area=135 Å^2 at gas/liquid expanded transition, 495
---fluorescence micrographs are homogeneous for gas or liquid expanded regions, 495
---fluorescence micrographs show dark regions (liquid compressed) which grow and move closer, as area decreases, until dark regions touch and merge into solid compressed region, 496, 497
---gel area=45.5 Å^2/molecule, 499
---monolayers at air-water interface, 492, 493, 495
---phase-separated structures due to mutual electrostatic interactions in the high-density domains, 499
---pressure-area isotherms at different T, 492, 495
---pressure=8.5 mN/m, area=80 Å^2 at liquid expanded/liquid compressed transition, 495
---pressure=48 mN/m, area=41.5 Å^2 at liquid compressed/solid transition, 495
Dingle temperature, 236
7-N,N-dioctylaminophenyl-7,8,'-diapo-carotene- 8'-al
---all-trans form predominates, 20:1, 534
---area per molecule=56 Å^2, 535
---collapse pressure=18 mN/m, 535
---^{1}H and ^{13}C nuclear magnetic resonance spectra, 533, 534
---LB films, transfer ratio=0.8, 535
---pressure-area isotherm, 536
---synthesis, 532, 533
diode laser (GaAlAs, 830 nm), 482, 570
dioxygen trap, 514-516
DIP *see* pyranylidenes
dipalmitoyl-nitrobenzoxadiazol-phosphatidylethanolamine, 494
diphenylpicrylhydrazyl (DPPH)
---ESR of, 218
---as sterically hindered stable radical, 4
diphosphite (pop), $H_2P_2O_5^{--}$, 263
diphosphite linear-chain platinum complexes, 267
DIP(S)=dithiapyranylidene, 59, 62
---structure, 47
DIP(S)-TCNQ
---good conductor at 300 K, 62
---lamellar structure above 123°C, 62
dipyrrolopyrazine, 360
dipyrrolyl polymers, 359
---conductivity, 361
---redox potential, 361
discotic liquid crystal, 61
discrete valences
---in Cs_2TCNQ_3, 11
---prevent metallic conduction, 11
dispersion relation, 49
---illustrations, 2, 9, 50
disproportionation, 5
dithiapyrene, fatty acid derivative, 529, 530
dithieno[3,2-b::2',3'-d]thiophene
---oxidation potential (solution), 412
---single-double bond distance difference, 412
dithieno[3,4-b::3',4'-d]thiophene
---oxidation potential (solution), 412
---single-double bond distance difference, 412
dithieno[3,2-b::2',3'-d]thiophene:TCNQ complex
---single-double bond distance difference, 412

2,3-dithio-dipyrazine, graphite intercalate ($DTDP_xC_y$): unit cell data; staging C-C-C-DTDP-C-C-C; pressed pellet conductivity=0.189 S/cm, 285-286
dithiolates, 45
---1,2-dithiolate-type, structure, 46
---4,5-dithiolato-5-cyano-isothiazole, (dcit), structure, 156
---4,5-dithiolato-1,3-dithiol-2-thione (dmit), *or* 4,5-dimercapto-1,3-dithiol-2-thione, structure, 46, 156, 519
---iso-maleonitriledithiolate (i-mnt), structure, 211
---maleonitriledithiolate (mnt), structure, 211
---(TTF)Ni$(dmit)_2$, superconductor, 116
DMAP-C-Me (N,N-dimethylaminophenyl carbamate, methyl ester)
---crystal structure, 619
---oxidation potentials, 617
---structure, 616
DM-DCNQI *see* 2,5-dimethyl-dicyanoquinodiimine
DMEDT-BMT-TTF *see* dimethylethylene-dithiolo-bis(methyl-thio)TTF
$(DMEDT\text{-}BMT\text{-}TTF)_xIBr_2$
---black needles, 183
$(DMEDT\text{-}BMT\text{-}TTF)_x$TCNQ
---black needles, 183
DMET *see* dimethyl(diselena)ethylene-dithio(dithia)fulvalene
$(DMET)_2AuBr_2$
---crystal structure not columnar, resembles that of κ-$(BEDT\text{-}TTF)_2Cu(NCS)_2$, 225
---semiconductor, 150 K to 300 K; metal below 150 K; superconductor (T_c=1.9 K at 1 bar, or at high pressure), 224
---superconductor, 157
$(DMET)_2AuCl_2$
---conductivity similar to $Au(CN)_2^-$ salt, 224
$(DMET)_2Au(CN)_2$, 57
---crystal structure similar to that of $(TMTSF)_2X$ salts, 224, 225
---long SDW-superconductivity boundary, 224
---temperature-pressure phase diagram similar to that of $(TMTSF)_2X$ salts, 224
---resistive upturn at 28 K and 1 bar, superconductor at high pressure, 223
$(DMET)_2AuI_2$
---crystal structure, 225
---conductivity similar to $Au(CN)_2^-$ salt, 224
$(DMET)_2I_3$
---crystal structure resembles that of β-$(BEDT\text{-}TTF)_2X$ salts, 224-226
---metal to low temperature, superconductor, T_c=0.47 K at 1 bar, 145, 224
$(DMET)_2IBr_2$
---crystal structure, 225
---conductivity similar to I_3^- salt, 224
---superconductor, 145
$(DMET)_2X$ salts
---more 2-dimensional than the $(BEDT\text{-}TTF)_2X$ salts, 57
---seven organic superconductors: $Au(CN)_2^-$, $AuCl_2^-$, AuI_2^-, I_3^-, IBr_2^-, $AuBr_2^-$ (two forms), 57, 223
dmit *see* dithiolate, --- 4,5-dithiolato-1,3-dithiol-2-thione
DMTCNQ *see* dimethyltetracyanoquinodimethane
DMtTSF *see* dimethyltrimethylenetetraselenafulvalene
$(DMtTSF)_2X$ salts
---longitudinal conduction is coherent, 55
---metallic to 4 K, 55, 58, 59, 145
---no cooperative ground state at low T, 58
DMTTT *see* dimethyltetrathiotetracene
(DMTTT)Cl
---ESR: single anisotropic line, 194
(DMTTT)I
---conductivity (pressed pellet, 300 K)=0.001 S/cm, 194
---ESR: single anisotropic line, 194
$(DMTTT)_2I_3$
---dark green with golden luster, 194
---ESR: single symmetrical line, 194, 195
---synthesis, 194
DMtTTF *see* dimethyltrimethylenetetrathiafulvalene
$(DMtTTF)_2X$ salts
---conductor, 55
---Hubbard gap, 55
---longitudinal conduction is diffusional, 55
---Pauli paramagnetism, 55
---weakly dimerized, 55
2-docosylamino-5-nitropyridine
---effective nonlinear optical suceptibility 25% that of $LiNbO_3$, 601
---Y-type Langmuir-Blodgett multilayers oriented with dipoles in layer plane so that optical second harmonic generation (2 x 10^{-7} efficiency) occurs from 540 monolayers, 600
(N-docosylpyridinium)(TCNQ) (NDP^+ $TCNQ^-$)
---monolayer (Pockels-Langmuir) at air-water interface: plateau in pressure-area isotherm at 10-16 mN/m: irreversible 2D-3D phase transition causes bilayers to form: transfer as LB film (at 35 mN/m) shows (by EPR and linear di-chroism) TCNQ planes to be parallel to substrate: full charge-transfer insulator. Exposure of LB film to I_2 vapor incorporates I_3- in film, partially oxidizes TCNQ, and reorients TCNQ perpendicularly to substrate, 504
(N-docosylpyridinium)(TCNQ) I_x semiconductive LB films, 62, 520

(N-docosylpyridinium)(TCNQ) I_x
semiconductive LB films *(cont'd)*
---activation energy=0.15 eV (dc) 0.08 eV (microwave), 506, 520
---bulk conductivity=0.01 or 0.1 S/cm, 504, 520
---carrier mobility=1 cm^2/V s, 506
---conductivity low or zero for 1 to 4 monolayers (lack of overlap of conducting platelets), 506
---conductivity is sensitive to oxidizing or reducing atmosphere: gas sensor, 506
---stable, even in vacuo, for months, 506
---in-plane conductivity is isotropic, 506
---optical micrograph, 505
---structure, 505
(N-docosylpyridinum) $(TCNQ)_n$
semiconductive LB films (n=2), 520
---activation energy=0.01 eV (n=2), 520
---bulk conductivity=0.01 S/cm (n=2), 520
---conductivity drops by 50% on electron irradiation (3 kV, 0.05 C/m^2),but it can be restored by iodination, 506
---conductivity is sensitive to oxidizing or reducing atmosphere: gas sensor, 506
---less than 10 LB monolayers: insulator, 506
---semiconducting LB films without iodination (n=1.4 to 2), 506
---sensitivity to phosphine: 0.5 ppm, 506
(N-docosyltrimethylammonium)(TCNQ), LB films, 506
donors *see* electron donors
donor-acceptor pairs, 33
---degree of ionicity, 34
donor-pi-acceptor (D-π-A) molecules
---as artificial photsynthetic systems, 662
---for non-linear optics (β or $\chi^{(2)}$), 565
---for organic devices: R-Py-CH=C(CN)-ϕ-$C(CN)_2$, 649
---for organic devices: R-Qn-CH=C(CN)-ϕ-$C(CN)_2$, 649
donor-sigma-acceptor (D-σ-A) molecules as artificial photosynthetic systems, 662
donor-sigma-acceptor (D-σ-A) molecules for organic rectifiers
---DDOP-C-BHTCNQ, 614, 615
---DDOP-C-ENP, 614, 615
---BDDAP-C-BHTCNQ, 614, 615
---BDDAP-C-HETCNQ, 617, 618
---BDDAP-C-HMTCAQ, 614, 615
---BHAP-C-HMTCAQ, 614, 615
---MTDAP-C-ENP, 614, 615
---Py-C-BHTCNQ, 614, 615
---Py-C-HECTNQ, 617
---TDDOP-C-ENP, 614, 615
---TDDOP-C-HETCNQ, 614, 615
---TTF-C-BHTCNQ, 614, 615
doping level, polymer, 294, 298, 361
D-π-A *see* donor-pi-acceptor molecules
DPPH *see* diphenylpicrylhydrazyl
drain conductance, 637
drain current, 637
Drude absorption, 298
Drude model, 132
Drude-Lorentz model, 55
D-σ-A *see* donor-sigma-acceptor molecules
Durham/Graz synthesis of polyacetylene, 415
Durham synthesis of polyacetylene, 295

E_A *see* electron affinity (of acceptor)
EB *see* emeraldine base*and* poly(aniline)
EDS-DMPyz-TTF *see* ethylenediselenolo-dimethypyrazino-TTF
$(EDS\text{-}DMPyz\text{-}TTF)_xI_3$
---golden chips or bronze plates, 183
$(EDS\text{-}DMPyz\text{-}TTF)_xIBr_2$
---black plates or copper chips, 183
$(EDS\text{-}DMPyz\text{-}TTF)_2TCNQ$
---dark-blue plates, 183
$(EDS\text{-}DMPyz\text{-}TTF)_xAuI_2$
---black plates, 183
EDS-DSDTF *see* (ethylenediselenolo-diselena)dithiafulvalene
$(EDS\text{-}DTDSF)_xAuI_2$
---black microcrystals, 183
$(EDS\text{-}DSDTF)_xI_3$
---golden needles or black crystals, plates, 183
$(EDS\text{-}DSDTF)_xIBr_2$
---black needles, 183
$(EDS\text{-}DSDTF)_xTCNQ$
---black microcrystals, 183
EDS-DTDSF *see* (ethylenediselenolo-dithia)diselenafulvalene
$(EDS\text{-}DTDSF)_xI_3$
---bronze needles, 183
EDS-MDT-TTF *see* ethylenediselenolo-methylenedithiolo-TTF
$(EDS\text{-}MDT\text{-}TTF)_xI_3$: needles, 183
$(EDS\text{-}MDT\text{-}TTF)_xTCNQ$
---black microcrystals, 183
EDS-Pyz-TTF *see* ethylenediselenolo-pyrazino-TTF
$(EDS\text{-}Pyz\text{-}TTF)_xBr_3$: golden brown, 183
$(EDS\text{-}Pyz\text{-}TTF)_xI_3$: golden needles, 183
$(EDS\text{-}Pyz\text{-}TTF)_xIBr_2$
---brown microcrystals, 183
EDS-TTF *see* ethylenediselenolo-TTF
$(EDS\text{-}TTF)_xI_3$: black microcrystals, 183
$(EDS\text{-}TTF)_xIBr_2$
---golden microcrystal or black, golden plates, 183
$(EDS\text{-}TTF)_xTCNQ$
---black microcrystals, 183
EDT-DTDSF *see* (ethylenedithiolo-dithia)diselenafulvalene
$(EDT\text{-}DTDSF)_2IBr_2$
---black needles, plates, 183
---conductivity (1.35 K) > 10,000 S/cm, E166
---metallic down to 1.35 K, E166

EDT-Thz-TTF *see* ethylenedithiolo-thiazolo-TTF
(EDT-Thz-TTF) $_xBr_3$
---black microcrystals, 183
EDT-TTF *see* ethylenedithiolo-tetrathia-fulvalene; *see also* ethylenedithiolo-di(octadecylthio)-tetrathiafulvalene
$(EDT-TTF)_2Ag(CN)_2$
---black needles, 183
$(EDT-TTF)_2AuBr_2$
---black plates, metallic to low temperature, 183
$(EDT-TTF)_2IBr_2$
---conductivity (1.35 K)>10,000 S/cm, 156
---metallic down to 1.35 K, 156
$(EDT-TTF)_2Ni(dcit)_2$
---black microcrystals, 183
effective mass of charge carrier, 77, 294, 336
eicosanoic acid, LB films, 521
electric dipole moment operator, 24
electrical conductivity *see* conductivity, electrical
electrical polarization, 564, 608
electric susceptibility and non-linear optics, 564
electroactive polymer, 365
electrochemical polymerization, 359, 364, 654
electrochromic device, 654
electroluminescent diode
---galllium phosphide, enhanced by LB film, 476
electrolytic polymerization, 654
---polymer has lower redox potential than monomer, 654
electron acceptors (π), 6, 46, 613
---BHTCNQ, 613
---bromanil, 46
---chloranil, 10, 45, 46, 613
---DCNQI, 6, 45, 46
---DDQ, 613
---1,2-diheterolenes, 146
---1,2-dithiolates, 45, 46
---DMTCNQ, 6, 16
---DM-2,5-DCNQI, 16
---fluoranil, 46
---HETCNQ, 613
---HMTCAQ, 616
---pentacenediquinone, 645
---TCNQ, 4-6, 459, 46, 613
---$TCNQF_2$, 8
---$TCNQF_4$, 8
---TNAP, 6
electron affinity (of acceptor) (A_A or E_A), 10, 44, 611-613, 651
---chloranil, 613
---DDQ, 613
---TCNQ, 613
electron beam lithography, 474
electron correlation, 129
electron diffraction, 397, 398
electron donors (π), 6, 47
---BEDO-TTF, 145
---BEDS-TTF, 47
---BEDT-TSF, 47
---BEDT-TTF or ET, 6, 47, 145
---corbien, 529, 530
---DIET, 176
---DIMET, 47
---DMET, 47, 57, 145
---dithiapyrene, 529, 530
---DMtTSF, 55
---DMtTTF, 55, 145
---DMTTT, 191
---EDT-DSDTF, 145
---EDT-TTF, 145
---ET *see* BEDT-TTF
---fluoranthene, 211
---2-FTST, 191
---HMTSF, 6
---HMTTeF, 6
---HMTTF, 5, 6
---MDT *see* MDT-TTF
---MDT-TTF, 57, 145, 227
---metal phthalocyanines M(Pc), 6, 47
---metal porphyrins M(Por), 47
---perylene, 6, 47
---precursors to tetraheterafulvalenes *see* 2-thioxo-1,3-dithioles, 143-147
---pyranylidenes, 47
---pyrene, 47, 613
---TMDTDSF, 85, 86
---TMET, 176
---TMPD, 38, 613
---TMTSF, 6, 47, 145
---TMTTF, 6, 47
---TSF, 6, 47
---TTeF, 6, 47
---TTF, 5, 6, 47, 613
---TST, 6, 47
---TTeT, 6
---TTT, 6, 47, 191
electron-electron interactions, 129
---comparable to one-electron interactions, 23
electronic susceptibility, 30
electron-lattice interactions, 129
electron-molecular vibration coupling, 133
electron-phonon coupling parameter, 93, 133, 136
electron spin resonance (ESR)
---linewidths affected by transverse transfer integral, 54
---of α'-$(BEDT-TTF)_2Ag(CN)_2$, 170, 171
---of $α_t$-$(BEDT-TTF)_2I_3$, 101, 107
---of β-$(BEDT-TTF)_2AuI_2$, 170, 171
---of β''-$(BEDT-TTF)_2AuBr_2$, 170, 171
---of β'-$(BEDT-TTF)_2AuCl_2$, 170, 171
---of $β_d$'-$(BEDT-TTF)_2I_3$, 259
---of κ-$(BEDT-TTF)_2Ag(CN)_2.H_2O$, 170, 171
---of κ-$(BEDT-TTF-h_8)Cu(NCS)_2$, 80

electron spin resonance (ESR) *(cont'd)*
---of $(DIMET)_2SbF_6$, 467
---of (DMTTT)Cl, 194
---of (DMTTT)I, 194
---of $(DMTTT)_2I_3$, 194
---of $K_{0.3}MoO_3$, 279
---of metalloporphyrins, 217, 221
---of $(perylene)_2Au(i\text{-}mnt)_2$, 213
---of poly(aniline), emeraldine base, 339
---of poly(aniline), emeraldine salt, 331
---of poly(aniline) in zeolites (g=2.0034, linewidths 8-10 Gauss), 425
---of poly(pyrrole) in zeolites (g=2.0027, linewidths 8-10 Gauss), 425
---of $[Pt(en)_2][Pt(en)_2Cl_2](ClO_4)_4$, 274, 275
---of (TTT)Cl, 194
---of $(TTT)_2I_3$, 194
electro-optic effect (Pockels effect), 608
---measurement, 609
Elliott relation for spin relaxation in 2-D metals, 79, 172, 213
ellipsometry, 609
en *see* 1,2-diaminoethane
emeraldine base (EB) *see* poly(aniline)
emeraldine salt (ES) *see* poly(aniline)
energy bands
---one-dimensional, 2, 9, 49, 50
enzyme, as polymeric anionic dopant in conducting polymers (for catalysis, sensors), 654
EPR *see* electron spin resonance
ESR *see* electron spin resonance
ET *see* BEDT-TTF
1-ethyl-4-carbomethoxypyridinyl, 4
ethylenediamine *see* 1,2-diaminoethane
ethylenediselenolo-dimethypyrazino-TTF (EDS-DMPyz-TTF), 182
(ethylenediselenolo-diselena)-dithiafulvalene (EDS-DSDTF), 150
(ethylenediselonolo-dithia)diselenafulvalene (EDS-DTDSF), 150
(ethylenediselenolo-diselena)-(ethylenedithiolo-dithia)fulvalene, 150
ethylenediselenolo-methylenedithiolo-TTF (EDS-MDT-TTF), 182
ethylenediselenolo-pyrazino-TTF (EDS-Pyz-TTF), 182
ethylenediselenolo-TSF, 150
ethylenediselenolo-TTF (EDS-TTF), 150
(ethylenedithiolo-diselena)-(ethylenediselenolo-diselena) fulvalene, 150
(ethylenedithiolo-diselena)-(ethylenediselenolo-dithia)fulvalene, 150
(ethylenediselenolo-diselena)-(ethylenedithiolo-dithia)fulvalene, 150
(ethylenedithiolo-diselena)dithiafulvalene, 150
(ethylenedithiolo-dithia)-(ethylenediselenolo-diselena)fulvalene, 150
ethylenedithiolo-ethylenediselenolo-TTF, 150
ethylenedithiolo-tetraselenafulvalene, 150
ethylenedithiolo-tetrathiafulvalene (EDT-TTF), 145, 150
(ethylenedithiolo-dithia)diselenafulvalene (EDT-DTDSF), 145
[ethylenedithiolo-bis(hexadecylthio)-TTF] $(TCNQF_4)I_x$, LB film semiconductor, 520
---conductivity=0.25 S/cm, 520
ethylenedithiolo-bis(octadecylthio)-tetrathiafulvalene (EDT-TTF)
---films at air-water interface (Pockels-Langmuir) is unstable, 507
---structure, 505
[ethylenedithiolo-bis(octadecylthio)-tetrathiafulvalene]$(TCNQF_4)I_x$, LB film semiconductor, 507, 508, 520
---activation energy=0.08 eV, 520
---conductivity=0.25 S/cm, 520
---conductivity=0.05 S/cm after I_2 doping, 508
---IR spectrum: $TCNQF_4^-$ only, even after I_2 doping, 508
---monolayer films stable, LB films with few defects, 507
[ethylenedithiolo-bis(octadecylthio)-tetrathiafulvalene]$(TCNQF_4)I_x$, mixed 1:1 with ω-tricosenoic acid: LB film semiconductor, 508, 520
---activation energy=0.08 eV, 508
---conductivity=0.2 S/cm, 508
ethylenedithiolo-thiazolo-TTF (EDT-Thz-TTF), 182
ethylene-tetrathiolate (TT)
---coordination polymer with Ni (Ni-TT), Cu (Cu-TT), 111
---structure, 111
excitonic superconductivity model (Little), 53, 59, 85
excitons
---in conducting polymers, 415
---intramolecular, 43
extended Hubbard model, 51
see also Hubbard
---charge, spin excitations decoupled, 51
---comparison with Pariser-Parr-Pople molecular orbital theory, 446
---diagrammatic valence bond treatment (12 site-cluster): intradimer energy= 0.079-0.084 eV, interdimer energy= 0.014-0.016 eV, 467, 471
---diagrammatic VB for half-filled band: site-CDW, bond-CDW, and SDW instabilities, 443
---estimates of U and V for TTF, TMTTF, DIMET, BEDT-TTF, and TCNQ, 58
---half-filled system: V can drive ground state from bond-CDW (or, equivalently, SDW) to site-CDW, 446
---half-filled system: X keeps system in bond-CDW (or SDW) gound state, 446
---Hubbard gap, 51

extended Hubbard model *(cont'd)*
---intrasite Coulomb repulsion U, 25, 51, 490
---nearest-neighbor transfer integral t, 25, 51, 468
---nearest-neighbor electron-adjacent bond attraction X (<0), 442
---nearest-neighbor Coulomb repulsion V, 51, 442, 468
---one-electron site energy ε, 51
---strong-coupling limit, 51
---weak-coupling limit, 51
---zero-bandwidth and V=0 limit: Heisenberg spin hamiltonian, 51
extended Hückel
---calculations for $(TMTTF)_2$ dimer, 164-166
---method of Hoffmann, 49
---theory, applied to 1-D band, 49
---tight-binding band calculations, 72, 73
external reflection absorption spectroscopy, 402

Fabry-Perot etalon, 569
---shows dispersive bistability for poly(imide), 569
Fano structure in infrared spectrum, 507
Fano-type antiresonance
---in quasi-one-dimensional charge-transfer salts, 299
---none in polyacetylene, 299
1,2-FeNc (iron 1,2-naphthalocyanine)
---conductivity=4 x 10^{-9} S/cm, 460
2,3-FeNc (iron 2,3-naphthalocyanine)
---conductivity=4 x 10^{-5} S/cm, 460
FeOCl
---poly(pyrrole) layered intercalate, 421
---poly(thiophene) layered intercalate, 421
FePc (iron phthalocyanine)
---conductivity=4 x 10^{-9} S/cm, 460
FePhc (iron phenanthrenocyanine)
---conductivity=8 x 10^{-9} S/cm, 460
Fermi
---energy E_F, 2, 89
---gas, degenerate, 294
---glass, 294
---level, 2, 248
---liquid, 87
---surface, 16, 86, 88, 118, 130
---velocity, 248, 294
---wavevector k_F, 2, 8, 131, 248
---wavevector k_F and ionicity, 9, 50
fermion, regular vs. spinless, 133
ferrite, 661
3-(13-ferrocenyl,13-hydroxytridecyl) pyrrole (Fc-Py) and 3-hexadecylpyrrole (3HDP), (1:1 to 1:20 ratio) LB films formed as polymer with $FeCl_3$, 543
ferroelectric liquid crystals
---can be switched much faster than smectic or nematic liquid crystals, 476
---good pyroelectric coefficient, 551
---large dielectric loss, 551
ferromagnet, organic *see* organic ferromagnet
FET *see* transistor, field effect
FISDW *see* spin density wave state, induced by magnetic field
---caused by quasi-two-dimensional open-orbit Fermi surface, 245
---complex Hall effect, 245
---in $(TMTSF)_2ClO_4$, 245
---second-order transition between metal and SDW state, 245
---weak first-order transitions between SDW states, 245
para -fluoranil
---structure, 46
$(fluoranthene)_2PF_6$
---ESR: linewidth=0.010 gauss, 211
fluorescence microscopy of monolayers, 492
fluoro-gallio-octamethylporphyrin [Ga(OMP)F], EXAFS, 218
---EPR spectrum after iodination, 221
---partial oxidation with iodine, 220
---Raman spectrum after iodination, 221
fluoro-gallio-octaethylporphyrin [Ga(OEP)F], 217
---EPR spectrum after iodination, 221
---infrared spectrum, 220
---NMR spectrum, 219
---partial oxidation with iodine, 220
---Raman spectrum after iodination, 221
fluoro-gallio-tetraphenylporphyrin [Ga(TPP)F], 218
---EPR spectrum after iodination, 221
---infrared spectrum, 220
---partial oxidation with iodine, 220
---Raman spectrum after iodination, 221
fluoro-thallio-octaethylporphyrin [Tl(OEP)F], NMR of, 219
fluoro-thallio-tetraphenylporphyrin [Tl(TPP)F], NMR of, 219
---crystal structure, 219, 220
2-fluorotetraselenotetracene, 191
flux exclusion in superconductors, 115
Fourier transform infrared (FTIR) spectra *see* infrared spectra
fractional charge transfer *see* ionicity, partial
Franck-Condon factor, 29, 32, 242, 613, 614, 662
free-electron approximation, 2
free radical, stable, 3
Fresnel's equation, 607
Friedel-Crafts reaction, 381
Fröhlich mode, 5, 277
FTIR *see* Fourier transform infrared
2-FTST *see* 2-fluorotetraselenotetracene
$(2\text{-FTST})_2Br$
---conductivity rises smoothly from 300 K to 4.2 K, 191
furan, 448

Gallium aluminium arsenide, diode laser, 482, 570
gallium arsenide, 611, 661
gallium phosphide
---electroluminescence enhanced by 8 LB monolayers of ω-tricosanoic acid, 476, 477
gamma irradiation
---of poly(acrylonitrile), 435
---of poly(dimethyldiallyl ammonium chloride), 435
---of poly(ethylenimine), 435
---of poly(pyrrole), 435-439
---of poly(thiophene hexafluorosulfate), 435
---of potential graft co-polymer of pyrrole and 4-vinylpyridine, 435
Ga(OMP)F *see* fluoro-gallio-octamethylporphyrin
Ga(OEP)Cl *see* chloro-gallio-octaethylporphyrin
Ga(OEP)F *see* fluoro-gallio-octaethylporphyrin
Ga(OEP)I *see* iodo-gallio-octaethylporphyrin
Ga(OMP)F *see* fluoro-gallio-octamethyl-porphyrin
Ga(Por)X *see* halo-gallioporphyrin
Ga(TPP)F *see* fluoro-gallio-tetraphenyl-porphyrin
gel electrolyte, 639
gerade symmetry, 35
Ginzburg-Landau coherence length
---κ-$(BEDT\text{-}TTF)_2Cu(NCS)_2$, 77
---α_t-$(BEDT\text{-}TTF)_2I_3$, 102
glassy resistivity, 100
glass transition temperature, 310, 639
g-ology models, 51
---with bipolarons, 59
graphite, 175
---pressed pellet conductivity=0.752 S/cm, 287
---intercalated compounds, 285-287
---intercalate with 2,3-dithio-dipyrazine ($DTDP_xC_y$): unit cell data; staging C-C-C-DTDP-C-C-C; pressed pellet conductivity=0.189 S/cm, 285-287
---intercalate with perylene (Per_xC_y): unit cell data; staging C-C-Per-C-TT-C-C; pressed pellet conductivity=0.087 S/cm, 286, 287
---intercalate with tetracene ($TT_{1.0}C_{60}$): unit cell data; staging C-C-C-TT-C-C-TT-C-C-C; pressed pellet conductivity=0.106 S/cm, 286, 287
---intercalate with 1,3,10,12-tetrachloro-tetracene (TCT_xC_y): unit cell data; staging C-C-C-TCT-C-C-C; pressed pellet conductivity=0.084 S/cm, 285-287
---intercalate with tetracyanotetrathiafulvalene ($TCN\text{-}TTF_xC_y$): unit cell data; staging C-C-C-TCN-TTF-C-C-C; pressed pellet conductivity=0.288 S/cm, 286, 287
---organic intercalates mostly small-ring, 285
---staging in intercalates, 286

H-aggregate, 512, 513
half-filled
---band, dimer model, 137
---systems, 34, 129
Hall effect, quantized
---in $(TMTSF)_2PF_6$, 85, 87
---in $(TMTSF)_2ClO_4$, 88
halobacterium halobium, 663
halo-gallio-porphyrin [Ga(Por)X], 217
---Ga(OMP)Cl, 217
---Ga(OMP)F, 217, 219
---Ga(OMP)I, 218
---Ga(OEP)Cl, 219
---Ga(OEP)F, 219
---Ga(TPP)Cl, 219
---Ga(TPP)F, 219
halo-indio-porphyrin [In(Por)X], 217
---In(Por)F, NMR of, 219
halo-thallio-porphyrins [Tl(Por)X], 217
---Tl(TPP)F, 219
Hamiltonian
---eigenvalues, 24
---Heisenberg spin, 51
---extended Hubbard, 51, 443, 468
---Hubbard, 1, 3, 25, 442
---for electron transfer, 130
---for dimerized chain, 24, 132
harmonic generation
---second and third, 608
heat capacity, 550
---by microcalorimetry, 246
Heisenberg
---paramagnetism in $(DIMET)_2SbF_6$, 561
---spin hamiltonian, 51
---exchange integral, 51
Hellmann-Feynman theorem, 30
hemicyanine-nitrostilbene alternate-layer LB film: second-order electric susceptibility≈7 x 10^{-7} esu, 481
$(heptadecylTTF)I_x$, LB film, 520
(heptadecylDMTTF)(TCNQ) with (TMTTF)(hexadecylTCNQ) LB film semiconductor, 520
herringbone motif in TTF TCNQ, 12
Herzberg-Teller coupling, 29, 30, 242
HETCNQ *see* 2-hydroxyethoxy-7,7,8,8-tetracyanoquinodimethane
heterostructures, 573
hexachloroplatinate, anionic dopant in conducting polymers (for electro-chromism, dispersed noble metal, catalysis), 654
(hexadecanoylTTF) I_x, LB film semiconductor, 520

3-hexadecylpyrrole (3HDP), LB monolayer
---area per molecule=22 Å^2, 539
3-hexadecylpyrrole (3HDP) and pyrrole (Py), mixed in 1:500 ratio at air-water interface with $FeCl_3$ in water, form 1:1 copolymer film, 537, 540
---conductive LB films, 540
---electron diffraction: 7 spots, fit monoclinic lattice, a=3.28Å, b=7.0Å, γ=50° (poly(pyrrole) lattice), 541
(hexadecylTET-TTF)($TCNQF_4$)I_xLB film semiconductor, 520
hexamethoxytriphenylene-$TCNQF_4$
---triplet excitons when doped with AsF_5, 61
---possible ferrimagnet, 61
hexamethylenetetramine, complex with iodoform, second-order non-linear optical material, 565
hexamethylenetetraselenafulvalene (HMTSF)
---complexes *see* HMTSF
---ionization potential, 11
---structure, 6
hexamethylenetetrathiafulvalene (HMTTF)
---complexes *see* HMTTF
---good donor, 5
---ionization potential, 11
---structure, 6
hexamethylenetetratellurafulvalene (HMTTeF)
---complexes *see* HMTTeF
---structure, 6
highest occupied molecular orbital (HOMO), 2, 3, 43, 44
---and LUMO, 3, 44
---splitting by Peierls distortion, 7, 8
---splitting by Jahn-Teller distortion, 8
HMTCAQ *see* 1-hydroxymethyl-7,7,8,8-tetracyanoanthraquinodimethane
HMTSF *see* hexamethylenetetraselenafulvalene
HMTSF-TCNQ
---conductivity, 14, 16, 121
---conductivity max.(broad, 32 K)=7,000 S/cm, 16
---conductivity (300 K)=2000 S/cm, 16
---crystal structure, 121
---small on-site repulsion U_{eff}, 15
---interstack N..Se van der Waals contact, 120
---ionicity=0.74, 15, 16
---metallic down to 0.01 K, 15, 16, 120
---two-dimensional system, 15
---2 k_F scattering, 15
HMTSF-$TCNQF_4$
---4 k_F scattering, 15
---ionicity=1, 15
---large on-site repulsion U_{eff}, 15
---structure classification 8
HMTSF-TNAP
---conductivity max.(50 K)=20,000 S/cm, 16
---conductivity (300 K)=2900 S/cm, 16
---metal-to-insulator transition at 50 K, 16
---structure classification, 8
HMTTeF *see* hexamethylenetetratellurafulvalene
HMTTeF-DMTCNQ
---conductivity max. (83 K, broad)=1000 S/cm, 16
---conductivity(300 K)=460 S/cm, 16
HMTTeF-TCNQ
---conductivity max. (73 K, broad)=900 S/cm, 16
---conductivity (300 K)=550 S/cm, 16
HMTTF *see* hexamethylenetetrathiafulvalene
HMTTF-TCNQ
---conductivity max.(75 K)=2,000 S/cm, 16
---conductivity (300 K)=500 S/cm, 16
---ionicity=0.7, 16
---metal-to-insulator transition (75 K), 16
---structure classification, 8
HMTTF $TCNQF_4$
---structure classification, 8
1,2-H_2Nc (1,2-naphthalocyanine), 455
---structure, 456
---transition (calc) $15a_u$-->$15b_g$* ($16b_g$*) is 1.84 (1.97) eV, 458
---valence effective hamiltonian calculation, 456
---X-ray diffraction crystal structure, 456
2,3-H_2Nc (2,3-naphthalocyanine), 455
---HOMO ($6a_u$) energy=-5.93 eV, 457
---structure, 456
---transition (calc) $6a_u$-->$8b_{2g}$*,$8b_{3g}$* = 1.64 eV, 458
---valence effective hamiltonian calculation, 456
Hoffmann's extended Hückel theory, 49
Holstein effect, 134
HOMO *see* highest occupied molecular orbital
hopping distance, 366
hopping theory of conductivity, 365
hot electron effect, 256
H_2P (porphine)
---gas-phase ionization potential=7.1 eV, 457
H_2Pc (phthalocyanine), 455
---gas-phase ioniz. potential=6.41 eV, 457
---HOMO ($4a_u$) energy= -6.32 eV, 457
---photovoltaic cell, gap=0.14 eV, 561
---neutron diffraction crystal structure, 456
---structure, 456
---transition (calc) $4a_u$-->$6b_{2g}$* ($6b_{3g}$*) = 1.81 (1.98) eV agrees with expt, 458
---valence effective hamiltonian calculation, 456
H_2(Pc)I (phthalocyanine iodide)
---conductivity max.(15 K)=4000 S/cm, 16

H_2(Pc)I (phthalocyanine iodide) *(cont'd)*
---conductivity (300 K)=750 S/cm, 16, 455
---ionicity=0.33, 16
---metal-to-insulator transition 15 K, 16
H_2Phc (9,10-phenanthrenocyanine), 455
---structure, 456
---transition (calc) $8a_u$-->$10b_{2g}$*, $10b_{3g}$* =1.88 eV, 458
---valence effective hamiltonian calculation, 456
H_2TAP (tetraazaporphyrin), 455
---HOMO ($2a_u$) energy=-7.27 eV, 457
---structure, 456
---transition (calc) $2a_u$-->$4b_{2g}$* ($4b_{3g}$*)= 2.12 (2.30) eV agrees with exp. Q bands, 458
---valence effective hamiltonian calculation, 456
Hubbard
see also extended Hubbard
---dimer model: dielectric function, 27
---dimer model: dipole operator, 25
---dimer model: eigenvalues, 25
---dimer model: Herzberg-Teller coupling, 29
---dimer model: localized excitations, 27
---dimer model: oscillator strengths, 27
---dimer model: transition dipoles, 26
---gap, 51, 55
---gap in $(TMTTF)_2X$ salts, 56
---ground state: commensurate charge density wave (Peierls-Fröhlich), 52
---ground state: spin density wave (antiferromagnetic order below Néel temperature), 52
---ground state: singlet superconductor, 52
---ground state: triplet superconductor, 52
---hamiltonian, 1, 3, 25, 442
---model, half-filled, 130
---nearest-neighbor transfer energy (t), 3, 25, 441, 442
---on-site Coulomb repulsion energy (U), 3, 25, 441, 442
Hubbard-Peierls gap, 54
---in $(DIMET)_2SbF_6$, 54, 55
---in β-$(BEDT\text{-}TTF)_2AuCl_2$, 55
Hückel
---extended Hückel theory, 49
---molecular orbital theory, 3, 49, 469
---on-site energy, 3
---resonance integral, 1
hydrogen (H_2) sensor: microeletrochemical transistor, 627
2-hydroxyethoxy-7,7,8,8-tetracyano-quinodimethane (HETCNQ)
---crystal structure of methyl ester, 619
---reduction potentials, 617
---structure, 613
1-hydroxymethyl-7,7,8,8-tetracyano-anthraquinodimethane (HMTCAQ)
---reduction potentials, 617
---structure, 616
hyperpolarizabilities β and γ, 565
---calculations by dipole derivative or by sum-over-states method, 570
---calculations (*ab initio* level): strong dependence on diffuse orbitals used, on π electrons, on π delocalization; for non-resonant case γ >0, 571

I_D *see* ionization potential (of electron donor)
i-mnt *see* iso-maleonitriledithiolate
incommensurate
---phases, 49
---superlattice in (TTF)(TCNQ), 12
index of refraction, intensity-dependent, 579
indigo carmine, for anionic dopant in conducting polymers (for electrochromism, photoelectric conversion), 654
indio-porphyrin halides [In(Por)X], 217
---In(OMP)F, NMR, 219
inelastic electron tunneling spectroscopy, 574
---of metal | insulator | metal sandwiches at 4 K, 574
---pinholes in LB films give problems, 574
---second derivative of current versus voltage shows vibrational absorption by bonds in insulator, 574
---spectra from LB films of poly-4-BCMU, 575
---sub-monolayer sensitivity, 574
$In_{0.53}Ga_{0.47}As$, covered by 1 LB monolayer of cadmium arachidate:modulation-doped FET (cut-off frequency=19 GHz, transconductance=0.170 S/mm), 478
In(Por)X *see* indio-porphyrin halides
infrared spectra, 23, 345
---attenuated total reflectance, 573
---conductivity, 30, 134, 138, 241
---external reflection-absorption, 405
---internal reflection, 403
---of $Ag(2,5\text{-}DM\text{-}DCNQI)_2$, 199, 201
---of $Cu(2,5\text{-}DM\text{-}DCNQI)_2$, 199, 201
---of (N-docosylpyridinium)(TCNQ), LB film, before and after iodination, 507
---of (N-docosyltrimethylammonium) (TCNQ) LBfilm, after iodination, 507
---of ethylenedithiolo-bis-octadecylthio-tetrathiafulvalene:ω-tricosenoic acid (1:1), LB film, 532
---of M(Por)Cl, 218
---of M(Por)F, 218
---of M_2P-TCNQ (reflectance), 38
---of $M_2P\text{-}TCNQF_4$ (reflectance), 38
---of $Na(2,5\text{-}DM\text{-}DCNQI)_2$, 199, 201
---of (octadecylphenanthrolinium) $(TCNQ)_2$ with and without I_2 doping, 31
---of poly(acetylene), 298, 349, 350
---of poly(aniline), 423

infrared spectra *(cont'd)*
---of poly(aniline) encapsulated in $H_{46}Na_{10}$ Y-zeolite, 423
---of poly(3-methylthiophene), 423
---of poly(3-methylthiophene) encapsulated in $Cu_{15}Na_{26}$ Y-zeolite, 423
---of poly(3-octylthiophene), 371
---of poly(3-octylthiophene)/poly(ethylene) structure, 372
---of poly(*para* -phenylene), 385
---of poly(*para* -phenylene vinylene), 395
---of poly(pyrrole), 403, 404, 423, 430, 439
---of poly(pyrrole), γ-irradiated, 439
---of poly(pyrrole) encapsulated in $Fe_{12}Na_{32}$ Y-zeolite, 423
---of poly(pyrrole) intercalated in V_2O_5 xerogel, 430
---of (Rb)($TCNQF_4$), 37
---of (TMPD) ($TCNQF_4$), 39
---of $(TMTSF)_2ClO_4$ (conductivity), 136
---of $(TMTSF)_2PF_6$ (conductivity), 134, 136
---of $(TMTSF)_2ReO_4$(conductivity), 138
---of $(TMTTF)_2Br$ (conductivity), 136
---of $(TMTTF)_2PF_6$ (conductivity), 136
---of Y-zeolite $Na_{56}Y$, 423
---results: vibrational frequencies, integrated intensities, band shapes, 345
---vibronic modes affected by electron-molecular vibration coupling, 54
inhomogenous charge and spin distribution in organic metals, 12
insulators
---structure classification, 8
---Mott-Hubbard, 8
interband transition, 130
intercalation compounds
---are two-dimensional, 175
---layered copper oxides, 175
---of graphite, 175, 287-287
---of iron oxychloride, 421
---of molybdenum sulfide, 175
---of tantalum sulfide, 175, 289-291
---of titanium sulfide, 289-291
---of vanadium oxide, 421
intercalation aided by sonication, 285, 289
interchain coupling
---can suppress CDW and SDW, 13
---can preserve organic metal state, 13
---can enable organic superconductivity, 13
intermolecular distances
---alternating diads, 7, 48
---alternating tetrads, 7, 48, 137
---alternating triads, 7, 48
---regular, 7, 48
intermolecular overlap, 10
internal reflection spectroscopy, 403
intervalence band
---Pt^{II}-Pt^{IV}, 264
intervalence transitions
---in linear-chain platinum complexes, 264
---correlate with Pt^{II}-Pt^{IV} distance, 264
---change Pt^{IV}-Cl bond length, 264
intramolecular electron transfer rate
---decreases if Franck-Condon factor increases, 613, 662
---increases with I_D-A_A, 612, 662
---Marcus "inverted region" found, 612, 613, 662
intramolecular exciton, 43
inversion center, 35
iodine doping
---in alkylpyridinium TCNQ LB films, 503-505
---in M(Por)X, 217, 218, 220, 221
---in Pt complexes, 266
iodoform: hexamethylenetetramine complex, a second-order non-linear optical material, 565
iodoform: sulfur complex, a second-order non-linear optical material, 565
iodo-gallio-octamethylporphyrin [Ga(OMP)I], 218
iodo-gallio-octaethylporphyrin [Ga(OEP)I], 218
Ioffe-Regel rule, 54
ionicity (*or* degree of charge transfer)
---and bond lengths, 11
---and shifts in IR and Raman bands, 11
---depends on CT intensity, 344
---depends on vibronic frequency shifts, 34
---depends on Fermi wavevector, 9, 11, 50
---≈0.0 (almost neutral), 7, 34
---=0.5, 34
---≈1.0 (almost ionic), 7, 34, 44
---=2, 44
---fractional, 11
---measured from X-ray superlattice spots, 9
---of HMTSF TCNQ, 11
---of HMTTF TCNQ, 11
---of TSF TCNQ, 11
---of TTF TCNQ, 11
---partial, criterion for organic metals, 11
---partial, discrete valence (localized), 7, 46
---partial, mixed-valent (delocalized), 7, 45, 46
---partial, models, 10
---partial, no simple relation with ionization potential, 11
---partial, stabilized theoretically, 10
---related to I_D-E_A, 44
---table, 11, 16
ionization potential (of electron donor) I_D, 11, 44, 611-613, 651
---of HMTSF, 11
---of HMTTF, 11
---of pyrene, 613
---of TMPD, 613
---of TSF, 11
---of TTF, 11, 613
ion-radical salts *or* radical ion salts (RIS)
---formed by direct oxidation, 45

ion-radical salts *(cont'd)*
---formed by direct reduction, 45
---formed by metathesis, 45
---formed by electrochemical crystal growth, 45
---have low donor ionization potential, 43
---have high acceptor electron affinity, 43
---one-stack, 8
(IPDT-BMT-TTF *see* isopropylenedithiolo-bis(methylthio)TTF
$(IPDT\text{-}BMT\text{-}TTF)_xIBr_2$
---black microcrystals, 182
iron bathephenanthroline disulfonic acid, for anionic dopant in conducting polymers (for electrochromism, photoelectric conversion, electrogenerated chemiluminescence), 654
iron tetrasulfophthalocyanine, for anionic dopant in conducting polymers (for catalysis), 654
irreversible oxidation, 360, 365
iso-maleonitriledithiolate (i-mnt), 221
(iso-propylenediselenolo)-diselena)-(iso-propylenedithiolo-dithia)fulvalene, 150
(iso-propylenediselenolo-diselena)-dithiafulvalene, 150
iso-propylenediselenolo-TSF, 150
iso-propylenediselenolo-TTF, 150
isopropylenedithiolo-bis(methylthio)TTF (IPDT-BMT-TTF), 182
(iso-propylenedithiolo-diselena)-(iso-propylenediselenolo-dithia)fulvalene, 150
(iso-propylenedithiolo-diselena)dithiafulvalene, 150
(iso-propylenedithiolo)diselena)-(isopropylenediselenolo)diselena) fulvalene, 150
(iso-propylenedithiolo-dithia)-(iso-propylenediselenolo-diselena)fulvalene, 150
iso-propylenedithiolo-iso-propylenediselenolo-TTF, 150
iso-propylenedithiolo-TSF, 150
iso-propylenedithiolo-TTF, 150
isothiazolo-bis(methylseleno)-TSF, 150
isothiazolo-bis(methylseleno)-TTF, 150
isothiazolo-bis(methyl)-TSF, 150
isothiazolo-bis(methyl)-TTF, 150
isothiazolo-bis(methylthio)-TSF, 150
isothiazolo-bis(methylthio)-TTF, 150
isothiazolo-(dimethylethylene-diselenolo)-TSF, 150
isothiazolo-(dimethylethylene-diselenolo)-TTF, 150
isothiazolo-(dimethylethylene-dithiolo)-TSF, 150
isothiazolo-(dimethylethylene-dithiolo)-TTF, 150
(isothiazolo-diselena)-(bis(methylthio)-dithia)fulvalene, 150
(isothiazolo-diselena)-(dimethylethylenedithiolo-dithia)fulvalene, 150
(isothiazolo-diselena)-dithiafulvalene, 150
(isothiazolo-diselena)-(ethylenedithiolo-dithia)fulvalene, 150
(isothiazolo-diselena)-(iso-propylenedithiolo-dithia)fulvalene, 150
isothiazolo-diselena)-(methylenedithiolo-dithia)fulvalene, 150
(isothiazolo-diselena)-(propylenedithiolo-dithia)fulvalene, 150
(isothiazolo-dithia)-(bis(methyl)diselena)-fulvalene, 150
(isothiazolo-dithia)-(bis(methylseleno)-diselena)fulvalene, 150
(isothiazolo-dithia)-(dimethylethylene-diselenolo-diselena)fulvalene, 150
(isothiazolo-dithia)-(ethylenediselenolo-diselena)fulvalene, 150
(isothiazolo-dithia)-(iso-propylenediselenolo-diselena)fulvalene, 150
(isothiazolo-dithia)-(methylenediselenolo-diselena)fulvalene, 150
(isothiazolo-dithia)-(propylenediselenolo-diselena)fulvalene, 150
(isothiazolo-dithia)-diselenafulvalene, 150
isothiazolo-(ethylenediselenolo)-TSF, 150
isothiazolo-(ethylenediselenolo)-TTF, 150
isothiazolo-(ethylenedithiolo)-TSF, 150
isothiazolo-(ethylenedithiolo)-TTF, 150
isothiazolo-(iso-propylenediselenolo)-TSF, 150
isothiazolo-(iso-propylenediselenolo)-TTF, 150
isothiazolo-(iso-propylenedithiolo)-TSF, 150
isothiazolo-(iso-propylenedithiolo)-TTF, 150
isothiazolo-(methylenediselenolo)-TSF,150
isothiazolo-(methylenediselenolo)-TTF,150
isothiazolo-(methylenedithiolo)-TSF, 150
isothiazolo-(methylenedithiolo)-TTF, 150
isothiazolo-(propylenediselenolo)-TSF, 150
isothiazolo-(propylenediselenolo)-TTF, 150
isothiazolo-(propylenedithiolo)-TSF, 150
isothiazolo-(propylenedithiolo)-TTF, 150
isothiazolo-TSF, 150
isothiazolo-TTF, 150
isothiazolo-(vinylenediselenolo)-TSF, 150
isothiazolo-(vinylenedithiolo)-TTF, 150
isotope effect, 67, 75
ITO (indium-doped tin oxide glass), 655

J-aggregate, 482, 512, 513, 647
Jacobi elliptic function, 582
Jahn-Teller
---distortion, 8, 60
---theorem, 8
Jensen coupler, 581

KCP *see* potassium tetracyanoplatinate
Keggin polyanions ($PMo_{12}O_{40}^{---}$, $PW_{12}O_{40}^{---}$, $SiMo_{12}O_{40}^{--}$, $SiW_{12}O_{40}^{--}$), 186

Kerr effect, 564
"key fits the lock" (Kitaigorodsky), 10
k_F *see* Fermi wavevector
---2 k_F scattering *see* diffuse scattering
---4 k_F scattering *see* diffuse scattering
kinetics of crystallization, 7
kink-solitons, 271
Kitaigorodsky argument ("key fits the lock"), 10
$K_{0.3}MoO_3$, 277
---electron spin resonance of Mo^V defects at 41 K shows incommensurate and commensurate CDW lines, 279
---incommensurate CDW below 180 K, 278
---Peierls transition at 180 K, 278
---quasi-one-dimensional metal, 278
---static CDW seen by neutron and X-ray diffraction, 278
---structure: chains of MoO_6 octahedra, 278
K_2(i-mnt)$_2$, 212
$K_2[Ni(dmit)_2]_5$, semiconductor, 155
K_2NiF_4 structure, 123
Koopmans' theorem, 622
Korringa relation, 79
Kovacic synthesis, 375, 381
Kosower's radical, 4
$K_2[Pt^{II}(NH_3)Cl_3][Pt^{IV}(NH_3)Cl_5].2H_2O$, 264
$K_4[Pt^{II}I_4]$ $[Pt^{IV}I_6]$, 264
$K_2Pt(CN)_4Br_{0.30}.3H_2O$ (KCP) *see* potassium tetracyanoplatinate
$K_4[Pt_2(H_2P_2O_5)_4Br].3H_2O$
---band gap, 267
---conductivity, 263, 267
---mixed-valent (class IIIA), 269
---oxidation state=2.5, 269
---Raman spectrum, 269
$K_4[Pt_2(H_2P_2O_5)_4Cl].3H_2O$
---crystal structure, 257, 268
---platinum valences close to II,III, 269
---Raman spectrum, 268
---X-ray structure, 269
$K_4[Pt_2(H_2P_2O_5)_4I].3H_2O$
---mixed-valent (class IIIA), 269
---oxidation state=2.5, 269
---Raman spectrum, 269
---X-ray structure, 269
Kramers-Kronig analysis, 24, 133, 200, 387
Krogmann's salt *see* potassium tetracyanoplatinate
K^+TCNQ^-
---ionicity =1, but band isn't half-filled, 50
---Mott-Hubbard insulator: 1 electron per site, 51
---structure classification, 8

La_2CuO_4
---orthorhombic (Cmca) (room temp.), 125
---antiferromagnetic, Néel temp.=270 K, 126
---tetragonal (I4/mmm) K_2NiF_4 structure (above 540 K), 125
$La_{2-x}M^{II}{}_xCuO_{4-y}$
---K_2NiF_4 structure, 123
---layers of corner-linked CuO_6 octahedra, 123
---superconducts, T_c = 20-40 K, 123
Langmuir-Blodgett (LB) devices (P = passive; A = active)
---antibody-antigen recognition molecule imoobilized in LB film (A), 479
---anti-static sheets (P), 506
---chemical sensors excellent, but mechanical stability problematic (A), 478
---chemical sensor can use transistors or optical or acoustic transducers to transmit chemically sensed signal (A), 479
---electroluminescent diode enhancement (P), 476
---electron beam lithography (positive and negative resists; nanolithography) (P), 474
---ferroelectric liquid crystal alignment (P), 476
---few monolayers: surface-plasmon electrooptic modulators; optical sensors, guided-wave device, 576
---few monolayers: pyroelectric devices (A), 549
---few monolayers: organic rectifiers (A), 661
---gas sensors sofar disappointing (A), 478
---gas sensor (A), 506
---inorganic semiconductor sensor with LB film at its center (P), 478
---low-reflectance glass coatings (P), 473
---lubrication of magnetic and optical storage media (P), 474
---many (≈ 1000) monolayers: optical waveguides (but difficult to suppress domain formation and dielectric losses), 576
---mechanical filters (P), 473
---one monolayer: organic rectifiers (A), 612
---one monolayer: photodiode (A), 663
---optical write-once memory (A), 482
---photovoltaic enhancement (P), 476
---step thickness gauges (P), 473
---surface acoustic wave device coating (P), 476
---surface acoustic wave sensor: LB film (e.g. pyridinium TCNQ) on quartz or lithium niobate (A), 479, 500
---surface plasmon resonance (A), 479, 480
---transistors using LB multilayers as insulator (P), 478
---unlikely to displace silicon technology, 478

Langmuir-Blodgett (LB) films, 31, 61, 62, 473, 491, 503, 511, 519, 527, 531, 549, 557, 563, 573, 591, 605, 647
see also monolayer films at air-water interface *and also* Langmuir-Blodgett film devices
---alternate layers (superstructure), 481, 511, 598
---and second harmonic generation, 573, 591
---and third harmonic generation, 573
---as anti-static sheets, 506
---conducting LB films, 61, 62, 519-522
---devices *see* Langmuir-Blodgett devices
---diffraction: X-ray, neutron, electron, synchrotron radiation, 414
---for alternate-layer heterostructures, 573
---for superlattice structures, 573
---for unimolecular organic rectifiers, 612
---inelastic electron tunneling spectroscopy, 573, 574
---infra-red attenuated total reflectance, 573
---non-linear optical processes, 573
---of alkylammonium metal(dmit)$_2$ [alkyl=$R_m(CH3)_{4-m}$, R=$C_{10}H_{21}$, $C_{12}H_{25}$, $C_{14}H_{29}$, $C_{16}H_{33}$, $C_{18}H_{37}$, $C_{22}H_{45}$, m=1, 2,3, 4, metal=Ni, Au, Pd, Pt] oxidized with bromine or electrochemically, and mixed 1:1 with eicosanoic acid: semiconductors, 520-522
---of Z-β-(N-alkyl-4-pyridinium)-α-cyano-4-styryldicyanomethanide, where alkyl=hexadecyl, octyl, 649
---of Z-β-(N-alkyl-4-quinolinium)-α-cyano-4-styryldicyanomethanide, where alkyl= decyl, dodecyl, eicosyl, nonyl, heptyl, hexadecyl, hexyl, octadecyl, octyl, pentadecyl, tridecyl, tetradecyl, undecyl, 649
---of barium stearate, 474, 475
---of BDDAP-BHTCNQ, 614, 616
---of BDDAP-C-HMTCAQ, 614
---of behenic acid, 515, 516, 552
---of BHAP-C-HMTCAQ, 614
---of bis-[bis(heptadecyl)-octadecyl-oxy-orthophenanthroline]Fe(II)(NCS)$_2$, 516-518
---of bis(decyl)bis(methyl)ammonium Au(dmit)$_2$ Br_x: eicosanoic acid (1:1), 520, 521
---of bis(decyl)bis(methyl)ammonium Ni(dmit)$_2$ Br_x : eicosanoic acid (1:1), 520, 521
---of bis(decyl)bis(methyl)ammonium Pd(dmit)$_2$ Br_x, 521
---of bis(decyl)bis(methyl)ammonium Pt(dmit)$_2$ Br_x, 521
---of bis(docosyl)bis(methyl)ammonium Au(dmit)$_2$ Br_x: eicosanoic acid (1:1), 521
---of bis(docosyl)bis(methyl)ammonium Ni(dmit)$_2$ Br_x : eicosanoic acid (1:1), 521
---of bis(dodecyl)bis(methyl)ammonium Ni(dmit)$_2$ Br_x : eicosanoic acid (1:1), 521
---of [bis(heptadecyl)dimethylTTF] (TCNQ) and (tetramethylTTF)(hexadecylTCNQ), semiconductor, 520
---of bis(hexadecyl)bis(methyl)ammonium Ni(dmit)$_2$ Br_x : eicosanoic acid (1:1), 521
---of bis(octadecyl)bis(methyl)ammonium Au(dmit)$_2$ Br_x: eicosanoic acid (1:1), 521
---of bis(octadecyl)bis(methyl)ammonium Ni(dmit)$_2$ Br_x : eicosanoic acid (1:1), 521
---of [bis(octadecylthio)bis(methyl)TTF] (TCNQF$_4$)I_x, semiconductor, 520
---of bis(tetradecyl)bis(methyl)ammonium Au(dmit)$_2$ Br_x: eicosanoic acid (1:1), 521
---of bis(tetradecyl)bis(methyl)ammonium Ni(dmit)$_2$ Br_x : eicosanoic acid (1:1), 521
---of bixin, 535
---of cadmium arachidate, 558
---of chlorophyll **b**, 558
---of Co(II)-5,10,15,20-tetra[4-oxy-(2'-docosanoic acid)]-phenylporphyrin, 515, 516
---of Cu tetra-*tert* - butyl phthalocyanine, 561
---of DDOP-C-BHTCNQ, 614
---of DDOP-C-ENP, 614
---of diethyl alkoxypentacene diquinone dicarboxylate, 645
---of 7-N,N-dioctylaminophenyl-7,8'-diapocarotene-8'-al, 535
---of 2-docosylamino-5-nitropyridine, 599
---of N-docosylpyridinium Ni(dmit)$_2$ Br_x : eicosanoic acid (1:1), 521
---of (N-docosylpyridinium)(TCNQ) I_x semiconductor, 62, 504, 520
---of (N-docosylpyridinum)(TCNQ)$_2$ semiconductor, 504, 520
---of (N-docosyltrimethylammonium) (TCNQ), 505, 506
---of dye based on 1,4-naphthoquinone, 482
---of eicosanoic acid, 521
---of [(ethylenedithiolo)bis(hexadecylthio)-TTF] (TCNQF$_4$)I_x, semiconductor, 520
---of [(ethylenedithiolo)bis(octadecylthio)-TTF](TCNQF$_4$)I_x, semiconductor, 507, 520
---of 3-(13-ferrocenyl,13-hydroxytridecyl)-pyrrole + 3-hexadecylpyrrole, 538-541
---of 3-(13-ferrocenyl,13-hydroxytridecyl)-pyrrole:pyrrole (1:1 to 1:20), 539, 541-544
---of a hemicyanine-nitrostilbene (alternate layers), 481

Langmuir-Blodgett films *(cont'd)*
---of ω-tricosenoic acid-docosylamine alternate layers, 485
---of ω-tricosenoic acid- ruthenium (η^5-cyclopentadienyl)-bis(triphenylphosphine)-4-cyano-4"-*n* -pentyl-*p* -terphenyl hexafluorophosphate alternate layers, 485
---of trisdecylmethylammonium $Au(dmit)_2$, electro-oxidized metal, 520
---of TTF-C-BHTCNQ, 614
---of TTF TCNQ (?), semiconductor, 520
---ordering and conductivity, 528
---photodiode, 611,659, 663
---piezoelectricity is not significant, 551
---polarized infrared spectrum, 31
---pyroelectricity, 551-553
---quartz crystal microbalance: follow mass gained in monolayer deposition, 575
---Stark effect in, 597
---surface plasmon resonance in, 573, 605
---visible-ultraviolet spectroscopy, 573
---waveguide Raman study, 573
---X-ray diffraction study, 414
---X-type: acentric, 598
---Y-type: centrosymmetric, 598
---Z-type: acentric, 598, 614
---Z-type used for second harmonic generation, 598, 614
LARC-TPI (a poly(imide)), structure, 568
large-U limit of Hubbard model, 51, 133
$La_{2-x}Sr_xCuO_4$ (also $La_{2-x}Ba_xCuO_4$)
---antiferromagnetic, 126
---electrons near E_F more O(2p) than Cu(3d), 126
---extensive covalent Cu-O mixing, 126
---SDW at (1/2, 1/2, 0), 126
---x>0.06: Néel temperature vanishes, 126
---x=0.15: orthorhombic-tetragonal transition at 180 K, 125
---x=0.15: superconducts, T_c=36 K, 125
---x>0.20: only tetragonal phase, 125
---x>0.20: no superconductivity, 125
lattice
---dimerization, 9
---parameter, interstack, 136
---period, magnetic, 9
---softness, 93
layered compounds *see* two-dimensional structures
LB *see* Langmuir-Blodgett
lead, a superconductor, T_c=7.2 K, 116
LEB = leucoemeraldine base *see* poly(aniline)
Lee-Rice-Anderson theory, 251
leucoemeraldine base (LEB) *see* poly(aniline)
$LiCF_3SO_3$, electrolyte, 629
$LiNbO_3$
---good inorganic single crystal for second harmonic generation, 598
---second harmonic coefficient $\chi^{(2)}$ = 1.4 x 10^{-8} esu, 481
---used in surface acoustic wave device, 476, 479
Lindquist anions ($Mo_6O_{19}^{--}$, $W_6O_{19}^{--}$), 186
linear-chain complexes, 263
---halide bridge increases conductivity, 264
---intense colors, 264
---intervalence transitions, 264
---of nickel, 264, 268
---of palladium and platinum, 263, 264
---polarons and solitons in, 266
linear dichroism, 504
Li (pentacenediquinone semiquinone)
---conductivity (300 K)=1 S/cm, 646
liquid crystals
---ferroelectric, 476, 551
---nematic, 476
---smectic, 476
Li^+ sensor: WO_3-based microelectrochemical transistor, 631
$LiTaO_3$ (pyroelectric coefficient)/(dielectric constant) ratio = 4.1 $\mu C\ m^{-2}\ K^{-1}$, 485
(Li)(TCNQ)
---conductivity (300 K)=5 x 10^{-5} S/cm, 11
lithium niobate *see* $LiNbO_3$
lithography, electron beam, 474
$LiTi_2O_4$, superconductor above 10 K, 116
Little's excitonic superconductivity model, 53, 59, 85
Ln = lanthanides
$LnBa_2Cu_3O_{7-y}$ superconductors, 123
---all Ln yield superconductors (except Pr, Ce), 126
---anion-deficient perovskite, 123
---arrays of square-planar CuO_4, 123
---layers of square-pyramidal CuO_5, 123
---Ln(4f) orders antiferromagnetically far below T_c, 126
---replacing Cu with other 3d metals depresses T_c strongly, 127
---T_c=70-95 K, 123
$Ln_{2-x}M^{IV}{}_xCuO_{4-y}$ superconductors, 123
---derived from K_2NiF_4 structure, 123
---CuO_e rotated in alternate layers, 123
---loss of apical oxygens, 123
---T_c=20 K, 123
localized charges, 46
localization
---Anderson, 58, 156
---can be reduced by interchain charge transfer, 296
---in quasi-one-dimensional systems, 296, 340, 468
---length, 296, 366
---limits conductivity by multiple resonant scattering, 296
London penetration depth for α_t-(BEDT-TTF)$_2I_3$, 102
lone dimers, 7, 48
Lorentz factor, 164

lowest unoccupied molecular orbital (LUMO), 2, 4
lubrication by LB films, 474
LUMO *see* lowest unoccupied molecular orbital
lutetium(tetrasulfophthalocyanine)$_2$, for anionic dopant in conducting polymers (for electrochromism, photoelectric conversion, electro-generated chemi-luminescence), 654

Madelung energy (crystal electrostatic binding energy), 10, 49
magnetic lattice period, 9
magnetic transition *see* spin-Peierls transition
magnetoresistance
---in κ-(BEDT-TTF)$_2$Cu(NCS)$_2$, 233
---in β_H-(BEDT-TTF)$_2$I$_3$, 81, 87, 88
maleonitriledithiolate (mnt)
---salts, 205, 211
---structure, 211
manganese tetrasulfophthalocyanine, for anionic dopant in conducting polymers (for catalysis), 654
Marcus "inverted region" in electron transfer rates, 613, 662
McConnell-Hoffman-Metzger neutral/ionic classification, 10, 49
McConnell proposal for organic ferromagnet, 60
McMillan equation, 93
M(dmit)$_2$ *see* metal-bis(4,5-dithiolato-1,3-dithiol-2-thione)
MDS-Pyz-TTF *see* methylenediselenolo-pyrazino-TTF
(MDS-Pyz-TTF)$_x$Br$_3$
---brown chips, 183
(MDS-Pyz-TTF)$_x$I$_3$
---dark-copper, 183
(MDS-Pyz-TTF)$_x$IBr$_2$
---dark-copper, 183
MDS-TTF *see* methylenediselenolo-TTF
(MDS-TTF)$_2$AuI$_2$
---black plates, 183
---metal-to-insulator transition at 60 K, 156
(MDS-TTF)$_x$I$_3$
---copper plates, 183
(MDS-TTF)$_2$IBr$_2$
---bronze plates, 183
(MDS-TTF)$_x$TCNQ
---black needles, 183
MDT-DTDSF *see* methylenedithiolo-dithiadiselenafulvalene
(MDT-DTDSF)$_x$IBr$_2$
---bronze needles, 183
MDT-EDT-TTF *see* methylenedithiolo-ethylenedithiolo-TTF
(MDT-EDT-TTF)$_x$IBr$_2$
---bronze plates, 183
(MDT-EDT-TTF)$_x$I$_3$
---bronze-black plates, 183
MDT-TTF *see* methylenedithiolo-TTF
(MDT-TTF)$_2$AuBr$_2$
---black plates, metallic to low temperature, 183, 184
(MDT-TTF)$_2$AuI$_2$
---almost isostructural with κ-(BEDT-TTF)$_2$Cu(NCS)$_2$: isolated dimers of MDT-TTF, 157
---superconductor, T_c≈5 K, 156
(MDT-TTF)I$_{3+x}$, semiconductor, 156
(MDT-TTF)$_2$Ni(dcit)$_2$
---black microcrystals, metallic to low temperatures, 183
mean free path, 54, 296
---if equal to lattice spacing, then≈Fermi wavevector: conductivity=300-500 S/cm (between metallic and semiconductive), 54, 296
Medien *see* N-methyldiethylene-triamine
MEEP see poly(bis(2-(2-methoxyethoxy)-ethoxy)phosphazene)
Meissner effect
---in α_t-(BEDT-TTF)$_2$I$_3$, 101, 104
---in β_p-(BEDT-TTF)$_2$I$_3$, 108
---in κ-(BEDT-TTF)$_2$Cu(NCS)$_2$, 79, 98
MEM = methylethylmorpholinium
(MEM)(TCNQ)$_2$
---ionicity=0.5 (uniform), 48
---irregular stacks of tetramers, 48
---semiconductor, 48
mercury, superconductor, T_c=4.15 K, 115, 116
mercury cadmium telluride, night-vision device (liquid N$_2$-cooled), 482
metal bis(dithiolates), 46, 143, 205-209, 211-214
metal-bis(4,5-dithiolato-1,3-dithiol-2-thione) [M(dmit)$_2$], 46
metal dichalcogenides, 175, 289, 291
metallic conductors
---scattering at 2 k_F in, 9
---structure classification, 8
metalloporphyrins M(Por), 47, 217
---Ga(Por)F, 217
---In(Por)F, 217
---Tl(Por)F, 217
metal oxide bronzes
---A$_{0.3}$MoO$_3$ blue bronzes, 277
---layered copper oxides, 175
---FeOCl intercalates, 421
---V$_2$O$_5$ intercalation xerogels, 421, 427
---W bronzes, 421
metal phthalocyanine (MPc)
---good donor, 6
---structure, 6, 47
4,5-(methylacetate diseleno)-1,3-dithiole-2-thione, 181
methylcarboxyl-1,3-diselenole-2-selenone, 146, 147
methylcarboxyl-1,3-dithiole-2-thione, 146, 147

1'-methylcarboxy-4,5-methylenediseleno-lo-1,3-dithiole-2-thione, 147
1'-methylcarboxy-4,5-methylenedithiolo-1,3-dithiole-2-thione, 147
N-methyldiethylenetriamine (Medien), 264
methyl (2,4-dinitrophenyl)aminopropanoate: second harmonic susceptibility $\chi^{(2)} = 4 \times 10^{-8}$ esu, 481
methylene-bis(phosphinate) (pcp) $CH_2(PO_2H)_2^{--}$, 267
(methylenediselenolo-diselena)-dithiafulvalene, 150
(methylenedithiolo-diselena)-(methylenediselenolo-diselena) fulvalene, 150
(methylenediselenolo)diselena)-(methylenedithiolodithia)fulvalene, 150
methylenediselenolo-pyrazino-TTF (MDS-Pyz-TTF), 182
methylenediselenolo-TSF, 150
methylenediselenolo-TTF (MDS-TTF), 150
(methylenedithiolo-diselena)-(methylenediselenolo-dithia)fulvalene, 150
(methylenedithiolo-diselena)dithiafulvalene, 150
(methylenedithiolo-dithia)-(methylenediselenolo-diselena)fulvalene, 150
(methylenedithiolo-dithia)diselenafulvalene (MDT-DTDSF), 150
methylenedithiolo-ethylenedithiolo-TTF (MDT-EDT-TTF), 150
methylenedithiolo-methylenediselenolo-TTF, 150
methylenedithiolo-TSF, 150
methylenedithiolo-TTF (MDT-TTF), 150, 182
2-methyl-4-nitroaniline (MNA)
---non-linear coherent coupler built: MNA in propylene carbonate (medium), two optical fibers with single-mode cutoff at 590 nm; with HeNe probe laser (633 nm) and Ar^+ pump laser (515 nm): 10 ms switching, 587
---non-linear index of refraction $n_2 = 25 \times 10^{-11}$ esu (better than Si or p-nitroaniline), 582
---refractive index in different solvents as a function of temperature: easy to match index of silica optical fibers, 584
---second-order material, 582
---third-order properties good, 582
---used for NLO applications, 579, 580
---UV-visible spectrum, 585
N-methylphenazinium (NMP), 8
---structure, 15
3-methylpyrrole-4-carboxylic acid, octadecyl ester, mixed LB multilayer with octadecane (600 Y-type layers on ITO glass: changes transparent to red brown on oxidative electrolysis: pyrrole polymerization: $\sigma_{\parallel} = 0.1$S/cm, $\sigma_{\perp} = 10^{-11}$ S/cm), 655
microelectrochemical diode
---using tungsten oxide [conducts if reduced]and poly(3-methylthiophene) [conducts if oxidized] with poly(bis(2-(2-methoxy-ethoxy)ethoxy)-phosphazene) (MEEP) doped with $LiCF_3SO_3$ as ionic (solid-state) electrolyte, 632
microelectrochemical transistor
---using poly(3-methylthiophene) as electroactive polymer, and either poly(ethyleneoxide), poly(vinylalcohol), or poly(bis(2-(2-methoxyethoxy)ethoxy)-phosphazene) (MEEP) doped with $LiCF_3SO_3$ as ionic (solid-state) electrolytes, 627-631
---using N,N,N',N'-tetramethyl-para-phenylenediamine as electroactive material and poly(bis(2-(2-methoxy-ethoxy)ethoxy)-phosphazene) (MEEP) doped with $LiCF_3SO_3$ as ionic (solid-state) electrolyte, 633
---using tungsten oxide as electroactive material, and poly(bis(2-(2-methoxy-ethoxy)-ethoxy)-phosphazene) (MEEP) doped with $LiCF_3SO_3$ as polymeric ionic (solid-state) electrolyte, 631
MINDO/3 (Modified Intermediate Neglect of Differential Overlap, version 3) molecular orbital method, 468
mixed regular stack (donor-acceptor), 35
---symmetric mode Raman perturbed, 36
---transition to mixed dimerized, 38
---with rare intermediate ionicity, 37
mixed stacks, 7, 8
mixed-valence
---bipolarons (±2,0) for possible new organic metals or superconductors, 59
---complexes, Pt, 264
---dimer, 31
---inorganic salts (non-integral oxidation stete), 116
---ion-radical salts, 44
---Robin-Day class I (alternate stacks; neutral ground state), 53
---Robin-Day class II (segregated stacks, inequivalent sites, ordered charges), 53
---Robin-Day class IIIA (clusters), 53, 269
---Robin-Day class IIIB (delocalized metals), 11, 53
MNA *see* 2-methyl-4-nitroaniline
MNDO (Moderate Neglect of Differential Overlap) molecular orbital method, 348, 620-622
mnt *see* maleonitriledithiolate
MO *see* molecular orbital
mobility, 3, 294
molecular dynamics of 1-D polymers
---eigenvalue equation (finite case), 346
---harmonic oscillator approximation, 346
---infinite polymer: coupling between oscillators, with wavevector **k**, 346
---lattice dynamics: density of vibrational states, 346

molecular dynamics of 1-D polymers *(cont'd)*
---model oligomers for polymers: *ab initio* or MNDO calculations, 349
---normal vibrations: bond stretching, angle bending, torsion, 346
---optical theory of valence: instantaneous dipole is sum of bond dipoles, 347
---vibrational intensity related to square of dipole moment derivative, 347
---vibrational potentials easy for σ-bonded polymers, hard for π-bonded polymers (long-range interactions), 347, 348
molecular electronics, 421, 531
see also molecular engineering, molecular lines, molecular machinery, molecular wires, *and* molecule-based transistor
---actual bulk devices (organic phase-change switch; molecule-based transistor using conducting polymers, phase-change optical memories), 661
---actual bulk devices: microelectrochemical devices; chemical sensors for pH, O_2, H_2, Li^+, and Na^+ in solution, 627
---actual bulk devices: solid-state microelectrochemical transistors, using either conductive polymers, or WO_3, or TMPD as electroactive materials, and either poly(ethyleneoxide), poly(vinylalcohol), or poly(bis(2-(2-methoxyethoxy)ethoxy)phosphazene) (MEEP) doped with $LiCF_3SO_3$ as polymer ionic conductor electrolytes, 627-633
---actual zero-dimensional device (Langmuir-Blodgett photodiode), 663
---broad definition (organic conductors, organic superconductors; conducting polymers; intercalated graphites; intercalated layered compounds; inorganic linear-chain complexes; Langmuir-Blodgett multilayers), 659
---centers, 659
---devices, 491, 537, 653
---future OK if electron transfer is fast, 660
---narrow definition (truly unimolecular "zero-dimensional" devices), 659
---passive molecular connectors (carotenes, linked thiophenes, molecular spacers, inclusion compounds), 660
---proposed device: chemical transducer, 656, 657
---proposed zero-dimensional devices (Aviram-Ratner rectifier; Aviram-Ratner proton transfer switch; Mitani proton transfer switch; Carter soliton switch), 662, 663
---zero-dimensional device ultra small, 662
molecular engineering (for LB assemblies), 511, 555
---emulation of biological machineries, 514
---example: dioxygen trap: single penta-coordinate Fe(II) porphyrin + imidazole, 514, 515
---example: dioxygen trap: LB alternate-layer films of Co(II)-5,10,15,20-tetra[4-oxy-(2'docosanoic acid)]-phenylporphyrin and of N-octadecyl-imidazole (easier synthesis), 515, 516
---example: spin-transition complex: LB film of bis-[bis(heptadecyl)-octadecyl-oxy-orthophenanthroline]Fe(II) $(NCS)_2$, 516-518
---structural criteria: amphiphilicity; in-plane cross-linking; plane-to-plane connection; reactive site, 511, 512
---work-up by mixed or alternate LB monolayers, 513
---work-up by reactivity, 513, 514
molecular lines, 643, 660, 664
---conductivity of lithium pentacene semiquinone (300 K)=1 S/cmK, 646
---conductivity of tetramethylammonium pentacene semiquinone (300 K)=0.02 S/cm, 646
---crystal structure of tetrabutylammonium pentacenediquinone, 646
---Langmuir-Blodgett film of pentacene tetrone, 645
---Langmuir-Blodgett film of diethyl-alkoxypentacene diquinone dicarboxylate, 645
---scanning transmission electron microscopy, 645
---scanning tunneling microscopy, 645
---solid-state conductivity, 646
---synthesis of imide-linked Diels-Alder adduct polyacenequinones, 643, 644
molecular machinery, 514
molecular orbital (MO), 1, 49
---highest occupied (HOMO), 2, 49
---lowest unoccupied (LUMO), 2, 49
---method: *ab initio,* 348, 448, 570
---method: CNDO/2, 468
---method: extended Hückel theory, 49
---method: Hückel theory, 3, 49
---method: MINDO/3, 468
---method: MNDO, 620-622
---method: Pariser-Parr-Pople, 446
---method: valence effective hamiltonian, 327, 448, 456
molecular polarizability, 5
molecular wires (polyene chains), 512, 531, 660, 664
molecule-based transistor (using conducting polymers), 627, 664
---mechanism: the drain current, induced by a source-to-drain voltage, increases if the gate potential (electrochemical potential), with the passage of a gate (faradaic) current, switches polymer from neutral to doped conducting state; electrodes are Au or Pt on Si chip; small spacings increase device speed, 628

monodentate ligands
---ammonia, 266
---methylamine, 266
---ethylamine, 266
monolayer films at air-water interface (Pockels-Langmuir films)
see also Langmuir-Blodgett films
---amphiphilic molecules, 531
---area per molecule, 495-499, 531, 648
---compression until molecules are upright, 607
---fluorescence microscopy, 492-497
---Langmuir's studies, 491
---phases (two-dimensional): gas (G), liquid expanded (LE), liquid compressed (LC) (or gel-like), solid condensed (SC) (like smectic-b), 492, 493, 531
---phase transitions: first-order G/LE, maybe first-order LE/LC, second-order LC/SC, 493
---pressure = surface tension difference (subphase - interface film), 492
---pressure-area isotherm, 531, 616
---synchrotron X-ray diffraction, 491
---transfer to solid at constant film pressure: LB film, 531, 607
MOPAC computer program *see* MNDO
Mott insulator: K(TCNQ), 118
Mott-Hubbard
---band filling from -2 k_F to 2 k_F, 51
---becomes Wigner lattice, 51
---insulator, 8
---insulator, 4 k_F scattering in, 9
---K(TCNQ) (ρ = 1): 1 electron per site, 51
---semiconductor, 9
M(Pc) *see* metal phthalocyanine
M_2P *see* N,N'-dimethylphenazine
M_2P-TCNQ, 37
---infrared reflectance spectrum, 38
M_2P-$TCNQF_4$, 37
---infrared reflectance spectrum, 38
μSR *see* muon spin resonance
MTDAP-C-ENP (methyltricosi-11,13-diyne-amino-phenyl-carbamate of hydroxyethyl-para-nitrobenzene)
---LB films, 614
---polymerization, 614
---structure, 615
---Z-type monolayers, 614
Mulliken
---charge-transfer integral, 1
---charge-transfer theory, 1, 3, 43
---Wolfsberg-Helmoltz approximation, 49
multifrequency optical memory, 647
multimer stacking
---dimer, 46, 47
---trimer, 46, 47
---tetramer, 47
muon spin resonance (μSR), 81

Na^+ sensor: microelectrochemical transistor, 627
$Na(DM\text{-}DCNQI)_2$
---absorption peaks at 13,000 cm^{-1} (CT) 28,000 and 38,000 cm^{-1} (DCNQI), 200
---infrared spectrum : peaks at 4,000 and 3500 cm^{-1}; strong triplet at 2160, 2130, and 2105 cm^{-1}, 197
---optical reflectance spectrum, 197
---Raman spectrum : peaks at 2180 cm^{-1} (C≡N stretch), 1623 (C=C stretch), 1465 cm^{-1} (C≡N stretch), 202
---synthesis, 197
---visible spectrum, 197
Nafion (® DuPont), polymeric anionic dopant in conducting polymers (for mechanical strength, transport membrane), 654, 656
nanolithography, 474
1,4-naphthoquinone-based dye, LB film for write-once ablative optical memory, 482
narrow-band solids, 1
$Na_{0.40}V_2O_5.nH_2O$
---EPR, 430
---electron semiconductor, 432
NBP *see* N-butylphenazinium
(NBP)(TCNQ), 8
Nb_3Ge superconducts, T_c= 23.2 K, 116
$NbSe_3$, non-linear conductivity below Peierls transition, 277
Nb_3Sn superconducts, T_c= 18.1 K, 115, 116
NDP^+ $TCNQ^-$ *see* (N-docosylpyridinium)(TCNQ)
$NDTA^+TCNQ^-$ *see* (N-docosyltrimethyl-ammonium)(TCNQ)
near edge X-ray absorption fine structure (NEXAFS), 537, 544
neutral-to-ionic boundary
---discontinuous (theory), 10
---interface crossed (TTF-chloranil), 10
---interface found, 10
NEXAFS *see* near edge X-ray absorption fine structure
Ni-bis(4,5-dithiolato-1,3-dithiol-2-thione) [$Ni(dmit)_2$], 46
$[Ni^{II}(en)_2]$ $[Pt^{IV}(en)Cl_2][ClO_4]_4$, 264
$Ni^{III}(chxn)_2$ Br_3
---no Peierls distortion, 265
---resonance Raman spectrum, 265
niobium superconducts, T_c=9.2 K, 115, 116
$Ni(Pc)(BF_4)_{0.33}$ (nickel phthalocyanine tetrafluoroborate)
---conductivity max.(80 K)=4000 S/cm, 16
---conductivity (300 K)=1000 S/cm, 16
---ionicity=0.33, 16
---metal-to-insulator transition 80 K, 16
$Ni(Pc)(ClO_4)_{0.33}$ (nickel phthalocyanine perchlorate)
---conductivity max. (200 K, broad)=1000 S/cm, 16

$Ni(Pc)(ClO_4)_{0.33}$ *(cont'd)*
---conductivity (300 K)=700 S/cm, 16
---ionicity=0.33, 16
Ni(Pc)I (nickel phthalocyanine iodide)
---conductivity max. (25 K, broad)=5000 S/cm, 16
---conductivity (300 K)=550 S/cm, 16
---ionicity=0.33, 16
---metal to 4 K, 58, 59
---no cooperative ground state at low T, 58, 59
p-nitroaniline: third-order NLO material, 582
nitrostilbene: alternate-layer LB film with a hemicyanine, 481
nitroxide spin labels, 4
Ni-TTF-TT [nickel (IV) tetrathiafulvalene-tetrathiolate coordination polymer], 111
---conductivity (300 K)=40 S/cm, (4.2 K)= 10^{-1} S/cmK, 112
---Cu polymer less stable than Ni polymer, 111
---thermopower: metallic, 111
Ni-TT [nickel (IV) ethylene-tetrathiolate coordination polymer], 111
---conductivity (300 K)=2 S/cm, (≈ 20 K)= 10^{-6} S/cm, 112
---Cu polymer less stable than Ni polymer, 111
---thermopower: metallic, 111
NLO *see* non-linear optics
NMP *see* N-methylphenazinium
(NMP)(TCNQ)
---dipolar disorder, 15
---conductivity (300 K)=200-400 S/cm, 11
---ionicity=2/3, 15
---not metallic, 15
---solid solution with phenazine, 15
---structure classification, 8
---2 k_F streaks, 15
non-linear coherent coupler, 597
---two parallel waveguides spaced for evanescent wave overlap, and separated by a non-linear medium; used as optical switch (switching time < 1 ps at power 0.01-0.1 W desired), 581, 582
non-linear optics (NLO), 531, 591, 605
see also second-order NLO *and* third-order NLO devices
---as dielectric phenomena, 563
---degenerate four-wave mixing (DFWM), 576
---dielectric constant ε, 564
---difference frequency generation, 608
---electric polarization P, 564
---electro-optic (Pockels) effect, 608
---linear susceptibility $\chi^{(1)}$, 564
---materials: molecular, with van der Waals intermolecular interactions: non-linearity is molecular effect, with molecular β and γ, and can be tailored (organic crystals and polymers), 565
---materials: covalent and ionic bulk: non-linearity is bulk effect; inorganic quantum-well semiconductors, inorganic photorefractive crystals, 565
---non-linear coherent coupler, 581
---optical Kerr effect, 564
---parametric oscillation, 608
---phase velocity v, 564
---refractive index n, 564
---second-harmonic optical susceptibility of 2-docosylamino-5-nitropyridine (non-centrosymmetric Y-type Langmuir-Blodgett film) 25% that of $LiNbO_3$, 601
---second harmonic generation, 591, 608
---second-order susceptibility $\chi^{(2)}$, 564, 608
---third-order susceptibility $\chi^{(3)}$, 564, 608
---third-order susceptibility $\chi^{(3)}$ depends strongly on π conjugation length, 576
---third-order susceptibility $\chi^{(3)}$, resonance-enhanced, measurable by DFWM even for sub-monolayers, 576
---sum frequency generation, 608
---third harmonic generation, 573, 576, 608
non-stoichiometric (Berthollitic) compounds, 7, 49
nuclear magnetic resonance spectroscopy
---of 7-N,N-dioctylaminophenyl-7,8'-diapocarotene-8'-al, 533-535
---of Ga(OEP)Cl (^1H), 218
---of Ga(OEP)F (^1H), 218
---of Ga(OEP)F (^1H), 218
---of poly(aniline), 325, 326
---of $Rb_{0.3}MoO_3$, 278, 281
---of Tl(OEP)F, 219
---of Tl(TPP)F, 218
---of $(TMTSF)_2ClO_4$, $(TMTSF)_2PF_6$, 87
Nucleopore membranes
---poly(pyrrole) fibrils incorporated, 421
nucleoside, anionic dopant in conducting polymers (for catalysis, sensor), 654
number operator, 3
Nylon-66, 311

Octadecylamine, intercalate with TaS_2, superconductor, T_c=4 K, 290
(octadecyl-BEDT-TTF)$(ICl)_x$, LB film, 520
(octadecyldimethylsulfonium)(TCNQ) I_x, LB film semiconductor, 506,520
---reorganization of TCNQ on I_2 exposure, 506
N-octadecyl,3-methylpyrrole-4-carboxylic acid, mixed LB multilayer with octadecane, 655
(octadecylphenanthrolinium) $(TCNQ)_2$
---iodine-doped, 31
---LB film, IR spectrum, 31
---charge-transfer peak, 31
(N-octadecylpyridinium) (TCNQ), LB film semiconductor, 520

(N-octadecylpyridinium) $(TCNQ)_2$, LB film semiconductor, 520
(octadecylTET-TTF)$(TCNQF_4)I_x$, LB film semiconductor, 520
(octadecyltrimethylphosphonium)(TCNQ) I_x, LB film semiconductor, 506, 520
---reorganization of TCNQ on I_2 exposure, 506
(octadecyltrimethylammonium) nickel $(dmit)_2$
---mixed 1:1 with eicosanoic acid and oxidized with Br_2, Langmuir-Blodgett film, conductivity (300 K)=0.11 S/cm, 521
---mixed 1:1 with eicosanoic acid and electro-oxidized, Langmuir-Blodgett film, conductivity (300 K)=0.9 S/cm, 521
(octadecyltrimethylammonium) $Ni(pdt)_2$, LB film semiconductor, 520
(octadecyltrimethylphosphonium)(TCNQ) I_x, LB film semiconductor, 520
(N-[p-(p-octylphenylazo)phenyloxy]-dodecylpyridinium)$(TCNQ)_2$, LB film photoswitchable semiconductor, 520, 522-524
---trans-to-cis isomerization with 365 nm light, 522, 523
---cis-to-trans isomerization with 436 nm light, 522, 523
---conductivity switches (high/low) with isomerization, 523, 524
(octyltetrathiotetracene)I_x, LB film semiconductor, 520
odd-alternant hydrocarbons, 4
off-site transfer energy (t), 3, 34, 51, 52, 54, 131, 132, 442, 468
one-dimensional energy bands
---tight-binding approximation, 49
---Hubbard and extended Hubbard models, 51
---deviations: transverse interactions, 52
one-stack ion-radical salt, 5
on-site Coulomb repulsion energy (U), 3, 8
---compared with bandwidth (4 t), 8, 9
open-shell
---anion, 3
---cation, 3
---ion, stable, 3
operator
---annihilation (destruction), 3, 51, 131, 441, 442, 468
---creation, 3, 51, 131, 441, 442, 468
---number, 3, 131, 441, 442, 468
optical *see also* NLO devices
---ablative memory, LB film (dye based on 1,4-naphthoquinone), 482
---computing, 563, 579
---gap, 132
---phase-change memory (phthalocyanine derivative), 661
---phase-change memory (J-aggregate of cyanine dye in mixture of dioctyldimethyl ammonium chloride and methyl stearate), 482
---phase-change switch (CuTCNQ, AgTCNQ), 661
---pseudo-gap, 132
---quasi-waveguide, 569
---reflectivity affected by longitudinal transfer integral, 54
---switching depends on intensity-dependent index of refraction: phase shifts, changes in coupling angles, or in cavity resonance, 569
---switching, 579
---switching studied in Fabry-Perot etalon, and in planar waveguides, 569
---waveguide by LB films: domain and dielectric loss problems, 576
---waveguide by self-assembling chemical systems more promising, 577
---waveguide device candidates: poly(diacetylenes); methylnitroaniline in e.g. poly(methylmethacrylate), 580
---waveguide device needs: high third-order non-linearity, ultra-fast response, and excellent other physical properties (transparency, no scattering centers, uniformity, optically flat surfaces, chemical and environmental stability, processability), 579
---write-one-read-many (WORM) memory, 482, 661
optical conductivity *see* conductivity, optical
optoelectronics *see* photonics
organic acceptors *see* electron acceptors
organic conducting polymers
---review, 293-300, 303-313, 317-333, 335-343
organic devices (electronic or photonic) *see also* photonics *and* molecular electronics
---diodes *see* organic rectifiers
---Langmuir-Blodgett *see* Langmuir-Blodgett devices
---molecule-based transistor, 627-633
---optical memory, 482, 661
---phase-change switch, 661
---photodiode, 663
---photovoltaic, 557
---polymer field-effect transistor (polyFET), 635-670, 661
---rectifiers *see* organic rectifiers
---review, 611-624, 661-663
organic donor-inorganic acceptor salts
---with Lindquist anions $Mo_6O_{19}^{--}$, $W_6O_{19}^{--}$, 186
---with Keggin polyanions $PMo_{12}O_{40}^{---}$, $PW_{12}O_{40}^{---}$, $SiMo_{12}O_{40}^{--}$, $SiW_{12}O_{40}^{--}$, 186
organic donor-acceptor complexes *see* charge-transfer complexes

pentacene (pen)
---intercalate with TaS_2 (pen_xTaS_2): unit cell data; pressed pellet conductivity (300 K) ≈ 0.01 S/cm; weak paramagnetism; superconductor, T_c=7.02 K at 5 G (quenched at 350 G), 290
---intercalate with TiS_2 (pen_xTiS_2): unit cell data; pressed pellet conductivity (300 K)≈0.01 S/cm; weak paramagnetism, 290
pentacenediquinone semiquinone
---Li salt, conductivity, 646
---TMA salt, conductivity, 646
---TMA salt, crystal structure, 646
pentacene tetrone, LB film, 645
PEO *see* poly(ethylene oxide)
perfect diamagnetism in superconductors, 115
periodicity, 2
permittivity
---complex, 553, 607
---relative, 550, 551
perovskite-based compounds, 116
---barium bismuthates, 116
---cuprate high-temperature superconductors, 59, 123
---tungsten bronzes, 116
perovskite, molecular, 185
per *see* perylene
permanganate, anionic dopant in conducting polymers (for electrochromism, dispersed noble metal, catalysis), 654
perruthenate, anionic dopant in conducting polymers (for electrochromism, dispersed noble metal, catalysis), 654
perylene (per)
---good donor, 5
---intercalate with graphite (per_xC_y): unit cell; staging C-C-Per-C-TT-C-C; pressed pellet conductivity=0.087 S/cm, 285-287
---intercalate with TaS_2 (per_xTaS_2): unit cell data; pressed pellet conductivity (300 K)≈0.01 S/cm; weak paramagnetism, 290
---intercalate with TiS_2 (per_xTiS_2): unit cell data; pressed pellet conductivity (300 K)≈0.01 S/cm; weak paramagnetism, 290
---structure, 6, 47
$(perylene)_3I_2$
---alternate stacking, neutral, 48
---diamagnetic insulator, 53
---Robin-Day class I mixed-valent, 53
$(perylene)_2AsF_6.y(CH_2Cl_2)$
---ESR: very narrow linewidths, 211
---metal-to-insulator transition ≈ 200 K, 211
$(perylene)_2AsF_6$.y(tetrahydrofuran)
---ESR: very narrow linewidths, 211
---metal-to-insulator transition ≈ 200 K, 211
$(perylene)_2PF_6.y(CH_2Cl_2)$
---ESR: very narrow linewidths, 211
---metal-to-insulator transition ≈200 K, 211
$(perylene)_2PF_6$.y(tetrahydrofuran)
---ESR: very narrow linewidths, 211
---metal-to-insulator transition ≈200 K, 211
$(perylene)_2(PF_6)_{1.1}$
---conductivity max.(200 K)=1000 S/cm, 16
---conductivity (300 K)=900 S/cm, 16
---ionicity=0.550, 16
---metal-to-insulator transition 200 K, 16
$(perylene)_2Au(i\text{-}mnt)_2$
---synthesis, 212
---conductivity (300 K)=1,000 S/cm; metallic to ≈ 15 K, 212
---ESR susceptibility T-independent (50 < T < 300 K); g = 2.0032-2.0026; linewidth (<0.5 Gauss): T-dependence does not obey Elliott relation: metal-like, 213
---thermopower positive: holes; bandwidth 0.67 eV, 212
α-$(perylene)_2Au(mnt)_2$
---conductivity, room-temperature, 208
---crystallizes in space group $P2_1/c$, 205
---diamagnetic Au ion, 205
---different structural instabilities than Pd or Pt compounds, 205
---ESR linewidth=0.5 Gauss, 211
---hole carriers on perylene, 205
---made by electrocrystallization, or by oxidation of perylene with iodine, 205
---metal-to-insulator (spin-Peierls) transition at low T, 205
---thermopower, 207, 208
α-$(perylene)_2Co(mnt)_2$
---conductivity, room-temperature, 208
---made by electrocrystallization, or by oxidation of perylene with iodine, 205
---metal above 73 K; insulating gap=0.060 eV, 205
---thermopower, 207, 208
β-$(perylene)_2Co(mnt)_2$
---conductivity, room-temperature, 208
---made by electrocrystallization, or by oxidation of perylene with iodine, 205
---semiconductor, gap=0.090 eV, 208
---thermopower, 207, 208
$(perylene)_2Co(mnt).0.5CH_2Cl_2$
---conductivity, room-temperature, 208
---half-filled band, 209
---made by electrocrystallization, or by oxidation of perylene with iodine, 205
---metal above 272-277 K, 209
---two insulator phases, gaps=0.350 eV, 0.300 eV, 208
---thermopower, 207, 208
α-$(perylene)_2Cu(mnt)_2$
---conductivity, room-temperature, 208
---hole carriers on perylene, 205
---made by electrocrystallization, or by oxidation of perylene with iodine, 205

α-(perylene)$_2$Cu(mnt)$_2$ *(cont'd)*
---metal above about 60 K, 206
---thermopower, 207, 208
β-(perylene)$_2$Cu(mnt)$_2$
---conductivity, room-temperature, 208
---hole carriers on perylene, 205
---made by electrocrystallization, or by oxidation of perylene with iodine, 205
---semiconductor, gap=0.090 eV, 208
---thermopower, 207, 208
α-(perylene)$_2$Fe(mnt)$_2$
---conductivity, room-temperature, 208
---made by electrocrystallization, 205
---made by oxidation of perylene with iodine, 205
---metal above 58 K; insulating gap=0.050 eV, 205
---thermopower, 207, 208
α-(perylene)$_2$Ni(mnt)$_2$
---conductivity, room-temperature, 208
---hole carriers on perylene, 205
---made by electrocrystallization, or by oxidation of perylene with iodine, 205
---metal above 25 K; insulating gap≈ 0.015 eV, 205
---thermopower, 207, 208
β-(perylene)$_2$Ni(mnt)$_2$
---conductivity, room-temperature, 208
---hole carriers on perylene, 205
---made by electrocrystallization, or by oxidation of perylene with iodine, 205
---semiconductor, gap=0.102 eV, 208
α-(perylene)$_2$Pd(mnt)$_2$
---conductivity, room-temperature, 208
---crystallizes in space group P2$_1$/c, 205
---hole carriers on perylene, 205
---localized magnetic moments (S=1/2) on Pd(mnt), 205, 211
---made by oxidation of perylene with iodine, 205
---metal-to-insulator (spin-Peierls) transition at 28 K, 205
---thermopower, 207, 208
(perylene)$_2$Pd(mnt).0.5CH$_2$Cl$_2$
---conductivity, room-temperature, 208
---half-filled band, 229
---poor semiconductor, 229
α-(perylene)$_2$Pt (mnt)$_2$
---conductivity, room-temperature, 208
---crystallizes in space group P2$_1$/c, 205
---hole carriers on perylene, 205
---localized magnetic moments (S=1/2) on Pt(mnt)$_2$, 205, 211
---made by electrocrystallization, or by oxidation of perylene with iodine, 205
---metal-to-insulator (spin-Peierls) transition at 7 K, insulating gap≈0.010 eV, 205
---thermopower, 207, 208
(perylene)$_2$M(maleonitriledithiolate)$_2$ [per$_2$M(mnt)$_2$]
---M=Au, Cu, Ni, Pd, Pt, 205
pFBS [bis-(methylene-*para* -fluorobenzenesulfonate)-diacetylene], 407
---crystals: monoclinic, space group P2$_1$/c, 2 molecules/cell, 407
---electric permittivity tensor ε at 1 kHz, 80-333 K, as function of polymerization, in 2 directions: only small changes in ε_2, ε_3, 407
---polymer: no phase transitions 4-300 K, 407
PFV *see* poly(furylene vinylene)
phase diagram
---pressure-temperature for (TMTTF)$_2$X and (TMTSF)$_2$X salts, 86, 163
---temperature-magnetic field for (TMTSF)$_2$ClO$_4$, 246
phase-sensitive detection, 594
phase transition
---metal-to-insulator, 9
---non-structural: SDW (antiferomagnetic ordering), 53
---non-structural: superconductive, 53
---structural: electronic Peierls (CDW) 2k$_F$, 9, 12, 53
---structural: spin-Peierls at 4 k$_F$, 53
---temperature is affected by transverse electrostatic interactions, 53, 54
Ph-C-BHTCNQ (phenyl-carbamate of 2-bromo,5-hydroethoxy,7,7,8,8-tetracyanoquinodimethane), crystal structure, 620
phenalenium radical, 4
9,10-phenanthrenocyanine (H$_2$Phc), 455
phenazine, structure, 15
pheophytin, 662
phonon-assisted variable-range hopping, 294
phonon dispersion curve, for C-C skeletal stretch in polyenes: positive curvature, 348
phonons, 129
phonon propagator, 30
phosphine sensor: LB films of (N-docosylpyridinum) (TCNQ)$_2$ (0.5 ppm), 506
phospholipids
---L-α-dimyristoyl phosphatidic acid, 492
---monolayer films, 492
---dipalmitoyl-nitrobenzoxadiazol-phosphatidylethanolamine, 494
phosphotungstate, anionic dopant in conducting polymers (for electrochromism, dispersed noble metal, catalysis), 654
photochemical hole burning, 484, 592
photochemical reaction center, bacterial, 592, 662
photochromic LB film, 647
photochromism in N-salicylideneaniline, 663
photoinduced
---absorption spectroscopy, 341
---bleaching, 342

photoluminescence of poly(*para* -phenylenevinylene), 394, 395
photonics (*synonym:* optoelectronics)
---advantages: processing speed, three-dimensional connectivitry (neural networking), no electromagnetic interference, compatibility with fiber-optics communications, 563
---applications: fiber-optics, 563
---applications: optical computing, 563
---non-linear devices: frequency converters, all-optical switches, optical logic, optical memory storage, optical limiter, 563
---non-linear processes *see* non-linear optics
---second-order devices: second harmonic generators, electro-optic spatial light modulators, electro-optic switches, 566
---synonyms: light-wave technology, optical circuitry, 563
---third-order devices: optical signal processors, optical communications devices, 566
---use photons, not electrons for infomation transfer, 563
---using LB films, 478
photopolymerization, 664
photorefractive crystals, 565
photoresists, 474
photosynthesis, 661, 662
photovoltaic sandwich cells
---chlorophyll **a**, chlorophyll **b**, and Zn porphyrin derivative, 557
---Al | 1 LB monolayer of Cd arachidate | 44 LB monolayers of chlorphyll **b** | Ag: power conversion=0.002-0.026%, decreases with light intensity, gap=0.19 eV, high internal resistance (mainly at Al | Cd arachidate junctions), 558
---n-GaP | Au diode, 558
---SnO_2 | CdS | x-H_2(Pc) | Au: gap=0.14 eV, 561
pH sensor: WO_3-based microelectrochemical transistor, 631
phthalocyanine (H_2Pc), 217, 455-460
---β-phase: crystal structure, 462, 466
---metallophthalocyanine salts *see* Ni(Pc), Cu(Pc); *see also* metal phthalocyanines
---salts *see under* H_2(Pc)
---stacking model calculations using pair potential function ($R^{-1,-4,-6,-12}$). Dimer: minimum energy (-236 kJ/mol) is for face-to-face slipped geometry; energy for face-to-face geometry (no slipping)-5 kJ/mol; dimer binding energies greater if charges on dimer increase, from (0,0: -236 kJ./mol) to (1/3,1/3: -243.6 kJ/mol) to (0,1: -276.1 kJ/mol). For two stacks of n-mers (n=3 to 10), side-by-side geometry is favored, 461-466
---used as phase-change optical memory, 661
π-electron delocalization, 5, 351
piezoelectricity, 551
planar molecules with delocalized π orbitals, 5
planar optical waveguides, 569
plasma edge, 134
plasma frequency, 133
---affected by longitudinal transfer integral, 54
plasmon, 605
plastoquinone, 662
platinum linear-chain complexes, 263- 269, 271-276
---chlorine-doping, 273, 275
---iodine doping, 266
---intervalence band: bond stretching, 264
---luminescence, 266
---polaron pair formation, 272
---Raman spectra, 263
PMMA *see* poly(methylmethacrylate)
pn *see* 1,2-diaminopropane
PNB =pernigraline base, *see* poly(aniline)
Pockels effect, 608
Pockels-Langmuir films *see* monolayers at air-water interface
polarizability, molecular, 5
polarization, electrical, expansion in powers of the electric field, 481, 564, 608
polarization factor, 164
polarized infrared reflectance spectra, 132, 133
polaronic metal, 299
polarons
---cannot be distinguished from solitons or bipolarons in IR spectra of doped poly(acetylene), 356
---in conducting polymers, 411, 415
---bond-order formalism, 341
---electron polaron, 299
---hole polaron, 299
---Holstein formalism, 341
---in linear-chain Pt complexes, 266
---in poly(aniline) at low protonation (positive or neutral polaron), 340
---in poly(aniline) encapsulated in Y-zeolite, 425
---in poly(pyrrole) encapsulated in Y-zeolite and mordenite, 425
poling, 551, 565
poly(acenequinones), LB films, 643, 645
poly(acetylene) $(CH)_x$ (PA)
---anisotropic optical absorption, 373
---atomic point charge (IR int.) q_H=0.13, 352
---bond alternation proved by IR, Raman, 335, 350, 351
---bond length (IR freq.) d_{C-H}=1.090, 352
---charge-conjugation symmetric, 335
---*cis* --> *trans* isomerization: IR: 735 --> 743 cm^{-1}, 1013 -->1008 cm^{-1}, 348, 349

polyacetylene *(cont'd)*
---conductivity of 10^5 S/cm achieved, 293
---conjugation (π-electron delocalization), 351
---doped: 3 new IR bands (ν_1 strong, ν_2 weak, ν_3 strong and very broad); the ν_i show dispersion, isotope shift, no shift with n or p doping, 355
---doping: charge-transfer between chain and dopant, and Raman-active A_g modes become strongly IR-active, 355
---doping: IR spectrum cannot distinguish between solitons, polarons, and bipolarons, 356
---doping (high, e.g. 12% I): a sort of phase transition occurs, 355
---doping with iodine to form conductor, 305
---Durham/Graz synthesis, 415
---Durham synthesis, 295
---effective conjugation force constant, from Raman spectrum, 353
---effective conjugation force constant, from MNDO calculations, 354
---electronic energy, dimerized *versus* undimerized *trans* PA, 351
---electronic structure not that of simple metal, 298
---force constants, 375
---high carrier concentration, 294
---highly oriented *cis* form, Raman spectrum and intensity ratios for 909, 1250, and 1520 cm^{-1} bands and for *trans* band at 1100 cm^{-1}, 387
---highly oriented *cis* form, orientation-dependent complex index of refraction, 387
---highly oriented, from polymerization in a liquid crystal, 369
---highly oriented, from stretching, 369
---infrared intensity parameters, 352
---infra-red active vibrational modes, 298
---infrared spectrum (*cis* and *trans*), 350
---infrared spectrum of stretched*trans* film: bond moment for C-H stretch tilts away from the 1-D polymer plane, 351
---interchain transfer energy =0.03 eV, 297
---intrachain transfer energy=2.5-3.0 eV, 295
---intrinsic conductivity may be 2 x 10^6 S/cm, 4 times that of copper, 293, 295
---is polaronic metal, 299
---no Fano-like antiresonances, 299
---Peierls gap, 335
---perdeuterated form, 355
---photoexcited: PA is fast decay time photoconductor, 357
---photoexcited: 3 new IR bands, ν_1, ν_2 (both identical to bands in doped PA) and ν_3 (lower than doping-induced band, and sensitive to conjugation length), 357
---photoinduced infrared absorption, 341
---π-electron delocalization (conjugation), 351
---pseudo-gap=0.2 eV, 298
---Raman dispersion due to varying conjugation lengths in sample, 352
---Raman spectrum (resonance-enhanced because π-electron delocalization decreases bandgaps), favors "σ-π dimerized" structure of *trans* PA, 351
---soliton "switches" (proposed), 659, 663
---strong T-dependence of intrinsic conductivity predicted, 295
---structure, 335, 351
---structure (*trans*): orthorhombic cell with 2 chains, 350
---synthesis, 295, 415, 448
---temperature dependence of conductivity not metallic: imperfection-limited, 293
---temp.-indep. Pauli susceptibility, 298
---transient photoconductivity: fast (ps) decay $\approx$ time$^{1.0}$, independent of temperature (down to 10 K), then slow (μs), decay $\approx$ time$^{0.5}$, which is strongly temperature-dependent: (maybe due to enhanced probability of electron-hole separation onto different chains), 415
---2 k_F phonon frequency 0.12 eV, 295
---valence effective hamiltonian calculation (all-*trans*): bandwidth 6.5 eV, ionization potential 4.7-4.8 eV, 451, 452
---vibrational modes (undimerized *trans* , point group D_{2h}): $2A_g(R) + 2B_{2g}(R) + 1B_{1u}(IR)+1B_{3u}(IR)$ in-plane, and $1B_{1g}(R)+1B_{1u}(IR)$ out-of-plane; (dimerized *trans* , point group D_{2h}): $4A_g(R)+2B_u(IR)$ in-plane, and $1B_g(R) +1A_u(IR)$ out-of-plane, 351
poly(acrylonitrile), gamma irradiation, 435
poly(3-alkylthiophene)
---lower carrier concentration than in poly(thiophene), 294
---*see* poly(3-methylthiophene)
---*see* poly(3-octylthiophene)
poly(amic acid): optical quasi-waveguide propagates signal over 5 cm, 569
poly(amides), 311
poly(aniline)
[principal forms (acronym, conducting (C) or insulating(I), average oxidation state OS) are:
leucoemeraldine base (LEB, I, OS=0),
emeraldine base (EB, I, OS=0.5),
pernigraniline base (PNB, I, OS=1),
emeraldine salt (ES, C, OS=0.5),
fully oxidized salt (FOS, I, OS=1)], 303, 320, 335, 655
---four principal forms: fully reduced leucoemeraldine base; partially oxidized emeraldine base; oxidized and fully protonated emeraldine salt; fully oxidized bipolaronic copolymer salt, 320

poly(aniline) *(cont'd)*
---band structure, 327, 328, 337, 338
---charge-conjugation symmetry: none, 336
---chemical oxidation of LEB to ES: Cl_2 in CCl_4; $NOPF_6$, $FeCl_3$, or $SnCl_4$ in organic solvents; O_2, or H_2O_2 in aqueous acid, 306, 307
---conductivity of ES depends on pH, 305
---conductivity=1-5 S/cm for hydrochloride ES, 305
---dedoping ES to EB with alkali, 305
---dedoping ES-I yields amorphous EB-I; solvent extraction of EB-I yields 50% crystalline, orthorhombic EB-II; redoping EB-II with HCl yields orthorhombic ES-II, 309, 338
---deposited on interdigitated Au electrodes: molecule-based transistor, 661
---dielectric constant, 340, 341
---EB: black-blue powder (dark blue by transmitted light) with coppery glint, made by treating ES with ammonium hydroxide, 304
---EB: broad absorption at 2 eV (benzenoid -to-quinoid exciton); π-π^* gap (phenyl rings) at 3.8 eV, 336, 337, 341
---EB: free-standing copper-colored films, 309
---EB: crystalline (coherence length 100Å) when carefully compensated, 319
---EB: ^{13}C and ^{15}N NMR, 304
---EB: heated to 300°C: cross-linked electroactive polymer, 310
---EB: 1 quinoid ring per 4 C_6H_4 rings, 318
---EB: photobleaching of 1.8 (or 2) eV and 3.8 eV absorptions, 320, 341
---EB: photo-induced absorptions at 0.9 (?), 1.4 and 3.0 eV (polaron), 320, 341
---EB: magnetic susceptibility, 320
---EB: optical absorptions at 2 and 3.9 eV, 320, 324, 327
---EB: soluble in H_2SO_4 and N-methyl-pyrrolidone, 309
---EB: stretch-oriented fibers: tensile strength 330 MPa, 311
---EB: stretch-oriented films (400%) made by heating above glass transition temperature (110°C), 310
---EB: VEH band structure, 327, 328
---EB-I: dc conductivity=10^{-10} S/cm, 340
---EB-I: max. molar mass=325,000, 309
---electrochemical oxidation LEB-->ES, 306
---emeraldine base (EB) or poly-(phenyleneamineimine)$(C_6H_{4.5}N)_x$ (insulating form), 318, 336
---emeraldine salt (ES) $(C_6H_5N)_xA_y$ (A=anion) (conducting form), 336
---ES: absorptions at 1.0 eV and 3.0 eV (intrachain), at 1.5 eV (interchain) (the 2 eV and 4 eV peaks indicate incomplete protonation), 323, 327, 328
---ES formed at pH 0.15: single broad absorption at 1.0-1.4 eV, 324
---ES formed in hydrochloric, sulfuric, alkylsulfonic acids, 305
---ES formed in H_2SO_4: is 2-sulfonate, 305
---ES formed with non-protonic acids can add groups to imine nitrogen, 306
---ES: high carrier concentration, 318
---ES (.HCl) forms not a bipolaronic diamagnet, but a Pauli paramagnet with separated polarons, 306
---ES (.HCl): flexible, lustrous purple-blue films, conductivity=1-4 S/cm, 309
---ES (.HCl): partly crystalline black-green solid (green in transmitted light), made from aniline in HCl with ammonium peroxydisulfate, 304
---ES in solution is spinless, 339
---ES: low-viscosity fraction: not fully metallic, 322
---ES: more crystalline (coherence length 100 Å) if anion A=HSO_4^- than if A=ClO_4^-,Cl^-, 319
---ES: polaron absorptions at 1.4 and 2.9 eV, 341
---ES: polaronic metal, 318
---ES: protonation mainly at imine, also at amine sites, 304
---ES: spin-cast from H_2SO_4, 320
---ES: stretch-oriented fibers (HCl-doped): tensile strength 185 MPa, conductivity 8 S/cm, 311, 340
---ES: temperature-independent metallic Pauli susceptibility (above 125 K) only in well-ordered samples, 327, 329
---ES: VEH band structure, 327, 328
---ES-I: conductivity limited by charging energy between metallic islands, 340
---ES-I: crystalline, monoclinic (metallic islands 100 Å apart), 338-340
---ES-II (8% H^+): conductivity=10^{-6} S/cm, 340
---encapsulated in mordenite, 422
---encapsulated in zeolite Y, 422
---encapsulted in zeolite, EPR (g=2.0034, linewidth 8-10 Gauss), 425
---encapsulated in zeolite: insulator, 425
---energy gap in PNB may be Peierls gap, 336
---formed by adsorption polymeriz., 304
---formed by electrochem. oxidation, 304
---FOS: all phenyl rings are equivalent, 326
---FOS: quinonoid-benzenoid bipolaronic copolymer, 318, 320
---FOS: stable in H_2SO_4, 325
---FOS: VEH band structure, 327, 328
---fully oxidized octamer, 327
---fully oxidized salt (FOS): absorptions at 2.2 eV (leucoemeraldine HOMO to bipolaron) and 4.3 eV (π-π^*), 329
---fully oxidized tetramer, 327

poly(aniline) *(cont'd)*
---inherent average viscosity 1.0 dL/g, 321
---inherent viscosity 1.6 dL/g implies molar mass=15,000-60,000, 321
---in sulfuric acid solution: "liquid metal",
---layered intercalate in $V_2O_5.nH_2O$: interlayer spacing 13.94 Å, n-type metal-like or n-type semiconductor, broad EPR (d^1 V^{IV} centers), 429
---leucoemeraldine base (LEB) or poly-(phenyleneamine) $(C_6H_5N)_x$ (insulating form), 336
---LEB and PNB, when mixed, form EB, 308
---LEB: band energy versus twist angle, 337, 338
---LEB: first synthesized in 1910, 308
---LEB: insulator with π-π* gap at 3.0 eV (onset), 3.7-3.8 eV (peak), 324
---LEB: off-white powder, 308
---LEB: phenyl ring π-π* gap 3.8 eV, 318, 337, 338, 341
---LEB: valence effective hamiltonian band structure, 327, 328
---long chains needed to get intrinsic spectrum, 327
---may be polaronic metal, 323
---metallic phase: crystalline islands in emeraldine salt, 336
---microwave absorption, 337
---mid-infrared spectrum, 423
---nonlinear optical effects, 337
---oxidation of LEB to ES partially depopulates the valence band, 337
---oxidation of LEB to PNB halves occupancy of valence band, 337
---oxidation of ES: bipolaronic insulator, 318
---partially soluble in dimethylformamide, dimethylsulfoxide, tetrahydrofuran, 320
---pernigraniline base (PNB) or poly-(phenyleneimine) $(C_6H_4N)_x$ (insulating form), 336
---PNB synthesized impure in 1910, pure in 1989, 308
---PNB may undergo Peierls dimerization, 342
---PNB: dark purple, microcrystals or copper-colored films, 308
---photoinduced infrared absorption, 341
---photoinduced optical spectra, 336, 341
---possible erasable optical storage device, based on long-lived photoinduced trap-charge states, 337, 342
---redoping EB-II with HCl first dopes amorphous regions, then, in ES-II the Pauli paramagnetism returns, 339
---ring torsion effects, 336, 337
---self-localization of charge, 337
---soluble in sulfuric acid (dopant), N-methylpyrrolidone, methylsulfonic acid, trifluromethylsulfonic acid, 317
---soluble form (SPAN-ES), 337
---solution in H_2SO_4: magnetic susceptibility 100 times less than in solid: possible Peierls superlattice forms, 331
---spin-casting, 321
---spinless bipolaron (polaron) is stable for short (long) ES segments, 339
---structures, 336
---substituted (ring): CH_3, C_2H_5, OCH_3, OC_2H_5 groups, 311, 312
---substituted (at nitrogen): CH_3, 312
---sulfonated form (SPAN-ES) is self-doped and water-soluble, 337
---synthesis, 320, 321
---temperature-independent Pauli susceptibility inferred, 298
---tetramer salt of aniline is spinless, with bipolaron state, 339
---visible spectrum, 424
---visible peaks at <1.8, 3.1 eV: polarons, 425
poly(aniline-2-sulfonate), emeraldine salt form (SPAN-ES) $(C_6H_4NSO_3)_x$, structure, 337
poly-3-BCMU (a diacetylene polymer containing the butoxycarboxylmethyl urethan substitutent)
---monolayer yellow-to-bilayer-blue transition, 575
poly-3-BCMU (a diacetylene polymer containing the butoxycarboxylmethyl urethan substitutent)
---blue form is most conjugated, 575
---structure, 568
poly-4-BCMU (a diacetylene polymer containing the butoxycarboxylmethyl urethan substitutent), 574
---monolayer yellow-to-bilayer-red transition, 575
---structure, 568
---third-order harmonic generation increases with monolayer film compression, 576
---third-order non-linear susceptibility $\chi^{(3)} \approx 3 \times 10^{-10}$ esu (red form), $\approx 2.5 \times 10^{-11}$ esu (yellow form), 568
---yellow (coil) form ($\lambda_{max} = 460$ nm) has higher C=C Raman frequency (by surface plasmon technique) than the bilayer red (extended-rod, more π-conjugated) form ($\lambda_{max} = 530$ nm), 575
poly-9-BCMU (a diacetylene polymer containing the butoxycarboxylmethyl urethan substitutent), 575
---monolayer film transferred as LB film, then u.v. polymerized: most order (λ_{max}= 655 nm), 575
---monolayer film u.v. polymerized at air-water interface, then transferred as LB film: intermediate order (λ_{max}= 645 nm), 575

poly-9-BCMU *(cont'd)*
---solution-cast film, then u.v. polymerized: least order (λ_{max}= 635 nm), 575
---structure, 568
poly(bis(2-(2-methoxyethoxy)ethoxy)-phosphazene) (MEEP) doped 4:1 with $LiCF_3SO_3$: solid electolyte for micro-electrochemical transistor, 627
poly(bis(pyrrolo)benzene), 361
poly(2,2'-bithiophene)
---layered intercalate in $V_2O_5.nH_2O$: layer spacing 14.26 Å, p-type metal above 210 K, broad EPR (d^1 V^{IV} centers), 429
poly(2-butylpyridinium) I_y + perfluorooctanoic acid, LB film semiconductor, 520
poly(3-butylthiophene),
---$NOPF_6$-doped, LB film semiconductor, 521
---used in field-effect transistor, 636
poly(carbonate), 661
poly(3,4-cycloalkylpyrrole), 363
poly(3,4-cycloalkylthiophene), 363
poly(cyclohexa-1,3-dienyl-5,6-diacetate), 381
poly(diacetylene), 664
see also diacetylenes
---anisotropy, 407
---dielectric properties, 407
---LB monolayer-yellow --> LB bilayer blue: most conjugated structure, 575
---organic non-linear optical solid, 580
---poly-3-BCMU, 575
---poly-4-BCMU, 573
---poly-9-BCMU, 575
---poly-pFBS [bis-(methylene-*para* -fluorobenzenesulfonate)-poly(diacetylene)], 407
---poly-pTS [bis-(methylene-*para* -toluenesulfonate)-poly(diacetylene)], 407
---solid-state polymerization, 407
---ultraviolet light can polymerize LB monolayers, 575
poly(3,4-dibutylpyrrole)I_x + perfluoro-octanoic acid, LB film semiconductor, 520
poly(2,6-dimethylaniline)
---layered intercalate in $V_2O_5.nH_2O$: layer spacing 14.55 Å, broad EPR (d^1 V^{IV} centers), 429
poly(dimethyldiallyl ammonium chloride), gamma irradiation, 435
poly(dipyrrolyl), 359
---conductivity, 361
poly(dithieno[3,4-d:3',4'-b]thiophene)
---molecular modelling; eight residues (10° bent) surrounded by perchlorates, 413
---single X-ray diffraction line (d=3.53 Å), 413
poly(3-dodecylthiophene)
---degenerate four-wave mixing at 602 nm for LB film: fast decay, resonant $\chi^{(3)} \approx$ 1.1 x 10^{-9} esu, 576
---LB films, 575
---mass-to-absorbance ratio linear up to tenth monolayer deposited, 575
polyene chains
---as molecular wires, 531
---donor and acceptor-substituted, 531
---solitons in donor-polyene-acceptor, 531
poly(ethylene) $(C_2H_4)_x$
---atomic pont charge (from IR intensities) q_H = 0.04, 352
---bond length (from IR frequency) d_{C-H}=1.096, 352
---film support, 369
---structure: orthorhombic unit cell with 2 chains/cell, 350
---when chain-aligned by gel-spinning and tensile drawing is one of strongest materials, 295
poly(ethyleneoxide) (PEO): solid electrolyte in microelectrochemical transistor, 627
poly(ethyleneterephthalate) film support, 369
poly(ethylenimine), gamma irradiation, 435
poly-[3-(13-ferrocenyl-13-hydroxytridecyl)pyrrole + 3-hexadecylpyrrole]
---cyclic voltammogram (ITO glass): ferrocene/ferricinium couple present, 546
---from mixtures polymerized with $FeCl_3$ at air-water interface, 542
---NEXAFS of 4-layer LB films: C-H in film plane (290 eV), C-C along normal (295 eV), 543, 544
---pressure-area isotherms show need of excess 3-hexadecylpyrrole, 542
---proposed structure, 545
polyFET *see* polymer field-effect transistor
poly(furan) $(C_4H_4O)_x$, 655
---valence effective hamiltonian calculation: HOMO bandwidth=3.7 eV, bandgap= 3.1 eV, ionization potential=4.9 eV, electron affinity=1.8 eV, 451, 452
poly(2,5-furylene vinylene) $(C_6H_4O)_x$
---conductivity=36 S/cm for acceptor-doped polymer, 448
---optical bandgap=1.76 eV, 452
---resonance Raman spectrum, 379
---structure, 448
---valence effective hamiltonian calculation (all-*trans*): HOMO bandwidth=2.92 eV, bandgap=2.15 eV, ionization potential= 4.64 eV, electron affinity=2.49 eV, 448, 452
---vinyl group decreases calculated ionization potential and bandgap, 453
poly(3-hexadecylpyrrole + pyrrole) *see* poly(pyrrole + 3-hexadecylpyrrole)
poly[3-hexadecylpyrrole + 3-(13-ferrocenyl-13-hydroxytridecyl)pyrrole] *see* poly[3-(13-ferrocenyl-13-hydroxytridecyl)pyrrole + 3-hexadecylpyrrole]
poly(3-hexylthiophene)
---glass transition temperature 340 K, 639

poly(3-hexylthiophene) *(cont'd)*
---used in field-effect transistor, 635
---work function 4.7 eV, 636
poly(imide), 476, 568, 569
poly(indole), 655
polymer-based devices *see* polymer-based microelectrochemical transistor *and* polymer field-effect transistor
---using viologen, ferrocene, and quinone, 633
polymer-based microelectrochemical transistor: the drain current, induced by a source-to-drain voltage, increases if the gate potential (electrochemical potential), with the passage of a gate (faradaic) current, switches polymer from neutral to doped conducting state; electrodes are Au or Pt on Si chip; small spacings increase device speed, 628
polymer field-effect transistor, 635-640
---limited by low mobility (10^{-3} to 10^{-6} cm^2/V s at 300 K) and low maximum. frequency, 640
---mobility (due to phonon-assisted hopping, then polarons and bipolarons) increases up to glass transition, then decreases, 639
---normally-on state at 300 K, normally off after heat treatment at 430 K, 63, 637
---poly(3-butylthiophene) (p-type, 700 Å) on Si | SiO_2 wafer (3000 Å), 636
---poly(3-hexylthiophene) (p-type, 700Å) on Si | SiO_2 wafer (3000 Å), 636
---poly(3-methylthiophene), on Au or Pt microfabricated electrodes (70 μm long, 2.4 μm wide, 1.4 mm apart, 0.1 μm high) on 3 mm x 3 mm Si chip, with added Ag external reference and counter electrodes, either exposed to liquid electrolyte, or coated with MEEP/ $LiCF_3SO_3$ (4:1 or 5;1) solid polymer electrolyte: in both cases, similar cyclic voltammograms and drain currents at gate voltages where poly-3-methylthiophene becomes conducting; electrochemical signals amplified, but at low frequency, 628-631
---poly(3-octylthiophene) (p-type, 700Å) on Si | SiO_2 wafer (3000 Å), 636
---use Au source and drain electrodes, 636, 637
---use nearly undoped polymer to avoid dopant diffusion, 635
polymerization
---by chemical oxidation, 364, 369, 654
---electrochemical, 364, 369, 654, 655
---oxidative cationic, 381
---using sensitizer and sacrificial oxidant, 655, 656
polymer, redox, 632
poly(methylmethacrylate) (PMMA), substituted, LB film, as photoresist, 474
poly(N-methylpyrrole), 361
---layered intercalate in $V_2O_5.nH_2O$: layer spacing 14.30 Å, n-type semiconductor, broad EPR (d^1 V^{IV} centers), 429
poly(3-methylthiophene), 365
---copolymer with pyrrole, grown by electropolymerization with alternate layers of poly(pyrrole), 655, 656
---encapsulated in mordenite, 422
---encapsulated in zeolite: insulator, 429
---encapsulated in zeolite Y, 422
---mid-infrared spectrum, 423
---in microelectrochemical diode, using tungsten oxide [conducts if reduced] and poly(3-methylthiophene)[conducts if oxidized] with poly(bis(2-(2-methoxy-ethoxy)ethoxy)-phosphazene) (MEEP) as ionic (solid-state)electrolytes, 632
---in polymer-based microelectrochemical transistor: the drain current, induced by a source-to-drain voltage, increases if the gate potential (electrochemical potential), with the passage of a gate (faradaic) current, switches polymer from neutral to doped conducting state; electrodes are Au or Pt on Si chip; small spacings increase speed, 628
---in polymer-based microelectrochemical transistor: deposited on Au or Pt micro-fabricated electrodes (70 μm long, 2.4 μm wide, 1.4 μm apart, 0.1 μm high) on 3 mm x 3 mm Si chip, with added Ag external reference and counter electrodes, either exposed to liquid electrolyte, or coated with MEEP/ $LiCF_3SO_3$ (4:1 or 5;1) solid polymer electrolyte: in both cases, similar cyclic voltammograms and drain currents at gate voltages where poly-3-methylthiophene becomes conducting; amplification of electrochemical signals, but at low frequency, 629-631
---visible peak at 2 eV: bipolaron, 425
---visible spectrum, 424
poly(octadecyl 4-methylpyrrolium-3-carboxylate perchlorate), LB film semiconductor, 520
poly(octadecyl 4-methylpyrole-3-carboxylate) I_x, LB film semiconductor, 520
---conductivity (in plane)=0.1 S/cm, 508, 520
---conductivity anisotropy=10^{10}, 508
poly(2-octadecyl-oxy-aniline) I_x, LB film semiconductor, 520
poly-[(N-octadecylpyrrole) + $FeCl_3$ + (pyrrole)], LB film semiconductor, 520
poly(3-octylthiophene), 369
---polarized infrared spectra, 371
---polarized visible spectra, 372
---grown using $FeCl_3$ catalyst, 369
---stretch-oriented, 369
---stretch-oriented: longer chains orient preferentially, 373

poly(3-octylthiophene) *(cont'd)*
---stretch-oriented and $FeCl_3$-doped: metallic, 373
---used in polymer field-effect transistor, 636
poly(*ortho* -toluidine)
---is not formed in V_2O_5 xerogel, 429
---emeraldine base form (POT-EB), 337
---increased localization and ring torsion angles (relative to poly(aniline)), 341
poly(*para* -phenylene) $(C_6H_4)_x$ (PPP), 375
---AsF_5 - doped, infrared spectrum, 383, 384
---AsF_5 - doped, resonance Raman spectrum, 376
---benzenoid structure, 378
---^{13}C NMR spectra, 384
---conductivity undoped (=10^{-14} S/cm) and doped (=3.9 S/cm at 9.2 wt% Fe), 386
---$FeCl_3$-doped, infrared spectrum
---high carrier concentration, 294
---infrared and Raman analysis, 357
---PF_5 - doped, infrared spectrum, 383
---polarons at low doping, 383
---quinoid bipolarons at high doping, 383
---resonance Raman spectrum of, 376
---synthesis, 447
---synthesis from benzene (Kovacic reaction): 15-oligomer chains, 373, 381
---synthesis from 5,6-dihydroxycyclohexa-1,3-diene (Ballard): high molecular weight, 381
---synthesis from para-dihalobenzenes (Ullmann reaction): 10-12 oligomers, 381
---vibrations, in-plane and out-of-plane, 376
poly(*para* -phenylene benzobisoxazole) (PBO)
---structure, 568
---third-order susceptibility $\chi^{(3)} \approx 10^{-11}$ esu, 568
poly(*para* -phenylene benzobisthiazole) (PBT)
---structure, 568
---third-order susceptibility $\chi^{(3)} \approx 10^{-11}$ esu, 568
poly(*para* -phenylene oxide) (PPO)
---structure, 319
poly(*para* -phenylene sulfide) (PPS)
---resonance Raman spectrum, 379
---structure, 319
poly(*para* -phenylene terephthalamide)
---structure, 319
poly(*para* -phenylene vinylene) $(C_8H_6)_x$ (PPV) two forms: A, B with different conjugation lengths, 393
---alternation of bond lengths, 378
---B has lower bandgap than A, 396
---high carrier concentration, 294
---highly oriented, from stretching film grown on In-Sn oxide coated poly(ethyleneterephthalate) films, 369
---infrared spectra: A has extra line at 2920 cm^{-1}, 396
---neutral polaron-exciton states, 417
---photoluminescence: A lines 0.1 eV lower in energy than B lines, 395
---resonance Raman spectrum, 378
---SO_3-doped, LB film semiconductor, 520
---stretch-oriented film (draw ratio=11) 600-1100 Å thick, shows electron diffraction and X-ray diffraction (glued layers): cell constants: a=8.02 Å, b=6.07 Å, c=6.4 Å (chain repeat distance), α=123°, plane group pgg, 397, 398
---structure, 568
---synthesis, 415, 447
---synthesis of A: heating the sulfonium electrolyte precursor polymer in vacuo at 300°C, 393
---synthesis of B: heat tetrahydrothiophenium precursor polymer *in vacuo* at 200-300°C, 393, 394
---third-order susceptibility $\chi^{(3)} \approx 4 \times 10^{-10}$ esu, 568
---transient photoconductivity: fast (ps) decay $\approx$ time$^{0.9}$, independent of temperature (down to 10 K) then slow, bimolecular decay $\approx$ time$^{0.6}$, which is strongly temperature-dependent, 415
---ultraviolet and visible spectra: A lines 0.1 eV lower in energy than B, 394
poly(perdeutero-*para* -phenylene) $(C_6D_4)_x$, 375
---AsF_5-doped, resonance Raman spectrum 376
---Kovacic synthesis, 375
---quinoid structure, 378
---resonance Raman spectrum, 376
---vibrations, in-and out-of-plane, 376
poly(phenyl acetylene) $(C_8H_5)_x$
---structure, 568
---third-order susceptibility $\chi^{(3)} = 5 \times 10^{-11}$ esu, 568
poly(phenyleneamine) *see* polyaniline: leucoemeraldine base
poly(phenyleneamineimine) *see* polyaniline: emeraldine base
poly(phenyleneimine) *see* polyaniline: pernigraline base
poly(pyrrole), $(C_4H_3N)_x$ (PPy), 366, 655
---bond alternation, 335
---charge-conjugation symmetric, 335
---charge excitations: positive and negative polarons and bipolarons, 335
---copolymer with 3-methylthiophene, grown by electropolymerization with alternate layers of poly(pyrrole), 655, 656
---crystal structure: experimental, 541
---crystal structure: proposed, 541
---electron affinity 0.3 eV, 452
---encapsulated in mordenite, 422

poly(pyrrole) *(cont'd)*
---encapsulated in zeolite A, 422
---encapsulated in zeolite, electron spin resonance (g=2.0027, linewidth 8-10 Gauss), 425
---encapsulated in zeolite: insulator, 425
---encapsulated in zeolite Y, 422
---FTIR, 439
---gamma irradiation (^{60}Co) of BF_4^- salt: FTIR shows new bands at 1080 and 1130 cm^{-1} (C=O-O-C stretch), 436, 439
---incoporated in nucleopore membranes, 421
---infrared and Raman analysis, 357
---*in situ* spectroscopy during doping, 401
---layered intercalate in FeOCl, 421
---layered intercalate in $V_2O_5.nH_2O$: black interlayer spacing=16.60 Å, degenerate p-type semiconductor, broad EPR (d^1 V^{IV} centers), 421, 429
---mid-infrared spectrum, 423
---oxygen binds irreversibly upon gamma irradiation, 439
---Peierls gap, 335
---perchlorate: IR bands at 1546, 1475, 1446, 1385, 1288, 1217,1038, 924, and 792 cm^{-1} (polymer); broad band above 1800 cm^{-1} ((bi)polaron); bands at 1110 and 624 cm^{-1} (perchlorate), 402, 405
---perchlorate: EPR signal broadens (spin-flip relaxation?) reversibly with exposure to oxygen, 437, 439
---polymerization using $Ru(bpy)_3^{++}$ as sensitizer and $Co(NH_3)_5Cl^{++}$ as sacrificial oxidant, 656
---structure, 335
---synthesis (tetrafluoroborate), 436
---tetrafluoroborate: EPR signal broadens reversibly with exposure to oxygen, 437, 439
---tetrafluoroborate, as ^{60}Co gamma-irradiated electrode in CH_3CN: ferrocene oxidation is not affected (even by later exposure to air); benzoquinone reduction is not affected *in situ*, but changes irreversibly with exposure to air (oxygen bound to polymer), 436-439
---tosylate on graphite: scanning tunneling microscopy, 645
---valence effective hamiltonian calculation: HOMO bandwidth=3.8 eV, bandgap= 3.6 eV, ionization potential=3.9 eV, electron affinity=0.3 eV, 451, 452
---visible peaks at 2.0, 2.7 eV: bipolarons, 425
---visible spectrum, 424
poly(pyrrole + 3-hexadecylpyrrole)
---from 500:1 mixture at air-water interface, polymerizedwith $FeCl_3$: 1:1 copolymer monolayer film, 541
---electron diffraction structure: monoclinic (7 spots) fits poly(pyrrole) model, 541
---NEXAFS spectrum, 543
poly(2,5-pyrrylene vinylene) $(C_6H_5O)_x$
---never synthesized, 448
---structure, 448
---valence effective hamiltonian calculation (all-*trans*): HOMO bandwidth=2.81 eV, bandgap=2.48 eV, ionization potential= 4.20 eV, electron affinity=1.72 eV, 451, 452
---vinyl group increases calculated ionization potential, but decreases bandgap, 453
polyquinones as molecular lines, 643
poly(sulfurnitride) $(SN)_x$
---superconducts, T_c=0.35 K, 116
poly(2,5-thienylene vinylene) $(C_6H_4S)_x$
---conductivity (I_2-doped form, 300 K)=60 S/cm, 447
---resonance Raman spectrum, 379
---structure, 448
---syntheses, 447
---valence effective hamiltonian calculation (all-*trans*): HOMO bandwidth=2.18 eV, LUMO bandwidth=2.30 eV, bandgap= 1.90 eV, ionization potential=5.04 eV, electron affinity=3.14 eV, 452
---vinyl group decreases calculated ionization potential and bandgap, 453
poly(thiophene), $(C_4H_2S)_x$ (PT), 365, 655
---bond alternation, 335
---charge-conjugation symmetric, 335
---charge excitations: positive and negative polarons and bipolarons, 335
---encapsulated in mordenite, 422
---encapsulated in zeolite A, 422
---encapsulated in zeolite: insulator, 425
---encapsulated in zeolite Y, 422
---gamma irradiation, 435
---infrared and Raman analysis, 357
---ionization potential=5.4 eV, 451
---layered intercalate in FeOCl, 421
---layered intercalate in V_2O_5, 421
---optical bandgap=2.0 eV, 451
---Peierls gap, 335
---photoinduced infrared absorption, 341
---structure, 335, 568
---temperature-indep. Pauli susceptibility, 298
---third-order non-linear susceptibility $\chi^{(3)} \approx 10^{-9}$ esu, 568
---valence effective hamiltonian calculation: HOMO bandwidth=2.1-2.6 eV, bandgap =2.2 eV, ionization potential=5.4 eV, electron affinity=3.2 eV, 451, 452
poly(vinyl alcohol) (PVA), 594
---solid electrolyte for microelectrochemical transistor, 627
microelectrochemical transistor based on poly(3-methylthiophene), 628, 629
poly(vinylidene fluoride)
---good pyroelectric coefficient, 551

poly(vinylidene fluoride) *(cont'd)*
---large piezoelectric activity, 551
pop *see* diphosphite
porphine (H_2P), 457
porphyrin (Por), 47, 217
---*see also* metalloporphyrins
potassium poly(vinylsulfate (PVSK), polymeric anionic dopant in conducting polymers (for mechanical strength, transport membrane), 654
potassium tetracyanoplatinate, $K_2Pt(CN)_4Br_{0.30}.3H_2O$ (KCP, Krogmann's salt)
---conductivity of, 263
---Peierls transition in, 57
PPP *see* poly(*para* -phenylene)
PPS *see* poly(*para* -phenylene sulfide)
PPV *see* poly(*para* -phenylene-vinylene)
profilometer, Dektak, 321
1,2-propanediamine *see* 1,2-diaminopropane
(propylenediselenolo-diselena)-dithiafulvalene, 150
(propylenediselenolo-diselena)-(propylenedithiolo-dithia) fulvalene, 150
propylenediselenolo-TSF, 150
propylenediselenolo-TTF (PDS-TTF), 150
propylenedithiolo-bis(methylthio)TTF (PDT-BMT-TTF), 182
(propylenedithiolo-diselena)dithiafulvalene, 150
(propylenedithiolo-diselena)-(propylenediselenolo-diathia)fulvalene, 150
(propylenedithiolo-dithia)-(propylenediselenolo-diselena)fulvalene, 150
(propylenedithiolo-diselena)-(propylenediselenolo-diselena)fulvalene, 150
propylenedithiolo-propylenediselenolo-TTF, 150
propylenedithiolo-TSF, 150
propylenedithiolo-TTF, 150
proton transfer switch (proposed), 659, 663
PSSNa *see* sodium poly(styrene sulfonate)
Pt-bis(4,5-dithiolato-1,3-dithiol-2-thione) $[Pt(dmit)_2]$, 46
$[Pt^{II}(Medien)I]$ $[Pt^{IV}(Medien)I_3]$ I_2, 264
$[Pt^{II}(en)Cl_2]$ $[Pt^{IV}(en)Cl_2](ClO_4)_4$, 271
---chlorine doping, 273
---crystal structure, 272
---defects strongly localized, 272
---doubly degenerate Peierls state, 271
---ESR spectra of irradiated crystals, 273, 274
---example of ..-X-M^{IV}-X-M^{II}-X-M^{IV}-..., 271
---photoinduced polaron pairs, 272, 273
---possible bipolarons (positive, negative), 272
---possible kink-solitons (neutral, positive, negative), 272
---possible polarons (positive, negative), 272
---resonant Raman spectrum, 272
---soliton formation has energy barrier, 276
$[Pt^{II}(en)Cl_2]$ $[Pt^{IV}(en)Cl_4]$, 264
$[Pt^{II}(C_2H_5NH_2)_4][Pt^{IV}(C_2H_5NH_2)_4Br_2]$ $Br_4.4\ H_2O$ *see* Reihlen's green salt
$[Pt^{II}(C_2H_5NH_2)_4][Pt^{IV}(C_2H_5NH_2)_4Cl_2]$ $Cl_4.4\ H_2O$ *see* Wolffram's red salt
$[Pt^{II}(pn)_2]$ $[Pt^{IV}(pn)_2Br_2][Cu_3Br_5]_2$
---resonance Raman spectrum, 265
pTS [bis-(methylene-*para* -toluene-sulfonate)-diacetylene], 407
---crystal monoclinic, space group $P2_1/c$, 2 molecules per cell, 407
---electric permittivity tensor ε at 1 kHz, 80-333 K, as function of polymerization: ε_2 most polymerization and temperature-sensitive: maximum at 220 K (ε_1, ε_3 change < 2%), 407
---phase transition at $\approx$ 200 K to incommensurate phase, and at 160 K to low-T phase ($P2_1/c$, 4 molecules/cell), 407
---polymer: one phase transition at 190 K to low-T phase (isomorphous to that of monomer), 407
PTV *see* poly(thienylene vinylene)
purple bacteria, 592, 662
push-pull substituent effect, 4
PVA *see* poly(vinyl alcohol)
PVSK *see* potassium poly(vinylsulfate)
Py *see* pyrene
Py-C-BHTCNQ (1-pyrenyl-carbamate of 2-bromo-5-hydroxyethoxy-7,7,8,8-tetracyanoquinodimethane)
---LB films, 614
---rectification attempts, 622, 623
---structure, 615
Py-C-HETCNQ (1-pyrenyl-carbamate of 2-hydroxyethoxy-7,7,8,8-tetracyanoquinodimethane)
---oxidation and reduction potential, 617
---structure, 615
pyranylidenes (DIP), good donors, structure, 47
pyrazino-bis(methylseleno)-TSF, 150
pyrazino-bis(methylseleno)-TTF, 150
pyrazino-bis(methylthio)-TSF, 150
pyrazino-bis(methylthio)-TTF, 150
pyrazino-bis(methyl)-TSF, 150
pyrazino-bis(methyl)-TTF, 150
pyrazino-(dimethylethylenediselenolo)-TSF, 150
pyrazino-(dimethylethylenediselenolo)-TTF, 150
pyrazino-(dimethylethylenedithiolo)-TSF, 150
pyrazino-(dimethylethylenedithiolo)-TTF, 150
(pyrazino-diselena)-(bis(methylthio)-dithia)fulvalene, 150
(pyrazino-diselena)-(dimethylethylene-dithiolo-dithia)fulvalene, 150

(pyrazino-diselena)-dithiafulvalene, 150
(pyrazino-diselena)-(ethylenedithiolo)
-dithia)fulvalene, 150
(pyrazino-diselena)-(iso-propylenedithio-lo)-dithia)fulvalene, 150
(pyrazino-diselena)-(methylenedithiolo-dithia)fulvalene, 150
(pyrazino-diselena)-(propylenedithiolo-dithia)fulvalene, 150
4,5-pyrazino-1,3-diselenole-2-thione, 147
4,5-pyrazino-1,3-diselenole-2-thione, 147
(pyrazino-dithia)-(bis(methyl)diselena)ful-valene, 150
(pyrazino-dithia)-(bis(methylseleno)-diselena)fulvalene, 150
(pyrazino-dithia)-(dimethylethylene-diselenolo)-diselena)fulvalene, 150
(pyrazino-dithia)-diselenafulvalene, 150
(pyrazino-dithia)-(ethylenediselenolo)
-diselena)fulvalene, 150
(pyrazino-dithia)-(iso-propylenediseleno-lo)-diselenselena)fulvalene, 150
(pyrazino-dithia)-(methylenediselenolo)
-diselena)fulvalene, 150
(pyrazino-dithia)-(propylenediselenolo)
-diselena)fulvalene, 150
4,5-pyrazino-1,3-dithiole-2-thione, 147
pyrazino-(ethylenediselenolo)-TSF, 150
pyrazino-(ethylenediselenolo)-TTF, 150
pyrazino-(ethylenedithiolo)-TSF, 150
pyrazino-(ethylenedithiolo)-TTF, 150
---3:1 salt with I_3^-: metallic, metal-to-insulator transition at ≈60 K, 155
---2:1 salt with BF_4^-: metallic, metal-to-insulator transition at ≈180 K, 155
pyrazino-(isopropylenediselenolo)-TSF, 150
pyrazino-(iso-propylenediselenolo)-TTF, 150
pyrazino-(iso-propylenedithiolo)-TSF, 150
pyrazino-(iso-propylenedithiolo)-TTF, 150
pyrazino-(methylenediselenolo)-TSF, 150
pyrazino-(methylenediselenolo)-TTF, 150
pyrazino-(methylenedithiolo)-TSF, 150
pyrazino-(methylenedithiolo)-TTF, 150
pyrazino-(propylenediselenolo)-TSF, 150
pyrazino-(propylenediselenolo)-TTF, 150
pyrazino-(propylenedithiolo)-TSF, 150
pyrazino-(propylenedithiolo)-TTF, 150
pyrazino-TSF, 150
pyrazino-TTF, 150
pyrazino-(vinylenediselenolo)-TSF, 150
pyrazino-(vinylenedithiolo)-TTF, 150
pyrene (Py)
---good donor, 47
---ionization potential=7.41 eV, 613
---structure, 47, 613
pyrene-σ-viologen-σ-ferrocene, an organic unimolecular photodiode, 611, 659, 663
pyridinium zwitterion, 647
pyridino-bis(methylseleno)-TSF, 150
pyridino-bis(methylseleno)-TTF, 150
pyridino-bis(methylthio)-TSF, 150
pyridino-bis(methylthio)-TTF, 150
pyridino-bis(methyl)-TSF, 150
pyridino-bis(methyl)-TTF, 150
pyridino-(dimethylethylenediselenolo)-TSF, 150
pyridino-(dimethylethylenediselenolo)-TTF, 150
pyridino-(dimethylethylenedithiolo)-TSF, 150
pyridino-(dimethylethylenedithiolo)-TTF, 150
(pyridino-diselena)-(bis(methylthio)-dithia)fulvalene, 150
(pyridino-diselena)-(dimethylethylene-dithiolo-dithia)fulvalene, 150
(pyridino-diselena)-dithiafulvalene, 150
(pyridino-diselena)-(ethylenedithiolo-dithia)fulvalene, 150
(pyridino-diselena)-(iso-propylenedithio-lo-dithia)fulvalene, 150
(pyridino-diselena)-(methylenedithiolo-dithia)fulvalene, 150
(pyridino-diselena)-(propylenedithiolo-dithia)fulvalene, 150
(pyridino-dithia)-(bis(methyl)-diselena)ful-valene, 150
(pyridino-dithia)-(bis(methylseleno)-diselena)fulvalene, 150
(pyridino-dithia)-(dimethylethylene-diselenolo)-diselena)fulvalene, 150
(pyridino-dithia)-(ethylenediselenolo)-diselena)fulvalene, 150
(pyridino-dithia)-(iso-propylenediseleno-lo)-diselena)fulvalene, 150
(pyridino-dithia)-(methylenediselenolo)-diselena)fulvalene, 150
(pyridino-dithia)-(propylenediselenolo)-diselena)fulvalene, 150
(pyridino-dithia)-diselenafulvalene, 150
4,5-pyridino-1,3-dithiole-2-thione, 147
pyridino-(ethylenediselenolo)-TSF, 150
pyridino-(ethylenediselenolo)-TTF, 150
pyridino-(ethylenedithiolo)-TSF, 150
pyridino-(ethylenedithiolo)-TTF, 150
pyridino-(iso-propylenediselenolo)-TSF, 150
pyridino-(iso-propylenediselenolo)-TTF, 150
pyridino-(iso-propylenedithiolo)-TSF, 150
pyridino-(iso-propylenedithiolo)-TTF, 150
pyridino-(methylenediselenolo)-TSF, 150
pyridino-(methylenediselenolo)-TTF, 150
pyridino-(methylenedithiolo)-TSF, 150
pyridino-(methylenedithiolo)-TTF, 150
pyridino-(propylenediselenolo)-TSF, 150
pyridino-(propylenedithiolo)-TSF, 150
pyridino-(propylenediselenolo)-TTF, 150
pyridino-(propylenedithiolo)-TTF, 150
pyridino-TSF, 150
pyridino-TTF, 150

pyridino-(vinylenedithiolo)-TTF, 150
pyridino-(vinylenediselenolo)-TSF, 150
pyroelectric coefficient /dielectric constant ratio (μC m^{-2} K^{-1})
---of acid/amine LB films=1.3 (P only), 552
---of azo compound, LB film (X-type)= 1.2, 485
---of biphenyl/ester liquid crystal =1, 485
---of liquid crystal Z-type LB film =0.2, 485
---of lithium tantalate crystal =4.1, 485
---of polysiloxane liquid crystal =1.2, 485
---of poly(vinyldenefluoride) polymer=3, 485
---of PZFNTU (ceramic $PbZrO_3$-$PbTiO_3$-$PbFeNb_{0.5}O_{1.5}UO_3$) =1.3, 485
---of ruthenium (η^5-cyclopentadienyl) (bistriphenylphosphine)-4-cyano-4"-*n* -pentyl- *para* -terphenyl hexafluorophosphate (RuTCP), 552
---of $Sr_{0.6}Ba_{0.4}Nb_2O_6$ =1.4, 485
---of ω-tricosanoic acid-docosyamine LB superlattice film =0.7, 485
---of ω-tricosanoic acid-RuTCP LB superlattice film =0.7, 485
---of triglycine sulfate crystal =6.0, 485
pyroelectric effect, 549
---competing piezoelectricity can cause microphony problems, 551
---in materials with temperature-dependent spontaneous electric polarization, 549
---use in infra-red and thermal imaging, 549
pyroelectrics
---ceramic: high pyroelectric coefficient, high permittivity, 550
---inorganic single crystal: high pyroelectric coefficient, high permittivity, 550
---polymeric: good pyroelectric coefficient, large piezoelectricity, 551
---ferroelectric liquid crystals: good pyroelectric coefficient, large dielectric loss, 551
---Langmuir-Blodgett films: smaller pyroelectric coefficient and permittivity, low loss, no piezoelectricity, 551
pyrrole, structure, 448
pyrrole and 3-hexadecylpyyrole (500:1 ratio): LB polymer with $FeCl_3$, 538, 541
---area per molecule=22 Å^2, 539
PZFNTU (ceramic $PbZrO_3$+$PbTiO_3$+$PbFeNb_{0.5}O_{1.5}UO_3$) (pyroelectric coefficient)/(dielectric constant) ratio=1.3 μC m^{-2} K^{-1}, 485

Quadridentate ligand
---1,4,8,11-tetraazacyclotetradecane (cyclam), 264
quantized Hall effect *see* Hall effect
quantized nesting model
---interference between cyclotron period and SDW periodicity, 248
quantum-well
---heterostructures, 293
---semiconductors: non-linear optical devices, 565
quarter-filled bands, 129
quarter-filled band, tetramer model, 137
quartz, in surface acoustic wave device, 476
quasi-one-dimensional mixed-valent ion-radical salts (1 site/cation)
---semiconductors, strongly dimerized, activation energy (Hubbard-Peierls gap), Heisenberg paramagnets (e.g. $(DIMET)_2MF_6$, β-$(BEDT\text{-}TTF)_2AuCl_2$) 54, 55
---conducting salts, weakly dimerized, small activation energy (Hubbard gap), enhanced Pauli paramagnetism (e.g. $(TMTTF)_2X$, $(DMtTTF)_2X$), 55
---metals (e.g. $(TMTSF)_2X$, $DMtTSF)_2X$, $(TST)_2I$), 55, 56
quaterthienyl, unit cell, 413
quinolinium
---structure, 15
---TCNQ, 1:2 salt, conductivity (300 K) =100-200 S/cm, 11
---zwitterion, 647

Radical ion salts (RIS) *see* ion-radical salts
Raman spectra (*see also* resonance Raman spectra), 23, 202, 239, 264, 345, 387
---affected by electron-molecular vibration coupling, 54
---dispersion for poly(acetylene), 352
---modes for tetramer, 137
---of $Ag(2,5\text{-}DM\text{-}DCNQI)_2$, 202
---of β_d'-$(BEDT\text{-}TTF)_2I_3$, 259
---of $Cu(2,5\text{-}DM\text{-}DCNQI)_2$, 202
---of 2,5-DM-DCNQI, 202
---of $Na(2,5\text{-}DM\text{-}DCNQI)_2$, 202
---of poly(acetylene), 349-352
---of $(TMTSF)_2ReO_4$, 139
---results: vibrational frequencies, integrated intensities, band shapes, 345
---surface plasmon wave, 573
---waveguide, 573
$Rb_{0.3}MoO_3$, 277
---incommensurate CDW below 180 K, 278
---^{87}Rb nuclear magnetic resonance spectrum shows incommensurate CDW, 279
---Peierls transition at 180 K, 278
---propagating voltage pulses (sliding CDW), 283
---quasi-one-dimensional metal, 278
---sliding CDW: motional narrowing of ^{87}Rb NMR, excess nonlinear current in applied electrical field, 280
---static CDW seen by neutron and X-ray diffraction, 278

$Rb_{0.3}MoO_3$ *(cont'd)*
---structure: chains of MoO_6 octahedra, 278
(Rb)(TCNQ)
---regular stacking, 48
---ionicity=1, 8, 48
---insulator, 8, 48
(Rb)($TCNQF_4$), 36
---infrared spectrum, 37
rectifiers
---organic *see* organic rectifiers
---p-n junction, 635, 661
---Schottky barrier, 635
---Schottky barrier: p-doped polymer needs a low-work function metal (e.g. Al), which is chemically reactive, 635
redox potential
see reduction potential
---depends on solvent, electrolyte, electrode, 11
---difference and criteria for organic metals, 12
---of BDDAP-C-HETCNQ, 617
---of benzoquinone on Pt/poly(pyrrole), 456
---of BHTCNQ, 617
---of DDOP-C-ENP, 617
---of DDOP-C-BHTCNQ, 617
---of dipyrrolyl monomers, 361
---of dipyrrolyl polymers, 361
---of DMAP-C-ME, 617
---of ferrocene on Pt/poly(pyrrole), 455
---of HETCNQ, 617
---of HMTCAQ, 617
---of Py-C-HETCNQ, 617
---of TCNQ, 617
---of TDDOP-C-ENP, 617
---of TDDOP-C-HETCNQ, 617
---solution, 11
redox cycles, reversible, 365
reduction potential, half-wave, 651
reflectance, polarized, at normal incidence, 133
reflection spectroscopy
---external reflection absorption, 402
---internal, 403
refractive index, 24, 609
---and non-linear optics, 564
Reihlen's green salt, $[Pt^{II}(C_2H_5NH_2)_4][Pt^{IV}(C_2H_5NH_2)_4Br_2]Br_4.4\,H_2O$, 263
---conductivity of, 263
regular stack, mixed (donor-acceptor), 7, 8, 35
---symmetric mode Raman perturbed, 36
regular stack, segregated, 7, 8, 35
---symmetric mode Raman unperturbed, 35
---single phonon branch, 35
resistivity, electrical (*see also* conductivity, electrical), 14, 55, 56, 76, 81, 98, 100, 101, 102, 104-106, 108, 110, 112, 113, 119, 121, 125, 157, 163, 178, 179, 206, 208, 212, 224, 229, 231, 232, 254, 255, 259, 260, 366, 431, 522
resonance integral (Hückel), 1
resonance Raman spectroscopy, 32
---affected by electron-lattice phonon coupling, 54
---Albrecht model, 32
---enhanced by non-Condon transition, 32
---of κ-(BEDT-TTF)$_2$Cu(NCS)$_2$, 240, 241
---of α-(BEDT-TTF)$_2I_3$, 243
---of α_t-(BEDT-TTF)$_2I_3$, 107, 242
---of $[Ga(OEP)FI_x]_n$, 221
---of $[Ga(OMP)FI_x]_n$, 221
---of $[Ga(TPP)FI_x]_n$, 221
---of halogen-bridged mixed-valence Pt complexes, 264
---of $K_4[Pt_2(H_2P_2O_5)_4Cl.3H_2O$, 268
---of $Ni^{III}(chxn)_2\,Br_3$, 265
---of poly(acetylene) (*cis*), 387
---of poly(*para* -phenylene), 376
---of poly(*para* -phenylene), AsF_5-doped, 376
---of poly(*para* -phenylene sulfide), 378
---of poly(*para*-pheneylene vinylene), 378
---of poly(perdeutero*para* -phenylene), 376
---of poly(perdeutero-*para* -phenylene), AsF_5-doped, 376
---of poly(thienylenevinylene), 379
---of poly(furylenevinylene), 379
---of$[Pt^{II}(pn)_2][Pt^{IV}(pn)_2Br_2][Cu_3Br_5]_2$, 265
---of (TTF)Br powder, 33
retinal, cis-trans isomerization, 663
Rhodobacter sphaeroides , 592
Rhodopseudomonas viridis , 592
Rietveld analysis, 414
RIS *see* ion-radical salts
Robin-Day classes of mixed valence *see* mixed valence
rose bengal, for anionic dopant in conducting polymers (for electrochromism, photoelectric conversion), 654
RuCTP *see* ruthenium (η^5-cyclopentadienyl) (bistriphenylphosphine)- 4-cyano -4"-*n* -pentyl- *para* -terphenyl hexafluorophosphate
ruthenium bathephenanthroline disulfonic acid, anionic dopant in conducting polymers (for electrochromism, photoelectric conversion, electrogenerated chemiluminescence), 654
Rumer diagrams, 468, 470
ruthenium (η^5-cyclopentadienyl) bis(triphenylphosphine)- 4-cyano-4"-*n* -dodecyloxy-*para* -phenyl hexafluorophosphate, 483
---pyroelectric effect, 483
---second harmonic generation, 483
ruthenium (η^5-cyclopentadienyl) bis(triphenylphosphine)- 4-cyano-4"-*n* -dodecyloxy- *para* -biphenyl hexafluorophosphate, 483
---pyroelectric effect, 483
---second harmonic generation, 483

ruthenium (η^5-cyclopentadienyl) bis-(triphenylphosphine)- 4-cyano-4"-*n* -pentyl- *para* -terphenyl hexafluorophosphate (RuTCP), 605
---absorption coefficient k=0.08, 609
---ellipsometry, 609
---layer thickness 3.0 nm, 609
---LB films, 551, 605, 609
---permittivity 2.34 + 0.25i, 609
---pyroelectric coefficient=0.6 μC m^{-2} K^{-1}, 552
---refractive index n=1.53, 609
---second harmonic generation, 483
---structure, 609
---surface plasmon resonance, 605, 610
ruthenium (η^5-cyclopentadienyl) bis(triphenylphosphine)- 4-cyano-4"-*n* -tridecyloxy- *para* -bisphenylenevinylene hexafluorophosphate, 483
---pyroelectric effect, 483
---second harmonic generation, 483
ruthenium (η^5-cyclopentadienyl) bis(triphenylphosphine)- 4-cyano-4"-*n* -tridecyloxy- *para* -diphenylacetylene hexafluorophosphate, 483
---pyroelectric effect, 483
---second harmonic generation, 483
ruthenium tris-bipyridyl [$Ru(bpy)_3^{++}$] (Creutz-Taube ion)
---electron transfer to methylviologen, 662
---sensitizer for pyrrole polymerization, 656

S-aggregate, 513
Saito-Ferraris criteria for organic metals, 12
N-salicylideneaniline, keto-enol switch, 663
SbF_6^- anion: F atoms more negatively charged than in AsF_6^- or PF_6^- anions, 167
SbI_3, complex with sulfur, a second-order non-linear optical material, 565
scanning transmission electron microscopy (STEM), 645
scanning tunneling microscopy (STM)
---of κ-$(BEDT)_2Cu(NCS)_2$, 81
---of benzene co-adsorbed with CO on Rh, 645
---of DNA, 645
---of ortho-quinone-alkyl bridge-catechol, 623
---of BDDAP-C-BHTCNQ, 623, 624
---of poly(pyrrole) tosylate on highly oriented pyrolitic graphite, 645
scattering
---Bragg, 8
---diffuse, 9
---Umklapp, 8
Schottky barrier, 635
---increased by adsorption of LB films, 476
SDW *see* spin density wave
second harmonic
---generation, 608
---modulation of Stark spectrum of bacterial reaction center, 594
second-order non-linear optical materials
---bulk needs non-zero second-order electric susceptibility $\chi^{(2)}$, 565
---bulk must be non-centrosymmetric: acentric crystals, poled polymers, side-chain liquid crystal polymers, X-type, Z-type, or alternate-layer LB films, 565
---example: donor-phenyl-acceptor, 565
---example: CHI_3: sulfur complex, 565
---example: SbI_3: sulfur complex, 565
---example: CHI_3: hexamenthylenetetramine complex, 565
---Langmuir-Blodgett films promising, but film thicknesses near wavelength of light are inconvenient, 482
---molecules need large second-order hyperpolarizability β, 565
---organics give $\chi^{(2)}$ values orders of magnitude larger than inorganics; for second harmonic generation, crystal quality remains a problem; for electro-optic applications, poled guest-host polymers OK, except for stability, 570
---second harmonic susceptibility $\chi^{(2)} \approx 1.4 \times 10^{-8}$ esu for $LiNbO_3$, 481
---second harmonic susceptibility $\chi^{(2)} \approx 4 \times 10^{-8}$ esu for methyl(2,4-dinitrophenyl)aminopropanoate, 481
---second harmonic susceptibility $\chi^{(2)} \approx 7 \times 10^{-7}$ esu for a hemicyanine-nitrostilbene alternate-layer LB film, 481
Seeman-Bohlin focussing, 414
segregated stacks, 7, 35
---criterion for organic metals, 11
semiconductor
---Mott-Hubbard, 8
---Peierls, 9
semimetal, 5
sensor, 506, 654
shape mode in infrared spectrum, 356, 357
Shubnikov-de Haas oscillations
---in κ-$(BEDT\text{-}TTF)_2Cu(NCS)_2$, 67, 77, 78, 230, 233
---in κ-$(BEDT\text{-}TTF)_2KHg(SCN)_4$, 81, 82
---in β_H-$(BEDT\text{-}TTF)_2I_3$, 85, 88
---in $(TMTSF)_2PF_6$, 246
silanized metal electrode, 664
silanized molecule attached to metal, 664
silica (SiO_2), 478
---all-optical switch, 581
silicon
---technology, 579, 611, 661
---third-order NLO material, 583
SiO_2 *see* silica
$Si(Pc)(OSiMePhOH)_2$
---degenerate four-wave mixing at 602 nm: first-order and second-order (exciton-exciton) decay; resonant $\chi^{(3)} \approx 10^{-9}$ esu, 576

Si(Pc)(OSiMePhOH)$_2$ *(cont'd)*
---good LB films, 576
---soluble in common solvents, 576
simple (unary) stoichiometry, 7
small-U approximation in Hubbard model, 51
(SN)$_x$ *see* poly(sulfurnitride)
sodium poly(styrene sulfonate) (PSSNa), polymeric anionic dopant in conducting polymers (for mechanical strength, transport membrane), 654
solar cell, 654
sol-gel derived materials, 427
solid electrolytes: β-aluminas, Nafion®, gels, polymers (PVA, PEO, MEEP), 627, 629
soliton band (infrared): misleading term, 356
solitons
---cannot be distinguished from polarons or bipolarons in IR spectra of doped poly(acetylene), 356
---in conducting polymers, 411, 415, 567
---in optical excitation and charge transfer in donor-polyene-acceptor molecules, 531
---in linear-chain platinum complexes, 266
---kink-solitons in halogen-bridged Pt complexes, 271
---potential switches, 531, 659, 663
solution oxidation potential, 11
solution reduction potential, 11
sonication, to help intercalation reactions, 285, 289
Soret band, 220
specific heat, electronic
---affected by longitudinal transfer integral, 54
---measured for (TMTSF)$_2$ClO$_4$ is between 30% and 400% of BCS value, 247
spin density wave (SDW), 9
---accompanied by periodic charge density, 252
---basic interaction is electron-electron (*versus* electron-phonon for CDW), 251
---can carry current (predicted effects: nonlinear current-voltage, frequency-dependent conductivity, coupling to rf field, hysteresis, memory effects), 252, 253
---coexists with bond-CDW in half-filled band, 446
---dc conductivity should increase above a finite threshold field, 253
---due to spin-Peierls transition (2 k_F), 9
---dynamics may resemble CDW (in Lee-Rice-Anderson theory), 251
---induced by magnetic field (FISDW), 245
---is an antiferromagnetic state, 52, 126
---model systems: some (TMTSF)$_2$X salts have transition temperature≈10 K, 251
---SDW ground state exists below a temperature related to the single-particle gap by the BCS expression, 251
---transition to superconductive state by pressure in (TMTSF)$_2$X salts, 126
---transition to superconductive state by band filling in the copper oxides, 126
---translational mode can carry current, and would lead to Fröhlich superconductivity, except that motion is pinned, 251
spinel structure of LiTi$_2$O$_4$, 116
spin-Peierls transition, 9
---in alkali TCNQ salts, 9
---in Wurster's blue perchlorate, 9
spiropyran, 647
SPR *see* surface plasmon resonance
SQUID (superconducting quantum interference device), 289
Sr$_{0.6}$Ba$_{0.4}$Nb$_2$O$_6$ (pyroelectric coefficient)/(dielectric constant) ratio=1.4 μC m^{-2} K^{-1}, 485
SrTiO$_{3-x}$, a superconductor, T_c<1 K, 116
stable free radicals
---by high spin density on heteroatoms, 4
---by steric hindrance, 4
---by push-pull substituent effect, 4
---from odd-alternant hydrocarbons, 4
stable
---anion (cation) radical of good accceptor (donor), 5
---dianion (dication) of good acceptor (donor), 5
---neutral acceptor (donor), 5
stacking mode
---alternate, 48
---deformed, 48
---eclipsed, 12
---herring-bone, 12
---irregular, 48
---lone dimers, 7
---mixed stacks, 7
---orthogonal multimers, 48
---regular, 48
---ring over external bond, 12
---segregated stacks, 7
stacks
---one-stack ion-radical salt, 7
---two-stack charge transfer crystal, 7
Stark spectroscopy, 593
---of reaction center proteins in *Rhodobacter sphaeroides* , 594-596
---of reaction center proteins in *Rhodopseudomonas viridis,* 594-596
---of Langmuir-Blodgett films, 597
STEM *see* scanning transmission electron microscopy
step thickness gauge, LB film, 473
steric hindrance, 4
STM *see* scanning tunneling microscopy
stoichiometry
---berthollitic, 7
---complex, 7

stoichiometry *(cont'd)*
---unary or simple, 7
structure
---refinement, full-matrix least-squares, 164, 210
---reliability factor, 164
---solution, Patterson heavy-atom method, 210
---solution, direct method, 164
sulfur, complex with iodoform, a second-order non-linear optical material, 565
sulfur, complex with SbI_3, a second-order non-linear optical material, 565
sum frequency generation, 608
sum rule, 130
superconductivity
---collective ground state of Fermi gas, 115
---Cooper pairs, 52
---defeated by too narrow bandwidths, 115
---defeated by too weak vibronic interactions, 115
---discovered in 1910, 115
---excitonic, 53, 59, 85
---flux exclusion, 115
---Josephson-coupled layered two-dimensional, 97, 102, 110
---Little model (excitonic), 53, 59, 85
---metallic state a prerequisite, 115
---perfect diamagnetism, 115
---singlet and triplet, 52
superconductivity in inorganic oxides and in organic crystals, 115
---comparative history, 67, 68, 94, 95, 117
---competition from CDW and SDW states, 115
---electron correlations important, 126
---Josephson-coupled layered two-dimensional superconductivity, 97, 102, 110
---lattice instabilities may be important, 124
---layered 2-D superconductivity in both, 98
---magnetic states near superconducting states, 126
---narrow-band systems, 126
---precondition: metallic state, 115
---short Ginzburg-Landau coherence lengths, 102, 110
superconductors, inorganic
---$BaPb_{1-x}Bi_xO_3$ (T_c >10 K), 116
---$Bi_2Sr_2Ca_{n-1}Cu_nO_{2n+4}$ class (T_c=80-110 K), 115
---discovered in 1910, 115
---high-temperature classes: CuO_2 sheets, 124
---high-temperature classes: CuO_4 chains no longer important, 124
---high-temperature classes: sheet Cu coordination number =4, 5, or 6, 124
---high-temperature classes: tetragonal symmetry often broken, 124
---high-temperature classes: very two-dimensional, 124
---history: critical temperature versus year of discovery, 67, 68, 94, 95, 117
---Josephson-coupled layered two-dimensional, 97, 102
---$La_{2-x}M^{II}_xCuO_{4-y}$ class (T_c=20-40 K), 123
---lead (T_c=7.2 K), 116
---$LnBa_2Cu_3O_{7-y}$ class (T_c=70-95 K), 123
---$Ln_{2-x}M^{IV}_xCuO_{4-y}$ class (T_c=20 K), 123
---$LiTi_2O_4$ (T_c>10 K), 116
---mercury (T_c=4.15 K), 115
---niobium (T_c=9.2 K), 115
---Nb_3Ge (T_c=23.2 K), 115
---Nb_3Sn (T_c=18.1 K), 115
---poly(sulfurnitride) (T_c=0.35 K), 116
---$Tl_2Ba_2Ca_{n-1}Cu_nO_{2n+4}$ class (T_c=80-125 K), 123
superconductivity, organic
---best found at border with SDW phase (not so in high-T_c cuprates), 86
---discovered in 1980, 85
---five families: TMTSF (7 members), BEDT-TTF (≈13 members), DMET (7), MDT-TTF (1), $M(dmit)_2$ (4), 67
---history: critical temperature versus year of discovery, 67, 68, 117
---in Bechgaard salts $(TMTSF)_2X$, 16, 67, 116
---in $(BEDT\text{-}TTF)_2X$ *or* $(ET)_2X$ salts, 18, 67, 116
---in BEDT-TTF family: four anion types: (1) tetrahedral anions (ReO_4^-), (2) linear trihalides (I_3^-, IBr_2^-) or metal halides (AuI_2-), (3) clusters ($Cl^-.H_2O$) or polymers ($Hg_{2.78}Cl_8$, $Hg_{2.89}Br_8$) (4) boomerang-shaped anions in zigzag 1-D polymer ($Cu(NCS)_2$), 67, 72
---in $(BEDT\text{-}TTF)_4Hg_3Cl_8$, 227
---in $(BEDT\text{-}TTF)_4Hg_3Br_8$, 227
---in $(BEDT\text{-}TTF)_3Cl_2.2H_2O$, 116
---in β-$(BEDT\text{-}TTF)_2AuI_2$, 18
---in κ-$(BEDT\text{-}TTF)_2Cu(NCS)_2$, 118
---in β-$(BEDT\text{-}TTF)_2I_3$, 18, 145
---in β-$(BEDT\text{-}TTF)_2IBr_2$, 18
---in $(BEDT\text{-}TTF)_2ReO_4$, 18
---in $(DMET)_2X$ salts, 67, 145, 223
---in $(MDT\text{-}TTF)_2AuI_2$, 57, 67, 145, 156, 227
---in $(octadecylamine)_xTaS_2$, 290
---in $(pentacene)_xTaS_2$, 290
---in $(TMA)[Ni(dmit)_2]_2$, 58, 145, 156, 519
---in $(TMTSF)_2ClO_4$, 57, 118, 145
---in $(TMTSF)_2PF_6$, 118, 145
---in $(TMTSF)_2X$ and $(BEDT\text{-}TTF)_2X$ families, 16-18, 67-83, 85-89, 91-95, 97-110
---in $(TTF)[Ni(dmit)_2]_2$, 58, 116, 519
---in $(TTF)[Pd(dmit)_2]_2$, 58, 519
---Josephson-coupled layered two-dimensional, 98, 110
---pairing mechanism an open question, 85

superconductivity, organic *(cont'd)*
---phonon mediation unclear, 85
---possible if CDW and SDW suppressed, 13, 120
---possible if lattice is not too one-dimensional, 120
---possible in organic polymers if ordered and symmetric areas are of size less than coherence length (=10-30Å) if Josephson coupling between layered sites is sufficient and electron-phonon mechanism applies, 110
---singlet versus triplet pairing unclear, 85
---structure classification, 8
superlattice
---design by Langmuir-Blodgett films, 573
---incommensurate, in TTF TCNQ, 12
---in LB films, 607
---X-ray reflections, 11
surface acoustic wave device
---lithium niobate, 476
---quartz, 476
---with LB film coating, 476, 477
surface plasmons in silver, 605, 606
surface plasmon resonance (SPR), 605
---reflectometer, 607
surface plasmon wave Raman spectroscopy, 573
susceptibility *see* electric susceptibility *or* magnetic susceptibility
Su-Schrieffer-Heeger theory, 441
synchrotron radiation, 414

TaS_3, 277
TaS_2, 289
---intercalation compound with octadecylamine superconducts, T_c=4 K, 290
---intercalation compound with pentacene (Pen_xTaS_2): unit cell data, pressed pellet conductivity (300 K)≈ 0.01 S/cm; superconductor T_c=7.02 K at 5 G (quenched at 350 G), 290
---intercalation compound with perylene (Per_xTaS_2): unit cell; pressed pellet conductivity (300 K)≈ 0.01 S/cm; weak paramagnetism, 290
---intercalation compound with tetracene (TT_xTaS_2): unit cell; pressed pellet conductivity (300 K)≈ 0.01 S/cm; weak paramagnetism, 290
---intercalation compound with 1,3,10,12-tetrachlorotetracene (TCT_xTaS_2): unit cell; pressed pellet conductivity (300 K)≈ 0.01 S/cm; weak paramagnetism, 290
T_c *see* critical temperature, metal-to-superconductor transition
TCNQ *see* 7,7,8,8-tetracyanoquinodimethane
$TCNQF_2$ *see* 2,5-difluoro-7,7,8,8-tetracyanoquinodimethane
$TCNQF_4$ *see* 2,3,5,6-tetrafluoro-7,7,8,8-tetracyanoquinodimethane
TDDOP-C-ENP (tris-(dodecyloxy)-phenyl carbamate of ethoxy-nitrobenzene)
---LB films: Z-type, 614
---oxidation and reduction potentials, 617
---structure, 615
TDDOP-C-HETCNQ (tris-(dodecyloxy)-phenyl carbamate of 2-hydroxyethoxy 7,7,8,8-tetracyanoquinodimethane)
---infrared spectra of multilayers, 622
---LB films: Z-type, 614
---oxidation and reduction potentials, 617
---second harmonic generation not seen, 614
---structure, 615
TEA *see* triethylammonium
$(TEA)(TCNQ)_2$
---conductivity, 119
---commensurate CDW, 118
---dimers $(TCNQ)_2^-$, 118
---ionicity=0.5 (uniform), 48
---irregular stacks of tetramers, 48
---semiconductor, 48
$(TEA)_2(TTF)_6(PW_{12}O_{40})$
---conductivity (300 K)=0.03 S/cm, 186
---crystal structure, 187
---four anions and four independent TTF form a channel, 189
---one optical CT band at 4000 cm^{-1} (shoulder at 5500 cm^{-1}), 189
---semiconductor, 1 unpaired electron per anion, 189
---TTF^+ donor stacks eclipsed, 189
---unit cell and space group, 186
temperature
---critical (metal-to-superconductor) T_c, *see* critical temperature
---interchain (perpendicular) coupling $T_\perp$, 14
---mean-field T_{MF}, 14
---transition, metal-to-insulator T_{MI}, 12, 17, 18
terdentate ligands
---diethylenetriamine (dien), 264
---N-methyldiethylenetriamine (Medien), 264
2,2,2-tet *see* tris(ethylene)tetramine
2,3,2-tet *see* bis(ethylene)propylenetetramine
3,2,3-tet *see* bis(propylene)ethylenetetramine
3,3,3-tet *see* tris(propylene)tetramine
tetralkylammonium $Au(dmit)_2$, 62
---Langmuir-Blodgett films, 62
---metallic to 4 K, 62
---possible antiferromagnetic fluctuations below 20 K, 62
1,4,7,10-tetraazacyclotridecane ($[13]aneN_4$), 266
1,4,8,11-tetraazacyclotetradecane ($[14]aneN_4$ or cyclam), 264

1,4,8,12-tetraazacyclopentadecane ([15]aneN_4), 266
(tetrabutylammonium)$_2$ zinc dithiolate, 177
tetracene (TT)
---intercalate with graphite ($TT_{1.0}C_{60}$): staging C-C-C-TT-C-C-TT-C-C-C; unit cell; pressed pellet conductivity=0.106 S/cm, 286, 287
---intercalate with TaS_2 (TT_xTaS_2): unit cell; pressed pellet conductivity (300 K)≈0.01 S/cm; weak paramagnetism, 290
---intercalate with TiS_2 (TT_xTiS_2): unit cell; pressed pellet conductivity (300 K)≈0.01 S/cm; weak paramagnetism, 290
tetrachloroaurate, anionic dopant in conducting polymers (for electrochromism, dispersed noble metal, catalysis), 654
1,3,10,12-tetrachlorotetracene (TCT)
---intercalate with graphite (TCT_xC_y): unit cell; staging C-C-C-TCT-C-C-C; pressed pellet conductivity=0.084 S/cm, 286, 287
---intercalate with TaS_2 (TCT_xTaS_2): unit cell; pressed pellet conductivity (300 K)≈0.01 S/cm; weak paramagnetism, 290
---intercalate with TiS_2 (TCT_xTiS_2): unit cell; pressed pellet conductivity (300 K)≈0.01 S/cm; weak paramagnetism, 290
tetracyanonaphthoquinodimethane (TNAP)
---good acceptor, 5
---HMTSF complex (1:1), 16
---structure, 6
7,7,8,8-tetracyanoquinodimethane (TCNQ)
---anthracene complex (1:1), 8
---complete charge transfer in salts, 118
---complex salts (1:2, 2:3, etc), 118
---Cs salt (2:3), $(Cs)_2(TCNQ)_3$, 8
---DBTTF complex (1:1), 8
---derivatives *see* $TCNQF_4$, DMTCNQ
---N,N'-dimethylphenazine complex, 37
---dithieno[3,2-b::2',3'-d]thiophene complex, 412
---electron affinity=2.8 eV, 613
---estimated Hubbard U (solution dimer)= 1.3 eV, 58
---estimated Hubbard U(solid)=1.4 eV, 58
---estimated Hubbard V(solid)=0.4 eV, 58
---good acceptor, 5
---HMTSF complex (1:1), 16
---HMTTeF complex (1:1), 16
---HMTTF complex (1:1), 8, 16
---(K)(TCNQ) (Mott insulator), 118
---Li salt (1:1), (Li)(TCNQ), 11, 647
---MEM salt (1:2), 48
---NBP salt (1:1), 8
---NMP salt (1:1), 8, 11
---(octadecylphenanthrolinium^{++}) $(TCNQ^-)_2$ salt, 31
---partial charge transfer in compounds, 118
---quinolinium salt (1:2), 11
---Rb salt (1:1), (Rb)(TCNQ), 8, 48
---simple (1:1) salts, 8, 118
---structure, 6, 46, 612
---TEA salt (1:2), 48
---TMPD complex (1:1), 8
---TMTSF complex (black) (1:1), 8, 16
---TMTSF complex (red)(1:1), 8
---TMTTF complex (1:1), 8
---triphenylmethylphosphonium salt (1:1), 11
---triphenylmethylphosphonium salt (1:2), 11
---TSF complex (1:1), 8
---TTeF complex (1:1), 15, 16
---TTF complex (1:1), 8, 12-15
tetracyano-tetrathiafulvalene, intercalate with graphite (TCN-TTF_xC_y): unit cell data; staging C-C-C-TCN-TTF-C-C-C; pressed pellet conductivity=0.288 S/cm, 286, 287
tetra(decyl)ammonium $Ni(dmit)_2$
---mixed 1:1 with eicosanoic acid and oxidized with Br_2, LB film, conductivity (300 K)=1.6 S/cm, 521
---mixed 1:1 with eicosanoic acid and electro-oxidized, LB film, conductivity (300 K)=0.012 S/cm, 521
tetradecylpyridinium $Ni(dmit)_2$
---mixed 1:1 with eicosanoic acid and oxidized with Br_2, LB film, conductivity (300 K)=1.5 S/cm, 521
---mixed 1:1 with eicosanoic acid and oxidized with Br_2, LB film, conductivity (300 K)=1.2 S/cm, 521
2,3,5,6-tetrafluoro-7,7,8,8-tetracyanoquinodimethane ($TCNQF_4$)
---HMTSF complex, 8, 15
---HMTTF complex, 8
---Rb salt (1:1) $(Rb)(TCNQF_4)$, 36
---N,N'-dimethylphenazine complex, 37
---TMPD complex, IR spectrum, 39
tetrakis-butyl-phenoxy-phthalocyanine, 575
tetrakis(methylseleno)-diselenadithiafulvalene, 150
tetrakis(methylseleno)-TSF, 150
tetrakis(methylseleno)-TTF, 150
tetrakis(methylthio)-dithiadiselenafulvalene, 150
tetrakis(methylthio)-TTF, 150
tetramethyl-bis-(ethylenedithiolene)-tetrathiafulvalene (TMET), synthesis, 176, 177
tetramethyldithiadiselenafulvalene (TMDTDSF), 85, 86
tetramethylmanganese chloride (TMMC), Mn^{++} magnetic chains, 61

N,N,N',N'-tetramethyl-*para*--phenylene-diamine (TMPD)
---complexes and salts *see* TMPD
---electroactive material in microelectrochemical transistor, using poly(bis(2-(2-methoxy- ethoxy)ethoxy)phosphazene) as the solid-state electrolyte, doped with $LiCF_3SO_3$: diffusion coefficient of $TMPD^+ = 2 \times 10^{-8}$ cm²/s, 633
---ionization potential=6.25 eV, 613
---structure, 613
tetramethyltetraselenafulvalene (TMTSF), 134
---complexes *see* TMTSF
---crystal structure, 134
---infrared spectrum, 134
---structure, 47
tetramethyltetrathiafulvalene (TMTTF)
---structure, 47
---complexes *see* TMTTF
---estimated Hubbard U(solid)=1.3 eV, 58
---estimated Hubbard V(solid)=0.3-0.4 eV, 58
tetraselenafulvalene (TSF)
---complexes *see* TSF
---derivatives *see* TMTSF, HMTSF
---ionization potential, 11
---salts with cluster anions [$Mo_6Cl_{14}^{--}$, $Re_6Se_5Cl_9^-$, $Ta_6Cl_{18}^{n-}$], 185
---structure, 6, 47
tetraselenatetracene (TST)
---complexes *see* TST
---salts *see* TST
---structure, 6, 47
tetratellurafulvalene (TTeF)
---complexes *see* TTeF
---derivatives *see* HMTTeF
---structure, 6, 47
tetratelluratetracene (TTeT)
---complexes *see* TTeT
---structure, 6
tetrathiafulvalene (TTF), 4, 6
---as 7π-7π system, 4
---chalcogenide analogs: TSF, TTeF: the oxidation potentials decrease (S-->Se) then increase (Se-->Te), 146
---complexes *see* TTF
---derivatives *see* BEDT-TTF, DBTTF, TMTTF, HMTTF, MDT-TTF, DMET, etc.
---derivatives as π donors are better (alkyl, cycloalkyl, thioalkyl groups) worse (trifluoromethyl, nitrile, ester groups; π-delocalization increased by benzene, vinylenedithio, pyrazine, and pyridine rings), 146
---estimated Hubbard U(solution dimer) =1.7 eV, 58
---estimated Hubbard U(solid)=1.7 eV, 58
---estimated Hubbard V(solid)=0.6 eV, 58
---fatty acid derivative, LB film: I_2 doping increases conductivity, 528
---fatty acid derivatives, 528-530
---intercalated in $V_2O_5.nH_2O$ xerogel, 427
---intermolecular contacts can increase the dimensionality, 146
---ionization potential=6.83 eV, 11, 613
---precursor: 2-thioxo-1,3-dithiole (*or* 1,3-dithiolo-2-thione), 146
---salts with cluster anions [$Mo_6Cl_{14}^{--}$, $Re_6Se_5Cl_9^-$, $Ta_6Cl_{18}^{n-}$], 185
---structure, 6, 47, 612
tetrathiafulvalene-1-thia-heptadecanoic acid, 529, 530
tetrathiafulvalene-tetrathiolate (TTF-TT)
---coordination polymer with Ni (Ni-TTF-TT), and Cu (Cu-TTF-TT), 111-113
---structure, 112
tetrathiafulvalene-vinyl-undecanoic acid
---crystal structure, 530
---synthesis, 529, 530
tetrathiafulvalene-vinyl-hexadecanoic acid, 528, 530
tetrathiotetracene (TTT)
---complexes *see* TTT
---good donor, 5
---radical cation in glass: ESR g-tensor, 194
---salts *see* TTT
---structure, 6, 47
TGS *see* triglycine sulfate
theories for 1-dimensional systems
---diagrammatic valence-bond for finite rings (4,6,8,10 sites), half-filled band, 444
---electron-electron correlations, 441
---Su-Schrieffer-Heeger, 441
---tight-binding 1-electron 1-dimensional theory, expanded in powers n of the overlap integral S. (1) For long-range intersite potential: n=1: extended Hubbard model; n>1: CT enters but two-electron part is S^n-invariant. (2) For delta-function intersite potential: n=1, S>0.2: simple Hubbard model; for smaller S, electron-adjacent bond attraction X is a complication, 442
---valence bond for half-filled band: site-CDW, bond-CDW and SDW instabilities, 443
thermal imaging, 484
---night vision using mercury cadmium telluride, 484
---using pyroelectric devices, 484
thermodynamic stability, 7
---and covalent bonding, 49
---and crystallization kinetics, 7
---and Madelung energy, 9, 49
---of TTF-TCNQ, 7, 9
---of $(BEDT\text{-}TTF)_2X$ salts, 7
thermoelectric power *or* thermopower
---affected by longitudinal transfer integral, 54

thermoelectric power *(cont'd)*
---in κ-(BEDT-TTF)$_2$Cu(NCS)$_2$, 67
---in α-(BEDT-TTF)$_2$I$_3$, 261
---in β-(BEDT-TTF)$_2$I$_3$, 261
---in β_d'-(BEDT-TTF)$_2$I$_3$, 258, 259, 261
---in λ_d-(BEDT-TTF)$_2$I$_3$, 260, 261
---in (DIET)$_2$(BF$_4$)$_3$, 179
---in (Per)$_2$Au(i-mnt)$_2$, 213
---in (Per)$_x$[M(mnt)$_2$] salts (M=Au,Co, Cu,Fe, Ni, Pd, Pt), 207-209
thiazolo-TTF (Thz-TTF), 182
thin films, 605, 635
thiophene
---molecular structure, 448
---oxidation potential (solution), 412
---single-double bond distance difference, 412
thiophthene
---oxidation potential (solution), 412
---single-double bond distance difference, 412
2-thioxo-1,3-dithioles (*also* 1,3-dithiole -2-thiones; precursors to tetrathiafulvalenes and metal 1,2-dithiolenes), 182
---alkyl,N-alkyl-pyrazolo-, 144
---alkylthiazolo-, 144
---alkyl-thioalkyl-dihydrothiolo-, 144
---5-cyano-iso-thiazolo-, 144
---cyano-pyrazino-, 144
---cyano-thioalkyl-, 144
---dialkyl-, 144
---dialkyl-N-alkyl-pyrrolo-, 144
---dialkyl-benzo-, 144
---dialkyl-dihydrothiolo-, 144
---dialkyl-pyrazino-, 144
---dialkyl-pyridino-, 144
---dialkylpyridino-pyrazino-, 144
---dialkylpyrazino-dithiolo-, 144
---dialkylpyridino-dithiolo-, 144
---dialkyl-thiolo-, 144
---dicyano-, 144
---dithioalkyl-, 144
---dithioalkyl-benzo-, 144
---dithioalkyl-ethylenedithiolo-, 144
---naphtho-, 144
---thiadiazolo-, 144
---thioalkyl-, 144
---thioxo-dithiolo-, 144
---thioxo-dithiepino-, 144
third harmonic generation, 576, 608
---strongly dependent on π conjugation length, 576
third-order non-linear optical materials
---all-optical waveguide devices, 579
---bulk needs non-zero third-order electric susceptibility $\chi^{(3)}$, 566
---bulk must be centrosymmetric: conjugated polymers best, 566
---example: CS$_2$, 582
---example: 2-methyl-4-nitroaniline, 583
---example: p-nitroaniline, 583
---example: poly(diacetylene), 566
---example: poly(*para* -phenylene-vinylene), 565
---example: poly(thiophene), 566
---example: silicon, 583
---molecules need large third-order hyperpolarizability γ, 566
---organic liquids based on 2-methyl-4-nitroaniline, 579
---oriented, well-packed polymers better than amorphous bulk, 566
---π-delocalization better for γ in thiophene oligomers than in benzene oligomers, 566
---strongly dependent on π–conjugation, 576
---studied by time-resolved degenerate four-wave mixing, 566
---waveguide device needs: high third-order non-linearity (non-linear response is n_2-I/$\alpha\lambda$), ultra-fast response, and excellent other physical properties (linear refractive index, for matching to glass optical fibers of index 1.45; transparency; no scattering centers; uniformity; optically flat surfaces; chemical and environmental stability; processability), 579, 580
Thz-TTF *see* thiazolo-TTF
(Thz-TTF)$_x$(TCNQ), brown microcrystals, 183
tight-binding
---chain, regular, 129
---band, 130, 133
---approximation, 1-dimensional, 49
time-resolved spectroscopy, 592
tin octaethylporphyrin dihalide, 220
TIR *see* total internal reflection
TiS$_2$, 289
---intercalation compound with pentacene (Pen$_x$TiS$_2$): unit cell; pressed pellet conductivity (300 K)$\approx$ 0.01 S/cm, 290
---intercalation compound with perylene (Per$_x$TiS$_2$): unit cell; pressed pellet conductivity (300 K) $\approx$ 0.01 S/cm; weak paramagnetism, 290
---intercalation compound with tetracene (TT$_x$TiS$_2$): unit cell; pressed pellet conductivity (300 K)$\approx$ 0.01 S/cm; weak paramagnetism, 290
---intercalation compound with 1,3,10,12-tetrachlorotetracene (TCT$_x$TiS$_2$): unit cell; pressed pellet conductivity (300 K)$\approx$ 0.01 S/cm; weak paramagnetism, 290
Tl$_{0.3}$MoO$_3$, 277
Tl(Por)X see halo-thallio-porphyrin
Tl(TPP)F *see* fluoro-thallio-tetraphenylporphyrin
Tl$_2$Ba$_2$Ca$_{n-1}$Cu$_n$O$_{2n+4}$, 123
---layers M$_2$O$_2$ with NaCl structure, 123
---1-4 layers of CuO$_5$ square pyramids, 123

$Tl_2Ba_2Ca_{n-1}Cu_nO_{2n+4}$ *(cont'd)*
---superconducts, T_c=80-125 K, 123
TMA = tetramethylammonium, 48
(TMA)[Ni(dmit)$_2$]$_2$, 145, 156
---conductor, 48
---ionicity=0.5 (uniform), 48
---irregular stacks of dimers, 48
---superconductor, T_c=5 K at 7 kbar, 58, 145, 156
(TMA)(pentacenediquinone semiquinone)
---conductivity (300 K)=0.02 S/cm, 646
TMDTDSF *see* tetramethyldithiadiselenafulvalene
(TMDTDSF)$_2$PF$_6$, 86
---antiferromagnetic resonance below 7 K, 86
---no superconductivity down to 0.4 K, 86
---pressure-temperature phase diagram, 86
---spin-Peierls state below 19 K (ESR), 86
---SDW state below 7 K, 86
TMET *see* tetramethyl-bis-(ethylene-dithiolene)-tetrathiafulvalene
TMMC *see* tetramethyl manganese chloride
TMPD *see* N,N,N',N'-tetramethyl-*para*--phenylenediamine
$TMPD^+ ClO_4^-$ (Wurster's blue perchlorate)
---structure classification, 8
---spin-Peierls transition in, 9
---2 k_F SDW, 9
(TMPD)(TCNQ)
---structure classification, 8
(TMPD)(TCNQF$_4$), 38
---dimerized TCNQF$_4$ chains, 38
---dimerized TMPD chains below 180 K, 38
---fully ionic, 38
---hopping integral, 38
---infrared spectrum, 39
---magnetic gap, 38
---on-site parameter U, 38
TMTSF *see* tetramethyltetraselenafulvalene
(TMTSF)$_2$AsF$_6$, 16
---SDW, 118
(TMTSF)$_2$BF$_4$, 16
(TMTSF)$_2$ClO$_4$, 16, 135, 145
---complex low-temperature Hall effect, 245
---conductivity max.(6 K, broad)=50,000 S/cm, 16
---conductivity (300 K)=650 S/cm, 16
---crystal structure, 17, 120
---dielectric constant, 135
---dimerization gap, 135
---electron-phonon coupling parameter, 135
---field-induced SDW state, 246
---heat capacity as a function of magnetic field by ac microcalorimetry: 30 to 400% of BCS prediction, 246, 247
---in pressure-temperature phase diagram, 86
---intrastack transfer integral=0.1 eV, 120
---magneto-caloric effect, 248
---mean-field correlation amplitude, 135
---no charge localization above T_c, 85
---quantized Hall effect, 88
---quantized nesting model of oscillating specific heat, 247, 248
---re-entrance of metallic phase at high magnetic field in B-T phase diagram: competition between different SDW parameters, 246
---resistivity, 119
---specific heat, 247, 248
---superconductor, T_c=1.2 K at 1 bar, 57, 86
---thermal coefficient of magnetization, 248
---transfer integral, 135
---transfer integral ratios are small, 120
---2 k_F spin fluctuations affect ^{77}Se nuclear relaxation rates below 30 K, 87
(TMTSF)(DMTCNQ)
---conductivity max.(42 K)=5,000 S/cm, 16
---conductivity (300 K)=500 S/cm, 16
---ionicity=0.5, 15, 16
---metal-to-insulator transition (42 K), 16
---large effective on-site repulsion, 15
---4 k_F scattering, 15
(TMTSF)$_3$[Ni(CN)$_4$]
---conductivity (300 K)=0.43 S/cm, 186
---unit cell and space group, 186
(TMTSF)$_x$Ni(dcit)$_2$, black microcrystals, 183
(TMTSF)$_2$NO$_3$, 16
---field-dependent conductivity in SDW state: threshold field =40 mV/cm at 4.2 K, and T-independent below 5.5 K, 253
---SDW state below T_c=11 K, 253
---single-particle gap=16 K, 253
(TMTSF)$_2$PF$_6$, 132, 134, 145
---anomalous Shubnikov-de Haas effect under pressure, 246
---bandwidth=1 eV, 134
---collective mode in frequency-dependent conductivity (pinning freq. =30 GHz), 253
---conductivity max.(18 K, broad)=70,000 S/cm, 16
---conductivity (300 K)=550 S/cm, 16
---dielectric constant=2.00, 135
---dimerization gap=0.066 eV, 134, 135
---electron-phonon coupling parameter, 135
---field-dependent conductivity in SDW state: threshold field is 7.5 mV/cm at 4.2 K, is T-independent below 5.75 K, increases close to T_c = 11.5 K, and increases greatly with impurities, 253
---ionicity=0.50, 16
---in pressure-temperature phase diagram, 86
---IR conductivity spectra, 134, 136
---mean-field correlation amplitude, 135

$(TMTSF)_2PF_6$ *(cont'd)*
---pressure suppresses SDW transition, 126
---pressure brings superconduction, 126
---quantized Hall effect, 85, 87, 88
---resistivity, 119
---SDW ground state below 10 K (low P), 118
---SDW state below T_c=11.5 K, 253
---single-particle gap=32 K, 253
---structure classification, 8
---superconducts, T_c=0.9 K at 10 kbar, 18, 67
---superconducts, T_c=1 K at 8 kbar, 120
---transfer integral=0.25 eV, 135
---2 k_F spin fluctuations affect ^{77}Se nuclear relaxation rates below 100 K, 87
$(TMTSF)_3[Pt(CN)_4]$
---conductivity (300 K)=0.25 S/cm, 186
---unit cell and space group, 186
$(TMTSF)_3(PW_{12}O_{40})$
---fully ionized TMTSF trimers, 189
---insulator at 300 K, 186, 189
---one optical band at 11,000 cm^{-1} (no CT), 189
---unit cell and space group, 186
$(TMTSF)_2ReO_4$, 18
---hidden spin density wave, 85
---ordering of non-centrosymmetric ReO_4^- ions needed for superconductivity, 86
---pressure-temperature phase diagram, 86
---semiconductor state at intermediate T, 86
---SDW semiconducting state below 15 K at 1 bar, 86
---superconductor above 10 bar, 86
$(TMTSF)_2SbF_6$, 16
(TMTSF)(TCNQ) (black)
---conductivity max.(61 K)=7,000 S/cm, 16
---conductivity (300 K)=1200 S/cm, 16
---ionicity=0.57, 16
---metal-to-insulator transition (61 K), 16
---resistivity, 14, 120
---structure classification, 8
(TMTSF)(TCNQ) (red)
---structure classification, 8
$(TMTSF)_3(Wo_6O_{19}).2C_3H_7ON$
---conductivity (300 K)=0.03 S/cm, 186
---crystal structure, 187
---trimers, criss-cross intrastack overlap, 188
---two optical absorptions (CT and other), 188
---unit cell and space group, 186
$(TMTSF)_2X$ salts, ("Bechgaard salts")
---almost two-dimensional, 18
---bandwidth=1.4 eV, 16
---correlated narrow-gap semiconductor, 141
---doping with "S" for "Se" lowers T_c, 18
---Fermi surface: slightly warped planes, 18
---4 k_F charge localization, 141
---half-filled valence band, 118
---ionicity=0.5 (uniform), 48
---irregular stacks of dimers, 48
---logitudinal conduction is coherent, 55
---magnetic SDW suppressed by pressure, 17
---metallic conductor, 48, 55
---metal-to-insulator transition temperature linear with tetrahedral anion size, 17, 18
---no nesting of Fermi planes, 18
---non-linear conductivity in SDW state, 251
---octahedral anions never order at low T, 18
---ordering of non-centrosymmetrical X, 52
---Pauli-type paramagnetism, 53
---pressure-temperature phase diagram: 1-D conductor, 4 k_F localization, spin-Peierls, SDW, 2-D(3-D) Fermi liquid, superconductor, 85, 86, 163
---precursor CDW not found, 17
---pressure induces superconductivity, 16
---Robin-Day class IIIB, 53
---SDW ground state at low temperature, 118
---slight dimerization, 118
---superconductors at low temperature, 18, 53
---triclinic, 118
TMTTF *see* tetramethyltetrathiafulvalene
(TMTTF)(octadecylTCNQ), LB films, metal?, 520
---conductivity=0.1 S/cm (if > 8 monolayers), 506
---instability in pressure-area isotherm below 15-20 mN/m, 506
---stronger film, with better conductivity, and transfer ratio, when mixing 1:1:1 TMTTF : octadecylTCNQ : ω-tricosenoic acid; used as antistatic agent, 506
(TMTTF)(tetradecylTCNQ), LB films, metal?, 520
$(TMTTF)_3(tetradecylTCNQ)_2$
---conductive Langmuir-Blodgett films, 62
---possible Peierls fluctuations at 150 K, 62
$(TMTTF)_2AsF_6$
---conductivity, 163
---crystal structure at 4 K, 164
---dimerization decrease (300 K to 4 K), 166
---electronic dimerization, 164
---temperature of maximum conductivity, 164
---transfer integrals (interchain, intrachain) from extended Hückel calculations for TMTTF dimers at 4 K and 300 K, 164-166
$(TMTTF)_2Br$, 132, 135
---antiferromagnetic ground state, 57
---dielectric constant=2.00, 135
---dimerization gap=0.086 eV, 135

(TMTTF)$_2$Br *(cont'd)*
---electron-phonon coupling parameter, 135
---in pressure-temperature phase diagram, 86
---mean-field correlation amplitude, 135
---Néel temperature=14 K, 57
---optical gap, 56
---room-temperature conductivity, 56
---3-D spin-Peierls state with no lattice distortion below 15-19 K, 85
---transfer integral=0.20 eV, 135
---weak charge localization below 100 K, 85
α-(TMTTF)$_2$IO$_4$
---optical gap, 56
---room-temperature conductivity, 56
β-(TMTTF)$_2$IO$_4$
---optical gap, 56
---room-temperature conductivity, 56
(TMTTF)$_2$[Ni(CN)$_4$]
---insulator at 300 K, 186
---unit cell and space group, 186
(TMTTF)$_3$[Ni(CN)$_4$]
---conductivity (300 K)=0.55 S/cm, 186
---unit cell and space group, 186
(TMTTF)$_2$PF$_6$, 132, 135
---antiferromagnet at 15 K and 15 kbar, 163
---conductivity, 163
---crystal structure at 4 K, 164
---dielectric constant=2.00, 135
---dimerization decreases (300 K to 4 K), 166
---dimerization gap=0.034 eV, 135
---electronic dimerization, 164
---in pressure-temperature phase diagram, 86
---1-D quantum antiferromagnet below 200 K, 85
---3-D spin-Peierls state and lattice distortion below 15-19 K at 1 bar, 85, 163
---strong charge localization below 200 K, 85
---temperature of maximum conductivity 164
---transfer integral=0.17 eV, 135
---transfer integrals (interchain, intrachain) from extended Hückel calculations for TMTTF dimers at 4 K and 300 K, 164-166
(TMTTF)$_2$ReO$_4$, 137
---charge-transfer transitions, 139, 140
---infrared conductivity, 138
---intersite transfer integrals t=0.173 eV, t'=0.127 eV, 139
---intratetramer site energy difference Δ= 0.055 eV, 139
---off-site energy V=0.14 eV, 139
---on-site energy U=0.6 eV, 139
---optical gap, 56
---Raman spectrum, 139
---room-temperature conductivity, 56
β-(TMTTF)$_2$ReSe$_5$Cl$_9$
---optical gap, 56
---room-temperature conductivity, 56
(TMTTF)$_2$SbF$_6$
---antiferromagnet at 6 K and 1 bar, 163-164
---conductivity, 163
---conductivity anomaly at 154 K, 164
---crystal structure at 135 K, 164-167
---crystal structure, room-temperature, 164
---dimerization: small decrease (300 K to 135 K), 166
---interchain interactions weaken (300 K to 135 K), 166
---lattice stiffens at 135 K (strong anion-cation interactions), 166
---optical gap, 56
---rigid-body spherical coordinates at 135 K and 300 K, 165
---room-temperature conductivity, 56
---transfer integrals (interchain, intrachain) from extended Hückel calculations for TMTTF dimers at 135 K and 300 K, 164-166
(TMTTF)$_2$SCN
---optical gap, 56
---room-temperature conductivity, 56
(TMTTF) (TCNQ)
---conductivity, 14
---conductivity max.(60 K)=5,000 S/cm, 18
---conductivity (300 K)=350 S/cm, 16
---ionicity=0.65, 16
---one-dimensional, 14
---structure classification, 8
(TMTTF)$_{1.3}$(TCNQ)$_2$, structure classification, 8
(TMTTF)$_2$X salts, 129
---close to spin-Peierls-antiferromagnetic state boundary for X=octahedral anion, 163
---correlated narrow-gap semiconductor, 141
---Hubbard gap, 55
---ionicity=0.5, 48
---irregular stacks of dimers, 48
---longitudinal conduction is diffusional, 55
---magnetic state at low T, 53
---metallic conductor, 48, 55
---metal-to-insulator transition temperature is linear with tetrahedral anion size, 17
---ordering of non-centrosymmetrical X, 52
---Pauli-type paramagnetism, 53, 55
---pressure-temperature phase diagram: 1-D conductor, 4 k_F localization, spin-Peierls, SDW, 2-D(3-D) Fermi liquid, superconductor, 85, 86, 163
---Robin-Day class IIIB, 53
---weakly dimerized, 55
tn *see* 1,3-diaminopropane
TNAP *see* tetracyanonaphthoquinodimethane

topotactic polymerization, 664
total internal reflection (TIR), 605
Torrance
---classification of ion-radical salts, 44, 118
---criteria for organic metals, 12
---proposal for organic ferromagnets, 60
transfer integral, 3, 51, 54, 129
---calculated values, 135, 139, 166
---half the dimer splitting, 49
---longitudinal (intrachain) $t_{||}$, 51
---longitudinal: affects specific heat, optical reflectivity, and thermopower, 54
---one-quarter the bandwidth, 3
---transverse (interchain) $t_{\perp}$, 52
---transverse: affects electrical conductivity anistropy, ESR linewidths, 54
transistor
---molecule-based (poly(aniline)) field-effect transistor, 661
---polymer field-effect (poly(3-hexylthiophene)), 661
---using LB multilayers as the insulator, 478
---microelectrochemical transistor based on poly(ethyleneoxide) and poly(vinyl alcohol), 627
---microelectrochemical transistor based on polymer ionic conductor poly(bis-(2-(2-methoxyethoxy)ethoxy)phosphazene) (MEEP), 627
transition *see also* phase transition
---matrix element, 24
---temperature, metal-to-insulator T_{MI}, 9
trans -stilbene, 378
transverse interactions in 1-dimension
---due to counterions, 52
---due to intrachain electron correlations, 52
ω-tricosenoic acid (ωTA)
---LB films, 476
---structure, 505
triethylammonium $(TCNQ)_2$, resistivity, 119
triglycine sulfate (TGS) (pyroelectric coefficient)/(dielectric constant) ratio $=6.0\ \mu C\ m^{-2}\ K^{-1}$, 485
trimethyl-dithiapyrene-1-thia-dodecanoic acid, 529, 530
trimethyl-TTF-vinyl-undecanoic acid, 529
trimethyl-TTF-1-thia-dodecanoic acid, 529
trimethyl-TTF-1-thia-heptadecanoic acid, 529
triphenylene-based conducting discotic liquid crystals, 61
triphenylmethylphosphonium (TCNQ)
---1:1 salt, low conductivity, 11
---1:2 salt, higher conductivity, 11
triplet superconductor (not yet found), 52, 53, 59
tris(decyl)methylammonium $Au(dmit)_2$
---mixed 1:1 with eicosanoic acid and oxidized with Br_2, LB film, conductivity (300 K)=15 S/cm, 521, 522
---mixed 1:1 with eicosanoic acid and electro-oxidized, LB film (>20 monolayers), conductivity (300 K)=33 S/cm, conduction: (i) macroscopically metallic 300-200 K, (ii) semiconductive (gap= 0.002 eV) 200-50 K, (iii) variable-range hopping below 50 K, 508, 521, 522
tris(decyl)methylammonium $Ni(dmit)_2$
---mixed 1:1 with eicosanoic acid and oxidized with Br_2, LB film, conductivity (300 K)=1.5 S/cm, 521
---mixed 1:1 with eicosanoic acid and electro-oxidized, LB film, conductivity (300 K)=1.4 S/cm, 521
tris(ethylene)tetramine (2,2,2-tet), 266
tris(tetradecyl)methylammonium $Au(dmit)_2$
---mixed 1:1 with eicosanoic acid and oxidized with Br_2, LB film, conductivity (300 K)=5.4 S/cm, 521
---mixed 1:1 with eicosanoic acid and electro-oxidized, LB film, conductivity (300 K)=19 S/cm, 521
tris(tetradecyl)methylammonium $Ni(dmit)_2$
---mixed 1:1 with eicosanoic acid and oxidized with Br_2, LB film, conductivity (300 K)=1.3 S/cm, 521
---mixed 1:1 with eicosanoic acid and electro-oxidized, LB film, conductivity (300 K)=0.87 S/cm, 521
tris(hexadecyl)methylammonium $Au(dmit)_2$
---mixed 1:1 with eicosanoic acid and oxidized with Br_2, LB film, conductivity (300 K)=2.6 S/cm, 521
---mixed 1:1 with eicosanoic acid and electro-oxidized$_2$, LB film, conductivity (300 K)=0.46 S/cm, 521
tris(octadecyl)methylammonium $Au(dmit)_2$
---mixed 1:1 with eicosanoic acid and oxidized with Br_2, LB film, conductivity (300 K)=1.4 S/cm, 521
---mixed 1:1 with eicosanoic acid and electro-oxidized, LB film, conductivity (300 K)=0.12 S/cm, 521
tris(propylene)tetramine (3,3,3-tet), 266
TSeF (= TSF) *see* tetraselenafulvalene
TSF *see* tetraselenafulvalene
(TSF)(TCNQ)
---conductivity, 15
---conductivity max.(40 K)=10,000 S/cm, 16
---conductivity (300 K)=800 S/cm, 16
---ionicity=0.63, 11, 16
---metal-to-insulator transition at 40 K, 16
---solid solution with TTF TCNQ, 12
---structure classification, 8
TST *see* tetraselenatetracene
$(TST)_2Cl$
---conductivity max.(26 K, broad)=10,000 S/cm, 16
---conductivity (300 K)=2100 S/cm, 16
---ionicity 0.50, 16

$(TST)_2I$
---conductivity max.(35 K, broad)=10,000 S/cm, 16
---conductivity (300 K)=1500-2000 S/cm, 16, 59
---ionicity=0.50, 16
---longitudinal conduction is coherent, 55
---metal to 4 K, 48, 55, 58, 59
---no cooperative ground state at low T, 59
---regular stacking, 48
TT *see* ethylene-tetrathiolate
TTeF *see* tetratellurafulvalene
(TTeF)(TCNQ)
---conductivity max.(<4 K)=25,000 S/cm, 16
---conductivity (300 K)=1800 S/cm, 16
---increased interchain interactions, 15
---ionicity=0.71, 16
---metal-to-insulator transition (< 4 K), 16
TTeT *see* tetratelluratetracene
TTF *see* tetrathiafulvalene
$(TTF)_3(BF_4)_2$
---chage-transfer bands, 53
---insulator, 48
---ionicity=0 and 1 (non-uniform), 48
---irregular stacks of trimers, 48
---paramagnetic, 53
---Robin-Day class II, 53
(TTF)(Br)
---electronic absorption spectrum at 12 K, 33
---resonance Raman spectrum, 33
---symmetric dimer in, 33
$(TTF)(Br)_{0.7}$
---regular stacking, 48
---ionicity <1, 48
---metal, 48
TTF-C-BHTCNQ (tetrathiafulvalenyl-carbamate of 2-bromo-5-hydroxy-ethoxy-7,7,8,8-tetracyanoquinodi-methane)
---difficult to purify; two forms, 614
---LB films (neutral form), 614
---structure, 615
(TTF)(chloranil)
---neutral-to-ionic transition, 10
---structure classification, 8
$(TTF)_6(MnCl_3)_3(MnCl_2)_3.13H_2O$
---coupling between Mn^{++} chains and conduction electrons sought, 61
---antiferromagnetic ordering ≈ 1 K, 61
$(TTF)_3(Mo_6O_{19})$
---conductivity=0.0014 S/cm at 300 K, 186
---crystal structure, 188
---trimers, criss-cross intrastack overlap, 188
---two optical absorptions (CT and other), 188
---unit cell and space group, 186
$(TTF)_xNi(dcit)_2$, black microcrystals, 183
$(TTF)Ni(dmit)_2$ semiconductor, 155
$(TTF)[Ni(dmit)_2]_2$ superconducts, T_c=1.6 K at 6 kbar, 58, 116, 145
$(TTF)[Pd(dmit)_2]_2$ superconducts, 58
$(TTF)_5[Pt(CN)_4].2CH_3CN$
---acetonitrile positions disordered, 186
---crystal structure, 187
---mixed-valence isolated clusters, 186
---pentamers $(TTF)_5^{+4}$ with middle TTF^0, 186
---two CT bands (5700 and 11000 cm^{-1}), 188
---unit cell and space group, 186
$(TTF)_3(Re_6Se_5Cl_9)(Cl)$, a "molecular perovskite", 185
(TTF)(TCNQ)
---bandwidth on TCNQ chain 4t=0.45 eV, 12
---bandwidth on TTF chain 4t=0.20 eV, 12
---CDW on TCNQ chains below 160 K, 12
---CDW coupling between TCNQ chains below 54 K, 12, 118
---CDW on TTF chains below 49 K, 12, 118
---conductivity, 12-14
---conductivity anisotropy, 12
---conductivity max.(59 K)=20,000 S/cm, 16
---conductivity (300 K)=500 S/cm, 16
---crystal structure, 13
---incommensurate superlattice, 12
---intermediate on-site repulsion U_{eff}, 15
---interstack coupling transition at 38 K, 118
---interweaving herringbone motif, 12
---intrastack transfer integral =0.1 eV, 120
---ionicity=0.59, 11, 12, 16
---LB films (?), 520
---metal-insulator transition (59-60 K), 12, 16
---Peierls distortion complete below 38 K, 12, 118
---resistivity, 119
---series (TTF)(TCNQ),(TMTTF)(TCNQ), (TMTSF)(TCNQ), (HMTSF)(TCNQ), 14
---series (TTF)(TCNQ), (TSF)(TCNQ), (TTeF)(TCNQ), 15
---sliding CDW, 277
---solid solution with (TSF)(TCNQ), 12
---structure classification, 8
---thermodynamic stability, 7
---transfer integral ratios large, 120
---2 k_F and 4 k_F scattering, 15
$(TTF)(TCNQF_2)$
---structure classification, 8
$(TTF)_3(SnMe_2Cl_4)$
---ionicity=0 and 1 (non-uniform), 48
---orthogonal trimers, 48
---semiconductor, 48
TTF-TT *see* tetrathiafulvalene-tetrathiolate
$(TTF)_3(W_6O_{19})$
---conductivity (300 K)=0.0005 S/cm, 186

$(TTF)_3(W_6O_{19})$ *(cont'd)*
---crystal structure, 187
---strong temperature dependence of optical spectra, 188
---trimers, criss-cross intrastack overlap, 188
---two optical absorptions (CT and other), 188
---unit cell and space group, 186
TTT *see* tetrathiotetracene
(TTT)(Cl)
---ESR: g-tensor, 194
---synthesis, 194
(TTT)(I), conductivity (300 K), 194
$(TTT)_2I_3$
---conductivity decreases dramatically below 35 K (Peierls transition?), 191
---conductivity max.($\approx$60 K, broad)=3,000 S/cm, 16
---conductivity (300 K)=1000 S/cm, 16, 191
---ESR: single almost symmetrical line, 194
---ionicity=0.50, 16, 48
---metal, 48
---regular stacking, 48
---synthesis, 194
two-dimensional structures
---$(BEDT\text{-}TTF)_2X$ salts, 18, 57, 70, 85, 98, 122, 176
---intercalation compounds of graphite, 175, 285
---intercalation compounds of FeOCl, 421
---intercalation compounds of MoS_2, 175
---intercalation compounds of TaS_2, 175, 289
---intercalation compounds of TiS_2, 289
---intercalation compounds of V_2O_5, 421
---Langmuir-Blodgett films *q.v.*
---layered copper oxides, 175
---metal dichalcogenides, 289
tungsten bronzes, 116
tungsten oxide *see* WO_3
tunneling, electron *see also* intramolecular electron transfer
---through bonds, 613, 662

Ullmann reaction, 381
ultraviolet-visible (UV-Vis) spectra *see* visible-ultraviolet spectra
Umklapp scattering
---at 2 k_F in metals, 8
---at 2 k_F in Peierls semiconductors, 9
---at 4 k_F in Mott-Hubbard semiconductors, 9
---term g_3, 86
unary (simple) stoichiometry, 7
UV-Vis *see* visible-ultraviolet spectra

Valence band, 2
---half-filled in $(TMTSF)_2X$, 118
valence bond
---method, diagrammatic, 467
---method, finite rings, 130
---theory, 49, 442
valence effective hamiltonian (VEH)
---band energy of poly(aniline), 327
---calculations for H_2TAP, H_2Pc, 2,3-H_2Nc, 1,2-H_2Nc, H_2Phc, 456
---calculations for poly(2,5-thienylene vinylene), poly(2,5-furylene vinylene), poly(2,5-pyrrylene vinylene), 448
valency (*see also* ionicity)
---almost ionic, 7
---almost neutral, 7
---discrete valences, 7
---divalent, 44, 45
---mixed-valent, 7
---monovalent, 45
---polyvalent, 45
vanadium oxide $(V_2O_5).nH_2O$ xerogel
---layered electron semiconductor, 427
---intercalants: alcohols, alkali ions, alkylamines, benzidine, 427
---intercalant: poly(pyrrole), 421
---intercalant: poly(thiophene), 421
---intercalant: sulfoxides, TTF, 427
---intercalative polymerization makes d^1 V^{IV} centers: V_2O_5 becomes a bronze, 429
---interlayer spacing 11.55 Å (pristine), 427
van der Pauw method for conductivity, 218
van der Waals area, end-on, 649
van der Waals energy
---charge-dependent contributions, 10
---weak effect on conducting polymers, 346
van Hove singularity in density of states, 89
VB *see* valence bond
VDS-BMT-TTF *see* (vinylenediselenolo)-(bis(methylthiolo)-TTF
$(VDS\text{-}BMT\text{-}TTF)_xI_3$, black microcrystals, 145
$(VDS\text{-}BMT\text{-}TTF)_xIBr_2$, black microcrystals, 145
VDT-BMT-TTF *see* vinylenedithiolo-bis(methylthio)-TTF
$(VDT\text{-}BMT\text{-}TTF)_xIBr_2$, black microcrystals, 183
VDT-TTF *see* vinylenedithiolo-TTF
$(VDT\text{-}TTF)_xAuI_2$, black needles, 183
$(VDT\text{-}TTF)_xIBr_2$, black needles, plates, metallic to low temperature, 183, 184
$(VDT\text{-}TTF)_xNi(dcit)_2$, black microcrystals, 183
VEH *see* valence effective hamiltonian
(vinylenediselenolo-diselena)-dithiafulvalene, 150
(vinylenediselenolo-diselena)-(vinylenedithiolo-dithia)fulvalene, 150
(vinylenediselenolo)(bis(methylthiolo)-TTF (VDS-BMT-TTF), 182
vinylenediselenolo-TSF, 150

vinylenediselenolo-TTF, 150
vinylenedithiolo-bis(methylthio)-TTF (VDT-BMT-TTF), 182
(vinylenedithiolo-diselena)dithiafulvalene, 150
(vinylenedithiolo)diselena)-(vinylenediselenolo)diselena) fulvalene, 150
(vinylenedithiolo-diselena)-(vinylenediselenolo-dithia)fulvalene, 150
(vinylenedithiolo-dithia)-(vinylenediselenolo-diselena)fulvalene, 150
vinylenedithiolo-TSF, 150
vinylenedithiolo-TTF (VDT-TTF), 150
vinylenedithiolo-vinylenediselenolo-TTF, 150
2-vinylfuran, *ab initio* calculation, 448
2-vinylpyrrole, *ab initio* calculation, 448
2-vinylthiophene, *ab initio* calculation, 448
vibrational progressions, 29
V_2O_5 see vanadium oxide
visible-ultraviolet (Vis-UV) spectra
---of $Ag(2,5\text{-}DM\text{-}DCNQI)_2$, 199
---of $(BEDT\text{-}TTF)_4Pt(CN)_4$, 188
---of $Cu(2,5\text{-}DM\text{-}DCNQI)_2$, 199
---of M(Por)X, 220
---of N-hexadecyl-4-pyridinium-tricyanoquinodimethanide (LB film), 648
---of N-hexadecyl-4-quinolinium-tricyanoquinodimethanide (LB film), 648
---of 2-methyl-4-nitroaniline, 585
---of mordenite Na_8Mor, 424
---of $Na(2,5\text{-}DM\text{-}DCNQI)_2$, 199
---of N-octyl-4-pyridinium-tricyanoquinodimethanide (solution), 648
---of (N-[p-(p-octylphenylazo)-phenyloxy]-dodecylpyridinium)$(TCNQ)_2$, 523
---of poly(aniline), emeraldine salt, 322-324
---of poly(aniline) encapsulated in $H_{46}Na_{10}$ Y-zeolite, 424
---of poly(aniline), leucoemeraldine and emeraldine base, 325
---of poly(aniline), fully oxidized salt, 326
---of poly(3-methylthiophene) encapsulated in $Cu_{15}Na_{26}$ Y-zeolite, 424
---of poly(3-octylthiophene)/poly(ethylene), 372
---of poly(*para* -phenylene vinylene), 395
---of poly(pyrrole) encapsulated in Fe_3Na_2 mordenite, 424
---of poly(pyrrole) encapsulated in $Fe_{12}Na_{32}$ Y-zeolite, 424
---of $[Pt(en)_2][Pt(en)_2Cl_2](ClO_4)_2$, 273, 275
---of Qy region of reaction center of *Rhodobacter sphaeroides*, 595
---of Qy region of reaction center of *Rhodopseudomonas viridis*, 595
---of $(TMTSF)_3Pt(CN)_4$, 188
---of $(TMTSF)_3W_6O_{19}.2DMF$, 188
---of $(TTF)_5Pt(CN)_4.2CH_3CN$, 188
---of $(TTF)_3Mo_6O_{19}$, 189
---of $(TTF)_3W_6O_{19}$, 189
---of zeolite $Na_{56}Y$, 424

Wannier functions, 49
W bronzes *see* tungsten bronzes
Wigner lattice, 51
---generalization of the Mott-Hubbard insulator, 51
---in liquid-compressed phase of L-α-dimyristoyl phosphatidic acid monolayers at air water interface, 491, 499
---period depends on ionicity, 51
Wittig reaction, 533
WO_3
---intercalates H^+, Li^+, Na^+ on reduction, and becomes conducting, 631
---use in microelectrochemical transistor sensitive to pH and Li^+: "off" when neutral, "on" when reduced, 631
---use in microelectrochemical diode [conducts if reduced] with poly(3-methylthiophene) [conducts if oxidized] with poly(bis(2-(2-methoxyethoxy)-ethoxy)-phosphazene) (MEEP) as ionic (solid-state) electrolyte, 632
---wide-gap semiconductor, 631
Wolffram's red salt (WR), $[Pt^{II}(C_2H_5NH_2)_4][Pt^{IV}(C_2H_5NH_2)_4Cl_2]Cl_4.4\,H_2O$, 263
work function, 558
---aluminum, 613, 635
---gold, 613, 636
---platinum, 613
---poly(3-hexylthiophene), 636
WR *see* Wolffram's red salt
Wudl proposal for organic ferromagnet, 60
Wurster's blue perchlorate *see* $TMPD^+ClO_4^-$

X-type LB multilayers, 598
xerogel, 427
X-ray photoelectron spectroscopy (XPS), 197
X-ray crystal structure *see* crystal structure
X-ray diffraction, 285, 289, 397, 398
X-ray scattering, diffuse *see* diffuse scattering

Yang's radical, 4
$YBa_2Cu_3O_6$
---linear Cu-O-Cu chains [Cu^I ($3d^{10}$)], 126
---magnetic moments in CuO_2 planes, 124
---two-dimensional antiferromagnet, 124
$YBa_2Cu_3O_{7-x}$
---anion-deficient perovskite, with layers of square-pyramidal CuO_5 and with chains of square-planar CuO_4, 123
---as x grows, Néel temp. and magnetic moments fall, 126
---as x grows, superconductivity sets in, 126

---Ginzburg-Landau coherence length is anisotropic=11.6 Å (< **c**) along **c**, =23-35 Å in **ab** plane, 77
---superconductivity around 90 K, 123
Y-type LB multilayers, 598

Zener breakdown, 255
zeolite types: mordenite (MOR), zeolite A, zeolite Y, 421
zeolite A encapsulates
---color changes (to blue and green), 422
---$Cu_8Na_{80}A$ with poly(pyrrole), 422
---$Cu_8Na_{80}A$ with poly(thiophene), 422
zeolite (mordenite) encapsulates
---color changes (to blue and green), 422
---$Cu_{2.5}Na_3MOR$ with poly(pyrrole), 422
---$Cu_{2.5}Na_3MOR$ with poly(thiophene), 422
---$Cu_{2.5}Na_3MOR$ with poly(3-methylthiophene), 422
---Fe_3Na_2MOR with poly(pyrrole), 422
---Fe_3Na_2MOR with poly(thiophene), 422
---Fe_3Na_2MOR with poly(3-methylthiophene), 422
---H_8MOR with poly(aniline), 422
---Na_8MOR with poly(aniline), 422
---Na_8MOR with poly(pyrrole), 422
---Na_8MOR with poly(thiophene), 422
---Na_8MOR with poly(3-methylthiophene), 422
zeolite Y encapsulates
---color changes (to blue and green), 422
---$Cu_{15}Na_{26}Y$ with poly(pyrrole), 422
---$Cu_{15}Na_{26}Y$ with poly(thiophene), 422
---$Cu_{15}Na_{26}Y$ with poly(3-methylthiophene), 422
---$Fe_{12}Na_{32}Y$ with poly(pyrrole), 422
---$Fe_{12}Na_{32}Y$ with poly(thiophene), 422
---$Fe_{12}Na_{32}Y$ with poly(3-methylthiophene), 422
---$H_{46}Na_{10}Y$ with poly(aniline), 422
---$Na_{56}Y$ with poly(aniline), 422
---$Na_{56}Y$ with poly(pyrrole), 422
---$Na_{56}Y$ with poly(thiophene), 422
zinc tetra(4-sulfophenyl)porphyrin, for anionic dopant in conducting polymers (for photoelectric conversion), 654
Z-type LB multilayers, 598, 614